Norbert Wiener: Collected Works
Volume II

Mathematicians of Our Time

Gian-Carlo Rota, series editor

Paul Erdös: The Art of Counting
edited by Joel Spencer [5]

Einar Hille: Classical Analysis and Functional Analysis
Selected Papers of Einar Hille
edited by Robert R. Kallman [11]

Mark Kac: Probability, Number Theory, and Statistical Physics
Selected Papers
edited by K. Baclawski and M. D. Donsker [14]

Charles Loewner: Theory of Continuous Groups
notes by Harley Flanders and Murray H. Protter [1]

Percy Alexander MacMahon: Collected Papers
Volume I
Combinatorics
edited by George E. Andrews [13]

George Pólya: Collected Papers
Volume I
Singularities of Analytic Functions
edited by R. P. Boas [7]

George Pólya: Collected Papers
Volume II
Location of Zeros
edited by R. P. Boas [8]

Collected Papers of Hans Rademacher
Volume I
edited by Emil Grosswald [3]

Collected Papers of Hans Rademacher
Volume II
edited by Emil Grosswald [4]

Stanislaw Ulam: Selected Works
Volume I
Sets, Numbers, and Universes
edited by W. A. Bayer, J. Mycielski, and G.-C. Rota [9]

Norbert Wiener: Collected Works
Volume I
Mathematical Philosophy and Foundations; Potential Theory; Brownian Movement, Wiener Integrals, Ergodic and Chaos Theories; Turbulence and Statistical Mechanics
edited by P. Masani [10]

Norbert Wiener: Collected Works
Volume II
Generalized Harmonic Analysis and Tauberian Theory; Classical Harmonic and Complex Analysis
edited by P. Masani [15]

Oscar Zariski: Collected Papers
Volume I
Foundations of Algebraic Geometry and Resolution of Singularities
edited by H. Hironaka and D. Mumford [2]

Oscar Zariski: Collected Papers
Volume II
Holomorphic Functions and Linear Systems
edited by M. Artin and D. Mumford [6]

Oscar Zariski: Collected Papers
Volume III
Topology of Curves and Surfaces, and Special Topics in the Theory of Algebraic Varieties
edited by M. Artin and B. Mazur [12]

Norbert Wiener

Collected Works with Commentaries

Volume II

Generalized Harmonic Analysis and Tauberian Theory;
Classical Harmonic and Complex Analysis

edited by
P. Masani

The MIT Press
Cambridge, Massachusetts, and London, England

This book was set in Hong Kong
by Asco Trade Typesetting Limited,
printed and bound by The Murray Printing Company
in the United States of America.

Library of Congress Cataloging in Publication Data

Wiener, Norbert, 1894–1964.
Generalized harmonic analysis and Tauberian theory.

(His Collected works; v. 2) (Mathematicians of our time; 15)
Bibliography: p.
Includes index.
1. Harmonic analysis—Addresses, essays, lectures. 2. Tauberian theorems—Addresses, essays, lectures. I. Masani, Pesi Rustom. II. Title. III. Series: Mathematicians of our time; 11.
QA3.W53 vol. 2 [QA403] 510′.8s[515′.2433] 79-14418
ISBN 0-262-23092-5

Commentators for Volume II

R. A. Askey, Professor of Mathematics, University of Wisconsin, Madison

P. T. Bateman, Professor of Mathematics, University of Illinois

V. E. Beneš, Mathematics and Mechanics Research Center, Bell Laboratories, Murray Hill, N.J.

C. A. Berenstein, Associate Professor of Mathematics, University of Maryland

R. H. Cameron, Professor Emeritus, University of Minnesota

H. G. Diamond, Professor of Mathematics, University of Illinois

W. H. Fuchs, Professor of Mathematics, Cornell University

R. P. Gosselin, Professor of Mathematics, University of Connecticut

E. Hewitt, Professor of Mathematics, University of Washington

T. Hida, Professor of Mathematics, Nagoya University

F. Holland, Lecturer in Mathematics, University College, Cork, Ireland

S. Koizumi, Professor of Mathematics, Keio University, Yokohama

P. J. Koosis, Professor of Mathematics, University of California at Los Angeles

J. Korevaar, Professor of Mathematics, University of Amsterdam

K.-S. Lau, Assistant Professor of Mathematics, University of Pittsburgh

S. Mandelbrojt, Honorary Professor of Mathematics, College de France, Paris

P. Masani, University Professor of Mathematics, University of Pittsburgh

J. W. Neuberger, Professor of Mathematics, North Texas University

H. R. Pitt, Former Vice-Chancellor, Reading University, England

J. I. Richards, Professor of Mathematics, University of Minnesota

W. Rudin, Vilas Research Professor of Mathematics, University of Wisconsin

L. Schwartz, Professor of Mathematics, Ecole Polytechnique, Paris

G. Strang, Professor of Mathematics, Massachusetts Institute of Technology

J. L. Walsh, *Perkins Professor of Mathematics, Harvard University

D. V. Widder, Emeritus Professor of Mathematics, Harvard University

---

*Deceased.

Norbert Wiener (1894–1964). Photograph taken around 1935.

Contents

(Bracketed numbers are from the Bibliography.)

ID
Generalized Harmonic Analysis and Tauberian Theory

IE
Classical Harmonic and Complex Analysis

Orthogonal developments, quasi-analyticity, gap theorems, and Fourier transforms in the complex domain

Preface

Norbert Wiener's profound contributions encompassed not only several areas of mathematics but mathematical philosophy, relativity and quantum mechanics, communication engineering, the physiology of the heart and the nervous system, brain wave encephalography, sensory prosthesis—the field now widely known as cybernetics. His scholarly work also included incisive social, educational, and literary essays. The object of these *Collected Works* is not only the reproduction of *all* his scholarly publications other than books but the placement of the overwhelming bulk of these in the context of present-day research by means of commentaries, tracing both their genesis and aftermath, written by eminent contemporary scholars.

The papers are organized by coherence of subject matter rather than by chronology, so that Wiener's intellectual evolution may stand out. Volume I included the three important categories of papers representative of his early work and its pursuance in later years. Volume II deals with generalized and classical harmonic analysis and Tauberian theory. Its organization is described in the section Introduction and Acknowledgments (p. 3). Volume III will cover the Hopf-Wiener equation, prediction and filtering, as well as relativity and quantum mechanics. Finally, Volume IV will include his philosophical, cybernetical, and sociological papers, commentaries, and reviews.

This volume, like the previous one and the ones to follow, owes a great deal to a great many people. I have thanked the commentators for this volume for their very valuable contributions in appropriate places in the section Introduction and Acknowledgments. But it remains to mention several other individuals and institutions without whose help this publication would have been severely handicapped.

The Indiana University Research Foundation provided grants-in-aid of publication during 1970–1972, and I wish to thank them very much, especially its then Director Harry G. Day for encouragement. The rather complicated secretarial work involved in the organization of these volumes has been carried out over the years by Theresa Klassen and Rosemarie Dawson at Indiana University, and by Susan Black, Mildred Kosanovich, and Alice Chapman at the University of Pittsburgh. I am most grateful for their devoted services. Finally, I thank Mrs. Margaret Wiener for providing me with photographs of Professor Wiener.

P. Masani
Pittsburgh, Pennsylvania
April 1979

Norbert Wiener: Collected Works
Volume II

The Classification of Wiener's Papers

Introduction and Acknowledgments

P. Masani

In these Works we have classified Wiener's publications into six broad categories labeled I, . . . , VI, and have divided the larger categories into subcategories labeled A, B, C, ... This classification has been motivated by practical editorial considerations, but we hope that it reflects to some extent the natural stages in Wiener's intellectual evolution.

The publications are indexed by the official year of their appearance. The internal ordering of the publications appearing in a given year is not chronological but according to the categories mentioned in the last paragraph: a run down the list a, b, c, . . . for a particular year entails a run down the list of categories IA, IB, . . . , usually with omissions and repetitions, as will be apparent from a glance at the right-hand columns in the bibliography (pp. 9–28). For instance, [33d] means "the *d*th paper, according to category, which appeared in a journal marked 1933."

The objective in these collected works is not only the reproduction of all papers in categories I through V, but the placement of the overwhelming bulk of these papers in the context of modern research. This is done by means of commentaries written by contemporary scholars on individual papers or on batches of related ones. The commentaries, long or short, review the papers and trace their genesis, as well as describe their impact on future research. They form a very valuable part of these Works, and to all commentators go our very sincere thanks for the time and energy they have so generously dedicated.

In our ordering of the reprints of papers and the accompanying comments we have eschewed chronology in favor of coherence of subject matter and conceptual development. We proceed to explain the specific groupings made and to record gratefully the services rendered by individual commentators.

Wiener's initial publications (1913–1922 approx.) are in the realm of mathematical philosophy and foundations, and have been categorized IA in our classification. The next stage of his research (1920–1925 approx.) pertains primarily to the Brownian motion and potential theory, papers in which areas we have categorized as IC and IB. All papers in the categories IA, IB, and IC, including those appearing after 1925, have been published along with comments in Volume I of these collected works, and we have there expressed our deep gratitude to the commentators involved.

The present volume covers the next stage of Wiener's work begun in the middle twenties, which has to do with his remarkable synthesis of stochastic ideas with generalized harmonic analysis and Tauberian theory, and also with his work of a more classical analytic nature done during the thirties.

Prompted by his engineering colleagues at MIT, Wiener became interested in the harmonic analysis of the current at a point along a telephone line. Assuming that the telephone conversation lasts forever, the current f, as a function of time, is neither periodic nor in L_2, for "it must be able to carry everything from a groan to a squeak" [56g], p. 75. The physicist A. Schuster, who was interested in the same type of problem in optics, had formulated a concept of correlation. By recasting this concept Wiener was able to associate with f his now famous (positive-definite) autocorrelation function ϕ and its (monotone increasing) spectral distribution S. It seems reasonable on physical grounds that for the telephone current and other chaotic signals f (e.g. sunlight, which is due to the random emission of billions of frequencies by solar atoms), S should be absolutely continuous. But it was difficult to define such an f mathematically. Wiener proved that when a time-invariant linear filter is subjected to a Brownian motion input, its output is almost invariably such an f. Today, after the Birkhoff Ergodic Theorem, the proof is straight forward. But Wiener had to resort to very hard analysis, in which a certain equality was crucial. This equality was "Tauberian" in nature as he learned from A. E. Ingham in 1926. In this way Wiener was able to open up a very interesting connection between (i) the Fourier analysis of haphazard signals, (ii) the Brownian motion, (iii) Tauberian theory. His work in this direction has revolutionized both the theory of stochastic processes and Tauberian theory, and has had a profound impact on communication theory. We have placed the papers in this area in category ID, entitled "Generalized Harmonic Analysis and Tauberian Theory."

The papers on Generalized Harmonic Analysis fall into two groups: the earlier ones in which Brownian motion is not considered, and the later ones where it is. We are indebted to Dr. V. Beneš for comments on the first group comprising [23f], [24f], and [25c, d] and to Professor S. Koizumi for comments on the second group comprising [26a, b] and [28a] and on the climactic memoir [30a].† We must also record here my own comments on [30a], and on the paper [27a] in which Wiener partially extended the theory (from the reals) to the group of integers.

Wiener's engineering colleagues pressed him for a satisfactory treatment of the Heaviside calculus. This led to the paper [26c] on the operational calculus, which foreshadows the modern theory of distributions by its extensive use of generalized Fourier transformations and certain other techniques, and also introduces new ideas such as those of causal and anti-causal operators. We are

† Additional comments by Sir Dennis Gabor on sections of [30a] dealing with optics will appear in Volume III of these collected works.

grateful to Professor L. Schwartz for his interesting comments on this paper. As he points out, Wiener's definition of generalized solution in [26c] supersedes several later definitions by others. Some of the ideas in [26c] are exposed in more readable terms in [29c], an appendix to an engineering treatise by Vanniver Bush. Wiener also treated difference equations from the standpoint of generalized harmonic analysis in the paper [25c], for comments on which we thank Professor G. Strang.

The almost periodic functions form an important subclass of those amenable to generalized harmonic analysis cf. [30a, Ch. V]. Wiener had written the papers [27b] and [29a] concerning these functions. We are thankful to Dr. F. Holland and Dr. C. Berenstein for comments on these papers and for pointing out an error in a proof in [27b]. (It is rectifiable as Dr. Holland indicates, and the results are correct.) The paper [29a] revolves around the idea of a "properly unitary" transformation. This rather strange idea is now beginning to bear fruit as Professor K.-S. Lau shows in his very interesting comments on [29a], for which we are again grateful.

The papers in ID mentioned above belong to Generalized Harmonic Analysis. Among the papers remaining in ID are those on Tauberian theory, [27c], [28b], [29b] culminating in the memoir [32a, b], with [33b] following in its wake. This work has had profound ramifications in different directions, and we felt the need to seek the assistance of more than one commentator. Our deep gratitude goes to both Dr. H. R. Pitt and Professor J. Korevaar for their comprehensive but differing commentaries on these papers. In 1932 Haar had already begun his studies of locally compact Abelian groups. Although the duality theory was not fully understood, Wiener and Paley sensed the possibilities of extending Wiener's ideas to such groups. Their rather incomplete results appear in the abstract [32f] and the paper [33e], for comments on which we thank Professor E. Hewitt.

An important offshoot of the work on generalized harmonic analysis and Tauberian theory are the seminal joint papers [38b, c] and [39c, d] with H. R. Pitt; A. Wintner and R. H. Cameron on the Fourier-Stieltjes transforms of measures. For comments on these we are grateful to Professor W. Rudin, especially for his revelation of a rather large lacuna in [39c] subsequently eradicated by R. Salem. For very valuable further comments on [38b, c] our thanks go again to Professor E. Hewitt, and for those on [39d] to its co-author Professor R. H. Cameron and his colleague Dr. I. Richards. We also thank Professor Rudin for commenting on the paper [33d] with R. C. Young on the translates of a bounded measure.

Wiener also applied his Tauberian theory in more classical directions. The papers [37a] and [38d] with W. T. Martin deal with Taylor series of functions of smooth growth, for comments on which we are thankful to Professor W. Fuchs. But his more concerted attempt in the classical direction was in the realm of number theory. From conversations with Professor Robert Schmidt in 1927, he had realized that his work on Tauberian theory had a bearing on the Prime Number Theorem and related questions in analytic number theory. His work in this direction, centering around the Wiener-Ikehara Theorem, ([32a], p. 44, Thm. XVII), appears in the papers [39e], [50a], [56a] and [57b, c]. We are very grateful to Professors P. Bateman and H. G. Diamond for their extremely careful and critical survey of this material, in which several lacunae appearing in the papers are cleared up.

In our classification we have separated the categories ID and IE. In doing so we have tried to draw a line between generalized analysis on the one hand and classical and complex analysis on the other. Papers on topics such as gap and density theorems and quasi-analytic functions have been placed in IE. The dividing line between ID and IE is of course rather hazy.

The papers in the category IE present some editorial difficulties. In the first place, the cohesiveness marking the earlier categories is reduced by the occurrence of several topics only tenuously related either logically or chronologically. This makes it hard to order the papers. A case in point: should the seven notes on Fourier transforms appearing in 1933 precede or follow the papers on gap theory commencing in 1927 and running into 1936? We have settled such questions as best we could. Next, several important contributions of Wiener in the more classical realms of analysis appear in his renowned book with R. A. E. C. Paley [34d], which is not reproduced in these collected works and is therefore outside the purview of direct comment. But the book lurks in the background of several papers under consideration, either as precursor or beneficiary. Consequently, there is an unfortunate discontinuity in our commentary. Luckily, a more complete but brief comment, embracing the book [34d], is easily accessible in the excellent article by Professor J. P. Kahane: "Norbert Wiener et l'analyse de Fourier," *Bull. Amer. Math. Soc.* 72(1966), no. 1, part II, 42–47.

Wiener's venture in classical analysis began with his joint paper [22d] on the equivalence of orthogonal developments, for a comment on which our thanks go to its co-author, the late Professor J. L. Walsh. Next the subject of quasi-analyticity engaged Wiener's attention. His first paper [25e] was written before the appearance of Carleman's treatise on the subject (1926), and dealt with a class of functions given by cosine series in the spirit of de la Vallee Poussin. The advent of Carleman's book brought the subject closer to Wiener's own point of view of generalized harmonic analysis. His more definitive researches on quasi-analyticity appear in Note I in [33e], in the book [34d,

Ch. I], with Paley, and the later paper [47a] with S. Mandelbrojt. We are grateful to Professor J. W. Neuberger for his comments on both [25e] and [33e].

Wiener's attention then shifted to gap theorems. In this area fall the papers [27e], [34b], [35a], [36b–e], and to an extent the paper [36a]. We are extremely grateful to Professor P. J. Koosis for organizing these widely scattered papers, for his very thorough survey of them in which several obscurities are cleared up, and for tracing their influence on the continuing activity in this area. Of the papers just mentioned, [36b] and [36c] are joint works with S. Mandelbrojt as is the much later paper [47a]. We would like to thank Professor Mandelbrojt for his specific comments on these three papers.

Turning to the miscellaneous papers falling in category IE, Hermite functions and polynomials have played a key role in certain aspects of Wiener's work such as his theory of homogeneous chaos. In the early paper [29d] Wiener relates these functions to Fourier developments of fractional order. We thank Professor T. Hida for his comments on this work. Next we should mention the seven notes [33e] on Fourier transforms written with R. A. E. C. Paley during their very important but short lived collaboration, 1932–1933. The main thrust is the use of the Fourier integral in the complex domain. These researches were however systematized and incorporated in the book [34d]. We have not therefore included any comments on these notes, apart from the first on quasi-analyticity, reviewed by Professor J. W. Neuberger, and the second on conjugate functions, reviewed by Professor R. P. Gosselin. To both of them go our thanks. In [38f] Wiener and D. V. Widder discuss inversion of the Laplace transform of a function in $L_2\ (0, \infty)$, for a brief comment on which we thank Professor Widder. Finally, there is the interesting paper [42a] with G. Pólya on the oscillation of the derivatives of a periodic function, for comments on which thanks are due to Professor R. Askey.

During the period in which the papers we have been reporting on were written, occurred the historic collaboration between Wiener and Eberhard Hopf, which ushered in the Hopf-Wiener integral equation (1931) and marked a shift in Wiener's activity. This work and its ramifications in prediction and filtering will be covered in Volume III of these collected works. From about 1926 Wiener, in collaboration with Professor Max Born (a visitor to MIT) and his colleague Professor D. J. Struik, began researches in quantum mechanics and relativity theory. These too will be the subject of Volume III.

Bibliography of Norbert Wiener

(Entries preceded by an asterisk are reprinted in this volume)

	Title	Category
[13a]	*On a method of rearranging the positive integers in a series of ordinal numbers greater than that of any given fundamental sequence of omegas*, Messenger of Math. 43(1913), 97–105.	IA
[14a]	*A simplification of the logic of relations*, Proc. Cambridge Philos. Soc. 17(1914), 387–390.	IA
[14b]	*A contribution to the theory of relative position*, Proc. Cambridge Philos. Soc. 17(1914), 441–449.	IA
[14c]	*The highest good*, J. Phil. Psych. and Sci. Method 11(1914), 512–520.	IIA
[14d]	*Relativism*, J. Phil. Psych. and Sci. Method 11(1914), 561–577.	IIA
[15a]	*Studies in synthetic logic*, Proc. Cambridge Philos. Soc. 18(1915), 14–28.	IA
[15b]	*Is mathematical certainty absolute?*, J. Phil. Psych. and Sci. Method 12(1915), 568–574.	IA
[16a]	*Mr. Lewis and implication*, J. Phil. Psych. and Sci. Method 13(1916), 656–662.	IA
[16b]	*The shortest line dividing an area in a given ratio*, Proc. Cambridge Philos. Soc. 18(1916), 56–58.	I,I
[16c]	Review of Cassius J. Keyser, *Science and Region: the Rational and the Superrational*, J. Phil. Psych. and Sci. Method 13(1916), 273–277.	IVA
[16d]	Review of A. A. Robb, *A Theory of Time and Space*, J. Phil. Psych. and Sci. Method 13(1916), 611–613.	IVA
[17a]	*Certain formal invariances in Boolean algebras*, Trans. Amer. Math. Soc. 18(1917), 65–72.	IA
[17b]	Review of C. J. Keyser, *The Human Worth of Rigorous Thinking*, J. Phil. Psych. and Sci. Method 14(1917), 356–361.	IVA

[18a] Review of Edward V. Huntington, *The Continuum and Other Types of Serial Order*, J. Phil. Psych. and Sci. Method 15(1918), 78–80. IVA

[20a] *Bilinear operations generating all operations rational in a domain Ω*, Ann. of Math. 21(1920), 157–165. IA

[20b] *A set of postulates for fields*, Trans. Amer. Math. Soc. 21(1920), 237–246. IA

[20c] *Certain iterative characteristics of bilinear operations*, Bull. Amer. Math. Soc. 27(1920), 6–10. IA

[20d] *Certain iterative properties of bilinear operations*, G. R. Strasbourg Math. Congress, 1920, 176–178. IA

[20e] *On the theory of sets of points in terms of continuous transformations*, G. R. Strasbourg Math. Congress, 1920, 312–315. IA

[20f] *The mean of a functional of arbitrary elements*, Ann. of Math. (2) 22(1920), 66–72. IC

[20g] Review of C. I. Lewis, *A Survey of Symbolic Logic*, J. Phil. Psych. and Sci. Method 17(1920), 78–79. IVA

[21a] *A new theory of measurement: A study in the logic of mathematics*, Proc. London Math. Soc. 19(1921), 181–205. IA

[21b] *The isomorphisms of complex algebra*, Bull. Amer. Math. Soc. 27(1921), 443–445. IA

[21c] *The average of an analytic functional*, Proc. Nat. Acad. Sci. U.S.A. 7(1921), 253–260. IC

[21d] *The average of an analytic functional and the Brownian movement*, Proc. Nat. Acad. Sci. U.S.A. 7(1921), 294–298. IC

[21e] *A new vector method in integral equations* (with F. L. Hitchcock), J. Math. and Phys. 1(1921), 1–20. I,I

[22a] *The relation of space and geometry to experience*, Monist 32(1922), 12–60, 200–247, 364–394. IA

[22b] *The group of the linear continuum*, Proc. London Math. Soc. 20(1922), 329–346. IA

[22c] *Limit in terms of continuous transformation*, Bull. Soc. Math. France 50(1922), 119–134. IA

*[22d] *The equivalence of expansions in terms of orthogonal functions* (with J. L. Walsh), J. Math. and Phys. 1(1922), 103–122. IE

[22e] *A new type of integral expansion*, J. Math. and Phys. 1(1922), 167–176. I,I

[23a] *On the nature of mathematical thinking*, Austral. J. Psych. and Phil. 1(1923), 268–272. IA

[23b] *Nets and the Dirichlet problem* (with H. B. Phillips), J. Math. and Phys. 2(1923), 105–124. (Reprinted in [64f].) IB

[23c] *Discontinuous boundary conditions and the Dirichlet problem*, Trans. Amer. Math. Soc. 25(1923), 307–314. IB

[23d] *Differential-space*, J. Math. and Phys. 2(1923), 131–174. (Reprinted in [64f].) IC

[23e] *O szeregach* $\sum_1^\infty (\pm 1/n)$*–Note on the series* $\sum_1^\infty (\pm 1/n)$, Bull. Acad. Polon. Ser. A, 13(1923), 87–90. IC

*[23f] *Note on a new type of summability*, Amer. J. Math. 45(1923), 83–86. ID

[23g] *Note on a paper of M. Banach*, Fund. Math. 4(1923), 136–143. I,I

[24a] *Certain notions in potential theory*, J. Math. and Phys. 3(1924), 24–51. IB

[24b] *The Dirichlet problem*, J. Math. and Phys. 3(1924), 127–146. (Reprinted in [64f].) IB

[24c] *Une condition nécessaire et suffisante de possibilité pour le problème de Dirichlet*, C. R. Acad. Sci. Paris 178(1924), 1050–1054. IB

[24d] *The average value of a functional*, Proc. London Math. Soc. 22(1924), 454–467. IC

[24e] *Un problème de probabilités denombrables*, Bull. Soc. Math. France 11(1924), 569–578. IC

*[24f] *The quadratic variation of a function and its Fourier coefficients*, J. Math. and Phys. 3(1924), 72–94. ID

[24g] Review of four books on space, Rudolf Carnap's *Der Raum: Ein Beitrag zur Wissenschaftslehre*, E. Study's *Mathematik und Physik: eine erkenntnistheoretische Untersuchung* and *Die realistische* IVA

Weltansicht und die Lehre vom Raume: zweite Auglage; erster Teil, and Hermann Weyl's *Mathematische Analyse des Raum-problems. Vorlesungen gehalten in Barcelona und Madrid*, Bull. Amer. Math. Soc. 30(1924), 258–262.

[24h] Review of E. Study, *Denken und Darstellung: Logik und Werte; Dingliches und Menschliches in Mathematik und Naturwissenschaften*, Bull. Amer. Math. Soc. 30(1924), 277. IVA

[24i] *In memory of Joseph Lipka*, J. Math. and Phys. 3(1924), 63–65. IVB

[25a] *Note on a paper of O. Perron*, J. Math. and Phys. 4(1925), 21–32. IB

*[25b] *The solution of a difference equation by trigonometrical integrals*, J. Math. and Phys. 4(1925), 153–163. ID

*[25c] *On the representation of functions by trigonometrical integrals*, Math. Z. 24(1925), 575–616. ID

*[25d] *Verallgemeinerte trigonometrische Entwicklungen*, Göttingen Nachr. (1925), 151–158. ID

*[25e] *Note on quasi-analytic functions*, J. Math. and Phys. 4(1925), 193–199. IE

[25f] *A contribution to the theory of interpolation*, Ann. of Math. (2) 26(1965), 212–216. I,I

*[26a] *The harmonic analysis of irregular motion*, J. Math and Phys. 5(1926), 99–121. ID

*[26b] *The harmonic analysis of irregular motion* (Second Paper), J. Math. and Phys. 5(1926), 158–189. ID

*[26c] *The operational calculus*, Math. Ann. 95(1926), 557–584. ID

[26d] *A new formulation of the laws of quantization of periodic and aperiodic phenomena* (with M. Born), J. Math and Phys. 5(1926), 84–98. IH

[26e] *Eine neue Formulierung der Quantengesetze für periodische und nicht periodische Vorgänge* (with M. Born), Z. Physik 36(1926), 174–187. IH

[26f] *Analytic approximations to topological transformations* (with P. Franklin), Trans. Amer. Math. Soc. 28(1926), 762–785. I,I

*[27a] *The spectrum of an array and its application to the study of the transla-* ID

tion properties of a simple class of arithmetical functions, Part I, J. Math. and Phys. 6(1927), 145–157. (Part II: *On the translation of a simple class of arithmetical functions*, by K. Mahler, *ibid.*, pp. 158–163.)

*[27b] *A new definition of almost periodic functions*, Ann. of Math. (2) 28(1927), 365–367. ID

*[27c] *On a theorem of Bochner and Hardy*, J. London Math. Soc. 2(1927), 118–123. ID

*[27d] *Une méthode nouvelle pour la démonstration des théorèmes de M. Tauber*, C. R. Acad. Sci. Paris 184(1927), 793–795. ID

*[27e] *On the closure of certain assemblages of trigonometrical functions*, Proc. Nat. Acad. Sci., U.S.A. 13(1927), 27–29. IE

[27f] *Quantum theory and gravitational relativity* (with D. J. Struik), Nature 119(1927), 853–854. IH

[27g] *A relativistic theory of quanta* (with D. J. Struik), J. Math and Phys. 7(1927), 1–23. IH

[27h] *Sur la théorie relativiste des quanta* (with D. J. Struik), C. R. Acad. Sci. Paris 185(1927), 42–44. IH

[27i] *Sur la théorie relativiste des quanta* (Note), C. R. Acad. Sci. Paris 185(1927), 184–185. IH

[27j] *Laplacians and continuous linear functionals*, Acta Sci. Math. (Szeged) 3(1927), 7–16. I,I

[27k] *Une généralisation des fonctions à variation bornée*, C. R. Acad. Sci. Paris 185(1927), 65–67. I,I

*[28a] *The spectrum of an arbitrary function*, Proc. London Math. Soc. (2) 27(1928), 483–496. ID

*[28b] *A new method in Tauberian theorems*, J. Math. and Phys. 7(1928), 161–184. ID

[28c] *The fifth dimension in relativistic quantum theory* (with D. J. Struik), Proc. Nat. Acad. Sci. U.S.A. 14(1928), 262–268. IH

[28d] *Coherency matrices and quantum theory*, J. Math. and Phys. 7(1928), 109–125. IH

*[29a] *Harmonic analysis and group theory*, J. Math. and Phys. 8(1929), 148–154. ID

*[29b] *A type of Tauberian theorem applying to Fourier series*, Proc. London Math. Soc. (20) 30(1929), 1–8. ID

*[29c] *Fourier analysis and asymptotic series*. Appendix to V. Bush, *Operational Circuit Analysis*, New York, John Wiley, 1929, 366–379. ID

*[29d] *Hermitian polynomials and Fourier analysis*, J. Math. and Phys. 8(1929), 70–73. IE

[29e] *Harmonic analysis and the quantum theory*, J. Franklin Inst. 207(1929), 525–534. IH

[29f] *On the spherically symmetrical statical field in Einstein's unified theory of electricity and gravitation* (with M. S. Vallarta), Proc. Nat. Acad. Sci. U.S.A. 15(1929), 353–356. IH

[29g] *On the spherically symmetrical statical field in Einstein's unified theory: A correction* (with M. S. Vallarta), Proc. Nat. Acad. Sci. U.S.A. 15(1929), 802–804. IH

[29h] *Mathematics and art (Fundamental identities in the emotional aspects of each)*, Tech. Rev. 32(1929), 129–132, 160, 162. IIA

[29i] *Einsteiniana (Facts and fancies about Dr. Einstein's famous theory)*, Tech. Rev. 32(1929), 403–404. IIIE

[29j] *Murder and mathematics*, Tech. Rev. 32(1929), 271–272. IVA

*[30a] *Generalized harmonic analysis*, Acta Math. 55(1930), 117–258. (Reprinted in [64f] and [66b].) ID

[30b] Review of A. Eddington's *Science and the Unseen World*, Tech. Rev. 33(1930), 150. IVA

*[31a] *Über eine Klasse singulärer Integralgleichungen* (with E. Hopf), Sitzber. Preuss. Akad. Wiss. Berlin, Kl. Math. Phys. Tech., 1931, pp. 696–706. (Reprinted in [64f].) IF

[31b] *A new deduction of the Gaussian distribution*, J. Math. and Phys. 10(1931), 284–288. I,I

[31c] *Reports from Cambridge—1931*, Tech. Rev. 34(1931), 82–83, 131, 218, 220. IIIE

*[32a] *Tauberian theorems*, Ann. of Math. 33(1932), 1–100. (Reprinted in [64f] and [66b].) ID

*[32b] *A note on Tauberian theorems*, Ann. of Math. 33(1932), 787. ID

[32c] *Back to Leibniz! (Physics reoccupies an abandoned position)*, Tech. Rev. 34(1932), 201–203, 222, 224. IIA

[32d] *Reports from Cambridge—1932*, Tech. Rev. 34(1932), 62, 74. IIIE

[32e] Review of A. S. Besicovitch, *Almost Periodic Functions*, Math. Gaz. 16(1932), 275–277. IVA

*[32f] *Analytic properties of the characters of infinite Abelian groups* (with R. E. A. C. Paley), Abstract, Int'l. Math. Congr., Zürich, 1932, 95. V

[33a] *Notes on random functions* (with R. E. A. C. Paley and A. Zygmund), Math. Z. 37(1933), 647–668. IC

*[33b] *A one-sided Tauberian theorem*, Math. Z. 36(1933), 787–789. ID

*[33c] *Characters of Abelian groups* (with R. E. A. C. Paley), Proc. Nat. Acad. Sci. U.S.A. 19(1933), 253–257. ID

*[33d] *The total variation of* $g(x+h)-g(x)$ (with R. C. Young), Trans. Amer. Math. Soc. 35(1933), 327–340. ID

*[33e] *Notes on the theory and application of Fourier transforms* (with R. E. A. C. Paley) I, II, Trans. Amer. Math. Soc. 35 (1933), 348–355; III, IV, V, VI, VII, Trans. Amer. Math. Soc. 35(1933), 761–791. IE

[33f] *Putting matter to work (The search for cheapter power)*, Tech. Rev. 35(1933), 47–49, 70, 72. IIIA

[33g] Review of Harald Bohr, *Fastperiodische Funktionen*, Math. Gaz. 17(1933), 54. IVA

[33h] *R. E. A. C. Paley—In Memoriam, Jan. 7, 1907–Apr. 7, 1933*, Bull. Amer. Math. Soc. 39(1933), 476. IVB

[33i] *The Fourier Integral and Certain of Its Applications*, Cambridge University Press, New York, 1933; reprint, Dover, New York, 1959; review by E. C. Titchmarsh in Math. Gaz. 17(1933), 129. VI

[34a] *Random functions*, J. Math. and Phys. 14(1934), 17–23. IC

*[34b] *A class of gap theorems*, Ann. Scuola Norm. Sup. Pisa, E(1934–1936), 1–6. IE

[34c] *Quantum mechanics, Haldane, and Leibniz*, Philos. Sci. 1(1934), 479–482. IIA

[34d] *Fourier Transforms in the Complex Domain* (with R. E. A. C. Paley), Amer. Math. Soc. Colloq. Publ. 19, Amer. Math. Soc., Providence, R. I., 1934. VI

*[35a] *Fabry's gap theorem*, Sci. Repts. of Nat'l. Tsing Hua Univ., Ser. A, 3(1935), 239–245. IE

[35b] *Limitations of science (The holiday fallacy and a response to the suggestion that scientists become sociologists)*, Tech. Rev. 37(1935), 255–256, 268, 270, 272. IIIA

[35c] *The student agitator (Is he accepting radicalism as an opiate?)* (with Carl Bridenbaugh), Tech. Rev. 37(1935), 310–312, 344, 346. IIIC

[35d] *Mathematics in American secondary schools*, J. Math. Assoc. Japan for Secondary Education (Tokyo) 17(1935), 1–5. IIIC

[35e] *The closure of Bessel functions*, Abstract 66, Bull. Amer. Math. Soc. 41(1935), 35. V

*[36a] *A theorem of Carleman*, Sci. Repts. of Nat'l. Tsing Hua Univ., Ser. A, 3(1936), 291–298. IE

*[36b] *Sur les séries de Fourier lacunaires. Théorèmes directs* (with S. Mandelbrojt), C. R. Acad. Sci. Paris 203(1936), 34–36. IE

*[36c] *Séries de Fourier lacunaires. Théorèmes inverses* (with S. Mandelbrojt), C. R. Acad. Sci. Paris 203(1936), 233–234. IE

*[36d] *Gap theorems*, C. R. de Congr. Int'l. des Math., 1936, 284–296. IE

*[36e] *A Tauberian gap theorem of Hardy and Littlewood*, J. Chinese Math. Soc. 1(1936), 15–22. IE

[36f] *Notes on the Kron theory of tensors in electrical machinery*, J. Electr. Engrg., China 7(1936), 277–291. I,I

[36g] *The role of the observer*, Philos. Sci. 3(1936), 307–319. IIA

*[37a] *Taylor's series of entire functions of smooth growth* (with W. T. Martin), Duke Math. J. 3(1937), 213–223. ID

[37b] *Random Waring's theorems*, Abstract (with N. Levinson), Science 85(1937), 439. V

[38a] *The homogeneous chaos*, Amer. J. Math. 60(1938), 897–936. (Reprinted in [64f].) IC

*[38b] *On absolutely convergent Fourier-Stieltjes transforms* (with H. R. Pitt), Duke Math. J. 4(1938), 420–440. ID

*[38c] *Fourier-Stieltjes transforms and singular infinite convolutions* (with A. Wintner), Amer. J. Math. 60(1938), 513–522. ID

*[38d] *Taylor's series of functions of smooth growth in the unit circle* (with W. T. Martin), Duke Math. J. 4(1938), 384–392. ID

*[38e] *The historical background of harmonic analysis*, Amer. Math. Soc. Semicentennial Publications Vol. II, Semicentennial Addresses, Amer. Math. Soc., Providence, R. I., 1938, 513–522. ID

*[38f] *Remarks on the classical inversion formula for the Laplace integral* (with D. V. Widder), Bull. Amer. Math. Soc. 44 1938), 573–575. IE

[38g] *The decline of cookbook engineering*, Tech. Rev. 40(1938), 23. IIIE

[38h] Review of L. Hogben, *Science for the Citizen*, Tech. Rev. 40(1938), 66–67. IVA

[39a] *The ergodic theorem*, Duke Math. J. 5(1939), 1–18. (Reprinted in [64f].) IC

[39b] *The use of statistical theory in the study of turbulence*, Proc. 5th Int'l. Congr. of Applied Mechanics, Sept. 12–16, 1938, Wiley, New York, 1939, 356–358. IC

*[39c] *On singular distributions* (with A. Wintner), J. Math. and Phys. 17(1939), 233–246. ID

*[39d] *Convergence properties of analytic functions of Fourier-Stieltjes transforms* (with R. H. Cameron), Trans. Amer. Math. Soc. 46(1939), 97–109; Math. Rev. 1(1940), 13; rev. 400.

*[39e] *A generalization of Ikehara's theorem* (with H. R. Pitt), J. Math. and Phys. 17(1939), 247–258. ID

[39f] Review of Roger Burlingame, *March of the Iron Men*, Tech. Rev. 41(1939), 115. IVA

[39g] Review of W. George, *The Scientist in Action*, Tech. Rev. 41(1939), 202. IVA

[39h] *A new method in statistical mechanics*, Abstract 133 (with B. McMillan), Bull. Amer. Math. Soc. 45(1939), 234; Science 90(1939), 410–411. V

[40a] Review of M. Fukamiya, *On dominated ergodic theorems in* $L_p(p \geqslant 1)$, Math. Rev. 1(1940), 148. IVA

[40b] Review of M. Fukamiya, *The Lipschitz condition of random function*, Math Rev. 1(1940), 149. IVA

[40c] Review of Th. De Donder, *L'énergétique déduite de la méchanique statistique générale*, Math. Rev. 1(1940), 192. IVA

[40d] *A canonical series for symmetric functions in statistical mechanics*, Abstract 133, Bull. Amer. Math. Soc. 46(1940), 57. V

[41a] *Harmonic analysis and ergodic theory* (with A. Wintner), Amer. J. Math. 63(1941), 415–426; Math. Rev. 2(1941), 319. IC

[41b] *On the ergodic dynamics of almost periodic systems* (with A. Wintner), Amer. J. Math. 63(1941), 794–824; Math. Rev. 4(1943), 15. IC

*[42a] *On the oscillation of the derivatives of a periodic function* (with G. Pólya), Trans. Amer. Math. Soc. 52(1942), 249–256. IE

[43a] *The discrete chaos* (with A. Wintner), Amer. J. Math. 65(1943), 279–298; Math. Rev. 4(1943), 220. IC

[43b] *Behavior, purpose, and teleology* (with A. Rosenblueth and J. Bigelow), Philos. Sci. 10(1943), 18–24. IIA

[45a] *La teoria de la estrapolacion estadistica*, Bol. Soc. Mat. Mexicana 2(1945), 37–45; Math. Rev. 7(1946), 461. IG

[45b] *The role of models in science* (with A. Rosenblueth), Philos. Sci. 12(1945), 316–322. IIA

[46a] *A generalization of the Wiener-Hopf integral equation* (with A. E. Heins), Proc. Nat. Acad. Sci. U.S.A. 32(1946), 98–101; Math. Rev. 8(1947), 29. IF

[46b] *The mathematical formulation of the problem of conduction of impulses in a network of connected excitable elements, specifically in cardiac* IIB

muscle (with A. Rosenblueth), Arch. Inst. Cardiol. Méxicana 16(1946), 205–265; Bol. Soc. Mat. Mexicana 2(1945), 37–42; Math. Rev.

*[47a] *Sur les fonctions indéfiniment dérivables sur une demidroite* (with S. Mandelbrojt), C. R. Acad. Sci. Paris 225(1947), 978–980; Math. Rev. 9(1948), 230. IE

[47b] *A scientist rebels*, Atlantic Monthly 179(1946), 46; Bill. Atomic Scientists 3(1947), 31. IIIB

[48a] *Time, communication and the nervous system*, Ann. New York Acad. Sci. 50(1948), 197–220; Math. Rev. 10(1949), 133. IIB

[48b] *Cybernetics*, Scientific American 179(1948), 14–18. IIB

[48c] *An account of the spike potential of axons* (with A. Rosenblueth, W. Pitts, J. Garcia Ramos, and the assistance of F. Webster), J. of Cellular and Comparative Physiol. 32(1948), 275–318. IIB

[48d] *A rebellious scientist after two years*, Bull. Atomic Scientists 4(1948), 338–339. IIIB

[48e] Review of L. Infeld, *Whom the Gods Love. The Story of Evariste Galois*, Scripta Math. 14(1948), 273–274. IVA

[48f] *Cybernetics, or Control and Communication in the Animal and the Machine*, Actualités Sci. Ind., no. 1053; Hermann et Cie., Paris; The MIT Press, Cambridge, Mass., and Wiley, New York, 1948; Math. Rev. 9(1948), 598. Partly reprinted as Rigidity and learning: ants and men, in *Classics in Biology (A Course of Selected Reading by Authorities)*, Philosophical Library, New York, 1960, pp. 205–213. VI

[49a] *Sur la théorie de la prévision statistique et du filtrage des ondes*, Analyse Harmonique, Colloques Internationaux du CNRS, No. 15, pp. 67–74. Centre National de la Recherche Scientifique, Paris, 1949; Math. Rev. 11(1950), 376. IG

[49b] *A statistical analysis of synaptic excitation* (with A. Rosenblueth, W. Pitts, and J. Garcia Ramos), J. of Cellular and Comparative Physiol. 34(1949), 173–205. IIB

[49c] *A new concept of communication engineering*, Electronics 22(1949), 74–77. IIB

[49d] *Sound communication with the deaf*, Philos. Sci. 16(1949), 260–262. IIB

[49e] *Some problems in sensory prosynthesis* (with J. Wiesner and L. Levine), Science 110(1949), 512. IIB

[49f] *Obituary—Godfrey Harold Hardy, 1877–1947*, Bull. Amer. Math. Soc. 55(1949), 72–77. IVB

[49g] *Extrapolation, Interpolation, and Smoothing of Stationary Time Series with Engineering Applications*, The MIT Press, Cambridge, Mass.; Wiley, New York; Chapman & Hall, London, 1949; paperback edition with the title *Time Series*, The MIT Press, 1964; Math. Rev. 11(1950), 118. VI

*[50a] *Some prime-number consequences of the Ikehara theorem* (with L. Geller), Acta Sci. Math. (Szeged) 12(1950), 25–28, Leopoldo Fejer et Frederico Riesz LXX annos natis dedicatus, Pars B; Math. Rev. 11(1950), 644; Math. Rev. 12(1951), 1002. ID

[50b] *Comprehensive view of prediction theory*, Proceedings of the International Congress of Mathematicians, Cambridge, Mass., 1950, vol. 2, pp. 308–321; Amer. Math. Soc., Providence, R. I., 1952, Expository lecture; Math. Rev. 13(1952), 477. IG

[50c] *Some maxims for biologists and psychologists*, Dialectica 4(1950), 186–191. IIA

[50d] *Purposeful and non-purposeful behavior* (with A. Rosenblueth), Philos. Sci. 17(1950), 318–326. IIA

[50e] *Cybernetics*, Bull. Amer. Acad. Arts and Sci. 3(1950), 2–4. IIB

[50f] *Speech, language, and learning*, J. Acoust. Soc. Amer. 22(1950), 696–697. IIB

[50g] *Entropy and information*, Proc. Sympos. Appl. Math., vol. 2, Amer. Math. Soc., Providence, R. I., 1950, p. 89; Math. Rev. 11(1950), 305. IIB

[50h] *Too big for private enterprise*, Nation 170(1950), 496–497; IIIA

[50i] *Too damn close*, Atlantic 186(1950), 50–52. IIIA

[50j] *The Human Use of Human Beings*, Houghton Mifflin, Boston, 1950; paperback edition by Doubleday, Anchor, Garden VI

City, N.Y., 1954. Chapter 3 reprinted in *Classics in Biology (A Course of Selected Reading by Authorities)*, Philosophical Library, New York, 1960, pp. 205–213.

[50k] *The brain* (short story), Tech. Eng. News 31(1950), 14–15, 33–34, 44, 50. (reprinted in paperback anthology, *Cross-roads in Time*, ed. Groff Conklin, Doubleday, Garden City, N.Y., 1953.) VI

[51a] *Problems of sensory prosthesis*, Bull. Amer. Math. Soc. 57(1951), 27–35. (Reprinted in [64f].) IIB

[51b] *Homeostasis in the individual and society*, J. Franklin Inst. 251(1951), 65–68. (Reprinted in [64f].) IIB

[52a] *Cybernetics (Light and Maxwell's demon)*, Scientia (Italy) 87(1952), 233–235. IIB

[52b] *The miracle of the broom closet* (short story), Tech. Eng. News 33(1952), 18–19, 50. (Reprinted in the Magazine of Fantasy and Science Fiction, ed. Anthony Boucher, February, 1954, pp. 59–63.) VI

[53a] *Optics and the theory of stochastic processes*, J. Opt. Soc. Amer. 43(1953), 225–228; Math. Rev. 17(1956), 33. IG

[53b] *A new form for the statistical postulate of quantum mechanics* (with A. Siegel), Phys. Rev. 91(1953), 1551–1560; Math. Rev. 15(1954), 273. IH

[53c] *Distributions quantiques dans l'espace différentiel pour les fonctions d'ondes dépendant du spin* (with A. Siegel), C. R. Acad. Sci. Paris 237(1953), 1640–1642; Math. Rev. 15(1954), 490. IH

[53d] *Les machines à calculer et la forme (Gestalt)*, *Les machines à calculer et la pensee humaine*, Colloques Internationaux du Centre National de la Recherche Scientifique, Paris, 1953, pp. 461–463; Math. Rev. 16 (1955), 529. IIB

[53e] *The concept of homeostasis in medicine*, Transactions and Studies of the College of Physicians of Philadelphia (4) 20(1953), No. 3, 87–93. IIB

[53f] *Problems of organization*, Bull. Menninger Clinic 17(1953), 130–138. IIB

[53g] *The future of automatic machinery*, Mech. Engrg. 75 (1953), 130–132. IIB

[53h] *Ex-prodigy: My Childhood and Youth*, Simon and Schuster, New York, 1953; The MIT Press, Cambridge, Mass., 1965 (paperback edition also by The MIT Press); Math. Rev. 15(1954), 277. VI

[54a] *Men, machines, and the world about*, in *Medicine and Science*, New York Academy of Medicine and Science, ed. I. Galderston, International Universities Press, New York, 1954, pp. 13–28. IIB

[54b] *Conspiracy of conformists*, Nation 178(1954), 375. IIIE

[55a] *Nonlinear prediction and dynamics*, Proc. Third Berkeley Symposium on Mathematical Statistics and Probability, University of California Press, Berkeley, Calif., 1954/5, pp. 247–252; Math. Rev. 18(1957), 949. IG

[55b] *On the factorization of matrices*, Comment. Math. Helv. 29(1955), 97–111; Math. Rev. 16(1955), 921. IG

[55c] *The differential-space theory of quantum systems* (with A. Siegel), Nuovo Cimento (10) 2(1955), 982–1003, No. 4, Suppl. IH

[55d] *Thermodynamics of the message*, in *Neurochemistry*, ed. K. E. C. Elliott, Thomas, Springfield, 1955, pp. 844–849. IIB

[55e] *Time and organization*, Second Fawley Foundation Lecture, University of Southampton, 1955, pp. 1–16. IIB

*[56a] *On a local L^2-variant of Ikehara's theorem* (with A. Wintner), Rev. Math. Cuyana 2(1956), 53–59. ID

[56b] *The theory of prediction*, in *Modern Mathematics for the Engineer*, ed. E. F. Beckenbach, McGraw-Hill, New York, 1956, pp. 165–187. IG

[56c] "*Theory of Measurement*" *in differential-space quantum theory* (with A. Siegel), Phys. Rev. 101(1956), 429–432. IH

[56d] *Pure patterns in a natural world*, in *The New Landscape in Art and Science*, ed. G. Kepes, Paul Theobald and Co., Chicago, 1956, pp. 274–276. IIB

[56e] *Brain waves and the interferometer*, J. Phys. Soc. Japan 18(1956), 499–507. IIB

[56f] *Moral reflections of a mathematician*, Bull. Atomic Scientists 12(1956), 53–57. (Reprinted from [56g].) IIIB

[56g] *I Am a Mathematician. The Later Life of a Prodigy*, Doubleday, Garden City, New York, 1956; paperback edition by The MIT Press, 1964; Math. Rev. 17(1956), 1037. VI

[57a] *The definition and ergodic properties of the stochastic adjoint of a unitary transformation* (with E. J. Akutowicz), Rend. Circ. Mat. Palermo (2) 6(1957), 205–217, Addendum, 349; Math. Rev. 20(1959), rev. 4328. IC

*[57b] *Notes on Pólya's and Turán's hypotheses concerning Liouville's factor* (with A. Wintner), Rend. Circ. Mat. Palermo (2) 6(1957), 240–248; Math. Rev. 20(1959), rev. 5759. ID

*[57c] *On the non-vanishing of Euler products* (with A. Wintner), Amer. J. Math. 79(1957), 801–808. ID

[57d] *The prediction theory of multivariate stochastic processes*, Part I (with P. Masani), Acta Math. 98(1957), 111–150; Math. Rev. 20(1959), rev. 4323. IG

[57e] *Rhythms in physiology with particular reference to encephalography*, Proceedings of the Rudolf Virchow Medical Society in the City of New York, vol. 16, 1957, pp. 109–124. IIB

[57f] *The role of the mathematician in a materialistic culture (A scientist's dilemma in a materialistic world)*, Columbia Engineering Quarterly, Proceedings of the Second Combined Plan Conference, Arden House, October 6–9, 1957, pp. 22–24. IIIA

[57g] *The role of the small cultural college in education of the scientists*, a speech given at Wabash College, Indiana, October 10, 1957. IIIC

[58a] *Logique, probabilité et méthode des sciences physiques*, in *La Méthode dans les Sciences Modernes*, Editions Science et Industrie, ed. Francois Le Lionnais, Paris, 1958, pp. 111–112. IH

[58b] *The prediction theory of multivariate stochastic processes*, Part II (with P. Masani), Acta Math. 99(1958), 93–137; Math. Rev. 20(1959), rev. 4325. IG

[58c] *Random time* (with A. Wintner), Nature 181(1958), 561–562. IIB

[58d] *Sur la prévision lineáire des processus stochastiques vectoriels à densité spectrale bornée.* I (with P. Masani), C. R. Acad. Sci. Paris 246(1958), 1492–1495; Math. Rev. 20(1959), rev. 4324a. IG

[58e] *Sur la prévision linéaire des processus stochastiques vectoriels à densité spectrale bornée.* II (with P. Masani), C. R. Acad. Sci. Paris 246(1958), 1655–1656; Math. Rev. 20(1959), rev. 4324b. IG

[58f] *My connection with cybernetics. Its origins and its future*, Cybernetica (Belgium) 1(1958), 1–14. IIB

[58g] *Time and the science of organization*, Part II, Scientia 93(1958), 225–230. IIB

[58h] *Science: The megabuck era*, New Republic 138(1958), 10–11. IIIA

[58i] *Nonlinear Problems in Random Theory*, The MIT Press, Cambridge, Mass., and Wiley, New York, 1958; paperback edition, The MIT Press, 1966. VI

[59a] *A factorization of positive Hermitian matrices* (with E. J. Akutowicz), J. Math. Mech. 8(1959), 111–120. IG

[59b] *Nonlinear prediction* (with P. Masani), in *Probability and Statistics*, The Harald Cramér Volume, ed. U. Grenander, Stockholm, 1959, 190–212. IG

[59c] *On bivariate stationary processes and the factorization of matrix-valued functions* (with P. Masani), Teor, Verojatnost. i Primenen. 4(1959), 322–331. (English transl. Theor. Probability App. 4(1959), 300–308.) IG

[59d] *Man and the machine* (Interview with N. Wiener), Challenge (The Magazine of Economic Affairs) 7(1959), 36–41. IIIA

[59e] *The Tempter* (novel), Random House, New York, 1959. VI

[60a] *The application of physics to medicine*, in *Medicine and Other Disciplines*, New York Academy of Medicine, ed. I. Galderston, International Universities Press, 1960, pp. 41–57. IIB

[60b] *The brain and the machine* (Summary of an address), in *Dimensions of Mind*, ed. S. Hook, Collier Books, 1960, (Proceedings of Third Annual New York Univ. Institute of Philosophy held on May 15–16, 1959), pp. 113–117. IIB

[60c] *Kybernetik*, Contribution to *Wörterbuch der Soziologie*, F. Enke Verlag, Stuttgart, 1960, pp. 620–622. IIB

[60d] *Some moral and technical consequences of automation*, Science 131(1960), 1355–1358. IIIA

[60e] *The duty of the intellectual*, Tech. Rev. 62(1960), 26–27; reprinted almost in entirety in *The grand privilege*, Sat. Rev. 43(1960), 51–52; also in Technion 18(1961), 86–87—"A professor tells what a professor is." IIIB

[60f] Preface to *Cybernetics of Natural Systems*, by D. Stanley-Jones, Pergamon Press, London, 1960, pp. v–viii. IVA

[61a] *Über Informationstheorie*, Naturwissenschaften 48(1961), 174–176. IIB

[61b] *Science and society*, Voprosy Filosofii (1961), No. 7, 117–122; reprinted in Estratto Rivista Methodos 13(1961), 1–8, and in Tech. Rev. 63(1961), 49–52. Excerpts in Science 138(1962), 651. IIB

[61c] *Cybernetics*, Second edition of [48f] (revisions and two additional chapters), The MIT Press and Wiley, New York, 1961; paperback edition, The MIT Press, 1965. VI

[62a] A verbal Contribution to Proc. of the International Symposium on the Application of Automatic Control in Prosthetics Design, August 27–31, 1962, Opatija, Yugoslavia, pp. 132–133. IIB

[62b] *The mathematics of self-organizing systems*, in *Recent Developments in Information and Decision Processes*, MacMillan, New York, 1962, pp. 1–21. IIB

[63a] *Random theory in classical phase space and quantum mechanics* (with Giacomo Della Riccia), Proc. Internat. Conference on Functional Analysis, Massachusetts Institute of Technology, Cambridge, Mass., June 9–13, 1963; *Analysis in Function Space*, The MIT Press, Cambridge, Mass., 1964, pp. 3–14. IH

[63b] *Introduction to neurocybernetics* (with J. P. Schadé) and *Epilogue*, in *Progress in Brain Research*, vol. 2 of Nerve, Brain and Memory Models, Elsevier Publishing Co., Amsterdam, 1963, pp. 1–7, 264–268. IIB

[63c] *The lonely nationalism of Rudyard Kipling* (with K. Deutsch), Yale Rev. 52(1963), 499–517. IIID

[64a] *On the oscillations of nonlinear systems*, Proc. Symposium on Stochastic Models in Medicine and Biology, Mathematics Research Center, U.S. Army, June 12–14, 1963, ed. John Gurland, University of Wisconsin Press, Madison, Wisconsin, 1964, pp. 167–177. IG

[64b] *Dynamical systems in physics and biology*, Contribution to series "Fundamental Science in 1984," The New Scientist (London) 21(1964), 211–212. IIB

[64c] *Machines smarter than men?* (Interview with N. Wiener), U.S. News and World Rept. 56(1964), 84–86; abbreviated in Reader's Digest 84(1964), 121–124. IIIA

[64d] *Intellectual honesty and the contemporary scientist* (Transcript of talk given to Hillel Group at Massachusetts Institute of Technology), Tech. Rev. 66(1964), 17–18, 44–45, 47. IIIA

[64e] *God, Golem, Inc.—A Comment on Certain Points Where Cybernetics Impinges on Religion*, The MIT Press, Cambridge, Mass., 1964; paperback edition, The MIT Press, 1966. VI

[64f] *Selected Papers of Norbert Wiener* with expository papers by Y. W. Lee, Norman Levinson, and W. T. Martin, The MIT Press, Cambridge, Mass., 1964. VI

[65a] L'homme et la machine, Proc. Colloques Philosophiques Internationaux de Royaumont, July, 1962; *Le concept d'information dans la science contemporaine*, Gauthier-Villars, Paris, 1965, pp. 99–132. IIB

[66a] *Wave mechanics in classical phase space, Brownian motion, and quantum theory* (with G. Della Riccia), J. Math. Phys. 7(1966), 1372–1383. IH

[66b] *Differential Space, Quantum Systems and Prediction* (with A. Siegel, B. Rankin, W. T. Martin), The MIT Press, Cambridge, Mass., 1966. VI

[66c] *Generalized Harmonic Analysis and Tauberian Theorems* (paperback edition of [30a] and [32a]), The MIT Press, Cambridge, Mass., 1966. VI

Supplementary Bibliography

[18b] *Aesthetics*, in Encyclopedia Americana, 1918–20 edition, vol. I, 198–203. IIA

[18c] *Algebra, definitions and fundamental concepts*, in Encyclopedia Americana, 1918–20 edition, vol. I, 381–385. I,I

[18d] *Alphabet*, in Encyclopedia Americana, 1918–20 edition, vol. I, 435–438. IIA

[18e] *Animals, chemical sense in*, in Encyclopedia Americana, 1918–20 edition, vol. I, 704. IIA

[18f] *Apperception*, in Encyclopedia Americana, 1918–20 edition, vol. II, 82–83. IIA

[18g] *Category*, in Encyclopedia Americana, 1918–20 edition, vol. VI, 49. IIA

[18h] *Dualism*, in Encyclopedia Americana, 1918–20 edition, vol. IX, 367. IIA

[18i] *Duty*, in Encyclopedia Americana, 1918–20 edition, vol. IX, 440–441. IIA

[18j] *Ecstasy*, in Encyclopedia Americana, 1918–20 edition, vol. IX, 570. IIA

[19a] *Geometry, non-euclidean*, in Encyclopedia Americana, 1918–20 edition, vol. XII, 463–467. I,I

[19b] *Induction, in logic*, in Encyclopedia Americana, 1918–20 edition, vol. XV, 70–73. IIA

[19c] *Infinity*, in Encyclopedia Americana, 1918–20 edition, vol. XV, 120–122. IIA

[19d] *Meaning*, in Encyclopedia Americana, 1918–20 edition, vol. XVIII, 478–479. IIA

[19e] *Mechanism and vitalism*, in Encyclopedia Americana, 1918–20 edition, vol. XVIII, 527–528. IIA

[19f] *Metaphysics*, in Encyclopedia Americana, 1918–20 edition, vol. XVIII, 707–710. IIA

[19g] *Pessimism*, in Encyclopedia Americana, 1918–20 edition, vol. XXI, 654 IIA

[19h] *Postulates*, in Encyclopedia Americana, 1918–20 edition, vol. XXII, 437–438. IIA

[20h] *Soul*, in Encyclopedia Americana, 1918–20 edition, vol. XXV, 268–271. IIA

[20i] *Substance*, in Encyclopedia Americana, 1918–20 edition, vol. XXV, 775–776. IIA

[20j] *Universals*, in Encyclopedia Americana, 1918–20 edition, vol. XXVII, 572–573. IIA

[49h] Review of Philipp Frank, *Modern Science and its Philosophy*, New York Times, Book Review, August 14, 1949, sec. 7, p. 3. IVA

[52c] *Cybernetics*, in Encyclopedia Americana Annual 1952 edition, 187–188. IIB

[65b] *Perspectives in cybernetics*, Progress in Brain Research 17 (1965), 399–404. IIB

ID

Generalized Harmonic Analysis and Tauberian Theory

NOTE ON A NEW TYPE OF SUMMABILITY.

By Norbert Wiener.

We have

$$\frac{x^2}{4} - \frac{\pi x}{2} + \frac{\pi^2}{6} = \cos x + \frac{\cos 2x}{4} + \cdots + \frac{\cos nx}{n^2} + \cdots. \tag{1}$$

From this it may be deduced at once that

$$\frac{\pi x - x^2}{2} = \sin^2 x + \frac{\sin^2 2x}{4} + \cdots + \frac{\sin^2 nx}{n^2} + \cdots, \tag{2}$$

whence

$$1 - \frac{1}{\pi n} = \frac{2n}{\pi}\left\{ \sin^2\frac{1}{n} + \frac{1}{4}\sin^2\frac{2}{n} + \cdots + \frac{1}{k^2}\sin^2\frac{k}{n} + \cdots \right\}. \tag{3}$$

Moreover,

$$\lim_{n\to\infty} \frac{2n}{\pi k^2}\sin^2\frac{k}{n} = 0. \tag{4}$$

This suggests that given a series $\sum_1^\infty a_n$, we may consider

$$T(a_n) = \lim_{n\to\infty} \frac{2n}{\pi}\left\{ a_1 \sin^2\frac{1}{n} + \frac{a_1 + a_2}{4}\sin^2\frac{2}{n} + \cdots + \frac{a_1 + \cdots + a_k}{k^2}\sin^2\frac{k}{n} + \cdots \right\} \tag{5}$$

as a generalized sum of the a_n's. This definition is of the sort called linear and regular by Carmichael* and Hurwitz,† since (3) and (4) hold, and series (3) is a series of positive terms. Hence it evaluates every convergent series correctly.

Let us now apply our method of summation to the Fourier series

$$\frac{1}{2\pi}\int_{-\pi}^{\pi} f(x)dx + \sum_1^\infty \frac{1}{\pi}\int_{-\pi}^{\pi} f(x)\cos n(x - x')dx. \tag{6}$$

The nth partial sum of this is, as is well known,

$$\frac{1}{2\pi}\int_{-\pi}^{\pi} f(x)\,\frac{\sin (2n+1)\dfrac{x-x'}{2}}{\sin\dfrac{x-x'}{2}}\,dx. \tag{7}$$

* *Bull. Am. Math. Soc.*, Vol. 25 (1918–19), p. 97.
† Ibid., Vol. 28 (1922), p. 20.

Hence we get for (5)

$$\lim_{n\to\infty} \frac{2n}{\pi} \sum_1^\infty \frac{1}{2\pi k^2} \sin^2 \frac{k}{n} \int_{-\pi}^{\pi} f(x) \frac{\sin (2n+1)\dfrac{x-x'}{2}}{\sin \dfrac{x-x'}{2}} dx. \tag{8}$$

This leads us to consider the series

$$\begin{aligned}
&\frac{2n}{\pi} \sum_{k=1}^\infty \frac{1}{2\pi k^2} \sin^2 \frac{k}{n} \frac{\sin (2k+1)u}{\sin u} \\
&= \frac{2n}{\pi} \sum_{k=1}^\infty \frac{1}{4\pi k^2} \left(1 - \cos \frac{2k}{n}\right) (\sin 2ku \cot u + \cos 2ku) \\
&= \frac{2n}{\pi} \sum_{k=1}^\infty \frac{1}{4\pi k^2} \left\{ \cot u \left[\sin 2ku - \sin 2ku \cos \frac{2k}{n} \right] \right. \\
&\qquad \left. + \cos 2ku - \cos 2ku \cos \frac{2k}{n} \right\} \\
&= \frac{2n}{\pi} \sum_{k=1}^\infty \frac{1}{4\pi k^2} \left\{ \cot u \left[\sin 2ku - \frac{1}{2} \sin 2k\left(u + \frac{1}{n}\right) \right.\right. \\
&\qquad \left. - \frac{1}{2} \sin 2k \left(u - \frac{1}{n}\right) \right] \\
&\qquad \left. + \cos 2ku - \frac{1}{2} \cos 2k\left(u + \frac{1}{n}\right) - \frac{1}{2} \cos 2k \left(u - \frac{1}{n}\right) \right\} \qquad (9) \\
&= \frac{n}{4\pi^2} \cot u [- 2 \int_0^u \log (4 \sin^2 x) dx \\
&\qquad + \int_0^{u+(1/n)} \log (4 \sin^2 x) dx + \int_0^{u-(1/n)} \log (4 \sin^2 x) dx] \\
&\qquad + \frac{n}{4\pi^2} \left[2\left(u^2 - \pi u + \frac{\pi^2}{6}\right) - \left(\left(u + \frac{1}{n}\right)^2 \right.\right. \\
&\qquad \left.\left. - \pi\left(u + \frac{1}{n}\right) + \frac{\pi^2}{6}\right) - \left(\left(u - \frac{1}{n}\right)^2 - \pi \left(u - \frac{1}{n}\right) + \frac{\pi^2}{6} \right) \right] \\
&= \frac{n}{4\pi^2} \cot u [\int_u^{u+(1/n)} \log (4 \sin^2 x) dx \\
&\qquad + \int_u^{u-(1/n)} \log (4 \sin^2 x) dx] - \frac{1}{2\pi^2 n} \\
&= \frac{1}{4\pi^2} \cot u \log \frac{\sin^2 \xi_1}{\sin^2 \xi_2} - \frac{1}{2\pi^2 n},
\end{aligned}$$

where ξ_1 lies between u and $u + \frac{1}{n}$, ξ_2 between $u - \frac{1}{n}$ and u, and $0 < u < \pi$. As n increases, expression (9) converges uniformly to 0 over $(\epsilon, \pi - \epsilon)$. Moreover, in the neighborhood of $u = 0$, expression (9) is positive, while it never has a negative value less than $\frac{1}{2\pi^2 n}$. Series (9) is even in u.

The partial sums of series (9) are subject to the formal transformation

$$\frac{2n}{\pi}\sum_1^m \frac{1}{2\pi k^2}\sin^2\frac{k}{n}\frac{\sin(2k+1)u}{\sin u}$$
$$= \frac{2n}{\pi}\left\{\frac{1}{2\pi}\sin^2\frac{1}{n}+\frac{1}{\pi}\sin^2\frac{1}{n}\cos 2u\right.$$
$$+\frac{1}{8\pi}\sin^2\frac{2}{n}+\frac{1}{4\pi}\sin^2\frac{2}{n}\cos 2u+\frac{1}{4\pi}\sin^2\frac{2}{n}\cos 4u$$
$$+\frac{1}{18\pi}\sin^2\frac{3}{n}+\frac{1}{9\pi}\sin^2\frac{3}{n}\cos 2u$$
$$+\frac{1}{9\pi}\sin^2\frac{3}{n}\cos 4u+\frac{1}{9\pi}\sin^2\frac{3}{n}\cos 6u \tag{10}$$
$$+\cdots\cdots\cdots\cdots\cdots\cdots\cdots$$
$$+\frac{1}{2\pi m^2}\sin^2\frac{m}{n}+\frac{1}{\pi m^2}\sin^2\frac{m}{n}\cos 2u$$
$$\left.+\cdots+\frac{1}{\pi m^2}\sin^2\frac{m}{n}\cos 2mu\right\}$$
$$= \frac{2n}{\pi}\left\{\sum_1^m \frac{1}{2\pi k^2}\sin^2\frac{k}{n}+\cos 2u\sum_1^m\frac{1}{\pi k^2}\sin^2\frac{k}{n}\right.$$
$$\left.+\cos 4u\sum_2^m\frac{1}{\pi k^2}\sin^2\frac{k}{n}+\cdots+\cos 2mu\frac{1}{\pi m^2}\sin^2\frac{m}{n}\right\}.$$

This suggests an investigation of the series

$$\frac{2n}{\pi}\left\{\sum_{k=1}^\infty \frac{1}{2\pi k^2}\sin^2\frac{k}{n}+\sum_{j=1}^\infty\cos 2ju\sum_{k=j}^\infty\frac{1}{\pi k^2}\sin^2\frac{k}{n}\right\}. \tag{11}$$

The sum of the squares of the coefficients of (11) may be shown to converge, since the mth coefficient is of the order of $1/m$. Since the coefficients of (10) are positive, less than those of (11) and convergent to those of (11), it follows that the sum of (10) converges in the mean to the sum of (11) with increasing m. Hence series (9) can be integrated term by term when multiplied by any summable function of summable square. Consequently if $f(x)$ is a summable function of summable square, we may write for (8)

$$\lim_{n\to\infty}\int_{-\pi}^{\pi} f(x)\frac{2n}{\pi}\sum_1^\infty\frac{1}{2\pi k^2}\sin^2\frac{k}{n}\frac{\sin(2k+1)\dfrac{x-x'}{2}}{\sin\dfrac{x-x'}{2}}\,dx. \tag{12}$$

Let us write (12) in the form $\lim\limits_{n\to\infty}\int_{-\pi}^{\pi} f(x)G_n(x,x')dx$. We have already shown that over $(-\pi, x'-\epsilon)$ and $(x'+\epsilon, \pi)$,

$$\lim_{n\to\infty} G_n(x,x') = 0 \tag{13}$$

uniformly in x, and that $G_n(x, x') + \frac{1}{2\pi^2 n}$ is positive. Moreover,

$$\begin{aligned}\int_{-\pi}^{\pi} G_n(x, x')dx &= \int_{-\pi}^{\pi} \frac{2n}{\pi}\left\{\sum_1^\infty \frac{1}{2\pi k^2}\sin^2\frac{k}{n}\right.\\ &\qquad \left. + \sum_{j=1}^\infty \cos j(x - x') \sum_{k=j}^\infty \frac{1}{\pi k^2}\sin^2\frac{k}{n}\right\} dx \\ &= \frac{2n}{\pi}\sum_1^\infty \frac{1}{k^2}\sin^2\frac{k}{n} \\ &= 1 - \frac{1}{\pi n}.\end{aligned} \tag{14}$$

From all these facts it follows that if $f(x)$ is continuous at $x = x'$,

$$\begin{aligned}\lim_{n\to\infty}\int_{-\pi}^{\pi} f(x)G_n(x, x')dx &= \lim_{n\to\infty}\int_{-\pi}^{\pi} f(x)\left[G_n(x, x') + \frac{1}{2\pi^2 n}\right]dx \\ &= \lim_{n\to\infty}\int_{-\pi}^{x'-\epsilon} f(x)\left[G_n(x, x') + \frac{1}{2\pi^2 n}\right]dx \\ &\qquad + \lim_{n\to\infty}\int_{x'-\epsilon}^{x'+\epsilon} f(x)\left[G_n(x, x') + \frac{1}{2\pi^2 n}\right]dx \\ &\qquad + \lim_{n\to\infty}\int_{x'+\epsilon}^{\pi} f(x)\left[G_n(x, x') + \frac{1}{2\pi^2 n}\right]dx \\ &= F\lim_{n\to\infty}\int_{x'-\epsilon}^{x'+\epsilon}\left[G_n(x, x') + \frac{1}{2\pi^2 n}\right]dx \\ &= F\lim_{n\to\infty}\int_{-\pi}^{\pi} G_n(x, x')dx = F,\end{aligned} \tag{15}$$

F being some quantity lying between the upper and the lower bounds of $f(x)$ over the interval $(x' - \epsilon, x' + \epsilon)$. Hence

$$\lim_{n\to\infty}\int_{-\pi}^{\pi} f(x)G_n(x, x')dx = f(x'). \tag{16}$$

By an obvious modification of this proof, if $f(x' + 0)$ and $f(x' - 0)$ exist,

$$\lim_{n\to\infty}\int_{-\pi}^{\pi} f(x)G_n(x, x')dx = \tfrac{1}{2}[f(x' + 0) + f(x' - 0)]. \tag{17}$$

In other words, at any point at which a summable function f of summable square is continuous, its Fourier series may be correctly summed by the method of this paper, and at any point at which f has determinate limits to the right and to the left, its Fourier series is summable to the mean of these values. Since a function is bounded in the neighborhood of a point of continuity, and since the contributions to the partial sums of a Fourier series at a point x' due to values of f for arguments outside $(x' - \epsilon, x' + \epsilon)$ converge uniformly to zero, the condition that f be of summable square is unessential.

MASSACHUSETTS INSTITUTE OF TECHNOLOGY,
November 28, 1922.

Comments on [23f]

V. Beneš

The regular method of summation $\Sigma a_{ij} b_j$ defined by

$$a_{nm} = \frac{2n}{\pi} \sum_{k=m}^{\infty} \frac{1}{k^2} \sin^2 \frac{k}{n}$$

sums the Fourier coefficients of an integrable function $f(\cdot)$ to $f(x)$ if x is a point of continuity and to $1/2[f(x+0) - f(x-0)]$ if $f(\cdot)$ has determinate limits at x from both sides. This result is the basis of paper [24f].†

† *Editor's note.* For the format of $f(x + 0)$ etc., see Wiener's [24f].

THE QUADRATIC VARIATION OF A FUNCTION AND ITS FOURIER COEFFICIENTS

By NORBERT WIENER

At the end of an article[1] in which he gives the first example of a function of period 2π, everywhere continuous, and of limited total variation, with Fourier coefficients a_n and b_n which do not fulfil at once both conditions

$$a_n = o(1/n), \qquad b_n = o(1/n), \tag{1}$$

F. Riesz says, " In conclusion I should like to propound the following further questions which seem to me more difficult of attack: Is there a continuous function of period 2π and of limited total variation for which the sequences $\{na_n\}$, $\{nb_n\}$ converge, and at least one of the two limits is distinct from 0? Is there a discontinuous function of limited total variation for which $a_n = o(1/n)$, $b_n = o(1/n)$? The two questions are closely connected."

These questions are discussed in a paper by L. Neder,[2] who points out, in accordance with a remark of Landau, that the second question is trivial unless we restrict our question to functions with other than removable discontinuities. Under this restriction, he answers both of Riesz's questions in the negative.

Neder's results may be put together into the following theorem: the necessary and sufficient condition that a function of limited total variation and of period 2π, for which $\lim na_n$ and $\lim nb_n$ exist, should be everywhere continuous, except for removable discontinuities, is that

$$\lim na_n = 0, \quad \lim nb_n = 0. \tag{2}$$

In this paper a larger question will be discussed, the answer to which completely includes Neder's theorem. We shall show

[1] Ueber die Fourierkoeffizienten einer stetigen Function von beschränkter Schwankung, *Math. Zeitschr.* 2 (1918), p. 312.

[2] Ueber die Fourierkoeffizienten der Funktionen beschränkter Schwankung, *Math. Zeitschr.* 6 (1920), p. 270.

that if $f(x)$ is a function of limited variation with period 2π and is represented by the Fourier series

$$a_0/2+\sum_1^\infty(a_n \cos nx+b_n \sin nx),$$

which has for its complete set of discontinuities over the interval $0\leq x<2\pi$ the numbers $\{\xi_k\}$, then

$$\lim_{n\to\infty} 2n\sum_1^\infty(a_m{}^2+b_m{}^2)\sin^2 m\pi/n$$

exists and

$$=\sum[f(\xi_k+0)-f(\xi_k-0)]^2. \tag{3}$$

From this it may be concluded at once that a necessary and sufficient condition that $f(x)$ should be everywhere continuous, apart from removable discontinuities, is that

$$\lim_{n\to\infty} 2n\sum_1^\infty(a_m{}^2+b_m{}^2)\sin^2 m\pi/n=0. \tag{4}$$

Since we shall show that if $\lim m^2(a_m{}^2+b_m{}^2)$ exists,

$$\pi^2 \lim_{m\to\infty} m^2(a_m{}^2+b_m{}^2)=\lim_{u\to\infty} 2n\sum_1^\infty(a_m{}^2+b_m{}^2)\sin^2 m\pi/n, \tag{5}$$

Neder's theorem may be deduced at once.

In the course of this demonstration we shall introduce the notion of the quadratic variation of a function, a precise analogue to the total variation. We shall orient this and related notions with reference to one another and to Lipschitz conditions of various orders. We shall show that if a function is of limited quadratic variation over any interval, its Fourier series converges everywhere within that interval, and is uniformly convergent over any interior interval if the function is continuous over the first interval. We shall finally parallel these results in the theory of trigonometric interpolation.

§1. **The Higher Variations of a Function.** Let $f(x)$ be a single-valued, real function of the real variable x, and let it have the period 2π. Then if we write $V_\epsilon(f)$ for the upper bound of

$$|f(x_2)-f(x_1)|+|f(x_3)-f(x_2)|+\dots+|f(x_n)-f(x_{n-1})|+|f(x_1)-f(x_n)|$$

for all sets of x_k's such that

$$0 \leq x_1 < x_2 < \ . \ . \ . \ < x_n \leq 2\pi,$$

while

$$x_k - x_{k-1} < \epsilon, \qquad (k = 2, 3, \ . \ . \ . \ , n)$$
$$x_1 + 2\pi - x_n < \epsilon, \tag{6}$$

we may define the total variation of $f(x)$ over the interval $(0, 2\pi)$ as

$$V(f) = \lim_{\epsilon \to 0} V_\epsilon(f).$$

It is understood here and in all analogous cases later that ∞ will be regarded as a possible limit or upper bound, so that $V_\epsilon(f)$ and $V(f)$ admit it as a value. There is no difficulty in showing the agreement between our definition of the total variation of a function and that more usual.

In the same way, we shall write $V_\epsilon^{(\nu)}(f)$ for the upper bound of

$$|f(x_2) - f(x_1)|^\nu + |f(x_3) - f(x_2)|^\nu + \ . \ . \ .$$
$$+ |f(x_n) - f(x_{n-1})|^\nu + |f(x_1) - f(x_n)|^\nu,$$

where the x_k's are subject to the same restrictions as in the case of $V_\epsilon(f)$. We shall then define the νth total variation of $f(x)$ as

$$V^{(\nu)}(f) = \lim_{\epsilon \to 0} V_\epsilon^{(\nu)}(f). \tag{7}$$

It is easy to show that if O_m is the oscillation of $f(x)$ in the interval (x_m, x_{m+1}), and O_n is the oscillation of $f(x)$ in the interval $(x_n, x_1 + 2\pi)$, then

$$V^{(\nu)}(f) = \lim_{\epsilon \to 0} \limsup \sum_1^n O_m{}^\nu, \tag{8}$$

where the lim sup is taken over all the sets of O_m's corresponding to sequences of x_k's satisfying (6). It is obvious on inspection that

$$\lim_{\epsilon \to 0} \limsup \sum_1^n O_m{}^\nu \geq V^{(\nu)}(f), \tag{9}$$

since the oscillation of a function in an interval is at least as great as the difference of the values of the function at the ends of the interval. On the other hand, $\Sigma O_m{}^\nu$ is the upper bound of

$$|f(\xi_1) - f(\eta_1)|^\nu + |f(\xi_2) - f(\eta_2)|^\nu +$$
$$+ |f(\xi_{n-1}) - f(\eta_{n-1})|^\nu + |f(\xi_n) - f(\eta_n)|^\nu,$$

where

$$x_k \leq \xi_k \leq x_{k+1}, \qquad x_k \leq \eta_k \leq x_{k+1}, \qquad (k=1, 2, \ldots, n-1)$$
$$x_n \leq \xi_n \leq 2\pi + x_1, \qquad x_n \leq \eta_n \leq 2\pi + x_1.$$

It results that

$$\Sigma O_m^{\nu} \leq V_\epsilon^{(\nu)}(f), \tag{10}$$

which in combination with (9) yields (8) almost at once.

If $V^{(\nu)}(f)$ is bounded, $f(x)$ can have no discontinuities not of the first kind. Otherwise we shall have some point ξ at which either

$$\bar{f}(\xi+0) > \underline{f}(\xi+0),$$

or

$$\bar{f}(\xi-0) > \underline{f}(\xi-0),$$

or

$$\bar{f}(\xi\pm 0) = \infty,$$

or

$$\underline{f}(\xi\pm 0) = -\infty.$$

In the last two cases, it is possible, given any A, to find in any neighborhood of ξ numbers ξ_1 and ξ_2 such that $f(\xi_2)-f(\xi_1) > A$. Hence $V^{(\nu)}(f) \geq A^{\nu}$, and is therefore infinite. Excluding these, in the first case in any neighborhood to the right of ξ there is a number ξ_1 such that

$$|\bar{f}(\xi+0)-f(\xi_1)| < \eta.$$

Between ξ and ξ_1 there is a number ξ_2 such that

$$|\underline{f}(\xi+0)-f(\xi_2)| < \eta.$$

Between ξ and ξ_2 there is a number ξ_3 such that

$$|\bar{f}(\xi+0)-f(\xi_3)| < \eta,$$

and so on indefinitely. Let us suppose that $2\eta < \bar{f}(\xi+0)-\underline{f}(\xi+0)$, and that $\xi_1 - \xi < \epsilon$. Then

$$\begin{aligned} V_\epsilon^{(\nu)}(f) &\geq \sum_1^n |f(\xi_{n+1})-f(\xi_n)|^{\nu} \\ &\geq n\{\bar{f}(\xi+0)-\underline{f}(\xi+0)-2\eta\}^{\nu}. \end{aligned} \tag{11}$$

Since n may have any value whatever, independently of η and ϵ, we find that $V_\epsilon^{(\nu)}(f)$ is infinite for all ϵ, so that $V^{(\nu)}(f)$ is infinite.

An argument along identical lines excludes the possibility that $f(\xi-0)$ does not have a definite finite value. Thus $f(x)$ has only discontinuities of the first kind, which by a familiar theorem must form a denumerable set. Here and throughout this paper we shall call these discontinuities $\xi_1, \xi_2, \ldots, \xi_k, \ldots$, modulo 2π. For the sake of simplicity of notation we shall confine ourselves throughout the remainder of the paper to functions such that for every k, $f(\xi_k)$ lies between $f(\xi_k-0)$ and $f(\xi_k+0)$. The modifications to be made in our argument in the contrary case will be obvious.

Let ϵ be any number less than the least distance (modulo 2π) between any tw oof the points ξ_k $(k \leq m)$, and such that if $\epsilon_1 \leq \epsilon$, then for all $k \leq m$,

$$|f(\xi_k-\epsilon_1)-f(\xi_k-0)|<\eta;$$

$$|f(\xi_k+\epsilon_1)-f(\xi_k+0)|<\eta.$$

Then if $x_1, \ldots, x_n=x_0$ is any set of numbers such that the residue modulo 2π of x_i-x_{i-1} lies between 0 and ϵ for all i, we have

$$\sum_1^n |f(x_i)-f(x_{i-1})|^\nu \geq \sum_1^m |f(\xi_k+0)-f(\xi_k-0)-2\eta|^\nu.$$

Since an ϵ can be found for any η, no matter how small, and since if ϵ is associated with η, we may associate with η any smaller number ϵ', it follows that no matter how the x_i's are chosen for each value of ϵ,

$$\lim_{\epsilon\to 0} \sum_1^n |f(x_i)-f(x_{i-1})|^\nu \geq \sum_1^m |f(\xi_k+0)-f(\xi_k-0)|^\nu.$$

Since this is true for all m, we have

$$\lim_{\epsilon\to 0} \sum_1^n |f(x_i)-f(x_{i-1})|^\nu \geq \sum_1^\infty |f(\xi_k+0)-f(\xi_k-0)|^\nu. \quad (12)$$

§2. **Relations between Variations of the νth and of Lower Orders.** By Taylor's Theorem with Remainder, if ν is a positive number greater than or equal to 2,

$$v^\nu = u^\nu + \nu(v-u)\{u+\theta(v-u)\}^{\nu-1},$$

where θ is some number between 0 and 1, inclusive. It follows that if u and v are both positive,

$$|v^\nu - u^\nu| \leq \nu(v-u)(v^{\nu-1}+u^{\nu-1}).$$

From this we may deduce at once that if H is the maximum fluctuation made by $f(x)-g(x)$ in an interval of length ϵ, then

$$|V_\epsilon^{(\nu)}(f) - V_\epsilon^{(\nu)}(g)| \leq \nu H(V_\epsilon^{(\nu-1)}(f) + V_\epsilon^{(\nu-1)}(g)). \tag{13}$$

Let $f(x)$ be a function of limited $(\nu-1)$st variation V, with discontinuities $\{\xi_k\}$. Let the saltus of f at ξ_k be a non-increasing function of k. Let $|f(\xi_n+0)-f(\xi_n-0)|<\eta$. Let $g(\xi_\nu+0)=f(\xi_\nu+0)$ and $g(\xi_\nu-0)=f(\xi_\nu-0)$ $[\nu=1, 2, \ldots, n]$, and let $g(x)$ be defined elsewhere by linear interpolation between these values. Since $V^{(\nu-1)}(g)\leq V^{(\nu-1)}(f)$, it follows that for sufficiently small values of ϵ,

$$|V_\epsilon^{(\nu)}(f) - V_\epsilon^{(\nu)}(g)| \leq 4\nu\eta V. \tag{14}$$

That is,

$$\left|V_\epsilon^{(\nu)}(f) - \sum_1^n |f(\xi_k+0)-f(\xi_k-0)|^\nu\right| \leq 4\nu\eta V.$$

From this it may be deduced at once that

$$\begin{aligned}
&\lim_{n\to\infty} \left|V^{(\nu)}(f) - \sum_1^\infty |f(\xi_k+0)-f(\xi_k-0)|^\nu\right| \\
&\leq \lim_{n\to\infty} \left\{\sum_{n+1}^\infty |f(\xi_k+0)-f(\xi_k-0)|^\nu + 4\nu\eta V\right\} \\
&\leq \lim_{\eta\to 0} \eta V(4\nu+1) = 0.
\end{aligned} \tag{15}$$

Hence

$$V^{(\nu)}(f) = \sum_1^\infty |f(\xi_k+0)-f(\xi_k-0)|^\nu. \tag{16}$$

An immediate consequence of this is that a continuous function of bounded $(\nu-1)$st variation is of νth variation 0. However, this results admits of considerable extension, since if $\nu_1<\nu$,

$$V^{(\nu)}(f) \leq (\limsup |f(x+0)-f(x-0)|)^{\nu-\nu_1} V^{(\nu_1)}(f). \tag{17}$$

Thus if $f(x)$ is of period 2π and everywhere continuous, and is of limited ν_1st variation, it is of νth variation 0 for $\nu>\nu_1$.

§3. Special Theorems Relating to the Quadratic Variation. Clearly

$$V^{(2)}_{\frac{2\pi}{n}}(f) \geq \left[f\left(u+\frac{2\pi}{n}\right)-f(u)\right]^2+\left[f\left(u+\frac{4\pi}{n}\right)-f\left(u+\frac{2\pi}{n}\right)\right]^2+\dots$$

$$+\left[f(u+2\pi)-f\left(u+2\pi\left(\frac{n-1}{n}\right)\right)\right]^2. \tag{18}$$

Now let $f(x)$ be summable and of summable square between 0 and 2π. Then there will be a Fourier series

$$a_0/2+\sum_1^\infty(a_n\cos nx+b_n\sin nx)$$

which converges in the mean to $f(x)$ over $(0, 2\pi)$. Then

$$f\left(u+\frac{2\pi}{n}\right)-f(u)\sim\sum_1^\infty\left\{a_m\left[\cos m\left(u+\frac{2\pi}{n}\right)-\cos mu\right]\right.$$

$$\left.+b_m\left[\sin m\left(u+\frac{2\pi}{n}\right)-\sin mu\right]\right\}$$

$$\sim\sum_1^\infty\left[2b_m\cos m\left(u+\frac{\pi}{n}\right)-2a_m\sin m\left(u+\frac{\pi}{n}\right)\right]\sin\frac{m\pi}{n}. \tag{19}$$

Hence

$$\int_0^{2\pi}\left[f\left(u+\frac{2\pi}{n}\right)-f(u)\right]^2du=4\pi\sum_1^\infty(a_m{}^2+b_m{}^2)\sin^2\frac{m\pi}{n}. \tag{20}$$

Therefore by (18),

$$V^{(2)}_{\frac{2\pi}{n}}(f)\geq\frac{n}{2\pi}\int_0^{\frac{2\pi}{n}}\left\{\left[f\left(u+\frac{2\pi}{n}\right)-f(u)\right]^2+\dots\right.$$

$$\left.+\left[f(u+2\pi)-f\left(u+2\pi\left(\frac{n-1}{n}\right)\right)\right]^2\right\}du$$

$$=\frac{n}{2\pi}\int_0^{2\pi}\left[f\left(u+\frac{2\pi}{n}\right)-f(u)\right]^2du$$

$$=2n\sum_1^\infty(a_m{}^2+b_m{}^2)\sin^2\frac{m\pi}{n}. \tag{21}$$

Combining this with (16), we see that if $f(x)$ is of limited total variation,

$$\overline{\lim_{n\to\infty}}\,2n\sum_1^\infty(a_m{}^2+b_m{}^2)\sin^2\frac{m\pi}{n}\leq\sum_1^\infty[f(\xi_k+0)-f(\xi_k-0)]^2. \tag{22}$$

The expression on the left hand of this inequality we shall term $Q(f)$. Moreover, since an integral lies between the extremes of the integrand, there is a u such that

$$\left[f\left(u+\frac{2\pi}{n}\right)-f(u)\right]^2+\ .\ .\ .\ +\left[f(u+2\pi)-f\left(u+2\pi\left(\frac{n-1}{n}\right)\right)\right]^2$$

$$\leq\frac{n}{2\pi}\int_0^{2\pi}\left[f\left(u+\frac{2\pi}{n}\right)-f(u)\right]^2du. \tag{23}$$

It follows from (12), (21), and (23) that

$$\varlimsup_{n\to\infty} 2n\sum_1^\infty(a_m{}^2+b_m{}^2)\sin^2\frac{m\pi}{n}\geq\sum_1^\infty[f(\xi_k+0)-f(\xi_k-0)]^2. \tag{24}$$

(22) and (24) together give:

$$\lim_{n\to\infty} 2n\sum_1^\infty(a_m{}^2+b_m{}^2)\sin^2\frac{m\pi}{n}=\sum_1^\infty[f(\xi_k+0)-f(\xi_k-0)]^2. \tag{3}$$

If $f(x)$ is everyhere continuous except for removable singularities, which are nugatory in so far as the Fourier coefficients are concerned, (4) follows at once.

The author has already shown in a paper forthcoming in the *American Journal of Mathematics* that if s_m is the mth partial sum of a series, the limit

$$\lim_{n\to\infty}\frac{2n}{\pi^2}\sum_1^\infty\frac{s_m}{m^2}\sin^2\frac{m\pi}{n}$$

may be regarded as a generalized sum of this series which is linear and regular in the sense of W. A. Hurwitz.[8] We shall develop some of the results of that paper in order to prove (5).

Let us assume that for all m

$$a_m{}^2+b_m{}^2\geq A/m^2.$$

Then for all n,

$$\sum_1^\infty[f(\xi_k+0)-f(\xi_k-0)]^2\geq 2An\sum_1^\infty\frac{1}{m^2}\sin^2\frac{m\pi}{n}$$

$$=An\sum_1^\infty\frac{\left(\cos 0-\cos\frac{2m\pi}{n}\right)}{m^2}$$

$$=A\pi^2(1-1/n). \tag{25}$$

[8] Bull. Am. Math. Soc., Vol. 28 (1922), p. 18.

A similar result holds with the sign of inequality reversed. Moreover, no batch of terms at the beginning of the a_m's and b_m's is of any effect in determining $\lim_{n\to\infty} 2n\sum_1^\infty (a^2_m+b_m^2)\sin^2 m\pi/n$. Hence

$$\underline{\lim}_{m\to\infty} \pi^2 m^2(a_m^2+b_m^2) \leq \sum_1^\infty [f(\xi_k+0)-f(\xi_k-0)]^2 \leq \overline{\lim}_{m\to\infty} \pi^2 m^2(a_m^2+b_m^2. \tag{26}$$

(5) follows at once.

Again, if $0 < z \leq \pi/2$, $\sin z \geq 2z/\pi$. Therefore

$$\begin{aligned}\sum_1^\infty [f(\xi_k+0)-f(\xi_k-0)]^2 &\geq 4n\sum_1^\infty (a_m^2+b_m^2)\sin^2 m\pi/2n \\ &\geq 4n\sum_1^n (a_m^2+b_m^2)\sin^2 m\pi/2n \\ &\geq 4n\sum_1^n (a_m^2+b_m^2)m^2/n^2 \\ &= \frac{4}{n}\sum_1^n m^2(a_m^2+b_m^2).\end{aligned} \tag{27}$$

Therefore, if $f(x)$ is a continuous function of limited total variation and period 2π, the Cesàro limit of first order of $m^2(a_m^2+b_m^2)$ is 0.

Now let $f(x)$ be a function of limited total variation and period 2π, and let

$$\lim_{n\to\infty}\sum_1^n \frac{m^2}{n}(a_m^2+b_m^2)=0.$$

By a well-known theorem concerning functions of limited total variation,

$$a_n=0(1/n), \qquad b_n=0(1/n).^4$$

Let the upper bound of $m^2(a_m^2+b_m^2)$ be A. Then

$$\begin{aligned}2n\sum_1^\infty (a_m^2+b_m^2)\sin^2\frac{m\pi}{n} &\leq 2n\sum_1^{np}(a_m^2+b_m^2)\frac{m^2\pi^2}{n^2}+2n\sum_{np+1}^\infty (a_m^2+b_m^2) \\ &\leq \frac{2\pi^2}{n}\sum_1^{np} m^2(a_m^2+b_m^2)+2nA\sum_{np+1}^\infty \frac{1}{m^2} \\ &\leq \frac{2\pi^2}{n}\sum_1^{np} m^2(a_m^2+b_m^2)+\frac{2A}{p}.\end{aligned} \tag{28}$$

[4] Cf. Riesz, loc. cit.

If we let n become infinite, we get:

$$\lim_{n\to\infty} 2n\sum_1^\infty (a_m^2+b_m^2)\sin^2\frac{m\pi}{n}\leq\frac{2A}{p}. \tag{29}$$

Since p is arbitrary, (3) yields us immediately:

$$\sum_1^\infty [f(\xi_k+0)-f(\xi_k-0)]^2=0. \tag{30}$$

Joining this result with that of the last paragraph, we obtain the highly important theorem: *if $f(x)$ is of limited total variation and of period 2π, the necessary and sufficient condition that it be continuous is that*

$$\lim_{n\to\infty}\frac{1}{n}\sum_\infty^n m^2(a_m^2+b_m^2)=0. \tag{31}$$

§4. **Higher Variations and Lipschitz Conditions.** We shall say that $f(x)$ satisfies a Lipschitz condition of order $1/\nu$ at $x=\xi$ if whenever ϵ is a number less in modulus than some positive E, there is an A such that

$$|f(\xi+\epsilon)-f(\xi)|\leq A\epsilon^{1/\nu}. \tag{32}$$

on the other hand, we shall say that $f(x)$ satisfies a Lipschitz condition of order $1/\nu$ uniformly over (a, b) if there is an A such that for all x and y on the closed interval (a, b)

$$|f(x)-f(y)|\leq A|x-y|^{1/\nu}. \tag{33}$$

It is clear that if $f(x)$ is a function of period 2π satisfying (33) over an interval containing in its interior $(0, 2\pi)$, then

$$V^{(\nu)}(f)\leq 2\pi A.$$

Again, let $f(x)$ be such that $V^{(\nu_1)}(f)$ is finite for some $\nu_1<\nu$. Then some $V_\epsilon^{(\nu_1)}(f)$, and hence every $V_\epsilon^{(\nu_1)}(f)$, will be finite. Let $B=V_{2\pi}^{(\nu_1)}(f)$. Divide the interval $(0, 2\pi)$ into consecutive halves, fourths, . . . , $1/2^n$ths. Let the kth part of length $2\pi/2^n$ be I_{kn}, and let the fluctuation of f on the closed interval I_{kn} be F_{kn}. Then

$$B\geq\sum_{k=1}^{2n} F_{kn}^{\nu_1}. \tag{34}$$

From this point on we shall use the symbol $[x]$ for the largest integer not greater than x. It results from (34) that over not more than $1+[B/\alpha_n^{\nu_1}]$ of the I_{kn}'s can the fluctuation of f exceed α_n. Let

$$\alpha_n = C(2\pi/2^n)^{1/\nu},$$

where C and ν are independent of n. Then the total length L_n of the I_{kn}'s for which

$$F_{k-1,n} > C(2\pi/2^n)^{1/\nu},\ F_{kn} > C(2\pi/2^n)^{1/\nu},\ \text{or}\ F_{k+1,n} > C(2\pi/2^n)^{1/\nu}\ {}^{5}$$

is not more than

$$3(2\pi/2^n)\{[B/(C^{\nu_1}(2\pi/2^n)^{\nu_1/\nu})]+1\}.$$

Hence for sufficiently large values of n,

$$L_n < 6B(2\pi)^{1-\nu_1/\nu}/\{2^{1-\nu_1/\nu}\}^n C^{\nu_1}. \tag{35}$$

It follows that there is a number N such that

$$\sum_N^\infty L_n < 6B(2\pi)^{1-\nu_1/\nu}\sum_N^\infty\{2^{\nu_1/\nu-1}\}^n/C^{\nu_1}$$
$$< D/C^{\nu_1}, \tag{36}$$

where D is a constant independent of C.

As a consequence, there is a set of points S_C of measure at least $2\pi - D/C^{\nu_1}$ with the property that if ξ belongs to S_C, then for any integer n from a certain N on, ξ lies in the middle third of an interval of length $6\pi/2^n$, over which the fluctuation of f is not greater than $3C(2\pi/2^n)^{1/\nu}$. In other words, if $\epsilon < 2\pi/2^N$, we have

$$|f(\xi+\epsilon)-f(\xi)| \leq 3C(2\epsilon)^{1/\nu}$$
$$\leq 3C2^{1/\nu}\epsilon^{1/\nu}. \tag{37}$$

That is, at any point in any S_C, $f(x)$ satisfies a Lipschitz condition of order $1/\nu$. However, ΣS_C is a measurable set with a measure $\geq 2\pi - D/C^{\nu_1}$, whatever C may be. Thus if $\nu_1 > 0$, ΣS_C is of measure 2π, and its complement is a set of zero measure. Consequently, if $V^{(\nu_1)}(f)$ is bounded and $0 < \nu_1 < \nu$, $f(x)$ satisfies a Lipschitz condition of order $1/\nu$ except over a set of zero measure. From this the further important conclusion may be drawn that

[5] Here $F_{0,n} = F_{2^n,n}$.

the Fourier series of a function of limited νth variation converges except at a set of points of zero measure.

§5. **The Convergence of the Fourier Series of a Function of Limited Quadratic Variation.** Let $s_n(x)$ be the nth partial sum of the Fourier series of $f(x)$, and let $\sigma_n(x)$ be $(s_1+\ \ldots\ +s_n)/n$, the nth approximation to the first Cesàro sum of the Fourier series of f. Then by the Schwarz inequality,

$$|s_{n+1}(x)-\sigma_n(x)| = \{ (a_1 \cos x+b_1 \sin x)+2(a_2 \cos 2x+b_2 \sin 2x) + \ldots + n(a_n \cos nx+b_n \sin nx) \}/n$$

$$\leq \sqrt{\begin{matrix}(a_1{}^2+b_1{}^2+4a_2{}^2+4b_2{}^2+\ \ldots\ +n^2a_n{}^2+n^2b_n{}^2)\times \\ (\cos^2 x+\sin^2 x+\ \ldots\ +\cos^2 nx+\sin^2 nx)/n^2\end{matrix}}$$

$$= \sqrt{\sum_1^n \frac{\nu^2}{n}(a_\nu{}^2+b_\nu{}^2)}$$

$$\leq \sqrt{4n\sum_1^\infty (a_\nu{}^2+b_\nu{}^2)\sin^2\frac{m\pi}{2n}}$$

$$= \sqrt{\frac{n}{\pi}\int_0^{2\pi}\left\{f\left(x+\frac{\pi}{n}\right)-f(x)\right\}^2 dx}. \tag{38}$$

Hence if $f(x)$ has a Fourier series summable $(C1)$ at any point (which will always be true if $V^{(2)}(f)\neq\infty$ and $f(x)$ has consequently only singularities of the first kind), and if $Q(f)=0$, $f(x)$ will converge at the point where the Fourier series is summable. Indeed, it will converge uniformly over any interval over which the Fourier series is uniformly summable, and hence over any interval contained in an interval over which $f(x)$ is continuous. Furthermore, if the Fourier series of $f(x)$ is summable at $x=x_1$ and $Q(f)$ is bounded, the nth partial sum of the Fourier series of $f(x)$ at x_1 lies in the interval

$$\left(f(x_1)-\sqrt{Q(f)}-\epsilon_n,\ f(x_1)+\sqrt{Q(f)}+\epsilon_n\right)$$

where ϵ_n tends to 0 with $1/n$, and tends to 0 uniformly over any interval over which $f(x_1)$ is uniformly summable.

It is well known that the convergence and uniform convergence properties of a Fourier series are entirely local properties of the function it represents. In more precise language, if $f(x)$ and $g(x)$ are two summable functions coinciding over the interval (a, b), and if (c, d) is an interval interior to (a, b), then the Fourier series for $f-g$ converges uniformly to 0 over (c, d). Therefore if $f(x)$ is represented by a Fourier series summable at x_1, the nth sum of the Fourier series for $f(x)$ at x_1 lies in the interval

$$\left(f(x_1)-\sqrt{Q(g)}-\epsilon_n,\ f(x_1)+\sqrt{Q(g)}+\epsilon_n\right)$$

where ϵ_n vanishes with $1/n$, and vanishes uniformly over any interval interior to (c, d) over which the Fourier series for $f(x_1)$ is uniformly summable.

Let $f(x)$ have no discontinuities of the second kind. Let us then put:

$$g(x)=f(x); \qquad (a<x<b)$$

$$g(x)=f(b-0)+\frac{x-b}{2\pi+a-b}f(a+0); \quad (b\leq x\leq 2\pi+a)$$

$$g(x+2\pi)=g(x).$$

We shall formulate the definitions:

$$\overset{b}{\underset{a}{Q}}(f)=Q(g),\quad \overset{b}{\underset{a}{V_\epsilon^{(2)}}}(f)=V_\epsilon^{(2)}(g),\ \text{and}\ \overset{b}{\underset{a}{V^{(2)}}}(f)=V^{(2)}(g). \tag{39}$$

From these it immediately follows that if $0<a<c<b<2\pi$,

$$\overset{b}{\underset{a}{Q}}(f)\leq\overset{c}{\underset{a}{Q}}(f)+\overset{b}{\underset{c}{Q}}(f)+\{f(c+0)-f(c-0)\}^2, \tag{40}$$

and

$$\overset{b}{\underset{a}{V^{(2)}}}(f)=\overset{c}{\underset{a}{V^{(2)}}}(f)+\overset{b}{\underset{c}{V^{(2)}}}(f)+\{f(c+0)-f(c-0)\}^2. \tag{41}$$

Further obvious relations are

$$\overset{b}{\underset{a}{V^{(2)}}}(f)\geq\overset{b}{\underset{a}{Q}}(f) \tag{42}$$

and

$$\overset{b}{\underset{a}{Q}}(f)\leq Q(f),\quad \overset{b}{\underset{a}{V^{(2)}}}(f)\leq V^{(2)}(f). \tag{43}$$

Among the corollaries of these statements are:

(a) $\overset{x}{\underset{a}{V^{(2)}}}(f)$ is an increasing function of x, and as such has only a denumerable set of discontinuities, provided that $V^{(2)}(f)$ exists;

(*b*) If ξ is a point of continuity of $\overset{x}{\underset{a}{V^{(2)}}}(f)$, $\overset{b_1}{\underset{a_1}{V^{(2)}}}(f)$ and hence $\overset{b_1}{\underset{a_1}{Q}}(f)$ can be made arbitrarily small for some interval (a_1, b_1) containing ξ, so that the Fourier series of $f(x)$ must converge to $f(x)$ at ξ.

It remains to discuss the continuity of $\overset{x}{\underset{a}{V^{(2)}}}(f)$. We shall show that if $\overset{b}{\underset{a}{V^{(2)}}}(f)$ is finite, $\overset{x}{\underset{a}{V^{(2)}}}(f)$ is continuous within (a, b) and the Fourier series of $f(x)$ converges to $f(x)$ except at the discontinuities of $f(x)$. For let $\overset{x}{\underset{a}{V^{(2)}}}(f)$ be discontinuous at ξ within (a, b), while $f(x)$ is continuous. It follows that no matter how small ϵ is, $\overset{\xi+\epsilon}{\underset{\xi-\epsilon}{V^{(2)}}}(f)$ remains greater than some positive number A. That is, given any positive ϵ and η, it is possible to find a sequence of numbers $\{x_k\}$ such that

$$\xi-\epsilon\leq x_1\leq \;.\;.\;.\; \leq x_n\leq\xi+\epsilon;$$
$$|x_k-x_{k-1}|<\eta; \quad (k=2, \;.\;.\;.\;, n)$$
$$\sum_2^n\{f(x_k)-f(x_{k-1})\}^2>A.$$

Clearly ξ will lie in the interior of some interval (x_{j-1}, x_{j+1}). Since $f(x)$ is continuous at ξ, we can make $\{f(x_j)-f(x_{j-1})\}^2 + \{f(x_{j+1})-f(x_j)\}^2$ as small as we choose by taking η sufficiently small. Thus we can make

$$\sum_2^{j-1}\{f(x_k)-f(x_{k-1})\}^2+\sum_{j+2}^{n}\{f(x_k)-f(x_{k-1})\}^2>A.$$

Repeating this process, we can find a sequence of numbers $\{y_k\}$ such that

$$x_{j-1}\leq y_1 \leq \;.\;.\;.\; \leq y_m\leq x_{j+1};$$
$$y_{h-1}\leq \xi \leq y_{h+1},$$

and

$$\sum_2^{h-1}\{f(y_k)-f(y_{k-1})\}^2+\sum_{h+2}^{m}\{f(y_k)-f(y_{k-1})\}^2>A.$$

This means that $\overset{b_1}{\underset{a_1}{V_\epsilon^{(2)}}}(f) > 2A$, where ϵ is arbitrarily small, and hence that $\overset{b_1}{\underset{a_1}{V^{(2)}}}(f) \geq 2A$. In the same way it may be shown that for any N, $\overset{b_1}{\underset{a_1}{V^{(2)}}}(f) \geq NA$ and hence $\overset{b_1}{\underset{a_1}{V^{(2)}}}(f) = \infty$. However, this contradicts our initial assumption.

It follows that if $\overset{b_1}{\underset{a_1}{V^{(2)}}}(f)$ is bounded and (a, b) is an interval interior to (a_1, b_1), $\overset{x}{\underset{a}{V^{(2)}}}(f)$ is *uniformly* continuous over (a,b). That is, given any small quantity θ, (a, b) may be included in a finite number of overlapping intervals (α_e, β_e), over each of which

$$\overset{\beta_e}{\underset{\alpha_e}{V^{(2)}}}(f) < \theta.$$

Since there is only a finite set of these, it follows that the range of oscillation of the nth sum of the Fourier series for $f(x)$ is included in the interval

$$(f(x) - \sqrt{\theta} - \epsilon_n,\quad f(x) + \sqrt{\theta} + \epsilon_n),$$

where ϵ_n vanishes uniformly with $1/n$ over (a, b). We may conclude directly from this the uniform convergence of the Fourier series for $f(x)$ over (a, b) to the value of the function.

If $\overset{b}{\underset{a}{V^{(2)}}}(f)$ is bounded and $f(x)$ is discontinuous at $\xi[a < \xi < b]$, the Fourier series of f converges to $\frac{1}{2}\{f(\xi+0) + f(\xi-0)]$ at ξ. To show this, it is only necessary to remark that $f(x)$ has only a discontinuity of the first kind at ξ. Let $g(x)$ be defined as follows:

$$\begin{aligned} & g(\xi) = 0; \\ & g(x) = f(x) - f(\xi - 0); \qquad a < x < \xi \\ & g(x) = f(x) - f(\xi + 0); \qquad \xi < x < b \\ & g(x) = f(x) \text{ otherwise.} \end{aligned}$$

Clearly $g(x)$ is continuous at $x = \xi$, and

$$\overset{b}{\underset{a}{V^{(2)}}}(g) \leq \overset{b}{\underset{a}{V^{(2)}}}(f) + 2\{f(\xi+0) - f(\xi - 0\}^2.$$

Hence the Fourier series of $g(x)$ converges to 0 for $x = \xi$. Moreover, $f(x)-g(x)$ is a function of limited total variation, and its Fourier series converges to

$$\tfrac{1}{2}\{f(\xi+0)-g(\xi+0)+f(\xi-0)-g(\xi-0)\}$$
$$=\tfrac{1}{2}\{f(\xi+0)+f(\xi-0)\}$$

for $x=\xi$. Adding these series term by term, we obtain our result.

§6. **The Relation between $Q(f)$ and $V^{(2)}(f)$.** The close relation between $Q(f)$ and $V^{(2)}(f)$ might lead us to believe that the boundedness of the former is a sufficient condition for the boundedness of the latter. That this is not the case will be shown by the following example:

$$f(x)=\sin\log x; \qquad 0<x\leq 2\pi$$
$$f(x+2\pi)=f(x).$$

Since $f(+0)$ is indeterminate, $V^{(2)}(f)$ cannot be finite. On the other hand,

$$Q(f)=\overline{\lim_{n\to\infty}}\frac{n}{2\pi}\left\{\int_0^{2\pi-\frac{2\pi}{n}}\left\{\sin\log\left(u+\frac{2\pi}{n}\right)-\sin\log u\right\}^2du\right.$$
$$\left.+\int_{2\pi-\frac{2\pi}{n}}^{2\pi}\left\{\sin\log\left(u+\frac{2\pi}{n}-2\pi\right)-\sin\log u\right\}^2du\right\}$$
$$\leq\overline{\lim_{n\to\infty}}\frac{n}{2\pi}\left\{\int_0^{\frac{2\pi}{n}}4\,du\right.$$
$$+\int_{\frac{2\pi}{n}}^{2\pi-\frac{2\pi}{n}}\left\{2\cos\log\left[u\left(u+\frac{2\pi}{n}\right)\right]\sin\log\left(1+\frac{2\pi}{nu}\right)\right\}^2du$$
$$\left.+\int_{2\pi-\frac{2\pi}{n}}^{2\pi}4\,du\right\}$$
$$\leq 8+\overline{\lim_{n\to\infty}}\frac{2n}{\pi}\int_{\frac{2\pi}{n}}^{2\pi-\frac{2\pi}{n}}\sin\log\left(1+\frac{2\pi}{nu}\right)du$$
$$\leq 8+\overline{\lim_{n\to\infty}}\frac{2}{\pi}\int_{2\pi}^{2\pi(n-1)}\left[\log\left(1+\frac{2\pi}{x}\right)\right]^2dx$$
$$\leq 8+\frac{2}{\pi}\int_{2\pi}^{\infty}\frac{4\pi^2}{x^2}dx=12.$$

§7. **An Expression Related to** $Q(f)$. If $Q(f)$ is bounded, the expression

$$\frac{\int_0^\infty 2n \sum_{m=1}^\infty (a_m^2+b_m^2)\sin^2 \frac{m\pi}{n} \cdot \frac{e^{-\frac{k}{n}}}{n^3} dn}{\int_0^\infty \frac{e^{-\frac{k}{n}}}{n^3} dn}$$

exists for every value of k, and has the upper bound of its ultimate oscillation not greater than $Q(f)$. That is,

$$Q(f) \geq \overline{\lim_{k\to\infty}} \frac{\int_0^\infty 2\sum_1^\infty (a_m^2+b_m^2) \sin^2 m\pi\xi \, e^{-\xi k} d\xi}{\int_0^\infty \xi e^{-\xi k} d\xi}$$

$$= \overline{\lim_{k\to\infty}} \sum_1^\infty \frac{(a_m^2+b_m^2)m^2\pi^2 k}{k^2+m^2\pi^2} \tag{44}$$

It follows that

$$Q(f) \geq \overline{\lim_{\mu\to\infty}} \sum_\mu^\infty \frac{(a_m^2+b_m^2)m^2\pi^3\mu}{\pi^2(m^2+\mu^2)} \geq \overline{\lim_{\mu\to\infty}} \sum_\mu^\infty \frac{\pi\mu}{2}(a_m^2+b_m^2), \tag{45}$$

this giving us a measure of the rapidity of the convergence in the mean of the Fourier series of a function with bounded Q. What is more, we can proceed directly from the series of the Poisson integral to a proof of the convergence of a Fourier series of a function of limited quadratic variation. We have

$$\left| a_0/2 + \sum_1^\infty \left(1-\frac{1}{n}\right)^m (a_m \cos mx + b_m \sin mx) \right.$$

$$\left. - \left\{ a_0/2 + \sum_1^n (a_m \cos mx + b_m \sin mx \right\} \right|$$

$$\leq \left| \sum_1^n \frac{m}{n} (a_m \cos mx + b_m \sin mx) \right|$$

$$+ \left| \sum_{m=n+1}^\infty \left(1-\frac{1}{n}\right)^m (a_m \cos mx + b_m \sin mx) \right|$$

$$\leq\sqrt{\sum_1^n\frac{m^2}{n^2}(a_m{}^2+b_m{}^2)\sum_1^n(\sin^2 mx+\cos^2 mx)}$$

$$+\sqrt{\sum_{n+1}^\infty(a_m{}^2+b_m{}^2)\sum_{n+1}^\infty\left(1-\frac{1}{n}\right)^{2m}(\sin^2 mx+\cos^2 mx)}$$

$$\leq\sqrt{\sum_1^n\frac{m^2}{n}(a_m{}^2+b_m{}^2)}+\sqrt{\frac{\left(1-\frac{1}{n}\right)^{2(n+1)}}{1-\left(1-\frac{1}{n}\right)^2}\sum_{n+1}^\infty(a_m{}^2+b_m{}^2)}$$

$$\leq\sqrt{\sum_1^n\frac{m^2}{n}(a_m{}^2+b_m{}^2)}+\sqrt{(n+1)\sum_{n+1}^\infty(a_m{}^2+b_m{}^2)}.$$

We have already seen that the first radical has the upper bound of its final oscillation no greater than $Q\{f\}$, while we have just shown that the second radical has the upper bound of its final oscillation no greater than $\frac{2}{\pi}Q\{f\}$. Hence the difference between $\sum_1^\infty\left(1-\frac{1}{n}\right)^m(a_m\cos mx+b_m\sin mx)$ and the nth partial sum of the Fourier series of $f(x)$ has a final oscillation not exceeding $Q\{f\}(1+2/\pi)$.

From this point we may proceed with an argument precisely parallel to that of §5.

§8. **Quadratic Variation and Trigonometric Interpolation.** Let $f(x)$ be a bounded function of x of period 2π, and let

$$\alpha_{nk}=\frac{2}{2n+1}\sum_1^{2n+1}f\left(\frac{2j\pi}{2n+1}\right)\cos\frac{2jk\pi}{2n+1},$$

$$\beta_{nk}=\frac{2}{2n+1}\sum_1^{2n+1}f\left(\frac{2j\pi}{2n+1}\right)\sin\frac{2jk\pi}{2n+1}.$$

Then

$$f_n(x)=\alpha_{n0}/2+\sum_{k=1}^n(\alpha_{nk}\cos kx+\beta_{nk}\sin kx)$$

will be the only trigonometric polynomial of degree n which will coincide with $f(x)$ for every x of the form $\frac{2k\pi}{2n+1}$. We shall show that if $f(x)$ is of limited quadratic variation, $f_n(x)$ converges everywhere to $(f(x+0)+f(x-0))/2$, and converges uniformly over every interval interior to the one over which $f(x)$ is continuous.

We make use of the auxiliary function

$$\begin{aligned}\phi_n(x) &= \alpha_{n0}/2+\sum_{k=1}^{n-1}\frac{n-k}{n}(\alpha_{nk}\cos kx+\beta_{nk}\sin kx)\\ &= \frac{2}{2n+1}\sum_{j=1}^{2n+1} f\left(\frac{2j\pi}{2n+1}\right)\\ &\qquad\times\left[1/2+\sum_{k=1}^{n-1}\frac{n-k}{n}\left(\cos\frac{2jk\pi}{2n+1}\cos kx-\sin\frac{2jk\pi}{2n+1}\sin kx\right)\right]\\ &= \frac{2}{n(2n+1)}\sum_{j=1}^{2n+1} f\left(\frac{2j\pi}{2n+1}\right)\left[n/2+\sum_{k=1}^{n-1}(n-k)\cos k\left(\frac{2j\pi}{2n+1}-x\right)\right]\\ &= \frac{1}{n(2n+1)}\sum_{j=1}^{2n+1} f\left(\frac{2j\pi}{2n+1}\right)\frac{\sin^2\left[\frac{n}{2}\left(\frac{2j\pi}{2n+1}-x\right)\right]}{\sin^2\left[\frac{1}{2}\left(\frac{2j\pi}{2n+1}-x\right)\right]}.\end{aligned} \tag{46}$$

This function, which is a slight modification of one employed by D. Jackson,[6] represents a weighted average of the quantities $f\left(\frac{2j\pi}{2n+1}\right)$, in which those outside any specified neighborhood of x tend to zero weight with increasing n, and in which the weighting of the points to the right of x becomes essentially the same as the weighting of those to the left. Jackson's arguments may be used without essential change, and we see that $\phi_n(x)$ converges to $\frac{1}{2}(f(x+0)+f(x-0))$ wherever this latter quantity exists and converges uniformly to $f(x)$ over any interval interior to one over which $f(x)$ is continuous, provided only that $f(x)$ is bounded, as we have assumed.

[6] A Formula of Trigonometric Interpolation, *Rend. Circ. Math. Palermo* 37 (1914), pp. 371–375.

Now,

$$|\phi_n(x)-f_n(x)|=\Big|\frac{k}{n}\sum_1^n(\alpha_{nk}\cos kx+\beta_{nk}\sin kx)\Big|$$

$$\leq\sqrt{\frac{1}{n^2}\sum_1^n k^2(\alpha_{nk}^2+\beta_{nk}^2)\sum_1^n(\cos^2 kx+\sin^2 kx)}$$

$$=\sqrt{\frac{1}{n}\sum_1^n k^2(\alpha_{nk}^2+\beta_{nk}^2)}. \tag{47}$$

Now, if u is any number of the form $\dfrac{2k\pi}{2n+1}$,

$$f\left(u+\frac{2\pi}{2n+1}\right)-f(u)=f_n\left(u+\frac{2\pi}{2n+1}\right)-f_n(u)$$

$$=\sum_1^n\left\{\alpha_{nm}\left[\cos m\left(u+\frac{2\pi}{2n+1}\right)-\cos mu\right]\right.$$

$$\left.+\beta_{nm}\left[\sin m\left(u+\frac{2\pi}{2n+1}\right)-\sin mu\right]\right\}$$

$$=\sum_1^n\left[2\beta_{nm}\cos m\left(u+\frac{\pi}{2n+1}\right)\right.$$

$$\left.-2\alpha_{nm}\sin m\left(u+\frac{\pi}{2n+1}\right)\right]\sin\frac{m\pi}{2n+1}.$$

By formulæ familiar in the theory of trigonometric sums and trigonometric interpolation, we obtain:

$$\sum_1^{2n+1}\left\{f\left(\frac{2(j+1)\pi}{2n+1}\right)-f\left(\frac{2j\pi}{2n+1}\right)\right\}^2$$

$$=\sum_1^n 2(2n+1)(\alpha_{nm}^2+\beta_{nm}^2)\sin^2\frac{m\pi}{2n+1}$$

$$\geq\sum_1^n\frac{8m^2(\alpha_{nm}^2+\beta_{nm}^2)}{2n+1}. \tag{48}$$

If we combine (47) and (48), we see that

$$|\phi_n(x)-f_n(x)|\leq\sqrt{\frac{2n+1}{8n}\sum_1^{2n+1}\left\{f\left(\frac{2(j+1)\pi}{2n+1}\right)-f\left(\frac{2j\pi}{2n+1}\right)\right\}^2}$$

$$\leq\sqrt{\frac{2n+1}{8n}V^{(2)}_{\frac{2\pi}{2n+1}}(f)}. \tag{49}$$

From this point on, the argument exactly parallels that following (38).

§9. **Step-Functions and Quadratic Variation.** Let $f(x)$ be a function of period 2π. Then

$$\left[\frac{1}{b}\int_a^{a+b} f(x)dx-\frac{1}{b}\int_{a+b}^{a+2b} f(x)dx\right]^2=\left[\frac{1}{b}\int_a^{a+b}[f(x)-f(x+b)]dx\right]^2$$
$$\leq \frac{1}{b}\int_a^{a+b}[f(x)-f(x+b)]^2dx$$

Hence

$$\sum_1^n \left\{\frac{n}{2\pi}\int_{u+\frac{2\pi k}{n}}^{u+\frac{2\pi(k+1)}{n}} f(x)dx-\frac{n}{2\pi}\int_{u+\frac{2\pi(k+1)}{n}}^{u+\frac{2\pi(k+2)}{n}} f(x)dx\right\}^2$$
$$\leq\frac{n}{2\pi}\sum_1^n\int_{u+\frac{2\pi k}{n}}^{u+\frac{2\pi(k+1)}{n}}\left[f(x)-f\left(x+\frac{2\pi}{n}\right)\right]^2dx$$
$$=\frac{n}{2\pi}\int_0^{2\pi}\left[f(x)-f\left(x+\frac{2\pi}{n}\right)\right]^2dx.$$

From this we deduce at once that

$$\overline{\lim_{n\to\infty}}\sum_1^n \left\{\frac{n}{2\pi}\int_{u+\frac{2\pi k}{n}}^{u+\frac{2\pi(k+1)}{n}} f(x)dx-\frac{n}{2\pi}\int_{u+\frac{2\pi(k+1)}{n}}^{u+\frac{2\pi(k+2)}{n}} f(x)dx\right\}^2\leq Q(f).$$

Let us put the definition:

$$f_n(x)=\frac{n}{2\pi}\int_{u+\frac{2\pi k}{n}}^{u+\frac{2\pi(k+1)}{n}} f(\xi)d\xi \quad \left[u+\frac{2\pi k}{n}\leq x<u+\frac{2\pi(k+1)}{n}\right].$$

Then

$$\int_0^{2\pi}[f_n(x)-f_{2n}(x)]^2dx$$
$$\leq\frac{\pi}{n}\sum_1^{2n}\left\{\frac{n}{\pi}\int_{u+\frac{\pi k}{n}}^{u+\frac{\pi(k+1)}{n}} f(x)dx-\frac{n}{\pi}\int_{u+\frac{\pi(k+1)}{n}}^{u+\frac{\pi(k+2)}{n}} f(x)dx\right\}^2,$$

since

$$\begin{cases} \left|\frac{n}{2\pi}\int_{u+\frac{\pi k}{n}}^{u+\frac{\pi(k+2)}{n}} f(x)dx - \frac{n}{\pi}\int_{u+\frac{\pi k}{n}}^{u+\frac{\pi(k+1)}{n}} f(x)dx\right| \\ \leq \left|\frac{n}{\pi}\int_{u+\frac{\pi(k+1)}{n}}^{u+\frac{\pi(k+2)}{n}} f(x)dx - \frac{n}{\pi}\int_{u+\frac{\pi k}{n}}^{u+\frac{\pi(k+1)}{n}} f(x)dx\right|, \\ \left|\frac{n}{2\pi}\int_{u+\frac{\pi k}{n}}^{u+\frac{\pi(k+2)}{n}} f(x)dx - \frac{n}{\pi}\int_{u+\frac{\pi(k+1)}{n}}^{u+\frac{\pi(k+2)}{n}} f(x)dx\right| \\ \leq \left|\frac{n}{\pi}\int_{u+\frac{\pi(k+1)}{n}}^{u+\frac{\pi(k+2)}{n}} f(x)dx - \frac{n}{\pi}\int_{u+\frac{\pi k}{n}}^{u+\frac{\pi(k+1)}{n}} f(x)dx\right|. \end{cases}$$

This gives us the result that

$$\overline{\lim_{n\to\infty}}\ \frac{n}{\pi}\int_0^{2\pi}[f_n(x)-f_{2n}(x)]^2dx = Q(f).$$

Let $\{w_n(x)\}$ be a set of normal, orthogonal functions over $(0, 2\pi)$, all less than or equal to A in maximum modulus, and such that the partial sum to n terms of the w-series for $f(x)$ is always $f_n(x)$ whenever n is a power of 2. A set of essentially this character has been given by Walsh[7] for the interval (0, 1). Then if $2^m < k < 2^{m+1}$ and $a_j = \int_0^{2\pi} f(\xi)w_j(\xi)d\xi$,

$$\begin{aligned} \left|\sum_{2^m+1}^{k} a_j w_j(x)\right| &\leq A\left|\sum_{2^m+1}^{k} a_j\right| \\ &\leq A\sqrt{(k-2^m)\sum_{2^m+1}^{k} a_j^2} \\ &\leq A\sqrt{2^m\sum_{2^m+1}^{2^{m+1}} a_j^2} \\ &= A\sqrt{2^m\int_0^{2\pi}[f_{2^m}(x)-f_{2^{m+1}}(x)]^2dx} \\ &\leq A\sqrt{\pi Q(f)+\epsilon} \end{aligned}$$

for any positive ϵ, for a sufficiently large m.

[7] American Journal of Math., V. 45 (1923), pp. 5–24.

Let $f(x)$ be a continuous function. Then over the interval over which it is continuous — possibly only a single point —

$$\max |f(x)-f_n(x)| \to 0.$$

Furthermore, Walsh has shown in the paper already mentioned that the convergence and uniform convergence properties of the w-series of a function over an interval depend only on the properties of the function over any including interval. Hence if $f(x)$ is summable over $(0, 2\pi)$ an argument precisely paralleling that of §5 will show that the w-series of $f(x)$ converges at every point at which $f(x)$ is continuous and uniformly over any interval interior to one over which $f(x)$ is continuous, provided only that $V^{(2)}(f)$ is finite. Moreover, the discussion which Walsh has given of the w-series of functions constant except for a single discontinuity will enable us to show that if $V^{(2)}(f)$ is finite and $f(x)$ is summable, the w-series of f converges at every point which is a binarily rational multiple of π.

Comments on [24f]

V. Beneš

The method of summation of [23f] is used to show that if a function on $[0, 2\pi)$ is of bounded variation, has $\{\xi_k\}$ for its set of discontinuities, and a_m, b_m for Fourier coefficients, then

$$\lim_{n\to\infty} 2n \sum_{m=1}^{\infty} (a_m^2 + b_m^2) \sin^2 \frac{m\pi}{n}$$

exists and equals $\Sigma_k[f(\xi_k + 0) - f(\xi_k - 0)]$. It then follows that such a function is continuous iff the Cesàro limit of $m^2(a_m^2 + b_m^2)$ is zero. Connections with higher (e.g., quadratic) variations and with Lipschitz conditions are established. In particular, the Fourier series of a function of limited quadratic variation converges everywhere, and uniformly on any set interior to an interval of continuity.

On the representation of functions by trigonometrical integrals.

By

Norbert Wiener, of Cambridge, Massachusetts.

Table of Contents.

Introduction.

At present the mathematical literature contains a number of distinct classes of functions capable of being represented by a trigonometric sum or integral. For example, if $f(x)$ is of period 2π, and is summable and of summable square over a period, we may write

$$f(x) \sim \frac{1}{2\pi}\int_0^{2\pi} f(\xi)\,d\xi$$

$$+\frac{1}{\pi}\sum_{n=1}^{\infty}\left\{\cos nx\int_0^{2\pi} f(\xi)\cos n\xi\,d\xi + \sin n\xi\int_0^{2\pi} f(\xi)\sin n\xi\,d\xi\right\}.$$

Here the sign $\sim$ is used to convey the fact that the series converges in the mean to $f(x)$ over the interval $(0, 2\pi)$ — that is, that

$$\lim_{m\to\infty}\int_0^{2\pi}\left\{f(x) - \left[\frac{1}{2\pi}\int_0^{2\pi} f(\xi)\,d\xi\right.\right.$$

$$\left.\left.+\frac{1}{\pi}\sum_{n=1}^{m}\left\{\cos nx\int_0^{2\pi} f(\xi)\cos n\xi\,d\xi + \sin nx\int_0^{2\pi} f(\xi)\sin n\xi\,d\xi\right\}\right]\right\}^2 dx = 0.$$

Again, if $f(x)$ is summable and of summable square over the entire interval $(-\infty, \infty)$, it may be proved[1]) that

$$\frac{1}{\pi}\int_{-A}^{A} f(\xi)\cos\lambda\xi\, d\xi \quad \text{and} \quad \frac{1}{\pi}\int_{-A}^{A} f(\xi)\sin\lambda\xi\, d\xi$$

converge in the mean over the interval $(-\infty, \infty)$ to functions of λ which may be termed, respectively,

$$\frac{1}{\pi}\int_{-\infty}^{\infty} f(\xi)\cos\lambda\xi\, d\xi \quad \text{and} \quad \frac{1}{\pi}\int_{-\infty}^{\infty} f(\xi)\sin\lambda\xi\, d\xi,$$

as A becomes infinite. It may then be demonstrated that

$$\frac{1}{\pi}\int_{0}^{A}\cos\lambda x\, d\lambda\int_{-\infty}^{\infty} f(\xi)\cos\lambda\xi\, d\xi + \frac{1}{\pi}\int_{0}^{A}\sin\lambda x\, d\lambda\int_{-\infty}^{\infty} f(\xi)\sin\lambda\xi\, d\xi$$

converges in the mean to $f(x)$ over the interval $(-\infty, \infty)$ as A becomes infinite.

We have thus determined two classes of functions, the one consisting of periodic functions, and the other of functions summable and of summable square over an infinite range, both of which, in very different senses, consist of functions with a representation of the form

$$\int_{0}^{\infty}[\cos\alpha x\, d\gamma(\alpha) + \sin\alpha x\, d\delta(\alpha)]$$

over the whole infinite X-axis. To these we must adjoin the almost periodic functions of H. Bohr[2]). It will be useful to recall his definitions and some of his cardinal theorems. His definitions read as follows: *Let $f(x)$ be a function of the real variable x, which is well-defined and continuous over the entire interval* $(-\infty < x < \infty)$. [For the sake of convenience in formulating his theorems, Bohr allows $f(x)$ to be of the form $u(x) + iv(x)$, where u and v are separately continuous real functions of the real variable x.] *If for a given* $\varepsilon > 0$, *the real number* $\tau = \tau(f, \varepsilon)$ *satisfies the equation*

$$|f(x+\tau) - f(x)| \leqq \varepsilon$$

[1]) This theorem may be deduced without difficulty from one proved by M. Plancherel, *Contribution à l'étude de la représentation d'une fonction arbitraire par des intégrales définies, Rendiconti di Palermo* **30** (1910), p. 330.

[2]) H. Bohr, *Zur Theorie der fast periodischen Funktionen. I. Eine Verallgemeinerung der Fourierreihen. Acta Mathematica* **45** (1924), pp. 29–127. Cf. W. L. Hart, *On trigonometric series, Annals of Mathematics* (2), **18** (1916), pp. 99–104.

for all x, τ is termed a translation number (Verschiebungszahl) of the function $f(x)$, pertaining to the given number ε. A function $f(x)$ continuous for every real x is termed almost periodic if to every positive ε there corresponds a length $l = l(\varepsilon) > 0$ such that every interval $\alpha < x < \beta$ of length $\beta - \alpha = l$ contains at least one translation number $\tau = \tau(f, \varepsilon)$.

Bohr's fundamental theorem reads as follows: *To every almost periodic function $f(x)$ there belongs a uniquely determined sequence of real numbers $\lambda_1, \lambda_2, \ldots, \lambda_n, \ldots$ and a series of the form $\sum a_n e^{i\lambda_n x}$, the "Fourier series" of the function $f(x)$. This Fourier series determines the function $f(x)$ uniquely, and converges in the mean to $f(x)$, in the sense that the expression*

$$S_N = \lim_{T\to\infty} \frac{1}{T}\int_0^T \left| f(x) - \sum_{n=1}^{N} a_n e^{i\lambda_n x} \right|^2 dx$$

converges to 0 as N becomes infinite.

We shall devote this paper to the study of the representation of a function $f(x)$ by an integral of the form

$$\int_0^\infty [\cos\alpha x \, d\gamma(\alpha) + \sin\alpha x \, d\delta(\alpha)]$$

under hypotheses of a very general nature. We shall obtain a generalization of periodicity, which does not involve continuity, as that of Bohr does. We shall obtain a theorem from which Bohr's fundamental theorem follows as an immediate corollary. In the proof of this theorem, unlike Bohr, we shall not employ arithmetical methods. To express more precisely what we intend to do, we shall have need of a few definitions. These are given below.

Definition I. *We shall say that $f(x)$ is nearly bounded if*

$$\int_\xi^{\xi+h} [f(x)]^2 \, dx$$

is bounded for all ξ, h remaining fixed. It is manifest that the particular value of h chosen is of no moment in this definition, for if a function is nearly bounded for one value of h, it is nearly bounded for any other value of h.

Definition II. *We shall say that $\varphi(x, y)$ converges almost in the mean to $f(x)$ over the interval (a, b) if*

$$\lim_{\xi\to\infty} \int_\xi^{\xi+h} dy \int_a^b [\varphi(x, y) - f(x)]^2 dx = 0.$$

We shall represent this fact by the equation

$$\underset{y\to\infty}{\text{l. a. m.}}\, \varphi(x, y) = f(x).$$

If $\varphi(x, y)$ *converges in the mean to* $f(x)$ *as* $y \to \infty$, *we shall write*

$$\underset{y\to\infty}{\text{l. i. m.}}\, \varphi(x, y) = f(x).$$

Definition III. *We shall write*

$$\overset{b}{\underset{a}{Q}}\{f(x)\} = \lim_{h\to 0} \frac{1}{2h} \int_a^b [f(x+h) - f(x-h)]^2 dx$$

if this latter limit exists. In any case, we shall write

$$\overline{\overset{b}{\underset{a}{Q}}}\{f(x)\} = \overline{\lim_{h\to 0}} \frac{1}{2h} \int_a^b [f(x+h) - f(x-h)]^2 dx.$$

$\underline{Q}$ receives a like definition.

Definition IV. *We shall write*

$$M\{f(x)\} = \lim_{T\to\infty} \frac{1}{2T} \int_{-T}^{T} f(x)\,dx$$

if this latter limit exists.

Definition V. *If* $f(x)$ *is nearly bounded and*

$$\lim_{T\to\infty} \frac{1}{2T} \int_{\xi-T}^{\xi+T} f(x) f(x+h)\,dx$$

exists uniformly in ξ *and* h *as* T *becomes infinite, we shall say that* $f(x)$ *is well-behaved.*

Definition VI. *We shall write*

$$C(f; h) = \lim_{T\to\infty} \frac{1}{2T} \int_{-T}^{T} f\left(x - \frac{h}{2}\right) f\left(x + \frac{h}{2}\right) dx.$$

Definition VII. *We shall understand by*

$$\overset{A}{\underset{0}{I}}\, \varphi(x)\,d\psi(x)$$

the expression

$$\varphi(A)\psi(A) - \int_0^A \psi(x)\varphi'(x)\,dx.$$

Definition VIII. *If $f(x)$ is a summable function with summable square, we shall call $\tau = \tau(f, \varepsilon)$ a displacement number pertaining to the positive number ε if for all ξ,*

$$\int_{\xi}^{\xi+1} [f(x+\tau) - f(x)]^2 dx \leqq \varepsilon.$$

The summable function $f(x)$ with square summable over any finite range will then be termed pseudoperiodic if to every positive ε there corresponds a length $l = l(\varepsilon) > 0$ such that every interval $\alpha < x < \beta$ of length $\beta - \alpha = l$ contains at least one displacement number $\tau = \tau(f, \varepsilon)$.

The main theorems which we shall prove are the following.

Theorem I. *If $f(x)$ is nearly bounded,*

$$\gamma(\alpha) = \underset{T\to\infty}{\text{l. i. m.}} \frac{1}{\pi} \int_{-T}^{T} f(\lambda) \frac{\sin \alpha\lambda}{\lambda} d\lambda$$

and

$$\delta(\alpha) = \underset{T\to\infty}{\text{l. i. m.}} \frac{1}{\pi} \int_{-T}^{T} f(\lambda) \frac{e^{-\lambda^2} - \cos \alpha\lambda}{\lambda} d\lambda$$

exist over any bounded interval of α whatever[3]).

Theorem II. *If we define $\gamma(\alpha)$ and $\delta(\alpha)$ as in Theorem I, we shall have*

$$f(x) = \underset{T\to\infty}{\text{l. a. m.}} [\overset{T}{\underset{0}{I}} \cos \alpha x \, d\gamma(\alpha) + \overset{T}{\underset{0}{I}} \sin \alpha x \, d\delta(\alpha)]$$

over any finite range of x whatever.

Theorem III. *If $f(x)$ is nearly bounded, $\gamma(\alpha)$ and $\delta(\alpha)$ being interpreted as above,*

$$\frac{1}{2h} \int_{-\infty}^{\infty} \{[\gamma(\alpha+h) - \gamma(\alpha-h)]^2 + [\delta(\alpha+h) - \delta(\alpha-h)]^2\} d\alpha.$$

exists for every h, and equals

$$\frac{2}{\pi h} \int_{-\infty}^{\infty} [f(\lambda)]^2 \frac{\sin^2 h\lambda}{\lambda^2} d\lambda.$$

It follows that if $M\{[f(x)]^2\}$ exists, so do $\overset{\infty}{\underset{-\infty}{Q}} \{\gamma(\alpha)\}$ and $\overset{\infty}{\underset{-\infty}{Q}} \{\delta(\alpha)\}$, while

$$\overset{\infty}{\underset{-\infty}{Q}} \{\gamma(\alpha)\} + \overset{\infty}{\underset{-\infty}{Q}} \{\delta(\alpha)\} = 2M\{[f(x)]^2\}.$$

---

[3]) Cf. H. Hahn: *Über Fouriersche Reihen und Integrale, Bericht über die Jahres-(versammlung zu Innsbruck, Jahresbericht der deutschen Mathematikervereinigung,* **33** 1925), S. 1 07.

37*

Theorem IV. *If $f(x)$ is well-behaved, $C(f;h)$ is a continuous even function of h which assumes its maximum value for $h=0$.*

Theorem V. *If $f(x)$ is well-behaved, and*

$$f_k(x) = \frac{1}{\pi k} \int_{-\infty}^{\infty} f(\xi) \left(\frac{\sin k \frac{x-\xi}{2}}{x-\xi} \right)^2 d\xi,$$

then $M\{[f_k(x)]^2\}$ exists for every positive K, and

$$M\{[f(x)]^2\} = \lim_{k \to \infty} M\{[f_k(x)]^2\}.$$

Moreover,

$$M\{[A f_k(x) + B f_l(x)]^2\}$$

always exists.

Theorem VI. *If $f(x)$ is well-behaved, if $f_k(x)$ is defined as above, and if $\gamma_k(\alpha)$ and $\delta_k(\alpha)$ are defined in terms of $f_k(x)$ in the same way in which $\gamma(\alpha)$ and $\delta(\alpha)$ were defined in terms of $f(x)$, then*

$$\gamma_k(\alpha) = \int_0^{\alpha} \left(1 - \frac{\alpha}{k}\right) d\gamma(\alpha) \qquad [k > \alpha \geqq 0];$$

$$\gamma_k(\alpha) = \int_0^{k} \left(1 - \frac{\alpha}{k}\right) d\gamma(\alpha) \qquad [\alpha \geqq k]$$

and

$$\delta_k(\alpha) = \int_0^{\alpha} \left(1 - \frac{\alpha}{k}\right) d\delta(\alpha) + C \qquad [k > \alpha \geqq 0];$$

$$\delta_k(\alpha) = \int_0^{k} \left(1 - \frac{\alpha}{k}\right) d\delta(\alpha) + C \qquad [\alpha \geqq k].$$

Theorem VII. *If*

$$\psi(\alpha) = \int_0^{\alpha} F(\alpha)\, d\varphi(\alpha) \qquad [-\tau - \varepsilon < \alpha < \tau + \varepsilon];$$

where $F(\alpha)$ is a bounded summable function satisfying the Lipschitz condition

$$|F(\alpha + h) - F(\alpha)| < M|h|$$

for all α and h, if φ is summable and of summable square, and if $\underset{-T}{\overset{T}{Q}}\{\varphi(\alpha)\}$ and $\underset{-T}{\overset{T}{Q}}\{\psi(\alpha)\}$ exist [T finite], then

$$\underset{-T}{\overset{T}{Q}}\{\psi(\alpha)\} = \lim_{h \to 0} \frac{1}{2h} \int_{-T}^{T} \{F(\alpha)[\varphi(\alpha + h) - \varphi(\alpha - h)]\}^2 d\alpha.$$

Theorem VIII. *Let $f(x)$ be a well-behaved function, and let $\gamma_k(\alpha)$ and $\delta_k(\alpha)$ be defined as above. We shall write*

$$R\{f(x); k\} = \lim_{\varepsilon \to 0} \left[\overset{\infty}{\underset{-\infty}{Q}} \left\{ \frac{k\gamma_k(\alpha) - (k-\varepsilon)\gamma_{k-\varepsilon}(\alpha)}{\varepsilon} \right\} + \overset{\infty}{\underset{-\infty}{Q}} \left\{ \frac{k\delta_k(\alpha) - (k-\varepsilon)\delta_{k-\varepsilon}(\alpha)}{\varepsilon} \right\} \right];$$

$$[k > 0]$$

$$R\{f(x); 0\} = 0.$$

Then $R\{f; \alpha\}$ is an increasing bounded non-negative function of α, such that

$$C(f; h) = \frac{1}{2} \int_0^\infty \cos \alpha h \, d R\{f; \alpha\}.$$

Theorem IX. *Let $f(x)$ be a well-behaved function. Then we may write $C(f; h) = F(h) + G(h)$, where $F(h)$ is almost periodic, and*

$$M\{[G(h)]^2\} = 0.$$

We shall be able to write

$$F(h) = \int_0^\infty \cos \alpha h \, d\Phi(\alpha)$$

and

$$G(h) = \int_0^\infty \cos \alpha h \, d\Psi(\alpha),$$

where $\Psi(\alpha)$ is a bounded continuous increasing function such that $\Psi(0) = 0$, while $\Phi(\alpha)$ is a bounded increasing function which has zero derivative except at a denumerable set of points. If $f(x)$, besides being well-behaved, is pseudoperiodic, $G(h)$ vanishes, and $C(f; h)$ is almost periodic.

Theorem X. *Let $f_1(x), \ldots, f_n(x)$ be a finite set of pseudoperiodic functions. Then given any positive ε, there is a number L such that in any interval $(a, a+L)$ there is a displacement number τ pertaining to ε and common to all the functions $f_1, \ldots, f_n$.*

Theorem XI. *If every function of the sequence $\{f_n(x)\}$ is pseudoperiodic, and if $f(x)$ is a nearly bounded function such that*

$$\lim_{n \to \infty} \int_\xi^{\xi+1} [f_n(x) - f(x)]^2 dx = 0$$

uniformly in ξ, then $f(x)$ is itself pseudoperiodic.

Theorem XII. *If $f(x)$ and $g(x)$ are pseudoperiodic, so are $c_1 f(x) + c_2 g(x)$, $f(x) \cos kx$, and $f(x) \sin kx$.*

Theorem XIII. *If $f(x)$ is pseudoperiodic, it is well-behaved.*

Theorem XIV. *If $f(x)$ is pseudoperiodic, the expression*

$$\frac{1}{T}\int_{\xi}^{\xi+T} f(x)\,dx$$

approaches a definite limit uniformly in ξ as $T\to\infty$.

Theorem XV. *Let $f(x)$ be a pseudoperiodic function. Let $\lambda_1, \ldots, \lambda_N$ be an arbitrary finite set of distinct positive numbers, and let $a_0, \ldots, a_N$, $b_1, \ldots, b_N$ be an arbitrary set of real numbers. We then have*

$$M\left\{\left[f(x) - \sum_1^N (a_n \cos\lambda_n x + b_n \sin\lambda_n x) - a_0\right]^2\right\}$$

$$= M\{[f(x)]^2\} - 2\sum_1^N\left\{[M\{f(x)\cos\lambda_n x\}]^2 + [M\{f(x)\sin\lambda_n x\}]^2\right\}$$

$$-[M\{f(x)\}]^2 + 2\sum_1^N\left\{\left[\frac{a_n}{2} - M\{f(x)\cos\lambda_n x\}\right]^2\right\}$$

$$+\left[\frac{b_n}{2} - M\{f(x)\sin\lambda_n x\}\right]^2 + [a_0 - M\{f(x)\}]^2.$$

Theorem XVI. *If $f(x)$ is a pseudoperiodic function, we have*

$$[M\{f(x)\}]^2 = \frac{1}{2}R\{f; +0\}.$$

Theorem XVII. *If $f(x)$ is a pseudoperiodic function, we have*

$$2\left\{[M\{f(x)\sin\lambda x\}]^2 + [M\{f(x)\cos\lambda x\}]^2\right\}$$

$$= \frac{1}{2}R\{f; \lambda+0\} - \frac{1}{2}R\{f; \lambda-0\}.$$

Theorem XVIII. *If $f(x)$ is a pseudoperiodic function, there is only a denumerable or finite set of values $\Lambda_n > 0$ of λ for which either*

$$M\{f(x)\sin\lambda x\} \neq 0$$

or

$$M\{f(x)\cos\lambda x\} \neq 0.$$

We have

$$M\{[f(x)]^2\} = [M\{f(x)\}]^2 + 2\sum_1^\infty\left\{[M\{f(x)\cos\Lambda_n x\}]^2 + [M\{f(x)\sin\Lambda_n x\}]^2\right\}.$$

Theorem XIX. *Using the notation of the last theorem, we have*

$$\lim_{N\to\infty} M\left\{\left[f(x) - M\{f(x)\} - 2\sum_1^N M\{f(\xi)\cos\Lambda_n(\xi - x)\}\right]^2\right\} = 0,$$

whenever $f(x)$ is pseudoperiodic.

Theorem XX. *Let $f(x)$ be pseudoperiodic. If a and $b > a$ are any two finite numbers, and if $\Lambda_1', \Lambda_2', \ldots, \Lambda_n', \ldots$ are the complete set of Λ_n's over the interval $a < \lambda < b$, let*

$$\sum |M\{f(x)\cos \Lambda_n' x\}|$$

and

$$\sum |M\{f(x)\sin \Lambda_n' x\}|$$

be finite. Then

$$\left.\begin{aligned} \gamma(\alpha) &= M\{f(x)\} + 2\sum M\{f(x)\cos \Lambda_n'' x\}; \\ \delta(\alpha) &= \qquad C + 2\sum M\{f(x)\sin \Lambda_n'' x\}; \end{aligned}\right\} \qquad [\alpha > 0]$$

where the set $\{\Lambda_n''\}$ consists of those positive Λ_n's not exceeding α.

Theorem XXI. *If $f(x)$ is pseudoperiodic, and the Λ_n's are distributed in the manner indicated by the hypothesis of Theorem XX, then over any interval (a, b) we have*

$$\lim_{\beta\to\infty} \int_\beta^{\beta+1} d\alpha \int_a^b \left\{f(x) - M\{f(\xi)\} - 2\sum M\{f(\xi)\cos \Lambda_n''(\xi - x)\}\right\}^2 dx = 0.$$

In particular, if the Λ_n's form a discrete set over any finite range, and are arranged in order of magnitude, we can pick out a sequence of values of N for which

$$\lim_{N\to\infty} \int_a^b \left\{f(x) - M\{f(\xi)\} - 2\sum_1^N M\{f(\xi)\cos \Lambda_n(\xi - x)\}\right\}^2 dx = 0.$$

Theorem XXII. *Let $\gamma(\alpha)$ and $\delta(\alpha)$ be any two functions of limited total variation over every finite range, assuming only a denumerable set of values, defined for all positive arguments, and vanishing at the origin. Furthermore, let*

$$\sum_0^\infty \left\{\left[\int_n^{n+1} |d\gamma(\alpha)|\right]^2 + \left[\int_n^{n+1} |d\delta(\alpha)|\right]^2\right\}$$

converge, where $\int_a^b |d\varphi(\alpha)|$ stands for the total variation of $\varphi(\alpha)$ over the interval (a, b). Then

$$\operatorname*{l.i.m.}_{T\to\infty} \left[\int_0^T \cos \alpha x\, d\gamma(\alpha) + \int_0^T \sin \alpha x\, d\delta(\alpha)\right]$$

exists, and is pseudoperiodic.

Theorem XXII is a generalization of the Riesz-Fischer theorem. It is intimately connected with the fact that every periodic function which is summable and of summable square is pseudoper dic. It is manifest that not all such periodic functions are almost periodic in the sense of Bohr, for Bohr's functions are of necessity continuous. The author of the present paper has not yet proved that the condition for pseudoperiodicity in this theorem is necessary as well as sufficient, and indeed does not believe that such is the case. It seems to him, however, to point definitely in the direction in which we must proceed if we ever are to obtain such a necessary and sufficient condition. It also seems to him that the prospects of obtaining such a condition are much better in the case of pseudoperiodicity than in that of almost-periodicity, where the purely adventitious notion of continuity plays such a prominent role.

The cardinal notion in this paper is that of $\overset{b}{\underset{a}{Q}}\{f(x)\}$, and the cardinal theorem is Theorem III. This notion and a like theorem have already been used by the author in the discussion of Fourier series[4]).

Another notion which merits especial mention is that of $C(f; h)$. The expression $C(f; h)/C(f; 0)$ represents the coefficient of correlation of $f(x)$ with the function which is obtained from it by a translation of h. This notion has already been employed by Mr. G. I. Taylor in a very important paper on *Diffusion by continuous movements*[5]). It is the opinion of the author of the present paper that this notion, together with the allied notion of well-behaved function, is destined to have important applications in fields remote from that of pseudoperiodic functions[6]).

§ 1.

The representation of a nearly bounded function by a trigonometrical integral

Let $f(x)$ be a nearly bounded function. Let T be any number larger than 2. Then

$$\int_{-T}^{T} f(\lambda)\frac{\sin\alpha\lambda}{\lambda}\,d\lambda = \int_{-T}^{-2} f(\lambda)\frac{\sin\alpha\lambda}{\lambda}\,d\lambda + \int_{-2}^{2} f(\lambda)\frac{\sin\alpha\lambda}{\lambda}\,d\lambda + \int_{2}^{T} f(\lambda)\frac{\sin\alpha\lambda}{\lambda}\,d\lambda$$

$$= \int_{-T}^{T} g(\lambda)\sin\alpha\lambda\,d\lambda + \int_{-2}^{2} f(\lambda)\frac{\sin\alpha\lambda}{\lambda}\,d\lambda\,.$$

---

[4]) *The quadratic variation of a function and its Fourier coefficients*, *Journal of Mathematics and Physics of the Massachusetts Institute of Technology* 3, pp. 72–94.

[5]) *Proc. Lond. Math. Soc.* (2), 20 (1920), pp. 196–212.

[6]) The author has recently received a note from Professor Bohr, to the effect that pseudoperiodicity together with certain allied notions have been treated by Stefanoff in a paper soon to appear in the *Annalen* July 21, 1925).

where $g(x) = 0 \ (-2 < x < 2)$ and $g(x) = f(x)/x$ otherwise. If N is any integer greater than 2, we shall have

$$\int_{-N}^{N} [g(\lambda)]^2 d\lambda = \int_{2}^{3} \frac{[f(\lambda)]^2}{\lambda^2} d\lambda + \int_{3}^{4} \frac{[f(\lambda)]^2}{\lambda^2} d\lambda + \dots + \int_{N-1}^{N} \frac{[f(\lambda)]^2}{\lambda^2} d\lambda$$

$$+ \int_{-3}^{-2} \frac{[f(\lambda)]^2}{\lambda^2} d\lambda + \dots + \int_{-N}^{1-N} \frac{[f(\lambda)]^2}{\lambda^2} d\lambda$$

$$\leqq \left[\frac{2}{2^2} + \frac{2}{3^2} + \dots + \frac{2}{(N-1)^2}\right] \max \int_{\xi}^{\xi+1} [f(\lambda)]^2 d\lambda \leqq 2 \max \int_{\xi}^{\xi+1} [f(\lambda)]^2 d\lambda .$$

Hence $\int_{-\infty}^{\infty} [g(\lambda)]^2 d\lambda$ is finite. Now, Plancherel has substantially proved in the paper which we have cited that if $\int_{-\infty}^{\infty} [g(\lambda)]^2 d\lambda$ is finite,

$$\int_{-T}^{T} g(\lambda) \sin \alpha\lambda \, d\lambda$$

converges in the mean over $(-\infty, \infty)$ to a limit which is defined except for a set of values of α of zero measure. Hence

$$\gamma(\alpha) = \underset{T\to\infty}{\text{l. i. m.}} \int_{-T}^{T} \frac{f(\lambda) \sin \alpha\lambda}{\lambda} d\lambda$$

exists, where the l. i. m. is taken over any finite range of α. A similar proof applies in the case of the existence of

$$\delta(\alpha) = \underset{T\to\infty}{\text{l. i. m.}} \int_{-T}^{T} f(\lambda) \frac{e^{-\lambda^2} - \cos \alpha\lambda}{\lambda} d\lambda .$$

We have thus established Theorem I. As to Theorem II, what we wish to prove is that

$$\lim_{\xi\to\infty} \int_{\xi}^{\xi+1} dy \int_{a}^{b} \Big[f(x) - \cos yx\,\gamma(y) - \sin yx\,\delta(y)$$

$$+ x \int_{0}^{y} \sin \alpha x\,\gamma(\alpha) d\alpha - x \int_{0}^{y} \cos \alpha x\,\delta(\alpha) d\alpha \Big]^2 dx = 0 .$$

Now,

$$f(x) - \cos yx\,\gamma(y) - \sin yx\,\delta(y) + x \int_{0}^{y} \sin \alpha x\,\gamma(\alpha) d\alpha - x \int_{0}^{y} \cos \alpha x\,\delta(\alpha) d\alpha \Big]^2 dx$$

$$= \int_{-b}^{b} \Big\{ f(x) - \underset{T\to\infty}{\text{l. i. m.}} \Big[\cos yx \int_{-T}^{T} f(\lambda) \frac{\sin \lambda y}{\lambda} d\lambda + \sin yx \int_{-T}^{T} f(\lambda) \frac{e^{-\lambda^2} - \cos \lambda y}{\lambda} d\lambda \Big]$$

$$+ \frac{x}{\pi} \int_{0}^{y} \underset{T\to\infty}{\text{l. i. m.}} \Big[\sin \alpha x \int_{-T}^{T} f(\lambda) \frac{\sin \lambda\alpha}{\lambda} d\lambda - \cos \alpha x \int_{-T}^{T} f(\lambda) \frac{e^{-\lambda^2} - \cos \lambda\alpha}{\lambda} d\lambda \Big] d\alpha \Big\}^2 dx$$

$$= \int_{-b}^{b} \Big\{ f(x) - \underset{T\to\infty}{\text{l. i. m.}} \frac{1}{\pi} \int_{-T}^{T} \frac{f(\lambda)}{\lambda} [\sin y(\lambda - x) + e^{-\lambda^2} \sin yx]\, d\lambda$$

$$+ \frac{x}{\pi} \int_0^y \Big[\underset{T\to\infty}{\text{l. i. m.}} \int_{-T}^{T} \frac{f(\lambda)}{\lambda} [\cos\alpha(\lambda - x) - e^{-\lambda^2} \cos \alpha x]\, d\lambda \Big] d\alpha \Big\}^2 dx$$

$$= \int_{-b}^{b} \Big\{ f(x) - \underset{T\to\infty}{\text{l. i. m.}} \frac{1}{\pi} \int_{-T}^{T} \frac{f(\lambda)}{\lambda} [\sin y(\lambda - x) + e^{-\lambda^2} \sin yx]\, d\lambda$$

$$+ \frac{x}{\pi} \lim_{T\to\infty} \int_{-T}^{T} \frac{f(\lambda)}{\lambda} \Big[\frac{\sin y(\lambda - x)}{\lambda - x} - \frac{e^{-\lambda^2} \sin yx}{x} \Big] d\lambda \Big\}^2 dx$$

$$= \int_{-b}^{b} \Big\{ f(x) - \underset{T\to\infty}{\text{l. i. m.}} \frac{1}{\pi} \int_{-T}^{T} f(\lambda) \frac{\sin y(\lambda - x)}{\lambda - x} d\lambda \Big\}^2 dx.$$

Hence Theorem II will be established if we can show that whenever $f(x)$ is almost bounded,

$$\lim_{\xi\to\infty} \int_{\xi}^{\xi+1} dy \int_{-b}^{b} \Big\{ f(x) - \underset{T\to\infty}{\text{l. i. m.}} \frac{1}{\pi} \int_{-T}^{T} f(\lambda) \frac{\sin y(\lambda - x)}{\lambda - x} d\lambda \Big\}^2 dx = 0.$$

Now,

$$\int_{\xi}^{\xi+1} dy \int_{-b}^{b} \Big\{ f(x) - \underset{T\to\infty}{\text{l. i. m.}} \frac{1}{\pi} \int_{-T}^{T} f(\lambda) \frac{\sin y(\lambda - x)}{\lambda - x} d\lambda \Big\}^2 dx$$

$$\leq 2 \int_{\xi}^{\xi+1} dy \int_{-b}^{b} \Big\{ f(x) - \frac{1}{\pi} \int_{-T_1}^{T_1} f(\lambda) \frac{\sin y(\lambda - x)}{\lambda - x} d\lambda \Big\}^2 dx \qquad [T_1 > b]$$

$$+ 4 \int_{\xi}^{\xi+1} dy \int_{-b}^{b} \Big\{ \underset{T\to\infty}{\text{l. i. m.}} \frac{1}{\pi} \int_{-T}^{-T_1} f(\lambda) \frac{\sin y(\lambda - x)}{\lambda - x} d\lambda \Big\}^2 dx$$

$$+ 4 \int_{\xi}^{\xi+1} dy \int_{-b}^{b} \Big\{ \underset{T\to\infty}{\text{l. i. m.}} \frac{1}{\pi} \int_{T_1}^{T} f(\lambda) \frac{\sin y(\lambda - x)}{\lambda - x} d\lambda \Big\}^2 dx$$

$$= 2 \int_{\xi}^{\xi+1} dy \int_{-b}^{b} \Big\{ f(x) - \frac{1}{\pi} \int_{-T_1}^{T_1} f(\lambda) \frac{\sin y(\lambda - x)}{\lambda - x} d\lambda \Big\}^2 dx$$

$$+ \frac{4}{\pi^2} \int_{-b}^{b} \Big\{ \lim_{T\to\infty} \int_{\xi}^{\xi+1} dy \Big\{ \int_{-T}^{-T_1} f(\lambda) \frac{\sin y(\lambda - x)}{\lambda - x} d\lambda \Big\}^2 \Big\} dx$$

$$+ \frac{4}{\pi^2} \int_{-b}^{b} \Big\{ \lim_{T\to\infty} \int_{\xi}^{\xi+1} dy \Big\{ \int_{T_1}^{T} f(\lambda) \frac{\sin y(\lambda - x)}{\lambda - x} d\lambda \Big\}^2 \Big\} dx$$

$$= 2\int_{\xi}^{\xi+1} dy \int_{-b}^{b} \left\{ f(x) - \frac{1}{\pi} \int_{-T_1}^{T_1} f(\lambda) \frac{\sin y(\lambda - x)}{\lambda - x} d\lambda \right\}^2 dx$$

$$+ \frac{4}{\pi^2} \int_{-b}^{b} dx \left\{ \int_{T_1}^{T} \frac{f(\lambda)}{\lambda - x} d\lambda \int_{T_1}^{T} \frac{f(\mu)}{\mu - x} d\mu \int_{\xi}^{\xi+1} \sin y(\lambda - x) \sin y(\mu - x) dy \right.$$

$$\left. + \int_{-T}^{-T_1} \frac{f(\lambda)}{\lambda - x} d\lambda \int_{-T}^{-T_1} \frac{f(\mu)}{\mu - x} d\mu \int_{\xi}^{\xi+1} \sin y(\lambda - x) \sin y(\mu - x) dy \right\}$$

$$= 2\int_{\xi}^{\xi+1} dy \left\{ \int_{-b}^{b} f(x) - \frac{1}{\pi} \int_{-T_1}^{T_1} f(\lambda) \frac{\sin y(\lambda - x)}{\lambda - x} d\lambda \right\}^2 dx$$

$$+ \frac{4}{\pi^2} \int_{-b}^{b} dx \left\{ \int_{T_1}^{T} \frac{f(\lambda)}{\lambda - x} d\lambda \int_{T_1}^{T} \frac{f(\mu)}{\mu - x} d\mu \left\{ \frac{\cos\left(\xi + \frac{1}{2}\right)(\lambda - \mu) \sin \frac{\lambda - \mu}{2}}{\lambda - \mu} \right.\right.$$

$$\left. - \frac{\cos\left(\xi + \frac{1}{2}\right)(\lambda + \mu - 2x) \sin \frac{\lambda + \mu - 2x}{2}}{\lambda + \mu - 2x} \right\}$$

$$+ \int_{-T}^{-T_1} \frac{f(\lambda)}{\mu - \lambda} d\lambda \int_{-T}^{-T_1} \frac{f(\mu)}{\mu - x} d\mu \left\{ \frac{\cos\left(\xi + \frac{1}{2}\right)(\lambda - \mu) \sin \frac{\lambda - \mu}{2}}{\lambda - \mu} \right.$$

$$\left.\left. - \frac{\cos\left(\xi + \frac{1}{2}\right)(\lambda + \mu - 2x) \sin \frac{\lambda + \mu - 2x}{2}}{\lambda + \mu - 2x} \right\}\right\}$$

$$\leqq 2\int_{\xi}^{\xi+1} dy \int_{-b}^{b} \left\{ f(x) - \frac{1}{\pi} \int_{-T_1}^{T_1} f(\lambda) \frac{\sin y(\lambda - x)}{\lambda - x} d\lambda \right\}^2 dx$$

$$+ \frac{8b}{\pi^2} \max_{x \leqq b} \int_{T_1}^{T} \frac{|f(\lambda)|}{\lambda - x} d\lambda \int_{T_1}^{T} \frac{|f(\mu)|}{\mu - x} d\mu \left\{ \frac{1}{1 + |\lambda - \mu|} + \frac{1}{1 + (\lambda + \mu - 2x)} \right\}$$

$$+ \frac{8b}{\pi^2} \max_{x \leqq b} \int_{T_1}^{T} \frac{|f(-\lambda)|}{\mu - x} d\mu \int_{T_1}^{T} \frac{|f(-\mu)|}{\mu - x} d\mu \left\{ \frac{1}{1 + |\lambda - \mu|} + \frac{1}{1 + (\lambda + \mu - 2x)} \right\}$$

$$\leqq 2\int_{\xi}^{\xi+1} dy \int_{-b}^{b} \left\{ f(x) - \frac{1}{\pi} \int_{-T_1}^{T_1} f(\lambda) \frac{\sin y(\lambda - x)}{\lambda - x} d\lambda \right\}^2 dx$$

$$+ \frac{8b}{\pi^2} \sum_{j=0}^{\infty} \sum_{k=0}^{\infty} \int_{T_1+j}^{T_1+j+1} \frac{|f(\lambda)|}{\lambda - b} d\lambda \int_{T_1+k}^{T_1+k+1} \frac{|f(\mu)|}{\mu - b} \left\{ \frac{1}{1 + |\lambda - \mu|} + \frac{1}{1 + (\lambda + \mu - 2b)} \right\}$$

$$+\frac{8b}{\pi^2}\sum_{j=0}^{\infty}\sum_{k=0}^{\infty}\int_{T_1+j}^{T_1+j+1}\frac{|f(-\lambda)|}{\lambda-b}d\lambda\int_{T_1+k}^{T_1+k+1}\frac{|f(-\mu)|}{\mu-b}\left\{\frac{1}{1+|\lambda-\mu|}+\frac{1}{1+(\lambda+\mu-2b)}\right\}$$

$$\leqq 2\int_{\xi}^{\xi+1}dy\int_{-b}^{b}\left\{f(x)-\frac{1}{\pi}\int_{-T_1}^{T_1}f(\lambda)\frac{\sin y(\lambda-x)}{\lambda-x}d\lambda\right\}^2dx$$

$$+\frac{16b}{\pi^2}\max\int_{\xi}^{\xi+1}[f(x)]^2dx\sum_{j=0}^{\infty}\sum_{k=0}^{\infty}\frac{1}{T_1+j-b}\cdot\frac{1}{T_1+k-b}\cdot\left\{\frac{1}{1+|j-k|}\right.$$

$$\left.+\frac{1}{1+2T_1+j+k-2b}\right\}$$

$$\leqq 2\int_{\xi}^{\xi+1}dy\int_{-b}^{b}\left\{f(x)-\frac{1}{\pi}\int_{-T_1}^{T_1}f(\lambda)\frac{\sin y(\lambda-x)}{\lambda-x}d\lambda\right\}^2dx$$

$$+\frac{16b}{\pi^2}\max\int_{\xi}^{\xi+1}[f(x)]^2dx\left\{\sum_{j=0}^{\infty}\sum_{k=0}^{\infty}\frac{1}{(T_1+j-b)(T_1+k-b)(1+j-k)}+\sum_{j=0}^{\infty}\left(\frac{1}{T_1+j-b}\right.\right.$$

$$\left.+\sum_{j=0}^{\infty}\sum_{k=0}^{\infty}\frac{1}{(T_1+j-b)(T_1+k-b)(1+2T_1+j+k-2b)}\right\}$$

$$\leqq 2\int_{\xi}^{\xi+1}dy\int_{-b}^{b}\left\{f(x)-\frac{1}{\pi}\int_{-T_1}^{T_1}f(\lambda)\frac{\sin y(\lambda-x)}{\lambda-x}d\lambda\right\}^2dx$$

$$+\frac{16b}{\pi^2}\max\int_{\xi}^{\xi+1}[f(x)]^2dx\left\{2\int_{T_1-b}^{\infty}\frac{du}{u-1}\int_{T_1-b}^{v}\frac{dv}{(v-1)(1+u-v)}\right.$$

$$\left.+\int_{T_1-b}^{\infty}\frac{du}{(u-1)^2}+\int_{T_1-b}^{\infty}\frac{du}{u-1}\int_{T_1-b}^{\infty}\frac{dv}{(v-1)(u+v-1)}\right\}$$

$$\leqq 2\int_{\xi}^{\xi+1}dy\int_{-b}^{b}\left\{f(x)-\frac{1}{\pi}\int_{-T_1}^{T_1}f(\lambda)\frac{\sin y(\lambda-x)}{\lambda-x}d\lambda\right\}^2dx$$

$$+\frac{16b}{\pi^2}\max\int_{\xi}^{\xi+1}[f(x)]^2dx\left\{4\int_{T_1-b-1}^{\infty}\frac{du}{u(u+1)}\log(u+1)+\frac{1}{T_1-b-1}\int_{T_1-b-1}^{\infty}\frac{du}{u^2}\log(u+1\right.$$

for all sufficiently large values of T_1. Since the infinite integrals thus obtained are convergent, we can choose T_1 so large that

$$\frac{16b}{\pi^2}\max\int_{\xi}^{\xi+1}[f(x)]^2dx\left\{4\int_{T_1-b-1}^{\infty}\frac{\log(u+1)\,du}{u(u+1)}+\frac{1}{T_1-b-1}\right.$$

$$\left.+\int_{T_1-b-1}^{\infty}\frac{\log(u+1)\,du}{u^2}\right\}<\varepsilon/2$$

for all values of ξ, f being nearly bounded. The expression

$$\frac{1}{\pi}\int_{-T_1}^{T_1} f(\lambda)\frac{\sin y(\lambda-x)}{\lambda-x}d\lambda$$

is everywhere less in absolute value than some expression

$$\frac{A}{B+|\lambda|}$$

and so is of summable square from $-\infty$ to ∞. Let

$$g(x)=f(x); \qquad [-T_1 \leqq x \leqq T_1]$$
$$g(x)=0. \qquad [|x|>T_1]$$

Then both $g(x)$ and

$$h(x)=\frac{1}{\pi}\int_{-T_1}^{T_1} f(\lambda)\frac{\sin y(\lambda-x)}{\lambda-x}d\lambda$$

may be developed in Fourier integrals convergent in the mean, since they are both summable and of summable square over the interval $(-\infty,\infty)$. We shall have

$$\underset{A\to\infty}{\text{l. i. m.}}\int_0^A h(x)\sin ux\,dx = \underset{A\to\infty}{\text{l. i. m.}}\frac{1}{\pi}\int_0^A \sin ux\,dx\int_{-T_1}^{T_1} f(\lambda)\,d\lambda\frac{\sin y(\lambda-x)}{\lambda-x}\,d\lambda$$

$$=\underset{A\to\infty}{\text{l. i. m.}}\frac{1}{\pi}\int_{-T_1}^{T_1} f(\lambda)\,d\lambda\int_0^A\frac{\sin ux\sin y(\lambda-x)}{\lambda-x}dx = \frac{1}{\pi}\int_{-T_1}^{T_1} f(\lambda)\,d\lambda\int_0^\infty\frac{\sin ux\sin y(\lambda-x)}{\lambda-x}dx$$

$$=-\frac{1}{\pi}\int_{-T_1}^{T_1} f(\lambda)\,d\lambda\int_{\lambda}^{-\infty}\sin u(\lambda-v)\frac{\sin yv}{v}\,dv$$

$$=\frac{1}{\pi}\int_{-T_1}^{T_1} f(\lambda)\,d\lambda\int_{-\infty}^{\lambda}(\sin u\lambda\cos uv-\cos u\lambda\sin uv)\frac{\sin yv}{v}\,dv$$

$$=\frac{1}{\pi}\int_{-T_1}^{T_1} f(\lambda)\,d\lambda\left[\frac{\sin u\lambda}{2}\int_{-\infty}^{\lambda}\frac{\sin(y+u)v+\sin(y+u)v}{v}dv\right.$$

$$\left.-\frac{\cos u\lambda}{2}\int_{-\infty}^{\lambda}\frac{\cos(y-u)v-\cos((y+u)v}{v}dv\right]$$

$$\begin{cases} =\int_0^\infty g(x)\sin ux\,dx; & [u<y] \\ =\frac{1}{2}\int_0^\infty g(x)\sin ux\,dx; & [u=y] \\ =0 & [u>y[\end{cases}$$

Similarly

$$\underset{A\to\infty}{\text{l. i. m.}} \int_0^A h(x) \cos ux\, dx = \int_0^\infty g(x) \cos ux\, dx; \qquad [u < y]$$

$$= \frac{1}{2}\int_0^\infty g(x) \cos ux\, dx; \qquad [u = y]$$

$$= 0. \qquad [u > y]$$

Is thus follows that as y increases indefinitely, the integrals of the squares of the sine and cosine Fourier transforms of $g(x) - h(x)$ tend to zero. Since, however, the integral of the square of a function vanishes with those of its Fourier transforms, it follows that

$$\lim_{y\to\infty} \int_{-\infty}^{\infty} \left\{ g(x) - \frac{1}{\pi} \int_{-T_1}^{T_1} f(\lambda) \frac{\sin y(\lambda - x)}{\lambda - x} d\lambda \right\}^2 dx = 0.$$

Hence Ξ may be chosen so large that for $\xi \geqq \Xi$,

$$2\int_{\xi}^{\xi+1} dy \int_{-b}^{b} \left\{ f(x) - \frac{1}{\pi} \int_{-T_1}^{T_1} f(\lambda) \frac{\sin y(\lambda - x)}{\lambda - x} d\lambda \right\}^2 dx < \varepsilon/2.$$

We shall then have for $\xi \geqq \Xi$,

$$\int_{\xi}^{\xi+1} dy \int_{-b}^{b} \left\{ f(x) - \underset{T\to\infty}{\text{l. i. m.}} \frac{1}{\pi} \int_{-T}^{T} f(\lambda) \frac{\sin y(\lambda - x)}{\lambda - x} d\lambda \right\}^2 dx < \varepsilon.$$

This completes the proof of Theorem II.

§ 2.

The mean of the square of a nearly bounded function.

Let us now turn to Theorem III. We have

$$\frac{1}{2h} \int_{-\infty}^{\infty} \left\{ [\gamma(\alpha + h) - \gamma(\alpha - h)]^2 + [\delta(\alpha + h) - \delta(\alpha - h)]^2 \right\} dh$$

$$= \frac{1}{2h} \int_{-\infty}^{\infty} \left\{ \left[\frac{2}{\pi} \underset{T\to\infty}{\text{l. i. m.}} \int_{-T}^{T} f(\lambda) \frac{\sin \alpha\lambda \cos h\lambda}{\lambda} d\lambda \right]^2 \right.$$

$$\left. + \left[\frac{2}{\pi} \underset{T\to\infty}{\text{l. i. m.}} \int_{-T}^{T} f(\lambda) \frac{\sin \alpha\lambda \sin h\lambda}{\lambda} d\lambda \right]^2 \right\} dh$$

$$= \frac{1}{2h}\int_{-\infty}^{\infty}\left\{\left[\frac{2}{\pi}\,\underset{T\to\infty}{\mathrm{l.\,i.\,m.}}\int_0^T [f(\lambda)+f(-\lambda)]\frac{\sin\alpha\lambda\cos h\lambda}{\lambda}\,d\lambda\right]^2\right.$$

$$\left.+\left[\frac{2}{\pi}\,\underset{T\to\infty}{\mathrm{l.\,i.\,m.}}\int_0^T [f(\lambda)-f(-\lambda)]\frac{\sin\alpha\lambda\sin h\lambda}{\lambda}\,d\lambda\right]^2\right\}dh$$

$$= \frac{1}{2h}\left[\frac{2}{\pi}\int_0^{\infty}[f(\lambda)+f(-\lambda)]^2\frac{\sin^2\alpha\lambda}{\lambda^2}\,d\lambda+\frac{2}{\pi}\int_0^{\infty}[f(\lambda)-f(-\lambda)]^2\frac{\sin^2\alpha\lambda}{\lambda^2}\,d\lambda\right]$$

$$= \frac{2}{\pi h}\int_0^{\infty}\left\{[f(\lambda)]^2+[f(-\lambda)]^2\right\}\frac{\sin^2\alpha\lambda}{\lambda^2}\,d\lambda = \frac{2}{\pi h}\int_{-\infty}^{\infty}[f(\lambda)]^2\frac{\sin^2\alpha\lambda}{\lambda^2}\,d\lambda.$$

Moreover,

$$\frac{2}{\pi h}\int_{-\infty}^{\infty}[f(\lambda)]^2\frac{\sin^2 h\lambda}{\lambda^2}\,d\lambda = \frac{2}{\pi h}\int_{-\frac{N\pi}{h}}^{\frac{N\pi}{h}}[f(\lambda)]^2\frac{\sin^2 h\lambda}{\lambda^2}\,d\lambda$$

$$+\frac{2}{\pi h}\int_{-\infty}^{-\frac{N\pi}{h}}[f(\lambda)]^2\frac{\sin^2 h\lambda}{\lambda^2}+\frac{2}{\pi h}\int_{\frac{N\pi}{h}}^{\infty}[f(\lambda)]^2\frac{\sin^2 h\lambda}{\lambda^2}\,d\lambda$$

$$= \frac{2}{\pi h}\int_{-\frac{N\pi}{h}}^{\frac{N\pi}{h}}[f(\lambda)]^2\frac{\sin^2 h\lambda}{\lambda^2}\,d\lambda+\frac{4\vartheta}{\pi h}\int_{\frac{N\pi}{h}}^{\infty}\left\{[f(\lambda)]^2+[f(-\lambda)]^2\right\}\frac{d\lambda}{\lambda^2}\qquad [|\vartheta|\leqq 1]$$

$$= \frac{2}{\pi h}\int_{-\frac{N\pi}{h}}^{\frac{N\pi}{h}}[f(\lambda)]^2\frac{\sin^2 h\lambda}{\lambda^2}\,d\lambda+\frac{8\vartheta_1}{\pi h}\,\frac{\max\int_{\xi}^{\xi+1}[f(\lambda)]^2}{\frac{N\pi}{h}-1}\,d\lambda\qquad [|\vartheta_1|\leqq 1]$$

$$= \frac{2}{\pi h}\int_{-\frac{N\pi}{h}}^{\frac{N\pi}{h}}[f(\lambda)]^2\frac{\sin^2 h\lambda}{\lambda^2}\,d\lambda+\frac{8\vartheta_2}{N\pi^2-1}\max\int_{\xi}^{\xi+1}[f(\lambda)]^2\,d\lambda\qquad [|\vartheta_2|\leqq 1]$$

for all sufficiently small values of h. Let it be noted that the second term of this sum vanishes with increasing N, irrespective of what happens to h. Furthermore,

$$\frac{2}{\pi h}\int_{-\frac{N\pi}{h}}^{\frac{N\pi}{h}}[f(\lambda)]^2\frac{\sin^2 h\lambda}{\lambda^2}\,d\lambda$$

$$=\int_0^{\frac{N\pi}{h}}\frac{8}{\pi h\xi^2}[\sin^2 h\xi-h\xi\sin h\xi\cos h\xi]\,d\xi\,\frac{1}{2\xi}\int_{-\xi}^{\xi}[f(\lambda)]^2\,d\lambda$$

$$=\int_0^{N\pi}\frac{8}{\pi u^2}[\sin^2 u-u\sin u\cos u]\,du\,\frac{h}{2u}\int_{-\frac{u}{h}}^{\frac{u}{h}}[f(\lambda)]^2\,d\lambda$$

$$=M\{[f(x)]^2\}\int_0^{N\pi}\frac{8}{\pi u^2}[\sin^2 u-u\sin u\cos u]\,du$$
$$+\int_0^{\varepsilon}\frac{8}{\pi u^2}[\sin^2 u-u\sin u\cos u]\,du\left\{\frac{h}{2u}\int_{-\frac{u}{h}}^{\frac{u}{h}}[f(\lambda)]^2\,d\lambda-M\{f^2\}\right\}$$
$$+\int_{\varepsilon}^{N\pi}\frac{8}{\pi u^2}[\sin^2 u-u\sin u\cos u]\,du\left\{\frac{h}{2u}\int_{-\frac{u}{h}}^{\frac{u}{h}}[f(\lambda)]^2\,d\lambda-M\{f^2\}\right\}$$

$$=M\{f^2\}\int_0^{N\pi}\frac{8}{\pi u^2}[\sin^2 u-u\sin u\cos u]\,du$$
$$+\frac{2}{\pi h}\int_{-\frac{\varepsilon}{h}}^{\frac{\varepsilon}{h}}[f(\lambda)]^2\frac{\sin^2 h\lambda}{\lambda^2}\,d\lambda+\frac{8\vartheta_3\varepsilon}{\pi}M\{f^2\}\qquad[|\vartheta_3|\leqq 1]$$
$$+\int_{\varepsilon}^{N\pi}\frac{8}{\pi u^2}[\sin^2 u-u\sin u\cos u]\,du\left\{\frac{h}{2u}\int_{-\frac{u}{h}}^{\frac{u}{h}}[f(\lambda)]^2\,d\lambda-M\{f^2\}\right\}$$

$$=M\{f^2\}\int_0^{N\pi}\frac{8}{\pi u^2}[\sin^2 u-u\sin u\cos u]\,du$$
$$+\frac{16\vartheta_4\varepsilon}{\pi}\max\int_{\xi}^{\xi+1}[f(\lambda)]^2\,d\lambda$$
$$+\int_{\varepsilon}^{N\pi}\frac{8}{\pi u^2}[\sin^2 u-u\sin u\cos u\,du]\left\{\frac{h}{2u}\int_{-\frac{u}{h}}^{\frac{u}{h}}[f(\lambda)]^2\,d\lambda-M\{f^2\}\right\}\qquad[|\vartheta_4|\leqq 1]$$

for all sufficiently small h, in case $M\{f^2\}$ exists.

It may be demonstrated without difficulty that

$$\int_0^\infty \frac{8}{\pi u^2}[\sin^2 u - u \sin u \cos u]\, du = 2.$$

Let η be any positive quantity. Let N be any positive integer so large that

$$\left| M\{f^2\}\left[\int_0^{N\pi} \frac{8}{\pi u^2}[\sin^2 u - u \sin u \cos u]\, du - 2\right]\right| < \frac{\eta}{4}$$

and

$$\frac{8}{N\pi^2 - 1}\max \int_0^{\xi+1} [f(\lambda)]^2\, d\lambda < \frac{\eta}{2}.$$

Let ε be a positive number so chosen that

$$\frac{16\,\varepsilon}{\pi}\max \int_\xi^{\xi+1} [f(\lambda)]^2\, d\lambda < \frac{\eta}{8}.$$

Finally, let H be chosen so that if $h \leqq H$,

$$\left|\frac{h}{2\varepsilon}\int_{-\frac{\varepsilon}{h}}^{\frac{\varepsilon}{h}} [f(\lambda)]^2\, d\lambda - M\{f^2\}\right| < \frac{\eta}{256\,N}.$$

For this and all smaller values of h, we shall have

$$\left|\frac{2}{\pi h}\int_{-\infty}^{\infty} [f(\lambda)]^2 \frac{\sin^2 h\lambda}{\lambda^2}\, d\lambda - 2\,M\{f^2\}\right| < \frac{\eta}{2} + \frac{\eta}{4} + \frac{\eta}{8} + \frac{\eta}{16} < \eta.$$

Since η is arbitrary, this proves that

$$\overset{\infty}{\underset{\infty}{Q}}\{\gamma(\alpha)\} + \overset{\infty}{\underset{-\infty}{Q}}\{\delta(\alpha)\} = 2\,M\{f^2\}.$$

§ 3.

Well-behaved functions.

Let us now turn to Theorems IV, V, and VI. To begin with

$$\frac{1}{2T}\int_{-T}^{T} f\left(\xi - \frac{h}{2}\right) f\left(\xi + \frac{h}{2}\right) d\xi$$

is a continuous function of h over any finite range if only $f(x)$ is summable and of summable square over every finite range. This may be seen by a consideration of the Fourier development of $f(x)$ over any interval containing $(-T, T)$. It will be seen that

$$\frac{1}{2T}\int_{-T}^{T} f\left(\xi - \frac{h}{2}\right) f\left(\xi + \frac{h}{2}\right) d\xi - \frac{1}{2T}\int_{-T}^{T} f\left(\xi - \frac{h+\varepsilon}{2}\right) f\left(\xi + \frac{h+\varepsilon}{2}\right) d\xi$$

will be expressed by a series which will be uniformly dominated by a convergent series in the squares of the **Fourier** coefficients of $f(x)$, while the individual terms of the series will all tend toward 0 with ε. Hence

$$\lim_{\varepsilon \to 0} \frac{1}{2T}\int_{-T}^{T} f\left(\xi - \frac{h+\varepsilon}{2}\right) f\left(\xi + \frac{h+\varepsilon}{2}\right) d\xi = \frac{1}{2T}\int_{-T}^{T} f\left(\xi - \frac{h}{2}\right) f\left(\xi + \frac{h}{2}\right) d\xi,$$

which is what we wish to establish.

We also have

$$\frac{1}{2T}\int_{-T}^{T} f\left(\xi - \frac{h}{2}\right) f\left(\xi + \frac{h}{2}\right) d\xi$$
$$\leqq \sqrt{\frac{1}{2T}\int_{-T}^{T}\left[f\left(\xi - \frac{h}{2}\right)\right]^2 d\xi \, \frac{1}{2T}\int_{-T}^{T}\left[f\left(\xi + \frac{h}{2}\right)\right]^2 d\xi}.$$

Hence $\frac{1}{2T}\int_{-T}^{T} f\left(\xi - \frac{h}{2}\right) f\left(\xi + \frac{h}{2}\right) d\xi$ is bounded, if $f(x)$ is nearly bounded.

If $f(x)$ is well-behaved, $C(f;h)$ is a uniform limit of

$$\frac{1}{2T}\int_{-T}^{T} f\left(\xi - \frac{h}{2}\right) f\left(\xi + \frac{h}{2}\right) d\xi,$$

and being a uniform limit of continuous functions, is continuous. Moreover,

$$\frac{1}{2T}\int_{-T}^{T}\left[f\left(\xi - \frac{h}{2}\right)\right]^2 d\xi = \frac{1}{2T}\int_{-T-\frac{h}{2}}^{T-\frac{h}{2}} [f(\xi)]^2 d\xi$$
$$= \frac{1}{2T}\int_{-T}^{T} [f(\xi)]^2 d\xi + \frac{1}{2T}\int_{-T-\frac{h}{2}}^{-T} [f(\xi)]^2 d\xi - \frac{1}{2T}\int_{T-\frac{h}{2}}^{T} [f(\xi)]^2 d\xi$$
$$= \frac{1}{2T}\int_{-T}^{T} [f(\xi)]^2 d\xi + \frac{\vartheta}{2}\max\int_{\eta}^{\eta+\frac{h}{2}} [f(\xi)]^2 d\xi \qquad [|\vartheta| \leqq 1].$$

A similar relation will hold for $\frac{1}{2T}\int_{-T}^{T}\left[f\left(\xi + \frac{h}{2}\right)\right]^2 d\xi$. Hence if $f(x)$ is well-behaved

$$C(f;0) = \lim_{T\to\infty} \frac{1}{2T}\int_{-T}^{T}\left[f\left(\xi - \frac{h}{2}\right)\right]^2 d\xi = \lim_{T\to\infty} \frac{1}{2T}\int_{-T}^{T}\left[f\left(\xi + \frac{h}{2}\right)\right] d\xi.$$

It follows that

$$C(f;h) \leqq C(f;0).$$

An argument of the same type will prove that

$$C(f;-h) = C(f;h).$$

This completes the proof of Theorem IV.

As to Theorem V, we see that

$$\frac{1}{2T}\int_{-T}^{T}[f_k(x)]^2\,dx = \frac{1}{2T}\int_{-T}^{T}\left\{\frac{1}{\pi k}\int_{-\infty}^{\infty} f(\xi)\left[\frac{\sin k\dfrac{x-\xi}{2}}{x-\xi}\right]^2 d\xi\right.$$

$$= \frac{1}{2T}\int_{-T}^{T} dx\left\{\frac{1}{2\pi k}\int_{-\infty}^{\infty} f(x+2u)\frac{\sin^2 ku}{u^2}\,du\right\}\left\{\frac{1}{2\pi k}\int_{-\infty}^{\infty} f(x+2v)\frac{\sin^2 kv}{v^2}\,dv\right\}$$

$$= \frac{1}{4\pi^2k^2}\int_{-\infty}^{\infty}\frac{\sin^2 ku}{u^2}\,du\int_{-\infty}^{\infty}\frac{\sin^2 kv}{v^2}\,dv\left\{\frac{1}{2T}\int_{-T}^{T} f(x+2u)f(x+2v)\,dx\right\}.$$

By an argument similar to that which we employed in the case of Theorem IV, it follows that $M\{f_k^2\}$ exists, and that

$$M\{f_k^2\} = \frac{1}{4\pi^2k^2}\int_{-\infty}^{\infty}\frac{\sin^2 ku}{u^2}\,du\int_{-\infty}^{\infty}\frac{\sin^2 kv}{v^2}\,dv\, C(f;2u-2v).$$

Hence

$$M\{f^2\} - M\{f_k^2\}$$

$$= \frac{1}{4\pi^2k^2}\int_{-\infty}^{\infty}\frac{\sin^2 ku}{u^2}\,du\int_{-\infty}^{\infty}\frac{\sin^2 kv}{v^2}\,dv\,[C(f;0) - C(f;2u-2v)]$$

$$\leqq \frac{1}{\pi k}\int^{\infty}\frac{\sin^2 ku}{u^2}\,du\, M\{f^2\}$$

$$- \frac{1}{4\pi^2k^2}\int_{-\varepsilon}^{\varepsilon}\frac{\sin^2 ku}{u^2}\,du\int_{-\varepsilon}^{\varepsilon}\frac{\sin^2 kv}{v^2}\,dv\,[C(f;0) - C(f;2u-2v)].$$

On account of the continuity of C, the second term in this sum may be made less in absolute value than $\eta/2$ by making ε sufficiently small. The first term can then be made less than $\eta/2$ by taking k sufficiently large. Hence for sufficiently large values of k,

$$|M\{f^2\} - M\{f_k^2\}| < \eta,$$

and

$$\lim_{k\to\infty} M\{f_k^2\} = \{f^2\}.$$

38*

The proof of the existence of

$$M\{[Af_k(x)+Bf_1(x)]^2\}$$

follows essentially similar lines.

We now come to Theorem VI. Let us put

$$\Gamma_k(\alpha)=\int_0^\alpha\left(1-\frac{\alpha}{k}\right)d\gamma(\alpha) \qquad [k>\alpha>0];$$

$$\Gamma_k(\alpha)=\int_0^k\left(1-\frac{\alpha}{k}\right)d\gamma(\alpha) \qquad [\alpha\geqq k]$$

and

$$\Delta_k(\alpha)=\int_0^\alpha\left(1-\frac{\alpha}{k}\right)d\delta(\alpha) \qquad [k>\alpha>0];$$

$$\Delta_k(\alpha)=\int_0^k\left(1-\frac{\alpha}{k}\right)d\delta(\alpha) \qquad [\alpha\geqq k].$$

We shall then have

$$\underset{T\to\infty}{\text{l. a. m.}}\left[\int_0^T\cos\alpha x\,d\Gamma_k(\alpha)+\int_0^T\sin\alpha x\,d\Delta_k(\alpha)\right]$$

$$=\int_0^k\cos\alpha x\,d\Gamma_k(\alpha)+\int_0^k\sin\alpha x\,d\Delta_k(\alpha)$$

$$=\cos kx\int_0^k\left(1-\frac{\alpha}{k}\right)d\gamma(\alpha)+\sin kx\int_0^k\left(1-\frac{\alpha}{k}\right)d\delta(\alpha)$$

$$+\lambda\int_0^k\sin\alpha x\,d\alpha\int_0^\alpha\left(1-\frac{\alpha}{k}\right)d\gamma(\alpha)-x\int_0^k\cos\alpha x\,d\alpha\int_0^\alpha\left(1-\frac{\alpha}{k}\right)d\delta(\alpha)$$

$$=\frac{\cos kx\int_0^k\gamma(\alpha)\,d\alpha+\sin kx\int_0^\alpha\delta(\alpha)\,d\alpha}{k}$$

$$+x\int_0^k\left\{\sin\alpha x\left[\left(\frac{1-\alpha}{k}\right)\gamma(\alpha)+\frac{1}{k}\int_0^\alpha\gamma(\alpha)d\alpha\right]\right.$$

$$\left.-\cos\alpha x\left[\left(1-\frac{\alpha}{k}\right)\delta(\alpha)+\frac{1}{k}\int_0^\alpha\delta(\alpha)\,d\alpha\right]\right\}d\alpha$$

$$=x\int_0^k\left\{\sin\alpha x\left(1-\frac{\alpha}{k}\right)\gamma(\alpha)-\cos\alpha x\left(1-\frac{\alpha}{k}\right)\delta(\alpha)\right\}d\alpha$$

$$-\frac{1}{k}\int_0^k[(\cos\alpha x\,\gamma(\alpha)+\sin\alpha x\,\delta(\alpha)]\,d\alpha$$

$$= \int_0^k \Big\{ \Big[x \sin\alpha x \Big(1 - \frac{\alpha}{k}\Big) - \frac{\cos\alpha x}{k} \Big] \operatorname*{l.i.m.}_{T\to\infty} \frac{1}{\pi} \int_{-T}^{T} f(\lambda) \frac{\sin\alpha\lambda}{\lambda} d\lambda$$

$$- \Big[x \cos\alpha x \Big(1 - \frac{\alpha}{k}\Big) + \frac{\sin\alpha x}{k} \Big] \operatorname*{l.i.m.}_{T\to\infty} \frac{1}{\pi} \int_{-T}^{T} f(\lambda) \frac{e^{-\lambda^2} - \cos\alpha\lambda}{\lambda} d\lambda \Big\} d\alpha$$

$$= \frac{1}{\pi} \int_{-\infty}^{\infty} \frac{f(\lambda)}{\lambda} d\lambda \int_0^k \Big\{ \Big[x \sin\alpha x \Big(1 - \frac{\alpha}{k}\Big) - \frac{\cos\alpha x}{k} \Big] \sin\alpha\lambda$$

$$- \Big[x \cos\alpha x \Big(1 - \frac{\alpha}{k}\Big) + \frac{\sin\alpha x}{k} \Big] (e^{-\lambda^2} - \cos\alpha\lambda) \Big\} d\alpha$$

$$= \frac{1}{\pi} \int_{-\infty}^{\infty} \frac{f(\lambda)}{\lambda} d\lambda \int_0^k \Big\{ x \Big(1 - \frac{\alpha}{k}\Big) \cos\alpha(x - \lambda) + \frac{\sin\alpha(x-\lambda)}{k}$$

$$- e^{-\lambda^2} \Big[x \Big(1 - \frac{\alpha}{k}\Big) \cos\alpha x + \frac{\sin\alpha x}{k} \Big] \Big\} d\alpha$$

$$= \frac{1}{\pi k} \int_{-\infty}^{\infty} f(\lambda) \frac{1 - \cos k(x - \lambda)}{(x - \lambda)} d\lambda = f_k(x).$$

Since

$$f_k(x) = \mathop{I}_0^k \cos\alpha x \, d\Gamma_k(\alpha) + \mathop{I}_0^k \sin\alpha x \, d\Delta_k(\alpha),$$

we have

$$\gamma_k(\beta) = \operatorname*{l.i.m.}_{T\to\infty} \frac{1}{\pi} \int_{-T}^{T} \Big[\mathop{I}_0^k \cos\alpha\lambda \, d\Gamma_k(\alpha) + \mathop{I}_0^k \sin\alpha\lambda \, d\Delta_k(\alpha) \Big] \frac{\sin\beta\lambda}{\lambda} d\lambda$$

$$= \operatorname*{l.i.m.}_{T\to\infty} \frac{1}{\pi} \int_{-T}^{T} \frac{\sin\beta\lambda}{\lambda} d\lambda \Big[\cos\lambda k \, \Gamma_k(k) + \lambda \int_0^k \Gamma_k(\alpha) \sin\alpha\lambda \, d\alpha \Big]$$

$$= \frac{\Gamma_k(k)}{\pi} \int_{-\infty}^{\infty} \frac{\sin\beta\lambda \cos\lambda k}{\lambda} d\lambda + \operatorname*{l.i.m.}_{T\to\infty} \frac{1}{\pi} \int_{-T}^{T} \sin\beta\lambda \, d\lambda \int_0^k \Gamma_k(\alpha) \sin\alpha\lambda \, d\alpha$$

$$\begin{cases} = \dfrac{\Gamma_k(k)}{2\pi} \displaystyle\int_{-\infty}^{\infty} \frac{\sin(\beta+k)\lambda + \sin(\beta-k)\lambda}{\lambda} d\lambda + \Gamma_k(\beta) & [\beta < k] \\ = \dfrac{\Gamma_k(k)}{2\pi} \displaystyle\int_{-\infty}^{\infty} \frac{\sin(\beta+k)\lambda + \sin(\beta-k)\lambda}{\lambda} d\lambda + \Gamma_k(\beta)/2 & [\beta = k] \\ = \dfrac{\Gamma_k(k)}{2\pi} \displaystyle\int_{-\infty}^{\infty} \frac{\sin(\beta+k)\lambda + \sin(\beta-k)\lambda}{\lambda} d\lambda & [\beta > k] \end{cases}$$

$$\begin{cases} = \Gamma_k(\beta) & [\beta < k] \\ = \Gamma_k(k)/2 + \Gamma_k(\beta)/2 & [\beta = k] \\ = \Gamma_k(k) & [\beta > k] \end{cases}$$

$$= \Gamma_k(\beta).$$

By a precisely similar proof, we can show that

$$\Delta_k(\beta) = \delta_k(\beta) + C,$$

thus completing the demonstration of Theorem VI.

§ 4.

The operator $\overset{T}{\underset{-T}{Q}}\{\varphi(\alpha)\}$.

Let us now devote our attention to Theorems VII, VIII, and IX. If

$$\psi(\alpha) = \int_0^\alpha F(\alpha)\, d\varphi(\alpha)$$

we shall have

$$\overset{T}{\underset{-T}{Q}}\{\psi(\alpha)\} = \lim_{h\to 0}\frac{1}{2h}\int_{-T}^{T}\left[\int_0^{\alpha+h} F(\alpha)\,d\varphi(\alpha) - \int_0^{\alpha-h} F(\alpha)\,d\varphi(\alpha)\right]^2 d\alpha$$

$$= \lim_{h\to 0}\frac{1}{2h}\int_{-T}^{T}\left\{F(\alpha+h)\,\varphi(\alpha+h) - F(\alpha-h)\,\varphi(\alpha-h) - \int_{\alpha-h}^{\alpha+h} F'(\beta)\,\varphi(\beta)\,d\beta\right\}^2 d\alpha$$

$$= \lim_{h\to 0}\frac{1}{2h}\int_{-T}^{T}\Big\{F(\alpha+h)[\varphi(\alpha+h) - \varphi(\alpha-h)] + \varphi(\alpha-h)[F(\alpha+h) - F(\alpha-h)$$

$$-\int_{\alpha-h}^{\alpha+h} F'(\beta)\,\varphi(\beta)\,d\beta\Big\}^2 d\alpha$$

in case F satisfies the Lipschitz condition in the hypothesis of Theorem VII. Hence under this hypothesis,

$$\frac{1}{2h}\int_{-T}^{T}\{\varphi(\alpha-h)[F(\alpha+h) - F(\alpha-h)]\}^2 d\alpha \leq M^2 h\int_{-T}^{T}[\varphi(\alpha-h)]^2 d\alpha,$$

and

$$\frac{1}{2h}\int_{-T}^{T}\left[\int_{\alpha-h}^{\alpha+h} F'(\beta)\,\varphi(\beta)\,d\beta\right]^2 d\alpha \leq \frac{M^2 h}{2}\int_{-T}^{T}\left[\int_{\alpha-h}^{\alpha+h}\varphi(\beta)\,d\beta\right]^2 d\alpha$$

$$\leq \frac{M^2}{2}\int_{-T}^{T}\left[\int_{-T-h}^{T+h}[\varphi(\beta)]^2 d\beta\int_{\alpha-h}^{\alpha+h} d\beta\right] d\alpha$$

$$\leq 2M^2 h\, T\int_{-T-h}^{T+h}[\varphi(\beta)]^2 d\beta.$$

Thus by an elementary use of the Schwarz inequality, it results that under the hypothesis of Theorem VII,

$$\overset{T}{\underset{-T}{Q}}\{\psi(\alpha)\} = \lim_{h\to 0}\frac{1}{2h}\int_{-T}^{T}\{F(\alpha)[\varphi(\alpha+h)-\varphi(\alpha-h)]\}^2\,d\alpha.$$

Let us apply this in particular to the case where

$$\begin{aligned} F(\alpha) &= 1 & &[0 \leqq |\alpha| \leqq k-\varepsilon];\\ F(\alpha) &= (k-\alpha)/\varepsilon & &[k-\varepsilon \leqq |\alpha| \leqq k];\\ F(\alpha) &= 0 & &[k \leqq \alpha]. \end{aligned}$$

We shall then have

$$\overset{\infty}{\underset{-\infty}{Q}}\left\{\frac{k\gamma_k(\alpha)-(k-\varepsilon)\gamma_{k-\varepsilon}(\alpha)}{\varepsilon}\right\} + \overset{\infty}{\underset{-\infty}{Q}}\left\{\frac{k\delta_k(\alpha)-(k-\varepsilon)\delta_{k-\varepsilon}(\alpha)}{\varepsilon}\right\}$$

$$= \overset{k+0}{\underset{-k-0}{Q}}\left\{\int_0^{\alpha} F(\alpha)\,d\gamma(\alpha)\right\} + \overset{k+0}{\underset{-k-0}{Q}}\left\{\int_0^{\alpha} F(\alpha)\,d\delta(\alpha)\right\}$$

$$= \lim_{h\to 0}\frac{1}{2h}\int_{-k-0}^{k+0}\{F(\alpha)\}^2\{[\gamma(\alpha+h)-\gamma(\alpha-h)]^2+[\delta(\alpha+h)-\delta(\alpha-h)]^2\}\,d\alpha.$$

Hence

$$\overset{\overline{k-\varepsilon}}{\underset{-k+\varepsilon}{Q}}\{\gamma(\alpha)\} + \overset{\overline{k-\varepsilon}}{\underset{-k+\varepsilon}{Q}}\{\delta(\alpha)\} \leqq R\{f(x);k\} \leqq \overset{k+\varepsilon}{\underset{\underline{-k-\varepsilon}}{Q}}\{\gamma(\alpha)\} + \overset{k+\varepsilon}{\underset{\underline{-k-\varepsilon}}{Q}}\{\delta(\alpha)\}.$$

Since $R\{f(x);k\}$ and

$$\frac{1}{2h}\int_{-k}^{k}\{[\gamma(\alpha+h)-\gamma(\alpha-h)]^2+[\delta(\alpha+h)-\delta(\alpha-h)]^2\}\,d\alpha$$

are both increasing functions of k, we may conclude from this that $R\{f(x);k\} =$

$$\underset{h\to 0}{\text{l. i. m.}}\frac{1}{2h}\int_{-k}^{k}\{[\gamma(\alpha+h)-\gamma(\alpha-h)]^2+[\delta(\alpha+h)-\delta(\alpha-h)]^2\}\,d\alpha.$$

Furthermore, since

$$\lim_{k\to\infty}\overset{\infty}{\underset{-\infty}{Q}}\{\gamma_k(\alpha)\} = \overset{\infty}{\underset{-\infty}{Q}}\{\gamma(\alpha)\}; \qquad \lim_{k\to\infty}\overset{\infty}{\underset{-\infty}{Q}}\{\delta_k(\alpha)\} = \overset{\infty}{\underset{-\infty}{Q}}\{\delta(\alpha)\}$$

we may conclude that

$$\lim_{k\to\infty} R\{f(x);k\} = \overset{\infty}{\underset{-\infty}{Q}}\{\gamma(\alpha)\} + \overset{\infty}{\underset{-\infty}{Q}}\{\delta(\alpha)\}.$$

We have

$$\frac{1}{2}\int_0^\infty \cos\alpha h\, dR\{f;\alpha) = \lim_{n\to\infty}\frac{1}{2}\int_0^{(n+\frac{1}{2})\pi}\cos\alpha h\, dR\{f;\alpha\} =$$

$$\lim_{n\to\infty}\frac{h}{2}\int_0^{(n+\frac{1}{2})\pi} R\{f;\alpha\}\sin\alpha h\, d\alpha = \lim_{n\to\infty}\frac{h}{2}\int_0^{(n+\frac{1}{2})\pi}\sin kh$$

$$\left\{\underset{j\to 0}{\text{l. i. m.}}\frac{1}{2j}\int_{-k}^{k}\{[\gamma(\beta+j)-\gamma(\beta-j)]^2+[\delta(\beta+j)-\delta(\beta-j)]^2\}d\beta\right\}dk =$$

$$\lim_{n\to\infty}\lim_{j\to 0}\frac{h}{4j}\int_0^{(n+\frac{1}{2})\pi}\sin kh$$

$$\left\{\int_{-k}^{k}\{[\gamma(\beta+j)-\gamma(\beta-j)]^2+[\delta(\beta+j)-\delta(\beta-j)]^2\}d\beta\right\}dk$$

$$=\lim_{n\to\infty}\lim_{j\to 0}\frac{h}{4j}\int_{-(n+\frac{1}{2})\pi}^{(n+\frac{1}{2})\pi}\cos\alpha h$$

$$\{[\gamma(\alpha+j)-\gamma(\alpha-j)]^2+[\delta(\alpha+j)-\delta(\alpha-j)]^2\}d\alpha$$

$$=\lim_{n\to\infty}\lim_{j\to 0}\frac{1}{4j}\int_{-(n+\frac{1}{2})\pi}^{(n+\frac{1}{2})\pi}\left[\cos^2\frac{\alpha h}{2}-\sin^2\frac{\alpha h}{2}\right]$$

$$\{[\gamma(\alpha+j)-\gamma(\alpha-j)]^2+[\delta(\alpha+j)-\delta(\alpha-j)]^2\}d\alpha$$

$$=\lim_{j\to 0}\frac{1}{4j}\int_{-\infty}^{\infty}\left[\cos^2\frac{\alpha h}{2}-\sin^2\frac{\alpha h}{2}\right]$$

$$\{\gamma(\alpha+j)-\gamma(\alpha-j)]^2+[\delta(\alpha+j)-\delta(\alpha-j)]^2\}d\alpha.$$

On the other hand

$$C(f;h)=\lim_{T\to\infty}\frac{1}{2T}\int_{-T}^{T}f\left(x-\frac{h}{2}\right)f\left(x+\frac{h}{2}\right)dx$$

$$=\lim_{T\to\infty}\frac{1}{8T}\int_{-T}^{T}\left\{\left[f\left(x+\frac{h}{2}\right)+f\left(x-\frac{h}{2}\right)\right]^2-\left[f\left(x+\frac{h}{2}\right)-f\left(x-\frac{h}{2}\right)\right]^2\right\}dx$$

$$=\frac{1}{4}\left\{M\left\{\left[f\left(x+\frac{h}{2}\right)+f\left(x-\frac{h}{2}\right)\right]^2\right\}-M\left\{\left[f\left(x+\frac{h}{2}\right)-f\left(x-\frac{h}{2}\right)\right]^2\right\}\right\}.$$

If we take

$$f(x)=\underset{T\to\infty}{\text{l. a. m.}}\left[\overset{T}{\underset{0}{I}}\cos\alpha x\, d\gamma(\alpha)+\overset{T}{\underset{0}{I}}\sin\alpha x\, d\delta(\alpha)\right],$$

we see that

$$\frac{f\left(x+\frac{h}{2}\right)+f\left(x-\frac{h}{2}\right)}{2} = \underset{T\to\infty}{\text{l. a. m.}} \left| \int_0^T \left[\frac{\cos\alpha\left(x+\frac{h}{2}\right)+\cos\alpha\left(x-\frac{h}{2}\right)}{2}\right] d\gamma(\alpha) + \int_0^T \left[\frac{\sin\alpha\left(x+\frac{h}{2}\right)+\sin\alpha\left(x-\frac{h}{2}\right)}{2}\right] d\delta(\alpha)\right|$$

$$= \underset{T\to\infty}{\text{l. a. m.}} \left[\int_0^T \cos\alpha x \cos\frac{\alpha h}{2}\, d\gamma(\alpha) + \int_0^T \sin\alpha x \cos\frac{\alpha h}{2}\, d\delta(\alpha)\right].$$

This leads us to suspect that

$$\text{l. i. m.}\ \frac{1}{\pi}\int_{-T}^{T} \frac{f\left(\lambda+\frac{h}{2}\right)+f\left(\lambda-\frac{h}{2}\right)}{2}\,\frac{\sin\alpha\lambda}{\lambda}\, d\lambda = \int_0^\alpha \cos\frac{\alpha h}{2}\, d\gamma(\alpha)$$

$$= \cos\frac{\alpha h}{2}\gamma(\alpha) + \frac{h}{2}\int_0^\alpha \gamma(\beta)\sin\frac{\beta h}{2}\, d\beta$$

$$= \underset{T\to\infty}{\text{l. i. m.}}\ \frac{1}{\pi}\int_{-T}^{T} f(\lambda)\,\frac{\sin\alpha\lambda\cos\frac{\alpha h}{2}}{\lambda}\, d\lambda + \lim_{T\to\infty}\frac{h}{2\pi}\int_0^\alpha \sin\frac{\beta h}{2}\, d\beta \int_{-T}^{T} f(\lambda)\frac{\sin\alpha\lambda}{\lambda}\, d\lambda$$

$$= \underset{T\to\infty}{\text{l. i. m.}}\ \frac{1}{\pi}\int_{-T}^{T} \frac{f(\lambda)}{\lambda}\, d\lambda \left[\sin\alpha\lambda\cos\frac{\alpha h}{2} + \frac{h}{2}\int_0^\alpha \sin\frac{\beta h}{2}\sin\beta\lambda\, d\beta\right]$$

$$= \underset{T\to\infty}{\text{l. i. m.}}\ \frac{1}{\pi}\int_{-T}^{T} \frac{f(\lambda)}{\lambda}\, d\lambda \left\{\sin\alpha\lambda\cos\frac{\alpha h}{2} + \frac{h}{4}\left|\frac{\sin\alpha\left(\lambda-\frac{h}{2}\right)}{\lambda-\frac{h}{2}} - \frac{\sin\alpha\left(\lambda+\frac{h}{2}\right)}{\lambda+\frac{h}{2}}\right|\right\}.$$

This may be put in the form

$$\underset{T\to\infty}{\text{l. i. m.}}\ \frac{1}{2\pi}\int_{-T}^{T} \frac{f(\lambda)}{\lambda}\, d\lambda \left\{\sin\alpha\left(\lambda-\frac{h}{2}\right)\left[1+\frac{\frac{h}{2}}{\lambda-\frac{h}{2}}\right] + \sin\alpha\left(\lambda+\frac{h}{2}\right)\left[1-\frac{\frac{h}{2}}{\lambda+\frac{h}{2}}\right]\right\}$$

$$= \underset{T\to\infty}{\text{l. i. m.}}\ \frac{1}{2\pi}\int_{-T}^{T} f(\lambda)\, d\lambda \left|\frac{\sin\alpha\left(\lambda-\frac{h}{2}\right)}{\lambda-\frac{h}{2}} + \frac{\sin\alpha\left(\lambda+\frac{h}{2}\right)}{\lambda+\frac{h}{2}}\right|$$

$$= \underset{T\to\infty}{\text{l. i. m.}}\ \frac{1}{2\pi}\int_{-T}^{T} \left[f\left(\lambda+\frac{h}{2}\right)+f\left(\lambda-\frac{h}{2}\right)\right]\frac{\sin\alpha\lambda}{\lambda}\, d\lambda.$$

This proves our conjecture. Similarly,

$$\underset{T\to\infty}{\mathrm{l.i.m.}}\frac{1}{\pi}\int_{-T}^{T}\frac{f\left(\lambda+\frac{h}{2}\right)+f\left(\lambda-\frac{h}{2}\right)}{2}\,\frac{e^{-\lambda^2}-\cos\alpha\lambda}{\lambda}\,d\lambda=\int_0^{\alpha}\cos\frac{\alpha h}{2}\,d\delta(\alpha)+C;$$

$$\underset{T\to\infty}{\mathrm{l.i.m.}}\frac{1}{\pi}\int_{-T}^{T}\frac{f\left(\lambda+\frac{h}{2}\right)-f\left(\lambda-\frac{h}{2}\right)}{2}\,\frac{\sin\alpha\lambda}{\lambda}\,d\lambda=-\int_0^{\alpha}\sin\frac{\alpha h}{2}\,d\gamma(\alpha);$$

$$\underset{T\to\infty}{\mathrm{l.i.m.}}\frac{1}{\pi}\int_{-T}^{T}\frac{f\left(\lambda+\frac{h}{2}\right)-f\left(\lambda-\frac{h}{2}\right)}{2}\,\frac{e^{-\lambda^2}-\cos\alpha\lambda}{\lambda}\,d\lambda=\int_0^{\alpha}\sin\frac{\alpha h}{2}\,d\delta(\alpha)+C_1.$$

Combining these with Theorem III, we may conclude that

$$C(f;h)=\frac{1}{2}\left\{\overset{\infty}{\underset{-\infty}{Q}}\left\{\int_0^{\alpha}\cos\frac{\alpha h}{2}\,d\gamma(\alpha)\right\}+\overset{\infty}{\underset{-\infty}{Q}}\left\{\int_0^{\alpha}\cos\frac{\alpha h}{2}\,d\delta(\alpha)\right\}\right\}$$

$$-\frac{1}{2}\left\{\overset{\infty}{\underset{-\infty}{Q}}\left\{\int_0^{\alpha}\sin\frac{\alpha h}{2}\,d\gamma(\alpha)\right\}+\overset{\infty}{\underset{-\infty}{Q}}\left\{\int_0^{\alpha}\sin\frac{\alpha h}{2}\,d\delta(\alpha)\right\}\right\}$$

$$=\lim_{j\to 0}\frac{1}{4j}\int_{-\infty}^{\infty}\left\{\left\{\int_0^{\alpha+j}\cos\frac{\alpha h}{2}\,d\gamma(\alpha)-\int_0^{\alpha-j}\cos\frac{\alpha h}{2}\,d\gamma(\alpha)\right\}^2\right.$$

$$+\left\{\int_0^{\alpha+j}\cos\frac{\alpha h}{2}\,d\delta(\alpha)-\int_0^{\alpha-j}\cos\frac{\alpha h}{2}\,d\delta(\alpha)\right\}^2$$

$$\left.-\left\{\int_0^{\alpha-j}\sin\frac{\alpha h}{2}\,d\gamma(\alpha)-\int_0^{\alpha-j}\sin\frac{\alpha h}{2}\,d\gamma(\alpha)\right\}^2-\left\{\int_0^{\alpha+j}\sin\frac{\alpha h}{2}\,d\delta(\alpha)-\int_0^{\alpha-j}\sin\frac{\alpha h}{2}\,d\delta(\alpha)\right\}^2\right\}$$

$$=\lim_{j\to 0}\frac{1}{4j}\left\{\int_{-T}^{T}\left[\cos^2\frac{\alpha h}{2}-\sin^2\frac{\alpha h}{2}\right]\right.$$

$$\{[\gamma(\alpha+j)-\gamma(\alpha-j)]^2+[\delta(\alpha+j)-\delta(\alpha-j)]^2\}\,d\alpha$$

$$\left.+\vartheta\left[\int_{-\infty}^{-T}+\int_{T}^{\infty}\right]\{[\gamma(\alpha+j)-\gamma(\alpha-j)]^2+[\delta(\alpha+j)-\delta(\alpha-j)]^2\}\,d\alpha\right\},$$

ϑ being less in absolute value than 2. The proof of this is in every essential respect similar to the proof of Theorem VII. Furthermore, we can deduce from the fact that

$$\lim_{k\to\infty}\overset{\infty}{\underset{-\infty}{Q}}\{\gamma_k(\alpha)\}=\overset{\infty}{\underset{-\infty}{Q}}\{\gamma(\alpha)\};\qquad \lim\overset{\infty}{\underset{-\infty}{Q}}\{\delta_k(\alpha)\}=\overset{\infty}{\underset{-\infty}{Q}}\{\delta(\alpha)\},$$

that T can be chosen so large that for all sufficiently small values of j

$$2\left[\int_{-\infty}^{-T}+\int_{T}^{\infty}\right]\{[\gamma(\alpha+j)-\gamma(\alpha-j)]^2+[\delta(\alpha+j)-\delta(\alpha-j)]^2\}\,d\alpha<\varepsilon.$$

Hence

$$C(f;h)=\lim_{j\to 0}\frac{1}{4j}\int_{-\infty}^{\infty}\left[\cos^2\frac{\alpha h}{2}-\sin^2\frac{\alpha h}{2}\right]\{[\gamma(\alpha+j)-\gamma(\alpha-j)]^2$$

$$+[\delta(\alpha+j)-\delta(\alpha-j)]^2\}\,d\alpha=\frac{1}{2}\int_0^{\infty}\cos\alpha h\,dR\{f;\alpha)\}.$$

We have already verified the boundedness of R. Its increasing character is obvious from Theorem VII. Thus Theorem VIII is established. Since R is increasing and bounded, it is continuous except for a denumerable set of finite jumps with finite sum. We may thus put

$$R\{f;\alpha\}=2\,\Phi(\alpha)+2\,\Psi(\alpha),$$

as in the statement of Theorem IX, where Ψ will be continuous, and Φ will be constant except at the jumps of R. Let us now consider

$$F(h)=\int_0^{\infty}\cos\alpha h\,d\Phi(\alpha).$$

The integral here may be taken as a Stieltjes integral in the classical sense, so that $F(h)$ reduces to an expression of the type

$$\sum A_j\cos B_j h,$$

where $\sum A_j$ converges, and all the A's are positive. As Bohr has shown, $F(h)$ will then be almost periodic.

On the other hand, let us consider

$$G(h)=\int_0^{\infty}\cos\alpha h\,d\Psi(\alpha).$$

It may be verified without difficulty that

$$\overset{\infty}{\underset{-\infty}{Q}}\{\Psi(\alpha)\}=0.$$

It may also be verified that

$$\underset{T\to\infty}{\mathrm{l.i.m.}}\frac{1}{\pi}\int_{-T}^{T}G(\lambda)\frac{\sin\alpha\lambda}{\lambda}\,d\lambda=\Psi(\alpha);\qquad \underset{T\to\infty}{\mathrm{l.i.m.}}\frac{1}{\pi}\int_{-T}^{T}G(\lambda)\frac{e^{-\lambda^2}-\cos\alpha\lambda}{\lambda}\,d\lambda=0.$$

To begin with,

$$\frac{1}{\pi}\int_{-T}^{T}G(\lambda)\frac{\sin\alpha\lambda}{\lambda}\,d\lambda=\frac{1}{\pi}\lim_{H\to\infty}\int_{-T}^{T}\frac{\sin\alpha\lambda}{\lambda}\,d\lambda\int_0^{H}\cos\beta\lambda\,d\Psi(\beta)$$

$$=\frac{1}{\pi}\lim_{H\to\infty}\int_0^{H}d\Psi(\beta)\int_{-T}^{T}\frac{\sin\alpha\lambda\cos\beta\lambda}{\lambda}\,d\lambda$$

$$=\frac{1}{\pi}\lim_{H\to\infty}\left[\Psi(H)\int_{-T}^{T}\frac{\sin\alpha\lambda\cos H\lambda}{\lambda}d\lambda+\int_0^H\Psi(\beta)\,d\beta\int_{-T}^{T}\sin\alpha\lambda\sin\beta\lambda\,d\lambda\right]$$

$$=\frac{1}{\pi}\lim_{H\to\infty}\left[\Psi(H)\int_{-T}^{T}\frac{\sin\alpha\lambda\cos H\lambda}{\lambda}d\lambda+\int_0^H\Psi(\beta)\left[\frac{\sin(\alpha-\beta)T}{\alpha-\beta}-\frac{\sin(\alpha+\beta)T}{\alpha+\beta}\right]d\beta\right]$$

$$=\frac{1}{\pi}\left[\Psi(\infty)\lim_{H\to\infty}\int_{-T}^{T}\frac{\sin\alpha\lambda\cos H\lambda}{\lambda}d\lambda+\int_0^\infty[\Psi(u+\alpha)-\Psi(u-\alpha)]\frac{\sin uT}{u}du\right]$$

$$=\frac{1}{\pi}\int_0^\infty[\Psi(u+\alpha)-\Psi(u-\alpha)]\frac{\sin uT}{u}du.$$

Hence

$$\underset{T\to\infty}{\text{l. i. m.}}\ \frac{1}{\pi}\int_{-T}^{T}G(\lambda)\frac{\sin\alpha\lambda}{\lambda}d\lambda=\lim_{T\to\infty}\frac{1}{\pi}\int_0^\infty[\Psi(u+\alpha)-\Psi(u-\alpha)]\frac{\sin uT}{u}du$$

$$=\frac{\Psi(\alpha)-\Psi(-\alpha)}{2}=\Psi(\alpha).$$

The rest of the proof results from the oddness of Ψ.

The thus see that

$$M\{[G(h]^2\}=0.$$

Moreover,

$$C(f;h)=F(h)+G(h).$$

Hence if $C(f;h)$ is almost periodic, G is almost periodic and vanishes identically, and

$$C(f;h)=F(h)=\sum A_j\cos B_j h.$$

Now, if τ is a displacement number pertaining to ε for $f(x)$, we have

$$\left|\frac{1}{2T}\int_{-T}^{T}f\left(x+\tau-\frac{h}{2}\right)f\left(x+\tau+\frac{h}{2}\right)dx-\frac{1}{2T}\int_{-T}^{T}f\left(x-\frac{h}{2}\right)f\left(x+\frac{h}{2}\right)dx\right|$$

$$\leqq\frac{1}{8T}\left\{\left|\int_{-T}^{T}\left[\left(f\left(x+\tau+\frac{h}{2}\right)+f\left(x+\tau-\frac{h}{2}\right)\right)^2-\left(f\left(x+\frac{h}{2}\right)+f\left(x-\frac{h}{2}\right)\right)^2\right]dx\right|\right.$$

$$\left.+\left|\int_{-T}^{T}\left[\left(f\left(x+\tau+\frac{h}{2}\right)-f\left(x+\tau-\frac{h}{2}\right)\right)^2-\left(f\left(x+\frac{h}{2}\right)-f\left(x-\frac{h}{2}\right)\right)^2\right]dx\right|\right\}$$

$$\leqq\frac{1}{8T}\left\{\sqrt{\int_{-T}^{T}\left[f\left(x+\tau+\frac{h}{2}\right)+f\left(x+\tau-\frac{h}{2}\right)-f\left(x+\frac{h}{2}\right)-f\left(x-\frac{h}{2}\right)\right]^2dx}\right.$$

$$\overline{\int_{-T}^{T}\left[f\left(x+\tau+\frac{h}{2}\right)+f\left(x+\tau-\frac{h}{2}\right)+f\left(x+\frac{h}{2}\right)+f\left(x-\frac{h}{2}\right)\right]^2dx}$$

$$+\sqrt{\int_{-T}^{T}\left[f\left(x+\tau+\frac{h}{2}\right)-f\left(x+\tau-\frac{h}{2}\right)-f\left(x+\frac{h}{2}\right)+f\left(x-\frac{h}{2}\right)\right]^2 dx}$$

$$\left.\overline{\int_{-T}^{T}\left[f\left(x+\tau+\frac{h}{2}\right)-f\left(x+\tau-\frac{h}{2}\right)+f\left(x+\frac{h}{2}\right)-f\left(x-\frac{h}{2}\right)\right]^2 dx}\right\}$$

$$\leq \frac{4N}{T}\sqrt{\varepsilon \max \int_{\xi}^{\xi+1}[f(x)]^2 dx},$$

where N is the integer next greater than $2T$. Thus τ is a translation number of

$$C_T(f;h)=\frac{1}{2T}\int_{-T}^{T} f\left(x-\frac{h}{2}\right)f\left(x+\frac{h}{2}\right)dx$$

pertaining to

$$\frac{4N}{T}\sqrt{\varepsilon \max \int_{\xi}^{\xi+1}[f(x)]^2 dx}.$$

It results that if $f(x)$ is pseudoperiodic, $C_T(f;h)$ is almost periodic. However, if f is well-behaved, we have

$$C(f;h)=\lim_{T\to\infty} C_T(f;h)$$

uniformly. Hence by a theorem of Bohr, $C(f;h)$ is almost periodic, and reduces to $F(h)$. That is, $R\{f;\alpha\}$ is an increasing function of α which is constant except in the neighborhood of its denumerable set of discontinuities.

§ 5.

Pseudoperiodic functions.

Let us now turn to the study of pseudoperiodic functions. Our work here will parallel closely the work of Bohr on almost periodic functions. To begin with, we wish to prove the lemma that if τ is a displacement number of $f(x)$ pertaining to ε, then there is a quantity A depending only on f and ε, such that if $|\sigma-\tau|\leq A$, then σ is a displacement number of $f(x)$ pertaining to 2ε. As may be seen without difficulty, it is enough to prove that

$$\int_{\xi}^{\xi+1}[f(x+\lambda)-f(x-\lambda)]^2 dx$$

tends to 0 uniformly in ξ as λ tends to 0. Now, express $f(x)$ over the interval (a,b) $[a>\xi+\lambda+1>\xi-\lambda>b]$ by the Fourier series

$$\frac{\alpha_0}{2}+\sum_{1}^{\infty}\left[\alpha_n \cos n\left(\frac{a-b}{\pi}\right)x+\beta_n \sin n\left(\frac{a-b}{\pi}\right)x\right].$$

We have

$$\int_{\xi}^{\xi+1} [f(x+\lambda) - f(x-\lambda)]^2\, dx \leq \frac{4}{a-b} \sum_{1}^{\infty} (\alpha_n^2 + \beta_n^2) \sin^2 n \left(\frac{\alpha-\beta}{\pi}\right) \lambda.$$

Hence $\int_{\xi}^{\xi+1} [f(x+\lambda) - f(x-\lambda)]^2 dx$ tends uniformly to 0 over the interval (a, b).

I now wish to prove that if η is any positive number, there is a positive number ε such that if $0 < \lambda < \varepsilon$, then whatever ξ_1 may be, we have

$$\int_{\xi_1}^{\xi_1+1} [f(x+\lambda) - f(x-\lambda)]^2 dx < \eta$$

provided $f(x)$ is pseudoperiodic. To begin with, let

$$L = l\left(\frac{\eta^2}{32 \max \int_{\eta}^{\eta+1} [f(x)]^2 dx}\right)$$

in accordance with Definition VIII. There will then be a value of ξ in the interval $(0, L)$ such that

$$\int_{\xi_1}^{\xi_1+1} [f(x+\lambda) - f(x-\lambda)]^2 dx - \int_{\xi}^{\xi+1} [f(x+\lambda) - f(x-\lambda)]^2 dx$$

$$= \int_{\xi}^{\xi+1} \{[f(x+\lambda+\xi_1-\xi) - f(x-\lambda+\xi_1-\xi)]^2 - [f(x+\lambda) - f(x-\lambda)]^2\} dx$$

$$\leq 4 \sqrt{\int_{\xi}^{\xi+1} [f(x+\lambda+\xi_1-\xi) - f(x-\lambda+\xi_1-\xi) - f(x+\lambda) + f(x-\lambda)]^2 dx \max \int_{\eta}^{\eta+1} [f(x)]^2 dx}$$

$$\leq 8 \sqrt{\frac{\eta^2}{32 \max \int_{\eta}^{\eta+1} [f(x)]^2 dx} \max \int_{\eta}^{\eta+1} [f(x)]^2 dx} = \eta/2.$$

We may then choose ε so small that if $0 < \lambda < \varepsilon$,

$$\int_{\xi}^{\xi+1} [f(x+\lambda) - f(x+\lambda)]^2 dx < \eta/2$$

uniformly over the interval $(0, L)$. We shall thus have

$$\int_{\xi_1}^{\xi_1+1} [f(x+\lambda) - f(x-\lambda)]^2 dx < \eta.$$

We are now in a position to prove Theorem X. Let $l_1(u)$, $l_2(u)$, ... be expressions bearing to f_1, f_2, ... the same relation which $l(u)$ does to $f(u)$ in Definition VIII. Let A_n be a quantity such that if τ is a displacement number of $f_n(x)$ pertaining to ε, and $|\sigma - \tau| < A_n$, then σ is a displacement number of $f_n(x)$ pertaining to 2ε. Let A be any quantity greater than $l_1(\varepsilon), \ldots, l_n(\varepsilon)$, and let B be any positive quantity less than each $A_k/3$. Consider space of n dimensions divided up into cubes of side A, and also into smaller cubes with the same orientation and with side less than B and a proper part of A. Let some vertex of a cube coincide in the two systems of subdivision. Each large cube will then contain at least one small cube consisting entirely of points whose coordinates are respectively displacement numbers for $f_1, f_2, \ldots$ pertaining to ε. It is manifest that if we consider the infinite array of large cubes possessing as their vertex with least coordinates a point of the form $(hA, hA, \ldots, hA)$, some two of them will have their small cube in the same position. Let the two cubes correspond in the manner indicated to $(hA, \ldots, hA)$ and $(HA, \ldots, HA)$ respectively. Then $(H-h)A$, being the difference between two displacement numbers pertaining to 2ε for each f_n, is itself a displacement number corresponding to 8ε.

We now come to Theorem XI. Let the number ε be given arbitrarily. We choose the number $N = N(\varepsilon)$ so large that the inequality

$$\int_{\xi}^{\xi+1} [f(x) - f_n(x)]^2 dx \leqq \varepsilon/12$$

holds for all ξ and all $n \geqq N$. We then have for every τ the obvious inequality,

$$\int_{\xi}^{\xi+1} [f(x+\tau) - f(x)]^2 dx \leqq 3 \int_{\xi}^{\xi+1} [f_n(x+\tau) - f_n(x)]^2 dx + \varepsilon/2.$$

Hence every displacement number of $f_n(x)$ pertaining to $\varepsilon/6$ is manifestly a displacement number of $f(n)$ pertaining to ε. From this fact the proof follows at once.

Theorem XII is an immediate corollary of Theorem I. Theorems XIII and XIV are proved in essentially the same way, which we shall exemplify only in the case of Theorem XIII. That a pseudoperiodic function is nearly bounded is obvious. Let L be a number so large that every interval $(a, a+L)$ contains at least one displacement number of $f(x)$ pertaining to ε. We can of course find such a number if f is pseudoperiodic. Let τ be a displacement number of $f(x)$ exceeding L. Then given any interval $(n\tau, (n+1)\tau)$, it may be divided into two parts $(n\tau, n\tau+\alpha)$ and $(n\tau+\alpha, (n+1)\tau)$, in such a way that the first of these may be put into correspondence with the interval $(\tau-\alpha, \tau)$ by a translation with displacement number σ belonging to ε, while the second may be put into correspondence with the interval $(0, \tau-\alpha)$ by the displacement number $\tau+\sigma$ pertaining to 4ε. Hence

$$\int_{n\tau}^{(n+1)\tau}[f(x)]^2\,dx-\int_0^{\tau}[f(x)]^2\,dx$$

$$=\int_{\tau-\alpha}^{\tau}\{[f(x+\sigma)]^2-[f(x)]^2\}\,dx+\int_0^{\tau-\alpha}\{[f(x+\sigma+\tau)]^2-[f(x)]^2\}\,dx$$

$$\leqq\sqrt{\int_{\tau-\alpha}^{\tau}[f(x+\sigma)-f(x)]^2\,dx\int_{\tau-\alpha}^{\tau}[f(x+\sigma)+f(x)]^2\,dx}$$

$$+\sqrt{\int_0^{\tau-\alpha}[f(x+\sigma+\tau)-f(x)]^2\,dx\int_0^{\tau-\alpha}[f(x+\sigma+\tau)+f(x)]^2\,dx}$$

$$\leqq 8T\sqrt{\varepsilon\max\int_{\xi}^{\xi+1}[f(x)]^2\,dx},$$

where T is the integer next larger than τ. Remembering that if τ is a displacement number pertaining to ε for f, it pertains to 4ε for $f(x+h)+f(x-h)$, we see that

$$\left|\frac{1}{(n-m)\tau}\int_{m\tau}^{n\tau}[f(x+h)+f(x-h)]^2\,dx-\frac{1}{\tau}\int_0^{\tau}[f(x+h)+f(x-h)]^2\,dx\right|$$

$$\leqq\frac{32\,T}{\tau}\sqrt{\varepsilon\max\int_{\xi}^{\xi+1}[f(x)]^2\,dx}$$

Therefore

$$\left|\frac{1}{b-a}\int_a^b[f(x+h)+f(x-h)]^2\,dx-\frac{1}{\tau}\int_0^{\tau}[f(x+h)+f(x-h)]^2\,dx\right|$$

$$\leqq\frac{8}{b-a}\max\int_{\xi}^{\xi+1}[f(x)]^2\,dx+\frac{32\,T}{\tau}\sqrt{\varepsilon\max\int_{\xi}^{\xi+1}[f(x)]^2\,dx}$$

Hence for all sufficiently large values of τ and $b-a$, we have

$$\left|\frac{1}{b-a}\int_a^b[f(x+h)+f(x-h)]^2\,dx-\frac{1}{\tau}\int_0^{\tau}[f(x+h)+f(x-h)]^2\,dx\right|$$

$$\leqq 100\sqrt{\varepsilon\max\int_{\xi}^{\xi+1}[f(x)]^2\,dx}\,.$$

Similarly,

$$\left|\frac{1}{b-a}\int_a^b [f(x+h)-f(x-h)]^2\,dx - \frac{1}{\tau}\int_0^\tau [f(x+h)-f(x-h)]^2\,dx\right|$$
$$\leq 100\sqrt{\varepsilon \max \int_\xi^{\xi+1} [f(x)]^2\,dx}\,.$$

Hence

$$\left|\frac{1}{b-a}\int_a^b f(x+h)f(x-h)\,dx - \frac{1}{\tau}\int_0^\tau f(x+h)f(x-h)\,dx\right|$$
$$\leq 50\sqrt{\varepsilon \max \int_\xi^{\xi+1} [f(x)]^2\,dx}\,.$$

Since ε can be made arbitrarily small by taking τ sufficiently large, we see that

$$\lim_{\xi\to\infty} \frac{1}{\xi}\int_a^{a+\xi} f(x+h)f(x-h)\,dx$$

exists uniformly in h and ξ. Thus f is well-behaved.

Theorem XV is Bohr's Satz IX, with his almost periodic functions replaced by pseudoperiodic functions, and all statements made in terms of reals rather than of complex numbers. It is essentially a form of stating the Bessel's inequality, and may be proved without difficulty by squaring out and taking the mean as indicated.

Theorem XVI is one of the key theorems of the present paper. We have

$$\frac{1}{2}R\{f;+0\} = \lim_{k\to 0}\frac{1}{2}\int_0^k \left(1-\frac{\alpha}{k}\right)^2 dR\{f;\alpha\}$$

$$= \lim_{k\to 0}\frac{1}{k}\int_0^k \left(1-\frac{\alpha}{k}\right) R\{f;\alpha\}\,d\alpha$$

$$= \lim_{k\to 0}\lim_{h\to 0}\frac{1}{k}\int_0^k \left(1-\frac{\alpha}{k}\right)\frac{d\alpha}{2h}\int_{-\alpha}^{\alpha}\{[\gamma(u+h)-\gamma(u-h)]^2$$
$$+[\delta(u+h)-\delta(u-h)]^2\}\,du$$

$$= \lim_{k\to 0}\lim_{h\to 0}\frac{1}{2h}\int_0^k \left(1-\frac{\alpha}{k}\right)^2\{[\gamma(\alpha+h)-\gamma(\alpha-h)]^2$$
$$+[\delta(\alpha+h)-\delta(\alpha-h)]^2\}\,d\alpha$$

$$= \lim_{k \to 0} \frac{1}{2} \left[\overset{k}{\underset{-k}{Q}} \{\gamma_k(\alpha)\} + \overset{k}{\underset{-k}{Q}} \{\delta_k(\alpha)\} \right] = \lim_{k \to 0} \frac{1}{2} \left[\overset{\infty}{\underset{-\infty}{Q}} \{\gamma_k(\alpha)\} + \overset{\infty}{\underset{-\infty}{Q}} \{\delta_k(\alpha)\} \right]$$

$$= \lim_{k \to 0} M\{[f_k(x)]^2\} = \lim_{k \to 0} M\left\{\left[\frac{1}{\pi k} \int_{-\infty}^{\infty} f(\xi) \left(\frac{\sin k \frac{x-\xi}{2}}{x-\xi}\right)^2 d\xi\right]^2\right\}.$$

Since

$$\lim_{T \to \infty} \frac{1}{T} \int_{\xi}^{\xi+T} f(x)\, dx = M\{f(x)\}$$

uniformly in ξ, an argument essentially like that used in the demonstration of Theorem III will show that

$$\lim_{k \to 0} \frac{1}{\pi k} \int_{-\infty}^{\infty} f(\xi) \left(\frac{\sin k \frac{x-\xi}{2}}{x-\xi}\right)^2 d\xi = M\{f(x)\}$$

uniformly in x. Hence

$$[M\{f(x)\}]^2 = \frac{1}{2} R\{f; +0\}.$$

We now turn to Theorem XVII. The function $F(x) = f(x) \sin \lambda_1 x$ has belonging to it a $\Gamma_1(\alpha)$ and $\Delta_1(\alpha)$ bearing to it the same relation which $\gamma(\alpha)$ and $\delta(\alpha)$ bear to $f(x)$, and determined by the equations

$$\Gamma_1(\alpha) = \underset{T \to \infty}{\text{l. i. m.}} \frac{1}{\pi} \int_{-T}^{T} f(\lambda) \frac{\sin \lambda_1 \lambda \sin \alpha \lambda}{\lambda} d\lambda = \frac{1}{2} (\delta(\alpha + \lambda_1) - \delta(\alpha - \lambda_1));$$

$$\Delta_1(\alpha) = \underset{T \to \infty}{\text{l. i. m.}} \frac{1}{\pi} \int_{-T}^{T} f(\lambda) \sin \lambda_1 \lambda \frac{e^{-\lambda^2} - \cos \alpha \lambda}{\lambda} d\lambda$$

$$= C - \frac{1}{2} (\gamma(\alpha + \lambda_1) - \gamma(\alpha - \lambda_1)).$$

Similarly, to the function $G(x) = f(x) \cos \lambda_1 x$ will correspond

$$\Gamma_2(\alpha) = \underset{T \to \infty}{\text{l. i. m.}} \frac{1}{\pi} \int_{-T}^{T} f(\lambda) \frac{\cos \lambda_1 \lambda \sin \alpha \lambda}{\lambda} d\lambda = \frac{1}{2} (\gamma(\alpha + \lambda_1) + \gamma(\alpha - \lambda_1));$$

$$\Delta_2(\alpha) = \underset{T \to \infty}{\text{l. i. m.}} \int_{-T}^{T} f(\lambda) \cos \lambda_1 \lambda \frac{e^{-\lambda^2} - \cos \alpha \lambda}{\lambda} d\lambda$$

$$= C_1 + \frac{1}{2} (\delta(\alpha + \lambda_1) + \delta(\alpha - \lambda_1)).$$

We have

$$[M\{f(x)\sin\lambda_1 x\}]^2+[M\{f(x)\cos\lambda_1 x\}]^2$$

$$=\frac{1}{2}R\{F;+0\}+\frac{1}{2}R\{G;+0\}$$

$$=\frac{1}{2}\lim_{\varepsilon\to 0}\left\{\overset{\infty}{\underset{-\infty}{Q}}\{\Gamma_{1,\varepsilon}(\alpha)\}+\overset{\infty}{\underset{-\infty}{Q}}\{\varDelta_{1,\varepsilon}(\alpha)\}+\overset{\infty}{\underset{-\infty}{Q}}\{\Gamma_{2,\varepsilon}(\alpha)\}+\overset{\infty}{\underset{-\infty}{Q}}\varDelta_{2,\varepsilon}(\alpha)\}\right\}$$

$$=\frac{1}{4}\lim_{\varepsilon\to 0}\lim_{h\to 0}\frac{1}{2h}\int_{-\infty}^{\infty}\{[\Gamma_{1,\varepsilon}(\alpha+h)-\Gamma_{1,\varepsilon}(\alpha-h)]^2$$
$$+[\varDelta_{1,\varepsilon}(\alpha+h)-\varDelta_{1,\varepsilon}(\alpha-h)]^2+[\Gamma_{2,\varepsilon}(\alpha+h)-\Gamma_{2,\varepsilon}(\alpha-h)]^2$$
$$+[\varDelta_{2,\varepsilon}(\alpha+h)-\varDelta_{2,\varepsilon}(\alpha-h)]^2\}\,d\alpha$$

$$=\frac{1}{8}\lim_{\varepsilon\to 0}\lim_{h\to 0}\frac{1}{2h}\int_{-\infty}^{\infty}[\Phi(\alpha)]^2\{[\gamma(\alpha+h+\lambda_1)-\gamma(\alpha-h+\lambda_1)]^2$$
$$+[\gamma(\alpha+h-\lambda_1)-\gamma(\alpha-h-\lambda_1)]^2+[\delta(\alpha+h+\lambda_1)-\delta(\alpha-h+\lambda_1)]^2$$
$$+[\delta(\alpha+h-\lambda_1)-\delta(\alpha-h-\lambda_1)]^2\}\,d\alpha,$$

where

$$\Phi(\alpha)=1-\frac{|\alpha|}{\varepsilon}\qquad[|\alpha|<\varepsilon];$$

$$\Phi(\alpha)=0\qquad[|\alpha|\geqq\varepsilon].$$

Then by an argument exactly similar to that which we used in the last theorem, we can show that

$$2\{[M\{f(x)\sin\lambda_1 x\}]^2+[M\{f(x)\cos\lambda_1 x\}]^2\}$$

$$=\frac{1}{2}\lim_{\varepsilon\to 0}\int_{-\infty}^{\infty}[\Phi(\alpha)]^2 dR\{f;\lambda_1+\alpha\}=\frac{1}{2}R\{f;\lambda_1+0\}-\frac{1}{2}R\{f;\lambda_1-0\}.$$

This proves Theorem XVII. As to Theorem XVIII, the denumerability of the set of $\varLambda_n$'s follows at once from Theorem XV. It results from Theorem IX that

$$\lim_{\alpha\to\infty}R\{f;\alpha\}=R\{f;+0\}+\sum_{1}^{\infty}\{R\{f;\varLambda_n+0\}-R\{f;\varLambda_n-0\}\}.$$

Hence

$$M\{[f(x)]^2\}=\frac{1}{2}\lim_{\alpha\to\infty}R\{f;\alpha\}$$

$$=[M\{f(x)\}]^2+2\sum_{1}^{\infty}\{[M\{f(x)\sin\varLambda_k x\}]^2+[M\{f(x)\cos\varLambda_k x\}]^2\}.$$

On combining this with Theorem XV, we immediately btain Theorem XIX.

39*

Theorem **XX** may be verified by direct computation in the case where all the A_n's lie in a finite interval, for then the series

$$M\{f(\xi)\} + 2\sum_{1}^{\infty} M\{f(\xi)\cos A_n(\xi - x)\}$$

in uniformly and absolutely convergent to $f(x)$, and can be integrated termwise over any interval. Hence under any circumstances consistent with the hypothesis of Theorem **XX**, we have

$$\gamma_k(\alpha) = M\{f_k(x)\} + 2\sum M\{f_k(x)\cos A_n'' x\}$$

$$= M\{f(x)\} + 2\sum{}' M\left\{\frac{1}{\pi k}\int_{-\infty}^{\infty} f(x-u)\cos A_n'' x \frac{\sin^2\frac{ku}{2}}{u^2}\,du\right\}$$

$$= M\{f(x)\} + 2\sum{}' M\left\{\frac{1}{\pi k}\int_{-\infty}^{\infty} f(x-u)\cos A_n''(x-u)\cos A_n'' \frac{\sin^2\frac{ku}{2}}{u^2}\,du\right.$$

$$\left. + \frac{1}{\pi k}\int_{-\infty}^{\infty} f(x-u)\sin A_n''(x-u)\sin A_n'' u\frac{\sin^2\frac{ku}{2}}{u^2}\,du\right\}$$

$$= M\{f(x)\} + 2\sum M\left\{\frac{1}{2\pi k}\int_{-\infty}^{\infty} f(x-u)\cos A_n''(x-u)\frac{\cos A_n'' u\,(1-\cos ku)}{u^2}\,du\right\}$$

$$+ \frac{2}{\pi k}\lim_{T\to\infty}\frac{1}{T}\int_{\xi}^{\xi+T} dx\int_{-\infty}^{\infty} f(x-u)\sin A_n''(x-u)\sin A_n'' u\frac{\sin^2\frac{ku}{2}}{u^2}\,du$$

$$= M\{f(x)\} + 2\sum{}' M\left\{\frac{1}{2\pi k}\int_{-\infty}^{\infty} f(x-u)\cos A_n''(x-u)\right.$$

$$\left.\frac{\cos A_n'' u - \frac{1}{2}\cos(A_n''+k)u - \frac{1}{2}\cos(A_n''-k)u}{u^2}\,du\right\}$$

$$+ \frac{2}{\pi k}\int_{-\infty}^{\infty} M\{f(x)\sin A_n'' x\}\sin A_n'' u\frac{\sin^2\frac{ku}{2}}{u^2}\,du$$

$$= M\{f(x)\} + \sum\left\{-\frac{2A_n''}{K}M\{f(x)\cos A_n'' x\} + \frac{A_n''+K}{K}M\{f(x)\cos A_n'' x\}\right.$$

$$\begin{cases} + \dfrac{K - A_n''}{K} M\{f(x)\cos A_n'' x\} & [K > A_n''] \\ + \dfrac{A_n'' - K}{K} M\{f(x)\cos A_n'' x\} & [A_n'' > K]. \end{cases}$$

Hence

$$\gamma_k(\alpha) = M\{f(x)\} + 2\sum \frac{K-\Lambda_n''}{K} M\{f(x)\cos \Lambda_n'' x\},$$

where the Σ is taken over all the Λ_n'' 's less than K. It is here assumed that K is so chosen that no Λ_n'' equals K.

If we then put

$$\gamma_1(\alpha) = M\{f(x)\} + 2\sum M\{f(x)\cos \Lambda_n'' x\},$$

we see that

$$0 = \int_0^k \left(1-\frac{\alpha}{k}\right) d(\gamma(\alpha) - \gamma_1(\alpha)) = \frac{1}{k}\int_0^k [\gamma(\alpha) - \gamma_1(\alpha)]\, d\alpha$$

for all except a denumerable set of positive values of K. Hence $\gamma(\alpha)$ and $\gamma_1(\alpha)$ can only differ at a set of points of zero measure, and as $\gamma(\alpha)$ is only determined with the exception of an arbitrary set of points of zero measure, we may write $\gamma_1(\alpha) = \gamma(\alpha)$. The argument in the case of $\delta(\alpha)$ is exactly similar.

With these γ 's and δ 's, the expression

$$\int_0^\alpha \cos \alpha x\, d\gamma(\alpha) + \int_0^\alpha \sin \alpha x\, d\delta(x)$$

reduces to

$$\int_0^\alpha \cos \alpha x\, d\gamma(\alpha) + \int_0^c \sin \alpha x\, d\delta(\alpha)$$

$$= M\{f(\xi)\} - 2\sum M\{f(\xi)\cos \Lambda_n''(\xi - x)\}.$$

Since we have already proved in Theorem II that this converges almost in the mean over any finite interval to $f(x)$, Theorem XXI follows at once.

We finally come to Theorem XXII. The existence and pseudoperiodic character of

$$\int_0^T \cos \alpha x\, d\gamma(\alpha) + \int_0^T \sin \alpha x\, d\delta(\alpha)$$

are obvious, since this represents an absolutely and uniformly convergent series of pseudoperiodic functions. We hence have only to prove that

$$\lim_{T\to\infty} \limsup_{\xi} \limsup_{t_2 > t_1 > T} \int_\xi^{\xi+1} \left\{\int_{t_1}^{t_2} \cos \alpha x\, d\gamma(\alpha) + \int_{t_1}^{t_2} \sin \alpha x\, d\delta(\alpha)\right\}^2 dx = 0$$

in order to show that the functions

$$\int_0^{t_1} \cos \alpha x\, d\gamma(\alpha) + \int_0^{t_1} \sin \alpha x\, d\delta(\alpha)$$

answer to the f_n's of Theorem XI, and that

$$\underset{T\to\infty}{\text{l. i. m.}} \left[\int_0^T \cos \alpha x \, d\gamma(\alpha) + \int_0^T \sin \alpha x \, d\delta(\alpha)\right],$$

which must exist, by the Riesz-Fischer theorem, is pseudoperiodic.

Let

$$\varphi(x) = \int_{t_1}^{t_2} \cos \alpha x \, d\gamma(\alpha) + \int_{t_1}^{t_2} \sin \alpha x \, d\delta(\alpha).$$

Then the γ and δ corresponding to $\varphi(x+\eta)$ will be

$$\int_{t_1}^{\alpha} \cos \alpha \eta \, d\gamma(\alpha) + \int_{t_1}^{\alpha} \sin \alpha \eta \, d\delta(\alpha)$$

and

$$\int_{t_1}^{\alpha} \cos \alpha \eta \, d\delta(\alpha) - \int_{t_1}^{\alpha} \sin \alpha \eta \, d\gamma(\alpha)$$

for $t_1 < \alpha < t_2$. We shall have by Theorem III

$$\frac{2}{\pi}\int_{-\infty}^{\infty} [\varphi(\lambda+\eta)]^2 \frac{\sin^2\lambda}{\lambda^2} d\lambda \leqq \int_{t_1-1}^{t_2+1} \left[\int_{\alpha-1}^{\alpha+1} |d\gamma(\alpha)| + \int_{\alpha-1}^{\alpha+1} |d\delta(\alpha)|\right]^2 d\alpha$$

$$\leqq 4 \sum_N^{\infty} \left\{\left[\int_n^{n+1} |d\gamma(\alpha)|\right]^2 + \left[\int_n^{n+1} |d\delta(\alpha)|\right]^2\right\},$$

where N is the integer next less than $T-1$. Hence under the hypothesis of Theorem XXII, we have

$$\lim_{T\to\infty} \int_{-\infty}^{\infty} [\varphi(\lambda+\eta)]^2 \frac{\sin^2\lambda}{\lambda^2} \delta\lambda = 0$$

uniformly in η. Moreover,

$$\int_{\xi}^{\xi+1} \left\{\int_{t_1}^{t_2} \cos \alpha x \, d\gamma(\alpha) + \int_{t_1}^{t_2} \sin \alpha x \, d\delta(\alpha)\right\}^2 dx = \int_0^1 [\varphi(\lambda-\xi)]^2 d\lambda$$

$$\leqq \csc^2 1 \int_{-\infty}^{\infty} |\varphi(\lambda-\xi)]^2 \frac{\sin^2\lambda}{\lambda^2} d\lambda.$$

Hence

$$\lim_{T\to\infty} \limsup_{\xi} \limsup_{t_2>t_1>t} \int_{\xi}^{\xi+1} \left\{\int_{t_1}^{t_2} \cos \alpha x \, d\gamma(\alpha) + \int_{t_1}^{t_2} \sin \alpha x \, d\delta(\alpha)\right\}^2 dx = 0,$$

and Theorem XXII is established.

(Eingegangen am 15. Januar 1925.)

Appendix.

In a letter to the author from Professor R. Courant, the question has been raised whether the methods developed in this paper in any way facilitate the handling of the arithmetical part of Bohr's work. The author wishes to take advantage of the opportunity presented by the correction of the proof to append a brief sketch of the manner in which his methods may be used to derive Bohr's theorem concerning the absolute convergence of the Fourier series of an almost periodic function in the case where there are no linear relations among the frequencies. In this proof, as in that recently given by Bohr[7]), the theorem that the formal product of the Fourier series of two almost periodic functions is the Fourier series of the product is essential. In the present note, this theorem is taken in the more general form in which almost periodic functions are replaced by pseudoperiodic functions. As Bohr showed in his *Acta* paper, this theorem is equivalent to his fundamental theorem, and his elementary proof still holds when pseudoperiodic functions are being considered.

Using polynomial approximations, it is easy to show on the basis of this theorem that if $\Phi(x)$ is an almost periodic real function, the function $\overline{\Phi(x)}$ which is 1 when $\Phi(x)$ is at least half its maximum value and which is 0 otherwise is pseudoperiodic, and is of the form $\sum_1^\infty A_n e^{i\lambda_n x}$, where each λ_n is linearly dependent on a finite number of the corresponding exponents for $\Phi(x)$. Let $f(x) = \sum_1^\infty B_n e^{i\Lambda_n x}$, and let us suppose — what is no essential restriction — that no Λ_n vanishes. It is easy to see then that $\Re\{B_1 e^{i\Lambda_1 x}\}, \ldots, \Re\{B_{k-1} e^{i\Lambda_{k-1} x}\}, \Re\{\sum_1^\infty B_n e^{i\Lambda_n x}\}$ all assume positive maxima, and are hence positive at the points where $\overline{\Re\{B_1 e^{i\Lambda_1 x}\}}, \ldots, \overline{\Re\{B_{k-1} e^{i\Lambda_{k-1} x}\}}, \overline{\Re\{\sum_{n=k}^\infty B_n e^{i\Lambda_n x}\}}$ are respectively 1. Consider the function

$$\left\{\prod_{m=1}^{k-1} \overline{\Re\{B_m e^{i\Lambda_m x}\}}\right\} \overline{\Re\{\sum_{n=k}^{\infty} B_n e^{i\Lambda_n x}\}}.$$

This cannot vanish identically, for — with the exception of the exponent 0 — no exponent can be formed more than once by a combination of the

---

[7]) *Einige Sätze über Fourierreihen fastperiodischer Funktionen*, *Math. Zeitschr.* **23** (1925), S. 38—45.

exponents of the different factors. Thus some non-zero exponent will appear with a non-zero coefficient. Hence there is some quantity ξ for which at the same time

$$\Re\{B_1 e^{i\Lambda_1\xi}\} \geqq |B_1|/2;$$

$$\cdots\cdots\cdots\cdots$$

$$\Re\{B_{k-1} e^{i\Lambda_{k-1}\xi}\} \geqq |B_{k-1}|/2;$$

$$\Re\{\sum_k^\infty B_n e^{i\Lambda_n\xi}\} > 0.$$

Hence

$$2\max\Re\{f(x)\} > \sum_1^{k-1}|B_n|.$$

Thus the Fourier series for $f(x)$ converges absolutely.

(Eingegangen am 23. Mai 1925.)

Comments on [25c]

V. Beneš

This work presages the second phrase of Wiener's generalized harmonic analysis, which culminates in the Acta paper [30a], by introducing for nearly bounded functions $f(\cdot)$, the "integrated" spectral functions

$$\gamma(\alpha) = \underset{T\to\infty}{\text{l.i.m.}}\,\frac{1}{\pi}\int_{-T}^{T} f(\lambda)\,\frac{\sin\alpha\lambda}{\lambda}\,d\lambda$$

$$\delta(\alpha) = \underset{T\to\infty}{\text{l.i.m.}}\,\frac{1}{\pi}\int_{-T}^{T} f(\lambda)\,\frac{e^{-\lambda^2} - \cos\alpha\lambda}{\lambda}\,d\lambda.$$

These functions are of bounded *quadratic* variation whenever

$$\lim_{T\to\infty}\frac{1}{2T}\int_{-T}^{T} |f(\lambda)|^2\,d\lambda$$

exists. The paper also introduces a new generalization of periodic functions, the pseudo-periodic, and shows that these have a denumerable spectrum. Subsequently, Paley and Wiener [34d] proved that $f(\cdot)$ is pseudo-periodic, iff it is almost periodic in the sense of Stepanoff and has a regular spectrum, i.e., $\inf|\lambda_m - \lambda_n| > 0$. The study of these functions culminated in the explicit calculation of the pseudo-period by J. P. Kahane[1] as 2π times the outer uniform density of the spectral numbers λ_n.

Reference

1. J. P. Kahane, Ann. Inst. Fourier, Grenoble 7(1957), 293–314. (Math. Rev. 21(1960), no. 1489.)

Verallgemeinerte trigonometrische Entwicklungen.

Von

Norbert Wiener aus Boston U. S. A.

Vorgelegt von R. Courant in der Sitzung vom 23. Oktober 1925.

Seit vielen Jahren haben die Physiker sich mit der Frage nach der Zerlegung von weißem Licht in periodische Komponenten beschäftigt. Die Möglichkeit einer solchen Zerlegung ist seit dem frühesten Beginn der Wellentheorie einleuchtend gewesen, und vielleicht vermuten die Physiker allgemein, daß die vollständige Theorie sich in der Literatur vorfindet. Doch zeigt uns schon eine kurze Überlegung, daß eine solche Theorie, wenn sie überhaupt wirklich existiert, von neuestem Datum sein muß; denn der berühmte Streit zwischen Poincaré [1]) auf der einen Seite und Gouy [2]), Schuster [3]) und Lord Rayleigh [4]) auf der andern über die Beschaffenheit des weißen Lichtes wäre unmöglich gewesen, wenn Klarheit über den präzisen mathematischen Sinn bestanden hätte, in welchem die allgemeine, beschränkte, nicht periodische Funktion in ihre Komponenten zerlegt werden kann. Es ist wohl bekannt, daß weißes Licht Interferenzerscheinungen bei einem Wegunterschiede von mehreren Zentimetern zeigt, vorausgesetzt, daß man es durch ein Spektrometer von genügender Auflösungskraft beobachtet. Poincaré hat diese Tatsache für einen sicheren Beweis dafür gehalten, daß die Ätherschwingungen, die weißes Licht bilden, in hohem Grade periodisch seien. Die anderen oben erwähnten Autoren wiesen jedoch darauf hin, daß die Periodizität von dem auflösenden Instrument allein abhängig sein könnte, ohne irgend eine Abhängigkeit von der Natur der aufgelösten Wellen. Dies bewiesen sie dadurch, daß sie weißes Licht in eine unregelmäßige

---

1) C. R., 120, p. 1895.

2) Journ. de Phys., 5, p. 354, 1886.

3) C. R., 120, pp. 915, 987.

4) Artikel über „Wave Theory", Encyclopaedia Britannica.

Folge von Ätherimpulsen auflösten, deren jeder durch ein Fouriersches Integral analysiert werden konnte.

Dabei wird die Zerlegung von weißem Licht in zwei Schritten ausgeführt. Es ist gar nicht klar, ob die Impulse wirklich physikalisch existieren oder bloß mathematische Konstruktionen sind. Diese Theorie ist weder direkt, noch einfach, noch elegant. Sie ist jedoch nützlich, in Ermangelung einer passenden Analyse. Es gibt schließlich noch eine indirekte Auflösungsmethode. Diese Methode, mit welcher die Namen von Gouy und von Planck verknüpft sind, stellt die elektrische Verschiebung über eine lange, jedoch endliche Zeit durch eine Fouriersche Reihe und infolgedessen durch ein Linienspektrum dar. Wenn das Ausgangs-Zeitintervall mehr und mehr verlängert wird, rücken die Linien immer näher aneinander, und die Energieverteilung — oder besser die Leistungsverteilung — nähert sich der in einem kontinuierlichen Spektrum stattfindenden Verteilung.

Es ist leicht zu sehen, daß weder das Fouriersche Integral noch die Fouriersche Reihe das richtige Instrument für die Behandlung weißen Lichtes ist, wenngleich sich diese in beide Rahmen zwängen läßt. Dieselbe Schwierigkeit erhebt sich in den Theorien der Geräusche, der Erdbeben- und Meereswellen. Es handelt sich in allen diesen Fällen um Funktionen, die erstens nichtperiodisch sind und außerdem nicht gegen Null streben, wenigstens nicht innerhalb der Intervalle, die für uns in Frage kommen. Die erste Tatsache verhindert uns, sie vermittels einer Fourierschen Reihe zu zerlegen, die zweite daran, daß wir zu diesem Zweck ein Fouriersches Integral benutzen. Trotzdem zeigen diese Funktionen eine gewisse Regelmäßigkeit. Das Ziel der vorliegenden Arbeit ist, diese Regelmäßigkeit zu charakterisieren und das geeignete Werkzeug zur Zerlegung dieser Funktionen zu schmieden.

Der spezielle Fall eines reinen Linienspektrums ist bereits in der Literatur genügend behandelt worden. Rein formale Ansätze für die Theorie wurden 1897 von Schuster in seiner sogenannten „periodogram analysis" konstruiert; die wirkliche Bestimmung der Funktionenklasse, für welche diese Methode anwendbar ist, und die strenge Rechtfertigung der Methode ergibt sich aber erst aus den Arbeiten von Harald Bohr [5]).

Bohr nennt eine Zahl τ eine „Verschiebungszahl", die zu

5) H. Bohr, Zur Theorie der fast periodischen Funktionen. I. Eine Verallgemeinerung der Theorie der Fourierreihen. Acta Mathematica 45 (1924), pp. 29—127.

einer positiven Zahl ε und der stetigen Funktion $f(x)$ einer reellen Variablen gehört, wenn für alle x die Ungleichung

$$|f(x+\tau)-f(x)| \leqq \varepsilon$$

gilt. Wenn dann für irgend ein vorgegebenes ε eine positive Zahl $l(\varepsilon)$ existiert, derart, daß für alle a die Funktion $f(x)$ im Intervalle $(a, a+l(\varepsilon))$ eine Verschiebungszahl besitzt, die zu l gehört, wird $f(x)$ „fast periodisch" genannt. Jede fast periodische Funktion ist umkehrbar eindeutig verknüpft mit einer Reihe der Form

$$\sum_{1}^{\infty} A_n e^{i A_n x},$$

wo die Anzahl der Glieder abzählbar ist. Die Art dieser Verknüpfung wird gegeben durch eine Verallgemeinerung des geläufigen Begriffes der Konvergenz im Mittel.

Die in der Theorie der allgemeinen Banden- und Linienspektra vorkommenden Funktionen erfüllen gleichfalls gewisse Regularitäts- oder Gleichmäßigkeitsbedingungen, die ihre harmonische Zerlegung erleichtern. Ich werde hier auf die weitest erreichbaren Bedingungen dieser Art nicht eingehen. Jedoch wird die zu diskutierende Klasse den in Betracht kommenden Typus von Gleichmäßigkeitsbedingungen klar erkennen lassen. Sie ist allerdings viel zu beschränkt, und ich habe keinen Beweis, daß sie irgend einen Fall eines kontinuierlichen Spektrums enthält. Doch bietet die dafür notwendige Erweiterung keine wirklichen Schwierigkeiten.

Sei $f(x)$ eine stetige Funktion von x; dann stellt bis auf einen konstanten Faktor

$$\varphi(T, A, t) = \frac{1}{T}\int_{A}^{A+T} f(x) f(x+t)\, dx$$

eine Art von Korrelationskoeffizient von f mit sich selbst bei einer Verschiebung t dar. Wenn irgend eine Gleichmäßigkeit in der Unregelmäßigkeit von $f(x)$ existiert, dann sollte $\varphi(T, A, t)$ für große Werte von T merkbar unabhängig von A sein. Wir nennen $f(x)$ „brauchbar", wenn $\lim_{T\to\infty} \Phi(T, A, t)$ gleichmäßig in A und t existiert. Es ist leicht zu sehen, daß dann

$$\lim_{T\to\infty} \varphi(T, A, t) = C(f; t)$$

unabhängig von A ist. Die Größe $C(f; t)$ stellt einen wichtigen

Begriff dar, der in das Studium der Wärmeausbreitung durch kontinuierliche Bewegungen von G. I. Taylor[6]) eingeführt worden ist.

Wir wenden uns nun zu der genaueren formelmäßigen Ausführung. Wir verfolgen zunächst einige Ideen von Hahn[7]) über die Fourierentwicklung der allgemeinen, beschränkten Funktion $f(x)$. Offenbar sollte diese Entwicklung wenigstens formal ein Stieltjesches Integral

$$\int_0^\infty \cos x\alpha \, d\gamma(\alpha) + \int_0^\infty \sin x\alpha \, d\delta(\alpha) \tag{1}$$

sein, denn sie muß die Fouriersche Reihe

$$\sum (A_n \cos nx + B_n \sin nx)$$

und das Fouriersche Integral

$$\int_0^\infty \cos x\alpha \, \varphi(\alpha) \, d\alpha + \int_0^\infty \sin x\alpha \, \psi(\alpha) \, d\alpha$$

enthalten. Formal hätten wir, in Analogie mit den Fourierschen Integralformeln

$$\varphi(\alpha) = \frac{1}{\pi} \int_{-\infty}^\infty f(x) \cos \alpha x \, dx; \qquad \psi(\alpha) = \frac{1}{\pi} \int_{-\infty}^\infty f(x) \sin \alpha x \, dx,$$

die Formeln

$$\left\{ \begin{aligned} \gamma(\alpha) &= \frac{1}{\pi} \int d\alpha \int_{-\infty}^\infty f(x) \cos \alpha x \, dx \\ &= \frac{1}{\pi} \int_{-\infty}^\infty f(x) \frac{\sin \alpha x}{x} \, dx; \\ \delta(\alpha) &= \frac{1}{\pi} \int d\alpha \int_{-\infty}^\infty f(x) \sin \alpha x \, dx \\ &= \frac{1}{\pi} \int_{-\infty}^\infty f(x) \frac{e^{-x^2} - \cos \alpha x}{x} \, dx, \end{aligned} \right. \tag{2}$$

wo e^{-x^2} eingeführt wird, um das Verschwinden des Zählers im letzten Integral für $x = 0$ zu sichern, ohne das Verhalten von $\int_A^\infty f(x) \frac{\cos \alpha x}{x} dx$ zu stören.

---

6) Proc. Lond. Math. Soc. (2) 20 (1920), pp. 196—212.

7) Bericht über die Jahresversammlung zu Innsbruck, Jahresbericht der deutschen Mathematikervereinigung 33 (1925), p. 107.

Die Formeln (1) und (2) gelten in unserer Theorie für irgend eine stetige, beschränkte Funktion $f(x)$, vorausgesetzt, daß die vorliegenden Integrale richtig verallgemeinert werden. Erstens müssen die ins Unendliche erstreckten Integrale, die γ und δ darstellen, als Limites im Mittel von Integralen über irgend einen endlichen Bereich aufgefaßt werden. Die Integrale in (1) verlangen eine speziellere Erklärung. Wenn wir eine „brauchbare" Funktion $f(x)$ betrachten, für welche $C(f; t)$ nicht gleich Null ist, und wenn γ und δ stetig sind — diese beiden Bedingungen darf man mit Recht als erfüllt betrachten, wenn es sich um Lichtspektra handelt —, dann ist es unmöglich, daß γ und δ von beschränkter totaler Variation sind. Das Stieltjesche Integral ist deshalb für unsere Zwecke ganz unzulänglich. Aus dieser Schwierigkeit hilft uns die Tatsache, daß Kosinus und Sinus differenzierbare Funktionen sind. Wir definieren mit Hilfe der Teilintegration

$$\int_0^A \cos x\alpha \, d\gamma(\alpha) + \int_0^A \sin x\alpha \, d\delta(\alpha) \tag{3}$$

$$= \cos xA\,\gamma(A) + \sin xA\,\delta(A) + x\left[\int_0^A \gamma(\alpha)\sin x\alpha\, d\alpha - \int_0^A \delta(\alpha)\cos x\alpha\, d\alpha\right].$$

Dennoch wird es im allgemeinen nicht der Fall sein, daß dieser Ausdruck gegen irgend einen Limes im Mittel konvergiert, wenn A unendlich wird. Tatsächlich gilt jedoch für irgend ein Intervall (a, b) die Gleichung

$$\lim_{\xi\to\infty}\int_\xi^{\xi+1} dA \int_a^b \left[f(x) - \int_0^A \cos x\alpha\, d\gamma(\alpha) - \int_0^A \sin x\alpha\, d\delta(\alpha)\right]^2 dx = 0.$$

Um diese Gleichung wörtlich auszudrücken, werden wir sagen, daß der Ausdruck (3) „fast" im Mittel gegen $f(x)$ konvergiert, wenn A unendlich wird.

Die Funktionen $\gamma(\alpha)$ bzw. $\delta(\alpha)$ drücken die totalen Amplituden von allen Kosinus- bzw. Sinuskomponenten von $f(x)$ mit Frequenzen bis zu α aus. In einem gewissen Sinne bestimmen sie das Spektrum von $f(x)$. Dieser Sinn ist nicht der des Physikers. Sein Spektrum ist vor allem ein Leistungsspektrum; Phasenverhältnisse kommen nicht zum Vorschein, und die Stärke einer Spektrallinie wird bestimmt durch das Quadrat ihrer Amplitude. Um zu diesem Begriff überzugehen, definieren wir

$$\overset{b}{\underset{a}{Q}}\{\varphi(\alpha)\} = \lim_{h\to 0} \frac{1}{2h}\int_a^b [\varphi(\alpha+h) - \varphi(\alpha-h)]^2\, d\alpha.$$

Es ist leicht zu sehen, daß, wenn φ eine Funktion von beschränkter totaler Variation im Intervalle (a, b) ist, $\overset{b}{\underset{a}{Q}} \{\varphi(\alpha)\}$ die Summe der Quadrate der Sprünge von φ repräsentiert. Im allgemeinen werden wir Q, wenn es überhaupt existiert, als Summe der Quadrate der „unendlich kleinen Sprünge", die φ im Intervalle (a, b) macht, zusammen mit der Summe der Quadrate der tatsächlichen Sprünge von φ betrachten. Dies ist natürlich nur eine Redewendung, wird aber zur Erläuterung des Folgenden nützlich sein.

Gewisse Schwierigkeiten in der Bestimmung von $\overset{K}{\underset{-K}{Q}} \{\gamma(\alpha)\} + \overset{K}{\underset{-K}{Q}} \{\delta(\alpha)\}$ entstehen in den Punkten K und $-K$ aus der Tatsache, daß die Definition von Q sich ein wenig über diese Punkte hinaus erstreckt; aber sie können mittels einer geringen Veränderung der Definition beseitigt werden. Bei „brauchbarem" $f(x)$ existiert dieses verallgemeinerte $\overset{K}{\underset{-K}{Q}} \{\gamma(\alpha)\} + \overset{K}{\underset{-K}{Q}} \{\delta(\alpha)\}$ immer und wird in der Form $R\{f; K\}$ geschrieben. Es ist leicht zu sehen, daß, sofern $\gamma(\alpha)$ und $\delta(\alpha)$ die Summen der Amplituden von allen f-Komponenten darstellen, deren Frequenzen zwischen 0 und $\frac{\alpha}{2\pi}$ liegen, $R\{f; \alpha\}$ die Summe der Quadrate der Amplituden von allen solchen Komponenten repräsentiert. Das heißt, $R\{f; \alpha\}$ ist eine zunehmende Funktion von α, die stetig bzw. sprungweise zunimmt, wenn f ein Banden- bzw. Linien-Spektrum hat.

Die wichtigsten Theoreme, die ich gefunden habe, sind, daß, wenn f „brauchbar" ist,

$$C(f; h) = \tfrac{1}{2} \int_0^\infty \cos \alpha h \, dR\{f; \alpha\}$$

ist, wo das Integral jetzt ein gewöhnliches Stieltjesches Integral bedeutet, und daß entsprechend

$$R\{f; \alpha\} = \lim_{T \to \infty} \frac{2}{\pi} \int_{-T}^{T} C\{f; h\} \frac{\sin \alpha h}{h} dh$$

ist. Die letzte Formel gestattet uns, direkt von C zu der Leistungsverteilung im Spektrum überzugehen. Da wir C unmittelbar aus f erhalten können, haben wir eine Methode, die sich rechnerisch zur Bestimmung des Leistungsspektrums einer gegebenen Kurve benutzen läßt. Diese Methode geht viel weiter als die „periodogram analysis", welche das Linienspektrum einer Funktion gibt, aber das Restglied im Falle eines kontinuierlichen Spektrums vernachlässigt.

Ein wichtiger Hilfssatz bei dem Beweise der vorangehenden Formeln ist folgender:

$f(x)$ sei irgend eine beschränkte Funktion, so daß

$$\lim_{T\to\infty} \frac{1}{T}\int_{-T}^{T} [f(x)]^2\,dx$$

existiert. Dann existieren $\overset{\infty}{\underset{-\infty}{Q}} \{\gamma(\alpha)\}$ und $\overset{\infty}{\underset{-\infty}{Q}} \{\delta(\alpha)\}$, und zwar ist

$$\overset{\infty}{\underset{-\infty}{Q}} \{\gamma(\alpha)\} + \overset{\infty}{\underset{-\infty}{Q}} \{\delta(\alpha)\} = \lim_{T\to\infty} \frac{1}{T}\int_{-T}^{T} [f(x)]^2\,dx.$$

Diesen Hilfssatz, der eine Verallgemeinerung des Parsevalschen Satzes ist, hat der Verfasser [8]) benutzt als ein Lemma für einen Beweis der Bohrschen Verallgemeinerung des Parsevalschen Satzes auf fast periodische Funktionen. Diese Arbeit, welche die strengen Entwicklungen zu den Sätzen der vorstehenden Note enthält, wurde erst geschrieben, nachdem der Verfasser die Bohrsche Arbeit kennen gelernt hatte. Es ist von Interesse, daß dieser Beweis des Bohrschen Fundamentalsatzes keine arithmetischen Betrachtungen enthält.

Die Beschaffenheit der Funktionen $\gamma(\alpha)$, für welche $\overset{\alpha}{\underset{0}{Q}} \{\gamma(\alpha)\}$ endlich, stetig und nicht identisch Null ist, kann vielleicht durch die folgenden Überlegungen anschaulich gemacht werden. Wenn man voraussetzt, daß die Einsteinsche Behandlung der Brownschen Bewegung vollständig exakt und nicht nur angenähert ist, so folgt daraus, wie leicht zu zeigen ist, daß die als Funktion der Zeit aufgetragene Projektion der Verschiebung eines Teilchens auf die x-Achse (mit der Wahrscheinlichkeit 1) eine stetige Funktion mit einem beschränkten, stetigen Q wird und daher nirgends von beschränkter totaler Variation ist.

Noch andere Berührungspunkte existieren zwischen der Theorie der verallgemeinerten trigonometrischen Entwicklungen und der Theorie der Brownschen Bewegung. Schuster behauptet, daß dann, wenn ein Ätherteilchen sich in einer gänzlich unregelmäßigen Weise bewegt — d. h. wie ein Teilchen, das dem Brownschen

---

8) Math. Zeitschrift 24 (1925), p. 575—616.

Bewegungsgesetze unterworfen ist —, die Verteilung der Leistung auf die verschiedenen Spektralfrequenzen gleichmäßig ist. Die Schustersche Behauptung streng zu beweisen ist gar nicht schwierig, wenn man die Definitionen der vorliegenden Arbeit etwas verallgemeinert. Hierzu ist eine Theorie der Integration im Funktionenraum nötig. Die erforderliche Theorie findet sich in einer Reihe von Arbeiten, die ich früher geschrieben habe.

Bis jetzt sind nur beschränkte Funktionen erwähnt worden. Aber auch gewisse Klassen von nicht beschränkten Funktionen sind einer verallgemeinerten trigonometrischen Analyse zugänglich. Diese Methode hat nützliche Resultate im Bereiche des Heavisideschen Operationskalküls [9]) und der Differenzengleichungen [10]) geliefert.

9) Vgl. meine Arbeit in der Mathematischen Zeitschrift.

10) Journ. Math. Phys. Mass. Inst. Technol. May, 1925.

Comments on [25d]

V. Beneš

This is an informal presentation of generalized harmonic analysis as seen by Wiener in 1925. It starts by illustrating the need for generalization by citing the problem of decomposition of light in physics. Bohr's theory of almost periodic functions is described, and then Wiener's own early harmonic analysis paper [25c].

THE HARMONIC ANALYSIS OF IRREGULAR MOTION

By Norbert Wiener

The author has already developed the theory of the harmonic analysis of a very general class of functions which neither have properties of periodicity, as do those to which the method of Fourier series applies, nor of vanishing at infinity, as is essential for the application of the classical forms of the Fourier integral.[1] The notions there employed are closely akin to those which had been previously introduced by Hahn,[2] although Hahn's discussion is strictly confined to ordinary convergence and to Stieltjes integrals of the classical sort, whereas the present author is primarily interested in convergence in the mean and in theorems of the Hurwitz and Riesz-Fischer order, and has found the classical Stieltjes integral inadequate to his needs. In a recent paper presented to the Academy of Sciences at Vienna,[3] of which Professor Hahn has been kind enough to send me a proof-sheet, integrals of the form

$$\lim_{\mu\to\infty}\frac{1}{\mu\pi}\int_0^\mu d\lambda\int_0^\lambda\left(\cos\mu x\frac{d^2\Phi_2(\tau)}{d\tau}+\sin\mu x\frac{d^2\Psi_2(\tau)}{d\tau}\right)$$

are introduced, where

$$\Phi_2(\mu)=2\int_{-\infty}^{\infty}f(x)\,\frac{\sin^2\frac{\mu}{2}x}{x^2}\,dx;\ \Psi_2(\mu)=\int_{-1}^{1}f(x)\,\frac{\mu x-\sin\mu x}{x^2}\,dx$$
$$-\int_1^\infty f(x)\,\frac{\sin\mu x}{x^2}\,dx-\int_{-\infty}^{-1}f(x)\,\frac{\sin\mu x}{x^2}\,dx.$$

The present author has been compelled in the course of his own researches to make use of integrals of similar form. It therefore seems to him an appropriate time to communicate the results of these researches.

The theory developed by the author in his *Zeitschrift* paper has one important defect. While the theory embodied in that paper

[1] Mathematische Zeitschrift, Bd. 24, Heft 3, pp. 575–616.

[2] H. Hahn: Über Fouriersche Reihen und Integrale, Bericht über die Jahresversammlung zu Innsbruck, Jahresbericht der deutschen Mathematiker-vereinigung, 33 (1925), p. 107.

[3] Jahrgang 1925, Nr. 25, Sitzung vom 3. Dezember, 1925.

has at least a formal niche for functions with continuous spectra, in all the more detailed developments of that paper, a certain hypothesis of uniformity plays an important rôle. This is that the functions discussed should be well-behaved. To the present time, the author has been unable either to produce a single example of a well-behaved function with a continuous spectrum, or to demonstrate that no such functions exist. In the theory here expounded, the necessary existence proofs may be filled in with the aid of the theory of probabilities, aided, it is true, by the Zermelo principle of selection. These existence proofs, however, will be reserved for a later communication.

Included in the range of ideas treated in this paper are the theory of white light, of noise, of the periodogram, and of the so-called Schrotteffekt. Inasmuch, however, as probability considerations play a large part in most of these topics, the author will put off a detailed discussion of them to a later paper.

§1. The Coefficient-Functions of a Trigonometrical Development.

Let $f(t)$ be a summable function defined for all real arguments, and let

$$\lim_{A\to\infty}\frac{1}{2A}\int_{-A}^{A}[f(t)]^2dt=M\left\{[f(t)]^2\right\}$$

exist. Then

$$\int_1^\infty\frac{dA}{A^3}\int_{-A}^{A}[f(t)]^2dt=\int_{A=1}^{A=\infty}d\left(-\frac{1}{2A^2}\right)\int_{-A}^{A}[f(t)]^2dt$$

$$=\frac{1}{2}\int_{-1}^{1}[f(t)]^2dt+\int_1^\infty\frac{[f(A)]^2+[f(-A)]^2}{2A^2}dA \tag{1}$$

exists. Now,

$$\left[\int_{B_1}^{B_2}f(t)\frac{1-\cos ut}{t^2}dt\right]^2\leq 4\int_{B_1}^{B_2}\frac{[f(t)]^2}{t^2}dt\int_{B_1}^{B_2}\frac{dt}{t^2}.$$

$$\leq\left[\frac{4}{B_1}-\frac{4}{B_2}\right]\int_{B_1}^{B_2}\frac{[f(t)]^2}{t^2}.$$

It follows that

$$\Gamma(u)=\frac{1}{\pi}\int_{-\infty}^{\infty}f(t)\frac{1-\cos ut}{t^2}dt \tag{2}$$

exists, and indeed exists as a limit uniform in u. A similar proof will show the existence of

$$\Delta(u) = \frac{1}{\pi}\int_{-\infty}^{\infty} f(t)\, \frac{ute^{-t^2} - \sin ut}{t^2}\, dt \tag{3}$$

uniformly in u.

We shall here introduce the following

Lemma. *Let $g(x)$ be a function summable and of summable square over the range $(-a, a)$. Let*

$$\frac{1}{4h^2}\int_{-a}^{a} [g(x+h) - g(x-h)]^2 dx$$

be bounded as h approaches 0. Then there is a function $g'(x)$ such that

$$\int_{-a}^{a} [g(x) - g(0) - \int_0^x g'(y)dy]^2 = 0. \tag{4}$$

To prove this, let us expand $g(x)$ in the Fourier series

$$g(x) = \frac{a_0}{2} + \sum_1^{\infty}\left[a_n \cos\frac{n\pi x}{a} + b_n \sin\frac{n\pi x}{a}\right].$$

We shall have

$$\frac{1}{4h^2}\int_{-a}^{a} [g(x+h) - g(x-h)]^2 dx = \frac{a}{h^2}\sum_1^{\infty}[a_n^2 + b_n^2]\sin^2\frac{n\pi h}{a}.$$

Hence

$$\sum_1^m n^2[a_n^2 + b_n^2]$$

is bounded, and by the Riesz-Fischer theorem, there is a function to which

$$\frac{\pi}{a}\sum_1^{\infty} n\left[b_n \cos\frac{n\pi x}{a} - a_n \sin\frac{n\pi x}{a}\right]$$

converges in the mean. This fulfills the requirements for $g'(x)$.

It may be proved without difficulty that $\Gamma'(u)$ and $\Delta'(u)$, defined in this sense, exist over any finite range of u. It is readily seen that the definitions for two overlapping ranges are then consistent. To prove the existence of $\Gamma'(u)$, we have only to note that

$$\frac{1}{4h^2}\int_{-a}^{a} du \left\{ \frac{1}{\pi}\int_{-\infty}^{\infty} f(t)\,\frac{1-\cos(u+h)t}{t^2}\,dt \right.$$

$$\left. -\frac{1}{\pi}\int_{-\infty}^{\infty} f(t)\,\frac{1-\cos(u-h)t}{t^2}\,dt \right\}^2 dx$$

$$=\frac{1}{4h^2}\int_{-a}^{a} du \left\{ \frac{1}{\pi}\int_{-\infty}^{\infty} f(t)\frac{2\sin ut\sin ht}{t^2}dt \right\}^2$$

$$\leq\frac{1}{2h^2}\int_{-\infty}^{\infty} du \left\{ \frac{1}{\pi}\left[\int_{-\infty}^{-A}+\int_{A}^{\infty}\right] f(t)\frac{2\sin ut\sin ht}{t^2}\,dt \right\}^2$$

$$+\frac{1}{2h^2}\int_{-a}^{a} du \left\{ \frac{1}{\pi}\int_{-A}^{A} f(t)\,\frac{2\sin ut\sin ht}{t^2}\,dt \right\}^2$$

$$\leq\frac{1}{\pi h^2}\left[\int_{-\infty}^{-A}+\int_{A}^{\infty}\right][f(t)]^2\,\frac{\sin^2 ht}{t^4}\,dt$$

$$+\frac{1}{2}\int_{-a}^{a} du \left\{ \frac{1}{\pi}\int_{-A}^{A} f(t)\,\frac{2\sin ut}{t}\,dt \right\}^2$$

$$\leq\frac{1}{2\pi}\left[\int_{-\infty}^{-A}+\int_{A}^{\infty}\right]\frac{[f(t)]^2}{t^2}dt+\frac{1}{2}\int_{-a}^{a} du \left\{ \frac{1}{\pi}\int_{-A}^{A} f(t)\frac{2\sin ut}{t}\,dt \right\}^2.$$

As this is independent of h, the existence of $\Gamma'(u)$ follows at once. A similar proof demonstrates the existence of $\Delta(u)$.

We have

$$\frac{1}{h^3}\int_{-\infty}^{\infty} [f(t)]^2\,\frac{\sin^4 ht}{t^4}\,dt=\frac{1}{h^3}\int_{t=0}^{t=\infty}\frac{\sin^4 ht}{t^4}\,d\int_{-t}^{t}[f(\xi)]^2 d\xi$$

$$=-\frac{1}{h^3}\int_{0}^{\infty}\left\{\int_{-t}^{t}[f(\xi)]^2 d\xi\right\}\frac{d}{dt}\left[\frac{\sin^4 ht}{t^4}\right]dt$$

$$=-\frac{1}{h^3}\int_{0}^{\infty}\left\{\frac{1}{2t}\int_{-t}^{t}[f(\xi)]^2 d\xi\right\}\frac{8\sin^3 ht(ht\cos ht-\sin ht)}{t^4}\,dt. \tag{5}$$

It may be verified immediately that

$$\int_{0}^{\infty}\frac{8\sin^3 x(x\cos x-\sin x)}{x^4}\,dx=-\int_{-\infty}^{\infty}\frac{\sin^4 u}{u^4}\,du=-\frac{2\pi}{3} \tag{6}$$

converges absolutely. Hence

$$\lim_{h\to 0}\frac{1}{h^3}\int_{-\infty}^{\infty}[f(t)]^2\frac{\sin^4 ht}{t^4}dt=\frac{2\pi}{3}\lim_{A\to\infty}\frac{1}{2A}\int_{-A}^{A}[f(\xi)]^2d\xi. \tag{7}$$

It follows at once from the properties of the Fourier transformation that

$$\begin{aligned}
&\frac{1}{h^3}\int_{-\infty}^{\infty}[f(t)]^2\frac{\sin^4 ht}{t^4}dt\\
&=\frac{2}{\pi h^3}\int_0^{\infty}\left\{\left[\int_{-\infty}^{\infty}f(t)\frac{\sin^2 ht\sin tu}{t^2}dt\right]^2\right.\\
&\qquad\left.+\left[\int_{-\infty}^{\infty}f(t)\frac{\sin^2 ht\cos tu}{t^2}dt\right]^2\right\}du\\
&=\frac{2}{\pi h^3}\int_0^{\infty}\left\{\left[\int_{-\infty}^{\infty}f(t)\frac{(1-\cos 2ht)\sin tu}{2t^2}dt\right]^2\right.\\
&\qquad\left.+\left[\int_{-\infty}^{\infty}f(t)\frac{(1-\cos 2\,ht)\cos tu}{2t^2}dt\right]^2\right\}du\\
&=\frac{2}{\pi h^3}\int_0^{\infty}\left\{\left[\int_{-\infty}^{\infty}f(t)\frac{\sin(t+2h)u+\sin(t-2h)u-2\sin tu}{4t^2}dt\right]^2\right.\\
&\qquad\left.+\left[\int_{-\infty}^{\infty}f(t)\frac{\cos(t+2h)u+\cos(t-2h)u-2\cos tu}{4t^2\cdot}dt\right]^2\right\}du\\
&=\frac{\pi}{8h^3}\int_0^{\infty}\{[\Delta(u+2h)+\Delta(u-2h)-2\Delta(u)]^2+[\Gamma(u+2h)\\
&\qquad+\Gamma(u-2h)-2\Gamma(u)]^2\}\,du.
\end{aligned} \tag{8}$$

As an immediate consequence, we have

$$\begin{aligned}
&\lim_{\epsilon\to 0}\frac{1}{\epsilon^3}\int_0^{\infty}\{[\Delta(u+\epsilon)+\Delta(u-\epsilon)-2\Delta(u)]^2\\
&\qquad+[\Gamma(u+\epsilon)+\Gamma(u-\epsilon)-2\Gamma(u)]^2\}\,du\\
&=\frac{2}{3}\lim_{A\to\infty}\frac{1}{2A}\int_{-A}^{A}[f(\xi)]^2d\xi=\frac{2}{3}M\{[f(\xi)]^2\},
\end{aligned} \tag{9}$$

in the sense that the former of these two limits exists whenever the latter does, and is identical with it.

There is another interesting property of $\Gamma(u)$ and $\Delta(u)$. Let us interpret

$$\int_0^A \cos ut\, d\Gamma'(u) + \int_0^A \sin ut\, d\Delta'(u)$$

as the result which we obtain by integrating by parts as indicated —that is, as

$$\cos At\,\Gamma'(A) + \sin At\,\Delta'(A) + t\int_0^A \sin ut\, d\Gamma(u) - t\int_0^A \cos ut\, d\,\Delta(u)$$

$$= \cos At\,\Gamma'(A) + \sin At\,\Delta'(A) + t[\sin At\,\Gamma(A) - \cos At\,\Delta(A)\}$$

$$-t^2\left[\int_0^A \cos ut\,\Gamma(u)du + \int_0^A \sin ut\,\Delta(u)du\right]. \tag{10}$$

We then have

$$\int_0^A \cos ut\, d\Gamma'(u) + \int_0^A \sin ut\, d\,\Delta'(u)$$

$$= \underset{h\to 0}{\text{limit in mean}} \left\{ \frac{1}{\pi h}\int_{-\infty}^{\infty} f(\tau)\,\frac{\cos At \sin A\tau \sin h\tau}{\tau^2}\,d\tau \right.$$

$$\left. - \frac{1}{\pi h}\int_{-\infty}^{\infty} f(\tau)\frac{\sin At[\cos A\tau \sin h\tau - h\tau e^{-\tau^2}]}{\tau^2}\,d\tau \right\}$$

$$+t\left\{ \frac{1}{\pi}\int_{-\infty}^{\infty} f(\tau)\,\frac{\sin At(1-\cos A\tau)}{\tau^2}\,d\tau \right.$$

$$\left. - \frac{1}{\pi}\int_{-\infty}^{\infty} f(\tau)\,\frac{\cos At(A\tau e^{-\tau^2} - \sin A\tau)}{\tau^2}\,d\tau \right\}$$

$$-t^2\int_0^A du \left\{ \frac{1}{\pi}\int_{-\infty}^{\infty} f(\tau)\,\frac{\cos ut\,(1-\cos u\tau)}{\tau^2}\,d\tau \right.$$

$$\left. + \frac{1}{\pi}\int_{-\infty}^{\infty} f(\tau)\,\frac{\sin ut\,(u\tau e^{-2\tau^2} - \sin u\tau)}{\tau^2}\,d\tau \right\}. \tag{11}$$

It is easy to verify that, in the last integral, the conditions for integrating under the sign of integration are fulfilled, and that

$$t^2\int_0^A du\left\{\frac{1}{\pi}\int_{-\infty}^{\infty} f(\tau)\,\frac{\cos ut(1-\cos u\tau)}{\tau^2}d\tau\right.$$
$$\left.+\frac{1}{\pi}\int_{-\infty}^{\infty} f(\tau)\,\frac{\sin ut(u\tau e^{-\tau^2}-\sin u\tau)}{\tau^2}\,d\tau\right\}$$
$$=\frac{t}{\pi}\int_{-\infty}^{\infty} f(\tau)\frac{\sin At}{\tau^2}\,d\tau-\frac{t^2}{2\pi}\int_{-\infty}^{\infty} f(\tau)\,\frac{\sin A(t+\tau)}{\tau^2\,(t+\tau)}d\tau$$
$$-\frac{t^2}{2\pi}\int_{-\infty}^{\infty} f(\tau)\,\frac{\sin A(\tau-t)}{\tau^2(\tau+t)}\,d\tau$$
$$+\frac{1}{\pi}\int_{-\infty}^{\infty}\frac{f(\tau)\sin At\,\tau e^{-\tau^2}}{\tau^2}\,d\tau-\frac{t}{\pi}\int_{-\infty}^{\infty}\frac{Af(\tau)\cos At\,\tau e^{-\tau^2}}{\tau^2}\,d\tau$$
$$+\frac{t^2}{2\pi}\int_{-\infty}^{\infty} f(\tau)\frac{\sin A(t+\tau)}{\tau^2\,(t+\tau)}d\tau-\frac{t^2}{2\pi}\int_{-\infty}^{\infty} f(\tau)\frac{\sin A(\tau-t)}{\tau^2(\tau-t)}\,d\tau.$$

Hence

$$\int_0^A \cos ut\; d\Gamma'(u)+\int_0^A \sin ut\; d\,\Delta'(u)$$
$$=\underset{h\to 0}{\text{l. i. m.}}\frac{1}{\pi}\int_{-\infty}^{\infty} f(\tau)\frac{\sin A(\tau-t)}{\tau^2}\left[\frac{\sin h\tau}{h}+\frac{\tau t}{\tau-t}\right]d\tau$$
$$=\frac{1}{\pi}\int_{-B}^{B} f(\tau)\,\frac{\sin A(\tau-t)}{\tau-t}\,d\tau+\frac{1}{\pi}\left[\int_{-\infty}^{-B}+\int_{B}^{\infty}\right]\frac{t\sin A(\tau-t)}{\tau(\tau-t)}\,d\tau$$
$$+\underset{h\to 0}{\text{l. i. m.}}\frac{1}{\pi}\left[\int_{-\infty}^{-B}+\int_{B}^{\infty}\right] f(\tau)\frac{\sin A(\tau-t)\,\sin h\tau}{h\tau^2}\,d\tau. \tag{12}$$

Now let us consider only such values of t as lie in the range $(-B+a,\ B-a)$. We have

$$\underset{A\to\infty}{\text{l. i. m.}}\ \frac{1}{\pi}\int_{-B}^{B} f(\tau)\,\frac{\sin A(\tau-t)}{\tau-t}\,d\tau=f(t), \tag{13}$$

and

$$\lim_{A\to\infty}\ \frac{1}{\pi}\left[\int_{\infty}^{-B}+\int_{B}^{\infty}\right] f(\tau)\,\frac{t\,\sin A(\tau-t)}{\tau(\tau-t)}\,d\tau=0 \tag{14}$$

uniformly in t over any finite range. By the well-known theorem of Plancherel on the Fourier transformation, we know that

$$\frac{1}{\pi}\left[\int_{-C}^{-B}+\int_{B}^{C}\right]f(\tau)\frac{\sin A\ (\tau-t)}{\tau}\,d\tau$$

converges in the mean in A and t as C becomes infinite to some function $G(A, t)$, since

$$\left[\int_{-\infty}^{-B}+\int_{B}^{\infty}\right]\frac{[f(\tau)]^2}{\tau^2}\,d\tau$$

is bounded. Let us put

$$\begin{aligned}\frac{1}{\pi}\left[\int_{-\infty}^{-B}+\int_{B}^{\infty}\right]&f(\tau)\frac{\sin A(\tau-t)\sin h\tau}{h\tau^2}\,d\tau\\ &=\frac{1}{\pi}\left[\int_{-\infty}^{-C}+\int_{C}^{\infty}\right]f(\tau)\frac{\sin A(\tau-t)\sin h\tau}{h\tau^2}\,d\tau\\ &+\frac{1}{\pi}\left[\int_{-C}^{-B}+\int_{B}^{C}\right]f(\tau)\frac{\sin A(\tau-t)\sin h\tau}{h\tau^2}\,d\tau.\end{aligned}$$

It is easy to see that

$$\begin{aligned}\lim_{h\to 0}\frac{1}{\pi}\left[\int_{-C}^{-B}+\int_{B}^{C}\right]&f(\tau)\frac{\sin A(\tau-t)\sin h\tau}{h\tau^2}\,d\tau\\ &=\frac{1}{\pi}\left[\int_{-C}^{-B}+\int_{B}^{C}\right]f(\tau)\frac{\sin A(\tau-t)}{\tau}\,d\tau\end{aligned}$$

uniformly over any range of A and t. Moreover,

$$\begin{aligned}&\int_0^{\alpha}\left\{\frac{1}{\pi}\left[\int_{-\infty}^{-C}+\int_{C}^{\infty}\right]f(\tau)\frac{\sin A(\tau-t)\sin h\tau}{h\tau^2}\,d\tau\right\}^2 dA,\\ &=\left|\frac{1}{\pi^2}\left[\int_{\infty}^{-C}+\int_{C}^{\infty}\right]f(\tau)\frac{\sin h\tau}{h\tau^2}\,d\tau\left[\int_{-\infty}^{-C}+\int_{C}^{\infty}\right]f(\sigma)\frac{\sin h\sigma}{h\sigma^2}\,d\sigma\right.\\ &\qquad\left.\cdot\frac{1}{2}\left[\frac{\sin\alpha(\tau-\sigma)}{\tau-\sigma}-\frac{\sin\alpha(\tau+\sigma-2t)}{\tau+\sigma-2t}\right]\right|\\ &\leq\frac{1}{\pi^2}\left[\int_{-\infty}^{-C}+\int_{C}^{\infty}\right]d\tau\left[\int_{-\infty}^{-C}+\int_{C}^{\infty}\right]\left|\frac{f(\tau)f(\sigma)}{\tau\sigma}\right|\\ &\qquad\left[\left|\frac{\sin\alpha(\tau-\sigma)}{\tau-\sigma}\right|+\left|\frac{\sin\alpha(\tau+\sigma-2t)}{\tau+\sigma-2t}\right|\right]d\sigma\end{aligned}$$

if the latter integral exists. This will be the case when

$$\left[\int_{-\infty}^{-C}+\int_{C}^{\infty}\right]\frac{[f(\tau)]^2}{\tau^{4/3}}\,d\tau;$$

$$\left[\int_{\infty}^{-C}+\int_{C}^{\infty}\right]d\tau\left[\int_{-\infty}^{-C}+\int_{C}^{\infty}\right]\frac{\sin^2\alpha(\tau-\sigma)}{(\tau-\sigma)^2(\tau\sigma)^{2/3}}\,d\sigma;$$

$$\left[\int_{\infty}^{-C}+\int_{C}^{\infty}\right]d\tau\left[\int_{-\infty}^{-C}+\int_{C}^{\infty}\right]\frac{\sin^2\alpha(\tau+\sigma-2t)}{(\tau+\sigma-2t)^2(\tau\sigma)^{2/3}}\,d\sigma$$

all exist. The existence of the first of these integrals follows from the existence of

$$\lim_{A\to\infty}\frac{1}{2A}\int_{A}^{A}[f(t)]^2dt$$

by the fact that

$$\int_1^{\infty}\frac{dA}{A^{7/3}}\int_{A}^{A}[f(t)]^2\,dt=\int_{A=1}^{A=\infty}d\left(\frac{-3}{4A^{4/3}}\right)\int_{-A}^{A}[f(t)]^2dt$$

$$=\frac{3}{4}\int_{-1}^{1}[f(t)]^2+\frac{3}{4}\int_1^{\infty}\left\{[f(A)]^2+[f(-A)]\right\}^2A^{-4/3}dA.$$

As to the second integral, we have

$$\left|\int_C^{C_1}d\tau\int_C^{C_1}\frac{\sin^2\alpha(\tau-\sigma)}{(\tau-\sigma)^2(\tau\sigma)^{2/3}}\,d\sigma\right|\leq\frac{1}{2}\int_0^{2C_1}\frac{\sin^2\alpha v}{v^2}dv\int_{v+2C}^{2C_1}\frac{du}{(u^2-v^2)^{2/3}}$$

$$+\frac{1}{2}\int_{-2C_1}^{0}\frac{\sin^2\alpha v}{v^2}dv\int_{2C-v}^{2C_1}\frac{du}{(u^2-v^2)^{2/3}}$$

$$\leq\int_0^{\infty}\frac{\sin^2\alpha v}{v^2}dv\int_{2C}^{\infty}\frac{dw}{w^2}.$$

The existence of the other parts of the second integral, as well as the existence of the third integral, and the fact that it has for any finite range of t an upper bound vanishing as $c\to\infty$, may be proved in the same manner.

We may readily conclude that

$$\underset{C\to\infty}{\text{l. i. m.}}\,\frac{1}{\pi}\left[\int_{-\infty}^{-C}+\int_{C}^{\infty}\right]f(\tau)\,\frac{\sin A(\tau-t)\sin h\tau}{h\tau^2}\,d\tau=0$$

in the variable A uniformly in t and h. Hence

$$\underset{C\to\infty}{\text{l. i. m.}} \quad \underset{h\to 0}{\text{l. i. m.}} \frac{1}{\pi}\left[\int_{-\infty}^{-C}+\int_{C}^{\infty}\right] f(\tau)\frac{\sin A(\tau-t)\sin h\tau}{h\tau^2}d\tau=0.$$

Therefore

$$\underset{C\to\infty}{\text{l. i. m.}} \quad \underset{h\to 0}{\text{l. i. m.}} \frac{1}{\pi}\left[\int_{-C}^{-B}+\int_{B}^{C}\right] f(\tau)\frac{\sin A(\tau-t)\sin h\tau}{h\tau^2}d\tau$$

$$=\underset{h\to 0}{\text{l. i. m.}} \frac{1}{\pi}\left[\int_{-\infty}^{-B}+\int_{B}^{\infty}\right] f(\tau)\frac{\sin A(\tau-t)\sin h\tau}{h\tau^2}d\tau.$$

That is,

$$\underset{C\to\infty}{\text{l. i. m.}} \frac{1}{\pi}\left[\int_{-C}^{-B}+\int^{C}\right] f(\tau)\frac{\sin A(\tau-t)}{\tau}d\tau$$

$$=\underset{h\to 0}{\text{l. i. m.}} \frac{1}{\pi}\left[\int_{-\infty}^{-B}+\int_{B}^{\infty}\right] f(\tau)\frac{\sin A(\tau-t)\sin h\tau}{h\tau^2}d\tau.$$

By the use of a now familiar theorem concerning Fourier transforms,

$$\int_{-\infty}^{\infty}\left\{\frac{1}{\pi}\left[\int_{-C}^{-B}+\int_{B}^{C}\right] f(\tau)\frac{\sin A(\tau-t)}{\tau}d\tau\right\}^2 dA$$

$$=\frac{1}{\pi}\left[\int_{-C}^{-B}+\int_{B}^{C}\right]\frac{[f(u+t)]^2}{[u+t)]^2}du. \tag{15}$$

Hence

$$\int_{-\infty}^{\infty}\left\{\underset{h\to 0}{\text{l. i. m.}} \frac{1}{\pi}\left[\int_{-\infty}^{-B}+\int_{B}^{\infty}\right] f(\tau)\frac{\sin A(\tau-t)\sin h\tau}{h\tau^2}d\tau\right\}^2 dA$$

$$=\frac{1}{\pi}\left[\int_{-\infty}^{-B}+\int_{B}^{\infty}\right]\frac{[f(u+t)]^2}{[u+t]^2}du. \tag{16}$$

Since we know that this latter integral exists, we have

$$\lim_{A\to\infty}\int_{A}^{A+1}du\left\{\underset{h\to 0}{\text{l.i.m.}} \frac{1}{\pi}\left[\int_{-\infty}^{-B}+\int_{B}^{\infty}\right] f(\tau)\frac{\sin u(\tau-t)\sin h\tau}{h\tau^2}d\tau\right\}^2 = . \tag{17}$$

If we combine this with propositions (12), (13), and (14), we see that

$$\lim_{A\to\infty}\int_A^{A+1}\left\{f(t)-\int_0^{w}\cos ut\, d\Gamma'(u)-\int_0^{w}\sin ut\, d\Delta'(u)\right\}^2 dw=0. \tag{18}$$

In the language of the author's previous paper,

$$\int_0^A \cos ut\, d\Gamma'(u)+\int_0^A \sin ut\, d\Delta'(u)$$

converges almost in the mean to $f(t)$ as A becomes infinite. We thus see that there is a certain reciprocality in the relation between $f(t)$ and the pair of functions $\Gamma'(u)$ and $\Delta'(u)$, inasmuch as each is enough to determine the other except at a set of points of measure 0.

As the ideas of this section are a little hard to picture, it may be worth while to illustrate them by a simple example. Let

$$f(t)=\sum_1^m(a_n\cos\alpha_n t+b_n\sin\alpha_n t) \qquad (\text{no } \alpha_n=0).$$

Then

$$\Gamma'(u)=\sum_{\alpha_n<u}a_n;$$

$$\Delta'(u)=\sum_{\alpha_n<u}b_n+C_1;$$

$$\Gamma(u)=\sum_{\alpha_n<u}a_n(u-\alpha_n)$$

$$\Delta(u)=\sum_{\alpha_n<u}b_n(u-\alpha_n)+C_1u+C_2.$$

If ϵ is sufficiently small, we have in the neighborhood of α_n

$$\begin{aligned}\Gamma(u+\epsilon)+\Gamma(u-\epsilon)-2\Gamma(u)&=a_n(u+\epsilon-\alpha_n); && [\alpha_n-\epsilon<u<\alpha_n]\\ &=a_n(\epsilon-u+\alpha_n); && [\alpha_n<u<\alpha_n+\epsilon]\\ &=0. && [\,|u-\alpha_n|>\epsilon\,]\end{aligned}$$

Hence

$$\frac{1}{\epsilon^3}\int_{a_n-\epsilon}^{a_n+\epsilon}[\Gamma(u+\epsilon)+\Gamma(u-\epsilon)-2\Gamma(u)]^2du$$

$$=\frac{a_n^2}{\epsilon^3}\left[\int_{a_n-\epsilon}^{a_n}[u+\epsilon-a_n]^2du+\int_{a_n}^{n+\epsilon}[\epsilon-u+a_n]^2du\right]$$

$$=\frac{2a_n^2}{\epsilon^3}\int_0^{\epsilon}u^2du=\frac{2a_n^2}{3}.$$

We thus see that except for a constant factor,

$$\lim_{\epsilon\to 0}\frac{1}{\epsilon^3}\int_0^{\infty}\{[\Delta(u+\epsilon)+\Delta(u-\epsilon)-2\Delta(u)]^2$$
$$+[\Gamma(u+\epsilon)+\Gamma(u-\epsilon)-2\Gamma(u)]^2\}du$$

will represent in this case

$$\sum_1^m(a_n^2+b_n^2)$$

or the sum of the squares of the jumps of $\Gamma'(u)$ and $\Delta'(u)$.

§2. The Coefficient-Functions of the Trigonometrical Development of the Transforms of a Given Function.

Now let $\Phi(x)$ be any even function which is summable and of summable square over the entire interval $(-\infty, \infty)$ and which is $0(1/x^3)$ at infinity. Then

$$g(t)=\int_{-\infty}^{\infty}f(\tau)\Phi(t-\tau)d\tau \tag{19}$$

exists. We retain the supposition of the previous section, that

$$\lim_{A\to\infty}\frac{1}{2A}\int_{-A}^{A}[f(t)]^2dt$$

exists. We shall have

$$\int_{-A}^{A}[g(t)]^2dt=\int_{-A}^{A}dt\left[\int_{-\infty}^{\infty}f(\tau)\Phi(t-\tau)d\tau\right]^2$$

$$=\int_{-A}^{A}dt\left[\int_{-\infty}^{\infty}f(t-u)\Phi(u)du\right]^2$$

$$=\int_{-\infty}^{\infty}\Phi(u)du\int_{-\infty}^{\infty}\Phi(v)dv\int_{-A}^{A}f(t-u)f(t-v)dt, \tag{20}$$

inasmuch as we can invert the order of integration, all the integrals concerned being uniformly and absolutely convergent. Hence

$$\int_{-A}^{A}[g(t)]^2dt \leq \int_{-\infty}^{\infty}|\Phi(u)|du\int_{-\infty}^{\infty}|\Phi(v)|\,dv\sqrt{\int_{-A}^{A}[f(t-u)]^2dt\int_{-A}^{A}[f(t-v)]^2dt}$$

$$\leq K\int_{-\infty}^{\infty}|\,\Phi(u)\,|du\int_{-\infty}^{\infty}|\,\Phi(v)\,|(A+w)dv, \tag{21}$$

where w stands for the larger of the quantities, $|u|$ and $|v|$. It follows that

$$\frac{1}{2A}\int_{-A}^{A}[g(t)]^2dt$$

is bounded. This is sufficient to prove the existence of $\gamma(u)$ and $\delta(u)$, which bear the same relation to $g(t)$ as $\Gamma(u)$ and $\Delta(u)$ do to $f(u)$. The methods of our first section also lead us to

$$\frac{1}{\pi}\lim_{h\to 0}\frac{1}{h^3}\int_{-\infty}^{\infty}[g(t)]^2\frac{\sin^4 ht}{t^4}dt=\lim_{\epsilon\to 0}\frac{1}{\epsilon^3}\int_0^{\infty}\Big\{[\delta(u+\epsilon)+\delta(u-\epsilon)-2\delta(u)]^2$$

$$+[\gamma(u+\epsilon)+\gamma(u-\epsilon)-2\gamma(u)]^2\Big\}du \tag{22}$$

if the former limit exists. We have

$$\gamma(u)=\frac{1}{\pi}\int_{-\infty}^{\infty}g(t)\frac{1-\cos ut}{t^2}dt=\frac{1}{\pi}\int_{-\infty}^{\infty}\frac{1-\cos ut}{t^2}\,dt\int_{-\infty}^{\infty}f(t)\Phi(t-\tau)d\tau$$

$$=\frac{1}{\pi}\int_{-\infty}^{\infty}f(\tau)d\tau\int_{-\infty}^{\infty}\Phi(t-\tau)\frac{1-\cos ut}{t^2}dt$$

$$=\frac{1}{\pi}\int_{-\infty}^{\infty}f(\tau)d\tau\int_{-\infty}^{\infty}\Phi(t-\tau)\int_0^u dv\int_0^v \cos wt\,dw$$

$$=\frac{1}{\pi}\int_{-\infty}^{\infty}f(\tau)d\tau\int_0^u dv\int_0^v \cos w\tau\,dw\int_{-\infty}^{\infty}\Phi(x)\cos wx\,dx$$

$$=\frac{1}{\pi}\int_{-\infty}^{\infty}f(\tau)d\tau\int_0^u dv\int_0^v\phi(w)\cos w\tau\,dw, \tag{23}$$

where

$$\phi(w)=\int_{-\infty}^{\infty}\Phi(x)\cos wx\,dx. \tag{24}$$

Similarly,

$$\delta(u) = C + \frac{1}{\pi}\int_{-\infty}^{\infty} f(\tau)d\tau \int_0^u dv \int_0^v \phi(w) \sin w\tau \, dw. \quad (25)$$

We have

$$\gamma(u+\epsilon)+\gamma(u-\epsilon)-2\gamma(u)$$

$$= \frac{1}{\pi}\int_{-\infty}^{\infty} f(\tau)d\tau \left[\int_u^{u+\epsilon} - \int_{u-\epsilon}^{u}\right] dv \int_0^v \phi(w) \cos w\tau \, dw$$

$$= \frac{1}{\pi}\int_{-\infty}^{\infty} f(\tau)d\tau \int_0^{\epsilon} d\xi \int_0^{u+\xi} \phi(w) \cos w\tau \, dw$$

$$- \frac{1}{\pi}\int_{-\infty}^{\infty} f(\tau)d\tau \int_0^{\epsilon} d\xi \int_0^{u+\xi-\epsilon} \phi(w) \cos w\tau \, dw$$

$$= \frac{1}{\pi}\int_{-\infty}^{\infty} f(\tau)d\tau \int_0^{\epsilon} d\xi \int_{-\epsilon}^{0} \phi(u+\xi+\eta) \cos (u+\xi+\eta)\tau \, d\eta$$

$$= \frac{1}{\pi}\int_{-\infty}^{\infty} f(\tau)d\tau \int_0^{\epsilon} d\xi \left[\frac{\sin (u+\xi)\tau}{\tau} \phi(u+\xi) - \frac{\sin(u+\xi-\epsilon)\tau}{\tau} \phi(u+\xi-\epsilon) - \int_{-\epsilon}^{0} \frac{\sin (u+\xi+\eta)\tau}{\tau} \phi'(u+\xi+\eta)d\eta\right]$$

$$= \frac{1}{\pi}\int_{-\infty}^{\infty} f(\tau)d\tau \int_0^{\epsilon} d\xi \left\{ \left[\frac{\sin (u+\xi)\tau}{\tau} - \frac{\sin (u+\xi-\epsilon)\tau}{\tau}\right] \phi(u+\xi) + \frac{\sin (u+\xi-\epsilon)\tau}{\tau} (\phi(u+\xi)-\phi(u+\xi-\epsilon)) + \left[\frac{\cos (u+\xi+\eta)\tau}{\tau^2} \phi'(u+\xi+\eta)\right]_{-\epsilon}^{0} - \int_{-\epsilon}^{0} \frac{\cos (u+\xi+\eta)\tau}{\tau^2} \phi''(u+\xi+\eta)\, d\eta \right\}$$

$$
\begin{aligned}
&= \frac{1}{\pi}\int_{-\infty}^{\infty} f(\tau)d\tau \left\{ \left[-\frac{\cos(u+\epsilon)\tau}{\tau^2}+\frac{2\cos u\tau}{\tau^2}-\frac{\cos(u-\epsilon)\tau}{\tau^2}\right]\phi(u)\right.\\
&\quad+\left[-\frac{\cos(u+\epsilon)\tau}{\tau^2}+\frac{\cos u\tau}{\tau^2}\right]\left[\phi(u+\epsilon)-\phi(u)\right]\\
&\quad+\left[-\frac{\cos u\tau}{\tau^2}+\frac{\cos(u-\epsilon)\tau}{\tau^2}\right]\left[\phi(u+\epsilon)-\phi(u)\right]\\
&\quad+\frac{\cos(u-\epsilon)\tau}{\tau^2}\left[2\phi(u)-\phi(u-\epsilon)-\phi(u+\epsilon)\right]\\
&\quad+\int_0^{\epsilon}\frac{\cos(u+\xi-\epsilon)\tau}{\tau^2}\left[\phi'(u+\xi)-\phi'(u+\xi-\epsilon)\right]d\xi\\
&\quad+\int_0^{\epsilon}\frac{\cos(u+\xi)\tau}{\tau^2}\left[\phi'(u+\xi)-\phi'(u+\xi-\epsilon)\right]d\xi\\
&\quad+\int_0^{\epsilon}\frac{\cos(u+\xi)\tau-\cos(u+\xi-\epsilon)\tau}{\tau^2}\phi'(u+\xi-\epsilon)d\xi\\
&\quad\left.-\int_0^{\epsilon}d\xi\int_{-\epsilon}^{0}\frac{\cos(u+\xi+\eta)\tau}{\tau^2}\phi''(u+\xi+\eta)d\eta\right\}. \qquad (26)
\end{aligned}
$$

It follows from this and the Schwarz inequality, together with the existence of $\Gamma'(u)$, that

$$
\begin{aligned}
&\int_A^B \Big\{ \gamma(u+\epsilon)+\gamma(u-\epsilon)-2\gamma(u)\\
&\qquad -\phi(u)\left[\Gamma(u+\epsilon)+\Gamma(u-\epsilon)-2\Gamma(u)\right]\Big\}^2 du = O(\epsilon^4). \qquad (27)
\end{aligned}
$$

Since

$$\int_A^B \left[\Gamma(u+\epsilon)+\Gamma(u-\epsilon)-2\Gamma(u)\right]^2 du = O(\epsilon^3), \qquad (28)$$

we may readily conclude that

$$
\begin{aligned}
&\int_A^B \left[\phi(u)\right]^2\left[\Gamma(u+\epsilon)+\Gamma(u-\epsilon)-2\Gamma(u)\right]^2 du\\
&\qquad -\int_A^B\left[\gamma(u+\epsilon)+\gamma(u-\epsilon)-2\gamma(u)\right]^2 du = o(\epsilon^3). \qquad (29)
\end{aligned}
$$

A similar result holds when we substitute Δ for Γ and δ for γ.

There is a particular class of functions ϕ which is of some importance. Let $\phi_{N\eta}(u)$ be 1 from $u=0$ to $u=N$ and 0 from $u=N+\eta$ onwards. In the intermediate interval from N to $N+\eta$ let $\phi_{N\eta}(u)$ assume values ranging from 1 to 0 in such a manner that $\phi_{N\eta}(u)$ will everywhere possess continuous first, second and third derivatives. It is easy to see that if u, $u-\epsilon$, and $u+\epsilon$ are all greater than $N+\eta$, and if $\phi(u)=\phi_{N\eta}(u)$

$$\gamma(u+\epsilon)+\gamma(u-\epsilon)-2\gamma(u)$$
$$=\frac{2}{\pi}\int_{-\infty}^{\infty}f(\tau)d\tau\int_0^{\epsilon}d\xi\int_{-\epsilon}^{0}\cos(u+\xi+\xi)\tau\ \phi_{N\eta}(u+\xi+\xi)d\xi=0. \quad (30)$$

Similarly, $$\delta(u+\epsilon)+\delta(u-\epsilon)-2\delta(u)=0. \quad (31)$$

Accordingly, if

$$\frac{3}{2\pi}\lim_{h\to 0}\frac{1}{h^3}\int_{-\infty}^{\infty}[g(t)]^2\frac{\sin^4 ht}{t^4}dt$$

exists, and in particular, if

$$\lim_{A\to\infty}\frac{1}{2A}\int_{-A}^{A}[g(\xi)]^2d\xi$$

exists, we have

$$\frac{3}{2\pi}\lim_{h\to 0}\frac{1}{h^3}\int_{-\infty}^{\infty}[g(t)]^2\frac{\sin^4 ht}{t^4}dt$$
$$=\frac{3}{2}\lim_{\epsilon\to 0}\frac{1}{\epsilon^3}\int_0^{\infty}[\phi(u)]^2\{[\Delta(u+\epsilon)+\Delta(u-\epsilon)-2\Delta(u)]^2$$
$$+[\Gamma(u+\epsilon)+\Gamma(u-\epsilon)-2\Gamma(u)]^2\}du. \quad (32)$$

Let us write $g_{N\eta}(t)$ for the $g(t)$ corresponding to $\phi_{N\eta}(u)$. From this stage on we make the following assumption: *the limit*

$$\lim_{h\to 0}\frac{1}{h^3}\int_{-\infty}^{\infty}[g_{N\eta}(t)]^2\frac{\sin^4 ht}{t^4}dt \quad (33)$$

shall exist for an everywhere dense set of N's from 0 to infinity, and for each of a denumerable set of η's containing terms of arbitrarily small size. Manifestly if $N_1>N+\eta$, we shall have

$$\lim_{h\to 0}\frac{1}{h^3}\int_{-\infty}^{\infty}[g_{N_1\eta_1}(t)]^2\frac{\sin^4 ht}{t^4}dt$$
$$\geq\lim_{h\to 0}\frac{1}{h^3}\int_{-\infty}^{\infty}[g_{N\eta}(t)]^2\frac{\sin^4 ht}{t^4}dt. \quad (34)$$

Thus the values of

$$\frac{3}{\pi} \lim_{h\to 0} \frac{1}{h^3} \int_{-\infty}^{\infty} [g_{N\eta}(t)]^2 \frac{\sin^4 ht}{t^4} dt$$

for $N+\eta<L$ will have a finite upper bound, which we shall call $R\{f;\ L\}$. This will be a non-decreasing function of L, and will hence be continuous except for a denumerable set of finite jumps. We put $R\{f;0\}=0$.

Let $\psi(u)$ be any even function of u with continuous first, second and third derivatives, which, together with its first, second and third derivatives, is summable and of summable square. If $Y>Z>0$, we have

$$\begin{aligned} 3 \overline{\lim_{\epsilon\to 0}} \frac{1}{\epsilon^3} \int_Y^Z & [\psi(u)]^2 \{ [\Delta(u+\epsilon)+\Delta(u-\epsilon)-2\Delta(u)]^2 \\ & +[\Gamma(u+\epsilon)+\Gamma(u-\epsilon)-2\Gamma(u)]^2\} du \\ & \leq \max_{Y\leq v\leq Z} [\psi(v)]^2 [R\{f;Z+\eta\}-R\{f;\ Y-\eta\}]. \end{aligned} \tag{35}$$

Similarly,

$$\begin{aligned} 3 \underline{\lim_{\epsilon\to 0}} \frac{1}{\epsilon^3} \int_Y^Z & [\psi(u)]^2 \{ [\Delta(u+\epsilon)+\Delta(u-\epsilon)-2\Delta(u)]^2 \\ & +[\Gamma(u+\epsilon)+\Gamma(u-\epsilon)-2\Gamma(u)]^2\} du \\ & \geq \min_{Y\leq v\leq Z} [\psi(v)]^2 [R\{f;Z-\eta\}-R\{f;\ Y+\eta\}] \end{aligned} \tag{36}$$

f Y and Z are poi nts of continuity of R, we have

$$\begin{aligned} 3 \overline{\lim_{\epsilon\to 0}} \frac{1}{\epsilon^3} \int_Y^Z & [\psi(u)]^2 \{ [\Delta(u+\epsilon)+\Delta(u-\epsilon)-2\Delta(u)]^2 \\ & +[\Gamma(u+\epsilon)+\Gamma(u-\epsilon)-2\Gamma(u)]^2\} du \\ & \leq \max [\psi(v)]^2 [R\{f;Z\}-R\{f;\ Y\}] \end{aligned} \tag{37}$$

and

$$\begin{aligned} 3 \underline{\lim_{\epsilon\to 0}} \frac{1}{\epsilon^3} \int_Y^Z & [\psi(u)]^2 \{ [\Delta(u+\epsilon)+\Delta(u-\epsilon)-2\Delta(u)]^2 \\ & +[\Gamma(u+\epsilon)+\Gamma(u-\epsilon)-2\Gamma(u)]^2\} du. \\ & \geq \min [\psi(v)]^2 [R\{f;Z\}-R\{f;\ Y\}]. \end{aligned} \tag{38}$$

It then results from the uniform continuity of $\psi(u)$ that

$$\int_0^\infty [\psi(u)]^2 dR\{f;u\} = 3\lim_{\epsilon\to 0}\frac{1}{\epsilon^3}\int_0^\infty [\psi(u)]^2\{[\Delta(u+\epsilon)+\Delta(u-\epsilon)$$
$$-2\Delta(u)]^2+[\Gamma(u+\epsilon)+\Gamma(u-\epsilon)-2\Gamma(u)]^2\}du$$
$$=\frac{3}{\pi}\lim_{h\to 0}\frac{1}{h^3}\int_{-\infty}^\infty \frac{\sin^4 ht}{t^4}\left[\frac{1}{\pi}\int_{-\infty}^\infty f(\tau)d\tau\int_0^\infty \psi(u)\cos u(t-\tau)du\right]^2 dt. \tag{39}$$

Let us consider the functions

$$F_1(t)=f(t+w)+f(t-w),\quad F_2(t)=f(t+w)-f(t-w).$$

Let us put

$$\Psi(t)=\frac{1}{\pi}\int_0^\infty \psi(u)\cos ut\,du;$$
$$G_1(t)=\int_{-\infty}^\infty F_1(t)\Psi(t-\tau)d\tau;$$
$$G_2(t)=\int_{\infty}^\infty F_2(t)\Psi(t-\tau)d\tau;$$
$$g(t)=\int_{-\infty}^\infty f(t)\Psi(t-\tau)d\tau.$$

We have

$$G_1(t)=\int_{-\infty}^\infty f(t)[\Psi(t-\tau+w)+\Psi(t-\tau-w)]d\tau$$
$$=\int_{-\infty}^\infty f(t)\left\{\frac{1}{\pi}\int_0^\infty \psi(x)[\cos(t-\tau+w)x+\cos(t-\tau-w)x]dx\right\}dt$$
$$=\int_{-\infty}^\infty f(t)\left[\frac{2}{\pi}\int_0^\infty \psi(x)\cos wx\cos(t-\tau)x\,dx\right]dt. \tag{40}$$

Accordingly,

$$4\int_0^\infty \cos^2 wu[\psi(u)]^2 dR\{f;u\} = \frac{3}{\pi}\lim_{h\to 0}\frac{1}{h^3}\int_{-\infty}^\infty [G_1(t)]^2\frac{\sin^4 ht}{t^4}dt.$$

Similarly

$$4\int_0^\infty \sin^2 wu[\psi(u)]^2 dR\{f;u\} = \frac{3}{\pi}\lim_{h\to 0}\frac{1}{h^3}\int_{-\infty}^\infty [G_2(t)]^2\frac{\sin^4 ht}{t^4}dt.$$

Hence by subtraction,

$$\int_0^\infty \cos 2wu[\psi(u)]^2 dR\{f;u\} = \frac{3}{\pi}\lim_{h\to 0}\frac{1}{h^3}\int_{-\infty}^\infty g(t+w)g(t-w)\frac{\sin^4 ht}{t^4}dt.$$

If we now let $\psi(u)$ approach 1, while it never exceeds 1 in absolute value,

$$\int_0^\infty \cos 2wu \, dR\{f;u\} = \lim_{\psi(u)\to 1} \frac{3}{\pi} \lim_{h\to 0} \frac{1}{h^3} \int_{-\infty}^\infty g(t+w)g(t-w) \frac{\sin^4 ht}{t^4} dt.$$

Now, (41)

$$\frac{1}{h^3}\int_{-\infty}^\infty \frac{\sin^4 ht}{t^4}[g(t+w)g(t-w)-f(t+w)f(t-w)]\,dt$$

$$= \frac{1}{4h^3}\int_{-\infty}^\infty \frac{\sin^4 ht}{t^4}\{[g(t+w)+g(t-w)]^2-[g(t+w)-g(t-w)]^2$$
$$-[f(t+w)+f(t-w)]^2+[f(t+w)-f(t-w)]^2\}\,dt$$

$$= \frac{1}{4h^3}\int_{-\infty}^\infty \frac{\sin^4 ht}{t^4}[g(t+w)+g(t-w)-f(t+w)-f(t-w)]$$
$$[g(t+w)+g(t-w)+f(t+w)+f(t-w)]dt$$
$$-\frac{1}{4h^3}\int_{-\infty}^\infty \frac{\sin^4 ht}{t^4}[g(t+w)-g(t-w)-f(t+w)+f(t-w)]$$
$$[g(t+w)-g(t-w)+f(t+w)-f(t-w)]\,dt$$

$$\leq \frac{1}{4h^3}\sqrt{\int_{-\infty}^\infty \frac{\sin^4 ht}{t^4}[g(t+w)+g(t-w)-f(t+w)-f(t-w)]^2dt}$$
$$\overline{\int_{-\infty}^\infty \frac{\sin^4 ht}{t^4}[g(t+w)+g(t-w)+f(t+w)+f(t-w)]^2dt}$$
$$+\frac{1}{4h^3}\sqrt{\int_{-\infty}^\infty \frac{\sin^4 ht}{t^4}[g(t+w)-g(t-w)-f(t+w)+f(t-w)]^2dt}$$
$$\overline{\int_{\infty}^\infty \frac{\sin^4 ht}{t^4}[g(t+w)-g(t-w)+f(t+w)-f(t-w)]^2dt}$$

$$\leq \frac{2}{h^3}\sqrt{2\int_{-\infty}^\infty \frac{\sin^4 ht}{t^4}[g(t)-f(t)]^2dt \int_{-\infty}^\infty \frac{\sin^4 ht}{t^4}[[g(t)]^2+[f(t)]^2]dt}$$

$$\leq \frac{4}{h^3}\sqrt{\int_{-\infty}^\infty \frac{\sin^4 ht}{t^4}[g(t)-f(t)]^2dt \int_{-\infty}^\infty \frac{\sin^4 ht}{t^4}[f(t)]^2dt}$$

$$\leq \frac{4}{h^3}\sqrt{\frac{\pi}{8}\int_0^\infty \{[\Delta_1(u+2h)+\Delta_1(u-2h)-2\Delta_1(u)]^2}$$
$$\overline{+[\Gamma_1(u+2h)+\Gamma_1(u-2h)-2\Gamma_1(u)]^2\}\,du \int_{-\infty}^\infty \frac{\sin^4 ht}{t^4}[f(t)]^2dt,}$$

(42)

where

$$\Gamma_1(u) = \Gamma(u) - \gamma(u), \quad \Delta_1(u) = \Delta(u) - \delta(u),$$

and $\psi(u)$ replaces $\phi(u)$ in the definition of $\gamma(u)$ and $\delta(u)$.

Let us give to $\psi(u)$ the particular value $\phi_{N1}(u)$, and let us write $\Phi_{N1}(t)$ for $\Psi(t)$. We have

$$\Gamma_1(u) = \frac{1}{\pi}\int_{-\infty}^{\infty} f(t)dt \left\{ \frac{1-\cos ut}{t^2} - \int_0^u dv \int_0^v \phi_{N1}(w) \cos wt\, dw \right\}. \tag{43}$$

If we denote this particular value of Γ_1 by Γ_{1N}, we may readily conclude that

$$\begin{cases} \Gamma_{1N}(u+2h) + \Gamma_{1N}(u-2h) - 2\Gamma_{1N}(u) = 0 \quad \text{if } u+2h < N; \\ \Gamma_{1N}(u+2h) + \Gamma_{1N}(u-2h) - 2\Gamma_{1N}(u) = \Gamma_{1N}(u+2h) + \Gamma_{1N}(u-2h) \\ \qquad -2\Gamma_{1N}(u) - \Gamma_{1,N+2}(u+2h) - \Gamma_{1,N+2}(u-2h) + 2\Gamma_{1,N+2}(u) \\ \qquad\qquad \text{if } u+2h < N+2; \\ \Gamma_{1N}(u+2h) + \Gamma_{1N}(u-2h) - 2\Gamma_{1N}(u) = \Gamma(u+2h) + \Gamma(u-2h) \\ \qquad -2\Gamma(u) \quad \text{if } u-2h > N+1. \end{cases}$$

With an analogous definition of Δ_{1N}, similar results hold with reference to Δ and its allied functions. As a consequence,

$$\begin{aligned} &\lim_{h\to 0} \frac{\pi}{8h^3}\int_0^{\infty} \{[\Delta_{1N}(u+2h) + \Delta_{1N}(u-2h) - 2\Delta_{1N}(u)]^2 \\ &\qquad + [\Gamma_{1N}(u+2h) + \Gamma_{1N}(u-2h) - 2\Gamma_{1N}(u)]^2\}\, du \\ &\leq \lim_{h\to 0} \frac{\pi}{8h^3}\int_0^{\infty} \{[\Delta(u+2h) + \Delta(u-2h) - 2\Delta(u)]^2 \\ &\qquad + [\Gamma(u+2h) + \Gamma(u-2h) - 2\Gamma(u)]^2\}\, du \\ &\qquad - \frac{\pi}{3}\int_0^{\infty} [\phi_{1N}(u)]^2 dR\{f; u\} \\ &\qquad + \frac{\pi}{3}\int_0^{\infty} [\phi_{1,N+2}(u) - \phi_{1N}(u)]^2 dR\{f; u\}. \end{aligned}$$

We see at once that if

$$\lim_{h\to 0}\frac{\pi}{8h^3}\int_0^\infty \{[\Delta(u+2h)+\Delta(u-2h)-2\Delta(u)]^2$$

$$+[\Gamma(u+2h)+\Gamma(u-2h)-2\Gamma(u)]^2\}du=\frac{\pi}{3}\int_0^\infty dR\{f;u\},$$

we have

$$\lim_{N\to\infty}\lim_{h\to 0}\int_{-\infty}^\infty \frac{\sin^4 ht}{t^4}[g_{1N}(t+w)g_{1N}(t-w)-f(t+w)f(t-w)]dt=0.$$

Under this hypothesis,

$$\int_0^\infty \cos 2wu\, dR\{f;u\}=\frac{3}{\pi}\lim_{h\to 0}\frac{1}{h^3}\int_{-\infty}^\infty \frac{\sin^4 ht}{t^4}f(t+w)f(t-w)dt. \tag{44}$$

That is, if

$$\lim_{A\to\infty}\frac{1}{2A}\int_{-A}^{A} f(t+w)f(t-w)dt$$

exists, we have

$$\frac{1}{2}\int_0^\infty \cos 2\,wu\, dR\{f;u\}=\lim_{A\to\infty}\frac{1}{2A}\int_{-A}^{A} f(t+w)f(t-w)dt.$$

Let us now collect together our hypotheses. These are:
(1) The limit

$$\lim_{A\to\infty}\frac{1}{2A}\int_{-\infty}^\infty [f(t)]^2 dt$$

exists; (2)

$$\lim_{h\to}\frac{1}{h^3}\int_{-\infty}^\infty [g_{N_\eta}(t)]^2\frac{\sin^4 ht}{t^4}dt$$

exists for an everywhere dense set of values of N, and for a denumerable set of values of η containing arbitrarily small values; (3)

$$\lim_{N\to\infty}\lim_{h\to 0}\frac{1}{h^3}\int_{-\infty}^\infty [g_{N_\eta}(t)]^2\frac{\sin^4 ht}{t^4}dt$$

$$=\lim_{h\to 0}\frac{1}{h^3}\int_{-\infty}^\infty [f(t)]^2\frac{\sin^4 ht}{t^4}dt.$$

It will be seen that these hypotheses involve directly only a denumerable set of transforms of $f(t)$, which is exceedingly important for the application of considerations from the theory of probability. This is the case because the probability that any one of a denumerable set of events of zero probability shall occur is itself zero. The hypotheses of the author's previous paper involve assertions of uniformity which comprise a non-denumerable infinity of discrete cases, and are far more difficult to tie up with probability theory.

Under these hypotheses we have, as we have just shown,

$$\int_0^\infty \cos 2\,wu\; dR\{f;u\} = \frac{3}{\pi} \lim_{h\to 0} \frac{1}{h^3}\int_{-\infty}^{\infty} \frac{\sin^4 ht}{t^4} f(t+w)f(t-w)dt,$$

where

$$R\{f,L\} = \limsup_{N+\eta<L} \frac{3}{\pi} \lim_{h\to 0} \frac{1}{h^3}\int_{-\infty}^{\infty} [g_{N_\eta}(t)]^2 \frac{\sin^4 ht}{t^4}\, dt.$$

Now let us consider the general case of an expression

$$\int_0^\infty \cos \alpha\, u\; dH(u) = K(\alpha),$$

where $H(u)$ is an odd non-decreasing function of u, $H(0)=0$, and where $H(\infty) = \lim_{u\to+\infty} H(u)$ is finite. We have

$$\frac{2}{\pi}\int_0^A K(\alpha)\frac{\sin \alpha v}{\alpha}\,d\alpha = \frac{2}{\pi}\int_0^A \frac{\sin \alpha v}{\alpha}\,d\alpha \int_0^\infty \cos \alpha u\; dH(u)$$

$$= \lim_{B\to\infty} \frac{2}{\pi}\int_0^A \frac{\sin \alpha v}{\alpha}\,d\alpha \left\{ \int_0^B \alpha \sin \alpha u\, H(u)du + H(B)\cos \alpha B \right\}$$

$$= \lim_{B\to\infty} \frac{2}{\pi}\int_0^A \sin \alpha v\, d\alpha \int_0^B \sin \alpha u\, H(u)du$$

$$= \lim_{B\to\infty} \frac{2}{\pi}\int_0^B H(u)du \int_0^A \sin \alpha v \sin \alpha u\, d\alpha$$

$$= \frac{1}{\pi}\int_0^\infty H(u) \left[\frac{\sin A(u-v)}{u-v} - \frac{\sin A(u+v)}{u+v}\right] du$$

$$= \frac{1}{\pi}\int_{-\infty}^\infty H(w+v)\frac{\sin Aw}{w}dw. \tag{45}$$

Hence, since $H(u)$ is of limited total variation, we have by Dirichlet's formula

$$H(v)=\frac{2}{\pi}\int_0^\infty K(\alpha)\frac{\sin \alpha v}{\alpha}\,d\alpha. \tag{46}$$

In particular,

$$R\{f;u\} = \frac{2}{\pi}\int_0^\infty \left\{\frac{3}{\pi}\lim_{h\to 0}\frac{1}{h^3}\int_{-\infty}^{\infty}\frac{\sin^4 ht}{t^4} f\left(t+\frac{\alpha}{2}\right) f\left(t-\frac{\alpha}{2}\right) dt\right\}\frac{\sin \alpha u}{\alpha}\,d\alpha. \tag{47}$$

THE HARMONIC ANALYSIS OF IRREGULAR MOTION

Second Paper

By Norbert Wiener

Table of Contents

Introduction

In a previous paper by the same name, also published in this Journal, the author gave a method of analyzing into its harmonic components a function neither periodic, as are the functions representable by Fourier series, nor asymptotically zero, as are the functions for which the Fourier integral is the appropriate development. The conditions to be fulfilled by a function $f(t)$ possessing such an analysis are:

(*a*) that its mean square value shall exist;

(*b*) that the mean square values of a certain denumerable set of its transforms

$$\int_{-\infty}^{\infty} \Phi(t-\tau)f(\tau)d\tau \qquad [\Phi \text{ even}]$$

shall exist;

(*c*) that the mean square value of the transform shall approach the mean square value of $f(t)$ as

$$\frac{2}{\pi}\int_0^\infty \Phi(t) \cos t\, u\, dt$$

approaches 1.

It is the purpose of the present paper to exhibit a function $f(x)$ of this sort, which in addition has in the language of the previous paper a continuous spectrum. In this way, we shall not only supplement the results already obtained by an existence theory, but shall also contribute to the rigorous mathematical theory of the disturbance of a linear oscillating system by haphazard disturbances, as exemplified by the theory of tube noises in radiotelephony. Let it be noted that our existence theory is dependent throughout on the multiplicative axiom of Zermelo. We shall assume this axiom without explicit reference in all that follows.

§1. The distribution of functions over an infinite range.

In an earlier paper,[1] the author has discussed the distribution of a class of functions defined for arguments between $t=0$ and $t=1$, which have as a class the following properties:

(*a*) The change in the value of a function between $t=t_1$ and $t=t_2$ is independent of the value of the function at and before $t=t_1$;

(*b*) The change in question has a Gaussian distribution;

(*c*) The parameter of this distribution depends only on t_2-t_1.

We shall call a class of functions with the probability distribution indicated in that paper a set *normally distributed* over the time-interval (0, 1). We wish here to generalize the notion of a normal distribution to functions defined over the time-interval $(-\infty, \infty)$.

Let $f(t)$ be a function of a normally distributed class over the interval (0, 1). Let us form the function

$$f\left(\frac{1}{\pi}\tan^{-1}t+\frac{1}{2}\right)$$

[1] Differential-Space, this Journal, Vol. 2 (1923), pp. 131–174.

This will be defined over the whole infinite set of values of t, and will furthermore satisfy conditions (a) and (b). It will not, however, satisfy condition (c), since the interval (t_1, t_2) will be drawn out into the interval $(-\cot \pi t_1, -\cot \pi t_2)$.

We can, however, compensate for this change in the range of arguments of the function by a corresponding change in the values of the function. We should perhaps expect that $g(t)$ would be uniformly distributed if

$$g(t) = \int_0^t \frac{1}{\frac{d}{dt}\left(\frac{1}{\pi}\tan^{-1}t + \frac{1}{2}\right)}\, df\left(\frac{1}{\pi}\tan^{-1}t + \frac{1}{2}\right)$$

$$= \pi \int^t (1+t^2)\, df\left(\frac{1}{\pi}\tan^{-1}t + \frac{1}{2}\right).$$

It must be remembered, however, that in such a haphazard motion as we are discussing, the change of g is proportional for a small interval of time, not to the change in t, but roughly to the change in the square root of t. This would give us formally

$$g(t) = \sqrt{\pi} \int_0^t \sqrt{1+t^2}\, df\left(\left(\frac{1}{\pi}\right) \tan^{-1}t + \frac{1}{2}\right).$$

Since, however, f is practically never of limited total variation, we must change this definition into a form where it makes sense. To do this, we integrate formally by parts, obtaining.

$$g(t) = \sqrt{\pi} \int_{-\infty}^{\infty} f\left(\left(\frac{1}{\pi}\right)\tan^{-1}\xi + \frac{1}{2}\right) d\phi_t(\xi)$$

$$= \sqrt{\pi} \int_0^1 f(u)\, d\psi_t(u), \tag{1}$$

where

$$\left.\begin{array}{ll} \phi_t(\xi) = \sqrt{1+\xi^2} & [0 \leq \xi \leq t]; \\ \phi_t(\xi) = 0 & [0 > \xi \text{ or } t < \xi]; \end{array}\right\} [t>0]$$

$$\left.\begin{array}{ll} \psi_t(u) = \csc \pi u & \left[\frac{1}{2} \leq u \leq \frac{\tan^{-1}t}{\pi} + \frac{1}{2}\right]; \\ \psi_t(u) = 0 & \left[\frac{1}{2} \geq u \text{ or } u \geq \frac{\tan^{-1}t}{\pi} + \frac{1}{2}\right]; \end{array}\right\} [t>0]$$

$$\left.\begin{array}{ll}\phi_t(\xi)=\sqrt{1+\xi^2} & [t\leq\xi\leq 0];\\ \phi_t(\xi)=0 & [\xi>0 \text{ or } \xi<t];\end{array}\right\}[t<0]$$

$$\left.\begin{array}{ll}\psi_t(u)=\csc\pi u & \left[\frac{1}{2}\geq u\geq\frac{\tan^{-1}t}{\pi}+\frac{1}{2}\right];\\ \psi_t(u)=0 & \left[\frac{1}{2}\leq u \text{ or } u\leq\frac{\tan^{-1}t}{\pi}+\frac{1}{2}\right];\end{array}\right\}[t<0]$$

$$g(0)=0. \tag{2}$$

Using the results of an earlier paper,[2] we obtain for the average value of $g(t_1)\,g(t_2)$,

$$\frac{\pi r}{2}\int_0^1 d\psi_{t_1}(u)\int_0^u v\,d\psi_{t_2}(v)+\frac{\pi^2 r}{2}\int_0^1 d\psi_{t_2}(u)\int_0^u v\,d\psi_{t_1}(v)$$

$$=-\frac{\pi r}{2}\int_0^1 u[\psi_{t_1}(u)d\psi_{t_2}(u)+\psi_{t_2}(u)d\psi_{t_1}(u)]$$

$$=-\frac{\pi r}{2}\int_0^1 u\,d[\psi_{t_1}(u)\psi_{t_2}(u)]$$

$$=\frac{\pi r}{2}\int_0^1 \psi_{t_1}(u)\psi_{t_2}(u)du$$

$$\left\{\begin{array}{lll}=\frac{\pi r}{2}\int_{1/2}^{\frac{1}{\pi}\tan^{-1}t_1+1/2}\csc^2\pi u\,du; & & \left[\begin{array}{l}0<t_1\leq t_2\\ \text{or } 0>t_1\geq t_2\end{array}\right]\\ =\frac{\pi r}{2}\int_{1/2}^{\frac{1}{\pi}\tan^{-1}t_2+1/2}\csc^2\pi u\,du; & & \left[\begin{array}{l}0<t_2\leq t_1\\ \text{or } 0>t_2\geq t_1\end{array}\right]\\ =0 & ; & [t_1t_2\leq 0]\end{array}\right.$$

$$\left\{\begin{array}{lll}= rt_1/2 & ; & \left[\begin{array}{l}0<t_1\leq t_2\\ \text{or } 0>t_1\geq t_2\end{array}\right]\\ = rt_2/2 & ; & \left[\begin{array}{l}0<t_2\leq t_1\\ \text{or } 0>t_2\geq t_1\end{array}\right]\\ =0 & . & [t_1\,t_2\leq 0]\end{array}\right. \tag{3}$$

[2] Cf. *The Average of an Analytic Function*, Proc. Nat. Acad. Sci. 1921, p. 256.

Here r is a parameter determining the mean square variation of $f(t)$ over a given time.

We thus obtain the same formal expression for the mean of $g(t_1)g(t_2)$ as for the mean of $f(t_1)f(t_2)$, which is as it should be. We furthermore see that the mean of $g(t_1)(g(t_2)-g(t_1))$ is 0, thus showing that there is at least a linear independence between $g(t_1)$ and $g(t_2)-g(t_1)$. The mean square value of $g(t_2)-g(t_1)$ is always $r|t_2-t_1|/2$, which is in harmony with condition (c).

§2. **Independent linear functionals.**

In the author's earlier paper, the notion of independent linear functionals was introduced. It was there shown that if

$$\int_x^1 K(x)dx \qquad \text{and} \qquad \int_x^1 Q(x)dx$$

are normal and orthogonal, and $\phi(u)$ is bounded and uniformly continuous, the mean of

$$\phi\left(\int_0^1 K(x)f(x)dx, \quad \int_0^1 Q(x)f(x)dx\right)$$

with respect to f is the same as the mean of

$$\phi\left(\int_0^1 K(x)f_1(x)dx \quad \int_0^1 Q(x)f_2(x)dx\right)$$

with respect to f_1 and f_2. This notion may be extended to a set containing any finite number of independent functionals. It may also be extended to linear functionals of the form

$$\int_0^1 f(x)d\,\alpha(x)$$

Without any fundamental change, the argument of the author's earlier paper will show that if

$$\alpha(1)-\alpha(x) \qquad \text{and} \qquad \beta(1)-\beta(x)$$

are normal and orthogonal, and ϕ is bounded and uniformly continuous, the mean of

$$\phi\left(\int_0^1 f(x)d\,\alpha(x), \quad \int_0^1 f(x)d\,\beta(x)\right)$$

with respect to f is the same as the mean of

$$\phi\left(\int_0^1 f_1(x)d\,\alpha(x),\ \int_0^1 f_2(x)d\,\beta(x)\right)$$

with respect to f_1 and f_2. The extension to any finite set of independent linear functionals also holds. It follows at once that the mean of

$$\phi(g(t_2)-g(t_1),\quad g(t_1)-g(t_0)) \qquad [t_0<t_1<t_2]$$

is the same as the mean of

$$\phi(g_1(t_2)-g_1(t_1),\quad g_2(t_1)-g_2(t_0))$$

with respect to the two normally distributed functions $f_1(t)$ and $f_2(t)$ on which g_1 and g_2 respectively depend. This is in accordance with (a).

Let us now consider two expressions of the form

$$\int_{-\infty}^{\infty} K(x)g(x)dx, \qquad \int_{-\infty}^{\infty} Q(x)g(x)dx,$$

where $K(x)\sqrt{1+x^2}$ and $Q(x)\sqrt{1+x^2}$ are absolutely integrable over the whole infinite line. We have (except in the case where $g(x)\neq O(x)$ at infinity, which is of 0 probability)

$$\begin{aligned}\int_{-\infty}^{\infty} K(x)g(x)dx &= \sqrt{\pi}\int_{\infty}^{\infty} K(x)dx\int_0^1 f(u)d\psi_x(u)\\ &= \sqrt{\pi}\int_0^1 f(u)\,d_u\int_{-\infty}^{\infty} K(x)\psi_x(u)dx\\ &= \sqrt{\pi}\int_{1/2}^1 f(u)d_u\int_{-\cot\pi u}^{\infty} K(x)\csc \pi u\; dx\\ &\quad -\sqrt{\pi}\int_0^{1/2} f(u)d_u\int_{-\infty}^{-\cot \pi u} K(x)\csc \pi u\; dx.\end{aligned}$$

Now, we have

$$\pi\int_0^1 du\int_{-\infty}^{\infty}K(x)[\psi_x(1)-\psi_x(u)]dx\int_{-\infty}^{\infty}Q(x)[\psi_x(1)-\psi_x(u)]dx$$

$$=\pi\int_{1/2}^1 du\int_{-\cot\pi u}^{\infty}K(x)\csc\pi u\,dx\int_{-\cot\pi u}^{\infty}Q(y)\csc\pi u\,dy$$

$$+\pi\int_0^{1/2} du\int_{-\infty}^{-\cot\pi u}K(x)\csc\pi u\,dx\int_{-\infty}^{-\cot\pi u}Q(y)\csc\pi u\,dy$$

$$=\int_{-\infty}^0 dw\int_w^{\infty}K(x)dx\int_w^{\infty}Q(y)dy$$

$$+\int_0^{\infty} dw\int_{-\infty}^{w}K(x)dx\int_{-\infty}^{w}Q(y)dy. \tag{4}$$

That is, we have the following theorem: *let* $K(x)\sqrt{1+x^2}$ and $Q(x)\sqrt{1+x^2}$ *be absolutely integrable over the whole infinite line, and let*

$$\int_{-\infty}^{\infty}K(x)dx=\int_{-\infty}^{\infty}Q(x)dx=0.$$

Let

$$\int_{-\infty}^{\infty}dw\int_{-\infty}^{w}K(x)dx\int_{-\infty}^{w}Q(y)dy=0;$$

$$\int_{-\infty}^{\infty}dw\int_{-\infty}^{w}K(x)dx\int_{-\infty}^{w}K(y)dy=\int_{-\infty}^{\infty}dw\int_{-\infty}^{w}Q(x)dx\int_{-\infty}^{w}Q(y)dy=1.$$

Then

$$\int_{-\infty}^{\infty}K(x)g(x)dx \quad \textit{and} \quad \int_{-\infty}^{\infty}Q(x)g(x)dx$$

will be independent functionals, with mean square values r^2. A similar result will hold with regard to any finite set of such functionals.

§3. The existence of the mean square of the transform of a function.

Now let us consider the expression

$$\frac{1}{2A}\int_{-A}^{A}\left[\int_{-\infty}^{\infty}\Phi(\tau-t)g(\tau)d\tau\right]^2 dt \tag{5}$$

where

$$\Phi(\tau)=O(\tau^{-3}) \text{ at } \pm\infty; \quad \int_{-\infty}^{\infty}\Phi(\tau)d\tau=0. \tag{6}$$

We have

$$\frac{1}{2A}\int_{-A}^{A}\left[\int_{-\infty}^{\infty}\Phi(\tau-t)g(\tau)d\tau\right]^2 dt$$
$$=\int_{-\infty}^{\infty}g(\sigma)d\sigma\int_{-\infty}^{\infty}g(\tau)d\tau\left[\frac{1}{2A}\int_{-A}^{A}\Phi(\sigma-t)\Phi(\tau-t)dt\right]$$
$$=\int_{-\infty}^{\infty}g(\sigma)d\sigma\int_{-\infty}^{\infty}g(\tau)F_A(\sigma,\tau)d\tau. \tag{7}$$

Let us put

$$G_A(\sigma,\tau)=\int_{-\infty}^{\sigma}d\alpha\int_{-\infty}^{\tau}F_A(\alpha,\beta)d\beta. \tag{8}$$

Let us now consider the Fredholm equation

$$u(\sigma)=\lambda\int_{-\infty}^{\infty}G_A(\sigma,\tau)u(\tau)d\tau \tag{9}$$

This will be satisfied for a denumerable set of values λ_n of λ, and for a denumerable set of corresponding functions $u_n(\tau)$. These can be chosen in such a way that

$$\int_{-\infty}^{\infty}u_m(\tau)u_n(\tau)d\tau=\delta_{mn}. \tag{10}$$

We shall have at least formally

$$G_A(\sigma,\tau)=\sum_1^{\infty}\frac{u_n(\sigma)u_n(\tau)}{\lambda_n} \tag{11}$$

and hence again formally

$$F_A(\sigma,\tau)=\sum_1^{\infty}\frac{u_n'(\sigma)u_n'(\tau)}{\lambda_n}. \tag{12}$$

If this latter formula holds, it follows that

$$\int_{-\infty}^{\infty}g(\sigma)d\sigma\int_{-\infty}^{\infty}g(\tau)F_A(\sigma,\tau)d\tau$$
$$=\sum_1^{\infty}\frac{1}{\lambda_n}\left[\int_{-\infty}^{\infty}g(\sigma)u_n'(\sigma)d\sigma\right]^2. \tag{13}$$

Then

$$\frac{1}{2A}\int_{-A}^{A}\left[\int_{-\infty}^{\infty}\Phi(\tau-t)d\tau\int_{-\infty}^{\tau}u_m(\alpha)d\alpha\right]^2 dt$$

$$=\sum_1^\infty \frac{1}{\lambda_n}\left[\int_{-\infty}^{\infty}\left(\int_{-\infty}^{\sigma}u_m(\alpha)d\alpha\right)u_n'(\sigma)d\sigma\right]^2$$

$$=\sum_1^\infty \frac{1}{\lambda_n}\left[u_n(\infty)\int_{-\infty}^{\infty}u_m(\alpha)\,d\alpha-\int_{-\infty}^{\infty}u_m(\sigma)u_n(\sigma)d\sigma\right]^2=\frac{1}{\lambda_m}. \tag{14}$$

That is, every λ_m is positive. In this proof we have used the obvious consequences of our hypotheses, that $G_A(\sigma, \tau)$ is $O\left(\frac{1}{\sigma^2+\tau^2}\right)$ at $\sigma, \tau=\pm\infty$, and that $u_n(\infty)=0$. From the latter fact, assuming that our bilinear formula holds, it results that we have reduced

$$\int_{-\infty}^{\infty}g(\sigma)d\sigma\int_{-\infty}^{\infty}g(\tau)F_A(\sigma, \tau)d\tau$$

to a sum of squares of independent linear functionals.

We have

$$\int_{-\infty}^{\infty}[u_n(\sigma)]^2 d\lambda=\lambda_n\int_{-\infty}^{\infty}u_n(\sigma)d\sigma\int_{-\infty}^{\infty}G_A(\sigma, \tau)u(\tau)d\tau$$

$$=\frac{\lambda_n}{2A}\int_{-\infty}^{\infty}u_n(\sigma)d\sigma\int_{-\infty}^{\infty}u_n(\tau)d\tau\int_{-A}^{A}dt\int_{-\infty}^{\sigma}\Phi(\alpha-t)d\alpha\int_{-\infty}^{\tau}\Phi(\beta-t)d\beta$$

$$=\frac{\lambda_n}{2A}\int_{-\infty}^{\infty}d\mu\int_{-\infty}^{\infty}u_n(\sigma)u_n(\sigma+\mu)d\sigma\int_{-A}^{A}dt\int_{-\infty}^{\sigma}\Phi(\alpha-t)d\alpha$$

$$\int_{-\infty}^{\sigma+\mu}\Phi(\beta-t)d\beta$$

$$\leq\frac{\lambda_n}{2A}\int_{-\infty}^{\infty}[u_n(\sigma)]^2 d\sigma\int_{-\infty}^{\infty}d\mu\left|\int_{-A+\sigma}^{A+\sigma}dt\int_{-\infty}^{0}\Phi(\alpha-t)d\alpha\int_{-\infty}^{\mu}\Phi(\beta-t)d\beta\right|$$

$$\leq\frac{\lambda_n}{2A}\int_{-\infty}^{\infty}[u_n(\sigma)]^2 d\sigma\int_{-\infty}^{\infty}d\mu\int_{-\infty}^{\infty}dt\,\frac{C}{1+t^2}\cdot\frac{C}{1+(\mu-t)^2} \tag{15}$$

where C is some constant. Inasmuch as the latter integral converges, we have a quantity R such that $\lambda_n \geq RA$.

We have now the task of establishing (13) rigorously. Owing to the continuity of the functionals involved, it is enough to establish (13) for differentiable functions $g(\alpha)$ which are $O(\sqrt{\alpha})$ at infinity. Under these conditions, we see on integrating by parts that (13) is equivalent to

$$\int_{-\infty}^{\infty} g'(\sigma)d\sigma \int_{-\infty}^{\infty} g'(\tau)G_A(\sigma, \tau)d\tau = \sum_1^{\infty} \frac{1}{\lambda_n}\left[\int_{-\infty}^{\infty} g'(\sigma)u_n(\sigma)d\sigma\right]^2. \tag{16}$$

This is, however, an immediate consequence of (11), provided the latter formula holds uniformly, or indeed, if this latter formula holds in the sense of convergence in the mean. On the hypothesis that all the λ_n's are positive, the analogue of (11) has been proved by Mercer[3] to hold uniformly and absolutely in the case where the interval of integration is finite. If the interval is infinite, our hypotheses (6) allow us to transform it into a finite interval in such a manner that Mercer's theorem holds everywhere, except possibly in the neighborhood of the end points. If we replace uniform convergence by convergence in the mean, this latter difficulty is obviated.

The expression

$$\sum_1^{N} \frac{1}{\lambda_n}\left[\int_{-\infty}^{\infty} g(\sigma)u_n'(\sigma)d\sigma\right]^2$$

will have a mean value. It is true that the square of a number is not a uniformly continuous function of that number over an infinite range, but it is the upper bound of an increasing sequence of such functions, and hence[4] will have a mean value. The computation of this value is easy: it will be

$$r^2\sum_1^{N} \frac{1}{\lambda_n},$$

[3] T. Mercer. Functions of Positive and Negative Type and their Connection with the Theory of Integral Equations. Phil. Trans. Royal. Soc. (A) Vol. 209, pp. 415–446. 1909.

[4] Cf. Differential-Space, §§9 and 10; The Average of an Analytic Functional.

in accordance with our fundamental theorem concerning independent functionals. If we use again our fundamental theorem concerning increasing sequences of functionals, to the effect that if we have an increasing sequence of functionals whose means approach a limit, this will be the mean of the limit functional of the sequence, we see that the mean of

$$\frac{1}{2A}\int_{-A}^{A}\left[\int_{-\infty}^{\infty}\Phi(\tau-t)g(\tau)d\tau\right]^{2}dt$$

will be

$$\begin{aligned}
r^2\sum_{1}^{\infty}\frac{1}{\lambda_n} &= r^2\int_{-\infty}^{\infty}G_A(\sigma,\sigma)d\sigma \\
&= \frac{r^2}{2A}\int_{-\infty}^{\infty}d\sigma\int_{-\infty}^{\sigma}d\alpha\int_{-\infty}^{\sigma}d\beta\int_{-A}^{A}\Phi(\alpha-t)\Phi(\beta-t)dt \\
&= \frac{r^2}{2A}\int_{-A}^{A}dt\int_{-\infty}^{\infty}d\sigma\left[\int_{-\infty}^{\sigma}\Phi(\alpha-t)d\alpha\right]^2 \\
&= r^2\int_{-\infty}^{\infty}d\sigma\left[\int_{-\infty}^{\sigma}\Phi(\alpha)d\alpha\right]^2. \qquad (17)
\end{aligned}$$

It will be noted that this is independent of A.

An argument of the same type will show the existence of the mean of

$$\left\{\frac{1}{2A}\int_{-A}^{A}\left[\int_{-\infty}^{\infty}\Phi(\tau-t)g(\tau)d\tau\right]^{2}dt\right\}^{2}.$$

This may be written

$$\left\{\sum_{1}^{\infty}\frac{1}{\lambda_n}\left[\int_{-\infty}^{\infty}g(\sigma)u_n'(\sigma)d\sigma\right]^2\right\}^2$$

$$= \lim_{N\to\infty}\sum_{j=1}^{N}\sum_{k=1}^{N}\frac{1}{\lambda_j\lambda_k}\left[\int_{-\infty}^{\infty}g(\sigma)u_j'(\sigma)d\sigma\right]^2\left[\int_{-\infty}^{\infty}g(\sigma)u_k'(\sigma)d\sigma\right]^2. \qquad (18)$$

The means of the several terms in this sum have already been evaluated by the author,[5] yielding us the result for the mean of this expression

$$\left[r^2\sum_1^\infty \frac{1}{\lambda_n}\right]^2+2r^4\sum_1^\infty \frac{1}{\lambda_n^2}\,. \tag{19}$$

Using the distributive property of the mean, we see that

$$\left\{\frac{1}{2A}\int_{-A}^{A}\left[\int_{-\infty}^{\infty}\Phi(\tau-t)g(\tau)d\tau\right]^2 dt-r^2\sum_1^\infty \frac{1}{\lambda_n}\right\}^2$$

$$=\left\{\frac{1}{2A}\int_{-A}^{A}\left[\int_{-\infty}^{\infty}\Phi(\tau-t)g(\tau)d\tau\right]^2 dt\right\}^2$$

$$-2r^2\sum_1^\infty \frac{1}{\lambda_n}\cdot\frac{1}{2A}\int_{-A}^{A}\left[\int_{-\infty}^{\infty}\Phi(\tau-t)g(\tau)d\tau\right]^2 dt$$

$$+r^4\left[\sum_1^\infty \frac{1}{\lambda_n}\right]^2 \tag{20}$$

has the mean

$$2r^4\sum_1^\infty \frac{1}{\lambda_n^2}\leq\frac{2r^4}{RA}\int_{-\infty}^{\infty}d\sigma\left[\int_{-\infty}^{\sigma}\Phi(\alpha)d\alpha\right]^2=B/A \tag{21}$$

where B is some constant. In a similar manner, we may obtain for the mean of

$$\left\{\frac{1}{2A}\int_{-A}^{A}\left[\int_{-\infty}^{\infty}\Phi(\tau-t)g(\tau)d\tau\right]^2 dt-r^2\sum_1^\infty \frac{1}{\lambda_n}\right\}^4$$

the expression

$$48r^8\sum_1^\infty \frac{1}{\lambda_n^4}+24r^8\left[\sum_1^\infty \frac{1}{\lambda_n^2}\right]^2 \tag{22}$$

which is less than some expression C/A^2. The argument that

$$\text{mean}\left\{\frac{1}{2A}\int_{-A}^{A}\left[\int_{-\infty}^{\infty}\Phi(\tau-t)g(\tau)d\tau\right]^2 dt-r^2\sum_1^\infty 1/\lambda_n\right\}^4$$

$$=\lim_{N\to\infty}\text{mean}\left\{\sum_1^N \frac{1}{\lambda_n}\left[\int_{-\infty}^{\infty}g(\sigma)u_n'(\sigma)d\sigma\right]^2-r^2\sum_1^N 1/\lambda_n\right\}^4$$

[5] The Average of an Analytic Functional. Also Differential-Space, §9.

follows exactly the same lines as in the second degree case. Let us use Σ' to denote a sum for several *distinct* subscripts. We have

$$\text{mean}\left\{\sum_1^N \frac{1}{\lambda_n}\left[\left(\int_{-\infty}^{\infty} g(\sigma)u_n'(\sigma)d\sigma\right)^2 - r^2\right]\right\}^4$$

$$= \text{mean}\sum_1^N \frac{1}{\lambda_n^4}\left[\left(\int_{-\infty}^{\infty} g(\sigma)u_n'(\sigma)d\sigma\right)^2 - r^2\right]^4$$

$$+4\ \text{mean}\sum_{m,n=1}^{N}{}' \frac{1}{\lambda_m^3\lambda_n}\left[\left(\int_{-\infty}^{\infty} g(\sigma)u_m'(\sigma)d\sigma\right)^2 - r^2\right]^3 \left[\left(\int_{-\infty}^{\infty} g(\sigma)u_n'(\sigma)d\sigma\right)^2 - r^2\right]$$

$$+\ 6\ \text{mean}\sum_{m,n=1}^{N}{}' \frac{1}{\lambda_m^2\lambda_n^2}\left[\left(\int_{-\infty}^{\infty} g(\sigma)u_m'(\sigma)d\sigma\right)^2 - r^2\right]^2 \left[\left(\int_{-\infty}^{\infty} g(\sigma)u_n'(\sigma)d\sigma\right)^2 - r^2\right]^2$$

$$+12\ \text{mean}\sum_{m,n,p=1}^{N}{}' \frac{1}{\lambda_m^2\lambda_n\lambda_p}\left[\left(\int_{-\infty}^{\infty} g(\sigma)u_m'(\sigma)d\sigma\right)^2 - r^2\right]^2 \left[\left(\int_{-\infty}^{\infty} g(\sigma)u_n'(\sigma)d\sigma\right)^2 - r^2\right] \left[\left(\int_{-\infty}^{\infty} g(\sigma)u_p'(\sigma)d\sigma\right)^2 - r^2\right]$$

$$+24\ \text{mean}\sum_{m,n,p,q=1}^{N}{}' \frac{1}{\lambda_m\lambda_n\lambda_p\lambda_q}\left[\left(\int_{-\infty}^{\infty} g(\sigma)u_m'(\sigma)d\sigma\right)^2 - r^2\right] \left[\left(\int_{-\infty}^{\infty} g(\sigma)u_n'(\sigma)d\sigma\right)^2 - r^2\right] \cdot\left[\left(\int_{-\infty}^{\infty} g(\sigma)u_p'(\sigma)d\sigma\right)^2 - r^2\right] \left[\left(\int_{-\infty}^{\infty} g(\sigma)u_q'(\sigma)d\sigma\right)^2 - r^2\right]$$

$$=\sum_1^N \frac{1}{\lambda_n^4}\left\{\text{mean}\left(\int_{\infty}^{\infty} g(\sigma)u_n'(\sigma)d\sigma\right)^8\right.$$

$$-4\,r^2\,\text{mean}\left(\int_{-\infty}^{\infty} g(\sigma)u_n'(\sigma)d\sigma\right)^6$$

$$+6\,r^4\,\text{mean}\left(\int_{-\infty}^{\infty} g(\sigma)u_n'(\sigma)d\sigma\right)^4$$

$$\left.-4\,r^6\,\text{mean}\left(\int_{-\infty}^{\infty} g(\sigma)u_n'(\sigma)d\sigma\right)^2+r^8\right\}$$

$$+6\sum_{m,\,n=1}^{N}{}' \frac{1}{\lambda_m^2\lambda_n^2}\left\{\text{mean}\left(\int_{-\infty}^{\infty} g(\sigma)u_m'(\sigma)d\sigma\right)^4\right.$$

$$\left.-2\,r^2\,\text{mean}\left(\int_{-\infty}^{\infty} g(\sigma)u_m'(\sigma)d\sigma\right)^2+r^4\right\}$$

$$\left\{\text{mean}\left(\int_{-\infty}^{\infty} g(\sigma)u_n'(\sigma)d\sigma\right)^4\right.$$

$$\left.-2\,r^2\,\text{mean}\left(\int_{-\infty}^{\infty} g(\sigma)u_n'(\sigma)d\sigma\right)^2+r^4\right\}$$

$$=\sum_1^N \frac{r^8}{\lambda_n^4}\frac{1}{\sqrt{2\pi}}\int_{-\infty}^{\infty}(\xi^8-4\xi^6+6\xi^4-4\xi^2+1)e^{-\frac{\xi^2}{2}}d\xi$$

$$+6\sum_{m,n=1}^{N}{}' \frac{r^8}{\lambda_m^2\lambda_n^2}\left[\frac{1}{\sqrt{2\pi}}\int_{-\infty}^{\infty}(\xi^4-2\xi^2+1)e^{-\frac{\xi^2}{2}}d\xi\right]^2$$

$$=60\,r^8\sum_1^N\frac{1}{\lambda_n^4}+24\,r^8\sum_{m,\,n=1}^{N}{}'\frac{1}{\lambda_m^2\lambda_n^2}$$

$$=48\,r^8\sum_1^N\frac{1}{\lambda_n^4}+24\,r^8\left[\sum_1^N\frac{1}{\lambda_n^2}\right]^2.$$

Let us now consider a functional of f which is P^4 whenever

$$\left\{\frac{1}{2A}\int_{-A}^{A}\left[\int_{-\infty}^{\infty}\Phi(\tau-t)g(\tau)d\tau\right]^2dt-r^2\sum_1^{\infty}\frac{1}{\lambda_n}\right\}^4\geq P^4 \qquad (23)$$

and which is zero otherwise. It is clear that this functional never exceeds

$$\left\{\frac{1}{2A}\int_{-A}^{A}\left[\int_{-\infty}^{\infty}\Phi(\tau-t)g(\tau)d\tau\right]^2 dt-r^2\sum_1^\infty\frac{1}{\lambda_n}\right\}^4 \tag{24}$$

Hence the lower bound of the mean of larger functionals, or, as we call it, the upper mean of this functional, cannot exceed C/A^2. We term the upper mean of the functional which is 1 on a given class of functions and 0 elsewhere, the outer probability that a function belongs to this class. It follows that the outer probability that a function $g(\tau)$ has the property (23) is not greater than C/A^2P^4. There is, moreover, a theorem due to the fact that our mean is a Daniell integral,[6] to the effect that the outer probability of the disjunction of two events cannot exceed the sum of their outer probabilities.

It follows that the outer probability that some expression of the form

$$\left|\frac{1}{2\alpha}\int_{-\alpha}^{\alpha}\left[\int_{-\infty}^{\infty}\Phi(\tau-t)g(\tau)d\tau\right]^2 dt-r^2\sum_1^\infty\frac{1}{\lambda_n}\right| \qquad [A\leq\alpha\leq A+1]$$

should exceed $3P$ cannot exceed the combined outer probabilities that

$$\left|\frac{1}{2(A+1)}\int_{-A-1}^{A+1}\left[\int_{-\infty}^{\infty}\Phi(\tau-t)g(\tau)d\tau\right]^2 dt-r^2\sum_1^\infty\frac{1}{\lambda_n}\right|\geq P;$$

$$\int_{A}^{A+1}\left[\int_{-\infty}^{\infty}\Phi(\tau-t)g(\tau)d\tau\right]^2 dt\geq 2P(A+1);$$

$$\int_{-A-1}^{A}\left[\int_{-\infty}^{\infty}\Phi(\tau-t)g(\tau)d\tau\right]^2 dt\geq 2P(A+1); \tag{25}$$

and hence does not exceed some expression S/A^2. To prove this, it is only necessary to carry through an argument concerning the second and third integrals here given analogous to that carried through in the case of the first, and to show that their mean

[6] P. J. Daniell, A General Form of Integral. Annals of Mathematics, Series 2, Vol. 19, pp. 279–294.

squares exist and have an upper bound independent of A. This independence follows at once from the fact that

$$\int_{A}^{A+1}\left[\int_{-\infty}^{\infty}\Phi(\tau-t)g(\tau)d\tau\right]^{2}dt=\int_{0}^{1}\left[\int_{-\infty}^{\infty}\Phi(\tau-t)g(\tau-A)d\tau\right]^{2}dt \tag{26}$$

combined with the fact that

$$\int_{-\infty}^{\infty}\Phi(\alpha)d\alpha=0$$

sȯ that

$$\int_{-\infty}^{\infty}\Phi(\tau-t)g(\tau)d\tau$$

only involves the differences between values of g.

It follows that the chance that some

$$\left|\frac{1}{2\alpha}\int_{-\alpha}^{\alpha}\left[\int_{-\infty}^{\infty}\Phi(\tau-t)g(\tau)\,d\tau\right]^{2}dt-r^{2}\sum_{1}^{\infty}\frac{1}{\lambda_n}\right|>3P$$

for $\alpha\geq N$ is less than or equal to

$$S\left(\frac{1}{N^2}+\frac{1}{(N+1)^2}+\frac{1}{(N+2)^2}+\ldots\right)\leq\frac{S}{N} \tag{27}$$

and hence can be made arbitrarily small by taking N large enough. Hence the outer chance that there is no N such that

$$\left|\frac{1}{2\alpha}\int_{-\alpha}^{\alpha}\left[\int_{-\infty}^{\infty}\Phi(\tau-t)g(\tau)d\tau\right]^{2}dt-r^{2}\sum_{1}^{\infty}\frac{1}{\lambda_n}\right|\leq 3P$$

for all $\alpha\geq N$ is 0. This results from certain theorems of Daniell concerning the integrability of the limit of a monotone sequence of functionals. Since the inner chance is also 0, we see that the expression

$$\overline{\lim_{\alpha\to\infty}}\left|\frac{1}{2\alpha}\int_{-\alpha}^{\alpha}\left[\int_{-\infty}^{\infty}\Phi(\tau-t)g(\tau)d\tau\right]^{2}dt-r^{2}\sum_{1}^{\infty}\frac{1}{\lambda_n}\right|\leq 3P \tag{28}$$

except for a set of cases of probability zero.

If we remember that the probability of the disjunction of a

denumerable set of events does not exceed the sum of their probabilities, we see that

$$\lim_{a\to\infty}\frac{1}{2a}\int_{-a}^{a}\left[\int_{-\infty}^{\infty}\Phi(\tau-t)g(\tau)d\tau\right]^2 dt = r^2\int_{-\infty}^{\infty}d\sigma\left[\int_{-\infty}^{\sigma}\Phi(\alpha)d\alpha\right]^2 \quad (29)$$

except for a set of cases of probability zero.

§4. The possibility of the harmonic analysis of an irregular motion.

It further follows from Daniell's theorems — which involve Zermelo's axiom in their proof — that the null set has measure zero, and that hence any set of functions of probability different from zero contains functions. Hence we get the following theorem: *given any denumerable set of functions* $\Phi_n(\tau)$ *satisfying conditions* (6), *the relations*

$$\lim_{a\to\infty}\frac{1}{2a}\int_{-a}^{a}\left[\int_{-\infty}^{\infty}\Phi_n(\tau-t)g(\tau)d\tau\right]^2 dt = r^2\int_{-\infty}^{\infty}d\sigma\left[\int_{-\infty}^{\sigma}\Phi_n(\alpha)d\alpha\right]^2 \quad (30)$$

will all hold except in a set of cases of zero probability, and will hence hold in a non-null set of cases.

Let us consider the particular set of cases

$$\Phi_1(\tau) = \Phi(\tau);$$

$$\Phi_k(\tau) = \int_{-\infty}^{\infty}\Phi(\sigma)\Psi_{k-1}(\tau-\sigma)d\sigma; \qquad [k>1]$$

where

$$\Psi_k(\sigma) = \Psi_k(-\sigma) = \frac{2}{\pi}\int_0^{\infty}\Psi_k(u)\cos u\sigma\, du,$$

and

$$\psi_k(u) = 1; \qquad [0\leq u\leq D_k]$$

$$0\leq\psi_k(u)\leq 1; \qquad [D_k\leq u\leq D_k+\epsilon_k]$$

$$\psi_k(u) = 0; \qquad [D_k+\epsilon_k\leq u]$$

while all the functions $\psi_k(u)$ possess at every point derivatives of all orders. We suppose the D_k's to form an everywhere dense set, and the ϵ_k's for each D_k to assume arbitrarily small values.

We may verify at once that if conditions (6) hold for $\Phi_1(\tau)$, they hold for all Φ_k's. If we put

$$\psi(u)=\int_{-\infty}^{\infty}\Phi(\tau)e^{iu\tau}d\tau, \tag{31}$$

we get

$$\begin{aligned}\int_{-\infty}^{\infty}\Phi_k(\tau)e^{iu\tau}d\tau &= \int_{-\infty}^{\infty}e^{iu\tau}d\tau\int_{-\infty}^{\infty}\Phi(\sigma)\Psi_{k-1}(\tau-\sigma)d\sigma\\ &=\int_{-\infty}^{\infty}\Phi(\sigma)e^{iu\tau}d\sigma\int_{-\infty}^{\infty}\Psi_{k-1}(\tau-\sigma)e^{iu(\tau-\sigma)}\,d\tau\\ &=\psi(u)\psi_{k-1}(u).\end{aligned}$$

Hence

$$\Phi_1(\tau)=\frac{1}{2\pi}\int_{-\infty}^{\infty}\psi(u)e^{-iu\tau}du;$$

$$\Phi_k(\tau)=\frac{1}{2\pi}\int_{-\infty}^{\infty}\psi(u)\psi_{k-1}(u)e^{-iu\tau}d\,u. \qquad [k>0]$$

Upon integration, we find that

$$\int_{-\infty}^{\sigma}\Phi_1(\alpha)d\alpha=-\frac{1}{2\pi}\int_{-\infty}^{\infty}\frac{\psi(u)}{u}e^{-iu\sigma}du$$

$$\int_{-\infty}^{\sigma}\Phi_k(\alpha)d\alpha=-\frac{1}{2\pi}\int_{-\infty}^{\infty}\frac{\psi(u)}{u}\psi_{k-1}(u)e^{-iu\sigma}du. \qquad [k>0]$$

It is to be noted that it follows from (6) that $\psi(u)=O(u)$ at the origin. Consequently,

$$\int_{-\infty}^{\infty}d\sigma\int_{-\infty}^{\sigma}\left[\Phi_1(\alpha)d\alpha\right]^2=\frac{1}{2\pi}\int_{-\infty}^{\infty}\frac{[\psi(u)]^2}{u^2}du;$$

$$\int_{-\infty}^{\infty}d\sigma\left[\int_{-\infty}^{\sigma}\Phi_k(\alpha)d\alpha\right]^2=\frac{1}{2\pi}\int_{-\infty}^{\infty}\frac{[\psi(u)]^2[\psi_{k-1}(u)]^2}{u^2}du. \tag{32}$$

We thus get

$$\lim_{D_{k-1}\to\infty}\int_{\infty}^{\infty}d\sigma\left[\int_{-\infty}^{\sigma}\Phi_k(\alpha)d\alpha\right]^2=\int_{-\infty}^{\infty}d\sigma\left[\int_{-\infty}^{\sigma}\Phi_1(\alpha)d\alpha\right]^2. \tag{33}$$

We have now established that except in a set of cases of zero probability,

$$h(t)=\int_{-\infty}^{\infty}\Phi(\tau-t)g(\tau)d\tau$$

possesses a generalized trigonometric analysis in the sense of the paper to which this is a sequel, for we have demonstrated that conditions (*a*), (*b*), and (*c*) of the introduction to this paper are fulfilled.

§5. The spectrum of an irregular motion.

We showed in the previous paper that if

$$\Gamma(u)=\frac{1}{\pi}\int_{-\infty}^{\infty}h(t)\frac{1-\cos ut}{t^2}dt;$$

$$\Delta(u)=\frac{1}{\pi}\int_{-\infty}^{\infty}h(t)\frac{ut\,e^{-t^2}-\sin ut}{t^2}dt;$$

$$H_k(t)=\int_{-\infty}^{\infty}\Phi_k(\tau-t)g(\tau)d\tau;$$

we have

$$\lim_{A\to\infty}\frac{1}{2A}\int_{-A}^{A}[H_k(t)]^2dt=\frac{3}{2}\lim_{\epsilon\to 0}\frac{1}{\epsilon^3}\int_0^{\infty}[\psi_{k-1}(u)]^2$$

$$\left\{[\Delta(u+\epsilon)+\Delta(u-\epsilon)-2\Delta(u)]^2+[\Gamma(u+\epsilon)+\Gamma(u-\epsilon)-2\Gamma(u)]^2\right\}du. \tag{34}$$

Hence except for a set of cases of zero probability,

$$\frac{3}{2}\lim_{\epsilon\to 0}\frac{1}{\epsilon^3}\int_0^{\infty}[\psi_{k-1}(u)]^2\{[\Delta(u+\epsilon)+\Delta(u-\epsilon)-2\Delta(u)]^2$$
$$+[\Gamma(u+\epsilon)+\Gamma(u-\epsilon)-2\Gamma(u)]^2\}du$$
$$=r^2\int_{-\infty}^{\infty}d\sigma\left[\int_{-\infty}^{\sigma}\Phi_k(\alpha)d\alpha\right]^2=\frac{r^2}{2\pi}\int_{-\infty}^{\infty}\frac{[\psi_{k-1}(u)]^2[\psi(u)]^2}{u^2}du. \tag{35}$$

We therefore have

$$\limsup_{D_k+\epsilon_k\le U}\frac{3}{2}\lim_{\epsilon\to 0}\frac{1}{\epsilon^3}\int_0^{\infty}[\psi_k(u)]^2$$

$$\{[\Delta(u+\epsilon)+\Delta(u-\epsilon)-2\Delta(u)]^2+[\Gamma(u+\epsilon)+\Gamma(u-\epsilon)-2\Gamma(u)]^2\}du$$
$$=\frac{r^2}{2\pi}\int_{-U}^{U}\frac{[\psi(u)]^2}{u^2}du.$$

In the preceding paper, we use the notation,

$$\limsup_{D_k+\epsilon_k\leq U} 3\lim_{\epsilon\to 0}\frac{1}{\epsilon^3}\int_0^\infty [\psi_k(u)]^2$$
$$\left\{[\Delta(u+\epsilon)+\Delta(u-\epsilon)-2\Delta(u)]^2+[\Gamma(u+\epsilon)+[(u-\epsilon)-2\Gamma(u)]^2\right\}du$$
$$=R\{h;U\}.$$

It follows that

$$R\{h;u\}=\frac{r^2}{\pi}\int_{-u+0}^{u-0}\frac{[\psi(u)]^2}{u^2}\,du. \tag{36}$$

In the preceding paper, we prove that

$$C(h;\xi)=\lim_{\epsilon\to 0}\frac{3}{2\pi\epsilon^3}\int_{-\infty}^{\infty}h\left(t-\frac{\xi}{2}\right)h\left(t+\frac{\xi}{2}\right)\frac{\sin^4\epsilon t}{t^4}\,dt$$
$$=\frac{1}{2}\int_0^\infty\cos\alpha\xi\,dR\{h;\alpha\}.$$

This becomes

$$C(h;\xi)=\frac{r^2}{2\pi}\int_{-\infty}^{\infty}\cos\alpha\,\xi\,\frac{[\psi(\alpha)]^2}{\alpha^2}\,d\alpha$$
$$=\frac{r^2}{2\pi}\int_{-\infty}^{\infty}\left[\cos^2\frac{\alpha\xi}{2}-\sin^2\frac{\alpha\xi}{2}\right]\frac{[\psi(\alpha)]^2}{\alpha^2}\,d\alpha$$
$$=\frac{r^2}{2\pi}\int_0^\infty\left[\cos^2\frac{\alpha\xi}{2}-\sin^2\frac{\alpha\xi}{2}\right]$$
$$\frac{[\psi(\alpha)+\psi(-\alpha)]^2+[\psi(\alpha)-\psi(-\alpha)]^2}{\alpha^2}\,d\alpha$$
$$=\frac{r^2}{4}\int_0^\infty\left\{\left[\int_0^\infty\cos\frac{\alpha\xi}{2}\,\frac{\psi(\alpha)+\psi(-\alpha)}{2\alpha}\cos\alpha x\,d\alpha\right]^2\right.$$
$$-\left[\int_0^\infty\sin\frac{\alpha\xi}{2}\,\frac{\psi(\alpha)+\psi(-\alpha)}{2\alpha}\sin\alpha x\,d\alpha\right]^2$$
$$+\left[\int_0^\infty\cos\frac{\alpha\xi}{2}\,\frac{\psi(\alpha)-\psi(-\alpha)}{2\alpha}\sin\alpha x\,d\alpha\right]^2$$
$$\left.-\left[\int^\infty\sin\frac{\alpha\xi}{2}\,\frac{\psi(\alpha)-\psi(-\alpha)}{2\alpha}\cos\alpha x\,d\alpha\right]^2\right\}dx$$

$$= \frac{r^2}{4}\int_0^\infty \left\{ \left[\int_{-\infty}^{\infty} \frac{\cos \alpha\left(x+\frac{\xi}{2}\right)+\cos \alpha\left(x-\frac{\xi}{2}\right)}{2\alpha}\psi(\alpha)d\alpha\right]^2 \right.$$

$$- \left[\int_{-\infty}^{\infty} \frac{\cos \alpha\left(x+\frac{\xi}{2}\right)-\cos \alpha\left(x-\frac{\xi}{2}\right)}{2\alpha}\psi(\alpha)d\alpha\right]^2$$

$$+ \left[\int_{-\infty}^{\infty} \frac{\sin \alpha\left(x+\frac{\xi}{2}\right)+\sin \alpha\left(x-\frac{\xi}{2}\right)}{2\alpha}\psi(\alpha)d\alpha\right]^2$$

$$\left. - \left[\int_{-\infty}^{\infty} \frac{\sin \alpha\left(x+\frac{\xi}{2}\right)-\sin \alpha\left(x-\frac{\xi}{2}\right)}{2\alpha}\psi(\alpha)d\alpha\right]^2 \right\} dx$$

$$= \frac{r^2}{4}\int_0^\infty dx \int_{-\infty}^{\infty} \frac{\psi(\alpha)}{\alpha}\cos \alpha\left(x+\frac{\xi}{2}\right)d\alpha \int_{-\infty}^{\infty} \frac{\psi(\beta)}{\beta}\cos \beta\left(x-\frac{\xi}{2}\right)d\beta$$

$$+ \frac{r^2}{4}\int_0^\infty dx \int_{-\infty}^{\infty} \frac{\psi(\alpha)}{\alpha}\sin \alpha\left(x+\frac{\xi}{2}\right)d\alpha \int_{-\infty}^{\infty} \frac{\psi(\beta)}{\beta}\sin \beta\left(x-\frac{\xi}{2}\right)d\beta$$

$$= \frac{r^2}{4}\int_0^\infty dx \int_{-\infty}^{\infty} \frac{\psi(\alpha)}{\alpha}d\alpha \int_{-\infty}^{\infty} \frac{\psi(\beta)}{\beta}d\beta \cos\left[(\alpha+\beta)x+(\alpha-\beta)\frac{\xi}{2}\right]$$

$$= \frac{r^2}{8}\int_0^\infty dx \int_{-\infty}^{\infty} \frac{\psi(\alpha)}{\alpha}d\alpha$$

$$\int_{-\infty}^{\infty} \frac{\psi(\beta)}{\beta}d\beta \left[e^{i\left[(\alpha+\beta)x+(\alpha-\beta)\frac{\xi}{2}\right]} + e^{-i\left[(\alpha+\beta)x+(\alpha-\beta)\frac{\xi}{2}\right]}\right]. \quad (37)$$

We now refer back to formula (31), which when inverted yields us

$$\Phi(\tau) = \frac{1}{2\pi}\int_{-\infty}^{\infty} \psi(u)e^{-iu\tau}du.$$

By integrating both sides of this expression, we obtain

$$\int_{-\infty}^{\tau} \Phi(\tau)d\tau = -\frac{1}{2\pi}\int_{-\infty}^{\infty} \frac{\psi(u)}{u}e^{-iu\tau}\,du. \quad (38)$$

If we substitute this result into (37), we get

$$C(h;\xi)=\frac{\pi^2}{2}r^2\int_0^\infty dx\left[\int_{-\infty}^{-x-\frac{\xi}{2}}\Phi(\tau_1)d\tau_1\int_{-\infty}^{-x+\frac{\xi}{2}}\Phi(\tau_2)d\tau_2\right.$$

$$\left.+\int_{-\infty}^{x+\frac{\xi}{2}}\Phi(\tau_3)d\tau_3\int_{-\infty}^{x-\frac{\xi}{2}}\Phi(\tau_4)d\tau_4\right]$$

$$=\frac{\pi^2}{2}r^2\int_{-\infty}^{\infty}dx\int_{-\infty}^{x+\frac{\xi}{2}}\Phi(\tau_1)d\tau_1\int_{-\infty}^{x-\frac{\xi}{2}}\Phi(\tau_2)d\tau_2\,. \tag{39}$$

That is, the mean of $h\left(t-\frac{\xi}{2}\right)h\left(t+\frac{\xi}{2}\right)$ differs only by a constant factor from the integral of $\int_{-\infty}^{x+\frac{\xi}{2}}\Phi(\tau_1)d\tau_1\int_{-\infty}^{x-\frac{\xi}{2}}\Phi(\tau_2)d\tau_2$.

We may regard $R\{h;u\}$ as a measure of the sum of the squares of the coefficients of the terms in the trigonometric expansion of $h(t)$ up to the term corresponding to frequency $u/2\pi$. We may regard $h(t)$ as the velocity of a motion in which the displacement is

$$\eta(t)=\int_{-\infty}^{\infty}g(\tau)d\tau\int_{-\infty}^{\tau-t}\Phi(\sigma)d\sigma.$$

We then see that $R\{h;u\}$ is proportional to the total power in the spectrum of this motion, up to the term of frequency $u/2\pi$. By (32), this only differs by a constant factor from the integral of the square of the amplitudes in the spectrum of $\int_{-\infty}^{\tau}\Phi(\sigma)d\sigma$ over the same range of frequencies. We may put this more expressively if we say that the power-spectrum of a haphazard succession of impulses of a given form is the same as the energy-spectrum of one of these impulses, or again if we say that the power-spectrum of the response of a linearly resonant system to a succession of haphazard sharp impulses is the same as the energy-spectrum of the response of the system to a single such impulse. The enunciation of this as a physical fact is not new, but the rigorous mathematical proof is new, to the best of the author's knowledge and belief. A full discussion is to be found in the next section.

While the function $g(t)$ has no power-spectrum, in the sense here defined, we may regard a motion with displacement $x=g(t)$ as having the distribution of power in its spectrum corresponding to the limit of the distribution of power in the spectrum of a motion with displacement

$$G(t)=\int_{-\infty}^{\infty}g(\tau)\Psi_k(t-\tau)d\tau$$

or with velocity

$$G'(t)=\int_{-\infty}^{\infty}g(\tau)\Psi_k'(t-\tau)d\tau.$$

We have, however,

$$R\left\{G';u\right\}=\frac{r^2}{\pi}\int_{-u}^{u}[\psi_k(u)]^2du=\frac{2ur^2}{\pi}\qquad[D_k\geq u]$$

and

$$\lim_{D_k\to\infty}R\left\{G';u\right\}=\frac{2ur^2}{\pi}.\tag{40}$$

That is, in a generalized sense, the power-spectrum of the function $g(t)$ may be regarded as uniformly distributed in frequency, in all except an assemblage of cases of zero probability.

§6. The Schroteffekt.

A linear system such as a resonator may be regarded as generating a linear transformation of one function, representing the impressed displacement, into another function, representing the emergent displacement. For example, the horn of the phonograph generates a transformation of the displacement of the diaphragm into the displacement of the air at a specified place. Precisely similar cases occur in electrical systems.

In all linear systems occurring in practice, there are frictional losses of energy. If these did not exist, the response of the system to impressed periodic displacements of certain frequencies would tend to infinity with increasing time. Such unlimited resonance is however never found.

The linear operations generated by resonators are said to be operators of the closed cycle. This means that the result of applying one of these operators to a given function is not affected by a displacement of the origin in time. In symbols, if I is such an operator,

$$I\{f(t+\tau)\} = [I\{f(\sigma)\}]_{\sigma=t+\tau}. \tag{41}$$

All resonators found in practice respond only slightly to very high frequencies. Such resonators have characteristic operators of the form

$$I\{f(t)\} = \int_{-\infty}^{\infty} F(\tau-t)f(\tau)d\tau.$$

If such a resonator is subjected to a displacement $g(t)$, the velocity corresponding to the emergent displacement is

$$h(t) = -\int_{-\infty}^{\infty} F'(\tau-t)g(\tau)d\tau,$$

provided F and F' are summable and of summable square over the whole line. If in addition $F'(\tau)$ is $O(-\tau^3)$ at $\pm\infty$, condition (6) is fulfilled for F', and we have almost always

$$R\{h; u\} = \frac{r^2}{\pi}\int_{-u}^{u}\left[\int_{-\infty}^{\infty} F(\tau)e^{iu\tau}d\tau\right]^2 du, \tag{42}$$

provided $g(t)$ has a normal distribution.

As a particular application: let us consider the response of an acoustical resonator to a noise consisting of a series of sharp impulses to the air, such as those produced by the grains of sand in a sand-blast, which are (1) short in duration as compared with the period of an audible sound, (2) so frequent that many occur within the period of an audible sound, (3) occurring at random intervals, and (4) such that the quantities determining the character of this random distribution do not vary with the time. If we make a harmonic analysis of the output of the resonator, in the sense of the paper to which this is a sequel, we shall find that the value of the R of this function for a given frequency is proportional to the energy of the response of the oscillator to a sinusoidal disturbance of this frequency and of unit energy. This property may be employed in the calibration of phonodeiks, in which the output function is permanently recorded as a curve, which may be analyzed by the methods we have given.

Another example is formed by the so-called Schroteffekt of Schottky.[7] The space current in a vacuum tube is carried by electrons, and hence can not be continuous. Its deviations from steadiness are of such a nature as to make the quantity of electricity passing in a given time a function with the same sort of distribution as $g(t)$. If the tube is connected to a linear system of inductances, capacities, and resistances, the irregularities of the space current will produce oscillations of the system. These can be amplified to the point where they are audible when rectified, and have been observed. The question is as to the distribution in frequency of the electric power in some member of the oscillating system. Our results show that it is proportional to the power in the same member of the system when the space current is replaced by a sinusoidal current of the given frequency and of unit power. We shall not here discuss the factor of proportionality in question, which has been adequately treated in the papers referred to, and which gives a valuable means of measuring the electronic charge.

The observation that the space current is uniformly distributed with respect to power over the different frequencies was made by Schottky, although the means of rigorously proving this were not at hand. Fry's paper is in large measure a polemic against the idea of making a harmonic analysis of an irregular motion. He points out that the notion of the amplitude of a component of the spectrum of the electron stream does not generalize properly when the number of electrons is taken to be very large, and draws from that the incorrect conclusion that the notion of power — he says " energy " — will not generalize.

[7]Cf. W. Schottky, Ann. der Phys., 57, 541–67 (1918); 68, 157–76 (1922); T. C. Fry, Jour. Frank. Inst., 199, 203–221 (1925).

§7. **White light and interference.**

It was first pointed out by Gouy[8] and later emp'·-sized by Schuster[9] that white light — that is, light with a uniform power distribution over its spectrum — may be regarded as being made up of a succession of perfectly irregular impulses. This is completely confirmed by the results of our last section. The author wishes, however, to take this opportunity to point out that neither Gouy nor Schuster have given a clear and unambiguous definition of what is meant by " a succession of perfectly irregular impulses," and that to the best of the author's knowledge, no one has yet exhibited a function possessing a continuous spectrum in the sense of a spectrum with a continuous power distribution over the range of frequencies. The Fourier series and the analogous series for Bohr's " almost periodic functions " represent nothing but line spectra. The continuous spectra of the Fourier integral have a continuous distribution of energy over the spectrum, but represent processes of zero power. They are the spectra of pulses, of transient phenomena, whereas it is highly desirable to have a theory of light with a continuous spectrum as a steady-state phenomenon. The present paper, therefore, forms a necessary part of the existence theory of physical optics.

The author wishes to point out its application to the theory of coherent and incoherent sources. Let $g_1(t)$ and $g_2(t)$ be two independently varying functions, each with a normal distribution of the type discussed in this paper. Consider

$$\left[\int_{-\infty}^{\infty}\Phi(\tau-t)[g_1(\tau)+g_2(\tau)]d\tau\right]^2$$
$$=\left[\int_{-\infty}^{\infty}\Phi(\tau-t)g_1(\tau)d\tau\right]^2+\left[\int_{-\infty}^{\infty}\Phi(\tau-t)g_2(\tau)d\tau\right]^2$$
$$+2\int_{-\infty}^{\infty}\Phi(\tau-t)g_1(\tau)d\tau\int_{-\infty}^{\infty}\Phi(\tau-t)g_2(\tau)d\tau. \tag{43}$$

The mean of the last term of this with respect to the time is

$$\frac{1}{A}\int_{-A}^{A}dt\int_{-\infty}^{\infty}\Phi(\tau-t)g_1(\tau)d\tau\int_{-\infty}^{\infty}\Phi(\tau-t)g_2(\tau)d\tau. \tag{44}$$

If we make use of our theorems on the average of analytic function-

[8] Journal de Physique, Vol. V., p. 354 (1886).
[9] Schuster and Nicholson, Theory of Optics (Third Edition), p. 333.

als, we see that the mean square value of this, taken separately with respect to g_1 and g_2, is

$$\frac{r^2}{A^2}\int_{-A}^{A} ds \int_{-A}^{A} dt \left[\int_0^{\infty} \Phi(\sigma - s)\, d\sigma \int^{\sigma} \Phi(\tau - t)\tau d\tau \right.$$
$$\left. - \int_{\infty}^{0} \Phi(\sigma - s) d\sigma \int_{\sigma}^{0} \Phi(\tau - t)\tau d\tau \right]^2 = O(1/A^2). \tag{45}$$

An argument such as that which we have developed in §3 will now show that in all except a set of cases of zero probability,

$$\lim_{A\to\infty} \frac{1}{A}\int_{-A}^{A} dt \int_{-\infty}^{\infty} \Phi(\tau - t) g_1(\tau) d\tau \int_{-\infty}^{\infty} \Phi(\tau - t) g_2(\tau) d\tau = 0.$$

It follows that in all except a set of cases of zero probability,

$$M\left[\int_{-\infty}^{\infty} \Phi(\tau - t)[g_1(\tau) + g_2(\tau)] d\tau\right]^2$$
$$= M\left[\int_{-\infty}^{\infty} \Phi(\tau - t) g_1(\tau) d\tau\right]^2 + M\left[\int_{-\infty}^{\infty} \Phi(\tau - t) g_2(\tau) d\tau\right]^2. \tag{46}$$

That is, if we have two independent sources of light, even though they may have identical power spectra, it is almost certain that their energies will be additive, and that they cannot interfere. On the other hand, we have

$$M\left[\int_{-\infty}^{\infty} \Phi(\tau - t)[g(\tau) + g(\tau)] d\tau\right]^2 = 4M\left[\int_{-\infty}^{\infty} \Phi(\tau - t) g(\tau) d\tau\right]^2;$$
$$M\left[\int_{-\infty}^{\infty} \Phi(\tau - t)[g(\tau) - g(\tau)] d\tau\right]^2 = 0; \tag{47}$$

thus showing the possibility of interference between light from non-independent sources.

We shall say that two functions $h_1(t)$ and $h_2(t)$ are *coherent* at frequency $u/2\pi$ if there exist quantities α and $\rho(\alpha)$ such that

$$0 \neq \rho(\alpha) = \left| \lim_{\epsilon\to 0} \frac{\left\{ \begin{array}{l} R\{h_1(t+\alpha) + h_2(t), u+\epsilon\} \\ \qquad - R\{h_1(t+\alpha) + h_2(t), u-\epsilon\} \\ - R\{h_1(t+\alpha) - h_2(t), u+\epsilon\} \\ \qquad + R\{h_1(t+\alpha) - h_2(t), u-\epsilon\} \end{array} \right\}}{4\{R\{h_1(t+\alpha), u+\epsilon\} - R\{h_1(t+\alpha), u-\epsilon\}\}^{1/2} \cdot \{R\{h_2(t), u+\epsilon\} - R\{h_2(t), u-\epsilon\}\}^{1/2}} \right| \tag{48}$$

$\max_\alpha \rho(\alpha)$ will then be said to be the measure of their coherence, and the corresponding angle α their phase angle. If we consider two components of the electric vector of light at right angles to one another and to the direction of its propagation, and represent them by $h_1(t)$ and $h_2(t)$, respectively, we shall call $100 \max_\alpha \rho(\alpha)$ the *percentage of polarization* of the light at frequency u. If $\max_\alpha \rho(\alpha)=1$, the directions making with the direction of oscillation of $h_1(t)$ angles θ satisfying the conditions

$$\sin 2\theta = \frac{-l_1^2 \sin 2u\,\alpha}{l_1^2+l_2^2-2l_1l_2 \cos 2\,u\alpha} \tag{49}$$

where

$$l_1^2 = \lim_{\epsilon\to 0} \frac{R\{h_1, u+\epsilon\} - R\{h_1, u-\epsilon\}}{2\epsilon};$$

$$l_2^2 = \lim_{\epsilon\to 0} \frac{R\{h_2, u+\epsilon\} - R\{h_2, u-\epsilon\}}{2\epsilon}; \tag{50}$$

are the directions of the axes of the elliptical polarization of our light, and the quantities

$$l_1 \sin\theta + l_2 \cos\theta;$$

$$l_1 \cos\theta - l_2 \sin\theta; \tag{51}$$

are these axes themselves. It is, of course, understood that all these are taken with respect to light of frequency $u/2\pi$. We shall not go into the question of the existence of the limits in question under general hypotheses, but it is a simple matter to produce cases where they exist.

§8. Taylor's theory of irregular motion.

In a paper entitled, "Diffusion by Continuous Movement,"[10] G. I. Taylor developed a theory of irregular motion depending on the idea of the correlation between a function and itself under a displacement, which we have borrowed from him in the present paper. He also introduces the notion of the coefficient of correlation between two derivatives of a function. His paper does not aim at a high mathematical rigor, and the precise conditions for the validity of his theorems are not stated with all desirable defi-

[10] G. I. Taylor, Proc. London Math. Soc., Ser. 2, 20, pp. 196–212 (1920).

niteness. He appears to confine himself, however, to analytic functions.

It will be the purpose of the present section to prove his results under fairly general hypotheses. We shall then discuss the harmonic analysis of the various derivatives of the functions of our last paper. We have.

$$\frac{1}{h^3}\int_{-\infty}^{\infty} f(t)f'(t)\frac{\sin^4 ht}{t^4}\,dt$$
$$=\left\{\frac{1}{h^3}[f(t)]^2\frac{\sin^4 ht}{t^4}\right\}_{-\infty}^{\infty}-\frac{1}{h^3}\int_{-\infty}^{\infty} f'(t)f(t)\frac{\sin^4 ht}{t^4}\,dt$$
$$-\frac{1}{h^3}\int_{-\infty}^{\infty}[f(t)]^2\frac{d}{dt}\left(\frac{\sin^4 ht}{t^4}\right)dt.$$

Consequently

$$\frac{1}{h^3}\int_{-\infty}^{\infty} f(t)f'(t)\frac{\sin^4 ht}{t^4}\,dt=-\frac{1}{2h^3}\int_{-\infty}^{\infty}[f(t)]^2\frac{d}{dt}\left(\frac{\sin^4 ht}{t^4}\right)dt$$
$$=-\frac{h}{2}\int_{-\infty}^{\infty}[f(u/h)]^2\frac{d}{du}\frac{\sin^4 u}{u^4}\,du. \tag{52}$$

Hence

$$\left|\frac{1}{h^3}\int_{-\infty}^{\infty} f(t)f'(t)\frac{\sin^4 ht}{t^4}\,dt\right|\leq 2h\max\left|\frac{u\sin u\cos u-\sin^2 u}{u^3}\right|$$
$$\cdot\int_{-\infty}^{\infty}[f(u/h)]^2\frac{\sin^2 u}{u^2}\,du. \tag{53}$$

It follows that if

$$\frac{1}{h}\int_{-\infty}^{\infty}[f(u)]^2\frac{\sin^2 hu}{u^2}\,du$$

is bounded,

$$\lim_{h\to 0}\frac{1}{h^3}\int_{-\infty}^{\infty} f(t)f'(t)\frac{\sin^4 ht}{t^4}\,dt=0. \tag{54}$$

In general, if

$$\frac{1}{h}\int_{-\infty}^{\infty}[f^{(k)}(u)]^2\frac{\sin^2 hu}{u^2}\,du \qquad [k=1,2,\ldots,n]$$

exists, we have

$$\lim_{h\to 0}\frac{1}{h^3}\int_{-\infty}^{\infty}[f(t)f^{(n)}(t)+f^{(k)}(t)+f^{(n-k)}(t)\,(-1)^k]\frac{\sin^4 ht}{t^4}\,dt=0. \tag{55}$$

That is, the mean value of $f^{(k)}(t)f^{(n-k)}(t)$ is 0 if n is odd, the average of $[f^{(n/2)}(t)]^2$ if $k-n/2$ is divisible by 2, and the negative average of $[f^{(n/2)}(t)]^2$ otherwise. This is essentially the theorem of G. I. Taylor. Taylor then goes on to formulate the Maclaurin expansion of what is essentially

$$\lim_{h\to 0}\frac{3}{2\pi h^3}\int_{-\infty}^{\infty} f(t)f(t+\xi)\frac{\sin^4 ht}{t^4}dt$$

in powers of ξ. This cannot be justified without much more stringent restrictions on f, and is not employed by him in his further work.

It is a matter of a certain amount of interest to ascertain a set of conditions under which $f(t)$ and $f'(t)$ can be simultaneously given a harmonic analysis. For this we shall need to employ certain auxiliary functions. Let $\Phi^{(1)}(t)$ be an odd summable function of summable square which is $O(1/t^3)$ at infinity. Let $f(t)$ be a function which fulfills the hypotheses necessary for a harmonic analysis in the sense of our previous paper. Let us consider the function

$$g_1(t)=\int_{-\infty}^{\infty}\Phi^{(1)}(\tau-t)f(\tau)d\tau.$$

In our previous paper we showed that if $\Phi^{(2)}(t)$ was an even function, summable and of summable square, which is $O(1/t^3)$ at infinity, and if

$$g_2(t)=\int_{-\infty}^{\infty}\Phi^{(2)}\,(\tau-t)f(\tau)d\tau$$

then g_2 is harmonically analyzable and

$$R\{g_2;u\}=\int_0^{u-0}\left[\int_{-\infty}^{\infty}\Phi^{(2)}(\xi)\cos u\,\xi\,d\xi\right]^2 dR\{f;u\}. \tag{56}$$

Precisely similar methods show that g_1 is harmonically analyzable and that

$$R\{g_1;u\}=\int_0^{u-0}\left[\int_{-\infty}^{\infty}\Phi^{(1)}(\xi)\sin u\,\xi d\xi\right]^2 dR\{f;u\}. \tag{57}$$

Now let $M\{[f(t)]^2\}$ and $M\{[f'(t)]^2\}$ exist, and let

$$\lim_{h\to 0}\frac{1}{h^3}\int_{-\infty}^{\infty}\left[\int_{-\infty}^{\infty}f(\tau)\Psi_k(t-\tau)d\tau\right]^2\frac{\sin^4 ht}{t^4}dt$$

exist for an everywhere dense set of D_k's and for arbitrarily small ϵ's. The arguments of the previous paper will show that

$$\frac{3}{\pi}\lim_{h\to 0}\frac{1}{h^3}\int_{-\infty}^{\infty}\left[\int_{-\infty}^{\infty}f'(\tau)\Psi_k(t-\tau)d\tau\right]^2\frac{\sin^4 ht}{t^4}dt$$

$$=\frac{3}{\pi}\lim_{h\to 0}\frac{1}{h^3}\int_{-\infty}^{\infty}\left[\int_{-\infty}^{\infty}f(\tau)\Psi_k'(t-\tau)d\tau\right]^2\frac{\sin^4 ht}{t^2}dt$$

will exist, and will in fact equal

$$\int^{\infty}\frac{[\Psi_k(u)]^2}{u^2}dR\{f;u\} \tag{58}$$

Let us make the additional assumption that

$$M\{[f'(t)]^2\}$$
$$=\lim_{D_k\to\infty}\frac{3}{2\pi}\lim_{h\to 0}\frac{1}{h^3}\int_{\infty}\left[\int_{-\infty}^{\infty}f'(\tau)\Psi_k(t-\tau)d\tau\right]^2\frac{\sin^4 ht}{t^4}dt. \tag{59}$$

The function $f'(t)$ will then have a harmonic analysis in the sense of the previous paper. Now over any finite range

$$f(t)-\int_{-\infty}^{\infty}f(\tau)\Psi_k(t-\tau)d\tau=\underset{D_l\to\infty}{\text{l. m.}}\int_{-\infty}^{\infty}f(\sigma)[\Psi_k(t-\sigma)-\Psi_k(t-\sigma)]d\sigma$$

$$=\underset{D_l\to\infty}{\text{l. m.}}\int_{-\infty}^{\infty}f'(\tau)d\tau\int_{-\infty}^{\tau}[\Psi_l(t-\sigma)-\Psi_k(t-\sigma)]d\sigma$$

$$=\underset{D_l\to\infty}{\text{l. m.}}\frac{2}{\pi}\int_{-\infty}^{\infty}f'(\tau)d\tau\int_0^{\infty}\cos u(t-\tau)\frac{\psi_l(u)-\psi_k(u)}{u}du$$

$$=\frac{2}{\pi}\int_{-\infty}^{\infty}f'(\tau)d\tau\int_0^{\infty}\cos u(t-\tau)\frac{1-\psi_k(u)}{u}du \tag{60}$$

except possibly over a set of zero measure. The function

$$\frac{2}{\pi}\int_0^{\infty}\cos u\tau\frac{1-\psi_k(u)}{u}du$$

is summable and of summable square, and is $O(1/\tau^3)$ at infinity. Hence

$$\frac{3}{\pi}\lim_{h\to 0}\frac{1}{h^3}\int_{-\infty}^{\infty}\left[f(t)-\int_{-\infty}^{\infty}f(\tau)\Psi_k(t-\tau)d\tau\right]^2\frac{\sin^4 ht}{t^4}\,dt$$

exists, and equals

$$\int_{-\infty}^{\infty}\frac{[1-\psi_k(u)]^2}{u^2}dR\{f;u\}. \tag{61}$$

It follows that

$$\lim_{D_k\to\infty}\frac{3}{\pi}\lim_{h\to 0}\frac{1}{h^3}\int_{-\infty}^{\infty}\left[f(t)-\int_{-\infty}^{\infty}f(\tau)\Psi_k(t-\tau)d\tau\right]^2\frac{\sin^4 ht}{t^4}\,dt=0$$

and by the application of a form of the Schwarz inequality,

$$M\{[f(t)]^2\}=\lim_{D_k\to\infty}\frac{3}{2\pi}\lim_{h\to 0}\int_{-\infty}^{\infty}\left[\int_{-\infty}^{\infty}f(\tau)\Psi_k(t-\tau)d\tau\right]^2\frac{\sin^4 ht}{t^4}\,dt. \tag{62}$$

That is, $f(t)$ also has a harmonic analysis in the sense of the previous paper. In general, if $M\{[f(t)]^2\}$ and $M\{[f^n(t)]^2\}$ exist, and every

$$\lim_{h\to 0}\frac{1}{h^3}\int_{-\infty}^{\infty}\left[\int_{-\infty}^{\infty}f(\tau)\Psi_k(t-\tau)d\tau\right]^2\frac{\sin^4 ht}{t^4}\,dt$$

then

$$\lim_{h\to 0}\frac{1}{h^3}\int_{-\infty}^{\infty}\left[\int_{-\infty}^{\infty}f^{(j)}(\tau)\Psi_k(t-\tau)d\tau\right]^2\frac{\sin^4 ht}{t^4}\,dt \quad [j=1,2,\ldots \tag{63}$$

will exist. If in addition $M\{[f^{(j)}(t)]^2\}$ $[j=1, 2, \ldots, n]$ exist and

$$M\{[f^{(n)}(t)]^2\}=\lim_{D_k\to\infty}\frac{3}{2\pi}\lim_{h\to 0}\frac{1}{h^3}\int_{-\infty}^{\infty}[f^{(n)}(\tau)\Psi_k(t-\tau)d\tau]^2\frac{\sin^4 ht}{t^4}\,dt, \tag{64}$$

then every $f^{(j)}(t)$ $[j=1, \ldots, n]$ will have a harmonic analysis in the sense of the previous paper, and indeed

$$R\{f^{(j)};u\}=\int_0^{u-0}\frac{1}{u^{2j}}dR\{f;u\}. \tag{65}$$

THE SPECTRUM OF AN ARBITRARY FUNCTION

By NORBERT WIENER.

[Received 1 December, 1926.—Read 9 December, 1926.]

Introduction.

The present article is the outcome of a series of investigations concerning the generalizations of the Fourier integral which the author has published* in the course of the last few years. In these the leading motives have been to find a bond of union between Bohr's theory of almost periodic functions† and the Plancherel theory of the Fourier integral, and to include in this united theory an adequate mathematical account of such continuous spectra as that of white light.

A somewhat similar problem has been attacked by Hahn‡, who was, however, primarily interested in questions of ordinary convergence, while the present author concerns himself rather with theorems of the Hurwitz-Riesz-Fischer type. Our methods differ considerably.

The previous papers of the author suffered, on the one hand, from the presence of a number of superfluous conditions imposed on the functions considered, and, on the other, from a certain complication in the methods

---

* N. Wiener, "On the representation of functions by trigonometrical integrals", *Math. Zeitschrift*, 24 (1925), 575-617; "The harmonic analysis of irregular motion", *J. Math. and Phys., Massachusetts Institute of Technology*, 1926.

† H. Bohr, "Zur Theorie der fastperiodischen Funktionen I, II, III", *Acta Math.*, 45-47 (1924-26).

‡ H. Hahn, "Über eine Verallgemeinerung des Fourierschen Integralformel", *Acta Math.*, 49 (1926), 301-353.

of proof. He therefore was glad to accept an invitation from Professors Hilbert and Courant in Göttingen to develop his theories in a course of lectures at the university. In the preparation of this course he received much valuable aid from Mr. K. Mahler. The necessary trip to Europe was made possible by the grant of a Guggenheim Fellowship. To the trustees of the Guggenheim Foundation and to his Göttingen colleagues he wishes to tender his most sincere thanks.

It was the intention of the author to develop the spectrum theory of this paper in book form. However, the methods have so simplified themselves in the course of his recent work that this procedure now seems to him inadvisable, if not impossible. Moreover, the theory is too young to be set down in more definitive form than that of a comprehensive article.

A recent article of Weyl* develops a new method of attacking the Bohr analogue of the Hurwitz-Parseval theorem for almost periodic functions. Like the present paper, and like an earlier investigation of Schur†, Weyl's paper depends on the study of the function

$$g(t) = \lim_{T\to\infty} \frac{1}{2T}\int_{-T}^{T} f(x)\bar{f}(x-t)\,dx$$

derived from an almost periodic function $f(x)$. Weyl's method differs from that of this paper in that Weyl considers the trigonometrical components of an almost periodic function as arranged in the order of magnitude of their coefficients, while we choose the order of magnitude of their exponents. The Weyl theory is primarily based on the Schmidt-Hilbert theory of integral equations; the present theory has its roots in the Plancherel theory of the Fourier integral and the Hurwitz-Parseval-Riesz-Fischer theory of Fourier series.

1. *The spectrum of a trigonometrical polynomial.*

Let us consider the complex trigonometrical polynomial

$$f(t) = \sum_{1}^{n} A_k e^{i\Lambda_k t}$$

* H. Weyl, "Beweis des Fundamentalsatzes in der Theorie der fastperiodischen Funktionen" (aus einem schreiben an Herrn Harald Bohr), *Sitzungsber. der preussischen Akademie der. Wiss. (Phys.-math. Klasse)*, 18, 19 (1926), 211–214. See also "Integralgleichungen und fastperiodische Funktionen", *Math. Annalen*, 97 (1926), 338–356.

† "Über die Entwicklung einer gegebenen Funktion nach den Eigenfunktionen eines positiv definiten Kerns", *Schwarz-Festschrift* (1914), 392–409 (404).

in the variable t. The individual term of $f(t)$ has a *frequency* characterized by Λ_k, an *amplitude* determined by $|A_k|$, and a *phase*, given by the argument of A_k. If we consider t as the time, and $f(t)$ as the displacement of a particle at time t, as in a sound vibration, the motion defined by $f(t)$ will have a certain spectrum. Now, in light spectra, as given by a prism or grating, and in the analogous sound spectra, amplitude and frequency play a definite rôle, but phase is neglected. A spectrum, indeed, is determined by a distribution of amplitude in frequency. It is thus a matter of some interest to find a mathematical expression depending on f, which shall determine the frequencies and amplitudes of the components of f, and shall be determined by them, without any reference to their phases.

The function

$$g(t) = \lim_{X \to \infty} \frac{1}{2X} \int_{-X}^{X} f(x) \bar{f}(x-t)\, dx$$

(where the bar denotes the conjugate) has the desired property. It is readily seen that

$$g(t) = \sum_{1}^{n} |A_k|^2 e^{i\Lambda_k t}.$$

The function $g(t)$ determines and is determined by the function

$$R(u) = \frac{1}{\pi} \int_{-\infty}^{\infty} g(t) \frac{\sin ut}{t}\, dt.$$

It is easily seen that, if $\Lambda_k \neq 0$,

$$\frac{1}{\pi} \int_{-\infty}^{\infty} e^{i\Lambda_k t} \frac{\sin ut}{t}\, dt = \begin{cases} -1 & (u < -|\Lambda_k|) \\ -\frac{1}{2} & (u = -|\Lambda_k|) \\ 0 & (-|\Lambda_k| < u < |\Lambda_k|) \\ \frac{1}{2} & (u = |\Lambda_k|) \\ 1 & (u > |\Lambda_k|) \end{cases}$$

and it is well known that

$$\frac{1}{\pi} \int_{-\infty}^{\infty} \frac{\sin ut}{t}\, dt = \operatorname{sgn} u.$$

Hence $R(u)$ is an increasing step-function which has a jump at $u = |\Lambda_k|$. If we group the Λ_k's into pairs of corresponding negative and positive values, and if $\Lambda_m > 0$, $\Lambda_n = -\Lambda_m$, then the jump of R at Λ_m is $|A_m|^2 + |A_n|^2$. Here, of course, A_m or A_n may vanish. $R(0)$ is zero, and $R(+0)$ is $|A_0|^2$. R is moreover an odd function.

The function R gives us a direct picture of the spectrum of f. If $f(t)$ is taken as the velocity of a moving particle, $R(u_2)-R(u_1)$ may, in general, be considered to be proportional to the power of those frequencies in the vibration of the particle which are contained between u_2 and u_1.

2. *The spectrum of an arbitrary function of one variable.*

In the definition of g and R, no use whatever is made of the harmonic resolution of f as a trigonometrical polynomial. It seems possible, therefore, to extend the definition of the intensity of the spectrum of f over the region (u_1, u_2) as $R(u_2)-R(u_1)$ to functions other than trigonometrical polynomials. Let us now investigate the conditions under which this can actually be done.

We shall make the assumption (A) that

$$g(t) = \lim_{T\to\infty} \frac{1}{2T}\int_{-T}^{T} f(x)\bar{f}(x-t)\,dx$$

exists for every t. By the Schwarz inequality,

$$\left|\int_{-T}^{T} f(x)\bar{f}(x-t)\,dx\right|^2 \leqslant \left[\int_{-T}^{T}|f(x)|^2\,dx\int_{-T}^{T}|f(x-t)|^2\,dx\right].$$

We have

$$\frac{1}{2T}\int_{-T}^{T}|f(x-t)|^2\,dx = \frac{1}{2T}\int_{-T-t}^{T-t}|f(x)|^2\,dx$$

$$= \frac{1}{2T}\int_{-T}^{T}|f(x)|^2\,dx + \frac{1}{2T}\int_{-T-t}^{-T}|f(x)|^2\,dx - \frac{1}{2T}\int_{T-t}^{T}|f(x)|^2\,dx.$$

Now

$$\frac{1}{2T}\int_{-T-t}^{-T}|f(x)|^2\,dx \leqslant \frac{1}{2T}\int_{-T-t}^{T+t}|f(x)|^2\,dx - \frac{1}{2T}\int_{-T}^{T}|f(x)|^2\,dx$$

$$= \left(1+\frac{t}{T}\right)\frac{1}{2(T+t)}\int_{-T-t}^{T+t}|f(x)|^2\,dx - \frac{1}{2T}\int_{-T}^{T}|f(x)|^2\,dx.$$

Hence
$$\lim_{T\to\infty}\frac{1}{2T}\int_{-T-t}^{-T}|f(x)|^2\,dx$$

$$\leqslant \lim_{T\to\infty}\left(1+\frac{t}{T}\right)\lim_{T\to\infty}\frac{1}{2(T+t)}\int_{-T-t}^{T+t}|f(x)|^2\,dx - \lim_{T\to\infty}\frac{1}{2T}\int_{-T}^{T}|f(x)|^2\,dx = 0.$$

Similarly,
$$\lim_{T\to\infty}\frac{1}{2T}\int_{T}^{T+t}|f(x)|^2\,dx = 0.$$

Hence $$\lim_{T\to\infty} \frac{1}{2T}\int_{-T}^{T} |f(x-t)|^2\,dx = \lim_{T\to\infty} \frac{1}{2T}\int_{-T}^{T} |f(x)|^2\,dx.$$

From this we readily conclude that $|g(t)| \leqslant g(0)$. The function $g(t)$ is thus bounded. Indeed, we can go further, and assert that $g(t)$ is the limit of a uniformly bounded sequence of summable functions, and hence is summable over any finite range. From this it follows readily that $g(t)/t$ is summable over any finite range which does not include the origin and it is of summable absolute square from $-\infty$ to $-\epsilon$, and from ϵ to ∞.

We now wish to introduce the chief theorems of the Plancherel theory of the Fourier transform*. We shall translate this into its complex form. If $f(x)$ is a complex function of the real variable x, summable over any finite range, and such that

$$\int_{-\infty}^{\infty} |f(x)|^2\,dx$$

exists, then $$F(u) = \underset{A\to\infty}{\text{l.i.m.}}\ \frac{1}{\sqrt{(2\pi)}}\int_{-A}^{A} f(x)\,e^{iux}\,dx$$

exists, and is summable over any finite range, and

$$\int_{-\infty}^{\infty} |F(u)|^2\,du$$

exists and is equal to $$\int_{-\infty}^{\infty} |f(x)|^2\,dx.$$

Furthermore, $$f(x) = \underset{A\to\infty}{\text{l.i.m.}}\ \frac{1}{\sqrt{(2\pi)}}\int_{-A}^{A} F(u)\,e^{-iux}\,du.$$

Here l.i.m. stands for "limit in mean" over $(-\infty, \infty)$.

Since $g(t)/t$ is summable over any finite range and of summable absolute square over $(-\infty, -\epsilon)$ and (ϵ, ∞), there exists

$$\underset{N\to\infty}{\text{l.i.m.}}\ \frac{1}{\pi}\left[\int_{-N}^{-\epsilon} + \int_{\epsilon}^{N}\right] g(t)\frac{\sin ut}{t}\,dt$$

over the whole u-line. Thus there exists as a limit in the mean over any finite segment of the u-line

$$R(u) = \underset{N\to\infty}{\text{l.i.m.}}\ \frac{1}{\pi}\int_{-N}^{N} g(t)\,\frac{\sin ut}{t}\,dt.$$

* Hobson, *The theory of functions of a real variable*, 2nd edition, 2 (1926), 742–752.

We now wish to prove that $R(u)$ is a non-decreasing function. This is essential if we are to consider $R(u_2)-R(u_1)$ as the total spectral power of f in the region (u_1, u_2). To prove this, we introduce the auxiliary functions

$$f_P(t)=f(t)\ \ [|t|\leqslant P];\quad f_P(t)=0\ \ [|t|>P];$$

$$g_P(t)=\frac{1}{2P}\int_{-\infty}^{\infty} f_P(x)\bar{f}_P(x-t)\,dx\,;$$

$$R_P(u)=\frac{1}{\pi}\int_{-\infty}^{\infty} g_P(t)\frac{\sin ut}{t}\,dt.$$

The integrals here are infinite integrals only in appearance; it is easily seen that we are justified in writing

$$\begin{aligned}\frac{dR_P(u)}{du}&=\frac{1}{\pi}\int_{-\infty}^{\infty} g_P(t)\cos ut\,dt\\&=\frac{1}{2\pi P}\int_{-\infty}^{\infty}\cos ut\,dt\int_{-\infty}^{\infty} f_P(x)\bar{f}_P(x-t)\,dx\\&=\frac{1}{2\pi P}\int_{-\infty}^{\infty} f_P(x)\,dx\int_{-\infty}^{\infty}\bar{f}_P(x-t)\cos ut\,dt\\&=\frac{1}{2\pi P}\int_{-\infty}^{\infty} f_P(x)\,dx\int_{-\infty}^{\infty}\bar{f}_P(\xi)\cos u(x-\xi)\,d\xi\\&=\frac{1}{2\pi P}\left\{\left|\int_{-\infty}^{\infty} f_P(x)\cos ux\,dx\right|^2+\left|\int_{-\infty}^{\infty} f_P(x)\sin ux\,dx\right|^2\right\}\\&\geqslant 0.\end{aligned}$$

Hence $R_P(u)$ is a monotone non-decreasing function of u for each P.

The function $g_P(t)$ converges boundedly to $g(t)$ as $P\to\infty$. Hence over the intervals $(-\infty, -\epsilon)$ and (ϵ, ∞), $g_P(t)/t$ converges in the mean to $g(t)/t$. It follows from Plancherel's theory of the Fourier transform that

$$\underset{P\to\infty}{\text{l.i.m.}}\,\frac{1}{\pi}\left[\int_{-\infty}^{-\epsilon}+\int_{\epsilon}^{\infty}\right]g_P(t)\frac{\sin ut}{t}\,dt=\underset{N\to\infty}{\text{l.i.m.}}\,\frac{1}{\pi}\left[\int_{-N}^{-\epsilon}+\int_{\epsilon}^{N}\right]g(t)\frac{\sin ut}{t}\,dt.$$

From this we readily conclude that over any finite range

$$R(u)=\underset{P\to\infty}{\text{l.i.m.}}\,R_P(u),$$

We see at once, since every R_P is non-decreasing, and since therefore $\frac{1}{2\eta}\int_{u-\eta}^{u+\eta}R_P(\lambda)\,d\lambda$ is monotone, that its limit, $\frac{1}{2\eta}\int_{u-\eta}^{u+\eta}R(\lambda)\,d\lambda$ is likewise monotone. By a well known theorem, however*,

$$\lim_{\eta \to 0} \frac{1}{2\eta} \int_{u-\eta}^{u+\eta} R(\lambda)\, d\lambda = R(u)$$

almost everywhere. If we define $R(u)$ as this limit wherever this limit exists, and choose the values of $R(u)$ suitably at all other points, it is easy to see that we can so determine $R(u)$ that it shall be monotone non-decreasing. It will always be possible to put $R(0) = 0$.

Let us choose $R(u)$ in this manner. Let us put

$$\rho(u) = R(u) - \frac{1}{\pi} \int_{-M}^{M} g(t) \frac{\sin ut}{t}\, dt.$$

The function $\rho(u)$ is of limited total variation over every finite interval, and by a familiar theorem†, its Fourier integral converges everywhere to $\frac{1}{2}\{\rho(u+0)+\rho(u-0)\}$. It is easy to conclude from these properties that, as an ordinary infinite integral,

$$\frac{R(u+0)+R(u-0)}{2} = \frac{1}{\pi} \int_{-\infty}^{\infty} g(t) \frac{\sin ut}{t}\, dt.$$

Inasmuch as $R(u)$ is of limited total variation, it can have at most a denumerable set of discontinuities, all of the first kind. This is of fundamental importance when we came to the study of almost periodic functions.

3. *The total spectral intensity.*

It is manifest that $\lim_{u \to \infty} R(u)$, or, as we shall write it, $R(\infty)$, if it exists, is a measure of the total spectral intensity of f. It is therefore a matter of some interest to determine, first, whether $R(\infty)$ is finite, and, secondly, what its relation with $f(t)$ may be.

We have, as an appeal to the Plancherel theory will show,

$$\frac{1}{S} \int_0^S R(u)\, du = \frac{1}{\pi S} \int_{-\infty}^{\infty} \frac{g(t)}{t}\, dt \int_0^S \sin ut\, du$$

$$= \frac{1}{\pi S} \int_{-\infty}^{\infty} g(t) \frac{1-\cos St}{t^2}\, dt.$$

* Hobson, 1, 541.
† Hobson, 2, 741.

Since R is monotone increasing, $R(\infty)$ will exist when, and only when,

$$\lim_{S\to\infty} \frac{1}{S}\int_0^S R(u)\,du$$

exists, and will assume the same value. Manifestly, however,

$$\frac{1}{S}\int_0^S R(u)\,du \leqslant \max|g|\frac{1}{\pi S}\int_{-\infty}^{\infty} \frac{1-\cos St}{t^2}\,dt$$

$$= g(0).$$

Thus $R(\infty)$ always exists. If $g(t)$ is continuous at $t=0$, an argument of the Fejér type will show that

$$R(\infty) = \lim_{S\to\infty} \frac{1}{S}\int_0^S R(u)\,du = g(0)$$

$$= \lim_{T\to\infty} \frac{1}{2T}\int_{-T}^{T} |f(t)|^2\,dt.$$

We thus see that, under this assumption, the mean square of the modulus of f gives the total spectral intensity.

We shall call the assumption that $g(t)$ is continuous at $t=0$ assumption (B). As an example of a function for which it is not fulfilled, we may put $f(t) = \sin t^2$. Here

$$g(t) = 0 \quad [t \neq 0];$$

$$g(0) = \tfrac{1}{2}.$$

We see that R vanishes identically, and $R(\infty) \neq g(0)$.

4. *The spectrum of an almost periodic function.*

In this and the following two paragraphs we shall confine our attention to the almost periodic functions of Bohr. A continuous function $f(t)$ is said to have a *translation number* τ pertaining to the quantity ϵ if, for every value of t,

$$|f(t+\tau)-f(t)| \leqslant \epsilon.$$

The function $f(t)$ is said to be *almost periodic* if, for every positive ϵ, there exists a positive number l_ϵ, such that every interval $(a, a+l_\epsilon)$ contains at least one translation number τ pertaining to ϵ. The following facts are capable of an *elementary* proof : the sum or product of a finite number of almost periodic functions is almost periodic, and so is any constant multiple of an almost periodic function; the limit of a uniformly convergent series of almost periodic functions is almost periodic; almost periodic functions are uniformly continuous. If $f(t)$ is almost periodic,

$$M\{f\} = \lim_{T\to\infty} \frac{1}{T}\int_z^{z+T} f(t)\,dt$$

exists as a uniform limit in z. The sum of a uniformly convergent series of sines and cosines of multiples of t is almost periodic. If $f(t)$ is almost periodic, $g(t) = M_x\{f(x)\bar{f}(x-t)\}$ is also almost periodic. If $f(t)$ is almost periodic and is greater than or equal to zero, and $M\{f\} = 0$, $f(t)$ is identically zero. We now wish to derive from these facts a proof of Bohr's analogue to Parseval's theorem.

Let $f(t)$ be an almost periodic function; then our conditions (A) and (B) are fulfilled, as we may easily show. Let us form $g(t)$ and $R(u)$ as before. Let $R(+0) = a_0$, and let the discontinuities of R for positive arguments, at most a denumerable set, be at $u = \lambda_1, \lambda_2, \ldots$. Let

$$a_n = R(\lambda_n+0)-R(\lambda_n-0).$$

All the coefficients a_n are positive, and

$$\sum_0^\infty a_k \leqslant R(\infty) = M\{|f|^2\}.$$

Let us form the function

$$\gamma(t) = \mathfrak{R}\,g(t)-\sum_0^\infty a_k \cos \lambda_k t,$$

where $\mathfrak{R}$ denotes the real part of the function. We have, as a simple computation will show,

$$\frac{2}{\pi}\int_0^\infty \gamma(t)\frac{\sin ut}{t}\,dt = R(u)-R_1(u),$$

where the odd function $R_1(u)$ consists of the sum of all the jumps of R with abscissae less than u, together with half the jump (if any) with abscissa u. That is

$$R_2(u) = \frac{2}{\pi}\int_0^\infty \gamma(t)\frac{\sin ut}{t}\,dt$$

is a continuous function of limited total variation, say V.

Consider
$$\frac{1}{2\epsilon}\int_{-\infty}^{\infty}|\Phi(u+\epsilon)-\Phi(u-\epsilon)|\,du.$$

This may be written

$$\frac{1}{2\epsilon}\sum_{-\infty}^{\infty}\int_{2k\epsilon}^{2(k+1)\epsilon}|\Phi(u+\epsilon)-\Phi(u-\epsilon)|\,du$$
$$=\frac{1}{2\epsilon}\int_{0}^{2\epsilon}\sum_{-\infty}^{\infty}|\Phi\{u+(2k+1)\epsilon\}-\Phi\{u+(2k-1)\epsilon\}|\,du$$
$$\leqslant \max_{u}\sum_{-\infty}^{\infty}|\Phi\{u+(2k+1)\epsilon\}-\Phi\{u+(2k-1)\epsilon\}|,$$

and this does not exceed the total variation of Φ over $(-\infty,\infty)$.

Similarly,
$$\frac{1}{2\epsilon}\int_{A}^{\infty}|\Phi(u+\epsilon)-\Phi(u-\epsilon)|\,du$$

does not exceed the total variation of Φ over $(A-\epsilon,\infty)$, and

$$\frac{1}{2\epsilon}\int_{-\infty}^{-A}|\Phi(u+\epsilon)-\Phi(u-\epsilon)|\,du$$

does not exceed the total variation of Φ over $(-\infty,-A+\epsilon)$.

We have

$$\frac{1}{2\epsilon}\int_{-\infty}^{\infty}[R_2(u+\epsilon)-R_2(u-\epsilon)]^2\,du$$
$$\leqslant \operatorname*{Max}_{A>u>-A}|R_2(u+\epsilon)-R_2(u-\epsilon)|\frac{1}{2\epsilon}\int_{-\infty}^{\infty}|R_2(u+\epsilon)-R_2(u-\epsilon)|\,du$$
$$+\operatorname{Max}|R_2(u_1)-R_2(u_2)|\frac{1}{2\epsilon}\left[\int_{A}^{\infty}+\int_{-\infty}^{-A}\right]|R_2(u+\epsilon)-R_2(u-\epsilon)|\,du.$$

The second term contains as a factor a quantity not exceeding the total variation of R_2 over $(A-\epsilon,\infty)$ and $(-\infty,\epsilon-A)$, taken together, and can be made arbitrarily small for all values of ϵ less than some fixed quantity by taking A large enough. When this is done, the first term, which is less than or equal to

$$V.\max_{A>u>-A}|R_2(u+\epsilon)-R_2(u-\epsilon)|,$$

can be made arbitrarily small by taking ϵ small enough. It follows, from the Fourier transform theory, that

$$0 = \lim_{\epsilon \to 0} \frac{1}{2\epsilon} \int_{-\infty}^{\infty} [R_2(u+\epsilon) - R_2(u-\epsilon)]^2 \, du$$

$$= \lim_{\epsilon \to 0} \frac{1}{2\epsilon} \int_{-\infty}^{\infty} \left[\frac{4}{\pi} \int_0^\infty \gamma(t) \frac{\sin \epsilon t}{t} \cos ut \, dt \right]^2 du$$

$$= \lim_{\epsilon \to 0} \frac{4}{\pi\epsilon} \int_{-\infty}^{\infty} [\gamma(t)]^2 \frac{\sin^2 \epsilon t}{t} \, dt.$$

Now, $\dfrac{\sin^2 \epsilon t}{\epsilon^2 t^2} > \dfrac{4}{\pi^2}$ for $|\epsilon t| \leqslant \dfrac{\pi}{2}$. Hence

$$0 = \lim_{\epsilon \to 0} \frac{16\epsilon}{\pi^3} \int_{-(\pi/2\epsilon)}^{\pi/2\epsilon} [\gamma(t)]^2 \, dt.$$

It follows that
$$M\{|\gamma(t)|^2\} = 0.$$

However, $\mathfrak{R} g(t)$ and $\sum_0^\infty a_k \cos \lambda_k t$ are both almost periodic, and hence the same is true of $\gamma(t)$. It follows that $\gamma(t)$ is identically zero, that

$$\mathfrak{R} g(t) = \sum_0^\infty a_k \cos \lambda_k t,$$

and that
$$R(u) = R_1(u).$$

Since $R_1(u)$ increases solely by jumps, we have

$$M\{|f(t)|^2\} = R(\infty) = R_1(\infty) = \sum_0^\infty a_k.$$

In physical language, the entire power of the spectrum of an almost periodic function is contained in a denumerable set of lines.

5. *The Parseval theorem for almost periodic functions.*

We now wish to seek for a means of expressing the various coefficients a_k directly in terms of f. We have

$$\mathfrak{R} g(t) = \sum_0^\infty a_k \cos \lambda_k t,$$

where this series converges uniformly and absolutely. Accordingly,

$$M\{\mathfrak{R} g(t) \cos \lambda_j t\} = \tfrac{1}{2} a_j \quad [j \neq 0];$$

$$M\{\mathfrak{R} g(t)\} = a_0.$$

However, since it follows at once from the definition of $g(t)$ that $g(-t)=\bar{g}(t)$,

$$M\{g(t)\cos\lambda_j t\} = \tfrac{1}{2}a_j.$$

That is, because of the uniform existence of means of almost periodic functions,

$$\begin{aligned}
a_j &= 2\lim_{T\to\infty}\frac{1}{2T}\int_{-T}^{T}\cos\lambda_j t\,dt\lim_{U\to\infty}\frac{1}{2U}\int_{-U}^{U}f(x)\bar{f}(x-t)\,dx\\
&= 2\lim_{U\to\infty}\frac{1}{2U}\int_{-U}^{U}f(x)\,dx\lim_{T\to\infty}\frac{1}{2T}\int_{-T}^{T}\bar{f}(x-t)\cos\lambda_j t\,dt\\
&= 2\lim_{U\to\infty}\frac{1}{2U}\int_{-U}^{U}f(x)\,dx\lim_{V\to\infty}\frac{1}{2V}\int_{-V}^{V}\bar{f}(y)\cos\lambda_j(x-y)\,dy\\
&= 2\lim_{U\to\infty}\frac{1}{2U}\int_{-U}^{U}f(x)\cos\lambda_j x\,dx\lim_{V\to\infty}\frac{1}{2V}\int_{-V}^{V}\bar{f}(y)\cos\lambda_j y\,dy\\
&\qquad+2\lim_{U\to\infty}\frac{1}{2U}\int_{-U}^{U}f(x)\sin\lambda_j x\,dx\lim_{V\to\infty}\frac{1}{2V}\int_{-V}^{V}\bar{f}(y)\sin\lambda_j y\,dy\\
&= 2|M\{f(x)\cos\lambda_j x\}|^2+2|M\{f(x)\sin\lambda_j x\}|^2\\
&= \tfrac{1}{2}|M\{f(x)(e^{i\lambda_j x}+e^{-i\lambda_j x})\}|^2+\tfrac{1}{2}|M\{f(x)(e^{i\lambda_j x}-e^{-i\lambda_j x})\}|^2\\
&= |M\{f(x)e^{i\lambda_j x}\}|^2+|M\{f(x)e^{-i\lambda_j x}\}|^2 \quad [j\neq 0].
\end{aligned}$$

Similarly, $$a_0 = |M\{f(x)\}|^2.$$

Hence

$$M\{|f(t)|^2\} = \sum_0^\infty a_k = |M\{f(x)\}|^2+\sum_1^\infty |M\{f(x)e^{i\lambda_j x}\}|^2+|M\{f(x)e^{-i\lambda_j x}\}|^2.$$

This is a precise analogue to the Hurwitz-Parseval theorem for periodic functions, and is the well known fundamental theorem of Bohr.

6. *The Weierstrass theorem for almost periodic functions*.*

Let $f(z)$ be almost periodic. Consider

$$\phi(x) = \max_t |f(x+t)-f(t)|.$$

* The method of the present section has much resemblance to that used by Weyl in the paper already cited. The function $\phi(x)$ is S. Bochner's "Verschiebungsfunktion", [S. Bochner. "Beiträge zur Theorie der fastperiodischen Funktionen", *Math. Annalen*, 96 (1926), 119–147].

We have

$$|\phi(x+\tau)-\phi(x)| \leqslant \max_t \Big| |f(x+t+\tau)-f(t+\tau)| - |f(x+t)-f(t)| \Big|$$
$$\leqslant \max_t |f(x+t+\tau)-f(x+t)| + \max |f(t+\tau)-f(t)|$$
$$\leqslant 2 \max_t |f(t+\tau)-f(t)|.$$

Hence any displacement number for f pertaining to ϵ is a displacement number for ϕ pertaining to 2ϵ, and $\phi(x)$ is almost periodic.

It is easy to show that any polynomial in an almost periodic function is almost periodic, and hence by Weierstrass's theorem and the fact that the uniform limit of an almost periodic function is almost periodic, that any continuous function of an almost periodic function is almost periodic. Let $H(u)$ be a continuous non-negative function of u which is 1 when $0 \leqslant u \leqslant \frac{1}{2}\epsilon$, between 1 and 0 when $\frac{1}{2}\epsilon \leqslant u \leqslant \epsilon$, and 0 when $u \geqslant \epsilon$. Let

$$\psi_\epsilon(x) = \frac{H[\phi(x)]}{MH[\phi(x)]}.$$

Since $H[\phi(x)]$ is somewhere positive, and it is everywhere non-negative and almost periodic, it is easy to show that $MH[\phi(x)]$ cannot vanish. The function $\psi_\epsilon(x)$ is almost periodic.

Let
$$f_\epsilon(x) = M[f(t)\,\psi_\epsilon(x-t)].$$

The existence and almost periodic character of $f_\epsilon(x)$ are proved without difficulty. The definition of ψ_ϵ ensures that

$$|f(x)-f(t)| \leqslant \epsilon$$

if $\psi_\epsilon(x-t) \neq 0$. Hence, since $f_\epsilon(x)$ is a mean of these values of $f(t)$,

$$\max_x |f(x)-f_\epsilon(x)| \leqslant \epsilon.$$

Similarly, if
$$f^{(\epsilon)}(x) = M[f_\epsilon(t)\,\psi_\epsilon(x-t)],$$

$f^{(\epsilon)}$ exists and is almost periodic, and

$$\max_x |f^{(\epsilon)}(x)-f_\epsilon(x)| \leqslant \epsilon.$$

Hence
$$\max_x |f^{(\epsilon)}(x)-f(x)| \leqslant 2\epsilon.$$

We have
$$f^{(\epsilon)}(x) = \underset{t}{M}\Big[\psi_\epsilon(x-t)\,\underset{\tau}{M}[f(\tau)\,\psi_\epsilon(t-\tau)]\Big].$$

Bearing in mind the uniformity properties of almost periodic functions, we have

$$f^{(\epsilon)}(x) = \underset{\tau}{M}\Big[f(\tau)\underset{t}{M}[\psi_\epsilon(x-t)\,\psi_\epsilon(t-\tau)]\Big].$$

However, by § 4,

$$\underset{t}{M}[\psi_\epsilon(x-t)\,\psi_\epsilon(t-\tau)] = \sum_0^\infty a_k \cos\lambda_k(x-\tau),$$

where all the coefficients a_k are positive, and $\sum_0^\infty a_k$ converges. Hence

$$f^{(\epsilon)}(x) = M\Big[f(\tau)\sum_0^\infty a_k \cos\lambda_k(x-\tau)\Big].$$

Since
$$\left|\frac{1}{2T}\int_{-T}^{T} f(\tau)\cos\lambda_k(x-\tau)\,d\tau\right| \leqslant \max|f(\tau)|,$$

it follows that we can invert the order of M and Σ, and that

$$f^{(\epsilon)}(x) = \sum_0^\infty a_k\Big\{\cos\lambda_k x M\{f(\tau)\cos\lambda_k\tau\} + \sin\lambda_k x M\{f(\tau)\sin\lambda_k\tau\}\Big\}.$$

Inasmuch as
$$\left|M\left\{f(\tau)\begin{matrix}\sin\\ \cos\end{matrix}\lambda_k\tau\right\}\right| \leqslant |f(\tau)|,$$

$f^{(\epsilon)}(x)$ is the sum of an absolutely convergent series of trigonometric terms. That is to say, we can choose N so large that

$$\max_x\left|f^{(\epsilon)}(x) - \sum_1^N a_k\Big\{\cos\lambda_k x M\{f(\tau)\cos\lambda_k\tau\} + \sin\lambda_k x M\{f(\tau)\sin\lambda_k\tau\}\Big\}\right| \leqslant \epsilon,$$

and hence that

$$\max_x\left|f(x) - \sum_1^N a_k\Big\{\cos\lambda_k x M\{f(\tau)\cos\lambda_k\tau\} + \sin\lambda_k x M\{f(\tau)\sin\lambda_k\tau\}\Big\}\right| \leqslant 3\epsilon.$$

In other words, we have proved Bohr's approximation theorem, to the effect that it it is possible to approximate uniformly to any desired degree of accuracy to an almost periodic function by means of trigonometrical polynomials.

Comments on [26a, b], [28a]

S. Koizumi

In these papers Wiener dispenses with the "hypothesis of uniformity" imposed on $f(t)$ in his work during the earlier period (1923–1925). He now imposes on $f(t)$ three hypotheses, viz. (1), (2) and (3) in [26a], p. 119, and demonstrates the possibility of a generalized harmonic analysis of $f(t)$. These hypotheses involve only a denumerable set of transforms of $f(t)$, which is a great advance over his earlier work and exceedingly important for a proper linkage to the theory of probability. It now becomes possible to apply the analysis to Brownian motion. It is proved that the Brownian paths have the same spectral distribution except for a set of paths of measure zero, and this is continuous. The method of proof depends on calculating the mean square of transforms of $f(t)$, which is a kind of energy integral.

The three hypotheses (1), (2), and (3) in [26a] are transformed into new ones (a), (b) and (c) in [26b]. These sets of hypotheses are equivalent, as he shows with the aid of a kind of Tauberian theorem (cf. (7) in [26a], p. 103).

Next Wiener devotes himself to the simplification and refinement of the theory. Guided by the work of H. Weyl on almost periodic functions, he examines the correlation function of trigonometrical polynomials. Let $f(t) = \Sigma A_k e^{i\Lambda_k t}$ be given. Then its correlation function $g(t) = \Sigma |A_k|^2 e^{i\Lambda_k t}$, and its spectral distribution $R(u)$ is an increasing step function and has a jump at $u = |\Lambda_k|$. Thus he arrives at the two hypotheses (A) and (B) in [28a], pp. 486, 490, under which it is proved that $R(u)$ can be chosen so as to be monotone non-decreasing and bounded. He applies his new method to Bohr's almost periodic functions and shows that fundamental results such as Parseval's equality, Weierstrass' approximation theorem are deducible from his general theory. Hypotheses (1) and (2), or (a) and (b), correspond to that of (A). Furthermore if (A) is assumed, then (B) is equivalent to (3) and (c). It should be noted that if we define the generalized Fourier transform (GFT) $\sigma(u)$ of the correlation function g *by*

$$\sigma(u) = \frac{1}{2\pi}\int_{-1}^{1} g(x)\frac{e^{iux}-1}{ix}\,dx$$

$$+ \underset{A\to\infty}{\text{l.i.m.}}\,\frac{1}{2\pi}\left[\int_{1}^{A} + \int_{-A}^{-1}\right] g(x)\frac{e^{iux}}{ix}\,dx,$$

then we have

$$R(u) = \tfrac{1}{2}\{\sigma(u) - \sigma(-u)\}.$$

In later work Wiener was able to define the GFT of f itself analogously, replacing $g(x)$ by $f(x)$, and he denoted it by $s(u)$. See [33i], Chapter IV, §21, 22, 23.

The results on Brownian motion have had a deep influence in the theory of probability and related topics. With the aid of Birkhoff's ergodic theorem, Paley and Wiener showed that there is an associated flow which is measure preserving and ergodic. See [34d], chapters IX and X. This enables us to give a simple proof.

GENERALIZED HARMONIC ANALYSIS.[1]

By

NORBERT WIENER

of Cambridge, Mass., U. S. A.

Reprinted from *Acta Math.*, *55*, 1930, pp. 117–258. (Courtesy of the Institut Mittag-Leffler.)

Introduction.

Generalized harmonic analysis represents the culmination and combination of a number of very diverse mathematical movements. The theory of almost periodic functions finds its precursors in the theory of Dirichlet series, and in the quasiperiodic functions of Bohl and Esclangon. These latter, in turn, are an answer to the demands of the theory of orbits in celestial mechanics; the former take their origin in the analytic theory of numbers. Quite independent of the regions of thought just enumerated, we have the order of ideas associated with the names of Lord Rayleigh, of Gouy, and above all, of Sir Arthur Schuster; these writers concerned themselves with the problems of white light, of noise, of coherent and incoherent sources. More particularly, Schuster was able to point out the close analogy between the problems of the harmonic analysis of light and the statistical analysis of hidden periods in such scientific data as are common in meteorology and astronomy, and developed the extremely valuable theory of the periodogram. The work of G. I. Taylor on diffusion represents another valuable anticipation of theories here developed, from the standpoint of an applied mathematician of the British school, with preoccupations much the same as those of Schuster.

The work of Hahn seems to have a much more definitely pure-mathematics motivation. To the pure mathematician in general, however, and the worker in real function theory in particular, we owe, not so much the setting of our problem, as the chief tool in its attack: the famous theorem of Plancherel, the proof of which Titchmarsh has extended and improved.

It may seem a little strange to the reader that the present paper should contain yet another proof of this much proved theorem. In view, however, of the centralness of the Plancherel theorem in all that is to follow, and more expecially of the fact that the proof here given furnishes an excellent introduction to the meaning and motivation of our proofs in more complicated cases, it has seemed worth while to prove the Plancherel theorem in full.

The germs of the generalized harmonic analysis of this paper are already in the work of Schuster, but only the germs. To make the Schuster theory assume a form suitable for extension and generalization, a radical recasting is necessary. This recasting brings out the fact that the expression

$$\varphi(x) = \lim_{T\to\infty} \frac{1}{2T} \int_{-T}^{T} f(x+t)\bar{f}(t)\,dt \tag{0.1}$$

plays a fundamental part in Schuster's theory, as does also

$$S(u) = \frac{1}{2\pi} \int_{-\infty}^{\infty} \varphi(x) \frac{e^{iux}-1}{ix}\,dx. \tag{0.2}$$

Accordingly, section 3 is devoted to the independent study of these two expressions, and to the definition of $S(u)$ under appropriate assumption as the *spectrum* of $f(x)$.

There are some interesting relations between the total spectral intensity of $f(x)$ as represented by $S(u)$ and the other expressions of the theory. Some of these demand for their proper appreciation a mode of connecting various weighted means of a positive quantity. The appropriate tool for this purpose is the general theory of Tauberian theorems developed by the author and applied to these problems by Mr. S. B. Littauer.

These latter Tauberian theorems enable us to correlate the mean square of the modulus of a function and the »quadratic variation» of a related function which determines its harmonic analysis. The theory of harmonic analysis here indicated has been extended by Bochner to cover the case of very general functions behaving algebraically at infinity. A somewhat similar theory is due to Hahn, who is, however, more interested in questions of ordinary convergence than in those clustering about the Parseval theorem.

The theory of generalized harmonic analysis is itself capable of extension in very varied directions. Mr. A. C. Berry has recently developed a vectorial extension of the theory to n dimensions, while on the other hand, the author himself has extended the theory to cover the simultaneous harmonic analysis of a set of functions and the notions of coherent and incoherent sources of light. A third extension depends on the replacement of the translation group, fundamental in all harmonic analysis, by another group.

To prove that the theory is not vacuous and trivial, it is of importance to give examples of different types of spectra. We do this, both by direct methods, and by methods involving an infinite series of choices between alternatives of equal probability. The latter method, of course, involves the assumption of the

Zermelo axiom: on the other hand, it yields a most interesting probability theory of spectra. This theory may be developed to cover the case where the infinite sequence of choices is replaced by a haphazard motion of the type known as Brownian.

The spectrum theory of the present paper has as one very special application the theory of almost periodic functions. It is not difficult to prove that the spectrum of such a function contains a discrete set of lines and no continuous part, and to deduce from this, Bohr's form of the Parseval theorem. The transition from the Parseval theorem to the Weierstrassian theorem that it is possible to approximate uniformly to any almost periodic function by a sequence of trigonometrical polynomials follows essentially lines laid down by Weyl, though it differs somewhat in detail.

Besides the well-known generalizations of almost periodic functions due to Stepanoff, Besicovitch, Weyl, and the author, there is the little explored field of extensions of almost periodic functions containing a parameter. These have been used by Mr. C. F. Muckenhoupt to prove the closure of the set of the Eigenfunktionen of certain linear vibrating systems. This is one of the few applications of almost periodic functions of a fairly general type to definite mathematicophysical problems. Our last section is devoted to this, and to related matters.

CHAPTER I.

1. Plancherel's theorem.

Plancherel's theorem reads as follows: *Let $f(x)$ be quadratically summable over $(-\infty, \infty)$ in the sense of Lebesgue — that is, let it be measurable, and let*

$$\int_{-\infty}^{\infty} |f(x)|^2 \, dx \tag{1.01}$$

exist and be finite. (i.e. $f \subset L_2$). *Then*

$$g(u) = \underset{A\to\infty}{\text{l.i.m.}} \frac{1}{\sqrt{2\pi}} \int_{-A}^{A} f(x) e^{iux} \, dx \tag{1.02}$$

(*where* l.i.m. *stands for »limit in the mean») will exist, and*

$$f(x) = \underset{A\to\infty}{\text{l.i.m.}} \frac{1}{\sqrt{2\pi}} \int_{-A}^{A} g(u) e^{iux}\, du. \tag{1.03}$$

$g(u)$ is known as the »Fourier transform» of $f(x)$. To prove this, let us put

$$f_A(x) = \begin{cases} f(x) & \text{if } |x| < A, \\ 0 & \text{if } |x| \geq A. \end{cases} \tag{1.04}$$

Let us represent $f_A(x)$ over $(-2A, 2A)$ by the Fourier series

$$f_A(x) \sim \sum_{-\infty}^{\infty} a_n e^{\frac{in\pi x}{2A}}. \tag{1.05}$$

Then

$$\int_{-\infty}^{\infty} f_A(x+\xi)\bar{f}_A(\xi)\, d\xi = \int_{-2A}^{2A} f_A(x+\xi)\bar{f}_A(\xi)\, d\xi$$

$$= \sum_{-\infty}^{\infty} a_n \bar{a}_n \int_{-2A}^{2A} e^{\frac{in\pi(x+\xi)}{2A}} e^{-\frac{in\pi\xi}{2A}}\, d\xi$$

$$= \sum_{-\infty}^{\infty} 4A\, |a_n|^2 e^{\frac{in\pi x}{2A}}. \tag{1.06}$$

This series of equations merits several comments. First, the infinite integrals which appear are infinite in appearance only, as the integrand vanishes beyond a certain point. Secondly, the period chosen for the Fourier representation of $f_A(x)$ is twice the length of the interval over which $f_A(x)$ may differ from 0, so that one period of $f_A(x+\xi)$ may overlap not more than one corresponding period of $\bar{f}_A(\xi)$. Third, the function $\int_{-\infty}^{\infty} f_A(x+\xi)\bar{f}_A(\xi)\, d\xi$ has a Fourier development which possesses only positive coefficients, and is absolutely and uniformly convergent, as follows at once from the Hurwitz theorem. The positiveness of the Fourier coefficients of this function forms the point of departure for the greater part of the present paper.

16—29764. *Acta mathematica.* 55. Imprimé le 7 avril 1930.

It follows at once that

$$\int_{-\infty}^{\infty} f_A(x+\xi) f_A(\xi)\, d\xi = \lim_{N\to\infty} \frac{1}{4A} \int_{-2A}^{2A} \frac{\sin \frac{2N+1}{4A}\pi(x-y)}{\sin \frac{\pi(x-y)}{4A}}\, dy \int_{-\infty}^{\infty} f_A(y+\xi) \bar{f}_A(\xi)\, d\xi. \quad (1.07)$$

However, Lebesgue's fundamental theorem on the Fourier coefficients, to the effect that they always tend to zero, yields us

$$\lim_{N\to\infty} \frac{1}{4A} \int_{-2A}^{2A} \sin \frac{2N+1}{4A}\pi(x-y) \left[\frac{1}{\frac{\pi(x-y)}{4A}} - \frac{1}{\sin \frac{\pi(x-y)}{4A}} \right] \cdot$$

$$dy \int_{-\infty}^{\infty} f_A(y+\xi) \bar{f}_A(\xi)\, d\xi = 0. \quad (1.08)$$

Combining these two relations, we see that

$$\int_{-A}^{A} |f(x)|^2 dx = \int_{-\infty}^{\infty} |f_A(x)|^2\, dx$$

$$= \lim_{N\to\infty} \frac{1}{\pi} \int_{-2A}^{2A} \frac{\sin \frac{2N+1}{4A}\pi y}{y}\, dy \int_{-\infty}^{\infty} f_A(y+\xi) \bar{f}_A(\xi)\, d\xi$$

$$= \lim_{N\to\infty} \frac{1}{\pi} \int_{-\infty}^{\infty} dy \int_{0}^{\frac{2N+1}{4A}\pi} \cos uy\, du \int_{-\infty}^{\infty} f_A(y+\xi) \bar{f}_A(\xi)\, d\xi$$

$$= \frac{1}{2\pi} \int_{-\infty}^{\infty} du \int_{-\infty}^{\infty} f_A(\eta)\, d\eta \int_{-\infty}^{\infty} \bar{f}_A(\xi)\, e^{iu(\eta-\xi)}\, d\xi$$

$$= \frac{1}{2\pi} \int_{-\infty}^{\infty} du \left| \int_{-A}^{A} f(\eta)\, e^{iu\eta}\, d\eta \right|^2. \quad (1.09)$$

The inversions of the order of integration are here justified by the fact that all the infinite limits are merely apparent, and are introduced to simplify the formal work of inversion. If we replace $f_A(x)$ by the function $f_B(x)-f_A(x)$ which has essentially the same properties, we see that

$$\int_{-B}^{B} |f(x)|^2\, dx - \int_{-A}^{A} |f(x)|^2\, dx = \frac{1}{2\pi}\int_{-\infty}^{\infty} du \left| \int_{-B}^{B} f(\eta)\, e^{iu\eta}\, d\eta - \int_{-A}^{A} f(\eta) e^{iu\eta}\, d\eta \right|^2. \tag{1.10}$$

In case $\int_{-\infty}^{\infty} |f(x)|^2\, dx$ exists,

$$\lim_{B,A\to\infty} \int_{-\infty}^{\infty} du \left| \int_{-B}^{B} f(\eta)\, e^{iu\eta}\, d\eta - \int_{-A}^{A} f(\eta)\, e^{iu\eta}\, d\eta \right|^2 = 0 \tag{1.11}$$

and we may use Weyl's lemma to the Riesz-Fischer theorem to prove that

$$g(u) = \underset{A\to\infty}{\mathrm{l.i.m.}} \frac{1}{\sqrt{2\pi}} \int_{-A}^{A} f(\eta)\, e^{iu\eta}\, d\eta \tag{1.12}$$

exists, and is »quadratically summable». Combining this definition of $g(u)$ with (1.09), we see that

$$\int_{-\infty}^{\infty} |g(u)|^2\, du = \int_{-\infty}^{\infty} |f(x)|^2\, dx. \tag{1.13}$$

That is, the integral of the square of the modulus of a function is invariant under a Fourier transformation.

To complete the proof of Plancherel's theorem, it is merely necessary to show that for functions $f(x)$ of some closed set,

$$f(x) = \underset{A\to\infty}{\mathrm{l.i.m.}} \frac{1}{\sqrt{2\pi}} \int_{-A}^{A} g(u)\, e^{-iux}\, du. \tag{1.14}$$

A particular choice of $f(x)$ is the following:

$$f(x) = \begin{cases} 0; & [x < \alpha] \\ 1; & [\alpha < x < \beta] \\ 0; & [\beta < x]. \end{cases} \tag{1.15}$$

Here

$$g(u) = \frac{1}{\sqrt{2\pi}} \int_{\alpha}^{\beta} e^{iux}\, dx = \frac{e^{iu\beta} - e^{iu\alpha}}{iu\sqrt{2\pi}}. \tag{1. 16}$$

Hence

$$\begin{aligned}
\underset{A\to\infty}{\text{l.i.m.}} \frac{1}{\sqrt{2\pi}} \int_{-A}^{A} g(u) e^{-iux}\, du &= \underset{A\to\infty}{\text{l.i.m.}} \frac{1}{2\pi} \int_{-A}^{A} \frac{e^{iu(\beta-x)} - e^{iu(\alpha-x)}}{iu}\, du \\
&= \underset{A\to\infty}{\text{l.i.m.}} \frac{1}{2\pi} \int_{-A}^{A} \frac{\sin u(\beta-x) - \sin u(\alpha-x)}{u}\, du \\
&= \frac{1}{2} [\operatorname{sgn}(\beta-x) - \operatorname{sgn}(\alpha-x)] \\
&= f(x)
\end{aligned} \tag{1. 17}$$

except possibly at the two points α and β, a set of zero measure. This completes the proof of Plancherel's theorem.

Plancherel states this theorem somewhat differently. He essentially defines $g(u)$ as

$$g(u) = \frac{d}{du} \frac{1}{\sqrt{2\pi}} \int_{-\infty}^{\infty} d\eta \int_{0}^{u} f(\eta)\, e^{iu\eta}\, dv. \tag{1. 18}$$

If we retain our definition, it follows from an elementary use of the Schwarz inequality that

$$\int_{0}^{u} g(v)\, dv = \frac{1}{\sqrt{2\pi}} \int_{-\infty}^{\infty} d\eta \int_{0}^{u} f(\eta)\, e^{iv\eta}\, dv. \tag{1. 19}$$

To see this, les us reflect that

$$\begin{aligned}
\left| \int_{0}^{u} g(v)\, dv - \frac{1}{\sqrt{2\pi}} \int_{-A}^{A} d\eta \int_{0}^{u} f(\eta) e^{iv\eta}\, dv \right| &= \lim_{B\to\infty} \left| \frac{1}{\sqrt{2\pi}} \int_{0}^{u} dv \left[\int_{A}^{B} + \int_{-B}^{-A} \right] f(\eta)\, e^{iv\eta}\, d\eta \right| \\
&\leq \lim_{B\to\infty} \left\{ \frac{1}{2\pi} \int_{0}^{u} dv \int_{0}^{u} \left| \left[\int_{A}^{B} + \int_{-B}^{-A} \right] f(\eta)\, e^{iv\eta}\, d\eta \right|^{2} dv \right\}^{1/2}
\end{aligned}$$

$$\leq u \lim_{B\to\infty} \left\{ \frac{1}{2\pi} \int_{-\infty} \left| \left[\int_A^B + \int_{-B}^{-A} \right] f(\eta)\, e^{iu\eta}\, d\eta \right|^2 dv \right\}^{1/2}$$

$$= u \left\{ \left[\int_A^\infty + \int_{-\infty}^{-A} \right] |f(\eta)|^2\, d\eta \right\}^{1/2},$$

and since $\int_{-\infty}^{\infty} |f(\eta)|^2\, d\eta$ is finite, it follows that

$$\lim_{A\to\infty} \left| \int_0^u g(v)\, dv - \frac{1}{\sqrt{2\pi}} \int_{-A}^{A} d\eta \int_0^u f(\eta)\, e^{iv\eta}\, dv \right| = 0.$$

From this (1. 19) follows at once. Since a summable function is almost everywhere the derivative of its integral, the transition to Plancherel's form of the definition is immediate.

It follows at once from Plancherel's theorem that if $f_1(x)$ and $f_2(x)$ are quadratically summable,

$$F_1(u) = \underset{A\to\infty}{\text{l.i.m.}} \frac{1}{\sqrt{2\pi}} \int_{-A}^{A} f_1(x)\, e^{iux}\, dx \tag{1. 20}$$

and

$$F_2(u) = \underset{A\to\infty}{\text{l.i.m.}} \frac{1}{\sqrt{2\pi}} \int_{-A}^{A} f_2(x)\, e^{iux}\, dx \tag{1. 21}$$

exist, and that

$$\int_{-\infty}^{\infty} |F_1(u) \pm F_2(u)|^2\, du = \int_{-\infty}^{\infty} |f_1(x) \pm f_2(x)|^2\, dx \tag{1. 22}$$

and

$$\int_{-\infty}^{\infty} |F_1(u) \pm iF_2(u)|^2\, du = \int_{-\infty}^{\infty} |f_1(x) \pm if_2(x)|^2\, dx. \tag{1. 23}$$

Combining the last four formulae with one another, we have

$$\int_{-\infty}^{\infty} F_1(u)\,\overline{F}_2(u)\,du = \int_{-\infty}^{\infty} f_1(x)\overline{f}_2(x)\,dx. \tag{1.24}$$

This we may know as the Parseval theorem for the Fourier integral. Since

$$\overline{F}_2(-u) = \underset{A\to\infty}{\text{l.i.m.}}\ \frac{1}{\sqrt{2\pi}} \int_{-A}^{A} \overline{f}_2(x)\,e^{iux}\,dx \tag{1.25}$$

we may deduce at once that

$$\int_{-\infty}^{\infty} F_1(u)\,F_2(-u)\,du = \int_{-\infty}^{\infty} f_1(x)f_2(x)\,dx. \tag{1.26}$$

Since furthermore

$$\overline{F}_2(v-u) = \underset{A\to\infty}{\text{l.i.m.}}\ \frac{1}{\sqrt{2\pi}} \int_{-A}^{B} \overline{f}_2(x)\,e^{ivx}\,e^{iux}\,dx \tag{1.27}$$

it follows that

$$\int_{-\infty}^{\infty} F_1(u)\,F_2(v-u)\,du = \int_{-\infty}^{\infty} f_1(x)f_2(x)\,e^{ivx}\,dx. \tag{1.28}$$

As a consequence, if $f_1(x)f_2(x)$ is quadratically summable, its Fourier transform is

$$\sqrt{2\pi}\int_{-\infty}^{\infty} F_1(u)\,F_2(v-u)\,du. \tag{1.29}$$

This theorem lies at the basis of the whole operational calculus.

2. Schuster's periodogram analysis.

The two theories of harmonic analysis embodied in the classical Fourier series development and the theory of Plancherel do not exhaust the possibilities of harmonic analysis. The Fourier series is restricted to the very special class of periodic functions, while the Plancherel theory is restricted to functions which are quadratically summable, and hence tend on the average to zero as their argument tends to infinity. Neither is adequate for the treatment of a ray of white light which is supposed to endure for an indefinite time. Nevertheless, the physicists who first were faced with the problem of analyzing white light into its components had to employ one or the other of these tools. Gouy accordingly represented white light by a Fourier series, the period of which he allowed to grow without limit, and by focussing his attention on the average values of the energies concerned, he was able to arrive at results in agreement with the experiments. Lord Rayleigh on the other hand, achieved much the same purpose by using the Fourier integral, and what we now should call Plancherel's theorem. In both cases one is astonished by the skill with which the authors use clumsy and unsuitable tools to obtain the right results, and one is led to admire the unfailing heuristic insight of the true physicist.

The net outcome of the work of these writers was to dispel the idea that white light consist in some physical, supermathematical way of homogeneous monochromatic vibrations. Schuster in particular, was led to the conclusion that when white light is analyzed by a grating, the monochromatic components are created by the grating rather than selected by it. Thus a great stimulus was given to the investigation of the sense in which any phenomenon may be said to contain hidden periodic components. The successful completion of this investigation is also due to Schuster.

Schuster sums up his conclusions as follows[2]: »Let y be a function of t, such that its values are regulated by some law of probability, not necessarily the exponential one, but acting in such a manner that if a large number of t be chosen at random, there will always be a definite fraction of that number depending on t_1 only, which lie between t_1 and t_1+T, where T is any given time interval.

»Writing

$$A=\int_{t_1}^{t_1+T} y\cos\varkappa t\,dt \quad\text{and}\quad B=\int_{t_1}^{t_1+T} y\sin\varkappa t\,dt,$$

and forming

$$R=\sqrt{A^2+B^2},$$

the quantity R will, with increasing values of T, fluctuate about some mean value, which increases proportionally to $\sqrt{T}$ provided T is taken sufficiently large.

»If this theorem is taken in conjunction with the two following well-known propositions,

(1) If $y = \cos \varkappa t$, R will, apart from periodical terms, increase proportionally to T;

(2) If $y = \cos \lambda t$, λ being different from $\varkappa$, the quantity R will fluctuate about a constant value;

it is seen that we have means at our disposal to separate any true periodicity of a variable from among its irregular changes, provided we can extend the time limits sufficiently. . . . The application of the theory of probability to the investigation of what may be called »hidden» periodicities . . . may be further extended . . .»

While Schuster's statement is perhaps not in all respects clear, it contains the germs of all subsequent generalizations of harmonic analysis. First among these is the emphasis on the notion of the *mean*. The operator which yields

$$A_1 = \lim_{T \to \infty} \frac{1}{T} \int_{t_1}^{t_1+T} y \cos \varkappa t \, dt \text{ or } B_1 = \lim_{T \to \infty} \frac{1}{T} \int_{t_1}^{t_1+T} y \sin \varkappa t \, dt \tag{2. 01}$$

annihilates all functions $y(t)$ made up in a purely fortuitous or haphazard manner, as well as all trigonometrical functions other than $\cos \varkappa t$ or $\sin \varkappa t$, respectively. Hence we may take A_1 and B_1 to indicate the amounts of $\cos \varkappa t$ or $\sin \varkappa t$ contained in y. As a simultaneous indication of these two quantities, neglecting phase, Schuster takes $\sqrt{A_1^2 + B_1^2}$ which he supersedes in his later papers by the somewhat simpler expression $A_1^2 + B_1^2$.

It is possible to lend a certain plausibility to this later choice of Schuster as contrasted with his earlier, by considering the expression

$$\varphi(x) = \lim_{T \to \infty} \frac{1}{2T} \int_{-T}^{T} f(x+t) \bar{f}(t) \, dt. \tag{2. 02}$$

If

$$f(t) = \sum_1^N a_n e^{i\lambda_n t}, \tag{2. 03}$$

we have

$$\varphi(x) = \sum_1^N |a_n|^2 e^{i\lambda_n x}. \tag{2. 04}$$

Accordingly,

$$\varrho(\lambda_n) = |a_n|^2 = \lim_{T \to \infty} \frac{1}{2T} \int_{-T}^{T} \varphi(x) e^{-i\lambda_n x} dx. \tag{2.05}$$

This function $\varphi(t)$ differs from $f(t)$ in that every emplitude of a trigonometric term in $f(t)$ is replaced by the square of its modulus.

The expression $|a_n|^2$ is necessarily positive. It is, however, unobservable in any actual case, as we only have a finite interval of time at our disposal. Let it be noted that if we put

$$\varphi'_T(x) = \frac{1}{2T} \int_{-T}^{T} f(x+t) \bar{f}(t) \, dt \tag{2.06}$$

and

$$\varrho'_T(\lambda_n) = \frac{1}{2T} \int_{-T}^{T} \varphi'_T(x) e^{-i\lambda_n x} dx, \tag{2.07}$$

it is *not* necessarily true that ϱ'_T is non-negative. On the other hand, if we put

$$f_A(t) = f(t) \, [|t| < A]; \; f_A(t) = 0 \text{ otherwise} \tag{2.08}$$

and

$$\varphi_A(x) = \frac{1}{2A} \int_{-\infty}^{\infty} f_A(x+t) \bar{f}_A(t) \, dt \tag{2.09}$$

then

$$\varrho_A(\lambda_n) = \frac{1}{2A} \int_{-\infty}^{\infty} \varphi_A(x) e^{-i\lambda_n x} dx = \frac{1}{4A^2} \left| \int_{-\infty}^{\infty} f_A(t) e^{-i\lambda_n t} dt \right|^2 \geq 0. \tag{2.10}$$

This suggests an improved method of treating the approximate periodogram of a function under observation for a finite time.

The periodogram of a function — that is, the graph of the discontinuous function $\varrho(\lambda_n)$ or its approximate continuous analyses $\varrho_A(\lambda_n)$ — contains but a small amount of the information which the complete graph of the original function is able to yield. Not only do we deliberately discard all phase relations, but a large part of the original function — often the most interesting and important part — is thrown away as the aperiodic residue. The chief reason for this that any measure for a continuous spectral density becomes infinite at a spectral line, while any measure for the intensity of a spectral line becomes zero over the continuous spectrum.

This is a difficulty, however, which has had to be faced in many other branches of mathematics and physics. Impulses and forces are treated side by side in mechanics, although they have no common unit. We are familiar in potential theory with distributions of charge containing point, line, and surface distributions, as well as continuous volume distributions. The basic theory of all these problems is that of the Stieltjes integral.

Let us put

$$S(u) = \frac{1}{2\pi} \int_{-\infty}^{\infty} \varphi(x) \frac{e^{ixu} - 1}{ix} dx. \tag{2. 11}$$

Here the term 1 is introduced to cancel the singularity which we should otherwise find for $x=0$. We have formally and heuristically

$$\begin{aligned} S(u+0) - S(u-0) &= \lim_{\varepsilon \to 0} \int_{-\infty}^{\infty} \varphi(x) \frac{e^{ix(u+\varepsilon)} - e^{ix(u-\varepsilon)}}{2\pi i x} dx \\ &= \lim_{\varepsilon \to 0} \frac{1}{\pi} \int_{-\infty}^{\infty} \varphi(x) e^{iux} \frac{\sin \varepsilon x}{x} dx \\ &= \lim_{\eta \to 0} \frac{1}{\pi \eta} \int_{0}^{\eta} d\varepsilon \int_{-\infty}^{\infty} \varphi(x) e^{iux} \frac{\sin \varepsilon x}{x} dx \\ &= \lim_{\eta \to 0} \frac{\eta}{\pi} \int_{-\infty}^{\infty} \varphi(x) e^{iux} \frac{1 - \cos \eta x}{\eta^2 x^2} dx \\ &= \lim_{T \to \infty} \frac{1}{\pi T} \int_{-\infty}^{\infty} \varphi(x) e^{iux} \frac{T^2 \left(1 - \cos \frac{x}{T}\right)}{x^2} dx. \end{aligned} \tag{2. 12}$$

Now,

$$\frac{T^2 \left(1 - \cos \frac{x}{T}\right)}{x^2} \tag{2. 13}$$

is a positive function assuming the value 1/2 for $x=0$, with a graph with a scale in the x direction proportional to T, and with a finite integral. Hence, it does not seem amiss to consider $S(u+0)-S(u-0)$ except for a constant factor, as the same expression as $\varrho(u)$. We shall later verify this fact in more detail and with more rigor. On the other hand, again formally,

$$S'(u) = \frac{1}{2\pi} \int_{-\infty}^{\infty} \varphi(x) e^{ixu} \, dx. \tag{2. 14}$$

Thus in case $\varphi(x)$ is of too small an order of magnitude to possess a line spectrum, $S(u)$ still has a significance. We shall interpret its derivative as meaning the density of the continuous portion of the spectrum of $f(t)$.

The graph of $S(u)$ shall be called the *integrated periodogram* of $f(t)$. We shall show later that under very general conditions, it may be so chosen as to be a monotone non-decreasing curve. The amount of rise of this curve between the arguments indicates the total intensity of the part of the spectrum lying between the frequencies. This shift of our attention from the periodogram itself to the integrated periodogram, which is monotone but not necessarily everywhere differentiable, is as we have said of the same nature as the shift from $g(x)$ in

$$\int f(x) \, g(x) \, dx \tag{2. 15}$$

to $\alpha(x)$ in

$$\int f(x) \, d\alpha(x). \tag{2. 16}$$

I wish to remark in passing that the formulae for the integrated periodogram are at least as convenient for computational purposes as the formulae of the Schuster analysis, that the monotony of the intergrated periodogram avoids the possibility of overlooking important periods by an insufficient search, while it gives an immediate indication of empty parts of the spectrum which need no further exploration; and that the computation of $\varphi(x)$ and $S(u)$ may be performed by such instruments as the product integraph of V. Bush. I also wish to call attention to a practical study of these modified periodogram methods by Mr. G. W. Kenrick of the Massachusetts Institute of Technology.

CHAPTER II.

3. The spectrum of an arbitrary function of a single variable.

The present section is devoted to the rigorous delimitation and demonstration of the theorems heuristically indicated in section 2. Let $f(t)$ be a measurable function such that

$$\varphi(x)=\lim_{T\to\infty}\frac{1}{2T}\int_{-T}^{T} f(x+t)\bar{f}(t)\,dt \tag{3.01}$$

exists for every x. This is the sole assumption necessary in the present section. By the Schwarz inequality

$$|\varphi(x)|\leq \lim_{T\to\infty}\frac{1}{2T}\left\{\int_{-T}^{T}|f(x+t)|^2\,dt\int_{-T}^{T}|f(t)|^2\,dt\right\}^{1/2}. \tag{3.02}$$

It follows from this that $\varphi(x)$ is bounded. To show this, it is only necessary to prove that

$$\lim_{T\to\infty}\frac{1}{2T}\int_{-T}^{T}|f(x+t)|^2\,dt=\lim_{T\to\infty}\frac{1}{2T}\int_{-T}^{T}|f(t)|^2\,dt=\varphi(0). \tag{3.03}$$

We have

$$\left|\frac{1}{2T}\int_{-T}^{T}|f(x+t)|^2\,dt-\frac{1}{2T}\int_{-T}^{T}|f(t)|^2\,dt\right|$$

$$=\left|\frac{1}{2T}\int_{T}^{T+x}|f(t)|^2\,dt-\frac{1}{2T}\int_{-T}^{-T+x}|f(t)|^2\,dt\right|$$

$$\leq\left|\frac{1}{2T}\int_{T}^{T+x}|f(t)|^2\,dt+\frac{1}{2T}\int_{-T}^{-T+x}|f(t)|^2\,dt\right|$$

$$\leq\left|\frac{1}{2T}\int_{-T-x}^{T+x}|f(t)|^2\,dt-\frac{1}{2T}\int_{-T+x}^{T-x}|f(t)|^2\,dt\right|$$

$$=\left|\left(1+\frac{x}{T}\right)\frac{1}{2(T+x)}\int_{-T-x}^{T+x}|f(t)|^2\,dt-\left(1-\frac{x}{T}\right)\frac{1}{2(T-x)}\int_{-T+x}^{T-x}|f(t)|^2\,dt\right|. \tag{3.04}$$

134

Hence

$$\overline{\lim_{T\to\infty}}\left|\frac{1}{2T}\int_{-T}^{T}|f(x+t)|^2\,dt-\frac{1}{2T}\int_{-T}^{T}|f(t)|^2\,dt\right|$$

$$\leq\lim_{T\to\infty}\left|\left(1+\frac{x}{T}\right)\frac{1}{2(T+x)}\int_{-T-x}^{T+x}|f(t)|^2\,dt-\left(1-\frac{x}{T}\right)\frac{1}{2(T-x)}\int_{-T+x}^{T-x}|f(t)|^2\,dt\right|=0. \tag{3.05}$$

Therefore

$$|\varphi(x)|\leq\varphi(0), \tag{3.06}$$

and $\varphi(x)$ is bounded.

As before, we put

$$\varphi_A(x)=\frac{1}{2A}\int_{-\infty}^{\infty}f_A(x+t)\bar{f}_A(t)\,dt. \tag{3.07}$$

By the Schwarz inequality

$$|\varphi_A(x)|\leq\frac{1}{2A}\sqrt{\int_{-\infty}^{\infty}|f_A(x+t)|^2\,dt\int_{-\infty}^{\infty}|f_A(t)|^2\,dt}$$

$$=\frac{1}{2A}\int_{-\infty}^{\infty}|f_A(t)|^2\,dt, \tag{3.08}$$

and $\varphi_A(x)$ is uniformly bounded in x and A for all values of A larger than some given value. Furthermore, if $x>0$,

$$\varphi_A(x)=\frac{1}{2A}\int_{-A}^{A-x}f(x+t)\bar{f}(t)\,dt$$

$$=\frac{1}{2A}\int_{-A}^{A}f(x+t)\bar{f}(t)\,dt-\frac{1}{2A}\int_{A-x}^{A}f(x+t)\bar{f}(t)\,dt. \tag{3.09}$$

We shall have a similar formula in case x is negative. We have furthermore

$$\left|\frac{1}{2A}\int_{A-x}^{A} f(x+t)\bar{f}(t)\,dt\right| \le \frac{1}{2A}\sqrt{\int_{A}^{A+x} |f(t)|^2\,dt \int_{A-x}^{A} |f(t)|^2\,dt}$$

$$\le \frac{1}{2A}\left|\int_{A-x}^{A+x} |f(t)|^2\,dt\right|. \tag{3.10}$$

Since

$$\lim_{A\to\infty} \frac{1}{2A}\left|\int_{A-x}^{A+x} |f(t)|^2\,dt\right| = 0, \tag{3.11}$$

it follows at once that

$$\lim_{A\to\infty} \varphi_A(x) = \lim_{A\to\infty} \frac{1}{2A}\int_{-A}^{A} f(x+t)\bar{f}(t)\,dt = \varphi(x). \tag{3.12}$$

Thus $\varphi(x)$ is the limit of a uniformly bounded sequence of measurable functions, and is measurable. Since it is also bounded, it is quadratically summable over any finite range, while $\varphi(x)/x$ is quadratically summable over any range excluding the origin. It is, moreover, easy to prove that

$$\varphi(x) = \underset{A\to\infty}{\text{l.i.m.}}\ \varphi_A(x) \tag{3.13}$$

over any finite range, and that

$$\frac{\varphi(x)}{x} = \underset{A\to\infty}{\text{l.i.m.}}\ \frac{\varphi_A(x)}{x} \tag{3.14}$$

over any range excluding the origin. Hence,

$$\varphi(x)\frac{\sin\mu x}{x} = \underset{A\to\infty}{\text{l.i.m.}}\ \varphi_A(x)\frac{\sin\mu x}{x}. \tag{3.15}$$

In as much as the Fourier transformation leaves invariant the integral of the square of the modulus of a function, and hence leaves invariant all properties of convergence in the mean,

$$\underset{N\to\infty}{\text{l.i.m.}} \int_{-N}^{N} \varphi(x)\frac{\sin\mu x}{x}e^{iux}\,dx = \underset{A\to\infty}{\text{l.i.m.}} \int_{-\infty}^{\infty} \varphi_A(x)\frac{\sin\mu x}{x}e^{iux}\,dx$$

$$= \underset{A\to\infty}{\mathrm{l.i.m.}} \int_{-\infty}^{\infty} \varphi_A(x)\, e^{iux}\, dx \int_0^{\mu} \cos \xi x\, d\xi$$

$$= \frac{1}{2} \underset{A\to\infty}{\mathrm{l.i.m.}} \int_0^{\mu} d\xi \int_{-\infty}^{\infty} \varphi_A(x) [e^{i(u+\xi)x} + e^{i(u-\xi)x}]\, dx. \tag{3.16}$$

The inversion of the order of integration is justified as usual by the fact that the infinite integral is only apparently infinite. This, let me remark parenthetically, is the case also in the next set of formulae.

The last expression is the limit in the mean of a real non-negative quantity, for

$$\int_{-\infty}^{\infty} \varphi_A(x) e^{i\nu x}\, dx = \frac{1}{2A} \int_{-\infty}^{\infty} e^{i\nu x}\, dx \int_{-\infty}^{\infty} f_A(x+t) \bar{f}_A(t)\, dt$$

$$= \frac{1}{2A} \int_{-\infty}^{\infty} \bar{f}_A(t)\, dt \int_{-\infty}^{\infty} f_A(x+t)\, e^{i\nu x}\, dx$$

$$= \frac{1}{2A} \int_{-\infty}^{\infty} \bar{f}_A(t)\, dt \int_{-\infty}^{\infty} f_A(w)\, e^{i\nu(w-t)}\, dw$$

$$= \frac{1}{2A} \left| \int_{-\infty}^{\infty} f_A(w)\, e^{i\nu w}\, dw \right|^2 \geq 0. \tag{3.17}$$

The limit in the mean of a function is determined with the exception of a set of points of zero measure, but the limit in the mean of a non-negative function may always be so chosen as to be non-negative. If we make this choice,

$$\underset{N\to\infty}{\mathrm{l.i.m.}} \int_{-N}^{N} \varphi(x) \frac{\sin \mu x}{x} e^{iux}\, dx \geq 0. \tag{3.18}$$

The expression

$$\sigma_1(u) = \underset{A\to\infty}{\mathrm{l.i.m.}} \frac{1}{2\pi} \left[\int_1^A + \int_{-A}^{-1} \right] \varphi(x) \frac{e^{ixu}}{ix}\, dx \tag{3.19}$$

exists, as the Fourier transform of a quadratically summable function. Moreover,

$$\sigma_2(u) = \frac{1}{2\pi} \int_{-1}^{1} \varphi(x) \frac{e^{ixu} - 1}{ix} dx \tag{3. 20}$$

exists. If we put

$$\sigma(u) = \sigma_1(u) + \sigma_2(u) \tag{3. 21}$$

we have

$$\sigma(u+\mu) - \sigma(u-\mu) = \underset{A\to\infty}{\text{l.i.m.}} \frac{1}{\pi} \int_{-A}^{A} \varphi(x) \frac{\sin \mu x}{x} e^{iux} dx \geq 0. \tag{3. 22}$$

Of course, when we say that a limit in the mean is non-negative, we merely mean that it can be so chosen. Thus the expression $\sigma(u)$ is monotone, or at least can be so chosen, for example, by putting

$$\sigma(u) = \frac{d}{du} \int_0^u \sigma(u)\, du \tag{3. 23}$$

at every point where the latter expression is defined. Here we introduce (3. 23), because $\sigma(u)$ is now almost everywhere the limit of the difference quotient of $\int_0^u \sigma(u)\, du$, namely, $\frac{1}{2\varepsilon} \int_{u-\varepsilon}^{u+\varepsilon} \sigma(u)\, du$, which is monotone as a consequence of (3. 22). Thus, except at a set of zero measure, $\sigma(u)$ is the limit, not merely the limit in the mean, of a set of monotone functions, and is monotone. Elsewhere, at a set of zero measure, we put

$$\sigma(u) = \frac{1}{2} [\sigma(u+0) + \sigma(u-0)]. \tag{3. 24}$$

It follows that $\sigma(u+\mu) - \sigma(u-\mu)$ is of limited total variation over any finite interval. We shall show in the next paragraph that

$$\lim_{\mu\to\infty} [\sigma(u+\mu) - \sigma(u-\mu)]$$

is finite, and that hence $\sigma(u+\mu) - \sigma(u-\mu)$ is of limited total variation over $(-\infty, \infty)$. It is moreover, quadratically summable, as the Fourier transform of a quadratically summable function. It tends to 0 as $u \to \pm\infty$ and hence, by a theorem of Hobson[3], we have

$$\frac{1}{2}[\sigma(u+0+\mu)-\sigma(u+0-\mu)+\sigma(u-0+\mu)-\sigma(u-0-\mu)]$$

$$= \frac{1}{\pi}\int_{-\infty}^{\infty} \varphi(x)\frac{\sin \mu x}{x} e^{iux}\,dx. \tag{3.25}$$

In particular, if $u=\mu=v/2$,

$$\frac{1}{2}[\sigma(v+0)+\sigma(v-0)] - \frac{1}{2}[\sigma(+0)+\sigma(-0)] = \frac{1}{2\pi}\int_{-\infty}^{\infty} \varphi(x)\frac{e^{ivx}-1}{ix}\,dx. \tag{3.26}$$

If therefore we define

$$S(u) = \frac{1}{2\pi}\int_{-\infty}^{\infty} \varphi(x)\frac{e^{iux}-1}{ix}\,dx, \tag{3.27}$$

$S(u)$ will exist, and

$$S(u)-\sigma(u) = \text{constant}. \tag{3.28}$$

4. The total spectral intensity.

It is manifest that $\lim_{\mu\to\infty}[S(u+\mu)-S(u-\mu)]$, or as we shall write it, $S(\infty)-$ $-S(-\infty)$, if it exists, is a measure of the total spectral intensity of $f(x)$. We shall prove that this quantity exists and is finite.

We have

$$\frac{1}{A}\int_0^A [\sigma(u+\mu)-\sigma(u-\mu)]\,d\mu = \frac{1}{\pi A}\int_0^A d\mu \operatorname*{l.i.m.}_{B\to\infty}\int_{-B}^{B} \varphi(x)\frac{\sin \mu x}{x} e^{iux}\,dx. \tag{4.01}$$

The limit in the mean is here taken with u as the fundamental variable, and with μ as parameter. It is not difficult to deduce from the boundedness of

$$\int_0^A d\mu \int_{-\infty}^{\infty} |\varphi(x)|^2 \frac{\sin^2 \mu x}{x^2}\,dx$$

that we may invert the order of integration, and get

18—29764. *Acta mathematica.* 55. Imprimé le 7 avril 1930.

$$\frac{1}{A}\int_0^A [\sigma(u+\mu)-\sigma(u-\mu)]\,d\mu = \frac{1}{\pi A}\ \underset{B\to\infty}{\mathrm{l.i.m.}}\int_{-B}^{B}\varphi(x)\frac{1-\cos Ax}{x^2}e^{iux}\,dx. \tag{4.02}$$

To show this, let us remark that

$$\int_{-\infty}^{\infty}\left|\frac{1}{\pi A}\int_0^A d\mu\ \underset{B\to\infty}{\mathrm{l.i.m.}}\int_{-B}^{B}\varphi(x)\frac{\sin\mu x}{x}e^{iux}\,dx-\frac{1}{\pi A}\int_0^A d\mu\int_{-C}^{C}\varphi(x)\frac{\sin\mu x}{x}e^{iux}\,dx\right|^2 du$$

$$=\frac{1}{\pi^2A^2}\int_{-\infty}^{\infty}\left|\int_0^A d\mu\ \underset{B\to\infty}{\mathrm{l.i.m.}}\left[\int_C^B+\int_{-B}^{-C}\right]\varphi(x)\frac{\sin\mu x}{x}e^{iux}\,dx\right|^2 du$$

$$\leq\frac{1}{\pi^2A^2}\int_{-\infty}^{\infty}du\int_0^A d\mu_1\int_0^A\left|\underset{B\to\infty}{\mathrm{l.i.m.}}\left[\int_C^B+\int_{-B}^{-C}\right]\varphi(x)\frac{\sin\mu x}{x}e^{iux}\,dx\right|^2 d\mu$$

$$=\frac{1}{\pi^2A}\int_0^A d\mu\int_{-\infty}^{\infty}du\left|\underset{B\to\infty}{\mathrm{l.i.m.}}\left[\int_C^B+\int_{-B}^{-C}\right]\varphi(x)\frac{\sin\mu x}{x}e^{iux}\,dx\right|^2$$

$$=\frac{2}{\pi A}\int_0^A d\mu\left[\int_C^{\infty}+\int_{-\infty}^{-C}\right]|\varphi(x)|^2\frac{\sin^2\mu x}{x^2}\,dx. \tag{4.03}$$

Inasmuch as this latter expression tends to 0 with increasing C,

$$\frac{1}{\pi A}\int_0^A d\mu\ \underset{B\to\infty}{\mathrm{l.i.m.}}\int_{-B}^{B}\varphi(x)\frac{\sin\mu x}{x}e^{iux}\,dx=\underset{C\to\infty}{\mathrm{l.i.m.}}\ \frac{1}{\pi A}\int_0^A d\mu\int_{-C}^{C}\varphi(x)\frac{\sin\mu x}{x}e^{iux}\,dx$$

$$=\underset{B\to\infty}{\mathrm{l.i.m.}}\ \frac{1}{\pi A}\int_{-B}^{B}\varphi(x)\frac{e^{iux}}{x}\,dx\int_0^A\sin\mu x\,d\mu$$

$$=\frac{1}{\pi A}\ \underset{B\to\infty}{\mathrm{l.i.m.}}\int_{-B}^{B}\varphi(x)\frac{1-\cos Ax}{x^2}e^{iux}\,dx, \tag{4.04}$$

thus proving our statement.

Our limit in the mean may be replaced by an ordinary limit, as this limit exists, owing to the boundedness of $\varphi(x)$. Therefore

$$\lim_{A\to\infty} \frac{1}{A} \int_0^A [\sigma(u+\mu)-\sigma(u-\mu)]\,d\mu = \lim_{A\to\infty} \frac{1}{\pi A} \int_{-\infty}^{\infty} \varphi(x)\, \frac{1-\cos Ax}{x^2}\, e^{iux}\, dx. \qquad (4.05)$$

It follows from the monotony of $\sigma(u+\mu)-\sigma(u-\mu)$ in μ that we may write

$$\lim_{\mu\to\infty} [\sigma(u+\mu)-o(u-\mu)] = \lim_{A\to\infty} \frac{1}{\pi} \int_{-\infty}^{\infty} \varphi\left(\frac{x}{A}\right) e^{iu\frac{x}{A}} \frac{1-\cos x}{x^2}\, dx$$

$$\leq \max \left|\varphi\left(\frac{x}{A}\right)\right| \frac{1}{\pi} \int_{-\infty}^{\infty} \frac{1-\cos x}{x^2}\, dx$$

$$= \varphi(0). \qquad (4.06)$$

This yields us the existence of $\sigma(\infty)-\sigma(-\infty)$ and hence, according to the last paragraph, of $S(\infty)-S(-\infty)$. We have

$$S(\infty)-S(-\infty) = \lim_{A\to\infty} \frac{1}{\pi} \int_{-\infty}^{\infty} \varphi\left(\frac{x}{A}\right) \frac{1-\cos x}{x^2}\, dx$$

$$= \varphi(0) + \lim_{A\to\infty} \frac{1}{\pi} \int_{-\infty}^{\infty} \left[\varphi\left(\frac{x}{A}\right) - \varphi(0)\right] \frac{1-\cos x}{x^2}\, dx. \qquad (4.07)$$

Hence for sufficiently large A

$$|S(\infty)-S(-\infty)-\varphi(0)|$$

$$\leq \frac{2}{\pi} \max |\varphi(\xi)| \left[\int_{A^{1/2}}^{\infty} + \int_{-\infty}^{-A^{1/2}}\right] \frac{1-\cos x}{x^2}\, dx + \max_{|\xi|<A^{-1/2}} |\varphi(\xi)-\varphi(0)| + \varepsilon. \qquad (4.08)$$

Since A is arbitrary,

$$|S(\infty)-S(-\infty)-\varphi(0)| \leq \overline{\lim_{|\xi|\to 0}} |\varphi(\xi)-\varphi(0)|. \qquad (4.09)$$

In case $\varphi(x)$ is continuous at the origin,

$$\varphi(0) = S(\infty)-S(-\infty). \qquad (4.10)$$

However, $\varphi(x)$ need not be continuous at the origin, even if $f(t)$ is everywhere continuous. Thus let $f(t)=\sin t^2$.

Then

$$\varphi(0)=\lim_{T\to\infty}\frac{1}{2T}\int_{-T}^{T}\sin^2 t^2\,dt$$

$$=\lim_{T\to\infty}\frac{1}{T}\int_{0}^{T}\frac{1-\cos 2t^2}{2}\,dt$$

$$=\frac{1}{2}-\lim_{T\to\infty}\frac{\sqrt{2}}{T}\int_{0}^{T\sqrt{2}}\cos u^2\,du$$

$$=\frac{1}{2}. \tag{4.11}$$

since $\int_0^\infty \cos u^2\,du$ is a Fresnel integral, and equals $\frac{1}{2}\sqrt{\frac{\pi}{2}}$.

On the other hand, if $x\neq 0$, we have

$$\varphi(x)=\lim_{T\to\infty}\frac{1}{2T}\int_{-T}^{T}\sin(t+x)^2\sin t^2\,dt$$

$$=\lim_{T\to\infty}\frac{1}{4T}\int_{-T}^{T}[-\cos(2t^2+2tx+x^2)+\cos(2tx+x^2)]\,dt.$$

The second part of this mean obviously vanishes. Hence

$$\varphi(x)=-\lim_{T\to\infty}\frac{1}{4T}\int_{-T}^{T}\cos(2t^2+2tx+x^2)\,dt$$

$$=-\lim_{U\to\infty}\frac{1}{2U}\int_{0}^{U}\cos\left(u^2+\frac{x^2}{2}\right)du$$

$$= 0 \left[\cos \frac{x^2}{2} \int_0^\infty \cos u^2 \, du - \sin \frac{x^2}{2} \int_0^\infty \sin u^2 \, du \right]$$

$$= 0. \tag{4.12}$$

since $\int_0^\infty \cos u^2 \, du = \int_0^\infty \sin u^2 \, du = \frac{1}{2} \sqrt{\frac{\pi}{2}}$.

Thus $\varphi(x)$ vanishes almost everywhere, $S(u)$ vanishes identically, and

$$\varphi(0) \neq S(\infty) - S(-\infty). \tag{4.13}$$

5. Tauberian theorems and spectral intensity.

In a recent paper, the author has proved the following general Tauberian theorem: *Let $M_1(x)$ and $M_2(x)$ be two functions bounded over every range $(\varepsilon, 1/\varepsilon)$, which are $O\left(\frac{1}{x(\log x)^2}\right)$ at 0 and ∞. Let $M_1(x)$ be measurable and non-negative, and let*

$$\int_0^\infty M_1(x) \, x^{iu} \, dx \neq 0. \qquad [-\infty < u < \infty] \tag{5.01}$$

Let $M_2(x)$ be continuous, except for a finite number of finite jumps. Let $f(x)$ be a measurable function bounded below and such that

(a) $$\lim_{\lambda \to 0 [\infty]} \int_0^\infty f(\lambda x) \, M_1(x) \, dx = A \int_0^\infty M_1(x) \, dx;$$

(b) $$\int_0^\infty f(\lambda x) \, M_1(x) \, dx \text{ is bounded.} \qquad [0 < \lambda < \infty]$$

Then

$$\lim_{\lambda \to 0 [\infty]} \int_0^\infty f(\lambda x) \, M_2(x) \, dx = A \int_0^\infty M_2(x) \, dx. \tag{5.02}$$

Here ∞ is put into brackets to indicate that at these points it may be consistently substituted for 0. There is manifestly no restriction in assuming

$f(x)$ non-negative, as the theorem, if true for a given $f(x)$, is unchanged as to its validity by the addition to $f(x)$ of a constant. The theorem assumes a more understandable form under the transformations

$$\left.\begin{array}{l} x = e^{\xi}; \quad \lambda = e^{-\eta}; \\ e^{\xi} M_1(e^{\xi}) = N_1(\xi); \quad e^{\xi} M_2(e^{\xi}) = N_2(\xi). \end{array}\right\} \tag{5.03}$$

It then becomes: *Let* $N_1(\xi)$ *and* $N_2(\xi)$ *be two bounded functions which are* $O(\xi^{-2})$ *at* $\pm\infty$. *Let* $N_1(\xi)$ *be measurable and non-negative, and let*

$$\int_{-\infty}^{\infty} N_1(\xi)\, e^{iu\xi}\, d\xi \neq 0. \qquad [-\infty < u < \infty] \tag{5.04}$$

Let $N_2(\xi)$ *be continuous, except for a finite number of finite jumps. Let* $g(\xi)$ *be a non-negative measurable function such that*

(a) $$\lim_{\eta \to \infty} \int_{-\infty}^{\infty} g(\eta - \xi)\, N_1(\xi)\, d\xi = A \int_{-\infty}^{\infty} N_1(\xi)\, d\xi;$$

(b) $$\int_{-\infty}^{\infty} g(\eta - \xi)\, N_1(\xi)\, d\xi \text{ is bounded.} \qquad [-\infty < \eta < \infty]$$

Then

(c) $$\lim_{\eta \to \infty} \int_{-\infty}^{\infty} g(\eta - \xi)\, N_2(\xi)\, d\xi = A \int_{-\infty}^{\infty} N_2(\xi)\, d\xi. \tag{5.05}$$

The proof proceeds as follows: We shall symbolize by C the class of all functions $N_2(\xi)$, bounded and $O(\xi^{-2})$ at $\pm\infty$, and continuous except for a finite number of finite jumps, for which (c) is a consequence of (a) and (b) for all non-negative measurable functions $g(\xi)$. Among the functions in C are all functions $N_2(\xi)$ of the form

$$N_2(\xi) = \int_{-\infty}^{\infty} N_1(\eta)\, R(\eta - \xi)\, d\eta \tag{5.06}$$

for which $\int_{-\infty}^{\infty} |R(\eta)|\, d\eta$ converges, inasmuch as the double integral

$$\int_{-\infty}^{\infty} R(\zeta)\,d\zeta \int_{-\infty}^{\infty} g(\eta-\xi-\zeta)\,N_1(\xi)\,d\xi$$

is absolutely convergent, so that

$$\int_{-\infty}^{\infty} N_2(\xi)\,g(\eta-\xi)\,d\xi = \int_{-\infty}^{\infty} g(\eta-\xi)\,d\xi \int_{-\infty}^{\infty} N_1(\zeta+\xi)\,R(\zeta)\,d\zeta$$

$$= \int_{-\infty}^{\infty} R(\zeta)\,d\zeta \int_{-\infty}^{\infty} g(\eta-\xi-\zeta)\,N_1(\xi)\,d\xi, \tag{5.07}$$

and

$$\lim_{\eta\to\infty} \int_{-\infty}^{\infty} N_2(\xi)\,g(\eta-\xi)\,d\xi = \lim_{\eta\to\infty} \int_{-\infty}^{\infty} R(\zeta)\,d\zeta \int_{-\infty}^{\infty} g(\eta-\xi-\zeta)\,N_1(\xi)\,d\xi$$

$$= A \int_{-\infty}^{\infty} R(\zeta)\,d\zeta \int_{-\infty}^{\infty} N_1(\xi)\,d\xi = A \int_{-\infty}^{\infty} N_2(\xi)\,d\xi. \tag{5.08}$$

A particular example of such a function is furnished by

$$N_2(\xi) = \int_{-B}^{B} \nu_2(u)\,e^{-iu\xi}\,du \tag{5.09}$$

where $\nu_2(u)$ is continuous over $(-B, B)$, while its first derivative is continuous except for a finite number of discontinuities of the first kind, and

$$\nu_2(B) = \nu_2(-B) = 0. \tag{5.10}$$

To prove this, let us reflect that $N_1(\xi)$ and $N_2(\xi)$ are quadratically summable by assumption, and that

$$\nu_1(u) = \int_{-\infty}^{\infty} N_1(\xi)\,e^{iu\xi}\,d\xi \tag{5.11}$$

exists, as well as

$$\nu'_1(u) = \operatorname*{l.i.m.}_{E\to\infty} \int_{-E}^{E} i\xi N_1(\xi)\,e^{iu\xi}\,d\xi. \tag{5.12}$$

By our hypothesis (5.04)

$$\nu_1(u) \neq 0. \qquad [-\infty < u < \infty] \tag{5.13}$$

Let us put

$$\mu(u) = \nu_2(u)/\nu_1(u). \tag{5.14}$$

Inasmuch as $\mu(u)$ is absolutely continuous, its derivative may be computed by the rules, and

$$\mu'(u) = \frac{\nu_1(u)\,\nu'_2(u) - \nu_2(u)\,\nu'_1(u)}{[\nu_1(u)]^2}. \tag{5.15}$$

I now say that we shall have

$$R(\zeta) = \int_{-B}^{B} \mu(u)\, e^{iu\zeta}\, du$$

$$= \frac{1}{i\zeta} \int_{-B}^{B} \mu'(u)\, e^{iu\zeta}\, du. \tag{5.16}$$

Inasmuch as

$$\int_{-B}^{B} \mu'(u)\, e^{iu\zeta}\, du \tag{5.17}$$

is quadratically summable,

$$\int_{-\infty}^{\infty} |R(\zeta)|\, d\zeta \tag{5.18}$$

exists. Since the integrals involved converge absolutely,

$$\int_{-\infty}^{\infty} N_1(\eta)\, R(\eta - \xi)\, d\eta = \int_{-\infty}^{\infty} N_1(\eta)\, d\eta \int_{-B}^{B} \mu(u)\, e^{iu(\eta-\xi)}\, du$$

$$= \int_{-B}^{B} \mu(u)\, e^{-iu\xi}\, du \int_{-\infty}^{\infty} N_1(\eta) e^{iu\eta}\, d\eta$$

$$= \int_{-B}^{B} \mu(u)\, \nu_1(u)\, e^{-iu\xi}\, du$$

$$= \int_{-B}^{B} \nu_2(u)\, e^{-iu\xi}\, du = N_2(\xi). \tag{5.19}$$

This justifies our evaluation of $R(\zeta)$, and proves that $N_2(\xi)$ belongs to C. The following are particular cases which may be probed to belong to C in this manner:

$$T_B(\xi) = \int_{-B}^{B} [e^{-|u|} - e^{-B}] e^{-i\xi u} du$$

$$= \int_0^B [e^{-u} - e^{-B}] \cos \xi u \, du$$

$$= \frac{1}{\xi} \int_0^B e^{-u} \sin \xi u \, du$$

$$= \frac{1 - e^{-B} \cos B\xi}{1 + \xi^2} - \frac{e^{-B} \sin B\xi}{\xi(1 + \xi^2)}. \tag{5.20}$$

Again,

$$Q_B(\xi) = \frac{1}{\pi} \int_{-B}^{B} \left(1 - \frac{|u|}{B}\right) \frac{\sin u}{u} e^{-iu\xi} du$$

$$= \int_{B(\xi-1)}^{B(\xi+1)} \frac{1 - \cos z}{\pi z^2} dz$$

$$= \frac{1}{\pi B(\xi - 1)} - \frac{1}{\pi B(\xi + 1)} + O\left(\frac{1}{B^2 \xi^2}\right). \qquad [B\xi \to \pm \infty] \tag{5.21}$$

If we already know certain members of the class C, we may obtain new members of the class in the following manner: Let $V(\xi)$ be a function continuous, except for a finite number of finite jumps, such that, when any positive ξ is given, we can find two members of C, $V_1(\xi)$ and $V_2(\xi)$, such that

$$V_1(\xi) \leq V(\xi) \leq V_2(\xi), \tag{5.22}$$

while

$$\int_{-\infty}^{\infty} [V_2(\xi) - V_1(\xi)] \, d\xi < \varepsilon. \tag{5.23}$$

Then $V(\xi)$ itself belongs to C. For

19—29764. *Acta mathematica*. 55. Imprimé le 7 avril 1930.

$$\overline{\lim_{\eta\to\infty}}\left|\int_{-\infty}^{\infty} g(\eta-\xi)\, V(\xi)\, d\xi - A\int_{-\infty}^{\infty} V(\xi)\, d\xi\right|$$

$$\leq \overline{\lim_{\eta\to\infty}}\left|\int_{-\infty}^{\infty} g(\eta-\xi)\, V_1(\xi)\, d\xi - A\int_{-\infty}^{\infty} V_2(\xi)\, d\xi\right|$$

$$+ \overline{\lim_{\eta\to\infty}}\left|\int_{-\infty}^{\infty} g(\eta-\xi)\, V_2(\xi)\, d\xi - A\int_{-\infty}^{\infty} V_1(\xi)\, d\xi\right|$$

$$< 2A\varepsilon. \tag{5.24}$$

and since ε is arbitrarily small, this limit is 0. Furthermore, any linear combination of a finite number of members of C belongs to C.

As a particular case, we have

$$[1+(1+B)e^{-B}]^{-1}\, T_B(\xi) < \frac{1}{1+\xi^2} < [1-(1-B)\, e^{-B}]^{-1}\, T_B(\xi). \tag{5.25}$$

Inasmuch as

$$\lim_{B\to\infty}(1+B)\, e^{-B} = 0, \tag{5.26}$$

it follows at once that $\frac{1}{1+\xi^2}$ belongs to C. An exactly similar proof will show that the same thing holds of

$$\frac{p}{(\xi-q)^2-r^2}.$$

Again,

$$\lim_{A\to\infty} Q_A(\xi) = \operatorname{sgn}(\xi+1) - \operatorname{sgn}(\xi-1) = V(\xi), \tag{5.27}$$

and this convergence is uniform except in the neighborhood of ± 1 while we always have for $B > 0$

$$Q_A(\xi) < \frac{1}{B(\xi^2+1)} \text{ over } (1+\eta,\, \infty) \text{ and } (-\infty,\, -1-\eta) \quad [A \text{ large enough}] \tag{5.28}$$

Furthermore,

$$\int_{-\infty}^{\infty} Q_A(\xi)\, d\xi = 4. \tag{5.29}$$

Let us put

$$\left.\begin{aligned}(1+\eta)\,Q_A(\xi(1-\eta)) &= V_2(\xi);\\(1-\eta)\,Q_A(\xi(1+\eta)) - \frac{1}{B(\xi^2+1)} &= V_1(\xi).\end{aligned}\right\} \tag{5.30}$$

We can so determine a large A when η and B are given, that for that A and all larger ones,

$$V_2(\xi) \geq V(\xi) \geq V_1(\xi). \tag{5.22}$$

We have

$$\int_{-\infty}^{\infty} [V_2(\xi) - V_1(\xi)]\,d\xi = 2\left[\frac{1+\eta}{1-\eta} - \frac{1-\eta}{1+\eta}\right] + \frac{\pi}{B} \tag{5.31}$$

which we may make arbitrarily small. Hence

$$\operatorname{sgn}(\xi+1) - \operatorname{sgn}(\xi-1)$$

belongs to C. As an immediate consequence, since

$$\operatorname{sgn}(\xi+\alpha) - \operatorname{sgn}(\xi+\beta)$$

may be shown by the same means to belong to C, any step function vanishing for large positive and negative arguments belongs to C, and hence any function continuous except for a finite set of discontinuities of the first kind, and vanishing outside of a finite interval, since the latter function may be penned in between two step functions enclosing an arbitrarily small area.

Now let $N_2(\xi)$ be a bounded function which is $O(\xi^{-2})$ at $\pm\infty$, and which is continuous except for a finite number of finite jumps. Let

$$|N_2(\xi)| < P/(\xi^2+1), \tag{5.32}$$

for all ξ. We put

$$V_1(\xi) = \begin{cases} N_2(\xi); & [|\xi| < M] \\ -P/(\xi^2+1); & [|\xi| \geq M] \end{cases} \tag{5.33}$$

$$V_2(\xi) = \begin{cases} N_2(\xi); & [|\xi| < M] \\ P/(\xi^2+1). & [|\xi| \geq M] \end{cases} \tag{5.34}$$

The functions $V_1(\xi)$ and $V_2(\xi)$ are sums of functions of C and functions of the form $\pm P/(\xi^2+1)$ which also belong to C. Hence, they themselves belong to C. We have

$$V_2(\xi) > N_2(\xi) > V_1(\xi) \tag{5.35}$$

and

$$\int_{-\infty}^{\infty} [V_2(\xi) - V_1(\xi)]\, d\xi < 2P[\pi - \tan^{-1} M] \tag{5.36}$$

which we may make as small as we like. Hence $N_2(\xi)$ belongs to C. This concludes the proof of our generalized Tauberian theorem.

As a corollary of our Tauberian theorem, Mr. S. B. Littauer has given a proof of the following theorem of Jacob: *If $f(t)$ is a measurable function, integrable in every finite interval of* $(0, \infty)$ *and if for some given* α $(0 \leq \alpha < 1)$

(a) $$\frac{1}{T^{1-\alpha}} \int_0^T |f(t)|\, dt < B \text{ for every } T;$$

(b) $$\lim_{T\to\infty} \frac{1}{T^{1-\alpha}} \int_0^T f(t)\, dt = A;$$

then

(c) $$\lim_{\varepsilon\to 0} \frac{\varepsilon^{1-\alpha}}{\gamma_\alpha \pi} \int_0^\infty f(t) \left(\frac{\sin \varepsilon t}{\varepsilon t}\right)^2 dt = A,$$

where

$$\gamma_\alpha = \frac{2^{\alpha-1}(1-\alpha)}{\Gamma(2+\alpha)\cos\frac{\pi\alpha}{2}}.$$

Furthermore, if $f(t)$ is measurable and non-negative (*or bounded below*), (c) *implies* (b).

The particular case of this theorem where $\alpha=0$ had already been treated by Bochner, Hardy, and the present author.

In all theorems of this type, there is a close relation between the theorem which one obtains by letting λ become infinite and that which one obtains by letting λ become 0. This is to be explained by the fact that the general Tauberian theorem assumes a perfectly symmetric form when we make the substitutions

$$x=e^\xi;\ \lambda=e^\eta;\ M_1(x)=N_1(\xi)\,e^\xi;\ M_2(x)=N_2(\xi)\,e^\xi. \tag{5.03}$$

If we take

$$\left.\begin{aligned} M_1(x)\,[\text{or } M_2(x)] &= 1 \text{ if } 0 < x < 1;\ = 0 \text{ otherwise;} \\ M_2(x)\,[\text{or } M_1(x)] &= \frac{4\sin^2 x/2}{x^2} \end{aligned}\right\} \tag{5.37}$$

150

in our general Tauberian theorem, since $\int_0^\infty M_1(x)\, e^{iux}\, dx \neq 0$ and $\int_0^\infty M_2(x)\, e^{iux}\, dx \neq 0$, we may deduce the conclusions of the theorem. We thus get the following result: *Let*

$$|f(x)| < B. \qquad [0 < x < \infty]$$

Then the two propositions,

(a)
$$\lim_{T\to 0} \frac{1}{T} \int_0^T f(x)\, dx = A$$

and

(b)
$$\lim_{T\to 0} \frac{2}{\pi} \int_0^\infty f(Tx) \frac{1-\cos x}{x^2}\, dx = A$$

are equivalent.

In the particular case where $f(x)$ is replaced by $\frac{1}{2}[\varphi(x)+\varphi(-x)]$, we see that

$$A = \lim_{r\to\infty} \frac{1}{\pi} \int_{-\infty}^{\infty} \varphi\left(\frac{x}{r}\right) \frac{1-\cos x}{x^2}\, dx \tag{5.38}$$

implies, and is implied by

$$A = \lim_{\varepsilon\to 0} \frac{1}{2\varepsilon} \int_{-\varepsilon}^{\varepsilon} \varphi(x)\, dx. \tag{5.39}$$

We have (see (3.27))

$$\int_{-\infty}^{\infty} e^{-i\lambda u}\, dS(u) = \lim_{\mu\to 0} \frac{1}{2\pi\mu} \int_{-\infty}^{\infty} e^{-i\lambda u}\, du \operatorname*{l.i.m.}_{A\to\infty} \int_{-A}^{A} \varphi(x) \frac{\sin \mu x}{x}\, e^{iux}\, dx$$

$$= \lim_{\mu\to 0} \frac{1}{\mu\lambda} \varphi(\lambda) \sin \mu\lambda = \varphi(\lambda), \tag{5.40}$$

except possibly at a set of points of zero measure. To see this, it is only necessary to reflect that it follows from the definition of the Stieltjes integral that if $\alpha(x)$ is of limited total variation over $(-\infty, \infty)$,

$$\int_{-\infty}^{\infty} f(x)\, d\alpha(x) = \lim_{\varepsilon\to 0} \sum_{-\infty}^{\infty} f(u+2n\varepsilon)[\alpha(u+(2n+1)\varepsilon)-\alpha(u+(2n-1)\varepsilon)]$$

$$= \lim_{\varepsilon \to 0} \frac{1}{2\varepsilon} \int_{-\varepsilon}^{\varepsilon} du \sum_{-\infty}^{\infty} f(u+2n\varepsilon)\,[\alpha(u+(2n+1)\varepsilon)-\alpha(u+(2n-1)\varepsilon)]$$

$$= \lim_{\varepsilon \to 0} \frac{1}{2\varepsilon} \int_{-\varepsilon}^{\varepsilon} f(u)\,[\alpha(u+\varepsilon)-\alpha(u-\varepsilon)]\,du. \tag{5.41}$$

Let us put

$$\Phi(\lambda) = \int_{-\infty}^{\infty} e^{-i\lambda u}\, d\,S(u). \tag{5.42}$$

This function will be defined for all real arguments, and we shall have

$$\Phi(\lambda+\varepsilon)-\Phi(\lambda) = -2\,i \int_{-\infty}^{\infty} e^{-iu\left(\lambda+\frac{\varepsilon}{2}\right)} \sin\frac{u\,\varepsilon}{2}\, d\,S(u). \tag{5.43}$$

Since the function $\sin u\varepsilon/2$ is uniformly bounded, and tends to 0 over every finite range of u as $\varepsilon \to 0$, while $e^{-iu\left(\lambda+\frac{\varepsilon}{2}\right)}$ has modulus 1, it follows that $\Phi(\lambda+\varepsilon)-\Phi(\lambda)$ is less than the sum of two terms, one of which is the total variation of $S(u)$ over a region receding towards infinity, while the other is less than the total variation of $S(u)$ multiplied by a factor tending to 0. Hence

$$\lim_{\varepsilon \to 0} [\Phi(\lambda+\varepsilon)-\Phi(\lambda)] = 0. \tag{5.44}$$

Thus the function $\Phi(\lambda)$ is continuous, and indeed, this proof shows it to be uniformly continuous. Hence

$$\lim_{\varepsilon \to 0} \frac{1}{2\varepsilon} \int_{x-\varepsilon}^{x+\varepsilon} \varphi(\xi)\,d\xi = \lim_{\varepsilon \to 0} \frac{1}{2\varepsilon} \int_{x-\varepsilon}^{x+\varepsilon} \Phi(\xi)\,d\xi = \Phi(x). \tag{5.45}$$

This gives another proof that

$$\lim_{\varepsilon \to 0} \frac{1}{2\varepsilon} \int_{-\varepsilon}^{\varepsilon} \varphi(\xi)\,d\xi = S(\infty)-S(-\infty), \tag{5.46}$$

and indeed proves considerably more.

It is thus possible to dispense with Tauberian theorems for this part of the theory. There is another point, however, where they play a more essential rôle. That is in the study of the generalized Fourier transform of a function. Let $\lim_{T\to\infty} \frac{1}{2T}\int_{-T}^{T} |f(x)|^2\,dx$ exist. Then

$$\left[\int_1^T + \int_{-T}^{-1}\right]\frac{|f(x)|^2}{x^2}\,dx = \int_1^T \frac{|f(x)|^2 + |f(-x)|^2}{x^2}\,dx$$

$$= \int_1^T \frac{1}{x^2}\,d\int_0^x [|f(\xi)|^2 + |f(-\xi)|^2]\,d\xi$$

$$= \frac{1}{T^2}\int_{-T}^{T}|f(x)|^2\,dx - \int_{-1}^{1}|f(x)|^2\,dx + \int_1^T \frac{2\,dx}{x^3}\int_{-x}^{x}|f(\xi)|^2\,d\xi$$

$$= O(1) + \int_1^T 2\,O(1)\frac{dx}{x^2}$$

$$= O(1). \tag{5.47}$$

Consequently

$$\left[\int_1^\infty + \int_{-\infty}^{-1}\right]\frac{|f(x)|^2}{x^2}\,dx \tag{5.48}$$

exists. It follows from this that

$$\psi_\mu(u) = \frac{1}{\pi}\operatorname*{l.i.m.}_{A\to\infty}\int_{-A}^{A} f(x)\frac{\sin \mu x}{x}\,e^{iux}\,dx \tag{5.49}$$

exists, and that

$$\psi_\mu(u) = s(u+\mu) - s(u-\mu), \tag{5.50}$$

where

$$s(u) = \frac{1}{2\pi}\int_{-1}^{1} f(x)\frac{e^{iux}-1}{ix}\,dx + \frac{1}{2\pi}\operatorname*{l.i.m.}_{A\to\infty}\left[\int_1^A + \int_{-A}^{-1}\right]\frac{f(x)\,e^{iux}}{ix}\,dx. \tag{5.51}$$

$s(u)$ has a somewhat artificial appearance, due to the fact that it is necessary to avoid the consequences of the vanishing of the denominator at the origin.

We shall see later, however, that we always actually work with $\psi_\mu(u)$ rather than with $s(u)$.

As a result of the Plancherel theory,

$$\frac{1}{2\mu}\int_{-\infty}^{\infty} |s(u+\mu)-s(u-\mu)|^2\,du = \frac{1}{\pi\mu}\int_{-\infty}^{\infty} |f(x)|^2 \frac{\sin^2 \mu x}{x^2}\,dx. \tag{5.52}$$

It follows from this by an immediate application of the Tauberian theorem associated with the names of Bochner, Hardy, Jacob, Littauer, and the author, and already proved in this section, that

$$\lim_{\mu\to 0}\frac{1}{2\mu}\int_{-\infty}^{\infty} |s(u+\mu)-s(u-\mu)|^2\,du = \lim_{T\to\infty}\frac{1}{2T}\int_{-T}^{T} |f(x)|^2\,dx. \tag{5.53}$$

The meaning of (5.53) is that *if* $f(x)$ *is quadratically summable over every finite range, and* $f(x)/x$ *is quadratically summable over any infinite range excluding the origin, then if either side of* (5.53) *exists, the other side exists and assumes the same value.*

This formula is worthy of some detailed attention. If $s(u)$ is of limited total variation, we shall always have

$$\frac{1}{2\mu}\int_{-\infty}^{\infty} |s(u+\mu)-s(u-\mu)|\,du \le V(s). \tag{5.54}$$

Accordingly, if in addition

$$\lim_{T\to\infty}\frac{1}{2T}\int_{-T}^{T} |f(x)|^2\,dx \neq 0, \tag{5.55}$$

the function $s(u)$ cannot be uniformly continuous. Again, if

$$s(u) = A_n, \quad [\lambda_n < u < \lambda_{n+1}] \tag{5.56}$$

we shall have

$$\lim_{\mu\to 0}\frac{1}{2\mu}\int_{-\infty}^{\infty} |s(u+\mu)-s(u-\mu)|^2\,du = \Sigma\,|A_{n+1}-A_n|^2, \tag{5.57}$$

so that $\lim_{T\to\infty} \frac{1}{2T}\int_{-T}^{T} |f(x)|^2\, dx$ represents the sum of the squares of the moduli of the jumps of $s(u)$. Let it be noted that if $f(x)$ is a periodic function with the period 2π,

$$\lim_{T\to\infty} \frac{1}{2T}\int_{-T}^{T} |f(x)|^2\, dx = \frac{1}{2\pi}\int_{-\pi}^{\pi} |f(x)|^2\, dx, \tag{5.58}$$

while if

$$a_n = \frac{1}{2\pi}\int_{-\pi}^{\pi} f(x)\, e^{inx}\, dx, \tag{5.59}$$

then

$$\psi_\mu(u) = s(u+\mu) - s(u-\mu) = \sum_{[u-\mu]+1}^{[u+\mu]} a_n. \tag{5.60}$$

Thus our formula (5.53) is a generalization of the Parseval formula for the Fourier series, though it is not a direct generalization of the Parseval formula for the Fourier integral. For the Fourier integral,

$$\lim_{\mu\to 0} \frac{1}{2\mu}\int_{-\infty}^{\infty} |s(u+\mu) - s(u-\mu)|^2\, du = 0, \tag{5.61}$$

although $s(u)$ exists, and indeed becomes the integral of the Fourier transform of $f(x)$. In this case, $s(u)$ is of limited total variation over every finite interval.

6. Bochner's generalization of harmonic analysis.

The study of the function $s(u)$ and its generalizations was first undertaken by Hahn, although, as we shall see later, on a basis insufficiently general to cover the needs of physics. The present author developed the theory for functions $f(x)$ with a finite mean square modulus, but the complete generalization of the theory is due to Bochner.

We have so far been interested in the problem of proceeding from $f(x)$ to $s(x)$. The question now arises, can we go backward, and determine $f(x)$ from $s(x)$? We should formally expect

20—29764. *Acta mathematica* 55. Imprimé le 8 avril 1930.

$$f(x) = \int_{-\infty}^{\infty} e^{-ixu}\, ds(u), \tag{6.01}$$

though the integral in question cannot be an ordinary Stieltjes integral, as $s(u)$ is not in general of limited total variation.

We may, however, develop this integral by a formal integration by parts, and we get

$$\int_{-A}^{A} e^{-ixu}\, ds(u) = e^{-iAx} s(A) - e^{iAx} s(-A) + ix \int_{-A}^{A} s(u)\, e^{-ixu}\, du$$

$$= e^{-iAx}\left[\frac{1}{2\pi}\int_{-1}^{1} f(\xi)\frac{e^{iA\xi}-1}{i\xi}\, d\xi + \underset{B\to\infty}{\text{l.i.m.}}\ \frac{1}{2\pi}\left[\int_{-B}^{-1} + \int_{1}^{B}\right] f(\xi)\frac{e^{iA\xi}}{i\xi}\, d\xi\right]$$

$$-e^{iAx}\left[\frac{1}{2\pi}\int_{-1}^{1} f(\xi)\frac{e^{-iA\xi}-1}{i\xi}\, d\xi + \underset{B\to\infty}{\text{l.i.m.}}\ \frac{1}{2\pi}\left[\int_{-B}^{-1} + \int_{1}^{B}\right] f(\xi)\frac{e^{-iA\xi}}{i\xi}\, d\xi\right]$$

$$+ix\int_{-A}^{A} e^{-ixu}\, du\left[\frac{1}{2\pi}\int_{-1}^{1} f(\xi)\frac{e^{iu\xi}-1}{i\xi}\, d\xi + \underset{B\to\infty}{\text{l.i.m.}}\ \frac{1}{2\pi}\left[\int_{-B}^{-1} + \int_{1}^{B}\right]\frac{f(\xi)e^{iux}}{i\xi}\, d\xi\right]$$

$$= \underset{B\to\infty}{\text{l.i.m.}}\ \frac{1}{2\pi}\int_{-B}^{B}\frac{f(\xi)}{i\xi}\left[e^{iA(\xi-x)} - e^{-iA(\xi-x)} + ix\int_{-A}^{A} e^{iu(\xi-x)}\, du\right] d\xi$$

$$= \underset{B\to\infty}{\text{l.i.m.}}\ \frac{1}{\pi}\int_{-B}^{B} f(\xi)\frac{\sin A(\xi-x)}{\xi-x}\, d\xi. \tag{6.02}$$

Even this expression fails in general to converge in the mean as $A\to\infty$. A natural device to choose to compel the desired convergence in the mean is to replace this integral by its Cesaro sum, and to investigate the behavior of

$$\frac{1}{D}\int_{0}^{D} dA \int_{-A}^{A} e^{-ixu}\, ds(u) = \frac{1}{\pi D}\int_{-\infty}^{\infty} f(\xi)\frac{1-\cos D(\xi-x)}{(\xi-x)^2}\, d\xi\,. \tag{6.03}$$

This is the familiar Fejér expression for the partial Cesaro sum of a Fourier series, at least in form. The classical Fejér argument will prove that at any point of continuity of $f(x)$ we shall have

$$f(x) = \lim_{D\to\infty} \frac{1}{\pi D} \int_{-N}^{N} f(\xi) \frac{1-\cos D(\xi-x)}{(\xi-x)^2} d\xi, \tag{6.04}$$

and indeed, that this will in any case be true almost everywhere. To proceed to

$$\lim_{D\to\infty} \frac{1}{D} \int_0^D dA \int_{-A}^{A} e^{-ixu}\, d\, s(u) = f(x) \tag{6.05}$$

requires only the reflection that $\left[\int_{-\infty}^{-1} + \int_{1}^{\infty}\right] \frac{f(t)}{t^2}\, dt$ converges.

In a manner similar to that in which we have proved

$$\int_{-A}^{A} e^{-ixu}\, ds(u) = \underset{B\to\infty}{\text{l.i.m.}} \frac{1}{\pi} \int_{-B}^{B} f(\xi) \frac{\sin A(\xi-x)}{\xi-x} d\xi \tag{6.02}$$

we may show that

$$\int_{P}^{A} e^{-ixu}\, ds(u) = \underset{B\to\infty}{\text{l.i.m.}} \frac{1}{2\pi i} \int_{-B}^{B} f(\xi+x) \frac{e^{iA\xi}-e^{iP\xi}}{\xi} d\xi \tag{6.06}$$

as a function of A and P. Thus, except for an additive constant, $\int_P^A e^{-ixu}\, ds(u)$ bears to $f(x+\xi)$ the relation which $s(A)$ bears to $f(\xi)$. Similarly, to

$$\left.\begin{aligned} &f(x) \pm f(t+x) \quad \text{there corresponds} \int_P^A (1 \pm e^{-itu})\, ds(u) \\ \text{and to}\quad &f(x) \pm if(t+x) \quad \text{there corresponds} \int_P^A (1 \pm ie^{-itu})\, d\, s(u). \end{aligned}\right\} \tag{6.07}$$

By an obvious linear combination of the formulae relating to these four functions separately (cf. (1.24)), we obtain

$$\lim_{\varepsilon\to 0}\frac{1}{2\varepsilon}\int_{-\infty}^{\infty}\left[\int_{u-\varepsilon}^{u+\varepsilon} e^{-itv}\,ds(v)\right]\left[\bar{s}(u+\varepsilon)-s(u-\varepsilon)\right]du$$

$$=\lim_{T\to\infty}\frac{1}{2T}\int_{-T}^{T} f(x+t)\bar{f}(x)\,dx=\varphi(t). \tag{6.08}$$

Here, by definition,

$$\int_{u-\varepsilon}^{u+\varepsilon} e^{-itv}\,ds(v)=e^{-it(u+\varepsilon)}\,s(u+\varepsilon)-e^{-it(u-\varepsilon)}\,s(u-\varepsilon)+it\int_{u-\varepsilon}^{u+\varepsilon} s(v)\,e^{-itv}\,dv$$

$$=e^{-iut}\left(s(u+\varepsilon)-s(u-\varepsilon)\right)$$

$$+\,it\left\{\int_{u}^{u+\varepsilon}\left[s(v)-s(u+\varepsilon)\right]e^{-itv}\,dv+\int_{u-\varepsilon}^{u}\left[s(v)-s(u-\varepsilon)\right]e^{itv}\,dv\right\}. \tag{6.09}$$

However, we have

$$\frac{1}{\varepsilon}\left[\int_{u-\varepsilon}^{u+\varepsilon} e^{-itv}\,ds(v)-e^{-iut}\left(s(u+\varepsilon)-s(u-\varepsilon)\right)\right]$$

$$=\underset{B\to\infty}{\mathrm{l.i.m.}}\ \frac{1}{2\pi i\varepsilon}\int_{-B}^{B} f(\xi+t)\left[\frac{e^{i(u+\varepsilon)\xi}-e^{i(u-\varepsilon)\xi}}{\xi}-e^{-iut}\,\frac{e^{i(u+\varepsilon)(\xi+t)}-e^{i(u-\varepsilon)(\xi+t)}}{\xi+t}\right]d\xi$$

$$=\underset{B\to\infty}{\mathrm{l.i.m.}}\ \frac{1}{2\pi i\varepsilon}\int_{-B}^{B} f(\xi+t)\left\{e^{i(u+\varepsilon)\xi}\left(\frac{1}{\xi}-\frac{e^{i\varepsilon t}}{\xi+t}\right)-e^{i(u-\varepsilon)\xi}\left(\frac{1}{\xi}-\frac{e^{-i\varepsilon t}}{\xi+t}\right)\right\}d\xi$$

$$=\underset{B\to\infty}{\mathrm{l.i.m.}}\ \frac{1}{\pi}\int_{-B}^{B} f(\xi+t)\,e^{iu\xi}\left[\frac{\sin\varepsilon\xi}{\varepsilon\xi}-\frac{\sin\varepsilon(\xi+t)}{\varepsilon(\xi+t)}\right]d\xi. \tag{6.10}$$

Now,

$$\left|\frac{\sin\varepsilon\xi}{\varepsilon\xi}-\frac{\sin\varepsilon(\xi+t)}{\varepsilon(\xi+t)}\right|$$

$$=\left|\frac{-2\xi\sin\frac{\varepsilon t}{2}\cos\varepsilon\left(\xi+\frac{t}{2}\right)+t\sin\varepsilon\xi}{\varepsilon\xi(\xi+t)}\right|\le\left|\frac{2t}{\xi+t}\right|\text{ and also }\le 2. \tag{6.11}$$

Thus, by the use of a Fourier transformation, it follows at once that

$$\int_{-\infty}^{\infty} \frac{1}{\varepsilon^2} \left| \int_{u-\varepsilon}^{u+\varepsilon} e^{-itv}\, ds(v) - e^{-iut}(s(u+\varepsilon)-s(u-\varepsilon)) \right|^2 du = O(1). \tag{6.12}$$

Inasmuch as we may readily show that

$$\lim_{\varepsilon \to 0} \int_{-\infty}^{\infty} |s(u+\varepsilon)-s(u-\varepsilon)|^2\, du = 0, \tag{6.13}$$

and since

$$\varphi(t) = \lim_{\varepsilon \to 0} \frac{1}{2\varepsilon} \int_{-\infty}^{\infty} e^{-itu} |s(u+\varepsilon)-s(u-\varepsilon)|^2\, du$$

$$+ \lim_{\varepsilon \to 0} \frac{1}{2\varepsilon} \int_{-\infty}^{\infty} \left[\int_{u-\varepsilon}^{u+\varepsilon} e^{-itv}\, ds(v) - e^{-iut}(s(u+\varepsilon)-s(u-\varepsilon)) \right] [\bar{s}(u+\varepsilon)-\bar{s}(u-\varepsilon)]\, du, \tag{6.14}$$

it follows by an elementary use of the Schwarz inequality that

$$\varphi(t) = \lim_{\varepsilon \to 0} \frac{1}{2\varepsilon} \int_{-\infty}^{\infty} e^{-itu} |s(u+\varepsilon)-s(u-\varepsilon)|^2\, du. \tag{6.15}$$

This formula holds in the same sense as (5. 53) *for each t independently.*

We may deduce from (6. 06) the analogous formula

$$\int_{-\lambda}^{\lambda} dy \int_{P}^{A} e^{-iyu}\, ds(u) = \int_{P}^{A} \frac{2 \sin \lambda u}{u}\, ds(u) \tag{6.16}$$

by an easily justified inversion of the order of integration. Furthermore,

$$\int_{-\lambda}^{\lambda} dy \operatorname*{l.i.m.}_{B \to \infty} \frac{1}{2\pi i} \int_{-B}^{B} f(\xi+y) \frac{e^{iA\xi}-e^{iP\xi}}{\xi}\, d\xi$$

$$= \operatorname*{l.i.m.}_{B \to \infty} \frac{1}{2\pi i} \int_{-B}^{B} \frac{e^{iA\xi}-e^{iP\xi}}{\xi}\, d\xi \int_{-\lambda}^{\lambda} f(\xi+y)\, dy, \tag{6.17}$$

as may be deduced from the fact that

$$\int_N^\infty \frac{|f(\xi+y)|^2}{\xi^2}\,d\xi \quad \text{and} \quad \int_{-\infty}^{-N} \frac{|f(\xi+y)|^2}{\xi^2}\,d\xi$$

tend uniformly to 0 with increasing N for all y in $(-\lambda, \lambda)$. Thus to

$$\int_{-\lambda}^{\lambda} f(x+y)\,dy \quad \text{there corresponds} \quad \int_P^A \frac{2\sin\lambda u}{u}\,ds(u)$$

in the same sense in which to $f(x)$ there corresponds $s(A)$. It follows from (5.53) that

$$\lim_{\mu\to 0}\frac{1}{2\mu}\int_{-\infty}^{\infty}\left|\int_{A-\mu}^{A+\mu}\frac{2\sin\lambda u}{u}\,ds(u)\right|^2 dA = \lim_{T\to\infty}\frac{1}{2T}\int_{-T}^{T}\left|\int_{-\lambda}^{\lambda} f(x+y)\,dy\right|^2 dx. \tag{6.18}$$

As in (6.09),

$$\int_{A-\mu}^{A+\mu}\frac{2\sin\lambda u}{u}\,ds(u) = \frac{2\sin\lambda(A+\mu)}{A+\mu}\,s(A+\mu) - \frac{2\sin\lambda(A-\mu)}{A-\mu}\,s(A-\mu)$$

$$-\int_{A-\mu}^{A+\mu} s(u)\,\frac{d}{du}\left(\frac{2\sin\lambda u}{u}\right)du, \tag{6.19}$$

and as in (6.10),

$$\frac{1}{\mu}\int_{A-\mu}^{A+\mu}\frac{2\sin\lambda u}{u}\,ds(u) - \frac{2\sin\lambda A}{A\mu}\,[s(A+\mu)-s(A-\mu)]$$

$$= \frac{1}{\mu}\int_{A-\mu}^{A+\mu} ds(u)\int_{-\lambda}^{\lambda} e^{-i\sigma u}\,d\sigma - 2\int_{-\lambda}^{\lambda} e^{-i\sigma A}\,d\sigma\,[s(A+\mu)-s(A-\mu)]$$

$$= \int_{-\lambda}^{\lambda} d\sigma\,\underset{B\to\infty}{\text{l.i.m.}}\,\frac{1}{\pi}\int_{-B}^{B} f(\xi+\sigma)\,e^{iA\sigma}\left[\frac{\sin\mu\xi}{\mu\xi} - \frac{\sin\mu(\xi+\sigma)}{\mu(\xi+\sigma)}\right]d\xi \tag{6.20}$$

because (6.10) holds uniformly over a finite range of σ, because of the fact that

$$\left| \frac{\sin \mu \xi}{\mu \xi} - \frac{\sin \mu(\xi + \sigma)}{\mu(\xi - \sigma)} \right| = O(\xi^{-1}) \tag{6. 21}$$

uniformly over such a range. Hence, as in (6. 12), we may show that

$$\int_{-\infty}^{\infty} \frac{1}{\mu^2} \left| \int_{A-\mu}^{A+\mu} \frac{2 \sin \lambda u}{u} \, ds(u) - \frac{2 \sin \lambda A}{A} [s(A+\mu) - s(A-\mu)] \right|^2 dA = O(1); \tag{6. 22}$$

and by an argument exactly similar to that leading to (6. 15), we obtain

$$\lim_{T \to \infty} \frac{1}{2T} \int_{-T}^{T} \left| \int_{-\lambda}^{\lambda} f(x+y) \, dy \right|^2 dx = \lim_{\mu \to 0} \frac{1}{2\mu} \int_{-\infty}^{\infty} \frac{4 \sin^2 \lambda u}{u^2} | s(u+\mu) - s(u-\mu) |^2 du \tag{6. 23}$$

for each λ independently, in the same sense as that in which we have proved (5. 53) *and* (6. 15).

We now proceed to the part of the theory that is specifically Bochner's. We wish to discuss the harmonic analysis of functions which are no longer bounded on the average (for so we may interpret the finiteness of $\lim_{T \to \infty} \frac{1}{2T} \int_{-T}^{T} |f(x)|^2 dx$), but instead have on the average an algebraic rate of growth as the argument proceeds to $\pm \infty$. That is, we assume the existence of

$$\lim_{T \to \infty} \frac{1}{2T^n} \int_{-T}^{T} |f(x)|^2 \, dx \tag{6. 24}$$

and we shall take n to be a positive odd integer. By arguments following identically the lines laid down for $n=1$, we show that

$$\left[\int_1^T + \int_{-T}^{-1} \right] \frac{|f(x)|^2}{x^{n+1}} \, dx = O(1), \tag{6. 25}$$

and hence that

$$\left[\int_1^{\infty} + \int_{-\infty}^{-1} \right] \frac{|f(x)|^2}{x^{n+1}} \, dx \tag{6. 26}$$

exists. We can then show the existence of

$$s(u) = s_1(u) + s_2(u)$$

$$= \frac{1}{2\pi} \int_{-1}^{1} f(x) \left[\left(\int_0^u \right)^{\frac{n+1}{2}} e^{ivx}\, dv \right] dx + \frac{1}{2\pi} \operatorname*{l.i.m.}_{A \to \infty} \left[\int_1^A + \int_{-A}^{-1} \right] f(x) \frac{e^{iux}}{(ix)^{\frac{n+1}{2}}}\, dx. \qquad (6.27)$$

If we now put

$$\Delta_\mu^{\frac{n+1}{2}}(s) = s\left(u + \mu\left(\frac{n+1}{2}\right)\right) - ns\left(u + \mu\left(\frac{n-3}{2}\right)\right) + \frac{n(n-1)}{2!} s\left(u + \mu\left(\frac{n-7}{2}\right)\right)$$

$$- \cdots \pm s\left(u - \mu\left(\frac{n+1}{2}\right)\right)$$

$$= \frac{2^{\frac{n+1}{2}}}{2\pi} \operatorname*{l.i.m.}_{A \to \infty} \int_{-A}^{A} f(x)\, e^{iux} \frac{\sin^{\frac{n+1}{2}} \mu x}{x^{\frac{n+1}{2}}}\, dx, \qquad (6.28)$$

we can show by a Plancherel argument that

$$\frac{1}{\mu} \int_{-\infty}^{\infty} \left| \Delta_\mu^{\frac{n+1}{2}}(s(u)) \right|^2 du = \frac{2^n}{\pi\mu} \int_{-\infty}^{\infty} |f(x)|^2 \frac{\sin^{n+1} \mu x}{x^{n+1}}\, dx. \qquad (6.29)$$

It is easy to show that for any function $F(x)$,

$$\frac{1}{T^n} \int_0^T F(x)\, dx = \frac{1}{T^n} \int_0^T \frac{F(x)}{x^{n-1}} x^{n-1}\, dx = \int_0^1 \frac{F(Ty)}{(Ty)^{n-1}} y^{n-1}\, dy. \qquad (6.30)$$

Inasmuch as

$$\int_0^1 y^{n-1}\, dy = 1/n, \qquad (6.31)$$

if we put

$$F(x)/nx^{n-1} = G(x), \qquad (6.32)$$

the above integral is a mean of G, in the sense in which means enter into our general Tauberian theorem. Furthermore,

$$\frac{2^{n+1}}{\pi\mu} \int_0^\infty F(x) \frac{\sin^{n+1} \mu x}{x^{n+1}}\, dx = \frac{2^{n+1} n}{\pi\mu} \int_0^\infty G(x) \frac{\sin^{n+1} \mu x}{x^2}\, dx$$

$$= \frac{2^{n+1} n}{\pi} \int_0^\infty G(y/\mu) \frac{\sin^{n+1} y}{y^2}\, dy \qquad (6.33)$$

and if we put

$$\frac{2^{n+1}n}{\pi}\int_0^\infty \frac{\sin^{n+1}y}{y^2}\,dy = P_n, \tag{6.34}$$

then

$$\frac{2^{n+1}}{\pi\mu P_n}\int_0^\infty F(x)\frac{\sin^{n+1}\mu x}{x^{n+1}}\,dx \tag{6.35}$$

is another mean of the quantity G. If we set

$$M_1(x)\,[\text{or } M_2(x)] = \begin{cases} n x^{n-1}; & [0 < x < 1] \\ 0 & ; \; [x \geq 1] \end{cases} \qquad M_2(x)\,[\text{or } M_1(x)] = \frac{2^{n+1}n}{\pi P_n}\frac{\sin^{n+1}x}{x^2}; \tag{6.36}$$

then

$$\int_0^\infty M_1(x)\,x^{iu}\,dx = \frac{n}{n+iu} \neq 0. \tag{6.37}$$

As to

$$\int_0^\infty M_2(x)\,x^{iu}\,dx,$$

we have

$$\int_0^\infty \sin^{n+1}x\,x^{iu-2}\,dx$$

$$= \int_0^\infty \frac{(n+1)!}{2^{n-1}}\left[\frac{\sin^2 x}{\frac{n-1}{2}!\,\frac{n+3}{2}!} - \frac{\sin^2 2x}{\frac{n-3}{2}!\,\frac{n+5}{2}!} + \frac{\sin^2 3x}{\frac{n-5}{2}!\,\frac{n+7}{2}!} - \cdots\right] x^{iu-2}\,dx.$$

Hence

$$\int_0^\infty M_2(x)\,x^{iu}\,dx = \frac{4n(n+1)!}{\pi P_n}\cdot$$

$$\left[\frac{1}{\frac{n-1}{2}!\,\frac{n+3}{2}!} - \frac{2^{iu-1}}{\frac{n-3}{2}!\,\frac{n+5}{2}!} + \frac{3^{iu-1}}{\frac{n-5}{2}!\,\frac{n+7}{2}!} - \cdots\right]\int_0^\infty \sin^2 x\,x^{iu-2}\,dx. \tag{6.38}$$

We have already seen that

$$\int_0^\infty \sin^2 x \, x^{iu-2}\, dx \neq 0. \qquad (6.39)$$

Thus the question of the possibility that

$$\int_0^\infty M_2(x)\, x^{iu}\, dx$$

should vanish depends on the possibility of the vanishing of

$$\frac{1}{\frac{n-1}{2}!\,\frac{n+3}{2}!} - \frac{2^{iu-1}}{\frac{n-3}{2}!\,\frac{n+5}{2}!} + \frac{3^{iu-1}}{\frac{n-5}{2}!\,\frac{n+7}{2}!} - \cdots. \qquad (6.40)$$

It is easy enough to prove that this cannot vanish for $n=1, 3, 5, 7, 9, 11$ but the author has not yet been able to produce a proof in the general case.

In case this expression does not vanish, we may apply our Tauberian theorem, replacing $f(x)$ by $|f(x)|^2+|f(-x)|^2$. We shall assume to begin with that $f(x)$ vanishes in the neighbourhood of the origin. Then

$$\lim_{T\to\infty} \frac{1}{2T^n} \int_{-T}^{T} |f(x)|^2\, dx = A \qquad (6.41)$$

and

$$\lim_{\mu\to 0} \frac{1}{P_n \mu} \int_{-\infty}^{\infty} \left| \Delta_\mu^{\frac{n+1}{2}} s(u) \right|^2 du = \lim_{\mu\to 0} \frac{2^n}{P_n \pi \mu} \int_{-\infty}^{\infty} |f(x)|^2 \frac{\sin^{n+1} \mu x}{x^{n+1}}\, dx \qquad (6.42)$$

are equivalent.

We have put $f(x)=0$ in the neighbourhood of the origin to be sure of the boundedness of

$$\frac{1}{2T^n} \int_{-T}^{T} |f(x)|^2\, dx \quad \text{and} \quad \frac{2^n}{P_n \pi \mu} \int_{-\infty}^{\infty} |f(x)|^2 \frac{\sin^{n+1} \mu x}{x^{n+1}}\, dx. \qquad (6.43)$$

In any case,

$$\lim_{T\to\infty} \frac{1}{2T^n} \int_{-T}^{T} |f(x)|^2\, dx$$

will not be changed if $f(x)$ is made to vanish in this neighbourhood. Moreover,

$$\frac{1}{\mu}\int_{-B}^{B} |f(x)|^2 \frac{\sin^{n+1}\mu x}{x^{n+1}}\,dx \leq \mu^n \int_{-B}^{B} |f(x)|^2\,dx. \tag{6. 44}$$

Thus

$$\lim_{\mu\to 0} \frac{1}{P_n \mu}\int_{-\infty}^{\infty} \left| \Delta_\mu^{\frac{n+1}{2}} s(u) \right| du$$

will not be changed either, and we may always write:

$$\lim_{\mu\to 0} \frac{1}{P_n \mu}\int_{-\infty}^{\infty} \left| \Delta_\mu^{\frac{n+1}{2}} s(u) \right|^2 du = \lim_{T\to\infty} \frac{1}{2T^n}\int_{-T}^{T} |f(x)|^2\,dx. \tag{6. 45}$$

We now come to the problem of returning from $s(u)$ to $f(x)$. We should expect formally

$$f(x) = \int_{-\infty}^{\infty} e^{-iux}\, d^{\frac{n+1}{2}} s(u)/du^{\frac{n-1}{2}} \tag{6. 46}$$

Again, we need to interpret

$$\int_{-\infty}^{\infty} e^{-iux}\, d^{\frac{n+1}{2}} s(u)/du^{\frac{n-1}{2}}$$

by an integration by parts, and again some form of summation is necessary to get a more manageable expression. One method is the following: Let us replace

$$\int_{-\infty}^{\infty} e^{-iux}\, d^{\frac{n+1}{2}} s(u)/du^{\frac{n-1}{2}}$$

by

$$\int_{-\infty}^{\infty} e^{-\lambda u^2 - iux}\, d^{\frac{n+1}{2}} s(u)/du^{\frac{n-1}{2}} \tag{6. 47}$$

and let us investigate its behavior as $\lambda \to 0$. This expression is to be interpreted by $\frac{n+1}{2}$ formal integrations by parts, which convert it to

$$(-1)^{\frac{n+1}{2}} \int_{-\infty}^{\infty} s_2(u) d^{\frac{n+1}{2}} (e^{-\lambda u^2 - iux}) / d u^{\frac{n+1}{2}} du. \tag{6.48}$$

It then becomes obvious that this expression will have the same limit in the mean as

$$\int_{-B}^{B} s_2(u) (ix)^{\frac{n+1}{2}} e^{-iux} du \qquad [B \to \infty] \tag{6.49}$$

It here will, however, be $f(x)$ if $|x| > A$, and will be 0 otherwise. To see this, we need only compare $\frac{d^{\frac{n+1}{2}}}{dx^{\frac{n+1}{2}}}(e^{-\lambda u^2 - iux})$ with $\frac{d^{\frac{n+1}{2}}}{dx^{\frac{n+1}{2}}} e^{-iux}$. A similar result holds for $s_1(u)$, and the final result is that over any finite interval not including the origin,

$$f(x) = \underset{\lambda \to 0}{\mathrm{l.i.m.}} \int_{-\infty}^{\infty} e^{-\lambda u^2 - iux} d^{\frac{n+1}{2}} s(u) / d u^{\frac{n-1}{2}} \tag{6.50}$$

7. The Hahn generalization of harmonic analysis.

Up to the present we have concerned ourselves rather with questions of convergence in the mean than of ordinary convergence. Retaining our previous notation, in the case where $n = 1$, we may raise the further questions: when does

$$\int_{-A}^{A} e^{-iux} ds(u)$$

exist for all u as an ordinary Stieltjes integral, rather than a generalized Stieltjes integral such as we have treated in the last section? When does

$$\lim_{A \to \infty} \int_{-A}^{A} e^{-iux} ds(u)$$

converge in the ordinary sense? These questions furnish the vital link between the generalized harmonic analysis of Hahn, and that developed here. In the second problem of this section, it is at present not likely that we can obtain conditions that are both necessary and sufficient. The necessary and sufficient answer to the first question is manifestly that $s(u)$ should be of limited total variation over any finite interval. We have

$$s(v)-s(u) = \underset{B\to\infty}{\text{l.i.m.}} \frac{1}{2\pi}\int_{-B}^{B} f(x)\frac{e^{ivx}-e^{iux}}{ix}\,dx$$

$$= \underset{B\to\infty}{\text{l.i.m.}} \frac{1}{2\pi}\int_{-B}^{B} \frac{e^{ivx}-e^{iux}}{ix}\,d\int_{0}^{x} f(\xi)\,d\xi$$

$$= \underset{B\to\infty}{\text{l.i.m.}} \frac{1}{2\pi}\left\{\frac{e^{ivB}-e^{iuB}}{iB}\int_{0}^{B} f(\xi)\,d\xi + \frac{e^{-ivB}-e^{-iuB}}{iB}\int_{0}^{-B} f(\xi)\,d\xi\right.$$

$$\left. + i\int_{-B}^{B}\left[\frac{ivx-1}{x^2}e^{ivx} - \frac{iux-1}{x^2}e^{iux}\right]dx\int_{0}^{x} f(\xi)\,d\xi\right\}. \qquad (7.01)$$

If now we assume that

$$F(x) = \int_{0}^{x} f(\xi)\,d\xi = O(x^{-1}), \qquad (7.02)$$

it follows that

$$s(v)-s(u) = i\int_{-\infty}^{\infty} F(x)\left[\frac{ivx-1}{x^2}e^{ivx} - \frac{iux-1}{x^2}e^{iux}\right]dx. \qquad (7.03)$$

Inasmuch as by our assumptions

$$\int_{1}^{\infty}\frac{F(x)}{x^2}e^{iux}\,dx \quad \text{and} \quad \int_{1}^{\infty}\frac{F(x)}{x}e^{iux}\,dx$$

converge and represent functions with quadratically summable derivatives, $s(u)$ is of limited total variation over any finite region. Thus (7.02) is a sufficient condition for the answer to our first question to be in the affirmative.

Let us now suppose our first question affirmatively answered. The formal integration by parts by which we have defined $\int_{-A}^{A} e^{-iux}\,ds(u)$ in the general case may now be carried out, and all our quasi-Stieltjes integrals become ordinary Stieltjes integrals. If, on the other hand, $s(u)$ is not of limited total variation, Hahn adopts a generalized definition of the Stieltjes integral identical in content with that here given. In either case, we have already seen that

$$\int_{-A}^{A} e^{-ixu}\, ds(u) = \underset{B\to\infty}{\text{l.i.m.}} \frac{1}{\pi}\int_{-B}^{B} f(\xi)\frac{\sin A(\xi-x)}{\xi-x}\, d\xi, \tag{7.04}$$

and that

$$f(x) = \lim_{A\to\infty}\frac{1}{2\pi A}\int_{-\infty}^{\infty} f(\xi)\frac{1-\cos 2A(\xi-x)}{(\xi-x)^2}\, d\xi. \tag{7.05}$$

Moreover,

$$\underset{B\to\infty}{\text{l.i.m.}}\frac{1}{2\pi A}\int_{-B}^{B} f(\xi)\frac{1-\cos 2A(\xi-x)-2A(\xi-x)\sin A(\xi-x)}{(\xi-x)^2}\, d\xi$$

$$= \frac{1}{2\pi}\int_{-\infty}^{\infty} f\left(\xi+\frac{w}{A}\right)\frac{1-\cos 2w-2w\sin w}{w^2}\, dw$$

$$= \frac{1}{2\pi}\int_{-\infty}^{\infty}\left[f\left(\xi+\frac{w}{A}\right)-f(\xi)\right]\frac{1-\cos 2w-2w\sin w}{w^2}\, dw$$

$$+\frac{f(\xi)}{2\pi}\int_{-\infty}^{\infty}\frac{1-\cos 2w-2w\sin w}{w^2}\, dw$$

$$= \frac{1}{2\pi}\int_{-D}^{D}\left[f\left(\xi+\frac{w}{A}\right)-f(\xi)\right]\frac{1-\cos 2w-2w\sin w}{w^2}\, dw$$

$$+\frac{1}{2\pi}\left[\int_{-\infty}^{-D}+\int_{D}^{\infty}\right]\left[f\left(\xi+\frac{w}{A}\right)-f(\xi)\right]\frac{1-\cos 2w-2w\sin w}{w^2}\, dw. \tag{7.06}$$

Thus a sufficient condition for

$$\int_{-\infty}^{\infty} e^{-ixu}\, ds(u) = f(x) \tag{7.07}$$

is that $f(x)$ should satisfy locally one of the sufficient conditions for the convergence of a Fourier series, and that

$$\left[\int_{-\infty}^{-1} + \int_{1}^{\infty}\right]\left|\frac{f(w)}{w}\right| dw \text{ converge}$$

or what is the same, that

$$\int_{-\infty}^{\infty}\left|\frac{f(w)}{\sqrt{1+w^2}}\right| dw \qquad (7.08)$$

exist. This condition thus constitutes a sufficient solution of the Hahn problem. To such a function we may add any function with a convergent Fourier series, without destroying the fact that it solves Hahn's problem.

Another condition under which (7.07) holds is that $f(x)$ should be of the form

$$g(x) \cdot h(x)$$

where $g(x)$ is periodic, and $h(x)$ bounded and monotone at $\pm\infty$, and that $f(x)$ should satisfy one of the sufficient conditions for the convergence of a Fourier series to its function. As in the previous case, the proof of his assertion made by Hahn depends simply on the fact that the second term of the last line of (7.06) will then vanish with increasing D, while the first term is asymptotically equal to the difference between the Cesaro and ordinary partial sums of the Fourier integral of the function $f(x)$ mutilated by being made to vanish outside the interval $(-D, D)$. It then follows from the Fejér theorem and the fact that the conditions for the convergence of such a Fourier integral and of a Fourier series are the same, that Hahn's theorem holds.

The class of functions for which $s(u)$ exists as an ordinary Stieltjes integral is too narrow to cover the physically interesting cases of continuous spectra. To see this, let $f(x)$ be a function for which $\varphi(x)$ and $S(u)$ exist, and let

$$f^{(\eta)}(x) = \frac{1}{2\eta}\int_{-\eta}^{\eta} f(x+\xi)\, d\xi. \qquad (7.09)$$

Let us suppose that

$$\varphi^{(\eta)}(x) = \lim_{T\to\infty} \frac{1}{2T} \int_{-T}^{T} f^{(\eta)}(x+t)\bar{f}^{(\eta)}(t)\,dt \tag{7.10}$$

exists for every x, and hence that

$$S^{(\eta)}(u) = \frac{1}{2\pi} \int_{-\infty}^{\infty} \varphi^{(\eta)}(x) \frac{e^{iux}-1}{ix}\,dx \tag{7.11}$$

exists. Then

$$\varphi^{(\eta)}(x) = \lim_{T\to\infty} \frac{1}{2T}\frac{1}{4\eta^2} \int_{-\eta}^{\eta} d\xi \int_{-\eta}^{\eta} d\zeta \int_{-T}^{T} f(x+t+\xi)\bar{f}(t+\zeta)\,dt$$

$$= \frac{1}{4\eta^2} \int_{-\eta}^{\eta} d\xi \int_{-\eta}^{\eta} d\zeta\, \varphi(x+\xi-\zeta)$$

$$= \frac{1}{4\eta^2} \int_{-\eta}^{\eta} d\xi \int_{-\eta-\xi}^{\eta-\xi} \varphi(x-y)\,dy$$

$$= \frac{1}{4\eta^2}\left\{\eta \int_{-2\eta}^{0} \varphi(x-y)\,dy + \eta \int_{0}^{2\eta} \varphi(x-y)\,dy - \int_{-\eta}^{\eta} \xi[\varphi(x+\eta+\xi)-\varphi(x-\eta+\xi)]\,d\xi\right\}$$

$$= \frac{1}{4\eta^2} \int_{-2\eta}^{2\eta} \varphi(x-y)(2\eta-|y|)\,dy; \tag{7.12}$$

$$S^{(\eta)}(u+\mu) - S^{(\eta)}(u-\mu) = \frac{1}{\pi} \operatorname*{l.i.m.}_{A\to\infty} \int_{-A}^{A} \varphi^{(\eta)}(x) \frac{\sin \mu x}{x} e^{iux}\,dx$$

$$= \operatorname*{l.i.m.}_{A\to\infty} \frac{1}{4\pi\eta^2} \int_{-2\eta}^{2\eta} (2\eta-|y|)\,dy \int_{-A-y}^{A-y} \varphi(x-y) \frac{\sin \mu x}{x} e^{iux}\,dx$$

$$= \operatorname*{l.i.m.}_{A\to\infty} \frac{1}{4\pi\eta^2} \int_{-A}^{A} \varphi(x)\,dx \int_{-2\eta}^{2\eta} (2\eta-|y|) \frac{\sin \mu(x+y)}{x+y} e^{iu(x+y)}\,dy$$

$$= \underset{A\to\infty}{\text{l.i.m.}} \frac{1}{8\pi\eta^2}\int_{-A}^{A}\varphi(x)\,dx\int_{u-\mu}^{u+\mu}dw\int_{-2\eta}^{2\eta}(2\eta-|y|)\,e^{iw(x+y)}\,dy$$

$$= \underset{A\to\infty}{\text{l.i.m.}} \frac{1}{4\pi\eta^2}\int_{-A}^{A}\varphi(x)\,dx\int_{u-\mu}^{u+\mu}\frac{e^{iwx}(1-\cos 2w\eta)}{w^2}\,dw$$

$$= \underset{A\to\infty}{\text{l.i.m.}} \frac{1}{4\pi\eta^2}\int_{u-\mu}^{u+\mu}\frac{1-\cos 2w\eta}{w^2}\,dw\int_{-A}^{A}\varphi(x)\frac{e^{iwx}-1}{ix}\,dx$$

$$= \frac{1}{4\pi\eta^2}\int_{u-\mu}^{u+\mu}\frac{1-\cos 2w\eta}{w^2}\,dw\int_{-\infty}^{\infty}\varphi(x)\frac{e^{iwx}-1}{ix}\,dx$$

$$= \frac{1}{2\eta^2}\int_{u-\mu}^{u+\mu}\frac{1-\cos 2w\eta}{w^2}\,d\,S(w) = \int_{u-\mu}^{u+\mu}\frac{\sin^2\eta w}{\eta^2 w^2}\,d\,S(w); \qquad (7.13)$$

$$s^{(\eta)}(u+\varepsilon)-s^{(\eta)}(u-\varepsilon) = \underset{A\to\infty}{\text{l.i.m.}} \frac{1}{\pi}\int_{-A}^{A}f^{(\eta)}(x)\frac{\sin\varepsilon x}{x}e^{iux}\,dx$$

$$= \underset{A\to\infty}{\text{l.i.m.}} \frac{1}{2\pi\eta}\int_{-\eta}^{\eta}d\xi\int_{-A}^{A}f(x+\xi)\frac{\sin\varepsilon x}{x}e^{iux}\,dx$$

$$= \frac{1}{2\eta}\int_{-\eta}^{\eta}d\xi\int_{u-\varepsilon}^{u+\varepsilon}e^{-i\xi u}\,ds(u) \;\text{[by (6.06)]}. \qquad (7.14)$$

Although this is not in general an ordinary Stieltjes integral, we may integrate by parts and then invert the order of integration, and thus obtain

$$s^{(\eta)}(u+\varepsilon)-s^{(\eta)}(u-\varepsilon) = \frac{1}{2\eta}\int_{u-\varepsilon}^{u+\varepsilon}ds(u)\int_{-\eta}^{\eta}e^{-i\xi u}\,d\xi = -\int_{u-\varepsilon}^{u+\varepsilon}\frac{\sin\eta u}{\eta u}\,ds(u). \qquad (7.15)$$

Hence, by (5.53) and (6.23),

$$\varphi^{(\eta)}(0) = \lim_{T\to\infty}\frac{1}{2T}\int_{-T}^{T}|f^{(\eta)}(x)|^2\,dx = \lim_{\varepsilon\to 0}\frac{1}{2\varepsilon}\int_{-\infty}^{\infty}\frac{\sin^2\eta u}{\eta^2 u^2}|s(u+\varepsilon)-s(u-\varepsilon)|^2\,du. \qquad (7.16)$$

22—29764. *Acta mathematica.* 55. Imprimé le 14 avril 1930.

By (7. 14),

$$\varphi^{(\eta)}(0) = \frac{1}{4\eta^2}\int_{-2\eta}^{2\eta}\varphi(\xi)(2\eta-|\xi|)\,d\xi$$

$$= \frac{1}{4\eta^2}\int_0^{2\eta} 2\lambda d\lambda\left[\frac{1}{2\lambda}\int_{-\lambda}^{\lambda}\varphi(\xi)\,d\xi\right], \tag{7. 17}$$

which in combination with (5. 47) leads us to

$$\lim_{\eta\to 0}\varphi^{(\eta)}(0) = S(\infty)-S(-\infty). \tag{7. 18}$$

In other words,

$$\lim_{\eta\to 0}\lim_{\varepsilon\to 0}\frac{1}{2\varepsilon}\int_{-\infty}^{\infty}\frac{\sin^2\eta u}{\eta^2 u^2}|s(u+\varepsilon)-s(u-\varepsilon)|^2\,du = S(\infty)-S(-\infty). \tag{7. 19}$$

If we now assume that

$$\lim_{\eta\to 0}\varphi^{(\eta)}(0) = \varphi(0), \tag{7. 20}$$

it follows from (7. 21), (7. 22), and (5. 54) that

$$\lim_{\eta\to 0}\lim_{\varepsilon\to 0}\frac{1}{2\varepsilon}\int_{-\infty}^{\infty}\left[1-\frac{\sin^2\eta u}{\eta^2 u^2}\right]|s(u+\varepsilon)-s(u-\varepsilon)|^2 du = 0, \tag{7. 21}$$

or since $\sin\eta u/\eta u$ tends to 1 as $\eta\to 0$

$$\lim_{A\to\infty}\lim_{\varepsilon\to 0}\frac{1}{2\varepsilon}\left[\int_A^{\infty}+\int_{-\infty}^{-A}\right]|s(u+\varepsilon)-s(u-\varepsilon)|^2 du = 0. \tag{7. 22}$$

If $\varphi(x)$ is a continuous function, not only shall we have (7. 20) as a consequence of (7. 17), but all the functions $\varphi^{(\eta)}(x)$ will be continuous, as follows from (7. 12). Then, by (6. 15), we shall have

$$\lim_{\varepsilon\to 0}\frac{1}{2\varepsilon}\int_{-\infty}^{\infty}e^{-iux}|s(u+\varepsilon)-s(u-\varepsilon)|^2\,du = \int_{-\infty}^{\infty}e^{-iux}\,d\,S(u). \tag{7. 23}$$

It follows at once that if $P(u)$ is a trigonometrical polynomial, or the uniform limit of such a polynomial,

$$\lim_{\varepsilon \to 0} \frac{1}{2\varepsilon} \int_{-\infty}^{\infty} P(u) \, | s(u+\varepsilon) - s(u-\varepsilon) |^2 \, du = \int_{-\infty}^{\infty} P(u) \, dS(u). \tag{7.24}$$

For the transition to the case where $P(u)$ is a uniform limit of a polynomial we need only make appeal to the boundedness of

$$\lim_{\varepsilon \to 0} \frac{1}{2\varepsilon} \int_{-\infty}^{\infty} | s(u+\varepsilon) - s(u-\varepsilon) |^2 \, du. \tag{7.25}$$

Hence, by Fejér's theorem, $P(u)$ may be any continuous periodic function, and because of (7.22), any continuous function differing from 0 only over a finite range, as the change in the left side of this expression due to making $P(u)$ artificially periodic tends to 0 as the period increases.

It follows at once that under the hypotheses:

(a) $\varphi^{(\eta)}(x)$ exists for every x and η;

(b) $\varphi(x)$ is continuous;

if

$$\lim_{\varepsilon \to 0} \frac{1}{2\varepsilon} \int_{\alpha}^{\beta} | s(u+\varepsilon) - s(u-\varepsilon) |^2 \, du = 0, \tag{7.26}$$

over any interval, $S(u)$ is a constant over any interior interval. Thus if $s(u)$ is of limited total variation over any finite interval, and is continuous, $S(u)$ reduces to a constant, and the spectrum of $f(x)$ vanishes. In other words, the very natural hypotheses (a) and (b) are inconsistent with the existence of a continuous spectrum, provided $s(u)$ is of limited total variation. To see this, we need only notice that almost everywhere

$$S(u_2) - S(u_1) = \lim_{\varepsilon \to 0} \frac{1}{2\varepsilon} \int_{u_1}^{u_2} | s(u+\varepsilon) - s(u-\varepsilon) |^2 \, du$$

$$= \text{sum of squares of jumps of } S(u) \text{ between } u_1 \text{ and } u_2. \tag{7.27}$$

Thus the expansion of $f(x)$ in an ordinary Stieltjes integral is not adequate to the discussion of such continuous spectra as occur in physics, inasmuch, as in these cases, as we shall see in section 13, conditions analogous to (a) and (b) are fulfilled.

CHAPTER III.

8. Harmonic analysis in more than one dimension.

The elementary function of harmonic analysis in one dimension is e^{iux}. In n dimensions, this is replaced by

$$e^{i(u_1x_1+\cdots+u_nx_n)}$$

which we may write vectorially

$$e^{i(U\cdot X)},$$

where the vector X represents the argument of the function to be analyzed, and the vector U the vectorial frequency. If we keep the term $(U.X)$ invariant, and X varies cogrediently, U varies contragrediently. Thus the familiar duality relation between Fourier transforms is intimately connected with the point-plane duality of geometry. This is why the relation between position-coordinates and momentum-coordinates in modern quantum physics appears as a Fourier duality, while the same relation appears in the theory of relativity as the relation between a certain cogredient tensor and a certain contragredient tensor.

Practically the whole generalized theory of harmonic analysis so far developed is susceptible to a generalization to n dimensions. This generalization has been carried out by Mr. A. C. Berry, who has been kind enough to furnish me with the following summary of his Harvard doctoral thesis.

It is necessary to introduce certain notations at once and to make a few preliminary remarks. Let there be given a real n-dimensional space and, in it, some fixed reference point, or origin, O. If X is an arbitrary point of this space, the symbol X shall be used to denote not only this point, but also any real n-dimensional vector equivalent to the directed line segment OX. Let $f(X)$ be any complex, measurable function defined for all such real arguments X. Hereafter it will be assumed that all functions with which we start satisfy these requirements. If R is any measurable point-set, then

$$\int_R f(X)\,d\,V_X \tag{8.01}$$

shall mean the n-dimensional volume integral, in the sense of Lebesgue, of $f(X)$ taken over the region R. Since, in general, n-dimensional »spheres» will be employed as regions of integration, it will be convenient to use the notation $(r;\,X)$ to signify a sphere of radius r and having its center at X. The vector interpretation of X enables us to write

$$\int_{(r;\,Y)} f(X)\,d\,V_X = \int_{(r;\,O)} f(X+Y)\,d\,V_X. \tag{8.02}$$

The »volume» or measure, of an n-dimensional sphere of radius r is known to be

$$\frac{\pi^{\frac{n}{2}}\, r^n}{\Gamma\left(\frac{n}{2}+1\right)} \tag{8.03}$$

which quantity, hereafter, shall be denoted by the symbol $v(r)$. Thus the average of a function $f(x)$ over a sphere of radius r about the point Y is

$$\frac{1}{v(r)}\int_{(r;\,Y)} f(X)\,d\,V_X. \tag{8.04}$$

Corresponding to the theorem, in one dimension, that a function is almost everywhere the derivative of its integral, there is here the fact that, for almost every Y,

$$\lim_{r\to 0}\frac{1}{v(r)}\int_{(r;\,Y)} f(X)\,d\,V_X = \lim_{r\to 0}\frac{1}{v(r)}\int_{(r;\,O)} f(X+Y)\,d\,V_X = f(Y). \tag{8.05}$$

For any positive integer m it readily follows that, almost everywhere,

$$\begin{aligned}
&\lim_{r\to 0}\frac{1}{[v(r)]^m}\int_{(r;\,Y)} d\,V_{X_m}\cdots\int_{(r;\,X_3)} d\,V_{X_2}\int_{(r;\,X_2)} f(X_1)\,d\,V_{X_1}\\
&=\lim_{r\to 0}\frac{1}{[v(r)]^m}\int_{(r;\,O)} d\,V_{X_m}\cdots\int_{(r;\,O)} d\,V_{X_2}\int_{(r;\,O)} f(X_1+X_2+\cdots+X_m+Y)\,d\,V_{X_1}\\
&=f(Y).
\end{aligned} \tag{8.06}$$

The classical Stieltjes integral, for functions of a single variable,

$$\int_{-\infty}^{\infty} f(x)\, d\varphi(x),$$

may be defined under suitable conditions as the following limit:

$$\lim_{r\to 0} \frac{1}{2r} \int_{-\infty}^{\infty} f(x)\{\varphi(x+r)-\varphi(x-r)\}\, dx. \tag{8. 07}$$

If one denotes by $(r;\, x)$ that interval of length $2r$ which has its center at the point x and if one constructs the following function of an interval:

$$M(r;\, x) = \varphi(x+r)-\varphi(x-r), \tag{8. 08}$$

then this limit may be written in the form

$$\lim_{r\to 0} \frac{1}{2r} \int_{-\infty}^{\infty} f(x)\, M(r;\, x)\, dx, \tag{8. 09}$$

which may be called the Stieltjes integral of the point-function f with respect to the point-set-function M. Proceding by analogy we shall call the expression

$$\lim_{r\to 0} \frac{1}{v(r)} \int_{\infty} f(X)\, M(r;\, X)\, d\, V_X, \tag{8. 10}$$

when it exists, the Stieltjes integral of the n-dimensional point-function f with respect to the region-function M, and shall denote it by the symbol

$$\int_{\infty} f(X)\, d_X M. \tag{8. 11}$$

It will be necessary to introduce a generalization of this integral in order to handle certain region-functions M to appear later. If the expression

$$\lim_{r\to 0} \frac{1}{[v(r)]^m} \int_{\infty} f(X)\, M(r;\, X)\, d\, V_X \tag{8. 12}$$

exists it shall be called the mth Stieltjes integral of f with respect to M and will be denoted by

$$\int_{\infty} f(X)\, d_X^m M. \tag{8.13}$$

A function $f(X)$ shall be said to be quadratically summable over all space, or q. s. over ∞, provided it satisfies the requirements laid down above and provided that

$$\int_{\infty} |f(X)|^2\, d V_X \tag{8.14}$$

exists. A one-parameter family of functions $f_n(X)$, each of which is q. s. over ∞, is said to converge in mean to a function $f(X)$, also q. s. over ∞, as $n \to \infty$, if

$$\lim_{\lambda \to \infty} \int_{\infty} |f(X) - f_n(X)|^2\, d V_X = 0. \tag{8.15}$$

We are now in a position to discuss the harmonic analysis of a given function $f(X)$. The fundamental theorem is, of course, Plancherel's theorem on the Fourier transform. In n dimensions this reads as follows:

If $f(X)$ is q. s. over ∞, there exists its Fourier transform

$$g(U) = \underset{R \to \infty}{\text{l.i.m.}} \frac{1}{(2\pi)^{n/2}} \int_{R} f(X)\, e^{i(X \cdot U)}\, d V_X, \tag{8.16}$$

where the expanding region R is selected from an arbitrary one-parameter family of regions which are such that ultimately R covers and continues to cover almost any given point of space. All such transforms are equivalent; i.e. any pair of such functions can differ at most on a set of zero measure. Furthermore, $g(U)$ is q. s. over ∞ and satisfies the equality

$$\int_{\infty} |g(U)|^2\, d V_U = \int_{\infty} |f(X)|^2\, d V_X. \tag{8.17}$$

For any integer $m \geq 1$, there exists as an absolutely convergent integral

$$\int_{(r;\,0)} d\,V_{T_m} \cdots \int_{(r;\,0)} d\,V_{T_2} \int_{(r;\,0)} g(T_1 + T_2 + \cdots + T_m + U)\, d\,V_{T_1}$$

$$= \frac{1}{(2\,\pi)^{n/2}} \int_{\infty} f(X) \left\{ \int_{(r;\,0)} e^{i(X \cdot T)}\, d\,V_T \right\}^m e^{i(X \cdot U)}\, d\,V_X, \tag{8. 18}$$

which yields the following explicit formula for $g(U)$:

$$g(U) = \lim_{r \to 0} \frac{1}{(2\,\pi)^{n/2}} \cdot \frac{1}{[v(r)]^m} \int_{\infty} f(X) \left\{ \int_{(r;\,0)} e^{i(X \cdot T)}\, d\,V_T \right\}^m e^{i(X \cdot U)}\, d\,V_X; \quad m = 1, 2, \ldots; \tag{8. 19}$$

this limit existing for almost every U. Conversely, $g(U)$ possesses a Fourier transform and this is $f(-X)$:

$$f(X) = \underset{R \to \infty}{\text{l.i.m.}} \frac{1}{(2\,\pi)^{n/2}} \int_{R} g(U)\, e^{-i(X \cdot U)}\, d\,V_U. \tag{8. 20}$$

An equivalent statement is that $\bar{f}(X)$ is the Fourier transform of $\bar{g}(U)$. As above, we may write explicitly, for almost every X,

$$f(X) = \lim_{r \to 0} \frac{1}{(2\,\pi)^{n/2}} \cdot \frac{1}{[v(r)]^m} \int_{\infty} g(U) \left\{ \int_{(r;\,0)} e^{-i(U \cdot T)}\, d\,V_T \right\}^m e^{-i(X \cdot U)}\, d\,V_U; \quad m = 1, 2, \ldots. \tag{8. 21}$$

Finally, if $f_1(X)$ and $f_2(X)$ are each q. s. over ∞, and if $g_1(U)$ and $g_2(U)$ are their respective Fourier transforms, then

$$\int_{\infty} f_1(X)\, g_2(X)\, d\,V_X = \int_{\infty} f_2(U)\, g_1(U)\, d\,V_U. \tag{8. 22}$$

We are thus possessed of a harmonic analysis of any quadratically summable function. For a given such function $f(X)$ this consists in associating with almost every vector frequency U a complex amplitude $g(U)$. It is suggestive to imagine this $g(U)$ as the density of a complex mass distribution in the U space. The converse procedure by which we rebuild the given $f(X)$ from this mass distribution may be described formally as follows: We begin by multiplying the density of this distribution by the factor,

$$e^{-i(X \cdot U)}$$

thus altering the complex phase of the density in an simply periodic fashion. This done, we calculate the total mass of the resulting distribution and find it to be $f(X)$.

Now, there exists a very important class of functions to which exactly this harmonic analysis cannot be given, namely the n-tuply periodic functions. If such a function $f(X)$ be also q. s. over an arbitrary finite region, it is known to possess an n-tuple Fourier series representation:

$$f(X) \sim \Sigma a_k e^{i(X \cdot U_k)}; \tag{8. 23}$$

where the summation is effected for all points U_k which are vertices of a certain rectangular network. This analysis can also be interpreted as a mass distribution. Here, however, the spread is not continuous but consists of masses a_k concentrated at the corresponding frequency points U_k. Yet the process by which $f(X)$ is reobtained is again that of calculating the total mass of a distribution.

If one seeks a uniform method of treating these two types of mass spreads, one naturally is led to construct a region-function: the total mass in an arbitrary region. Knowledge of this function is equivalent to knowledge of the particular distribution in question. The advantage derived in employing it is that it is of the same order of magnitude for the various types of spreads whereas the densities are not. We shall see that this region-function can readily be calculated. To effect a return to the given function $f(X)$ we shall employ the Stieltjes integral which will simultaneously correctly modify the complex phases of the spread and determine the total resulting mass. Let us note how easily all this is carried out.

However, we can handle at once a more general problem. Let $f(X)$ be a function such that, for some integer m, the product

$$f(X)\left\{\int\limits_{(r;\,O)} e^{i(X \cdot T)}\, d\, V_T\right\}^m \tag{8. 24}$$

is q. s. over ∞. Let the Fourier transform of this product be denoted by

$$M^{(m)}(r;\; U) = \underset{R \to \infty}{\text{l.i.m.}} \frac{1}{(2\,\pi)^{n/2}} \int\limits_{r} f(X)\left\{\int\limits_{(r;\,O)} e^{i(X \cdot T)}\, d\, V_T\right\}^m e^{i(X \cdot U)}\, d\, V_X. \tag{8. 25}$$

Hence

$$f(X)\left\{\int_{(r;\,O)} e^{i(X\cdot T)}\,d\,V_T\right\}^m = \underset{R\to\infty}{\mathrm{l.i.m.}}\ \frac{1}{(2\,\pi)^{n/2}}\int_R e^{-i(X\cdot U)}\,M^{(m)}(r;\,U)\,d\,V_U. \quad (8.\ 26)$$

Integrating the above expression we obtain the result:

$$\int_{(s;\,Y)} f(X)\left\{\int_{(r;\,O)} e^{i(X\cdot T)}\,d\,V_T\right\}^m d\,V_X$$

$$= \frac{1}{(2\,\pi)^{n/2}}\int_\infty\left\{\int_{(s;\,O)} e^{-i(X\cdot U)}\,d\,V_X\right\}e^{-i(Y\cdot U)}\,M^{(m)}(r;\,U)\,d\,V_U. \quad (8.\ 27)$$

Now since

$$\lim_{r\to 0}\frac{1}{[v(r)]^m}\left\{\int_{(r;\,O)} e^{i(X\cdot T)}\,d\,V_T\right\}^m = 1, \quad (8.\ 28)$$

and since $f(X)$ is summable, it follows that

$$\int_{(s;\,Y)} f(X)\,d\,V_X = \lim_{r\to 0}\frac{1}{(2\,\pi)^{n/2}}\frac{1}{[v(r)]^m}\int_\infty\left\{\int_{(s;\,O)} e^{-i(X\cdot U)}\,d\,V_X\right\}e^{-i(Y\cdot U)}\,M^{(m)}(r;\,U)\,d\,V_U. \quad (8.\ 29)$$

The right hand side is precisely an mth order Stieltjes integral. We have, then,

$$\int_{(s;\,Y)} f(X)\,d\,V_X = \frac{1}{(2\,\pi)^{n/2}}\int_\infty\left\{\int_{(s;\,O)} e^{-i(X\cdot U)}\,d\,V_X\right\}e^{-i(Y\cdot U)}\,d_U^m\,M^{(m)}, \quad (8.\ 30)$$

and, therefore, for almost every X,

$$f(X) = \lim_{s\to 0}\frac{1}{(2\,\pi)^{n/2}}\frac{1}{v(r)}\int_\infty\left\{\int_{(s;\,O)} e^{-i(T\cdot U)}\,d\,V_T\right\}e^{-i(X\cdot U)}\,d_U^m\,M^{(m)}. \quad (8.\ 31)$$

That the function $M^{(m)}(r;\,U)$ constitutes a harmonic analysis of the given $f(X)$ is established by the following considerations. It is a matter of simple calculation to show that if $f(X)$ is itself q. s. over ∞, $M^{(1)}(r;\,U)$ exists and has for its value the total mass in the sphere $(r;\,U)$ of a distribution of density $g(U)$. It is similarly easy to show that if $f(X)$ is possessed of a Fourier series development, $M^{(1)}(r;\,U)$ exists as a limit in the mean and is equal to the sum of those Fourier coefficients a_k which correspond to frequency points U lying in $(r;\,U)$. As we have seen above, $M^{(1)}(r;\,U)$ determines the mass distributions which constitute the harmonic analysis of $f(X)$. In the same fashion $M^{(2)}(r; U)$ determines $M^{(1)}(r; U)$; etc.

While, then, we are justified in considering that $M^{(m)}(r; U)$ is a harmonic analysis of $f(X)$, we must yet determine for what class of functions $f(X)$ the region-function M will exist at least for some integer m. It can be shown that the function

$$\int_{(r;\,O)} e^{i(X \cdot T)}\, d\, V_T \tag{8.32}$$

is bounded for all X and is of the order of $|X|^{-\frac{n+1}{2}}$ for large values of $|X|$. This at once establishes the fact that if, for some positive p, the quotient

$$\frac{f(X)}{1+|X|^r} \tag{8.33}$$

is q. s. over ∞, then a value of m can be determined for which $M^{(m)}$ will exist. Furthermore, a generalization to n dimensions of a Tauberian theorem such as given by Jacob readily shows that if $f(X)$ is q. s. over all finite regions and is such that for some positive p the expression

$$\frac{1}{[v(r)]^p} \int_{(r;\,O)} |f(X)|^2\, d\, V_X \tag{8.34}$$

is bounded for sufficiently large values of r, then it too belongs to the class in question. An n-tuply periodic function is included in this last type of function $f(X)$. Essentially, then, the functions which we can harmonically analyze are those which are »algebraic on the average» as $|X| \to \infty$.

When we come to the study of the energy spectrum of a given function $f(X)$ we again subject the function to a harmonic analysis but not in such great detail as before. In the corresponding mass spread we are no longer interested in the complex phase of the various masses and densities present, but solely in their absolute values, or, more precisely, in the squares of these absolute values. Obviously, again, there will be different orders of density, or, as we may say, different orders of energy in the component oscillations. For the functions of the class described above there will exist for some value of $k \geq 0$ the limit.

$$\varphi(X) = \lim_{r \to \infty} \frac{1}{[v(r)]^k} \int_{(r;\,O)} f(T) \bar{f}(T-X)\, d\, V_T. \tag{8.35}$$

The harmonic analysis of this latter function will show that its mass distribution consists of the squares of the absolute values of the highest order densities that appear in the distribution corresponding to $f(X)$. Terms of lower energy level do not appear. Thus, the total mass of the distribution corresponding to $\varphi(X)$ does not necessarily coincide with although it will never exceed the total energy which could be calculated from the spread associated with $f(X)$.

Precisely as in the one-dimensional case it can be shown that

$$|\varphi(X)| \leq \varphi(0) = \lim_{r \to \infty} \frac{1}{[v(r)]^k} \int_{(r;\, 0)} |f(T)|^2 \, d\, V_T. \tag{8.36}$$

From this boundedness it follows that the product

$$\varphi(X) \left\{ \int_{(r;\, 0)} e^{i(X \cdot T)} \, d\, V_T \right\} \tag{8.37}$$

is q. s. over ∞, and hence that there exists $M^{(1)}(r;\, U)$ formed with respect to $\varphi(X)$. This function is at present only defined as a limit in the mean. It is desired to show that it can be so defined that for any given U it will be a monotonic non-decreasing function of r. This is carried out as when $n=1$ by a series of limiting-processes which do not alter any monotonic properties. To begin with one shows that if

$$\left.\begin{aligned} \varphi_s(X) &= \frac{1}{[v(s)]^k} \int_{\infty} f_s(T) \bar{f}_s(T - X) \, d\, V_T; \\ f_s(T) &= \begin{cases} f(T); & [T \text{ in } (s;\, 0)] \\ 0\,; & [T \text{ elsewhere}] \end{cases} \end{aligned}\right\} \tag{8.38}$$

then

$$\varphi(X) = \lim_{s \to \infty} \varphi_s(X). \tag{8.39}$$

Furthermore one notes that

$$\int_{\infty} \varphi_s(X) \, e^{i(X \cdot Y)} \, d\, V_X = \frac{1}{[v(s)]^k} \left| \int_{\infty} f_s(T) \, e^{i(T \cdot Y)} \, d\, V_T \right|^2 \geq 0, \tag{8.40}$$

precisely as in the one-dimensional case. From this it follows that the function

$$M_s(r;\, U) = \frac{1}{(2\,\pi)^n} \int\limits_{\infty} \varphi_s(X) \left\{ \int\limits_{(r;\, O)} e^{i\,(X \cdot T)}\, d\, V_T \right\} e^{i\,(X \cdot U)}\, d\, V_X$$

$$= \frac{1}{(2\,\pi)^n} \int\limits_{(r;\, U)} d\, V_T \int\limits_{\infty} \varphi_s(X)\, e^{i(X \cdot T)}\, d\, V_X \qquad (8.\,41)$$

for given U is monotone non-decreasing in r. It is readily seen that

$$\int\limits_{(r';\, U')} M(r;\, U)\, d\, V_U = \lim_{S \to \infty} \int\limits_{(r';\, U')} M_s(r;\, U)\, d\, V_U, \qquad (8.\,42)$$

and hence that for given r' and U' the integral on the left has the same monotone property with respect to r. Finally since, for almost every U',

$$M(r;\, U') = \lim_{r' \to 0} \frac{1}{v(r')} \int\limits_{(r';\, U')} M(r;\, U)\, d\, V_U, \qquad (8.\,43)$$

we see that, in so far as it is thus defined, $M(r;\, U)$ has the desired property. Since the point-set on which M is not defined constitutes at most a set of zero measure, it is a simple matter to define M for all r and U so that for each U will be a monotonic non-decreasing function of r.

The total spectral intensity of $f(X)$, or more accurately, the total spectral intensity of those components which are associated with the maximum energy level, is given by the limit

$$\lim_{r \to \infty} M(r;\, U), \qquad (8.\,44)$$

if this exists. One fairly readily shows that

$$M(r;\, U) \leq \varphi(0), \qquad (8.\,45)$$

and, because of the monotony, hence that the limit in question exists and lies between 0 and $\varphi(0)$. The details of the argument whereby one next shows that

$$\lim_{r \to \infty} M(r;\, U) = \lim_{r' \to 0} \frac{1}{v(r')} \int\limits_{(r';\, O)} \varphi(X)\, d\, V_X \leq \varphi(0) \qquad (8.\,46)$$

will be omitted here to avoid too much complication.

9. Coherency matrices.

The spectrum theory of our earlier sections is a theory of the spectrum of an individual function. There are, however, many phenomena intimately connected with harmonic analysis which refer to several functions considered simultaneously. Chief among these are the phenomena of coherency and incoherency, of interference, and of polarization.

It is known to every beginner in physics that two rays of light from the same source may interfere: that is, they may be superimposed to form a darkness, or else a light more intense than is ordinarily formed by two rays of light of their respective intensities. On the other hand, two rays of light from independent sources or from different parts of the same source never exhibit this phenomenon. The former rays are said to be *coherent*, and the latter to be *incoherent*. Although it is mathematically impossible for two truly sinusoidal oscillations to be incoherent, even the most purely monochromatic light which we can sensibly produce never coheres with similar light from another source.

The physicist's explanation of incoherency is the following: the interference pattern produced by two sources of light depends on their relative displacement in phase. Now, the relative phase of two sensibly monochromatic sources of light is able to assume all possible values, and since light probably consists in a series of approximately sinusoidal trains of oscillations each lasting but a small portion of a millionth of a second, this relative phase assumes in any sensible interval all possible values with a uniform distribution which averages out light and dark bands into a sensibly uniform illumination.

This explanation of incoherency is unquestionable adequate to account for the phenomenon which it was invented to explain. Nevertheless, it is desirable to have a theory of coherency and incoherency which does not postulate a hypothetical set of constituent harmonic trains of oscillations, which at any rate must become merged in the general electromagnetic oscillation constituting the light. The present section is devoted to the development of a theory of coherency which is as direct as the theory of this paper concerning the harmonic analysis of a single function, and indeed forms a natural extension of the latter.

In interference, the components of the electromagnetic field of the constituent light rays combine additively. Accordingly, the theory of coherency and interference must be a theory of the harmonic analysis of all functions which can be obtained from a given set by linear combination. Let us see what the outlines of this theory are.

We start from a class of functions, $f_k(t)$, in general complex, and defined for all real arguments between $-\infty$ and ∞. For the present we shall assume this class of functions to be finite, although this restriction is not essential. Let

$$f(t)=a_1 f_1(t)+a_2 f_2(t)+\cdots+a_n f_n(t) \tag{9.01}$$

be the general linear combination of functions of the set. We shall have

$$\varphi(\tau)=\lim_{T\to\infty}\frac{1}{2T}\int_{-T}^{T} f(t+\tau)\bar{f}(t)\,dt=\sum_{j,k=1}^{n}\sum a_j\bar{a}_k \lim_{T\to\infty}\frac{1}{2T}\int_{-T}^{T} f_j(t+\tau)\bar{f}_k(t)\,dt \tag{9.02}$$

in case the latter limits exist. The necessary and sufficient condition for this to exist for all linear combinations $f(t)$ of functions of the set is that

$$\varphi_{jk}(\tau)=\lim_{T\to\infty}\frac{1}{2T}\int_{-T}^{T} f_j(t+\tau)\bar{f}_k(t)\,dt \tag{9.03}$$

should exist for every j, k, and τ. Then

$$\varphi(\tau)=\sum_j\sum_k a_j\bar{a}_k\varphi_{jk}(\tau). \tag{9.04}$$

Again, we shall have formally

$$S(u)=\frac{1}{2\pi}\int_{-\infty}^{\infty}\varphi(\tau)\frac{e^{iu\tau}-1}{i\tau}\,d\tau=\sum_j\sum_k a_j\bar{a}_k\frac{1}{2\pi}\int_{-\infty}^{\infty}\varphi_{jk}(\tau)\frac{e^{iu\tau}-1}{i\tau}\,d\tau, \tag{9.05}$$

where we may write

$$S_{jk}(u)=\frac{1}{2\pi}\int_{-\infty}^{\infty}\varphi_{jk}(\tau)\frac{e^{iu\tau}-1}{i\tau}\,d\tau. \tag{9.06}$$

If $\varphi_{jk}(\tau)$ exists for every j, k, and τ, we may readily show that each $S_{jk}(u)$ exists. Clearly

$$\varphi_{kj}(\tau)=\lim_{T\to\infty}\frac{1}{2T}\int_{-T}^{T} f_k(t+\tau)\bar{f}_j(t)\,dt=\lim_{T\to\infty}\frac{1}{2T}\int_{-T}^{T}\bar{f}_j(t-\tau)f_k(t)\,dt=\bar{\varphi}_{jk}(-\tau) \tag{9.07}$$

and

$$S_{kj}(u) = \frac{1}{2\pi}\int_{-\infty}^{\infty} \bar{\varphi}_{jk}(-\tau)\frac{e^{iu\tau}-1}{i\tau}\,d\tau = \frac{1}{2\pi}\int_{-\infty}^{\infty} \bar{\varphi}_{jk}(\tau)\frac{e^{-iu\tau}-1}{-i\tau}\,d\tau = \bar{S}_{jk}(u), \qquad (9.08)$$

so that the matrix

$$\| S_{jk}(u) \|$$

is Hermitian. This matrix determines the spectra of all possible linear combinations of $f_1(t), \ldots, f_n(t)$. Since it determines the precise coherency relations of the functions in question, we shall call it the coherency matrix.

Let us subject the functions $f_k(t)$ to the linear transformation

$$g_j(t) = \sum_k a_{jk} f_k(t). \qquad (9.09)$$

Then

$$\psi_{jk}(\tau) = \lim_{T\to\infty}\frac{1}{2T}\int_{-T}^{T} g_j(t+\tau)\,\bar{g}_k(t)\,dt = \sum_l \sum_m a_{jl}\bar{a}_{km}\,\varphi_{lm}(\tau) \qquad (9.10)$$

and

$$T_{jk}(u) = \frac{1}{2\pi}\int_{-\infty}^{\infty} \psi_{jk}(\tau)\frac{e^{iu\tau}-1}{i\tau}\,d\tau = \sum_l \sum_m a_{jl}\bar{a}_{km}\,S_{lm}(u). \qquad (9.11)$$

Thus the new coherency matrix $\| T_{jk}(u) \|$ may be written

$$M_1 \cdot \| S_{jk}(u) \| \cdot M_2, \qquad (9.12)$$

where

$$M_1 = \| a_{jk} \|, \qquad (9.13)$$

and

$$M_2 = \| \bar{a}_{kj} \|. \qquad (9.14)$$

In case the transformation with matrix M_1 has the property

$$M_1 \cdot M_2 = M_2 \cdot M_1 = 1, \qquad (9.15)$$

it is said to be *unitary*. For such transformations,

$$\| T_{jk} \| = M_1 \cdot \| S_{jk} \| \cdot M_1^{-1}. \qquad (9.16)$$

A matrix $\| a_{jk} \|$ is said to be in diagonal form if all the terms a_{jk} for which $j \neq k$ are identically zero. By a theorem of Weyl[5], every Hermitian matrix may be transformed into a diagonal matrix by a unitary transformation. Since we may regard a diagonal matrix as representing a set of completely incoherent phenomena, this transformation is of fundamental importance in the characterization of the state of coherency of the functions determining the matrix. Together with the numerical values of the diagonal elements of the diagonal matrix, it indeed consitutes a complete characterization of the state of coherency of the original function. In the case the values of the diagonal elements are distinct, this characterization is indeed to be carried through in but a single way.

The production of incoherent functions is a simple matter, when we have once settled the existence theory of functions with given types of spectra. Let $f_1(t)$ be any bounded function such that $\varphi(\tau)$ and consequently $S(u)$ exists. Then if

$$f_2(t) = f_1(t)\, e^{i\sqrt{|t|}}, \tag{9.17}$$

we have

$$\lim_{T\to\infty} \frac{1}{2T}\int_{-T}^{T} f_2(t+\tau)\bar{f}_2(t)\,dt = \lim_{T\to\infty} \frac{1}{2T}\int_{-T}^{T} f_1(t+\tau)\bar{f}_1(t)\,e^{i(\sqrt{|t+\tau|}-\sqrt{|t|})}\,dt$$

$$= \lim_{T\to\infty} \frac{1}{2T}\int_{-T}^{T} f_1(t+\tau)\bar{f}_1(t)\,\exp\left(i\frac{|t+\tau|-|t|}{\sqrt{|t+\tau|}+\sqrt{|t|}}\right)dt, \tag{9.18}$$

and hence, since $\lim\limits_{t\to\pm\infty} \exp\left(i\dfrac{|t+\tau|-|t|}{\sqrt{|t+\tau|}+\sqrt{|t|}}\right) = 1$,

$$\varphi_{22}(\tau) = \varphi_{11}(\tau) \tag{9.19}$$

and

$$S_{22}(u) = S_{11}(u). \tag{9.20}$$

On the other hand, if for example $f(t) = e^{i\lambda t}$,

$$\left.\begin{aligned} \varphi_{12}(\tau) &= \lim_{T\to\infty} \frac{1}{2T}\int_{-T}^{T} f_1(t+\tau)\bar{f}_1(t)\,e^{-i\sqrt{|t|}}\,dt = 0; \\ S_{12}(u) &= S_{21}(u) = 0. \end{aligned}\right\} \tag{9.21}$$

Thus the coherency matrix of $f_1(t)$ and $f_2(t)$ is

24 — 29764. *Acta mathematica*. 55. Imprimé le 14 avril 1930.

$$S_{11}(u)\begin{Vmatrix} 1 & 0 \\ 0 & 1 \end{Vmatrix}. \tag{9.22}$$

The coherency matrix of $\sqrt{2}f_1(t)$ and 0 is

$$S_{11}(u)\begin{Vmatrix} 2 & 0 \\ 0 & 0 \end{Vmatrix}; \tag{9.23}$$

that of $f_1(t)$ and $f_1(t)$ is

$$S_{11}(u)\begin{Vmatrix} 1 & 1 \\ 1 & 1 \end{Vmatrix}; \tag{9.24}$$

that of $f_1(t)$ and $if_1(t)$ is

$$S_{11}(u)\begin{Vmatrix} 1 & -i \\ i & 1 \end{Vmatrix}. \tag{9.25}$$

Let it be noted that the coherency matrices of real functions are in general complex. Thus if

$$f_2(t) = f_1(t+\lambda), \tag{9.26}$$

we have

$$\left.\begin{aligned} \varphi_{22}(\tau) &= \varphi_{11}(\tau); \\ \varphi_{12}(\tau) &= \lim_{T\to\infty}\frac{1}{2T}\int_{-T}^{T} f_1(t+\lambda+\tau)f_1(t)\,dt \\ &= \varphi_{11}(\tau+\lambda); \end{aligned}\right\} \tag{9.27}$$

and hence

$$\left.\begin{aligned} &S_{22}(u) = S_{11}(u); \\ &S_{12}(u) = \int_{-\infty}^{\infty}\varphi_{11}(\tau+\lambda)\frac{e^{iu\tau}-1}{i\tau}\,d\tau = \int_{-\infty}^{u} e^{-iv\lambda}\,dS_{11}(v) + S_{12}(-\infty) \end{aligned}\right\} \tag{9.28}$$

giving the coherency matrix with derivative

$$S_{11}^{1}(u)\begin{Vmatrix} 1 & e^{-iv\lambda} \\ e^{iv\lambda} & 1 \end{Vmatrix}. \tag{9.29}$$

In optics, coherency is generally considered for light of one particular frequency. From that standpoint, the coherency of a set of functions $f_1(t), \ldots, f_n(t)$

with continuous differentiable spectra may be regarded as determined for frequency u by the matrix

$$\left\| \begin{matrix} S^1_{11}(u), & \ldots, & S^1_{1n}(u) \\ \cdot & \cdot & \cdot \\ \cdot & \cdot & \cdot \\ S^1_{n1}(u), & \ldots, & S^1_{nn}(u) \end{matrix} \right\|; \tag{9.30}$$

or in the case of functions with line spectra, by

$$\left\| \begin{matrix} S_{11}(u+0)-S_{11}(u-0), & \ldots, & S_{1n}(u+0)-S_{1n}(u-0) \\ \cdot & \cdot & \cdot \\ \cdot & \cdot & \cdot \\ S_{n1}(u+0)-S_{n1}(u-0), & \ldots, & S_{nn}(u+0)-S_{nn}(u-0) \end{matrix} \right\|. \tag{9.31}$$

We may regard these matrices in a secondary sense as coherency matrices.

Coherency matrices of two functions are of particular interest in connection with the characterization of the state of polarization of light. As everyone knows, this characterization is identically the characterization of the state of coherency between two components of the electric vector at right angles to one another. With this interpretation, matrix (9.22) represents unpolarized light, matrix (9.23) light polarized completely in one plane, while

$$\left\| \begin{matrix} 0 & 0 \\ 0 & 2 \end{matrix} \right\| \tag{9.32}$$

represents light completely polarized in a plane perpendicular to the first. Matrix (9.24) and matrix

$$\left\| \begin{matrix} 1 & -1 \\ -1 & 1 \end{matrix} \right\| \tag{9.33}$$

represent light polarized respectively at 45° and at 135° to the first direction. Matrix (9.25) and matrix

$$\left\| \begin{matrix} 1 & i \\ -i & 1 \end{matrix} \right\| \tag{9.34}$$

represent respectively light polarized circularly in a counter-clockwise and a clockwise direction.

When the matrix of completely polarized light, whether linearly, elliptically, or circularly polarized, is brought into diagonal form by a linear unitary transformation, the resulting diagonal matrix will have only one element distinct from 0. On the other hand, completely unpolarized light has the diagonal terms equal. This suggests as a measure of the amount of polarization of the diagonal matrix

$$\left\|\begin{matrix} a & 0 \\ 0 & b \end{matrix}\right\|, \tag{9.35}$$

or of any other matrix equivalent to it under a unitary transformation, the quantity

$$a - b. \tag{9.36}$$

If we subtract from our original diagonal matrix the completely incoherent matrix

$$\left\|\begin{matrix} \frac{a+b}{2} & 0 \\ 0 & \frac{a+b}{2} \end{matrix}\right\|, \tag{9.37}$$

which is invariant under every unitary transformation, we get the matrix

$$\left\|\begin{matrix} \frac{a-b}{2} & 0 \\ 0 & \frac{b-a}{2} \end{matrix}\right\|, \tag{9.38}$$

which may be regarded as a representative of the quantity $a-b$. This suggests that given any coherency matrix

$$\left\|\begin{matrix} A & B+Ci \\ B+Ci & D \end{matrix}\right\|, \tag{9.39}$$

we may take $A+D$ to represent the intensity of the corresponding light, and the matrix

$$\left\|\begin{matrix} \frac{A-D}{2} & B-Ci \\ B+Ci & \frac{D-A}{2} \end{matrix}\right\| \tag{9.40}$$

as its polarization. Thus horizontal polarization is represented by the matrix

$$\begin{Vmatrix} 1 & 0 \\ 0 & -1 \end{Vmatrix}; \tag{9.41}$$

polarization at 45° by the matrix

$$\begin{Vmatrix} 0 & 1 \\ 1 & 0 \end{Vmatrix}; \tag{9.42}$$

and circular polarization by the matrix

$$\begin{Vmatrix} 0 & i \\ -i & 0 \end{Vmatrix}. \tag{9.43}$$

These are the same matrices which Jordan, Dirac and Weyl have employed to such advantage in the theory of quanta.

Since the most general Hermitian matrix of the second order may be written

$$\begin{Vmatrix} \alpha+\beta & \gamma+\delta i \\ \gamma-\delta i & \alpha-\beta \end{Vmatrix} = \alpha \begin{Vmatrix} 1 & 0 \\ 0 & 1 \end{Vmatrix} + \beta \begin{Vmatrix} 1 & 0 \\ 0 & -1 \end{Vmatrix} + \gamma \begin{Vmatrix} 0 & 1 \\ 1 & 0 \end{Vmatrix} + \delta \begin{Vmatrix} 0 & i \\ -i & 0 \end{Vmatrix}, \tag{9.44}$$

it appears that all light may characterized as to its state of polarization by given the total amount of light it contains, the excess of the amount polarized at 0° over that polarized at 90°, the excess of the amount polarized at 45° over that polarized at 135° and the excess of that polarized circularly to the right over that polarized circularly to the left. This characterization is complete and univocal. The total intensity of the light may be read off any sort of a photometer. The excess of light polarized horizontally over that polarized vertically may be determined by a doubly refracting crystal in one orientation, and the excess of light polarized at 45° over that polarized at 135° by the same crystal in another orientation. It is possible furthermore to devise an instrument which will read off the amount of circular polarization in the light in question. The three latter instruments possess some very remarkable group properties with respect to one another. Either portion of the light emerging from any one of the instruments will behave towards the other two exactly like unpolarized light. Rotation of the plane of polarization of the light through 45° will change the reading of the first of the last three instruments into that of the second, and the reading of the second into minus that of the first, leaving the reading of the third unchanged. There are precisely analogous unitary transformations interchanging any other pair of the three instruments, leaving the reading of the remaining instrument untouched. These transformations together with their powers and the identical transformation form a group.

A fact concerning polarized light which is so apparently obvious that it is generally regarded as not needing any proof is that all light is a case or limiting case of partially elliptically polarized light. It is nevertheless desirable to prove this statement. Completely elliptically polarized light with the coordinate axes as principal axes has a coherency matrix of the form

$$\left\| \begin{matrix} A^2 & -ABi \\ ABi & B^2 \end{matrix} \right\|; \tag{9.45}$$

and hence partially polarized light with the same principal axes has a coherency matrix of the form

$$P = \left\| \begin{matrix} A^2+D^2 & -ABi \\ ABi & B^2+D^2 \end{matrix} \right\|. \tag{9.46}$$

We wish to show that the general coherency matrix

$$M = \left\| \begin{matrix} \alpha & \gamma-\delta i \\ \gamma+\delta i & \beta \end{matrix} \right\| \tag{9.47}$$

may be transformed into this form by a real unitary transformation in such a way that

$$T \, . \, M \, . \, T^{-1} = P.$$

To do this, we need only put

$$T = \left\| \begin{matrix} \cos\varphi & \sin\varphi \\ \sin\varphi & -\cos\varphi \end{matrix} \right\|; \text{ where } \tan 2\varphi = \frac{2\gamma}{\alpha-\beta}. \tag{9.48}$$

Thus the axes of polarization of M are 1 and 2 directions when we replace $f_1(t)$ and $f_2(t)$ by

$$\left. \begin{aligned} g_1(t) &= f_1(t)\cos\varphi + f_2(t)\sin\varphi; \\ g_2(t) &= -f_2(t)\cos\varphi + f_1(t)\sin\varphi; \end{aligned} \right\} \tag{9.49}$$

the »lengths» of these axes are respectively

$$\left.\begin{aligned} A &= \left\{\frac{1}{2}\left[(\alpha-\beta)^2+4(\gamma^2+\delta^2)\right]^{1/2}+\frac{1}{2}\left[(\alpha-\beta)^2+4\gamma^2\right]^{1/2}\right\}^{1/2}; \\ B &= \left\{\frac{1}{2}\left[(\alpha-\beta)^2+4(\gamma^2+\delta^2)\right]^{1/2}-\frac{1}{2}\left[(\alpha-\beta)^2+4\gamma^2\right]^{1/2}\right\}^{1/2}; \end{aligned}\right\} \tag{9.50}$$

and the percentage of polarization

$$100\left(1-\frac{4(\gamma^2+\delta^2)}{(\alpha+\beta)^2}\right)^{1/2}. \tag{9.51}$$

The connection between coherency matrices and optical instruments, which we have already mentioned in the case of polarized light, is of far more general applicability. An optical instrument is a method, linear in electric and magnetic field vectors, of transforming a light input into a light output. In general, this transformation, in the language of Volterra[5], belongs to the group of the closed cycle with respect to the time, in the sense that it is independent of the position of our initial instant in time. Such a transformation leaves a simple harmonic input still simply harmonic in the time, although in general with a shift in phase.

An example of an optical instrument is a microscope. This may be regarded as a means of making an electromagnetic disturbance in the image plane depend linearly on a given electromagnetic disturbance in the object plane. Telescopes, spectroscopes, Nicol prisms, etc., serve as further examples of optical instruments in the sense in question. Among these, a particularly interesting ideal type is the conservative optical instrument, in which the power of the input and the power of the output are identical. This power depends quadratically on the electric and magnetic vectors, so that a conservative optical instrument has a quadratic invariant for the corresponding transformations. When only terms of the one frequency of e^{iut} are considered, this quadratic positive invariant becomes Hermitian, and has essentially the same properties as the expression

$$x_1\bar{x}_1 + x_2\bar{x}_2 + \cdots + x_n\bar{x}_n \tag{9.52}$$

which is invariant under all unitary transformations of $x_1, \ldots, x_n$. Thus the theory of the group of unitary transformations is physically applicable, not only in quantum mechanics, where Weyl has already employed it so successfully, but even in classical optics. It is the conviction of the author that this analogy is not merely an accident, but is due to a deep-lying connection between the two theories.

In quantum mechanics, while all the terms of a matrix enter in an essential way into its transformation theory, only diagonal terms are given an immediate physical significance. This is also in precise accord with the optical situation. Every optical observation ends with the measurement of an energy or power, either by direct bolometric or thermometric means, or by the observations of a visual intensity or the blackening of a photographic plate. Every such observation means the more or less complete determination of some diagonal term. The non-diagonal terms of a coherency matrix of light only have signifiance in so far as they enable us to predict the energies or intensities which the light will show after having been subjected to some linear transformation or optical instrument. This fact that new diagonal terms after a transformation cannot be read off from the old diagonal terms before a transformation, without the intervention of non-diagonal terms, is the optical analogue for the principle of indetermination in quantum theory, according to which observations on the momentum of an electron alone cannot yield a single value if its position is known, and vice versa. The statement that every observation of an electron affects its properties has the following analogy: if two optical instruments are arranged in series, the taking of a reading from the first will involve the interposition of a ground-glass screen or photographic plate between the two, and such a plate will destroy the phase relations of the coherency matrix of the emitted light, replacing it by the diagonal matrix with the same diagonal terms. Thus the observation of the output of the first instrument alters the output of the second. In this case, the possibility of taking part of the output of the first instrument for reading by a thinly silvered mirror warns us not to try to push the analogy with quantum theory too far.

Coherency matrices form a close analogue to the correlation matrices long familiar in statistical theory. If we have a set of n observations $x^{(1)}, x^{(2)}, \ldots, x^{(n)}$ all made together, and this set of observations is repeated on the occasions 1, 2, ..., m thus yielding sets $x_1^{(1)}, \ldots, x_1^{(n)}; x_2^{(1)}, \ldots, x_2^{(n)}; \ldots; \ldots; x_m^{(1)}, \ldots, x_m^{(n)}$, the correlation matrix of these observations is

$$\left\| \begin{array}{cccc} \sum_1^m (x_k^{(1)})^2, & \sum_1^m x_k^{(1)} x_k^{(2)}, & \cdots\cdots, & \sum_1^m x_k^{(1)} x_k^{(n)} \\ \sum_1^m x_k^{(2)} x_k^{(1)}, & \sum_1^m (x_k^{(2)})^2, & \cdots\cdots, & \sum_1^m x_k^{(2)} x_k^{(n)} \\ \cdots & \cdots & \cdots & \cdots \\ \cdots & \cdots & \cdots & \cdots \\ \sum_1^m x_k^{(n)} x_k^{(1)}, & \sum_1^m x_k^{(n)} x_k^{(2)}, & \cdots\cdots, & \sum_1^m (x_k^{(n)})^2 \end{array} \right\| . \tag{9.53}$$

This symmetrical matrix represents the entire amount and kind of linear relationship to be observed between the different observations in question. The further analysis of the information yielded by a correlation matrix depends on the nature of the data to be analysed. Thus if the two observations of each set are the x and y coordinates of the position of a shot on a target, the rotations of the x and y axes have a concrete geometrical meaning, and the question of reducing the matrix to diagonal form by a rotation of axes is the significant question of determining the ellipse which best represents the distribution of holes in the target. On the other hand, if the quantities whose correlation we are investigating are the price of wheat x in dollars per bushel and the marriage rate y, rotations have no significance, as there is no common scale, while on the other hand, the significant information yielded by the matrix must be invariant under the transformations

$$\left. \begin{array}{l} x_1 = kx; \\ y_1 = ly. \end{array} \right\} \tag{9.54}$$

The so-called coefficients of correlation and of partial correlation and the lines of regression of x on y and of y on x have this type of invariance.

Correlation matrices and their derived quantities are the tool for the statistical analysis of what is known as frequency series, series of data where no such variable as the time enters as a parameter. In the study of meteorology, of business cycles, and of many other phenomena of interest to the statistician, on the other hand, we must discuss time series, where the relations of the data in time are essential. The proper analysis of these has long been a moot point among statisticians and economists. As far as it is linear relationships which we are seeking for, it is only reasonable to suppose that coherency matrices

25—29764. *Acta mathematica*. 55. Imprimé le 14 avril 1930.

must play the same rôle for time series which correlation matrices play for frequency series. In statistical work, the group of transformations which will most frequently be permissible is as before

$$\begin{aligned} g_1(t) &= A f_1(t); \\ g_2(t) &= B f_2(t). \end{aligned} \qquad [A \text{ and } B \text{ real}] \qquad (9.55)$$

Under this group, the significant invariants of the Hermitian matrix

$$\begin{Vmatrix} S^1_{11}(u) & S^1_{12}(u) \\ S^1_{21}(u) & S^1_{22}(u) \end{Vmatrix} \qquad (9.56)$$

are

$$r(u) = S^1_{12}(u)\,[S^1_{11}(u)\, S^1_{22}(u)]^{-1/2}; \qquad (9.57)$$

which we may call the coefficient of coherency of f_1 and f_2 for frequency u, and

$$\sigma_1(u) = \frac{S^1_{12}(u)\sqrt{S^1_{11}(u)}}{S^1_{22}(u)} \text{ and } \sigma_2(u) = \frac{S^1_{21}(u)\sqrt{S^1_{22}(u)}}{S^1_{11}(u)}, \qquad (9.58)$$

the coefficients of regression respectively of f_1 on f_2 and of f_2 on f_1. The modulus of the coefficient of coherency represents the amount of linear coherency between $f_1(t)$ and $f_2(t)$ and the argument the phase-lag of this coherency. The coefficients of regression determine in addition the relative scale for equivalent changes of f_1 and f_2.

The computation of coefficients of coherency and of regression is to be done in the steps indicated in their definition. In the case where only a finite set of real data are at our disposal, distributed at equal unit intervals from 0 to n, say $x_0, x_1, \ldots, x_n$ and $y_0, y_1, \ldots, y_n$, the steps of our computation are:

$$\left.\begin{aligned} (\varphi_k)_{11} &= \frac{1}{n}\sum_0^{n-k} x_j\, x_{j+k}; \\ (\varphi_k)_{12} &= \frac{1}{n}\sum_0^{n-k} x_j\, y_{j+k}; \\ (\varphi_k)_{21} &= \frac{1}{n}\sum_0^{n-k} y_j\, x_{j+k}; \\ (\varphi_k)_{22} &= \frac{1}{n}\sum_0^{n-k} y_i\, y_{j+k}; \end{aligned}\right\} \qquad [0 \le k \le n] \qquad (9.59)$$

$$r(u)=\frac{\sum_{0}^{n}\{[(\varphi_k)_{21}+(\varphi_k)_{12}]\cos ku+i\,[(\varphi_k)_{21}-(\varphi_k)_{12}]\sin ku\}-(\varphi_0)_{12}/2}{2\left[\sum_{0}^{n}(\varphi_k)_{11}\cos ku-(\varphi_0)_{11}/2\right]^{1/2}\left[\sum_{0}^{n}(\varphi_k)_{22}\cos ku-(\varphi_0)_{22}/2\right]^{1/2}}; \qquad (9.60)$$

$$\left.\begin{aligned}\sigma_1(u) &= r(u)\frac{\left[\sum_{0}^{n}(\varphi_k)_{11}\cos ku-(\varphi_0)_{11}/2\right]^{1/2}}{\left[\sum_{0}^{n}(\varphi_k)_{22}\cos ku-(\varphi_0)_{22}/2\right]^{1/2}};\\ \sigma_2(u) &= \bar{r}(u)\frac{\left[\sum_{0}^{n}(\varphi_k)_{22}\cos ku-(\varphi_0)_{22}/2\right]^{1/2}}{\left[\sum_{0}^{n}(\varphi_k)_{11}\cos ku-(\varphi_0)_{11}/2\right]^{1/2}}.\end{aligned}\right\} \qquad (9.61)$$

In case we have at our disposal methods for performing the Fourier transformation, we may compute these coefficients directly from graphs. Several devices for this purpose are now being developed in the laboratory of Professor V. Bush in the Department of Electrical Engineering of the Massachusetts Institut of Technology.

10. **Harmonic analysis and transformation groups.**

Inasmuch, as the theory of Fourier series forms a special chapter in the theory of expansions in general sets of normal and orthogonal functions, it is reasonable to expect that the theory developed in the present paper is but a special chapter in the theory of general orthogonal developments. An attempt to translate the present theory into more general terms, however, incurs at once somewhat serious difficulties. This is due to the fact that the theory of the Fourier series involves only one fundamental Hermitian form,

$$\frac{1}{2\pi}\int_{-\pi}^{\pi} f(x)\bar{f}(x)\,dx; \qquad (10.01)$$

the closely related theory of the Fourier integral involves only the analogous form

$$\int_{-\infty}^{\infty} f(x)\bar{f}(x)\,dx; \qquad (10.\ 02)$$

while the present paper involves besides this latter form the singular quadratic form

$$M_x(f(x)) = \lim_{T\to\infty} \frac{1}{2T} \int_{-T}^{T} f(x)\bar{f}(x)\,dx. \qquad (10.\ 03)$$

The forms (10. 02) and (10. 03) are quite independent of one another in their formal properties, but the complex exponentials e^{iux} stand in close relation to both of them: to (10. 02) because if $a < b$, $c < d$,

$$\int_{-\infty}^{\infty} dx \int_a^b e^{iux}\,du \int_c^d e^{-ivx}\,dv = 2\pi \text{ [length common to } (a, b) \text{ and } (c, d)]; \qquad (10.\ 04)$$

and to (10. 03) because

$$\lim_{T\to\infty} \frac{1}{2T} \int_{-T}^{T} e^{iux} e^{-ivx}\,dx = \begin{cases} 0; & [u \neq v] \\ 1 & [u = v]. \end{cases} \qquad (10.\ 05)$$

In the classical Plancherel theory, only the first form is in evidence; in the Bohr theory of almost periodic functions, only the second; in the theory of the present paper, the two play an equal rôle.

Weyl has developed in some detail the relations between the theory of unitary groups and the theory of periodic and almost periodic functions. The groups which he introduces are one parameter groups of linear functional transformations leaving (10. 03) invariant. The Weyl theory is manifestly not susceptible to an extension to more general forms of harmonic analysis, unless a way is found to take cognizance of the invariance of (10. 02) as well. This is the purpose of the present section.

Let us restrict the functions $f(x)$ which we discuss in the present section to those for which

$$\int_{-\infty}^{\infty} \frac{|f(x)|^2}{1 + x^2}\,dx \qquad (10.\ 06)$$

is finite. Let

$$f_A(x)=\begin{cases} 0; & [|x|>A] \\ f(x); & [|x|\le A] \end{cases} \tag{10.07}$$

and let s_A be the transformation leading from $f(x)$ to $f_A(x)$. Let T be a transformation which is linear, and with an inverse, and is defined for all functions $f(x)$ subject to the finiteness of (10.06). Let T preserve (10.02) invariant, and in case (10.03) is finite, let

$$\int_{-\infty}^{\infty} |(Ts_A - s_A T)f(x)|^2\, dx = o(A). \tag{10.08}$$

Then since

$$\lim_{A\to\infty} \frac{1}{2}\int_{-\infty}^{\infty} \left|\frac{Ts_A f(x)}{\sqrt{A}}\right|^2 dx = M_x(|f(x)|^2), \tag{10.09}$$

and

$$\lim_{A\to\infty} \int_{-\infty}^{\infty} \left|\frac{Ts_A f(x)}{\sqrt{A}} - \frac{s_A Tf(x)}{\sqrt{A}}\right|^2 dx = 0, \tag{10.10}$$

it follows that

$$M_x(|Tf(x)|^2) = \lim_{A\to\infty} \frac{1}{2}\int_{-\infty}^{\infty} \left|\frac{s_A Tf(x)}{\sqrt{A}}\right|^2 dx = M_x(|f(x)|^2). \tag{10.11}$$

The transformations T form a group. If T_1 and T_2 are of this form,

$$\begin{aligned} &\int_{-\infty}^{\infty} |(T_1T_2s_A - s_AT_1T_2)f(x)|^2\, dx \\ &= \int_{-\infty}^{\infty} |[(T_1T_2s_A - T_1s_AT_2) + (T_1s_AT_2 - s_AT_1T_2)]f(x)|^2\, dx \\ &\le 2\int_{-\infty}^{\infty} |T_1T_2s_A - T_1s_AT_2)f(x)|^2\, dx + 2\int_{-\infty}^{\infty} |(T_1s_AT_2 - s_AT_1T_2)f(x)|^2\, dx \\ &= 2\int_{-\infty}^{\infty} |(T_2s_A - s_AT_2)f(x)|^2\, dx + 2\int_{-\infty}^{\infty} |(T_1s_A - s_AT_1)T_2f(x)|^2\, dx. \end{aligned} \tag{10.12}$$

Furthermore,

$$\int_{-\infty}^{\infty} |(T^{-1}s_A - s_A T^{-1}) f(x)|^2 \, dx = \int_{-\infty}^{\infty} |(s_A - T s_A T^{-1}) f(x)|^2 \, dx$$

$$= \int_{-\infty}^{\infty} |(T s_A - s_A T) T^{-1} f(x)|^2 \, dx. \qquad (10.\ 13)$$

Thus the product of two transformations satisfying (10. 08) and the inverse of a transformation satisfying (10. 08) likewise satisfy (10. 08).

An example of a transformation satisfying (10. 08) is

$$T^{\lambda} f(x) = f(x + \lambda), \qquad (10.\ 14)$$

for

$$\int_{-\infty}^{\infty} |(T^{\lambda} s_A - s_A T^{\lambda}) f(x)|^2 dx$$

$$\leq \int_{0}^{\lambda} [|f(x - A - \lambda)|^2 + |f(x + A - \lambda)|^2] \, dx = o(A). \qquad (10.\ 15)$$

If T satisfies (10. 08) and (10. 02) is invariant under it, we shall call it *properly unitary*. Let us consider a one parameter group consisting of all properly unitary transformations U^{λ}, where

$$U^{\lambda+\mu} = U^{\lambda} U^{\mu}. \qquad (10.\ 16)$$

Let $f(x)$ be such that

$$\varphi(t) = M_x [(U^t f(x)) \bar{f}(x)] \qquad (10.\ 17)$$

exists for every t. Clearly, by the Schwarz inequality,

$$\varphi(t) \leq [M_x(|U^t f(x)|^2) M_x(|f(x)|^2)]^{1/2}$$

$$= \left[\varphi(0) \lim_{A \to \infty} \frac{1}{2A} \int_{-\infty}^{\infty} |s_A U^t f(x)|^2 \, dx\right]^{1/2}$$

$$= \left[\varphi(0) \lim_{A \to \infty} \frac{1}{2A} \int_{-\infty}^{\infty} |U^t s_A f(x)|^2 \, dx\right]^{1/2}$$

$$= \varphi(0). \qquad (10.\ 18)$$

It follows that $\varphi(t)$ is a bounded function, and that

$$S(u_1, u_2) = \underset{B\to\infty}{\text{l.i.m.}} \frac{1}{\pi} \int_{-B}^{B} \varphi(t) \frac{\sin(u_2 - u_1)t/2}{t} e^{i\left(\frac{u_1+u_2}{2}\right)t} dt \qquad (10.19)$$

exists when $u_2 - u_1$ is given as a quadratically summable function of $u_2 + u_1$. For this one need only apply Plancherel's theorem.

Let us put

$$\varphi_A(t) = \frac{1}{2A} \int_{-\infty}^{\infty} [U^t s_A f(x)] \overline{s_A f(x)}\, dx. \qquad (10.20)$$

If condition (10.17) holds for every t, we have

$$\varphi(t) = \lim_{A\to\infty} \frac{1}{2A} \int_{-\infty}^{\infty} [s_A U^t f(x)] \overline{s_A f(x)}\, dx, \qquad (10.21)$$

which we may readily reduce to the form

$$\varphi(t) = \lim_{A\to\infty} \varphi_A(t) \qquad (10.22)$$

by means of (10.08). We may easily prove $\varphi_A(t)$ not to exceed $\varphi_A(0)$ and hence to be uniformly bounded in A and t, for

$$\lim_{A\to\infty} \varphi_A(0) = \varphi(0). \qquad (10.23)$$

Let us now introduce a new assumption concerning U^x. *Let the transformation W taking $f(x)$ into $U^x f(a)$ (a fixed) preserve* (10.02) *invariant.* Than if

$$\psi(x) = W s_A f(x) \qquad (10.24)$$

we have, by the new assumption and (10.20),

$$\varphi_A(t) = \frac{1}{2A} \int_{-\infty}^{\infty} U^t (U^x s_A f(a)) \overline{U^x s_A f(a)}\, dx$$

$$= \frac{1}{2A} \int_{-\infty}^{\infty} \psi(x+t) \overline{\psi}(x)\, dx. \qquad (10.25)$$

As we may readily see (cf. (1.29)) $\varphi_A(t)$ is absolutely integrable over $(-\infty, \infty)$, and

$$P_A(u) = \frac{1}{\pi}\int_{-\infty}^{\infty} \varphi_A(t)\, e^{iut}\, dt \qquad (10.26)$$

exists, and is real and positive. Indeed, we might have replaced our new assumption by the assumption that

$$\frac{1}{\pi}\int_{-\infty}^{\infty} e^{iut}\, dt \int_{-\infty}^{\infty} [U^t f(x)] f(x)\, dx \qquad (10.27)$$

is positive-definite, and exists for every quadratically summable $f(x)$.

Thus it appears that

$$S_A(u_1, u_2) = \frac{2}{\pi}\int_{-\infty}^{\infty} \varphi_A(t) \frac{\sin (u_2 - u_1) t/2}{t} e^{i\left(\frac{u_1+u_2}{2}\right)t}\, dt \qquad (10.28)$$

exists, is monotone in u_1 and u_2, and has the property that

$$S_A(u_1, u_2) + S_A(u_2, u_3) = S_A(u_1, u_3). \qquad (10.29)$$

To see this, we need only reflect that

$$S_A(u_1, u_2) = \int_{u_1}^{u_2} P_A(u)\, du. \qquad (10.30)$$

Now

$$\varphi(t) \frac{\sin (u_2 - u_1) t/2}{t} = \underset{A\to\infty}{\text{l.i.m.}}\ \varphi_A(t) \frac{\sin (u_2 - u_1) t/2}{t}, \qquad (10.31)$$

so that

$$S(u_1, u_2) = \underset{A\to\infty}{\text{l.i.m.}}\ S_A(u_1, u_2). \qquad (10.32)$$

From this we may readily conclude that

$$S(u_1, u_2) + S(u_2, u_3) = S(u_1, u_3), \qquad (10.33)$$

and that $S(u_1, u_2)$ may be so defined as to be monotone in both arguments and increasing in u_2. $S(u_1, u_2)$ is the analogue to $S(u_2) - S(u_1)$ in our earlier sections.

We have

$$\frac{1}{u}\int_0^u S(-u,u)\,du = \frac{1}{\pi u}\int_{-\infty}^{\infty}\varphi(t)\frac{1-\cos ut}{t^2}\,dt. \tag{10. 34}$$

Hence

$$S(-\infty,\infty) = \lim_{u\to\infty}\frac{1}{\pi u}\int_{-\infty}^{\infty}\varphi(t)\frac{1-\cos ut}{t^2}\,dt. \tag{10. 35}$$

As in (5. 47)

$$S(-\infty,\infty) = \lim_{\varepsilon\to 0}\frac{1}{2\varepsilon}\int_{-\varepsilon}^{\varepsilon}\varphi(t)\,dt. \tag{10. 36}$$

We thus have arrived at a spectrum theory closely paralleling the theory here developed for trigonometric expansions. Thus for the general case of harmonic analysis, it is the group theory of transformations satisfying (10. 08) and (10. 24) which is important, rather than the recognized theory of unitary transformations.

Transformations U^t with the properties demanded in this section make their appearance in physics in connection with the Schrödinger equation, which often has its Eigenfunktionen also Eigenfunktionen of an operator analogous to U^t. A more specific instance of $U^t f(x)$ is

$$U^t f(x) = e^{iAt} f(x+t). \tag{10. 37}$$

CHAPTER IV.

11. Examples of functions with continuous spectra.

The theory of harmonic analysis which we have so far developed has as one of its purposes the analysis of phenomena such as white light, involving continuous spectra. We have not yet proved our theory to be adequate to this purpose, for we have not yet given a single instance of a continuous spectrum. This lacuna it is the purpose of the present section to fill. To this end, we shall only consider functions $f(t)$ which over every interval $(n, n+1)$, n being an integer, assume one of the two values, 1 and -1. For such a function, the existence of

26—29764. *Acta mathematica.* 55. Imprimé le 15 avril 1930.

$$\varphi(x) = \lim_{T\to\infty} \frac{1}{2T} \int_{-T}^{T} f(t+x) f(t)\, dt$$

for all arguments will follow from its existence for all integral arguments, inasmuch as, if x lies between n and $n+1$,

$$\frac{1}{2T} \int_{-T}^{T} f(t+x) f(t)\, dt = \frac{x-n}{2T} \int_{-T}^{T} f(t+n+1) f(t)\, dt$$
$$+ \frac{n+1-x}{2T} \int_{-T}^{T} f(t+n) f(t)\, dt, \qquad (11.01)$$

so that

$$\varphi(x) = \lim_{T\to\infty} \frac{1}{2T} \int_{-T}^{T} f(t+x) f(t)\, dt = (x-n)\,\varphi(n+1) + (n+1-x)\,\varphi(n). \qquad (11.02)$$

An example of a function of this sort is given below, where the sequence of signs represents the signs of $f(t)$ over the intervals both to the right and the left of the zero point:

$+, -;$

$+, +;\ +, -;\ -, +;\ -, -$ repeated twice

$+, +, +;\ +, +, -;\ +, -, +;\ +, -, -;\ -, +, +;\ -, +, -;$
$-, -, +;\ -, -, -$ repeated four times (11.03)

$+, +, +, +;\ +, +, +, -;\ +, +, -, +;$ etc. repeated eight times

. . . .

Each repetition of a row is here counted as a separate row. In each row, all the possible arrangements of n symbols which may be either $+$ or $-$ occur arranged in a regular order. Thus in each row, the possible arrangements of p pluses or minuses occur equally often, except for the modification incurred by the possibility that such an arrangement may overlap one of the major divisions indicated by a semicolon in the above table. These semicolons become more and more infrequent as we proceed to later rows in our table, and do not affect the asymptotic distribution of sequence of p signs.

Thus the possible sequences of p signs occur with a probability approaching $1/2^p$ at the end of a comported row. However, the ratio of the number of terms in a row to that in all previous rows approaches zero, so that the effect of stopping at some intermediate point in a row becomes negligible. In other words,

$$\lim \frac{(\text{number of repetitions of a particular sequence of } p \text{ terms in first } n)}{n} = 1/2^p. \qquad (11.04)$$

Hence

$$\lim_{T\to\infty} \frac{1}{2T}\int_{-T}^{T} f(t+m)f(t)\,dt = 0. \qquad [m = \text{an integer} \neq 0] \quad (11.05)$$

Inasmuch as obviously

$$\lim_{T\to\infty} \frac{1}{2T}\int_{-T}^{T} [f(t)]^2\,dt = 1, \qquad (11.06)$$

we see that

$$\varphi(\tau) = \begin{cases} 1-|\tau|; & [|\tau| \leq 1] \\ 0. & [|\tau| > 1] \end{cases} \qquad (11.07)$$

It follows that

$$S(u) = \frac{1}{2\pi}\int_{-\infty}^{\infty} \varphi(x)\frac{e^{iux}-1}{ix}\,dx = \frac{1}{\pi}\int_{0}^{1}(1-x)\frac{\sin ux}{x}\,dx = \frac{1}{\pi}\int_{0}^{u}\frac{1-\cos v}{v^2}\,dv. \qquad (11.08)$$

Thus the function $f(t)$ has a continuous spectrum with spectral density $\frac{1}{\pi}\frac{1-\cos u}{u^2}$.
This fact that the spectrum has a spectral density is an even more restrictive condition than the condition that it should be continuous.

Every monotone function is known to be the sum of three parts: a step function with a denumerable set of steps; a function which is the integral of its derivative; and a continuous function with a derivative almost everywhere zero. This latter type of function has been ignored as a possibility in spectrum analysis. With both line and continuous spectra we are familiar, but the physicists have not considered the possibility of a spectrum in which the total energy of a region tends to zero as the width of the region decreases, but not proportionally in the limit to the width of the region. Nevertheless, functions with a spectrum of this type do exist, as Mr. Kurt Mahler has proved. I am incorporating into this paper an extremely ingenious note of Mr. Mahler, already published in the Journal of Mathematics and Physics of the Massachusetts Institute of Technology, giving an example of this kind.

Let ξ be a simple q-th root of unity, q being any positive integer greater than 1. Let $\bar{\xi}$ be the conjugate complex number, so that

$$\xi\bar{\xi} = 1. \tag{11.09}$$

We define the arithmetical function $\varrho(n)$ by the functional equations

$$\left.\begin{array}{l} \varrho(0) = 1; \\ \varrho(qn + l) = \xi^l \varrho(n) \text{ for } \begin{bmatrix} l = 0, 1, 2, \ldots, q-1 \\ n = 0, 1, 2, \ldots \end{bmatrix} \end{array}\right\} \tag{11.10}$$

We have thus defined $\varrho(n)$ unambiguously for every positive integer n. We may write

$$\varrho(n) = \xi^{q(n)} \tag{11.11}$$

where $q(n)$ is the sum of the digits of n in the q-ary system of notation.

Our problem here is to give an asymptotic evaluation of

$$S_k(n) = \sum_{l=0}^{n-1} \varrho(l)\bar{\varrho}(l+k), \tag{11.12}$$

for arbitrary positive integral values of k and large values of n. If $k=0$, we have the obvious formula

$$S_0(n) = n. \tag{11.13}$$

We shall use this as a basis on which to determine

$$S_1(n) = \sum_{l=0}^{n-1} \varrho(l)\bar{\varrho}(l+1). \tag{11.14}$$

We may deduce at once from our fundamental equation (11.10) the functional equations of $S_1(n)$, namely

$$\begin{aligned} &S_1(0)=0, \\ &S_1(qn+l)=\bar{\xi}\, S_1(n)+((q-1)n+l)\,\bar{\xi}. \qquad [l=0, 1, \ldots, q-1] \end{aligned} \tag{11.15}$$

As is obvious, these equations determine $S_1(n)$ unambiguously.

We now see, however, that the series

$$\Sigma_1(n) = \bar{\xi}\left\{[n] - \left[\frac{n}{q}\right]\right\} + \bar{\xi}^2\left\{\left[\frac{n}{q}\right] - \left[\frac{n}{q^2}\right]\right\} + \bar{\xi}^3\left\{\left[\frac{n}{q^2}\right] - \left[\frac{n}{q^3}\right]\right\} + \cdots \qquad (11.16)$$

satisfies the same functional equations (11.15) as $S_1(n)$ and hence is identical with $S_1(n)$. We thus have

$$S_1(n) = \bar{\xi}\left\{[n] - \left[\frac{n}{q}\right]\right\} + \bar{\xi}^2\left\{\left[\frac{n}{q}\right] - \left[\frac{n}{q^2}\right]\right\} + \bar{\xi}^3\left\{\left[\frac{n}{q^2}\right] - \left[\frac{n}{q^3}\right]\right\} + \cdots. \qquad (11.17)$$

Now let

$$q^r \leq n < q^{r+1}. \qquad (11.18)$$

We see that

$$\begin{aligned} S_1(n) &= \bar{\xi}\left\{[n] - \left[\frac{n}{q}\right]\right\} + \bar{\xi}^2\left\{\left[\frac{n}{q}\right] - \left[\frac{n}{q^2}\right]\right\} + \cdots + \bar{\xi}^{r+1}\left\{\left[\frac{n}{q^r}\right] - \left[\frac{n}{q^{r+1}}\right]\right\} \\ &= n\bar{\xi}\left(1 - \frac{1}{q}\right)\left(1 + \frac{\bar{\xi}}{q} + \frac{\bar{\xi}^2}{q^2} + \cdots + \frac{\bar{\xi}^r}{q^r}\right) + O(r) \\ &= n\bar{\xi}\left(1 - \frac{1}{q}\right)\frac{1}{1 - \frac{\bar{\xi}}{q}} + O(1) + O(r), \end{aligned} \qquad (11.19)$$

or by (11.18)

$$S_1(n) = \frac{q-1}{q\xi - 1}\, n + O\,(\log n). \qquad (11.20)$$

Formulae (11.13) and (11.20) are only special cases of the corresponding formula for arbitrary k. We obtain this in the following manner.

Since

$$S_k(qn + l) = S_k(qn) + O(1), \qquad [l = 0, 1, 2, \ldots, q-1] \quad (11.21)$$

we need only consider $S_k(qn)$. For this we have the formula

$$S_{qK+\lambda}(qn)\,\bar{\xi}^{\lambda}\,\{(q-\lambda)\,S_K(n) + \lambda\,S_{K+1}(n)\}. \qquad (11.22)$$

We define a sequence $\sigma(k)$ by the functional equations

$$\begin{aligned} &\sigma(0) = 1 \\ &\sigma(qK + \lambda) = \bar{\xi}^{\lambda}\left(\frac{q-\lambda}{q}\,\sigma(K) + \frac{\lambda}{q}\,\sigma(K+1)\right). \end{aligned} \qquad (11.23)$$

Then it is always true that

$$S_k(n) = \sigma(k)n + O(\log n). \tag{11.24}$$

To begin with, we have proved this theorem for $k=0$ and $k=1$. Formula (11.23) shows, however, that we may prove (11.24) in general by a mathematical induction with respect to k.

$\sigma(k)$ is a very complicated arithmetical function. For small values of its argument ($K=0, 1, \cdots, q-1$, $\lambda=0, 1, \cdots, q-1$) we have

$$\left.\begin{aligned} \sigma(\lambda) &= \bar{\xi}^{\lambda}\frac{(q-\lambda)+(\lambda-1)\bar{\xi}}{q-\bar{\xi}}, \\ \sigma(Kq+\lambda) &= \bar{\xi}^{K+\lambda}\frac{(q+K)(q-\lambda)+((K-1)(q-\lambda)+(q-K-1)\lambda)\bar{\xi}+K\lambda\bar{\xi}^2}{q(q-\bar{\xi})}. \end{aligned}\right\} \tag{11.25}$$

It is natural to extend our definition of $\sigma(k)$ to negative values of k by the formula

$$\sigma(-k) = \overline{\sigma(k)}. \tag{11.26}$$

Formula (11.24) is then true for negative as well as for positive arguments.

It is natural to investigate the functions

$$T_k(n) = \sum_{0}^{n-1} \sigma(l)\,\bar{\sigma}(l+k) \tag{11.27}$$

which arise from σ in the same fashion as S_k arises from ϱ. We shall confine ourselves to the case

$$q=2;\ \xi=\bar{\xi}=-1.$$

We have here the equations

$$\left.\begin{aligned} \sigma(k) &= \bar{\sigma}(k); \\ \sigma(2k) &= \sigma(k); \\ \sigma(2k+1) &= -\frac{\sigma(k)+\sigma(k+1)}{2}. \end{aligned}\right\} \tag{11.28}$$

Hence we have the following formulae:

$$T_{2k}(2n) = \sum_{m=0}^{n-1} (\sigma(2m)\,\sigma(2m+2k) + \sigma(2m+1)\,\sigma(2m+2k+1)) \tag{11.29}$$

$$= \sum_{m=0}^{n-1} \left(\sigma(m)\,\sigma(m+k) + \frac{(\sigma(m)+\sigma(m+1))(\sigma(m+k)+\sigma(m+k+1))}{4}\right) \tag{11.30}$$

or

$$\left| T_{2k}(2n) - \frac{3}{2} T_k(n) - \frac{1}{4} T_{k-1}(n) - \frac{1}{4} T_{k+1}(n) \right| < \text{const.} \tag{11.31}$$

and further

$$T_{2k+1}(2n) = \sum_{m=0}^{n-1} (\sigma(2m)\,\sigma(2m+2k+1) + \sigma(2m+1)\,\sigma(2m+2k+2))$$

$$= -\sum_{m=0}^{n-1} \left(\sigma(m)\frac{\sigma(m+k)+\sigma(m+k+1)}{2} + \frac{\sigma(m)+\sigma(m+1)}{2}\sigma(m+k+1)\right) \tag{11.32}$$

or

$$| T_{2k+1}(2n) + T_k(n) + T_{k+1}(n) | < \text{const.} \tag{11.33}$$

The array

$$\{\ldots, \varrho(n), \ldots, \varrho(1), \varrho(0), \varrho(1), \ldots, \varrho(n), \ldots\} = \{\ldots, a_{-n}, \ldots, a_{-1}, a_0, a_1, \ldots, a_n, \ldots\}$$

defines a function

$$f(t) = \begin{cases} a_n \text{ if } |t-n| \leq 1/8 \\ 0 \text{ if for no } n, |t-n| \leq 1/8 \end{cases} \tag{11.34}$$

and

$$\varphi(\tau) = \lim_{T\to\infty} \frac{1}{2T} \int_{-T}^{T} f(t+\tau)\bar{f}(t)\,dt = \frac{1}{4}\,\sigma\left(\left[\tau + \frac{1}{4}\right]\right) Q\left(\tau - \left[\tau + \frac{1}{4}\right]\right), \tag{11.35}$$

where

$$Q(x) = \begin{cases} 1 - 4|x|; & \left[-\frac{1}{4} < x < \frac{1}{4}\right] \\ 0. & \left[\frac{1}{4} \leq x < \frac{3}{4}\right] \end{cases}$$

We have

$$S(u) = \frac{1}{2\pi} \int_{-\infty}^{\infty} \varphi(\tau) \frac{e^{iu\tau} - 1}{i\tau}\,d\tau, \tag{11.36}$$

where $S(u)$ is of limited total variation. Then

$$S(u+\varepsilon) - S(u-\varepsilon) = \frac{1}{\pi} \int_{-\infty}^{\infty} \varphi(\tau) \frac{\sin \varepsilon\tau}{\tau} e^{iu\tau}\,d\tau. \tag{11.37}$$

Hence by (6. 15)

$$\lim_{\varepsilon\to 0}\frac{1}{2\varepsilon}\int_{-\infty}^{\infty} e^{-i\nu u}\,|S(u+\varepsilon)-S(u-\varepsilon)|^2\,du = \lim_{T\to\infty}\frac{1}{2T}\int_{-T}^{T}\varphi(x+\nu)\,\overline{\varphi}(x)\,dx. \qquad (11.\ 38)$$

Hence if the finite or denumerable set of discontinuities of $S(u)$ are at u_1, u_2, ... and have values A_1, A_2, ..., respectively

$$\sum_1^{\infty}|A_k|^2 = \lim_{\varepsilon\to 0}\frac{1}{2\varepsilon}\int_{-\infty}^{\infty} e^{-i\nu u}\,|S(u+\varepsilon)-S(u-\varepsilon)|^2\,du =$$

$$= \lim_{T\to\infty}\frac{1}{2T}\int_{-T}^{T}\varphi(x+\nu)\,\overline{\varphi}(x)\,dx \qquad (11.\ 39)$$

and exists for every ν. However,

$$\lim_{T\to\infty}\frac{1}{2T}\int_{-T}^{T}\varphi(x+\nu)\,\overline{\varphi}(x)\,dx = \frac{1}{16}\lim_{n\to\infty}\frac{T_{\left[\nu+\frac{1}{2}\right]}(n)}{n}R\left(\nu-\left[\nu+\frac{1}{2}\right]\right), \qquad (11.\ 40)$$

where

$$R(x) = \int_{-\frac{1}{2}}^{\frac{1}{2}} Q(y)\,Q(x+y)\,dy, \qquad (11.\ 41)$$

so that

$$\lim_{n\to\infty}\frac{T_k(n)}{n}$$

exists for every k. If we put

$$\varpi(k) = \lim_{n\to\infty}\frac{T_k(n)}{n} = \lim_{n\to\infty}\frac{1}{n}\sum_{l=0}^{n-1}\sigma(l)\,\sigma(l+k) \qquad (11.\ 42)$$

we may conclude from our equations for T_k that

$$\left.\begin{aligned}\varpi(2k) &= \frac{\varpi(k-1)+6\,\varpi(k)+\varpi(k+1)}{8};\\ \varpi(2k+1) &= -\frac{\varpi(k)+\varpi(k+1)}{2}.\end{aligned}\right\} \qquad (11.\ 43)$$

It follows that if

$$\varpi(0) = 0, \quad \varpi(1) = 0 \qquad (11.44)$$

then for every k

$$\varpi(k) = 0. \qquad (11.45)$$

We now put $k=0$ in (11.43), remembering that

$$\varpi(-k) = \varpi(k). \qquad (11.46)$$

We obtain

$$\varpi(0) = \varpi(1) = 3\,\varpi(1). \qquad (11.47)$$

Hence $\varpi(0)=\varpi(1)=0$ and $\varpi(k)$ is identically 0. In other words,

$$T_k(n) = o(n). \qquad (11.48)$$

As $\sigma(2k)=\sigma(k)$, $\sigma(1)=-1/3$, we see that we cannot have

$$\lim_{k\to\infty} \sigma(k) = 0, \qquad (11.49)$$

and hence we cannot have

$$\lim_{\tau\to\infty} \varphi(\tau) = 0. \qquad (11.50)$$

It is thus impossible that

$$S(u) = \int_{-\infty}^{u} S'(v)\,dv + S(-\infty), \qquad (11.51)$$

for then we should have by (5.43), (5.46)

$$\lim_{\tau\to\infty} \varphi(\tau) = \lim_{\tau\to\infty} \int_{-\infty}^{\infty} S'(u)\,e^{-iu\tau}\,du = 0. \qquad (11.52)$$

On the other hand, as

$$\varpi(0) = 0,$$

we must have

$$\lim_{\varepsilon\to 0} \frac{1}{2\varepsilon} \int_{-\infty}^{\infty} |S(u+\varepsilon)-S(u-\varepsilon)|^2\,du = 0. \qquad (11.53)$$

It follows that $S(u)$ is a continuous function which we have already seen not to be the integral of its derivative. This theorem of Mahler thus leads to a new type of spectrum.

27 — 29764. *Acta mathematica.* 55. Imprimé le 15 avril 1930.

12. Spectra depending on an infinite sequence of choices.

In the last section we gave concrete, well defined examples of functions with various continuous types of spectra. The present and the following sections are devoted to the generation of such functions by methods which instead of always yielding functions with continuous spectra, *almost* always yield such functions. The distinction is precisely analogous to that between rational mechanics of the classical kind and statistical mechanics. Theoretically, all the molecules of gas in a vessel might group themselves in a specified half of its volume; practically, such a contingency is unthinkably improbable.

The notion of probability is a new element not occurring in classical mechanics, but essential in statistical mechanics. It applies to a class of contingent situations, and has the essential properties of a measure. So too the idea of »almost always» in harmonic analysis depends on some more or less concealed notion of measure. In the present and the ensuing sections, we shall assimilate this notion of measure to that of Lebesgue, so that »almost always» will translate into »except for a set of Lebesgue measure zero».

Consider a sequence of independent tosses of a coin. By a sequence, we mean a record such as, »heads, tails, heads, heads, tails » For such a finite sequence, the probability is 2^{-n}, where n is the number of tosses. That is, it is the same as the measure of the set of all the points on $(0, 1)$ with coordinates whose binary expansion begins $.10110$. This mapping immediately suggests a definition of probability for infinite sequences of tosses. The probability of any set of sequences of tosses is defined as the Lebesgue measure of the set of points whose binary representations correspond to sequences of tosses in the set, in such a manner that 1 corresponds to »heads» and 0 to »tails».[5] We can even represent sequences infinite in both directions by binary numbers in such a way as to define the probability of a set of sequences, by having recourse to some definite enumerative rearrangement of such a sequence.

If we have made »probability» a mere translation of »measure», »average» becomes the equivalent of »integral». We are accordingly able to use the entire body of theorems concerning the Lebesgue integral in the service of the calculus of probabilities.

We have not yet, however, correlated with our sequence of throws functions susceptible of a harmonic analysis. To do this, we take a certain zero point on a doubly infinite line to correspond with the zero point of our doubly infinite sequence of tosses, and if the nth toss is a head, we define $f(t)$ to be 1 for $n < t < n+1$; if a tail, to be -1. The question we wish to ask is: what is the probability distribution of spectral functions $S(u)$ for these functions $f(t)$?

We have, taking $f(t) = a_n$ for $n < t < n+1$,

$$\frac{1}{2n}\int_{-n}^{n} f(t+m)f(t)\,dt = \frac{1}{2n}(a_{m-n}\,a_{-n} + a_{m-n+1}\,a_{-n+1} + \cdots + a_{m+n-1}\,a_{n-1}). \quad (12.01)$$

Since the distribution of each a_n between negative and positive values is symmetrical and independent of that of every other,

$$\text{Average } \frac{1}{2n}\int_{-n}^{n} f(t+m)f(t)\,dt = \begin{cases} 0 \text{ if } m \neq 0; \\ 1 \text{ if } m = 0. \end{cases} \quad (12.02)$$

When $m=0$, this average is indeed identically 1. In every other case, when m is an integer,

$$\text{Average } \left\{\frac{1}{2n}\int_{-n}^{n} f(t+m)f(t)\,dt\right\}^2$$

$$= \text{Average } \frac{1}{4n^2}(a_{m-n}\,a_{-n} + \cdots + a_{m+n-1}\,a_{n-1})^2$$

$$= \text{Average } \frac{1}{4n^2}(a_{m-n}^2\,a_{-n}^2 + \cdots + a_{m+n-1}^2\,a_{n-1}^2)$$

$$= 1/2n, \quad (12.03)$$

since the averages of all the non-square terms vanish. Hence

$$\frac{1}{2n^2}\int_{-n^2}^{n^2} f(t+m)f(t)\,dt > A \quad (12.04)$$

over a set of values of $f(t)$ with total probability

$$\leq \frac{A}{2n^2}. \quad (12.05)$$

Since the sum of these latter quantities forms a convergent series, with remainder after n terms tending to 0 as n increases, we must have

$$\lim_{n\to\infty} \frac{1}{2n^2}\int_{-n^2}^{n^2} f(t+m)f(t)\,dt < A, \tag{12.06}$$

except in a set of cases of arbitrarily small and hence of zero probability. Hence except in a set of cases of zero probability,

$$\lim_{n\to\infty} \frac{1}{2n^2}\int_{-n^2}^{n^2} f(t+m)f(t)\,dt = 0. \tag{12.07}$$

Here the procedure to the limit runs through integral values of n. This can be generalized at once. Let P be bounded, and let

$$n^2 < T < (n+1)^2.$$

Then

$$\frac{1}{2T}\int_{-T}^{T} P\,d\lambda = \frac{n^2}{T}\frac{1}{2n^2}\int_{-n^2}^{n^2} P\,d\lambda + \frac{1}{2T}\left[\int_{n^2}^{T} + \int_{-T}^{-n^2}\right] P\,d\lambda$$

$$= \frac{n^2}{T}\frac{1}{2n^2}\int_{-n^2}^{n^2} P\,d\lambda + \frac{\vartheta}{2T}\left[\int_{-(n+1)^2}^{(n+1)^2} P\,d\lambda - \int_{-n^2}^{n^2} P\,d\lambda\right] \qquad [0 \leq \vartheta \leq 1]$$

$$= \frac{n^2}{T}\frac{1}{2n^2}\int_{-n^2}^{n^2} P\,d\lambda + \vartheta\left[\frac{(n+1)^2}{T}\cdot\frac{1}{2(n+1)^2}\int_{-(n+1)^2}^{(n+1)^2} P\,d\lambda - \frac{n^2}{T}\cdot\frac{1}{2n^2}\int_{-n^2}^{n^2} P\,d\lambda\right]. \tag{12.08}$$

Thus if

$$\lim_{n\to\infty} \frac{1}{2n^2}\int_{-n^2}^{n^2} P\,d\lambda = L,$$

then

$$\lim_{T\to\infty} \frac{1}{2T}\int_{-T}^{T} P\,d\lambda = L. \tag{12.09}$$

We thus see that in case m is an integer other than zero, we almost always have

$$\lim_{T\to\infty} \frac{1}{2T} \int_{-T}^{T} f(t+m) f(t)\, dt = 0. \qquad (12.10)$$

As in (11.07), we may conclude that

$$\varphi(x) = \begin{cases} 0; & (x > 1) \\ 1 - x; & (0 < x < 1) \\ 1 + x; & (-1 < x < 0) \\ 0; & (-1 < x) \end{cases} \qquad (12.11)$$

and that

$$S(u) = \frac{1}{2\pi} \int_{-\infty}^{\infty} \varphi(x) \frac{e^{iux} - 1}{ix}\, dx = \frac{1}{\pi} \int_{0}^{1} (1 - x) \frac{\sin ux}{x}\, dx = \frac{1}{\pi} \int_{0}^{u} \frac{1 - \cos v}{v^2}\, dv. \qquad (12.12)$$

These propositions are true, not always, but almost always. Thus a haphazard sequence of positive and negative rectangular impulses almost always has the spectral intensity

$$\frac{1}{\pi} \frac{1 - \cos v}{v^2} \qquad (12.13)$$

which is numerically identical to the square of the Fourier transform of a single rectangular impulse. To see this, we need only reflect that the Fourier transform of such an impulse is

$$\frac{1}{\sqrt{2\pi}} \int_{-\frac{1}{2}}^{\frac{1}{2}} e^{iux}\, dx = \sqrt{\frac{2}{\pi}} \frac{\sin u/2}{u} = \sqrt{\frac{1 - \cos u}{\pi u^2}}. \qquad (12.14)$$

It would not be a difficult task to generalize this remark to impulses of other than rectangular shape. The essential generalization to make this fact of physical interest is, however, to eliminate the equal spacing of the individual impulses, to reduce the sequence of impulses to such an irregularity as is found in the Brownian motion. This is the problem of our next section. The principal difficulty is that the fundamental Lebesgue measure to which we reduce our probabilities is not so obviously at hand. There is no continuous infinity of choices which bears an obvious analogy to that involved in building up a binary fraction. Nevertheless, the distribution involved in the time paths of particles subject to the Brownian motion can be reduced to a Lebesgue measure, certain functions connected with these paths can almost always be analysed harmonically, and their spectra will almost always have a certain fixed distribution of energy if frequency. In other words, the properties of the distributions and functions of this section furnish an excellent working model for those to be expected of the functions discussed in the next section.

13. Spectra and integration in function-space.

From the very beginning, spectrum theory and statistics have joined hands in the theory of white light. The apparent contradiction between the additive character of electromagnetic displacement in the Maxwell theory and the observed additive character of the quadratic light-intensities is on the surface of things irreconcilable. The credit for resolving this antinomy is due to Lord Rayleigh. He considers the resultant of a large number of vibrations of arbitrary phase, and shows that the mean intensity of their sum is actually additive. He says, »It is with this mean intensity only that we are concerned in ordinary photometry. A source of light, such as a candle or even a soda flame, may be regarded as composed of a very large number of luminous centres disposed throughout a very sensible space; and even though it be true that the intensity at a particular point of a screen illuminated by it and at a particular moment of time is a matter of chance, further processes of averaging must be gone through before anything is arrived at of which our senses could ordinarily take cognizance. In the smallest interval of time during which the eye could be impressed, there would be opportunity for any number of rearrangements of phase, due either to motions of the particles or to irregularities in their modes of vibration. And even if we suppose that each luminous centre was fixed, and emitted perfectly regular vibrations, the manner of composition and consequent intensity would vary rapidly from point to point of the screen, and in ordinary cases the mean illumination over the smallest appreciable area would correspond to a thorough averaging of the phase-relationships. In this way the idea of the intensity of a luminous source, independently of any questions of phase, is seen to be justified, and we may properly say that two candles are twice as bright as one.»

Thus Rayleigh's statistics of light is a statistics in which the quantities distributed are amplitudes of sinusoidal vibrations. Such a theory involves a preliminary harmonic analysis, perhaps of a somewhat vague nature, but definite enough to be useful in the hands of a competent physicist. There is an alternative approach to spectrum statistics, which is of a more direct nature. Imagine a resonator — say a sea-shell — struck by a purely chaotic sequence of acoustical impulses. It will yield a response which still has a statistical element in it, but in which the selective properties of the resonator will have accentuated certain frequencies at the expense of others. It seems on the surface of it highly plausible that the outputs of two such resonators will almost always be additive as to intensities rather than merely as to amplitudes.

»Chaos» and »almost always» — there lies an entire statistical theory behind these terms. The simplest phenomenon to which the name »chaos» may be applied with any propriety is that of the Brownian motion. Here a small particle is impelled by atomic collisions in such a way that its future is entirely independent of its past. If we consider the X-coordinate of such a particle, the probability that this should alter a given amount in a given time is independent (1) of the entire past history of the particle; (2) of the instant from which the given interval is measured; (3) whether we are considering changes that increase or changes that decrease it. If we make the assumption that the distribution of the changes of x over a given interval of time is Gaussian, it follows as Einstein has pointed out that the probability that after a time t the change in x should lie between x_1 and x_1+dx_1 is

$$\frac{1}{\sqrt{\pi c t}} e^{-x_1^2/ct} \, dx_1. \tag{13.01}$$

Here c is a constant which we may and shall reduce to 1 by a proper choice of units. The particular manner in which t enters results from the fact that

$$\frac{1}{\sqrt{\pi(t_1+t_2)}} e^{-\frac{x_1^2}{t_1+t_2}} = \frac{1}{\pi \sqrt{t_1 t_2}} \int_{-\infty}^{\infty} e^{-\frac{x^2}{t_1}} e^{-\frac{(x_1-x)^2}{t_2}} \, dx. \tag{13.02}$$

This fundamental identity is tantamount to the statement that the probability that x should have changed by an amount lying between x_1 and x_1+dx_1 after a time t_1+t_2 is the total compound probability that the change of x over time

t_1 should be anything at all, and that it should then migrate in a subsequent interval of length t_2 to a position between x_1 and $x_1 + dx_1$.

A quantity x whose changes are distributed after the manner just mentioned is said to have them normally or chaotically distributed. Of course, what really is distributed is the function $x(t)$ representing the successive values of x. (There is no essential restriction in supposing $x(0)=0$.) Thus the conception of a purely chaotic distribution really introduces a certain system of measure and consequently of integration into function-space. This gives us the clue to the statistical study of spectra. We determine the response of a resonator in terms of functionals of the incoming chaotic disturbance. We then ask, »What is the distribution of various quantities connected with the spectrum of this response, as determined by integrating these quantities over function-space with respect to the original chaos?» Let it be remarked that the theory of integration appropriate to this problem is that developed by the author in his »Differential-space», and not the earlier theory of Gâteaux, which forms the starting point of most researches in this direction.

Before we can attack these more difficult problems we must establish out theory of integration on a firm basis. To do this, we shall establish a correspondence between the set of all functions and the points on a segment of a line AB of unit length, and shall use this correspondence to define integration over function-space in terms of Lebesgue integration over the segment. Let me say in passing that in my previous attacks on this problem, I have made use of a somewhat generalized form of integration due to P. J. Daniell. This form of integration, at least in all cases yet given, may be mapped into a Lebesgue integration over a one-dimensional manifold by a transformation retaining measure invariant. In as much as the literature contains a much greater wealth of proved theorems concerning the Lebesgue integral than of theorems concerning the Daniell integral — although the latter are not particularly difficult to establish — it has seemed to me more desirable to employ the method of mapping. This method of mapping is an extension to infinitely many dimensions of a method due to Radon.

The method of mapping consists in making certain sets of functions $x(t)$, which we shall call »quasi-intervals», correspond to certain intervals of AB. Our quasi-intervals will be sets of all functions $x(t)$ defined for $0 \leq t \leq 1$ such that

$$\begin{gathered} x(0)=0; \\ x_{11}\leq x(t_1)\leq x_{12}; \\ x_{21}\leq x(t_2)\leq x_{22}; \\ x_{31}\leq x(t_3)\leq x_{32}; \\ \cdot\ \cdot\ \cdot\ \cdot\ \cdot\ \cdot\ \cdot \\ x_{n\,1}\leq x(t_n)\leq x_{n\,2}. \\ (0\leq t_1\leq t_2\leq t_3\leq\cdots\leq t_n\leq 1). \end{gathered} \tag{13.03}$$

By our definition of probability, the probability that $x(t)$ should lie in this quasi-interval is

$$\pi^{-n/2}[t_1(t_2-t_1)(t_3-t_2)\cdots(t_n-t_{n-1})]^{-1/2}\int_{x_{11}}^{x_{12}}d\xi_1\cdots\int_{x_{n\,1}}^{x_{n\,2}}d\xi_n\cdot$$

$$\cdot\exp\left[-\xi_1^2\,t_1^{-1}-\sum_2^n(\xi_k-\xi_{k-1})^2(t_k-t_{k-1})^{-1}\right]. \tag{13.04}$$

Clearly, if the class of all functions $x(t)$ be divided into a finite number of quasi-intervals — some of which then must contain infinite values of $x_{h\,1}$ or $x_{h\,2}$ — the sum of their probabilities will be 1.

The quasi-intervals with which we shall be more specially concerned are the quasi-intervals $I(n;k_1,k_2,\ldots,k_{2^n})$ for which

$$\left.\begin{aligned} t_h&=h\,2^{-n}; \qquad (1\leq h\leq 2^n) \\ x_{h\,1}&=\tan\,(k_h\pi\,2^{-n}); \\ x_{h\,2}&=\tan\,((k_h+1)\pi\,2^{-n}); \end{aligned}\right\} \tag{13.05}$$

where k_h is some integer between -2^{n-1} and $2^{n-1}-1$, inclusive. For the probability that $x(t)$ should lie in this interval let us write

$$P\{I(n;k_1,\ldots,k_{2^n})\}.$$

Let us notice that $I(n;k_1,\ldots,k_{2^n})$ is made of a finite number of quasi-intervals $I(n;l_1,\ldots,l_{2^{n+1}})$, and that the sum of the probabilities belonging to the latter gives the probability belonging to the former.

Let us now map the four quasi-intervals $I(1;k_1,k_2)$ on the segment AB, assigning to each in order an interval with length equal to its probability. Let

28—29764. *Acta mathematica.* 55. Imprimé le 15 avril 1930.

us map the quasi-intervals $I(2; k_1, k_2, k_3, k_4)$ into intervals of the segment AB, translating probability into length, and in such a manner that if $I(2; l_1, l_2, l_3, l_4)$ forms a portion of $I(1; k_1, k_2)$, their images stand in the same relation. If we keep this process up indefinitely, we shall have mapped all the quasi-intervals $I(n; k_1, k_2, \ldots, k_{2^n})$ into intervals of AB in such a way that probability is translated into length, and that the end-points of the intervals of AB form an everywhere dense set.

Up to this point, our mapping has mapped quasi-intervals on intervals. We wish to deduce from it a mapping of functions on points. As a lemma for this purpose, we shall show that the functions $x(t)$ for which for any t_1 and t_2 that are terminating binaries,

$$|x(t_1)-x(t_2)| \geq 40\, h\, |t_1-t_2|^{1/4} \tag{13.06}$$

may be enclosed in a denumerable set of quasi-intervals such that the sum of the probabilities of these quasi-intervals is $O(h^{-n})$ for any n.

To show this, let us represent t_1 as the binary fraction

$$0 \cdot a_1\, a_2\, a_3 \cdots a_n \cdots$$

and t_2 as the binary fraction

$$0 \cdot b_1\, b_2\, b_3 \cdots b_n \cdots.$$

Let t_3 be a number whose binary expansion may be made to agree with that of t_1 up to and including a_j and with that of t_2 up to and including b_j. We shall choose t_3 so that j is as large as possible, even though this may necessitate the use of an expression for t_3 ending in $111 \cdots$ to agree with the smaller of the quantities t_1 and t_2 and of a terminating expression for t_3 to agree with the larger. The interval from t_1 to t_3 will then be expressible in the form

$$0 \cdot 00 \cdots 0 c_{j+1}\, c_{j+2} \cdots,$$

where there are j consecutive 0's after the final point, and every c is 0 or 1. The interval from t_2 to t_3 may be expressed in a similar manner. In other words, the interval from t_1 to t_2 may be reduced to the sum of a denumerable set of intervals from terminating binaries to adjacent terminating binaries of the same number of figures, such that there are not more than two intervals in the set of magnitude 2^{-j-k} where k is any positive integer, and such that every interval is of one of these sizes.

Now clearly,

$$|t_2-t_1| \geq 2^{-1-j}. \tag{13.07}$$

Hence, if it is for particular values of t_2 and t_1 in question that (13.06) is fulfilled,

$$\begin{aligned} |x(t_1)-x(t_2)| &\geq 40\,h \cdot 2^{-(1+j)/4} \\ &> h \cdot 2^{1-(1+j)/4}/(1-2^{-1/4}) \\ &= 2\,h \sum_{k=j+1}^{\infty} 2^{-l/4}. \end{aligned} \tag{13.08}$$

If we now appeal to our analysis of the interval (t_1, t_2), we see that for *some* interval from $m \cdot 2^{-j-k}$ to $(m+1) \cdot 2^{-j-k}$, where m and k are integers and $0 \leq m < 2^{j+k}$, we shall have

$$\left| x\left(\frac{m}{2^{j+k}}\right) - x\left(\frac{m+1}{2^{j+k}}\right) \right| > h \cdot 2^{-(j+k)/4}. \tag{13.09}$$

Thus if for any pair of values t_1 and t_2 that are terminating binary fractions,

$$|x(t_1)-x(t_2)| \geq 40\,h\,|t_1-t_2|^{1/4}, \tag{13.10}$$

then for some integers m and i $(m < 2^i)$

$$|x(m\,2^{-i})-x((m+1)\,2^{-i})| > h \cdot 2^{-i/4}. \tag{13.11}$$

It merely remains to determine the measure of a denumerable set of our quasi-intervals $I(n; k_1, \ldots, k_{2^n})$ containing all the functions $x(t)$ for which, for some m and i, (13.11) holds.

To begin with, let m and i be fixed. Since our selected quasi-intervals ultimately divide the range of values of $x(m\,2^{-i})$ and of $x((m+1)2^{-i})$ to an arbitrary degree of fineness, there is no trouble in proving that the functions satisfying (13.11) may be enclosed in a finite set of selected quasi-intervals of total probability not exceeding

$$\varepsilon + \frac{2}{\sqrt{\pi 2^{-i}}} \int_{h \cdot 2^{-i/4}}^{\infty} e^{-\frac{x^2}{2^{-i}}}\,dx = \varepsilon + \frac{2}{\sqrt{\pi}} \int_{h \cdot 2^{i/4}}^{\infty} e^{-x^2}\,dx, \tag{13.12}$$

where ε is arbitrarily small. If we sum for m and i, we get as the sum of the probabilities of all our enclosing sets a quantity not exceeding

$$\eta + \sum_{i=1}^{\infty} \frac{2^{i+1}}{\sqrt{\pi}} \int_{h\cdot 2^{i/4}}^{\infty} e^{-x^2}\, dx$$

$$< \sum_{i=1}^{\infty} \frac{2^{i+1}}{\sqrt{\pi}} e^{-h\cdot 2^{i/4}}$$

$$< \sum_{i=1}^{\infty} \frac{2^{i+1-ni} h^{-4n}}{\sqrt{\pi}} \qquad (13.\ 13)$$

for sufficiently large h. As usual η represents an arbitrarily small quantity. Expression (13. 13) clearly can be made to vanish more rapidly than any given negative power of h as h becomes infinite.

Let us now reconsider our mapping. If we leave out the ends of our intervals, which form a denumerable set of measure 0, every point on AB is uniquely characterized by and uniquely characterizes an infinite sequence of intervals, each containing the next, and tending to 0 in length. If we reject a denumerable set of quasi-intervals of arbitrarily small total probability, the remaining quasi-intervals and portions of quasi-intervals all contain functions satisfying the condition of equicontinuity

$$|x(t_1)-x(t_2)| < 40\, h\, |t_1-t_2|^{1/4}, \qquad (13.\ 14)$$

so that if we modify AB by the removal of a set of points of arbitrarily small outer measure, as well as by the removal of the end-points of our intervals, every point of AB is characterized by a sequence of intervals closing down on it, by the succession of corresponding quasi-intervals, and by the uniquely determined function $x(t)$ common to this sequence of quasi-intervals and satisfying (13. 14). It follows at once that except for a set of points of zero measure, we have determined a unique mapping of the points of AB by functions satisfying (13. 14) for *some* h. Thus any functional of the latter functions determines a function on the line, which may be summable Lebesgue. In the latter case, we shall define the average of the functional as the Lebesgue integral of the corresponding function on AB.

Among the summable functionals are the expressions

$$P(x(t_1), x(t_2), \ldots, x(t_n)),$$

where P stands for a polynomial. This is readily seen to be the case when the expressions $t_1, \ldots, t_n$ are terminating binaries, and the extension to other values follows from the equicontinuity conditions we have already laid down. To see this, let us note that we have already given information enough to prove that the upper average (corresponding to upper integral) of

$$[\max |x(t)|]^n$$

is finite. This functional will, however, simultaneously dominate

$$P(x(t_1), x(t_2), \ldots, x(t_n))$$

in which we suppose P of the nth degree, and the set of approximating functionals

$$P(x(t_{11}), x(t_{12}), \ldots, x(t_{1n}));$$
$$\cdots\cdots\cdots\cdots$$
$$P(x(t_{m1}), x(t_{m2}), \ldots, x(t_{mn}));$$
$$\cdots\cdots\cdots\cdots$$

in which $t_{11}, \ldots, t_{mn}, \ldots$ are terminating binaries, and $\lim_{m\to\infty} t_{mn}=t_n$. That these functionals are actually approximating functionals results from the fact that

$$P(y_1, \ldots, y_n)$$

is continuous, and that $x(t)$ is almost always continuous. Now, there is a theorem to the effect that if a sequence of functions uniformly dominated by a Lebesgue summable function converges for each argument to a limit, and if the Lebesgue integrals of these functions converge to a limit, this limit is the Lebesgue integral of the limit function. This proves our theorem.

In case $t_1 \leq t_2 \leq \cdots \leq t_n$, the average of $P(x(t_1), \ldots, x(t_n))$ is readily seen to be

$$\pi^{-\frac{n}{2}}[t_1(t_2-t_1)\cdots(t_n-t_{n-1})]^{-1/2}\int_{-\infty}^{\infty} d\xi_1 \cdots \int_{-\infty}^{\infty} d\xi_n\, P(\xi_1, \ldots, \xi_n)\cdot$$
$$\cdot\exp\left[-\xi_1^2 t_1^{-1} - \sum_2^n (\xi_k-\xi_{k-1})^2(t_k-t_{k-1})^{-1}\right]. \qquad (13.15)$$

In particular, if $t_1 \leq t_2$,

$$\text{Average } (x(t_1)\,x(t_2)) = \frac{1}{\sqrt{\pi t_1(t_2-t_1)}}\int_{-\infty}^{\infty} d\xi_1 \int_{-\infty}^{\infty} \xi_1\,\xi_2\, e^{-\frac{\xi_1^2}{t_1}-\frac{(\xi_2-\xi_1)^2}{t_2-t_1}}\, d\xi_2$$

$$= t_1/2. \tag{13.16}$$

and if $t_1 \leqq t_2 \leqq t_3 \leq t_4$,

$$\text{Average } (x(t_1)\,x(t_2)\,x(t_3)\,x(t_4)) = \frac{1}{\pi^2\sqrt{t_1(t_2-t_1)(t_3-t_2)(t_4-t_3)}}\,\cdot$$

$$\cdot\int_{-\infty}^{\infty} d\xi_1 \int_{-\infty}^{\infty} d\xi_2 \int_{-\infty}^{\infty} d\xi_3 \int_{-\infty}^{\infty} d\xi_4\, \xi_1\xi_2\xi_3\xi_4 \exp\left(-\frac{\xi_1^2}{t_1}-\frac{(\xi_2-\xi_1)^2}{t_2-t_1}-\frac{(\xi_3-\xi_2)^2}{t_3-t_2}-\frac{(\xi_4-\xi_3)^2}{t_4-t_3}\right)$$

$$= \frac{t_1\,t_2}{2} + \frac{t_1\,t_3}{4}. \tag{13.17}$$

The expressions just given are absolutely summable. Accordingly, by the familiar rules for inverting the order of integration, if $\alpha_1(t)$, $\alpha_2(t)$, $\alpha_3(t)$, $\alpha_4(t)$ are of limited total variation over $(0, 1)$,

$$\text{Average} \int_0^1 x(t)\, d\alpha_1(t) \int_0^1 x(t)\, d\alpha_2(t) = \int_0^1 \frac{t}{2}\, d\alpha_1(t) \int_t^1 d\alpha_2(s) + \int_0^1 \frac{t}{2}\, d\alpha_2(t) \int_t^1 d\alpha_1(s)$$

$$= \int_0^1 \frac{t}{2}(\alpha_2(1)-\alpha_2(t))\, d\alpha_1(t) + \int_0^1 \frac{t}{2}(\alpha_1(1)-\alpha_1(t))\, d\alpha_2(t)$$

$$= -\int_0^1 \frac{t}{2}\, d[\alpha_1(1)-\alpha_1(t)][\alpha_2(1)-\alpha_2(t)]$$

$$= \frac{1}{2}\int_0^1 [\alpha_1(1)-\alpha_1(t)][\alpha_2(1)-\alpha_2(t)]\, dt; \tag{13.18}$$

$$\text{Average} \int_0^1 x(t)\, d\alpha_1(t) \int_0^1 x(t)\, d\alpha_2(t) \int_0^1 x(t)\, d\alpha_3(t) \int_0^1 x(t)\, d\alpha_4(t)$$

$$= \int_0^1 \frac{t}{2}\, d\alpha_1(t) \int_t^1 d\alpha_2(s) \int_s^1 \left(s + \frac{u}{2}\right) d\alpha_3(u) \int_u^1 d\alpha_4(v)$$

$$+\int_0^1 \frac{t}{2}\,d\alpha_1(t)\int_t^1 d\alpha_2(s)\int_s^1\left(s+\frac{u}{2}\right)d\alpha_4(u)\int_u^1 d\alpha_3(v)$$

+ 22 other terms representing different orders of α_1, α_2, α_3, α_4

$$=\frac{1}{2}\int_0^1 \frac{t}{2}\,d\alpha_1(t)\int_t^1 d\alpha_2(s)\int_s^1 \frac{u}{2}\,d\alpha_3(u)\int_u^1 d\alpha_4(v)$$

$$+\frac{1}{2}\int_0^1 \frac{t}{2}\,d\alpha_1(t)\int_t^1 \frac{u}{2}\,d\alpha_3(u)\int_u^1 d\alpha_2(s)\int_s^1 d\alpha_4(v)$$

$$+\frac{1}{2}\int_0^1 \frac{u}{2}\,d\alpha_3(u)\int_u^1 \frac{t}{2}\,d\alpha_1(t)\int_t^1 d\alpha_2(s)\int_s^1 d\alpha_4(v)$$

$$+\frac{1}{2}\int_0^1 \frac{u}{2}\,d\alpha_3(u)\int_u^1 d\alpha_4(v)\int_v^1 \frac{t}{2}\,d\alpha_1(t)\int_t^1 d\alpha_2(s)$$

$$+\frac{1}{2}\int_0^1 \frac{u}{2}\,d\alpha_3(u)\int_u^1 \frac{t}{2}\,d\alpha_1(t)\int_t^1 d\alpha_4(v)\int_v^1 d\alpha_2(s)$$

$$+\frac{1}{2}\int_0^1 \frac{t}{2}\,d\alpha_1(t)\int_t^1 \frac{u}{2}\,d\alpha_3(u)\int_u^1 d\alpha_4(v)\int_v^1 d\alpha_2(s)$$

+ all other terms representing different orders of α_1, α_2, α_3, α_4

$$=\frac{1}{2}\int_0^1 \frac{t}{2}\,d\alpha_1(t)\int_t^1 d\alpha_2(s)\int_0^1 \frac{u}{2}\,d\alpha_3(u)\int_u^1 d\alpha_4(v)+\text{etc.}$$

$$=\frac{1}{4}\int_0^1 [\alpha_1(1)-\alpha_1(t)][\alpha_2(1)-\alpha_2(t)]\,dt\int_0^1 [\alpha_3(1)-\alpha_3(s)][\alpha_4(1)-\alpha_4(s)]\,ds$$

$$+\frac{1}{4}\int_0^1 [\alpha_1(1)-\alpha_1(t)][\alpha_3(1)-\alpha_3(t)]\,dt\int_0^1 [\alpha_2(1)-\alpha_2(s)][\alpha_4(1)-\alpha_4(s)]\,ds$$

$$+\frac{1}{4}\int_0^1 [\alpha_1(1)-\alpha_1(t)][\alpha_4(1)-\alpha_4(t)]\,dt\int_0^1 [\alpha_2(1)-\alpha_2(s)][\alpha_3(1)-\alpha_3(s)]\,ds. \qquad (13.19)$$

The point of this last argument is that

$$\frac{t}{2}\left(s+\frac{u}{2}\right)$$

may be written

$$\frac{t}{2}\frac{s}{2}+\frac{s}{2}\frac{t}{2}+\frac{t}{2}\frac{u}{2}$$

and that we may then take advantage of the existence in our expression of all permutations of $\alpha_1, \alpha_2, \alpha_3, \alpha_4$ to relabel our variables s, t, u, v so as to add the terms of our expression together again in a more symmetrical way, and represent it as a sum of three products of integrals such as we have already evaluated.

Up to the present, we have been considering probabilities depending on a basis function $x(t)$ defined over $(0, 1)$. For the purposes of harmonic analysis, we wish to replace this by a basis function defined over $(-\infty, \infty)$. We may do this as follows: Let

$$\xi(\tau)=\sqrt{\pi}\left[\int_{\frac{1}{2}}^{\frac{1}{\pi}\cot^{-1}(-\tau)} x(t)d\csc\pi t-x\left(\frac{1}{\pi}\cot^{-1}(-\tau)\right)\sqrt{1+\tau^2}+x(1/2)\right]. \quad (13.\ 20)$$

As $x(t)$ is almost always bounded, this will almost always exist. Then

$$\text{Average }(\xi(\beta)-\xi(\alpha))^2=\frac{\pi}{2}\int_{\frac{1}{\pi}\cot^{-1}(-\alpha)}^{\frac{1}{\pi}\cot^{-1}(-\beta)}\csc^2\pi t\,dt=\frac{\beta-\alpha}{2}, \quad (13.\ 21)$$

in the case that $\beta>\alpha$. This is merely a particular case of (13. 18). In the case that (α, β) and (γ, δ) do not overlap, a similar argument will show that

$$\text{Average }(\xi(\beta)-\xi(\alpha))(\xi(\gamma)-\xi(\delta))=0. \quad (13.\ 22)$$

Thus $\xi(\tau)$ has essentially the same distribution properties as $x(t)$, but over $(-\infty, \infty)$ instead of $(0, 1)$.

We might, of course, have defined our distribution of $\xi(\tau)$ originally, without any recourse to that of $x(t)$. In any case, we should have had to make use of the fact that this distribution has certain equicontinuity properties, and these are somewhat easier to develop over a finite than over an infinite interval. The function $\xi(\tau)$ represents the result of a haphazard sequence of impulses, uniformly distributed in time, extending from $-\infty$ to ∞, in such a way that their zero value is taken to be at $\tau=0$. It is consequently immediately available for a harmonic analysis such as we discuss in this paper, while $x(t)$ is itself immediately adapted only for a Fourier series analysis.

Now let $\vartheta(\tau)$ represent the characteristic response in time of some resonator, the so-called indicial admittance. It may be real or complex, but we shall assume $\vartheta(\tau)\sqrt{1+\tau^2}$ to be of limited total variation over $(-\infty, \infty)$ and ϑ to be quadratically summable. As an immediate consequence of these assumptions,

$$\vartheta(-\infty)=\vartheta(\infty)=0. \tag{13.23}$$

We have

$$\int_{-\infty}^{\infty}\xi(\tau)\,d\vartheta(\tau)=\sqrt{\pi}\int_{-\infty}^{\infty}d\vartheta(\tau)\left[\int_{\frac{1}{2}}^{\frac{1}{\pi}\cot^{-1}(-\tau)}x(t)\,d\csc t-x\left(\frac{1}{\pi}\cot^{-1}(-\tau)\right)\sqrt{1+\tau^2}+x(1/2)\right]$$

$$=\sqrt{\pi}\int_{0}^{1}x(t)[\vartheta(-\cot\pi t)\,d\csc\pi t+\csc\pi t\,d\vartheta(-\cot\pi t)]$$

$$=\sqrt{\pi}\int_{0}^{1}x(t)\,d[\vartheta(-\cot\pi t)\csc\pi t]. \tag{13.24}$$

Hence if $\vartheta_1(\tau)$, $\vartheta_2(\tau)$, $\vartheta_3(\tau)$, and $\vartheta_4(\tau)$ satisfy the conditions we have laid down for $\vartheta(\tau)$,

$$\text{Average}\int_{-\infty}^{\infty}\xi(\tau)\,d\vartheta_1(\tau)\int_{-\infty}^{\infty}\xi(\tau)\,d\vartheta_2(\tau)=\frac{1}{2}\int_{0}^{1}\pi\csc^2\pi t\,\vartheta_1(-\cot\pi t)\vartheta_2(-\cot\pi t)\,dt$$

$$=\frac{1}{2}\int_{-\infty}^{\infty}\vartheta_1(\tau)\vartheta_2(\tau)\,d\tau; \tag{13.25}$$

29—29764. *Acta mathematica.* 55. Imprimé le 29 avril 1930.

$$\text{Average} \int_{-\infty}^{\infty} \xi(\tau)\,d\vartheta_1(\tau) \int_{-\infty}^{\infty} \xi(\tau)\,d\vartheta_2(\tau) \int_{-\infty}^{\infty} \xi(\tau)\,d\vartheta_3(\tau) \int_{-\infty}^{\infty} \xi(\tau)\,d\vartheta_4(\tau)$$

$$= \frac{1}{4} \int_{-\infty}^{\infty} \vartheta_1(\tau)\vartheta_2(\tau)\,d\tau \int_{-\infty}^{\infty} \vartheta_3(\tau)\vartheta_4(\tau)\,d\tau$$

$$+ \frac{1}{4} \int_{-\infty}^{\infty} \vartheta_1(\tau)\vartheta_3(\tau)\,d\tau \int_{-\infty}^{\infty} \vartheta_2(\tau)\vartheta_4(\tau)\,d\tau$$

$$+ \frac{1}{4} \int_{-\infty}^{\infty} \vartheta_1(\tau)\vartheta_4(\tau)\,d\tau \int_{-\infty}^{\infty} \vartheta_2(\tau)\vartheta_3(\tau)\,d\tau. \tag{13.26}$$

We thus have succeeded in generalizing our theorems concerning the averages of products of linear functionals to the case where the basis function has an infinite range.

We wish to apply these results to the harmonic analysis of an expression $\int_{-\infty}^{\infty} \xi(\tau)\,d\vartheta(\tau+\lambda)$. To do this, we must evaluate the following averages:

$$\text{Average} \frac{1}{2T} \int_{-T}^{T} d\lambda \int_{-\infty}^{\infty} \xi(\tau)\,d\vartheta(\tau+\lambda) \int_{-\infty}^{\infty} \xi(\sigma)\,d\bar{\vartheta}(\sigma+\lambda)$$

$$= \frac{1}{2T} \int_{-T}^{T} d\lambda \;\text{Average} \left[\int_{-\infty}^{\infty} \xi(\tau)\,d\vartheta(\tau+\lambda) \int_{-\infty}^{\infty} \xi(\sigma)\,d\bar{\vartheta}(\sigma+\lambda) \right]$$

$$= \frac{1}{4T} \int_{-T}^{T} d\lambda \int_{-\infty}^{\infty} |\vartheta(\tau+\lambda)|^2\,d\tau$$

$$= \frac{1}{2} \int_{-\infty}^{\infty} |\vartheta(\tau)|^2\,d\tau. \tag{13.27}$$

(Here as in what follows, the inversion of the operators $\frac{1}{2T}\int_{-T}^{T} d\tau$ and Average is permissable, since the integral to which our mapping process leads us is absolutely convergent.)

$$\text{Average}\left[\frac{1}{2T}\int_{-T}^{T}d\lambda\int_{-\infty}^{\infty}\xi(\tau)\,d\vartheta(\tau+\lambda)\int_{-\infty}^{\infty}\xi(\sigma)\,d\bar{\vartheta}(\sigma+\lambda)-\frac{1}{2}\int_{-\infty}^{\infty}|\vartheta(\tau)|^2\,d\tau\right]^2$$

$$= \text{Average}\,\frac{1}{4T^2}\int_{-T}^{T}d\lambda\int_{-T}^{T}d\mu\int_{-\infty}^{\infty}\xi(\tau)\,d\vartheta(\tau+\lambda)\int_{-\infty}^{\infty}\xi(\sigma)\,d\bar{\vartheta}(\sigma+\lambda)\int_{-\infty}^{\infty}\xi(\alpha)\,d\vartheta(\alpha+\mu)\cdot$$

$$\int_{-\infty}^{\infty}\xi(\beta)\,d\bar{\vartheta}(\beta+\mu)-\frac{1}{4}\left[\int_{-\infty}^{\infty}|\vartheta(\tau)|^2\,d\tau\right]^2$$

$$=\frac{1}{4T^2}\int_{-T}^{T}d\lambda\int_{-T}^{T}d\mu\,\frac{1}{4}\left\{\left|\int_{-\infty}^{\infty}\vartheta(\tau+\lambda)\bar{\vartheta}(\tau+\mu)\,d\tau\right|^2+\left|\int_{-\infty}^{\infty}\vartheta(\tau+\lambda)\vartheta(\tau+\mu)\,d\tau\right|^2\right\}$$

$$\leq\frac{1}{32T^2}\int_{-2T}^{2T}d\lambda\int_{-2T}^{2T}du\left\{\left|\int_{-\infty}^{\infty}\vartheta(\tau)\bar{\vartheta}(\tau+u)\,d\tau\right|^2+\left|\int_{-\infty}^{\infty}\vartheta(\tau)\vartheta(\tau+u)\,d\tau\right|^2\right\}$$

$$\leq\frac{1}{8T}\int_{-\infty}^{\infty}du\left\{\left|\int_{-\infty}^{\infty}\vartheta(\tau)\bar{\vartheta}(\tau+u)\,d\tau\right|^2+\left|\int_{-\infty}^{\infty}\vartheta(\tau)\vartheta(\tau+u)\,d\tau\right|^2\right\}. \qquad (13.28)$$

The function in the bracket in the last expression will be summable since $\vartheta(\tau)$ is, as we see from (1. 28).

It follows that for any positive number A,

$$\left|\frac{1}{2T}\int_{-T}^{T}d\lambda\left|\int_{-\infty}^{\infty}\xi(\tau)\,d\vartheta(\tau+\lambda)\right|^2-\frac{1}{2}\int_{-\infty}^{\infty}|\vartheta(\tau)|^2\,d\tau\right|<A \qquad (13.29)$$

except for a set of values of $x(t)$ not exceeding

$$\frac{1}{8TA^2}\int_{-\infty}^{\infty}du\left\{\left|\int_{-\infty}^{\infty}\vartheta(\tau)\bar{\vartheta}(\tau+u)\,d\tau\right|^2+\left|\int_{-\infty}^{\infty}\vartheta(\tau)\vartheta(\tau+u)\,d\tau\right|^2\right\} \qquad (13.30)$$

in outer measure. Let T now assume the successive values 1, 4, 9, ... Then the probability that (13. 29) fails to be satisfied for some T from $1/n^2$ on does not exceed the remainder of the convergent series

$$\frac{1}{8A^2}\int_{-\infty}^{\infty} du \left\{ \left| \int_{-\infty}^{\infty} \vartheta(\tau)\bar{\vartheta}(\tau+u)\,d\tau \right|^2 + \left| \int_{-\infty}^{\infty} \vartheta(\tau)\vartheta(\tau+u)\,d\tau \right|^2 \right\} \left[1 + \frac{1}{4} + \frac{1}{9} + \cdots \right]. \quad (13.31)$$

Inasmuch as this remainder is arbitrarily small, we almost always have

$$\overline{\lim_{n\to\infty}} \left| \frac{1}{2n^2}\int_{-n^2}^{n^2} d\lambda \left| \int_{-\infty}^{\infty} \xi(\tau)\,d\vartheta(\tau+\lambda) \right|^2 - \frac{1}{2}\int_{-\infty}^{\infty} |\vartheta(\tau)|^2\,d\tau \right| \leq A. \quad (13.32)$$

Since, however, A is an arbitrary positive quantity,

$$\lim_{n\to\infty} \left| \frac{1}{2n^2}\int_{-n^2}^{n^2} d\lambda \left| \int_{-\infty}^{\infty} \xi(\tau)\,d\vartheta(\tau+\lambda) \right|^2 - \frac{1}{2}\int_{-\infty}^{\infty} |\vartheta(\tau)|^2\,d\tau \right| = 0. \quad (13.33)$$

As in the preceding section, we may conclude that

$$\lim_{T\to\infty} \frac{1}{2T}\int_{-T}^{T} d\lambda \left| \int_{-\infty}^{\infty} \xi(\tau)\,d\vartheta(\tau+\lambda) \right|^2 = \frac{1}{2}\int_{-\infty}^{\infty} |\vartheta(\tau)|^2\,d\tau, \quad (13.34)$$

except in a set of cases of zero probability.

Let us now consider

$$\int_{-A}^{A} d\lambda \int_{-\infty}^{\infty} \xi(\tau)\,d\vartheta(\tau+\lambda) = \int_{-\infty}^{\infty} \xi(\tau)\,d\int_{-A}^{A} \vartheta(\tau+\lambda)\,d\lambda, \quad (13.35)$$

for rational values of A, $\vartheta(\tau)$ being subject to the conditions already laid down. Let us put

$$\int_{-\infty}^{\infty} \xi(\tau)\,d\vartheta(\tau+\lambda) = f(\lambda). \quad (13.36)$$

We have almost always, for any denumerable set of values of A, as for example, for all rational values of A,

$$\lim_{T\to\infty}\frac{1}{2T}\int_{-T}^{T}d\lambda\left|\int_{-A}^{A}f(\lambda+\mu)\,d\mu\right|^2=\frac{1}{2}\int_{-\infty}^{\infty}d\tau\left|\int_{-A}^{A}\vartheta(\tau+\mu)\,d\mu\right|^2. \qquad (13.37)$$

This results from the fact that the sum of a denumerable set of null sets is a null set. As before, let us put

$$s(u)=\frac{1}{2\pi}\int_{-1}^{1}f(x)\frac{e^{ixu}-1}{ix}\,dx+\frac{1}{2\pi}\operatorname*{l.i.m.}_{M\to\infty}\left[\int_{1}^{M}+\int_{-M}^{-1}\right]f(x)\frac{e^{ixu}}{ix}\,dx. \qquad (13.38)$$

It then follows from (13. 34) and (13. 37), with the help of (6. 23), that we shall almost always have

$$\lim_{\varepsilon\to 0}\frac{1}{2\varepsilon}\int_{-\infty}^{\infty}|s(u+\varepsilon)-s(u-\varepsilon)|^2\,du=\frac{1}{2}\int_{-\infty}^{\infty}|\vartheta(\tau)|^2\,d\tau, \qquad (13.39)$$

and (for all rational A)

$$\lim_{\varepsilon\to 0}\frac{1}{2\varepsilon}\int_{-\infty}^{\infty}\frac{4\sin^2 Au}{u^2}|s(u+\varepsilon)-s(u-\varepsilon)|^2\,du=\frac{1}{2}\int_{-\infty}^{\infty}d\tau\left|\int_{-A}^{A}\vartheta(\tau+\mu)\,d\mu\right|^2. \qquad (13.40)$$

Now let us put

$$\vartheta(\tau)=\frac{1}{\sqrt{2\pi}}\operatorname*{l.i.m.}_{M\to\infty}\int_{-M}^{M}\psi(u)e^{iu\tau}\,du. \qquad (13.41)$$

Then

$$\int_{-A}^{A}\vartheta(\tau+\mu)\,d\mu=\sqrt{\frac{2}{\pi}}\int_{-\infty}^{\infty}\psi(u)\frac{\sin Au}{u}e^{iu\tau}\,du. \qquad (13.42)$$

Thus if $\vartheta(\tau)\sqrt{1+\tau^2}$ is of limited total variation and $\vartheta(\tau)$ is quadratically summable, we almost always have for all rational A

$$\int_{-\infty}^{\infty}d\tau\left|\int_{-A}^{A}\vartheta(\tau+\mu)\,d\mu\right|^2=\int_{-\infty}^{\infty}|\psi(u)|^2\frac{4\sin^2 Au}{u^2}\,du. \qquad (13.43)$$

In other words, we almost always have for all rational A,

$$\lim_{\varepsilon\to 0}\int_{-\infty}^{\infty}\frac{\sin^2 Au}{u^2}\left\{\frac{1}{\varepsilon}|s(u+\varepsilon)-s(u-\varepsilon)|^2-|\psi(u)|^2\right\}du=0, \tag{13.44}$$

and

$$\lim_{\varepsilon\to 0}\int_{-\infty}^{\infty}\left\{\frac{1}{\varepsilon}|s(u+\varepsilon)-s(u-\varepsilon)|^2-|\psi(u)|^2\right\}du=0. \tag{13.45}$$

Thus we almost always have

$$\lim_{\varepsilon\to 0}\int_{-\infty}^{\infty}P(u)\left\{\frac{1}{\varepsilon}|s(u+\varepsilon)-s(u-\varepsilon)|^2-|\psi(u)|^2\right\}du=0, \tag{13.46}$$

in case

$$P(u)=\sum_{1}^{n}\frac{A_k\sin^2 N_k(u)}{u^2} \tag{13.47}$$

it follows from (13.45) that we may even replace $P(u)$ by

$$Q(u)=\underset{n\to\infty}{\text{uniform limit}}\ P_n(u) \tag{13.48}$$

where $P_n(u)$ is of the form given above for $P(u)$. Thus by the Weierstrass theorem, $Q(u)$ may be the quotient by u^2 of any continuous periodic function with any period. Since we can approximate by such a function Q to any continuous function vanishing at $\pm\infty$, our sole condition on Q may be replaced by

$$Q(u)=o(1) \quad \text{at} \quad u=\pm\infty. \tag{13.49}$$

Even this does not represent the utmost extension of our theorem. It follows at once by subtracting from 1 a Q vanishing outside of a finite range that

$$\lim_{N\to\infty}\ \overline{\lim_{\varepsilon\to 0}}\left[\int_{N}^{\infty}+\int_{-\infty}^{-N}\right]\left\{\frac{1}{\varepsilon}|s(u+\varepsilon)-s(u-\varepsilon)|^2-|\psi(u)|^2\right\}du=0. \tag{13.50}$$

Thus a bounded modification of Q for large arguments produces a decreasing effect as the range of modification recedes to infinity, and we have as our sole condition to be imposed on the continuous function Q that

$$Q(u)=O(1) \quad \text{at} \quad u=\pm\infty. \tag{13.51}$$

A case of peculiar importance is where $e^{-ivu}=Q(u)$. Here

$$\lim_{\varepsilon\to 0}\frac{1}{2\varepsilon}\int_{-\infty}^{\infty} e^{-i\nu u}\,|s(u+\varepsilon)-s(u-\varepsilon)|^2\,du = \frac{1}{2}\int_{-\infty}^{\infty} e^{-i\nu u}\,|\psi(u)|^2\,du. \qquad (13.52)$$

This exists for *every* ν for almost all $x(t)$. Thus by (6.15),

$$\varphi(\nu) = \frac{1}{2}\int_{-\infty}^{\infty} e^{-i\nu u}\,|\psi(u)|^2\,du, \qquad (13.53)$$

and $\varphi(\nu)$ exists for *every* ν for almost every $x(t)$. As in section 3, let us put

$$S(u) = \frac{1}{2\pi}\int_{-\infty}^{\infty} \varphi(x)\frac{e^{iux}-1}{ix}\,dx, \qquad (13.54)$$

so that we get by (5.42), (5.45)

$$\varphi(x) = \int_{-\infty}^{\infty} e^{-ixu}\,dS(u). \qquad (13.55)$$

We have already shown φ and S to exist on the assumption that $\varphi(\nu)$ exists for every ν. We have, by (13.53) and (13.55), for almost all $x(t)$

$$\int_{-\infty}^{\infty} e^{-i\nu u}\,d\left[S(u) - \frac{1}{2}\int_{-\infty}^{u}|\psi(v)|^2\,dv\right] \equiv 0. \qquad (13.56)$$

By processes now familiar (cf. (13.46)), we can replace $e^{-i\nu u}$ by functions Q which are merely continuous and bounded. We here make use of the absolute convergence of (13.56). Hence the average of $S(u) - \frac{1}{2}\int_{-\infty}^{u}|\psi(v)|^2\,dv$ vanishes over every interval, and

$$S(u) - \frac{1}{2}\int_{-\infty}^{u}|\psi(v)|^2\,dv \equiv 0; \qquad (13.57)$$

and consequently

$$S'(u) = \frac{1}{2}|\psi(u)|^2, \qquad (13.58)$$

except for a set of values of a zero measure. Inasmuch as $S'(u)$ is the spectral density of $f(x)$, we see that as a consequence of our assumptions that $\vartheta(\tau)$ is quadratically summable and of limited total variation when multiplied by $\sqrt{1+\tau^2}$ *the spectral density of* $\int_{-\infty}^{\infty} \xi(\tau)\,d\vartheta(\tau+\lambda)$ *is half the square of the modulus of the Fourier transform of* ϑ. Another way of phrasing this fact is: *if a linear resonator is stimulated by a uniformly haphazard sequence of impulses, each frequency responds with an amplitude proportional to that which it would have if stimulated by an impulse of that frequency and of unit energy.* An even simpler statement is: *the energy of a haphazard sequence of impulses is uniformly distributed in frequency.* This law of distribution bears a curious analogy to that predicted for white light by the incorrect Boltzmann law of radiation. The physical conditions which lead to this law of distribution of energy in frequency are that the sequence of impulses in question should be distributed over every interval of time in a Gaussian manner, that their past should not influence their future, that very many should occur over the smallest period of time to be investigated, and that the modulus of the Gaussian distribution of these impulses for a given time interval should depend only on the length of this interval. These conditions are approximately realized in the case of the Schroteffekt, where an electrical resonating circuit is set in vibration by the irregularities in the stream of electrons across a vacuum tube. It might also be realized in the ease of an acoustical system set in oscillation by such a noise as that of a sand blast. Theoretically this equipartition of energy might be used in the absolute calibration of acoustical instruments.

Just as the average of an expression depending on a single function $x(t)$ may be reduced a Lebesgue single integral, so a similar average depending on two independent functions $x(t)$ and $y(t)$ may be reduced to a Lebesgue double integral. On the assumption that $\vartheta_1(\tau)$ and $\vartheta_2(\tau)$ satisfy the conditions we have already laid down for $\vartheta(\tau)$ and that

$$\left.\begin{aligned} \xi(\tau) &= \sqrt{\pi}\int_{1/2}^{\frac{1}{\pi}\cot^{-1}(-\tau)} x(t)\,d\csc\pi t - x\left(\frac{1}{\pi}\cot^{-1}(-\tau)\right) + x(1/2);\\ \eta(\tau) &= \sqrt{\pi}\int_{1/2}^{\frac{1}{\pi}\cot^{-1}(-\tau)} y(t)\,d\csc\pi t - y\left(\frac{1}{\pi}\cot^{-1}(-\tau)\right) + y(1/2); \end{aligned}\right\} \tag{13.59}$$

$$\vartheta_{1,2}(\tau) = \frac{1}{\sqrt{2\pi}} \underset{M\to\infty}{\text{l.i.m.}} \int_{-M}^{M} \psi_{1,2}(u) e^{iu\tau} du \tag{13.60}$$

it is easy to prove by methods substantially identical with those already employed that

$$\lim_{T\to\infty} \frac{1}{2T} \int_{-T}^{T} \left| \int_{-\infty}^{\infty} \xi(\tau) d\vartheta_1(\tau+\lambda) + \int_{-\infty}^{\infty} \eta(\tau) d\vartheta_2(\tau+\lambda) \right|^2 d\lambda \tag{13.61}$$

almost always has a certain definite value. Inasmuch as a normal distribution for $x(t)$ implies the same for $-x(t)$,

$$\lim_{T\to\infty} \frac{1}{2T} \int_{-T}^{T} \left| -\int_{-\infty}^{\infty} \xi(\tau) d\vartheta_1(\tau+\lambda) + \int_{-\infty}^{\infty} \eta(\tau) d\vartheta_2(\tau+\lambda) \right|^2 d\lambda \tag{13.62}$$

almost always has the same value. Subtracting, we almost always have

$$\lim_{T\to\infty} \frac{1}{2T} \int_{-T}^{T} \Re \left[\int_{-\infty}^{\infty} \xi(\tau) d\vartheta_1(\tau+\lambda) \int_{-\infty}^{\infty} \eta(\tau) d\bar{\vartheta}_2(\tau+\lambda) \right] d\lambda = 0. \tag{13.63}$$

If we work in a similar manner with

$$\lim_{T\to\infty} \frac{1}{2T} \int_{-T}^{T} \left| \int_{-\infty}^{\infty} \xi(\tau) d\vartheta_1(\tau+\lambda) \pm i \int_{-\infty}^{\infty} \eta(\tau) d\vartheta_2(\tau+\lambda) \right|^2 d\lambda, \tag{13.64}$$

we see that almost always

$$\lim_{T\to\infty} \frac{1}{2T} \int_{-T}^{T} \Im \left[\int_{-\infty}^{\infty} \xi(\tau) d\vartheta_1(\tau+\lambda) \int_{-\infty}^{\infty} \eta(\tau) d\bar{\vartheta}_2(\tau+\lambda) \right] d\lambda = 0. \tag{13.65}$$

Hence almost always

$$\lim_{T\to\infty} \frac{1}{2T} \int_{-T}^{T} \left[\int_{-\infty}^{\infty} \xi(\tau) d\vartheta_1(\tau+\lambda) \int_{-\infty}^{\infty} \eta(\tau) d\bar{\vartheta}_2(\tau+\lambda) \right] d\lambda = 0 \tag{13.66}$$

30—29764. *Acta mathematica.* 55. Imprimé le 29 avril 1930.

and the coherency matrix of $\int_{-\infty}^{\infty} \xi(\tau) d\vartheta_1(\tau+\lambda)$ and $\int_{-\infty}^{\infty} \eta(\tau) d\vartheta_2(\tau+\lambda)$ is almost always

$$\left\| \begin{matrix} \frac{1}{2} | \psi_1(u) |^2 & 0 \\ 0 & \frac{1}{2} | \psi_2(u) |^2 \end{matrix} \right\| . \tag{13.67}$$

As a direct consequence, if the motion of a particle is independently haphazard in two directions at right angles, and if this motion influences a resonator with the same characteristics in the two directions, the coherency matrix of the motion of the resonator is unpolarized.

In the opinion of the author, the chief importance of this section is in showing in a systematic manner how the Lebesgue integral may be adapted to the needs of statistical mechanics. It is no new observation that sets of zero measure and sets of phenomena, not necessarily impossible, of probability zero, are in essence the same sort of thing. It is not, however, a particularly easy matter to translate any specific problem in statistical mechanics into its precise counterpart in the theory of integration. The author feels confident that methods closely resembling those here developed are destined to play a part in the statistical mechanics of the future, in such regions as those now being invaded by the theory of quanta.

CHAPTER V.

14. The spectrum of an almost periodic function.

The last paragraph was exclusively devoted to functions with continuous spectra; we now come to the most important known class of functions with spectra that are discrete. This is the class of almost periodic functions, the discovery of which is due to Harald Bohr. Let $f(x)$ be a continuous function, not necessarily real, defined for all real values of x between $-\infty$ and ∞. If ε is any positive quantity, Bohr defines τ_ε to be a translation number of $f(x)$ belonging to ε, in case for every real x,

$$|f(x+\tau_\varepsilon)-f(x)| \leq \varepsilon. \tag{14.01}$$

In case, whenever, ε is given, a quantity L_ε can be assigned, such that no interval $(a, a+L_\varepsilon)$ is free of translation numbers τ_ε belonging to ε, $f(x)$ is said to be *almost periodic.* Bohr's most fundamental theorem is: *the necessary and sufficient condition for a function $f(x)$ to be almost periodic is that for any positive quantity ε, there exist a finite set of complex numbers $A_1, A_2, \ldots, A_n$ and a set of real numbers $\Lambda_1, \Lambda_2, \ldots, \Lambda_n$, such that for all x*

$$\left| f(x) - \sum_1^n A_k e^{i\Lambda_k x} \right| < \varepsilon. \tag{14.02}$$

The next few sections of this paper are devoted to the proof of this theorem. In this proof we shall avail ourselves of the following theorems of Bohr concerning almost periodic functions, which are susceptible of a completely elementary proof:

Any finite set of almost periodic functions is *simultaneously* almost periodic, in the sense that for any ε, L_ε may be assigned for the whole set at once, in such a manner that in any interval $(a, a+L_\varepsilon)$, there is at least one translation number τ_ε, such that for every function $f(t)$ in the set, and every t,

$$|f(t+\tau_\varepsilon) - f(t)| \leq \varepsilon. \tag{14.03}$$

Hence any continuous function of a finite number of almost periodic functions yields an almost periodic function, as for example the sum or the product of a finite number of almost periodic functions. The limit of a uniformly convergent series or sequence of almost periodic functions is almost periodic. Every function that is periodic in the classical sense is almost periodic, and the same is true of

$$\Sigma A_n e^{i\Lambda_n t}, \tag{14.04}$$

in case $\Sigma|A_n|$ converges. Every almost periodic function is uniformly continuous. If $f(t)$ is almost periodic,

$$M\{f\} = \lim_{T\to\infty} \frac{1}{T} \int_z^{z+T} f(t)\,dt \tag{14.05}$$

exists as a uniform limit in z. If $f(t)$ is almost periodic, so is

$$\varphi(t) = M_x\{f(x+t)\bar{f}(x)\}. \tag{14.06}$$

(Here and later the symbol under the M indicates the variable on which the averaging is being done.) If $f(t)$ is a real non-negative almost periodic function, and $M\{f\}=0$, $f(t)$ is identically zero.

If $f(t)$ is almost periodic, then since $\varphi(t)$ is also almost periodic, it is continuous. Let us form

$$S(u)=\frac{1}{2\pi}\int_{-\infty}^{\infty}\varphi(x)\frac{e^{iux}-1}{ix}\,dx. \tag{14.07}$$

From theorems already established

$$\varphi(0)=S(\infty)-S(-\infty). \tag{14.08}$$

Let the discontinuities of $S(u)$ be at $u=\lambda_1, \lambda_2, \ldots$ These form a denumerable set, as $S(u)$ is of limited total variation, and indeed monotone. Let

$$a_n=S(\lambda_n+0)-S(\lambda_n-0). \tag{14.09}$$

All these coefficients a_n are positive, and

$$\sum_1^{\infty} a_n \leq S(\infty)-S(-\infty)=M\{|f|^2\}. \tag{14.10}$$

Let us form the function

$$\gamma(t)=\varphi(t)-\sum_0^{\infty} a_k e^{-i\lambda_k t}, \tag{14.11}$$

As a simple computation will show (cf. (4.05)),

$$\frac{1}{2\pi}\int_{-\infty}^{\infty}\gamma(x)\frac{e^{iux}-1}{ix}\,dx=S(u)-S_1(u), \tag{14.12}$$

where $S_1(u)$ consists of the sum of all the jumps of $S(u)$ with abscissae less than u, together with half the jump (if any) with abscissa u. Hence

$$S_2(u)=\frac{1}{2\pi}\int_{-\infty}^{\infty}\gamma(x)\frac{e^{iux}-1}{ix}\,dx \tag{14.13}$$

is a continuous function of limited total variation, say V. Let the total variation of $S_2(u)$ over the ranges $(-\infty, -B)$ and (B, ∞) be $V(B)$.

We have

$$\frac{1}{2\varepsilon}\int_{-\infty}^{\infty} | S_2(u+\varepsilon)-S_2(u-\varepsilon)|^2 \, du$$

$$= \frac{1}{2\varepsilon}\left[\int_{A}^{\infty} + \int_{-\infty}^{-A}\right] | S_2(u+\varepsilon)-S_2(u-\varepsilon)|^2 \, du + \frac{1}{2\varepsilon}\int_{-A}^{A} | S_2(u+\varepsilon)-S_2(u-\varepsilon)|^2 \, du$$

$$\leq \frac{\max | S_2(v)|}{\varepsilon}\left[\int_{A}^{\infty} + \int_{-\infty}^{-A}\right] | S_2(u+\varepsilon)-S_2(u-\varepsilon)| \, du$$

$$+ \max_{-A \leq v \leq A} | S_2(v+\varepsilon)-S_2(v-\varepsilon)| \frac{1}{2\varepsilon}\int_{-A}^{A} | S_2(u+\varepsilon)-S_2(u-\varepsilon)| \, du$$

$$\leq \frac{\max | S_2(v)|}{\varepsilon}\left[\int_{A-\varepsilon}^{A+\varepsilon} + \int_{A+\varepsilon}^{A+3\varepsilon} + \int_{A+3\varepsilon}^{A+5\varepsilon} + \cdots + \int_{-A-\varepsilon}^{-A+\varepsilon} + \int_{-A-3\varepsilon}^{-A-\varepsilon}\right.$$

$$\left. + \int_{-A-5\varepsilon}^{-A-3\varepsilon} + \cdots\right] | S_2(u+\varepsilon)-S_2(u-\varepsilon)| \, du + \max_{-A \leq v \leq A} | S_2(v+\varepsilon)-S_2(v-\varepsilon)| \frac{1}{2\varepsilon}$$

$$\left[\int_{-A-\varepsilon}^{-A+\varepsilon} + \int_{-A+\varepsilon}^{-A+3\varepsilon} + \cdots + \int_{-A+\left(\left[\frac{2A}{\varepsilon}\right]-2\right)\varepsilon}^{-A+\left[\frac{2A}{\varepsilon}\right]\varepsilon}\right] | S_2(u+\varepsilon)-S_2(u-\varepsilon)| \, du$$

$$= \frac{\max | S_2(v)|}{\varepsilon}\int_{-\varepsilon}^{\varepsilon}\{| S_2(u+A+\varepsilon)-S_2(u+A-\varepsilon)| + | S_2(u+A+3\varepsilon)$$

$$-S_2(u+A+\varepsilon)| + \cdots + | S_2(u-A+\varepsilon)-S_2(u-A+\varepsilon)| + | S_2(u-A-\varepsilon)$$

$$-S_2(u-A-3\varepsilon)| + \cdots\} \, du + \max_{-A \leq v \leq A} | S_2(v+\varepsilon)-S_2(v-\varepsilon)|$$

$$\cdot \frac{1}{2\varepsilon}\int_{-\varepsilon}^{\varepsilon}\Big\{| S_2(u-A+\varepsilon)-S_2(u-A-\varepsilon)| + | S_2(u-A+3\varepsilon)-S_2(u-A+\varepsilon)|$$

$$+\cdots+\left|S_2\left(u-A+\left[\frac{2A}{\varepsilon}\right]\varepsilon\right)-S_2\left(u-A+\left[\frac{2A}{\varepsilon}\right]\varepsilon-2\varepsilon\right)\right|\Bigg\}du$$

$$\leq 2\,V(A-2\varepsilon)\max|S_2(v)|+V\max_{-A\leq v\leq A}|S_2(v+\varepsilon)-S_2(v-\varepsilon)|. \qquad (14.\ 14)$$

Since the function $S(u)$ is uniformly continuous over any finite range, this gives us

$$\lim_{\varepsilon\to 0}\frac{1}{2\varepsilon}\int_{-\infty}^{\infty}|S_2(u+\varepsilon)-S_2(u-\varepsilon)|^2\,du\leq 2\,V(A-\eta)\max|S_2(v)|. \qquad (14.\ 15)$$

However, $V(A-\eta)$ tends to zero as A tends to infinity, and may be arbitrarily small. Hence

$$\lim_{\varepsilon\to 0}\frac{1}{2\varepsilon}\int_{-\infty}^{\infty}|S_2(u+\varepsilon)-S_2(u-\varepsilon)|^2\,du=0. \qquad (14.\ 16)$$

Applying (5. 53), we get

$$\lim_{T\to\infty}\frac{1}{2T}\int_{-T}^{T}|\gamma(t)|^2\,dt=0. \qquad (14.\ 17)$$

Since, however, $\gamma(t)$ and $|\gamma(t)|^2$ are almost periodic, we must have

$$\gamma(t)=0, \qquad (14.\ 18)$$

which yields us

$$S(u)=S_1(u) \qquad (14.\ 19)$$

and

$$\varphi(t)=\sum_{1}^{\infty}a_k e^{-i\lambda_k t}. \qquad (14.\ 20)$$

Thus $S(u)$ is a step function, and the spectrum of an almost periodic function is a pure line spectrum.

15. The Parseval theorem for almost periodic functions.

A further result is

$$M\{\varphi(t)\,e^{i\lambda_k t}\}=a_k. \qquad (15.\ 01)$$

If we remember the uniformity properties of the means of almost periodic functions, this yields

$$a_k = \lim_{T\to\infty} \frac{1}{2T} \int_{-T}^{T} e^{i\lambda_k t} dt \lim_{U\to\infty} \frac{1}{2U} \int_{-U}^{U} f(x+t)\bar{f}(x)\, dx$$

$$= \lim_{U\to\infty} \frac{1}{2U} \int_{-U}^{U} \bar{f}(x) e^{-i\lambda_k x}\, dx \lim_{T\to\infty} \frac{1}{2T} \int_{-T}^{T} f(x+t) e^{i\lambda_k (x+t)} dt$$

$$= \lim_{U\to\infty} \frac{1}{2U} \int_{-U}^{U} \bar{f}(x) e^{-i\lambda_k x}\, dx \lim_{T\to\infty} \frac{1}{2T} \int_{-T}^{T} f(y) e^{i\lambda_k y} dy$$

$$= |M\{f(x) e^{i\lambda_k x}\}|^2. \tag{15.02}$$

Hence

$$M\{|f|^2\} = \varphi(0) = S(\infty) - S(-\infty)$$

$$= S_1(\infty) - S_1(-\infty)$$

$$= \sum_0^\infty |M\{f(x)\, e^{i\lambda_k x}\}|^2. \tag{15.03}$$

This is a precise analogue to the Hurwitz-Parseval theorem for periodic functions, and is the well-known fundamental theorem of Bohr.

16. The Weierstrass theorem for almost periodic functions.

The present section 16 is devoted to the proof of the approximation theorem for almost periodic functions, which tells us that the necessary and sufficient condition for a function to be almost periodic is that it can be expressed as the uniform limit of a trigonometrical polynomial. The main idea of the present proof is due to Weyl, although the form of the argument is much changed from that on his paper. The essence of the proof is that harmonic analysis is not applied directly to the almost periodic function discussed, but to certain related functions derived from what Bochner calls the Verschiebungsfunktion of the given function. In the discussion of the many different extensions of almost periodic functions, there is a function in each case analogous to this Verschiebungsfunktion which is almost periodic in the strict Bohr sense. As we shall see in the next section, this enables us to carry over to these more general cases practically the entire Bohr approximation theorem redefined to suit each particular definition.

Let $f(t)$ be almost periodic. Consider

$$g(x) = \max_t |f(x+t) - f(t)|. \tag{16.01}$$

We have

$$\begin{aligned} |g(x+\tau) - g(x)| &\leq \max_t \big||f(x+t+\tau) - f(t)| - |f(x+t) - f(t)|\big| \\ &\leq \max_t |f(x+t+\tau) - f(x+t)| \\ &= \max_t |f(t+\tau) - f(t)|. \end{aligned} \tag{16.02}$$

Hence any translation number for $f(t)$ pertaining to ε is a translation number for $g(t)$ pertaining to ε, and $g(t)$ is almost periodic. It is this function which Bochner calls the Verschiebungsfunktion of $f(t)$.

We have already indicated the fact that any continuous function of an almost periodic function is almost periodic. Let $H_\varepsilon(U)$ be befined as follows:

$$H_\varepsilon(U) = \begin{cases} 1; & [0 \leq U \leq \varepsilon/2] \\ 2 - \dfrac{2U}{\varepsilon}; & [\varepsilon/2 \leq U \leq \varepsilon] \\ 0. & [\varepsilon \leq U] \end{cases} \tag{16.03}$$

Let

$$\psi_\varepsilon(x) = \frac{H_\varepsilon[g(x)]}{M_x H_\varepsilon[g(x)]}. \tag{16.04}$$

Since $H_\varepsilon[g(x)]$ is somewhere positive, and it is everywhere non negative and almost periodic, $M_x H_\varepsilon[g(x)]$ cannot vanish. Hence $\psi_\varepsilon(x)$ exists and is almost periodic.

Let

$$f_\varepsilon(x) = M_t\{f(t)\,\psi_\varepsilon(x-t)\}. \tag{16.05}$$

The existence and almost periodic character of $f_\varepsilon(x)$ are proved without difficulty. The definition of ψ_ε ensures that

$$|f(x) - f(t)| \leq \varepsilon \tag{16.06}$$

if $\psi_\varepsilon(x-t) \neq 0$. Hence, since $f_\varepsilon(x)$ is a mean of these values of $f(t)$,

$$\max_x |f(x) - f_\varepsilon(x)| \leq \varepsilon. \tag{16.07}$$

Similarly, if

$$f^{(\varepsilon)}(x) = M_t\{f_\varepsilon(t)\,\psi_\varepsilon(x-t)\}, \tag{16.08}$$

$f^{(\varepsilon)}(x)$ exists and is almost periodic, and

$$\max_x |f^{(\varepsilon)}(x)-f_\varepsilon(x)| \leq \varepsilon. \tag{16.09}$$

Hence, by (16.07) and (16.09),

$$\max_x |f^{(\varepsilon)}(x)-f(x)| \leq 2\,\varepsilon. \tag{16.10}$$

We have

$$f^{(\varepsilon)}(x)=M_t\{\psi_\varepsilon(x-t)\,M_\tau\{f(\tau)\,\psi_\varepsilon(t-\tau)\}\}. \tag{16.11}$$

Bearing in mind the uniformity properties of almost-periodic functions, we have

$$f^{(\varepsilon)}(x)=M_\tau\{f(\tau)\,M_t\{\psi_\varepsilon(x-t)\,\psi_\varepsilon(t-\tau)\}\}. \tag{16.12}$$

However, by (14.20),

$$M_t\{\psi_\varepsilon(x-t)\,\psi_\varepsilon(t-\tau)\} = \sum_1^\infty a_k e^{i\lambda_k(x-\tau)} \tag{16.13}$$

where all the coefficients a_k are positive, and $\sum_1^\infty a_k$ converges. Hence

$$f^{(\varepsilon)}(x) = M_\tau\left\{f(\tau)\sum_1^\infty a_k e^{i\lambda_k(x-\tau)}\right\}. \tag{16.14}$$

Since

$$\left|\frac{1}{2\,T}\int_{-T}^{T} f(\tau)\,e^{i\lambda_k(x-\tau)}\,d\tau\right| \leq \max |f(\tau)|, \tag{16.15}$$

it follows that we can invert the order of M and Σ, and that

$$f^{(\varepsilon)}(x) = \sum_1^\infty a_k\,e^{i\lambda_k x}\,M\{f(\tau)\,e^{-i\lambda_k\tau}\}. \tag{16.16}$$

Inasmuch as

$$|M\{f(\tau)\,e^{-i\lambda_k\tau}\}| \leq \max |f(\tau)|, \tag{16.17}$$

$f^{(\varepsilon)}(x)$ is the sum of a uniformly convergent series of trigonometric terms. That is to say, we can choose N so large that

31—29764. *Acta mathematica.* 55. Imprimé le 29 avril 1930.

$$\max_x \left| f^{(\varepsilon)}(x) - \sum_1^N a_k e^{i\lambda_k x} M\{f(\tau)e^{-i\lambda_k\tau}\} \right| \leq \varepsilon, \tag{16.18}$$

and hence that

$$\max_x \left| f(x) - \sum_1^N a_k e^{i\lambda_k x} M\{f(\tau)e^{-i\lambda_k\tau}\} \right| \leq 3\,\varepsilon. \tag{16.19}$$

In other words, we have proved Bohr's approximation theorem, to the effect that it is possible to approximate uniformly to any desired degree of accuracy to an almost periodic function by means of trigonometrical polynomials.

17. Certain generalizations of almost periodic functions.

It will be noticed that in the proof of the Weierstrass theorem for almost periodic functions, the spectrum of the function to be analyzed was not directly introduced, but rather that of the auxiliary function $\psi_\varepsilon(t)$. In many cases, when the function $f(t)$ is not almost periodic in the classical sense, an auxiliary function $\psi_\varepsilon(t)$ may be defined, which will be almost periodic in the classical sense, and which may be employed to establish the approximation theorem for $f(t)$, in whatever sense this theorem may hold. It would be possible in this manner to establish the approximation theorems for the almost periodic functions of the generalized types of Weyl, Besicovitch, Stepanoff, and others, but one example will suffice to show the power of the method, and to this we shall confine ourselves. This example, which is due to Mr. C. F. Muckenhoupt, is that of functions almost periodic in the mean.

We shall confine our attention to functions $f(x, t)$ defined over the range $(-\infty < t < \infty,\ x_0 \leq x \leq x_1)$, quadratically summable in x, and continuous in the mean in t in the sense that

$$\lim_{\tau\to 0} \int_{x_0}^{x_1} |f(x,t) - f(x,t+\tau)|^2\, dx = 0. \tag{17.01}$$

We shall say that τ_ε is a *translation number* of $f(x, t)$ pertaining to ε in case for all t,

$$\int_{x_0}^{x_1} |f(x, t+\tau_\varepsilon) - f(x,t)|^2\, dx < \varepsilon^2. \tag{17.02}$$

In case, given ε, we can always assign a finite quantity L_ε, such that each interval $(A, A+L_\varepsilon)$ contains at least one translation number τ_ε pertaining to ε, then $f(x, t)$ is said to be *almost periodic in the mean.* I case $f(x, t)$ is almost periodic in the mean, Mr. Muckenhoupt's theorem is:

Given any positive quantity ε there can be assigned a trigonometrical polynomial

$$P_\varepsilon(x, t) = \sum_1^N A_n(x)\, e^{i\Lambda_n t}, \tag{17.03}$$

such that

$$A_1(x),\ A_2(x),\ \ldots,\ A_n(x) \tag{17.04}$$

are all quadratically summable, and for all t,

$$\int_{x_0}^{x_1} |f(x, t) - P_\varepsilon(x, t)|^2\, dx < \varepsilon^2. \tag{17.05}$$

There are a number of elementary theorems which Mr. Muckenhoupt proves along lines not differing in any essential way from those followed by Bohr in the proof of the corresponding theorems for functions almost periodic in the original sense. Thus every function almost periodic in the mean is bounded in the mean, in the sense that

$$\int_{x_0}^{x_1} |f(x, t)|^2\, dx \tag{17.06}$$

is bounded; and is uniformly continuous in the mean, in the sense that

$$\lim_{\varepsilon \to 0} \max_{\tau < \varepsilon} \int_{x_0}^{x_1} |f(x, t) - f(x, t+\tau)|^2\, dx = 0. \tag{17.07}$$

In this and subsequent formulas, the maximum value indicated by »max» need not be actually attained. Any finite set of functions almost periodic in the mean are *simultaneously* almost periodic in the mean, in the sense that, given ε, an L_ε may be assigned in such a manner that every interval $(A, A+L_\varepsilon)$ contains at least one τ_ε which is a translation number pertaining to ε of all the functions of the set. Hence the sum of two or more functions almost periodic in the mean is almost periodic in the mean. Similarly, the product in the ordinary sense is almost periodic in the mean. The uniform limit in the mean of a set of functions almost periodic in the mean is itself almost periodic in the mean. If $f(x, t)$ is almost periodic in the mean,

$$\underset{T\to\infty}{\mathrm{l.i.m.}}\ \frac{1}{T}\int_a^{a+T} f(x,t)\,dt \tag{17. 08}$$

exists as a uniform limit in the mean in a, and is independent of a. We shall represent it by the symbol

$$M\{f(x,t)\}. \tag{17. 09}$$

Mr. Muckenhoupt now puts

$$g(t) = \max_z \int_{x_0}^{x_1} |f(x, t+z) - f(x, z)|^2\,dx. \tag{17. 10}$$

Clearly

$$|g(t+\tau) - g(t)| \leq \max_z \int_{x_0}^{x_1} \big||f(x, t+\tau+z) - f(x, z)|^2 - |f(x, t+z) - f(x, z)|^2\big|\,dx$$

$$\leq \max_z \left\{\int_{x_0}^{x_1} \big||f(x, t+\tau+z) - f(x, z)| + |f(x, t+z) - f(x, z)|\big|^2\,dx\right\}^{1/2}$$

$$\cdot \max_z \left\{\int_{x_0}^{x_1} \big||f(x, t+\tau+z) - f(x, z)| - |f(x, t+z) - f(x, z)|\big|^2\,dx\right\}^{1/2}. \tag{17. 11}$$

To evaluate this, let us consider the maximum of each of the integrals under the radical sign separately. The first does not exceed

$$16 \max_z \int_{x_0}^{x_1} |f(x, z)|^2\,dx; \tag{17. 12}$$

the second does not exceed

$$\max_z \int_{x_0}^{x_1} |f(x, \tau+z) - f(x, z)|^2\,dx. \tag{17. 13}$$

Hence we may write

$$|g(t+\tau)-g(t)| \leq 16\left[\max_z \int_{x_0}^{x_1} |f(x, z+\tau)-f(x, z)|^2 \, dx\right]^{1/2}$$

$$\cdot\left[\max_z \int_{x_0}^{x_1} |f(x, z)|^2 dx\right]^{1/2}. \qquad (17.\ 14)$$

If τ_ε is a translation number of $f(x, t)$ pertaining to ε, we have

$$|g(t+\tau_\varepsilon)-g(t)| \leq 16\,\varepsilon\left[\max_z \int_{x_0}^{x_1} |f(x, z)|^2 dx\right]^{1/2}, \qquad (17.\ 15)$$

so that any translation number of $f(x, t)$ pertaining to ε is a translation number of $g(t)$ pertaining to

$$16\,\varepsilon\left[\max_z \int_{x_0}^{x_1} |f(x, z)|^2 dx\right]^{1/2}. \qquad (17.\ 16)$$

Thus $g(t)$ is almost periodic in the classical sense, and is distinct from 0 unless $f(x, t)$ is independent of t, in the sense that $f(x, t_1)=f(x, t_2)$ almost everywhere.

As in the last section, let

$$H_\varepsilon(U)=\begin{cases} 1 \quad ; & [0 \leq U \leq \varepsilon/2] \\ 2-\dfrac{2\,U}{\varepsilon}; & [\varepsilon/2 \leq U \leq \ \varepsilon\ ] \\ 0 \quad ; & [\ \varepsilon\ \leq U \qquad] \end{cases} \qquad (17.\ 17)$$

and let

$$\psi_\varepsilon(t)=\frac{H_\varepsilon[g(t)]}{M H_\varepsilon[g(t)]}. \qquad (17.\ 18)$$

As before, $H_\varepsilon[g(t)]$ is dictinct from 0, and $\psi_\varepsilon(x)$ exists, and is almost periodic. As before, we put

$$f_\varepsilon(x, t) = M_\tau[f(x, \tau)\,\psi_\varepsilon(t-\tau)], \qquad (17.\ 19)$$

and

$$f^{(\varepsilon)}(x, t) = M_\tau[f(x, \tau)\,\psi_\varepsilon(t-\tau)]. \qquad (17.\ 20)$$

A proof precisely parallel to that of (16. 10) and (16. 12) shows that

$$\max_t \int_{x_0}^{x_1} |f(x, t) - f^{(\varepsilon)}(x, t)|^2 \, dx \leq 4\,\varepsilon. \tag{17.21}$$

and that

$$f^{(\varepsilon)}(x, t) = M_\sigma[f(x, \sigma)\, M_\tau[\psi_\varepsilon(t-\tau)\, \psi_\varepsilon(\tau-\sigma)]]. \tag{17.22}$$

As before,

$$M_\tau[\psi_\varepsilon(t-\tau)\, \psi_\varepsilon(\tau-\sigma)] = \sum_0^\infty a_k e^{i\lambda_k(t-\sigma)} \tag{17.23}$$

where all the a_k's are positive, and $\sum_0^\infty a_k$ converges. Hence

$$f^{(\varepsilon)}(x, t) = M_\sigma\left[f(x, \sigma) \sum_0^\infty a_k e^{i\lambda_k(t-\sigma)}\right]. \tag{17.24}$$

We have

$$\lim_{N\to\infty} \int_{x_0}^{x_1} \left| M_\tau\left[f(x, \sigma) \sum_N^\infty a_k e^{i\lambda_k(t-\sigma)}\right]\right|^2 dx$$

$$\leq \lim_{N\to\infty} \left[\sum_N^\infty a_k\right]^2 \max_\sigma \int_{x_0}^{x_1} |f(x, \sigma)|^2 \, dx = 0. \tag{17.25}$$

Hence since

$$\sum_0^\infty a_k$$

converges, we can invert the order of M and Σ in (17.25), and get

$$f^{(\varepsilon)}(x, t) = \underset{N\to\infty}{\text{l.i.m.}} \sum_0^n a_k e^{i\lambda_k t} \underset{\sigma}{M}[f(x, \sigma)\, e^{-i\lambda_k \sigma}]. \tag{17.26}$$

This convergence in the mean is uniform with respect to t. Combining (17.25) and (17.26), our theorem is proved.

This theorem has an interesting dynamical application. Really significant dynamical applications of almost periodic functions have been rather scarce, as no one has yet produced an example of an almost periodic function entering into a dynamical system with a finite number of degrees of freedom in which the frequencies or exponents are not linearly dependent (with rational coefficients) on a finite set of quantities. However, dynamical systems with an infinite number of degress of freedom are familiar enough in connection with boundary value problems, and in these, it is well known that the solution may involve an infinite linearly independent set of time frequencies. Mr. Muckenhoupt has succeeded in showing, under certain very general conditions, that the solution of such a problem is almost periodic in the mean with respect to the time, the space variables playing the rôle above assigned to x. In this proof, the existence of an integral invariant such as the energy is of the utmost importance, as is also the condition that when all the coordinates and velocities of the system are less in value than some given constant, the energy is also necessarily less than some constant.

Let us consider as an example a vibrating string, whose density and tensions are functions of position, but not of time. Let the mass density be $\mu(x)$ and the tension $T(x)$. The equation of motion is then

$$\frac{\partial}{\partial x}\left[T(x)\frac{\partial y}{\partial x}\right] = \mu(x)\frac{\partial^2 y}{\partial t^2}. \tag{17.27}$$

We consider the ends to be fixed, giving us

$$y(x_0) = y(x_1) = 0, \tag{17.28}$$

and we take T and μ, as is always physically the case, finite and positive. We shall also suppose them to have bounded derivatives of all orders.

Thus the total energy of the system is

$$E_0 = \frac{1}{2}\int_{x_0}^{x_1}\left[\mu(x)\left(\frac{\partial y}{\partial t}\right)^2 + T(x)\left(\frac{\partial y}{\partial x}\right)^2\right]dx. \tag{17.29}$$

If we assume density and tension independent of the time, we have

$$\frac{\partial E_0}{\partial t} = \int_{x_0}^{x_1}\left[\mu(x)\frac{\partial y}{\partial t}\frac{\partial^2 y}{\partial t^2} + T(x)\frac{\partial y}{\partial x}\frac{\partial^2 y}{\partial x\,\partial t}\right]dx, \tag{17.30}$$

or by (17.27),

$$\frac{\partial E_0}{\partial t} = \int_{x_0}^{x_1}\left\{\frac{\partial y}{\partial t}\frac{\partial}{\partial x}\left[T(x)\frac{\partial y}{\partial x}\right] + T(x)\frac{\partial y}{\partial x}\frac{\partial}{\partial x}\left(\frac{\partial y}{\partial t}\right)\right\}dx$$

$$= \left[T(x)\frac{\partial y}{\partial x}\frac{\partial y}{\partial t}\right]_{x_0}^{x_1} = 0. \tag{17.31}$$

Thus the total energy is invariant, as was to be expected from physical considerations.

Inasmuch as $\partial^n y/\partial t^n$ also satisfies equation (17. 27) and boundary conditions (17. 28), we see that all the expressions

$$E_n = \frac{1}{2}\int_{x_0}^{x_1}\left[\mu(x)\left(\frac{\partial^{n+1}y}{\partial t^{n+1}}\right)^2 + T(x)\left(\frac{\partial^{n+1}y}{\partial t\,\partial x^n}\right)^2\right]dx \tag{17. 32}$$

are invariants, at least if $y(x, t)$ is sufficiently often differentiable. We shall term E_n the $(n+1)$st energy of the system.

Let us now take $\partial y/\partial t$ and $\partial y/\partial x$ to be continuous, and let

$$E_0 \leq E; \quad \mu(x) \geq M; \quad T(x) \geq T.$$

Clearly $E_0 \geq 0$, and

$$\int_{x_0}^{x_1}\left(\frac{\partial y}{\partial t}\right)^2 dx \leq \frac{2E_0}{M} \leq \frac{2E}{M}. \tag{17. 33}$$

Similarly, $\int_{x_0}^{x_1}\left(\frac{\partial y}{\partial x}\right)^2 dx$ is bounded, provided only the first energy E_0 is finite. Furthermore, since

$$y^2 = \left[\int_{x_0}^{x}\frac{\partial y}{\partial x}\,dx\right]^2 \leq \int_{x_0}^{x}dx\int_{x_0}^{x}\left(\frac{\partial y}{\partial x}\right)^2 dx, \tag{17. 34}$$

by the Schwarz inequality, y is bounded. Similarly, if E_1 is also bounded,

$$\int_{x_0}^{x_1}\left(\frac{\partial^2 y}{\partial t^2}\right)^2 dx, \quad \int_{x_0}^{x_1}\left(\frac{\partial^2 y}{\partial x\,\partial t}\right)^2 dx, \quad \text{and} \quad \int_{x_0}^{x_1}\left(\frac{\partial^2 y}{\partial x^2}\right)^2 dx \tag{17. 35}$$

will be likewise; in the last case, as a result of (17. 27); if E_2 is also bounded,

$$\int_{x_0}^{x_1}\left(\frac{\partial^3 y}{\partial t^3}\right)^2 dx, \quad \int_{x_0}^{x_1}\left(\frac{\partial^3 y}{\partial t^2\,\partial x}\right)^2 dx, \quad \int_{x_0}^{x_1}\left(\frac{\partial^3 y}{\partial t\,\partial x^2}\right)^2 dx, \quad \text{and} \quad \int_{x_0}^{x_1}\left(\frac{\partial^3 y}{\partial x^3}\right)^2 dx \tag{17. 36}$$

will be, and so on indefinitely.

Let us now introduce

$$(y, y_1) = \left\{ \int_{x_0}^{x_1} \left[|y - y_1|^2 + \left| \frac{\partial y}{\partial x} - \frac{\partial y_1}{\partial x} \right|^2 + \left| \frac{\partial y}{\partial t} - \frac{\partial y_1}{\partial t} \right|^2 \right] dx \right\}^{1/2} \qquad (17.37)$$

as the distance between two functions, $y(x, t)$ and $y_1(x, t)$. If we write

$$y \sim \sum_{-\infty}^{\infty} A_n(t) e^{\frac{2\pi i n x}{x_1 - x_0}}; \quad y_1 \sim \sum_{-\infty}^{\infty} B_n(t) e^{\frac{2\pi i n x}{x_1 - x_0}};$$

we may approximate *uniformly* to (y, y_1) by

$$\int_{x_0}^{x_1} \left[\left| \sum_{-N}^{N} (A_n - B_n) e^{\frac{2\pi i n x}{x_1 - x_0}} \right|^2 \left(1 + \frac{4\pi n^2}{(x_1 - x_0)^2} \right) + \left| \sum_{-N}^{N} (A'_n - B'_n) e^{\frac{2\pi i n x}{x_1 - x_0}} \right|^2 dx \right] \qquad (17.38)$$

for all functions y and y_1 for which E_0 and E_1 are finite, since then

$$\left. \begin{aligned} \sum_{-\infty}^{\infty} n^4 |A_n(t) - B_n(t)|^2 &= \frac{x_1 - x_0}{(2\pi)^4} \int_{x_0}^{x_1} \left(\frac{\partial^2 y}{\partial t^2} \right)^2 dx; \\ \sum_{-\infty}^{\infty} n^2 |A'_n(t) - B'_n(t)|^2 &= \frac{x_1 - x_0}{(2\pi)^2} \int_{x_0}^{x_1} \left(\frac{\partial^2 y}{\partial x \partial t} \right)^2 dx \end{aligned} \right\} \qquad (17.39)$$

are uniformly bounded. Now, a bounded region in space of m dimensions may be divided into a finite number of compartments such that the distance between two points in the same compartment does not exceed ε. Hence we can divide the entire class of functions $y(x, t)$ for which E_0 and E_1 are finite into a finite number of classes such that the distance between two functions in the same class does not exceed ε.

Let us do this, and let us discard every class which is not actually represented by $y(x, t)$ for some value of t. Then we may assign a time-interval L'_ε within which $y(x, t)$ enters every class that it ever enters. Then, whatever τ may be, we may determine τ_1 between 0 and L'_ε such that

$$(y(x, \tau), \quad y(x, \tau_1))^2 < \varepsilon. \qquad (17.40)$$

Since

$$y(x, t) - y(x, t + \tau - \tau_1)$$

satisfies the differential equation (17. 27),

$$\mathfrak{E} = \frac{1}{2} \int_{x_0}^{x_1} \left[\mu(x) \left(\frac{\partial y(x, t)}{\partial t} - \frac{\partial y(x, t+\tau-\tau_1)}{\partial t} \right)^2 \right.$$

$$\left. + T(x) \left(\frac{\partial y(x, t)}{\partial x} - \frac{\partial y(x, t+\tau-\tau_1)}{\partial x} \right)^2 \right] dx \qquad (17.41)$$

is invariant, and since for $t = \tau_1$,

$$\mathfrak{E} \leq (\max \mu + \max T)\,(y(x, \tau_1),\ y(x, \tau))^2$$

$$< (\max \mu + \max T)\,\varepsilon, \qquad (17.42)$$

it follows that for all t,

$$\mathfrak{E} < (\max \mu + \max T)\varepsilon, \qquad (17.43)$$

and hence by (17. 34) and (17. 37)

$$|y(x, t) - y(x, t+\tau-\tau_1)| < \sqrt{2\,\frac{x_1 - x_0}{M}(\max \mu + \max T)\varepsilon}\,. \qquad (17.44)$$

Since for every τ, there is a value of τ_1 between 0 and L'_ε, there is a value of $\tau-\tau_1$ over every interval of length L'_ε. Thus $y(x, t)$ is an a almost periodic function taken with respect to the time, uniformly in x, and is *a fortiori* almost periodic in the mean, in case E_0 and E_1 are finite. It follows that we may so determine $A_1(x), \ldots, A_n(x); \Lambda_1, \ldots, \Lambda_n$ that for all t,

$$\int_{x_0}^{x_1} \left| y(x, t) - \sum_1^n A_k(x)\, e^{i\Lambda_k(t)} \right|^2 dx < \varepsilon. \qquad (17.45)$$

It is possible to go further than this, as Mr. Muckenhoupt has done, and to show that the method we have given for obtaining $f_\varepsilon(x, t)$, $f^{(\varepsilon)}(x, t)$, and $A_k(x)\, e^{i\Lambda_k(t)}$ assures us that all the functions

$$A_k(x)\, e^{i\Lambda_k t}$$

are solutions of the original differential equation, or that the functions

$$A_k(x)$$

are all solutions of the ordinary differential equation

$$\frac{d}{dx}(T(x)\,A'_k(x)) + \Lambda_k^2\, \mu(x)\, A_k(x) = 0; \qquad (17.46)$$

— that is, are what is known as Eigenfunktionen of the dynamical problem. This proof rests on the fact that each one of these functions may be obtained from its predecessor, and ultimately from $f(x, t)$, by a process of weighted averaging in the variable t which transforms every solution of a linear differential equation with coefficients constant with respect to the time into another solution of the same equation, or at least of the corresponding integral equation. Hence, if $y(x, 0) = F(x); \frac{\partial y(x, 0)}{\partial t} = 0$ is a possible set of initial conditions for the motion of the vibrating string. We may write

$$F(x) = \underset{n \to \infty}{\text{l.i.m.}} \sum_{1}^{n} A_k(x), \tag{17.47}$$

where the $A_k(x)$ are in general Eigenfunktionen of the problem that depend on n. Thus if the set of possible initial conditions of the string is closed, as we may show to be the case by direct methods, every quadratically summable function may be expanded in terms of a denumerable set of Eigenfunktionen, and the Eigenfunktionen may be shown to be a denumerable closed set.

The methods of Mr. Muckenhoupt are susceptible of extension to the treatment of a much wider class of Eigenfunktion problems, in any finite number of dimensions. The detail of this extension awaits further investigation.

Bibliography.

The works and papers covering the various theories belonging to general harmonic analysis fall into several imperfectly related categories. Among these are:

(1) The various papers written from the physical standpoint, with the explicit purpose of clearing up obscure points in the theories of interference, of coherency, and of polarization.

(2) Directly related to these, the various memoirs connecting with the Schuster theory of the periodogram.

(3) A group of memoirs preceding the Bohr theory of almost periodic functions, applying various extensions of the notion of periodicity in celestial mechanics and other similar fields.

(4) Papers written from the point of view of the mathematician, and dealing with trigonometric series not proceeding according to integral multiples of the argument.

(5) The Bohr theory of almost periodic functions, and papers directly inspired by it.

(6) Papers dealing with haphazard motion, and using ideas directly pertinent to generalized harmonic analysis.

(7) The Hahn direction of work, treating generalized harmonic analysis from the standpoint of ordinary convergence, rather than from that of convergence in the mean.

(8) The papers assuming essentially the standpoint of the present author, in whose work the generalizations of the Parseval theorem play the central rôle.

(9) Papers dealing rather with the rigorous theory of the Fourier integral itself than with its generalizations.

(10) Papers not dealing directly with generalized harmonic analysis, which it is desirable to cite for one reason or another.

In citing any paper, it will be indicated to which of these categories it belongs. Each paper will furthermore be quoted in the footnotes by the name of its author, together with an index number given in the bibliography.

A. C. BERRY 1. Doctoral dissertation, Harvard, 1929. Unpublished. (8)

—— 2. The Fourier transform theorem. Jour. Math. and Phys. Mass. Inst. Technology 8, 106—118. (1929). (8)

A. BESICOVITCH 1. Sur quelques points de la théorie des fonctions presque périodiques. C.R. 180, 394—397. (1925). (5)

—— 2. On Parseval's theorem for Dirichlet series. Proc. Lond. Math. Soc. 25, 25—34. (1926). (5)

—— 3. On generalized almost periodic functions. Proc. Lond. Math. Soc. 24, 495—512. (1926). (5)

A. BESICOVITCH and H. BOHR 1. Some remarks on generalizations of almost periodic functions. Danske Vidensk. Selskab. 8, No. 5. (1927). (5)

—— 2. On generalized almost periodic functions. Journ. Lond. Math. Soc. 3, 172—176. (1928). (5)

G. D. BIRKHOFF. Dynamical systems. Am. Math. Society, New York, 1927. Pp. 218—220. (5)

S. BOCHNER 1. Properties of Fourier series of almost periodic functions. Proc. Lond. Math. Soc. 26, 433. (1925). (5)

—— 2. Sur les fonctions presque-périodiques de Bohr. C.R. 180, 1156. (1925). (5)

—— 3. Beiträge zur Theorie der fastperiodischen Funktionen. Math. Ann. 96, 119—147. (1926). (5)

—— 4. Beiträge zur Theorie der fastperiodischen Funktionen. Math. Ann. 96, 383—409. (1926). (5)

—— 5. Über Fourierreihen fastperiodischer Funktionen. Berliner Sitzungsberichte, 26. (1926). (5)

—— 6. Konvergenzsätze für Fourierreihen grenzperiodischer Funktionen. Math. Zeitschr. 27, 187—211. (1927). (5)

—— 7. Darstellung reellvariabler und analytischer Funktionen durch verallgemeinerte Fourier- und Laplace-Integrale. Math. Ann. 97, 632—674. (1926—7). (8)

S. Bochner 8. Über gewisse Differential- und allgemeinere Gleichungen, deren Lösungen fastperiodisch sind. I. Teil. Der Existenzsatz. Math. Ann. 102, 489—504. (1929). (5)

S. Bochner and G. H. Hardy. Note on two theorems of Norbert Wiener. Jour. Lond. Math. Soc. 1, 240—242. (1926). (10)

P. Bohl. Über die Darstellung von Funktionen einer Variabeln durch trigonometrische Reihen mit mehreren einer Variabeln proportionalen Argumenten. (Dorpat, 1893). (3)

H. Bohr 1. Sur les fonctions presque périodiques. C.R. Oct. 22, 1923. (5)

—— 2. Sur l'approximation des fonctions presque périodiques par des sommes trigonométriques. C.R. Nov. 26, 1923. (5)

—— 3. Über eine quasi-periodische Eigenschaft Dirichletscher Reihen mit Anwendung auf die Dirichletschen L Funktionen. Math. Ann. 85, 115—122. (4)

—— 4. Zur Theorie der fastperiodischen Funktionen. Acta Math. 45, 29—127. (124). (5)

—— 5. Zur Theorie der fastperiodischen Funktionen. II. Acta Math. 46, 101—214. (1925). (5)

—— 6. Zur Theorie der fastperiodischen Funktionen. III. Acta Math. 47, 237—281. (1926). (5)

—— 7. Einige Sätze über Fourierreihen fastperiodischer Funktionen. Math. Ztschr. 23, 38—44. (1925). (5)

—— 8. Sur une classe de transcendantes entières. C.R. 181, 766. (1925). (5)

—— 9. Sur le théorème d'unicité dans la théorie des fonctions presque-périodiques. Bull. Sci. Math. 50, 1—7. (1926). (5)

—— 10. On the explicit determination of the upper limit of an almost periodic function. Jour. Lond. Math. Soc. 1. (1926). (5)

—— 11. Ein Satz über analytische Fortsetzung fastperiodischer Funktionen. Crelle, 157, 61—65. (5)

—— 12. En Klasse hele transcendente Funktioner. Mat. Tidsskrift, B. Aarg. 1926, 41—45. (5)

—— 13. Allgemeine Fourier- und Dirichlet-Entwicklungen. Abh. aus dem math. Sem. d. Hamburgischen Univ. 4, 366—374. (1926). (5)

—— 14. Fastperiodische Funktionen. Jahresb. d. D.M.V. 33, 25—41. (1925). (5)

—— 15. En Sætning om Fourierrækker for næstenperiodiske Funktioner. Mat. Tid. B. Aarg. 1924, 31—37. (5)

—— 16. Über die Verallgemeinerungen fastperiodischer Funktionen. Math. Ann. 99, 357—366. (1928). (5)

H. Bohr and O. Neugebauer. Über lineare Differentialgleichungen mit konstanten Koeffizienten und fastperiodischer rechter Seite. Gött. Sitzb. 1926, 1—13. (5)

E. Borel. Les probabilités dénombrables et leurs applications arithmétiques. Rend. di Palermo 27, 247—271. (1909). (10)

J. C. Burkill 1. The expression in Stieltjes integrals of the inversion formulae of Fourier and Hankel. Proc. Lond. Math. Soc. 25, 513—524. (1926). (7)

J. C. BURKILL 2. On Mellin's inversion formula. Proc. Camb. Phil. Soc. 23, 356—360. (1926). (7)

V. BUSH, F. D. GAGE, and H. R. STEWART. A continuous integraph. Jour. Franklin Institute. 63—84. (1927). (10)

V. BUSH and H. L. HAZEN. Integraph solution of differential equations. 575—615. (1927). (10)

CARSE and SHEARER. A course in Fourier's analysis and periodogram analysis for the mathematical laboratory. London, 1915. (2)

P. J. DANIELL 1. A general form of integral. Annals of Math. 19, 279—294. (1918). (10)

—— 2. Integrals in an infinite number of dimensions. Ann. of Math. 20, 281—288. (1919). (10)

—— 3. Further properties of the general integral. Ann. of Math. 21, 203—220. (1920). (10)

W. DORN. Fouriersche Integrale als Grenzwerte Fourierscher Reihen. Wiener Sitzungsber., 1926, 127—147. (7)

A. EINSTEIN. Zur Theorie der Brownschen Bewegung. Ann. der Phys. (4) 19, 372—381. (1906). (10)

E. ESCLANGON 1. Sur une extension de la notion de périodicité. C.R. 135, 891—894. (1902). (3)

—— 2. Sur les fonctions quasi périodiques moyennes, déduites d'une fonction quasi périodique. C.R. 157, 1389—1392. (1913). (3)

—— 3. Sur les intégrales quasi périodiques d'une équation différentielle linéaire. C.R. 160, 652—653. (1915). (3)

—— 4. Sur les intégrales quasi périodiques d'une équation différentielle linéaire. C.R. 161, 488—489. (1915). (3)

J. FAVARD 1. Sur les fonctions harmoniques presque-périodiques. C.R. 182, 757. (1926). (5)

—— 2. Sur les equations différentielles à coefficients presque-periodiques. Acta Math. 51, 31—81. (1927). (5)

P. FRANKLIN 1. Almost periodic recurrent motions. Math. Ztschr. 30, 325—331. (1929). (5)

—— 2. The elementary theory of almost periodic functions of two variables. Jour. Math. Phys. M.I. T. 5, 40—55. (1925). (5)

—— 3. The fundamental theorem of almost periodic functions of two variables. Jur. Math. Phys. M.I. T. 5, 201—237. (1926). (5)

—— 4. Classes of functions orthogonal on an infinite interval, having the power of the continuum. Jour. Math. Phys. M.I. T. 8, 74—79. (1929). (5)

—— 5. Approximation theorems for generalized almost periodic functions. Math. Ztschr. 29, 70—87. (1928). (5)

G. L. GOUY. Sur le mouvement lumineux. Journal de physique, 5, 354—362. (1886). (1)

H. HAHN 1. Über die Verallgemeinerung der Fourierschen Integralformel. Acta Math. 49, 301—353. (1926). (7)

H. HAHN 2. Über die Methode der arithmetischen Mittel in der Theorie der verallgemeinerten Fourierintegrale. Wiener Sitzungsber., 1925, 449—470. (7)

G. H. HARDY, A. E. INGHAM, and G. PÓLYA. Notes on moduli and mean values. Proc. Lond. Math. Soc. 27, 401—409. (1928). (10)

E. W. HOBSON. The theory of functions of a real variable and the theory of Fourier series. Vol. 2, second edition. Cambridge, 1927. (9)

S. IZUMI 1. Über die Summierbarkeit der Fourierschen Integralformel. Tôhoku Journal 30, 96—110. (1929). (7)

—— 2. On the Cahen-Mellin's Inversion Formula. Tôhoku Journal 30, 111—114. (1929). (7)

M. JACOB 1. Über ein Theorem von Bochner-Hardy-Wiener. Jour. Lond. Math. Soc. 3, 182—187. (1928). (10)

—— 2. Über den Eindeutigkeitssatz in der Theorie der verallgemeinerten trigonometrischen Integrale. Math. Ann. 100, 279—294. (1928). (7)

—— 3. Über den Eindeutigkeitssatz in der Theorie der trigonometrischen Integrale. Math. Ann. 97, 663—674. (1927). (9)

G. W. KENRICK 1. Doctoral dissertation, Mass. Inst. Technology, 1927. (8)

—— 2. The analysis of irregular motions with applications to the energy-frequency spectrum of static and of telegraph signals. Phil. Mag. (7) 7, 176—196. (1929). (8)

E. H. LINFOOT. Generalization of two theorems of H. Bohr. Jour. Lond. Math. Soc. 3, 177—182. (1928). (5)

S. B. LITTAUER. On a theorem of Jacob. Jour. Lond. Math. Soc. 4, 226—231. (1929). (10)

K. MAHLER. On the translation properties of a simple class of arithmetical functions. J. Math. Phys. Mass. Inst. Technology 6, 158—164. (1927). (8)

C. F. MUCKENHOUPT. Almost periodic functions and vibrating systems. Doctoral dissertation, Mass. Inst. Technology, 1929. J. Math. Phys. Mass. Inst. Technology 8, 163—200. (1929). (5)

M. PLANCHEREL 1. Contribution à l'étude de la répresentation d'une fonction arbitraire par des intégrales définies. Rendiconti di Palermo, 30, 289—335. (1910). (9)

—— 2. Sur la représentation d'une fonction arbitraire par une intégrale définie. C.R. 150, 318—321. (1910). (9)

—— 3. Sur la convergence et sur la sommation par les moyennes de Cesàro de $\lim_{z=\infty} \int_a^z f(x) \cos xy\, dx$. Math. Ann. 76, 315—326. (1915). (9)

H. POINCARÉ. Leçons sur la théorie mathématique de la lumière. Paris, 1889. (1)

S. POLLARD 1. The summation of a Fourier integral. Proc. Camb. Phil. Soc. 23, 373—382. (1926). (9)

—— 2. On Fourier's integral. Proc. Lond. Math. Soc. 26, 12—24. (1927). (9)

S. POLLARD 3. Identification of the coefficients in a trigonometrical integral. Proc. Lond. Math. Soc. 25, 451—468. (1926). (9)

A. PRINGSHEIM. Über neue Gültigkeitsbedingungen der Fourierschen Integralformel. Math. Ann. 68, 367—408. (1910). (9)

J. RADON, Theorie und Anwendungen der absolutadditiven Mengenfunktionen. Wien. Ber. 122, 1295—1438. (1913). (10)

Lord RAYLEIGH 1. On the resultant of a large number of vibrations of the same pitch and of arbitrary phase. Phil. Mag. 10, 73—78. (1880). (6)

—— 2. Wave theory of light. Ency. Britt., 1888. Cf. especially § 4. (1)

—— 3. On the character of the complete radiation at a given temperature. Phil. Mag. 27, 460. (1889). (1)

—— 4. Röntgen rays and ordinary light. Nature, 57, 607. (1898). (1)

—— 5. On the spectrum of an irregular disturbance. Phil. Mag. 5, 238—243. (1903). (6)

—— 6. Remarks concerning Fourier's theorem as applied to physical problems. Phil. Mag. 24, 864—869. (1912). (1)

—— 7. On the problem of random vibrations, and of random flights in one, two, or three dimensions. Phil. Mag. 37, 321—347. (1919). (6)

—— 8. On the resultant of a number of unit vibrations, whose phases are at random over a range not limited to an integral number of periods. Phil. Mag. 37, 498—515. (1919). (6)

F. RIESZ. Sur la formule d'inversion de Fourier. Acta litt. ac Sci. Univ. Hung. 3, 235—241. (1927). (9)

H. L. RIETZ. Mathematical statistics. Chicago, 1927. (10)

R. SCHMIDT 1. Über divergente Folgen und lineare Mittelbildungen. Math. Ztschr. 22, 89—152. (1925). (10)

—— 2. Über das Borelsche Summierungsverfahren. Schriften der Köningsberger gelehrten Gesellschaft, 1, 202—256. (1925). (10)

—— 3. Die trigonometrische Approximation für eine Klasse von verallgemeinerten fastperiodischen Funktionen. Math. Ann. 100, 334—356. (1928). (5)

I. SCHOENBERG 1. Über total monotone Folgen mit stetiger Belegungsfunktion. Mat. Ztschr. 30, 761—768. (1929). (8)

—— 2. Über die asymptotische Verteilung reeller Zahlen mod 1. Mat. Ztschr. 28, 177—200. (1928). (8)

A. SCHUSTER 1. On interference phenomena. Phil. Mag. 37, 509—545. (1894). (1)

—— 2. The periodogram of magnetic declination. Camb. Phil. Trans. 18, 108. (1899). (2)

—— 3. The periodogram and its optical analogy. Proc. Roy. Soc. 77, 136—140. (1906). (2)

—— 4. The theory of optics. London, 1904. (1)

—— 5. On lunar und solar periodicities of earthquakes. Proc. Roy. Soc. London 61, 455—465. (1897). (2)

—— 6. On hidden periodicities. Terrestrial Magnetism 3, 13. (1897). (2)

—— 7. The periodogram of magnetic declination. Trans. Camb. Phil. Soc. 18, 107—135. (1900). (2)

H. STEINHAUS. Les probabilités dénombrables et leur rapport à la théorie de la mésure. Fund. Math. 3, 286—310. (1923). (10)
W. STEPANOFF. Über einige Verallgemeinerungen der fastperiodischen Funktionen. Math. Ann. 90, 473—492. (1925). (5)
G. SZEGÖ. Zur Theorie der fastperiodischen Funktionen. Math. Ann. 96, 378—382. (1926). (5)
G. I. TAYLOR. Diffusion by continuous movements. Proc. Lond. Math. Soc. 20, 196—212. (1920). (6)
E. C. TITCHMARSH 1. A contribution to the theory of Fourier transforms. Proc. Lond. Math. Soc. 23, 279—289. (1924). (9)
—— 2. Recent advances in science mathematics. Science progress 91, 372—386. (1929). (5)
C. DE LA VALLÉE POUSSIN 1. Sur les fonctions presque périodiques de H. Bohr. Annales de la Societé Scentifique de Bruxelles, A, 47, 141. (1927). (5)
—— 2. Sur les fonctions presque périodiques de H. Bohr. Note complementaire et explicative. A.S.S. Bruxelles, A, 48, 56—57. (1928). (5)
T. VIJAYARAGHAVAN 1. A Tauberian theorem. Jour. Lond. Math. Soc. 1, 113—120. (1926). (10)
—— 2. A theorem concerning the summability of series by Borel's method. Proc. Lond. Math. Soc. 27, 316—326. (1928). (10)
V. VOLTERRA. Leçons sur les fonctions de lignes. Paris, 1913. (10)
J. D. WALSH. A generalization of the Fourier cosine series. Trans. Am. Math. Soc. 21, 101—116. (1920). (4)
H. WEYL 1. Integralgleichungen und fastperiodische Funktionen. Math. Ann. 97, 338—356. (1926). (5)
—— 2. Beweis des Fundamentalsatzes in der Theorie der fastperiodischen Funktionen. Berliner Sitzungsber. 1926, 211—214. (5)
—— 3. Quantenmechanik und Gruppentheorie. Ztschr. f. Physik 46, 1—46. (1927). (10)
N. WIENER 1. On the representation of functions by trigonometrical integrals. Math. Ztschr. 24, 575—617. (1925). (8)
—— 2. The harmonic analysis of irregular motion. J. Math. Phys. Mass. Inst. Technology 5, 99—122. (1925). (8)
—— 3. The harmonic analysis of irregular motion II. J.M.P.M.I.T. 5, 158—191. (1926). (8)
—— 4. The spectrum of an arbitrary function. Proc. Lond. Math. Soc. 27, 487—496. (1928). (8)
—— 5. Coherency matrices and quantum theory. J.M.P.M.I.T. 7, 109—125. (1928). (8)
—— 6. Harmonic analysis and the quantum theory. Jour. of Franklin Institute 207, 525—534. (1929). (8)
—— 7. The average of an analytic functional. Proc. Nat. Acad. Sci. 7, 253—260. (1921). (6)
—— 8. The average of an analytic functional and the Brownian motion. Proc. Nat. Acad. Sci. 7, 294—298. (1921). (6)

33—29764. *Acta mathematica.* 55. Imprimé le 9 mai 1930.

N. WIENER 9. The quadratic variation of a function and its Fourier coefficients. J.M.P.M.I.T. 3, 72—94. (1924). (8)

—— 10. Differential-space. J.M.P.M.I.T. 2, 131—174. (1923). (6)

—— 11. The average value of a functional. Proc. Lond. Math. Soc. 22, 454—467. (1922). (6)

—— 12. On a theorem of Bochner and Hardy. Jour. Lond. Math. Soc. 2, 118—123. (1927). (10)

—— 13. A new method in Tauberian theorems. J.M.P.M.I.T 7, 161—184. (1928). (10)

—— 14. The spectrum of an array and its application to the study of the translation properties of a simple class of arithmetical functions. J.M.P.M.I.T. 6, 145—157. (1927). (8)

—— 15. Harmonic analysis and group theory. J.M.P.M.I.T. 8, 148—154. (1929). (8)

—— 16. The operational calculus. Math. Ann. 95, 557—584. (1926). (8)

—— 17. Verallgemeinerte trigonometrische Entwicklungen. Gött. Nachrichten, 1925, 151—158. (8)

A. WINTNER. Spektraltheorie der unendlichen Matrizen. Leipzig, 1929. (5)

W. H. YOUNG. On non-harmonic Fourier series. Proc. Lond. Math. Soc. 18, 307—335. (1919). (4)

Footnotes.

1. In accordance with the listing adopted in the bibliography, the following references will give the background of the corresponding sections.

   1. PLANCHEREL 1, 2, 3; and in general, papers under rubric (9).
   2. Papers under rubric (2).
   3 and 4. WIENER 1, 2, 3, 4, 17.
   5. SCHMIDT 1, 2; VIJAYARAGHAVAN 1, 2; WIENER 13; JACOB 1.
   6. BOCHNER 7.
   7. HAHN 1. 2; DORN; JACOB 2, 3.
   8. BERRY 1, 2.
   9. POINCARÉ; WIENER 5, 6; WEYL 3; RIETZ.
   10. WIENER 15.
   11. MAHLER; WIENER 14.
   12. WIENER 14.
   13. WIENER 2, 3, 7, 8, 10, 11; EINSTEIN; TAYLOR; RAYLEIGH 1, 2, 3, 4, 5, 7, 8.
   14, 15, 16. Papers under rubric (5).
   17. MUCKENHOUPT.

2. SCHUSTER 5., p. 464.
3. HOBSON, § 492.
4. WEYL 3.
5. VOLTERRA, Ch. 7.
6. BOREL; STEINHAUS.

Commentary on the Memoire [30a] on Generalized Harmonic Analysis

S. Koizumi

The purpose of generalized harmonic analysis (GHA) is to give an account of phenomena which cannot be described either by Fourier series or classical Fourier integrals. Phenomena such as noise in tubes, white light, interference and polarization, turbulence, and especially the Brownian motion are typical examples. In this review of [30a] we will describe briefly each chapter or refer to earlier reviews for this purpose, but more importantly supply references to work bearing on or influenced by [30a].

Chapter I. The Plancherel theorem is the chief tool in attacking the problems of GHA. Wiener reproves it here in a way that serves as an excellent introduction to his theory. The original Plancherel theorem is very general. The special case of Fourier integral was treated by many authors, see, e.g., Titchmarsh, *Introduction to the Theory of Fourier Integrals.*[58]

Chapter II. Let us replace Wiener's $f(t)$ by a trigonometrical polynomial. This special case makes the GHA expecially easy to understand, and therefore to elucidate. Using the same notations as his, let

$$\phi(x) = \lim_{T\to\infty} \frac{1}{2T}\int_{-T}^{T} f(x+t)\bar{f}(t)\,dt$$

and

$$f(t) = \sum_{1}^{N} a_n e^{i\lambda_n t}, \quad \lambda_n \text{ real.}$$

Then

$$\phi(x) = \sum_{1}^{N} |a_n|^2 e^{i\lambda_n t}.$$

Let

$$S(u) = \frac{1}{2\pi}\int_{-\infty}^{\infty} \phi(x)\frac{e^{iux}-1}{ix}\,dx = \sum_{\lambda_n < u} |a_n|^2,$$

the last equality holding at points u of continuity of S. Then

$$\phi(x) = \int_{-\infty}^{\infty} e^{-iux}\, dS(u).$$

On the other hand, for the generalized Fourier transform (GFT) of f, defined by (5.51), viz.,

$$s(u) = \frac{1}{2\pi}\int_{-1}^{1} f(x)\frac{e^{iux}-1}{ix}\,dx + \frac{1}{2\pi}\underset{A\to\infty}{\mathrm{l.i.m.}}\left[\int_{1}^{A} + \int_{-A}^{-1}\right]\frac{f(x)e^{iux}}{ix}\,dx,$$

we find that

$$\begin{aligned} s(u+\varepsilon) - s(u-\varepsilon) &= \sum_{1}^{N} \frac{a_n}{2}\{\mathrm{sign}\,(\lambda_n - u + \varepsilon) - \mathrm{sign}\,(\lambda_n - u - \varepsilon)\} \\ &= \sum_{1}^{N} a_n c_n(u, \varepsilon), \end{aligned}$$

where $c_n\,(u, \varepsilon)$ denote the indicator function on the interval $(\lambda_n - \varepsilon, \lambda_n + \varepsilon)$. If we take ε so small that no two of these intervals overlap, then

$$\lim_{\varepsilon\to 0}\frac{1}{2\varepsilon}\int_{-\infty}^{\infty} |s(u+\varepsilon) - s(u-\varepsilon)|^2\, du = \sum_{1}^{N} |a_n|^2 = \lim_{T\to\infty}\frac{1}{2T}\int_{-T}^{T} |f(t)|^2\, dt$$

$$\lim_{\varepsilon\to 0}\frac{1}{2\varepsilon}\int_{-\infty}^{\infty} e^{iux}|s(u+\varepsilon) - s(u-\varepsilon)|^2\, du = \sum_{1}^{N} |a_n|^2\, e^{i\lambda_n x} = \phi(x)$$

$$\lim_{\varepsilon\to 0}\frac{1}{2\varepsilon}\int_{-\infty}^{u} |s(v+\varepsilon) - s(v-\varepsilon)|^2\, dv = \sum_{\lambda_n < u} |a_n|^2 = S(u),$$

at points u of continuity of S, cf., (5.53).

Another approach to this theory is as follows. For $f \in S'$, the correlation function $\phi(x)$ exists and is (i) continuous and bounded on $(-\infty, \infty)$, (ii) Hermitian symmetric, i.e. $\overline{\phi(-x)} = \phi(x)$, and (iii) for any function $w(x) \in L_1$

$$\begin{aligned} &\int_{-\infty}^{\infty}\int_{-\infty}^{\infty} \phi(x-y)w(x)\overline{w(y)}\,dx\,dy \\ &\quad = \lim_{T\to\infty}\frac{1}{2T}\int_{-T}^{T}\left|\int_{-\infty}^{\infty} f(t-x)w(x)\,dx\right|^2 dt \geqslant 0. \end{aligned}$$

Hence by Bochner's representation theorem for positive-definite functions, there exists a nondecreasing bounded function $S(u)$ such that

$$\phi(x) = \int_{-\infty}^{\infty} e^{-iux}\, dS(u).$$

Furthermore by the Lévy inversion formula

$$S(u) = \frac{1}{2\pi}\int_{-\infty}^{\infty} \phi(x)\frac{e^{iux} - 1}{ix}\, dx.$$

See Bochner,[8,9] Lévy,[43] Cramer,[19,20] and Kawata.[36]

The Wiener Formula (5.53) is deduced from the general theory of Tauberian theorems; see [28b] and [32a]. As for direct proof, see [27c] and also Littauer.[45]

For GHA in the complex domain, see Paley and Wiener [34d], Chapter VIII. Related to these topics, is the GHA of Hilbert transforms developed by Koizumi.[38–40]

To turn to the problem of representation of $f(x)$ by its generalized Fourier integral $s(u)$, it was treated by two different methods depending on the topologies used. The first uses the ordinary convergence topology; see Hahn,[25,26] Burkill,[17] Izumi,[29,30] and Koizumi and Suniochi.[27] The second uses the mean convergence topology; see [26a], [26b], and Bochner.[7] Some theorems in §6 and §7 could be obtained by the method developed in the preceeding three sections (3, 4, and 5); see [33i], Chapter IV, §21, 22, 23.

The demonstration that $f(x)$ can be represented by its generalized Fourier integral $s(u)$ is remarkable (§6). Let f belong to the class W_2, i.e., $f(x)$ be measurable and locally square summable and

$$\int_{-\infty}^{\infty} \frac{|f(x)|^2}{1 + x^2}\, dx < \infty.$$

Consider the representation

$$f(x) = \int_{-\infty}^{\infty} e^{-iux}\, ds(u),$$

where $s(u)$ is the generalized Fourier transform defined above, cf. (5.51). The $s(u)$ is in general not a function of bounded variation, and so the representation formula has to be obtained by a formal integration by parts, viz.,

$$\int_{-A}^{A} e^{-ixu}\, ds(u) = e^{-iAx} s(A) - e^{iAx} s(-A) + ix \int_{-A}^{A} e^{-ixu} s(u)\, du$$

$$= \underset{B\to\infty}{\text{l.i.m.}} \frac{1}{\pi} \int_{-B}^{B} f(t) \frac{\sin A(t-x)}{t-x}\, dt.$$

The following difficulties should be noted:

(i) $f(x)$ is not square integrable near infinity, and $1/(1 + x^2)$ plays the role of a convergence factor;

(ii) $f(x)e^{iux}/ix$ is not integrable near the origin, and the term $(e^{iux} - 1)$ cancels the singularity;

(iii) $s(u)$ is not in general differentiable, and the term e^{-iux} carries over this operation.

Analogous difficulties have appeared in many branches of mathematical science, for example in the Heaviside operational calculus, in the Dirac (δ) formulation of quantum mechanics, the weak derivatives in partial differential equation, etc.

Along with the big paper on Tauberian theorems [32a], GHA tremendously stimulated the study of the general theory of harmonic analysis, e.g., spectral analysis and spectral synthesis. See Paley and Wiener [32g], Carleman,[18] Beurling,[3,5,6] and Pollard.[51] See also Kahane.[34]

L. Schwartz was led to investigate these questions from the point of view of linear functionals, resulting in his *Theorie des Distributions.*[53] The distributions are defined as linear functionals on $\mathfrak{D}$, $\mathscr{E}$, and $\mathscr{S}$ etc. each of which is the set of sufficiently regular functions. The Dirac δ is such a distribution, being in fact the measure with unit mass concentrated at the origin. It is also the unit of convolution multiplication. The Heaviside operational calculus is then rigorously subsumed by the theory of convolution algebras. The operations such as derivatives and Fourier transforms are carried over by test functions. The finite parts of Hadamard in divergent integrals appearing in potential theory correspond to pseudo-functions. We refer to the book of Schwartz,[53] and for further developments to Gel'fand and Vilenkin,[24] and Ehrenpreis.[21]

Chapter III. This chapter is devoted to the generalizations and applications of GHA. Regarding this it will suffice to cite the excellent comment by Masani.[49] For references to §9 on coherency matrices, see [53a], Masani,[48] and Born and Wolf.[16] For references to §10 on transformation groups, see Weyl.[60]

Chapter IV. This chapter is devoted to show several types of spectra and the analysis of the Brownian motion. Again it will suffice to cite the excellent comments of Levinson,[42] Kac,[31] and Masani.[49]

For the later developments and applications of the Brownian motion, Wiener measure and related topics, see [66c], Lévy,[44] and Ito and Mckean.[28] See also Friedrichs and Shapiro et al.,[23] Martin and Segal,[47] Imamura, Meecham, and Siegel,[27] and Siegel and Imamura.[54]

Chapter V. This chapter is devoted to the study of almost periodic (a.p.) functions from the standpoint of GHA. Let $f(x)$ be uniformly almost periodic in the sense of H. Bohr. Then it is well known that the mean value

$$\lim_{T \to \infty} \frac{1}{2T} \int_{-T}^{T} f(x) e^{-i\lambda x} \, dx$$

exists, and vanishes except at most a denumerable number of λ. Let these be λ_n, and let

$$\lim_{T \to \infty} \frac{1}{2T} \int_{-T}^{T} f(x) e^{-i\lambda_n x} \, dx = \alpha_n .$$

We then associate with the almost periodic function f the Fourier series

$$f(x) \sim \sum \alpha_n e^{i\lambda_n x}.$$

The approximation of $f(x)$ by trigonometrical polynomials is now more difficult than with a purely periodic function f. This is because the set $\{\lambda_n\}$ may have a number of limiting points, and therefore the manner of summation of terms $\alpha_n e^{i\lambda_n x}$ is crucial. For the fundamental theorems of Weierstrass and Parseval for a.p. functions there are three different proofs depending on the order of summation of the terms:

(i) Arrangement of the terms $\alpha_n e^{i\lambda_n x}$ in an order depending on the arithmetical properties of the λ_n. See Bohr.[11]

(ii) Arrangement of these terms in the descending order of magnitude of the coefficients $|\alpha_n|$. See Weyl.[59]

(iii) Arrangement of these terms in the order of exponents λ_n. This is the only method of summation which is compatible with a unified treatment of almost periodic functions and functions with continuous spectra. It is the one Wiener develops.

The first half of this chapter is devoted to the deduction of fundamental theorems of a.p. functions from those of GHA proved earlier. See [25c], [28a], and [33i], Chapter IV, §24.

Generalizations of a.p. functions have been given by several authors: Wiener [25c], Stepanoff,[55, 56] Weyl,[59] and Besicovitch.[1]

Independently of Stepanoff, Wiener arrived at the general a.p. functions of the Stepanoff class S_2a.p. With Paley, he studied the problem of representation by trigonometrical series such as

$$\sum \frac{a_n}{\sqrt{2\pi}} e^{i\lambda_n x}, \quad |\lambda_n - n| \leqslant L < \frac{1}{\pi^2},$$

calling them *non-harmonic Fourier series.* The best result in this field was obtained by N. Levinson. Following this, they introduced a new class of a.p. functions called the *pseudo-periodic functions* (p.p.f.), and proved that the class of p.p.f. is equivalent to the subclass of S_2a.p. for which the exponents are so spaced that

$$\text{g.l.b.}\, |\lambda_m - \lambda_n| < 0.$$

See Paley and Wiener [34d], Chapter VII, p. 116, and Levinson.[41]

A.p. functions in the complex domain are also studied and related to the Riemann ζ-function. See Bohr and Jessen,[12, 13] and Paley and Wiener [34d], Chapter X.

The last part of Chapter V is devoted to the mean almost periodic functions introduced by Muckenhaupt.[50] Such functions can be applied to dynamical systems with infinitely many degrees of freedom. Wiener gives this application for a vibrating string whose density and tension depend on position but not on time. As further references we may cite Favard[22] and Stoker.[57]

The subject of a.p. functions has of course expanded considerably. For detailed earlier references see Bohr,[14] Besicovitch,[2] and Maak.[46] For more recent work on a.p. functions see Wiener and Wintner [41a] and [41b], Bohr and Folner,[15] Beurling,[4] Schwartz,[52] Kahane,[32, 33, 35] and Bochner and Izumi.[10] There are of course many other references.

References

1. A. S. Besicovitch, Proc. London Math. Soc. (2) 25(1926), 495–512.

2. A. S. Besicovitch, *Almost periodic functions*, Cambridge Univ. Press, 1932.

3. A. Beurling, Neuviene Congr. Math. Scandinaves (Hesingfors, 1939).

4. A. Beurling, C. R. Acad. Sci. Paris 225(1947), 326–328.

5. A. Beurling, Colloque Analyse Harmonique, Nancy (Paris, 1949), 9–30.

6. A. Beurling, Ann. Inst. Fourier 14(1964), 1–32.

7. S. Bochner, Math. Ann. 97(1926–7), 632–674.

8. S. Bochner, *Vorlesungen über Fouriersche Integrals*, §41, Leipzig 1932.

9. S. Bochner, Math. Ann. 108(1933), 378–410.

10. S. Bochner and S. Izumi, Ann. Math. 69(1959), 718–732.

11. H. Bohr, Acta Math. 45(1924), 29–127.

12. H. Bohr and B. Jessen, Acta Math. 54(1930), 1–35.

13. H. Bohr and B. Jessen, Acta Math. 58(1932), 1–55.

14. H. Bohr, *Fastperiodische Funktionene*, Springer, 1932.

15. H. Bohr and E. Folner, Acta Math. 76(1944–5), 31–155.

16. M. Born and J. Wolf, *Principles of Optics*, Pergamon, 1964, Chapter X.

17. J. C. Burkill, Proc. London Math. Soc. 25(1926), 513–524.

18. T. Carleman, *L'integrale de Fourier et questions qui s'y rattachent*, Uppsala, 1944.

19. H. Cramer, Ann. Math. 46(1939), 191–201.

20. H. Cramer, Ann. Inst. Math. 41(1940), 215–230.

21. L. Ehrenpreis, *Fourier analysis in several complex variables*, Pure and Applied Math. XVII (New York, 1969).

22. J. Favard, Acta Math. 51(1927), 31–81.

23. K. O. Friedrichs and H. N. Shapiro et al., *Integration of functionals*, New York Univ., 1957.

24. I. M. Gel'fand and N. Ya. Vilenkin, *Generalized functions, Vol. 4, Applications of harmonic analysis* (transl. from Russian), Academic Press, New York, 1964.

25. *H. Hahn, Über die Methode der arithmetischen Mittel in die Theorie der verallgemeinerten Fourierschen Integrale*, Sitzungsberichte der Akademie Wissenschaften in Wien, Mathematischen-Naturwissenschaftliche Klasse Section II, 134(1925), 449–470.

26. H. Hahn, Acta Math. 49(1926), 301–353.

27. T. Imamura, W. C. Meecham, and A. Siegel, J. Math. Phys. 6(1965), 695–706.

28. K. Ito and H. P. Mckean, *Diffusion processes and their sample paths*, Springer, 1969.

29. S. Izumi, Sci. Rep. Tohoku Univ. 23(1935), 880–906.

30. S. Izumi, Sci. Rep. Tohoku Univ. 23(1935), 907–919.

31. M. Kac, Bull. Amer. Math. Soc. 72(1966, special issue), 52–68.

32. J. P. Kahane, Ann. Inst. Fourier 7(1957), 293–314.

33. J. P. Kahane, Proc. International Symposium of Linear Analysis (Jerusalem, 1961), 274–281.

34. J. P. Kahane, *Algebras de convolucion de sucesiones, fonciones y medidas sumables*, Cursos y seminarios de matematica, Fas 6 (Univ. de Buenos Aires, 1961).

35. J. P. Kahane, Ann. Sci. Ec. Norm. Sup. 79(1962), 93–150.

36. T. Kawata, Ann. Inst. Math. 1(1950).

37. S. Koizumi and G. Suniochi, Tohoku Math. J. (2) 5(1954), 243–260.

38. S. Koizumi, Proc. Japan Acad. 35(1959), 323–328.

39. S. Koizumi, J. Faculty of Sci. Hokkaido Univ. 14(1959), 153–224.

40. S. Koizumi, J. Faculty of Sci. Hokkaido Univ. 15(1960), 93–130.

41. N. Levinson, *Gap and density theorems*, Amer. Math. Soc. Colloq. Publ. Vol. XXVI (1939), Chapter IV.

42. N. Levinson, Bull. Amer. Math. Soc. 72(1966, no. 1, part II), 1–32.

43. P. Lévy, *Calcul des probabilites* (Paris, 1925), 163–172.

44. P. Lévy, *Processus stochastiques et mouvement Brownian*, Paris, 1948.

45. S. B. Littauer, J. London Math. Soc. 4(1929), 226–231.

46. W. Maak, *Fastperiodische Funktionen*, Springer, 1950.

47. W. T. Martin and I. E. Segal, *Analysis in function space*, The MIT Press, Cambridge, Mass., 1964.

48. P. Masani, in *Probability and Statistics*, The Harald Cramer Volume, U. Grenander, ed., Wiley, New York, 1959, 175–189.

49. P. Masani, Bull. Amer. Math. Soc. 72(1966, no. 1, part II), 73–125.

50. C. F. Muckenhaupt, J. Math. Phys. MIT 8(1929), 163–200.

51. H. Pollard, Duke Math. J. 20(1953), 449–512.

52. L. Schwartz, Ann. Math. 48(1927), 857–929.

53. L. Schwartz, *Theorie des distributions*, Hermann, Paris, 1950.

54. A. Siegel and T. Imamura, J. Math. Phys. 6(1965), 707–721.

55. W. Stepanoff, C. R. Acad. Sci. Paris 181(1925), 90–92.

56. W. Stepanoff, Math. Ann. 95(1926), 473–498.

57. J. J. Stoker, *Nonlinear vibrations in mechanical and electrical systems*, Pure and Applied Math. II (New York, 1950), Chapter V, B, §13, Appendix II.

58. E. C. Titchmarsh, *Introduction to the theory of Fourier integrals*, Oxford, 1937.

59. H. Weyl, Math. Ann. 97(1926–7), 338–356.

60. H. Weyl, *The theory of groups and quantum mechanics* (transl. from German), Dover Publ., 1931.

Commentary on the Memoire [30a] on Generalized Harmonic Analysis

P. Masani

This important paper has yet to be interpreted and assimilated in the standard repertory of abstract harmonic analysis.

Hewitt and Ross {19: vol. II, p. 547}†

Our object is (1) to show that Wiener's Generalized Harmonic Analysis (GHA) can be assimilated in a widened form of functional analysis in which the notions of Hilbert space and subspace are extended to those of *convergence vector-graph* and *conditionally linear subspace*, and in which the concept of stationarity is studied in this wider graphical setting with the aid of Tauberian identities (§§A–C); (2) to report on the interesting ramifications of recent attempts to apply GHA to the theory of turbulence (§D); (3) to gauge the significance of GHA from the perspective of the history of the Calculus of Probabilities in its search for explications of the notion of "random" (§E). It is hoped that our account will dispel any opinion that GHA has been superceded by the theory of stochastic processes (SP) and that the memoir [30a] is only of historical interest.

This commentary will thus supplement the preceding one by Professor Koisumi, which stresses the genesis and immediate aftermath of [30a], its relation to cognate work by H. Weyl, H. Bohr, S. Bochner and E. C. Titshmarsh among others, and traces its influence during the next few decades.

A. The Banach graphs $(\mathscr{X}, \|\cdot\|, \mathrm{corr})$, $(\mathscr{X}, \|\cdot\|, \tau\mathrm{corr})$

Let $L_2^{loc}(\mathbb{R})$ be the class of all Borel measurable functions f on $\mathbb{R}$ to $\mathbb{C}$ such that for any compact subset A of $\mathbb{R}$, $f \cdot \chi_A$ is in $L_2(\mathbb{R})$. GHA is concerned with the Wiener classes $\mathscr{P}$, $\mathscr{S}$, $\mathscr{S}_c$ of functions $f \in L_2^{loc}(\mathbb{R})$ for which

$$
\text{(A.1)} \quad \begin{cases} \phi_{ff}(0) \underset{\mathrm{d}}{=} \lim\limits_{T\to\infty} \dfrac{1}{2T} \displaystyle\int_{-T}^{T} |f(s)|^2\, ds \quad \textit{exists}, \\ \forall t \in \mathbb{R}, \quad \phi_{ff}(t) \underset{\mathrm{d}}{=} \lim\limits_{T\to\infty} \dfrac{1}{2T} \displaystyle\int_{-T}^{T} f(s+t)\overline{f(s)}\, ds \quad \textit{exists}, \\ \forall t \in \mathbb{R}, \quad \phi_{ff}(t) \textit{ exists} \quad \& \quad \phi_{ff}(\cdot) \textit{ is continuous at } 0, \end{cases}
$$

respectively. $\phi_{ff}(\cdot)$ is called the (*auto-*) *covariance function* of f in $\mathscr{S}$, and

† Numbers in braces refer to the bibliography at the end of this commentary. Those in square brackets refer to the Wiener bibliography.

$\sqrt{\phi_{ff}}(0)$ the *mean total power* of f in $\mathscr{P}$. A major difficulty in assimilating GHA into modern functional analysis is that none of the classes $\mathscr{P}$, $\mathscr{S}$, $\mathscr{S}_c$ is closed under addition. Each is a non-linear subspace of the Marcinkiewicz vector space $\bar{\mathscr{X}}$ defined by

$$\text{(A.2)} \quad \begin{cases} \bar{\mathscr{X}} \underset{\mathrm{d}}{=} \{f: f \in L_2^{loc}(\mathbb{R}) \ \& \ \|f\| < \infty\}, \\ \textit{where} \\ \|f\| \underset{\mathrm{d}}{=} \overline{\lim_{T\to\infty}} \sqrt{\left\{\frac{1}{2T}\int_{-T}^{T} |f(s)|^2\, ds\right\}} \in [0, \infty]. \end{cases}$$

This $\bar{\mathscr{X}}$ becomes a Banach space under the norm $\|\cdot\|$ when we identify f, g in $\bar{\mathscr{X}}$ for which $\|f - g\| = 0$, as was shown by Marcinkiewicz {36} and independently by Bohr and Folner {11}.

The members of the Banach space $\bar{\mathscr{X}}$ are huge equivalence classes

$$\text{(A.3)} \qquad [f] = \{g: g \in \bar{\mathscr{X}} \ \& \ \|f - g\| = 0\}, \quad f \in \bar{\mathscr{X}}.$$

It is easy to show that if f is in $\mathscr{P}$ or $\mathscr{S}$ or $\mathscr{S}_c$, then so is each member of $[f]$. Functions in $\bar{\mathscr{X}} \setminus \mathscr{P}$ seem to be incapable of generalized harmonic analysis. So the Banach space $\bar{\mathscr{X}}$ is too large for the purposes of GHA, and some recent work by J. K. Lee[†] {28} suggests that this is also the case with the smallest Banach subspace of $\bar{\mathscr{X}}$ we can conceive for this purpose, viz. the (closed linear) subspace $\mathfrak{S}(\mathscr{P})$ spanned by $\mathscr{P}$ in $\bar{\mathscr{X}}$. The fact that $\bar{\mathscr{X}}$ and $\mathfrak{S}(\mathscr{P})$ are Banach spaces and therefore amenable to functional analysis thus has little import as far as GHA is concerned.

It is apparent that *in order to embrace GHA current functional analysis has to be widened*. The new ideas which have to be injected for this purpose are those of *vector-graph* and *conditional Banach space*, as was indicated in our 1966 abstract {41}, and outlined more recently in the paper {51}. In $\bar{\mathscr{X}}$ define two relations "*correlated*" and "*τ-correlated*", abbreviated "corr" and "τcorr" by: $\forall f, g \in \bar{\mathscr{X}}$,

$$\text{(A.4)}^{\ddagger} \quad \begin{cases} f \text{ corr } g \underset{\mathrm{d}}{\Leftrightarrow} ((f, g)) \underset{\mathrm{d}}{=} \lim_{T\to\infty} \dfrac{1}{2T}\displaystyle\int_{-T}^{T} f(t)\overline{g(t)}\,dt \quad \textit{exists}, \\ f \ \tau\text{corr } g \underset{\mathrm{d}}{\Leftrightarrow} \forall h \in \mathbb{R}, \quad \tau_h f \text{ corr } g \\ \qquad \textit{i.e.} \Leftrightarrow \forall h \in \mathbb{R}, \quad \phi_{fg}(h) \underset{\mathrm{d}}{=} ((\tau_h f, g)) \quad \textit{exists}. \end{cases}$$

[†] Not to be confused with Wiener's distinguished engineering colleague and close associate Professor Y. W. Lee.

[‡] $\tau_h f$ is the h-translation of f, cf. (A. 30) below.

$\phi_{gf}(\cdot)$, when it exists, is called the (*cross-*) *covariance function* of f and g. Since corr & τcorr $\subseteq \bar{\mathscr{X}} \times \bar{\mathscr{X}}$, it follows that ($\bar{\mathscr{X}}$, corr) and ($\bar{\mathscr{X}}$, τcorr) are graphs in the current sense, cf. Berge {5}.

Both the relations corr and τcorr are non-reflexive and non-transitive on $\bar{\mathscr{X}}$. Moreover, corr is symmetric on $\bar{\mathscr{X}}$, but it is not yet known if τcorr is symmetric on $\bar{\mathscr{X}}$. The possible pathology that it is non-symmetric can be bypassed by restricting τcorr to the smaller space:

(A.5)† $$\mathscr{X} \underset{\mathrm{d}}{=} \{f : f \in \bar{\mathscr{X}} \ \& \ \forall a > 0, \ \lim_{T \to \pm\infty} \frac{1}{T} \int_T^{T+a} |f(t)|^2 \, dt = 0\}.$$

We can show that

(A.6) $$\mathscr{S}_c \underset{\neq}{\subset} \mathscr{S} \underset{\neq}{\subset} \mathscr{P} \underset{\neq}{\subset} \mathscr{X} \underset{\neq}{\subset} \bar{\mathscr{X}}.$$

We have $f \in \bar{\mathscr{X}} \setminus \mathscr{X}$ and $g \in \mathscr{X} \setminus \mathscr{P}$, when for instance

$$f(t) \underset{\mathrm{d}}{=} \sum_{k=0}^{\infty} 2^{k/2} \chi_{[2^k, 2^k+1]}(|t|), \quad t \in \mathbb{R},$$

and

$$g(t) \underset{\mathrm{d}}{=} \sum_{k=1}^{\infty} \frac{2^{k/2}}{\sqrt{k}} \chi_{[2^k, 2^k+k]}(|t|), \quad t \in \mathbb{R}.$$

For a $\phi \in \mathscr{S} \setminus \mathscr{S}_c$ see [33i, p. 151 (iii)]. An example of $\psi \in \mathscr{P} \setminus \mathscr{S}$ is too technical for inclusion here.

The fundamental properties of the space $\mathscr{X}$ are stated in the following theorem:

A.7 Thm.

(a) $f \in \mathscr{X}$ & $g \in [f] \Rightarrow g \in \mathscr{X}$;

(b) $\mathscr{X}$ *is a translation-invariant* (*closed, linear*) *subspace of the Banach space* $\bar{\mathscr{X}}$;

(c) $\forall f, g \in \mathscr{X}$ & $\forall h \in \mathbb{R}$,

$$f \text{ corr } g \Rightarrow \tau_h f \text{ corr } \tau_h g \ \& \ ((\tau_h f, \tau_h g)) = ((f, g))$$
$$\tau_h f \text{ corr } g \Leftrightarrow f \text{ corr } \tau_{-h} g$$
$$\tau_h f \text{ corr } g \Rightarrow ((\tau_h f, g)) = ((f, \tau_{-h} g));$$

(d) $\forall f, g \in \mathscr{X}, \quad f \, \tau\text{corr } g \Rightarrow g \, \tau\text{corr } f \ \& \ \phi_{gf}(t) = \overline{\phi_{fg}(-t)}$;

(e) $Rstr._{\mathscr{X}}$ τcorr *is symmetric on* $\mathscr{X}$.

Since in GHA we are only interested in $\mathscr{P}$ and its subsets $\mathscr{S}$ and $\mathscr{S}_c$, it

† In the notation of Bertrandias {6}, $\bar{\mathscr{X}} = \mathfrak{M}_2(\mathbb{R})$ and $\mathscr{X} = \mathfrak{M}_{2r}(\mathbb{R})$, the r standing for "regular", the term he uses to describe functions in $\mathscr{X}$.

is reasonable in the light of (A.6) to treat $\mathscr{X}$, and not the larger $\bar{\mathscr{X}}$, as our overall Banach space, and to restrict the domains and ranges of corr and τcorr to $\mathscr{X}$. Then by A.7(e) τcorr becomes symmetric, and it can in fact be shown that both corr *and* τcorr satisfy all the conditions (a) (i)–(iv) imposed on $\mathfrak{c}$ in the following definition:

A.8 Def.

(a) $(\mathscr{X}, |\cdot|, \mathfrak{c})$ is called a *Banach graph* over $\mathbb{F}$ ($\mathbb{F} = \mathbb{R}$ or $\mathbb{C}$), iff

(i) $\mathscr{X}$ is a Banach space over $\mathbb{F}$ with norm $|\cdot|$

(ii) $\mathfrak{c} \subseteq \mathscr{X} \times \mathscr{X}$ is a symmetric relation whose domain and range are $\mathscr{X}$

(iii) $f, g \in \mathscr{X}$ & $f\mathfrak{c}g \Rightarrow \forall c \in \mathbb{F},\quad cf\mathfrak{c}g$

$f_1, f_2, g \in \mathscr{X},\quad f_1\mathfrak{c}g$ & $f_2\mathfrak{c}g \Rightarrow f_1 + f_2\mathfrak{c}g$

(iv) $\forall n \geqslant 1,\quad f_n, g_n \in \mathscr{X},\quad f_n\mathfrak{c}g_n$ & $f_n \to f,\quad g_n \to g$, as $n \to \infty$

$\Rightarrow f\mathfrak{c}g.$

(b) $(\mathscr{H}, (\cdot, -), \mathfrak{c})$ is called a *Hilbert graph* over $\mathbb{F}$, iff

(i) $\mathscr{H}$ is a Hilbert space over $\mathbb{F}$ with inner product $(\cdot, -)$

(ii) with the norm $|\cdot| \underset{\mathrm{d}}{=} \sqrt{(\cdot, \cdot)}$ of $\mathscr{H}$, $(\mathscr{H}, |\cdot|, \mathfrak{c})$ is a Banach graph over $\mathbb{F}$ in the sense of (a).

As just stated, we can show that

(A.9) $\quad (\mathscr{X}, \|\cdot\|, \text{corr})$ & $(\mathscr{X}, \|\cdot\|, \tau\text{corr})$ *are Banach graphs over* $\mathbb{C}$.

In order to determine other properties of these graphs, it is necessary to introduce some concepts and results (A.10–A.16) of the general theory of vector-graphs.

A.10 Def. Let $(\mathscr{X}, |\cdot|, \mathfrak{c})$ be a Banach graph over $\mathbb{F}$, and $\emptyset \neq A, B \subseteq \mathscr{X}$. Then

(a) $A\mathfrak{c}B \underset{\mathrm{d}}{\Leftrightarrow} \forall f \in A$ & $\forall g \in B, \quad f\mathfrak{c}g$;

(b)† $A^{\mathfrak{c}} = \mathfrak{c}(A) \underset{\mathrm{d}}{=} \{g : g \in \mathscr{X}$ & $\forall f \in A, g\mathfrak{c}f\}$,

$A^{\mathfrak{c}}$ is called the *correlate of* A;

(c) A is called *self-correlated* $\Leftrightarrow A\mathfrak{c}A$;

(d) $\mathscr{X}_{\mathfrak{c}} = \{f : f \in \mathscr{X}$ & $f\mathfrak{c}f\}$.

$\mathscr{X}_{\mathfrak{c}}$, the *set of self-correlated vectors*, need not be self-correlated. Rstr. ${}_{\mathscr{X}_{\mathfrak{c}}}\mathfrak{c}$ is of course reflexive. It follows from A.8 that

(A.11) $\quad \begin{cases} \mathfrak{c} \textit{ is a closed, homogeneous, (many-valued) relation on } \mathscr{X} \textit{ onto } \mathscr{X} \\ \ni \mathfrak{c}^{-1} = \mathfrak{c}. \end{cases}$

† As with $A^{\perp}$ in Hilbert spaces, the notation $A^{\mathfrak{c}}$ is more convenient than the more logical notation $\mathfrak{c}(A)$.

The relation $\mathfrak{c}$ need not, however, be additive. Thus *in general* $\mathfrak{c}$ *is a non-linear relation.* It also follows from A.8 and A.10(d) that

$$\text{(A.12)} \quad \begin{cases} \text{(a) } \mathscr{X}_{\mathfrak{c}} \text{ is a closed, homogeneous subset of } \mathscr{X} \\ \text{(b) } \mathfrak{c} \text{ is a linear relation} \Rightarrow \mathscr{X}_{\mathfrak{c}} \text{ is a linear manifold in } \mathscr{X}. \end{cases}$$

The following triviality and lemma show that $\mathfrak{c}$ and $A^{\mathfrak{c}}$ have many of the properties of the relation $\perp$ and the set $A^{\perp}$ in a Hilbert space.

A.13 Triv. *Let* $(\mathscr{X}, |\cdot|, \mathfrak{c})$ *be a Banach graph over* $\mathbb{F}$, *and* $\emptyset \neq A, B \subseteq \mathscr{X}$. *Then*

(a) $0 \in \mathscr{X}^{\mathfrak{c}}$;

(b) $A\mathfrak{c}B$, $\emptyset \neq A_0 \subseteq A$ & $\emptyset \neq B_0 \subseteq B \Rightarrow A_0\mathfrak{c}B_0$;

(c) $A \subseteq B \Rightarrow B^{\mathfrak{c}} \supseteq A^{\mathfrak{c}}$ & $A^{\mathfrak{cc}} \subseteq B^{\mathfrak{cc}}$;

(d) $A \subseteq B^{\mathfrak{c}} \Leftrightarrow A\mathfrak{c}B \Leftrightarrow B \subseteq A^{\mathfrak{c}}$;

(e) A *is self-corr* $\Leftrightarrow A \subseteq A^{\mathfrak{c}} \cap \mathscr{X}_{\mathfrak{c}}$.

A.14 Lma. *Let* $\mathscr{X}, |\cdot|, \mathfrak{c}, A, B$ *be as in* A.8. *Then*

(a) $A\mathfrak{c}B \Rightarrow \mathfrak{S}(A)\mathfrak{c}\mathfrak{S}(B)$;

(b) $A^{\mathfrak{c}}$ = *a (closed linear) subspace of* $\mathscr{X}$;

(c) $A^{\mathfrak{c}} = \{\mathfrak{S}(A)\}^{\mathfrak{c}}$;

(d) $A \subseteq \mathfrak{S}(A) \subseteq A^{\mathfrak{cc}}, \quad A^{\mathfrak{ccc}} = A^{\mathfrak{c}}$;

(e)† A *is a correlate, i.e.* $\exists B \subseteq \mathscr{X} \ni A = B^{\mathfrak{c}} \Leftrightarrow A = A^{\mathfrak{cc}}$;

(f) $A\mathfrak{c}A \Leftrightarrow \mathfrak{S}(A)\mathfrak{c}\mathfrak{S}(A) \Rightarrow \mathfrak{S}(A) \subseteq \mathscr{X}_{\mathfrak{c}}$;

(g) $A\mathfrak{c}A$ & $A^{\mathfrak{c}}\mathfrak{c}A^{\mathfrak{c}} \Leftrightarrow A^{\mathfrak{c}} = A^{\mathfrak{cc}}$.

Many of the naturally occurring subsets of $(\mathscr{X}, |\cdot|, \mathfrak{c})$ are not closed under addition and therefore are non-linear; but they are "conditionally" linear in the following sense:

A.15 Def. Let $(\mathscr{X}, |\cdot|, \mathfrak{c})$ be a Banach graph over $\mathbb{F}$. Then

(a) $\mathfrak{M}$ is called a *conditionally linear* (or $\mathfrak{c}$-*linear*) *manifold* of $\mathscr{X}$, iff

$$\emptyset \neq \mathfrak{M} \subseteq \mathscr{X}; \quad \forall c \in \mathbb{F}, \quad c \cdot \mathfrak{M} \subseteq \mathfrak{M};$$
$$f, g \in \mathfrak{M} \ \& \ f\mathfrak{c}g \Rightarrow f + g \in \mathfrak{M};$$

(b) $\mathfrak{M}$ is called a *conditional subspace* (or $\mathfrak{c}$-*subspace*) of $\mathscr{X}$, iff $\mathfrak{M}$ is a conditionally linear manifold of $\mathscr{X}$ and $\mathfrak{M}$ is (sequentially) closed.

(c) An operator T on a $\mathfrak{c}$-linear manifold $\mathfrak{M}$ is called (*conditional-*) *linear*, iff $\forall f, g \in \mathfrak{M}$ & $\forall a, b \in \mathbb{F}$,

$$af + bf \in \mathfrak{M} \Rightarrow T(af + bg) = aT(f) + bT(g).$$

A.16 Triv. *Let* $(\mathscr{X}, |\cdot|, \mathfrak{c})$ *be a Banach graph over* $\mathbb{F}$. *Then*

(a) $\mathscr{X}$ *is a* $\mathfrak{c}$-*subspace of* $\mathscr{X}$;

(b) $\mathfrak{M}$ *is a* $\mathfrak{c}$-*linear manifold* & $\mathfrak{M}\mathfrak{c}\mathfrak{M} \Rightarrow \mathfrak{M}$ *is a linear manifold in* $\mathscr{X}_{\mathfrak{c}}$;

---

† In words, A is a correlate iff A is the correlate of its correlate.

(c) $\mathfrak{M}$ *is a* $\mathfrak{c}$*-subspace &* $\mathfrak{M}\mathfrak{c}\mathfrak{M} \Rightarrow \mathfrak{M}$ *is a subspace of* $\mathscr{X}_{\mathfrak{c}}$;

(d) $\mathfrak{M}, \mathfrak{N}$ *are* $\mathfrak{c}$*-linear manifolds* $\Rightarrow \mathfrak{M} \cap \mathfrak{N}$ *is a* $\mathfrak{c}$*-linear manifold.*

We are now ready to revert to the specific Banach graphs ($\mathscr{X}$, $\|\cdot\|$, corr) and ($\mathscr{X}$, $\|\cdot\|$, τcorr) defined by (A.5), (A.4), which concern GHA. From A.14(b) we see that

(A.17) $$\begin{cases} \forall A \subseteq \mathscr{X},\ A^{\text{corr}},\ A^{\tau\text{corr}} \textit{ are (closed linear) subspaces of } \mathscr{X}; \\ \qquad A^{\tau\text{corr}} \subseteq A^{\text{corr}}. \end{cases}$$

Also by the definitions of $\mathscr{P}$, $\mathscr{S}$ and $\mathscr{X}_{\mathfrak{c}}$, cf. (A.1) and A.10(d),

(A.18) $$\mathscr{X}_{\text{corr}} = \mathscr{P}, \quad \mathscr{X}_{\tau\text{corr}} = \mathscr{S};$$

and by A.16(a),

(A.19) $$\begin{cases} \mathscr{P} \textit{ is a corr-(linear) subspace of } \mathscr{X} \\ \mathscr{S} \textit{ is a } \tau\textit{corr-(linear) subspace of } \mathscr{X}. \end{cases}$$

But we know that the spaces $\mathscr{P}$ and $\mathscr{S}$ are non-linear, and so it follows from A.12(b) that

(A.20) *corr &* τ*corr are non-linear relations on* $\mathscr{X}$ *onto* $\mathscr{X}$.

The graphs ($\mathscr{X}$, $\|\cdot\|$, corr), ($\mathscr{X}$, $\|\cdot\|$, τcorr) of course possess several properties, not derivable from the general theory, which depend on the specific manner in which corr and τcorr are introduced, viz. an inner-product-like kernel $((\cdot, \cdot))$ is defined by means of a limit for all pairs $(f; g)$ such that g corr f, and a function $\phi_{gf}(\cdot)$ is so defined for each pair $(f; g)$ such that g τcorr f.† Before turning to these additional properties, it is again necessary to introduce a concept from the general theory of Banach graphs:

A.21 Def. Let ($\mathscr{X}$, $|\cdot|$, $\mathfrak{c}$) be a Banach graph over $\mathbb{F}$. Then

(a) a kernel $k(\cdot, \cdot)$ is called a *conditionally* (briefly, $\mathfrak{c}$-) *sesquilinear functional for* $\mathscr{X}$, iff its domain is $\mathfrak{c} \subseteq \mathscr{X} \times \mathscr{X}$ and its range is in $\mathbb{F}$, and $\forall f, g \in \mathscr{X}$,

$$k(\cdot, g) \text{ is a linear functional on } \{g\}^{\mathfrak{c}},$$
$$k(f, \cdot) \text{ is a semi-linear functional on } \{f\}^{\mathfrak{c}};$$

(b) a $\mathfrak{c}$-sesquilinear functional $\phi(\cdot, \cdot)$ is called *Hermitian*, iff $\forall (f; g) \in \mathfrak{c}$, $k(g; f) = \overline{k(f; g)}$; it is called *non-negative*, iff $\forall f \in \mathscr{X}_{\mathfrak{c}}$, $k(f; f) \geqslant 0$, and is called *positive*, iff $\forall f \in \mathscr{X} \setminus \{0\}$, $k(f; f) > 0$.

It is easy to check that

---

† Very recently Dr. K.-S. Lau {25′} has proved the extremely interesting result that every function f in $\mathscr{P}$ of norm $\|f\| = 1$ is an extreme point of the closed unit ball of the Marcinkiewicz space $\bar{\mathscr{X}}$.

(A.22) $\begin{cases} ((\cdot, -)) \textit{ is a corr-sesquilinear, Hermitian, positive functional for the} \\ \textit{Banach graph } (\mathscr{X}, \|\cdot\|, corr); \\ \forall x \in \mathscr{P}, ((x, x)) = \|x\|^2, \textit{ where } \|\cdot\| \textit{ is the norm of } \mathscr{X}. \end{cases}$

From A.16(c), (A.22) and the definition (A.1) of $((\cdot, -))$ as the limit as $T \to \infty$, of genuine inner products $((\cdot, -))_T$, viz.

(A.22′) $$((f, g))_T \underset{\mathrm{d}}{=} \frac{1}{2T} \int_{-T}^{T} f(t)\overline{g(t)}\, dt,$$

it is a straightforward exercise to establish the following theorem:

A.23 Thm. (a) *The following conditions are equivalent:*
(α) *$\mathfrak{M}$ is a linear manifold $\subseteq \mathscr{P}$,*
(β) *$\mathfrak{M}$ is a corr-linear manifold $\subseteq \mathscr{X}$ & $\mathfrak{M}$ corr $\mathfrak{M}$.*

(b) *If $\emptyset \neq A \subseteq \mathscr{X}$ & A corr A, then $\mathscr{H} \underset{\mathrm{d}}{=} \mathfrak{S}(A)$ is a Hilbert space over $\mathbb{F}$ with the inner product $Rstr._{\mathscr{H} \times \mathscr{H}}((\cdot, -))$, and $\mathscr{H} \subseteq \mathscr{P}$.*

(c) *Both* (a) *and* (b) *hold with τcorr and $\mathscr{S}$ substituted for corr and $\mathscr{P}$, respectively.*

Thm. A.23(b) marks an important stage in the development of the theory, for it permits an easy extension to the Banach graphs $(\mathscr{X}, \|\cdot\|, \text{corr})$ and $(\mathscr{X}, \|\cdot\|, \tau\text{corr})$ of the fruitful concepts of a stationary curve in Hilbert space, and of a Hilbert-space-valued orthogonally scattered measure, cf. {46: 2.8} and {42: 1.2}:

A.24 Def. (a) We call $x(\cdot)$ a *continuous stationary curve in $\mathscr{X}$*, iff (i) $x(\cdot)$ is a continuous function on $\mathbb{R}$ to $\mathscr{X}$, and (ii) $\forall s, t, h \in \mathbb{R}$,

$$x(s) \text{ corr } x(t) \quad \& \quad ((x(s + h), x(t + h))) = ((x(s), x(t))).$$

(b) The *covariance function* $\gamma_x(\cdot)$ of $x(\cdot)$ is defined by

$$\gamma_x(t) \underset{\mathrm{d}}{=} ((x(t), x(0))), \quad t \in \mathbb{R}.$$

(c) The *subspace of* $x(\cdot)$ in $\mathscr{X}$ is defined by

$$\mathfrak{M}_x \underset{\mathrm{d}}{=} \mathfrak{S}\{x(t): t \in \mathbb{R}\}.$$

A.25 Def. (a) We call $\xi(\cdot)$ a *$\mathscr{X}$-valued countably additive, orthogonally scattered* (c.a.o.s.) *measure on a ring $\mathscr{R}$ with control measure μ*, iff (i) μ is a c.a. measure on $\mathscr{R}$ to $\mathbb{R}_{0+}$, and (ii) $\xi(\cdot)$ is a function on $\mathscr{R}$ to $\mathscr{X}$ such that

$$\forall A, B \in \mathscr{R}, \quad \xi(A) \text{ corr } \xi(B) \quad \& \quad ((\xi(A), \xi(B))) = \mu(A \cap B).$$

(b) The *subspace of* $\xi(\cdot)$ is defined by $\mathfrak{M}_\xi \underset{\mathrm{d}}{=} \mathfrak{S}\{\xi(A): A \in \mathscr{R}\}$.

Let $x(\cdot)$ and $\xi(\cdot)$ be as in Defs. A.24 and A.25. Then the ranges of $x(\cdot)$ and $\xi(\cdot)$ are self-correlated. Hence by Thm. A.23(b)

(A.26) $\begin{cases} \mathfrak{M}_x \,\&\, \mathfrak{M}_\xi \textit{ are Hilbert spaces contained in } \mathscr{P}, \textit{ having the inner products} \\ Rstr._{\mathfrak{M}_x \times \mathfrak{M}_x}((\cdot, -)), \quad Rstr._{\mathfrak{M}_\xi \times \mathfrak{M}_\xi}((\cdot, -)), \textit{ respectively.} \end{cases}$

Clearly $x(\cdot)$ is a genuine stationary curve in the Hilbert space $\mathfrak{M}_x$, and $\xi(\cdot)$ is a genuine $\mathfrak{M}_\xi$-valued c.a.o.s. measure. Thus, *the entire theory of Hilbert-space-valued stationary curves and Hilbert-space-valued c.a.o.s. measures carries over verbatim to the Banach graph* $(\mathscr{X}, \|\cdot\|, corr)$.

It follows, cf. e.g. {52: (3.4)}, that any continuous stationary curve $x(\cdot)$ in $\mathscr{X}$ possesses a strongly continuous *propagator* (or *shift*) *group* $(U_t : t \in \mathbb{R})$ of unitary operators U_t on the Hilbert space $\mathfrak{M}_x$ onto $\mathfrak{M}_x$ such that

$$\text{(A.27)} \qquad \forall s, t \in \mathbb{R}, \quad U_s\{x(t)\} = x(s+t).$$

Letting $E(\cdot)$ be the spectral measure of this group on $Bl(\mathbb{R})$ and $\xi_x(\cdot) \underset{\mathrm{d}}{=} E(\cdot)x(0)$ & $\mu_x(\cdot) \underset{\mathrm{d}}{=} \|\xi_x(\cdot)\|^2$, it follows cf. [57d: §7][†] that

$$\text{(A.28)} \qquad \begin{cases} \forall t \in \mathbb{R}, \quad U_t = \displaystyle\int_{\mathbb{R}} e^{-it\lambda} E(d\lambda), \\ \qquad x(t) = \displaystyle\int_{\mathbb{R}} e^{-it\lambda} \xi_x(d\lambda), \quad \gamma_x(t) = \displaystyle\int_{\mathbb{R}} e^{-it\lambda} \mu_x(d\lambda). \end{cases}$$

We naturally retain the usual names *spectral vector measure of* $x(\cdot)$ and *spectral distribution of* $x(\cdot)$ for the measures $\xi_x(\cdot)$ and $\mu_x(\cdot)$.

Conversely, it follows, cf. e.g. {46: 3.3}, that the FS transform of any bounded $\mathscr{X}$-valued c.a.o.s. measure on $Bl(\mathbb{R})$ is a $\mathscr{X}$-valued continuous stationary curve $x(\cdot)$ in $\mathscr{X}$, and $\mathfrak{M}_x = \mathfrak{M}_\xi$. We also have the Lévy inversion formulae, cf. {49: 1.17 & 1.24}: for any compact subset C *of* $\mathbb{R}$,

$$\text{(A.29)} \qquad \begin{cases} \mu_x(C) = \displaystyle\lim_{h \to 0+} \frac{1}{\sqrt{2\pi}} \int_{\mathbb{R}} \bar{\eta}_{C+[-h,h]}(t) \cdot \frac{\sin ht}{ht} \cdot \gamma_x(t)\, dt \\ \xi_x(C) = \displaystyle\lim_{h \to 0+} \frac{1}{\sqrt{2\pi}} \int_{\mathbb{R}} \bar{\eta}_{C+[-h,h]}(t) \cdot \frac{\sin ht}{ht} \cdot x(t)\, dt, \end{cases}$$

where the second integration is Bochner[‡] in the Hilbert space $\mathfrak{M}_x$, and where

$$\eta_B(t) \underset{\mathrm{d}}{=} \frac{1}{\sqrt{2\pi}} \int_B e^{-it\lambda}\, d\lambda,$$

[†] In writing $e^{-it\lambda}$ rather than $e^{it\lambda}$, we are departing from [30a] and [33i] but conforming to Wiener's later writings on prediction theory, e.g. [57d: §7]. We shall often refer to the book [33i], wherein much of the contents of [30a] appear in more cogent form.

[‡] And not just Pettis, since the range of the integrand, being a continuous image of $\mathbb{R}$, is separable in $\mathscr{X}$.

for any Borel subset B of $\mathbb{R}$ of finite Lebesgue measure.

Now let $f \in \mathscr{X}$ and let $x_f(t) \underset{\mathrm{d}}{=} \tau_t(f)$, $t \in \mathbb{R}$, where τ_t is translation through t, i.e.

$$\text{(A.30)} \qquad \forall f \in \mathbb{F}^{\mathbb{R}} \quad \& \quad \forall s, t \in \mathbb{R}, \quad (\tau_t f)(s) \underset{\mathrm{d}}{=} f(s + t).$$

Then $x_f(\cdot)$ is a curve in $\mathscr{X}$. It is easy to prove that

$$\text{(A.31)} \begin{cases} x_f(\cdot) \text{ is a stationary curve in } \mathscr{X} \text{ with propagator } (\tau_t \colon t \in \mathbb{R}) \Leftrightarrow f \in \mathscr{S}_c; \\ \forall f \in \mathscr{S}_c;\ \gamma_{x_f}(\cdot) = \phi_{ff}(\cdot), \quad \& \quad \mathfrak{M}_{x_f} = \mathfrak{S}\{\tau_t(f) \colon t \in \mathbb{R}\}. \end{cases}$$

This shows that $\forall f \in \mathscr{S}_c$, $\phi_{ff}(\cdot)$ *is a continuous positive-definite (CPD) function on* $\mathbb{R}$ *to* $\mathbb{C}$, a result which can of course be proved ab initio, cf. {3: App. 1}.†

We have been concerned so far wfth the *time-domain analysis* of functions in $\mathscr{X}$, especially of functions in $\mathscr{S}_c$, $\mathscr{S}$ and $\mathscr{P}$. We must now turn to their *spectral domain analysis*, which is the real purpose of GHA. Let $f \in \mathscr{S}_c$. It follows from (A.31) and (A.28) that there is a spectral measure $E(\cdot)$ for $\mathfrak{M}_{x_f}$ such that

$$\text{(A. 32)} \quad \begin{cases} \forall t \in \mathbb{R}, \quad \tau_t = \displaystyle\int_{\mathbb{R}} e^{-it\lambda} E(d\lambda), \quad x_f(t) = \int_{\mathbb{R}} e^{-it\lambda} \xi_f(d\lambda). \\ \qquad\qquad \phi_{ff}(t) = \displaystyle\int_{\mathbb{R}} e^{-it\lambda} \mu_x(d\lambda). \end{cases}$$

This shows at once that *the control measure* $\mu_f(\cdot)$ *of the c.a.o.s. measure* $\xi_f(\cdot)$ *is the (usual) spectral distribution* $\mu_{ff}(\cdot)$ *of* $f(\cdot)$.

The spectral distribution $\mu_{ff}(\cdot)$ and the $\mathscr{X}$-valued c.a.o.s. measure $\xi_f(\cdot)$ can be recovered by means of the Lévy inversion formula (A.29) in which $x(t) \underset{\mathrm{d}}{=} \tau_t(f)$; thus

$$\text{(A.33)} \quad \begin{cases} \mu_{ff}(C) = \displaystyle\lim_{h \to 0+} \frac{1}{\sqrt{2\pi}} \int_{\mathbb{R}} \bar{\eta}_{C+[-h,h]}(t) \cdot \frac{\sin ht}{ht} \cdot \phi_{ff}(t)\,dt, \\ \xi_f(C) = \displaystyle\lim_{h \to 0+} \frac{1}{\sqrt{2\pi}} \int_{\mathbb{R}} \bar{\eta}_{C+[-h,h]}(t) \cdot \frac{\sin ht}{ht} \cdot \tau_t(f)\,dt, \end{cases}$$

where C is any compact subset of $\mathbb{R}$, and the second integration is Bochner in the Hilbert space $\mathfrak{M}_f \underset{\mathrm{d}}{=} \mathfrak{S}\{\tau_t f : t \in \mathbb{R}\} \subset \mathscr{X}$.

† It is a sad anamoly that even in papers written after the advent of Bochner's Thm. in 1933, Wiener did not mention that $\phi_{ff}(\cdot)$ is PD, but continued giving substandard proofs instead of the much clearer standard ones that Bochner's Thm. allows.

Results such as (A.32) and (A.33) constitute, however, only the elementary part of the GHA of f. Wiener subjected f to a much deeper kind of analysis than that above. He observed that his functions f are all square integrable with respect to the Cauchy distribution over $\mathbb{R}$, cf. [33i: Thm.20]. Indeed, he showed that

$$\text{(A.34)} \qquad \bar{\mathscr{X}} \subseteq L_{2,c}(\mathbb{R}), \quad \textit{where } c(B) \underset{\mathrm{d}}{=} \frac{1}{\pi}\int_B \frac{dt}{1+t^2}, \quad B \in Bl(\mathbb{R}).$$

This enabled Wiener to bring to bear his famous *Tauberian identity* [30a: p. 148(a)–(c)], or [33i: (20.09), (20.10)], and to associate with any f in $\mathscr{X}$ a "*generalized Fourier transform*" $s_f(\cdot)$ in the space of functions of bounded quadratic variation. It is this deeper aspect of GHA which is the hardest to assimilate in the framework of current ideas, and an easy expedient, adopted by most critics, is simply to ignore it. In §§B, C we shall report on this part of GHA in the light of the headway made towards its assimilation by the writer and by Dr. J. K. Lee in his doctoral dissertation {28} at Indiana University in 1971, which has unfortunately remained unpublished until now cf. {25″}.†

B. The dual Banach graph $(\tilde{\mathscr{X}}, \|\cdot\|_V, V)$ in Hilbert space

Generalized harmonic analysis over $\mathbb{R}$, like classical harmonic analysis over $\mathbb{R}$, is governed by the $L_2(\mathbb{R})$-valued basic c.a.o.s. measure $\eta_{(\cdot)}$, termed *Fourier-Plancherel* (*FP*) *measure*, which is defined in terms of the characters of $\mathbb{R}$ by‡

$$\text{(B.1)} \qquad \forall B \in Bl_0(\mathbb{R}), \quad \bar{\eta}_B(t) \underset{\mathrm{d}}{=} \frac{1}{\sqrt{2\pi}}\int_B e^{it\lambda}\,d\lambda.$$

This is a *stationary measure*, in the sense that its control measure is translation-invariant, i.e. Lebesgue; more fully

$$\text{(B.2)} \qquad \forall A, B \in Bl_0(\mathbb{R}), \quad (\bar{\eta}_A, \bar{\eta}_B)_2 = Leb(A \cap B),$$

where $(\cdot,\cdot)_2$ denotes the inner product for $L_2(\mathbb{R})$.

The measure $\bar{\eta}_{(\cdot)}$ remains crucial in the parts of GHA centering on the Wiener Tauberian Lma. [30a: p. 148(a)–(c)], [33i: (20.09), (20.10)]. To simplify and assimilate the abstruse parts of [30a] involving this lemma, it is necessary, however, to adopt an abstract standpoint, and to consider in place of $L_2(\mathbb{R})$ and $\eta_{(\cdot)}$ an arbitrary separable Hilbert space $\mathscr{H}$ over $\mathbb{F}$, and a $\mathscr{H}$-

---

†In the forthcoming joint paper {25″} the salient results of Lee's dissertation {28} are supplemented by very interesting new results obtained by Dr. K. -S. Lau.

‡$Bl_0(\mathbb{R}) \underset{\mathrm{d}}{=} \{B: B \in Bl(\mathbb{R})\ \&\ \text{Leb. } B < \infty\}$. Obviously, $Bl_0(\mathbb{R})$ is a sub δ-ring of $Bl(\mathbb{R})$.

valued (stationary) c.a.o.s. measure $\xi(\cdot)$ over $(\mathbb{R}, Bl_0(\mathbb{R}), \text{Leb.})$. Given $\mathscr{H}$, our Thm. 7.11 in {42} assures us of the existence of an infinitude of $\mathscr{H}$-valued c.a.o.s. measures over $(\mathbb{R}, Bl_0(\mathbb{R}), \text{Leb.})$, one corresponding to each o.n. basis of $\mathscr{H}$. We shall therefore work with the following general set-up:

B.3 Assumptions

(i) *$\mathscr{H}$ is a separable Hilbert space over $\mathbb{C}$;*

(ii) *ξ is a $\mathscr{H}$-valued basic c.a.o.s. measure over $(\mathbb{R}, Bl_0(\mathbb{R}), Leb.)$;*

(iii) *$\mathfrak{M}_\xi(\cdot)$ is the spatial measure of ξ, i.e.,*

$$\forall B \in Bl(\mathbb{R}), \quad \mathfrak{M}_\xi(B) \underset{\mathrm{d}}{=} \mathfrak{S}\{\xi(\Delta) \colon \Delta \in 2^B \cap Bl_0(\mathbb{R})\};$$

(iv) *$Q_\xi(\cdot)$ is the spatial spectral measure of ξ, i.e.*

$$\forall B \in Bl(\mathbb{R}), \quad Q_\xi(B) \underset{\mathrm{d}}{=} \textit{Projection on } \mathscr{H} \textit{ onto } \mathfrak{M}_\xi(B);$$

(v) $\forall \alpha, \beta \in \mathscr{H} \quad \& \quad \forall B \in Bl(\mathbb{R}), \qquad \mu_{\alpha\beta}(B) \underset{\mathrm{d}}{=} (Q_\xi(B)\alpha, Q_\xi(B)\beta)_{\mathscr{H}};$

(vi) $\forall \lambda \in \mathbb{R}, \quad V_\lambda \underset{\mathrm{d}}{=} \int_{\mathbb{R}} e^{i\lambda t} Q_\xi(dt);$

(vii) $\forall (a, b) \subset \mathbb{R}, \quad T_V(a, b) \underset{\mathrm{d}}{=} \dfrac{1}{\sqrt{2}} \left\{ V_b - V_a - \displaystyle\int_a^b V_\lambda \, d\lambda \right\};$

(viii) *$(U_t \colon t \in \mathbb{R})$ is the propagator-group of ξ, and $E(\cdot)$ is its spectral measure, i.e.*

$$\forall t \in \mathbb{R} \quad \& \quad \forall A \in Bl_0(\mathbb{R}), \quad U_t\{\xi(A)\} = \xi(A + t)$$

$$\forall t \in \mathbb{R}, \quad U_t = \int_{\mathbb{R}} e^{it\lambda} E(d\lambda).$$

In (viii) we are appealing to the known fact, cf. {42:7.5}, that a stationary c.a.o.s. measure ξ possesses a strongly continuous propagator group (also called *shift group*) $(U_t \colon t \in \mathbb{R})$ made up of unitary operators U_t on $\mathscr{H}$ onto $\mathscr{H}$.

The operator-valued measure $T_V(\cdot)$ on the pre-ring of half-open-closed intervals of $\mathbb{R}$ plays the same important role in the explicit formulation of GHA that it does in other facets of harmonic analysis in Hilbert space.† With its aid we define a conditional sesquilinear functional $((\cdot, \cdot))_{\mathbf{V}}$ for $\mathscr{H}$, and a corresponding correlation $\mathbf{V} \subseteq \mathscr{H} \times \mathscr{H}$ as follows:

B.4 Def. Let $\alpha, \beta \in \mathscr{H}$. We say that α, β are **V**-*correlated*, in symbols $\alpha \, \mathbf{V} \, \beta$, iff

$$((\alpha, \beta))_{\mathbf{V}} \underset{\mathrm{d}}{=} \lim_{h \to 0+} \frac{1}{2h} (T_V(-h, h](\alpha), T_V(-h, h](\beta))_{\mathscr{H}} \quad \text{exists.}$$

Clearly $(\mathscr{H}, \mathbf{V})$ is a vector graph. But $(\mathscr{H}, |\cdot|_{\mathscr{H}}, \mathbf{V})$ is not a Banach graph

† For instance, in the theory of the innovation process of a stationary stochastic process, the Wold decomposition of isometric semi-groups and Cooper's Thm., the fundamental theorem of scattering, helix-theory, infinitely decomposable probability measures, etc., cf. {52, 38, 43, 46, 47, 45}.

in the sense of Def. A.8(a), because the condition A.8(a), (iv), viz. $\alpha_n \mathbf{V} \beta_n$ & $\alpha_n \to \alpha$, $\beta_n \to \beta$ in $\mathscr{H} \Rightarrow \alpha \mathbf{V} \beta$, need not hold. To obtain a Banach graph we must define a linear manifold $\mathscr{Y}$ of $\mathscr{H}$ and a norm $\|\cdot\|_{\mathrm{V}}$ for $\mathscr{Y}$, which conforms to the relation **V**, just as we had previously to bring in the Marcinkiewicz linear manifold $\bar{\mathscr{X}}$ of the Hilbert space $L_{2,c}(\mathbb{R})$ and a norm $\|\cdot\|$ for $\bar{\mathscr{X}}$ which is in consonance with the Wiener correlation relation $((\cdot, \cdot))$ defined in (A.4). This we do in analogy with (A.2) as follows:

B.5 Def. We define

(a) $\forall \alpha \in \mathscr{H}, \quad \|\alpha\|_{\mathrm{V}} \underset{\mathrm{d}}{=} \overline{\lim_{h \to 0+}} \dfrac{1}{\sqrt{(2h)}} |T_V(-h, h](\alpha)|_{\mathscr{H}} \in [0, \infty].$

(b) $\qquad \mathscr{Y} \underset{\mathrm{d}}{=} \{\alpha : \alpha \in \mathscr{H} \ \& \ \|\alpha\|_{\mathrm{V}} < \infty\} \subseteq \mathscr{H}.$

It is easily seen that $(\mathscr{Y}, \|\cdot\|_{\mathrm{V}})$ is a normed vector space over $\mathbb{F}$, when vectors α, β in $\mathscr{Y}$ such that $\|\alpha - \beta\|_{\mathrm{V}} = 0$ are identified. But it can also be shown that it is complete and meets all the requirements of Def. A.8(a); i.e.

(B.6) $\qquad (\mathscr{Y}, \|\cdot\|_{\mathrm{V}}, \mathbf{V})$ *is a Banach graph* $\subseteq \mathscr{H}$.

Just as we defined $\mathscr{P} = \mathscr{X}_{\mathrm{corr}} \underset{\mathrm{d}}{=} \{f : f \in \mathscr{X} \ \& \ f \text{ corr } f\}$, cf. (A.1), A.10 and (A.18), so we now define

(B.7) $$\tilde{\mathscr{P}} \underset{\mathrm{d}}{=} \mathscr{Y}_{\mathrm{V}} \underset{\mathrm{d}}{=} \{\alpha : \alpha \in \mathscr{Y} \ \& \ \alpha \mathbf{V} \alpha\}.$$

Corresponding to (A.19) we now have

(B.8) $\qquad \tilde{\mathscr{P}}$ *is a* **V**-*linear subspace of* $\mathscr{Y}$.

Now there is a natural isometry on the Hilbert space $L_{2,c}(\mathbb{R})$ onto $L_2(\mathbb{R})$, cf. (A.34), viz.

(B.9) $$f(\cdot) \to f(\cdot) l_0^+(\cdot), \quad \textit{where} \quad l_0^+(t) \underset{\mathrm{d}}{=} \frac{1}{\sqrt{\pi}} \frac{1}{t + i}, \qquad t \in \mathbb{R}.^{\dagger}$$

It follows readily from B.3(ii) and the isomorphism theorem of c.a.o.s. integration {42: 5.9} that the correspondence

---

† $l_0^+(\cdot)$ is the 0th-term of the *Laguerre basis* $(l_n^+)_{n=0}^{\infty}$ of the space $L_2^{0+}(\mathbb{R})$ of boundary values of the Hardy class $H_2(\Delta_+)$, where $\Delta_+ \underset{\mathrm{d}}{=} \{z : z = u + iv \ \& \ v > 0\}$ is the upper half-plane, defined by

$$l_n^+(t) \underset{\mathrm{d}}{=} \frac{1}{\sqrt{\pi}} \frac{(t - i)^n}{(t + i)^{n+1}}, \quad n \in \mathbb{N}_{0+}, \quad t \in \mathbb{R}.$$

$$\text{(B.10)} \qquad \Sigma : f \to \alpha_f \underset{\mathrm{d}}{=} \int_{\mathbb{R}} f(t) l_0^+(t) \xi(dt)$$

is a unitary operator on $L_{2,c}(\mathbb{R})$ onto $\mathscr{H}$. Thus,

$$\forall f \in L_{2,c}(\mathbb{R}), \quad |\alpha_f|_{\mathscr{H}} = |f|_{2,c}.$$

The inverse Σ^{-1} of the isometry Σ is readily obtained from {42: 5.10} in the form

$$\text{(B.11)} \qquad \forall \alpha \in \mathscr{H}, f_\alpha \underset{\mathrm{d}}{=} \Sigma^{-1}(\alpha) = \frac{d v_\alpha}{dc}, \quad \textit{where } v_\alpha(\cdot) \underset{\mathrm{d}}{=} (\alpha, \xi(\cdot))_{\mathscr{H}}.$$

To get other equations for f_α, observe that from (B.10) and the cut-off property $Q_\xi(B)\{\xi(A)\} = \xi(B \cap A)$, which emerges from B.3(v), (iv), cf. {45: 10.25(b)}, we have

$$Q_\xi(B)\alpha = \int_B f_\alpha(t) l_0^+(t) \xi(dt).$$

From this and its analogue for β it follows, cf. B.3(v), that

$$\mu_{\alpha\beta}(B) = \int_B f_\alpha(t)\overline{f_\beta(t)} |l_0^+(t)|^2 \, dt = \int_B f_\alpha(t)\overline{f_\beta(t)} c(dt),$$

from which we get the equations: $\forall \alpha, \beta \in \mathscr{H}$,

$$\text{(B.12)} \qquad \begin{cases} f_\alpha(\cdot)\overline{f_\beta(\cdot)} = \dfrac{d\mu_{\alpha\beta}}{dc}(\cdot), \quad |f_\alpha(\cdot)|^2 = \dfrac{d\mu_{\alpha\alpha}}{dc}(\cdot), \quad \textit{a.e. Leb.} \\ f_\alpha(\cdot) = \dfrac{d\mu_{\alpha\alpha_1}}{dc}(\cdot), \quad \textit{a.e. Leb., where } \alpha_1 \underset{\mathrm{d}}{=} \Sigma(1) = \displaystyle\int_{\mathbb{R}} l_0^+(t)\xi(dt). \end{cases}$$

We now maintain that the restriction $\mathrm{Rstr.}_{\mathscr{P}}\, \Sigma$ of Σ to $\mathscr{P}$ is a conditional linear isometry on $\mathscr{P}$ onto $\tilde{\mathscr{P}}$ which carries the Marcinkiewicz norm $\|\cdot\|$ to the **V**-norm $\|\cdot\|_{\mathbf{V}}$, i.e. $\|\alpha_f\|_{\mathbf{V}} = \|f\|$. This rather unexpected result hinges on on two or three important identities. First, we have

$$\text{(B.13)} \qquad \begin{cases} \forall f, g \in L_{2,c}(\mathbb{R}) \;\&\; \forall h \in \mathbb{R}_+, \\ (T_V(-h, h](\alpha_f), T_V(-h, h](\alpha_g))_{\mathscr{H}} = \dfrac{2}{\pi} \displaystyle\int_{\mathbb{R}} \frac{\sin^2 ht}{t^2} f(t)\bar{g}(t)\, dt, \end{cases}$$

and in particular

$$\text{(B.14)} \qquad \begin{cases} \forall f \in L_{2,c}(\mathbb{R}) \quad \& \quad \forall h \in \mathbb{R}_+, \\ |T_V(-h, h](\alpha_f)|^2_{\mathscr{H}} = \dfrac{2}{\pi} \displaystyle\int_{\mathbb{R}} \dfrac{\sin^2 ht}{t^2} |f(t)|^2 \, dt. \end{cases}$$

Second, we have the Wiener Tauberian identity [30a: p. 148(a)–(c)], [33i: (20.09) & (20.10)], viz. $\forall f \in L_2^{\text{loc}}|(\mathbb{R})$,

$$\text{(B.15)} \quad \lim_{h \to 0+} \frac{1}{\pi h} \int_{\mathbb{R}} \frac{\sin^2 ht}{t^2} |f(t)|^2 \, dt = \lim_{T \to \infty} \frac{1}{2T} \int_{-T}^{T} |f(t)|^2 \, dt \in [0, \infty],$$

where it is to be understood that the existence of one limit implies that of the other. The proof of (B.13) rests on the spectral representation of $T_V(-h, h]$, viz.

$$T_V(-h, h] = \frac{1}{\sqrt{2\pi}} \int_{\mathbb{R}} (e^{iht} - e^{-iht}) \frac{1}{t} \cdot \frac{1}{l_0^+(t)} Q(dt),$$

which emerges from B.3(vii), (vi), cf. {46: 3.1}, and on the above-mentioned cut-off property: $Q_\xi(B)\{\xi(A)\} = \xi(B \cap A)$.

The combination of (B.13), (B.14) and (B.15) yields the fundamental result, alluded to earlier, that Rstr.$_{\mathscr{P}}$ Σ is a conditional-linear isometry on $\mathscr{P}$ onto $\tilde{\mathscr{P}}$:

B.16 Main Thm. *Rstr.$_{\mathscr{P}}$ Σ is a conditional-linear isometry on the corr-linear subspace $\mathscr{P}$ of $\mathscr{X}$ onto the* **V**-*linear subspace $\tilde{\mathscr{P}}$ of $\mathscr{Y}$. Hence, letting f, $g \in L_{2,c}(\mathbb{R})$, we have*
(a) $f \in \mathscr{P} \Leftrightarrow \alpha_f \in \tilde{\mathscr{P}}$,
(b) $f \in \mathscr{P} \Rightarrow \|\alpha_f\|_{\mathbf{V}} = \|f\|$,
(c) $\forall f, g \in \mathscr{P}$, f corr $g \Leftrightarrow \alpha_f$ **V** α_g,
(d) $f, g \in \mathscr{P}$ & f corr $g \Rightarrow ((\alpha_f, \alpha_g))_{\mathbf{V}} = ((f, g))$.

Part (c) is inferred from parts (a) and (b) by polarization, after appealing to the result (A.19) that $\mathscr{P}$ is a corr-linear manifold in $\mathscr{X}$.

We have left unanswered the question as to how Σ behaves on the linear manifold $\mathscr{X}$ of $L_{2,c}(\mathbb{R})$ under its own (complete) norm $\|\cdot\|$. On the basis of a generalization of an important theorem of Dr. J. K. Lee, {28: Thm. 4.30}, cf. {26′ Thm. 5.2}, we now claim that

(B.17) *Rstr.$_{\mathscr{X}}$ Σ is a one-one continuous linear operator on the Banach space $(\mathscr{X}, \|\cdot\|)$ into the Banach space $(\mathscr{Y}, \|\cdot\|_{\mathbf{V}})$ and has a closed range.*

The relations (B.12) immediately tell us what $\Sigma(\mathscr{X})$ is: define

(B.18) $$\tilde{\mathscr{X}} \underset{\mathrm{d}}{=} \{\alpha\colon \alpha \in \mathscr{Y} \ \&\ \forall a > 0,\ \lim_{T\to\infty} \frac{1}{T} \int_T^{T+a} t^2 \mu_{\alpha\alpha}(dt) = 0\};$$

then

(B.19) $$\Sigma^{-1}(\tilde{\mathscr{X}}) = \mathscr{X}, \quad \textit{i.e.}\ \Sigma(\mathscr{X}) = \tilde{\mathscr{X}}.$$

(B.17) shows that $\alpha \in \tilde{\mathscr{X}} \Rightarrow f_\alpha \in \mathscr{X}$; the rest of (B.19) emerges from the triviality that for any bounded complex-valued c.a. measure $\mu(\cdot)$ on $Bl(\mathbb{R})$ such that $\mu(\cdot) \ll c(\cdot)$,

$$\pi \int_B t^2 \mu(dt) = \int_B \frac{d\mu}{dc}(t)\,dt - \pi\mu(B), \quad B \in Bl(\mathbb{R}),$$

upon setting $B = [T, T + a]$, $\mu = \mu_{\alpha\alpha}$, and noting that by (B.12), $d\mu/dc = |f_\alpha(\cdot)|^2$ a.e., Leb., and that $\mu[T, T + a]/T \to 0$, as $T \to \infty$.

(B.10), B.16 and (B.17) yield the remarkable result that the unitary operator Σ on the Hilbert space $L_{2,c}(\mathbb{R})$ onto the Hilbert space $\mathscr{H}$ is not only continuous with respect to the entirely different norms $\|\cdot\|$ and $\|\cdot\|_{\mathbf{V}}$ over the linear manifolds $\mathscr{X}$ and $\mathscr{Y}$ of $L_{2,c}(\mathbb{R})$ and $\mathscr{H}$, but is in fact a conditional-linear isometry with respect to these new norms on $\mathscr{P}$ onto $\tilde{\mathscr{P}}$, as well as a "unitary" operator with respect to the corr-inner-product $((\cdot, \cdot))$ and the **V**-inner product $((\cdot, \cdot))_{\mathbf{V}}$.

Just as it is reasonable in the time-domain analysis to treat $\mathscr{X}$ rather than $\bar{\mathscr{X}}$ as the overall Banach space, cf. (A.5) et seq., so in the light of (B.19) it is reasonable to regard $\tilde{\mathscr{X}}$ rather than the larger $\mathscr{Y}$ as the overall Banach space for spectral analysis. We can therefore dispense with (B.6) and (B.8) in favor of the result:

(B.20) $$\begin{cases} (\tilde{\mathscr{X}}, \|\cdot\|_{\mathbf{V}}, \mathbf{V}) \textit{ is a Banach graph } \subseteq \mathscr{H}, \\ \tilde{\mathscr{P}} \textit{ is a } \mathbf{V}\textit{-linear subspace of } \tilde{\mathscr{X}}. \end{cases}$$

Since $L_{2,c}(\mathbb{R})$ and $\mathscr{X}$ are invariant under translations τ_a, $a \in \mathbb{R}$ and under multiplication by characters e_a, $a \in \mathbb{R}$, ($e_a(t) \underset{\mathrm{d}}{=} e^{iat}$), it is natural to ask what $\alpha_{e_a f}$ and $\alpha_{\tau_a(f)}$ are, for any $f \in L_{2,c}(\mathbb{R})$. It is easy to see that

(B.21) $$\forall a \in \mathbb{R} \ \&\ \forall f \in L_{2,c}(\mathbb{R}), \quad \alpha_{e_a f} = V_a(\alpha_f) \quad \textit{in } \mathscr{H}.$$

The same equality will of course prevail in the Banach space $(\mathscr{X}, \|\cdot\|_V)$, since $|\alpha|_{\mathscr{H}} = 0 \Rightarrow \|\alpha\|_V = 0$.

The issue is not quite as simple with regard to $\alpha_{\tau_a(f)}$. Since, cf. B.3(viii) and {45: 11.17(a)}, $U_a = \Sigma_\xi \cdot \tau_{-a} \cdot \Sigma_\xi^{-1}$, we have $\forall \phi \in L_2(\mathbb{R})$, $\Sigma_\xi\{\tau_a(\phi)\} = U_{-a}\{\Sigma_\xi(\phi)\}$ in $\mathscr{H}$. From this it is clear that we cannot expect the equality $\alpha_{\tau_a(f)} = U_{-a}(\alpha_f)$ to hold for $f \in L_{2,c}(\mathbb{R})$. It is a remarkable fact, however, that for $f \in \mathscr{X}$, the last equality holds in the Banach space $\tilde{\mathscr{X}}$. This follows from the result that

$$\forall f \in L_{2,c}(\mathbb{R}) \ \&\ \forall a \in \mathbb{R}, \quad \|\alpha_{\tau_a(f)} - U_{-a}(\alpha_f)\|_V = 0,$$

which is a fairly straightforward consequence of (B.13) and simple properties of $l_0^+(\cdot)$. Thus,

$$\text{(B.22)} \qquad \forall a \in \mathbb{R} \ \&\ \forall f \in \mathscr{X}, \quad \alpha_{\tau_a(f)} = U_{-a}(\alpha_f) \quad \textit{in } \tilde{\mathscr{X}}.$$

This relation enables us to express the auto-covariance $\phi_{ff}(\cdot)$ of an f in $\mathscr{S}$, cf. (A.1), in terms of α_f. Before turning to this question, it is convenient to obtain *alternative expressions for the* **V***-norm* $\|\cdot\|_V$ *and for the* **V***-inner product* $((\cdot, \cdot))_V$, and then to consider the intertwining relations between V_λ and U_t, cf. (B.3) (viii).

Adopt the notation:

$$\text{(B.23)} \qquad \forall h \in \mathbb{R}_+, \quad \Delta_V(h) \underset{\text{d}}{=} V_h - V_{-h}.$$

We then assert that

$$\text{(B.24)} \begin{cases} \text{(a)} \ \forall \alpha \in \mathscr{H}, \quad \|\alpha\|_V^2 = \overline{\lim_{h \to 0+}} \dfrac{1}{4h} |\Delta_V(h)(\alpha)|_{\mathscr{H}}^2 \in [0, \infty], \\ \text{(b)} \ \forall \alpha, \beta \in \mathscr{H}, \quad \alpha \mathbf{V} \beta \Leftrightarrow \lim_{h \to 0+} \dfrac{1}{4h} (\Delta_V(h)(\alpha), \ \Delta_V(h)(\beta))_{\mathscr{H}} \quad \textit{exists}, \\ \text{(c)} \ \forall \alpha, \beta \in \mathscr{H} \ni \alpha \mathbf{V} \beta, \quad ((\alpha, \beta))_V = \lim_{h \to 0+} \dfrac{1}{4h} (\Delta_V(h)(\alpha), \ \Delta_V(h)(\beta))_{\mathscr{H}}. \end{cases}$$

This follows from the observation that $T_V(-h, h]$ differs from $(1/\sqrt{2})\Delta_V(h)$ by $-(1/\sqrt{2}) \int_{-h}^{h} V_\lambda \, d\lambda$, cf. B.3(vii), and that

$$\operatorname{slim}_{h\to 0+} \frac{1}{2h}\int_{-h}^{h} V_\lambda \, d_\lambda = I \quad \& \text{ so } \quad \operatorname{slim}_{h\to 0+} \frac{1}{\sqrt{2h}}\int_{-h}^{h} V_\lambda \, d_\lambda = 0.$$

Thus we can rewrite the result B.16(b) in the form

$$\text{(B.25)} \quad \forall f \in \mathscr{P}, \quad \lim_{T\to\infty} \frac{1}{2T}\int_{-T}^{T} |f(t)|^2 \, dt = \lim_{h\to 0+} \frac{1}{4h} |V_h(\alpha_f) - V_{-h}(\alpha_f)|^2_{\mathscr{H}}.$$

Next, it follows at once from B.3(iv), (viii), cf. {45: 11.18(b)}, that†

$$\forall t \in \mathbb{R} \quad \& \quad \forall B \in Bl(\mathbb{R}), \quad Q_\xi(B + t) = U_t \cdot Q_\xi(B) U_t^*.$$

Consequently we have the *intertwining relations*

$$\forall t, \lambda \in \mathbb{R}, \quad V_\lambda U_t = e^{i\lambda t} U_t V_\lambda.$$

These relations in turn entail, with the commutator notation $[A, B] \underset{\mathrm{d}}{=} AB - BA$, that $\forall h \in \mathbb{R}_+$ & $\forall a \in \mathbb{R}$,

$$\text{(B.26)} \qquad \operatorname{slim}_{h\to 0+} \frac{1}{2h}[\Delta_V(h), U_a] = iaI, \quad \operatorname{slim}_{h\to 0+} \frac{1}{\sqrt{2h}}[\Delta_V(h), U_a] = 0.$$

We are now ready to assert the following theorem on the auto-covariance function ϕ_{ff}:

B.27 Thm. *Let $f \in \mathscr{S}$. Then $\forall t \in \mathbb{R}$,*

$$\phi_{ff}(t) = ((U_{-t}(\alpha_f), \alpha_f))_{\mathbf{V}} = \lim_{h\to 0+} \frac{1}{4h}(U_{-t}\Delta_V(h)(\alpha_f), \Delta_V(h)(\alpha_f))_{\mathscr{H}}.$$

The first equality follows at once from the definition (A.1) of $\phi_{ff}(\cdot)$, (B.22) and B.16(d); the second equality follows on expressing the **V**-inner product in the form (B.24) (c) and then exploiting the second equality in (B.26). Thm. B.27 suggests the definition

$$\text{(B.28)} \qquad \begin{cases} \tilde{\mathscr{S}} \underset{\mathrm{d}}{=} \{\alpha : \alpha \in \mathscr{H} \ \& \ \forall t \in \mathbb{R}, \ U_{-t}(\alpha) \mathbf{V} \ \alpha\}. \\ \tilde{\mathscr{S}}_c \underset{\mathrm{d}}{=} \{\alpha : \alpha \in \mathscr{S} \ \& \ \lim_{t\to 0} ((U_{-t}(\alpha), \alpha))_{\mathbf{V}} = \|\alpha\|^2_{\mathbf{V}}\}. \end{cases}$$

† In Mackey's terminology {34}, $Q(\cdot)$ is an "imprimitivity system" for the group $(U_t : t \in \mathbb{R})$.

It follows easily that

$$\text{(B.29)}\qquad \tilde{\mathscr{S}}_c \subset \tilde{\mathscr{S}} \subset \tilde{\mathscr{P}} \subset \tilde{\mathscr{X}}, \quad \Sigma(\mathscr{S}_c) = \tilde{\mathscr{S}}_c \;\&\; \Sigma(\mathscr{S}) = \tilde{\mathscr{S}}.$$

It is easy to express $\phi_{ff}(\cdot)$ in terms of the spectral measure $E(\cdot)$ of the propagator group $(U_t: t \in \mathbb{R})$ of ξ. For define

$$\text{(B.30)}\qquad \forall f \in L_{2,c}(\mathbb{R}) \;\&\; \forall h \in \mathbb{R}_+, \quad \nu_{f,h}(\cdot) \underset{\mathrm{d}}{=} |E(\cdot)\Delta_V(h)\alpha_f|^2_{\mathscr{H}};$$

it then readily follows from Thm. B.27 and B.3 (viii) that

$$\text{(B.31)}\qquad \forall f \in \mathscr{S} \;\&\; \forall t \in \mathbb{R}, \quad \phi_{ff}(t) = \lim_{h \to 0+} \frac{1}{4h} \int_{\mathbb{R}} e^{-it\lambda} \nu_{f,h}(d\lambda).$$

Now let $f \in \mathscr{S}$. By employing (B.31) in conjunction with the Fourier-Stieltjes integral version of the well-known Fejer kernel expression of the Cesaro sum of a Fourier-Stieltjes series, we get

$$\frac{1}{a}\int_{-a}^{a}\left(1 - \frac{|t|}{a}\right)\phi_{ff}(t)\,dt = \lim_{h \to 0+} \frac{1}{4h}\int_{\mathbb{R}} \left\{\frac{\sin(a\lambda/2)}{a\lambda/2}\right\}^2 \nu_{f,h}(d\lambda).$$

This yields after some complicated manipulation the following necessary and sufficient condition for an f in $\mathscr{S}$ to be in $\mathscr{S}_c$:

B.32 Cor. *Let* $f \in \mathscr{S}$. *Then* $f \in \mathscr{S}_c$, *iff*

$$\lim_{A \to \infty} \overline{\lim_{h \to 0+}} \frac{1}{4h} \nu_{f,h}(\mathbb{R} \setminus [-A, A]) = 0.$$

Finally let $f \in \mathscr{S}_c$, and let μ_{ff} be its spectral distribution, cf. (A.32). How are the measures $\nu_{f,h}$ and μ_{ff} related? First observe that by (B.31), $\forall t \in \mathbb{R}$,

$$\int_{\mathbb{R}} e^{-it\lambda} \mu_{ff}(d\lambda) = \lim_{h \to 0+} \frac{1}{4h}\int_{\mathbb{R}} e^{-it\lambda} \nu_{f,h}(d\lambda).$$

Hence for any generalized trigonometric polynomial $P(\cdot)$ on $\mathbb{R}$,

$$\int_{\mathbb{R}} P(\lambda)\mu_{ff}(d\lambda) = \lim_{h \to 0+} \frac{1}{4h}\int_{\mathbb{R}} P(\lambda)\nu_{f,h}(d\lambda).$$

Now given $(a, b] \subset \mathbb{R}$ there exists a uniformly bounded sequence $(\mathrm{P}_n)_1^\infty$ of generalized trigonometric polynomials that convergences to $\chi_{(a,b]}$ pointwise on $\mathbb{R}$. Substituting $P_n(\lambda)$ in the last equation, we get on applying the Lebesgue-bounded convergence theorem, and making a delicate interchange of limits on the RHS (which we need not justify here), the following interesting corollary:

B.33 Cor. $\forall f \in \mathscr{S}_c$, *the spectral distribution* μ_{ff} *of* f *and the measure* $\nu_{f,h}$, *cf.* (A.32) *and* (B.30), *are related by the limiting equation:*

$$\forall (a, b] \subset \mathbb{R}, \quad \mu_{ff}(a, b] = \lim_{h \to 0+} \frac{1}{4h} \nu_{f,h}(a, b].$$

C. The dual Banach graph $(\mathscr{X}, \|\cdot\|_\tau, \tau)$ in $L_2(\mathbb{R})$

We shall now apply the general theory developed in §B to the specific case considered by Wiener. For this we must take, cf. §B outset,

$$\text{(C.1)} \qquad \mathscr{H} \underset{\mathrm{d}}{=} L_2(\mathbb{R}) \ \& \ \forall B \in Bl_0(\mathbb{R}), \quad \xi(B) \underset{\mathrm{d}}{=} \bar{\eta}_B(\cdot),$$

where $\bar{\eta}_{(\cdot)}$ is the $L_{2,\mathbb{R}}$-valued FP measure defined in terms of the characters of $\mathbb{R}$ by (B.1). The condition B.3(i) is then met, and as (B.2) shows so is the condition B.3(ii).

Classical analysis now enables us to identify the mathematical entities introduced in the rest of B.3. The spatial spectral measure $Q(\cdot)$ of $\bar{\eta}$ is known to be convolution multiplication by $(1/\sqrt{2\pi})\bar{\eta}$, and the propagator group $U(\cdot)$ of ξ is known to consist of (ordinary) multiplication by the characters $e_t(\cdot)$ of $\mathbb{R}$, cf. {44: 4.2}. It follows that V_λ is the translation τ_λ and that $E(B)$ is just multiplication by the indicator-function $\chi_B(\cdot)$ of B. Thus in place of B.3 we now have

C.3 Assumptions

(a) $\forall B \in Bl(\mathbb{R}), \quad Q(B) = \overset{*}{M}_{\eta_B/\sqrt{2\pi}}$, *i.e.*

$$\forall \alpha(\cdot) \in L_2(\mathbb{R}), \quad Q(B)(\alpha) = \frac{1}{\sqrt{2\pi}} (\bar{\eta}_B * \phi)(\cdot);$$

(b) $\forall \alpha(\cdot), \beta(\cdot) \in L_2(\mathbb{R}) \ \& \ \forall B \in Bl(\mathbb{R})$,†

$$\mu_{\alpha\beta}(B) \underset{\mathrm{d}}{=} (Q(B)\alpha, Q(B)\beta)_2 = \frac{1}{2\pi} (\eta_B * \alpha, \eta_B * \beta)_2;$$

(c) $\forall \lambda \in \mathbb{R}, \quad V_\lambda \underset{\mathrm{d}}{=} \int_{\mathbb{R}} e^{it\lambda} Q(d\lambda) = \tau_\lambda;$

(d) $\forall (a, b] \subset \mathbb{R}, \quad T_V(a, b] = T_\tau(a, b] \underset{\mathrm{d}}{=} \frac{1}{\sqrt{2}} \{\tau_b - \tau_a - \int_a^b \tau_\lambda d\lambda\};$

† We denote the inner product and norm of $L_2(\mathbb{R})$ by $(\cdot, \cdot)_2$ and $|\cdot|_2$.

(e) $\forall t \in \mathbb{R}, \quad U_t = M_{e_t}, \quad$ *i.e.* $\forall \phi \in L_2(\mathbb{R}), \quad U_t(f) = e_t(\cdot)\phi(\cdot);$
(f) $\forall B \in Bl(\mathbb{R}), \quad E(B) = M_{\chi_B}, \quad$ *i.e.* $\forall \alpha(\cdot) \in L_2(\mathbb{R}), \quad E(B)(\alpha) = \chi_B(\cdot)\phi(\cdot).$

The important Defs. B.4, B.5, and (B.7) now give way to the following:

C.4 Def.

(a) Let $\alpha(\cdot), \beta(\cdot) \in L_2(\mathbb{R})$. We say that $\alpha(\cdot), \beta(\cdot)$ are *τ-correlated*, in symbols $\alpha\tau\beta$, iff

$$((\alpha, \beta))_\tau \underset{\mathrm{d}}{=} \lim_{h \to 0+} \frac{1}{2h}(T_\tau(-h, h](\alpha), T_\tau(-h, h](\beta))_2 \quad \text{exists}$$

(b) $\forall \alpha(\cdot) \in L_2(\mathbb{R}), \quad \|\alpha\|_\tau \underset{\mathrm{d}}{=} \overline{\lim_{h \to 0+}} \frac{1}{\sqrt{(2h)}} |T_\tau(-h, h](\alpha)|_2 \in [0, \infty]$

(c) $\mathscr{Y} \underset{\mathrm{d}}{=} \{\alpha(\cdot) : \alpha(\cdot) \in L_2(\mathbb{R}) \;\&\; \|\alpha\|_\tau < \infty\} \subseteq L_2(\mathbb{R})$

(d) $\tilde{\mathscr{P}} \underset{\mathrm{d}}{=} \mathscr{Y}_\tau \underset{\mathrm{d}}{=} \{\alpha(\cdot) : \alpha(\cdot) \in \mathscr{Y} \;\&\; \alpha(\cdot)\, \tau\, \alpha(\cdot)\}.$

Now, cf. {44: §5}, the correspondence

$$\sum_\eta : \quad \phi \to \int_{\mathbb{R}} \phi(t)\bar{\eta}_{dt}, \quad \phi \in L_2(\mathbb{R}),$$

is precisely the *indirect FP transformation on* $L_2(\mathbb{R})$. Denoting the indirect and direct FP transforms of ϕ in $L_2(\mathbb{R})$ by $\tilde{\phi}$ and $\hat{\phi}$,† it follows that the unitary operator Σ on $L_{2,c}(\mathbb{R})$ onto $L_2(\mathbb{R})$ defined in (B.10) and its inverse Σ^{-1} are now given by

$$\text{(C.5)} \qquad \begin{cases} \forall f \in L_{2,c}(\mathbb{R}), & \alpha_f(\cdot) \underset{\mathrm{d}}{=} \Sigma\{f(\cdot)\} \underset{\mathrm{d}}{=} \{f\, \ell_0^+\}^\sim(\cdot), \\ \forall \alpha \in L_2(\mathbb{R}), & f_\alpha(\cdot) \underset{\mathrm{d}}{=} \Sigma^{-1}\{\alpha(\cdot)\} \underset{\mathrm{d}}{=} \hat{\alpha}(\cdot)/\ell_0^+(\cdot). \end{cases}$$

Thus *the generalized Fourier transform (GFT)* $\alpha_f(\cdot)$ *of* f *in* $L_{2,c}(\mathbb{R})$ *is just the indirect FP transform of* $f \cdot \ell_0^+$, where $\ell_0^+(\cdot)$ is the 0th Laguerre function defined in (B.9). In view of (C.5), the relations in (B.12) governing the measure $\mu_{\alpha,\beta} \underset{\mathrm{d}}{=} (Q(\cdot)\alpha, Q(\cdot)\beta)_2$, take the simple form:

$$\text{(C.6)} \qquad \forall \alpha(\cdot), \beta(\cdot) \in L_2(\mathbb{R}), \quad \frac{d\mu_{\alpha\beta}}{d\,Leb}(\cdot) = \hat{\alpha}(\cdot)\overline{\hat{\beta}(\cdot)}.$$

---

† Usually written

$$\tilde{\phi}(u) = \underset{A \to \infty}{\text{l.i.m.}} \frac{1}{\sqrt{2\pi}} \int_{-A}^{A} \phi(t) e^{iut}\, dt$$

$$\hat{\phi}(u) = \underset{A \to \infty}{\text{l.i.m.}} \frac{1}{\sqrt{2\pi}} \int_{-A}^{A} \phi(t) e^{-iut}\, dt.$$

Our Main Thm. B.16 now reads:

C.7 Main Thm. *$Rstr._{\mathscr{P}}\Sigma$ is a conditional-linear isometry on the corr-linear subspace $\mathscr{P}$ of $\mathscr{X}$ onto the τ-linear subspace $\tilde{\mathscr{P}}$ of $\mathscr{Y}$. Hence, letting $f, g \in L_{2,c}(\mathbb{R})$, we have*

(a) $f \in \mathscr{P} \Leftrightarrow \alpha_f \in \tilde{\mathscr{P}}$,

(b) $\forall f \in \mathscr{P}, \quad \|\alpha_f\|_\tau = \|f\|$,

(c) $\forall f, g \in \mathscr{P}, \quad$ f corr $g \Leftrightarrow \alpha_f \tau \alpha_g$,

(d) $f, g \in \mathscr{P}$ & f corr $g \Rightarrow ((\alpha_f, \alpha_g))_\tau = ((f, g))$.

In view of (C.6), the definition (B.18) of the space $\tilde{\mathscr{X}}$ now takes the simple form

(C.8) $$\tilde{\mathscr{X}} \underset{\mathrm{d}}{=} \{\alpha : \alpha \in \mathscr{Y} \ \& \ \forall a > 0, \lim_{T \to \infty} \frac{1}{T} \int_T^{T+a} t^2 \mu_{\alpha\alpha}(dt) = 0\}.$$

By (B.19) this $\tilde{\mathscr{X}}$ is the image of the space $\mathscr{X}$ of (A.5) under the generalized Fourier transformation Σ:

(C.9) $$\tilde{\mathscr{X}} = \Sigma(\mathscr{X}), \qquad \mathscr{X} = \Sigma^{-1}(\tilde{\mathscr{X}}).$$

As in (B.20) we conclude

(C.10) $$\begin{cases} (\tilde{\mathscr{X}}, \|\cdot\|_\tau, \tau) \textit{ is a Banach graph} \subseteq L_2(\mathbb{R}) \\ \tilde{\mathscr{P}} \textit{ is a } \tau\textit{-linear subspace of } \tilde{\mathscr{X}}, \end{cases}$$

and as in §B we can discard $\mathscr{Y}$ in favour of $\tilde{\mathscr{X}}$ from here on.

From C.3(e) and (B.21), $\forall f \in L_{2,c}(\mathbb{R})$, $\alpha_{e_a f} = \tau_a(\alpha_f)$ in $L_2(\mathbb{R})$. This equality will of course prevail in the Banach space $\tilde{\mathscr{X}}$ as well. On the other hand, in general, $\alpha_{\tau_a(f)} \neq e_{-a} \cdot \alpha_f$ in $L_2(\mathbb{R})$; nevertheless we have as in (B.22)

(C.11) $$\forall a \in \mathbb{R} \quad \& \quad \forall f \in \mathscr{X}, \quad \alpha_{\tau_a(f)} = e_{-a}(\cdot)\alpha_f(\cdot) \text{ in } \tilde{\mathscr{X}}.$$

Now that $\Delta_\tau(h) \underset{\mathrm{d}}{=} \tau_h - \tau_{-h}$ and $\mathscr{H} = L_2(\mathbb{R})$, the relations (B.24) can be cast in the following more familiar form:

C.12 Lma. *Let $\alpha(\cdot), \beta(\cdot) \in L_2(\mathbb{R})$. Then*

(a) $$\|\alpha\|_\tau^2 = \overline{\lim_{h \to 0+}} \frac{1}{4h} \int_{\mathbb{R}} |\alpha(\lambda + h) - \alpha(\lambda - h)|^2 d\lambda \in [0, \infty];$$

(b) $$\alpha \tau \beta \Leftrightarrow \lim_{h \to 0+} \frac{1}{4h} \int_{\mathbb{R}} \{\alpha(\lambda + h) - \alpha(\lambda - h)\}\{\bar{\beta}(\lambda + h) - \bar{\beta}(\lambda - h)\} d\lambda$$

exists;

(c) $\alpha\tau\beta \Rightarrow ((\alpha, \beta))_\tau = \lim_{h\to 0+} \frac{1}{4h} \int_{\mathbb{R}} \{a(\lambda + h) - \alpha(\lambda - h)\}\{\bar{\beta}(\lambda + h) - \bar{\beta}(\lambda - h)\}\, d\lambda.$

The results (B.25) and B.27 now reduce to the following theorem connecting the mean total power of an f in $\mathscr{P}$ as well as the auto-covariance of an f in $\mathscr{S}$ to its GFT $\alpha_f(\cdot)$:

C.13 Thm. (a) $\forall f \in \mathscr{P}$,

$$\lim_{T\to\infty} \frac{1}{2T} \int_{-T}^{T} |f(t)|^2 = \lim_{h\to 0+} \frac{1}{4h} \int_{\mathbb{R}} |\alpha_f(\lambda + h) - \alpha_f(\lambda - h)|^2\, d\lambda;$$

i.e. the mean-square modulus of f equals the quadratic variation of $\alpha_f(\cdot)/\sqrt{2}$, (Wiener's terminology);

(b) $\forall f \in \mathscr{S}$, & $\forall t \in \mathbb{R}$,

$$\phi_{ff}(t) = \lim_{h\to 0+} \frac{1}{4h} \int_{\mathbb{R}} e^{-it\lambda} |\alpha_f(\lambda + h) - \alpha_f(\lambda - h)|^2\, d\lambda.$$

In view of C.3(e), the definitions (B.28) take the form

$$\text{(C.14)} \qquad \begin{cases} \tilde{\mathscr{S}} \underset{\mathrm{d}}{=} \{\alpha \colon \alpha \in L_2(\mathbb{R})\ \&\ \forall t \in \mathbb{R},\ e_{-t}(\cdot)\alpha(\cdot)\tau\alpha(\cdot)\} \\ \tilde{\mathscr{S}}_c \underset{\mathrm{d}}{=} \{\alpha \colon \alpha \in \mathscr{S}\ \&\ \lim_{t\to 0}((e_{-t}(\cdot)\alpha(\cdot), \alpha(\cdot)))_\tau = \|\alpha\|_\tau^2\}. \end{cases}$$

We have, cf. (B.29),

$$\text{(C.15)} \qquad \tilde{\mathscr{S}}_c \subset \tilde{\mathscr{S}} \subset \tilde{\mathscr{P}} \subset \tilde{\mathscr{X}}, \quad \Sigma(\mathscr{S}_c) = \tilde{\mathscr{S}}_c\ \&\ \Sigma(\mathscr{S}) = \tilde{\mathscr{S}}.$$

Finally, for the measure $\nu_{f,h}(\cdot)$ of (B.30) we now have

$$\text{(C.16)} \qquad \begin{cases} \forall f \in L_{2,c}(\mathbb{R}), \quad \forall h \in \mathbb{R}_+ \quad \& \quad \forall B \in Bl(\mathbb{R}), \\ \nu_{f,h}(B) \underset{\mathrm{d}}{=} |\chi_B \Delta_\tau(h)\alpha_f|_2^2 = \int_B |\alpha_f(\lambda + h) - \alpha_f(\lambda - h)|^2\, d\lambda. \end{cases}$$

Consequently, Cors. B.32 and B.33 reduce to the following:

C.17 Cor. (a) *Let $f \in \mathscr{S}$. Then $f \in \mathscr{S}_c$, iff*

$$\lim_{A\to\infty} \lim_{h\to 0+} \frac{1}{4h} \left(\int_{\mathbb{R}} - \int_{-A}^{A} \right) |\alpha_f(\lambda + h) - \alpha_f(\lambda - h)|^2\, d\lambda = 0.$$

(b) *$\forall f \in \mathscr{S}_c$, the spectral distribution μ_{ff} of f is related to its GFT $\alpha_f(\cdot)$ by*

$$\forall (a, b] \subset \mathbb{R}, \quad \mu_{ff}(a, b] = \lim_{h\to 0+} \frac{1}{4h} \int_a^b |\alpha_f(\lambda + h) - \alpha_f(\lambda - h)|^2\, d\lambda.$$

We must now find out how these results of ours tally with those obtained by Wiener for his GTF. Wiener defines the GFT of a function f in $L_{2,c}(\mathbb{R})$ by

(C.18) $$s(u) \underset{\mathrm{d}}{=} \underset{A\to\infty}{\mathrm{l.i.m.}} \frac{1}{\sqrt{(2\pi)}}\left(\int_{-A}^{-1} + \int_{1}^{A}\right)\frac{f(t)e^{iut}}{it}\,dt + \frac{1}{\sqrt{(2\pi)}}\int_{-1}^{1} f(t)\frac{e^{iut}-1}{it}\,dt,$$

cf. [33i: (20.06)] or [30a: (5.51)]. Although our GFT $\alpha_f(\cdot)$ defined in (C.5) is in $L_2(\mathbb{R})$ and Wiener's GFT $s_f(\cdot)$ is not, the fundamental equations they obey are basically the same, as a comparison of our C.13(a), C.13(b), C.17(a) and C.17(b) with Wiener's (20.13), (21.7), Thm. 28 and (21.26) in [33i] shows.† The slight differences between the two sets of equations are easily accounted for: whereas we have $1/\sqrt{\pi}(t+i)$ in our definition of $\ell_0^+(t)$, Wiener just has $1/t$ in his definition of $s(\cdot)$; Wiener's spectral distribution, moreover, is by definition $\sqrt{2\pi}$ times ours, cf. [33i: (23.07)] and (A.31).

Let us trivially modify Wiener's GFT by the insertion of $\sqrt{\pi}$ in the denominator, and denote it now by $s_f(\cdot)$ instead of $s(\cdot)$, so that

(C.18′) $$s_f(u) \underset{\mathrm{d}}{=} \underset{A\to\infty}{\mathrm{l.i.m.}} \frac{1}{\sqrt{(2\pi)}}\left(\int_{-A}^{-1} + \int_{1}^{A}\right)\frac{f(t)e^{iut}}{it\sqrt{\pi}}\,dt + \frac{1}{\sqrt{(2\pi)}}\int_{1}^{1} f(t)\frac{e^{iut}-1}{it\sqrt{\pi}}\,dt.$$

Then *the fundamental equations obeyed by* $s_f(\cdot)$ *are the very same as those satisfied by* $\alpha_f(\cdot)$. This strongly suggests a close connection between $\alpha_f(\cdot)$ and $s_f(\cdot)$. Such is indeed the case as we proceed to show.

The realization that the Fourier dual $\tilde{\mathscr{X}}$ of the space $\mathscr{X}$ of (A.5) could be construed as a linear manifold in $L_2(\mathbb{R})$ dawned on us only recently. In the late sixties, when Mr. J. K. Lee started working on his dissertation, we observed that the space $Q_2(\mathbb{R})$ of *functions of bounded quadratic variation*, i.e. of measurable functions ϕ on $\mathbb{R}$ to $\mathbb{C}$ for which

(C.19) $$\|\phi\|_Q \underset{\mathrm{d}}{=} \overline{\lim_{h\to 0+}} \sqrt{\left\{\frac{1}{2h}\int_{\mathbb{R}} |\phi(t+h)-\phi(t-h)|^2\,dt\right\}} < \infty,$$

is in a rather natural sense the "Fourier dual" of the Marcinkiewicz space

† The same results in [30a] are labeled (5.53), (6.15), (7.22) and (7.27).

$\mathfrak{M}_2(\mathbb{R})$, designated by $\bar{\mathscr{X}}$ in (A.2). Accordingly, we believed that the dual $\tilde{\mathscr{X}}$ of $\mathscr{X}$ had to be a subset of the (non-Hilbert) space $Q_2(\mathbb{R})$, a belief reinforced by the presence of the second term in Wiener's expression (C.18) for his GFT. This, somewhat off-the-mark, conceptualization led to some interesting research on the part of Mr. Lee.

The question as to whether the linear extension $\bar{U}$ to $\mathfrak{M}_2(\mathbb{R})$ of the (non-linear) Wiener isometry U on $\mathscr{P}$ is onto or merely into $Q_2(\mathbb{R})$ immediately raised the more basic question as to whether $Q_2(\mathbb{R})$ is complete under the norm $|\cdot|_Q$. This question, as well as the corresponding one for $Q_2(C)$, where C is the unit circumference in $\mathbb{C}$, were still open, a difficulty being that $|\cdot|_Q$ is not a "function semi-norm" in the sense of Luxemberg and Zaanan {33} and therefore unamenable to treatment by standard methods. Lee proved the completeness of $Q_2(C)$ by utilizing its inclusion in $L_2(C)$, and showing that a function in $L_2(C)$ could be lifted from each term (equivalence class) ϕ_{n_i} of a suitable subsequence $(\phi_{n_i})_{i=1}^{\infty}$ of any Cauchy sequence $(\phi_n)_1^{\infty}$ in $Q_2(C)$, and then combined to yield a convergent sequence $(\psi_k)_1^{\infty}$ in $L_2(C)$. The equivalence class ϕ in $Q_2(C)$ of its limit ψ was then shown to be the limit of $(\phi_n)_1^{\infty}$. For $Q_2(\mathbb{R})$, which is *not* a subset of $L_2(\mathbb{R})$, Lee appealed to the theory of helices in Hilbert space {46}. Let

$$\text{(C.20)} \qquad \forall \phi \in Q_2(\mathbb{R}) \quad \& \quad \forall t \in \mathbb{R}, \quad x_\phi(t) \underset{\mathrm{d}}{=} \tau_t(\phi) - \phi.$$

Then $x_\phi(\cdot)$ is a helix in the Hilbert space $L_2(\mathbb{R})$, which we shall call *the helix due to* ϕ. Its "average vector" is given by

$$\text{(C.21)} \quad \psi_\phi \underset{\mathrm{d}}{=} \sqrt{2}\int_0^\infty e^{-t}\{x_\phi(0) - x_\phi(t)\}\,dt = \sqrt{2}\int_0^\infty e^{-t}\{\phi - \tau_t(\phi)\}\,dt,$$

where the integration is Bochner in $L_2(\mathbb{R})$, cf. {46: 2.12}. Lee showed that $\psi_\phi/\sqrt{2}$ has the rather amazing property of not only being in the equivalence class of ϕ, but also of yielding for each Cauchy sequence $(\phi_n)_1^{\infty}$ in $Q_2(\mathbb{R})$ a subsequence $(n_i)_1^{\infty}$ and a sequence of bounded Borel sets $(B_i)_1^{\infty}$ such that, cf. C.3(a),

$$\psi \underset{\mathrm{d}}{=} \frac{1}{\sqrt{2}}\sum_{i=1}^{\infty} Q(B_i)(\psi_{\phi_{n_i}}) = \frac{1}{2\sqrt{\pi}}\sum_{i=1}^{\infty} \bar{\eta}_{B_i} * \psi_{\phi_{n_i}}$$

exists in $L_2(\mathbb{R})$ and yields an equivalence class ϕ in $Q_2(\mathbb{R})$, which is the $\|\cdot\|_Q$-limit of $(\phi_n)_1^{\infty}$.

In much the same way Mr. Lee was able to establish the completeness of the spaces $Q_p(\mathbb{R})$, where $p \in (1, \infty)$, which are defined like $Q_2(\mathbb{R})$ but with the L_p-norm instead of the L_2-norm in (C.19), by an appeal to the theory of helices in a Banach space {48} and the theory of semi-groups of operators therein. For $p = 1$ Lee's method fails, and the completeness of $Q_1(\mathbb{R})$ remains an open question.†

Reverting to the case $p = 2$, it is easy to check that for any ϕ in $Q_2(\mathbb{R})$, $|\phi|_Q$ is related to the τ-norm of the average vector in $L_2(\mathbb{R})$ of the helix $x_\phi(\cdot)$ due to ϕ, cf. (C.19)–(C.21), by the equation

$$\text{(C.22)} \qquad \forall \phi \in Q_2(\mathbb{R}), \quad \|\phi\|_Q = \|\psi_\phi/\sqrt{2}\|_\tau.$$

On the other hand we know from C.7(b) and [33i, (20.13)] that with the amended definition (C.18′) of $s_f(\cdot)$, $\forall f \in \mathscr{P}$, $\|\alpha_f\|_\tau = \|f\| = \|s_f\|_Q$. Hence

$$\forall f \in \mathscr{P}, \quad \|\alpha_f\|_\tau = \|\psi_{s_f}/\sqrt{2}\|_\tau.$$

This suggests that for all f in $\mathscr{P}$, and perhaps even for all $f \in L_{2,c}(\mathbb{R})$, $\alpha_f = c\psi_{s_f}/\sqrt{2}$, where c is a constant of norm one. This is indeed the case as the following theorem shows:

C.23 Thm. *Let (i) $f \in L_{2,c}(\mathbb{R})$, (ii) $\alpha_f(\cdot)$ be the GFT of f in $L_2(\mathbb{R})$ defined in* (C.5), *(iii) $s_f(\cdot)$ be Wiener's GFT of f in $Q_2(\mathbb{R})$ as amended in* (C.18′). *Then $\sqrt{2\alpha_f}(\cdot)/i$ is the average vector of the helix in $L_2(\mathbb{R})$ due to $s_f(\cdot)$.*

The constant i in the theorem could have been easily removed by trivially redefining $\ell_0^+(t)$ to be $1/\sqrt{\pi}(t + i)i$. The proof consists of establishing the equality

$$\int_0^\infty e^{-u}\{s_f - \tau_u(s_f)\}\, du = \frac{1}{i}(f\ell_0^+)^\sim = \frac{1}{i}\int_{\mathbb{R}} f(t)\ell_0^+(t)\bar{\eta}_{dt},$$

in which the first integral is Bochner in $L_2(\mathbb{R})$, while the second has a scalar integrand but a $L_2(\mathbb{R})$-valued c.a.o.s. measure. The crucial step is to change the order of iterated integrations of these two types, which result from the expansion of the LHS by means of (C.18′), cf. {42: 5.20}, and then to notice that $\ell_0^+(\cdot)i/\sqrt{2}$ is the indirect FP transform of $e^{-(\cdot)}\chi_{[0,\infty)}(\cdot)$.

Notwithstanding the elegant relationship revealed by Wiener between his $s_f(\cdot)$ and the auto-covariance and spectral distribution functions of f in

---

† The completeness has been proved very recently by Mr. R. Nelson, cf. his forthcoming paper {53′}.

$\mathscr{S}_c$, it would be best at this point in history to discard $s_f(\cdot)$ in favour of our $\alpha_f(\cdot)$. For this $\alpha_f(\cdot)$ has not only the advantage of bearing the same elegant relationships as $s_f(\cdot)$, but of having a very simple definition which, moreover, grants it membership in the Hilbert space $L_2(\mathbb{R})$. By contrast, the definition of $s_f(\cdot)$ has arbitrary features such as the termini ± 1 of the integrals, that though harmless are nonetheless artificial as Wiener himself noted [30a: pp. 151–152]. Finally, the choice of $\alpha_f(\cdot)$ results in much simpler proofs than Wiener's, witness our easy deduction of all the major results from general theorems concerning stationarity in Hilbert spaces (§B). These abstract proofs reveal in fact the enormous scope and generalizability of Wiener's ideas.

Wiener was aware that by omitting the second integral from his definition of $s_f(\cdot)$ he could have reduced some of its artificiality and bestowed it with membership in $L_2(\mathbb{R})$ without affecting its Q-norm and therefore its potency as the GFT of f. But with this omission† the clear nexus between $s_f(\cdot)$ and our GFT $\alpha_f(\cdot)$ via helix-theory (Thm. C.23) would be broken. It is a sign of Wiener's far-reaching wisdom that in work completed before the advent of either the theories of Hilbert spaces or helices he decided to retain the term despite its artificial appearance.

By establishing the prototype of Wiener's major results using essentially Hilbert space theoretic methods in the context of *conditionally sesquilinear* inner products, and by proving the Thm. C.23 linking these concepts with Wiener's own, we have fulfilled to a considerable extent the "assimilation" to which Professors Hewitt and Ross referred in the statement we quoted at the outset.

There is, however, a strange result mentioned in [30a] but left out from Wiener's later writings which we have been unable to assimilate, viz. the inversion formula:

$$f(t) = \int_{\mathbb{R}} e^{-it\lambda}\, ds_f(\lambda) \qquad [30a: (6.01)].$$

As Wiener explained, the formula is absurd if the integral is interpreted literally; it has to be interpreted as the Cesaro limit of a function obtained after a formal integration-by-parts, cf. [30a: (6.05)]. The equation bears resemblance to the first one in (A.32) for $f \in \mathscr{S}_c$. But we have not so far been able to link the GFT $s_f(\cdot)$, or even our own $\alpha_f(\cdot)$, with the spectral vector

† This omission is made in the treatize {29} on communication engineering by Professor Y. W. Lee.

measure $\xi_x(\cdot)$, taking values in the Hilbert space $\mathfrak{M}_f \underset{d}{=} \mathfrak{S}\{\tau_t f : t \in \mathbb{R}\} \subset \mathscr{X}$, which emerges quite naturally in §A.

On the other hand we now know more about one aspect of the GFT $s_f(\cdot)$ than Wiener did. The reviewer had surmised that the linear operator $T: f \to s_f(\cdot)$ on $L_{2,c}(\mathbb{R})$, which Wiener had shown to be an isometry on the corr-linear subspace $\mathscr{P}$, with respect to the norm $\| \ \|$ of f and the Q-norm $\| \ \|_Q$ of $s_f(\cdot)$, would retain this isometric property on the Banach space $\mathfrak{S}(\mathscr{P})$ spanned by $\mathscr{P}$. Dr. J. K. Lee {28: Ch. 4, §D} produced an f deep inside $\mathfrak{S}(\mathscr{P})$, viz. in $\mathfrak{S}(\mathscr{S}_c)$, disproving this. At the same time he was able to prove the following important results {28, Thm. 4.30}:

(a) *T is one-one on the Banach space $\mathfrak{M}_2(\mathbb{R})$, i.e. $\bar{\mathscr{X}}$, onto the Banach space $Q_2(\mathbb{R})$;*

(b) $T(\mathscr{P}) = \tilde{\mathscr{P}}$;

(c) *T and T^{-1} are continuous,* $|T| < 1.05$ & $|T^{-1}| < 1/0.67$.

The estimates in (c) are approximations to more profound estimates involving the roots of some transcendental equations, which are derived from a fine structure analysis of the kernel $k(\cdot)$ where $k(x) \underset{d}{=} (\sin^2 x)/x^2$, $x \in \mathbb{R}$. The exact values of $|T|$ and $|T^{-1}|$ are unknown. The writer would surmise that $|T| = 1$, i.e. T is a norm 1 contraction.†

D. GHA Turbulence and Monte Carlo

We shall now report on the demarcation of some interesting subclasses of the Wiener class $\mathscr{S}_c$ (A.1), which has ensued from the attempts made since the late fifties to utilize the ideas of GHA in the study of turbulence, in the belief that this would yield better results than those that had been obtained from the stochastic treatment. This work has brought to light interconnections between GHA and the 1916 work of H. Weyl on uniform distributions {60}, and with the Monte Carlo method. Its continuation has resulted in fruitful ideas such as that of a covariance functional applicable even to functions in $\bar{\mathscr{X}} \setminus \mathscr{S}_c$ and of asymptotic measure.

The turbulent movement of a particle in a fluid exhibits persistent local irregularities as well as a stable mean asymptotic behaviour, somewhat like the chaotic signals of white light which led Wiener to his GHA. But, as will be indicated in §E below, the quick appearance of the Birkhoff ergodic theorem and the Kolmogorov theory of SP's after the publication of [30a] created an intellectual climate favoring stochastic analysis rather than GHA. Thus most researchers, including Wiener, came to regard the turbulent movements of particles as the sample functions of an ergodic stationary

† This surmise has been disproved by Dr. K. -S. Lau in his joint paper {25″} with Dr. J. K. Lee.

random field. But this treatment has met with only partial success. Let $\mathbf{u} = \mathbf{u}(t, \mathbf{x})$ be the velocity vector field of a turbulent fluid. According to J. Bass {2: pp. 5, 6}, one of the difficulties of the stochastic treatment is that while the ergodic theorem gives a reasonable accounting for the mean asymptotic behavour of the function $\mathbf{u}(\cdot, \mathbf{x})$ on the time-domain $\mathbb{R}$, it fails to explain the highly irregular oscillatory behavior of $\mathbf{u}(\cdot, \mathbf{x})$ in-the-small, and especially to explain why, after perhaps some small-scale averaging, $\mathbf{u}(\cdot, \mathbf{x})$ should satisfy the non-linear Navier-Stokes equations. Bass felt that a more fruitful treatment might result from a nonstochastic approach along the lines of GHA by starting from certain functions in $\mathscr{S}_c$. He proposed such a theory in the late fifties, which we proceed to describe.

Unfortunately Bass does not clearly state that his functions belong to the class $\mathscr{S}_c$ which Wiener studied extensively in [30a], and this omission might leave the impression that such functions were found during the fifties. Actually they were discussed by Wiener in the mid-twenties, but with regard to white light and the schrot-effect rather than turbulence, cf. e.g. [26b]. Wiener's reason for staying away from turbulence during those early years is explained in [56g: pp. 36, 37]. It may therefore help in understanding Professor Bass's point of departure to recall briefly the factors that led Wiener towards GHA in the mid-twenties.

Optical origins of GHA. Newton's experiments on dispersion revealed the polychromatic nature of white light, and the question arose as to the kind of harmonic analysis that one could make of a white light signal $f(\cdot)$ on $\mathbb{R}$ at a fixed point of an optical medium. As indicated in our review {40: §1}, one could not regard f to be in $L_2(\mathbb{R})$, and this ruled out representation by the FP transform. Nor could one assume that f had a sharp line spectrum, and this ruled out representation by a generalized trigonometric polynomial or series. In short, our problem is to represent a function f, which is on the one hand so irregular that its spectrum is not made up of sharp lines alone, and on the other so lastingly vigorous that $\int_t^{t+h} |f(t)|^2\,dt \not\to 0$, as $t \to \infty$. Wiener, benefiting from the ideationally deep but mathematically nonrigorous deliberations of the physicists Rayleigh, Schuster and Taylor on the periodogram, was led by this problem to his GHA.

Around 1958 J. Bass {2} asked the same kind of question for a turbulent velocity-component function $u(\cdot, \mathbf{x})$ on $\mathbb{R}$. He too realized the inappropriateness of representing $u(\cdot, \mathbf{x})$ by either an FP integral or a generalized trigonometric series. But guided by H. Weyl's work {60} on uniformly distributed sequences, he was led to ask a rather specific question: *will the representation of $u(\cdot, x)$ be possible by means of series or integrals build from functions $e^{i\psi(\cdot)}$*

with certain $\mathbb{R}$-valued $\psi(\cdot)$ other than $\psi(t) = \lambda t$? In short, his idea was to see if the replacement of the $e^{i\lambda t}$ of classical harmonic analysis by $e^{i\psi(t)}$ for suitable real-valued $\psi(\cdot)$ would do the job. The answer turned out to be negative, but it suggested to Bass a wider and more appropriate class of functions.

The simplest choice of $\psi(\cdot)$ is of course a polynomial $P(\cdot)$ with real coefficients and degree exceeding 1. (For deg. $P = 1$ just gives us the rejected case $e^{i\lambda t}$.) But Bass showed that when deg. $P \geqslant 2$ and $f \underset{\mathrm{d}}{=} e^{iP(\cdot)}$ on $\mathbb{R}$, we have $\phi_{ff} = \chi_{\{0\}}$, i.e. $f \in \mathscr{S} \setminus \mathscr{S}_c$, cf. (A.1). The spectra of such functions f just have a sharp line at the frequency $\lambda = \infty$, and obviously they are inappropriate for representing the velocity component $u(\cdot, \mathbf{x})$.

Guided by a result of Van der Corput (cf. Donoghue {13: p. 199}) Bass next tried the step-function:

(D.1) $$f(\cdot) \underset{\mathrm{d}}{=} e^{iP([\cdot])} = \sum_{k\in\mathbb{N}} e^{iP(k)} \chi_{[k,k+1)}(\cdot),$$

where $[t] = \max \mathbb{N} \cap (-\infty, t]$, $t \in \mathbb{R}$.† For this f he found the following suggestive result:

D.2 Lma. *Let (i) $f(\cdot) \underset{\mathrm{d}}{=} e^{iP([\cdot])}$ on $\mathbb{R}$, where $P(\cdot)$ is a polynomial with real coefficients and deg. $P \geqslant 2$, (ii) $\forall n \in \mathbb{N}_+$, $\tau_n(f)$* corr *f. Then*

(a)‡ *$f \in \mathscr{S}_c$, and ϕ_{ff} has a polygonal graph with vertices at $(n, \phi_{ff}(n))$, $n \in \mathbb{N}$;*

(b) *when the coefficients of $P(\cdot) - P(0)$ are rational, f and ϕ_{ff} are periodic with period equal to the l.c.m. of the denominators of these coefficients in lowest terms;*

(c) *when a coefficient of $P(\cdot) - P(0)$ is irrational, we have*
$$f(\cdot) = e^{iP_1([\cdot])} e^{iP_2([\cdot])},$$
where $P_1(\cdot)$ is a polynomial such that $P_1(\cdot) - P_1(0)$ has rational coefficients, and $P_2(\cdot)$ is a polynomial with leading coefficient irrational.

The case (b) in which f is periodic is obviously of no interest in turbulence theory, and the same is true of the factor $e^{iP_1([\cdot])}$ in case (c). One is thus led to consider the pertinent case in which $P(\cdot)$ has an irrational leading coefficient. For this case Bass proved the following important theorem:

D.3 Thm. *Let $f(\cdot) = e^{iP([\cdot])}$ on $\mathbb{R}$, where $P(\cdot)$ is a polynomial with real coefficients, the leading one irrational, and deg. $P \geqslant 2$. Then $f \in \mathscr{S}_c$, and*

---

†The transition from $e^{ip(\cdot)}$ to $e^{iP([\cdot])}$ is somewhat like the quantization which leads from the harmonic oscillator of classical statistical mechanics to the harmonic oscillator of classical statistical mechanics to the harmonic oscillator of Planck's quantum theory.

‡The result that a step-function $f = \Sigma_{k=1}^{\infty} a_k \chi_{[k,k+1]}$, where $a_k = \pm 1$, for which $\tau_n(f)$ corr f, $\forall n \in \mathbb{N}$, belongs to $\mathscr{S}_c$ and $\phi_{ff}(\cdot)$ is as in D.3 (below) was known to Wiener, cf. [30a: (11.03)–(11.07)]. But Wiener does not seem to have ascribed any special significance to this result: to him its value lay in suggesting a method for constructing f having absolutely continuous spectral distributions, cf. [30a: §§11–13].

$$\forall t \in \mathbb{R}, \quad \phi_{ff}(t) \underset{\mathrm{d}}{=} (1 - |t|)\chi_{[-1,1]}(t).$$

The step-functions $f(\cdot)$ considered in Thm. D.3 are quite unlike those considered in D.2(b) and seem to have the persistency and high irregularity we are seeking. But the rather restricted "triangular" covariance function ϕ_{ff} they all have in common indicates that even they lack the degree of irregularity needed to represent $u(\cdot, \mathbf{x})$. They suggest, however, the requirement that $\phi_{ff}(t) \to 0$, as $t \to \pm\infty$, and the demarcation of the following proper subclass of $\mathscr{S}_c$:

(D.4) $$\mathscr{S}_{c0} \underset{\mathrm{d}}{=} \{f : f \in \mathscr{S}_c \ \& \lim_{t \to \infty} \phi_{ff}(t) = 0\}.$$

Bass calls functions of the form $f(\cdot) + c$, where $f \in \mathscr{S}_{c0}$ and $c \in \mathbb{C}$, *pseudo-aleatoires* or *pseudo-random* (PR); those in $\mathscr{S}_{c0}$ are called *centered PR functions*, cf. {3: p. 368}.†

Before we can conveniently deal with the properties of PR functions, we must mention the class of Besicovitch almost periodic functions, and recall some results of Wiener on Fourier-Stieltjes transforms [33i: Thm. 32], as well as a result of Van der Corput. Logically speaking, these considerations belong to our §A.

The closed linear subspace of the Marcinkiewicz space $\bar{\mathscr{X}}$, (A.2), spanned by the characters $e_\lambda(\cdot)$ of $\mathbb{R}$, $e_\lambda(t) \underset{\mathrm{d}}{=} e^{i\lambda t}$, is easily seen to be the class of AP functions, with $p = 2$, which Besicovitch introduced, cf. {7: pp. 78, 109–112}. Thus we may adopt the definition:

(D.5) $$\begin{cases} \hat{\mathbb{R}} \underset{\mathrm{d}}{=} \{e_\lambda : \lambda \in \mathbb{R}\}, \textit{the set of characters of } \mathbb{R}. \\ B_2\mathrm{AP} \underset{\mathrm{d}}{=} \mathfrak{S}(\hat{\mathbb{R}}) \subset \bar{\mathscr{X}}. \end{cases}$$

Next, adopting the notation

(D.6) $$\forall A \subseteq \bar{\mathscr{X}}, \quad A^\perp \underset{\mathrm{d}}{=} \{y : y \in A^{\mathrm{corr}} \ \& \ \forall x \in A, ((y, x)) = 0\},$$

we have

(D.7) $$\forall A \subseteq \bar{\mathscr{X}}, \quad A^\perp \textit{ is a (closed linear) subspace of } A^{\mathrm{corr}}.$$

It follows readily from A.23(c) that

---

†The reason for the qualification "centered" for f in $\mathscr{S}_{c0}$ is apparent from the result D.12(a).

$$\text{(D.8)} \qquad \begin{cases} B_2\text{AP } \textit{is a (non-separable) Hilbert subspace of } \mathscr{S}_c, \\ \& \qquad (B_2\text{AP})^{\text{corr}} = B_2\text{AP} + (B_2\text{AP})^{\perp}. \end{cases}$$

In this context Wiener's Thm. on FS transforms [33i: Thm. 32] may be rendered in the following widened form:

D.9 Wiener's Thm. *Let* (i) μ *be a bounded, complex-valued c.a. measure on* $Bl(\mathbb{R})$, (ii) $\tilde{\mu}$ *be the FS transform of* μ: $\tilde{\mu}(t) \underset{\mathrm{d}}{=} \int_{\mathbb{R}} e^{it\lambda}\mu(d\lambda)$, $t \in \mathbb{R}$. *Then*

(a) $\qquad \tilde{\mu} \in (B_2\text{AP})^{\text{corr}} \quad \& \quad \forall \lambda \in \mathbb{R}, \quad ((\tilde{\mu}, e_\lambda)) = \mu\{\lambda\};$

(b) $\qquad \tilde{\mu} \in \mathscr{S}_c \quad \& \quad \forall t \in \mathbb{R}, \quad \phi_{\tilde{\mu}\tilde{\mu}}(t) = \sum_{\lambda \in \mathbb{R}} \mu\{\lambda + t\}\bar{\mu}\{\lambda\},$

in particular,

$$\lim_{T\to\infty} \frac{1}{2T} \int_{-T}^{T} |\tilde{\mu}(t)|^2\, dt = \phi_{\tilde{\mu}\tilde{\mu}}(0) = \sum_{\lambda \in \mathbb{R}} |\mu\{\lambda\}|^2.$$

Since $\forall f \in \mathscr{S}_c$, $\phi_{ff}(t) = \tilde{\mu}_{ff}(-t)$, cf. (A.28) and (A.31), it follows from D.9(a) with $\lambda = 0$ that

$$\text{(D.10)} \qquad \forall f \in \mathscr{S}_c, \quad \lim_{T\to\infty} \frac{1}{2T} \int_{-T}^{T} \phi_{ff}(t)\, dt \quad \textit{exists} \quad \& \quad = \mu_{ff}\{0\}.$$

This brings us to the following result of Van der Corput (cf. {13: p. 194} for a proof):

$$\text{(D.11)} \qquad \forall f \in \mathscr{S}_c, \quad \overline{\lim_{T\to\infty}} \left| \frac{1}{2T} \int_{-T}^{T} f(t)\, dt \right|^2 \leqslant \lim_{T\to\infty} \frac{1}{2T} \int_{-T}^{T} \phi_{ff}(t)\, dt.$$

With the aid of these results one can establish the following properties of PR functions, cf. Bass {3: §11}.

D.12 Prop.

(a) $\mathscr{S}_{c0} \subseteq (B_2\text{AP})^{\perp}$, *i.e.* $\forall f \in \mathscr{S}_{c0}$ & $\forall \lambda \in \mathbb{R}$,

$$f \text{ corr } e_\lambda \quad \& \quad ((f, e_\lambda)) = 0,$$

in particular, an f *in* $\mathscr{S}_{c0}$ *has zero mean value:*

$$\lim_{T\to\infty} \frac{1}{2T} \int_{-T}^{T} f(t)\, dt = ((f, 1)) = 0;$$

(b) $\forall f \in \mathscr{S}_{c0}$, $\mu_{ff}(\cdot)$ *has no atoms.*

A rather superficial converse of (b) is provided by the triviality:

$$\text{(D.13)} \qquad f \in \mathscr{S}_c \quad \& \quad \mu_{ff} \ll \textit{Leb} \Rightarrow f \in \mathscr{S}_{c0},$$

an immediate consequence of the Riemann-Lebesgue Lemma. Since singular measures do not reveal themselves in physics, we may assume in the physical applications that for our centered PR function f, $\mu_{ff} \ll$ Leb.

PR functions that are real-valued and supported on $\mathbb{R}_{0+}$ have the following additional properties:

D.14 Cor. *Let* $f \in \mathscr{S}_{c0}$, *range* $f \subseteq \mathbb{R}$ *and supp* $f \subseteq \mathbb{R}_{0+}$. *Then*

(a) *either* f *is unbounded in every neighborhood of* ∞, *or* f *changes sign in every neighborhood of* ∞;

(b) $f(t)$ *oscillates* (*i.e. has no limit in* $[-\infty, \infty]$) *as* $t \to \infty$.

The properties listed in D.12 and in D.14 are what one would expect a turbulent velocity component to have, in addition of course to obvious ones such as smoothness. We are therefore tempted to advance the following as a tentative definition of "turbulent function":

D.15 Def. A $\mathbb{R}^3$-vector-valued function $\mathbf{u}(\cdot, \cdot)$ on $\mathbb{R}_{0+} x \mathbb{R}^3$ is called a *turbulent function*, iff

(i) its components and their first-order partial derivatives are continuous on $\mathbb{R}_{0+} x \mathbb{R}^3$, and $\mathbf{u}(\cdot, \cdot)$ satisfies the Navier-Stokes equations.

(ii) $\forall \mathbf{x} \in \mathbb{R}^3$, the three components of $\mathbf{u}(\cdot, \mathbf{x})$ are PR, i.e. of the form $c + f(\cdot)$, where $c \in \mathbb{R}$, $f \in \mathscr{S}_{c0}$, and moreover $\mu_{ff} \ll$ Leb.

The assumption of the absolute continuity of μ_{ff} in D.15(ii) is harmless. For we know from D.12(b) that μ_{ff} has no atoms, and from experience that singular distributions are not detectable in physics.

We are now confronted with the problem of generating functions in $\mathscr{S}_{c0}$ that, in the contrast to the step-functions obtained in D.3, meet the continuity and differentiability requirements of Def. D.15. For this purpose the following result of Bass on convolutions is very useful {3: §13}:

D.16 Thm. *Let* $K \in L_1(\mathbb{R})$, $f \in \mathscr{S}_{c0}$ *and* $\mu_{ff} \ll$ Leb. *Then*

(a) $g \underset{\mathrm{d}}{=} K * f \in \mathscr{S}_{c0}$,

(b) *if* K *is continuous, so is* g,

(c) *if* K *is differentiable, then so is* g, *and* $g' = K' * f$.

It is still too early to pronounce judgement on a theory of turbulence based on Def. D.15, cf. Bass {4: §§2–5}. But certain problems it poses deserve mention.

In the first place, in the stochastic approach different sample paths of the same ergodic SP $(f(t, \cdot): t \in \mathbb{R})$ on $(\Omega, \mathscr{B}, \mathscr{P})$ are always in $\mathscr{S}_c$ and are τ-correlated. Since, cf. (A.19), $\mathscr{S}_c$ is τcorr-linear, it follows by definition that their sum is in $\mathscr{S}_c$. The Superposition Principle thus prevails. But in the non-stochastic treatment we cannot conclude that if two such paths are in $\mathscr{S}_{c0}$, then their sum is in $\mathscr{S}_{c0}$. This is because unlike $\mathscr{S}_c$, $\mathscr{S}_{c0}$ is not τcorr-linear. This breach in the Principle of Superposition is avoided, however, by the absolute continuity requirement in D.15(ii), since, cf. {3: §14},

(D.17) $$f, g \in \mathscr{S}_c,\ f \tau\text{corr}\, g \ \&\ \mu_{ff}, \mu_{gg} \ll Leb. \Rightarrow f + g \in \mathscr{S}_{c0}.$$

The absolute continuity requirement thus seems to be indispensable.†

Secondly, it is not enough in a treatment of turbulence based on a definition such as D.15 to consider just the covariance function ϕ_{ff} and the spectral distribution μ_{ff} of the turbulent velocity component $f(\cdot)$ as Bass does, and ignore the GET $s_f(\cdot)$ or $\alpha_f(\cdot)$ of $f(\cdot)$. For as Wiener has stressed, cf., e.g. [49g: p. 6], the functions $\phi_{ff}(\cdot)$ and $\mu_{ff}(\cdot)$ throw away all information regarding the phase relationships of the frequency components of $f(\cdot)$, preserving only information as to their amplitudes. It would seem that the information of the phase relationships contained in the GFT would be germane to a study of local irregularities—a study unavoidable in a subject based on differential equations such as turbulence.

We shall now leave aside the difficult subject of turbulence and turn to an interesting outcome of Professor Bass's researches: the revelation of contacts between Wiener's GHA and H. Weyl's theory of uniformly distributed sequences (mod 1), {60}. Adopt the following notation and definition:

(D.18) $$\begin{cases} \forall t \in \mathbb{R}, \quad (t) \underset{\text{d}}{=} \textit{the fractional part of } t \\ \forall \mathbf{t} = (t_1, \ldots, t_q) \in \mathbb{R}^q, \quad (\mathbf{t}) \underset{\text{d}}{=} ((t_1), \ldots, (t_q)) \\ \forall \mathbf{s} = (s_i)_1^q \quad \& \quad \forall \mathbf{t} = (t_i)_1^q \in \mathbb{R}^q, \quad (\mathbf{s}, \mathbf{t}) \underset{\text{d}}{=} \sum_1^q s_i t_i. \end{cases}$$

D.19 Def. (a) A sequence $(t_n)_{n=1}^\infty$ in $[0, 1]^q$ is called *uniformly distributed*, iff $\forall$ subintervals J of $[0, 1]^q$,

$$\lim_{n\to\infty} \frac{1}{n} \sum_1^n \chi_J(\mathbf{t}_k) = \ell_q(J),$$

where ℓ_q is q-dimensional Lebesgue measure.

(b) A sequence $(\mathbf{t}_k)_{k=1}^\infty$ in $\mathbb{R}^q$ is called *uniformly distributed* (*mod* 1), iff the seq. $((\mathbf{t}_k))_{k=1}^\infty$ in $[0, 1]^q$ is uniformly distributed.

The following is Weyl's Thm. {60}.

D.20 Thm. *Let* $(\mathbf{t}_k)_1^\infty$ *be a sequence in* $[0, 1]^q$. *Then the following conditions are equivalent:*

(α) $(\mathbf{t}_k)_1^\infty$ *is uniformly distributed,*

(β) $\forall$ *Riemann-integrable* f *on* $[0, 1]^q$ *to* $\mathbb{C}$,

$$\lim_{n\to\infty} \frac{1}{n} \sum_1^n f(\mathbf{t}_k) = \int_{[0,1]^q} f(\mathbf{t}) \ell_q(d\mathbf{t}),$$

---

† What Professor Bass gains by not making it a part of the definition of a PR function is not clear to us.

$$(\gamma) \qquad \forall \mathbf{m} \in \mathbb{N}^q, \quad \lim_{n\to\infty} \frac{1}{n} \sum_1^n e^{2\pi i(\mathbf{m}, \mathbf{t}_k)} = \delta_{\mathbf{m}0}.$$

This theorem is deducible from its $q = 1$ version by appealing to the result that a necessary and sufficient condition for a sequence $(\mathbf{t}_n)$ in $\mathbb{R}^q$ to be uniformly distributed (mod 1) is that $\forall \mathbf{m} \in \mathbb{N}^q \setminus \{\mathbf{0}\}$, the sequence $((\mathbf{m}, \mathbf{t}_n))_{n=1}^{\infty}$ in $\mathbb{R}$ be uniformly distributed (mod 1). In order to state the connection between uniform distributions and Wienerian ideas we need one more definition:

D.21 Def. Let $(t_n)_1^{\infty}$ be a sequence in $\mathbb{R}$. Then

(a) $(t_n)_1^{\infty}$ is called *q-uniformly distributed* (*mod* 1), where $q \in \mathbb{N}_+$, iff the sequence $(\mathbf{t}_n)_{n=1}^{\infty}$ in $\mathbb{R}^q$, where

$$\mathbf{t}_n \underset{\mathrm{d}}{=} (t_n, t_{n+1}, \ldots, t_{n+q-1}),$$

is uniformly distributed (mod 1);

(b) $(t_n)_1^{\infty}$ is called *completely uniformly distributed* (*mod* 1), iff $\forall q \in \mathbb{N}_+$, $(t_n)_1^{\infty}$ is q-uniformly distributed (mod 1).

The nexus with GHA is then given by the following theorem of Bass {3: §12}, the proof of which hinges on Weyl's Thm. D.20:

D.22 Thm. *Let* (i) *the sequence* $(t_k)_1^{\infty}$ *in* $\mathbb{R}$ *to be* 2-*uniformly distributed* (*mod* 1), (ii) $\psi(\cdot)$ *be a Riemann-integrable function on* $[0, 1]$ *to* $\mathbb{C}$, (iii)

$$f(\cdot) \underset{\mathrm{d}}{=} \sum_{n=1}^{\infty} \psi((t_n))\chi_{[n-1,n)}(\cdot) - \left(\int_0^1 \psi(t)\,dt\right)\chi_{[0,\infty)}(\cdot) \text{ on } \mathbb{R}.$$

Then $f \in \mathscr{S}_{c0}$.

We do not have the space to elaborate on the obvious significance of D.20(β) for the Monte Carlo method, cf. Bass {3: §§2, 3}. For such applications it is useful to have at one's disposal a large collection of uniformly distributed sequences. They are provided by number-theoretic results such as the following:

D.23 Thm. (a) *Let* $q \in \mathbb{N}_+$, $t_1, \ldots, t_q$ *be irrational numbers, linearly independent over* $\mathbb{N}$, *and* $\mathbf{t} \underset{\mathrm{d}}{=} (t_1, \ldots, t_q)$. *Then the sequence* $(n\mathbf{t})_{n=1}^{\infty}$ *in* $\mathbb{R}^q$ *is uniformly distributed* (*mod* 1). *In particular,* $\forall$ *irrational numbers t, the sequence* $(nt)_{n=1}^{\infty}$ *in* $\mathbb{R}$ *is uniformly distributed* (*mod* 1).

(b) *Let t be transcendental, and* $\mathbf{t} \underset{\mathrm{d}}{=} (t, t^2, \ldots, t^q)$, *where* $q \in \mathbb{N}_+$. *Then the sequence* $(n\mathbf{t})_{n=1}^{\infty}$ *in* $\mathbb{R}^q$ *is uniformly distributed* (*mod* 1).

(c) *Let* $P(\cdot)$ *be a polynomial of degree r with real coefficients and leading coefficient irrational. Then* $\forall q \in \{1, 2, \ldots, r\}$ *the sequence* $(P(n))_{n=1}^{\infty}$ *in* $\mathbb{R}$ *is q-uniformly distributed.*

(d) *For almost all transcendental* $t > 1$, $(t^n)_{n=1}^{\infty}$ *is completely uniformly distributed* (*mod* 1).

D.23(a) is a fairly straightforward result, of which D.23(b) is an obvious consequence. For a proof of D.23(c) see Bass {3: pp. 360, 361}. The deeper result D.23(d) is due to J. N. Franklin {17}. For a specific method of constructing completely uniformly distributed sequences see {3: App.II}. Thms. D.22 and D.23(c) and (d) reveal an interesting link between GHA and number theory that calls for careful further study.

The ideas in [30a] received fresh impetus when in continuing the work of Bass, Dr. J. P. Bertrandias introduced the *covariance functional over any separable Banach subspace* $\mathscr{Y}$ *of* $\bar{\mathscr{X}}$ *due to any element* $g \in \bar{\mathscr{X}}$, and a certain fixed *associate* $\psi_{\mathscr{Y}}$ *of* $\mathscr{Y}$, and with their aid characterized some interesting Banach subspaces of almost periodic functions {6}. He dealt with the Marcinkiewicz space $\mathfrak{M}_p(\mathbb{R})$, $p \in [1, \infty)$. We shall discuss the $p = 2$ versions of just some of his results. Bertrandias's primary goal was to find a sufficiently small Banach space $\bar{\mathscr{S}}_{c0}$ containing the non-linear space $\mathscr{S}_{c0}$ of centered PR functions, cf. (D.4) and D.39(b) below, whose members share some of their properties. For reasons unknown to us he did not consider the smallest such space, viz. $\mathfrak{S}(\mathscr{S}_{c0})$. Instead, guided by the work of Bochner and Eberlein {9, 16} on AP functions, he brought to bear the concepts of *compactness* and the *weak topology* for the Marcinkiewicz Banach space $\bar{\mathscr{X}}$. Thereby he demarcated several Banach spaces of AP functions all within $\bar{\mathscr{X}}$, and characterized them in terms of his covariance functional. He defined $\bar{\mathscr{S}}_{c0}$ as a subspace of one of these Banach spaces.

The covariance functional of Bertrandias has its origin in the following basic result {6: p. 22}:

D.24 Lma. *Let* $\mathscr{Y}$ *be a separable Banach subspace of* $\bar{\mathscr{X}}$ *and* $g \in \bar{\mathscr{X}}$. *Then*

(a) $\exists$ *a strictly increasing sequence* $(n_k)_1^\infty$ *in* $\mathbb{N}_+ \ni$

$$\forall f \in \mathscr{Y}, \quad \ell_f \underset{\mathrm{d}}{=} \lim_{k \to \infty} ((f, g))_{n_k} \quad \textit{exists}, \quad \textit{cf.}\ (\mathrm{A}.22');$$

(b) *letting* $L_g = \{(f\,; \ell_f): f \in \mathscr{Y}\ \&\ \ell_f$ *is as in* (a)$\}$, *we have* $L_g \in \mathscr{Y}'$, *the dual space of* $\mathscr{Y}$, *and* $|L_g| \leqslant \|g\|$.

D.25 Def. Let $\mathscr{Y}$, g, L_g be as in the last Lma. Then

(a) L_g is called the *covariance functional on* $\mathscr{Y}$ *due to* g;

(b) $\mathscr{L}_{g,\mathscr{Y}} \underset{\mathrm{d}}{=} \{L_g : L_g$ is a covariance functional on $\mathscr{Y}$ due to $g\}$;

(c) $\mathscr{L}_{\mathscr{Y}} \underset{\mathrm{d}}{=} \bigcup \{\mathscr{L}_{g,\mathscr{Y}} : g \in \bar{\mathscr{X}}\ \&\ \|g\| \leqslant 1\}$.

It is clear that

$$(\mathrm{D}.26) \qquad \begin{cases} g \in \mathscr{Y}^{\mathrm{corr}}, \ \textit{iff}\ \ \mathscr{L}_{g,\mathscr{Y}}\ \textit{is a singleton}, \\ \& \\ \forall g \in \mathscr{Y}^{\mathrm{corr}}, \quad \mathscr{L}_{g,\mathscr{Y}} = \{\textit{Rstr.}_{\mathscr{Y}}((\cdot, g))\}, \quad \textit{cf.}\ (\mathrm{A}.4). \end{cases}$$

Thus for $g \in \mathscr{Y}^{\text{corr}}$ D.24 tells us nothing new. But for $g \notin \mathscr{Y}^{\text{corr}}$ it tells us that there still are collections of covariance functionals. These collections have moreover the interesting properties described in the following theorem, {6: pp. 23, 24, 34}.

D.27 Thm. *Let $\mathscr{Y}$ be a separable Banach subspace of $\bar{\mathscr{X}}$. Then*

(a) $\forall g \in \bar{\mathscr{X}}$, $\mathscr{L}_{g,\mathscr{Y}}$ *is a weak* compact subset of the closed ball* $\bar{N}(0, \|g\|)$ *of the dual space* $\mathscr{Y}'$ *with center* 0 *and radius* $\|g\|$;

(b) $\mathscr{L}_{\mathscr{Y}}$ *is a weak* compact subset of the closed unit ball* $\bar{N}(0, 1)$ *of* $\mathscr{Y}'$;

(c) $\mathscr{L}_{\mathscr{Y}} \supseteq$ *Extrm* $\bar{N}(0, 1)$, *where the last is the set of extreme points of* $\bar{N}(0, 1)$.

Another contribution of Bertrandias is his construction of a single function $\psi_{\mathscr{Y}}$ for any given separable Banach subspace $\mathscr{Y}$ of $\bar{\mathscr{X}}$, such that $\|\psi_{\mathscr{Y}}\| = 1$ and

$$\text{(D.28)} \qquad \forall f \in \mathscr{Y}, \quad \|f\| = \overline{\lim_{T \to \infty}} \, |((f, \psi_{\mathscr{Y}}))_T| = \overline{\lim_{T \to \infty}} \, real((f, \psi_{\mathscr{Y}}))_T,$$

cf. (A.22′). This $\psi_{\mathscr{Y}}$, called *the associate of* $\mathscr{Y}$, is defined stepwise by

$$\text{(D.29)} \qquad \psi_{\mathscr{Y}}(\cdot) = \sum_{k=1}^{\infty} \frac{f_{j(k)}(\cdot)}{\|f_{j(k)}(\cdot)\|} \chi_{-J_k \cup J_k}(\cdot) \quad on\ \mathbb{R},$$

where $(f_n)_1^{\infty}$ is a suitable everywhere-dense sequence in $\mathscr{Y}$, $j(\cdot)$ is a specific function on $\mathbb{N}_+$ to $\mathbb{N}_+$ and the J_k are certain subintervals of $\mathbb{R}_+$, the definitions of which are too technical to mention here, cf. {6: pp. 30, 31}.

The ideas in D.25 and in (D.29), (D.26) come to fruition when brought to bear on the space spanned by the translates of f:

$$\text{(D.30)} \qquad \forall f \in \bar{\mathscr{X}}, \quad \mathscr{Y}_f \underset{\text{d}}{=} \mathfrak{S}\{\tau_t(f) : t \in \mathbb{R}\},$$

{6: Ch. III}. $\mathscr{Y}_f$ is of course separable Banach subspace of $\bar{\mathscr{X}}$. Now with the aid of any covariance functional L_g on $\mathscr{Y}_f$, we can define a function $\Phi_{f,L_g}(\cdot)$ on $\mathbb{R}$ to $\mathbb{C}$ by:

$$\text{(D.31)} \qquad \forall f, g \in \bar{\mathscr{X}}, \forall L_g \in \mathscr{L}_{g,\mathscr{Y}_f} \quad \& \quad \forall t \in \mathbb{R}, \quad \Phi_{f,L_g}(t) \underset{\text{d}}{=} L_g\{\tau_t(f)\}.$$

Φ_{f,L_g} is called *a (cross-) covariance of* f *and* g, {6: p. 36}.

Let $f \in \bar{\mathscr{X}}$. When g τcorr f, it follows from (D.26) that there is exactly one such function Φ_{f,L_g}, viz. $\phi_{fg}(\cdot)$, cf. (A.4). But for other g in $\bar{\mathscr{X}}$ we have many covariances Φ_{f,L_g} and these yield two classes corresponding to the two classes defined in D.2.5(b) (c), viz.

$$\text{(D.32)} \qquad \begin{cases} \forall f, g \in \bar{\mathscr{X}}, & \Gamma_{f,g} \underset{\mathrm{d}}{=} \{\Phi_{f,L_g} : L_g \in \mathscr{L}_{g,\mathscr{Y}_f}\} \\ \forall f \in \bar{\mathscr{X}}, & \Gamma_f \underset{\mathrm{d}}{=} \cup \{\Gamma_{f,g} : g \in \bar{\mathscr{X}} \ \& \ \|g\| \leqslant 1\} \end{cases}$$

cf. {6: p. 38}.

The covariance Φ_{f,L_g}, like ϕ_{fg} when it exists, is in general discontinuous. To secure its continuity we have to take a "$\|\cdot\|$ continuous" and "regular" f, i.e. restrict f to the class $\mathscr{X}_c$ defined in

$$\text{(D.33)} \qquad \begin{cases} \bar{\mathscr{X}}_c \underset{\mathrm{d}}{=} \{f : f \in \bar{\mathscr{X}} \ \& \ \lim_{h \to 0} \tau_h(f) = f \ in \ \bar{\mathscr{X}}\}, \\ \mathscr{X}_c \underset{\mathrm{d}}{=} \mathscr{X} \cap \bar{\mathscr{X}}_c, \quad \mathscr{P}_c \underset{\mathrm{d}}{=} \mathscr{P} \cap \bar{\mathscr{X}}_c, \quad \mathscr{S}_c \underset{\mathrm{d}}{=} \mathscr{S} \cap \bar{\mathscr{X}}_c. \end{cases}$$

It is easy to show that $\bar{\mathscr{X}}_c$ and $\mathscr{X}_c$ are Banach subspaces of $\bar{\mathscr{X}}$ and of $\mathscr{X}$ respectively, cf. (A.5) and A.7(b), and that the $\mathscr{S}_c$ defined here equals the one defined by (A.1). Briefly, *$\bar{\mathscr{X}}_c$ is the Banach subspace of $\bar{\mathscr{X}}$ on which the group of isometries $(\tau_t : t \in \mathbb{R})$ is strongly continuous.* The result on the continuity of the covariance now reads, cf. {6: pp. 36, 37}:

$$\text{(D.34)} \qquad \begin{cases} \forall f \in \mathscr{X}_c, \quad \forall g \in \bar{\mathscr{X}} \ \& \ \forall L_g \in \mathscr{L}_{g,\mathscr{Y}_f}, \\ \Phi_{f,L_g}(\cdot) \ \textit{is weakly AP in Eberlein's sense.} \end{cases}$$

When for any $f \in \mathscr{X}_c$ we take $g = f$ in (D.31), we get *an auto-covariance function* $\Phi_{f,L_f}(\cdot)$. Any *such function is CPD on* $\mathbb{R}$, cf. {6: p. 81}. It is therefore the FS transform of a bounded, non-negative c.a. measure on $Bl(\mathbb{R})$, as well as the covariance function of a stationary curve in Hilbert space. Corresponding to the class Γ_{ff} of (D.32) we now have a class of bounded non-negative c.a. representing measures on $Bl(\mathbb{R})$, as well as a class of stationary curves in Hilbert space. A study of these measures and curves, i.e. a rudimentary GHA of his new concepts, is undertaken by Bertrandias {6: Ch.IX}. But as with Bass, the Tauberian considerations that Wiener applied are absent from his work. The more abstruse regions of his GHA thus remain unexplored.

To turn to the subspaces of AP functions which Bertrandias introduces, we recall a fundamental result in the theory of the uniformly almost periodic (UAP) functions of Bohr, {10}:

D.35 Thm. *The following conditions are equivalent:*

(α) *$f \in \mathfrak{S}(\hat{\mathbb{R}})$, where $\hat{\mathbb{R}} = \{e_\lambda : \lambda \in \mathbb{R}\}$ is the set of characters of $\mathbb{R}$, and where $\mathfrak{S}(\hat{\mathbb{R}})$ now denotes the (closed linear) subspace spanned by $\hat{\mathbb{R}}$ in the Banach space $C(\mathbb{R}, \mathbb{C})$ with the sup-norm;*

(β) *the orbit $\{\tau_t(f) : t \in \mathbb{R}\}$ is pre-compact in the Banach space $C(\mathbb{R}, \mathbb{C})$, cf. {9};*

(γ) $f \in UAP$.

This theorem establishes the equality of the class of *Bohr AP functions* with the class of what may be termed *Bohr-Bochner AP functions*. Bertrandias subjected the class B_2AP of Besicovitch AP functions (D.5) to analoguous consideration. For this it is necessary to define first the class, {6: p. 42},

$$\text{(D.36)} \qquad K \underset{\mathrm{d}}{=} \{f : f \in \bar{\mathscr{X}} \ \&\ \forall t \in \mathbb{R}, \tau_t(f) = f \ in\ \bar{\mathscr{X}}^{\dagger}\}$$

of so-called *pseudo-constants*, and $\forall A \subseteq \bar{\mathscr{X}}$ to define $\mathfrak{S}_K(A)$ as the closure in $\bar{\mathscr{X}}$ of the set of finite linear combinations of elements in A with *coefficients in* K (instead of $\mathbb{C}$). We then have the following interesting analogue of the last Thm. D.35

D.37 Thm. (a) *The following conditions are equivalent:*

(α) $$f \in \mathfrak{S}_K(\hat{\mathbb{R}});$$

(β) *the orbit* $\{\tau_t(f): t \in \mathbb{R}\}$ *is pre-compact in the Banach space* $\bar{\mathscr{X}}$. *Cf.* {6: *p.* 71, *Thm.III*}.

(b) B_2AP *is a proper subspace of* $\mathfrak{S}_K(\hat{\mathbb{R}})$. *Cf.* {6: *p.* 69}.

On the strength of this result we may in analogy with (D.5) define the class of "*Besicovitch-Bochner*" *AP functions* by

$$\text{(D.38)} \qquad \bar{B}_2\ \mathrm{AP} \underset{\mathrm{d}}{=} \mathfrak{S}_K(\hat{\mathbb{R}}).$$

Bertrandias also considered the analoguous "*Besicovitch-Eberlein*" *AP functions*, (E_2AP), and an important subspace thereof:

D.39 Def.[‡] (a) E_2AP is the class of all f in $\bar{\mathscr{X}}$ for which the orbit $\{\tau_t(f): t \in \mathbb{R}\}$ is pre-compact in the space $\bar{\mathscr{X}}$ with its weak topology, {6: p. 64}.

(b) $\mathscr{S}_{c0}$ is the class of all f in E_2AP, the weak closure of the translates which contains 0, i.e. such that

$$0 \in \mathrm{wcls}\,\{\tau_t(f): t \in \mathbb{R}\}.$$

Bertrandias proved, {6: pp. 76, 77, 79}, that $\mathscr{S}_{c0}$ is a Banach subspace of E_2AP and that

---

[†] That is, $\|\tau_t(f) - f\| = 0$. For instance, $f \in K$ when $f(t) \underset{\mathrm{d}}{=} e^{i \log|t|}$, $t \in \mathbb{R}$.

[‡] From our standpoint a more convenient definition would be $E_2\mathrm{AP} \underset{\mathrm{d}}{=} \mathfrak{S}_w(\hat{\mathbb{R}})$, where $\mathfrak{S}_w(\hat{\mathbb{R}})$ is the weak closure of the linear manifold spanned by $\hat{\mathbb{R}}$ in the space $\bar{\mathscr{X}}$. But we do not know if this definition is equivalent to the one given in D.39(a), which is Bertrandias's definition, {6: p. 64}. Thm. IV in {6: p. 78} shows that our D.39(b) is equivalent to Bertrandias's definition in {6: p. 74}.

$$
\text{(D.40)} \qquad \begin{cases} \mathscr{S}_{c0} \subsetneqq \bar{\mathscr{S}}_{c0} = (B_2\text{AP})^{\perp} \cap (E_2\text{AP}), \\ \& \\ E_2\text{AP} = \bar{\mathscr{S}}_{c0} + \bar{B}_2\text{AP} \qquad (\textit{Jacob's Thm.}\ \{20\}). \end{cases}
$$

According to him the members of the Banach space $\bar{\mathscr{S}}_{c0}$ retain the attributes of the centered PR functions required by Bass, and thus $\bar{\mathscr{S}}_{c0}$ serves as a more convenient substitute for the non-linear $\mathscr{S}_{c0}$.

Bertrandias was also able to characterize these different spaces in terms of sets of covariance functions defined in (D.31). He proved the following theorem, {6: pp. 70, 64, 74, 78}:

D.41 Thm. *Let* $f \in \mathscr{X}_c$. *Then*

(a) $f \in \bar{B}_2$ AP, *iff any one of the following sets*

$$\Gamma_{f,g}, \quad (\textit{any } g \in \bar{\mathscr{X}}), \quad \Gamma_{f,\psi_f}, \quad \Gamma_g$$

is a compact subset of the Banach space $C(\mathbb{R}, \mathbb{C})$;

(b) $f \in E_2$ AP, *iff any one of the sets mentioned in* (a) *is a compact subset of the space* $C(\mathbb{R}, \mathbb{C})$ *with its weak topology*;

(c) $f \in \bar{\mathscr{S}}_{c0}$, *iff any one of the sets mentioned in* (a) *is a compact subset of* $C(\mathbb{R}, \mathbb{C})$ *with its weak topology, and comprises only functions* $\Phi \ni \|\Phi\| = 0$.

Bass's work has been continued in other directions. We have space to mention just one of these. Say that a Borel measurable f on $\mathbb{R}$ to $\mathbb{R}$ has an *asymptotic distribution* $\nu_f(\cdot)$ on $Bl(\mathbb{R})$, iff $\forall B \in Bl(\mathbb{R})$, the set $f^{-1}(B)$ has an asymptotic Lebesgue density α, and define $\nu_f(B) \underset{\text{d}}{=} \alpha$. Pham Phu Hien and J. Couot have established a central limit theorem for the asymptotic density of convolutions built from a completely uniformly distributed sequence. (For references see Bass {3}.) This reads as follows:

D.42 Central Limit Thm. *Let* (i) *the sequence* $(a_n)_0^\infty$ *be completely uniformly distributed on* $[0, 1]$, (ii) ψ *be continuous on* $[0, 1]$ *to* $\mathbb{R}$, (iii) $f(\cdot)$ *be the even function on* $\mathbb{R}$ *such that* $f(\cdot) \underset{\text{d}}{=} \psi(a_{[\cdot]})$ *on* $\mathbb{R}_{0+}$, (iv) $\forall \lambda \in \mathbb{R}_+$ $g_\lambda(t) = \sqrt{\lambda \cdot f(\lambda t)}$, (v) $K(\cdot)$ *be a uniformly continuous function in* $L_1(\mathbb{R})$. *Then as* $\lambda \to \infty$, *the asymptotic distribution of* $K * g_\lambda$ *tends to the normal distribution with mean value* m *and standard deviation* σ *given by*

$$m = ((f, 1)) \int_{\mathbb{R}} K(t)\,dt, \quad \sigma^2 = \|f\|^2 \int_{\mathbb{R}} |K(t)|^2\,dt - m.$$

Dr. Couot has also developed an *ergodic theory* of uniform distributions. In this connection we should perhaps mention the cognate work of Kakutani and Slud {21, 56} on the uniform distribution resulting from a sequence of random subdivisions of $[0, 1]$.

To sum up, several interesting new ideas have been injected into GHA and new connections revealed, as a result of Bass's attempts to apply GHA to the difficult subject of turbulence.

E. The role of GHA in the quest for randomness

The calculus of probabilities is meant to deal with functions whose behavior is "random" in some sense. Most often the domain J of such a function f is a subset of $\mathbb{N}$ or $\mathbb{R}$. It has been the trend during the last 40 years to regard such an f as being the sample-function of a stochastic process parametrized on J.

Wiener himself was one of the great pioneers who helped usher in this trend towards stochastic processes by virtue of his researches on the Brownian movement [20f; 21c,d; 23d; 24d]. This research surpassed the 1912 contribution {1} of P. Bachelier by bringing to the forefront completely new attitudes. Perhaps the most remarkable of these were the emphasis on the construction of a probability measure over the infinite dimensional space $C[a, b]$ (the so-called *Wiener measure*), and on the definition of a L_2-vector-valued measure over $\mathbb{R}$ (now called an *orthogonally scattered measure*) which permitted the introduction of a *stochastic integration*, cf. {42}. The axiomatization of probability theory by Kolmogorov {43} and his proof of the existence of a stochastic process corresponding to an arbitrary stochastic family came only about 10 years later. The demarcation of the province of *stationary processes* by Khinchine {24} followed a year thereafter, and only then came the systematic theory of *processes of independent increments*, of which the Brownian motion is a special case, by P. Lévy, cf. {32}.

It is noteworthy that just before Kolmogorov began laying the foundations of the general theory of stochastic processes, by constructing (like Wiener) a probability measure over an infinite-dimensional space, Wiener himself was culminating work begun in 1924 on erratically behaving functions on $\mathbb{R}$ in the memoir [30a] under review, cf. §D outset. For fuller accounts of this development see the preceeding commentary of Professor Koisumi, my {40: §1} and Levinson {30; 31}.

During the thirties, however, Wiener's interests reverted from the study of individual chaotic functions to stochastic processes. This shift in interest was prompted to some extent by Kolmogorov's fundamental work, but its major thrust came from the appearance in 1931 of G. D. Birkhoff's individual ergodic theorem. According to this theorem, almost all the trajectories of a stationary process are members of the class defined in [30a] which we have called $\mathscr{S}_c$ in §A. But Wiener had shown in [30a: §13] that the output of a convolution filter fed with a Brownian input is also in $\mathscr{S}_c$. This led him to the study of the Brownian movement from the standpoint of ergodicity. We refer to his work on the ergodicity of the flow of the Brownian motion (more accurately, of white noise) with Paley in [34d] and to his work on the ergodic theorem and the homogeneous and discrete chaos in part with Wintner [38a; 39a; 41a, b; 43a], which have been reviewed in Volume I of these collected works.

Wiener attempted to apply the ideas advanced in the papers just cited to setting up statistical mechanics on a sound footing [39b, h; 40d], cf. the reviews in Volume I of these works, especially the survey by B. McMillan and G. Deem {53}. The Wiener program has so far proved to be impotent in dealing with turbulent flow and of limited value in the kinetic theory of gases, but it is still too early to gauge its potential in these and other areas. By contrast, the application of some of his other ideas, especially the ones nearest those of Khinchine and Kolmogorov, to the problem of anti-aircraft fire control with radar, which loomed large during the advent of World War II, proved exceedingly fruitful: it brought into being the theories of prediction and filtering. These theories placed communication engineering on a firm footing and set them moving on a proper course, with significant ramifications in physiology. This in turn led Wiener to the demarcation of a science of communication and control called *cybernetics*. These historic developments will be discussed systematically in Volumes III and IV of these collected works. Our motive for alluding to them here is to explain how it happened that most of Wiener's later work came to be based on stochastic processes, and why he had neither the time nor the inclination to pick up the loose ends of GHA.

The belief that the concepts of GHA have counterparts in the theory of stochastic processes, Wiener's own reliance on the Brownian motion SP to demonstrate the existence in $\mathscr{S}_c$ of functions having absolutely continuous spectral distributions, and his abandonment of GHA in favour of SP theory during the thirties are some of the factors which lend credence to a current opinion that [30a] marks the culmination of GHA and its supercession by the more advanced theory of processes. This view is voiced, for instance, by G. W. Mackey in his recent interesting essay {35: p. 217}.

Such an opinion is questionable, however, on at least two counts. In the first place, as the discussion in §§A–D shows, there are deep ideas in [30a] the full potential of which has barely been gauged. It would seem that the impact of GHA on pure mathematics will be increasingly felt as more and more of its aspects are assimilated into abstract analysis. Secondly, only the relatively straightforward ideas of GHA have found counterparts within the theory of stochastic processes. As far as the reviewer knows, cf. also Doob {14}, no such counterparts have as yet been found for the more abstruse parts of GHA centering on the Tauberian identity and the generalized Fourier transform. Thus [30a] marks more a beginning than an end of GHA.

But the attitude embodied in GHA has another significance, perhaps more lasting, which has gone unnoticed so far, viz. its bearing on the whole question of "randomness." To this we shall now turn.

Many probabilitists have felt that the concept randomness should be defined as directly as possible. From this standpoint it seems artificial to conceive of a chaotic function on a subset J of $\mathbb{R}$ or $\mathbb{N}$ as the sample function of a SP, i.e. as the section $f(\cdot, \omega)$ of a function $f(\cdot, \cdot)$ on $J \times \Omega$, where Ω is some arbitrary set over which a probability measure is assigned, and all the sections $f(t, \cdot)$, for $t \in J$, are assumed to be measurable on Ω. But the only notable attempt before 1930 to set up the probability calculus more directly was the theory of *Collectives* of R. von Mises in 1919, cf. {57, 58}. Here we should like to point out that a white light signal, whether conceived in terms of the classical electro-magnetic theory or the theory of quanta, appears intuitively to be an excellent example of a reproducible mass phenomenon which von Mises tried to conceptualize by the idea of a collective. Thus with a somewhat widened concept of Collective, *GHA could be viewed as the harmonic analysis of a special type of collectives, viz. those possessing covariance functions.* But the view just advanced was bȧrely heard during the ascendency of the theory of processes, i.e. before and during World War II. Wiener certainly did not subscribe to it, nor even allude to it.

But research in the theory of stochastic processes undertaken during the late forties and early fifties began revealing pathologies of a measure-theoretic nature, to circumvent which rather technical maneuvers had to be undertaken, cf. Blackwell {8}. There was no unique mode of maneuvering. Thus the theory began to lose some of its original charm and simplicity, and tended towards technical complexity and over-specialization. This rekindled the latent yearning for a more direct definition of random function over $\mathbb{N}$ or $\mathbb{R}$, without the aid of a complex notion of SP. We may cite the papers {54} of B. Rankin and the book of H. Furstenberg {16}, (cf. {39}) as indicative of this new trend. In both SP's appear, but random functions on $\mathbb{N}$, or $\mathbb{R}$, not dependent on ω, are also emphasized.

But in the fifties, unlike 1919 when von Mises began his work, much more powerful tools were available to the collective-minded probabilist. To understand these let us recall that Kolmogorov's construction of a probability measure in an infinite-dimensional space, and indeed many crucial constructions in functional and abstract analysis require an appeal to the *diagonalization procedure* and the *Principle of transfinite induction* of

G. Cantor.† Off hand it would appear that these transfinite ideas or their logical or dialectical offshoots should be no less necessary for a successful direct attack on the notion of randomness. The last few decades seem to confirm this view, as we proceed to show.

As just indicated, the enormous edifice of pure mathematics depends for its footing on Cantorian ideas. But this edifice by itself fails to convey the profound depth of these ideas. To sound their depth we must turn to the "diagonalization paradoxes" that accompanied their baptism, especially *the Russell and Richard paradoxes.* The attempt to eliminate the Russell paradox from the formalized system of logic and set theory put forward by G. Frege, in which he successfully embedded the vocabulary and postulates of arithmetic, resulted in the theory of types and the formulation of the system *Principia Mathematica* of Whitehead and Russell (1913)—a great watershed in the history of mathematical thought. Then in 1931 came the remarkable observation by K. Gödel that the syntax of arithmetic including the Richard sentence could be formulated in the Principia Mathematica, and that this sentence when made to talk about the system yields a proof of the incompleteness of the system.‡ This revealed the unimaginably enormous scope and beauty of pure mathematics: even an infinitude of deduction from n axioms, however large be n, will not reach all arithmetical truths, leave alone all mathematical truths.

These amazing discoveries naturally created a counter-tendency to demarcate mathematical territory comprising decidable statements exclusively, or even comprising only those statements which could be proved or disproved mechanically, i.e. for which a *decision procedure* was available. Using Cantorian and Gödelian ideas, Skolem, Post, Church, Kleene, Tarski and others proved a host of remarkable results, the description of which would be out of place in this commentary. The central idea of *recursion* running through the work of Whitehead, Russell and Gödel came in for intensive study, and a flourishing *theory of recursive functions* emerged. The notions conceived by recursive function theorists such as *computability, effectiveness, constructivity, Turing machine* and *algorithm* became the very

---

† Think of the proofs of the Tychonov Thm. in topology, the theorems on the existence of Haar measure over locally compact groups, of orthonormal bases in a Hilbert space and of maximal ideals in linear algebras, the Hahn-Banach and Krein-Milman Thms. in vector space theory, among others. They all appeal to basically Cantorian ideas, usually in the guise of Zermelo's Axiom of Choice or of the Maximality axioms for posets of Hausdorff or Zorn.

‡ Cf. H. Weyl {61: pp. 223–}, where his own cautious attitude towards the Richard paradox in the intuitionist framework is illuminatingly contrasted with Gödel's revolutionary utilization of it in the setting of the Principia Mathematica.

basis of the *theory of automata* in general and the *theory of computors* in particular. Thus, they became important ingredients of *cybernetics*, the theoretical propellant of our age of automation, on different aspects of which Wiener had much to say.

Attempts soon began to bring the ideas of recursive function theory to bear on the notion of collective, cf. Martin-Lof {37}. The von Mises definition contained an irregularity condition, the so-called "place-selection" postulate. Certain formulations of this postulate resulted in contradictions, but even after a more critical recursive formulation was provided, the existence of a collective remained in doubt till 1937, when A. Wald {59} supplied a proof by formalizing the postulate in a system of symbolic logic. In 1940 Church {12} took the decisive step of demanding that the selection functions appearing in the irregularity postulate be *effectively calculable* in the sense of recursive function theory. Then in 1963, Kolmogorov himself proposed the definition of a finite collective in which the allied concept of a *finite algorithm* is used in formulating the irregularity postulate {24}, and he suggested inquiry towards a possible reformulation of the foundations of probability theory on its basis.

It is worth noting that the existence of the *constructive* Church collective follows from a theorem of Doob, but that all existence proofs are necessarily *non-constructive*. Thus, Cantorian ideas and their offshoots, both constructive and non-constructive, enter essentially in the modern concept of collective.

None of the strict definitions of collective advanced during the last few decades meet all the requirements of "randomness" as this word is used in ordinary pre-scientific parlance. This suggests that these requirements are perhaps mutually inconsistent, and attempts at their mathematical explication are best abandoned. Instead, we should attempt to define several classes of chaotic functions (on $\mathbb{N}^q$ or $\mathbb{R}^q$ say, $q \geqslant 1$), exemplifying different types and degrees of randomness, on the basis of their covariance or other asymptotic behavior. Thus, one would define types of chaotic functions, such as stationary, ergodic, Markovian, stationary-Markovian and non-stationary.

Wiener's GHA would have a definite place in such a venture. For simplicity consider his discrete GHA [27a]. The properties of the auto-covariance function on $\mathbb{N}$, or of the spectral distribution on $[0, 2\pi]$, or of the GFT on $[0, 2\pi]$ could be conceived as demarcating different types of "randomness" within the subclass of "stationary, covariance-possessing" collectives. We cannot without serious investigation surmise the effectiveness of such a pursuit of randomness on the basis of Wienerian ideas. Nor can we say today if the ideas of recursive function theory will lend themselves to produce a more useful substitute, as far as the applications are concerned, than the current theory of stochastic processes. But the interesting dialectic movement we have traced shows that it is premature to regard [30a] as a dated paper and GHA as a passing and superceded phase of mathematical development.

In his encyclopedia article {25}, Kolmogorov writes

> Mathematical probability is an expression of the qualitatively distinctive connection between the random and the necessary. In the exposition of probability theory, those features of probability which at a given stage of scientific development were necessary for such development are formulated in the form of axioms. However, neither these axioms, nor the classical approach to probability, nor the statistical approach give an exhaustive definition of the real substance of the concept of probability; they are merely known approximations to its ever more complete disclosure.

Feeling the limitations of the current axiomatization, which was very largely his own handiwork, he has in {24} suggested a possible return to the ideas of von Mises. The integrity and wisdom reflected in this attitude should point the way towards the future. Side-by-side with the vigorous pursuit of the theory of stochastic processes, must coexist a more direct process-free inquiry of randomness of different classes of functions. If such a trend does indeed materialize, then Wiener's [30a], like his papers on the Brownian movement, will again have been ahead of its time, now by about three decades instead of one.

References

1. L. Bachelier, *Calcul des probabilities*, Gautier Villas, Paris, 1912.

2. J. Bass, *Les fonctions pseudo-aleatoires*, Memorial des Sciences Math., Gautier Villas, Paris, 1962.

3. J. Bass, *Stationary functions and their applications to the theory of turbulence, I*, J. Math. Anal. and Appl. 47(1974), 354–399.

4. J. Bass, *Stationary functions and their applications to the theory of turbulence, II*, J. Math. Anal. and Appl. 47(1974), 458–503.

5. C. Berge, *The theory of graphs and its applications*, Wiley, New York, 1962.

6. J. P. Bertrandias, *Espaces de fonctions bornees et continues en moyenne asymptotique d'ordre p*, Bull. Soc. Math. France (1966), Memoire 5.

7. A. S. Besicovitch, *Almost periodic functions*, Cambridge Univ. Press, 1932.

8. D. Blackwell, *On a class of probability spaces*, Proc. 3rd Berkeley Symp. on Math. Stat. and Prob. vol. II (1956), 1–6.

9. S. Bochner, *Beitrage zur Theorie der fastperiodischen Functionen, I, II*, Math. Ann. 96(1926), 119–147, 383.

10. H. Bohr, *Zur Theorie der fastperiodischen Functionen, I, II, III*, Acta Math. 45(1924), 29–127; 46(1925), 101–214; 47(1926), 237–281.

11. H. Bohr and E. Følner, *On some types of functional spaces*, Acta Math. 76(1944), 31–155.

12. A. Church, *On the concept of a random sequence*, Bull. Amer. Math. Soc. 46(1940), 130–135.

13. W. F. Donoghue, *Distributions and Fourier transforms*, Academic Press, New York, 1969.

14. J. L. Doob, *Times series and harmonic analysis*, Proc. 1st Berkeley Symp. on Math. Stat. and Prob. Univ. of Calif. Press, Berkeley, Calif., 1949, 303–343.

15. J. L. Doob, *Stochastic processes*, Wiley, New York; Chapman and Hall, London; 1953.

16. W. F. Eberlein, *Abstract ergodic theorems and weak almost periodic functions*, Trans. Amer. Math. Soc. 67(1949), 217–240.

17. J. N. Franklin, *Deterministic simulation of random processes*, Math. Comp. 17(1963).

18. H. Furstenberg, *Stationary processes and prediction theory*, Princeton Univ. Press, 1960.

19. E. Hewitt and K. Ross, *Abstract harmonic analysis*, vols. I, II, Springer Verlag, New York, 1963, 1970.

20. K. Jacobs, *Neuere Methoden und Ergebnisse der Ergodentheorie*, Springer Verlag, 1960.

21. S. Kakutani, *A problem of equidistribution on the unit interval* [0, 1], Measure Theory (Oberwolfach, 1975), 369–375. Edited by A. Bellow and D. Kolzow, Lecture Notes No. 541, Springer Verlag, New York, 1976.

22. A. Y. Khinchin, *Korrelationstheorie der stationaren stochastischen Prozesse*, Math. Ann. 109(1934), 604–615.

23. A. N. Kolmogorov, *Foundations of the theory of probability*, Chelsea, New York, 1950.

24. A. N. Kolmogorov, *On a table of random numbers*, Sankhya 25(1963), 369–376.

25. A. N. Kolmogorov, *Probability*, Great Soviet Ency. IV, 423–424, English translation of 3rd Ed., MacMillan, New York, 1970.

25.′ K.-S. Lau, "On the Banach spaces of functions with bounded upper means" (unpublished).

25.″ K.-S. Lau and J. K. Lee, *On generalized harmonic analysis*, Trans. Amer. Math Soc. (in press).

26. J. K. Lee, *On a class of functions in generalized harmonic analysis*, Abstract 674–106, Notices Amer. Math. Soc. 17(1970), 634.

27. J. K. Lee, *The completeness of the class of functions of bounded upper p-variation*, $1 < p < \infty$, Abstract 681-B5, Notices Amer. Math. Soc. 17(1970), 1057.

28. J. K. Lee, *On generalized harmonic analysis*, Thesis, Indiana University, 1971.

29. Y. W. Lee, *Statistical theory of communication*, Wiley, New York, 1960.

30. N. Levinson, *Introduction and some comments on Wiener's selected papers*, pp. 1–16 in Selected Papers of N. Wiener, The MIT Press, Cambridge, Mass., 1964.

31. N. Levinson, *Wiener's life*, Bull. Amer. Math. Soc. 72(1966), 1–32.

32. P. Lévy, *Theorie de l'addition des variables aleatoires*, Gautier Villars, Paris, 1937.

33. W. A. J. Luxemburg and A. C. Zaanen, *Notes on Banach function spaces, I*, Indag. Math. 25(1963), 135–147.

34. G. W. Mackey, *Induced representations of groups and quantum mechanics*, Benjamin, New York, 1968.

35. G. W. Mackey, *Ergodic theory and its significance for statistical mechanics and probability theory*, Adv. Math. 12(1974), 178–268.

36. J. Marcinkiewicz, *Une remarque sur les espaces de M. Besicovitch*, C. R. Acad. Sci. Paris 208(1939), 157–159.

37. P. Martin-Lof, *The literature on von Mises' Kollektivs revisited*, Theoria, part 1, 35(1969), 12–37.

38. P. Masani, *Isometric flows on Hilbert space*, Bull. Amer. Math. Soc. 68(1962), 624–632.

39. P. Masani, *Review of "Stationary processes and prediction" by H. Furstenberg*, Bull Amer. Math. Soc. 69(1963), 195–207.

40. P. Masani, *Wiener's contributions to generalized harmonic analysis, prediction theory and filter theory*, Bull. Amer. Math. Soc. 72(1966), 73–125.

41. P. Masani, *Graph-theoretic aspects of generalized harmonic analysis*, Abstract, Notices, Amer. Math. Soc. 14(1967), 407–408.

42. P. Masani, *Orthogonally scattered measures*, Adv. Math. 2(1968), 61–117. (Originally, Technical Report no. 378, Mathematics Research Center, University of Wisconsin, 1967).

43. P. Masani, *On the representation theorem of scattering*, Bull. Amer. Math. Soc. 74(1968), 618–624.

44. P. Masani, *Explicit form for the Fourier-Plancherel transform over locally compact Abelian groups*, in Abstract Spaces and Approximation, 162–182. Edited by P. L. Butzer and B. Sz.-Nagy, Birkhauser, Basel, 1969.

45. P. Masani, *Quasi-isometric measures and their applications*, Bull. Amer. Math. Soc. 76(1970), 427–528.

46. P. Masani, *On helixes in Hilbert space I*, Theor. Probability Appl., USSR, 17(1972), 3–20. (English Edition, SIAM 17(1972), 1–19).

47. P. Masani, *On infinitely decomposable probability distributions and helical varieties in Hilbert space*, in Multivariate Analysis (Proc. of the International Symposium held in Dayton, Ohio, June, 1972), 209–223. Edited by P. R. Krishnaiah, Academic Press, New York, 1973.

48. P. Masani, *On helixes in Banach spaces*, Sankhya 38, (1976), 1–27.

49. P. Masani, *Generalizations of Lévy's inversion theorem*, J. Multivariate Analysis 7(1977), 292–335.

50. P. Masani, *Measurability and Pettis integration in Hilbert space*, Crelle's J. 297(1978), 92–135.

51. P. Masani, *Vector graphs and conditional Banach spaces*, pp. 72–89 in "Linear Spaces and approximation" ed. by P. L. Butzer and B. Sz.-Nagy, Birkhauser, Basel, 1978.

52. P. Masani and J. Robertson, *The time-domain analysis of continuous-parameter, weakly stationary stochastic processes*, Pac. J. Math. 12(1962), 1361–1378.

53. B. McMillan and G. S. Deem, *The Wiener program in statistical physics*, pp. 654–671 in Norbert Wiener: Collected Works. Edited by P. Masani, The MIT Press, Cambridge, Mass., 1976.

53.′ R. Nelson, *The spaces of functions of finite upper p-variation*, Trans. Amer. Math Soc. 253(1979), July issue.

54. B. Rankin, *Computable probability spaces*, Acta. Math. 103(1960), 89–122.

55. J. B. Robertson, *Orthogonal decompositions of multivariate weakly stationary stochastic processes*, Canadian J. Math. 20(1968), 368–383.

56. E. V. Slud, *Entropy and maximal spacing for random partitions*, Zeit. Wahrscheinlikeitstheorie Verw. Geb. 41(1978), 341–352.

57. R. von Mises, *Grundlagen der Wahrschienlichkeitsrechnung*, Math. Zeit 5(1919), 52–99.

58. R. von Mises, *Mathematical theory of probability and statistics*, Academic Press, New York, 1963.

59. A. Wald, *Die Widerspruchsfreiheit des Kollektivbegriffes der Wahrscheinlichkeitsrechnung*, Ergebnisse eines Mathematischen Kolloquiums, 8(1937), 38–72.

60. H. Weyl, *Über die Ghechverteilung von Zahlen modulo eins*, Math. Ann. 77(1916), 313–352.

61. H. Weyl, *Philosophy of mathematics and natural science*, Princeton Univ. Press, 1949.

THE SPECTRUM OF AN ARRAY AND ITS APPLICATION TO THE STUDY OF THE TRANSLATION PROPERTIES OF A SIMPLE CLASS OF ARITHMETICAL FUNCTIONS

Part One

The Spectrum of an Array

By NORBERT WIENER

The purpose of the present paper is to extend the spectrum theory already developed by the author in a series of papers[1] to the harmonic analysis of functions only defined for a denumerable set of arguments — *arrays*, as we shall call them — and the application of this theory to the study of certain power series admitting the unit circle as an essential boundary.

An *array* A is a sequence of numbers a_n with indices n running from $-\infty$ to ∞. From the array A we can under certain circumstances form an array B according to the rule

$$b_n = \lim_{k\to\infty} \frac{1}{2k+1} \sum_{-k}^{k} a_j \bar{a}_{j-n},$$

where the bar denotes the operation of taking the conjugate. If b_n exists for every integral value of n, we shall call the array A *regular*.

We now form the function

$$A(x) \sim b_0(x) + \sum_{-\infty}^{\infty}{}' \frac{b_j e^{ijx}}{ij}.$$

It is easy to show by the use of the Schwartz inequality that $|b_j| \leq b_0$, and hence that $A(x)$ exists as a limit in the mean over any finite interval.

$A(x)$ is a monotone non-decreasing function of x. To see this let us put $\alpha_j = a_j[|j| \leq k]$; $\alpha_j = 0[|j| > k]$. Let

$$\alpha(x) = \sum_{-\infty}^{\infty} \alpha_j e^{ijx}.$$

[1] N. Wiener, On the Representation of Functions by Trigonometrical Integrals, Math. Ztschr. Vol. 24, pp. 575–616; The Harmonic Analysis of Irregular Motion, this *Journal*, February and March, 1926.

We have

$$|a(x)|^2 = \left[\sum_{-\infty}^{\infty} a_j e^{ijx}\right]\left[\sum_{-\infty}^{\infty} \bar{a}_k e^{-ikx}\right]$$

$$= \sum_{j=-\infty}^{\infty} a_j e^{ijx} \sum_{n=-\infty}^{\infty} \bar{a}_{j-n} e^{-i(j-n)x}$$

$$= \sum_{-\infty}^{\infty} e^{inx} \sum_{j=-\infty}^{\infty} a_j \bar{a}_{j-n}.$$

Thus if

$$\beta_n = \frac{1}{2k+1} \sum_{j=-\infty}^{\infty} a_j \bar{a}_{j-n},$$

we see that

$$A_k(x) = \beta_0 x + \sum_{-\infty}^{\infty}{}' \frac{\beta_j e^{ijx}}{ij}.$$

has a non-negative derivative, and is monotone.

We have however uniformly in j

$$\frac{b_j}{j} = \lim_{k\to\infty} \frac{\beta_j}{j},$$

while the whole set of β's is bounded in j and k. Hence over any finite interval

$$A(x) = \underset{k\to\infty}{\text{l. m.}} \quad A_k(x).$$

It follows readily that $A(x)$ may itself be so defined as to be monotone non-decreasing. We shall call $A(x)$ the *spectral function* of A. It always exists when A is regular. Since it is monotone, it follows from the theory of the Dirichlet conditions that

$$\frac{A(x+0)+A(x-0)}{2} = b_0 x + \sum_{-\infty}^{\infty}{}' \frac{b_j e^{ijx}}{ij}.$$

If A is regular,

$$b_0 = \lim_{k\to\infty} \frac{1}{2k+1} \sum_{j=-k}^{k} |a_j|^2$$

is bounded, and

$$\sum_{-k}^{k}{}' |a_j|^2/j^2 = \sum_{-k}^{k}{}' |a_j|^2 \int_j^\infty \frac{2}{x^3}\, dx$$

$$\leq \sum_{-k}^{k}{}' |a_j|^2 \sum_{l=j}^{\infty} \frac{2}{l^3}$$

$$\leq \sum_{l=1}^{\infty} \frac{2}{l^3} \sum_{j=-l}^{l} |a_j|^2$$

$$\leq \sum_{1}^{\infty} \frac{4}{l^2} \underset{m \geq 1}{\text{upper bound}} \frac{1}{2m} \sum_{-m}^{m} |a_j|^2.$$

This clearly has an upper bound independent of k, and thus

$$\sum_{-\infty}^{\infty} |a_j|^2/j^2$$

and hence

$$\phi(x) \sim a_0 x + \sum_{-\infty}^{\infty}{}' \frac{a_j e^{ijx}}{ij}$$

exists.

Let us assume that $a_0 = 0$. We have

$$\int_x^{x+h} \phi(\xi)\,d\xi - \int_{x-h}^{x} \phi(\xi)\,d\xi = -4 \sum_{-\infty}^{\infty} \frac{a_j e^{ijx}}{j^2} \sin^2 \frac{jh}{2} \cdot$$

Hence

$$\frac{1}{h^3} \int_0^{2\pi} \left| \int_x^{x+h} \phi(\xi)\,d\xi - \int_{x-h}^{x} \phi(\xi)\,d\xi \right|^2 dx = \frac{32\pi}{h^3} \sum_{-\infty}^{\infty} |a_j|^2 \frac{\sin^4 \dfrac{jh}{2}}{j^4}$$

$$= 32\pi \sum_{k=0}^{\infty} (2k+1)h \left[\frac{\sin^4 \dfrac{kh}{2}}{k^4 h^4} - \frac{\sin^4 \dfrac{(k+1)h}{2}}{(k+1)^4 h^4} \right] \frac{1}{2k+1} \sum_{j=-k}^{j=k} |a_j|^2.$$

It is easy to show that

$$32\pi\sum_{k=0}^{\infty}(2k+1)h\left|\frac{\sin^4\frac{kh}{2}}{k^4h^4}-\frac{\sin^4\frac{(k+1)h}{2}}{(k+1)^4h^4}\right|$$

is bounded as $h\to 0$, that

$$\lim_{h\to 0}\ (2k+1)h\left[\frac{\sin^4\frac{kh}{2}}{k^4h^4}-\frac{\sin^4\frac{(k+1)h}{2}}{(k+1)^4h^4}\right]=0,$$

and that

$$\lim_{h\to 0}\ 32\pi\sum_{k=0}^{\infty}(2k+1)h\left[\frac{\sin^4\frac{kh}{2}}{k^4h^4}-\frac{\sin^4\frac{(k+1)h}{2}}{(k+1)^4h^4}\right]$$

$$=\lim_{h\to 0}\frac{1}{h^3}\int_0^{2\pi}\left|\int_x^{x+h}(\pi-\xi)\operatorname{sgn}\xi\,d\xi-\int_{x-h}^{x}(\pi-\xi)\operatorname{sgn}\xi\,d\xi\right|^2dx=\frac{4}{3}\pi.$$

Hence an argument of the same type as that used in the proof of Féjer's theorem will show that

$$\lim_{h\to 0}\frac{3}{4\pi h^3}\int_0^{2\pi}\left|\int_x^{x+h}\phi(\xi)d\xi-\int_{x-h}^{x}\phi(\xi)d\xi\right|^2dx=b_0.$$

If A is regular, the arrays A_1 and A_2, respectively,

$$\cdots a_{-1}+a_{k-1},\ a_0+a_k,\ a_1+a_{k+1},\ \cdots$$

and

$$\cdots a_{-1}-a_{k-1},\ a_0-a_k,\ a_1-a_{k+1},\ \cdots$$

are also regular. We can without essential restriction suppose $a_0=a_{\pm k}=0$. The ϕ's pertaining to A_1 and A_2 are

$$\phi(x)\pm\phi(x)e^{-ikx}.$$

We thus have

$$\lim_{h\to 0}\frac{3}{4\pi h^3}\int_0^{2\pi}\left|\int_0^{x+h}[\phi(\xi)\pm\phi(\xi)e^{-ik\xi}]\,d\xi\right.$$

$$\left.-\int_{x-h}^{x}[\phi(\xi)\pm\phi(\xi)e^{-ik\xi}]\,d\xi\right|^2dx=2b_0\pm 2Rb_k.$$

Hence

$$Rb_k = \lim_{h\to 0} \frac{3}{4\pi h^3}\int_0^{2\pi} R\left\{\left[\int_x^{x+h}\phi(\xi)d\xi - \int_{x-h}^x \phi(\xi)d\xi\right]\right.$$
$$\left.\cdot \operatorname{Conj}\left[\int_x^{x+h}\phi(\xi)e^{-ik\xi}d\xi - \int_{x-h}^x \phi(\xi)e^{-ik\xi}d\xi\right]\right\} dx$$
$$= \lim_{h\to 0} \frac{3}{4\pi h^3}\int_0^{2\pi} \cos kx R\left\{\left[\int_x^{x+h}\phi(\xi)d\xi - \int_{x-h}^x \phi(\xi)d\xi\right]\right.$$
$$\left.\cdot \operatorname{Conj}\left[\int_x^{x+h}\phi(\xi)e^{ik(x-\xi)}d\xi - \int_{x-h}^x \phi(\xi)e^{ik(x-\xi)}d\xi\right]\right\} dx$$
$$= \lim_{h\to 0}\left\{\frac{3}{4\pi h^3}\int_0^{2\pi}\cos kx \left|\int_x^{x+h}\phi(\xi)d\xi - \int_{x-h}^x \phi(\xi)d\xi\right|^2 dx\right.$$
$$- \frac{3}{4\pi h^3}\int_0^{2\pi}\cos kx\ R\left\{\left[\int_x^{x+h}\phi(\xi)d\xi - \int_{x-h}^x \phi(\xi)d\xi\right]\right.$$
$$\cdot \operatorname{Conj}\left[\int_x^{x+h}\phi(\xi)[1-e^{ik(x-\xi)}]d\xi\right.$$
$$\left.\left.\left. - \int_{x-h}^x \phi(\xi)[1-e^{ik(x-\xi)}]d\xi\right]\right\} dx\right\}.$$

This last remainder term, by the Schwarz inequality, cannot exceed in modulus

$$\frac{3}{4\pi h^3}\sqrt{\int_0^{2\pi}\left|\int_x^{x+h}\phi(\xi)d\xi - \int_{x-h}^x \phi(\xi)d\xi\right|^2 dx}$$
$$\overline{\cdot \int_0^{2\pi}\left|\int_x^{x+h}\phi(\xi)\,[1-e^{ik(x-\xi)}]d\xi - \int_{x-h}^x \phi(\xi)[1-e^{ik(x-\xi)}]d\xi\right|^2 dx.}$$

We have, however,

$$\int_h^{x+h}\phi(\xi)\,[1-e^{ik(x-\xi)}]\,d\xi - \int_{x-h}^x \phi(\xi)[1-e^{ik(x-\xi)}]d\xi$$
$$= \sum_{-\infty}^{\infty}{}' \frac{a_j}{ij}\int_x^{x-h} e^{ij\xi}\,[1-e^{ik(x-\xi)}]d\xi - \int_x^{x-h} e^{ij\xi}[1-e^{ik(x-\xi)}]d\xi$$
$$= \sum_{-\infty}^{\infty}{}' \frac{2a_j}{ij}\, e^{ijx}\left(\frac{\cos jh}{ij} - \frac{\cos (j-k)h}{i(j-k)}\right).$$

Hence

$$\int_0^{2\pi}\left|\int_x^{x+h}\phi(\xi)[1-e^{ik(x-\xi)}]d\xi-\int_{x-h}^{x}\phi(\xi)[1-e^{ik(x-\xi)}]d\xi\right|^2dx$$

$$=8\pi\sum_{-\infty}^{\infty}{}'\frac{|a_j|^2}{j^2}\left(\frac{\cos jh-1}{j}-\frac{\cos(j-k)h-1}{j-k}\right)^2$$

$$\leq 8\pi k\sum_{-\infty}^{\infty}{}'\frac{|a_j|^2}{j^2}\max\left[\left|\frac{d}{dy}\left(\frac{\cos yh-1}{y}\right)\right|^2\right]$$

$$=O(h^4)\sum_{-}^{\infty}{}'\frac{|a_j|^2}{j^2}=O(h^4).$$

It follows at once that the remainder tends to zero with h, and that

$$\lim_{h\to 0}\frac{3}{4\pi h^3}\int_0^{2\pi}\cos kx\left|\int_x^{x+h}\phi(\xi)d\xi-\int_{x-h}^{x}\phi(\xi)d\xi\right|^2dx=Rb_k$$

$$=\frac{1}{2\pi}\int_0^{2\pi}\cos kx\, dA(x).$$

The restrictions imposed that $a_0=a_{\pm k}=0$ are not essential, and may readily be removed.

It follows that for any trigonometrical polynomial $P(x)$,

$$\lim_{h\to 0}\frac{3}{4\pi h^3}\int_0^{2\pi}P(x)\left|\int_x^{x+h}\phi(\xi)d\xi-\int_{x-h}^{x}\phi(\xi)d\xi\right|^2dx$$

$$=\frac{1}{2\pi}\int_0^{2\pi}P(x)dA(x),$$

and by the use of the Weierstrass approximation theorem and the fact that

$$\overline{\lim_{h\to 0}}\left|\frac{3}{4\pi h^3}\int_0^{2\pi}P(x)\left|\int_x^{x+h}\phi(\xi)d\xi-\int_{x-h}^{x}\phi(\xi)d\xi\right|^2dx\right|\leq\max|P(x)|\,b_0,$$

this theorem may be extended to all continuous functions $P(x)$ with period 2π.

Let $P(x)$ differ from 0 in and only in the interval (a, b), interior to $(0, 2\pi)$. Let the increase of A over every interval interior to (a, b) be different from 0. Then

$$\lim_{h\to 0}\frac{3}{4\pi h^3}\int_0^{2\pi}P(x)\left|\int_x^{x+h}\phi(\xi)d\xi-\int_{x-h}^{x}\phi(\xi)d\xi\right|^2dx>0.$$

By choosing P properly, we see that

$$\lim_{h\to 0}\frac{3}{4\pi h^3}\int_{a+\epsilon}^{b-\epsilon}\left|\int_x^{x+h}\phi(\xi)d\xi-\int_{x-h}^{x}\phi(\xi)d\xi\right|^2dx>0.$$

It is, however, easy to show that if ϕ possesses a bounded derivative in $(a+\epsilon, b-\epsilon)$,

$$\int_{a+\epsilon}^{b-\epsilon}\left|\int_x^{x+h}\phi(\xi)d\xi-\int_{x-h}^{x}\phi(\xi)d\xi\right|^2dx=O(h^4).$$

Hence *if* $A(x)$ *is properly increasing in every interval,* $\phi(x)$ *possesses a bounded derivative in no interval. This will in particular be the case if* b_0 *is the only* b_n *to differ from* 0.

An example of an array with this latter property is the array for which $a_n=a_{-n}$, while $a_1, a_2, \cdots$ have the values

1, −1

[1, 1; 1, −1; −1, 1; −1, −1] repeated twice

[1, 1, 1; 1, 1, −1; 1, −1, 1; 1, −1, −1; −1, 1, 1; −1, 1, −1; −1, −1, 1; −1, −1, −1] repeated four times

[1, 1, 1, 1; 1, 1, 1, −1; 1, 1, −1, 1; etc.] repeated eight times

. . . .

Here $b_0=1$; $b_{\pm 1}=b_{\pm 2}=\cdots=0$.

Again, let us consider an array for which $a_n=a_{-n}$ and where every a is ± 1. Let us make correspond to the array

$$a_0, a_1, a_2, \cdots$$

the binary number

$$\frac{a_0+1}{2}\ \frac{a_1+1}{2}\ \frac{a_2+1}{2}\cdots.$$

It is easy to show that for all such binary numbers excepting those of a set of measure 0, $b_0=1$; $b_{\pm 1}=b_{\pm 2}=\cdots=0$. We can interpret this as saying that the probability that

$$\pm 1,\ \pm 1,\ \cdots$$

determine a function $A(x)$ other than x itself is 0.

Let us consider a regular array A for which every $a_{-n}=0$, $a_0=0$, and let us consider the analytic function defined inside of the unit circle

$$\psi(x)=\sum_0^\infty \frac{a_n}{n} z^n = \sum_0^\infty \frac{a_n}{n} r^n e^{in\theta}.$$

Clearly as $r\to 1$, $\dfrac{\psi(re^{i\theta})}{i}$ converges in the mean to $\phi(\theta)$. Hence if $\psi(x)$ possesses an analytic continuation across any arc of the unit circle, $\phi(\theta)$ must be analytic over same arc. It follows that if $A(\theta)$ is properly increasing over any arc, $\psi(x)$ possesses the unit circle as an essential boundary. For example, the series

$$\sum \pm \frac{x^n}{n}$$

almost always has the unit circle as an essential boundary.

Let the array A be regular. We wish to prove that the array. B consisting of

$$(\cdots,\ Rb_n,\ \cdots,\ Rb_1,\ b_0,\ Rb_1,\ \cdots,\ Rb_n,\ \cdots)$$

is also regular. That is, we wish to prove that

$$C_n=\lim_{k\to\infty}\frac{1}{2k+1}\sum_{j=-k}^{j=k} R(b_{j+\frac{n}{2}}\bar{b}_{j-\frac{n}{2}})$$

exists for every n. Here j is summed over integral values if n is even, over values of the form $\nu+1/2$ if n is odd. This is equivalent to proving that

$$\lim_{k\to\infty}\frac{1}{2k+1}\sum_{j=-k}^{k}|b_{j+\frac{n}{2}}\pm b_{j-\frac{n}{2}}|^2$$

exists for every n. By a theorem proved by the author in a paper forthcoming in the *Journal of the London Mathematical Society* this is equivalent to proving that

$$\lim_{\epsilon \to 0} \frac{2}{\epsilon} \sum_{-\infty}^{\infty} \frac{|b_{j+\frac{n}{2}} \pm b_{j-\frac{n}{2}}|^2 \sin^2 j\epsilon}{j^2}$$

exists for every n, or that

$$\lim_{\epsilon \to 0} \frac{2}{\epsilon} \sum_{-\infty}^{\infty} R(b_{j+\frac{n}{2}} \bar{b}_{j-\frac{n}{2}}) \frac{\sin^2 j\epsilon}{j^2}$$

exists for every n. [Here we put $\sin^2 j\epsilon / j^2 = \epsilon^2$ if $j = 0$.] This again is equivalent to the statement that

$$\lim_{\epsilon \to 0} \frac{2}{\epsilon} \sum_{-\infty}^{\infty} R(b_{j+\frac{n}{2}} \bar{b}_{j-\frac{n}{2}}) \frac{\sin^2 j\epsilon \cos^2 \frac{n\epsilon}{2} - \cos^2 j\epsilon \sin^2 \frac{n\epsilon}{2}}{j^2 - \frac{n^2}{4}}$$

exists for every n, with proper definitions of the terms $j = \pm \frac{n}{2}$. This last expression is, however, merely another way of writing

$$\lim_{\epsilon \to 0} \frac{1}{2\epsilon} \int_0^{2\pi} \cos \frac{nx}{2} [A(x+\epsilon) - A(x-\epsilon)]^2 dx.$$

As $A(x)$ is a monotone function, the latter limit exists. It is the Stieltjes integral

$$\int_{-0}^{2\pi - 0} \cos \frac{nx}{2} \, d\psi(x),$$

where $\psi(x)$ is the sum of the squares of the jumps of $A(x)$ in the interval $(-0, x)$ if n is even, and in the interval $(+0, x)$, if n is odd.

The proof proceeds as follows: let $\Phi(x)$ be a step-function having as jumps all those jumps of $A(x)$ exceeding η in magnitude. Then

$$\left| \frac{1}{2\epsilon} \int_0^{2\pi} \cos \frac{nx}{2} [A(x+\epsilon) - A(x-\epsilon)]^2 dx - \frac{1}{2\epsilon} \int_0^{2\pi} \cos \frac{nx}{2} [\Phi(x+\epsilon) - \Phi(x-\epsilon)]^2 dx \right|$$

$$\leq \frac{1}{2\epsilon} \sqrt{\int_0^{2\pi} [A(x+\epsilon) - A(x-\epsilon) + \Phi(x+\epsilon) - \Phi(x-\epsilon)]^2 dx \cdot \int_0^{2\pi} [A(x+\epsilon) - A(x-\epsilon) - \Phi(x+\epsilon) - \Phi(x-\epsilon)]^2 dx}.$$

It can readily be shown by a use of the Heine-Borel theorem that ϵ can be taken so small that the second integral under the radical sign does not exceed $2\epsilon\eta A(2\pi)$. As the first integral is $O(\epsilon)$ uniformly in η, we have

$$\lim_{\epsilon\to 0}\left|\frac{1}{2\epsilon}\int_0^{2\pi}\cos\frac{nx}{2}[A(x+\epsilon)-A(x-\epsilon)]^2dx - \frac{1}{2\epsilon}\int_0^{2\pi}\cos\frac{nx}{2}[\Phi(x+\epsilon)-\Phi(x-\epsilon)]^2dx\right| < K\eta,$$

where K is independent of η.

We have, however,

$$\lim_{\epsilon\to 0}\frac{1}{2\epsilon}\int_0^{2\pi}\cos\frac{nx}{2}[\Phi(x+\epsilon)-\Phi(x-\epsilon)]^2dx = \int_{\pm 0}^{2\pi-0}\cos\frac{nx}{2}\,d\psi_\eta(x),$$

where $\psi_\eta(x)$ is the sum of the squares of the jumps of Φ in $(\pm 0, x)$. This latter integral differs from

$$\int_{\pm 0}^{2\pi-0}\cos\frac{nx}{2}\,d\psi(x)$$

by less than $\eta A(2\pi)$. Hence

$$\lim_{\epsilon\to 0}\left|\frac{1}{2\epsilon}\int_0^{2\pi}\cos\frac{nx}{2}[A(x+\epsilon)-A(x-\epsilon)]^2dx - \int_{\pm 0}^{2\pi-0}\cos\frac{nx}{2}\,d\psi(x)\right| \leq \eta(K+A(2\pi)).$$

In other words,

$$\lim_{\epsilon\to 0}\frac{1}{2\epsilon}\int_0^{2\pi}\cos\frac{nx}{2}[A(x+\epsilon)-A(x-\epsilon)]^2dx = \int_{\pm 0}^{2\pi-0}\cos\frac{nx}{2}\,d\psi(x),$$

and B is regular.

By a theorem of Fréchet, a monotone function $A(x)$ may conceivably contain three additive parts — a monotone step-function, a monotone function which is the integral of its derivative, and a monotone continuous function which has almost everywhere a derivative zero. The first type of function A is to be found in periodic arrays, for instance. The second type has already been illustrated. The necessary and sufficient condition for the first part to be lacking is that A be continuous, or what is the same thing, that $c_0=0$. If $A(x)$ contains only a part of the second type, we may write

$$A'(x)=\sum_{-\infty}^{\infty} b_j e^{ijx},$$

where A' is positive, integrable, and periodic. Hence, by a well-known theorem,

$$\lim_{j=\pm\infty} b_j=0.$$

If, therefore, we have at the same time

$$\begin{cases} c_0=0; \\ \lim\limits_{j=\pm\infty} b_j\neq 0 \quad \text{or} \quad \overline{\lim\limits_{j=\pm\infty}}\, b_j\neq 0, \end{cases}$$

we know that the third possible part of A, the monotone function with derivative almost everywhere zero, does not vanish. An example of an array of this sort is given by Mr. Mahler in the following paper.

We now wish to consider certain functions connected with an array, which in a certain sense possess a spectrum. Let $\mu(x)$ be any function which differs from 0 only over a finite interval, and is bounded. Let A be a regular array. Let us form the function

$$M(x)=\sum_{-\infty}^{\infty} a_k \mu(x-k).$$

We have

$$\begin{aligned}
\frac{1}{2T}\int_{-T}^{T} & M(x)\overline{M}(x-t)dx \\
&=\frac{1}{2T}\sum_{j=-\infty}^{\infty}\sum_{k=-\infty}^{\infty} a_j\bar{a}_k \int_{T}^{T}\mu(x-j)\bar{\mu}(x-k-t)dx \\
&=\frac{1}{2T}\sum_{j=-\infty}^{\infty}\sum_{l=-\infty}^{\infty} a_j\bar{a}_{j-l}\int_{-T}^{T}\mu(x-j)\bar{\mu}(x-j+l-t)dx \\
&=\sum_{l=-\infty}^{\infty}\sum_{j=-\infty}^{\infty}\frac{a_j\bar{a}_{j-l}}{2T}\int_{T-j}^{T-j}\mu(x)\bar{\mu}(x+l-t)dx.
\end{aligned}$$

It is easy to show that

$$\lim_{T\to\infty}\sum_{j=-\infty}^{\infty}\frac{a_j\bar{a}_{j-l}}{2T}\int_{-T-j}^{T-j}\mu(x)\bar{\mu}(x+l-t)dx=b_l\int_{-\infty}^{\infty}\mu(x)\bar{\mu}(x+l-t)dx.$$

Hence

$$\lim_{T\to\infty}\frac{1}{2T}\int_{-T}^{T}M(x)\bar{M}(x-t)dx=\sum_{l=-\infty}^{\infty}b_l\int_{-\infty}^{\infty}\mu(x)\bar{\mu}(x+l-t)dx.$$

This new function, which we shall call $g(t)$, is continuous in t. We have

$$\underset{A\to\infty}{\text{l.m.}}\ \frac{1}{\pi}\int_{-A}^{A}g(t)\frac{\sin ut}{t}dt$$

$$=\underset{A\to\infty}{\text{l.m.}}\sum_{l=-\infty}^{\infty}\frac{b_l}{\pi}\int_{-A}^{A}\frac{\sin ut}{t}dt\int_{-\infty}^{\infty}\mu(x)\bar{\mu}(x+l-t)dx$$

$$=\underset{N\to\infty}{\text{l.m.}}\sum_{l=-N}^{N}\frac{b_l}{\pi}\int_{-\infty}^{\infty}\frac{\sin ut}{t}dt\int_{-\infty}^{\infty}\mu(x)\bar{\mu}(x+l-t)dx$$

$$=\underset{N\to\infty}{\text{l.m.}}\int_{0}^{u}\sum_{l=-N}^{N}\frac{b_l}{\pi}\left\{\int_{-\infty}^{\infty}\cos ut\,dt\int_{-\infty}^{\infty}\mu(x)\bar{\mu}(x+l-t)dx\right\}du$$

$$=\underset{N\to\infty}{\text{l.m.}}\int_{0}^{u}\sum_{l=-N}^{N}\frac{b_l}{2\pi}\left\{\left|\int_{-\infty}^{\infty}\mu(x)e^{uix}dx\right|^2e^{uil}+\left|\int_{-\infty}^{\infty}\mu(x)e^{-uix}dx\right|^2e^{-uil}\right\}du$$

$$=\underset{N\to\infty}{\text{l.m.}}\int_{-u}^{u}\left|\int_{-\infty}^{\infty}\mu(x)e^{uix}dx\right|^2d\left\{\sum_{-N}^{N}{}'\frac{b_l}{\pi}\frac{e^{inl}}{il}+b_0u\right\}$$

$$=\int_{-u}^{u}\left|\int_{-\infty}^{\infty}\mu(x)e^{uix}\,dx\right|^2dA(u)=R(u).$$

This expression $R(u)$ is a monotone function, like A, and differs from A only in vertical scale, by the factor

$$\frac{\left|\int_{-\infty}^{\infty}\mu(x)e^{uix}dx\right|^2+\left|\int_{-\infty}^{\infty}\mu(x)e^{-uix}dx\right|^2}{2}.$$

As we have shown in the papers already cited $R(u)$ may be regarded as the spectral function of the function $M(x)$, in an analogous sense to that in which $A(x)$ may be regarded as the spectral function of the array A. We thus see how we may construct functions with continuous spectra, with or without a spectral density. This is important for the existence theory of white light.

Comments on [27a]

P. Masani

In this paper Wiener develops the generalized harmonic analysis of functions defined on the group $\mathbb{N}$ of integers along the lines of the theory worked out for the group $\mathbb{R}$ in [25c] and [26b]. For review purposes we shall refer instead to the more accessible and complete sources [30a] and [33i].

An array $a(\cdot)$, i.e. a complex-valued function on $\mathbb{N}$, is called *regular*, iff

$$\forall n \in \mathbb{N}, \qquad b(n) \underset{\mathrm{d}}{=} \lim_{k\to\infty} \frac{1}{2k+1} \sum_{j=-k}^{k} a(j)\overline{a(j-n)} \quad \text{exists.}$$

The regular arrays are thus the analogues of signals f on $\mathbb{R}$ of the class S in [33i, p. 150]. We may call $b(\cdot)$ the *covariance sequence* of $a(\cdot)$. As is easily seen, $b(\cdot)$ is positive definite on $\mathbb{N}$, and hence

$$b(n) = \frac{1}{\sqrt{(2\pi)}} \int_0^{2\pi} e^{-nix}\, dA(x),$$

where $A(\cdot)$ is a bounded, monotone increasing function on $[0, 2\pi]$, called the *spectral distribution* of $a(\cdot)$. As in the case of $\mathbb{R}$ Wiener does not introduce $A(\cdot)$ in this way but defines it by

$$\sqrt{(2\pi)} \cdot A(x) = b(0)\cdot x + \underset{k\to\infty}{\mathrm{l.i.m.}} \sum_{0<|j|\leqslant k} b(j)e^{ijx}/ij,$$

and then shows that $A(\cdot)$ has a realization that is monotone increasing.

In analogy with [33i, p. 138, Thm. 20 & (20.04)(20.05)] Wiener concludes that for a regular $a(\cdot)$

$$\sum_{j=-\infty}^{\infty} |a(j)|^2/j^2 < \infty,$$

and that

$$\sqrt{2\pi}\cdot \phi(x) \underset{\mathrm{d}}{=} a(0)\cdot x + \underset{k\to\infty}{\mathrm{l.i.m.}} \sum_{j=-k}^{k} a(j)e^{ijx}/ij$$

exists (p. 147). We may call $\phi(\cdot)$ the *generalized Fourier-transform* of $a(\cdot)$. He now proves that for all continuous functions $P(x)$ with period 2π,

$$\overline{\lim_{h\to 0}}\left|\frac{3}{4\pi h^3}\int_0^{2\pi} P(x)\left|\left(\int_x^{x+h} - \int_{x-h}^{x}\right)\phi(\xi)d\xi\right|^2 dx\right| \leqslant \max|P(x)|\cdot b(0).$$

Actually, the stronger and more important equality

$$\lim_{h\to 0}\frac{1}{4\pi h}\int_0^{2\pi} P(x)|\phi(x+h) - \phi(x-h)|^2 dx = \frac{1}{\sqrt{(2\pi)}}\int_0^{2\pi} P(x)dA(x)$$

prevails, this being the analogue of [30a, (7.24)]. But the inequality suffices to prove the following theorem (p. 151):

Theorem. *If for a regular array $a(\cdot)$, $A(\cdot)$ has no intervals of constancy, then $\phi(\cdot)$ cannot have a bounded derivative in any interval.*

Since for an uncorrelated array $a(\cdot)$, we have $b(j) = \delta_{j0}$ and $A(x) = \sqrt{(2\pi)}x$ on $[0, 2\pi]$, it follows that $\phi(\cdot)$ does not have a bounded derivative on any subinterval. Wiener points out that *almost all arrays $a(\cdot)$ such that $a(n) = \pm 1$ and $a(-n) = a(n)$ are uncorrelated*; here the probability measure is the transplantation of the Lebesgue measure over [0, 1] via the mapping that assigns to the real number $x = 0.\,x_1 x_2 \ldots$ (binary scale) the array $a(\cdot)$ for which $a(-n) = a(n) = 1, -1$ according as $x_n = 1, 0$. Thus almost all arrays $a(\cdot)$ have the absolutely continuous spectral distribution $A(\cdot)$, given by $A(x) = \sqrt{(2\pi)}x$ on $[0, 2\pi]$. On p. 151 Wiener gives the actual construction of such an array. In part II of the paper K. Mahler gives the construction of an array for which $A(\cdot)$ is a singular continuous function.

Next Wiener associates with an array $a(\cdot)$ supported on the set $\mathbb{N}_+$ of positive integers, the holomorphic function $\psi(\cdot)$ defined by

$$\psi(z) = \sum_{n=0}^{\infty} \{a(n)/n\} z^n, \quad |z| < 1.$$

From the last Theorem he easily deduces that *if $A(\cdot)$ has no intervals of constancy, then $\psi(\cdot)$ has the unit circumference as a natural boundary.*

Next, Wiener shows that *if the array $a(\cdot)$ is regular, so is the Re $a(\cdot)$*, (p. 152). To prove this surprisingly nontrivial result, Wiener appeals to the Tauberian lemma, that for $c(n) \geqslant 0$, $n \geqslant 0$, if one of the limits

$$\lim_{n\to\infty}\frac{1}{n+1}\sum_{k=0}^{n} c(n) \quad \& \quad \lim_{\varepsilon\to 0}\frac{2}{\varepsilon}\sum_{k=0}^{n} c(k)\frac{\sin^2 k\varepsilon}{k^2}$$

exists, then so does the other and the two are equal. This last is the discrete analogue of the more familiar result [33i, pp. 139, Thm. 21], and is alluded to in [27c, p. 123].

Finally, Wiener affects a transition from the discrete to the continuous by associating with any regular array $a(\cdot)$ a signal f on $\mathbb{R}$ defined by

$$(*) \qquad f(t) = \sum_{k=-\infty}^{\infty} a(k)\mu(t-k),$$

μ being any (measurable) bounded function on $\mathbb{R}$ with pre-compact support (p. 155). He shows that such an f has a continuous auto-correlation function $\phi(\cdot)$ on $\mathbb{R}$ given by

$$\phi(t) = \sum_{k=-\infty}^{\infty} b(k)(\mu^* \hat{\mu})(t-k),$$

where $\hat{\mu}(s) = \overline{\mu(-s)}$ and * denotes convolution, and that its spectral distribution $F(\cdot)$ is given by

$$F(\lambda) = 2\pi \int_{-\lambda}^{\lambda} |\tilde{\mu}(u)|^2 \, dA(u),$$

where $\tilde{\mu}$ is the (indirect) Fourier transform of $\mu(\cdot)$. Thus f is in the class S' of [33i, p. 150]. So the equation (*) provides a method of construction members of S' having spectral distributions F of a wide variety.

This paper is not definitive in that the analogues of several results for $\mathbb{R}$ have been left out. Also, Wiener seems to have been unaware of the utility of Herglotz's theorem on positive-definite functions. Nevertheless the paper is a valuable contribution. It demonstrates clearly how the haphazardness of $a(\cdot)$ entails the smoothness of the spectral distribution $A(\cdot)$ and the irregularity of the generalized Fourier transform $\phi(\cdot)$. It indicates how generalized harmonic analysis can contribute to the study of the sample sequences of stochastic sequences, and how the haphazardness of the Taylor coefficients of a function shapes its domain of holomorphy. It shows that a large class of signals on $\mathbb{R}$, amenable to generalized harmonic analysis, are weighted moving averages of haphazard impulses. All this has stimulated important new work, too copious to cite here, and the ideas employed have proved to be of lasting significance.

Only Part I of this paper is by Wiener. Part II is solely by K. Mahler, and therefore is not reprinted in these works. The circumstances that led to this form of collaboration, as kindly narrated to the reviewer by Professor Mahler, are as follows:

It fell on Mahler to act as Wiener's contact at Göttingen during the latter's visit in 1927. Wiener, who already had the results of Part I, posed the problem of finding an array $a(\cdot)$ with a continuous, purely singular spectral distribution $A(\cdot)$. Mahler was able to solve this problem in the light of his acquaintance with the following results:

Thm. *Let* $g(x) = \prod_{n=0}^{\infty} (1 - x^{2^n}) = \sum_{n=0}^{\infty} \rho(n)x^n$, *so that* $\rho(n) = 1, -1, -1, 1, -1, 1, \ldots$. *Then* $g(x)$ *is transcendental for algebraic numbers* x *such that* $0 < |x| < 1$.

Thm. $f(z) =: \sum_{n=1}^{\infty} \rho(n)n^{-z}$ *defines an entire function* f *having zeros at* $0, -1, -2, \ldots$.

Wiener translated Mahler's solution to the problem into English and sent it for publication to the MIT Journal, as Part II of this paper.

The Operational Calculus.

Von

Norbert Wiener in Cambridge (Mass. U. S. A.).

Table of Contents.

Introduction.

The operational calculus[1]) owes its inception to Leibniz, who was struck by the resemblance between the formula for the n-fold differentiation of a product and the formula for the nth power of a sum. Lagrange carried this analogy further, and set up a regular algorithm in which the differential operator plays the part of a sort of hypercomplex unit. It is to him, for example, that we owe the symbolic formula

$$f(x+h) = \left(e^{h\frac{d}{dx}}\right)(f(x)).$$

This operational calculus was further developed by many mathematicians, including Laplace, and more especially Boole, who employed it effectively

[1]) For a thorough historical account of the earlier phases of this subject, see S. Pincherle, Funktionaloperationen und -gleichungen, Enc. Math. Wiss. II A 11, or Équations et opérations fonctionnelles, Encyclopédie des sciences mathématiques pures et appliquées, tome II, vol. 5, fascicule 1.

in the well-known theory of the linear differential equation with constant coefficients. Other names which deserve particular mention in connection with the history of the theory are those of Arbogast and Oltramare.

The earlier work on the subject is of a purely formal character. The developments obtained are often of the nature of series, but there is little or no attempt to investigate their convergence. The fundamental notion of operator is but ill defined. For example, in the equation of Lagrange just cited, the operation on the left-hand side is applicable to any well-defined function whatever, while that on the right-hand side demands that $f(x)$ be analytic over a region bounded by two parallels to the axis of reals.

In recent years, the operational calculus has experienced a growth in two somewhat antithetical directions. On the one hand, we have the work of the rigorists, who start with a clear definition of an operator and of equations between operators. Chief among these are Pincherle [2]), who considers his operators as transformations which change power series into power series, and Volterra [3]), whose operators are first, such integral transformations as

$$\int_0^x f(\xi)\, F(\xi - x)\, d\xi,$$

and secondarily, such transformations as may be obtained by compounding these transformations with the inverse of a transformation such as

$$\frac{1}{(m-1)!}\int_0^x f(\xi)\,(\xi - x)^{m-1}\, d\xi.$$

On the other hand, we have the practical application of operational methods by O. Heaviside [4]) to the solution of the partial differential equations of the electric circuit. These two developments have not met.. The brilliant work of Heaviside is purely heuristic, devoid even of the pretense to mathematical rigor. Its operators apply to electric voltages and currents, which may be discontinuous and certainly need not be analytic. For example, the favorite *corpus vile* on which he tries out his operators is a function which vanishes to the left of the origin and is 1 to the right. This excludes any direct application of the methods of Pincherle. On the other hand, he makes much use of expansions of operators in in-

[2]) S. Pincherle, Atti R. Accad. Lincei Rendic. (5) **4**, 1 (1895), p. 142; Math. Annalen **49** (1897), p. 349.

[3]) V. Volterra, Leçons sur les fonctions de lignes, Paris, Gauthier-Villars (1913); Functions of Composition, The Rice Institute Pamphlet **7**, N° 4, October 1920.

[4]) O. Heaviside, On Operators in Physical Mathematics, Roy. Soc. Proc. **52** (1893), p. 504; **54** (1893), p. 105.

finite series of powers of d/dt, leading to asymptotic series of much value for computation. Such series are not treated by Volterra, whose method is indeed at its best in treating operators of what we may call the integral type.

Although Heaviside's developments have not been justified by the present state of the purely mathematical theory of operators, there is a great deal of what we may call experimental evidence of their validity, and they are very valuable to the electrical engineer. There are cases, however, where they lead to ambiguous or contradictory results. It has hence become important to put them on a sound mathematical basis or failing that, to establish heuristically criteria for the avoidance of contradiction. V. Bush[5]) has given a set of formal criteria of this type, while Carson[6]) has developed a theory of the Volterra type from the standpoint of the engineer, and has amplified it by a somewhat indirect justification of Heaviside's asymptotic series. It is here necessary to refer also to the work of Fry[7]), Berg[8]), and Bromwich[9]).

Nevertheless, a thoroughly satisfactory foundation for the Heaviside theory does not as yet exist. Carson[10]) himself says that his deduction of the asymptotic expansions 'entirely lacks the extraordinary simplicity and directness of the Heaviside solution'. Moreover, he establishes his asymptotic solution separately in each individual case, and does not make it the central feature of a general theory. The problem of obtaining a rigorous interpretation of Heaviside's theory which shall modify the form of his developments as little as possible is therefore still open[11]).

---

5) V. Bush, Note on Operational Calculus, Journ. Math. Phys. Massachusetts Institute of Technology **3** (1924), pp. 95–107.

6) J. R. Carson, Theory of the Transient Oscillations of Electrical Networks and Transmission Systems, Proc. Amer. Inst. Elec. Engineers 1919, pp. 345–427; The Heaviside Operational Calculus, The Bell System Technical Journal, November, 1922, pp. 1–13.

7) T. C. Fry, The Application of Modern Theories of Integration to the Solution of Differential Equations, Annals of Mathematics 22 (1920/21), pp. 182–211.

8) E. G. Berg, Heaviside's Operators in Engineering and Physics, Journ. Franklin Institute **198** (1924), pp. 647–702.

9) T. J. I'A. Bromwich, Normal Coordinates in Dynamical Systems, Proc. Lond. Math. Soc. **15** (1916), pp. 401–445.

10) Loc. cit. p. 409.

11) A justification of the method of Carson's second paper which is at once general in standpoint and rigorous is contained in a paper by Gronwall and Carson, to appear shortly in the Bull. Amer. Math. Soc. The precise conditions of validity of the methods there developed are in some respects more general, in others less general, than those of the present paper. Cf. also J. R. Carson, Electric Circuit Theory and the Operational Calculus, The Bell System Technical Journal, October, 1925, pp. 685–763.

There are several paths which seem to lead to this goal. Besides the Volterra theory of permutable kernels and the Pincherle theory of transformations of power series, the Laplace transformation[12]) and the Fourier integral appear to be promising tools. Like the theory of Pincherle, however, the theory of the Laplace transformation is directly applicable only to analytic functions. The Fourier integral, which may be regarded as derived from a complex form of the Laplace transformation, is not open to this objection. On the other hand, the functions to which the classical form of the Fourier integral applies are subject to very severe restrictions as to their behavior at infinity. These restrictions are far more severe than those to which Heaviside subjects the functions to which he applies his methods. We must therefore generalize the Fourier integral. This has been done in part by Fry in the paper already cited and in part by the author[13]). It is done to a still wider extent in the first section of the present paper.

When applied to the function e^{nit}, the operator $f(d/dt)$ is equivalent to the multiplier $f(ni)$. The result of applying a given operator to a given Fourier integral may naturally be conceived, then, as the multiplication of each e^{nit} term in the integral by a multiplier depending only on n. That is, the operator d/dt has no particular location in the complex plane, but ranges up and down the entire axis of imaginaries. It is thus too much to expect in general that any particular series expansion or other analytic representation of f should make $f(d/dt)$ converge when applied to a purely arbitrary function. In the present paper, the device is adopted of dissecting a function into a finite or infinite number of ranges of frequency, and of applying to each range the particular expansion of an operator yielding results convergent over that range.

It is by this method of dissection that Heaviside's asymptotic series are justified. If we apply an operator of a certain specified type to a function f of a certain specified type, we obtain a function g with a behavior at infinity essentially the same as that of the function g_λ obtained by applying our operator to the function f_λ which consists, in essence, of the components of f with frequencies within a certain restricted range. Thus if the operator is representable in a convergent series expansion, when applied to functions containing only frequencies withen the given range, we get an asymptotic expansion of $g_\lambda(x)$.

[12]) Cf. Probleme aus der Theorie der Wärmeleitung. Erste Mitteilung. F. Bernstein u. G. Doetsch. Zweite Mitteilung. G. Doetsch. Math. Zeitschr. 22 (1925), pp. 284—306.

[13]) N. Wiener, On the Representation of Functions by Trigonometrical Integrals, to appear in the Math. Zeitschrift; The Solution of a Difference Equation by Trigonometrical Integrals, Journ. Math. Phys. Massachusetts Institute of Technology, 4 (1925), pp. 153—164.

It may be said in passing that many of the apparent contradictions found by the electrical engineer in the work of Heaviside are due to a misunderstanding of the nature of an asymptotic expansion. Thus the asymptotic expansion of 0 for positive values of t is also an asymptotic expansion for e^{-t}. There is, however, another point where the theory of Heaviside falls into real contradictions. Both the Volterra theory of permutable kernels and the operational calculus here developed are in essence methods of determining the arbitrary constants in the solution of a linear differential equation of constant coefficients. In the operational calculus to which the Volterra theory leads, these are so determined that the result of applying an operator to a function vanishing to the left of a given point itself vanishes to the left of that point. In my theory, on the other hand, the result of applying an operator to a function behaving algebraically at $\pm\infty$ must itself behave algebraically — or at least, must not become exponentially infinite — at $\pm\infty$. These requirements may be in conflict. For example, the general solution of the differential equation

$$\frac{du}{dt} - u = \varphi(t),$$

where

$$\varphi(t) = 0 \qquad [t < 0];$$

$$\varphi(t) = 1 \qquad [t > 0]$$

is

$$u = \varphi(t)\,[e^t - 1] + c\,e^t.$$

The solution corresponding to an operational calculus of the Volterra-Carson type is

$$\varphi(t)\,[e^t - 1],$$

whereas the solution corresponding to an operational calculus of my type is

$$\varphi(t)\,[e^t - 1] - e^t.$$

Now, Heaviside's expansions in negative powers of d/dx correspond to an operational calculus of the Volterra-Carson type, while his expansions in positive powers of d/dx correspond to a theory such as that developed in the present paper. It is thus possible by following out his methods to make very real errors.

There is a large class of cases, however, where the determination of the constants of integration of a differential equation in accordance with the Volterra theory and that here developed are in complete harmony. In these cases, the operators of my theory depend only on the past of the functions to which they apply. I call such operators ***retrospective.*** In an electrical problem, all operators arising from a system with only positive resistances and leakages will be retrospective. All other operators correspond to electrical systems where there is at some point an unlimited input of energy and hence correspond only to such exceptional conditions as thermionic valves.

Our operators $f(d/dt)$ need special treatment in the case where they possess poles or branch-points on the axis of imaginaries, as well as when their behavior at infinity is too complicated. We shall discuss this, and shall investigate the degree of indeterminateness caused in the problem of inverting an operator by such singularities. We shall here confine ourselves to the simplest cases.

We shall finally investigate the application of our operators to the solution of partial differential equations.

1. The Trigonometric Expansion of a Function of Finite Order at Infinity.

Let $f(x)$ be a function which is summable and of summable square over every finite range, and let there be numbers A and k such that

$$f(x) \leqq A x^k$$

for every x of sufficient positive or negative magnitude. In general, $f(x)$ will not possess a Fourier integral expansion. There is nevertheless a type of extended Fourier integral by which $f(x)$ may be represented.

To establish this, we shall have need of a certain set of auxiliary functions. Let $\Phi(u)$ be a function which is defined over the interval (0, 1), which has at every point derivatives of all orders, which vanishes at 1 and equals 1 at 0, and which has its derivatives of every order vanish at 0 and at 1. We may, for example, put $\Phi(u) = \int_x^1 e^{\frac{1}{\xi^2-\xi}} d\xi \quad \int_0^1 e^{\frac{1}{\xi^2-\xi}} d\xi$. Let $\Psi(\xi, \lambda)$ be defined by the equation

$$\Psi(\xi, \lambda) = \frac{\sin \lambda \xi}{\pi \xi} + \frac{1}{\pi} \int_0^1 \Phi(u) \cos(u + \lambda) \xi \, du.$$

I then wish to prove that

$$\int_{-\infty}^{\infty} f(\xi) \Psi(\xi - x, \lambda) d\xi$$

exists, and converges in the mean to $f(x)$ over any finite range as $\lambda \to +\infty$.

To begin with,

$$|\Psi(\xi,\lambda)| = \left| \frac{\sin\lambda\xi}{\pi\xi} + \frac{\Phi(1)\sin(\lambda+1)\xi - \Phi(0)\sin\lambda\xi}{\pi\xi} - \frac{1}{\pi\xi}\int_0^1 \Phi'(u)\sin(u+\lambda)\xi\,du \right.$$

$$= \left| \frac{1}{\pi\xi}\int_0^1 \Phi'(u)\sin(u+\lambda)\xi\,du \right|$$

$$= \left| \frac{1}{\pi\xi^2}\int_0^1 \Phi''(u)\cos(u+\lambda)\xi\,du \right|$$

$$= \left| \frac{1}{\pi\xi^m}\int_0^1 \Phi^m(u) \begin{matrix}\cos\\ \sin\end{matrix} (u+\lambda)\xi\,du \right|$$

$$\leqq \frac{\int_0^1 |\Phi^{(m)}(u)|\,du}{\pi|\xi^m|}$$

Hence there is a ϑ such that $|\vartheta| \leqq 1$, and

$$\int_a^b dx \left\{ \int_{-\infty}^{\infty} f(\xi)\Psi(\xi-x,\lambda)\,d\xi - f(x) \right\}^2$$

$$= \int_a^b dx \left\{ \int_{-A}^{A} f(\xi)\Psi(\xi-x,\lambda)\,d\xi - f(x) \right.$$

$$\left. + \frac{\vartheta}{\pi}\int_0^1 |\Phi^{(k+2)}(u)|\,du \left\{ \left[\int_{-\infty}^{-A} + \int_A^{\infty} \right] \frac{f(\xi)\,d\xi}{(\xi-x)^{k+2}} \right\} \right\}^2$$

$$= \int_a^b dx \left\{ \int_{-A}^{A} f(\xi)\Psi(\xi-x,\lambda)\,d\xi - f(x) + \varepsilon(A) \right\}^2$$

$$\leqq \left[\sqrt{\int_a^b dx \left\{ \int_{-A}^{A} f(\xi)\Psi(\xi-x,\lambda)\,d\xi - f(x) \right\}^2} + \varepsilon(A)\sqrt{b-a} \right]^2 .$$

Here $\varepsilon(A)$ is some quantity depending on A alone, and vanishing as A increases without limit.

We shall therefore have proved the desired convergence in the mean if we can show that for all sufficiently large A,

$$\lim_{\lambda\to\infty} \int_a^b dx \left\{ \int_{-A}^{A} f(\xi)\Psi(\xi-x,\lambda)\,dx - f(x) \right\}^2 = 0 .$$

That is, it is enough to prove that if $g(x)$ vanishes for sufficiently large values of x,

$$\lim_{\lambda\to\infty}\int_{-\infty}^{\infty} dx\left\{\int_{-\infty}^{\infty} g(\xi)\,\Psi(\xi-x,\lambda)\,d\xi - g(x)\right\}^2 = 0.$$

By hypothesis, we may write

$$g(x) = \underset{B\to\infty}{\text{l. m.}}\int_{-B}^{B}\varphi(\xi)\,e^{i\xi x}\,d\xi.$$

in accordance with a theorem of Plancherel[14]. We shall then have

$$\int_{-\infty}^{\infty} g(\eta)\,\Psi(\eta-x,\lambda)\,d\eta = \int_{-\infty}^{\infty} g(x-\eta)\,\Psi(\eta,\lambda)\,d\eta$$

$$= \int_{-\infty}^{\infty}\left[\underset{B\to\infty}{\text{l. m.}}\int_{-B}^{\infty}\varphi(\xi)\,e^{i\xi(x-\eta)}\,d\xi\right]\left[\frac{\sin\lambda\eta}{\eta}+\frac{1}{\pi}\int_0^1 \Phi(u)\cos(u+\lambda)\,\eta\,du\right]d\eta$$

$$= \int_{-\infty}^{\infty}\varphi(\xi)\,e^{i\xi x}\,d\xi\int_{-\infty}^{\infty}\frac{e^{-i\eta\xi}}{\pi}\left[\frac{\sin\lambda\eta}{\eta}+\int_0^1\Phi(u)\cos(u+\lambda)\,\eta\,du\right]d\eta$$

$$= \frac{1}{2\pi}\int_{-\infty}^{\infty}\varphi(\xi)\,e^{i\xi x}\,d\xi\int_{-\infty}^{\infty}e^{-i\eta\xi}\left[\int_{-\infty}^{\infty}F_\lambda(u)\,e^{iu\eta}\,du\right]d\eta,$$

where

$$F_\lambda(u) = 1; \qquad [-\lambda \leqq u \leqq \lambda].$$
$$F_\lambda(u) = \Phi(u-\lambda); \qquad [\ \lambda \leqq u \leqq \lambda+1].$$
$$F_\lambda(u) = \Phi(\lambda-u); \qquad [-\lambda-1 \leqq u \leqq -\lambda].$$
$$F_\lambda(u) = 0; \qquad [\ |u| \geqq \lambda+1].$$

Therefore, by the theory of the Fourier integral,

$$\int_{-\infty}^{\infty} g(\eta)\,\Psi(\eta-x,\lambda)\,d\eta = \int_{-\infty}^{\infty}\varphi(\xi)\,F_\lambda(\xi)\,e^{i\xi x}\,d\xi,$$

and

$$\int_{-\infty}^{\infty} dx\left\{\int_{-\infty}^{\infty} g(\xi)\,\Psi(\xi-x,\lambda)\,d\xi - g(x)\right\}^2$$

$$= \int_{-\infty}^{\infty} dx\left[\underset{B\to\infty}{\text{l. m.}}\int_{-B}^{B}\varphi(\xi)\,[1-F_\lambda(\xi)]\,e^{i\xi x}\,d\xi\right]^2$$

$$= 2\pi\int_{-\infty}^{\infty}[\varphi(\xi)(1-F_\lambda(\xi))]^2\,d\xi.$$

[14]) M. Plancherel, Contribution à l'étud*e* de la représentation d'une fonction arbitraire par des intégrales définies, Rend. Palermo **30** (2nd sem. 1910) pp. 289–335.

From this it is easy to conclude that

$$\lim_{n\to\infty} \int_{-\infty}^{\infty} dx \left\{ \int_{-\infty}^{\infty} g(\xi)\,\Psi(\xi - x, \lambda)\, d\xi - g(x) \right\}^2 = 0,$$

and hence that

$$f_\lambda(x) = \int_{-\infty}^{\infty} f(\xi)\,\Psi(\xi - x, \lambda)\, d\xi$$

converges in the mean to $f(x)$ as λ becomes infinite.

It will be noticed that $f_\lambda(x)$ may be regarded as consisting of all the components of $f(x)$ with period not exceeding $2\pi\lambda$, together with a portion of the components of $f(x)$ with period exceeding $2\pi\lambda$ but not $2\pi(\lambda+1)$. We have here a method of dividing $f(x)$ into the sum of a number of functions, each of wich contains only harmonics restricted to a given range. This is the key-note of the theory which follows.

2. The Definition of a Non-Singular Operator.

Now let $H(u)$ be a function defined for all pure imaginary values of u, and assuming, in general, complex values. It will be natural to interpret $H(d/dx)\,\Psi(\xi - x, \lambda)$ as

$$G(\xi - x, \lambda) = \frac{1}{2\pi} \int_{-\infty}^{\infty} F_\lambda(u)\, H(-iu)\, e^{iu(\xi - x)}\, du.$$

Accordingly, the natural interpretation of

$$H(d/dx) \int_{-\infty}^{\infty} f(\xi)\,\Psi(\xi - x, \lambda)\, d\xi$$

will be

$$\int_{-\infty}^{\infty} f(\xi)\, G(\xi - x, \lambda)\, d\xi.$$

In case H possesses at every point of the closed interval $(-\lambda i - i, \lambda i + i)$ derivatives of all orders, we may show as in the last paragraph that $G(\xi - x, \lambda)$ is $o(x^{-k})$ at infinity for any positive k, and that

$$\int_{-\infty}^{\infty} f(\xi)\, G(\xi - x, \lambda)\, d\xi$$

will exist for such functions f as we have been considering. We shall have

$$\int_{-\infty}^{\infty} f(\xi)[G(\xi-x,\lambda_1)-G(\xi-x,\lambda)]\,d\xi \qquad [\lambda_1 \geqq \lambda+1 \geqq 1]$$

$$=\frac{1}{2\pi}\int_{-\infty}^{\infty} f(\xi)\,d\xi\left\{\left[\int_{-\lambda_1-1}^{-\lambda}+\int_{\lambda}^{\lambda_1+1}\right][F_{\lambda_1}(u)-F_{\lambda}(u)]\,H(-iu)\,e^{iu(\xi-x)}\,du\right\}$$

$$=\frac{1}{2\pi}\int_{x-1}^{x+1} f(\xi)\,d\xi\left\{\left[\int_{-\lambda_1-1}^{-\lambda}+\int_{\lambda}^{\lambda_1+1}\right][F_{\lambda_1}(u)-F_{\lambda}(u)]\,H(-iu)\,e^{iu(\xi-x)}\,du\right\}$$

$$+\frac{i^n}{2\pi}\left[\int_{-\infty}^{x-1}+\int_{x+1}^{\infty}\right]\frac{f(\xi)}{(\xi-x)^n}\,d\xi\left\{\left[\int_{-\lambda_1-1}^{-\lambda}+\int_{\lambda}^{\lambda_1+1}\right]\frac{d^n}{du^n}[(F_{\lambda_1}(u)-F_{\lambda}(u))\,H(-iu)]\,e^{iu(\xi-x)}\,du\right.$$

$$=\frac{1}{2\pi}\left[\int_{-\lambda_1-1}^{-\lambda}+\int_{\lambda}^{\lambda_1+1}\right]du\left\{[F_{\lambda_1}(u)-F_{\lambda}(u)]\,H(-iu)\int_{-1}^{1} f(x+\eta)\,e^{iu\eta}\,d\eta\right.$$

$$\left.+\frac{i^n\,d^n}{du^n}[(F_{\lambda_1}(u)-F_{\lambda}(u))\,H(-iu)]\left[\int_{-\infty}^{-1}+\int_{1}^{\infty}\right]\frac{f(x+\eta)}{\eta^n}\,e^{iu\eta}\,d\eta\right\}.$$

Therefore, by a ready extension of Plancherel's theorem on the Fourier transform, combined with the Schwarz inequality,

$$\left|\int_{-\infty}^{\infty} f(\xi)[G(\xi-x,\lambda_1)-G(\xi-x,\lambda)]\,d\xi\right|^2$$

$$\leqq\frac{1}{2\pi^2}\left\{\left|\left[\int_{-\lambda_1-1}^{-\lambda}+\int_{\lambda}^{\lambda_1+1}\right]du\Big[F_{\lambda_1}(u)-F_{\lambda}(u)\Big]H(-iu)\int_{-1}^{1} f(x+\eta)\,e^{iu\eta}\,d\eta\right|^2\right.$$

$$+\left|\left[\int_{-\lambda_1-1}^{-\lambda}+\int_{\lambda}^{\lambda_1+1}\right]du\left\{\frac{d^n}{du^n}\Big[\Big(F_{\lambda_1}(u)-F_{\lambda}(u)\Big)H(-iu)\Big]\left[\int_{-\infty}^{-1}+\int_{1}^{\infty}\right]\frac{f(x+\eta)}{\eta^n}\,e^{iu\eta}\,d\eta\right.\right.$$

$$\leqq\frac{1}{\pi^2}\left\{\operatorname*{Max}_{|v|\geqq\lambda}|H(v)|^2\left|\left[\int_{-\lambda_1-1}^{-\lambda}+\int_{\lambda}^{\lambda_1+1}\right]du\int_{-1}^{1} f(x+\eta)\,e^{iu\eta}\,d\eta\right|^2\right.$$

$$+\operatorname*{Max}_{|v|\geqq\lambda}\left|H^{(n)}(v)\right|^2\left|\left[\int_{-\lambda_1-1}^{-\lambda}+\int_{\lambda}^{\lambda_1+1}\right]du\left[\int_{-\infty}^{-1}+\int_{1}^{\infty}\right]\frac{f(x+\eta)}{\eta^n}\,e^{iu\eta}\,d\eta\right|^2$$

$$+\left|\left[\int_{-\lambda_1-1}^{-\lambda_1}+\int_{-\lambda-1}^{-\lambda}+\int_{\lambda}^{\lambda+1}+\int_{\lambda_1}^{\lambda_1+1}\right]du\sum_{h=1}^{n}\frac{d^h}{du^h}[F_{\lambda_1}(u)-F_{\lambda}(u)]\frac{d^{n-h}}{du^{n-h}}H(-u)\frac{n!}{h!\,(n-\right.$$

$$\left.\left.\left[\int_{-\infty}^{-1}+\int_{1}^{\infty}\right]\frac{f(x+\eta)}{\eta^n}\,e^{iu\eta}\,d\eta\right|^2\right\}.$$

It is now easy to show that if for all k, $\lim_{v \to \pm\infty} H^{(k)}(v) = 0$, then given any ε, λ can be chosen so large that for any λ_1, we shall have

$$\int_{-a}^{a} \left| \int_{-\infty}^{\infty} f(\xi) [G(\xi - x, \lambda_1) - G(\xi - x, \lambda)] \, d\xi \right|^2 dx$$

$$\leqq \frac{1}{\pi^2} \operatorname*{Max}_{|v| \geqq \lambda} |H(v)|^2 \int_{-\infty}^{a} \left| \left[\int_{-\lambda_1 - 1}^{-\lambda} + \int_{\lambda}^{\lambda_1 - 1} \right] du \int_{-1}^{1} f(x+\eta) e^{iu\eta} \, d\eta \right|^2 dx$$

$$+ \frac{1}{\pi^2} \operatorname*{Max}_{|v| \geqq \lambda} |H^{(n)}(v)|^2 \int_{-a}^{a} \left| \left[\int_{-\lambda_1 - 1}^{-\lambda} + \int_{\lambda}^{\lambda_1 + 1} \right] du \left[\int_{-\infty}^{-1} + \int_{1}^{\infty} \right] \frac{f(x+\eta)}{\eta^n} e^{iu\eta} d\eta \right| dx + \varepsilon$$

$$\leqq \frac{1}{\pi^2} \operatorname*{Max}_{|v| \geqq \lambda} |H(v)|^2 \int_{-a}^{a} \left\{ \left\{ \int_{\lambda}^{\lambda_1 + 1} du \int_{-1}^{1} (f(x+\eta) + f(x-\eta)) \cos u\eta \, d\eta \right\}^2 \right.$$

$$\left. + \left\{ \int_{\lambda}^{\lambda_1 + 1} du \int_{-1}^{1} (f(x+\eta) - f(x-\eta)) \sin u\eta \, d\eta \right\}^2 \right\} dx$$

$$+ \frac{1}{\pi^2} \operatorname*{Max}_{|v| \geqq \lambda} |H^{(n)}(v)|^2 \int_{-a}^{a} \left\{ \left\{ \int_{\lambda}^{\lambda_1 + 1} du \left[\int_{-\infty}^{-1} + \int_{1}^{\infty} \right] \frac{f(x+\eta) + f(x-\eta)}{\eta^n} \cos u\eta \, d\eta \right\}^2 \right.$$

$$\left. + \left\{ \int_{\lambda}^{\lambda_1 + 1} du \left[\int_{\infty}^{-1} + \int_{1}^{\infty} \right] \frac{f(x+\eta) - f(x-\eta)}{\eta^n} \sin u\eta \, d\eta \right\}^2 \right\} dx + \varepsilon.$$

Here n is taken to be even.

We shall consider n to be chosen so large that

$$f(x) = o(x^{n+2})$$

at infinity. I then say that all the integrals in this expression will tend to zero with increasing λ. For exemple, we shall have

$$I = \int_{-a}^{a} \left\{ \int_{\lambda}^{\lambda_1 + 1} du \int_{-1}^{1} (f(x+\eta) - f(x-\eta)) \cos u\eta \, d\eta \right\}^2 dx$$

$$\int^{+1} du \int_{\lambda}^{\lambda_1 + 1} dv \int_{-1}^{1} \cos u\xi \, d\xi \int_{-1}^{1} \cos v\eta \, d\eta \int_{-a}^{a} (f(x+\xi) + f(x-\xi))(f(x+\eta) + f(x-\eta)) \, dx.$$

Now, if $\psi(x, \eta)$ is any summable function of summable square, it is easy to prove by expanding it in a double Fourier series that

$$\int_{-1}^{1} \cos u\xi \, d\xi \int_{-1}^{1} \psi(\xi, \eta) \cos v\eta \, d\eta$$

is a function of summable square over an infinite range of both variables.

From this we may conclude that I vanishes with increasing λ, independently of λ_1.

The proof goes through in a similar manner for the other cases, and the factor η^n in the denominator guarantees that all the infinite integrals we obtain from our second integral shall be convergent. Hence if $H(v)$ is bounded at infinity and if all its derivatives vanish at infinity, we have

$$\lim_{\lambda \to \infty} \int_{-a}^{a} \left| \int_{-\infty}^{\infty} f(\xi)[G(\xi - x, \lambda_1) - G(\xi - x, \lambda)] d\xi \right|^2 dx = 0,$$

and

$$\underset{\lambda \to \infty}{\text{l. m.}} \int_{-\infty}^{\infty} f(\xi) G(\xi - x, \lambda) d\xi$$

exists. We shall term this $H(d/dx)f(x)$.

We shall have

$$\int_{-\infty}^{\infty} G_1(\xi - x, \lambda_1) d\xi \text{ l. m.} \int_{-\infty}^{\infty} f(\eta) G_2(\eta + \xi, \lambda_2) d\eta$$

$$= \lim_{\lambda_2 \to \infty} \int_{-\infty}^{\infty} f(\eta) d\eta \int_{-\infty}^{\infty} G_1(\xi - x, \lambda_1) G_2(\eta - \xi, \lambda_2) d\xi$$

$$= \lim_{\lambda_2 \to \infty} \int_{-\infty}^{\infty} f(\eta) d\eta \frac{1}{4\pi^2} \int_{-\infty}^{\infty} d\xi \int_{-\infty}^{\infty} F_{\lambda_1}(u) H_1(-iu) e^{iu(\xi - x)} du \int_{-\infty}^{\infty} F_{\lambda_2}(v) H_2(-iv) e^{iv(\eta -}$$

$$= \lim_{\lambda_2 \to \infty} \int_{-\infty}^{\infty} f(\eta) d\eta \frac{1}{2\pi} \int_{-\infty}^{\infty} F_{\lambda_1}(u) F_{\lambda_2}(u) H_1(-iu) H_2(-iu) e^{iu(\eta - x)} du$$

$$= \int_{-\infty}^{\infty} f(\eta) d\eta \frac{1}{2\pi} \int_{-\infty}^{\infty} F_{\lambda_1}(u) H_1(-iu) H_2(-iu) e^{iu(\eta - x)} du,$$

where H_1 and H_2 are functions everywhere infinitely differentiable, finite at infinity, and with all their derivatives finite at infinity, and G_1 and G_2 bear to them the same relation which G bears to H. If H satisfies this condition as to its differentiability and its order at infinity, we shall call $H(d/dx)$ a *non-singular operator*. It is clear that if $H_1(d/dx)$ and $H_2(d/dx)$ are non-singular, $H_1(d/dx) H_2(d/dx)$ will also be non-singular. We have just seen that the result of applying to $f(x)$ first $H_2(d/dx)$ and then $H_1(d/dx)$ is $H_1(d/dx) H_2(d/dx)$, so that the operational product of two non-singular operators is always a non-singular operator.

A particularly interesting non-singular operator is $F_\lambda(id/dx)$. By multiplying this with an arbitrary operator, we obtain that operator as applied to a given frequency range. The operator thus obtained may be non-singular even when the operator multiplied with $F_\lambda(id/dx)$ is singular. This is the case, for instance, with the operator d/dx itself.

In addition to the non-singular operators, there are other operators of the form $H(d/dx)$ which apply to all summable functions bounded over every finite region which are of finite order at infinity. For example,

$$e^{a d/dx} f_\lambda(x) = f_\lambda(x+a).$$

Hence if we define $e^{a d/dx} f(x)$ as the limit in the mean of $e^{a d/dx} f_\lambda(x)$ as λ becomes infinite, we have

$$e^{a d/dx} f(x) = f(x+a).$$

We can similarly be sure of the existence of

$$H(d/dx) f(x) = \underset{\lambda \to \infty}{\text{l. m.}}\, H(d/dx) f_\lambda(x),$$

where

$$H(d/dx) = e^{a d/dx} J(d/dx),$$

while $J(d/dx)$ is non-singular. An example in point is

$$e^{\sqrt{(d/dx+a)(d/dx+b)}}.$$

We shall call such operators *non-singular in the generalized sense.*

3. The Analytic Expansion of a Non-Singular Operator.

Let $H(u)$ be a function analytic within a circle of radius $r > 1$ about the origin. Let $\lambda < r - 1$. We shall have

$$H\left(\frac{d}{dx}\right)\Psi(\xi - x, \lambda) = \frac{1}{2\pi}\int_{-\infty}^{\infty} F_\lambda(u) H(-iu) e^{iu(\xi - x)}\, du$$

$$= \frac{1}{2\pi}\int_{-\infty}^{\infty} F_\lambda(u)\Big\{ H(0) - iu H'(0) + \dots + \frac{(-iu)^{n-1}}{(n-1)!} H^{(n-1)}(0)$$

$$+ \frac{(-i)^n}{(n-1)!}\int_0^u (u-\eta)^{n-1} H^{(n)}(-i\eta)\, d\eta \Big\} e^{iu(\xi-x)}\, du$$

$$= H(0)\Psi(\xi - x, \lambda) + H'(0)\frac{d\Psi(\xi - x, \lambda)}{dx} + \dots$$

$$+ \frac{H^{(n-1)}(0)}{(n-1)!}\frac{d^{n-1}\Psi(\xi - x, \lambda)}{dx^{n-1}} + \frac{\vartheta(\lambda+1)^{n+1}}{\pi(n-1)!}\max_{|\eta| \leqq \lambda+1} |H^{(n)}(-i\eta)| \qquad [|\vartheta| \leqq 1].$$

We may conclude at once from this that

$$\sum_n H^{(n)}(0)\frac{d^n \Psi}{dx^n}(\xi - x, \lambda)$$

converges uniformly to $H(d/dx)\Psi(\xi - x, \lambda)$ over any range, finite or infinite.

Again,

$$\frac{1}{2\pi}\int_{-\infty}^{\infty} F_\lambda(u)\frac{(-i)^n}{(n-1)!}\int_0^u (u-\eta)^{n-1} H^{(n)}(-i\eta)\,d\eta\, e^{iu(\xi-x)}\,du$$

$$= \frac{1}{2\pi(\xi-x)^k}\int_{-\infty}^{\infty} e^{iu(\xi-x)}\frac{d^k}{du^k}\left[F_\lambda'(u)\frac{(-i)^{n-k}}{(n-1)!}\int_0^u (u-\eta)^{n-1} H^{(n)}(-i\eta)\,d\eta\right]du$$

$$= \frac{1}{2\pi(\xi-x)^k}\int_{-\infty}^{\infty} e^{iu(\xi-x)}\left[\sum_{h=0}^{k}\frac{(-i)^{n-k}\,k!}{(n-1)!\,h!\,(k-h)!}F_\lambda^{(h)}(u)\right.$$

$$\left.\frac{d^{k-h}}{du^{k-h}}\int_0^u (u-\eta)^{n-1} H^{(n)}(-i\eta)\,d\eta\right]du$$

$$= \frac{1}{2\pi(\xi-x)^k}\int_{-\infty}^{\infty} e^{iu(\xi-x)}\left[\sum_{h=0}^{k}\frac{(-i)^{n-h}\,k!}{(n-k+h-1)!\,h!\,(k-h)!}F_\lambda^{(h)}(u)\right.$$

$$\left.\int_0^u (u-\eta)^{n-k+h-1} H^{(n-k+h)}(-i\eta)\,d\eta\right]du$$

$$= \frac{\theta_1}{\pi(\xi-x)^k}\sum_{h=0}^{k}(\lambda+1)^{n-k+h+1}\frac{k!\,F_\lambda^{(h)}(u)}{(n-k+h-1)!\,h!\,(k-h)!}\operatorname*{Max}_{|\eta|\leqq\lambda+1} H^{(n-k+h)}(-i\eta).$$

It follows that

$$\lim_{N\to\infty}(\xi-x)^k\left[\sum_0^N \frac{H^{(n)}(0)}{n!}\frac{d^n}{dx^n}\Psi(\xi-x,\lambda) - H\left(\frac{d}{dx}\right)\Psi(\xi-x,\lambda)\right]=0$$

uniformly in $\xi - x$, whatever k may be.

It results that in case f satisfies the conditions which we have laid down for it,

$$\sum_0^\infty \frac{H^{(n)}(0)}{n!}\frac{d^n}{dx^n}f_\lambda(x) = H\left(\frac{d}{dx}\right)f_\lambda(x)$$

uniformly over any finite range whatever. That is, in case f contains only components of sufficiently low frequency, if we apply to f an operator $H(d/dx)$, we may expand this operator in a Maclaurin series. In order to show this, it is enough to prove that

$$\frac{d}{dx}f_\lambda(x) = \int_{-\infty}^{\infty} f(\xi)\frac{d}{dx}\Psi(\xi-x,\lambda)\,d\xi,$$

with a similar theorem for the derivatives of higher order. Now,

$$\int_{-\infty}^{\infty} f(\xi)\frac{d}{dx}\Psi(\xi-x,\lambda)\,d\xi = \left[\int_{-A}^{A}+\int_{-\infty}^{-A}+\int_{A}^{\infty}\right]f(\xi)\frac{d}{dx}\Psi(\xi_1-x,\lambda)\,d\xi$$

$$= \lim_{A\to\infty}\frac{d}{dx}\int_{-A}^{A} f(\xi)\,\Psi(\xi-x,\lambda)\,d\xi$$

uniformly in x over any finite range of x. Now, it is easy to show that if $\varphi_k(x)$ converges uniformly to $\varphi(x)$ over some closed range of x, while $\varphi_k'(x)$ converges uniformly to $\psi(x)$, then within this range,

$$\psi(x) = \varphi'(x).$$

Thus

$$\int_{-\infty}^{\infty} f(\xi) \frac{d}{dx} \Psi(\xi - x, \lambda)\, d\xi = f_\lambda'(x).$$

In the case where $H(u)$ possesses a Taylor expansion valid within a circle of radius $r > 1$ about the point ia, a precisely similar argument will show that if $\lambda < r - 1$ and

$$g_\lambda(x) = \frac{1}{2\pi} \int_{-\infty}^{\infty} f(\xi)\, d\xi \left[\int_{a-\lambda-1}^{a-\lambda} \Phi(\lambda - a - u)\, e^{iu(\xi - x)}\, du + \int_{a-\lambda}^{a+\lambda} e^{iu(\xi - x)}\, du \right.$$

$$\left. + \int_{a+\lambda}^{a+\lambda+1} \Phi(u - a - \lambda)\, e^{iu(\xi - x)}\, du \right]$$

we shall have

$$H\left(\frac{d}{dx}\right) g_\lambda(x) = \sum_0^\infty \frac{H^{(n)}(ia)}{n!} \left(\frac{d}{dx} - ia\right)^n g_\lambda(x)$$

uniformly over any finite range of x.

We shall now consider Taylor expansions about the point at infinity. Let $H(u)$ possess such an expansion We can then put

$$H(u) = H(\infty) + \left(\frac{1}{u}\right) K\left(\frac{1}{u}\right),$$

where K is analytic in some circle about the origin of radius R^{-1}. Let λ be any number greater than R. We shall have

$$H\left(\frac{d}{dx}\right)[f(x) - f_\lambda(x)] = \frac{1}{2\pi} \underset{\mu\to\infty}{\text{l. m.}} \int_{-\infty}^{\infty} [f(\xi) - f_\lambda(\xi)]\, d\xi \int_{-\infty}^{\infty} F_\mu(u)\, H(-iu)\, e^{iu(\xi - x)}\, du$$

$$= \frac{1}{2\pi} \underset{\mu\to\infty}{\text{l. m.}} \int_{-\infty}^{\infty} f(\xi)\, d\xi \int_{-\infty}^{\infty} F_\mu(u)\, H(-iu)\, e^{iu(\xi - x)}\, du$$

$$- \frac{1}{4\pi^2} \underset{\mu\to\infty}{\text{l. m.}} \int_{-\infty}^{\infty} d\xi \int_{-\infty}^{\infty} f(\eta)\, d\eta \int_{-\infty}^{\infty} F_\lambda(v)\, e^{iv(\eta - \xi)}\, dv \int_{-\infty}^{\infty} F_\mu(u)\, H(-iu)\, e^{iu(\xi - x)}\, du$$

$$= \frac{1}{2\pi} \underset{\mu\to\infty}{\text{l. m.}} \int_{-\infty}^{\infty} f(\xi)\, d\xi \int_{-\infty}^{\infty} F_\mu(u)\, H(-iu)\, e^{iu(\xi - x)}\, du$$

$$- \frac{1}{2\pi} \underset{\mu\to\infty}{\text{l. m.}} \int_{-\infty}^{\infty} f(\eta)\, d\eta \int_{-\infty}^{\infty} F_\mu(u)\, F_\lambda(u)\, H(-iu)\, e^{iu(\eta - x)}\, du$$

37*

$$= \frac{1}{2\pi} \operatorname*{l.m.}_{\mu \to \infty} \int_{-\infty}^{\infty} f(\xi)\, d\xi \int_{-\infty}^{\infty} [F_\mu(u) - F_\lambda(u)]\, H(-iu)\, e^{iu(\xi - x)}\, du$$

$$= H(\infty)[f(x) - f_\lambda(x)] + \frac{1}{2\pi} \operatorname*{l.m.}_{\mu \to \infty} \int_{-\infty}^{\infty} f(\xi)\, d\xi \int_{-\infty}^{\infty} \frac{F_\mu(u) - F_\lambda(u)}{-iu} K\left(\frac{i}{u}\right) e^{iu(\xi - x)}\, du$$

$$= H(\infty)[f(x) - f_\lambda(x)] + \frac{1}{2\pi} \int_{-\infty}^{\infty} f(\xi)\, d\xi \int_{-\infty}^{\infty} \frac{i[1 - F_\lambda(u)]}{u} K\left(\frac{i}{u}\right) e^{iu(\xi - x)}\, du$$

$$= H(\infty)[f(x) - f_\lambda(x)]$$

$$+ \frac{1}{2\pi} \int_{-\infty}^{\infty} f(\xi)\, d\xi \int_{-\infty}^{\infty} \frac{i[1 - F_\lambda(u)]}{u} \left[K(0) + \frac{i}{u} K'(0) + \dots \right.$$

$$\left. + \frac{i^{n-1}}{u^{n-1}(n-1)!} K^{n-1}(0) + \frac{i^n}{(n-1)!} \int_u^{\infty} \left(\frac{1}{u} - \frac{1}{\eta}\right)^{n-1} K^{(n)}\left(\frac{i}{\eta}\right) \frac{d\eta}{\eta^2} \right] e^{iu(\xi - x)}\, du$$

$$= H(\infty)[f(x) - f_\lambda(x)] + K(0) \frac{1}{d/dx} [f(x) - f_\lambda(x)] + \dots$$

$$+ \frac{K^{(n-1)}(0)}{(d/dx)^n (n-1)} [f(x) - f_\lambda(x)] + \int_{-\infty}^{\infty} f(\xi)\, Q(\xi - x)\, d\xi,$$

where $\frac{1}{(d/dx)^n} [f(x) - f_\lambda(x)]$ signifies that solution of the differential equation

$$g^{(n)}(x) = f(x) - f_\lambda(x)$$

which has the property that $g_\mu(x)$ vanishes for sufficiently small values of μ, while

$$Q(w) = \frac{i^{n+1-k}}{2(n-1)!\, \pi w^k} \int_{-\infty}^{\infty} e^{iuw} \frac{d^k}{du^k} \left[\frac{1 - F_\lambda(u)}{u} \int_u^{\infty} \left(\frac{1}{u} - \frac{1}{\eta}\right)^{n-1} K^{(n)}\left(\frac{i}{\eta}\right) \frac{d\eta}{\eta^2} \right] du.$$

It is easy to establish that the analytic character of K makes

$$\lim_{n \to \infty} [w^k Q(w)] = 0$$

for all positive values of k. We thus establish the analytic expansion of an operator about the point at infinity.

4. Singular Operators with Poles.

Let $H(d/dx)$ be a non-singular operator without finite or infinite zeros on the axis of imaginaries. Then $H(d/dx)$ will be another such operator, which will be the inverse of the first. The question arises, has H any other inverses?

It has not. Suppose $H(d/dx)f(x) = H(d/dx)g(x)$, where f and g differ at more than a set of points of measure zero. Then $H(d/dx)(f(x) - g(x))$ is identically zero, except on such a set. Hence

$$f(x) - g(x) = \frac{1}{H(d/dx)} H\left(\frac{d}{dx}\right)(f(x) - g(x)) = \frac{1}{H(d/dx)} 0 = 0,$$

except at the most at a set of points of zero measure. This is however a contradiction.

We now turn to the case where $H(d/dx)$ has a finite number of zeros of finite orders. Let us suppose, for purposes of simplicity, that there is a zero of the nth order at the origin, and that no other zero is as near as $\lambda + 1$. We have

$$\begin{aligned} H\left(\frac{d}{dx}\right)f(x) &= H\left(\frac{d}{dx}\right)f_\lambda(x) + H\left(\frac{d}{dx}\right)(f(x) - f_\lambda(x)) \\ &= H_1\left(\frac{d}{dx}\right)f_\lambda(x) + H_2\left(\frac{d}{dx}\right)(f(x) - f_\lambda(x)), \end{aligned}$$

where H_1 equals H within a circle of radius exceeding $\lambda + 1$ about the origin, and defines a non-singular operator with no zeros on the imaginary axis outside this circle, except for a zero of order n at infinity, while H_2 equals H outside a circle of radius less than λ about the origin, and defines a non-singular operator with no zeros on the imaginary axis within this circle. It is easy to show that the equation

$$H\left(\frac{d}{dx}\right)f(x) = g(x)$$

reduces to the two simultaneous equations

$$H_1\left(\frac{d}{dx}\right)f_\lambda(x) = g_\lambda(x)$$

and

$$H_2\left(\frac{d}{dx}\right)(f(x) - f_\lambda(x)) = g(x) - g_\lambda(x).$$

We may without difficulty transform the first into the form

$$\frac{H_1(d/dx)}{(d/dx)^n}\left(\frac{d}{dx}\right)^n f_\lambda(x) = g_\lambda(x).$$

The solution of this may be achieved by the inversion of a non-singular operator without zeros, followed by an n-fold integration, which will in general introduce n constants of integration. The second equation will now involve the inversion of an operator with one less zero than H. Thus by a repetition of the same process, we ultimately reduce the inversion of our operator to the inversion of a non-singular operator without zeros in the finite portion of the plane. The inversion of an operator with an n-fold zero at infinity may similarly be reduced to an n-fold differentiation.

We shall represent the inverse of the non-singular operator $H(d/dx)$ by $1/H(d/dx)$. This will manifestly contain one arbitrary constant for every finite zero of H, counted with its multiplicity, when we apply it to an arbitrary function. The effect of a zero at infinity will be to restrict the class of functions to which $1/H(d/dx)$ can apply. We shall consider an operator containing both poles and zeros on the axis of imaginaries as the product of two operators, one with poles alone and the other with zeros alone. We can investigate the problem of inverting such an operator by our now-familiar method of dissecting each function to which we apply it into frequency ranges, each containing only a single singularity or zero, and replacing the operator over this range by one without other zeros or singularities.

We shall later investigate the question of removing the arbitrary character of the constants in the definition of singular operators.

5. Branch-Points and Fractional Differentiation and Integration.

Let us now consider operators such as $(d/dx)^{1/2}$, with branch points. We can limit ourselves to begin with to the case of an operator whose only notable point on the axis of imaginaries is a single branch-point of order n, where the sheets of the Riemann surface interchange in circular order. Within a certain distance of this point, our operator $H(d/dx)$ is determined by an analytic function, and every where except at the branch-point this is continuous, and has continuous derivatives of all orders. Our function $H(-iu)$ is bounded at infinity, whele all its derivatives vanish there. For purposes of simplicity, we shall only consider the case where the origin is the branch-point.

Let it be noted that our operator is singular, and accordingly has not yet received any precise definition. As before, we shall write

$$G(\xi - x, \lambda) = \frac{1}{2\pi} \int_{-\infty}^{\infty} F_\lambda(u) H(-iu) e^{iu(\xi - x)} du .$$

Since H is analytic within some circle about the origin, except at the center,

$$\frac{1}{2\pi} \int_{-a}^{a} F_\lambda(u) H(-iu) e^{iu(\xi - x)} du$$

for sufficiently small values of a may be regarded as an integral along some path in the complex plane other than the axis of imaginaries. The choice of this path depends on the choice we make of the sheets of the Riemann surface in which our function is taken to the right and the left of the origin. If the junction between these sheets is made trough positive real arguments of $H(u)$, our path of integration will pass above the origin; in the contrary case, below.

Let us now assume that our path of integration from $-\infty$ to ∞, as modified by our detour at the origin, passes above the origin, and let us call it C. We shall then have

$$G(\xi - x, \lambda) = \frac{1}{2\pi}\int_C F_\lambda(u) H(-iu) e^{iu(\xi - x)} du$$

$$= \frac{ik}{2\pi(\xi - x)^k}\int_C \frac{d^k}{du^k}[F_\lambda(u) H(-iu)] e^{iu(\xi - x)} du.$$

Hence if ε is any positive quantity,

$$|G(\xi - x, \lambda)| \leqq \frac{1}{2\pi(\xi - x)^k} \max_C \frac{d^k}{du^k}[F_\lambda(u) H(-iu)]$$

if $\xi - x \leqq 0$, and

$$|G(\xi - x, \lambda)| \leqq \frac{e^{\varepsilon(\xi - x)}}{2\pi(\xi - x)^k} \max_C \frac{d^k}{du^k}[F_\lambda(u) H(-iu)]$$

otherwise. Therefore, if $f(x)$ is any function which is $O(e^{-qx})$ to the right of the origin, q being greater than 0, and is $O(x^m)$ to the left,

$$\int_{-\infty}^{\infty} f(\xi) G(\xi - x, \lambda) d\xi,$$

exists. We shall term this

$$H\left(\frac{d}{dx}\right) f_\lambda(x).$$

This function will become only algebraically infinite at infinity. We may use the same methods as previously to show that

$$H\left(\frac{d}{dx}\right) f(x) = \underset{\lambda\to\infty}{\text{l. m.}}\; H\left(\frac{d}{dx}\right) f_\lambda(x)$$

exists. We shall call this, for obvious reasons, the backward value of $H(d/dx) f(x)$. The forward value of this symbol will correspond to a path of integration passing below the origin. It will apply to functions $f(x)$ which only become algebraically infinite for negative large values of x, and which become exponentially small for positive large values of x. In the applications of the operational calculus, x generally stands for the time, and $f(x)$ represents an impressed voltage or current. It is hence the backward values of our operators which have physical significance. We shall discuss this in more detail when we come to the theory of prospective and retrospective operators. Unless the contrary is stated, then, our operators with branch-points will be taken as backward.

As far as what we have said about the existence of $H(d/dx)f_\lambda(x)$, the properties of H for arguments greater than $\lambda+1$ in modulus are irrelevant. Thus if $f(x)$ is determined to have the value $(1+\operatorname{sgn} x)/2$, $(d/dx)^{1/2} f_\lambda(x)$ exists. We have:

$$\left(\frac{d}{dx}\right)^{1/2} f_\lambda(x) = \frac{1}{2\pi}\int_0^\infty d\xi \int_{-\lambda-1}^{\lambda+1} F_\lambda(u)\sqrt{-iu}\, e^{iu(\xi-x)}\,du$$

$$= \frac{1}{2\pi}\int_0^\infty d\xi \left[\int_0^{\lambda+1} F_\lambda(u)\frac{1-i}{\sqrt{2}}\sqrt{u}\, e^{iu(\xi-x)}\,du\right.$$

$$\left. + \int_0^{\lambda+1} F_\lambda(u)\frac{1+i}{\sqrt{2}}\sqrt{u}\, e^{iu(\xi-x)}\,du\right]$$

$$= \frac{\sqrt{2}}{2\pi}\int_0^\infty d\xi \int_0^{\lambda+1} F_\lambda(u)[\sin u(\xi-x)+\cos u(\xi-x)]\sqrt{u}\,du$$

$$= \lim_{H\to x}\frac{\sqrt{2}}{2\pi}\int_0^{\lambda+1} F_\lambda(u)\sqrt{u}\,\frac{-\cos u(H-x)+\cos ux+\sin u(H-x)+\sin ux}{u}\,du$$

$$= \frac{\sqrt{2}}{2\pi}\int_0^{\lambda+1} F_\lambda(u)\,\frac{\sin ux+\cos ux}{\sqrt{u}}\,du,$$

since $F_\lambda(u)/du$ is summable from 0 to $\lambda+1$, so that

$$\lim_{H\to x}\frac{\sqrt{2}}{2\pi}\int_0^{\lambda+1}\frac{F_\lambda(u)}{\sqrt{u}}\cos u(H-x)\,du = \lim_{H\to x}\frac{\sqrt{2}}{2\pi}\int_0^{\lambda+1}\frac{F_\lambda(u)}{\sqrt{u}}\sin u(H-x)\,du = 0.$$

Thus

$$\left(\frac{d}{dx}\right)^{1/2} f_\lambda(x) = \frac{\sqrt{2}}{2\pi}\int_0^\infty \frac{\sin ux+\cos ux}{\sqrt{u}}\,du - \frac{\sqrt{2}}{2\pi}\int_0^\infty \frac{[1-F_\lambda(u)][\sin ux+\cos ux]}{\sqrt{u}}\,du$$

$$\begin{cases} = \dfrac{1}{\sqrt{\pi x}} - \dfrac{\sqrt{2}}{2\pi}\displaystyle\int_0^\infty \frac{[1-F_\lambda(u)][\sin ux+\cos ux]}{\sqrt{u}}\,du; & [x>0] \\ = -\dfrac{\sqrt{2}}{2\pi}\displaystyle\int_0^\infty \frac{[1-F_\lambda(u)][\sin ux+\cos ux]}{\sqrt{u}}\,du. & [x<0] \end{cases}$$

Now,

$$\frac{\sqrt{2}}{2\pi}\int_0^\infty \frac{[1-F_\lambda(u)]\,[\sin ux+\cos ux]}{\sqrt{u}}\,du$$

$$=\pm\frac{\sqrt{2}}{2\pi x^n}\int_0^\infty (\sin ux \pm \cos ux)\frac{d^n}{du^n}\left(\frac{1-F_\lambda(u)}{\sqrt{u}}\right)du$$

$$=\frac{1}{x^n}\int_0^\infty \frac{(-iu)}{u^{n-1}}(\sin ux \pm \cos ux)\,du$$

where $G(u)$ is a continuous function which is $o(1/u)$ at infinity. We may readily conclude that

$$\frac{d^n}{dx^n}\frac{\sqrt{2}}{2\pi}\int_0^\infty \frac{[1-F_\lambda(u)]\,[\sin ux+\cos ux]}{\sqrt{u}}\,du$$

exists, and is $o(1/u^k)$ at infinity for every value of k. Thus we get *the* fundamental theorem that

$$\left(\frac{d}{dx}\right)^{n+1/2} f_\lambda(x)=\frac{1}{\sqrt{\pi}}\left(-\frac{1}{2}\right)\left(-\frac{3}{2}\right)\ldots\left(-n+\frac{3}{2}\right)x^{-n+1/2}+o(x^{-k})$$

for all values of k, as x becomes positively infinite, and

$$\left(\frac{d}{dx}\right)^{n+1/2} f_\lambda(x)=o(x^{-k})$$

for all values of k, as x becomes negatively infinite. Here $(d/dx)^{n+1/2}f_\lambda(x)$ is taken as the nth derivative of $(d/dx)^{1/2}f_\lambda(x)$. It is easy to show that this agrees with the direct definition of the derivative of order $n+1/2$.

In the same way, if $H(d/dx)$ is an operator which is non-singular except at the origin, and analytic about the origin, except for a finite branch-point at the origin, we may show that

$$H\left(\frac{d}{dx}\right)f(x)-H\left(\frac{d}{dx}\right)f_\lambda(x)=o(x^{-k})$$

at $\pm\infty$, $f(x)$ being given as above the interpretation $(1+\operatorname{sgn} x)/2$.

We now wish to discuss the definition of an operator with a branch point of finite order which is at the same time an infinity. As in the case of a pole, we can first isolate this singularity and then reduce our operator to a non-singular operator combined with an operator of the form $(d/dx-ai)^{-n}$. We shall only take up in detail the case of the operator $(d/dx)^{-1/2}$, but our methods will be of general applicability.

Let us consider a function $f(x)$ which is $O(e^{kx})$ at $-\infty$, k being positive, and which is at most algebraically infinite at $+\infty$. It will be natural to make the definition

$$\left(\frac{d}{dx}\right)^{-1/2} f(x) = \underset{\lambda\to\infty}{\text{l. m.}} \int_{-\infty}^{x} \left(\frac{d}{dx}\right)^{1/2} f_\lambda(x)\,dx.$$

We have, however,

$$\int_{-\infty}^{x} \left(\frac{d}{dx}\right)^{1/2} f_\lambda(x)\,dx = \frac{1}{2\pi}\int_{-\infty}^{x} dx \int_{-\infty}^{\infty} f(\xi)\,d\xi \int_{-\infty}^{\infty} F_\lambda(u)\sqrt{-iu}\,e^{iu(\xi-x)}\,du$$

$$= \frac{1}{2\pi}\lim_{A\to\infty}\int_{-\infty}^{x} dx \int_{-A}^{A} f(\xi)\,d\xi \int_{-\infty}^{\infty} F_\lambda(u)\sqrt{-iu}\,e^{iu(\xi-x)}\,du$$

$$= \frac{1}{2\pi}\int_{-\infty}^{\infty} f(\xi)\,d\xi \int_{-\infty}^{x} dx \int_{-\infty}^{\infty} F_\lambda(u)\sqrt{-iu}\,e^{iu(\xi-x)}\,du$$

$$= \frac{1}{2\pi}\int_{-\infty}^{\infty} f(\xi)\,d\xi \left[\int_{-\infty}^{\infty} \frac{F_\lambda(u)}{\sqrt{-iu}}\,e^{iu(\xi-\eta)}\,du\right]_{\eta=-\infty}^{\eta=x}$$

$$= \frac{1}{2\pi}\int_{-\infty}^{\infty} f(\xi)\,d\xi \int_{-\infty}^{\infty} \frac{F_\lambda(u)}{\sqrt{-iu}}\,e^{iu(\xi-x)}\,du$$

$$= \frac{1}{\sqrt{\pi}}\int_{-\infty}^{x} \frac{f(\xi)}{\sqrt{x-\xi}}\,d\xi - \frac{\sqrt{2}}{2\pi}\int_{-\infty}^{\infty} f(\xi)\,d\xi \int_{0}^{\infty} \frac{1-F_\lambda(u)}{\sqrt{u}}\,[\sin u\,(x-\xi) - \cos u\,(x-\xi)]\,du.$$

It may then be established without much difficulty that the second term in this sum converges in the mean to 0 over any finite range as λ increases without limit. Thus

$$\left(\frac{d}{dx}\right)^{-1/2} f(x) = \frac{1}{\sqrt{\pi}}\int_{-\infty}^{x} \frac{f(\xi)}{\sqrt{x-\xi}}\,d\xi.$$

As far as functions of the form $f_\lambda(x)$ are concerned, the operator $(d/dx)^{-1/2}$ here defined will be a true inverse to the operator $(d/dx)^{1/2}$. It is only for such functions that we have defined the derivative of order $1/2$. In the more general case, we shall define it as the inverse of $(d/dx)^{-1/2}$. The theory of the latter is well known, and it is known that it can have only one inverse on the range running from $-\infty$ up.

It is important to note that the operator $(d/dx)^{-1/2}$, and if it apply to f, the operator $(d/dx)^{1/2}$, are not affected by values of f for arguments greater than the point where we examine the function generated from f by the operator in question.

The operator $(d/dx)^{n+1/2}$ is to be conceived as the result of applying first $(d/dx)^{1/2}$ and then $(d/dx)^n$. Similar definitions are to be made for other improper fractional powers of the differentiator, negative as well as positive. If we confine our attention to functions f which are $O(e^{kx})$ at $-\infty$ and algebraically infinite or finite at $+\infty$ and let all the integrations in the definitions of our inverse operators be made from $-\infty$, we shall find that the laws of indices will always be obeyed.

6. Asymptotic Series.

Let $f(x)$ be $(1+\operatorname{sgn} x)/2$. Let $H(d/dx)$ be defined by a function $H(u)$ determined everywhere on the axis of imaginaries, non-singular, and such that

$$M(u) = \frac{H(u)}{\sqrt{u}}$$

is analytic within a circle of radius $r > \lambda + 1$ about the origin. Then

$$H\left(\frac{d}{dx}\right) f(x) = H\left(\frac{d}{dx}\right) f_\lambda(x) + o\left(\frac{1}{x^k}\right)$$

$$= M\left(\frac{d}{dx}\right)\left[\left(\frac{d}{dx}\right)^{1/2} f_\lambda(x)\right] + o\left(\frac{1}{x^k}\right)$$

$$= M(0)\left[\left(\frac{d}{dx}\right)^{1/2} f_\lambda(x)\right] + M'(0)\left[\left(\frac{d}{dx}\right)^{3/2} f_\lambda(x)\right] + M''(0)\left[\left(\frac{d}{dx}\right)^{5/2} f_\lambda(x)\right] + \cdots$$

$$+ \frac{M^{(k-1)}(0)}{(k-1)}\left[\left(\frac{d}{dx}\right)^{k-1/2} f_\lambda(x)\right] + N\left(\frac{d}{dx}\right)\left[\left(\frac{d}{dx}\right)^{k+1/2} f_\lambda(x)\right] + o\left(\frac{1}{x^k}\right)$$

$$= \frac{1}{\sqrt{\pi x}}\left[M(0) - \frac{1}{2}\frac{M'(0)}{x} + \frac{1}{2}\cdot\frac{3}{2}\frac{M''(0)}{x^2} - \cdots\right.$$

$$\left. + (-1)^{k-1}\left(\frac{1}{2}\cdot\frac{3}{2}\cdots\frac{2k-3}{2}\right)\frac{M^{(k-1)}(0)}{x^{k-1}}\right] + o\left(\frac{1}{x^k}\right)$$

$$+ \frac{1}{2\pi}\int_{-\infty}^{\infty}\left(\frac{d}{dx}\right)^{k+1/2} f_\lambda(\xi)\,d\xi \int_{-\infty}^{\infty} F_{\lambda+2}(u)\,N(-iu)\,e^{iu(\xi-x)}\,du$$

$$= \frac{1}{\sqrt{\pi x}}\left[M(0) - \cdots + (-1)^{k-1}\left(\frac{1}{2}\cdot\frac{3}{2}\cdots\frac{2k-3}{2}\right)\frac{M^{(k-1)}(0)}{x^{k-1}}\right] + o\left(\frac{1}{x^k}\right)$$

$$+ \frac{1}{2\pi}\int_{-\infty}^{\infty}\frac{(d/d\xi)^{1/2} f_\lambda(\xi)}{(x-\xi)^k}\int_{-\infty}^{\infty} F_{\lambda+2}(u)\,N(-iu)\,e^{iu(\xi-x)}\,du$$

$$= \frac{1}{\sqrt{\pi x}}\left[M(0) - \cdots + (-1)^{k-1}\left(\frac{1}{2}\cdot\frac{3}{2}\cdots\frac{2k-3}{2}\right)\frac{M^{(k-1)}(0)}{x^{k-1}}\right] + O\left(\frac{1}{x^k}\right).$$

Here $N(u)$ is a function without singularities on the imaginary axis between $-\lambda-2$ and $\lambda+2$. We have proved, that is, that

$$\frac{1}{\sqrt{\pi x}}\left[M(0)-\ldots+(-1)^{k-1}\left(\frac{1}{2}\cdot\frac{3}{2}\cdots\frac{2k-3}{2}\right)\frac{M^{k-1}(0)}{x^{k-1}}+\ldots\right]$$

is an asymptotic series for $H(d/dx)f(x)$.

7. Prospective and Retrospective Operators.

Let us now consider a non-singular operator $H(d/dx)$, where $H(u)$ is analytic in the entire right half-plane, and continuous in the right half-plane together with the axis of imaginaries. Furthermore, let $u^2 H(u)$ be bounded in the right half-plane. Then if $\xi > x$.

$$\left|\frac{1}{2\pi}\int_{-\infty}^{\infty} H(-iu)\, e^{iu(\xi-x)}\, du\right| = \left|\lim_{r\to\infty}\frac{1}{2\pi}\int_{-\frac{\pi}{2}}^{\frac{\pi}{2}} r H(re^{i\vartheta})\, e^{re^{i\vartheta}(x-\xi)}\, d\vartheta\right|$$

$$\leqq \frac{\max|uH(u)|}{2\pi}\lim_{r\to\infty}\int_{-\frac{\pi}{2}}^{\frac{\pi}{2}} e^{r\cos\vartheta(x-\xi)}\, d\vartheta = 0.$$

That is, $\frac{1}{2\pi}\int_{-\infty}^{x} H(-iu)\, e^{iyu}\, du$ vanishes for positive values of y. We then have, as we may easily show,

$$H(d/dx)f(x) = \frac{1}{2\pi}\int_{-\infty}^{x} f(\xi)\, d\xi \int_{-\infty}^{\infty} H(-iu)\, e^{iu(\xi-x)}\, du.$$

Consequently, if $f(x)$ vanishes for negative values of x, we have

$$H(d/dx)f(x) = \frac{1}{2\pi}\int_{0}^{x} f(\xi)\, d\xi \int_{-\infty}^{\infty} H(-iu)\, e^{iu(\xi-x)}\, du.$$

This is however an operator of the type discussed by Carson and Volterra, and when applied to a function $f(x)$ yields a function $g(x)$ which only depends on the values of f for arguments in the neighborhood of x and smaller arguments. Such an operator we call *retrospective.* Clearly, when applied to doubly differentiable functions, $(d/dx)^2 H(d/dx)$ is likewise retrospective. By hypothesis, however, $u^2 H(u)$ is bounded on the axis of imaginaries, so that the operator $(d/dx)^2 H(d/dx)$ is continuous, if we consider as the *écart* between two functions the square root of the integral between $-\infty$ and $+\infty$ of the square of their difference. From this we may conclude at once that $(d/dx)^2 H(d/dx)$ is retrospective in its most general application. In other words, an operator $H_1(d/dx)$ is retrospective if $H_1(u)$ is continuous and bounded in the right half-plane together with the axis of imaginaries, is analytic to the right of the axis, vanishes like u^2 at the origin, and is such that $H_1(d/dx)/(d/dx)^2$ is non-singular. The last two conditions are readily shown to be superfluous if H_1 is non-singular in the generalized sense. The presence of a finite set of branch-points or poles on the axis of imaginaries does not interfere with this retrospective character of $H_1(d/dx)$, provided that in all the integrals which arise from carrying out our operator, we let the lower bound of integration be $-\infty$. The sum of two retrospective operators is retrospective, as is the inverse of a retrospective operator, if it be properly chosen. There is an analogues theory of prospective operators, involving only the present and future of a function, not its past. Here the rôles of the right and the left half-planes are interchanged.

If an operator has singularities in the right half-plane, it will not in general be retrospective, as

$$\frac{1}{2\pi}\int_{-\infty}^{\infty} H(-iu)\, e^{iu(\xi-x)}\, du - \frac{1}{2\pi}\int_{-\frac{\pi}{2}}^{\frac{\pi}{2}} rH(re^{i\vartheta})\, e^{re^{i\vartheta}(x-\xi)}\, d\vartheta$$

will now represent the sum of the residues of

$$H(u)\, e^{u(x-\xi)}$$

within the region bounded by the axis of imaginaries and the semi-circle of radius r in the right half-plane with the origin as center. In such a case $H(d/dx)$ may represent an operator both in the Volterra-Carson theory and in that here developed, but it will not represent the same operator. Thus the present paper demands that

$$\frac{1}{d/dx-1}[1+\operatorname{sgn} x] = (e^x-1)(1+\operatorname{sgn} x) - 2e^x,$$

while the Volterra-Carson theory requires that

$$\frac{1}{d/dx-1}[1+\operatorname{sgn} x] = (e^x-1)(1+\operatorname{sgn} x).$$

In other words, the asymptotic expansions of Heaviside and his expansions in positive powers of t can only be depended on to yield the same results when we confine ourselves to operators analytic in the right half-plane. As a matter of fact, such operators correspond to problems arising from positive resistances only.

8. The Operational Solution of Partial Differential Equations.

Before we enter on this topic in detail, it is important to consider the nature of the solution of a partial differential equation. Let us consider the linear equation

$$A\frac{\partial^2 u}{\partial x^2}+B\frac{\partial^2 u}{\partial x\,\partial y}+C\frac{\partial^2 u}{\partial y^2}+D\frac{\partial u}{\partial x}+E\frac{\partial u}{\partial y}+Fu=0$$

where for simplicity's sake, we shall suppose that the coefficients have as many derivatives as we shall need in the work which follows. If u satisfies this equation, it must manifestly possess the various derivatives indicated in the equation. As is familiar, however, in the case of the equation of the vibrating string, there are cases where u must be regarded as a solution of our differential equation in a general sense without possessing all the orders of derivatives indicated in the equation, and indeed without being differentiable at all. It is a matter of some interest, therefore, to render precise the manner in which a non-differentiable function may satisfy in a generalized sense a differential equation.

Let $G(x, y)$ be a function positive and infinitely differentiable within a certain bounded polygonal region R of the XY plane, vanishing with its derivatives of all orders on the periphery of R, and zero outside R. Then there is a function $G_1(x, y)$ such that

$$\begin{aligned}\iint_R (Au_{xx}+Bu_{xy}+Cu_{yy}+Du_x+Eu_y+Fu)\,G(x, y)\,dx\,dy\\=\iint_R u(x, y)\,G_1(x, y)\,dx\,dy\end{aligned}$$

for all u with bounded summable derivatives of the first two orders, as we may show by integration by parts. Thus a necessary and sufficient condition for u to satisfy our differential equation almost everywhere is that

$$\iint_R u(x, y)\,G_1(x, y)\,dx\,dy=0$$

for every possible G (as the G's form a complete set over any region), and that u possess the requisite derivatives. We can thus regard a function orthogonal to every G_1 as satisfying our differential equation in a generalized sense. This sense is more general than that developed by Bôcher [15]), as not even the existence of first order derivatives is postulated. If a sequence of generalized solutions of our differential equation converges in the mean over a given area to a function $\Phi(x, y)$, it is now manifest

[15]) M. Bôcher, On harmonic functions in two dimensions, Proc. Amer. Acad. Sc. **41** (1905–1906); G. C. Evans, On the reduction of integro-differential equations, Trans. Amer. Math. Soc. **15** (1914), pp. 477–496.

that Φ is itself a generalized solution of this differential equation over this area.

Now let us discuss

$$e^{-x\sqrt{(d/dt+a)(d/dt+b)}} f_\lambda(t)$$

where a and b are positive, where f vanishes exponentially at $-\infty$, and is algebraically infinite at $+\infty$, and where x is positive.

$$e^{-x\sqrt{(d/dt+a)(d/dt+b)}}$$

is then a non-singular operator (in the generalized sense), and is retrospective. Let $|\lambda+1|<a$, $|\lambda+1|<b$. Then

$$e^{-x\sqrt{(d/dt+a)(d/dt+b)}} f_\lambda(t) = \sum_h \sum_j A_{hj} x^h f_\lambda^{(j)}(t),$$

where the methods of section 3 enable us to show that this double series is uniformly and absolutely convergent over any finite range. The coefficients are the same as those obtained in the expansion of

$$e^{-x\sqrt{(y+a)(y+b)}}$$

in a double series in x and y. Similarly, we have uniformly and absolutely

$$(d/dt+a)(d/dt+b)\, e^{-x\sqrt{(d/dt+a)(d/dt+b)}} f_\lambda(t) = \sum_h \sum_j B_{hj} x^h f_\lambda^{(j)}(t),$$

where

$$\sum_h \sum_j B_{hj} x^h y^j = (y+a)(y+b)\, e^{-x\sqrt{(y+a)(y+b)}} = \frac{\partial^2}{\partial x^2} \sum_h \sum_j A_{hj} x^h y^j$$
$$= \sum_h \sum_j h(h-1) A_{hj} x^{h-2} y^j.$$

Thus

$$B_{hj} = h(h-1) A_{hj},$$

from which we may conclude that

$$(d/dt+a)(d/dt+b)\, e^{-x\sqrt{(d/dt+a)(d/dt+b)}} f_\lambda(t)$$
$$= \frac{\partial^2}{\partial x^2} e^{-x\sqrt{(d/dt+a)(d/dt+b)}} f_\lambda(t).$$

By using in addition to expansions about the origin, expansions about other points on the axis of imaginaries, we may prove in a similar manner that for any value of λ, this last equation holds. We know, however, that for any given value of x,

$$e^{-x\sqrt{(d/dt+a)(d/dt+b)}} f(t) = \underset{\lambda\to\infty}{\text{l.m.}}\, e^{-x\sqrt{(d/dt+a)(d/dt+b)}} f_\lambda(t)$$

and there is no difficulty in showing that this holds uniformly over any finite range of x. Thus

$$u = e^{-x\sqrt{(d/dt+a)(d/dt+b)}} f(t)$$

satisfies in the generalized sense the differential equation

$$u_{tt} + (a+b)\,u_t + abu = u_{xx}.$$

If then it possesses the necessary derivatives, it satisfies this equation in the ordinary sense. When $x = 0$, it reduces to $f(t)$, and it depends retrospectively on $f(t)$. It may be shown, moreover, that if $f(t)$ is everywhere continuous, so is $u(x, t)$. We thus have completely solved the telegrapher's equation for a semi-infinite line with a given impressed voltage at the end.

Massachusetts Institute of Technology, April 6, 1925.

(Eingegangen am 20. 4. 1925.)

Comments on [26c]

L. Schwartz

Le but de ce papier de Wiener est de définir correctement des opérateurs de la forme $H\left(\frac{1}{i}\frac{d}{dx}\right)$, où H est une fonction sur $\mathbb{R}^{\dagger}$; ces opérateurs doivent opérer sur les fonctions complexes sur $\mathbb{R}$. Bien des auteurs antérieurs avaient étudié des opérateurs symboliques de ce type (le fameux calcul symbolique d'Heaviside date de 1893, celui de Pincherle de 1895)). Mais ces calculs manquaient généralement de rigueur ou de généralité. Wiener ici montre comment on peut définir $H\left(\frac{1}{i}\frac{d}{dx}\right)$ rigoureusement, de manière très générale, et en donne les principales propriétés et applications, notamment les développements asymptotiques. Bien entendu, aujourd'hui, les choses se traitent de manière bien plus simple et encore plus générale: si f est une *distribution tempérée*, et si H est une fonction C^{∞}, à croissance lente à l'infini ainsi que chacune de ses dérivées, $H \in \mathcal{O}_M$,[‡] alors $H\left(\frac{1}{i}\frac{d}{dx}\right)f$ est la distribution tempérée dont l'image de Fourier est $H(\xi)\hat{f}(\xi)$, où $\hat{f}$ est l'image de Fourier de f; autrement dit $H\left(\frac{1}{i}\frac{d}{dx}\right)f$ n'est autre que $L * f$, où $\mathscr{F}L = H$. En outre, on étudie aujourd'hui les opérateurs pseudo-différentiels, de la forme $H\left(x, \frac{1}{i}\frac{d}{dx}\right)$,[#] dont les applications à la théorie des équations aux dérivées partielles ne se comptent plus.

Wiener procède de la manière suivante. Il suppose d'abord que f est une fonction localement L^2 à croissance polynomiale à l'infini; alors $\hat{f}$ (qu'on ne peut pas encore définir correctement !) sera une somme de dérivées de fonctions de L^2. Soit $\hat{\alpha}$ une fonction convenable C^{∞} sur $\mathbb{R}$, égale à 1 pour $|\xi| \leqslant \lambda$, et à 0 pour $|\xi| \geqslant \lambda + 1$. Elle est l'image de Fourier d'une fonction α qui va servir de régularisante. En effet, $f * \alpha$ a un sens, et converge vers f dans L^2 local pour λ infini. Il est naturel de définir $H\left(\frac{1}{i}\frac{d}{dx}\right)(f * \alpha)$ par $f * H\left(\frac{1}{i}\frac{d}{dx}\right)\alpha$,

---

† Wiener écrit $H\left(\frac{d}{dx}\right)$, H fonction sur l'axe imaginaire; je change sa notation pour harmoniser avec les écriture actuelles.

‡ Voir L. Schwartz, *Théorie des distributions*, Hermann, Paris, 1966, p. 243.

\# Issus des opérateurs intégraux singuliers de Calderon et Zygmund, ils ont été introduits simultanément par Kohn, Nirenberg, Unterberger, Mme Bokobza, vers. 1964.

où $H\left(\frac{1}{i}\frac{d}{dx}\right)\alpha$ a pour image de Fourier $H\hat{\alpha}$.

Maintenant $H\hat{\alpha}$ est bien définie, elle est bien l'image de Fourier d'une fonction qu'on peut appeler $H\left(\frac{1}{i}\frac{d}{dx}\right)\alpha$, et, si H est bornée sur $\mathbb{R}$ ainsi que chacune de ses dérivées, on montre que $f * H\left(\frac{1}{i}\frac{d}{dx}\right)\alpha$ ainsi définie tend, dans L^2 local, vers une limite, qu'on peut définir comme étant $H\left(\frac{1}{i}\frac{d}{dx}\right)f$ (et qui est bien le $H\left(\frac{1}{i}\frac{d}{dx}\right) f$ défini aujourd'hui directement par la méthode des distributions). Ceci n'a pas nécessité pour Wiener de savoir définir l'image de Fourier $\hat{f}$. Il est alors aisé ensuite d'étudier les développements en série ou asymptotiques de $H\left(\frac{1}{i}\frac{d}{dx}\right)f$, moyennant des hypothèses sur H ou sur f. Au §4, Wiener étudie le cas où H possède des pôles; dans ce cas, si nous regardons le résultat en termes de distributions, la multiplication $\hat{f} \to H\hat{f}$ comporte une division (aux pôles de H, donc des indéterminations), et effectivement $H\left(\frac{1}{i}\frac{d}{dx}\right)f$, défini par Wiener, avec d'abord une dérivation puis la recherche d'une primitive, comporte ces indéterminations. Au §5, H est supposé posséder des points critiques. Par exemple, pour $H\left(\frac{1}{i}\frac{d}{dx}\right) = \left(\frac{d}{dx}\right)^{1/2}$, on retrouve les dérivations d'ordre "fractionnaire"; l'auteur retrouve ainsi les propriétés déjà bien connues de ces opérations. Au §7, il introduit les opérateurs rétrospectifs (et prospectifs); en physique, il sont essentiels, ce sont ceux qui peuvent s'écrire $f \to H\left(\frac{1}{i}\frac{d}{dx}\right)f$, où H est holomorphe dans le demi-plan inférieur, c'est-à-dire pour lesquels on exécute en fait une convolution de f, $f \to L * f$, avec une distribution $L = \mathscr{F}H$ tempérée à support dans la demi-droite positive.

Le §8 est intéressant. A part le fait qu'il applique les opérateurs symboliques précédents à des solutions d'équations aux dérivées partielles, Wiener réfléchit à la notion de solution généralisée d'une équation aux dérivées partielles; par exemple, une fonction de la forme $u(x + ct) + v(x - ct)$, où u et v sont des fonctions seulement continues, "doit" être regardée comme une solution généralisée de l'équation des cordes vibrantes. Il est amusant de remarquer que c'est exactement cette idée qui m'a poussé moi-même à introduire les distributions![†] Elle a tourmenté de nombreux mathématiciens, comme le montrent ces quelques pages. Or Wiener donne une très bonne définition d'une solution généralisée; j'en avais, dans mon livre sur les Distributions, attribué les premières définitions à Leray (1934), Sobolev (1936), Friedrichs (1939), Bochner (1946), la définition la plus générale étant celle de Bochner; or la définition de Wiener est la même que celle de Bochner, et date donc de ce mémoire, c'est-à-dire de 1926, elle est antérieure à toutes les autres.

[†] Voir l'introduction de mon livre sur les distributions p. 4.

APPENDIX B

FOURIER ANALYSIS AND ASYMPTOTIC SERIES

BY NORBERT WIENER

Integration. The integral between a and b of the continuous function $f(t)$ is defined in the usual way:

$$\int_a^b f(t)\,dt = \lim_{n\to\infty} \frac{b-a}{n} \sum_{k=1}^{n} f\left(a + \frac{k}{n}(b-a)\right). \qquad (1)$$

We then have:

if
$$f(t) = \phi_1(t) + \phi_2(t)$$
then
$$\int_a^b f(t)\,dt = \int_a^b \phi_1(t)\,dt + \int_a^b \phi_2(t)\,dt. \qquad (2)$$

This theorem cannot, however, be applied to infinite sums for if

$$f_m(t) = \phi_1(t) + \phi_2(t) + \cdots + \phi_m(t)$$
and
$$f(t) = \lim_{m\to\infty} f_m(t)$$

exists (even as a bounded limit) and

$$\lim_{m\to\infty} \int_a^b f_m(t)\,dt$$

exists, it does not follow that $f(t)$ has an integral in the sense already defined.

As an example, let $f(t) = 0$ except between $\frac{1}{n} - \frac{1}{n^4}$ and $\frac{1}{n} + \frac{1}{n^4}$, where

$$f(t) = n\left(1 - n^4\left|t - \frac{1}{n}\right|\right), \qquad n = 2, 3 \ldots.$$

366

Clearly $f(t)$ will not possess an integral in the sense already given. On the other hand, if

$$f_m(t) = f(t) \qquad \text{for } t > \frac{1}{m} + \frac{1}{m^4}$$

and 0 otherwise, then $f_m(t)$ will be continuous, and the conditions stated will hold.

Now, probabilities and integrals are related in the very closest possible way, and it is frequently useful in the theory of probabilities and elsewhere — in statistical mechanics, for example — to add an **infinite** set of integrals and to obtain a single integral. This obviously demands an extension of the notion of integral, and fortunately this extension can be carried through in just one way. In other words, there is just one operator $\int_a^b f(t)\,dt$ which yields for continuous functions the result already defined, and which has the property that if the set of functions $\{f_m(t)\}$ contains no function with absolute value exceeding A, and

$$\lim_{m\to\infty} f_m(t) = f(t), \qquad a \leq t \leq b$$

then

$$\lim_{m\to\infty} \int_a^b f_m(t)\,dt$$

exists and equals $\int_a^b f(t)\,dt$.

This definition yields, as far as bounded functions are concerned, the integral of Lebesgue, discovered by Lebesgue in 1900.[1] It is discussed in detail in such books as Vol. I of Hobson's Theory of Functions of a Real Variable, and is indeed given a quite different definition. We are, however, interested here in stating its properties rather than in deducing them.

[1] H. Lebesgue, Leçons sur l'Intégration, Paris, Gauthier-Villars.

If $f(t)$ is not bounded, let

$$f_{A,B}(t) \begin{cases} = f(t), & A \le f(t) \le B, \\ = A, & f(t) < A, \\ = B, & f(t) > B \end{cases} \tag{3}$$

If $\int_a^b f_{A,B}(t)\, dt$ approaches a unique limit as $A \to -\infty$ and $B \to \infty$ in any way, this limit is again called $\int_a^b f(t)\, dt$ and is known as the Lebesgue integral. In what follows all integrals are taken in the sense of Lebesgue.

2. **Fourier Coefficients.** Let $f(t)$ be any function (possibly complex valued) of a real variable t, defined for

$$-\pi \le t \le \pi,$$

possessing a Lebesgue integral, and such that the square of its modulus possesses a Lebesgue integral. Let us consider

$$I_n = \int_{-\pi}^{\pi} \left| f(t) - \sum_{-n}^{n} \frac{a_k}{\sqrt{2\pi}} \epsilon^{jkt} \right|^2 dt \tag{4}$$

which may be regarded in the least square sense as a measure of the amount by which

$$\sum_{-n}^{n} \frac{a_k}{\sqrt{2\pi}} \epsilon^{jkt}$$

fails to equal $f(t)$. We ask, what values of a_k minimize I? By elementary methods of calculus, we can show that these values are

$$a_k = \frac{1}{\sqrt{2\pi}} \int_{-\pi}^{\pi} f(t)\, \epsilon^{-jkt}\, dt. \tag{5}$$

These quantities are known as the **Fourier coefficients** of $f(t)$.

As we have just seen, if we replaced a_n and a_{-n} by zero instead of the values (5), we increase I_n or leave it unchanged. Hence if we give the a_k's the values in (5), we have

$$I_1 \ge I_2 \ge I_3 \cdots \ge I_n \ge \cdots \ge 0.$$

This suggests the possibility that

$$\lim_{n \to \infty} I_n = 0$$

and this is actually the case. The theorem to this effect is due to Hurwitz,[1] and takes his name.

It follows from the Hurwitz theorem that

$$\int_{-\pi}^{\pi} | f(t) |^2 \, dt = \sum_{-\infty}^{\infty} | a_k |^2 \tag{6}$$

or more generally, that if a_k is defined as above, while $g(t)$ satisfies the conditions laid down for $f(t)$ and

$$b_k = \int_{-\pi}^{\pi} g(t) \, \epsilon^{-jkt} \, dt,$$

then

$$\int_{-\pi}^{\pi} f(t) \, \bar{g}(t) \, dt = \sum_{-\infty}^{\infty} a_k \bar{b}_k, \tag{7}$$

the bar denoting the conjugate complex quantity of the quantity over which it is placed. Theorems (6) and (7) are known as the Parseval theorems.

An important question is whether Hurwitz's theorem possesses a valid inverse. It does not possess a valid inverse unless the integration involved is of the Lebesgue type. This inverse, known as the Riesz-Fischer theorem,[2] reads:

Let $\sum_{-\infty}^{\infty} | a_k |^2$ converge. Then there is a function $f(t)$, such that

$$\int_{-\pi}^{\pi} f(t) \, dt \quad \textbf{and} \quad \int_{-\pi}^{\pi} | f(t) |^2 \, dt$$

[1] Hurwitz, Math. Annalen, Vol. LVII, 1903, p. 426.

[2] F. Riesz, Sur les Systèmes orthogonaux de Fonctions, Comptes Rendus, Vol. 144, 1907, pp. 615–619.

E. Fischer, Sur la convergence en moyenne, *Ibid.*, pp. 1022–1024.

exist, and such that

$$\lim_{n\to\infty}\int_{-\pi}^{\pi}\left|f(t)-\sum_{-n}^{n}\frac{a_k}{\sqrt{2\pi}}\epsilon^{jkt}\right|^2 dt = 0. \tag{8}$$

Formula (5) will hold, of course.

The following remarks are in order as to the Riesz-Fischer theorem:

(*a*) It forms the chief justification for the use of the Lebesgue integral. No less general integral is adequate to the needs of harmonic analysis.

(*b*) The function $f(t)$ is not uniquely determined. It may be arbitrarily modified, for example, at all rational points. If, however, two functions $\phi_1(t)$ and $\phi_2(t)$ are considered as **equivalent** if

$$\int_{-\pi}^{\pi}|\phi_1(t)-\phi_2(t)|^2\,dt = 0, \tag{9}$$

then all functions $f(t)$ are equivalent. Two functions equivalent to a third function are equivalent to each other, and two equivalent **continuous** functions are identical.

3. **Fourier Duality.** Let it be noted that the relations already mentioned may be written in pairs as follows:

$a_k = \frac{1}{\sqrt{2\pi}}\int_{-\pi}^{\pi} f(t)\epsilon^{-jkt}\,dt$	$f(t)$ approached in least-square sense by partial sums of $\frac{1}{\sqrt{2\pi}}\sum a_k\,\epsilon^{jkt}$
$\int_{-\pi}^{\pi}\|f(t)\|^2\,dt$	$=\sum_{-\infty}^{\infty}\|a_k\|^2$
$\int_{-\pi}^{\pi} f(t)\bar{g}(t)\,dt$	$=\sum_{-\infty}^{\infty} a_k\bar{b}_k$

The expressions on the right correspond to those on the left under the following transformations:

Subscript k corresponds to argument t

a_k, b_k correspond to $f(t)$, $g(t)$

$$\sum_{-\infty}^{\infty} \text{ corresponds to } \int_{-\pi}^{\pi} dt$$

$$\epsilon^{jkt} \text{ corresponds to } \epsilon^{-jkt}$$

If we replace the interval $(-\pi, \pi)$ by the interval $(-A, A)$, it is easy to see that our duality takes the form:

$a(\mu) = \frac{1}{\sqrt{2\pi}} \int_{-A}^{A} f(t)\, \epsilon^{-j\mu t}\, dt$	$f(t)$ approached in least-square sense by partial sums of $\frac{1}{\sqrt{2\pi}} \sum_{\frac{A\mu}{\pi}=-\infty}^{\frac{A\mu}{\pi}=\infty} a(\mu)\epsilon^{j\mu t}\left(\frac{\pi}{A}\right)$
$\int_{-A}^{A} \lvert f(t) \rvert^2\, dt$	$= \sum_{\frac{A\mu}{\pi}=-\infty}^{\frac{A\mu}{\pi}=\infty} \lvert a(\mu) \rvert^2 \left(\frac{\pi}{A}\right)$
$\int_{-A}^{A} f(t)\bar{g}(t)\, dt$	$= \sum_{\frac{A\mu}{\pi}=-\infty}^{\frac{A\mu}{\pi}=\infty} a(\mu)\bar{b}(\mu) \left(\frac{\pi}{A}\right)$

where $\mu = \frac{k\pi}{A}$.

Note that

$$\mu \text{ corresponds to } t$$

$$\frac{\pi}{A} \sum_{\frac{A\mu}{\pi}=-\infty}^{\frac{A\mu}{\pi}=\infty} \text{ corresponds to } \int_{-A}^{A} dt$$

$$a(\mu),\, b(\mu) \text{ corresponds to } f(t),\, g(t)$$

$$\epsilon^{j\mu t} \text{ corresponds to } \epsilon^{-j\mu t}$$

The range of μ is from $-\infty$ to ∞ by steps of the size $\frac{\pi}{A}$. The range of t is continuously from $-A$ to A. As $A \to \infty$, these ranges both tend in different ways towards the continuous range $(-\infty, \infty)$. In the limit we get formally

$$a(\mu) \backsim \frac{1}{\sqrt{2\pi}} \int_{-\infty}^{\infty} f(t)\epsilon^{-j\mu t}\, dt \qquad f(t) \backsim \frac{1}{\sqrt{2\pi}} \int_{-\infty}^{\infty} a(\mu)\epsilon^{j\mu t}\, d\mu$$

$$\int_{-\infty}^{\infty} | f(t) |^2\, dt = \int_{-\infty}^{\infty} | a(\mu) |^2\, d\mu$$

$$\int_{-\infty}^{\infty} f(t)\bar{g}(t) = \int_{-\infty}^{\infty} a(\mu)\bar{b}(\mu)\, d\mu$$

We understand by $\backsim$ some correspondence later to be specified. Here perfect symmetry and reciprocality is obtained, barring the change of the sign of the exponent of $\epsilon^{j\mu t}$. How can the conditions for this equivalence be specified?

4. The Plancherel Theorem. In the theory of Fourier series we have seen that there is a certain duality between the Hurwitz theorem and the Riesz-Fischer theorem. Nevertheless the Hurwitz theorem may be treated on the basis of a theory of continuous functions alone, while the Riesz-Fischer theorem involves Lebesgue integration. This discrepancy is due to the asymmetry of the ordinary Fourier series theory, in which to a function of a real variable there corresponds that much simpler thing, an array of coefficients. In the case where t runs over an infinite range, the coefficients and the function assume the same limiting form — as did the tortoise and the hedgehog in Kipling's Just-So explanation of the origin of the armadillos. In this infinite case the quasi-Hurwitz theorem becomes identical with the quasi-Riesz-Fischer theorem, and involves the same apparatus of Lebesgue integration and least-square approximation which we have found to characterize the simple Riesz-Fischer theorem. This is perhaps the explanation of the much later growth of the infinite or Fourier integral theory, as opposed to that of Fourier series. The fundamental theorem is due to Plancherel[1] and reads as follows:

[1] M. Plancherel, Contribution à l'étude de la Représentation d'une fonction arbitraire par des integrales definies. Rend. di Palermo, Vol. 30, 1910, pp. 289–335.

Let $f(t)$ be a function such that

$$\int_{-A}^{A} f(t)\, dt$$

exists in the Lebesgue sense for every A, while

$$\lim_{A\to\infty} \int_{-A}^{A} | f(t) |^2\, dt$$

exists as a finite limit of a Lebesgue integral. Then there is a function $a(\mu)$ such that

$$\lim_{A\to\infty} \lim_{B\to\infty} \int_{-B}^{B} \left| a(\mu) - \frac{1}{\sqrt{2\pi}} \int_{-A}^{A} f(t)\epsilon^{-j\mu t}\, dt \right|^2 d\mu = 0. \quad (10)$$

The integrals

$$\int_{-A}^{A} a(\mu)\, d\mu \quad \text{and} \quad \int_{-A}^{A} | a(\mu) |^2\, d\mu$$

exist in the Lebesgue sense for every A, while

$$\lim_{A\to\infty} \int_{-A}^{A} | a(\mu) |^2\, d\mu = \lim_{A\to\infty} \int_{-A}^{A} | f(t) |^2\, dt \quad (11)$$

Furthermore,

$$\lim_{A\to\infty} \lim_{B\to\infty} \int_{-B}^{B} | f(t) - \frac{1}{\sqrt{2\pi}} \int_{-A}^{A} a(\mu)\epsilon^{j\mu t}\, d\mu |^2\, dt = 0. \quad (12)$$

Formulae (10) and (12) give us the proper interpretation for the symbol $\backsim$.

The function $a(\mu)$ is known as the **Fourier transform** of $f(t)$. It is so determined that if $a_1(\mu)$ is another possible Fourier transform of $f(t)$,

$$\int_{-A}^{A} | a_1(\mu) - a(\mu) |^2\, d\mu = 0 \quad (13)$$

for every A. A function can have only one continuous Fourier transform, if any.

The Parseval theorem for Fourier transforms reads as follows:

Let $f(t)$ and $g(t)$ fulfill the conditions laid down for $f(t)$ in the hypothesis to the Plancherel theorem, and let $a(\mu)$ and $b(\mu)$ be their Fourier transforms. Then

$$\lim_{A,B\to\infty}\int_{-A}^{B} f(t)\,\bar{g}(t)\,dt = \lim_{A,B\to\infty}\int_{-A}^{B} a(\mu)\bar{b}(\mu)\,d\mu. \qquad (14)$$

5. **Local and Asymptotic Properties.** In a generalized sense we may say that the Fourier transform of a function of period 2π is a function defined only for integral arguments. Periodicity is a property at infinity or an asymptotic property, while the property of being defined for only integral arguments is a local property, as it involves each integral argument separately. This is a crude illustration of a duality running through the whole theory of Fourier analysis.

Again, let us write formally for the $f(t)$ in terms of its Fourier transform $a(\mu)$

$$f(t) = \frac{1}{\sqrt{2\pi}}\int_{-\infty}^{\infty} a(\mu)\epsilon^{j\mu t}\,d\mu. \qquad (15)$$

On differentiating formally,

$$f'(t) = \frac{1}{\sqrt{2\pi}}\int_{-\infty}^{\infty} j\mu a(\mu)\epsilon^{j\mu t}\,d\mu. \qquad (16)$$

That is, the derivative of $f(t)$ is the Fourier transform of $j\mu a(\mu)$. The condition

$$\int_{-\infty}^{\infty} |\, f'(t)\,|^2\,dt \qquad \text{finite}$$

is primarily a local, differential condition on $f(t)$; the condition

$$\int_{-\infty}^{\infty} \mu^2\,|\, a(\mu)\,|^2\,d\mu \qquad \text{finite}$$

is an asymptotic condition. All of this can be made rigorous with the proper apparatus of assumptions.

More generally, we define $F\left(\frac{d}{dt}\right)f(t)$ as the function with the Fourier transform $F(j\mu)a(\mu)$, at least when F is a polynomial. While this is presented here formally, it may be put on a sound basis. This method furnishes one of the most useful approaches to the operatioual calculus.

6. **Retrospective or Historical Operators.** An interesting question is: What is the Fourier transform of a function vanishing for negative arguments? Such a transform will be represented as the least-square limit as $A \to \infty$ of

$$a(\mu) = \frac{1}{\sqrt{2\pi}} \int_0^A f(t)\epsilon^{-j\mu t}\,dt \tag{17}$$

where f possesses the conditions already laid down for it to possess a Fourier transform. The necessary and sufficient additional condition imposed on $a(\mu)$ reads as follows: No condition is imposed on the real and the pure imaginary parts of $a(\mu)$ separately, as may be seen on inspection, but together they must constitute the real and the imaginary parts of the boundary values of a function analytic in the lower half-plane. Of course this needs further mathematical specification to be made precise, but its general meaning can be seen from the fact that the integral

$$\frac{1}{\sqrt{2\pi}} \int_0^\infty f(t)\epsilon^{-\lambda t - j\mu t}\,dt$$

converges for $\lambda > 0$.

It follows readily from this that if $F(j\mu)$ is a function of μ admitting of a Fourier transform, analytic below the axis of reals and with certain restrictions as to its behavior at infinity, which in effect prevent the existence of a source there, $F(j\mu)a(\mu)$ enjoys the same property as does $a(\mu)$, of being the Fourier transform of a function vanishing to the left of the origin. That is, $F\left(\frac{d}{dt}\right)f(t)$ will vanish to the left of the origin. If $F(j\mu)$ is bounded in the lower half-plane, all necessary restrictions as to its behavior at infinity are fulfilled.

It is possible to reverse the argument and to deduce from this that such an operator $F\left(\frac{d}{dt}\right)$ only depends on the history of a function, inasmuch as two functions with the same history up to a given point have a difference vanishing up to this point.

7. The Separation of a Function into High- and Low-Frequency Ranges. If t is taken as the time in the above formulae, μ stands for an angular velocity. In the discussion of functions of time, from the standpoint of the theory of Fourier transforms, there are certain difficulties which belong peculiarly to the high-frequency range of the functions considered, and certain quite distinct difficulties which belong to the low-frequency range. In general it is the high-frequency terms in the harmonic analysis of a function which determine its properties of continuity, while the low-frequency terms are more concerned with its behavior at infinity. It is possible in many cases to separate a function into low-frequency and high-frequency parts and thus to secure a simplification of our problems.

As a useful tool in this separation let us introduce the notion of a non-analytic, infinitely differentiable function. An example is

$$\left.\begin{aligned} \phi_{A,B}(\mu) &= 1, \quad |\mu| < A \\ &= 0, \quad |\mu| > B \\ &= \frac{\displaystyle\int_A^\mu \epsilon^{\frac{1}{(\xi-A)(\xi-B)}}\,d\xi}{\displaystyle\int_A^B \epsilon^{\frac{1}{(\xi-A)(\xi-B)}}\,d\xi}, \quad A \leq |\mu| \leqq B \end{aligned}\right\} \tag{18}$$

We can represent any frequency function $a(\mu)$ as the sum of two parts; a low-frequency part, $a(\mu)\phi_{A,B}(\mu)$, and a high-frequency part, $a(\mu)\,(1 - \phi_{A,B}\,(\mu))$. If $a(\mu)$ is the Fourier transform of $f(t)$ and $\phi_{A,B}(\mu)$ is the Fourier transform of

$\Phi_{A,B}(t)$, this separation of a into two parts corresponds to the separation of $f(t)$ into the two parts:

$$f_1(t) = \frac{1}{\sqrt{2\pi}} \int_{-B}^{B} a(\mu)\phi_{A,B}(\mu)\epsilon^{j\mu t}\,d\mu \qquad (19)$$

$$= \frac{1}{\sqrt{2\pi}} \int_{-\infty}^{\infty} f(\tau)\Phi_{A,B}(\tau + t)\,d\tau \qquad (20)$$

and

$$f_2(t) = f(t) - f_1(t), \qquad (21)$$

which we may call the low- and high-frequency parts of $f(t)$, respectively. Now, it is easy to show that for any k

$$\lim_{t\to\pm\infty} \Phi_{A,B}(t)t^k = 0, \qquad (22)$$

while $\Phi_{A,B}$ is a continuous function with derivatives of all orders. Hence even if $f(t)$ does not possess a Fourier transform, but has with its square a Lebesgue integral over every finite range, and is of finite order at $\pm\infty$, the division into $f_1(t)$ and $f_2(t)$ will still hold valid.

8. **Asymptotic Series.** A case of particular interest is where

$$f(t) = 0, \quad t < 0$$
$$= 1, \quad t > 0$$

that is, the Heaviside *1* function. Here $f_1(t)$ will be of the form shown in Fig. 1 and will be analytic, while $f_2(t)$ will be

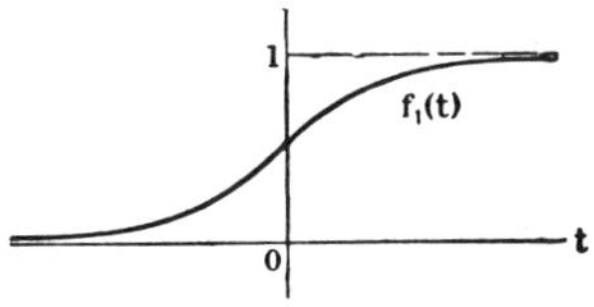

FIG. 1

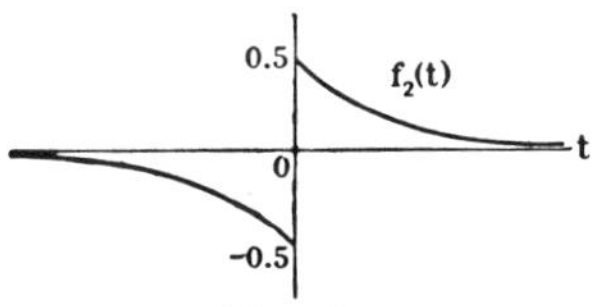

FIG. 2

of the form shown in Fig. 2 and will vanish at $\pm\infty$ faster than any negative power of t. This suggests that we may divide the problem of defining

$$F\left(\frac{d}{dt}\right)f(t)$$

into two parts, one for the high-frequency terms and one for the low-frequency terms, and separate those difficulties which come from the singularity of $f(t)$ at the origin from those which arrive from its behavior at infinity.

We shall consider the particular case where $F(j\mu)$ possesses the following properties:

(*1*) There is a neighborhood of the origin u within which $F(\mu^2)$ assumes the values of an analytic function of μ.

(*2*) $F(j\mu)$ is of the form *constant* $+ g(\mu)$ for $\mu =$ real, where $g(\mu)$ is a function with a Lebesgue integral over every finite range, and is such that

$$\lim_{A\to\infty} \int_{-A}^{A} |\, g(\mu)\,|^2\, d\mu$$

exists. All the successive derivatives of $F(j\mu)$ exist and are of the form $g(\mu)$, under the same conditions. We can then show that

(*a*) A and B may be so chosen that if we put

$$F(j\mu) = \sum a_n (j\mu)^{\frac{n}{2}}, \tag{23}$$

in the neighborhood of the origin, then the series

$$\sum a_n \left(\frac{d}{dt}\right)^{\frac{n}{2}} f_1(t)$$

converges, and has after $2\,k + 1$ terms a remainder which is of the order of magnitude of $t^{-(k+\frac{3}{2})}$ at infinity. Here we define

$$\left(\frac{d}{dt}\right)^{\frac{2k+1}{2}} h(t) = \left(\frac{d}{dt}\right)^{-\frac{1}{2}} \left(\frac{d}{dt}\right)^{k+1} h(t)$$

where

$$\left(\frac{d}{dt}\right)^{-\frac{1}{2}} \psi(t) = \frac{1}{\sqrt{\pi}} \int_{-\infty}^{t} \frac{\psi(\tau)}{\sqrt{t-\tau}}\, d\tau.$$

We can show that

$$(b)\quad \lim_{t\to\infty} \left(\frac{d}{dt}\right)^{\frac{2k+1}{2}} f_1(t) - \frac{(-\frac{1}{2})^k (1\cdot 3\cdot 5\cdots 2\,k-1)}{\sqrt{\pi}\, t^{\frac{2k+1}{2}}} = 0 \tag{24}$$

and that

$$\lim_{t\to\infty} \left(\frac{d}{dt}\right)^k f_1(t) = 0. \tag{25}$$

So far only condition (*1*) has come into consideration.
For $f_1(t)$, let us then put by definition

$$F\left(\frac{d}{dt}\right) f_1(t) = \sum a_n \left(\frac{d}{dt}\right)^{\frac{n}{2}} f_1(t). \tag{26}$$

As to $f_2(t)$, it has a Fourier transform, and we define $F\left(\frac{d}{dt}\right) f_2(t)$ as before. We can show that under our hypotheses,

$$(c) \qquad \lim_{t\to\infty} \quad t^k F\left(\frac{d}{dt}\right) f_2(t) = 0 \tag{27}$$

for all k.

If we put

$$F\left(\frac{d}{dt}\right) f(t) = F\left(\frac{d}{dt}\right) f_1(t) + F\left(\frac{d}{dt}\right) f_2(t) \tag{28}$$

we can show that this function is independent of the auxiliary numbers A and B chosen. Theorems (*a*), (*b*), and (*c*) then give

$$F\left(\frac{d}{dt}\right) f(t) - F(0) + \frac{a_1}{2\sqrt{\pi t}}$$

$$- \sum_1^n \frac{a_{2k+1}(-\frac{1}{2})^k (1 \cdot 3 \cdot 5 \cdot \cdot \cdot 2\,k - 1)}{\sqrt{\pi} t^{\frac{2\,k+1}{2}}}$$

= expression of order of magnitude of $t^{-\frac{2\,n+3}{2}}$ at $+\infty$. This is, however, what we mean when we say that the asymptotic development of

$$F\left(\frac{d}{dt}\right) f(t)$$

is

$$F(0) - \frac{a_1}{2\sqrt{\pi t}} t + \sum_1^\infty \frac{a_{2\,k+1}(-\frac{1}{2})^k (1 \cdot 3 \cdot 5 \cdot \cdot \cdot 2\,k - 1)}{\sqrt{\pi}\, t^{\frac{2\,k+1}{2}}}. \tag{29}$$

Comments on [29c]

P. Masani

Wiener's interest in the Heaviside operational calculus was stimulated by requests from his engineering colleagues to work out a satisfactory treatment. In this note, which appends Bush's engineering treatise, Wiener gives a brief exposition of the rigorization of the calculus obtained in his paper [26c]. The rigorization is accomplished by use of generalized Fourier transformation. In §§1–4 classical Fourier analysis is discussed informally, and in §§5–8 causal operations, the generalized Fourier transform, fractional derivatives and asymptotic series are discussed.

This paper is more readable than the much harder, pioneering and historically important but now superseded paper [26c].

THE SOLUTION OF A DIFFERENCE EQUATION BY TRIGONOMETRICAL INTEGRALS

By Norbert Wiener

Introduction

1. In a paper soon to appear,[1] the author has discussed the problem of the expansion of the general bounded or almost bounded function in a trigonometrical form. It is manifest that expansions of this type throw much light on the properties of the difference equation

$$f(x+1)-f(x)=\phi(x). \tag{1}$$

It is the purpose of this paper to develop this method of treatment to the extent at least of duplicating certain results which Nörlund[2] has reached by other methods. In the course of this investigation, methods will be developed which have possible applications in fields much wider than the calculus of finite differences, such as the calculus of operations and the methods of Heaviside for the solution of the partial differential equations of the electric circuit. I shall be content to indicate these here, leaving their elaboration to a subsequent paper.[3]

We shall put

$$\Theta(x)=0; \qquad [0\leq x\leq \tfrac{1}{2}]$$

$$\Theta(x)=\int_{\frac{1}{2}}^{x} e^{\frac{1}{6\xi^2-7\xi+2}}\,d\xi; \qquad [\tfrac{1}{2}\leq x\leq \tfrac{2}{3}]$$

$$\Theta(x)=\int_{\frac{1}{2}}^{\frac{2}{3}} e^{\frac{1}{6\xi^2-7\xi+2}}\,d\xi. \qquad [\tfrac{2}{3}\leq x]$$

We shall write

$$\Phi(x)=\frac{1}{\pi}\int_0^\infty\left[1-\frac{\Theta(\alpha)}{\Theta(1)}\right]\cos\alpha x\,d\alpha$$

and

$$\Psi(x)=\frac{1}{\pi}\int_0^\infty\frac{\Theta(\alpha)}{\alpha\Theta(1)}\sin\alpha x\,d\alpha.$$

[1] On the Representation of Functions by Trigonometrical Integrals, Mathematische Zeitschrift.

[2] N. E. Nörlund, Mémoire sur le calcul aux différences finies. Acta Mathematica, V. 44 (1923), pp. 71–212.

[3] The Operational Calculus, forthcoming in the Mathematische Annalen.

We shall then put

$$\phi_1(x)=\int_{-\infty}^{\infty}\phi(\xi)\Phi(\xi-x)d\xi$$

and

$$\phi_2(x)=\phi(x)-\phi_1(x).$$

If

$$f_1(x+1)-f_1(x)=\phi_1(x) \tag{2}$$

and

$$f_2(x+1)-f_2(x)=\phi_2(x), \tag{3}$$

we shall manifestly have

$$f_1(x+1)+f_2(x+1)-[f_1(x)+f_2(x)]=\phi_1(x)+\phi_2(x).$$

Nörlund points out that the difference equation (1) has the formal solutions

$$-\sum_{s=0}^{\infty}\phi(x+s)$$

and

$$\int_a^x\phi(z)dz+\sum_{\nu=1}^{\infty}\frac{B_\nu}{\nu!}\phi^{(\nu-1)}(x),$$

which, however, are generally divergent. In place of these, he develops the solution

$$\lim_{\eta\to 0}\left\{\int_a^{\infty}\phi(x)e^{-\eta x}\,dx-\sum_{s=0}^{\infty}\phi(x+s)e^{-\eta(x+s)}\right\},$$

which converges whenever:

(1) The function $\phi(x)$ possesses for all positive values of x a continuous derivative of order m which tends to 0 as x increases indefinitely; and

(2) The integral

$$\int_0^{\infty}\dot{B}_m(-z)\phi^{(m)}(x+\omega z)dz$$

converges uniformly over the interval (0, 1), where

$$\dot{B}_{2m}(x)=(-1)^{m+1}2(2m)!\sum_{k=1}^{\infty}\frac{\cos 2\pi kx}{(2\pi k)^{2m}};$$

$$\dot{B}_{2m+1}(x)=(-1)^{m+1}2(2m+1)!\sum_{k=1}^{\infty}\frac{\sin 2\pi kx}{(2\pi k)^{2m+1}}.$$

We shall show that under Nörlund's conditions on $\phi(x)$,

$$f_1(x) = \int_a^x \phi_1(z)dz + \sum_{\nu=1}^{\infty} \frac{B_\nu}{\nu!} \phi_1^{(\nu-1)}(x) \tag{4}$$

and

$$f_2(x) = -\sum_{s=0}^{\infty} \phi_2(x+s) \tag{5}$$

converge to functions satisfying respectively (2) and (3). We shall show, moreover, that the solution which we thus obtain is identically that of Nörlund.

The Properties of $\Phi(x)$ and $\Psi(x)$

2. $\Theta(\alpha)$ clearly possesses at every point derivatives of all orders. Thus integrating by parts,

$$\Phi(x) = \frac{\pm 1}{\pi\Theta(1)x^n} \int_0^{\frac{1}{2}} \begin{matrix}\sin\\ \cos\end{matrix}\, \alpha x \frac{d^n \Theta(\alpha)}{d\alpha^n} d\alpha$$

and

$$\Psi(x) = \pm \frac{1}{\pi x^n} \int_{\frac{1}{2}}^{\infty} \frac{d^n}{d\alpha^n} \left[\frac{\Theta(\alpha)}{\alpha\Theta(1)}\right] \begin{matrix}\sin\\ \cos\end{matrix}\, \alpha x \, d\alpha.$$

We hence see that whatever k may be, we have at $\pm\infty$

$$\Phi(x) = o(x^{-k}); \quad \Psi(x) = o(x^{-k}).$$

Similarly, $\Phi^{(n)}(x)$ and

$$\Psi^{(-n)}(x) = (-1)^n \int_x^{\infty} d\xi_1 \int_{\xi_1}^{\infty} d\xi_2 \ldots \int_{\xi_{n-1}}^{\infty} \Psi(\xi_n) d\xi$$

exist, and are $o(x^{-k})$ at infinity for all finite k.

Now let $\phi(x)$ be any function of finite order at $\pm\infty$. We have

$$\int_A^x \phi_1(x)dx = \int_A^x du \int_{-\infty}^{\infty} \phi(\xi)\Phi(\xi-u)d\xi$$

$$= \int_{-\infty}^{\infty} \phi(\xi)d\xi \int_A^x \Phi(\xi-u)du$$

$$= \int_{-\infty}^{\infty} \phi(\xi)d\xi \int_{\xi-x}^{\xi-A} \Phi(u)du$$

$$= \frac{1}{\pi}\int_{\infty}^{\infty} \phi(\xi)d\xi \int_{\xi-x}^{\xi-A} du \int_0^1 \left[1-\frac{\Theta(\alpha)}{\Theta(1)}\right] \cos \alpha u \, d\alpha$$

$$= \frac{1}{\pi}\int_{-\infty}^{\infty} \phi(\xi)d\xi \int_0^1 \left[1-\frac{\Theta(\alpha)}{\Theta(1)}\right] d\alpha \int_{\xi-x}^{\xi-A} \cos \alpha u \, du$$

$$= \frac{1}{\pi}\int_{-\infty}^{\infty} \phi(\xi)d\xi \int_0^{\infty} \left[1-\frac{\Theta(\alpha)}{\Theta(1)}\right] \frac{\sin \alpha(\xi-A)-\sin \alpha(\xi-x)}{\alpha} dx$$

$$= \int_{-\infty}^{\infty} \phi(\xi)d\xi[\Psi(\xi-x)-\Psi(\xi-A)] + \int_A^x \phi(\xi)d\xi.$$

Thus by differentiation,

$$\phi_1(x) - \frac{d}{dx}\int_{\infty}^{\infty} \phi(\xi)\Psi(\xi-x)d\xi = \phi(x).$$

Hence if $\phi(x)$ admits a derivative which is of finite order at $\pm\infty$

$$\phi_2(x) = -\frac{d}{dx}\int_{-\infty}^{\infty} \phi(\xi)\Psi(\xi-x)d\xi$$

$$= -\frac{d}{dx}\int_{-\infty}^{\infty} \phi(\xi)d\Psi^{(-1)}(\xi-x)$$

$$= \frac{d}{dx}\int_{-\infty}^{\infty} \phi'(\xi)\Psi^{(-1)}(\xi-x)d\xi$$

$$= \lim_{h\to 0}\int_{-\infty}^{\infty} \phi'(\xi)\frac{\Psi^{(-1)}(\xi-x-h)-\Psi^{(-1)}(\xi-x)}{h}d\xi$$

We have already seen that $u^n\Psi(u)$ has the Fourier integral expansion

$$u^n\Psi(u)=\frac{1}{\pi}\int_0^\infty \frac{\Theta(a)}{a\Theta(1)}u^n \sin au\, da$$
$$=\pm\frac{1}{\pi}\int_0^\infty \frac{d^n}{da^n}\left[\frac{\Theta(a)}{a\Theta(1)}\right]\begin{matrix}\sin\\ \cos\end{matrix}\, au\, da.$$

In a similar manner, we may prove that

$$u^n\left[\frac{\Psi^{(-1)}(u+h)-\Psi^{(-1)}(u)}{h}\right]=\pm\frac{1}{\pi}\int_0^\infty \frac{d^n}{da^n}\left[\frac{\Theta(a)\sin\frac{ah}{2}}{\frac{a^2h}{2}\Theta(1)}\right]\begin{matrix}\sin\\ \cos\end{matrix}\left(u+\frac{h}{2}\right)da.$$

There is no difficulty in establishing that

$$\frac{d^n}{da^n}\left[\frac{\Theta(a)}{a\Theta(1)}\right]=\underset{h\to 0}{l.m.}\left[\frac{\Theta(a)\sin\frac{ah}{2}}{\frac{a^2h}{2}\Theta(1)}\right]$$

where the *l.m.* is understood to extend to the entire postive half-line. Hence

$$u^n\Psi(u)=\underset{h\to 0}{l.m.}\; u^n\left[\frac{\Psi^{(-1)}(u+h)-\Psi^{(-1)}(u)}{h}\right],$$

where the *l.m.* refers to the limit in the mean from $-\infty$ to ∞. Thus if $\phi'(x)$ is of finite order at $\pm\infty$, we have

$$\phi_2(x)=\lim_{h\to 0}\int_{-\infty}^\infty \phi'(\xi)\frac{\Psi^{(-1)}(\xi-x-h)-\Psi^{(-1)}(\xi-x)}{h}d\xi$$
$$=-\int_{-\infty}^\infty \phi'(\xi)\Psi(\xi-x)d\xi.$$

Integrating by parts, we see that if $\phi^{(m)}(x)$ is of finite order at $\pm\infty$,

$$\phi_2(x)=-\int_{-\infty}^\infty \phi^{(m)}(\xi)\,\Psi^{(-m+1)}(\xi-x)d\xi.$$

From the standpoint of trigonometric representation, $\phi_1(x)$ represents those components of $\phi(x)$ with a frequency less than π, together with a portion of those components with a frequency between π and $4\pi/3$. On the other hand, $\phi_2(x)$ represents the remainder of those components with a frequency between π and $4\pi/3$, together with all those having a higher frequency.

The Low Frequency Components of $\phi(x)$.

3. Up to this point we have made no direct reference to the difference equation, and indeed our methods are useful in the treatment of a much wider class of problems concerning the operational method. We now turn to the difference equation (2) which we wish to prove to be satisfied by series (4), under the hypothesis that $\phi^{(m)}(x)$ is bounded.

We have

$$\begin{aligned}\phi_1{}^{(m+p)}(x) &= (-1)^{(m+p)}\int_{-\infty}^{\infty}\phi(\xi)\Phi^{(m+p)}(\xi-x)d\xi \\ &= (-1)^p\int_{-\infty}^{\infty}\phi^{(m)}(\xi)\Phi^{(p)}(\xi-x)d\xi.\end{aligned}$$

Now,

$$\begin{aligned}u^2\Phi^{(p)}(u) &= \pm\frac{u^2}{\pi}\int_0^\infty\left[1-\frac{\Theta(\alpha)}{\Theta(1)}\right]\alpha^p\,{\sin \atop \cos}\,\alpha u\,d\alpha \\ &= \pm\frac{1}{\pi}\int_0^\infty\frac{d^2}{d\alpha^2}\left\{\alpha^p\left[1-\frac{\Theta(\alpha)}{\Theta(1)}\right]\right\}{\sin \atop \cos}\,\alpha u\,du \\ &= \pm\frac{1}{\pi}\int_0^\infty\left\{p(p-1)\alpha^{p-2}\left[1-\frac{\Theta(\alpha)}{\Theta(1)}\right]-p\alpha^{p-1}\frac{\Theta'(\alpha)}{\Theta(1)}\right. \\ &\qquad\left.-\alpha^p\frac{\Theta''(\alpha)}{\Theta(1)}\right\}{\sin \atop \cos}\,\alpha u\,du\end{aligned}$$

Hence there is a constant K independent of p such that

$$|\Phi^{(p)}(u)|\leq\frac{K(\frac{2}{3})^p}{u^2}.$$

Similarly, there is an L independent of p such that

$$|\Phi^{(p)}(u)|\leq L(\tfrac{2}{3})^p$$

and an M depending only on ϕ such that for all positive p,

$$\phi_1{}^{(m+p)}(x)\leq M(\tfrac{2}{3})^p.$$

The series

$$\sum_{\nu=1}^{\infty} \frac{B_\nu}{\nu!} \xi^\nu$$

has the radius of convergence 1. Hence

$$f_1(x) = \int_a^x \phi_1(z)dz + \sum_{\nu=1}^{\infty} \frac{B_\nu}{\nu!} \phi_1{}^{(\nu-1)}(x)$$

converges absolutely and uniformly for any finite range of x. Thus

$$f_1(x+1) - f_1(x) = \int_x^{x+1} \phi_1(z)dz + \sum_{\nu=1}^{\infty} \frac{B_\nu}{\nu!} \left[\phi_1{}^{(\nu-1)}(x+1) - \phi_1{}^{(\nu-1)}(x)\right]$$

$$= \int_x^{x+1} \phi_1(z)dz - \sum_{\nu=1}^{\infty} \frac{B_\nu}{\nu!}\left[\phi_1{}^{(\nu)}(x) + \frac{\phi_1{}^{(\nu+1)}(x)}{2!} + \ldots + \frac{\phi_1{}^{(\nu+k)}(x)}{k!} + \ldots\right]$$

$$= \int_x^{x+1} \phi_1(z)dz - \sum_{k=1}^{\infty} \sum_{\nu=1}^{k} \frac{B_\nu \phi_1{}^{(k)}(x)}{\nu!\, k!}$$

$$= \int_x^{x+1} \phi_1(z) - \sum_{k=1}^{\infty} \frac{\phi_1{}^{(k)}(x)}{(k+1)!} = \phi_1(x).$$

The inversion of limits in this reasoning is justified by the fact that the double series which is summed is absolutely convergent.

In a similar manner, we may show that the series

$$F(x) = \int_x^{-a} \Phi(z)dz + \sum_{\nu=1}^{\infty} \frac{B_\nu}{\nu!}(-1)^{\nu-1}\Phi^{(\nu-1)}(x)$$

is convergent, and solves the difference equation

$$F(x-1) - F(x) = \Phi(x).$$

It is easy to verify that

$$f_1(x) = \int_{-\infty}^{\infty} \phi(\xi)\left[F(\xi-x) - \int_{\xi-x}^{-a} \Phi(z)dz\right]d\xi + \int_a^x \phi_1(z)dz. \qquad (6)$$

The Fourier integral expression for

$$F(x)-\int_x^{-a}\Phi(z)dz$$

is

$$\sum_{\nu=1}^{\infty}\frac{B^\nu}{2\pi(\nu!)}\int_0^\infty\left[1-\frac{\Theta(\alpha)}{\Theta(1)}\right]\left[(-i\alpha)^{\nu-1}e^{i\alpha x}+(i\alpha\nu)^{\nu-1}e^{-i\alpha x}\right]d\alpha$$

$$=\frac{1}{2\pi}\int_0^\infty\left[1-\frac{\Theta(\alpha)}{\Theta(1)}\right]\left\{\frac{e^{i\alpha x}}{e^{-i\alpha}-1}+\frac{e^{i\alpha x}}{i\alpha}+\frac{e^{-i\alpha x}}{e^{i\alpha}-1}-\frac{e^{-i\alpha x}}{i\alpha}\right\}d\alpha$$

$$=\frac{1}{\pi}\int_0^\infty\left[1-\frac{\Theta(\alpha)}{\Theta(1)}\right]\left\{\frac{\sin\alpha x}{\alpha}+\frac{\cos(x+1)\alpha-\cos\alpha x}{2-2\cos\alpha}\right\}d\alpha.$$

Nörlund's principal solution of (1) is that for which there exists

$$\lim_{x\to\infty} f^{(m)}(x).$$

Let

$$\lim_{x\to\infty}\phi^{(m-1)}(x)=0.$$

Differentiating expression (6) m times, we get

$$f_1^{(m)}(x)=\int_{-\infty}^{\infty}\phi^{(m-1)}(\xi)\frac{d}{dx}\left[F(\xi-x)-\int_{-x}^{-a}\Phi(z)dz\right]d\xi+\phi_1^{(m-1)}(x).$$

It is then easy to see that the solution which we have obtained for (2) is the principal solution.

The High Frequency Components of $\phi(x)$

4. Let us now subject our function $\phi(x)$ to the entire force of Nörlund's hypothesis. We shall have formally

$$f_2(x)=\sum_{s=0}^{\infty}\int_{-\infty}^{\infty}\phi^{(m)}(\xi)\Psi^{(-m+1)}(\xi-x-s)d\xi.$$

We wish to prove the convergence of this series.

To begin with, let us consider the function

$$K(u)=\sum_{s=-\infty}^{\infty}\Psi^{(-m+1)}(u-s)$$

This function exists, and is periodic with period 1. It is hence appropriate to consider its Fourier series.

We have

$$\int_0^1 K(u)\sin 2\pi nu du = \sum_{s=-\infty}^{\infty} \int_0^1 \Psi^{(-m+1)}(u-s)\sin 2\pi nu\, du$$
$$= \sum_{s=-\infty}^{\infty} \int_{-s}^{1-s} \Psi^{(-m+1)}(\nu)\sin 2\pi n\nu\, d\nu$$
$$= \int_{-\infty}^{\infty} \Psi^{(-m+1)}(\nu) \sin 2\pi n\nu\, d\nu$$
$$= \pm\frac{1}{\pi}\int_{-\infty}^{\infty} \sin 2\pi n\nu\, d\nu \int_0^{\infty} \frac{\Theta(\alpha)}{\alpha^m\Theta(1)} \begin{matrix}\sin\\ \cos\end{matrix}\, \alpha\nu\, d\alpha$$
$$\begin{cases} = \pm \dfrac{\Theta(2\pi n)}{(2\pi n)^m\Theta(1)} = \dfrac{\pm 1}{(2\pi n)^m}; & [m \text{ odd}] \\ = 0. & [m \text{ even}] \end{cases}$$

Again,

$$\int_0^1 K(u) \cos 2\pi nu\, du \begin{cases} = \pm \dfrac{1}{(2\pi n)^m}; & [m \text{ even}] \\ = 0. & [m \text{ odd}] \end{cases}$$

This means, however, that

$$K(u) = \frac{1}{n!} \dot{B}_n(u)$$

That is, Nörlund's condition tells us that

$$\int_{-\infty}^{\infty} \phi^m(\xi) \left[\sum_{-\infty}^{\infty} \Psi^{(-m+1)}(\xi - x - s)\right] d\xi$$

exists.

Now let us compare

$$U = \int_{\mu}^{\nu} \phi^{(m)}(\xi) \left[\sum_{-\infty}^{\infty} \Psi^{(-m+1)}(\xi - x - s)\right] d\xi$$

and

$$V = \int_{-\infty}^{\infty} \phi^{(m)}(\xi) \left[\sum_{\mu}^{\nu} \Psi^{(-m+1)}(\xi - x - s)\right] d\xi$$

We have

$$U-V=\int_{-\infty}^{\infty}\phi^{(m)}(\xi)Q(\xi-x)d\xi,$$

where

$$Q(u)=\sum_{s=\mu}^{\nu}\Psi^{(-m+1)}(u-s);\quad [u<\mu \text{ or } u>\nu)$$

$$Q(u)=\sum_{-\infty}^{\mu-1}\Psi^{(-m+1)}(u-s)+\sum_{\nu+1}^{\infty}\Psi^{(-m+1)}(u-s).\quad [\mu<u<\nu].$$

We have seen that

$$\Psi^{(-m+1)}(u)\leq\frac{R}{S+u^4}.$$

Thus

$$|Q(u)|\leq\sum_{t=0}^{\infty}\left[\frac{R}{S+[\,|u-\mu|+t]^4}+\frac{R}{S+[\,|u-\nu|+t]^4}\right]$$

$$\leq\frac{R_1}{S_1+|u-\mu|^3}+\frac{R_2}{S_2+|u-\nu|^3},$$

where R_1, R_2, S_1, and S_2 are properly chosen constants. If now $\phi^{(m)}(x)$ is a bounded function vanishing at $+\infty$, it follows that for any finite range of x, $U-V$ tends to 0 as μ and ν tend to $+\infty$ independently. Now, we already know that U tends to 0 as μ and ν tend to $+\infty$ independently. Thus if $\nu>\mu$, we have

$$0=\lim_{\mu\to\infty}\int_{-\infty}^{\infty}\phi^{(m)}(\xi)\left[\sum_{\mu}^{\nu}\Psi^{(-m+1)}(\xi-x-s)\right]d\xi$$

$$=\lim_{\mu\to\infty}\sum_{\mu}^{\nu}\int_{-\infty}^{\infty}\phi^{(m)}(\xi)\Psi^{(-m+1)}(\xi-x-s)d\xi.$$

Therefore

$$\sum_{s=0}^{\infty}\int_{-\infty}^{\infty}\phi^{(m)}(\xi)\Psi^{(-m+1)}(\xi-x-s)d\xi$$

converges. It manifestly satisfies equation (3), and as manifestly is the principal solution. Thus the principal solution of (1) is

$$f_1(x)+f_2(x)=\int_a^x \phi_1(z)dz+\sum_{\nu=1}^{\infty}\frac{B_\nu}{\nu!}\phi_1^{(\nu-1)}(x)-\sum_{s=0}^{\infty}\phi_2(x+s)$$

$$=\int_a^x dz\int_{-\infty}^{\infty}\phi(\xi)\Phi(\xi-z)d\xi$$

$$+\frac{1}{\pi}\int_{-\infty}^{\infty}\phi(\xi)d\xi\int_0^{\infty}\left[1-\frac{\Theta(\alpha)}{\Theta(1)}\right]\left\{\frac{\sin\alpha\,(\xi-x)}{\alpha}-\frac{\sin\alpha\,(\xi-x+\frac{1}{2})}{2\sin\frac{\alpha}{2}}\right\}d\alpha$$

$$+\sum_{s=0}^{\infty}\int_{-\infty}^{\infty}\phi^{(m)}(\xi)\Psi^{(-m+1)}(\xi-x-s)d\xi.$$

Comments on [25b]

G. Strang

This paper was written during the period when Wiener was determined to clarify, and thereby to justify, the operational calculus of Heaviside. Mathematically, the crucial point was always the same—to prove that certain expansions, whether into infinite series or integrals, actually converged. His favorite tool was of course the Fourier transform, and the slow decay at infinity of the transform of a delta-function, or even of a step-function, was the source of difficulty which provoked Wiener's powers of invention. (Elsewhere he refers to the step-function as Heaviside's favorite *corpus vile*.)

In the present paper he is looking for the principal solution of the simple difference equation $f(x+1) - f(x) = \phi(x)$. Nörlund had already pointed out the two formal solutions

$$f_1(x) = -\sum_0^\infty \phi(x+s) \quad \text{and} \quad f_2(x) = \int_a^x \phi(z)\,dz + \sum_1^\infty \frac{B_\nu}{\nu!}\phi^{(\nu-1)}(x).$$

Wiener's observation is this, that the first of these solutions tends to converge when ϕ is composed of high frequencies, and (because of the differentiations) the second behaves best in the opposite situation, when the Fourier transform of ϕ is concentrated in a low frequency band $|\xi| \leqslant$ constant. Inevitably, Wiener splits the given ϕ into $\phi_1 + \phi_2$ (the splitting itself is astonishingly complicated, and unlikely ever to become popular; integrands like exp $\{(6\xi^2 - 7\xi + 2)^{-1}\}$ appear without explanation), and the solution f is a sum of the corresponding $f_1 + f_2$.

In later publications he extends this technique, dissecting f even into an infinite number of frequency ranges—as in the Besov norms of today. The present paper is a comparatively small step in the development of his operational calculus.

A NEW DEFINITION OF ALMOST PERIODIC FUNCTIONS.

By Norbert Wiener.*

According to the definition of Harald Bohr,† a function $f(x)$ (possibly complex valued) of the real argument $x(-\infty < x < \infty)$ is said to have the translation number τ_ε pertaining to ε if

$$|f(x+\tau)-f(x)| \leqq \varepsilon$$

for all real x. If for every positive ε there is a finite number L_ε such that every interval $(A, A+L_\varepsilon)$ contains at least one translation number τ_ε pertaining to ε, the function $f(x)$ is said to be *almost periodic*. Bohr has shown‡ that the class of almost periodic functions is identical with the class of functions to which we may approximate to any desired degree of accuracy by a trigonometrical polynomial

$$\sum_1^N A_n e^{i\lambda_n x}.$$

Bochner§ has given an equivalent but distinct definition of the class of almost periodic functions. The present paper gives yet another definition of this class.

We shall call a real function $\varphi(x)$ *regular* if

$$\lim_{T\to\infty} \frac{1}{2T}\int_{-T}^{T} \varphi(x)\varphi(x-\tau)\,dx$$

exists for every real x. A complex function is regular if its real and pure imaginary parts are regular taken separately. Our theorem is: *if $f(x)$ is regular, and if for every ε there is a regular real function $\varphi_\varepsilon(x)$ such that*

$$g_\varepsilon(x) = \lim_{T\to\infty} \frac{1}{2T}\int_{-T}^{T} f(x-\xi)\varphi_\varepsilon(\xi)\,d\xi$$

exists for every ε while

$$|f(x)-g_\varepsilon(x)| < \varepsilon, \quad (-\infty < x < \infty)$$

---

* Received January 24, 1927.

† H. Bohr, Acta Math., vol. 45 (1924), pp. 29–127.

‡ H. Bohr, Acta Math., vol. 46 (1925), pp. 101–214.

§ S. Bochner, Math. Annalen, vol. 96 (1926), p. 143.

then $f(x)$ is almost periodic. Conversely, if $f(x)$ is almost periodic, it is regular, and there exists for every ε a function $\varphi_\varepsilon(x)$ with the property just indicated.

We may manifestly confine ourselves to the case where $f(x)$ is real. Our hypotheses concerning f and φ_ε demand the regularity of $f(x)+\varphi_\varepsilon(x)$ and $f(x)-\varphi_\varepsilon(x)$; hence $g_\varepsilon(\tau)$ is the difference of two expressions of the form

$$\lim_{T\to\infty} \frac{1}{2T}\int_{-T}^{T} \psi(x)\,\psi(x-\tau)\,dx.$$

The author has shown* that every such function is the sum of an almost periodic function and a function which is $o(1)$ at $\pm\infty$. Hence we may write

$$g_\varepsilon(x) = g_\varepsilon^{(1)}(x)+g_\varepsilon^{(2)}(x),$$

where $g_\varepsilon^{(1)}(x)$ is almost periodic and $g_\varepsilon^{(2)}(x) = o(1)$ at $\pm\infty$.

We have

$$g_\varepsilon(x)-g_\eta(x) = g_\varepsilon^{(1)}(x)-g_\eta^{(1)}(x)+(g_\varepsilon^{(2)}(x)-g_\eta^{(2)}(x)).$$

The pair of terms in the parenthesis is $o(1)$; the preceding pair is almost periodic. An almost periodic function comes arbitrarily near to its maximum value in any half line; hence

$$\max|g_\varepsilon^{(1)}(x)-g_\eta^{(2)}(x)| \leqq \max|g_\varepsilon(x)-g_\eta(x)|.$$

Hence

$$\lim_{\varepsilon\to 0} g_\varepsilon^{(1)}(x) = f^{(1)}(x)$$

exists as a uniform limit and is almost periodic.† Similarly

$$\lim_{\varepsilon\to 0} g_\varepsilon^{(2)}(x) = f^{(2)}(x)$$

exists as a uniform limit and is $o(1)$ at $\pm\infty$. Clearly

$$f(x) = f^{(1)}(x)+f^{(2)}(x).$$

Let us consider

$$\frac{1}{2T}\int_{-T}^{T} f^{(2)}(x-\xi)\,\varphi_\varepsilon(\xi)\,d\xi.$$

By the Schwarz inequality this cannot exceed

* In a paper soon forthcoming in the Proc. Lond. Math. Soc. Cf. N. Wiener, Math. Zeit., vol. 24, p. 581 (Theorem IX).

† Cf. the first cited paper of Harald Bohr.

$$\frac{1}{2\tau}\left[\int_{-\tau}^{\tau}|f^2(x-\xi)|^2\,d\xi\int_{-\tau}^{\tau}|\varphi_\varepsilon(\xi)|^2\,d\xi\right]^{\frac{1}{2}},$$

and, since

$$\lim_{\tau\to\infty}\frac{1}{2\tau}\int_{-\tau}^{\tau}|f^2(x-\xi)|^2\,d\xi = 0, \qquad \lim_{\tau\to\infty}\frac{1}{2\tau}\int_{-\tau}^{\tau}|\varphi_\varepsilon(\xi)|^2\,d\xi<\infty,$$

we have

$$\lim_{\tau\to\infty}\frac{1}{2\tau}\int_{-\tau}^{\tau}f^{(2)}(x-\xi)\,\varphi_\varepsilon(\xi)\,d\xi = 0,$$

and

$$g_\varepsilon(x) = \lim_{\tau\to\infty}\frac{1}{2\tau}\int_{-\tau}^{\tau}f^{(1)}(x-\xi)\,\varphi_\varepsilon(\xi)\,d\xi.$$

Since $f^{(1)}(x)$ is almost periodic, $g_\varepsilon(x)$ must be almost periodic, $g_\varepsilon^{(2)}(x)$ is identically 0, $f^{(2)}(x)$ vanishes, and $f(x)$ is almost periodic.

On the other hand, if $f(x)$ is almost periodic, it is clearly regular. By a theorem of Bochner,* $\varphi(x) = \max\limits_{\tau}|f(\tau+x)-f(\tau)|$ is almost periodic. Let $F_\xi(u)$ be a continuous non-negative function, defined for positive and zero u, not identically zero, and vanishing for $u>\varepsilon/2$. Let

$$\varphi_\varepsilon(u) = F_\varepsilon(\varphi(x))/\lim_{\tau\to\infty}\frac{1}{2\tau}\int_{-\tau}^{\tau}F_\varepsilon(\varphi(x))\,dx.$$

The author has proved elsewhere† the existence and almost periodic character of $\varphi_\varepsilon(u)$ as well as the fact that

$$\left|\lim_{\tau\to\infty}\frac{1}{2\tau}\int_{-\tau}^{\tau}f(x-\xi)\,\varphi_\varepsilon(\xi)\,d\xi-f(x)\right|<\varepsilon.$$

The converse part of our theorem is thus proved.

* Bochner, loc. cit., p. 136.

† In the forthcoming paper in the Proc. Lond. Math. Soc.

Comments on [27b]

F. Holland

Motivated by the methods he developed in [25c] and [28a] to prove, *inter alia*, Bohr's approximation theorem, to the effect that an almost periodic function is a uniform limit of trigonometric polynomials, Wiener presents here a new structural characterization of almost periodic functions.

In the course of proving the first part of the theorem in the paper, Wiener states that from a result proved in a "forthcoming" paper it follows that

$$g_\varepsilon(x) = g_\varepsilon^{(1)}(x) + g_\varepsilon^{(2)}(x),$$

where $g_\varepsilon^{(1)}(x)$ is almost periodic and $g_\varepsilon^{(2)}(x) = o(1)$ at $\pm\infty$. This result is incorrect. Indeed the example due to K. Mahler, reproduced in [30a, pp. 205–211]† disproves it. Consequently, the proof of the first part of the theorem is incorrect from this point on. It can be amended, however, by replacing Wiener's subsequent argument by the following one.

Thus $g_\varepsilon(t)$ is the difference of two continuous positive-definite functions and is therefore the Fourier transform of a finite measure. Letting ε tend to zero through a sequence of values, we conclude from W. Rudin's book,[1] pp. 32–34, that f is the Fourier transform of a finite measure μ, say. Decomposing μ into its discrete and continuous parts, we can write

$$f = f^{(1)} + f^{(2)},$$

where $f^{(1)}$ is almost periodic and

$$\lim_{\tau\to\infty} \frac{1}{2\tau} \int_{-\tau}^{\tau} |f^{(2)}(t)|^2\, dt = 0.$$

The condition on $f^{(2)}$ now forces

$$g_\varepsilon(x) = \lim_{\tau\to\infty} \frac{1}{2\tau} \int_{-\tau}^{\tau} f^{(1)}(x - t)\varphi_\varepsilon(t)\, dt.$$

Since $f^{(1)}$ is almost periodic, g_ε is almost periodic; and f, being the uniform limit of almost periodic functions, must be almost periodic.

---

† *Editor's note.* Cf. comments on [27a] for the origins of this work of K. Mahler.

Errata

p. 365, line 1:	Insert "continuous" before "function."
p. 365, line 4:	For τ read τ_ε.
p. 365, line 9:	Insert "uniformly" after "approximate".
p. 365, line 16, p. 366, line 7:	For $x - \tau$ read $x - t$.
p. 365, line 17:	For x read t.
p. 365, line 18:	Add to the hypotheses of the theorem that f and g_ε are continuous.
p. 365, line 22:	For ε read x.
p. 366, line 6:	For τ read t.
p. 366, line 18:	For (2) read (1).
p. 367, lines 1 and 3:	For $f^2(x - \xi)$ read $f^{(2)}(x - \xi)$.
p. 367, line 12:	For ξ read ε.
p. 367, line 14:	For u read x.

Reference

1. W. Rudin, *Fourier analysis on groups*, Interscience, New York, 1962.

Comments on [27b]

C. Berenstein

Let us recall the standard notation

$$M_t(f(x-t)g(t)) = \lim_{\tau\to\infty}\frac{1}{2\tau}\int_{-\tau}^{\tau} f(x-t)g(t)\,dt$$

whenever f, g are regular in Wiener's sense. The proof of the first part of the theorem in [27b] has a flaw, since it is not true that $g_\varepsilon(x) = M_\tau(f(x-\tau)\varphi_\varepsilon(\tau))$ can be written as $g_\varepsilon = g_\varepsilon^{(1)} + g_\varepsilon^{(2)}$, with $g_\varepsilon^{(1)}$ almost periodic and $g_\varepsilon^{(2)} = o(1)$ at $\pm\infty$. In fact, an example of K. Mahler's[1] in J. Math. & Phys.[1], shows that in general $g_\varepsilon^{(2)} \neq o(1)$ though $M_\tau(|g_\varepsilon^{(2)}(\tau)|^2) = 0$.[†] The proof can be corrected by reducing it to the case where f, φ_ε are Fourier transforms of measures of bounded variation, since in this case $g_\varepsilon^{(2)} = 0$. The second part of the theorem in this paper is correct, and it is the main idea in Wiener's proof of the Weierstrass approximation theorem for almost periodic functions (cf. [28a] and [33i], pp. 192–199). In fact, Wiener shows that if f is almost periodic, then for any $\varepsilon > 0$ there exists a function φ_ε; represented by the absolutely convergent series $\sum_{n=0}^{\infty} b_n e^{i\lambda_n x}$ $(\lambda_n \in \mathbb{R})$, such that $|f(x) - g_\varepsilon(x)| \leqslant \varepsilon$ for every $x \in \mathbb{R}$. In the last page of the above-mentioned book, Wiener gives a historical account of the different proofs of Weierstrass's approximation theorem.

Errata

In all instances:

for $\lim_{\tau\to\infty}\frac{1}{2\tau}\int_{-\tau}^{\tau}\varphi(x-\tau)\varphi(x)\,dx$ read $\lim_{\tau\to\infty}\frac{1}{2\tau}\int_{-\tau}^{\tau}\varphi(x-t)\varphi(t)\,dt$.

Reference

1. K. Mahler, J. Math. and Phys. 6(1927), 158–164.

[†] *Editor's note.* This paper of Mahler is Part II of [27a]. For its origins see comment on [27a].

HARMONIC ANALYSIS AND GROUP THEORY

By Norbert Wiener

Weyl [1] has developed in some detail the relations between the theory of unitary groups and the theory of periodic and almost periodic functions. The groups which he introduces are one-parameter groups of linear functional transformations which leave invariant the expression

$$\lim_{T\to\infty} \frac{1}{2T}\int_{-T}^{T} f(x)\bar{f}(x)dx \tag{1}$$

(which we shall abbreviate $M_x(f(x)\bar{f}(x))$. A similar theory of functions expansible in Fourier integral form would involve the invariancy of the integral quadratic form

$$\int_{-\infty}^{\infty} f(x)\bar{f}(x)dx. \tag{2}$$

The author [2] has shown that the theory of almost periodic functions may very satisfactorily be established on the basis of the Fourier integral. If this treatment is to be associated with the theory of groups, it becomes necessary to consider transformations which leave both (1) and (2) invariant.

If (1) exists, the author has shown [3] that

$$\left[\int_{A}^{\infty} + \int_{-\infty}^{-A}\right]\left|\frac{f(x)}{x}\right|^2 dx$$

is finite. In what follows, we shall consider only measurable functions fulfilling this condition, and such that

$$\int_{-A}^{A} |f(x)|^2 dx$$

is finite. All integrals are of course understood in the sense of Lebesgue.

[1] H. Weyl, "Integralgleichungen und fastperiodische Funktionen," *Math. Annalen.*, 97 (1926), 338–356.

[2] N. Wiener, "The Spectrum of an Arbitrary Function," *Proc. Lond. Math. Soc.*, Ser. 2, 27, 483–496.

[3] N. Wiener, "On the representation of Functions by Trigonometrical Integrals," *Math. Ztschr.* 24 (1925), 577.

Let T be a transformation defined for all functions of this class, and preserving (2) invariant for all the functions for which it is defined. Let S_A be the transformation leading from $f(x)$ to the function $f_A(x)$, where

$$f_A(x) = \begin{cases} 0 \text{ if } |x| > A \\ f(x) \text{ if } |x| \leq A \end{cases}$$

There will be certain transformations T keeping (2) invariant and such that for all functions f of our class for which $M_x(|f(x)|^2)$ exists,

$$\int_{-\infty}^{\infty} |(TS_A - S_A T)f(x)|^2 dx = o(A). \tag{3}$$

For such a transformation, we may readily deduce from

$$\lim_{A\to\infty} \frac{1}{2} \int_{-\infty}^{\infty} \left| \frac{TS_A f(x)}{\sqrt{A}} \right|^2 dx = M_x(|f(x)|^2)$$

and

$$\lim_{A\to\infty} \int_{-\infty}^{\infty} \left| \frac{TS_A f(x)}{\sqrt{A}} - \frac{S_A Tf(x)}{\sqrt{A}} \right|^2 dx = 0$$

that

$$M_x(|Tf(x)|^2) = \lim_{A\to\infty} \frac{1}{2} \int_{-\infty}^{\infty} \left| \frac{S_A Tf(x)}{\sqrt{A}} \right|^2 dx = M_x(|f(x)|^2).$$

These transformations form a group, for

$$\begin{aligned}
&\int_{-\infty}^{\infty} |(T_1T_2S_A - S_AT_1T_2)f(x)|^2 dx \\
&= \int_{-\infty}^{\infty} |[(T_1T_2S_A - T_1S_AT_2) + (T_1S_AT_2 - S_AT_1T_2)]f(x)|^2 dx \\
&\leq 2\int_{-\infty}^{\infty} |(T_1T_2S_A - T_1S_AT_2)f(x)|^2 dx \\
&\quad + 2\int_{-\infty}^{\infty} |(T_1S_AT_2 - S_AT_1T_2)f(x)|^2 dx \\
&= 2\int_{-\infty}^{\infty} |(T_2S_A - S_AT_2)f(x)|^2 dx \\
&\quad + 2\int_{-\infty}^{\infty} |(T_1S_A - S_AT_1)T_2f(x)|^2 dx.
\end{aligned}$$

Furthermore,

$$\int_{-\infty}^{\infty} |(T^{-1}S_A - S_A T^{-1})f(x)|^2 dx$$
$$= \int_{-\infty}^{\infty} |(S_A - TS_A T^{-1})f(x)|^2 dx$$
$$= \int_{-\infty}^{\infty} |(TS_A - S_A T)T^{-1}f(x)|^2 dx.$$

Thus the product of two transformations satisfying (3) and the inverse of a transformation satisfying (3) also satisfy (3).

An example of a transformation satisfying (3) is

$$T^\lambda f(x) = f(x + \lambda),$$

for

$$\int_{-\infty}^{\infty} |(T^\lambda S_A - S_A T^\lambda)f(x)|^2 dx$$
$$= \int_0^\lambda [|f(x - A - \lambda)|^2 + f(x + A - \lambda)|^2]dx = o(A).$$

We shall call a transformation satisfying (3) and keeping (2) invariant *properly unitary*. Let us consider the one-parameter group of properly unitary transformations U^λ, where

$$U^{\lambda+\mu} = U^\lambda U^\mu.$$

Let $f(x)$ be such that

$$g(t) = M_x[(U^t f(x))\bar{f}(x)]. \tag{4}$$

exists for every t. Clearly by the Schwarz inequality

$$g(t) \leq \sqrt{M_x(|U^t f(x)|^2) M_x(|f(x)|^2)}$$
$$= \sqrt{g(0) \lim_{A\to\infty} \frac{1}{2A} \int_{-\infty}^{\infty} |S_A U^t f(x)|^2 dx}$$
$$= \sqrt{g(0) \lim_{A\to\infty} \frac{1}{2A} \int_{-\infty}^{\infty} [U^t S_A f(x)|^2 dx}$$
$$= g(0).$$

It follows that $g(t)$ is a bounded function, and that

$$R(u_1, u_2) = \underset{B\to\infty}{\text{l.i.m.}} \frac{1}{\pi}\int_{-B}^{B} g(t)\frac{\sin(u_2 - u_1)^{t/2}}{t} e^{i(u_1+u_2/2)t}dt$$

exists when $(u_1 - u_2)/2$ is given as a quadratically summable function of $(u_1 + u_2)/2$. For this one only need apply Plancherel's theorem.[4]

Let us put

$$g_A(t) = \frac{1}{2A}\int_{-\infty}^{\infty} [U^t S_A f(x)]\overline{S_A f(x)}dx. \tag{5}$$

Under condition (4) for every t, we have

$$g(t) = \lim_{A\to\infty} \frac{1}{2A}\int_{-\infty}^{\infty} [S_A U^t f(x)]\overline{S_A f(x)}dx,$$

which one readily reduces to the form

$$g(t) = \lim_{A\to\infty} g_A(t)$$

by means of (3). We may easily prove $g_A(t)$ to be less than or equal to $g_A(0)$, and hence to be uniformly bounded in t and A, as $g_A(0) \to g(0)$ as $A \to \infty$.

Let us now introduce a new assumption concerning U^t. Let the transformation W taking $f(x)$ into $\varphi(a, t)$ (a fixed), where

$$\varphi(x, t) = U^t f(x)$$

preserve (2) invariant. Then if $\psi(x) = WS_A f(x)$,

$$g_A(t) = \frac{1}{2A}\int_{-\infty}^{\infty} \psi(x + t)\bar{\psi}(x)dx.$$

It follows readily that $g_A(t)$ is absolutely integrable over $(-\infty, \infty)$ and[5] that

$$P_A(u) = \frac{1}{\pi}\int_{-\infty}^{\infty} g_A(t)e^{iut}dt$$

[4] Hobson, "The Theory of Functions of a Real Variable," 2d edition 2 (1926), 742–752.

[5] N. Wiener, "The Spectrum of an Arbitrary Function," p. 483.

exists and is real and positive.[6] Thus

$$R_A(u_1, u_2) = \frac{1}{\pi}\int_{-\infty}^{\infty} g_A(t)\,\frac{\sin (u_2 - u_1)t/2}{t}\,e^{i(u_1+u_2/2)t}dt$$

exists, is monotone in u_1 and u_2, and has the property that

$$R_A(u_1, u_2) + R_A(u_2, u_3) = R_A(u_1, u_3).$$

To see this, we need only reflect that

$$R_A(u_1, u_2) = \int_{u_1}^{u_2} P_A(u)du.$$

Now

$$g(t)\,\frac{\sin (u_2 - u_1)t/2}{t} = \underset{A\to\infty}{\text{l.i.m.}}\; g_A(t)\,\frac{\sin (u_2 - u_1)t/2}{t},$$

so that

$$R(u_1, u_2) = \underset{A\to\infty}{\text{l.i.m.}}\; R_A(u_1, u_2).$$

From this we may readily conclude[7] that

$$R(u_1, u_2) + R(u_2, u_3) = R(u_1, u_3),$$

and that $R(u_1, u_2)$ may be so defined as to be monotone in both arguments and increasing in u_2.

We have

$$R_A(-u, u) = \frac{1}{\pi}\int_{-\infty}^{\infty} g_A(t)\,\frac{\sin ut}{t}\,dt.$$

Hence

$$\frac{1}{u}\int_0^u R_A(-u, u)du = \frac{1}{\pi}\int_{-\infty}^{\infty} g_A(t)\,\frac{1 - \cos ut}{t^2}\,dt.$$

Now,

$$R_A(-\infty, \infty) = \lim_{u\to\infty}\frac{1}{u}\int_0^u R_A(-u, u)du$$

$$= \lim_{A\to\infty}\frac{1}{\pi u}\int_{-\infty}^{\infty} g_A(t)\,\frac{1 - \cos ut}{t^2}\,dt = g_A(0).$$

[6] Indeed, we might assume directly that $\frac{1}{\pi}\int_{-\infty}^{\infty} e^{int}dt \int_{-\infty}^{\infty} [U^t f(x)]\bar{f}(x)dx$ is positive definite and exists for every quadratically summable $f(x)$.

[7] N. Wiener, loc. cit., p. 489.

by a Fejér theorem. Hence in the limit,

$$R(-\infty, \infty) \leq g(0).$$

In the same way, we can show[8] that

$$\frac{1}{u}\int_0^u R(-u, u)du = \frac{1}{\pi u}\int_{-\infty}^{\infty} g(t)\,\frac{1-\cos ut}{t^2}\,dt.$$

Hence

$$R(-\infty, \infty) = \lim_{u\to\infty}\frac{1}{\pi u}\int_{-\infty}^{\infty} g(t)\,\frac{1-\cos ut}{t^2}\,dt.$$

It follows by a Tauberian theorem[9] that

$$R(-\infty, \infty) = \lim_{\epsilon\to 0}\frac{1}{2\epsilon}\int_{-\epsilon}^{\epsilon} g(t)dt.$$

The theory of almost periodic functions may also be generalized to more general groups of transformations than the translation group. It is an interesting fact that the conditions on this group are somewhat different from those of spectrum theory. This is connected with the fact that, as Franklin has shown,[10] all of the theory of almost periodic functions and of their extensions follows by elementary means from the classical Bohr approximation theorem for almost periodic functions of the usual sort.

Let $\{T^\lambda\}$ be a set of linear transformations into itself of the class of all functions $f(x)$ of summable square modulus over any finite interval, and such that $\frac{1}{A}\int_{-A}^{A} |f(x)|^2 dx$ is bounded in A. Let

$$T^\lambda(T^\mu(f(x))) = T^{\lambda+\mu}(f(x)).$$

Let the transformations T^λ be unitary, and let

$$M_\lambda T^\mu(f(x, \lambda)) = T^\mu M_\lambda f(x, \lambda).$$

We shall say that the function $f(x)$ has a T transformation-number τ_ϵ with respect to ϵ when for all x and λ,

$$|T^{\tau_\epsilon} f(x) - f(x)| < \epsilon.$$

[8] Ibid., p. 489.
[9] Cf. N. Wiener, "A New Method in Tauberian theorems," this *Journal*, vol. 6 (October, 1928), p. 163.
[10] P. Franklin, *Math. Zeitschr.*, 29 (1928), pp. 70–86.

In case for every $\epsilon > 0$ there is an L_ϵ such that for every mod A, $(A, A + L_\epsilon)$ contains at least one T transformation-number τ_ϵ, and in case $T^\lambda f(x)$ is equally continuous in λ for all x we shall say that $f(x)$ is almost periodic with respect to the group $\{T^\lambda\}$.

The arguments of the author's previous paper[11] hold substantially without change to show that we can approximate uniformly in λ and x to $T^\lambda f(x)$ by polynomials of the form $\sum_1^N A_n(x)e^{i\Lambda_n\lambda}$. These polynomials may be obtained from $f(x)$ by a series of steps from f_{k-1} to f_k of the form

$$\begin{aligned} f_k(x, \lambda) &= M_\lambda\{T^\lambda f_{k-1}(x, 0)K(\lambda - \mu)\} \\ &= M_\lambda\{T^{\mu+\lambda} f_{k-1}(x, 0)K(\lambda)\} \\ &= T^\mu M_\lambda\{T^\lambda f_{k-1}(x, 0)K(\lambda)\}, \end{aligned}$$

and by a similar process, we can write

$$\begin{aligned} A_n(x)e^{i\Lambda_n\mu} &= M_\lambda\{T^\lambda[\sum_1^N A_n(x)]e^{-i\Lambda_n(\lambda-\mu)}\} \\ &= T^\mu M_\lambda\{T^\lambda[\sum_1^N A_n(x)]e^{-i\Lambda_n\lambda}\}. \end{aligned}$$

Letting $\mu = 0$,

$$A_n(x) = M_\lambda\{T^\lambda[\sum_1^N A_n(x)]e^{-i\Lambda_n\lambda}\},$$

and

$$A_n(x)e^{i\Lambda_n\mu} = T^\mu A_n(x),$$

or in particular,

$$TA_n(x) = e^{i\Lambda_n}A_n(x).$$

Thus $e^{i\Lambda_n}$ is an Eigenwert of T possessing $A_n(x)$ as an Eigenfunktion, and $f(x)$ may be approximated uniformly by the sum $\sum_1^N A_n(x)$ of Eigenfunktionen of the transformation T.

[11] "The Spectrum of an Arbitrary Function," 494–6.

Comments on [29a]

F. Holland

Let E denote the vector space of locally square summable functions f such that

$$\limsup_{\tau \to \infty} \frac{1}{2\tau} \int_{-\tau}^{\tau} |f(x)|^2 \, dx$$

is finite.

The note opens with a discussion of certain linear transformations T, called *properly unitary*, that map E into itself and from the definition of which it follows that

$$\lim_{\tau \to \infty} \frac{1}{2\tau} \int_{-\tau}^{\tau} |(Tf)(x)|^2 \, dx = \lim_{\tau \to \infty} \frac{1}{2\tau} \int_{-\tau}^{\tau} |f(x)|^2 \, dx,$$

whenever the right-hand limit exists. This is followed by an outline of arguments modeled on [28a] which, when coupled with Bochner's theorem on positive-definite functions, support the following assertion:

Let $\{U^\lambda: -\infty < \lambda < \infty\}$ be a group of properly unitary transformations on E, where U^0 is the identity and $U^{\lambda+\mu} = U^\lambda U^\mu$ and let f be such that

$$\Phi(t) = \lim_{\tau \to \infty} \frac{1}{2\tau} \int_{-\tau}^{\tau} (U^t f)(x) \, \overline{f(x)} \, dx$$

exists for every real t. Suppose also that for some fixed a the transformation W defined by

$$(Wg)(t) = (U^t g)(a)$$

maps $L^2(-\infty, \infty)$ isometrically into itself. Then Φ agrees almost everywhere with the Fourier transform of a finite positive measure.

The last part of the paper deals with the notion of almost periodicity of a function f relative to a group of transformations $\{T^\lambda: -\infty < \lambda < \infty\}$. Sufficient conditions are imposed on the action of the group and on the collection $\{T^\lambda f: -\infty < \lambda < \infty\}$ that imply $(T^\lambda f)(x)$ is the uniform limit of polynomials of the form $\sum_{k=1}^{n} A_k(x) e^{i\lambda_k \lambda}$, where $TA_k = e^{i\lambda_k} A_k$.

Apart from the section dealing with almost periodicity, the bulk of this article is reproduced with minor modifications in §10 of the memoir [30a].

Errata

p. 149, line 16: For "group" read "semigroup."

p. 149, line 17: The lower limit on the integral sign should be $-\infty$.

p. 150, line 6: Insert "invertible" before "transformation."

p. 150, line 11: should read

$$\leqslant \int_{-|\lambda|}^{|\lambda|} [|f(x - A)|^2 + |f(x + A)|^2]\,dx = o(A).$$

p. 151, line 2: The assumptions thus far imposed do not seem to bear out the implication made at this stage that g is measurable. The expression under the integral sign should read

$$g(t)\frac{\sin(u_2 - u_1)t/2}{t}e^{i(u_1+u_2)t/2}.$$

p. 151, line 3: The phrase "when $(u_1 - u_2)/2$ is given" should be placed at the end of the sentence.

p. 151, line 12: For $g_A(t)$ read $|g_A(t)|$.

p. 151, line 16: Here, and in the next line, $f(x)$ seems to refer to a general element, and not to the function under discussion.

p. 151, line 20: The assumptions do not seem to justify the claim that $g_A(t)$ is absolutely integrable over $(-\infty, \infty)$. To ensure this, it is sufficient to assume further either that W maps the space L_c^2 of quadratically summable functions with compact supports into itself or that the map

$$t \to \int_{-\infty}^{\infty} (U^t g)(x)\overline{g(x)}\,dx$$

belongs to $L(-\infty, \infty)$ whenever g belongs to L_c^2.

p. 151, line 22: For n read u.

p. 152, line 2: For $(u_1 + u_2/2)t$ read $(u_1 + u_2)t/2$.

p. 152, line 6: The equation should be

$$2R_A(u_1, u_2) = \int_{u_1}^{u_2} P_A(u)\,du.$$

p. 152, line 10: The meaning of this seems to be: for each fixed a

$$R(u, u + a) = \underset{A \to \infty}{\text{l.i.m}}\, R_A(u, u + a).$$

p. 152, line 21: For $A \to \infty$ read $u \to \infty$. Delete the period after $g_A(0)$.

p. 152: In the first line of the footnote replace e^{int} by e^{iut}.

p. 153, line 1: For Fejer read Fejér.

p. 153, line 23: Presumably, $f(x, \lambda) = (T^{\lambda}f)(x)$.

p. 153, line 26: The displayed inequality should read

$$|(T^{\lambda + \tau_\varepsilon}f)(x) - (T^{\lambda}f)(x)| < \varepsilon.$$

p. 154, line 2: For "mod A" read "real A."

p. 154, line 3: For "equally continuous" read "equicontinuous." Add that, for each λ, $T^{\lambda}f$ is a bounded function.

p. 154, line 11: For $f_k(x, \lambda)$ read $f_k(x, \mu)$.

Comments on [29a]

C. Berenstein

Wiener's idea of a systematic use of Fourier analysis and the spectrum of an arbitrary function, (cf. [28a]), leads him in this paper to point out a first step of a theory relating unitary representations with almost periodicity. A very clear introduction to the subject of group representations and its relation to almost periodicity can be found in Godement's papers.[1,2] On the other hand, Wiener proposes to study in particular transformations leaving invariant not only the mean-square modulus

$$\lim_{\tau \to \infty} \frac{1}{2\tau} \int_{-\tau}^{\tau} |f(x)|^2 \, dx$$

but also the L^2-norm $\int_{-\infty}^{\infty} |f(x)|^2 \, dx$. This particular suggestion has not been taken up. Section 7 of Mackey's survey paper has a thorough discussion of the relation between this work and probability theory.[3]

References

1. R. Godement, Trans. Amer. Math. Soc. 63(1948), 1–84.
2. R. Godement, Ann. of Math. (2) 53(1951), 68–124.
3. G. W. Mackey, Advances in Math. 12(1974), 178–268.

Comments on [29a]

Ka-Sing Lau

Guided by Weyl's work on unitary groups on almost periodic functions, Wiener introduced the notion of proper unitary transformation T on the class $\mathscr{P}$ of functions f of finite quadratic mean, i.e. such that

$$\lim_{A \to \infty} \frac{1}{2A} \int_{-A}^{A} |f(x)|^2 \, dx$$

exists in $\mathbb{R}$. A transformation T from $\mathscr{P}$ to $\mathscr{P}$ is called *proper unitary* if (i) when restricted to L^2, it is an L^2-isometry, (ii) for each $f \in \mathscr{P}$, $\int_{-\infty}^{\infty} |(TS_A - S_A T)f(x)|^2 \, dx = o(A)$, as $A \to \infty$, where S_A is the characteristic function of $[-A, A]$, $A > 0$.

A natural question to ask is whether Wiener's idea can be extended to operators whose restriction to L^2 is not an isometry. The main obstacle is that even elementary operators may fail to transform $\mathscr{P}$ into $\mathscr{P}$. Recently, attempts had been made to study Wiener's generalized harmonic analysis by embedding the nonlinear space $\mathscr{P}$ into a suitable Banach space, for example, the Marcinkiewicz space $\mathscr{M}^2(\mathbb{R})$, cf. {1}, {2}, {3}.[†] Let $\mathscr{X}$ be the closed subspace of regular functions in $\mathscr{M}^2(\mathbb{R})$ {3, A2, A5}. In these comments, we report how the concept of proper transformation can be extended to $\mathscr{X}$ and yields new results that allow us to define normal operators on $\mathscr{X}$ and obtain the spectral resolution theory on $\mathscr{X}$.

Let $\mathscr{T}$ denote the class of *proper operators* T from $\mathscr{X}$ into $\mathscr{X}$, namely, operators T such that

(i) the restriction of T to $L^2(\mathbb{R})$ is an L^2-bounded linear operator;

(ii) for $f \in \mathscr{X}$, $\int_{-\infty}^{\infty} |(TS_A - S_A T)f(x)|^2 \, dx = o(A)$, as $A \to \infty$;

(iii) for $f, g \in \mathscr{X}$, $\int_{-\infty}^{\infty} |(TS_A - S_A T)f(x)g(x)| \, dx = o(A)$, as $A \to \infty$.

The additional condition (iii) is introduced in order to consider the adjoints. It follows easily from (i), (ii) that $\|T\|_{\mathscr{X}} \leq \|T\|_{L^2}$. An important subclass of $\mathscr{T}$ is that of the convolution operators T_μ defined by

$$T_\mu f = \mu * f, \; f \in \mathscr{X},$$

[†] These numbers refer to the papers listed at the end of these comments.

where μ is a bounded countably additive regular Borel measure over $\mathbb{R}$ with $\int_{-\infty}^{\infty} |y| d|\mu|(y) < \infty$.

Let $f, g \in \mathscr{X}$ and let ν be a finitely additive regular Borel measure over $[1, \infty)$ which vanishes on bounded intervals. We define the "inner product" of f and g with respect to ν by

$$\langle f, g \rangle_\nu = \int_1^{\infty} \left(\frac{1}{2A} \int_{-A}^{A} f(x)\overline{g(x)}\, dx \right) d\nu(A).$$

Theorem. *For $T \in \mathscr{T}$, there exists a unique $T^* \in \tau$ such that*
(i) *$T^*|_{L^2}$ coincides with the L^2-adjoint of $T|_{L^2}$;*
(ii) *for any measure ν of the type described above,*

$$\langle T^* f, g \rangle_\nu = \langle f, Tg \rangle_\nu \ f, g \in \mathscr{X}.$$

By using this theorem, we are able to study the concept of adjoint and normal operators in the closure $\bar{\mathscr{T}}$ of $\mathscr{T}$ under the strong-operator topology. Spectral theorems on $\bar{\mathscr{T}}$ can then be obtained which are extensions of the well-known ones for normal operators on Hilbert spaces. The details will be shown in a subsequent paper.

References

1. K.-S. Lau and J. Lee, *On generalized harmonic analysis*, Tran. Amer. Math. Soc. (in press).
2. K.-S. Lau, *On the Banach spaces of functions with bounded upper means* (forthcoming).
3. P. Masani, *Commentary on the memoire [30a] on generalized harmonic analysis*, this volume, pp. 333–379.

ON A THEOREM OF BOCHNER AND HARDY

NORBERT WIENER*.

IN a paper† connected with my recent researches on general harmonic analysis, Bochner and Hardy have proved the following theorem.

THEOREM I. *If*

$$\frac{1}{T}\int_0^T |f(t)|\,dt$$

is bounded, and

$$\lim_{T\to\infty}\frac{1}{T}\int_0^T f(t)\,dt = A, \tag{1}$$

then

$$\lim_{\epsilon\to 0}\frac{2}{\pi\epsilon}\int_0^\infty f(t)\,\frac{\sin^2 \epsilon t}{t^2}\,dt = A. \tag{2}$$

We can and shall suppose A to be zero. This introduces no essential restriction, as we only replace $f(t)$ by $f(t)-A$. I shall here give an alternative proof of this theorem, which I shall call Theorem I, together with a proof of a partial converse of Theorem I.

THEOREM II. *If $f(t)$ is non-negative and* (2) *is true, then* (1) *is true.*

To prove Theorem I, we shall need the following lemma :

LEMMA A. *If, for all T,*

$$\frac{1}{T}\int_0^T |f(t)|\,dt < B,$$

then for all positive C, and all $D > C$,

$$\int_C^D \frac{|f(t)|}{t^2}\,dt \leqslant \frac{3B}{C}.$$

---

* Received 16 November, 1926; read 9 December, 1926.

† "Notes on two theorems of Norbert Wiener", *Journal London Math. Soc.*, 1 (1926), 240–4.

We have

$$\int_C^D \frac{|f(t)|}{t^2}\,dt = \int_C^D \frac{1}{t^2}\,d\left[\int_0^t |f(\xi)|\,d\xi\right]$$

$$= \frac{1}{D^2}\int_0^D |f(\xi)|\,d\xi - \frac{1}{C^2}\int_0^C |f(\xi)|\,d\xi + \int_C^D \frac{2}{t^3}\left[\int_0^t |f(\xi)|\,d\xi\right]dt$$

$$\leqslant \frac{B}{C} + B\int_C^\infty \frac{2}{t^2}\,dt = \frac{3B}{C}.$$

COROLLARY. *Under the same conditions*

$$\int_C^\infty \frac{|f(t)|}{t^2}\,dt$$

exists, and does not exceed $3B/C$; *and*

$$\frac{2}{\pi\epsilon}\int_0^\infty f(t)\,\frac{\sin^2 \epsilon t}{t^2}\,dt$$

exists, and

$$\left|\frac{2}{\pi\epsilon}\int_{N/\epsilon}^\infty f(t)\,\frac{\sin^2 \epsilon t}{t^2}\,dt\right| \leqslant \frac{6B}{\pi N}. \tag{3}$$

We can now prove Theorem I. Let us consider

$$\frac{2}{\pi\epsilon}\int_0^{N/\epsilon} f(t)\,\frac{\sin^2 \epsilon t}{t^2}\,dt.$$

On integration by parts, we see that this equals

$$\frac{2}{\pi}\int_0^N f\left(\frac{t}{\epsilon}\right)\frac{\sin^2 t}{t^2}\,dt = \frac{2}{\pi}\int_0^N \frac{\sin^2 t}{t^2}\,d\left[\int_0^t f\left(\frac{\xi}{\epsilon}\right)d\xi\right] \tag{4}$$

$$= \frac{2}{\pi}\,\frac{\sin^2 N}{N^2}\int_0^N f\left(\frac{\xi}{\epsilon}\right)d\xi$$

$$+\frac{2}{\pi}\int_0^N \left[\frac{2\sin^2 t}{t^3} - \frac{2\sin t\cos t}{t^2}\right]\left[\int_0^t f\left(\frac{\xi}{\epsilon}\right)d\xi\right]dt$$

$$= \frac{2}{\pi}\,\frac{\sin^2 N}{N^2}\int_0^N f\left(\frac{\xi}{\epsilon}\right)d\xi$$

$$+\frac{2}{\pi}\int_0^h \left[\frac{2\sin^2 t}{t^3} - \frac{2\sin t\cos t}{t^2}\right]\left[\int_0^t f\left(\frac{\xi}{\epsilon}\right)d\xi\right]dt$$

$$+\frac{2}{\pi}\int_h^N \left[\frac{2\sin^2 t}{t^3} - \frac{2\sin t\cos t}{t^2}\right]\left[\int_0^t f\left(\frac{\xi}{\epsilon}\right)d\xi\right]dt$$

Now let N be so large that $6B/\pi N < \frac{1}{2}\eta$. Having chosen N, let h be any positive number less than N such that

$$\frac{2B}{\pi}\int_0^h \left|\frac{2\sin^2 t}{t^2} - \frac{2\sin t\cos t}{t}\right| dt < \frac{1}{4}\eta.$$

Having chosen N and h, let us choose E positive and so small that, if $0 < \epsilon \leqslant E$,

$$\left|\frac{1}{h}\int_0^h f\left(\frac{\xi}{\epsilon}\right) d\xi\right| < \frac{\eta}{4} \Big/ \left[\frac{2}{\pi}\frac{\sin^2 N}{N} + \frac{2}{\pi}\int_h^N \left|\frac{2\sin^2 t}{t^2} - \frac{2\sin t\cos t}{t}\right| dt\right].$$

All these choices are manifestly possible. For $\epsilon \leqslant E$, if we add (3) and (4), we get

$$\frac{2}{\pi\epsilon}\int_0^\infty f(t)\frac{\sin^2 \epsilon t}{t^2} dt < \frac{1}{2}\eta + \frac{1}{4}\eta + \frac{1}{4}\eta = \eta.$$

Since, for every η, we can find an E with this property, we have

$$\lim_{\epsilon \to 0} \frac{2}{\pi\epsilon}\int_0^\infty f(t)\frac{\sin^2 \epsilon t}{t^2} dt = 0,$$

and Theorem I is proved.

Now as to Theorem II: we can, without essential restriction, let $f(t) = 0$ for sufficiently small positive values of t. We shall first prove

LEMMA B. *If*

$$\int_0^\infty \frac{|f(t)|}{t^2} dt$$

exists and (2) *is true, then*

$$\lim_{x \to 1-0} (1-x)\int_0^\infty f(t)x^t dt = A. \tag{5}$$

What we wish to show is essentially that

$$\lim_{\epsilon \to 0}\int_0^\infty f\left(\frac{t}{\epsilon}\right) e^{-t} dt = A.$$

It is enough to show that e^{-t} can be developed into an integral of the form

$$e^{-t} = \frac{2}{\pi}\int_0^\infty \Psi(u)\frac{\sin^2 ut}{ut^2} du,$$

where $\Psi(u) = O(u)$ at 0, $\int_0^\infty |\Psi(u)|du$ converges and $\int_0^\infty \Psi(u)du = 1$. In this case

$$\int_0^\infty f\left(\frac{t}{\epsilon}\right) e^{-t} dt = \frac{2}{\pi}\int_0^\infty f\left(\frac{t}{\epsilon}\right)\int_0^\infty \Psi(u)\frac{\sin^2 ut}{ut^2} du.$$

The double integral

$$\frac{2}{\pi}\int_0^\infty f\left(\frac{t}{\epsilon}\right)\int_0^\infty \Psi(u)\,\frac{\sin^2 ut}{ut^2}\,du$$

is absolutely convergent, and may be written

$$\int_0^\infty \Psi(u)\,du\,\frac{2}{\pi\epsilon u}\int_0^\infty f(t)\,\frac{\sin^2 \epsilon ut}{t^2}\,dt.$$

An argument of the Fejér type will now show that

$$\lim_{\epsilon\to 0}\int_0^\infty \Psi(u)\,du\,\frac{2}{\pi\epsilon u}\int_0^\infty f(t)\,\frac{\sin^2 \epsilon ut}{t^2}\,dt = \lim_{u\to 0}\frac{2}{\pi u}\int_0^\infty f(t)\,\frac{\sin^2 ut}{t^2}\,dt.$$

This, however, completes the proof.

It remains to discuss the development of e^{-t} in an expression of the form

$$\frac{2}{\pi}\int_0^\infty \Psi(u)\,\frac{\sin^2 ut}{ut^2}\,du,$$

or, what is the same thing, the development of t^2e^{-t} in an expression of the form

$$\frac{1}{\pi}\int_0^\infty \frac{\Psi(u)}{u}\,(1-\cos 2ut)\,du.$$

We have, however, by the theory of the Fourier integral,

$$t^2e^{-t} = \frac{2}{\pi}\int_0^\infty E(u)\cos ut\,du,$$

where $$E(u) = \int_0^\infty t^2e^{-t}\cos ut\,dt = \frac{2(1-3u^2)}{(1+u^2)^3}.$$

That is $$\Psi(u) = -4uE(2u) = \frac{8u(12u^2-1)}{(1+4u^2)^3} = O(u^{-3})$$

at ∞. Hence $\int_0^\infty |\Psi(u)|\,du$ exists, and we can verify directly that $\int_0^\infty \Psi(u)du = 1$. Thus our lemma is proved.

Let it be noted that, if $f(t)$ is non-negative, and (2) is true, then

$$\frac{2\epsilon}{\pi}\int_0^{\pi/2\epsilon} f(t)\,dt < \frac{4A}{\pi^2}+\eta$$

for all sufficiently small values of ϵ, and

$$\frac{1}{T}\int_0^T |f(t)|\,dt$$

is bounded. Hence, by Lemma A,

$$\int_0^\infty \frac{|f(t)|}{t^2}\,dt$$

exists. It follows that, *if* $f(t)$ *is non-negative, and* (2) *is true, then* (5) *is true.*

Hardy and Littlewood have proved* that if $a_1, a_2, \ldots$ are non-negative, and

$$\lim_{x\to 1-0}(1-x)\sum_1^\infty a_n x^n = A,$$

then
$$\lim_{n\to\infty}\frac{1}{n}\sum_1^n a_k = A.$$

Without essential change, the proof may be made to show that if $f(t)$ is non-negative and

$$\lim_{x\to 1-0}(1-x)\int_0^\infty f(t)\,x^t\,dt = A,$$

then
$$\lim_{T\to\infty}\frac{1}{T}\int_0^T f(t)\,dt = A.$$

This completes the proof of Theorem II.

It may similarly be shown that if $f(t)\geqslant 0$, and

(6)
$$\lim_{\epsilon\to 0}\frac{K_n}{\epsilon^{2n-1}}\int_0^\infty f(t)\,\frac{\sin^{2n}\epsilon t}{t^{2n}}\,dt = A \qquad (n = 1, 2, \ldots),$$

where
$$\frac{1}{K_n} = \int_0^\infty \frac{\sin^{2n} t}{t^{2n}}\,dt,$$

then (1) is true.

The converse theorem, that if

$$\int_0^\infty \frac{|f(t)|}{t^2}\,dt$$

* Hardy and Littlewood, *Proc. London Math. Soc.* (2), 13 (1914), 175.

exists, and (1) is true, then (6) is true, is trivial, except in the case $n = 1$, which follows from Theorem I. We thus see that the propositions

$$\lim_{T\to\infty} \frac{1}{T} \int_0^T |f(t)|\, dt = A$$

and

$$\lim_{\epsilon\to 0} \frac{K_n}{\epsilon^{2n-1}} \int_0^\infty |f(t)| \frac{\sin^{2n} \epsilon t}{t^{2n}}\, dt = A \quad (n = 1, 2, \ldots)$$

are entirely equivalent.

All the theorems of this paper have analogues for sums, which may be proved in exactly the same manner as for integrals. In particular, the propositions

$$\lim_{n\to\infty} \frac{1}{n} \sum_1^n |a_k| = A$$

and

$$\lim_{\epsilon\to 0} \sum_1^\infty |a_k| \frac{\sin^{2n} \epsilon t}{t^{2n}} = A \quad (n = 1, 2, \ldots)$$

are equivalent.

Comments on [27c]

H. R. Pitt

These theorems were proved before Wiener's general Tauberian theorem was established, and the proofs are elementary. In fact, the general theorem is not directly applicable without the stronger condition that $f(t)$ is bounded. There is, in fact, an Abelian element in Theorem 1 depending on the very simple and explicit properties of the Cesàro transformation

$$T^{-1}\int_0^T f(t)\,dt.$$

There is a similar Abelian argument in the proof of Theorem 2, where it is shown that (2) implies

$$\int_0^\infty f(t/\varepsilon)\,e^{-t}\,dt \to A \quad \text{as } \varepsilon \to 0,$$

through the explicit formula

$$e^{-t} = \frac{2}{\pi}\int_0^\infty \Psi(u)\frac{\sin^2 ut}{ut^2}\,du, \quad \int_0^\infty |\Psi(u)|\,du < \infty.$$

The appeal to the Hardy-Littlewood theorem in the final stage is genuinely Tauberian, in the sense that $f(t)$ non-negative (or bounded below, which is trivially equivalent), or some condition of roughly equivalent strength is essential. The condition $\int_0^T |f(t)|\,dt = O(T)$, for example, is not enough.

ANALYSE MATHÉMATIQUE. — *Une méthode nouvelle pour la démonstration de théorèmes de M. Tauber* Note (¹) de M. **Norbert Wiener**, présentée par M. Hadamard.

M. Tauber (²) a démontré que la formule

$$\lim_{x \to 1-0} \sum_1^\infty a_n X^n = A$$

exige la convergence de $\sum_1^\infty a_n$ vers A, sous l'hypothèse que

$$a_n = O\left(\frac{1}{n}\right).$$

MM. Hardy et Littlewood ont généralisé ce théorème dans plusieurs Notes (³). Ils appellent les lois qui résultent de cette généralisation *théorèmes de Tauber*. Leurs méthodes emploient l'intégration indéfiniment itérée et sont, en outre, très variées. M. Robert Schmidt (⁴) a appliqué la théorie des moments de Stieltjes à cette question, qu'il a systématisée.

J'emploie la théorie complexe analogue à la théorie des moments, c'est-à-dire la théorie de l'intégrale de Fourier, et je démontre le théorème suivant : *Soient* $M_1(x)$ *et* $M_2(x)$ *deux fonctions qui soient* $O\left[\frac{1}{x(\log x)^2}\right]$ *aussi bien à zéro qu'à l'infini et qui soient bornées sur chaque intervalle* $\left(\varepsilon, \frac{1}{\varepsilon}\right)$. *Supposons* $M_1(x)$ *partout positive ou nulle, mesurable, et telle que*

$$\int_0^\infty M_1(x)\, x^{iu}\, dx \neq 0$$

pour tout u réel. Supposons $M_2(x)$ *continue ou ayant des sauts finis en nombre fini. Soit* $\varphi(x)$ *une fonction mesurable et non négative telle que*

$$(a) \qquad \lim_{y \to 0} \int_0^\infty \varphi\left(\frac{x}{y}\right) M_1(x)\, dx = A \int_0^\infty M_1(x)\, dx$$

(¹) Séance du 21 mars 1927.

(²) *Monatsh. f. Math. u. Phys.*, 8, 1897, p. 273-277.

(³) Il convient aussi de mentionner à cet égard les résultats bien connus de M. Fatou. (Note de J. H.)

(⁴) *Math. Ztsch.*, 22, 1925, p. 89-152.

et

$$(b) \qquad \int_0^\infty \varphi\left(\frac{x}{y}\right) \mathbf{M}_1(x)\,dx \text{ est bornée} \qquad (0 < y < \infty).$$

Alors on a

$$(c) \qquad \lim_{y \to 0} \int_0^\infty \varphi\left(\frac{x}{y}\right) \mathbf{M}_2(x)\,dx = \mathbf{A} \int_0^\infty \mathbf{M}_2(x)\,dx.$$

La méthode générale que nous employons est l'introduction de variables nouvelles définies par les relations

$$x = e^\xi \quad \text{et} \quad y = e^\eta,$$

et l'analyse harmonique des fonctions

$$e^\xi \mathbf{M}_1(e^\xi) \quad \text{et} \quad e^\xi \mathbf{M}_2(e^\xi).$$

Nous déduisons de ce théorème le corollaire suivant : *Soit* $\mathrm{F}(x)$ *une fonction dérivable qui admette la valeur* 1 *pour* $x = 0$ *et qui s'annule à l'infini. Supposons qu'il existe un nombre* λ *plus grand que* 1 *et un nombre* $\mathfrak{z}$ *entre* 0 *et* 1 *tels que pour toute valeur de* x *suffisamment grande*

$$\mathrm{F}(\lambda x) < \mathfrak{z}\mathrm{F}(x).$$

Supposons $\mathrm{F}'(x)$ *bornée et partout négative ou nulle. Supposons*

$$\mathrm{G}(z) = \int_0^\infty \mathrm{F}(y)\, y^{z-1}\,dy$$

et son prolongement analytique analytiques dans le voisinage de $\mathcal{R}(z) = 0$, $z \neq 0$, *mais ayant un pôle d'ordre* 1 *pour* $z = 0$. *Supposons qu'il n'existe aucun zéro de* $\mathrm{G}(z)$ *sur* $\mathcal{R}(z) = 0$. *Soit* $\psi(x)$ *une fonction mesurable satisfaisant à la relation*

$$\psi(x) > -\mathrm{K}(x).$$

Alors la formule

$$\lim_{\varepsilon \to 0} \int_0^\infty \mathrm{F}(\varepsilon x)\,\psi(x)\,dx$$

exige que

$$\int_0^\infty \psi(x)\,dx = \mathrm{A}.$$

Ce théorème nous amène très directement à la plupart des théorèmes de Tauber déjà connus. Par exémple, nous démontrons sans aucune difficulté que si

$$a_n > -\frac{\mathrm{K}}{n}$$

et

$$\lim_{\varepsilon \to 0} \sum_{1}^{\infty} a_n Q^{-\varepsilon n} = A,$$

il suit que

$$\sum_{1}^{\infty} a_n = A.$$

La seule difficulté est l'étude des propriétés de $G(z)$, qui est la fonction $\zeta(z+1)\Gamma(z+1)$. Nous employons le théorème que $\zeta(z)$ ne s'annule pas sur la droite $\mathcal{R}(z) = 1$, et qu'elle est analytique sur cette droite sauf pour un pôle d'ordre 1 pour $z = 1$.

MM. Hardy et Littlewood (¹) ont démontré que le théorème que nous venons de donner se réduit par des procédés élémentaires au théorème classique sur la répartition des nombres premiers, qui donne $\frac{n}{\log n}$ comme la représentation asymptotique du nombre de nombres premiers qui ne dépassent pas n. La seule démonstration du théorème de Tauber déjà donnée suppose cette représentation comme connue. Nous avons donc une démonstration toute nouvelle du théorème des nombres premiers, qui ne suppose rien sur la croissance de la fonction zêta sur la droite $\mathcal{R}(z) = 1$.

Comments on [27d]

H. R. Pitt

The contents of this paper with proofs and full discussion are contained in the paper "A new method in Tauberian theorems" [28b], which comes later. But note that $O(1/n)$ should read $o(1/n)$ in line 4.

A NEW METHOD IN TAUBERIAN THEOREMS

By Norbert Wiener

The classical theorem of Abel states that if $\Sigma a_n = A$, then

$$\lim_{x \to 1-0} \Sigma a_n x^n = A. \tag{1}$$

The converse of this theorem is not unconditionally true.[1] Tauber[2] showed that if (1) is true and $\lim n a_n = 0$, it follows that $\Sigma a_n = A$. Hardy and Littlewood[3] have shown that (1) and the auxiliary condition that $n a_n > -K$ form a sufficient hypothesis for this conclusion. Theorems of this type, which involve some hypothesis on the rapidity of growth of the terms of a series, and under this assumption proceed from the given summability of the series by some particular method to its convergence in the ordinary sense, they term Tauberian theorems.

The methods of Hardy and Littlewood are difficult, and involve a repeated differentiation of arbitrarily high order. They are furthermore but little systematized, so that each particular theorem involves a considerable amount of manipulation peculiar to itself. To avoid this, Robert Schmidt[4] introduced the machinery of the theory of moments, and at the cost of a considerable amount of complication in the case of the theorems already proved, gave a general theory of Tauberian theorems adequate to deal with the majority of known cases.

There is however one particular Tauberian theorem which has proved refractory even to the improved tools of the Schmidt method. Hardy and Littlewood have shown[5] that the classical theorem of the distribution of primes, to the effect that the

1 Cf. $a_n = (-1)^n$.

2 A. Tauber, Ein Satz aus der Theorie der unendlichen Reihen. *Monatsch. f. Math. u. Phys.* 8 (1897), pp. 273–277.

3 Hardy and Littlewood, Tauberian Theorems concerning Power Series and Dirichlet's Series whose Coefficients are Positive. *Proc. Lond. Math. Soc.*, Series 2, 13 (1913), pp. 174–191.

4 R. Schmidt, Über divergente Folgen und lineare Mittelbildungen. *Math. Ztschr.* 22 (1925), pp. 89–152.

5 Hardy-Littlewood, New Proof of the Prime Number Theorem, *Quarterly Journal of Pure and Applied Math.*, 46 (1915), pp. 215–219.

number of primes less than n is asymptotically represented by $n/\log n$, is an elementary consequence of the Tauberian theorem to the effect that if $na_n > -K$ and

$$\lim_{w \to 0} \Sigma\, a_n \frac{nw\, e^{-nw}}{1-e^{-nw}} = A, \tag{2}$$

then $\Sigma a_n = A$. This theorem they are only able to prove by a recourse to the prime number theorem itself.

The present paper was suggested to the author by the investigations of Mr. Schmidt, to whom the author is also indebted to the suggestion to apply his method to the Tauberian theorem with the hypothesis (2). It bears the same relation to the Schmidt theory which the theory of the Fourier integral bears to the method of moments and the theory of the Laplace transformation. In other words, it involves the theory of complex moments. It is the author's opinion that it brings not merely a systematization, but a simplification, of a large region in the theory of Tauberian theorems. It is in particular adequate for a direct treatment of the case with hypothesis (2).

Hypotheses (1) and (2) are alike of the form

$$\lim_{w \to 0} \Sigma\, a_n N(nw) = A, \tag{3}$$

where $N(0)=1$. If we put $a(x) = \sum_{k=0}^{[x]} a_k$, (3)

$$\lim_{w \to 0} \int_0^\infty N(nw)\,da(n) = A. \tag{4}$$

If N vanishes sufficiently rapidly at infinity, we may integrate by parts and (4) becomes

$$\lim_{w \to 0} \int_0^\infty a(u/w)\, N'(u)\,du = -A. \tag{5}$$

On this hypothesis, together with a further hypothesis restricting the rate of growth of a, we wish to prove that

$$\lim_{u \to \infty} a(u) = A. \tag{6}$$

It will be seen that expression (5) is a mean of the values of $a(u)$ for large values of u. Our Tauberian theorems may then be put in the form: if a function has a mean value of a certain type for large values of its arguments, and if the rate of growth of the function is subject to certain restrictions, then the function has a limiting value at infinity. Closely related to theorems of this type are theorems which read as follows: if a function has a mean of a certain given type at infinity, and its growth is restricted in a certain given manner, then it has a mean of a certain other given type at infinity. We shall find it more convenient to prove a theorem of this latter type, and then to deduce from this theorems of a more obviously Tauberian nature.

Our key theorem reads as follows: *Let* $M_1(x)$ *and* $M_2(x)$ *be two real functions which are* $O\left(\frac{1}{x(\log x)^2}\right)$ *both at* 0 *and at* ∞, *and which are bounded over every interval* $(\epsilon, 1/\epsilon)$. *Let* $M_1(x)$ *be non-negative and measurable, and let*

$$\int_0^\infty M_1(x)x^{iu}dx \neq 0. \quad [u = \text{any real number}]$$

Let $M_2(x)$ *be continuous except for a finite number of finite jumps. Let* $\phi(x)$ *be a non-negative measurable function such that*

$$(a)\ \lim_{y\to 0}\int_0^\infty \phi\left(\frac{x}{y}\right)M_1(x)dx = A\int_0^\infty M_1(x)dx;$$

$$(b)\quad \int_0^\infty \phi\left(\frac{x}{y}\right)M_1(x)dx \text{ is bounded.} \quad [0<y<\infty]$$

Then

$$(c)\ \lim_{y\to 0}\int_0^\infty \phi\left(\frac{x}{y}\right)M_2(x)dx = A\int_0^\infty M_2(x)dx.$$

In order to prove this theorem, we shall introduce the transformations

$$x = e^\xi;\ y = e^\eta;\ xM_1(x) = N_1(\xi);\ xM_2(x) = N_2(\xi);\ \phi(x) = f(\xi).$$

Our theorem now assumes the form: *let* $N_1(\xi)$ *and* $N_2(\xi)$ *be two bounded functions which are* $O(1/\xi^2)$ *at* $\pm\infty$. *Let* $N_1(\xi)$ *be measurable and non-negative, and let*

$$\int_{-\infty}^{\infty} N_1(\xi)e^{iu\xi}\,d\xi \neq 0. \quad [-\infty < u < \infty].$$

let $N_2(\xi)$ *be continuous, except for a finite number of finite jumps. Let* $f(\xi)$ *be a non-negative measurable function such that*

$$(a)\ \lim_{\eta\to-\infty}\int_{-\infty}^{\infty} f(\xi-\eta)\,N_1(\xi)d\xi = A\int_{-\infty}^{\infty} N_1(\xi)d\xi;$$

$$(b)\ \int_{-\infty}^{\infty} f(\xi-\eta)\,N_1(\xi)d\xi \text{ is bounded.} \quad [-\infty < \eta < \infty].$$

Then

$$(c)\ \lim_{\eta\to-\infty}\int_{-\infty}^{\infty} f(\xi-\eta)\,N_2(\xi)d\xi = A\int_{-\infty}^{\infty} N_2(\xi)d\xi.$$

In order to prove the theorem in this form, let Φ be the class of all bounded measurable functions $N_2(\xi)$ for which (c) holds whenever $f(\xi)$ is a non-negative function satisfying (a) and (b). If $R(\xi)$ is an absolutely integrable function,

$$S(\xi) = \int_{-\infty}^{8} N_1(\xi-\zeta)R(\zeta)d\zeta$$

belongs to Φ. For we have

$$\int_{-\infty}^{\infty} f(\xi-\eta)S(\xi)d\xi = \int_{-\infty}^{\infty} f(\xi-\eta)d\xi \int_{-\infty}^{\infty} N_1(\xi-\zeta)R(\zeta)d\zeta$$
$$= \int_{-\infty}^{\infty} R(\zeta)d\zeta \int_{-\infty}^{\infty} f(\xi-\eta+\zeta)\,N_1(\xi)d\xi,$$

as the order of integration may be inverted owing to the absolute integrability of R and the non-negative character of f and N_1. Let us put

$$\int_{-\infty}^{\infty} f(\xi-\eta)\,N_1(\xi)d\xi - A\int_{-\infty}^{\infty} N_1(\xi)d\xi = \psi(\eta).$$

The function $\psi(\eta)$ is then a bounded function approaching 0 as η tends toward $-\infty$. We have

$$\left|\int_{-\infty}^{\infty} R(\zeta)\psi(\eta-\zeta)d\zeta\right| \leq \left|\int_{-\infty}^{B} R(\zeta)\psi(\eta-\zeta)d\zeta\right| + \left|\int_{B}^{\infty} R(\zeta)\psi(\eta-\zeta)d\zeta\right|$$
$$\leq \max |\psi(\eta)| \left|\int_{-\infty}^{B} |R(\zeta)|d\zeta\right.$$
$$+ \max_{\zeta \geq B} |\psi(\eta-\zeta)| \left|\int_{-\infty}^{\infty} |R(\zeta)|d\zeta.\right.$$

It follows at once that

$$\overline{\lim_{\eta\to-\infty}} \left|\int_{-\infty}^{\infty} R(\zeta)\psi(\eta-\zeta)d\zeta\right| \leq \max |\psi(\eta)| \left|\int_{-\infty}^{B} |R(\zeta)|d\zeta,\right.$$

and as B is arbitrary, that

$$\overline{\lim_{\eta\to-\infty}} \left|\int_{-\infty}^{B} R(\zeta)\psi(\eta-\zeta)d\zeta\right| = 0,$$

or as we may write it, that

$$\overline{\lim_{\eta\to-\infty}} \int_{-\infty}^{\infty} f(\xi-\eta)S(\xi)d\xi = A\int_{-\infty}^{\infty} R(\zeta)d\zeta \int_{-\infty}^{\infty} N_1(\xi)d\xi$$
$$= A\int_{-\infty}^{\infty} S(\xi)d\xi.$$

The boundedness of S follows from that of N_1 and the absolute integrability of R. Hence S belongs to Φ.

Let us put

$$\nu_1(u) = \int_{-\infty}^{\infty} N_1(\xi)e^{iu\xi}d\xi.$$

By the Plancherel theory of the Fourier transform,

$$\mu(u) = \underset{B\to\infty}{\text{limit in mean}} \int_{-B}^{B} i\xi N_1(\xi)e^{iu\xi}d\xi$$

exists as a measurable function the square of whose modulus is Lebesgue integrable from $-\infty$ to ∞.[6] We have

[6] M. Plancherel, Contribution à l'étude de la représentation d'une function arbitraire par des intégrales définies. *Rendiconti di Palermo*, 30 (1910), pp. 289–335.

$$\int_{\alpha}^{\beta} \mu(u)du = \int_{-\infty}^{\infty} i\xi N_1(\xi)\left[\int_{\alpha}^{\beta} e^{iu\xi}du\right]d\xi$$
$$= \nu_1(\beta) - \nu_1(\alpha).$$

The function $\nu_1(u)$ is thus a non-vanishing function which is an indefinite integral.

Let $S(\xi)$ be a bounded function which is $O(1/\xi^2)$ at $\pm\infty$. Let

$$\sigma(u)=\int_{-\infty}^{\infty} S(\xi)e^{iu\xi}d\xi,$$

and let

$$\tau(u)=\underset{B\to\infty}{\text{l. i. m.}}\int_{-B}^{B} i\xi S(\xi)e^{iu\xi}d\xi.$$

As above, we have

$$\int_{-\alpha}^{\beta}\tau(u)du=\sigma(\beta)-\sigma(\alpha).$$

Let us make the further restriction that

$$\tau(u)=0 \qquad [|u|\geq D].$$

I then say that $S(\xi)$ belongs to Φ.

The proof is as follows: consider

$$\omega(u)=\frac{\nu_1(u)\tau(u)-\mu(u)\sigma(u)}{[\nu_1(u)]^2}$$

It is easily seen that $\sigma(u)$ as well as $\tau(u)$ must vanish for $|u|\geq D$, as it is constant outside this range and is quadratically summable. Over the range $(-D, D)$, $\nu_1(u)$ and $\sigma(u)$ are bounded continuous functions, and $\nu_1(u)$ does not vanish. Hence $\omega(u)$ is a bounded measurable function vanishing outside $(-D, D)$, such that the square of its absolute value is Lebesgue integrable. We can take $D=\pi$ in what follows without any essential restriction of the argument.

Let $\nu_1(u)$ have the Fourier series $\sum_{-\infty}^{\infty} a_n e^{inu}$. Then $\mu(u)$ will have the Fourier series $\Sigma\, in\, a_n e^{inu}$. Similarly, let $\sigma(u)$ have the Fourier series $\Sigma b_n e^{inu}$, which means that $\tau(u)$ will have the Fourier series $\Sigma\, in\, b_n e^{inu}$. Let

$$\nu_1{}^N(u)=\sum_{-N}^{N} a_n e^{inu}\left(1-\frac{|n|}{N}\right);$$

$$\mu^N(u)=\sum_{-N}^{N} in\, a_n e^{inu}\left(1-\frac{|n|}{N}\right);$$

$$\sigma^N(u)=\sum_{-N}^{N} b_n e^{inu}\left(1-\frac{|n|}{N}\right);$$

$$\tau^N(u)=\sum_{-N}^{N} in\, b_n e^{inu}\left(1-\frac{|n|}{N}\right).$$

It is clear that

$$\int_{-\pi}^{\pi}\left|\omega(u)-\frac{\nu^N{}_1(u)\tau^N(u)-\mu^N(u)\sigma^N(u)}{[\nu^N{}_1(u)]^2}\right|^2 du$$

$$=\int_{-\pi}^{\pi}\left|\frac{\nu_1(u)\tau(u)[\nu^N{}_1(u)]^2-\mu(u)\sigma(u)[\nu^N{}_1(u)]^2-\nu^N{}_1(u)\tau^N(u)[\nu_1(u)]^2+\mu^N(u)\sigma^N(u)[\nu_1(u)]^2}{[\nu_1(u)]^2[\nu^N{}_1(u)]^2}\right|du$$

The functions $\nu^N{}_1(u)$, $\sigma^N(u)$ and $1/\nu^N{}_1(u)$ are all uniformly bounded in u and N and converge uniformly to the corresponding functions without the N. It can then be shown by a simple but tedious argument that

$$\underset{N\to\infty}{\mathrm{l.i.m.}}\frac{\nu^N{}_1(u)\tau^N(u)-\mu^N(u)\sigma^N(u)}{[\nu^N{}_1(u)]^2}=\omega(u).$$

Similarly,

$$\lim_{N\to\infty}\frac{\sigma^N(u)}{\nu^N{}_1(u)}=\frac{\sigma(u)}{\nu_1(u)}$$

uniformly in u. Since

$$\int_{\alpha}^{\beta}\frac{\nu^N{}_1(u)\tau^N(u)-\mu^N(u)\sigma^N(u)}{[\nu^N{}_1(u)]^2}du=\frac{\sigma^N(\beta)}{\nu^N{}_1(\beta)}-\frac{\sigma^N(\alpha)}{\nu^N{}_1(\alpha)}$$

we have

$$\int_{\alpha}^{\beta}\omega(u)du=\frac{\sigma(\beta)}{\nu_1(\beta)}-\frac{\sigma(\alpha)}{\nu_1(\alpha)}.$$

Let us put

$$R(\xi)=\frac{1}{2\pi}\int_{-\infty}^{\infty}\frac{\sigma(u)}{\nu_1(u)}e^{-iu\xi}du. \tag{7}$$

Then

$$S(\xi)=\int_{-\infty}^{\infty}N_1(\xi-\zeta)R(\zeta)d\zeta.$$

We also have

$$R(\xi)=\frac{1}{2\pi i\xi}\int_{-\infty}^{\infty}\omega(u)e^{-iu\xi}du.$$

Since $\int_{-\infty}^{\infty}\omega(u)e^{-iu\xi}du$ is quadratically summable, it follows by a simple use of the Schwarz inequality that

$$\int_{B}^{\infty}|R(\xi)|d\xi \text{ and } \int_{-\infty}^{-B}|R(\xi)|d\xi$$

are bounded. Combining this fact with the boundedness of R, which we deduce from (7), we see that $R(\xi)$ is absolutely integrable from $-\infty$ to ∞. It follows from our previous lemma that $S(\xi)$ belongs to Φ.

If we already know certain members of the class Φ, we may enlarge the class in the following manner. Let $S(\xi)$ be a function such that, if any positive ϵ is given, we can find two members of Φ, S_1 and S_2, such that $S_1(\xi)\leq S(\xi)\leq S_2(\xi)$, while

$$\int_{-\infty}^{\infty}[S_2(\xi)-S_1(\xi)]d\xi<\epsilon.$$

Then S itself belongs to Φ. For

$$\overline{\lim_{\eta\to-\infty}}\int_{-\infty}^{\infty}f(\xi-\eta)S(\xi)d\xi\leq\lim_{\eta\to-\infty}\int_{-\infty}^{\infty}f(\xi-\eta)S_2(\xi)d\xi$$

$$\leq A\left[\int_{-\infty}^{\infty}S(\xi)d\xi+\epsilon\right];$$

$$\underline{\lim}_{\eta\to-\infty}\int_{-\infty}^{\infty}f(\xi-\eta)S(\xi)d\xi\geq A\left[\int_{-\infty}^{\infty}S(\xi)d\xi-\epsilon\right].$$

Since ϵ is arbitrarily small

$$\lim_{\eta\to-}\int_{-\infty}^{\infty}f(\xi-\eta)S(\xi)d\xi=A\int_{-\infty}^{\infty}S(\xi)d\xi,$$

and S belongs to Φ.

As an example of a function S which can be proved in this manner to belong to Φ, let us take

$$S(\xi)=\frac{p}{(\xi-q)^2+r^2}.$$

We shall carry the proof through in the case $S(\xi)=1/(1+\xi^2)$; the method is general.

We have

$$S(\xi)=\frac{1}{2}\int_{-\infty}^{\infty}e^{-|u|}e^{i\xi u}du.$$

Let us consider the slightly modified function.

$$T(\xi)=\frac{1}{2}\int_{-B}^{B}[e^{-|u|}-e^{-B}]e^{i\xi u}du$$

which belongs to Φ as a result of our previous discussion. We have

$$\begin{aligned} T(\xi)&=\frac{1}{2\xi i}\int_{-B}^{B}[e^{-|u|}-e^{-B}]de^{i\xi u}\\ &=\frac{1}{2\xi i}\int_{0}^{B}e^{i\xi u}e^{-u}du-\frac{1}{2\xi i}\int_{-B}^{0}e^{i\xi u}e^{u}du\\ &=R\left(\frac{1}{\xi i}\int^{B}e^{u(i\xi-1)}du\right)=R\left\{\frac{1}{\xi i(\xi i-1)}[e^{B(i\xi-1)}-1]\right\}\\ &=\frac{1-e^{-B}\cos B\xi}{\xi^2+1}-\frac{e^{-B}\sin B\xi}{\xi(\xi^2+1)}. \end{aligned}$$

Hence

$$|T(\xi)-S(\xi)|<(1+B)e^{-B}S(\xi),$$

from which it follows that for large enough values of B,

$$[1+(1+B)e^{-B}]^{-1}T(\xi)<S(\xi)<[1-(1-B)e^{-B}]^{-1}T(\xi).$$

If we make use of the fact that

$$\lim_{B\to\infty}(1+B)e^{-B}=0,$$

we see that we may put

$$[1+(1+B)e^{-B}]^{-1}T(\xi)=S_1(\xi);\ [1-(1+B)e^{-B}]^{-1}T(\xi)=S_2(\xi),$$

and apply our last lemma. It follows at once that $S(\xi)$ belongs to Φ.

Another function which belongs to Φ is

$$Q_A(\xi)=2/\pi\int_0^A\left(1-\frac{u}{A}\right)\frac{\sin u\cos u\xi}{u}\,du.$$

We have

$$Q_A(\xi)=G(A(\xi+1))-G(A(\xi-1)),$$

where

$$G(\xi)=\frac{1}{\pi}\int_0^1\frac{(1-v)\sin v\,\xi}{v}\,dv.$$

Now,

$$G(0)=0;\quad G'(\xi)=\frac{1}{\pi}\int_0^1(1-v)\cos v\xi\,dv$$
$$=\frac{1-\cos\xi}{\pi\xi^2}.$$

Hence

$$G(\xi)=\int_0^\xi\frac{1-\cos z}{\pi z^2}\,dz$$
$$=\int_0^\infty\frac{1-\cos z}{\pi z^2}\,dz-\int_\xi^\infty\frac{dz}{\pi z^2}+\int_\xi^\infty\frac{\cos z}{\pi z^2}\,dz$$
$$=\frac{1}{2}-\frac{1}{\pi\xi}+O(1/\xi^2),$$

as may be seen by considering the graph of $\cos z/\pi z^2$. It follows that

$$Q_A(\xi)=\frac{1}{\pi A(\xi-1)}-\frac{1}{\pi A(\xi+1)}+O(1/A^2\xi^2)$$
$$=O(1/A\xi^2)$$

as $A\xi^2$ becomes positively infinite in any manner whatever, provided ξ remains larger than some constant greater than 1 (or less than some constant less than -1). Hence the function

$$Q_A(\xi) - \frac{1}{B(\xi^2+1)}$$

is negative over $(1+\eta, \infty)$ and $(-\infty, -1-\eta)$ for large enough values of A.

As A increases, $Q_A(\xi)$ tends uniformly to

$$\operatorname{sgn}(\xi+1) - \operatorname{sgn}(\xi-1)$$

except in the neighborhood of ± 1, while it always lies between 0 and 1. Furthermore,

$$\int_{-\infty}^{\infty} Q_A(\xi) d\xi = 2.$$

It follows that if we put $S(\xi) = \operatorname{sgn}(\xi+1) - \operatorname{sgn}(\xi-1)$, we may put

$$(1+\eta)Q_A(\xi(1-\eta)) = S_2(\xi);$$

$$(1-\eta)Q_A(\xi(1+\eta)) - \frac{1}{B(\xi^2+1)} = S_1(\xi);$$

for $Q_A(K\xi)$ contains only a limited range of frequencies, and so belongs to Φ, $\frac{1}{B(\xi^2+1)}$ belongs to Φ, the difference between two members of Φ belongs to Φ, and we can so choose η, B and A that

$$(1+\eta)Q_A(\xi(1-\eta)) > \operatorname{sgn}(\xi+1) - \operatorname{sgn}(\xi-1)$$
$$> (1-\eta)Q_A(\xi(1+\eta)) - \frac{1}{B(\xi^2+1)}, \quad (8)$$

while

$$\int_{-\infty}^{\infty} \left[(1+\eta)Q_A(\xi(1-\eta)) - (1-\eta)Q_A(\xi(1+\eta)) + \frac{1}{B(\xi^2+1)} \right] d\xi \quad (9)$$
$$= 2\left[\frac{1+\eta}{1-\eta} - \frac{1-\eta}{1+\eta}\right] + \frac{\pi}{B} < \epsilon.$$

To make this determination, we first choose η and B so that (9) holds, and then determine A in such a manner as to make (8) hold. It follows that sgn $(\xi+1)-$sgn $(\xi-1)$ belongs to Φ.

The same method of proof will show that any step function of a finite number of steps, vanishing for large positive and negative arguments, belongs to Φ, for the sum of a number of members of Φ belongs to Φ, and the argument in the case of any function of a single step proceeds exactly as in the case of sgn $(\xi+1)-$ sgn$(\xi-1)$. It follows that any function continuous except for a finite number of finite jumps, vanishing for large positive and negative arguments also belongs to Φ, for we can determine step-functions S_1 and S_2 for it.

Now let $N_2(\xi)$ be any bounded function which is $O(1/\xi^2)$ at $\pm\infty$, which is continuous except for a finite number of finite jumps. Let

$$|N_2(\xi)|<P/\xi^2+1$$

for all ξ. We put

$$\begin{aligned} S_1(\xi)&=N_2(\xi); & [|\xi|<M] \\ S_1(\xi)&=-P/\xi^2+1; & [|\xi|\geq M]. \\ S_2(\xi)&=N_2(\xi); & [|\xi|<M] \\ S_2(\xi)&=P/\xi^2+1. & [|\xi|\geq M] \end{aligned}$$

It is immediately clear that $N_2(\xi)$ belongs to Φ, as S_1 and S_2 are the sums of finite sets of functions in Φ. Our main theorem is thus proved.

It is worth while noting that our main theorem has at least a partial inverse which is valid. We shall suppose that $\int_0^\infty M_1(x)dx\neq 0$.

If $\int_0^\infty M_1(x)x^{iu}dx=0$, then let us put

$$\phi(x)=x^{iu}+x^{-iu}.$$

Then

$$(a)\quad \lim_{y\to 0}\int_0^\infty \phi\left(\frac{x}{y}\right)M_1(x)dx=A\int_0^\infty M_1(x)dx;\quad [A=0]$$

$$(b)\quad \int_0^\infty \phi\left(\frac{x}{y}\right)M_2(x)dx \text{ is bounded.}$$

Unless $M_2(x)$ satisfies the special condition

$$\int_0^\infty M_2{}^{(x)} x^{iu} dx = 0,$$

we shall *not* have

$$(c)\quad \lim_{y\to 0} \int_0^\infty \phi\left(\frac{x}{y}\right) M_2(x)dx = A \int_0^\infty M_2(x)dx.$$

It is only the restriction on the order of magnitude and the boundedness of M_1 which prevents us from regarding this as a perfect inverse of the fundamental theorem. Except for this, our theorem is a "best possible theorem," and no alternative method such as the method of moments can go substantially beyond it.

Our main theorem is only implicitly Tauberian. We wish to make the transition to the theorems of a more obviously Tauberian character. To this end, we put

$$M_2(x) = \begin{cases} 1; \ [x < T] \\ 0; \ [x \geq T] \end{cases}; \tag{10}$$

$$M_1(x) = -\int_x^\infty \frac{F'(y)}{y} dy.$$

$M_2(x)$ satisfies all the conditions necessary for the hypothesis of our theorem. If $F'(y)$ is bounded, and is $O(y^{-2-\theta})$ at infinity, we shall have

$$M_1(x) = O(\log x) \text{ at } 0;$$

$$M_1(x) = O(x^{-3-\theta}) \text{ at } \infty;$$

and *a fortiori*

$$M_1(x) = O\left(\frac{1}{x(\log x)^2}\right)$$

at 0 and ∞. The condition

$$\int_x^\infty M_1(x) x^{iu} dx \neq 0$$

may be written

$$H(u) = \int_0^\infty x^{iu}\, dx \int_x^\infty \frac{F'(y)}{y} dy \neq 0 \qquad (u \text{ real}).$$

$H(u)$ is an analytic function of u in the strip $0 > I(u) \geq -1$, and is continuous over the region $0 \geq I(u) \geq -1$. In the strip of analyticity,

$$H(u) = \int_0^\infty \frac{F'(y)}{y} dy \int_0^y x^{iu} dx$$

$$= \frac{1}{iu+1} \int_0^\infty F'(y) y^{iu} dy$$

$$= -\frac{iu}{iu+1} \int_0^\infty F(y) y^{iu-1} dy.$$

If then the function $\int_0^\infty F(y)\, y^{z-1} dy$ is analytic in the neighborhood of the line $R(z)=0$, with the exception of a pole of order 1 at $z=0$, and does not vanish on this line,

$$\int_0^\infty M_1(x) x^{iu} dx \neq 0.$$

The only remaining assumption concerning M_1 alone is that it is non-negative. This will be satisfied if $F'(x)$ is non-positive.

Examples of functions F satisfying our conditions are

$$F(x) = e^{-x};$$

$$F(x) = \frac{xe^{-x}}{1-e^{-x}}$$

In the first case,

$$\int_0^\infty F(y) y^{z-1} dy = \int_0^\infty e^{-y} y^{z-1} dy = \Gamma(z);$$

and in the second,

$$\int_0^\infty F(y) y^{z-1} dy = \int_0^\infty \frac{e^{-y}}{1-e^{-y}} y^z dy = \zeta(z+1)\, \Gamma(z+1).$$

For such functions F, if $\phi(x)$ is non-negative (or simply bounded below, as we may easily generalize), and

$$(a) \qquad -\lim_{\epsilon \to 0} \int_0^\infty \phi\left(\frac{x}{\epsilon}\right) \int_x^\infty \frac{F'(y)}{y} dy = AF(0);$$

(b) $$\int_0^\infty \phi\left(\frac{x}{\epsilon}\right)\int_x^\infty \frac{F'(y)}{y}dy \text{ is bounded,}$$

we have by our theorem

(c) $$\lim_{E\to\infty}\frac{1}{E}\int_\epsilon^E \phi(x)dx=A.$$

We may however write

$$-\int_0^\infty \phi\left(\frac{x}{\epsilon}\right)dx\int_x^\infty \frac{F'(y)}{y}dy=-\int_0^\infty \frac{F'(y)}{y}dy\int_0^y \phi\left(\frac{x}{\epsilon}\right)dx$$

$$=-\int_0^\infty F'(y)dy\frac{\epsilon}{y}\int_0^{y/\epsilon}\phi(\xi)d\xi.$$

Let us now introduce the supposition that we may express $\frac{1}{y}\int_0^y \phi(x)dx$ in the form

$$\frac{1}{y}\int_0^y \phi(x)dx=\int_0^y \psi(x)dx. \tag{11}$$

From the lower boundedness of $\phi(x)$ and the assumptions

(a′) $$AF(0)=\lim_{\epsilon\to 0} -\int_0^\infty F'(y)dy\int_0^{y/\epsilon}\psi(x)dx$$

(b′) $$\int_0^\infty F'(y)dy\int_0^{y/\epsilon}\psi(x)dx \text{ is bounded}$$

we can deduce by our theorem

(c′) $$\int_0^\infty \psi(x)dx=A.$$

Let $\psi(x)$ be any function summable over every finite interval and vanishing for arguments $\leq\eta$ and satisfying the relation

$$\psi(x)>-K/x.$$

Let us put

$$\phi(x)=x\psi(x)+\int_0^x \psi(\xi)d\xi.$$

We then see that (11) is valid. Proposition (a') takes the form,

$$AF(0) = -\lim_{\epsilon\to 0}\int_0^\infty F'(y)dy\int_0^{y/\epsilon}\psi(x)dx$$

$$= \lim_{\epsilon\to 0}\lim_{B\to\infty} - \int_0^B F'(y)dy\int_0^{y/\epsilon}\psi(x)dx$$

$$= \lim_{\epsilon\to 0}\lim_{B\to\infty}\int_0^{B/\epsilon}\left\{F(\epsilon x)-F(B)\right\}\psi(x)dx.$$

We shall show that we can deduce this from the assumption that

$$(a'') \qquad AF(0) = \lim_{\epsilon\to 0}\int_0^\infty F(\epsilon x)\psi(x)dx$$

together with the auxiliary assumption that $F(x)$ is of the order of magnitude of some negative power of x at infinity, and the assumptions which we have already made concerning F and ψ. Under these assumptions,

$$\int_\eta^\infty F(\epsilon x)\left[\psi(x)+\frac{K}{x}\right]dx = O(1) + K\int_{\eta\epsilon}^\infty F(\xi)\frac{d\xi}{\xi}$$

$$= O(\log \epsilon).$$

Since the expression $\psi(x)+\dfrac{K}{x}$ is non-negative, we can replace $F(x)$ by any smaller expression, and in particular by an expression equal to a constant from $x=0$ to $x=L>0$. Then

$$\int_0^{1/\epsilon}\left[\psi(x)+\frac{K}{x}\right]dx = O(\log \epsilon).$$

A repetition of our argument will give us

$$\int_0^{1/\epsilon}\psi(x)dx = O(\log \epsilon).$$

Hence

$$\lim_{B\to\infty} F(B)\int_0^{B/\epsilon}\psi(x)dx = 0,$$

and (a') is established as a consequence of (a'').

In order to prove proposition (b') as a consequence of (a''), let us notice that

$$\left|\int_0^\infty F'(y)dy\int_0^{y/\epsilon}\psi(x)dx\right| \leq C_1\left|\int_{\eta\epsilon}^\infty F'(y)(\log y - \log \epsilon)dy\right|$$
$$\leq C_1\left|-F(\eta\epsilon)(\log \eta\epsilon - \log \epsilon) - \int_{\eta\epsilon}^\infty \frac{F(y)}{y}dy\right|$$
$$\leq C_2$$

for all values of ϵ greater than a given value, C_1 and C_2 being properly chosen constants. In combination with (a'), this establishes (b').

The expression $x\psi(x)$ is bounded below. If then $\int_0^x \psi(\xi)d\xi$ is bounded below, (c') is valid whenever (a') is.

We wish to prove that under suitable hypotheses, the boundedness of $\int_0^x \psi(\xi)d\xi$ follows from that of $x\psi(x)$. We shall assume that $F(0)=1$. We define $M_2(x)$ as in (10). Then

$$\int_0^\infty [F(\epsilon x) - M_2(\epsilon x)]\psi(x)dx \leq \int_0^{T/\epsilon}\frac{K}{x}(1-F(\epsilon x))dx$$
$$+\mu\int_0^\infty \psi(x)[F(\epsilon x) - F(\lambda\epsilon x)]dx$$
$$+\mu\int_0^{T/\epsilon}\frac{K}{x}[F(\epsilon x) - F(\lambda\epsilon x)]dx$$
$$+\int_{T/\epsilon}^\infty \frac{K}{x}[(\mu-1)F(\epsilon x) - \mu F(\lambda\epsilon x)]dx,$$

provided for a $\mu>1$ and $\lambda>1$

$$F(x) \geq \frac{\mu}{\mu-1}F(\lambda x) \qquad [x>T]$$

Under this condition,

$$\int_0^\infty [F(\epsilon x) - M_2(\epsilon x)]\psi(x)dx$$
$$\leq K\left[\int_0^T \frac{1-F(x)}{x}dx + \mu\int^T \frac{F(x)-F(\lambda x)}{x}dx\right.$$
$$\left. + \int_T^\infty \frac{(\mu-1)F(x) - \mu F(\lambda x)}{x}dx\right] + \mu\int_0^\infty [F(\epsilon x) - F(\lambda\epsilon x)]\psi(x)dx.$$

It then follows from (a') and (b') that

$$\int_0^\infty M_2(\epsilon x)\psi(x)dx = \int_0^{1/\epsilon} \psi(x)dx$$

is bounded below.

To gather these results into our second fundamental theorem: *Let $F(x)$ be a differentiable function assuming the value 1 at 0 and tending to 0 at infinity. Let $F'(x)$ be nowhere positive, bounded, $O(x^{-2-\theta})$ at ∞, and let $F(x)$ be subject to the relation*

$$F(\lambda x) < \theta_1 F(x), \quad [x > T].$$

where $\lambda > 1$, $0 < \theta_1 < 1$. Let

$$\int_{0}^\infty F(y) y^{z-1} dy$$

be analytic in the neighborhood of the line $R(z) = 0$ with the exception of a pole of order 1 at $z = 0$, and let it have no zeros on this line. Let $\psi(x)$ be any measurable function (the restriction that it shall not vanish in the neighborhood of the origin is obviously unnecessary) satisfying the relation

$$\psi(x) > -K/x.$$

Then if

$$\lim_{\epsilon \to 0} \int_0^8 F(\epsilon x)\psi(x)dx = A$$

it follows that

$$\int_0^\infty \psi(x)dx = A.$$

Possible values of F are

$$F(x) = e^{-x};$$

$$F(x) = \frac{xe^{-x}}{1-e^{-x}}\ .$$

In particular, let

$$\psi(x) = a_n \qquad [n-1 < x \leq n]$$

where

$$a_n > -K/n.$$

Let us investigate the meaning of the condition

$$\lim_{\epsilon\to 0}\int_0^\infty F(\epsilon x)\psi(x)dx = A.$$

We have

$$\int^\infty F(\epsilon x)\psi(x)dx = \sum_1^\infty a_n\int_{n-1}^n F(\epsilon x)dx$$

$$= \sum_1^\infty a_nF(\epsilon n) + \sum_1^\infty\left(a_n+\frac{K}{n}\right)\left[\int_{n-1}^n F(\epsilon x)dx - F(\epsilon n)\right]$$

$$+\sum_1^\infty \frac{K}{n}\left[F(\epsilon n) - \int_{n-1}^n F(\epsilon x)dx\right].$$

By the law of the mean,

$$\int_{n-1}^n F(\epsilon x)dx - F(\epsilon n) < -\epsilon F'(\epsilon(n-\theta_n)). \quad [0\leq\theta_n\leq 1].$$

Let us introduce the new hypothesis,

$$-F'(x) < CF(x+1) < De^{-\Lambda x}. \tag{12}$$

It follows that

$$\left|\sum_1^\infty \frac{K}{n}\left[F(\epsilon n) - \int_{n-1}^n F(\epsilon x)dx\right]\right| \leq \sum_1^\infty \frac{K\epsilon De^{-\Lambda(n-1)\epsilon}}{n}$$

$$= K\epsilon De^{\Lambda\epsilon}\log(1-e^{-\Lambda\epsilon});$$

and that

$$\sum_1^\infty\left(a_n+\frac{K}{n}\right)\left[\int_{n-1}^n F(\epsilon x)dx - F(\epsilon n)\right] \leq C\epsilon\sum_1^\infty\left(a_n+\frac{K}{n}\right)F(\epsilon n)$$

$$\leq C\epsilon\sum_1^\infty a_nF(\epsilon n) + C\epsilon KD\sum_1^\infty\frac{e^{-\Lambda(\epsilon n-1)}}{n}$$

$$= C\epsilon\sum_1^\infty a_nF(\epsilon n) + C\epsilon KDe^{\Lambda}\log(1-e^{-\Lambda\epsilon}).$$

Hence if

$$\lim_{\epsilon\to 0}\sum_1^\infty a_nF(\epsilon n) = A,$$

we have

$$\lim_{\epsilon \to 0} \int_0^0 F(\epsilon x)\psi(x)dx = A,$$

and under the hypotheses of our last principal theorem,

$$\int_0^\infty \psi(x)dx = \sum_1^\infty a_n = A.$$

As particular examples of this theorem, let us suppose that

$$a_n > -K/n$$

and that either

$$\lim_{\epsilon \to 0} \sum_1^\infty a_n e^{-\epsilon n} = A$$

or

$$\lim_{\epsilon \to 0} \sum_1^\infty a_n \frac{\epsilon n e^{-\epsilon n}}{1 - e^{-\epsilon n}} = A.$$

We then have

$$\sum_1^\infty a_n = A.$$

In the Hardy-Littlewood terminology, we have proved the O_L Tauberian theorems for Abel and for Lambert summation. As we have previously said, the latter theorem is equivalent to the classical theorem on the asymptotic distribution of primes.

A third theorem on summability may be formulated as follows: *Let $G(u)$ be a differentiable function assuming the value 1 at $-\infty$ and tending to 0 at $+\infty$. Let $G'(u)$ be nowhere positive, less than Ce^u, and $O(e^{-(1+\theta)u})$ at $+\infty$. Let $G(u)$ be subiect to the relation*

$$G(u+l) < \theta_1 G(u) \qquad [x < T]$$

where $l < 0$; $0 \leq \theta_1 < 1$. Let

$$\int_{-\infty}^\infty G(u)e^{zu}du$$

be analytic in the neighborhood of the line $R(z) = 0$, with the exception of a pole of order 1 at $z = 0$, and let it have no zeros on this line.

Let $\chi(x)$ be any measurable function defined for all positive arguments and satisfying the relation

$$\chi(x) > -K/\sqrt{x}.$$

Then if

$$\lim_{N\to\infty}\int_{-\infty}^{\infty}\chi(x)G(\sqrt{x}-\sqrt{N})dx = A,$$

it follows that

$$\int_0^{\infty}\chi(x)dx = A.$$

The truth of this theorem follows from the fact that it may be obtained from the second theorem by the following equivalences:

$$G(\log x) = F(x);$$

$$2\chi[(\log x)^2]\,\frac{\log x}{x} = \psi(x).$$

A possible value of $G(x)$ is

$$\sqrt{\frac{2}{\pi}}\int_8^{\infty}e^{-2\xi^2}d\xi.$$

Here $\displaystyle\int_{-\infty}^{\infty}G(u)e^{zu}du = \frac{1}{z}e^{z^2/16}.$

This theorem is of importance in that it indicates a mode of bringing Borel summation under the methods of this paper. Vijayaraghavan[7] has given a proof that the condition

$$e^{-x}\,\Sigma\frac{s_n x^n}{n!}\to s \qquad \left(s_n = \sum_{r=0}^{n} a_r\right)$$

together with

$$s_n > -K/\sqrt{n} \qquad (O_L)$$

leads to the conclusion that

$$\sum_0^{\infty} a_n = s.$$

[7] T. Vijayaraghavan, A Theorem Concerning the Summability of Series by Borel's Method. *Proc. Lond. Math. Soc.*, Series 2, 27 (1928), pp. 316–326. Cf. also a paper by the same author in *Jour. Lond. Math. Soc.*, 1 (1926), pp. 113–120.

Indeed, he shows that (O_L) may be replaced by the weaker condition (s)

$$\lim (s_m - s_n) \geq 0 \quad [m>n;\ m,\ n \to \infty\ ;\ n^{-1/2}(m-n) \to 0].$$

due to R. Schmidt. On this assumption he shows by elementary means that $s_n = O(1)$ and that

$$\lim_{x\to\infty} \frac{1}{\sqrt{2\pi x}} \int_{-\infty}^{\infty} e^{-t^2/2x} f(t+x)\,dt = s,$$

where

$$f(y) = \begin{cases} s_n + (y-n)(s_{n+1}-s_n) & (n \leq y \leq n+1), \\ 0 & (y<0). \end{cases}$$

His argument can be used without substantial change to show that

$$\lim_{x\to\infty} \int_{-\infty}^{\infty} \frac{1}{\sqrt{2\pi t}} e^{-2(\sqrt{t}-\sqrt{x})^2} f(t)\,dt = s. \tag{A}$$

Instead of taking this step, Vijayaraghavan proceeds directly and in an elegant and ingenious manner to the Tauberian theorem for Borel summation itself. This is the only step in his paper which is in any intimate way concerned with the specific properties of the Borel coefficients. This step may be filled in — at least under the more stringent hypothesis (O_L) — by our third theorem. Indeed, it is not a matter of much difficulty to recast the argument in such a manner that this may be replaced by the more lenient hypothesis of Schmidt.

To show that Vijayaraghavan's assumptions imply (A) let us consider

$$\left| \frac{1}{\sqrt{2\pi x}} e^{-\frac{(t-x)^2}{2x}} - \frac{1}{\sqrt{2\pi t}} e^{-2(\sqrt{t}-\sqrt{x})^2} \right|,$$

putting

$$\sqrt{t} - \sqrt{x} = \lambda.$$

We have

$$\left| \frac{1}{\sqrt{2\pi x}} e^{-\frac{(t-x)^2}{2x}} - \frac{1}{\sqrt{2\pi t}} e^{-2(\sqrt{t}-\sqrt{x})^2} \right|$$

$$= \frac{1}{\sqrt{2\pi t}}\left|\left(1+\frac{\lambda}{\sqrt{x}}\right)e^{-\left(2\lambda^2+\frac{2\lambda^3}{\sqrt{x}}+\frac{\lambda^4}{2x}\right)}-e^{-2\lambda^2}\right|$$

$$\leq \frac{1}{\sqrt{2\pi t}}\left|\frac{\lambda}{\sqrt{x}}e^{-\lambda^2/2}+\left[\frac{2\lambda^2}{x}+\frac{\lambda^3}{x\sqrt{x}}\right]e^{-\lambda^2/2}\right|.$$

From this it follows that

$$\lim_{x\to\infty}\int_0^\infty |f(t)|\left|\frac{1}{\sqrt{2\pi x}}e^{-\frac{(t-x)^2}{2x}}-\frac{1}{\sqrt{2\pi t}}e^{-2(\sqrt{t}-\sqrt{x})^2}\right|dt$$

$$\leq \max|f(t)| \lim_{x\to\infty}\int_{-\infty}^\infty \sqrt{\frac{2}{\pi}}\left|\frac{\lambda}{\sqrt{x}}e^{-\lambda^2/2}+\left[\frac{2\lambda^2}{x}+\frac{\lambda^3}{x\sqrt{x}}\right]e^{-\lambda^2/2}\right|d\lambda$$

$$=0.$$

This proves formula (A).
Now, formula (A) may be written

$$s=\lim_{x\to\infty}\int_{-\infty}^\infty\frac{e^{-2(\sqrt{t}-\sqrt{x})^2}}{\sqrt{2\pi t}}\left[\int_0^t a_{[\xi]}d\xi\right]dt$$

$$=\lim_{x\to\infty}\int_{-\infty}^\infty a_{[\xi]}\int_\xi^\infty\frac{e^{-2(\sqrt{t}-\sqrt{x})^2}}{\sqrt{2\pi t}}\,dt$$

$$=\lim_{x\to\infty}\int_{-\infty}^\infty a_{[\xi]}\sqrt{\frac{2}{\pi}}\int_{\sqrt{\xi}-\sqrt{x}}^\infty e^{-2u^2}du.$$

This comes directly under our third theorem, for the $G(x)$ already discussed.

We may indeed say that practically every Tauberian theorem yet proved may be treated by the methods of this paper. In this connection, it is a matter of interest to compare the methods of Schmidt and of myself. In all cases, the types of summability handled either already are such as are treated in my second and third theorems, or may be reduced thereto by a further theorem, itself Tauberian, but of a looser character. This is the meaning of Schmidt's "gestrahlte Mittelbildungen," and of the formulae of Hardy and Littlewood invoked on pages 323 and 324 of Vijayaraghavan's paper on Borel summation.

The main Tauberian theorem — the one which puts the closest

restriction on the coefficients — is then of the type of the theorems of the present paper, save perhaps that conditions of the (O_L) type are replaced by those of the (s) type. Explicitly or implicitly, all three methods reduce this theorem to the solution of a singular integral equation

$$\int_{-\infty}^{\infty} f(x)\, K(y-x)dx = \phi(y)$$

of the first kind. Vijayaraghavan's attack depends upon the consideration of a finite or infinite set of equations which may be obtained from this by successive differentiation. Schmidt attacks this problem more systematically by the method of moments, which yields a general answer, but an answer not always easy of concrete application, as in the case of Lambert summation. The present paper uses the method of Fourier analysis, also of general bearing, and in the opinion of the author, simpler. The divergence of standpoints here reflected is but a reflection of that prevailing in the operational calculus at large. It is in that more general region that the relative merits of the different methods must be decided. The present author[8] has already given grounds for believing that the power of Fourier methods is here of the same order as that of methods depending on successive differentiation or on the theory of moments and of the Laplace transformation and has succeeded in using them to establish the asymptotic formulae of Heaviside. He is therefore inclined to look favorably upon them in this closely related field.

At all events, Tauberian theorems have undergone a remarkable change of status since the early days of pioneer work, in which four or five years of brilliant technique and persistent effort were often necessary to reduce an O condition to an o form. All three of the present methods are in essence reasonably simple, and at least two of them seem to be of sufficient generality for all our present needs. What the reason for this change may be is perhaps hard to say; the fairest explanation is perhaps that Hardy and Littlewood[9] had already succeeded in casting a large group of Tauberian theorems into the form of the equivalence of two means of positive quantities

[8] N. Wiener, The Operational Calculus. *Math. Ann.* 95 (1926), pp. 557–584.
[9] Loc. cit.

Comments on [28b]

H. R. Pitt

The paper contains the first formulation of Wiener's general Tauberian theorem, but in an incomplete and imperfect form. It is of considerable historical importance, but the actual results are superseded by his later work, particularly, in *Tauberian Theorems* [32a]. The main gap in relation to the final form is in the need for the order condition $O(\xi^{-2})$ at $\pm\infty$ for the two kernels N_1 and N_2. The other additional conditions are less serious and are needed because the condition $f(\xi) \geqslant 0$, which can readily be replaced by $f(\xi)$ bounded below, is weaker than the boundedness of $f(\xi)$ which the final version demands. In fact, if boundedness of $f(\xi)$ were assumed, the theorem could be stated more simply and would be equally effective. The order condition on the kernels does allow a different and easier approach from that needed in the final version, and a particularly simple proof of the theorem with the slightly stronger order condition $N(\xi)[1 + x^2]$ integrable was later given by Bochner.[1]

The second main theorem on the deduction of $\int_0^\infty \psi(x)\,dx = A$ from

$$\int_0^\infty F(\varepsilon x)\,\psi(x)\,dx \to A \quad \text{as } \varepsilon \to 0,$$

is deduced and is subject, of course, to the same unnecessarily restrictive conditions on the kernel $F(x)$; but even in this imperfect form the theorem is adequate to cover the classical special Tauberian theorem including, after an appropriate change of scale, that for Borel summation.

Reference

1. S. Bochner, *Umkehrsätze für allgemeine Limitierungsverfahren*, Berliner Sitz. (1933), 126–144.

PAPERS

PUBLISHED IN THE

PROCEEDINGS OF THE LONDON MATHEMATICAL SOCIETY

A TYPE OF TAUBERIAN THEOREM APPLYING TO FOURIER SERIES

By NORBERT WIENER.

[Received 1 January, 1929.—Read 10 January, 1929.]

The purpose of the present paper is to apply a method already developed by the author* to the proof of a theorem of Hardy and Littlewood† concerning the summability of Fourier series. Restated in an equivalent form for Fourier integrals, it reads: *Let* $f(x)$ *be a measurable function defined over* $(-\infty, \infty)$ *and zero outside* $(-A, A)$. *Let*

$$\phi(y) = \tfrac{1}{2}\{f(x+y)+f(x-y)-2C\}.$$

Then if we write B_m *for the proposition*

$$\lim_{\lambda\to\infty} \lambda \lim_{\epsilon\to 0} \int_\epsilon^{1/\lambda} \phi(y)(1-\lambda y)^m\, dy = 0$$

and C_m *for the proposition*

$$\lim_{\lambda\to\infty} \lambda \int_0^\infty \phi(y)\, dy \int_0^1 (1-z)^m \cos \lambda yz\, dz = 0,$$

the infinite integral being taken in the sense $\lim_{\epsilon\to 0,\, E\to\infty} \int_\epsilon^E$, *the truth of* B_m *for some m implies the truth of C for some n, and vice versa.* C_m asserts

---

* N. Wiener, "A new method in Tauberian theorems", *Journal Math. and Phys., Mass. Inst. of Tech.*, 7 (1928), 161–184.

† G. H. Hardy and J. E. Littlewood, "Fourier series and power series", *Proc. London Math. Soc.* (2), 21 (1922–23), xxxi–xxxii. Cf. also Hardy and Littlewood, "The equivalence of certain integral means", *ibid.*, 22 (1923–24), xl–xliii; "Solution of the Cesàro summability problem for power series and Fourier series", *Math. Zeitschrift*, 19 (1924), 67–96.

the m-th order Cesàro summability of the Fourier integral for $f(x)$ to the value C.

In proving this theorem, we shall use the following lemma :

LEMMA I. *Let $F(x)$ be a (real or complex) function of the real variable x, defined over $0 \leqslant x < \infty$ and summable over every interval $(0, A)$. Let $Q(x)$ be a function defined and measurable over $0 \leqslant x < \infty$ and not vanishing for $x > B$. Let $x^n Q(x)$ be of limited total variation over $(0, \infty)$, and let its reciprocal be of limited total variation over (B, ∞), so that $L = \lim_{x\to\infty} x^n Q(x)$ exists and is not zero. Let $\int_B^\infty F(x)\,Q(x)\,dx$ exist as the limit of a Lebesgue integral, and let us write*

$$g(y, \lambda) = \lambda \int_B^y F(x)\,Q(\lambda x)\,dx\,;$$

$$h(y) = \int_B^y F(x)\,x^{-n}\,dx.$$

Then $h(\infty)$ and $g(\infty, \lambda)$ exist, and

$$g(\infty, \lambda) = \lambda^{1-n}\left\{h(\infty)\,L - \int_B^\infty h(x)\,d\left[(\lambda x)^n\,Q(\lambda x)\right]\right\} = O(\lambda^{1-n}).$$

Proof. We have

$$h(N) = \int_B^N F(x)\,Q(x)\frac{dx}{x^n\,Q(x)} = \frac{g(N, 1)}{N^n Q(N)} - \int_B^N g(x, 1)\,d\left(\frac{1}{x^n\,Q(x)}\right).$$

Clearly $g(x, 1)$ is bounded. Going to the limit, $h(\infty)$ exists and

$$h(\infty) = \frac{g(\infty, 1)}{L} - \int_B^\infty g(x, 1)\,d\left(\frac{1}{x^n Q(x)}\right).$$

In the same way,

$$\begin{aligned} g(N, \lambda) &= \lambda \int_B^N F(x)\,x^{-n}\left\{x^n\,Q(\lambda x)\right\}\,dx \\ &= \lambda h(N)\,N^n\,Q(\lambda N) - \lambda \int_B^N h(x)\,d\left\{x^n\,Q(\lambda x)\right\} \\ &= \lambda^{1-n}\left\{h(N)(\lambda N)^n\,Q(\lambda N) - \int_B^N h(x)\,d\left[(\lambda x)^n\,Q(\lambda x)\right]\right\}. \end{aligned}$$

Since $h(x)$ is bounded, $g(\infty, \lambda)$ exists, and

$$g(\infty, \lambda) = \lambda^{1-n}\left\{h(\infty)\,L - \int_B^\infty h(x)\,d\left[(\lambda x)^n\,Q(\lambda x)\right]\right\}.$$

Thus, for $\lambda > 0$,

$$|g(\infty, \lambda)| \leqslant \lambda^{1-n}\left\{|L h(\infty)| + \max_{B \leqslant x < \infty} |h(x)| \int_0^\infty |d[x^n Q(x)]|\right\} = O(\lambda^{1-n}).$$

This completes our proof. We are now in a position to prove

LEMMA II. *If F and Q are subject to the restrictions of Lemma I, and if*

$$\int_0^\infty \lambda^{1-n} P(\lambda)\, d\lambda$$

is absolutely convergent, then

$$\int_0^\infty \lambda P(\lambda)\, d\lambda \int_B^\infty F(x)\, Q(\lambda x)\, dx = \int_0^\infty F(x)\, dx \int_0^\infty \lambda P(\lambda)\, Q(\lambda x)\, d\lambda.$$

Proof. The integral

$$\int_0^\infty \lambda P(\lambda)\, d\lambda \int_B^\infty F(x)\, Q(\lambda x)\, dx$$

may be arranged in the form

$$\begin{aligned}
\int_0^\infty & P(\lambda)\, g(\infty, \lambda)\, d\lambda \\
&= \int_0^\infty \lambda^{1-n} P(\lambda)\, d\lambda \left\{h(\infty) L - \int_B^\infty h(x)\, d[(\lambda x)^n Q(\lambda x)]\right\} \\
&= L h(\infty) \int_0^\infty \lambda^{1-n} P(\lambda)\, d\lambda - \int_B^\infty h(x)\, d\left[x^n \int_0^\infty \lambda P(\lambda)\, Q(\lambda x)\, d\lambda\right] \\
&= L h(\infty) \int_0^\infty \lambda^{1-n} P(\lambda)\, d\lambda - \lim_{N\to\infty} \left\{h(N) N^n \int_0^\infty \lambda P(\lambda)\, Q(\lambda N)\, d\lambda\right. \\
&\qquad\qquad \left. + \int_B^N \left[x^n \int_0^\infty \lambda P(\lambda)\, Q(\lambda x)\, d\lambda\right] dh(x)\right\}.
\end{aligned}$$

But $\lim_{N\to\infty} \left(h(N) N^n \int_0^\infty \lambda P(\lambda)\, Q(\lambda N)\, d\lambda\right)$

$$\begin{aligned}
&= \lim_{N\to\infty} \left(h(N) \int_0^\infty \lambda^{1-n} P(\lambda) (\lambda N)^n Q(\lambda N)\, d\lambda\right) \\
&= L h(\infty) \int_0^\infty \lambda^{1-n} P(\lambda)\, d\lambda.
\end{aligned}$$

Hence $\int_0^\infty P(\lambda)\, g(\infty, \lambda)\, d\lambda = \int_B^\infty F(x)\, dx \int_0^\infty \lambda P(\lambda)\, Q(\lambda x)\, d\lambda.$

B 2

Lemma II is of particular interest when

$$\lim_{\lambda\to\infty} \lambda \int_B^\infty F(x)\, Q(\lambda x)\, dx = M$$

and
$$P(\lambda) = \frac{1}{\mu} \Phi\left(\frac{\lambda}{\mu}\right).$$

Then our condition concerning P assumes the form that

$$\int_0^\infty \lambda^{1-n} \Phi(\lambda)\, d\lambda$$

is to converge absolutely. Let us put

$$\int_0^\infty \lambda \Phi(\lambda)\, Q(\lambda x)\, d\lambda = K(x).$$

Let us introduce the new assumption that

$$\int_0^\infty \Phi(\lambda)\, d\lambda$$

is to converge absolutely. We then have

$$\begin{aligned} M\int_0^\infty \Phi(\lambda)\, d\lambda &= \lim_{\mu\to 0} \int_0^\infty \Phi(\lambda)\, g(\infty, \lambda\mu)\, d\lambda \\ &= \lim_{\mu\to 0} \int_0^\infty \Phi(\lambda)\, d\lambda \int_B^\infty \lambda\mu F(x)\, Q(\lambda\mu x)\, dx \\ &= \lim_{\mu\to 0} \mu \int_B^\infty F(x)\, dx \int_0^\infty \lambda\Phi(\lambda)\, Q(\lambda\mu x)\, dx \\ &= \lim_{\mu\to 0} \mu \int_B^\infty F(x)\, K(\mu x)\, dx. \end{aligned}$$

All the integrals and limits in this last formula exist, as we may verify by means of Lemma I. The inversion of the order of integration is justified by the fact that it is carried out in an absolutely convergent double Stieltjes integral.

Hence, *if Q and F satisfy the condition which we have already laid down for them, and if the integral equation*

$$\int_0^\infty \lambda\Phi(\lambda)\, Q(\lambda x)\, d\lambda = K(x) \tag{I}$$

has a solution $\Phi(\lambda)$ for which

$$\int_0^\infty |\Phi(\lambda)|\, d\lambda, \quad \int_0^\infty \lambda^{1-n} |\Phi(\lambda)|\, d\lambda$$

are finite, then

$$\lim_{\mu\to 0} \mu \int_B^\infty F(x)\,K(\mu x)\,dx = M\int_0^\infty \Phi(\lambda)\,d\lambda.$$

This is a theorem of the type termed Tauberian* by Hardy and Littlewood.

Now let us put $\lambda = e^\xi$; $x = e^\eta$;

$$Q(x) = R(\eta);\quad \Phi(x) = S(\eta);\quad K(x) = T(\eta).$$

Then (I) assumes the form

$$\int_{-\infty}^\infty e^{2\xi}\,S(\xi)\,R(\xi+\eta)\,d\xi = T(\eta). \qquad (\mathrm{I}^1)$$

Now let us suppose that $e^\xi R(\xi)$ and $e^\xi T(\xi)$ are quadratically summable over $(-\infty, \infty)$. Let

$$r(u) = \underset{A\to\infty}{\mathrm{l.i.m.}} \int_{-A}^{A} e^\xi R(\xi)\,e^{i\xi u}\,d\xi = \underset{\epsilon\to 0}{\mathrm{l.i.m.}}\int_\epsilon^{1/\epsilon} Q(x)\,x^{iu}\,dx;$$

$$t(u) = \underset{A\to\infty}{\mathrm{l.i.m.}} \int_{-A}^{A} e^\xi T(\xi)\,e^{i\xi u}\,d\xi = \underset{\epsilon\to 0}{\mathrm{l.i.m.}}\int_\epsilon^{1/\epsilon} K(x)\,x^{iu}\,dx.$$

These limits, in the mean, exist by the Plancherel theory of the Fourier transforms, which we now assume *in toto*†. In case $t(u)/r(u)$ is quadratically summable, it follows from the Parseval-Plancherel theorem as applied to the Fourier transform of the product of two functions that

$$S(\xi) = \underset{A\to\infty}{\mathrm{l.i.m.}}\, e^{-\xi}\int_{-A}^{A}\left(\frac{t(-u)}{r(-u)}\right) e^{-i\xi u}\,du$$

is a possible solution of (I^1). This gives us

$$\Phi(x) = \underset{A\to\infty}{\mathrm{l.i.m.}}\int_{-A}^{A}\frac{t(-u)}{r(-u)}\,x^{-iu-1}\,du;$$

$$\frac{2\pi t(-u)}{r(-u)} = \underset{\epsilon\to 0}{\mathrm{l.i.m.}}\int_\epsilon^{1/\epsilon}\Phi(x)\,x^{iu}\,dx.$$

In a particular case, it may happen that

$$\int_0^\infty Q(x)\,x^{iu}\,dx,\quad \int_0^\infty K(x)\,x^{iu}\,dx$$

---

* Cf., *e.g.*, G. H. Hardy and J. E. Littlewood, "Tauberian theorems concerning power series and Dirichlet's series whose coefficients are positive", *Proc. London Math. Soc.* (2), 19 (1921), 174–191.

† M. Plancherel, "Contribution à l'étude de la représentation d'une fonction arbitraire par des intégrales définies", *Rend. di Palermo*, 30 (1910), 289–335.

exist and define analytic functions over the range $-B < \mathfrak{R}(w) < C$; $-\infty < \mathfrak{I}(w) < \infty$. Let us call the functions $\rho(w)$ and $\tau(w)$ respectively. Let the function $\tau(w)/\rho(w)$ be analytic over the range $\mu < \mathfrak{R}(w) < \nu$, including $\mathfrak{R}(w) = 0$. Let $\tau(w)/\rho(w)$ be uniformly $o(1)$ over this strip as $|\mathfrak{I}(w)| \to \infty$. By Cauchy's theorem,

$$S(\xi) = e^{-\xi}\int_{-\infty}^{\infty} \frac{\tau(-iu)}{\rho(-iu)} e^{-i\xi u}\,du = \int_{-\infty}^{\infty} \frac{\tau(w_0-iu)}{\rho(w_0-iu)} e^{\xi(w_0-iu-1)}\,du,$$

for any w_0 of the strip, provided that either integral exists. Indeed, the integrals in question need only exist as limits in the means. Thus, if

$$\tau(w_0-iu)/\rho(w_0-iu)$$

is quadratically summable, it is the Fourier transform of $S(\xi)e^{-\xi(w_0-1)}$, which is therefore also quadratically summable.

Let us assume that $\tau(w_0-iu)/\rho(w_0-iu)$ is quadratically summable in u for $\mu < \mathfrak{R}(w) < \nu$, and is uniformly $o(1)$ in this strip as $|u| \to \infty$. Since

$$\int_0^A |S(\xi)|e^{-\xi(w_0-1)}\,d\xi \leqq \sqrt{\left\{\int_0^A |S(\xi)e^{-\xi(w-1)}|^2\,d\xi \int_0^A e^{-\xi(w_0-w_1)}|^2\,d\xi\right\}},$$

where w_1 is a number of the strip less than w_0, we see that

$$\int_0^\infty |S(\xi)|e^{-\xi(w_0-1)}\,d\xi$$

exists for every w_0 in our strip. Similarly,

$$\int_{-\infty}^0 |S(\xi)|e^{-\xi(w_0-1)}\,d\xi$$

exists, and hence

$$\int_0^\infty |\Phi(x)|x^{-w_0}\,dx = \int_{-\infty}^\infty |S(\xi)|e^{-\xi(w_0-1)}\,d\xi$$

exists. Hence, if 0 and $1-n$ lie within the range $(-\nu, -\mu)$, the function Φ satisfies the condition laid down in the hypothesis of our Tauberian theorem. Assuming that $n > 1$, we may write the condition that 0 and $1-n$ are to lie within this range in the form $\mu < 0 < n-1 < \nu$. Thus μ may have any value less than 0, and ν any value greater than $n-1$. Combining these results, we obtain

LEMMA III. *Let $F(x)$ be a (real or complex) function, bounded for $0 \leqq x \leqq A$ for all A, and defined over $(0, \infty)$. Let $Q(x)$ be a function defined over $(0, \infty)$ and not vanishing for $x > B$. Let $x^n Q(x)$ be of* *limited total variation over* $(0, \infty)$, *and let its reciprocal be of limited total variation over* (B, ∞), *for a given* $n > 1$. *Let* $K(x)$ *be a measurable function defined over* $(0, \infty)$, *and let*

$$\int_0^\infty Q(x)\,dx = \int_0^\infty K(x)\,dx$$

(as will be the case when $\int_0^\infty \Phi(\lambda)d\lambda = 1$, *according to our earlier definitions). Let*

$$\rho(w) = \int_0^\infty Q(x)\,x^w\,dx \quad \textit{and} \quad \tau(w) = \int_0^\infty K(x)\,x^w\,dx$$

exist as functions such that $\tau(w)/\rho(w)$ *is analytic over*

$$\{-\eta < \mathfrak{R}(w) < n-1+\eta\},$$

where $\eta > 0$. *Let* $\tau(w)/\rho(w)$ *be* $o(1)$ *uniformly in* $\mathfrak{J}(w)$ *at* $+\infty$ *for* $\{-\eta < \mathfrak{R}(w) < n-1+\eta\}$, *and let it be quadratically summable over every ordinate in that interval. Let*

$$\lim_{\lambda\to 0}\int_A^\infty F(x)\,Q(\lambda x)\,dx = M.$$

Then

$$\lim_{\lambda\to 0}\int_A^\infty F(x)\,K(\lambda x)\,dx = M.$$

By the transformation $y = 1/x$, we get

LEMMA III[1]. *Let* $F(x)$ *be a (real or complex) function, bounded over* (ϵ, ∞) *for all* ϵ, *and defined over* $(0, \infty)$. *Let* $Q(y)$ *be a function defined over* $(0, \infty)$ *and not vanishing for* $x < \epsilon$. *Let* $y^{-m}Q(y)$ *be of limited total variation over* $(0, \infty)$, *and let its reciprocal be of limited total variation over* $(0, \epsilon)$, *for a given* $m > -1$. *Let* $K(y)$ *be a measurable function defined over* $(0, \infty)$, *and let*

$$\int_0^\infty Q(y)\,dy = \int_0^\infty K(y)\,dy.$$

Let
$$\rho'(w) = \int_0^\infty Q(y)\,y^w\,dy \quad \textit{and} \quad \tau'(w) = \int_0^\infty K(y)\,y^w\,dy$$

exist as functions such that $\tau'(w)/\rho'(w)$ *is analytic over*

$$\{-m-1-\eta < \mathfrak{R}(w) < \eta\}.$$

Let $\tau'(w)/\rho'(w)$ *be* $o(1)$ *uniformly in* $\mathfrak{J}(w)$ *at* $\pm\infty$ *for*

$$\{-m-1-\eta < \mathfrak{R}(w) < \eta\},$$

and let it be quadratically summable over every ordinate in that interval. Let

$$\lim_{\lambda\to\infty} \lambda \int_0^\epsilon F(y)\, Q(\lambda y)\, dy = M.$$

Then

$$\lim_{\lambda\to\infty} \lambda \int_0^\epsilon F(y)\, K(\lambda y)\, dy = M.$$

We now proceed to our main theorem. As $Q(y)$ or $K(y)$ we consider the functions

$$K_\mu(y) = \begin{cases} (\mu+1)(1-y)^\mu & [0 < y < 1], \\ 0 & [y > 1]; \end{cases}$$

$$K_\mu^1(y) = \frac{2}{\pi}\int_0^1 (1-z)^\mu \cos yz\, dz.$$

The quantity m is obviously 0*. If we write

$$\tau_\mu(w) = \int_0^\infty K_\mu(y)\, y^w\, dy, \quad \sigma_\mu(w) = \int_0^\infty K_\mu^1(y)\, y^w\, dy,$$

we have

$$\tau_\mu(w) = \frac{\Gamma(\mu+2)\,\Gamma(w+1)}{\Gamma(w+\mu+2)}, \quad \sigma_\mu(w) = \frac{\Gamma(\mu+1)}{\Gamma(\mu+1-w)\cos\frac{1}{2}w\pi}.$$

For any positive μ, these functions are free from zeros and singularities for all w, such that $\mathfrak{R}(w)$ lies in $(-1-\eta, \eta)$ for sufficiently small η, with the exception in all cases of a simple pole at -1. Thus $\sigma_\mu(w)/\tau_\nu(w)$ and $\tau_\mu(w)/\sigma_\nu(w)$ are alike free from singularities over $(-1-\eta, \eta)$. It is easy to see from the known asymptotic properties of the gamma function that

$$\left|\frac{\sigma_\mu(w)}{\tau_\nu(w)}\right| \sim \frac{2\Gamma(\mu+1)}{e\sqrt{2\pi}\,\Gamma(\nu+2)}\, \mathfrak{J}(w)^{\nu-\mu+\mathfrak{R}(w)+\frac{1}{2}}.$$

Over $-1-\eta < \mathfrak{R}(w) < \eta$ the exponent of $\mathfrak{J}(w)$ varies between $\nu-\mu-\frac{1}{2}-\eta$ and $\nu-\mu+\frac{1}{2}+\eta$. If both these numbers are greater than $\frac{1}{2}$, K_μ can serve for Q and K_ν^1 for K. In this case $\mu > \nu+1$. If both these numbers are greater than $\frac{1}{2}$, K_μ^1 can serve for Q and K_ν for K. In this case $\nu > \mu+1$. Thus, in the notation of the main theorem of our paper, B_m implies $C_{m+1+\eta}$ and C_m implies $B_{m+1+\eta}$. This is, however, sufficient to establish our main theorem.

* Cf. M. Jacob, "Über die Summierbarkeit von Fourierschen Reihen und Integralen", *Math. Zeitschrift*, 29 (1928), 21.

Comments on [29b]

H. R. Pitt

The theorem of this paper is a special one in the sense that it relates to averages of a function with respect to two special and explicit kernels, but the proof is obtained by the application of a theorem (Lemma III) about pairs of kernels of a general class. The theorem is not deducible from the general Tauberian theorem and depends on a quite different argument. It is, in fact, essentially Abelian in character since it establishes between the kernels $Q(x)$ and $K(x)$ a relationship of the form

$$K(x) = \int_0^\infty \lambda \Phi(\lambda) Q(\lambda x)\, d\lambda$$

with $\Phi(\lambda)$ and $\lambda^{1-n}\Phi(\lambda)$ both integrable in $(0, \infty)$. This is, of course, a stronger relationship than that between two kernels in a Tauberian theorem, and the reason why we should expect the required conditions on K and Q to be heavier. The condition is indeed of a different order in that the ratio

$$\int_0^\infty Q(x) x^w\, dx \Big/ \int_0^\infty K(x) x^w\, dx$$

must be analytic in the open strip $-\eta < R(w) < n - 1 + \eta$, whereas the appropriate conditions for a Tauberian theorem would simply be the non-vanishing of $\int K(x) x^w\, dx$ on a *line* $R(w) = 0$.

TAUBERIAN THEOREMS.*

By Norbert Wiener.

INTRODUCTION.

Numerous important branches of mathematics and physics concern themselves with the asymptotic behavior of functions for very large or very

* Received July 20, 1931.

small values of their arguments, or of certain parameters. Statistical mechanics is that branch of mechanics which concerns itself, not with an individual dynamical system of a finite number of degrees of freedom, but with the asymptotic behavior of dynamical systems as the number of degrees of freedom increases without limit. The analytical theory of numbers is likewise concerned with the behavior of assemblages of whole numbers as the size of these assemblages increases. To this domain of ideas belong asymptotic series in analysis, and a whole order of concepts clustering about the operational calculus of Heaviside and the Fourier series and integrals.

The Fourier integral is of peculiar interest in the study of asymptotic problems. If $f(x)$ is a function of Lebesgue class L_2, by a theorem of Plancherel,[1] there is related to it another function $g(u)$, also of L_2, defined by the equation

$$g(u) = \underset{A\to\infty}{\text{l. i. m.}} \frac{1}{\sqrt{2\pi}} \int_{-A}^{A} f(x)\, e^{iux}\, dx \tag{0.01}$$

(where l. i. m. stands for "limit in the mean"), such that

$$\int_{-\infty}^{\infty} |g(u)|^2\, du = \int_{-\infty}^{\infty} |f(x)|^2\, dx, \tag{0.02}$$

and

$$f(x) = \underset{A\to\infty}{\text{l. i. m.}} \frac{1}{\sqrt{2\pi}} \int_{-A}^{A} g(u)\, e^{-iux}\, du. \tag{0.03}$$

The functions $f(x)$ and $g(u)$ have the reciprocal relation, that asymptotic properties of each correspond to local properties of the other. They are known as Fourier transforms of one another. It is easy to see that $f(x+\lambda)$ and $g(u)\,e^{-iu\lambda}$ are likewise Fourier transforms of one another.

There is a large class of asymptotic problems in which the asymptotic property to be investigated is connected in some obvious and simple way with the entire class of functions $f(x+\lambda)$. Since the Fourier transforms of the functions of this class only differ by factors $e^{-iu\lambda}$, independent of the particular function $f(x)$, Fourier transformation is here a peculiarly useful tool. An example of such a problem is the investigation of the asymptotic behavior of the integral

$$\int_{-\infty}^{\infty} k(\lambda)\, f(x+\lambda)\, d\lambda. \tag{0.04}$$

Many problems which on first investigation do not appear to be concerned with integrals of the above sort may be put into such a form by an elementary transformation of variables. For example, if we put

[1] Cf. Plancherel (1), Titchmarsh (3), Mellin (1), Wiener (7). Cf. Bibliography on p. 94.

(0.05) $x = \log \xi, \quad \lambda = \log \mu, \quad f(x) = \varphi(\xi), \quad \mu k(\lambda) = Q(\mu),$

the integral in question assumes the form

(0.06) $$\int_0^\infty Q(\mu)\, \varphi(\xi\mu)\, d\mu.$$

Thus the study of the asymptotic properties of this integral also fall under those accessible through Fourier developments.

In 1925 a paper by Robert Schmidt[2] appeared in which the class of theorems known as Tauberian was brought into relation with the asymptotic properties of integrals of this type. Tauberian theorems gain their name from a theorem published by A. Tauber[3] in 1897, to the effect that if

(0.07) $$\lim_{x \to 1-0} \frac{1}{1-x} \sum_0^\infty a_n x^n = A,$$

and

(0.08) $$a_n = o(1/n),$$

then

(0.09) $$\sum_0^\infty a_n = A.$$

This is a conditioned converse of Abel's theorem, which stated that (0.07) follows from (0.09), without the mediation of any hypothesis such as (0.08). Such conditioned inverses of Abel's theorem, and of other analogous theorems which assert that the convergence of a series implies its summability by a certain method to the same sum, have been especially studied by G. H. Hardy and J. E. Littlewood, and have been termed by them *Tauberian.*

It is the service of Hardy and Littlewood[4] to have replaced hypothesis (0.08) by hypotheses of the form

(0.10) $$a_n = O\left(\frac{1}{n}\right),$$

or even of the form

(0.11) $$n a_n > -K.$$

The importance of these generalizations is scarcely to be exaggerated. They far exceed in significance Tauber's original theorem. The work of Hardy and Littlewood, unlike that of Tauber, makes very appreciable demands on analytical technique, and is capable, among other things, of supplying the gaps in Poisson's imperfect discussion of the convergence

[2] Schmidt (2).

[3] Tauber (1).

[4] Hardy (2), Littlewood (1), Hardy and Littlewood (4), (10), (13), (18) etc.

1*

of the Fourier series. For these reasons, I feel that it would be far more appropriate to term these theorems Hardy-Littlewood theorems, were it not that usage has sanctioned the other appelation.

As we said, Tauberian theorems reduce to theorems on the conditioned equivalence of the asymptotic values of certain averages. The auxiliary conditions of Hardy and Littlewood, together with certain more extended conditions given by Schmidt[5] and discussed by Vijayaraghavan[6] and Szász,[7] become restrictions on the magnitude of the function or mass distribution of which the average is to be taken. An even further reduction eliminates this function from all consideration, and finds the essence of the Tauberian theorem in the linear properties of the weighting functions used in determining the average in question. The linear properties which are of importance concern the closure of the set of functions derivable from a given weighting function by translation, and the first chapter of this monograph is devoted to the study of such closure properties. The second chapter is occupied with the formulation and proof of the fundamental Tauberian theorems concerning averages, and the third with the transformation of these theorems into a recognized Tauberian form.

Some of the most interesting applications of Tauberian theorems have been in the analytic theory of numbers. Here two different avenues of approach are possible. Hardy and Littlewood[8] have shown that the prime number theorem, to the effect that the number of primes less than n is asymptotically $n/\log n$, is entirely equivalent to the Tauberian theorem concerning Lambert series, that if

(0.12) $$\lim_{x\to 1-0} \sum_0^\infty \frac{n\, a_n\, x^n}{1-x^n} = A$$

and

(0.11) $$n\, a_n > -K,$$

then

(0.09) $$\sum_0^\infty a_n = A.$$

This theorem falls under the category of theorems demonstrable by the methods of the present paper, and involves no further information concerning the Riemann zeta function $\zeta(\sigma + \tau i)$ than that this must be free of zeros on the line $\sigma = 1$. This theorem was already contained in the author's first note on his method, dated 1928, and establishes the usefulness of Lambert series in the proof of the prime number theorem, which had

[5] Schmidt (1) and (2).

[6] Vijayaraghavan (1) and (2).

[7] Szász (1) and (2).

[8] Hardy and Littlewood (11).

frequently been questioned[9]. The proof was however needlessly complicated by the fact that after a good deal of labor was spent in transforming the author's average theorem into recognizable Tauberian form, this form did not appear to be especially direct in its relation to the prime number theorem. The present memoir gives a much more direct proof, starting from a theorem of the average type.

The other approach to the prime number theorem employs a theorem of S. Ikehara[10], which is itself a generalization of a theorem of Landau. Landau[11] showed that if $\sum a_n n^{-x}$ is a Dirichlet series with positive coefficients, representing a function $\psi(x)$ analytic on and to the right of $\Re(x) = 1$, except for a pole of order 1 at $x = 1$, for which the residue is A, and if the function in question is $O(|x|^k)$ for $\Re(x) \geqq 1$, then

$$\lim_{n\to\infty} \frac{1}{n} \sum_0^n a_l = A. \tag{0.13}$$

Ikehara proved by a Tauberian theorem that the condition that $\psi(x) = O(|x|^k)$ is inessential. By applying the theorem to $\psi(x) = -\zeta'(x)/\zeta(x)$ it is at once seen that the prime number theorem again follows from the fact that the Riemann zeta function has no zeros on the line $\Re(x) = 1$.

Chapter V is devoted to miscellaneous applications of Tauberian theorems. Among these perhaps the most important are to trigonometric developments and their summability. Here a theorem of Ramanujan[12] on Fourier transforms is of service. A particular application of Tauberian theorems to the criterion of Young[13] for the convergence of a Fourier series is due to Dr. Littauer[14], and is here discussed. We also take up the application of Tauberian theorems to asymptotic series and Wintner's[15] work.

There is a further field of application for Tauberian theorems where we need to introduce something like Robert Schmidt's[16] notion of "gestrahlte Matrizen". This we discuss in Chapter VI. In this category we find certain theorems of Hardy and Littlewood[17] related to Abel summation, as also the Tauberian theorems that concern themselves with Borel summation[18].

[9] Wiener (4).
[10] Ikehara (1).
[11] Landau (3).
[12] Hardy and Littlewood (8).
[13] Young (1).
[14] Littauer (2).
[15] Wintner (1).
[16] Schmidt (2).
[17] Cf. Hardy and Littlewood (10), (13), (18); Karamata (1), (2), (3), (4); Szász (1), (2); Doetsch (3); etc.
[18] Cf. Hardy and Littlewood (2), (3); Doetsch (4); Schmidt (1); Vijayaraghavan (1); Wiener (4).

In Chapter VII, we discuss certain theorems not strictly of a Tauberian nature, in that they involve no positiveness or boundedness condition. We here follow an earlier paper[19] in which the author proved the Hardy-Littlewood necessary and sufficient condition for the summability of a Fourier series. There our result was not a "best possible" one, as the work of Bosanquet[20] and Paley has demonstrated. At present, our method has succeeded in yielding their strict theorem.

Chapter VIII is devoted to the development of certain Tauberian theorems intimately connected with our generalized harmonic analysis[21]. These suggest certain new definitions of summability, and we discuss definitions of this sort.

The genesis of the present paper may deserve a word of comment. In the preparation of a previous investigation on generalized harmonic analysis, the author found himself obliged to make use of certain theorems communicated to him by Mr. A. E. Ingham and of a Tauberian nature. A correspondence arose with Professor Hardy, finally resulting in a number of papers of a Tauberian character. While interested in theorems of this type, the author had the work of Dr. Robert Schmidt brought to his notice, and at Dr. Schmidt's suggestion, began to search for a method of combining the trigonometric attack of his own earlier papers with the generality of the Schmidt standpoint. It was only by a radical change in the manner in which trigonometric methods were applied that this attempt succeeded. This change consisted in a logarithmic change of base before the introduction of the harmonic analysis. It was quite along lines contemplated by Schmidt, who correlated with his Tauberian theorems a certain moment problem. Schmidt, however, found in the problem of uniqueness the significant aspect of this moment problem. The present author looked for it rather in the problem of the existence of a solution together with certain associated problems of approximation. It was Dr. Schmidt himself who furnished the experimentum crucis which established the greater scope of the methods of this paper. He suggested that the Tauberian theorem for Lambert series had resisted his methods of proof, and that it might be desirable to try on it the edge of any new method.

The origination of a new method is but the prelude to a large amount of detailed application before its power can be judged or its limitations defined. In this task the author has had the valuable aid of his students, Drs. S. B. Littauer and S. Ikehara. He also wishes to express his gratitude for the criticisms of Professors Hardy and Tamarkin at more than one stage of his work.

[19] Cf. Hardy and Littlewood (12); Wiener (5).

[20] Bosanquet (1); Paley (1).

[21] Wiener (7).

CHAPTER I.

THE CLOSURE OF THE SET OF TRANSLATIONS OF A GIVEN FUNCTION.

1. **Closure in class L_2.** Let $f(x)$ be a function, real or complex, defined for all real arguments over $(-\infty, \infty)$. Let it belong to Lebesgue class L_2—that is, let it be measurable, and let

$$\int_{-\infty}^{\infty} |f(x)|^2\, dx \tag{1.01}$$

be finite. We shall term the class of all functions $f(x+\lambda)$ for all real values of λ the class of *translations of* $f(x)$. We wish to know when this class is closed or complete—that is, when it is possible, whenever a function $F(x)$ from L_2 is given, and any positive quantity ε, to find a function $F_1(x)$ of the form

$$F_1(x) = \sum_1^n A_k f(x+\lambda_k), \tag{1.02}$$

such that

$$\int_{-\infty}^{\infty} |F(x) - F_1(x)|^2\, dx \leqq \varepsilon. \tag{1.03}$$

The situation is governed by the following theorem:

THEOREM I. *The necessary and sufficient condition for the set of all translations of $f(x)$ to be closed is that the real zeros of its Fourier transform*

$$g(u) = \underset{A\to\infty}{\text{l. i. m.}} \frac{1}{\sqrt{2\pi}} \int_{-A}^{A} f(x)\, e^{iux}\, dx \tag{0.01}$$

should form a set of zero measure.

The existence of this Fourier transform results from a familiar theorem of Plancherel.[1] This theorem further requires that

$$\int_{-\infty}^{\infty} |g(u)|^2\, du = \int_{-\infty}^{\infty} |f(x)|^2\, dx \tag{0.02}$$

and that

$$f(x) = \underset{A\to\infty}{\text{l. i. m.}} \frac{1}{\sqrt{2\pi}} \int_{-A}^{A} g(u)\, e^{-iux}\, du. \tag{0.03}$$

The Fourier transform of $f(x+\lambda)$ is $e^{-iu\lambda} g(u)$.

On account of (0.02), properties of convergence in the mean are conserved under a Fourier transformation, for these concern merely the integral of the square of the modulus of the difference of two functions, which is equal to the integral of the square of the difference of their Fourier transforms. Accordingly, Theorem I is a corollary of:

LEMMA Ia. *If $g(u)$ is a function belonging to L_2, then the set of functions $e^{-iu\lambda}g(u)$ is closed when and only when the zeros of $g(u)$ form a set of zero measure.*

Since every zero of $g(u)$ is also a zero of $e^{-iu\lambda}g(u)$, the necessity of this condition is at once obvious, for if

$$G_1(u) = \sum_1^n A_k e^{-iu\lambda_k} g(u) \tag{1.04}$$

is any linear combination of the functions $e^{-iu\lambda_k}g(u)$, then

$$\int_A^B |G_1(u) - 1|^2 \, du \tag{1.05}$$

must equal or exceed the measure of the set of the points between A and B where $g(u)$ vanishes. That is, if the measure of this set is not zero, the function which is 1 between A and B and zero elsewhere cannot be a limit in the mean of functions $G_1(u)$.

Now as to the sufficiency of the condition in Lemma Ia: let

$$\varphi(u) = \begin{cases} 1 - \dfrac{|u|}{A}, & [|u| \leq A]; \\ 0, & [|u| > A]. \end{cases} \tag{1.06}$$

Then

$$\varphi(u) = \frac{1}{\pi A} \int_{-\infty}^{\infty} e^{-iu\mu} \frac{1 - \cos A\mu}{\mu^2} \, d\mu. \tag{1.07}$$

This is an absolutely convergent integral, and hence a limit of a sequence of polynomials

$$\sum A_n e^{-iu\lambda_n} \tag{1.08}$$

converging boundedly, and uniformly over every finite interval. Hence

$$e^{-i\frac{un\pi}{A}} \varphi(u) g(u) \tag{1.09}$$

is a limit in the mean of such functions as those in (1.04).

Now let $\psi(u)$ be a quadratically summable function satisfying the conditions

$$\int_{-A}^{A} \psi(u)\varphi(u)g(u) e^{\frac{-iun\pi}{A}} \, du = 0. \tag{1.10}$$

$$[n = \cdots, -2, -1, 0, 1, 2, \cdots]$$

Since $\psi(u)\varphi(u)g(u)$ is absolutely summable,[22] we shall have

$$\psi(u)\varphi(u)g(u) = 0 \tag{1.11}$$

[22] This follows from the Parseval theorem for Fourier integrals.

except over a set of measure zero.[23] If $g(u)$ does not vanish except over a set of measure zero, since $\varphi(u)$ has no zeros, $\psi(u)$ vanishes over $(-A, A)$ except over a set of measure zero. Thus if $g(u) \neq 0$, there is no quadratically summable function not almost everywhere zero, orthogonal to every function $e^{\frac{-iun\pi}{A}} \varphi(u) g(u)$ over $(-A, A)$. This means that every function of L_2 vanishing everywhere outside $(-A, A)$ is a limit in the mean of polynomials[24] in $e^{\frac{-iun\pi}{A}} \varphi(u) g(u)$ and hence in $e^{-iu\lambda_n} g(u)$. Every function of L_2 is however the limit in the mean of a sequence of functions of L_2 vanishing for large arguments. Thus Lemma Ia and Theorem I are established.

2. **Closure in class L_1.** The Lebesgue class L_1 consists of all measurable functions $f(x)$ which are absolutely integrable over $(-\infty, \infty)$. If $f(x)$ is such a function, its Fourier transform

(2.01) $$g(u) = \frac{1}{\sqrt{2\pi}} \int_{-\infty}^{\infty} f(x) e^{iux} dx$$

of course exists, and is bounded and continuous. As before, the Fourier transform of $f(x+\lambda)$ is $e^{-iu\lambda} g(u)$.

We shall say that a class C of functions $\varphi_\lambda(x)$ of L_1 is closed L_1 if, whenever $F(x)$ is a function of L_1 and ϵ is a positive number, there is a polynomial

(2.02) $$F_1(x) = \sum_1^n A_k \varphi_{\lambda_k}(x)$$

of functions of C, such that

(2.03) $$\int_{-\infty}^{\infty} |F(x) - F_1(x)| \, dx \leqq \epsilon.$$

We shall prove the following theorem:

THEOREM II. *If $f(x)$ is a function of L_1, a necessary and sufficient condition for the set of all translations of $f(x)$ to be closed L_1 is that its Fourier transform*

(2.01) $$g(u) = \frac{1}{\sqrt{2\pi}} \int_{-\infty}^{\infty} f(x) e^{iux} dx$$

should have no real zeros.

[23] This follows from the Riemann theorem on the unicity of the function with a given Fourier series.

[24] Here we make use of the familiar theorem that if any linear class of functions of L_2 is given, every function of L_2 is the sum of a function expressible in the mean by the functions of the class and a function orthogonal to all functions of the class.

The necessity of this condition is again obvious. If $g(u_1) = 0$, the same is true for the argument u_1 of the Fourier transform of

$$F_1(x) = \sum_1^n A_k f(x+\lambda_k). \tag{1.02}$$

Let $f_1(x)$ be a function of L_1 with a Fourier transform $g_1(x)$ for which

$$g_1(u) \neq 0.$$

Then

$$\begin{aligned}\int_{-\infty}^{\infty} |F_1(x)-f_1(x)|\, dx &\geqq \left|\int_{-\infty}^{\infty} (f_1(x)-F_1(x))\, e^{iu_1 x}\, dx\right| \\ &= \sqrt{2\pi}\, |g_1(u_1)| > 0.\end{aligned} \tag{2.04}$$

This proves that f_1 is not a limit in the mean (L_1) of any sequence of F_1's.

To prove the sufficiency of the condition of Theorem II, we shall have recourse to a sequence of lemmas.

Lemma IIa. *If*

$$f_1(x) = \sum_{-\infty}^{\infty} a_n e^{inx}, \qquad f_2(x) = \sum_{-\infty}^{\infty} b_n e^{inx}, \tag{2.05}$$

and

$$\sum_{-\infty}^{\infty} |a_n| = A < \infty, \qquad \sum_{-\infty}^{\infty} |b_n| = B < \infty, \tag{2.06}$$

and if

$$f_1(x)\, f_2(x) \sim \sum_{-\infty}^{\infty} c_n e^{inx} \tag{2.07}$$

then

$$\sum_{-\infty}^{\infty} |c_n| \leqq AB. \tag{2.08}$$

The proof is immediate, for

$$\begin{aligned}\sum_{-\infty}^{\infty} |c_n| &= \sum_{n=-\infty}^{\infty} \left|\sum_{k=-\infty}^{\infty} a_k b_{n-k}\right| \leqq \sum_{n=-\infty}^{\infty} \sum_{k=-\infty}^{\infty} |a_k|\, |b_{n-k}| \\ &= \sum_{-\infty}^{\infty} |a_n| \sum_{-\infty}^{\infty} |b_n| = AB.\end{aligned} \tag{2.09}$$

Lemma IIb. *If $f(x)$ is a function of period 2π, and if at every point y, there is a function $f_y(x)$, coincident with $f(x)$ over some interval $(y-\varepsilon_y, y+\varepsilon_y)$, of period 2π, and such that its Fourier series converges absolutely, then the Fourier series of $f(x)$ converges absolutely*[25].

[25] We shall say that a Fourier series converges absolutely when the sum of the moduli of the coefficients converges, the series being represented in complex exponential form. This is of course equivalent to the absolute convergence of the series in this form for any single real argument.

To prove this, let us reflect that by the Heine-Borel theorem, any period of $f(x)$ may be covered by a finite number of overlapping intervals $(y-\varepsilon_y, y+\varepsilon_y)$. Let such an interval be (A, B), let its neighbor to the left have its right-hand end-point at C, and let its neighbor to the right have its left-hand end-point at D. There will be no loss in generality if we suppose that $A<C<D<B$. Let us define $\varphi_y(x)$ by

$$\varphi_y(x) = \begin{cases} 0\ ; & [-\pi \leqq x < A] \\ \dfrac{x-A}{c-A}; & [\ A \leqq x < C] \\ 1\ ; & [\ C \leqq x < D] \\ \dfrac{B-x}{B-D}; & [\ D \leqq x < B] \\ 0\ . & [\ B \leqq x < \pi]. \end{cases} \tag{2.10}$$

It is easy to show that the Fourier series of $\varphi_y(x)$ converges absolutely. Thus the Fourier series of

$$\varphi_y(x)\, f(x) = \varphi_y(x)\, f_y(x) \tag{2.11}$$

converges absolutely, by Lemma IIa. However, if we add together the functions $\varphi_y(x)$ for a set of intervals completely covering a period of $f(x)$, the sum will be $f(x)$ itself. Thus $f(x)$ will be the sum of a finite number of functions with absolutely convergent Fourier series and must itself have an absolutely convergent Fourier series.

LEMMA IIc. *If*

$$f(x) = \sum_{-\infty}^{\infty} a_n e^{inx} \tag{2.12}$$

and

$$|a_0| > \sum_{-\infty}^{-1} |a_n| + \sum_{1}^{\infty} |a_n|, \tag{2.13}$$

then the Fourier series of $1/f(x)$ *converges absolutely.*

We may write

$$\begin{aligned} \frac{1}{f(x)} &= \frac{1}{a_0 + \sum_{-\infty}^{-1} a_n e^{inx} + \sum_{1}^{\infty} a_n e^{inx}} \\ &= \frac{1}{a_0}\left\{1 - \frac{1}{a_0}\left[\sum_{-\infty}^{-1} a_n e^{inx} + \sum_{1}^{\infty} a_n e^{inx}\right]\right. \\ &\qquad \left. + \frac{1}{a_0^2}\left[\sum_{-\infty}^{-1} a_n e^{inx} + \sum_{1}^{\infty} a_n e^{inx}\right]^2 - \cdots\right\}. \end{aligned} \tag{2.14}$$

That this is more than formally true, and that $1/f(x)$ is continuous, results from the fact that the geometrical progression

$$(2.15)\quad \frac{1}{|a_0|}\left\{1+\frac{1}{|a_0|}\left[\sum_{-\infty}^{-1}|a_n|+\sum_{1}^{\infty}|a_n|\right]+\frac{1}{a_0^2}\left[\sum_{-\infty}^{-1}|a_n|+\sum_{1}^{\infty}|a_n|\right]^2+\cdots\right\}$$

converges. Term by term, series (2.15) is greater than or equal to the sum of the sums of the moduli of the coefficients of the Fourier series of the successive terms of (2.14), and hence is greater than or equal to the sum of the moduli of the coefficients of the Fourier series of $1/f(x)$.

LEMMA IId. *Let*

$$(2.16)\qquad \sum_{-\infty}^{\infty}|a_n| = A < \infty,$$

and let

$$(2.17)\qquad f(x) = \sum_{-\infty}^{\infty} a_n e^{inx}.$$

Let

$$f(y) \neq 0.$$

Then there is a neighborhood $(y-\varepsilon,\ y+\varepsilon)$ *of* y *and a function* $f_y(x)$ *with absolutely convergent Fourier series such that over* $(y-\varepsilon,\ y+\varepsilon)$,

$$(2.18)\qquad f_y(x) = f(x)$$

and that the Fourier series of $f_y(x)$, *let us say*

$$(2.19)\qquad \sum_{-\infty}^{\infty} c_n e^{inx}$$

has the property that

$$(2.20)\qquad |c_0| > \sum_{-\infty}^{-1}|c_n| + \sum_{1}^{\infty}|c_n|.$$

There is manifestly no restriction in taking y to be 0. Let us introduce the auxiliary function

$$(2.201)\qquad \varphi_\varepsilon(x) = \begin{cases} 1 & ;\ [|x|<\varepsilon] \\ 2-\dfrac{|x|}{\varepsilon} & ;\ [\varepsilon \leqq |x| < 2\varepsilon] \\ 0 & .\ [2\varepsilon \leqq |x|]. \end{cases}$$

This will have the Fourier series

$$(2.202)\qquad \frac{3\varepsilon}{2\pi}+\sum_{1}^{\infty}\frac{e^{inx}+e^{-inx}}{\pi n^2 \varepsilon}(\cos \varepsilon n - \cos 2\varepsilon n).$$

If we represent the function $\{f(x)\,\varphi_\varepsilon(x)+f(0)\,[1-\varphi_\varepsilon(x)]\}$ by the Fourier series

$$\sum_{-\infty}^{\infty} c_n e^{inx},$$

we shall have

$$c_n = \frac{3\varepsilon}{2\pi} a_n + \sum_{k=1}^{\infty} \frac{a_{n-k} + a_{n+k}}{\pi k^2 \varepsilon} (\cos \varepsilon k - \cos 2\varepsilon k) - \sum_{-\infty}^{\infty} a_k \frac{\cos \varepsilon n - \cos 2\varepsilon n}{\pi n^2 \varepsilon}; \qquad [n \neq 0],$$

(2.203)

$$c_0 = a_0 + \sum_{1}^{\infty} (a_{-k} + a_k) \left[1 + \frac{\cos \varepsilon k - \cos 2\varepsilon k}{\pi k^2 \varepsilon} - \frac{3\varepsilon}{2\pi} \right].$$

Hence

(2.204) $$\lim_{\varepsilon \to 0} |c_0| = \left| \sum_{-\infty}^{\infty} a_k \right| = |f(0)| > 0,$$

and

$$\sum_{1}^{\infty} [|c_n| + |c_{-n}|]$$

(2.205)

$$\leq \sum_{m=-\infty}^{\infty} |a_m| \left\{ \sum_{n=-\infty}^{\infty}{}'' \left| \frac{\cos \varepsilon n - \cos 2\varepsilon n}{\pi n^2 \varepsilon} - \frac{\cos \varepsilon(m-n) - \cos 2\varepsilon(m-n)}{\pi (m-n)^2 \varepsilon} \right| + \left| \frac{\cos \varepsilon m - \cos 2\varepsilon m}{\pi m^2 \varepsilon} - \frac{3\varepsilon}{2\pi} \right| \right\}$$

$$= \sum_{-\infty}^{\infty} |a_m| A_m.$$

Here $\sum''$ means that the values 0 and m are omitted.

We have

(2.206) $$\left| \frac{\cos \varepsilon n - \cos 2\varepsilon n)}{\pi n^2 \varepsilon} \right| = \left| \frac{2 \sin \frac{\varepsilon n}{2} \sin \frac{3\varepsilon n}{2}}{\pi n^2 \varepsilon} \right| \leq \begin{cases} \frac{3\varepsilon}{2\pi}; \\ \frac{2}{\pi n^2 \varepsilon}; \end{cases}$$

and

$$\left| \frac{\cos \varepsilon n - \cos 2\varepsilon n}{\pi n^2 \varepsilon} - \frac{\cos \varepsilon (m-n) - \cos 2\varepsilon (m-n)}{\pi (m-n)^2 \varepsilon} \right|$$

(2.21) $$= \left| \int_{n-m}^{n} \frac{d}{dx} \left(\frac{\cos \varepsilon x - \cos 2\varepsilon x}{\pi x^2 \varepsilon} \right) dx \right|$$

$$\leq \frac{\varepsilon^2}{\pi} \left| \int_{n-m}^{n} \right| \max_y \left| \frac{d}{dy} \frac{\cos y - \cos 2y}{y^2} \right| dx \Big| \leq |m| \varepsilon^2 c.$$

This formula may be applied in the limiting form whenever $m = n$ or $m = 0$. It results from (2.205) and (2.206) that for all sufficiently small ε

(2.22) $$A_m \leq 4 \left[\frac{1}{\varepsilon} + 1 \right] \frac{3\varepsilon}{2\pi} + \frac{8}{\pi\varepsilon} \sum_{[1/\varepsilon]}^{\infty} n^{-2} < \frac{15}{\pi},$$

and from (2.205) and (2.21) that for all sufficiently small ε

$$A_m \leqq 2[\varepsilon^{-3/2}+1]\,|\,m\,|\,\varepsilon^2 c + \frac{8}{\pi\varepsilon}\sum_{[\varepsilon^{-3/2}]}^{\infty} n^{-2} \tag{2.23}$$
$$< \varepsilon^{1/2}[2\,|\,m\,|\,c+9/\pi].$$

Combining (2.205), (2.22) and (2.23), we get

$$\lim_{\varepsilon\to 0}\sum_{-\infty}^{\infty}|\,a_m\,|\,A_m = 0, \tag{2.24}$$

which with (2.204) yields us

$$\sum_{1}^{\infty}[|\,c_n\,|+|\,c_{-n}\,|] < |\,c_0\,|. \tag{2.25}$$

If we take

$$f_y(x) = f(x)\,\varphi_\varepsilon(x) + f(0)\,(1-\varphi_\varepsilon(x)), \tag{2.26}$$

Lemma IId is established.

As a direct consequence of Lemmas IIb, IIc, and IId we obtain:

LEMMA IIe. *If $f(x)$ is a function with an absolutely convergent Fourier series, which nowhere vanishes for real arguments, $1/f(x)$ has an absolutely convergent Fourier series.*

To prove this let us note that, in every neighborhood, by IId, $f(x)$ coincides with a function which by IIc has a reciprocal with an absolutely convergent Fourier series. By IIa, since in every neighborhood $1/f(x)$ coincides with a function with an absolutely convergent Fourier series, its Fourier series converges absolutely.

LEMMA IIf. *Let $f(x)$ be a function with an absolutely convergent Fourier series over the interval $(-\pi, \pi)$, and vanishing over neighborhoods including π and $-\pi$. Let $f(x)$ vanish everywhere outside $(-\pi, \pi)$. Let*

$$g(u) = \frac{1}{\sqrt{2\pi}}\int_{-\pi}^{\pi} f(x)\,e^{iux}\,dx. \tag{2.27}$$

Then the integral

$$\int_{-\infty}^{\infty}|\,g(u)\,|\,du \tag{2.28}$$

will converge. Conversely, if $f(x)$ vanishes over the region indicated, and (2.28) converges, its Fourier series will converge absolutely.

Let ε be such that $f(x)$ has no non-zero, values outside $(-\pi+2\varepsilon, \pi-2\varepsilon)$. Let us define $\varphi(x)$ by:

$$\varphi(x) = \begin{cases} 1 & ; \ [|x| < \pi - \varepsilon] \\ \dfrac{\pi - |x|}{\varepsilon} & ; \ [\pi - \varepsilon \leq |x| < \pi] \\ 0 & . \ [\pi \leq |x|] \end{cases} \tag{2.29}$$

Let

$$f(x) = \sum_{-\infty}^{\infty} a_n e^{inx}. \tag{2.17}$$

We shall have

$$\varphi(x) = \frac{1}{\pi} \int_{-\infty}^{\infty} e^{-iux} \frac{\cos u(\pi - \varepsilon) - \cos u\pi}{u^2 \varepsilon} du, \tag{2.30}$$

and hence

$$\begin{aligned} f(x) = f(x)\varphi(x) &= \frac{1}{\pi} \int_{-\infty}^{\infty} du \sum_{-\infty}^{\infty} a_n e^{-i(u-n)x} \frac{\cos u(\pi - \varepsilon) - \cos u\pi}{u^2 \varepsilon} \\ &= \frac{1}{\pi} \int_{-\infty}^{\infty} e^{iux} \left[\sum_{-\infty}^{\infty} a_n \frac{\cos(u+n)(\pi-\varepsilon) - \cos(u+n)\pi}{(u+n)^2 \varepsilon} \right] du, \end{aligned} \tag{2.31}$$

where the rearrangement is possible since both integral and sum form part of an absolutely convergent double summation process. Then

$$g(u) = \sqrt{\frac{2}{\pi}} \sum_{-\infty}^{\infty} a_n \frac{\cos(u+n)(\pi-\varepsilon) - \cos(u+n)\pi}{(u+n)^2 \varepsilon} \tag{2.32}$$

will satisfy the condition that (2.28) shall be finite, and equation (2.27).

To prove the converse part of Lemma IIf, let us notice that as a result of the convergence of (2.28),

$$\sum_{-\infty}^{\infty} \left| \int_{n-1/2}^{n+1/2} g(u)\, du \right| \tag{2.33}$$

converges. Now,

$$\begin{aligned} \int_{n-1/2}^{n+1/2} g(u)\, du &= \frac{1}{\sqrt{2\pi}} \int_{-\pi}^{\pi} f(x)\, dx \int_{n-1/2}^{n+1/2} e^{iux}\, du \\ &= \frac{1}{\sqrt{2\pi}} \int_{-\pi}^{\pi} f(x) \frac{2 \sin x/2}{x} e^{inx}\, dx. \end{aligned} \tag{2.34}$$

Thus the Fourier series of

$$f(x) \frac{2 \sin x/2}{x} \tag{2.35}$$

converges absolutely. Since the same is true of the Fourier series of

$$\frac{x}{2 \sin x/2} \varphi(x) \tag{2.36}$$

as may be determined by a simple computation, it follows from Lemma IIa that the same is true for the Fourier series of $f(x)$.

LEMMA IIg. *Let $f(x)$ be a function of class L_1. Then*

$$\lim_{\varepsilon\to 0}\int_{-\infty}^{\infty}|f(x+\varepsilon)-f(x)|\,dx = 0. \tag{2.37}$$

This is a well-known theorem and is proved in Hobson's *Theory of Functions of a Real Variable.*

LEMMA IIh. *Let $f(x)$ be a function of class L_1. Then if $\Phi(y)$ is any function of class L_1, differing from 0 only over $(-\varepsilon, \varepsilon)$,*

$$\begin{aligned}\int_{-\infty}^{\infty}\left|f(x)\int_{-\infty}^{\infty}\Phi(y)\,dy-\int_{-\infty}^{\infty}f(x+y)\,\Phi(y)\,dy\right|dx \\ \leq \int_{-\infty}^{\infty}|\Phi(y)|\,dy\,\max_{|z|\leq\varepsilon}\int_{-\infty}^{\infty}|f(x+z)-f(x)|\,dx.\end{aligned} \tag{2.38}$$

The proof of this lemma merely depends on the inversion of the order of integration in an absolutely convergent double integral.

LEMMA IIi. *Let $f(x)$ be a function of class L_1. Then*

$$\lim_{n\to\infty}\int_{-\infty}^{\infty}\left|f(x)-\frac{1}{\pi n}\int_{-\infty}^{\infty}f(x+y)\frac{\sin^2 ny}{y^2}\,dy\right|dx = 0. \tag{2.39}$$

This is a theorem of the Fejér type. It depends on the splitting of

$$\Phi_0(y) = \frac{\sin^2 ny}{y^2}$$

into the sum $\Phi_1(y)+\Phi_2(y)$ where

$$\Phi_1(y) = \begin{cases}\Phi_0(y)\,(1-|y|\,n^{1/2}); & [|y|\leq n^{-1/2}]\\ 0; & [|y|>n^{-1/2}].\end{cases} \tag{2.40}$$

Lemmas IIg and IIh enable us to show that

$$\lim_{n\to\infty}\int_{-\infty}^{\infty}\left|f(x)-\frac{1}{\pi n}\int_{-\infty}^{\infty}f(x+y)\,\Phi_1(y)\,dy\right|dx = 0 \tag{2.41}$$

while

$$\begin{aligned}\lim_{n\to\infty}\int_{-\infty}^{\infty}\left|\frac{1}{\pi n}\int_{-\infty}^{\infty}f(x+y)\,\Phi_2(y)\,dy\right|dx \\ \leq \int_{-\infty}^{\infty}|f(x)|\,dx\,\lim_{n\to\infty}\frac{1}{\pi n}\int_{-\infty}^{\infty}\Phi_2(y)\,dy.\end{aligned} \tag{2.42}$$

Now

$$\frac{1}{\pi n}\int_{-\infty}^{\infty} \Phi_2(y)\,dy$$

$$= \frac{1}{\pi n}\int_{-\infty}^{\infty} \frac{\sin^2 ny}{y^2}\,dy - \frac{1}{\pi n}\int_{-n^{-1/2}}^{n^{-1/2}} (1-|y|\,n^{1/2})\,\frac{\sin^2 ny}{y^2}\,dy$$

(2.421)
$$= \frac{1}{\pi}\int_{-\infty}^{\infty} \frac{\sin^2 z}{z^2}\,dz - \frac{1}{\pi}\int_{-n^{1/2}}^{n^{1/2}} (1-|z|\,n^{-1/2})\,\frac{\sin^2 z}{z^2}\,dz$$

$$= \frac{2}{\pi}\int_{n^{1/2}}^{\infty} \frac{\sin^2 z}{z^2}\,dz + \frac{2}{\pi n^{1/2}}\int_{0}^{n^{1/2}} \frac{\sin^2 z}{z}\,dz$$

$$= O(n^{-1/2}\log n).$$

Hence

(2.422)
$$\lim_{n\to\infty}\int_{-\infty}^{\infty}\left|\frac{1}{\pi n}\int_{-\infty}^{\infty} f(x+y)\,\Phi_2(y)\,dy\right| dx = 0,$$

and Lemma IIi is established.

We are now in a position to proceed to the proof of Theorem II. Let $F(x)$ be any function of L_1. By Lemma IIi, we can find a function $F_\eta(x)$ of the form

(2.43)
$$\frac{1}{\pi n}\int_{-\infty}^{\infty} F(x+y)\,\frac{\sin^2 ny}{y^2}\,dy,$$

such that

(2.44)
$$\int_{-\infty}^{\infty} |F(x)-F_\eta(x)|\,dx < \eta.$$

The Fourier transform $H_1(u)$ of this function $F_\eta(x)$ is

$$\frac{1}{\sqrt{2\pi}}\int_{-\infty}^{\infty} e^{iux}\,dx\,\frac{1}{\pi n}\int_{-\infty}^{\infty} F(x+y)\,\frac{\sin^2 ny}{y^2}\,dy$$

(2.45)
$$= \frac{1}{\sqrt{2\pi}}\int_{-\infty}^{\infty} F(x)\,e^{iux}\,dx\,\frac{1}{\pi n}\int_{-\infty}^{\infty}\frac{\sin^2 ny}{y^2}\,e^{-iuy}\,dy$$

$$= \begin{cases} \left(1-\dfrac{|u|}{2n}\right)\dfrac{1}{\sqrt{2\pi}}\displaystyle\int_{-\infty}^{\infty} F(x)\,e^{iux}\,dx; & [|u|<2n] \\ 0 \qquad ; & [|u|\geqq 2n] \end{cases}$$

and will vanish for all values of its argument u larger than $2n$ in modulus. The same will be the case with the Fourier transform $H_2(u)$ of

(2.46)
$$\frac{1}{2\pi n}\int_{-\infty}^{\infty} f(x+y)\,\frac{\sin^2 2ny}{y^2}\,dy$$

2

which will vanish for values of its argument greater in modulus than $4n$, and only for such values. If we expand these Fourier transforms in Fourier series over $(-8n, 8n)$, by Lemma IIf, these series converge absolutely.

By an argument substantially identical with that used in proving Lemma IIb, we see that in the neighborhood of every point of $(-3n, 3n)$, there is a function coinciding locally with $1/H_2(u)$, and with an absolutely convergent Fourier series. It follows from Lemma IIa that the same is true of $H_1(u)/H_2(u)$ over the same interval. Since 0 enjoys this property also, the function which is $H_1(u)/H_2(u)$ over $(-3n, 3n)$ and 0 over the rest of a period $(-8n, 8n)$ has an absolutely convergent Fourier series, and hence, by Lemma IIf, an absolutely convergent Fourier integral. That is, we may write

$$H_1(u) = H_2(u)\, H_3(u) \tag{2.47}$$

where $H_3(u)$ is of the form

$$H_3(u) = \int_{-\infty}^{\infty} \Phi(x)\, e^{iux}\, dx \tag{2.48}$$

and

$$\int_{-\infty}^{\infty} |\Phi(x)|\, dx \tag{2.49}$$

converges. We shall then have

$$\begin{aligned}
&\frac{1}{\sqrt{2\pi}} \int_{-\infty}^{\infty} F_\eta(x)\, e^{iux}\, dx \\
&\quad = \frac{1}{\sqrt{2\pi}} \int_{-\infty}^{\infty} \frac{e^{iux}}{2\pi n}\, dx \int_{-\infty}^{\infty} f(x+y) \frac{\sin^2 2ny}{y^2}\, dy \int_{-\infty}^{\infty} \Phi(z)\, e^{iuz}\, dz \\
&\quad = \frac{1}{\sqrt{2\pi}} \int_{-\infty}^{\infty} e^{iux}\, dx \\
&\qquad \times \left[\frac{1}{2\pi n} \int_{-\infty}^{\infty} f(x+\omega)\, d\omega \int_{-\infty}^{\infty} \frac{\sin^2 2ny}{y^2}\, \Phi(y-\omega)\, dy \right],
\end{aligned} \tag{2.50}$$

or

$$\begin{aligned}
&\int_{-\infty}^{\infty} e^{iux}\, dx \\
&\quad \times \left[F_\eta(x) - \frac{1}{2\pi n} \int_{-\infty}^{\infty} f(x+\omega)\, d\omega \int_{-\infty}^{\infty} \frac{\sin^2 2ny}{y^2}\, \Phi(y-\omega)\, dy \right] = 0.
\end{aligned} \tag{2.51}$$

Here we may replace e^{iux} by any function with an absolutely convergent Fourier integral, and hence by functions positive over an arbitrarily small region and zero everywhere else. This can only be if

$$F_\eta(x) = \frac{1}{2\pi n}\int_{-\infty}^{\infty} f(x+\omega)\,d\omega \int_{-\infty}^{\infty} \frac{\sin^2 2ny}{y^2}\,\Phi(y-\omega)\,dy \tag{2.52}$$
$$= \int_{-\infty}^{\infty} f(x+\omega)\,\Psi(\omega)\,d(\omega).$$

Here $\Psi(\omega)$ is absolutely integrable.

By Lemma IIh and Lemma IIg,

$$\lim_{n\to\infty}\int_{-\infty}^{\infty}\left|\int_{-\infty}^{\infty} f(x+\omega)\,\Psi(\omega)\,d\omega - \sum_{k=-n^2}^{n^2-1} f\left(x+\frac{k}{n}\right)\int_{\frac{k}{n}}^{\frac{k+1}{n}} \Psi(\omega)\,d\omega\right| dx = 0. \tag{2.53}$$

Combining this with (2.52) and Lemma IIi, we obtain Theorem II.

With only a slight modification of detail, we have proved:

THEOREM III. *Let $f_1(x)$ and $f_2(x)$ be two functions of L_1. Let*

$$g_1(u) = \frac{1}{\sqrt{2\pi}}\int_{-\infty}^{\infty} f_1(x)\,e^{iux}\,dx \tag{2.54}$$

and

$$g_2(u) = \frac{1}{\sqrt{2\pi}}\int_{-\infty}^{\infty} f_2(x)\,e^{iux}\,dx. \tag{2.55}$$

Let the set of points where $g_2(u) \neq 0$ consist only of inner points of the set where $g_1(u) \neq 0$. Then if $\epsilon > 0$ there is a polynomial

$$f_3(x) = \sum_1^N a_n f_1(x+\lambda_n) \tag{2.56}$$

such that

$$\int_{-\infty}^{\infty} |f_2(x) - f_3(x)|\,dx < \epsilon. \tag{2.57}$$

On the basis of Theorem III, we are in a position to prove the general:

THEOREM IV. *Let Σ be a class of functions of L_1. Then a necessary and sufficient condition for the class Σ_1, containing all functions $f(x+\lambda)$ whenever $f(x)$ belongs to Σ, to be closed L_1, is that there should be no real zero common to all the Fourier transforms of functions of Σ.*

The proof of the necessity of the condition of Theorem IV does not differ from the similar proof in the case of Theorem II. As to the sufficiency, it follows at once from Theorem III and the fact that the Fourier transform of a function of L_1 differs from 0 on an open set, that for every u, an ϵ may be assigned, such that

$$e^{iux}\frac{\sin^2 \epsilon x}{x^2} \tag{2.58}$$

is a function approximable L_1 by polynomials in functions of Σ_1. Let $(-U, U)$ be any interval of frequencies. By the Heine-Borel theorem, this may be overlaid with a finite number of overlapping intervals $U_n - \varepsilon_n$, $U_n + \varepsilon_n$. Let us form

$$\sum e^{iU_n x} \frac{\sin^2 \varepsilon_n x}{x^2} \tag{2.59}$$

for this set of intervals. This will be a function of L_1, approximable L_1 by polynomials in the function of Σ_1, whose Fourier transform may be shown to be

$$\sum \frac{\varepsilon_n}{4}\left(1-2\left|\frac{U_n+u}{\varepsilon_n}\right|\right)\left[\operatorname{sgn}\left(1-2\left|\frac{U_n+u}{\varepsilon_n}\right|\right)+1\right] \tag{2.60}$$

and to have no zeros over $(-U, U)$. Thus by Theorem III, every function of L_1 with a Fourier transform vanishing outside $(-U+n, U-n)$ will be approximable L_1 by polynomials in the functions of Σ_1. At this stage, the introduction of Lemma IIi serves to complete the proof of Theorem IV.

In Theorems II and IV, the necessary part is nearly trivial and all the difficulty resides in the proof of sufficiency. There are certain applications, however, where it is precisely the necessary part of the theorem that is significant. We may prove a function to have no zeros if it is the Fourier transform of a function of L_1 the set of whose translations is closed L_1. To establish this closure is indeed more than is needed: it is enough to produce for each u among the class of functions to which we may approximate L_1 by polynomials in the translations of our given functions, at least one function whose Fourier transform does not vanish for the argument u.

It may be shown if $F(x)$ is a non-negative function for which

$$\int_{-\infty}^{\infty} F(x)\,dx > A \tag{2.61}$$

and

$$\left[\int_B^{\infty} + \int_{-\infty}^{-B}\right] F(x)\,dx < \varepsilon A, \tag{2.62}$$

then

$$\left|\int_{-\infty}^{\infty} F(x)\,e^{iux}\,dx\right| > \left(\frac{1}{2}-\frac{3\varepsilon}{2}\right)A \qquad \left[|u| \leqq \frac{\pi}{3B}\right]. \tag{2.63}$$

This results from the fact that for $-B \leqq \lambda \leqq B$,

$$\Re\,(e^{iux}) = \cos ux \geqq \cos\frac{\pi}{3} = \frac{1}{2}.$$

Accordingly, if for every value of B and ε, we may find a function of this sort which may be approximated L_1 by polynomials in the translations of $f(x)$, the Fourier transform of $f(x)$ has no zeros. As in the case of Theorem IV, the function $f(x)$ may be replaced by the class Σ, and the class of translations of $f(x)$ by Σ_1. Then the condition that the Fourier transform of $f(x)$ shall have no zeros is to be replaced by the condition that there shall be no zero common to all the Fourier transforms of functions of Σ.

This method is a conceivable one for the investigation of the zeros of functions with known Fourier transforms, and might be applied to the study of the zeros of the Riemann zeta function. So far it has yielded no results. Under certain conditions, the Taylor series may be introduced to reduce the question of the closure of the set of translations of $f(x)$ to that of the closure of the set of functions

$$f^{(n)}(x). \tag{2.65}$$

This matter has not yet, however, been subjected to an adequate investigation.

3. A sub-class of L_1. In connection with Stieltjes distributions of mass, we shall have to consider "kernels" of class L_1, fulfilling a certain more restrictive condition. This condition on a measurable function $f(x)$ is that the series

$$\sum_{k=-\infty}^{\infty} \overline{\lim_{kA+B\leq x\leq(k+1)A+B}} |f(x)| \tag{3.01}$$

shall converge. We may easily verify that this condition is in fact independent of A and B, and that it can only be fulfilled by functions of L_1. With it goes a certain definition of the "distance" between two functions $f(x)$ and $g(x)$, this "distance" being defined as

$$(f(x), g(x)) = \sum_{-\infty}^{\infty} \overline{\lim_{k\leq x\leq k+1}} |f(x)-g(x)|. \tag{3.02}$$

This notion of "distance" has a very important difference from

$$\int_{-\infty}^{\infty} |f(x)-g(x)|^2 dx \tag{3.03}$$

and

$$\int_{-\infty}^{\infty} |f(x)-g(x)| dx \tag{3.04}$$

in that it is not true that if (3.01) is finite, then

$$\lim_{\varepsilon\to 0} \sum_{k=-\infty}^{\infty} \overline{\lim_{k\leq x\leq k+1}} |f(x+\varepsilon)-f(x)| = 0. \tag{3.05}$$

Accordingly, if we are to prove theorems analogous to those in the last paragraph without a radical change in method of proof, we must introduce a restriction which will make (3.05) hold. Otherwise we shall be unable to establish the analogue of Lemma IIg.

A condition of the desired sort is that $f(x)$ shall be continuous. It will then be uniformly continuous over any finite interval, and we shall be able, first to choose A so large that

$$\sum_{k=0}^{\infty}\left\{\max_{A+k\leq x\leq A+k+1}|f(x)|+\max_{-A-k-1\leq x\leq -A-k}|f(x)|\right\}<\varepsilon \tag{3.06}$$

and then to choose η so small that for $-A-1\leqq x\leqq A+1,\ 0<\eta_1<\eta$,

$$\max|f(x+\eta_1)-f(x)|<\frac{\varepsilon}{2A+2}. \tag{3.07}$$

It will follow that for $0<\eta_1<\eta$,

$$\sum_{k=-\infty}^{\infty}\max_{k\leq x\leq k+1}|f(x+\varepsilon)-f(x)|<3\varepsilon. \tag{3.08}$$

We shall call the class of all continuous functions for which (3.01) is finite, the class M_1, and we shall say that a class Σ of functions of M_1 is *closed* M_1, if whenever $F(x)$ belongs to M_1 and $\varepsilon>0$, there is a polynomial

$$F_1(x)=\sum_{h=1}^{n}A_h f_h(x) \tag{3.09}$$

in functions of the class Σ such that

$$\sum_{k=-\infty}^{\infty}\max_{k\leq x\leq k+1}|F(x)-F_1(x)|<\varepsilon. \tag{3.10}$$

The following theorem is valid:

THEOREM V. *Let $f(x)$ belong to L_1. Then a necessary and sufficient condition that whenever $f_1(x)$ is a function of M_1 and $\varepsilon>0$, there shall exist a function $\alpha(x)$ of M_1 such that*

$$\sum_{k=-\infty}^{\infty}\max_{k\leq x\leq k+1}\left|f_1(x)-\int_{-\infty}^{\infty}f(y)\,\alpha(x-y)\,dy\right|<\varepsilon \tag{3.11}$$

is that for no real u

$$\frac{1}{\sqrt{2\pi}}\int_{-\infty}^{\infty}f(x)\,e^{iux}\,dx=0. \tag{3.12}$$

The proof differs in no essential respect from that of Theorem II. Let us note that if $f_2(x)$ belongs to L_1 and

$$\frac{1}{\sqrt{2\pi}} \int_{-\infty}^{\infty} f_2(x) e^{iux} dx = 0, \qquad [|u| > A] \tag{3.13}$$

then $f_2(x)$ belongs to M_1. To see this, let us put

$$\frac{1}{\sqrt{2\pi}} \int_{-\infty}^{\infty} f_2(x) e^{-iux} dx = g_2(u); \tag{3.14}$$

$g_2(u)$ is bounded and differs from 0 only over a bounded range, and hence belongs to L_2. Thus

$$\underset{B\to\infty}{\text{l.i.m.}} \frac{1}{\sqrt{2\pi}} \int_{-B}^{B} g_2(u) e^{iux} du = f_3(x) \tag{3.15}$$

exists as a function of L_2. Further

$$g_2(u) = \underset{B\to\infty}{\text{l.i.m.}} \frac{1}{\sqrt{2\pi}} \int_{-B}^{B} f_3(x) e^{-iux} dx \tag{3.16}$$

exists and belongs to L_2. If we put

$$F(x) = \int_{-\infty}^{\infty} G(u)\, e^{iux} du \tag{3.17}$$

where G belongs both to L_2 and L_1, it follows readily from (3.14) and (3.16) that

$$\int_{-\infty}^{\infty} F(x)\, [f_2(x) - f_3(x)]\, dx = 0. \tag{3.18}$$

We can choose a set of F's vanishing outside a given finite interval and closed L_2 and hence L_1 over this interval. Then over this interval, which is arbitrary, f_2 and f_3 can differ at most on a null set. Thus f_2 belongs to L_2.

Now

$$g_2(u) = g_2(u)\, W(u) \tag{3.19}$$

where

$$W(u) = \begin{cases} 0 \;; & [|u| > 2A] \\ 2 - \dfrac{|u|}{A} \;; & [A < |u| < 2A] \\ 1 \;. & [|u| < A] \end{cases} \tag{3.20}$$

Thus by the Parseval theorem,

$$\begin{aligned} f_2(x) &= \frac{1}{2\pi} \int_{-\infty}^{\infty} f_2(y)\, dy \int_{-\infty}^{\infty} W(u)\, e^{iu(x-y)} du \\ &= \frac{1}{\pi A} \int_{-\infty}^{\infty} f_2(y) \frac{\cos A(x-y) - \cos 2A(x-y)}{(x-y)^2} dy. \end{aligned} \tag{3.21}$$

From this it follows that

$$(3.22)\quad \begin{aligned} \sum_{k=-\infty}^{\infty} \max_{k\leq x\leq k+1} |f_2(x)| &\leqq \frac{1}{\pi A}\int_{-\infty}^{\infty} |f_2(y)|\, dy \sum_{-\infty}^{\infty} O\left(\frac{1}{k-m}\right)^2 \\ &\leqq O(1)\int_{-\infty}^{\infty} |f_2(y)|\, dy \end{aligned}$$

and that $f_2(x)$, which obviously can be modified so as to be continuous, belongs to M_1.

It only remains to prove the analogue of Lemma IIi, and to show that every function of class M_1 may be approximated with any degree of accuracy in the M_1 sense by functions with Fourier transforms vanishing for large arguments. This follows exactly the lines suggested by IIi. We wish to show, that is, that if $f(x)$ belongs to M_1,

$$(3.221)\qquad \lim_{N\to\infty}\left(f(x),\, \frac{1}{\pi N}\int_{-\infty}^{\infty} f(x+y)\frac{\sin^2 Ny}{y^2}\, dy\right) \equiv 0.$$

As before, we put

$$(3.222)\qquad \Phi_1(y) = \begin{cases} \dfrac{\sin^2 Ny}{y^2}(1-|y|\, N^{1/2}); & [|y| \leqq N^{-1/2}] \\ 0 & [y > N^{-1/2}] \end{cases}$$

$$(3.223)\qquad \Phi_2(y) = \frac{\sin^2 Ny}{y^2} - \Phi_1(y).$$

The proof that

$$(3.224)\qquad \lim_{N\to\infty}\left(f(x),\, \frac{1}{\pi N}\int_{-\infty}^{\infty} f(x+y)\, \Phi_1(y)\, dx\right) = 0$$

follows exactly the lines of that of (2.41), for the analogue of Lemma IIg has already been shown to be true, while that of Lemma IIh is proved by the same arguments of absolute convergence as the lemma itself. Again,

$$(3.225)\quad \begin{aligned} &\lim_{N\to\infty} \sum_{k=-\infty}^{\infty} \max_{k\leq x\leq k+1} \left|\frac{1}{\pi N}\int_{-\infty}^{\infty} f(x+y)\, \Phi_2(y)\, dy\right| \\ &\leqq \sum_{k=-\infty}^{\infty} \max_{k\leq x\leq k+1} |f(x+y)| \lim_{N\to\infty} \frac{1}{\pi N}\int_{-\infty}^{\infty} \Phi_2(y)\, dy \\ &\leqq 2\sum_{k=-\infty}^{\infty} \max_{k\leq x\leq k+1} |f(x)| \lim_{N\to\infty} \frac{1}{\pi N}\int_{-\infty}^{\infty} \Phi_2(y)\, dy, \end{aligned}$$

and this, by (2.421), is zero. This completes the proof of (3.221).

An extension of Theorem V in the direction of Theorem IV is the following:

THEOREM VI. *Let Σ be a class of functions of L_1. Let Σ_1 be the class of all functions of the form*

$$\sum_1^N \int_{-\infty}^{\infty} f_n(y)\, \alpha_n(x+y)\, dy \tag{3.23}$$

where $f_1, \cdots, f_N$ belong to $\sum$ and $\alpha_1, \cdots, \alpha_N$ to M_1. Then Σ_1 is closed M_1 when and only when there is no real value of u which is a zero common to all the functions

$$\frac{1}{\sqrt{2\pi}} \int_{-\infty}^{\infty} f_n(x)\, e^{iux}\, dx \tag{3.24}$$

corresponding to functions $f_n(x)$ of class Σ.

As an easy corollary of Theorem V, using (3.10), we have:

THEOREM VII. *If $f(x)$ is a function of M_1, a necessary and sufficient condition for the set of its translations to be closed M_1 is that for no u should we have*

$$\frac{1}{\sqrt{2\pi}} \int_{-\infty}^{\infty} f(x)\, e^{iux}\, dx = 0. \tag{3.25}$$

CHAPTER II.

ASYMPTOTIC PROPERTIES OF AVERAGES[26].

4. Averages of bounded functions. Our fundamental theorem is:

THEOREM VIII. *Let $f(x)$ be a bounded measurable function, defined over $(-\infty, \infty)$. Let $K_1(x)$ be a function in L_1, and let*

$$\frac{1}{\sqrt{2\pi}} \int_{-\infty}^{\infty} K_1(x)\, e^{iux}\, dx \neq 0 \tag{4.01}$$

for every real u. Let

$$\lim_{x\to\infty} \int_{-\infty}^{\infty} f(\xi)\, K_1(\xi - x)\, d\xi = A \int_{-\infty}^{\infty} K_1(\xi)\, d\xi. \tag{4.02}$$

Then if $K_2(x)$ is any function in L_1,

$$\lim_{x\to\infty} \int_{-\infty}^{\infty} f(\xi)\, K_2(\xi - x)\, d\xi = A \int_{-\infty}^{\infty} K_2(\xi)\, d\xi. \tag{4.03}$$

Conversely, let $K_1(\xi)$ be a function of L_1, and let $\int_{-\infty}^{\infty} K_1(\xi)\, d\xi \neq 0$. Let (4.02) *imply* (4.03) *whenever $K_2(x)$ belongs to L_1 and $f(x)$ is bounded. Then* (4.01) *holds.*

[26] The emphasis here placed on averages is in the same order of ideas as was first introduced into the theory by Schmidt (1) (2).

As to the first part of Theorem VIII, it is clear that it is valid whenever $K_2(x)$ is of the form $\sum_{k=1}^{n} A_k K_1(x+\lambda_k)$. It is also clear that if

$$|f(x)| \leq B; \qquad \int_{-\infty}^{\infty} |K_2(x) - K_3(x)|\, dx \leq \varepsilon, \tag{4.04}$$

then

$$\left| \int_{-\infty}^{\infty} f(\xi)\, K_2(\xi - x)\, d\xi - \int_{-\infty}^{\infty} f(\xi)\, K_3(\xi - x)\, d\xi \right| \leq B\varepsilon.$$

An application of Theorem II completes the proof.

As to the second part of the proof, it is merely necessary to suppose that

$$\frac{1}{\sqrt{2\pi}} \int_{-\infty}^{\infty} K_1(x)\, e^{iu_1 x}\, dx = 0 \tag{4.05}$$

and to take

$$f(x) = e^{iu_1 x} \tag{4.06}$$

to obtain a contradiction.

Theorem VIII has an extension in the sense of Theorem III, in which $K_1(x)$ is replaced by a class of functions Σ_1 (4.02) holds for every function of the class, and there is no u for which for *every* function $K_1(x)$ of Σ_1

$$\frac{1}{\sqrt{2\pi}} \int_{-\infty}^{\infty} K_1(x)\, e^{iux}\, dx = 0. \tag{4.07}$$

The converse of this extended theorem is also valid.

5. **Averages of bounded Stieltjes distributions.** Here our theorem is:

THEOREM IX. *Let $f(x)$ be a function of limited total variation over every finite range, and let*

$$\int_{y}^{y+1} |d f(x)| \tag{5.01}$$

be bounded in y. Let $K_1(x)$ be a continuous function of L_1, and let

$$\sum_{k=-\infty}^{\infty} \overline{\lim_{k \leq x \leq k+1}} |K_1(x)| \tag{5.02}$$

converge. Let

$$\frac{1}{\sqrt{2\pi}} \int_{-\infty}^{\infty} K_1(x) e^{iux}\, dx \neq 0 \qquad [-\infty < u < \infty] \tag{4.01}$$

and let

$$\lim_{x \to \infty} \int_{-\infty}^{\infty} K_1(\xi - x)\, d f(\xi) = A \int_{-\infty}^{\infty} K_1(\xi)\, d\xi. \tag{5.03}$$

Then if $K_2(x)$ is any function of M_1,

$$\lim_{x \to \infty} \int_{-\infty}^{\infty} K_2(\xi - x)\, d f(\xi) = A \int_{-\infty}^{\infty} K_2(\xi)\, d\xi. \tag{5.04}$$

Conversely, let (5.02) *converge. Let* $\int_{-\infty}^{\infty} K_1(x)\,dx \neq 0$. *Let* (5.03) *imply* (5.04) *for every function of* K_2 *of* M_1 *and every* $f(x)$ *for which* (5.01) *is bounded. Then* (4.01) *holds.*

To prove this theorem we need the following elementary:

LEMMA IXa. *Let*

$$\int_{y-0}^{y+1+0} |df(x)| \leqq B \qquad [-\infty < y < \infty] \tag{5.05}$$

and let

$$\sum_{k=-\infty}^{\infty} \overline{\lim_{k \leq x \leq k+1}} |K_2(x) - K_3(x)| \leqq \epsilon. \tag{5.06}$$

Then

$$\left| \int_{-\infty}^{\infty} K_2(\xi - x)\,df(x) - \int_{-\infty}^{\infty} K_3(\xi - x)\,df(x) \right| \leqq B\epsilon. \tag{5.061}$$

With this lemma at our disposal, the proof of Theorem IX does not differ in any important respect from that of Theorem VIII. Of course, as in Theorem VI, we must replace the polynomials in translations of K_1 that figure in Theorem VIII by absolutely convergent integrals in these translations. The proof of the lemma itself is immediate.

The modification of Theorem IX with a hypothesis involving a whole class of kernels will be used later, so we shall formulate it as a separate theorem:

THEOREM X. *Let* $f(x)$ *be a function of limited total variation over every finite range, and let*

$$\int_{y}^{y+1} |df(x)| \tag{5.01}$$

be bounded in y. *Let* Σ *be a class of continuous functions of* L_1, *each one of which, for example* $K_1(x)$, *has the properties that*

$$\sum_{k=-\infty}^{\infty} \overline{\lim_{k \leq x \leq k+1}} |K_1(x)| \tag{5.02}$$

converges, and that

$$\lim_{x \to \infty} \int_{-\infty}^{\infty} K_1(\xi - x)\,df(\xi) = A \int_{-\infty}^{\infty} K_1(\xi)\,d\xi. \tag{5.03}$$

Let there be no u *which is a real zero for all the functions*

$$\frac{1}{\sqrt{2\pi}} \int_{-\infty}^{\infty} K_1(x)\,e^{iux}\,dx \tag{5.07}$$

for which $K_1(x)$ *belongs to* Σ. *Then if* $K_2(x)$ *is any function belonging to* M_1,

$$\lim_{x\to\infty}\int_{-\infty}^{\infty} K_2(\xi-x)\,df(\xi) = A\int_{-\infty}^{\infty} K_2(\xi)\,d\xi. \tag{5.04}$$

Conversely, if the class Σ *contains at least one function with non zero integral and has the property that each member of it is a member of* L_1 *and satisfies the condition that* (5.02) *shall converge, and if, whenever* $f(x)$ *satisfies the condition that* (5.01) *be bounded, and* (5.03) *for every member of* Σ, *then* (5.04) *holds for every function* $K_2(x)$ *belonging to* M_1, *then there is no* u *such that* (4.07) *holds for every function* $K_1(x)$ *belonging to* Σ.

6. **Averages of unilaterally bounded distributions and functions.** Let $K_1(x)$ not be equivalent to 0, and let it be a continuous function (or a function continuous except for a finite number of finite jumps) not identically zero and such that

$$K_1(x) \geqq 0, \quad \sum_{k=-\infty}^{\infty} \max_{k\leqq x\leqq k+1} K_1(x) < p \qquad (-\infty < x < \infty). \tag{6.01}$$

We shall say that the mass-distribution determined by $f(x)$ is *bounded below* when

$$\int_{y}^{y+1} |df(x)| - f(y+1) + f(y) \leqq N \qquad (-\infty < x < \infty) \tag{6.02}$$

and that it is *bounded above* when $-f(x)$ determines a mass-distribution bounded below. We shall prove:

LEMMA XIa. *Let the distribution corresponding to* $f(x)$ *be bounded below (or above). Let* (6.01) *hold, and let*

$$\overline{\int_{-\infty}^{\infty}} K_1(\xi-x)\,df(\xi) \leqq M; \quad (\geqq M); \qquad (-\infty < x < \infty). \tag{6.03}$$

Then there is a Q *such that*

$$\int_{y}^{y+1} |df(x)| \leqq Q \qquad (-\infty < y < \infty). \tag{6.04}$$

To prove this, let us notice that

$$\begin{aligned}\overline{\int_{-\infty}^{\infty}} K_1(\xi-x)\,d\int_{\xi_0}^{\xi}|df(y)| &\leqq \overline{\int_{-\infty}^{\infty}} K_1(\xi-x)\,d\left[\int_{\xi_0}^{\xi}|df(y)| - f(\xi)\right] \\ &+ \overline{\int_{-\infty}^{\infty}} K_1(\xi-x)\,df(\xi) \leqq N\sum_{k=-\infty}^{\infty}\max_{k\leqq x\leqq k+1} K(\xi) + M.\end{aligned} \tag{6.05}$$

Since K is "stückweise stetig", B, a and b exist such that

$$K_1(u) > B > 0 \qquad (a \leqq u \leqq b). \tag{6.06}$$

Hence

$$M + N\sum_{k=-\infty}^{\infty}\max_{k\leqq x\leqq k+1} K_1(\xi) \geqq B\int_{a+x}^{b+x} d\int_{\xi_0}^{\xi}|df(y)|, \tag{6.07}$$

or in other terms

$$\frac{1}{B}\left\{\left[\frac{1}{b-a}\right]+1\right\}\left\{M+N\sum_{k=-\infty}^{\infty}\max_{k\leq x\leq k+1}K_1(\xi)\right\} \tag{6.08}$$
$$\geqq \int_y^{y+1}|df(x)| \qquad (-\infty<y<\infty).$$

If we combine this lemma with Theorem X, we get:

THEOREM XI. *Let $f(x)$ be a function of limited total variation over every finite interval, for which*

$$\int_y^{y+1}|df(x)|-f(y+1)+f(y)\leqq N \qquad (-\infty<y<\infty). \tag{6.02}$$

Let Σ be a class of continuous functions K_1, for each of which

$$\sum_{k=-\infty}^{k=\infty}\max_{k\leqq x\leqq k+1}|K_1(x)| \tag{5.02}$$

converges, and let

$$\lim_{x\to\infty}\int_{-\infty}^{\infty}K_1(\xi-x)\,df(\xi)=A\int_{-\infty}^{\infty}K_1(\xi)\,d\xi \tag{5.03}$$

for every K_1 belonging to Σ. Let there be no real u for which every

$$\frac{1}{\sqrt{2\pi}}\int_{-\infty}^{\infty}K_1(x)\,e^{iux}\,dx \tag{5.07}$$

vanishes, for which K_1 belongs to Σ. Let $Q(x)$ be a continuous function belonging to Σ, for which

$$Q(x)\geqq 0, \qquad \left|\int_{-\infty}^{\infty}Q(\xi-x)\,df(x)\right|\leqq T \qquad (-\infty<x<\infty). \tag{6.09}$$

Then if $K_2(-x)$ is any function belonging to M_1,

$$\lim_{x\to\infty}\int_{-\infty}^{\infty}K_2(\xi-x)\,df(\xi)=A\int_{-\infty}^{\infty}K_2(\xi)\,d\xi. \tag{5.04}$$

We shall have frequent occasion to use Theorems X and XI in a form in which the infinite range of x is mapped on a semi-infinite range by an exponential transformation. This may happen in two ways: either 0 or ∞ for the new argument may correspond to $+\infty$ for x. In the first case, let us write:

$$\xi=-\log\lambda,\ f(\xi)=\int\lambda^{-1}\,d\varphi(\lambda),\ \ K_{1,2}(\xi)=\lambda N_{1,2}(\lambda),\ \ Q(\xi)=\lambda M(\lambda) \tag{6.10}$$

and in the second,

$$\xi=\log\lambda,\ f(\xi)=\int\lambda^{-1}\,d\varphi(\lambda),\ \ K_{1,2}(\xi)=\lambda N_{1,2}(\lambda),\ \ Q(\xi)=\lambda M(\lambda). \tag{6.11}$$

Then Theorem XI yields:

THEOREM XI′. *Let $\varphi(\lambda)$ be a function of limited total variation over every interval $(\varepsilon, 1/\varepsilon)$ where $0<\varepsilon<1$. Let $\varphi(0) = 0$ and let*

(6.12) $$\int_u^{2u} \lambda^{-1} |d\varphi(\lambda)| - \int_u^{2u} \lambda^{-1} d\varphi(\lambda) \leqq N \qquad (0<u<\infty).$$

Let Σ be a class of continuous functions $N_1(\lambda)$ for each of which

(6.13) $$\sum_{k=-\infty}^{\infty} \max_{2^k \leqq \lambda \leqq 2^{k+1}} \lambda |N_1(\lambda)|$$

converges, and let

(6.14) $$\lim_{\lambda\to 0(\infty)} \frac{1}{\lambda} \int_0^\infty N_1\left(\frac{\mu}{\lambda}\right) d\varphi(\mu) = A \int_0^\infty N_1(\mu)\, d\mu$$

for every $N_1(\lambda)$ belonging to Σ. Let there be no real u for which every

(6.15) $$\int_0^\infty N_1(\lambda)\, \lambda^{iu}\, d\lambda$$

vanishes, for which N_1 belongs to Σ. Let $M(\lambda)$ be a function belonging to Σ, for which

(6.16) $$M(\lambda) \geqq 0, \quad \left| \frac{1}{\lambda} \int_0^\infty M\left(\frac{\mu}{\lambda}\right) d\varphi(\mu) \right| \leqq \text{const. } for\ 0 \leqq \lambda < \infty.$$

Then if $N_2(\lambda)$ is any continuous function for which

(6.17) $$\sum_{k=-\infty}^{\infty} \max_{2^k \leqq \lambda \leqq 2^{k+1}} \lambda |N_2(\lambda)|$$

converges,

(6.18) $$\lim_{\lambda\to 0(\infty)} \frac{1}{\lambda} \int_0^\infty N_2\left(\frac{\mu}{\lambda}\right) d\varphi(\mu) = A \int_0^\infty N_2(\mu)\, d\mu.$$

In the case where

(6.19) $$\varphi(\lambda) = \int_{\lambda_0}^{\lambda} f(x)\, dx,$$

(6.12) will be satisfied if

(6.20) $$f(\lambda) > -\frac{N}{2 \log 2}.$$

If the other hypotheses of Theorem XI′ are satisfied, (6.18) will become

(6.21) $$\lim_{\lambda\to 0(\infty)} \frac{1}{\lambda} \int_0^\infty N_2\left(\frac{\mu}{\lambda}\right) f(\mu)\, d\mu = A \int_0^\infty N_2(\mu)\, d\mu.$$

As a particular admissible N_2, we may take

$$(6.22)\qquad N_2(\mu) = \begin{cases} 1 & ; \quad [0 \leqq \mu < 1] \\ 1 - \dfrac{\mu - 1}{\varepsilon}; & [1 \leqq \mu < 1 + \varepsilon] \\ 0 & . \quad [1 + \varepsilon \leqq \mu] \end{cases}$$

Thus (6.21) will assume the form

$$(6.23)\qquad \lim_{\lambda \to 0(\infty)} \frac{1}{\lambda} \left[\int_0^\lambda f(\mu)\, d\mu + \int_\lambda^{\lambda(1+\varepsilon)} \left(1 - \frac{\mu - \lambda}{\lambda \varepsilon}\right) f(\mu)\, d\mu \right] = A\left(1 + \frac{\varepsilon}{2}\right).$$

Now, by (6.20)

$$(6.24)\qquad \frac{1}{\lambda} \int_\lambda^{\lambda(1+\varepsilon)} \left(1 - \frac{\mu - \lambda}{\lambda \varepsilon}\right) f(\mu)\, d\mu \geqq \frac{1}{\lambda} \int_\lambda^{\lambda(1+\varepsilon)} \left(1 - \frac{\mu - \lambda}{\lambda \varepsilon}\right) \left(\frac{-N}{2 \log 2}\right) d\mu$$
$$= \frac{-N\varepsilon}{4 \log 2}.$$

Hence by (6.23)

$$(6.25)\qquad \overline{\lim_{\lambda \to 0(\infty)}} \frac{1}{\lambda} \int_0^\lambda f(\mu)\, d\mu \leqq A\left(1 + \frac{\varepsilon}{2}\right) + \frac{N\varepsilon}{4 \log 2}.$$

Since ε is arbitrarily small,

$$(6.25)\qquad \overline{\lim_{\lambda \to 0(\infty)}} \frac{1}{\lambda} \int_0^\lambda f(\mu)\, d\mu \leqq A.$$

Again, we may write (6.21) in the form

$$(6.26)\qquad \lim_{\lambda \to 0(\infty)} \frac{1}{\lambda(1+\varepsilon)} \left[\int_0^{\lambda(1+\varepsilon)} f(\mu)\, d\mu - \int_\lambda^{\lambda(1+\varepsilon)} \frac{\mu - \lambda}{\lambda \varepsilon} f(\mu)\, d\mu \right] = \frac{A(1+\varepsilon/2)}{1+\varepsilon}$$

from which we may conclude as above that

$$(6.27)\qquad \underline{\lim_{\lambda \to 0(\infty)}} \frac{1}{\lambda} \int_0^\lambda f(\mu)\, d\mu \geqq A.$$

Combining (6.25) and (6.27), we see that

$$(6.28)\qquad \lim_{\lambda \to 0(\infty)} \frac{1}{\lambda} \int_0^\lambda f(\mu)\, d\mu = A.$$

We thus get

THEOREM XI″. *Let $f(x)$ be a function bounded over every interval $(\varepsilon, 1/\varepsilon)$, where $0 < \varepsilon < 1$. Let*

$$(6.29)\qquad f(\lambda) > -K \ (or\ f(\lambda) < K)$$

for every argument. Let Σ be a class of continuous functions $N_1(\lambda)$ for each of which (6.13) *converges, and let*

$$(6.30)\qquad \lim_{\lambda\to 0(\infty)} \frac{1}{\lambda}\int_0^\infty N_1\left(\frac{\mu}{\lambda}\right) f(\mu)\,d\mu = A\int_0^\infty N_1(\mu)\,d\mu$$

for every $N_1(\lambda)$ *belonging to* Σ. *Let there be no real* u *for which every expression* (6.15) *vanishes with* N_1 *belonging to* Σ. *Let* $M(\lambda)$ *belong to* Σ, *and let*

$$(6.31)\qquad M(\lambda)\geqq 0,\ \frac{1}{\lambda}\int_0^\infty M\left(\frac{\mu}{\lambda}\right) f(\mu)\,d\mu \leqq \text{const. } for\ 0\leqq\lambda<\infty.$$

Then (6.28) *is true.*

Another case where a stricter conclusion than (6.18) may be drawn is where $\varphi(\lambda)$ is monotone. In this case, (6.12) is automatically satisfied. If we take $N_2(\mu)$ as in (6.22), we shall be able to write (6.18) in the form

$$(6.32)\qquad \lim_{\lambda\to 0(\infty)} \frac{1}{\lambda}\left[\varphi(\lambda)+\int_\lambda^{\lambda(1+\varepsilon)}\left(1-\frac{\mu-\lambda}{\lambda\varepsilon}\right)d\varphi(\mu)\right] = A\left(1+\frac{\varepsilon}{2}\right)$$

and hence

$$(6.33)\qquad \overline{\lim_{\lambda\to 0(\infty)}}\ \frac{\varphi(\lambda)}{\lambda}\leqq A.$$

From the analogue of (6.26) it follows that

$$(6.34)\qquad \underline{\lim_{\lambda\to 0(\infty)}}\ \frac{\varphi(\lambda)}{\lambda}\geqq A,$$

and hence that

$$(6.35)\qquad \varphi(\lambda)\sim\lambda A.$$

We thus obtain

THEOREM XI‴. *If in the hypothesis of Theorem* XI′, (6.12) *is replaced by the condition that* $\varphi(\lambda)$ *is monotone. Then* (6.35) *follows.*

It is even possible to weaken the hypothesis here given, and to replace (6.12) by the condition that

$$(6.36)\qquad \lim_{\varepsilon\to+0}\ \overline{\lim_{-\infty<\mu<\infty}}\left[\left|\int_\mu^{\mu(1+\varepsilon)}\lambda^{-1}\,|d\varphi(\lambda)| - \int_\mu^{\mu(1+\varepsilon)}\lambda^{-1}\,d\varphi(\lambda)\right|\right]=0$$

to establish (6.25).

CHAPTER III.

TAUBERIAN THEOREMS AND THE CONVERGENCE OF SERIES AND INTEGRALS.

7. The Hardy-Littlewood condition. We now enter upon the realm of ideas which has longest been associated with Tauberian theorems.[27] The class Σ of the last paragraph now consists of all functions of the form

[27] Hardy (2): Hardy and Littlewood (4), (10), (13) etc.: Littlewood (1): Landau (9), (4).

(7.01) $$M(x)-M(x+\lambda)$$

for which we have

(7.02) $$M(-\infty)=1, \quad M(\infty)=0, \quad \int_{-\infty}^{\infty} M(\xi-x)\, df(\xi) \quad \text{bounded.}$$

Condition (5.02) now asserts the convergence of

(7.03) $$\sum_{k=-\infty}^{\infty} \max_{k\leq x\leq k+1} |M(x)-M(x+\lambda)|$$

which will be automatically fulfilled if $M(x)$ is monotone and satisfies (7.02). We shall also suppose $M(x)$ continuous. Condition (5.07) becomes the condition that

(7.04) $$\frac{1}{\sqrt{2\pi}} \int_{-\infty}^{\infty} [M(x)-M(x+\lambda)]\, e^{iux}\, dx$$

shall not vanish for any real u for every λ. Let us suppose M to vanish exponentially at $+\infty$, and let

(7.05) $$\frac{1}{\sqrt{2\pi}} \int_{-\infty}^{\infty} M(x)\, e^{xz}\, dx = \varphi(z).$$

$\varphi(z)$ will clearly be analytic over some vertical strip to the right of the origin. We shall obviously have

(7.06) $$\frac{1}{\sqrt{2\pi}} \int_{-\infty}^{\infty} [M(x)-M(x+\lambda)]\, e^{xz}\, dz = \varphi(z)\,(1-e^{-\lambda z})$$

over this strip, and by analytic continuation, which we assume to be possible, the non-vanishing of (7.04) becomes the non-vanishing of $\varphi(z)$ over the imaginary axis.

If

(7.07) $$\lim_{x\to\infty} \int_{-\infty}^{\infty} M(\xi-x)\, df(\xi) = B,$$

(5.03) is satisfied for $A=0$. As is obviously permissible, let us put

(7.08) $$K_2(\xi) = \begin{cases} M(\xi)-1 \;; & (-\infty<\xi<0) \\ M(\xi)-1+\dfrac{\xi}{\varepsilon}\;; & (0\leq\xi<\varepsilon) \\ M(\xi) \;. & (\varepsilon\leq\xi<\infty) \end{cases}$$

Then (5.04) assumes the form

(7.09) $$\lim_{x\to\infty} \int_{-\infty}^{\infty} K_2(\xi-x)\, df(\xi) = 0,$$

which becomes

(7.10) $$\lim_{x\to\infty} \left[\int_{-\infty}^{x} df(\xi) + \int_{x}^{x+\varepsilon} \left(1-\frac{\xi-x}{\varepsilon}\right) df(\xi)\right] = B.$$

3

If we put

(7.11) $$f(-\infty) = 0$$

(5.04) assumes the form

(7.12) $$\lim_{x\to\infty} \frac{1}{\varepsilon} \int_x^{x+\varepsilon} f(\xi)\, d\xi = B.$$

Restating Theorem XI in this new form, we get

THEOREM XII. *Let $f(x)$ be a function of limited total variation over every range $(-\infty, A)$, let $f(-\infty) = 0$, and let*

(6.02) $$\int_y^{y+1} |df(x)| - f(y+1) + f(y) \leq N \qquad (-\infty < y < \infty).$$

Let $M(x)$ be a monotonely decreasing continuous function, such that

(7.13) $$M(-\infty) = 1, \quad M(x) = O(e^{-ux}) \text{ at } \infty \qquad (u > 0).$$

Let

(7.05) $$\frac{1}{\sqrt{2\pi}} \int_{-\infty}^{\infty} M(x)\, e^{xz}\, dx = \varphi(z)$$

over a strip to the right of the origin, and let $\varphi(z)$, continued analytically on to the imaginary axis, have no zeros there. (We assume the possibility of this continuation.) *Let*

(7.14) $$\int_{-\infty}^{\infty} M(\xi - x)\, df(\xi)$$

be bounded, and let

(7.07) $$\lim_{x\to\infty} \int_{-\infty}^{\infty} M(\xi - x)\, df(\xi) = B.$$

Then

(7.12) $$\lim_{x\to\infty} \frac{1}{\varepsilon} \int_x^{x+\varepsilon} f(\xi)\, d\xi = B.$$

Let us put

$$M(\log x) = N(x); \quad f(\log x) = F(x).$$

Theorem XII then becomes:

THEOREM XIII. *Let $F(x)$ be a function vanishing at the origin and of limited total variation over any finite range (including the origin), and let for some $A > 1$,*

(7.16) $$\int_y^{Ay} |dF(x)| - F(Ay) + F(y) \leq N \qquad (0 < y < \infty).$$

Let $N(x)$ be a monotonely decreasing continuous function, such that

(7.17) $$N(0) = 1, \qquad N(x) = O(x^{-\lambda}) \text{ at } \infty \qquad [\lambda > 0].$$

Let

(7.18) $$\frac{1}{\sqrt{2\pi}} \int_0^{\infty} N(x)\, x^{z-1}\, dx = \varphi(z)$$

over a strip to the right of the origin, and let $\varphi(z)$, continued analytically on to the imaginary axis, as we assume to be possible for all points but the origin, have no zeros there. Let

$$\int_0^\infty N\left(\frac{\xi}{x}\right) dF(\xi) \tag{7.19}$$

be bounded, and let

$$\lim_{x\to\infty} \int_0^\infty N\left(\frac{\xi}{x}\right) dF(\xi) = B. \tag{7.20}$$

Then if $\Lambda_1 > 1$,

$$\lim_{x\to\infty} \frac{1}{\log \Lambda_1} \int_x^{x\Lambda_1} F(\xi)\frac{d\xi}{\xi} = B. \tag{7.21}$$

A case of particular interest and importance is where

$$F(x) = \sum_{n=0}^{[x]} a_n. \tag{7.22}$$

Here

$$\begin{aligned} \int_y^{\Lambda y} |dF(x)| - F(\Lambda y) + F(y) &= \sum_{n=[y]+1}^{[\Lambda y]} (-a_n + |a_n|) \\ &\leq \begin{cases} \max 2na_n; \ (n \leq [\Lambda y]), \\ 0. \end{cases} \end{aligned} \tag{7.23}$$

if Λ is sufficiently near to 1. (7.23) is to be interpreted as meaning that the expression on the left hand side is less than or equal to the greater of the two expressions in brackets.

We thus obtain the

COROLLARY. *If $N(x)$ is subject to the conditions of Theorem* XIII, *if*

$$\int_0^\infty N\left(\frac{\xi}{x}\right) dF(\xi) = \sum_{n=0}^\infty a_n N\left(\frac{n}{x}\right) \tag{7.24}$$

is bounded, if

$$na_n < K \qquad (\text{or } na_n > -K) \tag{7.25}$$

for all n, and if

$$\lim_{x\to\infty} \sum_{n=0}^\infty a_n N\left(\frac{n}{x}\right) = B \tag{7.26}$$

then

$$\sum_0^\infty a_n = B. \tag{7.27}$$

Here instead of (7.27), what we directly prove is

$$\lim_{x\to\infty} \frac{1}{\log \Lambda_1} \int_x^{x\Lambda_1} \sum_{n=0}^{[\xi]} a_n \frac{d\xi}{\xi} = B. \tag{7.28}$$

3*

(7.27) may however be written in the form

$$\lim_{x\to\infty} \frac{1}{\log A_1} \int_x^{xA_1} \sum_{n=0}^{[x]} a_n \frac{d\xi}{\xi} = B. \tag{6.281}$$

Our conclusion (7.27) will then follow if we can show that we can so choose A_1, that

$$\begin{aligned} &\overline{\lim_{x\to\infty}} \frac{1}{\log A_1} \int_x^{xA_1} \sum_{n=[x]+1}^{[\xi]} a_n \frac{d\xi}{\xi} < \varepsilon, \\ &\overline{\lim_{x\to\infty}} \frac{1}{\log A_1} \int_x^{xA_1} \sum_{n=[\xi]+1}^{[xA_1]} a_n \frac{d\xi}{\xi} < \varepsilon. \end{aligned} \tag{7.29}$$

However,

$$\frac{1}{\log A_1} \int_x^{xA_1} \sum_{n=[x]+1}^{[\xi]} a_n \frac{d\xi}{\xi} \leqq \frac{1}{\log A_1} \int_x^{xA_1} \frac{K}{x}(xA_1 - x)\frac{d\xi}{\xi} = K(A_1 - 1). \tag{7.30}$$

Again,

$$\frac{1}{\log A_1} \int_x^{xA_1} \sum_{n=[\xi]+1}^{[A_1 x]} a_n \frac{d\xi}{\xi} \leqq K(A_1 - 1). \tag{7.31}$$

Since we may take A_1 as near to 1 as we wish, (7.29) is established, and (7.27) follows at once. This type of argument is to be found in the work of Szász.

8. **The Schmidt condition.**[28] This last corollary covers the work of Hardy and Littlewood on Tauberian theorems. An extension of their conditions is due to Robert Schmidt. To arrive at theorems of his type, let M be subject to the conditions stated in the hypothesis to Theorem XII; let us put

$$g(x) = f(x+\alpha) - f(x) \tag{8.01}$$

and let us assume that

$$-f(x) + |f(x)| < A(x + |x|) + c \qquad (A, c > 0). \tag{8.02}$$

As the integrals in question converge absolutely by (8.02) and (7.13), whenever they exist we may invert our order of integration and write:

$$\begin{aligned} \int_y^{y+\alpha} dx \int_{-\infty}^{\infty} M(\xi - x)\, df(\xi) &= \int_y^{y+\alpha} dx \int_{-\infty}^{\infty} M(\xi)\, df(\xi + x) \\ &= \int_{-\infty}^{\infty} M(\xi)\big(f(\xi+y+\alpha) - f(\xi+y)\big)\, d\xi \\ &= \int_{-\infty}^{\infty} M(\xi - y)\big(f(\xi+\alpha) - f(\xi)\big)\, d\xi \\ &= \int_{-\infty}^{\infty} M(\xi - y)\, g(\xi)\, d\xi. \end{aligned} \tag{8.03}$$

[28] Schmidt (1), (2). Throughout this section, the author has been strongly influenced by the methods of Vijayaraghavan and Szász.

Thus if $M(x)$ is subject to the conditions in the hypothesis to Theorem XII, if it is bounded above or below, and (7.07) is valid, and (7.14) bounded, whether (6.02) is true or not, it follows from Theorem XII itself on replacing $f(x)$ by $\frac{1}{\alpha}\int_{-\infty}^{x} g(\xi)\,d\xi$ that

$$(8.04)\qquad \begin{aligned} B &= \lim_{y\to\infty} \frac{1}{\alpha}\int_{y}^{y+\alpha} \frac{dx}{\alpha}\int_{-\infty}^{x} g(\xi)\,d\xi \\ &= \lim_{y\to\infty} \frac{1}{\alpha}\int_{y}^{y+\alpha} \frac{dx}{\alpha}\int_{x}^{x+\alpha} f(\xi)\,d\xi \\ &= \lim_{y\to\infty} \frac{1}{\alpha^2}\int_{-\alpha}^{\alpha} f(\xi+\alpha+y)\,(\alpha-|\xi|)\,d\xi. \end{aligned}$$

Let us now introduce Schmidt's notion of a "slowly decreasing" function. Let $f(u)$ have the property that when u and v run through a sequence of pairs of values for which

$$(8.05)\qquad v \geqq u, \qquad v-u\to 0$$

then

$$(8.06)\qquad \varliminf_{u\to\infty} (f(v)-f(u)) \geqq 0;$$

we shall call $f(u)$ *slowly decreasing*. (Schmidt treats sequences instead of functions, and on another scale, but the difference is unimportant.) Schmidt shows that we may write

$$(8.07)\qquad f(u) = f_1(u)+f_2(u)$$

where $f_2(u)$ is monotone increasing, and $f_1(u)$ satisfies the condition that whenever (8.05) is fulfilled, then

$$(8.08)\qquad \lim_{u\to\infty} (f_1(v)-f_1(u)) = 0.$$

He proves that there exists a function $T(\alpha)$ such that

$$(8.09)\qquad \varlimsup_{\substack{u\to\infty \\ u\leqq v\leqq u+\alpha}} |f_1(v)-f_1(u)| = T(\alpha)$$

and

$$(8.10)\qquad \lim_{\alpha\to 0} T(\alpha) = 0.$$

He proves that a number T exists, such that

$$(8.11)\qquad |f_1(v)-f_1(u)| \leqq T\cdot(v-u) \qquad (v>u+\alpha).$$

From this, it readily follows that if f is bounded near $-\infty$, (8.02) is valid and $g(x)$ is bounded below. Furthermore,

$$(8.12)\qquad \varlimsup_{x\to\infty}\left(f(x)-\frac{1}{\alpha^2}\int_{-\alpha}^{\alpha} f(\xi+\alpha+x)\,(\alpha-|\xi|)\,d\xi\right) \leqq T(2\alpha)$$

and

$$(8.13)\quad \underline{\lim}_{x\to\infty}\left(-\frac{1}{\alpha^2}\int_{-\alpha}^{\alpha} f(\xi+\alpha+x)\,(\alpha-|\xi|)\,d\xi+f(x+2\alpha)\right)\geqq -T(2\alpha);$$

going to the limit

$$(8.14)\qquad \underline{\lim}_{x\to\infty} f(x)\geqq -T(2\alpha)+B$$

and

$$(8.15)\qquad \overline{\lim}_{x\to\infty} f(x)\leqq B+T(2\alpha).$$

Then by (8.10), it follows that

$$(8.16)\qquad \underline{\lim}_{x\to\infty} f(x) = B.$$

This yields

THEOREM XIV. *In the hypothesis of Theorem* XII, *condition* (6.02) *may be replaced by the assumption that $f(x)$ is of limited total variation over any finite range, and is slowly decreasing, and the conclusion* (7.12) *may be replaced by* (8.16). Again, Theorem XIII becomes

THEOREM XV. *In the hypothesis to Theorem* XIII, *condition* (7.16) *may be replaced by the condition that if u and v run through a sequence for which*

$$(8.17)\qquad v\geqq u,\quad \frac{v}{u}\to 1$$

then

$$(8.18)\qquad \underline{\lim}_{u\to\infty}\,(F(v)-F(u))\geqq 0.$$

In the conclusion to Theorem XIII, (7.21) *may then be replaced by*

$$(8.161)\qquad \lim_{x\to\infty} F(x) = B.$$

In the corollary to Theorem XIII, (7.25) *may be replaced by the hypothesis that if*

$$(8.19)\qquad s_n = a_0+a_1+\cdots+a_n$$

and $p\to\infty$, $q\to\infty$ in such a manner that

$$(8.20)\qquad q\geqq p,\quad \frac{q}{p}\to 1$$

then

$$(8.21)\qquad \lim_{p\to\infty}\,(s_q-s_p)\geqq 0.$$

The conclusion of this corollary remains unchanged.

The hypothesis that (8.20) shall imply (8.21) is Schmidt's hypothesis concerning slowly decreasing sequences in unaltered form.

It will be noted that the proof here given for the Schmidt theorems does not essentially differ from the proofs given by Vijayaraghavan[29] and Szász[30] for theorems of these types. In both cases the transition is made to a Tauberian theorem of more standard type in which s_n is itself averaged, instead of appearing in the conclusion as the average of a unilaterally or bilaterally bounded mass distribution. In both cases, moreover, the Schmidt condition is again used to give a bilateral or unilateral estimate of the difference between s_n and its average.

CHAPTER IV.

TAUBERIAN THEOREMS AND PRIME NUMBER THEORY.

9. **Tauberian theorems and Lambert series.** One of the most important applications of Tauberian theorems is to the proof of the prime number theorem of Hadamard[31] and de la Vallée Poussin,[32] to the effect that the number of primes less than N is asymptotically $N/\log N$. The prime numbers bear an exceedingly close relation to series of the form

$$\sum a_n \frac{x^n}{1-x^n} \tag{9.01}$$

known as Lambert series, after their eighteenth century discoverer. Until recent times, however, all attempts to employ Lambert series effectively in the study of prime numbers had proved a failure, and indeed Knopp[33] has characterized one of these directions of attack as "verführerisch". Hardy and Littlewood[34] finally showed that the prime number theorem was equivalent to a Tauberian theorem concerning Lambert series, but did not succeed in establishing an autonomous proof of this theorem.

Our general Tauberian theorems suffice to furnish this autonomous proof, and indeed, the Tauberian theorem which we shall find it easiest to establish directly leads more directly to the prime number theorem than does the theorem of Hardy and Littlewood. The latter is also directly demonstrable by our methods. In both, the cardinal point in the proof is that the Riemann zeta function, $\zeta(x+iy)$, has no zeros on the line $x = 1$. The proof of this goes back to the first proofs of the prime number theorem, and has always been recognized as that property of the Riemann zeta

[29] Vijayaraghavan (1), (2).
[30] Szász (1), (2).
[31] Hadamard (1).
[32] de la Vallée Poussin (1).
[33] Knopp (3).
[34] Hardy and Littlewood (11); Hardy (6); Ananda-Rau (2).

function which is most central in the proofs of this theorem, but all earlier proofs had made some use of the behavior of the zeta function at infinity. These further properties now appear as inessential, and the non-vanishing of the function becomes the only non-elementary feature of the zeta function in question.

For example, the usual proof of the prime number theorem employs a lemma of Landau[35] to the following effect:

LANDAU'S LEMMA. *Let*

$$F(x) = \sum_{n=1}^{\infty} a_n n^{-x} \qquad [\Re(x) > 1] \tag{9.02}$$

and let

$$a_n \geqq 0 \qquad [n - 1, 2, \cdots]. \tag{9.03}$$

Let $F(x)$ *when analytically continued be without singularities on* $\Re(x) = 1$, *except for a pole of order one at* $x = 1$, *with principal part* $A/(x-1)$. *Let there be some* α *for which*

$$F(x) = O(|x|^{\alpha}) \tag{9.04}$$

in the right half-plane. Then

$$A = \lim_{n\to\infty} \frac{1}{n} \sum_{1}^{n} a_k. \tag{9.05}$$

In the following section, we shall show—following Ikehara— that condition (9.04) is not needed. When the lemma is used in the proof of the prime number theorem,

$$a_n = \Lambda_n \text{ (to be defined immediately)}; \qquad F(x) = -\zeta'(x)/\zeta(x) \tag{9.06}$$

and it will be seen that here too the only non-elementary property of the zeta function which we use is that it does not vanish on the line $\Re(x) = 1$.

To return to Lambert series, let $\Lambda(n)$ be the number-theoretic function determined by:

$\Lambda(p^k) = \log p$ if p is a prime and k is a positive integer;
$\Lambda(n) = 0$ if n is not of the form p^k.

Let $0 \leqq x < 1$; then

$$\sum_{1}^{\infty} \log m x^m = \sum_{m=1}^{\infty} x^m \sum_{n/m} \Lambda(n) = \sum_{n=1}^{\infty} \Lambda(n) \sum_{n/m} x^m = \sum_{n=1}^{\infty} \Lambda(n) \frac{x^n}{1-x^n}. \tag{9.07}$$

[35] Landau (1).

Hence

$$\sum_1^\infty A(n) \frac{x^n}{1-x^n} = \sum_1^\infty \log m \frac{x^m - x^{m+1}}{1-x}$$

(9.08)

$$= \frac{x}{1-x} \sum_1^\infty \log\left(1+\frac{1}{m}\right) x^m$$

$$= \frac{x}{1-x} \sum_1^\infty \left[\frac{1}{m} + O\left(\frac{1}{m^2}\right)\right] x^m$$

$$= \frac{x}{1-x} \left[\log \frac{1}{1-x} + \sum_1^\infty O\left(\frac{1}{m^2}\right) x^m\right].$$

If we put $x = e^{-\xi}$ and multiply by ξ we have

(9.09) $$\sum_1^\infty A(n) \frac{\xi e^{-n\xi}}{1-e^{-n\xi}} = \frac{\xi e^{-\xi}}{1-e^{-\xi}} \left[\sum_1^\infty O\left(\frac{1}{m^2}\right) e^{-m\xi} - \log(1-e^{-\xi})\right].$$

For $\xi > 0$, the derived series on both sides converge absolutely and uniformly, and we may differentiate (9.09) term by term. On multiplication by ξ this yields us

$$\xi \sum_1^\infty A(n) \frac{d}{dn\xi}\left(\frac{n\xi e^{-n\xi}}{e^{-n\xi}-1}\right)$$

(9.10)

$$= \frac{\xi e^{-\xi}(1-\xi-e^{-\xi})}{(1-e^{-\xi})^2}\left[\log(1-e^{-\xi}) - \sum_1^\infty O\left(\frac{1}{m^2}\right) e^{-m\xi}\right]$$

$$+ \frac{\xi^2 e^{-\xi}}{1-e^{-\xi}}\left[\frac{e^{-\xi}}{1-e^{-\xi}} - \sum_1^\infty O\left(\frac{1}{m}\right) e^{-m\xi}\right]$$

$$= O(\xi)\,[O(\log \xi) - O(1)] + [\xi + O(\xi^2)]\left[\frac{1}{\xi} + O(\log \xi)\right]$$

$$= 1 + O(\xi \log \xi)$$

as $\xi \to 0$. If we put

(9.11) $$N_1(u) = \frac{d}{du}\left(\frac{u e^{-u}}{e^{-u}-1}\right)$$

this leads us to

(9.12) $$\lim_{\xi\to 0} \xi \sum_1^\infty A(n)\, N_1(n\xi) = 1.$$

This is a particular case of (6.14), which forms part of the hypothesis to Theorem XI‴. $A(n)$ is positive, as the hypothesis of that theorem further demands. The function $N_1(u)$ may be written

(9.13) $$N_1(u) = \frac{d}{du}\left(\frac{u}{1-e^u}\right) = \frac{1+(u-1)e^u}{(1-e^u)^2} = \begin{cases} O(1) & \text{as } u\to 0; \\ O(u e^{-u}) & \text{as } u \to \infty. \end{cases}$$

Thus of that part of the hypothesis of Theorem XI‴ containing N_1, which is now both M and the whole class Σ, it only remains for us to verify the boundedness of

(9.14) $$\xi \sum_1^\infty \Lambda(n)\, N_1(n\xi)$$

which follows from the fact that there is a K such that for $\xi > \xi_0 > 0$

(9.15) $$\begin{aligned} \left| \xi \sum_1^\infty \Lambda(n)\, N_1(n\xi) \right| &\leq K \sum_1^\infty n(n\xi e^{-n\xi}) \\ &= K\xi^2 \frac{d^2}{d\xi^2} \frac{e^{-\xi}}{1-e^{-\xi}} \to 0 \quad \text{as} \quad \xi \to 0 \end{aligned}$$

and from (9.12), and to verify further that

(9.16) $$\int_0^\infty N_1(u)\, u^{ix}\, du \neq 0.$$

Now

$$\begin{aligned} &\int_0^\infty N_1(u)\, u^{ix}\, du \\ &= \int_0^\infty u^{ix}\, d\left(\frac{u}{1-e^u}\right) \\ &= \lim_{\lambda \to 0} \int_0^\infty u^{ix+\lambda}\, d\left(\frac{u}{1-e^u}\right) \\ &= -\lim_{\lambda \to 0} \int_0^\infty \frac{u}{1-e^u}\, d(u^{ix+\lambda}) \end{aligned}$$

(9.17) $$\begin{aligned} &= -\lim_{\lambda \to 0} (ix+\lambda) \int_0^\infty u^{ix+\lambda} \frac{du}{1-e^u} \\ &= \lim_{\lambda \to 0} (ix+\lambda) \int_0^\infty u^{ix+\lambda} (e^{-u} + e^{-2u} + \cdots)\, du \\ &= \lim_{\lambda \to 0} (ix+\lambda)\, \Gamma(ix+\lambda+1)\, (1 + 2^{-(ix+\lambda+1)} + 3^{-(ix+\lambda+1)} + \cdots) \\ &= \lim_{\lambda \to 0} (ix+\lambda)\, \Gamma(ix+\lambda+1)\, \zeta(ix+\lambda+1) \\ &= ix\, \Gamma(ix+1)\, \zeta(ix+1). \end{aligned}$$

Let us take over from the theory of the Riemann zeta function the following facts:

(a) that the Riemann zeta function $\zeta(z)$ is analytic on the line with real part 1, except for a pole of the first order with principal part $1/(z-1)$;

(b) that the Riemann zeta function has no zeros with real part 1.

It then follows that

(9.18) $$\int_0^\infty N_1(u)\, u^{ix}\, du \neq 0$$

for any real x, and that

(9.19) $$\int_0^\infty N_1(u)\,du = \lim_{\lambda\to 0} \lambda\,\Gamma(\lambda+1)\,\zeta(\lambda+1) = 1.$$

Thus it follows from (6.35) that

(9.22) $$\lim_{N\to\infty} \frac{1}{N}\sum_{n=1}^{N} \Lambda(n) = 1.$$

Let us put $\pi(u)$ for the number of primes less than or equal to u, and let us write

(9.23) $$\varpi(u) = \pi(u) + \tfrac{1}{2}\pi(u^{1/2}) + \tfrac{1}{3}\pi(u^{1/3}) + \cdots.$$

It is easy to prove by elementary means that

(9.24) $$\lim_{u\to\infty} \frac{\pi(u)}{u} = 0.$$

Moreover,

(9.25) $$\begin{aligned} |\varpi(u) - \pi(u)| &= \left| \tfrac{1}{2}\pi(u^{1/2}) + \cdots + \frac{\pi\left(u^{\frac{1}{[\log u/\log 2+1]}}\right)}{[\log u/\log 2]+1} \right| \\ &\leqq u^{1/2}([\log u/\log 2]+1) \\ &= O(u^{1/2}\log u). \end{aligned}$$

Thus

(9.26) $$\lim_{u\to\infty} \frac{\varpi(u)}{u} = 0.$$

We may write (9.22) in the form

(9.27) $$\lim_{N\to\infty} \frac{1}{N}\int_0^N \log u\, d\varpi(u) = 1.$$

On integrating by parts, we see that

(9.28) $$\lim_{N\to\infty} \left\{ \frac{\varpi(N)\log N}{N} - \frac{1}{N}\int_0^N \frac{\varpi(u)\,du}{u} \right\} = 1.$$

Now

(9.29) $$\lim_{N\to\infty} \frac{1}{N}\int_0^N \frac{\varpi(u)}{u}\,du = 0$$

because of (9.26). Thus

(9.30) $$\lim_{N\to\infty} \frac{\varpi(N)\log N}{N} = 1$$

or

(9.31) $$\varpi(N) \sim \frac{N}{\log N}.$$

From (9.25) it follows that

(9.32) $$\pi(N) \sim \frac{N}{\log N}.$$

This is the famous prime number theorem of de la Valleé Poussin and Hadamard.

10. Ikehara's Theorem. The Landau Theorem (XVI) received several successive generalizations at the hands of Landau himself, and of Hardy and Littlewood, perhaps the most general of which was indicated by Hardy and Littlewood[36] to be the following:

THEOREM XVI. *Let:*

(i) *the series* $\Sigma a_n \lambda_n^{-s}$ *be absolutely convergent for* $\Re(s) > \sigma_0 > 0$;

(ii) *the function* $F(s)$ *defined by the series be regular for* $\Re(s) > c$ *where* $0 < c \leqq \sigma_0$ *and continuous for* $\Re(s) \geqq c$, *except for a simple pole with residue* g *at* $s = c$;

(iii)

$$F(s) = O(e^{C|t|}) \tag{10.01}$$

for some finite C, *uniformly for* $\sigma \geqq c$;

(iv)

$$\frac{\lambda_n}{\lambda_{n-1}} \to 1; \tag{10.02}$$

(v) a_n *be real, and satisfy one of the inequalities*

$$a_n > -K \lambda_n^{c-1} (\lambda_n - \lambda_{n-1}), \quad a_n < K \lambda_n^{c-1} (\lambda_n - \lambda_{n-1}) \tag{10.03}$$

or complex, and of the form

$$O\{\lambda_n^{c-1} (\lambda_n - \lambda_{n-1})\}. \tag{10.04}$$

Then

$$A_n = a_1 + a_2 + \cdots + a_n \sim \frac{g \lambda_n^c}{c}. \tag{10.05}$$

The vital change between the Landau and the Hardy-Littlewood theorem is the looser form of (iii), which replaces a restriction of the form

$$F(s) = O(|t|^k). \tag{9.04}$$

Both these restrictions are inessential, and the true theorem is that of Ikehara[37], which reads as follows:

THEOREM XVII. *Let* $\alpha(x)$ *be a monotone increasing function, and let*

$$\int_{1+0}^{\infty} x^{-u} \, d\alpha(x) = f(u) \tag{10.06}$$

[36] Hardy and Littlewood (8).

[37] Ikehara (1').

converge for $\mathfrak{R}(u) > 1$. *Let*

$$f(u) - \frac{A}{u-1} = g(u) \tag{10.07}$$

converge uniformly to a finite limit as

$$\mathfrak{R}(u) \to 1 \tag{10.08}$$

over any finite interval of the line $\mathfrak{R}(u) = 1$. *Then*

$$A = \lim_{N\to\infty} \frac{\alpha(N)}{N}. \tag{10.09}$$

To prove this, let us put

$$\beta(\xi) = \alpha(e^\xi)\, e^{-\xi} + \int_0^\xi e^{-\xi}\,\alpha(e^\xi)\, d\xi - A\xi. \quad (\xi > 0). \tag{10.10}$$

Thus

$$d\beta(\xi) = e^{-\xi}\, d\alpha(e^\xi) - A\, d\xi. \tag{10.11}$$

Let us assume—what is no essential restriction—that

$$\beta(x) = \beta(+0), \quad (-\infty \leqq x \leqq 0). \tag{10.12}$$

Then (10.07) becomes

$$g(u) = \int_{-\infty}^{\infty} e^{(1-u)\xi}\, d\beta(\xi), \quad [\mathfrak{R}(u) > 1]. \tag{10.13}$$

What we wish to prove is that

$$\lim_{\eta\to\infty} \int_{-\infty}^{\eta} e^{\xi-\eta}\, d\beta(\xi) = 0 \tag{10.14}$$

which is equivalent to (10.09).

If $\varepsilon > 0$ and η is real, we have

$$\begin{aligned} &\int_{-B}^{B} \left(1 - \frac{|u|}{B}\right) g(iu+\varepsilon+1)\, e^{iu\eta}\, du \\ &\qquad = -\int_{-B}^{B} \left(1 - \frac{|u|}{B}\right) du \int_{-\infty}^{\infty} e^{iu(\eta-\xi)}\, e^{-\varepsilon\xi}\, d\beta(\xi). \end{aligned} \tag{10.15}$$

As this double integral is absolutely convergent, it becomes

$$\begin{aligned} &\int_{-\infty}^{\infty} e^{-\varepsilon\xi}\, d\beta(\xi) \int_{-B}^{B} \left(1 - \frac{|u|}{B}\right) e^{iu(\eta-\xi)}\, du \\ &\qquad = -\int_{-\infty}^{\infty} \frac{2(\cos B(\eta-\xi) - 1)}{B(\eta-\xi)^2}\, e^{-\varepsilon\xi}\, d\beta(\xi). \end{aligned} \tag{10.16}$$

Now,

$$\begin{aligned} &\lim_{\varepsilon\to 0} \int_{-B}^{B} \left(1 - \frac{|u|}{B}\right) g(iu+\varepsilon+1)\, e^{iu\eta}\, d\eta \\ &\qquad = \int_{-B}^{B} \left(1 - \frac{|u|}{B}\right) g(iu+1)\, e^{iu\eta}\, d\eta. \end{aligned} \tag{10.17}$$

Thus

$$\lim_{\varepsilon \to 0} \int_{-\infty}^{\infty} \frac{2(\cos B(\eta - \xi) - 1)\, e^{-\varepsilon\xi}}{B(\eta - \xi)^2}\, d\beta(\xi) \tag{10.18}$$

exists. Remembering that

$$\lim_{\varepsilon \to 0} \int_0^{\infty} \frac{2(\cos B(\eta - \xi) - 1)}{B(\eta - \xi)^2}\, e^{-\varepsilon\xi}\, d\xi = \int_0^{\infty} \frac{2(\cos B(\eta - \xi) - 1)}{B(\eta - \xi)^2}\, d\xi, \tag{10.19}$$

and that

$$\frac{2(\cos B(\eta - \xi) - 1)}{B(\eta - \xi)^2}\, e^{(-1-\varepsilon)\xi}\, d\alpha(e^{\xi}) \tag{10.20}$$

is a non-positive integrand, we have, by a theorem due to Bray[38] and fundamental in the theory of the Stieltjes integral

$$\lim_{\varepsilon \to 0} \int_0^{\infty} \frac{2(\cos B(\eta - \xi) - 1)}{B(\eta - \xi)^2}\, e^{(-1-\varepsilon)\xi}\, d\alpha(e^{\xi}) = \int_0^{\infty} \frac{2(\cos B(\eta - \xi) - 1)}{B(\eta - \xi)^2}\, e^{-\xi}\, d\alpha(e^{\xi}) \tag{10.21}$$

and hence

$$\lim_{\varepsilon \to 0} \int_{-\infty}^{\infty} \frac{2(\cos B(\eta - \xi) - 1)}{B(\eta - \xi)^2}\, e^{-\varepsilon\xi}\, d\beta(\xi) = \int_{-\infty}^{\infty} \frac{2(\cos B(\eta - \xi) - 1)}{B(\eta - \xi)^2}\, d\beta(\xi). \tag{10.22}$$

Moreover,

$$-\lim_{\eta \to \infty} \int_{-\infty}^{\infty} \frac{2(\cos B(\eta - \xi) - 1)}{B(\eta - \xi)^2}\, d\beta(\xi) = \lim_{\eta \to \infty} \int_{-B}^{B} \left(1 - \frac{|u|}{B}\right) g(iu + 1)\, e^{iu\eta}\, du = 0 \tag{10.23}$$

because $\left(1 - \frac{|u|}{B}\right) g(iu + 1)$ is summable over $(-B, B)$. Similarly

$$\int_{-\infty}^{\infty} \frac{2(\cos B(\eta - \xi) - 1)}{B(\eta - \xi)^2}\, d\beta(\xi) \tag{10.24}$$

may be proved to be bounded because of (10.23) and (10.12).

We know that

$$\int_n^{n+1} d\beta(\xi) > -A \tag{10.25}$$

[38] Bray (1).

and that all the other conditions of Theorem XI are satisfied, with the possible exception of the non-vanishing of (6.15). To see that this is also satisfied, we need only reflect that

$$\int_{-\infty}^{\infty} \frac{2(\cos B\eta - 1)}{B\eta^2} e^{iu\eta}\, d\eta = \frac{1}{2\pi}\left(1 - \frac{|u|}{B}\right) \qquad [|u| < B]. \tag{10.26}$$

Thus there are no zeros common to all these functions for all values of B, and the non-vanishing of (6.15) follows.

It will be observed that the full force of our Tauberian method is scarcely needed for this theorem. In Theorem XI, which is the critical part of the proof, the difficulty of proof is considerably lessened if $K_1(x)$ assumes such a special form as $2(\cos Bx - 1)$.

As a corollary of Theorem XVII, we may prove:

THEOREM XVIII. *Let $\gamma(x)$ be a monotone increasing function, and let*

$$\int_{1+0}^{\infty} x^{-u}\, d\gamma(x) = \varphi(u) \tag{10.27}$$

converge for $\Re(u) > 1$. Let

$$F(u) = e^{\varphi(u)}(u-1)^A \qquad [0 < A < \tfrac{1}{3}] \tag{10.28}$$

when continued analytically, be regular for $\Re(u) = 1$, and let it not vanish for $u = 1$. Then,

$$A = \lim_{N\to\infty} \frac{1}{N}\int_{1+0}^{N} \log x\, d\gamma(x). \tag{10.29}$$

To prove this, let us put

$$\begin{cases} \displaystyle\int_{1+0}^{x} \log \xi\, d\gamma(\xi) = \alpha(x); \\ -\varphi'(u) = f(u). \end{cases} \tag{10.30}$$

The theorem then reduces itself to Theorem XVII, provided we can establish that (10.07) approaches a finite limit as $\Re(u) \to 1$. From the regularity of (10.28), it follows that

$$\varphi(u) + A\log(u-1) \tag{10.31}$$

is regular for $\Re(u) = 1$, except for logarithmic singularities with *negative* infinities. It is also clear that there is no singularity for $u = 1$.

Now,

$$\Re(\varphi(u)) = \int_{1+0}^{\infty} x^{-\Re(u)} \cos(\Im(u)\log x)\, d\gamma(x) \tag{10.32}$$

so that

$$\Re(\varphi(1+\varepsilon+iv)) = \int_{1+0}^{\infty} x^{-(1+\varepsilon)} \cos(v\log x)\, d\gamma(x); \tag{10.33}$$

$$\Re(\varphi(1+\varepsilon+2iv)) = \int_{1+0}^{\infty} x^{-(1+\varepsilon)} \cos(2v\log x)\, d\gamma(x); \tag{10.34}$$

and

$$\varphi(1+\varepsilon) = \int_{1+0}^{\infty} x^{-(1+\varepsilon)}\, d\gamma(x). \tag{10.35}$$

Thus

$$\begin{aligned} &3\varphi(1+\varepsilon)+4\Re(\varphi(1+\varepsilon+iv))+\Re(\varphi(1+\varepsilon+2iv)) \\ &= \int_{1+0}^{\infty} x^{-(1+\varepsilon)}(3+4\cos v\log x+\cos(2v\log x))\, d\gamma(x) \end{aligned} \tag{10.36}$$

and since

$$\begin{aligned} 3+4\cos\varphi+\cos 2\varphi &= 3+4\cos\varphi+2\cos^2\varphi-1 \\ &= 2(1+\cos\varphi)^2 \\ &\geqq 0 \end{aligned} \tag{10.37}$$

it follows that

$$3\varphi(1+\varepsilon)+4\Re(\varphi(1+\varepsilon+iv))+\Re(\varphi(1+\varepsilon+2iv)) \geqq 0, \tag{10.38}$$

or

$$\Re(\varphi(1+\varepsilon+iv)) \geqq -\tfrac{3}{4}\varphi(1+\varepsilon)-\tfrac{1}{4}\Re(\varphi(1+\varepsilon+2iv)). \tag{10.39}$$

Thus

$$\begin{aligned} &\overline{\lim_{\varepsilon\to 0}} \frac{\Re(\varphi(1+\varepsilon+iv))}{\log\varepsilon} \\ &\leqq -\tfrac{3}{4}\lim_{\varepsilon\to 0}\frac{\varphi(1+\varepsilon)}{\log\varepsilon} - \tfrac{1}{4}\underline{\lim_{\varepsilon\to 0}}\frac{\Re(\varphi(1+\varepsilon+2iv))}{\log\varepsilon} \leqq \frac{3A}{4} < 1. \end{aligned} \tag{10.40}$$

On the other hand, if $1+iv$ is a logarithmic singularity of φ with a negative coefficient, it is a zero of $F(u)$ of integral order n, and

$$\overline{\lim_{\varepsilon\to 0}} \frac{\Re(\varphi(1+\varepsilon+iv))}{\log\varepsilon} = n > 1. \tag{10.41}$$

Thus $\varphi(u)+A\log(u-1)$ has no logarithmic singularities with negative coefficients, for $\Re(u)=1$, and hence is analytic throughout this whole line. Thus, by differentiation,

$$f(u) - \frac{A}{u-1} \tag{10.42}$$

is analytic on the line in question, and the finiteness of the limit of (10.07) is established.

It will be observed that our proof, which completes the demonstration of Theorem XVIII, follows closely the lines of Landau's proof[39] that the Riemann zeta function has no zeros on the line $\Re(u)=1$, and includes it

[39] Landau (2).

as a particular case. The prime number theorem itself is a particular case of Theorem XVIII. Let us put

(10.43) $$\gamma(x) = \varpi(x).$$
Then
(10.44) $$e^{\varphi(u)} = \zeta(u)$$

and the hypothesis of Theorem XVIII is manifestly satisfied. (10.29) then becomes

(9.27) $$1 = \lim_{N\to\infty} \frac{1}{N} \int_{1+0}^{N} \log x \, d\varpi(x)$$

which we have shown to be equivalent to the prime number theorem.

Thus we have repeatedly shown that the prime number theorem is basically Tauberian in nature. It might consequently be expected that the more refined theorems as to the distribution of the primes, based on Riemann's hypothesis as to the distribution of the zeros of the zeta function, and established by Hardy, Littlewood and others, might be easy to establish on a Tauberian basis. Such a formula as

(10.45) $$\lim_{\xi\to 0} \xi \sum_{1}^{\infty} (\Lambda(n)-1)\, n^{\alpha}\, (n^{-\alpha}\, \xi^{-\alpha}\, N_1(n\xi)) = 0 \quad \left(0 < \alpha < \frac{1}{2}\right)$$

which follows from (9.10) much as does (9.12), appears to lend a certain color to this view. Here the theorem to which this seems to lead is

(10.46) $$\lim_{N\to\infty} \frac{1}{N} \sum_{1}^{N} (\Lambda(n)-1)\, n^{\alpha} = 0,$$

and the condition of non-vanishing on the Fourier transform becomes the Riemann condition
(10.47) $$\zeta(ix+1-\alpha) \neq 0.$$

The author considers these hopes illusory and deceptive. Let it be noted that (10.45) does not form a satisfactory hypothesis to a Tauberian theorem until we have some hold on the boundedness of the mass distribution whose integral is

(10.48) $$M(u) = \sum_{1}^{[u]} (\Lambda(n)-1)\, n^{\alpha}.$$

Such information would already presuppose as much information as (10.45) can yield concerning all smaller values of α than occur in (10.46). In other words, Tauberian theorems merely transform a O into a o.

4

Another way of stating the same thing is to say that a Tauberian theorem always operates in the neighborhood of a single ordinate in the plane of the zeta function. This is because it depends on a division of the range of this function into near and remote parts, and because this division has validity in the theory of functions of a real variable, not in the theory of functions of a complex variable. On the other hand, the more refined properties of the distribution of primes depend on the behavior of the zeta function in the entire strip between ordinates $\frac{1}{2}$ and 1, inclusive, and can only be discussed with the aid of Cauchy's theorem.

Of course, no proof of the limitations of so vague a thing as a method has real mathematical cogency. At any time some super-Tauberian theorem may come to light and prove to be central in the utmost refinements of prime number theory. For the present, however, Tauberian theorems do not seem to lie on the main avenue of progress.

CHAPTER V.

SPECIAL APPLICATIONS OF TAUBERIAN THEOREMS.

11. **On the proof of special Tauberian theorems.** In the sequel, we shall show that the greater part of all known Tauberian theorems may be proved without great difficulty on the basis of the general theorems of the present paper. However, most of the particular theorems were proved in the first instance by entirely different methods. In individual cases, these methods are simpler and more direct than the general method here indicated. This is especially noticeable in the case of Karamata's[40] proof of the original Abel-Tauber theorem.

Being in possession of a general method, we may consider with advantage the particular methods and why they function. All Tauberian theorems of the type discussed in this paper are intimately related to the solution of an integral equation of the form

$$\int_{-\infty}^{\infty} F(y)\, G(x-y)\, dy = H(x). \tag{11.01}$$

The most direct and general method of solving such an equation is by the use of Fourier transforms. Nevertheless, there are many cases in which a repeated differentiation will reduce such an equation to a linear differential equation of finite order, and many more where the same repeated differentiation will lead to a differential equation of infinite order, but of manageable form. Thus it is appropriate in many cases to employ a technique

[40] Karamata (2), (3), (4).

of repeated differentiation, and this has been done by Hardy, Littlewood[41], and Vijayaraghavan,[42] though scarcely from an explicit consideration of the integral equations in question. So far, the successes of this method have been confined to cases where the analytic properties of the Fourier transform of F are extremely simple, and it has failed to throw any light on the Tauberian theorems of prime number theory.

The methods of Robert Schmidt[43] lie more along the lines of the present paper, in as much as he has seen the essential role played by the integral

$$\int_0^y K(x)\, x^u\, dx = R(u) \tag{11.02}$$

in the study of the kernel of a Tauberian theorem. However, he has devoted his attention to $R(u)$ for real integral arguments instead of for complex general arguments. Furthermore, Schmidt's general theorem concerns the unicity of the solution of his moment problem rather than its existence theory. As a consequence there is a wide gap between his general moment theorem and the particular Tauberian theorems which he obtains as corollaries. This gap he actually fills in in two cases, that of the Abel-Tauber theorem and that of the Borel-Tauber theorem, but he gives no general method by which it may be filled in in a new case. His actual procedure is closely allied to that of Hardy, Littlewood and Vijayaraghavan. Schmidt's chief service to the subject is in his great improvement of what may be called the auxiliary apparatus of the theory of Tauberian theorems, through his invention of the notions of "langsam abfallende Funktionen" and "gestrahlte Matrizen".

Karamata's elegant method leads to the study of the closure of a set of translations of a given function, and thus most closely approximates to that developed here. His function is

$$f(x) = e^x e^{-e^x} \tag{11.03}$$

and the problem is solved through Weierstrass' theory of polynomial approximation. The translations considered are accordingly those of the form

$$f(x + \log \mu). \tag{11.04}$$

Szász has carried further the study of this particular set of translations of a given function. This is a far more difficult study than that of the

[41] Cf. Littlewood (1).

[42] Vijayaraghavan (1), (2).

[43] Schmidt (1), (2).

closure of the complete set of translations, and here the general solution of the closure problem is not yet known to me.

12. Examples of kernels for which Tauberian theorems hold. Among the kernels admissible in the role of the $N(x)$ of Theorem XIII are:

(1) The Riesz kernels[44]

$$N(x) = \begin{cases} (1-x)^{\lambda}; & [0<x<1] \\ 0 \quad ; & [1<x] \end{cases} \quad [\lambda>0]. \tag{12.01}$$

(2) The Abel kernel

$$N(x) = e^{-x}; \tag{12.02}$$

(3) The kernel

$$N(x) = \frac{1}{1+x^2}, \tag{12.03}$$

corresponding to the method of summation of the series Σa_n, which gives as its partial sum the Abel average

$$\lambda \int_0^\infty f(x)\, e^{-\lambda x}\, dx \tag{12.04}$$

of the cosine series

$$f(x) = \sum a_n \cos nx. \tag{12.05}$$

In the respective cases we have

(1)

$$\int_0^\infty N(x)\, x^{z-1}\, dx = \frac{\Gamma(\lambda+1)\,\Gamma(z)}{\Gamma(\lambda+z+1)}; \tag{12.06}$$

(2)

$$\int_0^\infty N(x)\, x^{z-1}\, dx = \Gamma(z); \tag{12.07}$$

(3)

$$\int_0^\infty N(x)\, x^{z-1}\, dx = \frac{\pi}{2 \sin \frac{\pi z}{2}}. \tag{12.08}$$

None of these functions vanishes for purely imaginary values of z.

A possible $N_1(\mu)$ of Theorem XI''' is

$$N_1(\mu) = \begin{cases} (\lambda+1)(1-\mu)^{\lambda}; & [0<\mu<1] \\ 0 & [1<\mu] \end{cases} \quad [\lambda \geqq 0]. \tag{12.09}$$

Here

$$\int_0^\infty N_1(\mu)\, \mu^{iu}\, d\mu = \frac{\Gamma(\lambda+2)\,\Gamma(iu+1)}{\Gamma(iu+\lambda+2)} \neq 0 \quad \text{if } u \text{ is real.} \tag{12.10}$$

[44] Riesz (1).

13. A theorem of Ramanujan. As a lemma in our further work, we shall find it convenient to introduce a theorem formulated by Ramanujan[45] and first proved by Hardy and Titchmarsh,[46] although with a formulation somewhat different from that here given. The theorem is intrinsically interesting, and is perhaps worth presenting in some detail.

Let $f(x)$ be a function of L_2, defined over $(0, \infty)$. Then $e^{x/2}f(e^x)$ will also belong to L_2 over $(-\infty, \infty)$; for

$$(13.01) \qquad \int_{-\infty}^{\infty} |e^{x/2} f(e^x)|^2 \, dx = \int_{-0}^{\infty} |f(x)|^2 \, dx.$$

The Fourier transform of $e^{x/2} f(e^x)$ will be

$$(13.02) \qquad \begin{aligned} h(u) &= \sqrt{\frac{1}{2\pi}} \operatorname*{l.i.m.}_{A\to\infty} \int_{-A}^{A} e^{iux} e^{x/2} f(e^x) \, dx \\ &= \frac{1}{\sqrt{2\pi}} \operatorname*{l.i.m.}_{\varepsilon\to 0} \int_{\varepsilon}^{1/\varepsilon} f(x)\, x^{iu-1/2} \, dx. \end{aligned}$$

The sine transform of $f(x)$ will be

$$(13.03) \qquad g_1(y) = \sqrt{\frac{2}{\pi}} \operatorname*{l.i.m.}_{B\to\infty} \int_0^B f(x) \sin xy \, dx,$$

and its cosine transform,

$$(13.04) \qquad g_2(y) = \sqrt{\frac{2}{\pi}} \operatorname*{l.i.m.}_{B\to\infty} \int_0^B f(x) \cos xy \, dx.$$

Let us put

$$(13.05) \qquad k_1(u) = \frac{1}{\sqrt{2\pi}} \operatorname*{l.i.m.}_{\varepsilon\to 0} \int_{\varepsilon}^{1/\varepsilon} g_1(y)\, y^{iu-1/2} \, dy,$$

and

$$(13.06) \qquad k_2(u) = \frac{1}{\sqrt{2\pi}} \operatorname*{l.i.m.}_{\varepsilon\to 0} \int_{\varepsilon}^{1/\varepsilon} g_2(y)\, y^{iu-1/2} \, dy.$$

We have

$$(13.07) \qquad \begin{aligned} \frac{1}{z}\int_0^z g_1(y)\, dy &= \sqrt{\frac{2}{\pi}} \int_0^{\infty} f(x)\, dx \frac{1}{z} \int_0^z \sin xy \, dy \\ &= \sqrt{\frac{2}{\pi}} \int_0^{\infty} f(x) \frac{1-\cos xz}{xz} \, dx. \end{aligned}$$

Similarly

$$(13.08) \qquad \frac{1}{z}\int_0^z g_2(y)\, dy = \sqrt{\frac{2}{\pi}} \int_0^{\infty} f(x) \frac{\sin xz}{xz} \, dx.$$

---

[45] Hardy and Littlewood (8).

[46] Hardy and Titchmarsh (1).

However, on an exponential transformation, these become

$$(13.09)\quad \int_{-\infty}^{\zeta} e^{\frac{\eta-\xi}{2}} (e^{\eta/2} g_2(e^{\eta}))\, d\eta = \sqrt{\frac{2}{\pi}} \int_{-\infty}^{\infty} \sin (e^{\xi+\zeta})\, e^{-\frac{\xi+\zeta}{2}} (e^{\xi/2} f(e^{\xi}))\, d\xi,$$

and

$$(13.10)\quad \int_{-\infty}^{\zeta} e^{\frac{\eta-\zeta}{2}} (e^{\eta/2} g_1(e^{\eta}))\, d\eta = \sqrt{\frac{2}{\pi}} \int_{-\infty}^{\infty} (1-\cos(e^{\xi+\zeta}))\, e^{-\frac{\xi+\zeta}{2}} (e^{\xi/2} f(e^{\xi}))\, d\xi.$$

If we now make a Fourier transformation and make use of the Parseval theorem, it appears that

$$(13.11)\quad k_1(u) \int_0^{\infty} e^{-\eta/2} e^{iu\eta}\, d\eta = \sqrt{\frac{2}{\pi}}\, h(-u) \int_{-\infty}^{\infty} (1 - \cos e^{\xi})\, e^{-\xi/2} e^{iu\xi}\, d\xi,$$

and

$$(13.12)\quad k_2(u) \int_0^{\infty} e^{-\eta/2} e^{iu\eta}\, d\eta = \sqrt{\frac{2}{\pi}}\, h(-u) \int_{-\infty}^{\infty} \sin e^{\xi}\, e^{-\xi/2} e^{iu\xi}\, d\xi.$$

Thus

$$(13.13)\quad \begin{aligned} \frac{k_1(u)}{h(-u)} &= \frac{\sqrt{\frac{2}{\pi}} \int_0^{\infty} (1 - \cos x)\, x^{iu-3/2}\, dx}{\int_0^{\infty} e^{-\eta(1/2 - iu)}\, d\eta} \\ &= \sqrt{\frac{2}{\pi}} \sin \left(iu + \frac{1}{2}\right) \frac{\pi}{2}\, \Gamma\left(iu + \frac{1}{2}\right). \end{aligned}$$

Here we make use of the formulae:

$$(13.131)\quad \int_0^{\infty} x^{\nu} \sin x\, dx = \Gamma(\nu+1) \cos \frac{\nu\pi}{2} \qquad [-1 \geqq \Re(\nu) > -2]$$

and

$$(13.132)\quad \int_0^{\infty} x^{\nu} (1 - \cos x)\, dx = \Gamma(\nu+1) \sin \frac{\nu\pi}{2} \qquad [-1 > \Re(\nu) > -3].$$

Similarly

$$(13.14)\quad \frac{k_2(u)}{h(-u)} = \sqrt{\frac{2}{\pi}} \cos \left(iu + \frac{1}{2}\right) \frac{\pi}{2}\, \Gamma\left(iu + \frac{1}{2}\right).$$

The duality of the relation between $k_1(u)$ or $k_2(u)$ and $h(u)$ is shown by the familiar formulae of the gamma function,

$$(13.15)\quad \begin{aligned} &\frac{2}{\pi} \sin \left(iu + \frac{1}{2}\right) \frac{\pi}{2} \sin \left(-iu + \frac{1}{2}\right) \frac{\pi}{2}\, \Gamma\left(iu + \frac{1}{2}\right) \Gamma\left(-iu + \frac{1}{2}\right) \\ &= \frac{2}{\pi} \cos \left(iu + \frac{1}{2}\right) \frac{\pi}{2} \cos \left(-iu + \frac{1}{2}\right) \frac{\pi}{2}\, \Gamma\left(iu + \frac{1}{2}\right) \Gamma\left(-iu + \frac{1}{2}\right) \\ &= \frac{1}{\pi} \cos iu\pi\, \Gamma\left(iu + \frac{1}{2}\right) \Gamma\left(-iu + \frac{1}{2}\right) = 1. \end{aligned}$$

Thus

$$\left|\frac{k_2(u)}{h(-u)}\right| = 1, \quad \left|\frac{k_1(u)}{h(-u)}\right| = 1, \tag{13.16}$$

and the real zeros of $k_1(u)$, $k_2(u)$, and $h(-u)$ are the same. In other words, if the integrals in question exist, the zeros of

$$\begin{cases} \dfrac{1}{\sqrt{2\pi}} \displaystyle\int_0^\infty f(x)\, x^{-iu-1/2}\, dx, \\ \dfrac{1}{\sqrt{2\pi}} \displaystyle\int_0^\infty g_1(y)\, y^{iu-1/2}\, dy, \\ \text{and } \dfrac{1}{\sqrt{2\pi}} \displaystyle\int_0^\infty g_2(y)\, y^{iu-1/2}\, dy, \end{cases} \tag{13.17}$$

are the same. This is a very valuable way of determining new kernels whose Fourier transforms (on a logarithmic scale) have no zeros.

14. The summation of trigonometrical developments. Let $f(x)$ be an even function of class L_2, and, as above, let

$$g_2(y) = \sqrt{\frac{2}{\pi}} \operatorname*{l.i.m.}_{B\to\infty} \int_0^B f(x) \cos xy\, dx. \tag{13.04}$$

Let $K(x)$ belong to L_2 over $(0, \infty)$, and let us form

$$x \int_0^\infty g_2(w)\, K(wx)\, dw. \tag{14.01}$$

This will equal

$$\frac{1}{x} \int_0^\infty z f(z) \frac{x}{z}\, k\left(\frac{z}{x}\right) dz, \tag{14.02}$$

where

$$k(z) = \sqrt{\frac{2}{\pi}} \operatorname*{l.i.m.}_{B\to\infty} \int_0^B K(w) \cos wz\, dw; \tag{14.03}$$

for the cosine transform of $xK(wx)$ is $k\left(\dfrac{z}{x}\right)$. Thus an average of $g_2(w)$ at infinity will appear formally as an average of $z f(z)$ about the origin, and an average of $g_2(w)$ about the origin will appear as an average of $z f(z)$ at infinity. The kernels in the two cases will be $K(x)$ and $\dfrac{1}{x} k(x)$. Now by § 13, we have formally

$$\left|\sqrt{\frac{1}{2\pi}} \int_0^\infty \frac{k(x)}{x} x^{\lambda}\, dx\right| = \left|\sqrt{\frac{1}{2\pi}} \int_0^\infty K(x)\, x^{-\lambda}\, dx\right|, \tag{14.04}$$

when $\mathfrak{R}(\lambda) = \frac{1}{2}$. If

$$(14.05)\qquad \sqrt{\frac{1}{2\pi}}\int_0^\infty \frac{k(x)}{x}x^\lambda\,dx \quad\text{and}\quad \sqrt{\frac{1}{2\pi}}\int_0^\infty K(x)\,x^{-\lambda}\,dx$$

converge for $0 \leqq \mathfrak{R}(\lambda) \leqq \frac{1}{2}$, it then follows that

$$(14.06)\qquad \sqrt{\frac{1}{2\pi}}\int_0^\infty \frac{k(x)}{x}x^{iu}\,dx \neq 0 \qquad [-\infty < u < \infty]$$

when, and only when,

$$(14.07)\qquad \frac{1}{\sqrt{2\pi}}\int_0^\infty K(x)\,x^{iu}\,dx \neq 0 \qquad [-\infty < u < \infty].$$

Thus if $k(x)/x$ in the role of $N_1(x)$, and $N_2(x)$ satisfy the other conditions specified in the hypothesis of Theorem XI, and (14.07) holds, it is possible to infer from

$$(14.08)\qquad \lim_{x\to {0 \atop \infty}} x\int_0^\infty g_2(w)\,K(wx)\,dw = A\int_0^\infty K(w)\,dw$$

to the conclusion that

$$(14.09)\qquad \lim_{x\to {0 \atop \infty}} x\int_0^\infty z\,f(z)\,N_2(zx)\,dz = A\int_0^\infty N_2(z)\,dz.$$

The condition that f belongs to L_2 may in many cases be very considerably altered and relaxed.

It is possible to read the relation between (14.01) and (14.02) in the reverse direction, and to infer from

$$(14.10)\qquad \lim_{x\to\infty}\int_0^\infty f(z)\,k\left(\frac{z}{x}\right)dz = A\,k(0)$$

and other conditions completing the hypothesis of Theorem XI, to

$$(14.11)\qquad \lim_{x\to 0} x\int_0^\infty g_2(w)\,K_2(wx)\,dw = A\int_0^\infty K_z(w)\,dw.$$

The condition that

$$(14.12)\qquad \frac{1}{\sqrt{2\pi}}\int_0^\infty k(x)\,x^{z-1}\,dx$$

should determine a function over a strip to the right of the origin, which, when continued has no zeros over the imaginary axis, becomes the same condition for

$$(14.13)\qquad \frac{1}{\sqrt{2\pi}}\int_0^\infty K(x)\,x^{-z}\,dx.$$

We thus arrive at a whole class of theorems relating the partial sum of a Fourier development with the average of the function represented about some point. For example, it is easy to prove:

THEOREM XVIII. *If $f(x)$ is a non-negative function of class L_2 over $(-\pi, \pi)$, if it is summable at any point by a Riesz mean of any positive order, or by an Abel mean, to any value A, it is so summable to A by Riesz means of all positive orders, and by an Abel mean. A necessary and sufficient condition that this should take place at a given point x is that*

$$\lim_{\varepsilon \to 0} \frac{1}{2\varepsilon} \int_{x-\varepsilon}^{x+\varepsilon} f(x)\, dx = A. \tag{14.14}$$

This theorem is due to Hardy and Littlewood.

15. Young's criterion for the convergence of a Fourier series. We now come to a region in which S. B. Littauer has done work, which constitutes the theorems proved in the present section. Young has proved the following theorem: *for the Fourier series of the integrable function $f(u)$ to converge to s for $u = x$, it is sufficient that*

$$\varphi(t) \to 0 \tag{15.01}$$

and that

$$\int_0^t |d(u\,\varphi(u))| = O(t) \tag{15.02}$$

for small t, where

$$\varphi(t) = \frac{1}{2}\{f(x+t)+f(x-t)-2s\}. \tag{15.03}$$

Further work on this theorem has been done by Young, Pollard, Hardy and Littlewood. The chief theorem to which Hardy and Littlewood came was the following:

THEOREM XIX. *A necessary and sufficient condition that the Fourier series of the integrable function $f(u)$ be summable $(C, -1+\delta)$ for any positive δ is that*

$$\varphi_1(t) = o(t) \tag{15.04}$$

[*provided that*

$$\int_0^t |d(u\,\varphi(u))| = O(t), \qquad [0 \leq t < \pi]] \tag{15.05}$$

where

$$\varphi_1(t) = \int_0^t \varphi(u)\, du. \tag{15.06}$$

This theorem in its original form is not directly adaptable to proof by the methods of this paper, inasmuch as Cesàro summation of fractional order does not depend on a kernel of the form $N(nx)$. On the other hand, if the Cesàro summation is replaced by Riesz summation, the theorem is reduced to a particular case of theorems already proved. We shall here content ourselves with proving this related theorem, as applied to Fourier integrals rather than series, in the case where $f(x)$ differs from zero only over a finite range.[47] Let us write

$$t\,\varphi(t) = \psi(t). \tag{15.07}$$

Then (15.05) becomes

$$\int_0^t |d\,\psi(t)| = O(t), \qquad [0 \leqq t < \pi] \tag{15.08}$$

which yields the boundedness of

$$\int_\mu^{2\mu} \lambda^{-1} |d\,\psi(\lambda)| - \int_\mu^{2\mu} \lambda^{-1}\, d\,\psi(\lambda). \tag{15.09}$$

Furthermore

$$\begin{aligned} \varphi_1(t) &= \int_0^t \varphi(u)\,du = \int_0^t \psi(u)\frac{du}{u} \\ &= \int_0^t \frac{du}{u}\int_0^u d\,\psi(v) = \int_0^t d\,\psi(v)\int_v^t \frac{du}{u} \\ &= \int_0^t \log\frac{t}{v}\, d\,\psi(v). \end{aligned} \tag{15.10}$$

Thus (15.04) becomes

$$\lim_{t\to 0}\frac{1}{t}\int_0^t \log\frac{t}{v}\, d\,\psi(v) = 0. \tag{15.11}$$

We have

$$\int_0^1 \log\lambda\ \lambda^{iu}\, d\lambda = -(iu+1)^{-2} \neq 0. \tag{15.12}$$

This is similar in form to (6.15).

The Riesz sum of kth order of the Fourier series of $f(x)-s$ for $x=0$ is

$$\begin{aligned} &\lim_{\varepsilon\to 0}\frac{1}{\varepsilon}\int_0^\infty \frac{\varphi(t)\,dt}{\Gamma(1+k)}\int_0^1 (1-\lambda)^k\cos\frac{\lambda t}{\varepsilon}\,d\lambda \\ &= \lim_{\varepsilon\to 0}\frac{1}{\varepsilon}\int_0^\infty \frac{dt}{t\Gamma(1+k)}\int_0^t d\psi(u)\int_0^1 (1-\lambda)^k\cos\frac{\lambda t}{\varepsilon}\,d\lambda \\ &= \lim_{\varepsilon\to 0}\frac{1}{\varepsilon}\int_0^\infty d\psi(u)\int_u^\infty \frac{dt}{t\Gamma(1+k)}\int_0^1 (1-\lambda)^k\cos\frac{\lambda t}{\varepsilon}\,d\lambda \\ &= \lim_{\varepsilon\to 0}\frac{1}{\varepsilon}\int_0^\infty d\psi(u)\int_{u/\varepsilon}^\infty \frac{dt}{t\Gamma(k+1)}\int_0^1 (1-\lambda)^k\cos\lambda t\,d\lambda. \end{aligned} \tag{15.13}$$

[47] It is possible to make the transition to Theorem XIX, but several of the steps require some consideration. In particular, the equivalence of Riesz and Cesàro summation for orders > -1 was proved by Riesz (Proc. Lond. Math. Soc., 2 (22), 412–419 (1924).

Here if $k > -1$,

$$\int_0^\infty \left| \int_0^1 (1-\lambda)^k \cos \lambda \tau \, d\lambda \right| d\tau \tag{15.14}$$

is finite, and the inversion of integration in (15.13) is hence applied to an absolutely convergent integral, in view of (15.08).

We have

$$\int_0^\infty \omega^{iu} \, d\omega \int_\omega^\infty \frac{d\tau}{\tau \Gamma(1+k)} \int_0^1 (1-\lambda)^k \cos \lambda \tau \, d\lambda$$

$$= \frac{1}{iu+1} \int_0^\infty \frac{\omega^{iu}}{\Gamma(1+k)} \, d\omega \int_0^1 (1-\lambda)^k \cos \lambda \omega \, d\lambda$$

$$= \frac{1}{iu+1} \int_0^\infty \frac{\omega^{iu}}{\Gamma(1+k)} \, d\omega \int_{\lambda=0}^{\lambda=1} (1-\lambda)^k \frac{d \sin \lambda \omega}{\omega}$$

$$= \frac{1}{(iu+1)\,\Gamma(k)} \int_0^\infty \omega^{iu-1} \, d\omega \int_0^1 (1-\lambda)^{k-1} \sin \lambda \omega \, d\lambda$$

$$= \frac{1}{(iu+1)\,\Gamma(k)} \int_0^1 (1-\lambda)^{k-1} \, d\lambda \int_0^\infty \omega^{iu-1} \sin \lambda \omega \, d\omega$$

$$= \frac{1}{(iu+1)\,\Gamma(k)} \int_0^1 (1-\lambda)^{k-1} \lambda^{-iu} \, d\lambda \int_0^\infty z^{iu-1} \sin z \, dz \tag{15.15}$$

$$= \frac{1}{(iu+1)\,\Gamma(k)} \frac{\Gamma(k)\,\Gamma(1-iu)}{\Gamma(k+1-iu)} \Gamma(iu) \cos (iu-1) \frac{\pi}{2}$$

$$= \frac{\Gamma(1-iu)\,\Gamma(iu)}{(iu+1)\,\Gamma(k+1-iu)} \sin \frac{\pi i u}{2}$$

$$= \frac{\pi \sin \frac{\pi i u}{2}}{(iu+1)\,\Gamma(k+1-iu) \sin \pi i u}$$

$$= \frac{\pi}{2(iu+1)\,\Gamma(k+1-iu) \cos \frac{\pi i u}{2}}$$

$$\neq 0.$$

In proving this, we have made use of (13.131) and (13.132).

The other conditions of the hypothesis of Theorem XI′ are readily proved to be satisfied, and we have already shown that if (15.05) is satisfied, (15.11) *and*

$$\lim_{\varepsilon \to 0} \frac{1}{\varepsilon} \int_0^\infty \frac{\varphi(t)\,dt}{\Gamma(1+k)} \int_0^1 (1-\lambda)^k \cos \frac{\lambda t}{\varepsilon} \, d\lambda = 0, \tag{15.16}$$

are equivalent. Thus if (15.05) *is satisfied,* (15.04) *is completely equivalent to the statement that the Fourier series of* $f(x)$ *is summable to* s *by Riesz (or Cesàro) sums of any given order exceeding* -1.

16. Tauberian theorems and asymptotic series. Certain asymptotic problems arise in the discussion of the behavior of an integral

$$\int_a^b (g(t))^x \Phi(t)\, dt \tag{16.01}$$

for large values of x. By a change of variable, this problem may be reduced to the consideration of

$$\int_0^\infty e^{-xt} \Phi(t)\, dt. \tag{16.02}$$

In particular, let us discuss the situation which arises when $\nu > 0$ and

$$\lim_{x\to\infty} x^\nu \int_0^\infty e^{-xt} \Phi(t)\, dt = A, \tag{16.03}$$

which we may write

$$\lim_{x\to\infty} x \int_0^\infty (x\,t)^{\nu-1} e^{-xt} \frac{\Phi(t)}{t^{\nu-1}}\, dt = A. \tag{16.04}$$

If

$$x \int_0^\infty (x\,t)^{\nu-1} e^{-xt} \frac{\Phi(t)}{t^{\nu-1}}\, dt \tag{16.05}$$

is bounded and $\Phi(t)$ is positive, it will follow by Theorem XI″ that

$$\lim_{x\to\infty} x \int_0^{1/x} \frac{\Phi(t)}{t^{\nu-1}}\, dt = \frac{A}{\int_0^\infty t^{\nu-1} e^{-t}\, dt} = \frac{A}{\Gamma(\nu)}, \tag{16.06}$$

or that

$$\lim_{\varepsilon\to 0} \frac{1}{\varepsilon} \int_0^\varepsilon \frac{\Phi(t)}{t^{\nu-1}}\, dt = \frac{A}{\Gamma(\nu)}. \tag{16.07}$$

Again, it will follow that

$$\lim_{x\to\infty} x \int_{1/x}^\infty \frac{\Phi(t)}{t^{\nu-1}} \frac{dt}{x^2 t^2} = \frac{A}{\Gamma(\nu)}, \tag{16.08}$$

which yields

$$\lim_{\varepsilon\to 0} \varepsilon \int_0^\varepsilon t^{\nu-1} \Phi\left(\frac{1}{t}\right) dt = \frac{A}{\Gamma(\nu)}. \tag{16.09}$$

The condition that

$$\Gamma(iu+\nu) = \int_0^\infty e^{-t}\, t^{iu+\nu-1}\, dt \neq 0 \tag{16.10}$$

is obviously satisfied. Indeed the Tauberian theorems of this section (which are due to Ikehara[48]) differ only from those of the last section in that x tends to infinity instead of to 0.

If $\frac{\Phi(t)}{t^{\nu-1}}$ is bounded and tends to B at the origin and $\nu > 0$,

$$\lim_{x\to\infty} x \int_0^\infty (xt)^{\nu-1} e^{-xt} \left[\frac{\Phi(t)}{t^{\nu-1}} - B\right] dt = 0. \tag{16.11}$$

Then

$$\lim_{x\to\infty} x \int_0^\infty (xt)^{\nu-1} e^{-xt} \frac{\Phi(t)}{t^{\nu-1}}\, dt = B x^\nu \int_0^\infty t^{\nu-1} e^{-xt}\, dt = B\Gamma(\nu). \tag{16.12}$$

Thus

$$\lim_{x\to\infty} x^\nu \int_0^\infty e^{-xt}\, \Phi(t)\, dt = B\,\Gamma(\nu). \tag{16.13}$$

In particular, let there be a neighborhood of the origin in which $\Phi(t)$ is $\nu-1$ times differentiable, with a bounded derivative of the $(\nu-1)$st order. Let

$$\Phi(0) = 0; \quad \Phi'(0) = 0; \quad \cdots \quad \Phi^{(\nu-2)}(0) = 0; \quad \Phi^{(\nu-1)}(0) = A \tag{16.14}$$

and let $\frac{\Phi(t)}{t^{\nu-1}}$ be bounded outside the neighborhood in question. Then

$$\lim_{x\to\infty} x^\nu \int_0^\infty e^{-xt}\, \Phi(t)\, dt = A. \tag{16.03}$$

If now $\Phi(t) - A t^{\nu-1}$ has a bounded derivative of order $\mu - 1 > \nu - 1$ at the origin, and vanishes there with all its derivatives of order less than $\mu - 1$, and $\Psi(t)$ is a function which equals $\Phi(t) - \frac{A t^{\nu-1}}{\Gamma(\nu)}$ in some neighborhood of the origin, and for which $\frac{\Psi(t)}{t^{\mu-1}}$ is bounded and tends to A at the origin, we have

$$\lim_{x\to\infty} x^\mu \int_0^\infty e^{-xt}\, \Psi(t)\, dt = A. \tag{16.15}$$

By a repetition of this process, it is easy to show that if

$$\Phi(t) = \frac{A_1 t^{\nu_1 - 1}}{\Gamma(\nu_1)} + \frac{A_2 t^{\nu_2 - 1}}{\Gamma(\nu_2)} + \cdots + \frac{A_k t^{\nu_k - 1}}{\Gamma(\nu_k)} + \Psi(t), \tag{16.16}$$

if $\Psi(t)$ is bounded and $(\mu - 1)$ times differentiable with bounded $(\mu - 1)$st derivative in some neighborhood of the origin, if

[48] In a paper not yet published.

$$(16.17) \qquad \Psi(0) = 0; \quad \Psi'(0) = 0; \quad \cdots \quad \Psi^{(\mu-1)}(0) = 0,$$

and if $\mu > \nu_1, \mu > \nu_2, \cdots, \mu > \nu_k$, then

$$(16.18) \qquad \int_0^\infty e^{-xt}\,\Phi(t)\,dt = A_1 x^{-\nu_1} + A_2 x^{-\nu_2} + \cdots + A_k x^{-\nu_k} + O(x^{-\mu+\varepsilon})$$

as $x \to \infty$. This is an adequate basis for the theory of asymptotic expansions of Laplace integrals, and enables Wintner's work[49] on the subject to be greatly simplified.

CHAPTER VI.

KERNELS ALMOST OF THE CLOSED CYCLE.

17. **The reduction of kernels almost of the closed cycle to kernels of the closed cycle.** In the present section, we shall approach very close to the work of Robert Schmidt[50] in his discussion of "gestrahlte Mittelbildungen", although our terminology will be somewhat different. Up to this point, we have been discussing means of the form

$$(17.01) \qquad \int_{-\infty}^{\infty} K_1(\xi - x)\,f(\xi)\,d\xi$$

or

$$(17.02) \qquad \int_{-\infty}^{\infty} K_1(\xi - x)\,d\varphi(\xi),$$

or means only differing from these by a change of variable from ξ to a function of ξ. Let us now turn our attention to means of the form

$$(17.03) \qquad \int_{-\infty}^{\infty} K_1(\xi, x)\,d\varphi(\xi),$$

where

$$(17.04) \qquad K_1(\xi, x) = K_1(\xi - x) + K_1^*(\xi, x)$$

where $K_1(x)$ satisfies the condition that (5.02) converges and that (5.07) does not vanish, and where

$$(17.05) \qquad \lim_{x\to\infty} \sum_{-\infty}^{\infty} \max_{n \leq y \leq n+1} |K_1^*(y, x)| = 0.$$

Let $K_1(x, y)$ and $K_1(x)$ be continuous, and let

$$(17.06) \qquad \int_{-\infty}^{\infty} K_1(x, y)\,d\varphi(x) \quad \text{and} \quad \int_y^{y+1} |d\varphi(x)| - \varphi(y+1) + \varphi(y)$$

[49] Wintner (1).
[50] Schmidt (2).

be bounded. Let

$$\lim_{x\to\infty}\int_{-\infty}^{\infty} K_1(\xi, x)\, d\varphi(\xi) = A\int_{-\infty}^{\infty} K_1(\xi)\, d\xi. \tag{17.061}$$

The argument of Lemma XIa will need no substantial alteration to show that there is a Q such that

$$\int_{y}^{y+1} |d\varphi(x)| \leq Q. \tag{6.04}$$

It will then follow from (17.05) that

$$\lim_{x\to\infty}\int_{-\infty}^{\infty} K_1(\xi - x)\, d\varphi(\xi) = A\int_{-\infty}^{\infty} K_1(\xi)\, d\xi \tag{5.03}$$

and that

$$\int_{-\infty}^{\infty} K_1(\xi - x)\, d\varphi(\xi)$$

is bounded. Hence, by Theorem X, if K_2 is any function belonging to N_1,

$$\lim_{x\to\infty}\int_{-\infty}^{\infty} K_2(\xi - x)\, d\varphi(\xi) = A\int_{-\infty}^{\infty} K_2(\xi)\, d\xi. \tag{5.04}$$

In Chapter III, the kernel $K(\xi - x)$ is replaced by a kernel of the form

$$M(\xi + \alpha - x) - M(\xi - x) \tag{17.07}$$

where $M(x)$ is a monotone function for which

$$M(-\infty) = 1, \qquad M(x) = O(e^{-\mu x}) \text{ at } +\infty; \qquad [\mu > 0] \tag{7.13}$$

and for which the function

$$\frac{1}{\sqrt{2\pi}}\int_{-\infty}^{\infty} M(x)\, e^{xz}\, dx \tag{7.05}$$

which is defined over a strip to the right of the origin, when continued analytically on to the imaginary axis, has no zeros there. We can replace the kernel $M(\xi - x)$ by a kernel of the form

$$M(\xi, x) = M_1(\xi - x) + M_2(\xi, x) \tag{17.08}$$

where $M_1(x)$ satisfies the conditions we have already laid down for $M(x)$, and where

$$\lim_{x\to\infty}\int_{0}^{\infty} |dM_2(\xi, x)| = 0 \tag{17.09}$$

and

$$M_2(\xi, x) = O(e^{-\mu\xi}) \text{ at } \infty \text{ uniformly.} \tag{17.10}$$

Under this change, Theorem XIV still remains valid.

18. A Tauberian theorem of Hardy and Littlewood.[51] Hardy and Littlewood have proved the following theorem:

THEOREM XX. *Let* $f(x) = \sum a_n x^n$ *be a power series with positive coefficients, and let*

(18.01) $$f(x) \sim \frac{A}{(1-x)^\alpha} \quad (\alpha > 0,\ x \to 1).$$

Then

(18.02) $$\sum_1^n a_k \sim \frac{n^\alpha A}{\Gamma(\alpha+1)} \quad (n \sim \infty).$$

We may write (18.01)

(18.03) $$\sum a_n e^{-n\xi} \sim A\,\xi^{-\alpha}$$

or

(18.04) $$\lim_{\xi \to 0} \xi \sum \frac{a_n}{n^{\alpha-1}} (n\xi)^{\alpha-1} e^{-n\xi} = A.$$

This falls under (6.30), and

(18.05) $$\int_0^\infty \xi^{\alpha-1} e^{-\xi} \xi^{iu}\, d\xi = \Gamma(\alpha+iu) \neq 0.$$

Since $a_n/n^{\alpha-1} > 0$,

(18.06) $$\lim_{\xi \to 0} \xi \sum_1^{[1/\xi]} \frac{a_n}{n^{\alpha-1}} (n\xi)^{\alpha-1} = A \frac{\int_0^1 \xi^{\alpha-1}\, d\xi}{\int_0^\infty \xi^{\alpha-1} e^{-\xi}\, d\xi} = \frac{A}{\Gamma(\alpha+1)}$$

because of Theorem XI‴. Writing m for $1/\xi$, and letting m become infinite through integral values, we get

(18.07) $$\lim_{m \to \infty} \frac{1}{m^\alpha} \sum_1^m a_n = \frac{A}{\Gamma(\alpha+1)},$$

which is only another way of writing (18.02).

This however is not the most general theorem proved by Hardy and Littlewood in this connection. They show that if

(18.08) $$f(x) \sim \frac{A}{(1-x)^\alpha} L\left(\frac{1}{1-x}\right),$$

where

(18.09) $$L(u) = (\log u)^{\alpha_1} (\log\log u)^{\alpha_2} \cdots (\log^{(n)} u)^{\alpha_n}$$

and

(18.10) $$u^\alpha L(u) \neq O(1) \text{ at } \infty,$$

then

(18.11) $$\sum_1^n a_k \sim \frac{n^\alpha A\, L(n)}{\Gamma(\alpha+1)}.$$

[51] Hardy and Littlewood (4).

We may write our theorem to be proved in the form that if

$$\xi \sum_{1}^{\infty} \frac{a_n}{n^{\alpha-1}} (n\xi)^{\alpha-1} e^{-n\xi} \sim AL(\xi^{-1}), \tag{18.12}$$

then (18.11) follows. This we may again write

$$e^{-\mu} \sum_{1}^{\infty} \frac{a_n}{n^{\alpha-1}} e^{(\log n-\mu)(\alpha-1)} e^{-e^{\log n-\mu}} \sim AL(e^{\mu}) \tag{18.13}$$

as $\mu \to \infty$. Now, if $M(\xi)$ is a continuous function, defined over $(0, \infty)$, and asymptotic to $L(\xi)$ at ∞, we may show that:

$$\int_0^{\infty} e^{(u-\mu)\alpha} e^{-e^{(u-\mu)}} M(e^u)\,du \sim M(e^{\mu}) \int_{-\infty}^{\infty} e^u e^{-e^u}\,du \sim L(e^{\mu}). \tag{18.14}$$

Here $M(\xi)$ is introduced instead of $L(\xi)$ in order that we may have no trouble with finite singularities of $L(\xi)$.

To prove (18.14), let us reflect that

$$\begin{aligned} \int_0^{\infty} e^{(u-\mu)\alpha} e^{-e^{(u-\mu)}} M(e^u)\,du &= \int_0^{\mu-A} e^{(u-\mu)\alpha} e^{-e^{(u-\mu)}} M(e^u)\,du \\ &\quad + \int_{\mu-A}^{\mu+A} e^{(u-\mu)\alpha} e^{-e^{(u-\mu)}} M(e^u)\,du \\ &\quad + \int_{\mu+A}^{\infty} e^{(u-\mu)\alpha} e^{-e^{(u-\mu)}} M(e^u)\,du. \end{aligned} \tag{18.141}$$

Since $M(\xi)$ is asymptotic to $L(\xi)$, which is asymptotically increasing, the first integral is asymptotically less than

$$\begin{aligned} &L(e^{u}) \int_0^{\mu-A} e^{(u-\mu)\alpha} e^{-e^{(u-\mu)}}\,du \\ &\leqq L(e^{\xi}) \int_{-\infty}^{-A} e^u e^{-e^u}\,du \leqq L(e^{\mu})\, e^{-A}. \end{aligned} \tag{18.142}$$

As to the second integral, it is asymptotically

$$L(e^{\mu}) \int_{-A}^{A} e^{\alpha u} e^{-e^u}\,du, \tag{18.143}$$

while the third one is ultimately less than

$$\begin{aligned} &e^{A^2} e^{-e^A} \int_{\mu+A}^{\infty} e^{-(u-\mu)} L(e^u)\,du \\ &\leqq e^{A^2} e^{-e^A} e^{\mu} \int_{\mu}^{\infty} L(e^u)\, d(-e^{-u}) \\ &= e^{A^2} e^{-e^A} \left[L(e^{\mu}) + e^{\mu} \int_{\mu}^{\infty} e^{-u} \frac{d}{du} (L(e^u))\,du \right] \\ &\leqq e^{A^2} e^{-e^A} \left[L(e^{\mu}) + C e^{\mu} \int_{\mu}^{\infty} e^{-u} u^{\alpha_1 - 1/2}\,du \right]. \end{aligned} \tag{18.144}$$

5

Now,

$$
\begin{aligned}
\int_{\mu}^{\infty} e^{-u} u^{\alpha_1-1/2} du &= -\int_{\mu}^{\infty} u^{\alpha_1-1/2} d(e^{-u}) \\
&= e^{-\mu} \mu^{\alpha_1-1/2} + \int_{\mu}^{\infty} (\alpha_1 - \tfrac{1}{2}) e^{-u} u^{\alpha_1-3/2} du \\
&\leqq e^{-\mu} \left(\mu^{\alpha_1-1/2} + \int_{\mu}^{\infty} (\alpha_1 - \tfrac{1}{2}) u^{\alpha_1-3/2} du \right) \\
&\leqq 2 e^{-\mu} \mu^{\alpha_1-1/2}.
\end{aligned} \tag{18.145}
$$

Thus by (18.144)

$$
\int_{\mu+A}^{\infty} e^{(u-\mu)\alpha} e^{-e^{(u-\mu)}} M(e^u) du \leqq e^{A^2} e^{-e^A} [L(e^{\mu}) + o(L(e^{\mu}))]. \tag{18.146}
$$

Combining this with (18.141), (18.142) and (18.143), we see that

$$
\begin{aligned}
\overline{\lim_{\mu\to\infty}} \left| \int_0^{\infty} e^{(u-\mu)\alpha} e^{-e^{(u-\mu)}} M(e^u) du - L(e^{\mu}) \int_{-\infty}^{\infty} e^{u} e^{-e^{u}} du \right| \Big/ L(e^{\mu}) \\
\leqq \varphi(A)
\end{aligned} \tag{18.147}
$$

where $\varphi(A)$ vanishes as A becomes infinite. Since A is arbitrarily large, (18.14) follows.

Consequently, by (18.14), if we put

$$
d\psi(u) = \frac{1}{u} d \sum_{0}^{[u]} \frac{a_n}{n^{\alpha-1}} - A M(u), \tag{18.15}
$$

(18.12) becomes

$$
\int_0^{\infty} (w\xi)^{\alpha} e^{-w\xi} d\psi(w) \to 0. \tag{18.16}
$$

On the other hand, we may write (18.11) in the form

$$
\int_0^{1/\xi} (w\xi)^{\alpha} d\psi(w) \to 0. \tag{18.17}
$$

Both integrals, (18.16) and (18.17), converge absolutely. We wish to make the transition from (18.16) to (18.17).

If $\alpha_1 \leqq 1$, $\psi(w)$ is "slowly decreasing" in the sense of Schmidt, we can apply Theorem XV, and it appears that $\psi(\lambda) \to 0$ as $\lambda \to \infty$. From this it follows that

$$
\int_0^{1/\xi} (w\xi)^{\alpha} d\psi(w) = \psi\left(\frac{1}{\xi}\right) - \psi(0) - \alpha \int_0^{1/\xi} \xi^{\alpha} w^{\alpha-1} \psi(w) dw \to 0. \tag{18.18}
$$

We now come to the more general case where $\alpha_1 > 1$. Our hypothesis becomes

$$
e^{-\mu} \sum \frac{a_n}{n^{\alpha-1} (\log n)^{[\alpha_1]}} e^{(\log n-\mu)(\alpha-1)} e^{-e^{\log n-\mu}} \frac{\log n^{[\alpha_1]}}{\mu} \sim A \Lambda(e^{\mu}), \tag{18.19}
$$

where

$$\Lambda(u) = (\log u)^{\alpha_1-[\alpha_1]} (\log\log u)^{\alpha_2} \cdots (\log^{(n)} u)^{\alpha_n}. \tag{18.20}$$

The kernel of (18.19), in the sense of § 17, is

$$\begin{aligned} & e^{(\alpha-1)(y-x)} e^{-e^{y-x}} \left\{\frac{y}{x}\right\}^{[\alpha_1]} = e^{(\alpha-1)(y-x)} e^{-e^{y-1}} \left\{1 - \frac{x-y}{x}\right\}^{[\alpha_1]} \\ & + \left\{- [\alpha_1]\frac{x-y}{x} + \frac{[\alpha_1]\{[\alpha_1]-1\}}{2} \frac{(x-y)^2}{x^2} + \cdots + \frac{(x-y)}{x^{[\alpha_1]}}\right\}. \end{aligned} \tag{18.21}$$

It hence appears that this kernel is of the form indicated in (17.04) where K_1 and K_1^* satisfy the appropriate conditions. Thus if we put $M^*(\xi) \sim \Lambda(\xi)$ at ∞ and $M^*(\xi)$ is continuous, and if

$$d\psi^*(u) = \frac{1}{u} d \sum_0^{[u]} \frac{a_n}{n^{\alpha-1} (\log n)^{[\alpha_1]}} - A M^*(u), \tag{18.22}$$

it follows just as in the case where $\alpha_1 \leqq 1$ that

$$\int_0^{1/\xi} (w\xi)^\alpha \, d\psi^*(w) \to 0. \tag{18.23}$$

This however is only another way of writing (18.11).

19. The Tauberian theorem of Borel summation.[52] The Borel sum of the series with partial sums s_m is

$$\lim_{x\to\infty} e^{-x} \sum_0^\infty \frac{s_n x^n}{n!} = s. \tag{19.01}$$

Let us put $n = [u^2]$ and

$$\sum_1^n s_m (\sqrt{m} - \sqrt{m-1}) = f(u) \tag{19.02}$$

for $u > 0$. Let $f(u) = 0$ for negative u. Let us also put $x = y^2$. Then (19.01) becomes

$$\lim_{y\to\infty} \int_0^\infty \frac{e^{-y^2} y^{2u^2}}{\Gamma(u^2+1)} \frac{d f(u)}{u - \sqrt{u^2-1}} = s. \tag{19.03}$$

Now

$$\begin{aligned} \frac{e^{-y^2} y^{2u^2}}{\Gamma(u^2+1)(u-\sqrt{u^2-1})} &= \sqrt{\frac{2}{\pi}} e^{u^2-y^2} \left(\frac{y}{u}\right)^{2u^2} \left(1 + O\left(\frac{1}{u^2}\right)\right) \\ &= \sqrt{\frac{2}{\pi}} e^{-2(u-y)^2} e^{3u^2-4uy+y^2} \left(\frac{y}{u}\right)^{2u^2} \left(1 + O\left(\frac{1}{u^2}\right)\right). \end{aligned} \tag{19.04}$$

[52] For this section, cf. Schmidt (2); Vijayaraghavan (1).

Thus over the range $y - y^{1/6} \leqq u \leqq y + y^{1/6}$,

$$\frac{e^{-y^2} y^{2u^2}}{\Gamma(u^2+1)(u - \sqrt{u^2-1})} - \sqrt{\frac{2}{\pi}}\, e^{-2(u-y)^2} = \sqrt{\frac{2}{\pi}}\, e^{-2(u-y)^2} O\left(\frac{1}{y^2}\right) \tag{19.05}$$

because

$$e^{3u^2-4uy+y^2}\left(\frac{y}{u}\right)^{2u^2}$$

$$\begin{aligned} &= \exp\left\{3u^2 - 4uy + y^2 + 2u^2\left[\frac{y-u}{u} - \frac{(y-u)^2}{2u^2} + O\left(\frac{(y-u)^3}{u^3}\right)\right]\right\} \\ &= 1 + O\left(\frac{(y-u)^3}{u}\right) \\ &= 1 + O(y^{-1/2}). \end{aligned} \tag{19.06}$$

Moreover, it is always true that

$$\begin{aligned} \frac{e^{-y^2} y^{2u^2}}{\Gamma(u^2+1)(u - \sqrt{u^2-1})} &= \sqrt{\frac{2}{\pi}}\, e^{u^2-y^2}\left(\frac{y}{u}\right)^{2u^2}\left(1 + O\left(\frac{1}{u^2}\right)\right) \\ &\leqq \sqrt{\frac{2}{\pi}}\, e^{u^2-y^2}\, e^{\left(\frac{y-u}{u}\right)2u^2}\left(1 + O\left(\frac{1}{u^2}\right)\right) \\ &= \sqrt{\frac{2}{\pi}}\, e^{-(u-y)^2}\left(1 + O\left(\frac{1}{u^2}\right)\right). \end{aligned} \tag{19.07}$$

Thus the conditions (17.05) and (17.06) are satisfied if we put

$$\frac{e^{-y^2} y^{2u^2}}{\Gamma(u^2+1)(u - \sqrt{u^2-1})} = K_1(u, y) \tag{19.08}$$

and

$$\sqrt{\frac{2}{\pi}}\, e^{-2(u-y)^2} = K_1(u-y). \tag{19.09}$$

It follows at once that

$$\lim_{y\to\infty} \sqrt{\frac{2}{\pi}} \int_{-\infty}^{\infty} e^{-2(u-y)^2}\, df(u) = s \tag{19.10}$$

and that

$$\sqrt{\frac{2}{\pi}} \int_{-\infty}^{\infty} e^{-2(u-y)^2}\, df(u) \tag{19.101}$$

is bounded.

We now introduce again in its appropriate form Robert Schmidt's definition of a "langsam abfallende Folge". The sequence $\{s_m\}$ has this property if, whenever $q = q(p)$ $(p = 0, 1, \cdots)$ runs through such a sequence of indices that

$$q \geqq p \quad \text{and} \quad \frac{q-p}{\sqrt{p}} \to 0 \tag{19.11}$$

then

(19.12) $$\varliminf_{p\to\infty} (s_q - s_p) \geqq 0.$$

Schmidt proves by elementary methods that if $s_0, s_1, \cdots$ is such a sequence, then

(19.13) $$C(\Lambda) = \varliminf_{\substack{p\to\infty \\ p\leqq q\leqq p+\Lambda\sqrt{p}}} (s_q - s_p)$$

exists, and

(19.14) $$\lim_{\Lambda\to 0} C(\Lambda) = 0.$$

Furthermore, he shows that there exists a constant K, such that

(19.15) $$s_q - s_p \geqq -K(\sqrt{q} - \sqrt{p})$$

for all $p = 0, 1, \cdots$ and all $q \geqq p + \sqrt{p}$. Thus he shows that a K exists such that

(19.16) $$\frac{s_p}{\sqrt{p+1}} \geqq -K \qquad (p = 0, 1, \cdots).$$

Let the sequence s_m be "langsam abfallend" and let (19.01) hold. Still following Schmidt, we see that

(19.17) $$\begin{aligned} e^{-x}\sum_0^\infty \frac{s_n x^n}{n!} &= \sum_0^{[x]} \frac{e^{-x} s_n x^n}{n!} + \sum_{[x]+1}^{[x+\sqrt{x}]} \frac{e^{-x} s_n x^n}{n!} + \sum_{[x+\sqrt{x}]+1}^{\infty} \frac{e^{-x} s_n x^n}{n!} \\ &\geqq -K\sum_0^\infty \frac{e^{-x}\sqrt{n+1}\, x^n}{n!} + s_{[x]} \sum_{[x]+1}^{[x+\sqrt{x}]} \frac{e^{-x} x^n}{n!} - C(1) \sum_{[x]+1}^{[x+\sqrt{x}]} \frac{e^{-x} x^n}{n!}. \end{aligned}$$

If we build up the appropriate $f(u)$ as in (19.02), and make the transformation which led from (19.01) to (19.03), we get:

(19.18) $$e^{-x}\sum_0^\infty \frac{\sqrt{n+1}\, x^n}{n!} \leqq \int_0^\infty \frac{e^{-x} x^{u^2}}{\Gamma(u^2+1)} \frac{d(u^2)}{u - \sqrt{u^2-1}}.$$

Again, by (19.07)

(19.19) $$e^{-x}\sum_0^\infty \frac{\sqrt{n+1}\, x^n}{n!} \leqq \int_{-\infty}^\infty e^{-(u-\sqrt{x})^2} O(u)\, du = O(\sqrt{x}).$$

Obviously

(19.20) $$\sum_{[x]+1}^{[x+\sqrt{x}]} \frac{e^{-x} x^n}{n!} < 1.$$

On the other hand

$$\lim_{x\to\infty}\sum_{[x]+1}^{[x+\sqrt{x}]}\frac{e^{-x}x^n}{n!}>\lim_{x\to\infty}\frac{e^{-x}x^x}{e^{-x-\sqrt{x}}(x+\sqrt{x})^x\sqrt{2\pi(x+\sqrt{x})}}[\sqrt{x}+1]$$

(19.21)
$$=\lim_{x\to\infty}\frac{e^{\sqrt{x}}}{\left(1+\frac{1}{\sqrt{x}}\right)^x\sqrt{2\pi}}=\left(\frac{e}{2\pi}\right)^{1/2}.$$

Thus (19.17) yields

(19.22) $$\overline{\lim_{x\to\infty}}\, x^{-1/2}s_{[x]}\leqq\overline{\lim_{x\to\infty}}\sqrt{\frac{2\pi}{e}}\,[s x^{-1/2}+KO(1)+C(1)\,x^{-1/2}]=O(1),$$

which we may simplify and write

(19.23) $$s_{[x]}=O(x^{1/2}).$$

Let us now introduce the function

(19.24) $$g_\alpha(x)=\frac{1}{\alpha^2}\left[\int_{x+\alpha}^{x+2\alpha}-\int_x^{x+\alpha}\right]f(y)\,dy.$$

By (19.02),

(19.25) $$g_\alpha(x)=\frac{1}{\alpha^2}\left[\int_{(x+\alpha)^2}^{(x+2\alpha)^2}-\int_{x^2}^{(x+\alpha)^2}\right]\frac{dM}{2\sqrt{M}}\sum_1^{[M]}s_m(\sqrt{m}-\sqrt{m-1}).$$

Now,

(19.26)
$$\begin{aligned}\int_0^\nu\frac{dM}{2\sqrt{M}}\sum_1^{[M]}s_m(\sqrt{m}-\sqrt{m-1})&=\sum_1^{[\nu]}s_m(\sqrt{m}-\sqrt{m-1})\int_m^\nu\frac{dM}{2\sqrt{M}}\\&=\sum_1^{[\nu]}s_m(\sqrt{m}-\sqrt{m-1})(\sqrt{\nu}-\sqrt{m}).\end{aligned}$$

Substituting this in (19.25), and writing S for an average of the s_n's for $x^2<n<(x+2\alpha)^2+1$, we have

$$\begin{aligned}g_\alpha(x)&=\frac{1}{\alpha^2}\sum_1^{[(x+2\alpha)^2]}s_m(\sqrt{m}-\sqrt{m-1})(x+2\alpha-\sqrt{m})\\&\quad-\frac{2}{\alpha^2}\sum_1^{[(x+\alpha)^2]}s_m(\sqrt{m}-\sqrt{m-1})(x+\alpha-\sqrt{m})\\&\quad+\frac{1}{\alpha^2}\sum_1^{[x^2]}s_m(\sqrt{m}-\sqrt{m-1})(x-\sqrt{m})\\&=\frac{1}{\alpha^2}\left\{\sum_{[x^2]+1}^{[(x+\alpha)^2]}s_m(\sqrt{m}-\sqrt{m-1})(\sqrt{m}-x)\right.\\&\quad\left.+\sum_{[(x+\alpha)^2]+1}^{[(x+2\alpha)^2]}s_m(\sqrt{m}-\sqrt{m-1})(x+2\alpha-\sqrt{m})\right\}\end{aligned}$$

$$(19.27) \quad = \frac{S}{\alpha^2}\left\{\sum_{[x^2]+1}^{[(x+\alpha)^2]} (\sqrt{m}-\sqrt{m-1})(\sqrt{m}-x) + \sum_{[(x+\alpha)^2]+1}^{[(x+2\alpha)^2]} (\sqrt{m}-\sqrt{m-1})(x+2\alpha-\sqrt{m})\right\}$$

$$= \frac{S}{\alpha^2}\left[\int_{x+\alpha}^{x+2\alpha} - \int_{x}^{x+\alpha}\right] \sqrt{[x^2]}\,dx$$

$$= S\left\{1+\frac{1}{\alpha^2}\left[\int_{x+\alpha}^{x+2\alpha} - \int_{x}^{x+\alpha}\right] (\sqrt{[x^2]}-x)\,dx\right\}$$

$$= S\left\{1+O\left(\frac{1}{x^2}\right)+\frac{1}{\alpha^2}\left[\sum_{[x^2]}^{[(x+\alpha)^2]} - \sum_{[(x+\alpha)^2]}^{[(x+2\alpha)^2]}\right]\frac{(\sqrt{m+1}-\sqrt{m})^2}{2}\right\}$$

$$= S\left(1+O\left(\frac{1}{x^2}\right)\right).$$

It is indeed easy to show that as x increases, the weight of the later s_n's averaged in S increases at the expense of the weight of the earlier ones.

We may show that the condition that $\{s_m\}$ is "langsam abfallend" may be put in the form that if

$$(19.28) \qquad v \geqq u, \qquad v-u\to 0$$

then

$$(19.29) \qquad \underline{\lim}_{u\to\infty} (s_{[v^2]}-s_{[u^2]}) \equiv 0.$$

Hence by (19.27) and (19.23), we may readily show that

$$(19.30) \qquad \underline{\lim}_{u\to\infty} (g_\alpha(v)-g_\alpha(u)) \geqq 0.$$

Again remembering (19.23), which makes all the integrals in question absolutely convergent, we have

$$(19.31) \quad \begin{aligned} &\lim_{y\to\infty} \sqrt{\frac{2}{\pi}}\int_{-\infty}^{\infty} e^{-2(u-y)^2} g_\alpha(u)\,du \\ &= \lim_{y\to\infty} \sqrt{\frac{2}{\pi}}\int_{-\infty}^{\infty} e^{-2(u-y)^2}\,du\,\frac{1}{\alpha}\int_{u}^{u+\alpha}\frac{dv}{\alpha}\int_{v}^{v+\alpha} df(w) \\ &= \lim_{y\to\infty} \frac{1}{\alpha^2}\sqrt{\frac{2}{\pi}}\int_{y}^{y+\alpha} dv\int_{v}^{v+\alpha} dw\int_{-\infty}^{\infty} e^{-2(u-w)^2}\,df(u) \\ &= \lim_{y\to\infty} \sqrt{\frac{2}{\pi}}\int_{-\infty}^{\infty} e^{-2(u-y)^2}\,df(u) = s. \end{aligned}$$

An integration by parts yields:

$$(19.32) \qquad s = \lim_{y\to\infty}\sqrt{\frac{2}{\pi}}\int_{-\infty}^{\infty} d\,g_\alpha(v)\int_{v}^{\infty} e^{-2(u-y)^2}\,du.$$

Similar methods establish the boundedness of

$$\sqrt{\frac{2}{\pi}} \int_{-\infty}^{\infty} d\, g_\alpha(v) \int_v^{\infty} e^{-2(u-y)^2}\, du. \tag{19.33}$$

We may then apply Theorem XIV, taking

$$M(x) = \sqrt{\frac{2}{\pi}} \int_x^{\infty} e^{-2u^2}\, du. \tag{19.34}$$

In this case

$$\varphi(z) = \sqrt{\frac{1}{2\pi}}\, e^{z^2/8}. \tag{19.35}$$

Hence

$$\lim_{x\to\infty} g_\alpha(x) = s. \tag{19.36}$$

Let us now return to (19.27), making use of the existence of (19.13) and of (19.14). Then if x is sufficiently large,

$$S - s_{[x^2]} \geqq C(5\alpha) \tag{19.37}$$

and

$$s_{[(x+2\alpha)^2]} - S \geqq C(5\alpha). \tag{19.38}$$

It then follows from (19.27) and (19.36) that

$$\overline{\lim_{n\to\infty}}\, s_n \leqq s - C(5\alpha) \tag{19.39}$$

and

$$\underline{\lim_{n\to\infty}}\, s_n \geqq s + C(5\alpha). \tag{19.40}$$

Combining these, and remembering that $C(\Delta) \leqq 0$, we see that the following theorem holds:

THEOREM XXI. *Let* (19.01) *hold, and let* s_m *be "langsam abfallend". Then*

$$\lim_{n\to\infty} s_n = s. \tag{19.41}$$

CHAPTER VII.

A QUASI-TAUBERIAN THEOREM.

20. **The quasi-Tauberian theorem.** Up to the present point, all the Tauberian theorems we have discussed have involved some auxiliary condition of boundedness or positiveness or slow decrease. In the present chapter, we shall discuss a theorem without any such auxiliary condition as to the function averaged. The type of theorem is so fundamentally

different from that already discussed that the nomenclature, "Tauberian", seems to the author unfortunate. The theorems now to be discussed are in essence much closer to Abel's theorem than to Tauber's theorem.

By an integration by parts, we establish the following:

LEMMA XXIIa. *Let*

$$\int_B^C |F(x)|\,|df(x)| \tag{20.01}$$

exist for every $C > B$, *and let*

$$\int_B^\infty F(x)\,df(x) \tag{20.02}$$

exist as the limit of $\int_B^C F(x)\,df(x)$ *as* $C \to \infty$. *Let*

$$\int_B^C |G(x)|\,|df(x)| \tag{20.03}$$

exist for each $C > B$, *let* $G(x)$ *and* $F(x)$ *be continuous, and let*

$$\frac{G(x)}{F(x)} \to L \text{ as } x \to \infty; \qquad \int_D^B \left| d\,\frac{G(x)}{F(x)} \right| < \infty. \tag{20.04}$$

Then

$$\int_B^\infty G(x)\,df(x) \tag{20.05}$$

will exist, and

$$\int_B^\infty G(x)\,df(x) = L\int_B^\infty F(x)\,df(x) - \int_B^\infty d\left(\frac{G(x)}{F(x)}\right)\int_B^x F(\xi)\,df(\xi). \tag{20.06}$$

Now let us suppose that $f(x)$ is of limited total variation over every finite interval, that $K_1(x)$ is bounded and continuous, that

$$\int_{-\infty}^\infty |d(K_1(x)\,e^{-\lambda x})| < \text{const.}, \tag{20.061}$$

and that

$$\int_D^\infty K_1(-x)\,d\,f(x) \tag{20.07}$$

exists in the sense of (20.02). Let us suppose further that as $x \to -\infty$

$$K_1(x) \sim A_1 e^{\lambda x}, \qquad (\lambda > 0) \tag{20.08}$$

where $A_1 \neq 0$. Then

$$\int_{-\infty}^{-D} \left| d \frac{e^{\lambda x}}{K_1(x)} \right| < \text{const.}, \tag{20.081}$$

if D is sufficiently large that for $x \leqq -D$,

$$|e^{-\lambda x} K_1(x)| \geqq \text{const.} > 0. \tag{20.082}$$

As a particular case of (20.06), if B is large enough,

$$\begin{aligned} &\int_B^\infty e^{-\lambda x}\, d\, f(x) \\ &= \frac{1}{A_1} \int_B^\infty K_1(-x)\, d\, f(x) - \int_B^\infty d\left(\frac{e^{-\lambda x}}{K_1(-x)}\right) \int_B^x K_1(-\xi)\, d\, f(\xi) \end{aligned} \tag{20.083}$$

and hence

$$\int_B^\infty e^{-\lambda x}\, d\, f(x) \tag{20.084}$$

is bounded. A further application of (20.06) yields us

$$\begin{aligned} &\int_B^\infty K_1(y-x)\, d\, f(x) \\ &= A_1 e^{\lambda y} \int_B^\infty e^{-\lambda x}\, d\, f(x) - \int_B^\infty d(K_1(y-x)\, e^{\lambda x}) \int_B^x e^{-\lambda \xi}\, d\, f(\xi). \end{aligned} \tag{20.085}$$

From this we may conclude that

$$\int_0^\infty K_1(y-x)\, d\, f(x) = \begin{cases} O(e^{\lambda y}) & \text{as } y \to \infty, \\ O(1) & \text{as } y \to -\infty \end{cases} \tag{20.09}$$

and that

$$\lim_{B\to\infty} e^{-\lambda y} \int_B^\infty K_1(y-x)\, d\, f(x) = 0 \tag{20.10}$$

uniformly in y.

Now let $R(z)$ be a function for which

$$\int_{-\infty}^\infty e^{-\lambda z}\, |dR(z)| \tag{20.11}$$

and

$$\int_{-\infty}^\infty |dR(z)| \tag{20.12}$$

are finite. It follows from (20.10) that

$$\begin{aligned} &\int_{-\infty}^\infty dR(y-z) \int_0^\infty K_1(z-x)\, d\, f(x) \\ &= \lim_{B\to\infty} \int_{-\infty}^\infty dR(y-z) \int_0^B K_1(z-x)\, d\, f(x) \\ &= \lim_{B\to\infty} \int_0^B d\, f(x) \int_{-\infty}^\infty K_1(z-x)\, d\, R(y-z) \\ &= \int_0^\infty d\, f(x) \int_{-\infty}^\infty K_1(y-x-z)\, d\, R(z). \end{aligned} \tag{20.13}$$

Thus if

$$\lim_{y \to \infty} \int_0^\infty K_1(y-x)\, d\, f(x) = A \int_{-\infty}^\infty K_1(x)\, dx \tag{20.14}$$

then

$$\int_0^\infty K_1(y-x)\, d\, f(x) \tag{20.15}$$

will be bounded, by (20.09), and it follows that

$$\begin{aligned}
&\lim_{y \to \infty} \int_0^\infty df(x) \int_{-\infty}^\infty K_1(y-x-z)\, dR(z) \\
&= \lim_{y \to \infty} \int_{-\infty}^\infty dR(y-z) \int_{-\infty}^\infty K_1(z-x)\, df(x) \\
&= \int_{-\infty}^\infty dR(-z) \lim_{y \to \infty} \int_0^\infty K_1(z+y-x)\, df(x) \qquad (20.16) \\
&= A \int_{-\infty}^\infty K_1(x)\, dx \int_{-\infty}^\infty dR(-z) \\
&= A \int_{-\infty}^\infty dx \int_{-\infty}^\infty K_1(x-z)\, dR(z).
\end{aligned}$$

Hence if $K_1(x)$ *is bounded and continuous, if* $f(x)$ *is of limited total variation over every finite interval, if* (20.14), (20.061) *and* (20.08) *hold, if* (20.11) *and* (20.12) *are finite, and if*

$$K_2(x) = \int_{-\infty}^\infty K_1(x-z)\, dR(z), \tag{20.17}$$

it follows that

$$\lim_{y \to \infty} \int_0^\infty K_2(y-x)\, df(x) = A \int_{-\infty}^\infty K_2(x)\, dx. \tag{20.18}$$

This is a sufficiently important theorem to dignify by a number; we shall call it Theorem XXII.

A closely related proposition is:

THEOREM XXIII. *In the hypothesis of Theorem* XXII, *in case*

$$K_1(x) = 0, \qquad [x > 0], \tag{20.19}$$

we may replace the assumption of the finiteness of (20.11) *and* (20.12) *by that of the finiteness of*

$$\int_{-\infty}^\infty |dR(z)|. \tag{20.20}$$

The conclusion remains valid.

To prove this, let us reflect that if (20.08) holds as $x \to \infty$, since

$$\int_0^\infty K_1(y-x)\,df(x) = \int_y^\infty K_1(y-x)\,df(x), \tag{20.21}$$

it will follow from (20.06) that

$$\begin{aligned}\int_y^\infty e^{-\lambda x}\,df(x) &= \frac{e^{-\lambda y}}{A_1}\int_y^\infty K_1(y-x)\,df(x) \\ &\quad - \int_y^\infty d\left(\frac{e^{-\lambda x}}{K_1(y-x)}\right)\int_y^x K_1(y-\xi)\,df(\xi) \\ &= e^{-\lambda y}\,O(1).\end{aligned} \tag{20.22}$$

By a precisely similar argument,

$$\begin{aligned}&\int_B^\infty K_1(y-x)\,df(x) \\ &= A_1 e^{\lambda y}\int_B^\infty e^{-\lambda x}\,df(x) - \int_B^\infty d_x\left(e^{\lambda x}K_1(y-x)\right)\int_B^\infty e^{-\lambda\xi}\,df(\xi) \\ &= A_1 e^{\lambda(y-B)}O(1) - \int_B^\infty d_x\left(e^{\lambda x}K_1(y-x)\right)\left(e^{-\lambda B}O(1)+e^{-\lambda x}O(1)\right) \\ &= O\left(e^{\lambda(y-B)}\right) \quad \textit{as} \quad B\to\infty\end{aligned} \tag{20.23}$$

if $B > y$. Thus by (20.21), $\int_B^\infty K_1(y-x)\,df(x)$ is less than a function which is bounded and decreases monotonely to 0 as B becomes infinite. Consequently, by a theorem of Daniell on the Stieltjes integral,[53] in combination with the finiteness of (20.21),

$$\lim_{B\to\infty}\int_{-\infty}^\infty dR(y-z)\int_B^\infty K_1(z-x)\,df(x) = 0. \tag{20.24}$$

Thus the inversions of integration in (20.13) are again permissible.

Formally and heuristically, (20.17) is equivalent to

$$\frac{\int_{-\infty}^\infty K_2(x)\,e^{ux}\,dx}{\int_{-\infty}^\infty K_1(x)\,e^{ux}\,dx} = \int_{-\infty}^\infty e^{ux}\,dR(x). \tag{20.25}$$

If K_1 and K_2 are both $O(e^{-(\mu+\varepsilon)x})$ at $+\infty$ and $O(e^{+\varepsilon x})$ at $-\infty$,

$$k_2(u) = \int_{-\infty}^\infty K_2(x)\,e^{ux}\,dx \tag{20.26}$$

and

$$k_1(u) = \int_{-\infty}^\infty K_1(x)\,e^{ux}\,dx \tag{20.27}$$

[53] Daniell (1).

are both analytic over $-\varepsilon < \Re(u) < \varepsilon + \mu$. Formally,

$$R(x) = \frac{1}{2\pi i} \int_0^x d\xi \int_{-\infty}^{\infty} \frac{k_2(u)}{k_1(u)} e^{-u\xi} du, \tag{20.28}$$

the second integral being taken along any ordinate in the strip $-\varepsilon < \Re(u) < \varepsilon + \mu$.

Now let $K_1(x)$ *belong to* L_2. *Let* $k_2(u)/k_1(u)$ *be analytic over* $-\varepsilon \leqq \Re(u) \leqq \mu + \varepsilon$, *and let it be quadratically summable over every ordinate in that strip. Then*

$$e^{-\varepsilon x} \operatorname*{l.i.m.}_{A\to\infty} \int_{-A}^{A} \frac{k_2(iu)}{k_1(iu)} e^{-iux} du \tag{20.29}$$

and

$$e^{(\mu+\varepsilon)x} \operatorname*{l.i.m.}_{A\to\infty} \int_{-A}^{A} \frac{k_2(iu)}{k_1(iu)} e^{-iux} du \tag{20.30}$$

are quadratically summable. As a consequence,

$$\operatorname*{l.i.m.}_{A\to\infty} \int_{-A}^{A} \frac{k_2(iu)}{k_1(iu)} e^{-iux} du \tag{20.31}$$

and

$$e^{\mu x} \operatorname*{l.i.m.}_{A\to\infty} \int_{-A}^{A} \frac{k_2(iu)}{k_1(iu)} e^{-iux} du \tag{20.32}$$

are absolutely summable. If we assume R to be defined as in (20.28) and $\mu = \lambda$, we obtain:

THEOREM XXII′. *Let* $K_1(x)$ *be bounded and continuous. Let* $f(x)$ *be of limited total variation over every finite interval. Let* (20.14), (20.061) *and* (20.08) *hold. Let* $k_2(u)$ *and* $k_1(u)$ *be defined as in* (20.26) *and* (20.27), *respectively. Let* $K_1(x)$ *belong to* L_2. *Let* $k_2(u)/k_1(u)$ *be analytic over* $-\varepsilon \leqq \Re(u) \leqq \lambda + \varepsilon$, *and let it belong to* L_2 *over every ordinate in that strip. Then* (20.18) *follows.*

Similarly, we have:

THEOREM XXIII′. *In the hypothesis of Theorem* XXII′, *in case* $K_1(x)$ *vanishes for positive arguments, we may replace the strip* $-\varepsilon \leqq \Re(u) \leqq \lambda + \varepsilon$ *by the narrower strip* $-\varepsilon \leqq \Re(u) \leqq \varepsilon$.

21. Applications of the quasi-Tauberian theorem. If

$$K^{(m)}(x) = \begin{cases} (m+1)(1-e^x)^m e^x; & [x < 0] \\ 0 & [x > 0] \end{cases} \tag{21.01}$$

and

$$^{(m)}K(x) = \frac{2e^x}{\pi} \int_0^1 (1-z)^m \cos(z e^x)\, dz; \qquad [m > 0] \tag{21.02}$$

then (20.08) is clearly satisfied for $\lambda = 1$, and

(21.03) $$K_1(x) = K^{(m)}(x)$$

or

(21.04) $$K_1(x) = {}^{(m)}K(x).$$

If we put

(21.05) $$\tau_m(u) = \int_{-\infty}^{\infty} K^{(m)}(x)\, e^{ux}\, dx = \frac{\Gamma(m+2)\,\Gamma(u+1)}{\Gamma(u+m+2)},$$

and

(21.06) $$\sigma_m(u) = \int_{-\infty}^{\infty} {}^{(m)}K(x)\, e^{ux}\, dx = \frac{\Gamma(m+1)}{\Gamma(m+1-u)\cos\frac{\pi u}{2}},$$

then as $|\Im(u)| \to \infty$

(21.07) $$\left|\frac{\sigma_m(u)}{\tau_n(u)}\right| \sim \frac{2\,\Gamma(m+1)}{\sqrt{2\pi}\,\Gamma(n+2)\, e^{\Re(u)-m}} |\Im(u)|^{\Re(u)-m+n+1/2}.$$

Thus if $n > m$, then

(21.08) $$\lim_{y\to\infty} \int_0^{\infty} {}^{(m)}K(y-x)\, df(x) = A\int_0^{\infty} {}^{(m)}K(x)\, dx$$

will imply

(21.09) $$\lim_{y\to\infty} \int_0^{\infty} K^{(n)}(y-x)\, df(x) = A\int_{-\infty}^{\infty} K^{(n)}(x)\, dx,$$

while if $m > n+1$, (21.09) will imply (21.08).

With a little manipulation, already indicated by the authors in question, this result is seen to be equivalent to an important theorem of Hardy and Littlewood,[54] which gives the necessary and sufficient condition for the summability of Fourier series and integrals by Cesàro sums of some order. In the integral form, the theorem reads as follows: *Let $f(x)$ be a measurable function defined over $(-\infty, \infty)$, and zero outside $(-A, A)$. Let*

(21.10) $$\varphi(y) = \tfrac{1}{2}\{f(x+y)+f(x-y)-2s\}.$$

Then if we write B_m for the proposition

(21.11) $$\lim_{\lambda\to\infty} \lambda \lim_{\varepsilon\to 0} \int_{\varepsilon}^{1/\lambda} \varphi(y)\,(1-\lambda y)^{m-1}\, dy = 0,$$

and C_m for the proposition

(21.12) $$\lim_{\lambda\to\infty} \lambda \lim_{\varepsilon\to 0} \int_{\varepsilon}^{\infty} \varphi(y)\, dy \int_0^1 (1-z)^m \cos\lambda yz\, dz = 0,$$

B_m implies $C_{m+\varepsilon}$ for $m \geqq 1$, while C_m implies $B_{m+1+\varepsilon}$ for $m > 0$.

[54] Hardy and Littlewood (12).

While the general method of the present section has been developed in an earlier paper of the author,[55] his final results were not stated correctly. The correct result is due to Bosanquet[56] and Paley, who have shown it to be a "best possible" result in both directions. Their theorem also applies to $m > -1$, and is otherwise somewhat more general.

CHAPTER VIII.

TAUBERIAN THEOREMS AND SPECTRA.

22. A further type of asymptotic behavior. The O and o symbols do not exhaust the possible terms in which we may describe the behavior of a function at infinity. A proposition which may be regarded as in some wise a generalization of

$$\lim_{x\to\infty} f(x) = A \tag{22.01}$$

is

$$\lim_{B\to\infty} \frac{1}{B}\int_0^B |f(x) - A|^2\, dx = 0. \tag{22.02}$$

If $f(x)$ is bounded and measurable, (22.01) clearly implies (22.02), while (22.02) does not imply (22.01). Proposition (22.02) has a certain analogy to the different types of "strong convergence" to a limit which a function may exhibit: namely

$$\int_0^\infty |f(x) - A|\, dx \text{ converges} \tag{22.03}$$

and

$$\int_0^\infty |f(x) - A|^2 dx \text{ converges.} \tag{22.04}$$

The series analogues of the latter

$$\sum_1^\infty |s_n - A| \text{ converges} \tag{22.05}$$

and

$$\sum_1^\infty |s_n - A|^2 \text{ converges} \tag{22.06}$$

imply the ordinary convergence of s_n to A, but are not implied by it. On the other hand, neither (22.03) nor (22.04) is implied by ordinary convergence, although (22.04) implies (22.02). In contrast with propositions

[55] Wiener (5).

[56] Bosanquet (1), (2); Paley (1).

(22.03)—(22.06), which represent various types of "strong convergence", we shall express (22.02) in the usual language by saying that $f(x)$ is *strongly summable*[57] to A as $x \to \infty$. We shall say that A is a *sublimit* of $f(x)$, and shall write

$$\operatorname*{slm}_{x\to\infty} f(x) = A. \tag{22.07}$$

The sublimit of a function $f(x)$ differs markedly from the ordinary limit in that neither its existence nor its value are invariant if we replace x by a monotone function of x becoming infinite with x. On the other hand, the sublimit of $f(x)$ has a relation to the harmonic analysis of $f(x)$ far closer than does the ordinary limit.

Closely related to the notion of sublimit is that of subboundedness, which bears to the ordinary notion of boundedness much the same relation which that of sublimit does to the ordinary notion of limit. A function $f(x)$ is said to be *subbounded* if

$$\frac{1}{B}\int_0^B |f(x)|^2\,dx \tag{22.08}$$

is bounded. This juxtaposition of a notion of limit and a notion of boundedness suggests a generalized form of Tauberian theorem. To be specific, let us ask what conditions beyond the subboundedness of $f(x)$ are sufficient to make

$$\operatorname*{slm}_{x\to\infty} \int_0^\infty K_1(x-\xi)f(\xi)\,d\xi = A\int_{-\infty}^\infty K_1(\xi)\,d\xi \tag{22.09}$$

imply

$$\operatorname*{slm}_{x\to\infty} \int_0^\infty K_2(x-\xi)f(\xi)\,d\xi = A\int_{-\infty}^\infty K_2(\xi)\,d\xi. \tag{22.10}$$

This problem belongs to the range of ideas treated by the author in his work on generalized harmonic analysis. It gives a clearer picture of the real significance of Tauberian theorems. On the other hand, it does not at present offer an alternative approach to them, since Tauberian theorems of the type already discussed in this paper play an essential role in the establishment of a theory of generalized harmonic analysis. Results involving these theorems will be applied in this section.

To return to (22.09) and (22.10), no generality is lost by taking $A = 0$, as this simply amounts to replacing $f(x)$ by $f(x)-A$. Let us then take $A = 0$, and let us put $f(x) = 0$ for negative arguments, which

[57] Hardy and Littlewood (15).

is clearly permissible. We then wish to find a set of conditions which in conjunction with the boundedness for large B of

$$\frac{1}{2B}\int_{-B}^{B}|f(x)|^2\,dx \tag{22.11}$$

(or what is the same, with the fact that $\frac{f(x)}{1+|x|}$ belongs to L_2) and the proposition

$$\lim_{B\to\infty}\frac{1}{2B}\int_{-B}^{B}\left|\int_{-\infty}^{\infty}K_1(x-\xi)f(\xi)\,d\xi\right|^2dx = 0 \tag{22.12}$$

are sufficient to imply

$$\lim_{B\to\infty}\frac{1}{2B}\int_{-B}^{B}\left|\int_{-\infty}^{\infty}K_2(x-\xi)f(\xi)\,d\xi\right|^2d\xi = 0. \tag{22.13}$$

We shall prove the following:

THEOREM XXIV. *Let $f(x)$ be measurable, and let* (22.11) *be bounded. Let* (22.12) *be valid. Let K_1 and K_2 be measurable, and let*

$$\begin{gathered}\int_{-\infty}^{\infty}(1+|x|)^{3+\epsilon}\,|K_1(x)|^2\,dx<\infty\\ \left(\text{and hence}\int_{-\infty}^{\infty}(1+|x|)\,|K_1(x)|\,dx<\infty\right),\end{gathered} \tag{22.14}$$

and

$$\begin{gathered}\int_{-\infty}^{\infty}(1+|x|)^{3+\epsilon}\,|K_2(x)|^2\,dx<\infty\\ \left(\text{and hence}\int_{-\infty}^{\infty}(1+|x|)\,|K_2(x)|\,dx<\infty\right).\end{gathered} \tag{22.15}$$

Furthermore, let

$$\int_{-\infty}^{\infty}K_1(x)\,e^{iux}\,dx \neq 0 \tag{22.16}$$

for all real u. Then (22.13) *is true.*

For this theorem we shall need the following:

LEMMA XXIVa. *Let $f(x, u, \lambda)$ belong to L_2 as a function of u and be measurable in (x, u). Let $\int_{-\infty}^{\infty}|f(x, u, \lambda)|^2\,du$ be bounded, and let*

$$\int_{-\infty}^{\infty}|f(x, u, \lambda)-f(x, u)|^2\,du\to 0 \tag{22.17}$$

uniformly over any finite range of x as $\lambda\to\infty$. Let K be measurable, and let

6

$$(22.18) \qquad \int_{-\infty}^{\infty} |K(x)|\, dx$$

be finite. Let

$$(22.19) \qquad \int_{-\infty}^{\infty} K(x) f(x, u, \lambda)\, dx$$

exist for all λ. *Then*

$$(22.20) \quad \int_{-\infty}^{\infty} \left| \int_{-\infty}^{\infty} K(x) f(x, u, \lambda)\, dx - \int_{-\infty}^{\infty} K(x) f(x, u)\, dx \right|^2 du \to 0 \text{ as } \lambda \to \infty.$$

To prove this lemma, let us reflect that

$$(22.21) \quad \begin{aligned} &\left[\int_A^B |K(x)|\, dx\right]^2 \max_x \int_{-\infty}^{\infty} |f(x, u, \lambda) - f(x, u)|^2\, du \\ &\geqq \int_A^B |K(x)|\, dx \int_A^B |K(y)|\, dy \int_{-\infty}^{\infty} du\, |f(x, u, \lambda) - f(x, u)| \\ &\qquad\qquad \times |f(y, u, \lambda) - f(y, u)| \\ &= \int_{-\infty}^{\infty} du \int_A^B |K(x)|\, dx \int_A^B |K(y)|\, dy\, |f(x, u, \lambda) - f(x, u)| \\ &\qquad\qquad \times |f(y, u, \lambda) - f(y, u)| \\ &\geqq \int_{-\infty}^{\infty} du \left| \int_A^B K(x)\, [f(x, u, \lambda) - f(x, u)]\, dx \right|^2 \end{aligned}$$

By taking A large and B infinite, or $-B$ large and $-A$ infinite, we may make the left-hand member uniformly small. Over any finite range (A, B) the left-hand member tends to 0 as $\lambda \to \infty$. By combining these facts the lemma follows. It will be noted that it involves the existence almost everywhere of

$$(22.22) \qquad \int_{-\infty}^{\infty} K(x) f(x, u)\, dx.$$

We may now return to the proof of our main theorem. In a previous paper[58], the author has shown that

$$(22.23) \quad s(u) = \frac{1}{2\pi} \int_{-1}^{1} f(x) \frac{e^{iux} - 1}{ix}\, dx + \underset{A\to\infty}{\text{l.i.m.}} \frac{1}{\pi} \left[\int_1^A + \int_A^{-1}\right] f(x) \frac{e^{iux}}{ix}\, dx$$

exists. Moreover,

$$(22.24) \quad \begin{aligned} \int_{-\infty}^{\infty} |K_1(x-\xi) f(\xi)|\, d\xi &\leq \int_{-\infty}^{\infty} |K_1(\xi)|\, |f(x-\xi)|\, d\xi \\ &\leq \left[\int_{-\infty}^{\infty} (1+|\xi|)^2 |K_1(\xi)|^2\, d\xi \int_{-\infty}^{\infty} \frac{|f(x-\xi)|^2}{(1+|\xi|)^2}\, d\xi\right]^{1/2} \\ &< \infty, \end{aligned}$$

[58] Wiener (7).

so that

$$\frac{1}{2B}\int_{-B}^{B}\left|\int_{-\infty}^{\infty}K_1(x-\xi)f(\xi)\,d\xi\right|^2 dx$$

$$\leqq \frac{1}{2B}\int_{-B}^{B}dx\int_{-\infty}^{\infty}|K_1(\xi)|\,|f(x-\xi)|\,d\xi\int_{-\infty}^{\infty}|K_1(\eta)|\,|f(x-\eta)|\,d\eta$$

$$(22.25)\qquad = \frac{1}{2B}\int_{-\infty}^{\infty}|K_1(\xi)|\,d\xi\int_{-\infty}^{\infty}|K_1(\eta)|\,d\eta\int_{-B}^{B}|f(x-\xi)|\,|f(x-\eta)|\,dx$$

$$\leqq \frac{1}{2B}\int_{-\infty}^{\infty}|K_1(\xi)|\,d\xi\int_{-\infty}^{\infty}|K_1(\eta)|\,d\eta\int_{-B-|\xi|-|\eta|}^{B+|\xi|+|\eta|}|f(x)|^2\,dx$$

$$\leqq \frac{\text{const.}}{B}\int_{-\infty}^{\infty}|K_1(\xi)|\,d\xi\int_{-\infty}^{\infty}|K_1(\eta)|\,d\eta\,[B+|\xi|+|\eta|]$$

$$< \text{const.}$$

Thus the function

$$(22.26)\qquad f_1(x) = \int_{-\infty}^{\infty}K_1(x-\xi)f(\xi)\,d\xi$$

and the analogous function

$$(22.27)\qquad f_2(x) = \int_{-\infty}^{\infty}K_2(x-\xi)f(\xi)\,d\xi$$

define functions $s_1(u)$ and $s_2(u)$, corresponding to them in the same way in which $s(u)$ corresponds to $f(x)$.

In his previous paper,[58] the author has shown that

$$(22.28)\qquad s(u+\varepsilon)-s(u-\varepsilon) = \frac{1}{\pi}\operatorname*{l.i.m.}_{A\to\infty}\int_{-A}^{A}f(x)\frac{\sin\varepsilon x}{x}e^{iux}\,dx.$$

Similarly

$$(22.29)\qquad \begin{aligned} &s_1(u+\varepsilon)-s_1(u-\varepsilon)\\ &= \frac{1}{\pi}\operatorname*{l.i.m.}_{A\to\infty}\int_{-A}^{A}\frac{\sin\varepsilon x}{x}e^{iux}\,dx\int_{-\infty}^{\infty}K_1(\xi)f(x-\xi)\,d\xi.\end{aligned}$$

There is no trouble in modifying (22.28) into

$$(22.30)\qquad s(u+\varepsilon)-s(u-\varepsilon) = \frac{1}{\pi}\operatorname*{l.i.m.}_{A\to\infty}\int_{-A-\xi}^{A-\xi}f(x)\frac{\sin\varepsilon x}{x}e^{iux}\,dx.$$

Thus

$$s_1(u+\varepsilon)-s_1(u-\varepsilon)-(s(u+\varepsilon)-s(u-\varepsilon))\int_{-\infty}^{\infty}K_1(\xi)e^{iu\xi}\,d\xi$$

$$(22.31)\qquad \begin{aligned} &= \frac{1}{\pi}\Bigg[\operatorname*{l.i.m.}_{A\to\infty}\int_{-\infty}^{\infty}K_1(\xi)\,d\xi\int_{-A}^{A}f(x-\xi)\frac{\sin\varepsilon x}{x}e^{iux}\,dx\\ &\qquad -\int_{-\infty}^{\infty}K_1(\xi)\,d\xi\operatorname*{l.i.m.}_{A\to\infty}\int_{-A}^{A}f(x-\xi)\frac{\sin\varepsilon(x-\xi)}{x-\xi}e^{iux}\,dx\Bigg]\\ &= \frac{1}{\pi}\operatorname*{l.i.m.}_{A\to\infty}\int_{-\infty}^{\infty}K_1(\xi)\,d\xi\int_{-A}^{A}f(x-\xi)\left[\frac{\sin\varepsilon x}{x}-\frac{\sin\varepsilon(x-\xi)}{x-\xi}\right]e^{iux}\,dx.\end{aligned}$$

In the proof of this we have used our last lemma.

6*

Clearly

$$\left|\frac{\sin \varepsilon x}{\varepsilon x} - \frac{\sin \varepsilon (x-\xi)}{\varepsilon (x-\xi)}\right| \leqq 2. \tag{22.32}$$

Again,

$$\left|\frac{\sin \varepsilon x}{\varepsilon x} - \frac{\sin \varepsilon (x-\xi)}{\varepsilon (x-\xi)}\right| = \left|\int_{x-\xi}^{x} \frac{d}{dw}\left(\frac{\sin \varepsilon w}{\varepsilon w}\right) dw\right|$$

$$= \left|\int_{x-\xi}^{x} \frac{\varepsilon^2 w \cos \varepsilon w - \varepsilon \sin \varepsilon w}{\varepsilon^2 w^2} dw\right|$$

$$\leqq \varepsilon |\xi| \max_{w \operatorname{in}(x-\xi,\, x)} \left|\frac{\cos \varepsilon w}{\varepsilon w} - \frac{\sin \varepsilon w}{\varepsilon^2 w^2}\right| \tag{22.321}$$

$$\leqq 2\varepsilon |\xi| \max \frac{1}{|\varepsilon w|}$$

$$\leqq \left|\frac{8\xi}{|x|+|\xi|}\right|.$$

Thus and hence ly an application of (22.32)

$$\frac{1}{\varepsilon \xi} \int_{-A}^{A} f(x-\xi)\left[\frac{\sin \varepsilon x}{x} - \frac{\sin \varepsilon (x-\xi)}{x-\xi}\right] e^{iux}\, dx \tag{22.33}$$

converges in the mean uniformly in u as $A \to \infty$ over any finite range of ξ. Hence by our lemma, we may write

$$\begin{aligned} s_1(u+\varepsilon) - s_1(u-\varepsilon) - (s(u+\varepsilon) - s(u-\varepsilon)) \int_{-\infty}^{\infty} K_1(\xi)\, e^{iu\xi}\, d\xi \\ = \frac{1}{\pi} \int_{-\infty}^{\infty} K_1(\xi)\, d\xi \operatorname*{l.i.m.}_{A\to\infty} \int_{-A}^{A} f(x-\xi)\left[\frac{\sin \varepsilon x}{x} - \frac{\sin \varepsilon (x-\xi)}{x-\xi}\right] e^{iux}\, dx. \end{aligned} \tag{22.34}$$

By a further use of (22.321) and our lemma, since

$$\operatorname*{l.i.m.}_{\varepsilon\to 0} \frac{1}{\varepsilon^{1/2}\xi} \operatorname*{l.i.m.}_{A\to\infty} \frac{1}{\pi} \int_{-A}^{A} f(x-\xi)\left[\frac{\sin \varepsilon x}{x} - \frac{\sin \varepsilon (x-\xi)}{x-\xi}\right] e^{iux} dx = 0 \tag{22.35}$$

uniformly in ξ, it follows that

$$\begin{aligned} \lim_{\varepsilon\to 0} \frac{1}{\varepsilon} \int_{-\infty}^{\infty} \Big| s_1(u+\varepsilon) - s_1(u-\varepsilon) \\ - (s(u+\varepsilon) - s(u-\varepsilon)) \int_{-\infty}^{\infty} K_1(\xi)\, e^{iu\xi}\, d\xi \Big|^2 du = 0. \end{aligned} \tag{22.36}$$

From (22.12) and a theorem of the previous paper,[59] it follows that

$$\lim_{\varepsilon\to 0} \frac{1}{\varepsilon} \int_{-\infty}^{\infty} |s_1(u+\varepsilon) - s_1(u-\varepsilon)|^2\, du = 0. \tag{22.37}$$

[59] Wiener (7), (5.52).

Combining (22.36) and (22.37), we get

$$\text{(22.38)}\quad \lim_{\varepsilon\to 0}\frac{1}{\varepsilon}\int_{-\infty}^{\infty}\left|\int_{-\infty}^{\infty}K_1(\xi)\,e^{iu\xi}\,d\xi\right|^2 |s(u+\varepsilon)-s(u-\varepsilon)|^2\,du = 0.$$

Since over any finite range, by (22.16),

$$\text{(22.39)}\quad \left|\int_{-\infty}^{\infty}K_1(\xi)\,e^{iu\xi}\,d\xi\right| > \text{const.},$$

it results that for any finite C,

$$\text{(22.40)}\quad \lim_{\varepsilon\to 0}\frac{1}{\varepsilon}\int_{-C}^{C}|s(u+\varepsilon)-s(u-\varepsilon)|^2\,du = 0.$$

As in (22.36), we have

$$\text{(22.41)}\quad \lim_{\varepsilon\to 0}\frac{1}{\varepsilon}\int_{-\infty}^{\infty}\Big|s_2(u+\varepsilon)-s_2(u-\varepsilon) - (s(u+\varepsilon)-s(u-\varepsilon))\int_{-\infty}^{\infty}K_2(\xi)\,e^{iu\xi}\,d\xi\Big|^2 du = 0.$$

This yields us

$$\text{(22.42)}\quad \lim_{\varepsilon\to 0}\frac{1}{\varepsilon}\int_{-C}^{C}\Big|s_2(u+\varepsilon)-s_2(u-\varepsilon) - (s(u+\varepsilon)-s(u-\varepsilon))\int_{-\infty}^{\infty}K_2(\xi)\,e^{iu\xi}\,d\xi\Big|^2 du = 0.$$

Combining this with (22.40), we get

$$\text{(22.43)}\quad \lim_{\varepsilon\to 0}\frac{1}{\varepsilon}\int_{-C}^{C}\left|s_2(u+\varepsilon)-s_2(u-\varepsilon)\right|^2 du = 0.$$

By our previous paper,[60]

$$\text{(22.44)}\quad \frac{1}{\pi\varepsilon}\int_{-\infty}^{\infty}|f_2(x)|^2\frac{\sin^2\varepsilon x}{x^2}\,dx = \frac{1}{2\varepsilon}\int_{-\infty}^{\infty}|s_2(u+\varepsilon)-s_2(u-\varepsilon)|^2\,du,$$

and

$$\text{(22.45)}\quad \frac{1}{\pi\varepsilon}\int_{-\infty}^{\infty}\frac{\sin^2\varepsilon x}{x^2}\,dx\left|\frac{1}{2\lambda}\int_{-\lambda}^{\lambda}f_2(x+y)\,dy\right|^2 = \frac{1}{2\varepsilon}\int_{-\infty}^{\infty}\frac{\sin^2\lambda u}{\lambda^2u^2}|s_2(u+\varepsilon)-s_2(u-\varepsilon)|^2\,du.$$

[60] Wiener (7), (5.52).

Now

$$\overline{\lim_{\varepsilon\to 0}}\left|\frac{1}{\pi\varepsilon}\int_{-\infty}^{\infty}|f_2(x)|^2\frac{\sin^2\varepsilon x}{x^2}dx-\frac{1}{\pi\varepsilon}\int_{-\infty}^{\infty}\frac{\sin^2\varepsilon x}{x^2}dx\left|\frac{1}{2\lambda}\int_{-\lambda}^{\lambda}f_2(x+y)\,dy\right|^2\right|$$

$$\leqq \overline{\lim_{\varepsilon\to 0}}\frac{\text{const.}}{\varepsilon}\int_{-\infty}^{\infty}\left|f_2(x)-\frac{1}{2\lambda}\int_{-\lambda}^{\lambda}f_2(x+y)\,dy\right|^2\frac{\sin^2\varepsilon x}{x^2}dx$$

$$=\overline{\lim_{\varepsilon\to 0}}\frac{\text{const.}}{\varepsilon}\int_{-\infty}^{\infty}\left|\int_{-\infty}^{\infty}f(x-\xi)\right.$$

$$\left.\times\left[K_2(\xi)-\frac{1}{2\lambda}\int_{-\lambda}^{\lambda}K_2(\xi+y)\,dy\right]d\xi\right|^2\frac{\sin^2\varepsilon x}{x^2}dx$$

$$\leqq \overline{\lim_{B\to\infty}}\text{const.}\frac{1}{2B}\int_{-B}^{B}\left|\int_{-\infty}^{\infty}f(x-\xi)\right. \tag{22.46}$$

$$\left.\times\left[K_2(\xi)-\frac{1}{2\lambda}\int_{-\lambda}^{\lambda}K_2(\xi+y)\,dy\right]d\xi\right|^2dx$$

$$\leqq \text{const.}\overline{\lim_{B\to\infty}}\frac{1}{B}\int_{-\infty}^{\infty}\left|K_2(\xi)-\frac{1}{2\lambda}\int_{-\lambda}^{\lambda}K_2(\xi+y)\,dy\right|d\xi$$

$$\times\int_{-\infty}^{\infty}\left|K_2(\eta)-\frac{1}{2\lambda}\int_{-\lambda}^{\lambda}K_2(\eta+z)\,dz\right|d\eta\,(B+|\xi|+|\eta|)$$

$$\leqq \text{const.}\overline{\lim_{B\to\infty}}\left[\int_{-\infty}^{\infty}\left|K_2(\xi)-\frac{1}{2\lambda}\int_{-\lambda}^{\lambda}K_2(\xi+y)\,dy\right|\left(1+\frac{|\xi|}{B}\right)d\xi\right]^2,$$

where

$$\overline{\lim_{B\to\infty}}\int_{-\infty}^{\infty}\left|K_2(\xi)-\frac{1}{2\lambda}\int_{-\lambda}^{\lambda}K_2(\xi+y)\,dy\right|\left(1+\frac{|\xi|}{B}\right)d\xi$$

$$=\int_{-\infty}^{\infty}\left|K_2(\xi)-\frac{1}{2\lambda}\int_{-\lambda}^{\lambda}K_2(\xi+y)\,dy\right|d\xi. \tag{22.47}$$

Hence

$$\overline{\lim_{\varepsilon\to 0}}\left|\frac{1}{\pi\varepsilon}\int_{-\infty}^{\infty}|f_2(x)|^2\frac{\sin^2\varepsilon x}{x^2}dx-\frac{1}{\pi\varepsilon}\int_{-\infty}^{\infty}\frac{\sin^2\varepsilon x}{x^2}dx\left|\frac{1}{2\lambda}\int_{-\lambda}^{\lambda}f_2(x+y)\,dy\right|^2\right|$$

$$\leqq \text{const.}\left[\int_{-\infty}^{\infty}\left|K_2(\xi)-\frac{1}{2\lambda}\int_{-\lambda}^{\lambda}K_2(\xi+y)\,dy\right|d\xi\right]^2. \tag{22.48}$$

As we have seen in section 1,

$$\lim_{\lambda\to 0}\int_{-\infty}^{\infty}\left|K_2(\xi)-\frac{1}{2\lambda}\int_{-\lambda}^{\lambda}K_2(\xi+y)\,dy\right|d\xi=0, \tag{22.49}$$

from which it immediately follows that

$$\lim_{\lambda \to 0} \overline{\lim_{\varepsilon \to 0}} \left| \frac{1}{\pi \varepsilon} \int_{-\infty}^{\infty} |f_2(x)|^2 \frac{\sin^2 \varepsilon x}{x^2} dx - \frac{1}{\pi \varepsilon} \int_{-\infty}^{\infty} \frac{\sin^2 \varepsilon x}{x^2} dx \left| \frac{1}{2\lambda} \int_{-\lambda}^{\lambda} f_2(x+y)\, dy \right|^2 \right| = 0. \tag{22.50}$$

In combination with (22.44) and (22.45) this yields us

$$\lim_{\lambda \to 0} \overline{\lim_{\varepsilon \to 0}} \frac{1}{2\varepsilon} \int_{-\infty}^{\infty} \left(1 - \frac{\sin^2 \lambda u}{\lambda^2 u^2}\right) |s_2(u+\varepsilon) - s_2(u-\varepsilon)|^2 du = 0. \tag{22.51}$$

Since

$$1 - \frac{\sin^2 \lambda u}{\lambda^2 u^2} > \frac{1}{2} \quad \text{if} \quad u > \frac{2}{\lambda} \tag{22.52}$$

it results at once that

$$\lim_{C \to \infty} \overline{\lim_{\varepsilon \to 0}} \frac{1}{2\varepsilon} \left[\int_{C}^{\infty} + \int_{-\infty}^{-C} \right] |s_2(u+\varepsilon) - s_2(u-\varepsilon)|^2 du = 0. \tag{22.53}$$

Combining this with (22.43), we may readily conclude that

$$\lim_{\varepsilon \to 0} \frac{1}{2\varepsilon} \int_{-\infty}^{\infty} |s_2(u+\varepsilon) - s_2(u-\varepsilon)|^2 du = 0, \tag{22.54}$$

or in view of a theorem of our previous paper, that

$$\lim_{T \to \infty} \frac{1}{2T} \int_{-T}^{T} |f_2(x)|^2 dx = 0. \tag{22.13}$$

This establishes Theorem XXIV. It will be noted that this theorem has an immediate extension in the direction of Theorem XI′.

There is another Tauberian theorem concerning strong summability which we may discuss here. It is due to Hardy and Littlewood.[61] The particular case of it which most directly interests us is the following:

THEOREM XXV. *Let*

$$\frac{1}{\xi} \int_{0}^{\xi} |f(x)|^2 dx \tag{22.55}$$

be bounded, and let

$$\lim_{\varepsilon \to 0} \frac{1}{\varepsilon} \int_{0}^{\varepsilon} f(x)\, dx = 0. \tag{22.56}$$

[61] Hardy and Littlewood (16).

Then if

$$(22.57) \qquad s_n(0) = \frac{1}{\pi}\int_0^\pi f(x)\,\frac{\sin\frac{2n+1}{2}x}{\sin\frac{x}{2}}\,dx$$

it follows that

$$(22.58) \qquad \lim_{N\to\infty}\frac{1}{N}\sum_0^N |s_n(0)|^2 = 0.$$

In the proof of this theorem, it is easy to see that we may replace the conclusion (22.58) which we are to establish, by the equivalent

$$(22.59) \qquad \lim_{H\to\infty}\frac{1}{H}\int_0^H\left|\frac{2}{\pi}\int_0^\pi f(x)\,\frac{\sin ux}{x}\,dx\right|^2 du = 0.$$

By a Tauberian theorem of the type of Theorem XI, already established in a previous paper of the author,[58] it follows that (22.59) is completely equivalent to

$$(22.60) \qquad \lim_{\varepsilon\to 0}\frac{2}{\pi}\int_0^\infty \frac{\sin^2\varepsilon u}{\varepsilon u^2}\left|\frac{2}{\pi}\int_0^\pi f(x)\frac{\sin ux}{x}\,dx\right|^2 du.$$

On the assumption that $f(x) = 0$ if $x > \pi$, we have by the Plancherel theorem

$$\frac{2}{\pi}\int_0^\infty \frac{\sin^2\varepsilon u}{\varepsilon u^2}\left|\frac{2}{\pi}\int_0^\pi f(x)\frac{\sin ux}{x}\,dx\right|^2 du$$

$$= \frac{2}{\pi\varepsilon}\int_0^\infty\left|\frac{2}{\pi}\int_0^\pi f(x)\frac{\sin\varepsilon u\sin ux}{ux}\,dx\right|^2 du$$

$$= \frac{2}{\pi\varepsilon}\int_0^\infty\left|\frac{2}{\pi}\int_0^\pi \frac{\sin\varepsilon u\sin ux}{u}\,d\int_x^\infty\frac{f(\xi)}{\xi}\,d\xi\right|^2 du$$

$$= \frac{2}{\pi\varepsilon}\int_0^\infty\left|\frac{2}{\pi}\int_0^\infty \sin\varepsilon u\cos ux\,dx\int_x^\infty\frac{f(\xi)}{\xi}\,d\xi\right|^2 du$$

$$(22.61) \qquad = \frac{2}{\pi\varepsilon}\int_0^\infty\left|\frac{1}{\pi}\int_0^\infty (\sin u(x+\varepsilon)-\sin u(x-\varepsilon))\,dx\int_x^\infty\frac{f(\xi)}{\xi}d\xi\right|^2 du$$

$$= \frac{2}{\pi\varepsilon}\int_0^\infty\left|\frac{1}{2\pi}\int_{-\infty}^\infty (\sin u(x+\varepsilon)-\sin u(x-\varepsilon))\,dx\int_{|x|}^\infty\frac{f(\xi)}{\xi}d\xi\right|^2 du$$

$$= \frac{2}{\pi\varepsilon}\int_0^\infty\left|\frac{1}{2\pi}\int_{-\infty}^\infty \sin ux\,dx\int_{|x-\varepsilon|}^{|x+\varepsilon|}\frac{f(\xi)}{\xi}\,d\xi\right|^2 du$$

$$= \frac{1}{\pi^2\varepsilon}\int_0^\infty\left|\int_{|x-\varepsilon|}^{|x+\varepsilon|}\frac{f(\xi)}{\xi}\,d\xi\right|^2 dx.$$

Thus (22.59) and (22.58) are equivalent to

$$\lim_{\varepsilon\to 0}\frac{1}{\varepsilon}\int_0^\infty\left|\int_{|x-\varepsilon|}^{|x+\varepsilon|}\frac{f(\xi)}{\xi}\,d\xi\right|^2 dx = 0. \tag{22.62}$$

By the Schwarz inequality,

$$\begin{aligned}\frac{1}{\varepsilon}\left|\int_{|x-\varepsilon|}^{|x+\varepsilon|}\frac{f(\xi)}{\xi}\,d\xi\right|^2 &\leqq \frac{1}{\varepsilon}\int_{|x-\varepsilon|}^{|x+\varepsilon|}|f(\xi)|^2\,d\xi\int_{|x-\varepsilon|}^{|x+\varepsilon|}\frac{d\xi}{\xi^2}\\ &= \left(\frac{1}{|x-\varepsilon|}-\frac{1}{|x+\varepsilon|}\right)\frac{1}{\varepsilon}\int_{|x-\varepsilon|}^{|x+\varepsilon|}|f(\xi)|^2\,d\xi.\end{aligned} \tag{22.63}$$

Thus by the boundedness of (22.55) we have uniformly in ε for all sufficiently large values of N:

$$\begin{aligned}\frac{1}{\varepsilon}\int_{N\varepsilon}^\infty\left|\int_{|x-\varepsilon|}^{|x+\varepsilon|}\frac{f(\xi)}{\xi}\,d\xi\right|^2 dx &\leqq \int_{N\varepsilon}^\infty\frac{2\varepsilon}{x^2-\varepsilon^2}\,d\int_0^x\frac{d\eta}{\varepsilon}\int_{|\eta-\varepsilon|}^{|\eta+\varepsilon|}|f(\xi)|^2\,d\xi\\ &\leqq \int_{N\varepsilon}^\infty\frac{4\varepsilon x\,dx}{(x^2-\varepsilon^2)^2}\int_0^x\frac{d\eta}{\varepsilon}\int_{|\eta-\varepsilon|}^{|\eta+\varepsilon|}|f(\xi)|^2\,d\xi\\ &= \frac{1}{2}\int_{N\varepsilon}^\infty\frac{4\varepsilon x\,dx}{(x^2-\varepsilon^2)^2}\int_{-x-\varepsilon}^{x+\varepsilon}|f(\xi)|^2\,d\xi\int_{\xi-\varepsilon}^{\xi+\varepsilon}\frac{d\eta}{\varepsilon}\\ &\leqq \text{const.}\int_{N\varepsilon}^\infty\frac{4\varepsilon x^2\,dx}{(x^2-\varepsilon^2)^2}\\ &= \text{const.}\int_N^\infty\frac{4x^2\,dx}{(x^2-1)^2}.\end{aligned} \tag{22.64}$$

This can be made as small as we wish by taking N large enough. Moreover,

$$\begin{aligned}\frac{1}{\varepsilon}\int_0^{N\varepsilon}\left|\int_{|x-\varepsilon|}^{|x+\varepsilon|}\frac{f(\xi)}{\xi}\,d\xi\right|^2 &= \frac{1}{\varepsilon}\int_0^{N\varepsilon}\left|\int_{|x-\varepsilon|}^{|x+\varepsilon|}\frac{1}{\xi}\,d\int_0^\xi f(\eta)\,d\eta\right|^2 dx\\ &= \frac{1}{\varepsilon}\int_0^{N\varepsilon}\left|\left[\frac{1}{\xi}\int_0^\xi f(\eta)\,d\eta\right]_{|x-\varepsilon|}^{|x+\varepsilon|}+\int_{|x-\varepsilon|}^{|x+\varepsilon|}\frac{d\xi}{\xi^2}\int_0^\xi f(\eta)\,d\eta\right|^2 dx\\ &\leqq \frac{1}{\varepsilon}\int_0^{N\varepsilon}\left|A\left(1+\log\frac{|x+\varepsilon|}{|x-\varepsilon|}\right)\right|^2 dx\\ &= A^2\int_0^N\left(1+\log\frac{|x+1|}{|x-1|}\right)^2 dx,\end{aligned} \tag{22.65}$$

where

$$A > \left|\frac{1}{\xi}\int_0^\xi f(\eta)\,d\eta\right| \qquad (0<\xi\leqq|N+1|\varepsilon) \tag{22.651}$$

By (22.56), we may make A as small as we wish for a given N by taking ε small enough. Now let N be so large that for all ε,

$$(22.66)\qquad \frac{1}{\varepsilon}\int_{N\varepsilon}^{\infty}\left|\int_{|x-\varepsilon|}^{|x+\varepsilon|}\frac{f(\xi)}{\xi}\,d\xi\right|^{2}dx<\varepsilon_1/2,$$

and then let ε be so small that

$$(22.67)\qquad \frac{1}{\varepsilon}\int_{0}^{N\varepsilon}\left|\int_{|x-\varepsilon|}^{|x+\varepsilon|}\frac{f(\xi)}{\xi}\,d\xi\right|^{2}dx<\varepsilon_1/2.$$

It will then follow that for this and all smaller values of

$$(22.68)\qquad \frac{1}{\varepsilon}\int_{0}^{\infty}\left|\int_{|x-\varepsilon|}^{|x+\varepsilon|}\frac{f(\xi)}{\xi}\,d\xi\right|^{2}dx<\varepsilon_1.$$

Thus (22.62) follows, and Theorem XXV is established.

23. **Generalized types of summability.** The subject-matter of the last section leads us to interesting reflections on the notion of summability itself. The ordinary processes for summing a series are linear processes: that is, they consist in replacing the partial sums of a series by linear combinations of partial sums, and then investigating the ordinary limits of these linear combinations. This much of linearity must always remain in a definition of summability, that if two summable series are added term by term, the corresponding sums are also added, in the sense that the new sum-series will be summable to the sum of the sums corresponding to the individual series.

It is possible, however, to consider summability from another standpoint, from which linearity is not so obvious an attribute of the process. We may confine our attention to series, or rather to their sequences of partial sums, which are summable to zero. Thus a method of summability will sum to A the series whose partial sums are:

$$(23.01)\qquad s_1,\ s_2,\ \cdots,\ s_n,\ \cdots$$

if it will sum to zero the series whose partial sums are:

$$(23.02)\qquad s_1-A,\ s_2-A,\ \cdots,\ s_n-A,\ \cdots.$$

We may thus center our concept of summability about the notion of *null-sequence* or *null-function*. Whatever definition we choose for the class of such sequences or functions, this class should be closed additively, in the sense that the sum of any two members of the class must belong to the class. *It is not however essential that a member of the class should be characterized by the vanishing of some particular linear transform of that member.* In this sense, we are introducing a non-linear theory of summability.[62]

It is clearly a desideratum of a definition of a null-function that every function tending to a zero limit at infinity should be null. This requirement at once makes the corresponding definition of summability consistent with convergence and inclusive of it in scope. It is not satisfied by definitions of the type that assert that f is null if

$$\int_{-\infty}^{\infty} |f(x)|^p\, dx < \infty \tag{23.03}$$

but is satisfied by the definition that f is null if

$$\lim_{B\to\infty} \frac{1}{B}\int_{-B}^{B} |f(x)|^2\, dx = 0. \tag{23.04}$$

We have already seen that if $s(u)$ is defined as in (22.23), this last statement is equivalent to the assertion that

$$\lim_{\varepsilon\to 0} \frac{1}{2\varepsilon}\int_{-\infty}^{\infty} |s(u+\varepsilon)-s(u-\varepsilon)|^2\, du = 0. \tag{23.05}$$

This suggests an even looser definition of a null-function, according to which a function $f(x)$ is null if

$$\frac{1}{2B}\int_{-B}^{B} |f(x)|^2\, dx \tag{23.06}$$

is bounded, and if

$$\lim_{\varepsilon\to 0} \frac{1}{2\varepsilon}\int_{-A}^{A} |s(u+\varepsilon)-s(u-\varepsilon)|^2\, du = 0 \tag{23.07}$$

for all finite A, or even if (23.07) is true for *some* A greater than 0. The first of these definitions will make $f(x)$ null if (23.06) holds and

$$\int_{-\infty}^{\infty} K(x-\xi) f(\xi)\, d\xi \tag{23.08}$$

is null, provided that

$$\int_{-\infty}^{\infty} (1+|x|)^{3+\varepsilon}\, |K_1(x)|^2\, dx < \infty \tag{23.09}$$

and

$$\int_{-\infty}^{\infty} K(x)\, e^{iux}\, dx \neq 0, \qquad [-\infty < u < \infty]. \tag{23.10}$$

---

[62] As far as the author knows, strong summability of various sorts is the only example of a summability process of this sort now in the literature.

The second definition will have the same property, provided only (23.10) holds for all values of u in some neighborhood of 0. All this follows from the theory developed in the last section.

These two last definitions of summability therefore fit in well with the ordinary linear definitions of summability, which define the generalized limit of a function as an expression of the form

$$\lim_{x\to\infty} \frac{\int_{-\infty}^{\infty} K(x, \xi) f(\xi)\, d\xi}{\int_{-\infty}^{\infty} K(x, \xi)\, d\xi} \tag{23.11}$$

on an appropriate scale of measurement. They have the great advantage of not involving any reference to a particular kernel $K(x, \xi)$. They are, however, restricted to functions for which (23.06) is bounded.

If we confine our attention to functions $f(x)$ vanishing for negative arguments, this difficulty may be overcome in its turn. We now put

$$\sigma(z) = \frac{1}{2\pi} \int_{-\infty}^{\infty} f(x)\, e^{zx}\, dx \tag{23.12}$$

throughout the half-plane $\Re(z) > u_0$: Then, by analytic continuation if necessary, $\sigma(i\nu)$ may be defined in some cases on a section of the real axis containing the origin. We put

$$s(u) = \int_0^u \sigma(iw)\, dw \tag{23.121}$$

giving σ its boundary value along the axis of imaginaries. We now define nullity as in (23.06) and (23.07). If for all z with real part between $-\epsilon$ and $u_0+\epsilon$

$$\int_{-\infty}^{\infty} |K(x)|\, e^{zx}\, dx < \infty, \tag{23.13}$$

we have

$$\frac{1}{2\pi} \int_{-\infty}^{\infty} e^{zx}\, dx \int_{-\infty}^{\infty} K(x-\xi) f(\xi)\, d\xi = \sigma(z) \int_{-\infty}^{\infty} K(\xi)\, e^{z\xi}\, d\xi, \tag{23.14}$$

so that if $\int_{-\infty}^{\infty} K(\xi)\, e^{z\xi}\, d\xi$ is free from zeros and singularities in the left half plane, $f(x)$ and $\int_{-\infty}^{\infty} K(x-\xi) f(\xi)\, d\xi$ are null or not null simultaneously. With an appropriate definition of "analytic continuation", zeros and singularities not at the origin are of no concern, provided the function

$\int_{-\infty}^{\infty} K(\xi)\, e^{z\xi}\, d\xi$ is representible on a single-sheeted Riemann surface in the left half-plane.

It will be noted that all the definitions here given of the nullity of $f(x)$ depend on the scale chosen for x. If for all large x,

$$\frac{y'(x)}{y(A)} < \text{const. for } x < A \tag{23.15}$$

and

$$\frac{1}{2A}\int_{-A}^{A} |f(y(x))|^2\, dx \to 0 \tag{23.16}$$

then it at once follows that

$$\frac{1}{2B}\int_{-B}^{B} |f(x)|^2\, dx \to 0. \tag{23.17}$$

A similar study of the effect of a change of scale on nullity of other types would be of interest.

24. **Some unsolved problems.** In a piece of work of the ambitious length of the present, it is perhaps worth while to point out to the reader promising future directions of research. The following remarks may therefore not be amiss.

(1) The closure of the set of translations of a given function has been investigated in class L_1 and in class L_2. The methods of proof have been widely different in the two cases, but both results may be stated in a single formulation, that the set of translations of $f(x)$ is closed in the appropriate class when and only when the Fourier transform of $f(x)$, when properly defined and chosen, has no zeros. This formulation continues to constitute a reasonable proposition for L_p, where p is intermediate between 1 and 2, or even exterior to this interval. Is this proposition true? Certainly, for neither of the special cases already given is the method of proof extensible without serious modification. My own suspicion is that the general theorem is at least true for $1 \leq p \leq 2$.

(2) Obviously the power of Tauberian theorems in number theory has not been exhausted. Is there any Tauberian theorem which will reach from one complex ordinate to another, and enable us to handle the more refined forms of the prime number theorem?

(3) In particular, can we make a direct study of the closure of the set of all polynomials in the functions

$$e^{-\lambda(x-\xi)} \frac{d}{dx} \frac{e^{x-\xi}}{e^{e^{x-\xi}} - 1}, \tag{24.01}$$

and thus attack directly the problem of the zeros of the zeta function in the critical strip?

(4) In section 22, the conditions to which $K_1(x)$ and $K_2(x)$ are subjected in Theorem XXIV are probably needlessly stringent. What is the best possible theorem in this connection?

(5) The "quasi-Tauberian" theorem of section 21 has not yet been exploited to the full. In particular, in the discussion of the relations between Riesz summability, Cesàro summability, Hölder summability, and the like, it is extremely desirable to have theorems of this type in which the kernels are not of the form $K(x-y)$, but are in some sense *nearly* of this form. Here the cruder theorems, depending solely on the use of dominant functions, are probably not difficult to elicit. On the other hand, the more refined ones will almost certainly require the full armament of Carleman's theory of singular integral equations and of the modern von Neumann-Stone[63] calculus of operators. Indeed, the time will soon come when the entire Tauberian theory must be reconsidered from the point of view of this calculus.

BIBLIOGRAPHY.

The present list of memoirs, while it cannot in the nature of things be complete, represents an attempt to bring together in one place as much as possible of the literature of Tauberian theorems. Up to 1915, we have been able to draw on the bibliography of the Cambridge tract by Hardy and M. Riesz on The General Theory of Dirichlet Series, and up to about 1924, on Smail's History and Synopsis of the Theory of Summable Infinite Processes. The bibliography also contains certain memoirs mentioned in the notes which are not of themselves Tauberian.

K. Ananda-Rau.

(1) A note on a theorem of Mr. Hardy's. Proc. Lond. Math. Soc., (2) 17 (1918), 334–336.

(2) On Lambert's series. Proc. Lond. Math. Soc., (2) 19 (1921). 1–20.

(3) On the relation between the convergence of a series and its summability by Cesàro's means. Jour. Indian Math. Soc., 15 (1923–5), 264–268.

(4) On Dirichlet's series with positive coefficients. Rendiconti di Palermo, 54 (1930), 455–462.

S. Bochner and G. H. Hardy.

(1) Note on two theorems of Norbert Wiener. Jour. Lond. Math. Soc., 1 (1926), 240–242.

E. Bortolotti.

(1) Sulle condizioni di applicabilità e di coerenza dei processi di sommazione asintotica di algoritmi infiniti. Rend. del R. Ac. di Bologna, 31 (1926–7), 33–41.

[63] Von Neumann (1); Stone (1).

S. Bosanquet.

(1) The summability of Fourier series. Math. Gaz., 15 (1931), 293-296.
(2) On the summability of Fourier series. Proc. Lond. Math. Soc., (2) 31 (1930), 144-164.

L. S. Bosanquet and E. H. Linfoot.

(1) On the zero order summability of Fourier series. Jour. Lond. Math. Soc., 6 (1931), 117-126.
(2) Generalized means and the summability of Fourier series. Quarterly Journal (Oxford series), 2 (1931), 207-229.

H. E. Bray.

(1) Elementary properties of the Stieltjes integral. Ann. of Math., 20 (1919), 176-186.

T. J. I'a. Bromwich.

(1) An introduction to the theory of infinite series. London 1908. Cf. especially p. 251
(2) On the limits of certain infinite series and integrals. Math. Ann. 65, (1908), 350-369.

Anna Caldarera.

(1) Su talune estenzioni dei criteri di convergenza di Hardy e Landau. Note e memorie di mat. (Catania), 2 (1923), 77-98.

T. Carleman.

(1) A theorem concerning Fourier series. Proc. Lond. Math. Soc., (2) 21 (1923), 483-492.

H. S. Carslaw.

(1) Introduction to the theory of Fourier series and integrals. New edition London 1921. Cf. especially pp. 150-154, 234-243.

S. Chapman.

(1) On non-integral orders of summability of series and integrals. Proc. Lond. Math. Soc., (2) 9 (1910-11), 369-409. Cf. p. 374, second note.

M. Cipolla.

(1) Criteri di convergenza riducibili a quello di Hardy-Landau. Naples Rendiconti (3a), 27 (1921), 28-37.

P. J. Daniell.

(1) A general form of integral. Ann. of Math., 19 (1917), 279-294.

P. Dienes.

(1) Sur la sommabilité de la série de Taylor. C. R., 153 (1911), 802-805.

G. Doetsch.

(1) Ein Konvergenzkriterium für Integrale. Math. Ann., 82 (1920), 68-82.
(2) Die Integrodifferentialgleichungen vom Faltungstypus. Math. Ann., 89 (1923), 192-207.
(3) Sätze von Tauberschem Charakter im Gebiet der Laplace- und Stieltjes Transformation. Berliner Berichte, Phys.-Math. Klasse (1930), 144-157.
(4) Über den Zusammenhang zwischen Abelscher und Borelscher Summabilität. Math. Ann., 104 (1931), 403-414.

P. Fatou.

(1) Séries trigonométriques et séries de Taylor. Acta Math., 30 (1906), 335-400. (These, Paris 1907.)

L. Fejér.

(1) Fourierreihe und Potenzreihe. Monatshefte f. Math., 28 (1917), 64-76.

M. Fekete.

(1) Viszgálatok a Fourier sorokrol. Math. és Termész. Ért., 34 (1916).

M. Fujiwara.

(1) Über die Verallgemeinerung des Tauberschen Satzes auf Doppelreihen. Science reports of the Tôhoku national university, Sendai, Japan, 8 (1919), 43-50.
(2) Über summierbare Reihen und Integrale. Tôhoku math. jour., 15 (1919), 323-329.
(3) Ein Satz über die Borelsche Summation. Tôhoku math. jour., 17 (1920), 339-343.

J. Hadamard.

(1) Sur la distribution des zéros de la fonction $\zeta(s)$ et ses consequences arithmétiques. Bull. Soc. Math. France, 24 (1896), 199-220.

G. H. Hardy.

(1) Researches in the theory of divergent series and divergent integrals. Quarterly journal, 35 (1904), 22-66.
(2) Theorems relating to the summability and convergence of slowly oscillating series. Proc. Lond. Math. Soc., (2) 8 (1909), 301-320.
(3) On the multiplication of Dirichlet series. Proc. Lond. Math. Soc., (2) 10 (1912), 396-405.
(4) On the summability of Fourier's series. Proc. Lond. Math. Soc., (2) 12 (1913), 365-372.
(5) An extension of a theorem on oscillating series. Proc. Lond. Math. Soc., (2), 12 (1913), 174-180.
(6) Note on Lambert's series. Proc. Lond. Math. Soc., (2) 13 (1913), 192-198.
(7) The application of Abel's method of summation to Dirichlet series. Quarterly journal, 47 (1916), 176-192.
(8) The second theorem of consistency for summable series. Proc. Lond. Math. Soc., (2) 15 (1916), 72-88.
(9) On certain criteria for the convergence of the Fourier series of a continuous function. Messenger of mathematics, 49 (1920), 149-155.
(10) The summability of a Fourier series by logarithmic means. Quarterly Journal (Oxford series), 2 (1931), 107-113.

G. H. Hardy and J. E. Littlewood.

(1) Contributions to the arithmetic theory of series. Proc. Lond. Math. Soc., (2) 11 (1913), 411-478.
(2) The relations between Borel's and Cesàro's methods of summation. Proc. Lond. Math. Soc., (2) 11 (1913), 1-16.
(3) Sur la série de Fourier d'une fonction à carré sommable. C. R. April 28, 1913.
(4) Tauberian theorems concerning power series and Dirichlet's series whose coefficients are positive. Proc. Lond. Math. Soc., (2) 13 (1913), 174-191.
(5) Some theorems concerning Dirichlet's series. Mess. math., 43 (1914), 134-147.
(6) Theorems concerning the summability of series by Borel's exponential method. Rendiconti di Palermo, 41 (1916), 36-53.
(7) Sur la convergence des séries de Fourier et des séries de Taylor. C. R. Dec. 24, 1917.
(8) The Riemann zeta-function and the theory of the distribution of primes. Acta Math., 41 (1918), 119-196.
(9) On the Fourier series of a bounded function. Proc. Lond. Math. Soc., (2) 17 (1918), xiii-xv. (Abstract).
(10) Abel's theorem and its converse. Proc. Lond. Math. Soc., (2) 18 (1920), 205-235.

(11) On a Tauberian theorem for Lambert's series, and some fundamental theorems in the analytic theory of numbers. Proc. Lond. Math. Soc., (2) 19 (1921), 21-29.
(12) Solution of the Cesàro summability problem for power-series and Fourier series. Math. Ztschr., 19 (1923), 67-96.
(13) Abel's theorem and its converse (II). Proc. Lond. Math. Soc., (2) 22 (1924), 254-269.
(14) The allied series of a Fourier series. Proc. Lond. Math. Soc., 24 (1925), 211-246.
(15) The strong summability of Fourier series. Proc. Lond. Math. Soc., (2) 26 (1926), 273-286.
(16) Elementary theorems concerning power series with positive coefficients and moment-constants of positive functions. Crelle, 157 (1926), 141-158.
(17) A further note on the converse of Abel's theorem. Proc. Lond. Math. Soc., (2) 25 (1926), 219-236.
(18) Notes on the theory of series (VII): on Young's convergence criterion for Fourier series. Proc. Lond. Math. Soc., (2) 28 (1928) 301-311.
(19) Notes on the theory of series (XI): on Tauberian theorems. Proc. Lond. Math. Soc., (2) 30 (1929), 23-37.
(20) Notes on the theory of series (III): on the summability of the Fourier series of a nearly continuous function. Proc. Camb. Phil. Soc., 23 (1927), 681-684.
(21) The equivalence of certain integral means. Proc. Lond. Math. Soc. records, (2) 22 (1924), xl-xliii.

G. H. Hardy and M. Riesz.

(1) The general theory of Dirichlet series. Cambridge Tract in Mathematics and Mathematical Physics, No. 18. Cambridge 1915.

G. H. Hardy and E. C. Titchmarsh.

(1) Self-reciprocal functions. Quarterly journal (Oxford series), 1 (1930), 196-232.

W. A. Hurwitz.

(1) A trivial Tauberian theorem. Bull. Am. Math. Soc., 32 (1926), 77-82.

S. Ikehara.

(1) An extension of Landau's theorem in the analytic theory of numbers. Journal of Math. and Phys. of the Mass. Inst. of Technology, 10 (1931), 1-12.

S. Izumi.

(1) A generalization of Tauber's theorem. Proc. Imperial Acad. Japan, 5 (1929), 57-59.

M. Jacob.

(1) Über den Eindeutigkeitssatz in der Theorie der verallgemeinerten trigonometrischen Integrale. Math. Ann., 100 (1926-7), 278-294.
(2) Über ein Theorem von Bochner-Hardy-Wiener. Jour. Lond. Math. Soc., 3 (1928), 182-187.

J. Karamata.

(1) Sur le mode de croissance regulière des fonctions. Mathematica (Cluj, Roumania), 4 (1930), 38-53.
(2) Über die Hardy-Littlewoodschen Umkehrungen des Abelschen Stetigkeitssatzes. Math. Zeitschr., 32 (1930), 319-320.
(3) Neuer Beweis und Verallgemeinerung einiger Tauberian-Sätze. Math. Zeitschr., 33 (1931), 294-300.
(4) Neuer Beweis und Verallgemeinerung der Tauberschen Sätze, welche die Laplacesche und Stieltjessche Transformation betreffen. Crelle, 164 (1931), 27-40.

A. Kienast.

(1) Extension to other series of Abel's and Tauber's theorems on power series. Proc. Lond. Math. Soc., (2) 25 (1926), 45-52.

K. Knopp.

(1) Neuere Untersuchungen in der Theorie der divergenten Reihen. Jahresber. d. deutschen Mathematikervereinigung, 32 (1923), 43-67.

(2) Theorie und Anwendung der unendlichen Reihen. Berlin, Springer, 1922. Chap. XIII.

E. Landau.

(1) Über die Konvergenz einiger Klassen von unendlichen Reihen am Rande des Konvergenzgebietes. Monatsh. f. Math. und Phys., 18 (1907), 8-28.

(2) Handbuch der Verteilung der Primzahlen. Leipzig 1909, 2v.

(3) Über die Bedeutung einiger neuerer Grenzwertsätze der Herren Hardy und Axer. Prace mat.-Fiz., 21 (1910), 97-177.

(4) Über einen Satz des Herrn Littlewood. Rendiconti di Palermo, 35 (1913), 265-276.

(5) Ein neues Konvergenzkriterium für Integrale. Münch. Sitzber., 43 (1913), 461-467.

(6) Darstellung und Begründung einiger neuerer Ergebnisse der Funktionentheorie. Berlin 1916, 2nd ed., 1930.

(7) Sobre los números primos en progressión aritmética. Rev. Mat. Hispano-Americana, 4 (1923), 1-16, 33-44.

(8) Vorlesungen über Zahlentheorie. Leipzig 1927, 3v.

H. Lebesgue.

(1) Recherches sur la convergence des séries de Fourier. Math. Ann., 61 (1905), 251-280.

P. Lévy.

(1) Sur les conditions d'application et sur la régularité des procédés de sommation des séries divergentes. Bull. Soc. Math. de France, 54 (1926), 1-25.

S. B. Littauer.

(1) On a theorem of Jacob. Jour. Lond. Math. Soc., 4 (1929), 226-231.

(2) A new Tauberian theorem with application to the summability of Fourier series and integrals. Jour. Math. Phys. Mass. Insti. Tech., 8 (1928), 216-234.

J. E. Littlewood.

(1) On the converse of Abel's theorem on power series. Proc. Lond. Math. Soc., (2) 9 (1910), 434-448.

H. J. Mellin.

(1) Die Theorie der asymptotischen Reihen vom Standpunkte der reziproken Funktionen und Integrale. Ann. Ac. Sci. Fennicae, (A) 18 (1922), No. 4, 108 p.

G. Mignosi.

(1) Inversione d'un teorema sul rapporto delle medie (Cp) di due serie. Naples Rendiconti (3a), 27 (1921), 17-28.

L. Neder.

(1) Über Taubersche Bedingungen. Proc. Lond. Math. Soc., (2) 23 (1925), 172-184.

J. von Neumann.

(1) Allgemeine Eigenwerttheorie Hermitescher Funktionaloperatoren. Math. Ann., 102 (1930), 49-131.

R. E. A. C. Paley.

(1) On the Cesàro summability of Fourier series and allied series. Proc. Camb. Phil. Soc., 26 (1930), 173-203.

M. Plancherel.

(1) Contribution à l'étude de la répresentation d'une fonction arbitraire par des intégrales définies. Rendiconti di Palermo, 30 (1910), 289–335.

A. Pringsheim.

(1) Über das Verhalten von Potenzreihen auf dem Konvergenzkreise. Münch. Ber., 30 (1900), 43–100.
(2) Über die Divergenz gewisser Potenzreihen an der Konvergenzgrenze. M. Ber., 31 (1901), 505–524.
(3) Über eine Konvergenzbedingung für unendliche Reihen, die durch iterierte Mittelbildung reduzibel sind. M. Ber., 50 (1920), 275–284.

M. Riesz.

(1) Une methode de sommation équivalente à la methode des moyennes arithmétiques. C. R., 12 June 1911.
(2) Über einen Satz des Herrn Fatou. Crelle, 140 (1911) 89–99.

R. Schmidt.

(1) Über das Borelsche Summierungsverfahren. Schriften der Königsberger gelehrten Gesellschaft, 1 (1925), 202–256.
(2) Über divergente Folgen und lineare Mittelbildungen. M. Ztsch., 22 (1925), 89–152.

W. Schnee.

(1) Über Dirichletsche Reihen. Rend. di Palermo, 27 (1909), 87–116.

J. Schur.

(1) Über lineare Transformationen in der Theorie der unendlichen Reihen. Crelle, 151 (1921). 79–111.

L. L. Smail.

(1) History and synopsis of the theory of summable infinite processes. U. of Oregon publ., 2 (Feb. 1925), No. 8.

M. H. Stone.

(1) Linear transformations in Hilbert space III. Operational methods and group theory. Proc. Mat. Acad. Sci., 16 (1930), 172–175.

O. G. Sutton.

(1) On a theorem of Carleman. Proc. Lond. Math. Soc., (2) 23 (1925), xlvii–li.

O. Szász.

(1) Verallgemeinerung und neuer Beweis einiger Sätze Tauberischer Art. Münch. Ber., (1929), 325–340.
(2) Über einen Satz von Hardy und Littlewood. Berliner Berichte 1930, 4p.
(3) Über die Approximation stetiger Funktionen durch gegebene Funktionenfolgen. Math. Ann., 104 (1930), 155–160.
(4) Abel hatványsortételével kapcsolatos újabb vizsgálatokról. Mat. és Phys. Lapok., 36 (1929), 10–22.
(5) Über Sätze Tauberscher Art. Jahresb. d. Deutschen Mathematikervereinigung, 39 (1930). 28–31.
(6) Über einige Sätze von Hardy und Littlewood. Göttinger Nachrichten (1930). Fachgruppe I, 315–333.
(7) Verallgemeinerung eines Littlewoodschen Satzes über Potenzreihen. J. L. M. S., 3 (1930), 254–262.
(8) Über Dirichletsche Reihen an der Konvergenzgrenze. Rend. del Cong. Math. di Bologna, (1928), 269–276.

A. Tauber.

(1) Ein Satz aus der Theorie der unendlichen Reihen. Monatshefte f. Math., 8 (1897), 273–277.

E. C. Titchmarsh.

(1) A theorem on infinite products. Jour. Lond. Math. Soc., 1 (1926), 35–37.

(2) On integral functions with real negative zeros. Proc. Lond. Math. Soc., (2) 26 (1927), 185–200.

(3) A contribution to the theory of Fourier transforms. Proc. Lond. Math. Soc., (2) 23 (1924), 279–289.

C. de la Vallée Poussin.

(1) Récherches analytiques sur la théorie des nombres premiers. Première partie. Ann. Soc. Sci. Brux., 20: 2 (1896), 183–256.

S. Verblunsky.

(1) The convergence of singular integrals. Proc. Camb. Phil. Soc., 26 (1930), 312–322.

(2) On summable trigonometric series. Proc. Lond. Math. Soc., (2) 31 (1930), 370–386.

(3) On the limit of a function at a point. Proc. Lond. Math. Soc., (2) 32 (1931), 163–200.

(4) A uniqueness theorem for trigonometric series. Quarterly Journal (Oxford series), (2) (1931), 81–90.

T. Vijayaraghavan.

(1) A theorem concerning the summability of a series by Borel's method. Proc. Lond. Math. Soc., (2) 27 (1928), 316–326.

(2) A Tauberian theorem. Jour. Lond. Math. Soc., 1 (1926), 113–230.

N. Wiener.

(1) The harmonic analysis of irregular motion. Jour. Math. and Phys. Mass. Inst. Tech., 5 (1925), 99–122.

(2) Une methode nouvelle pour la démonstration des théorèmes de M. Tauber. C. R., 184 (1927), 793–795.

(3) On a theorem of Bochner and Hardy. Jour. Lond. Math. Soc., 2 (1927), 118-123.

(4) A new method in Tauberian theorems. Jour. Math. Phys. M. I. T., 7 (1928), 161–184.

(5) A type of Tauberian theorem applying to Fourier series. Proc. Lond. Math. Soc., (2) 30 (1929), 1–8.

(6) The spectrum of an arbitrary function. Proc. Lond. Math. Soc., (2) 27 (1928), 487–496.

(7) Generalized harmonic analysis. Acta. Math., 55 (1930), 117–258.

(8) The operational calculus. Math. Ann., 95 (1926), 557-584.

A. Wintner.

(1) Untersuchungen über Funktionen großer Zahlen. (Erste Mitteilung) M. Ztschr., 28 (1928), 416–429.

W. H. Young.

(1) On the determination of the summability of a function by means of its Fourier constants. Proc. Lond. Math. Soc., (2) 12 (1912), 71–88.

A NOTE ON TAUBERIAN THEOREMS.[1]

By Norbert Wiener.

The following is a list of corrections to my paper Tauberian Theorems in this volume of the Annals, pp. 1–100.

Page 3, formula (0.07). Read

$$\lim_{x\to 1-0} \sum_0^\infty a_n x^n = A$$

for $\lim_{x\to 1-0} \frac{1}{1-x} \sum_0^\infty a_n x^n = A$.

Page 4, formula (0.12). Read

$$\lim_{x\to 1-0} (1-x) \sum_0^\infty \frac{n a_n x^n}{1-x^n} = A$$

for $\lim_{x\to 1-0} \sum_0^\infty \frac{n a_n x^n}{1-x^n} = A$.

Page 20, line following formula (2.63). Read x for λ.

Page 24, formula (3.225), second line. Read

$$\leqq \sum_{k=-\infty}^{+\infty} \max_z \max_{k\leqq x\leqq k+1} |f(x+z)| \lim_{N\to\infty} \frac{1}{\pi N} \int_{-\infty}^{+\infty} \Phi_2(y)\, dy$$

for $\leqq \sum_{k=-\infty}^{+\infty} \max_{k\leqq x\leqq k+1} |f(x+y)| \lim_{N\to\infty} \frac{1}{\pi N} \int_{-\infty}^{+\infty} \Phi_2(y)\, dy$.

Page 28, line 9. Delete "not be equivalent to 0, and let it".

Page 40, formula (9.06). Read $A(n)$ for A_n.

Page 45, formula (10.15). In the second line, delete the minus sign before the integral.

Page 47, formula (10.26). Insert a — after the sign of equality.

Page 47, line 10. For $2(\cos Bx - 1)$ read $2(\cos Bx - 1)/Bx^2$.

Page 66, formula (18.15). Insert du after $M(u)$.

Page 67, formula (18.22). Insert du after $M^*(u)$.

Page 72, formula (19.35). Read $\varphi(z) = \frac{e^{z^2/8}}{z\sqrt{2}}$.

Page 81, line 4. Delete "the same", and insert in its place "implied by this boundedness".

Page 84. The phrase "and hence by an application of (22.32)" should be raised from the line following (22.321) to a position directly to the left of the last line of that formula.

Page 86, line 2. The expression on this line should be squared.

Page 86, formula (22.48). The expression in the first line of this formula should be squared.

Page 88, formula (22.60). The formula should end "= 0".

[1] Received April 18, 1932.

Commentary on the Memoire [32a, b] on Tauberian Theorems.

H. R. Pitt

This paper provides the first fully developed account of Wiener's general Tauberian theory with proofs of the general theorems in their final and best possible form and an account of the more notable applications to special cases.

The methods expounded in previous papers are brought together and shown to depend essentially on a single theorem (Theorem II) that the translations of a function of $L(-\infty, \infty)$ are closed in L, if and only if its Fourier transform has no zero. The proof is based on his earlier theorem that the reciprocal of a periodic function with absolutely convergent Fourier series also has an absolutely convergent Fourier series, if and only if it has no zero. Simple and more direct proofs are now available, but the theorem itself stays unchanged as the essential element in Tauberian theory. The generalizations of the main theorem in Chapter I of the paper are of great interest in themselves, but are not often needed in applications as it has generally been found much simpler and more straightforward to appeal to the main theorem.

In Chapter II, it is a short step from the closure theorems to the more familiar statement of the general Tauberian theorem in the form Theorem VIII. In the remainder of Chapter II and in Chapter III, there is an apparently parallel development for the Stieltjes form in which

$$\int_{-\infty}^{\infty} K_1(\xi - x)\,df(\xi) \to A \int_{-\infty}^{\infty} K_1(\xi)\,d\xi$$

implies

$$\int_{-\infty}^{\infty} K_2(\xi - x)\,df(\xi) \to A \int_{-\infty}^{\infty} K_2(\xi)\,d\xi$$

under certain conditions. The theorems are difficult and, in spite of their apparent generality, can be deduced without difficulty from Theorem VIII, or from a modified version of it in which $s(x) \to A$ as $x \to \infty$ is deduced directly from $\int_{-\infty}^{\infty} K(\xi - x)s(\xi)\,d\xi \to A$ under the conditions $\int_{-\infty}^{\infty} K(x)e^{iux}\,dx \neq 0$, $s(\xi)$ bounded and $s(\xi)$ satisfying a Tauberian condition such as

$\int_y^{y+1} |ds|$ bounded or $s'(\xi)$ bounded, or (as in Theorem XII) $\int_y^{y+1} |ds(x)| - s(y+1) + s(y)$ bounded. These are, of course, unnecessarily restrictive but on the right scale, the last being a one-sided condition covering the classical $na_n \geqslant -K$ condition of Hardy and Littlewood in the Abel case as well as later contributions by R. Schmidt, Szasz, Vijayaraghavan and others. In fact, the Wiener theorems in the Stieltjes form are not often used or needed in applications and, in spite of their power, it is generally easier and more straightforward to appeal directly to the standard (non-Stieltjes) form.

Wiener's proof of the Tauberian theorem for Lambert summation in Chapter IV is probably the finest example of the power of his method. It is, in fact, even more straightforward to derive it from his Theorem VIII than from XI''' as in the text, and it remains the only proof. The Tauberian theorem for Lambert summation was a landmark in the history of the prime number theorem (PNT), but the relationship between the two theorems is clearer now than at any time, since the Lambert Tauberian theorem is now known to be deeper than the PNT and has not so far been proved by the elementary methods which suffice for the PNT itself. It has a rather paradoxical relationship to Littlewood's Tauberian theorem for Abel summation. It is logically weaker in that Lambert summability implies Abel summability, but the proof of this Abelian result is more difficult than the PNT and has not so far been reduced to elementary terms.

The proof of Ikehara's theorem (XVII) is by an appeal to the general Tauberian theorem, and while this is interesting and perfectly valid, it must be said that it is not the *right* one in any logical sense and helps little towards an understanding of the nature of the theorem. In fact, it is less useful to regard Ikehara's theorem as Tauberian than to appreciate its relationship to the Riemann-Lebesgue theorem for Fourier coefficients.

Chapter V contains applications of the general Tauberian theorems to special transformations which are *exactly* of the convolution type, $\int k(x-y)\, s(y)\,dy$. It differs in this respect from Chapter VI in which the emphasis is on transformations that are only approximately of this kind. In other words, the transformation studied Chapter VI is $\int_{-\infty}^{\infty} k(x,y)s(y)\,dy$, or $\int_{-\infty}^{\infty} k(x,y)\, d\phi(y)$ in the Stieltjes form, where

$$k(x, y) = h(x - y) + h^*(x, y)$$

and h^* is small in some appropriate sense. The general theorem is based on the Stieltjes form and is powerful enough to deal with the Tauberian theorems for Borel summation and the Hardy-Littlewood theorem to the effect that

$$f(x) = \sum_0^\infty a_n x^n \sim A(1 - x)^{-\alpha}, \quad \alpha > 0, \quad a_n \geqslant 0, \quad x \to 1,$$

implies that $s_n \sim An^\alpha/\Gamma(\alpha + 1)$. It would seem, however, that both of these special theorems could be proved more directly without bringing in the Stieltjes form.

In Chapter VII, Wiener is quite right in describing XXII as Abelian, since it is formally parallel to the standard general Abelian theorem that $\int_{-\infty}^{\infty} k_1(x - y)s(y)\,dy \to 0$ implies that $\int_{-\infty}^{\infty} k_2(x - y)s(y)\,dy \to 0$, provided that

$$k_2(x) = \int_{-\infty}^{\infty} k_1(x - y)\gamma(y)\,dy,$$

k_1, and γ are integrable $(-\infty, \infty)$ and $s(y)$ is bounded. Theorem XXII is then one of many variants in which the integrability and order conditions are modified. The nature of the theorem comes out clearly in its application to the Abelian Hardy-Littlewood theorem on summability of Fourier series.

The main Theorem XXIV of Chapter VIII is a most interesting extension of the general Tauberian theorem to a different type of asymptotic behaviour in which the property

$$B^{-1} \int_{-B}^{B} |g(x)|^2\,dx \to 0 \quad \text{as } B \to \infty$$

takes the place of $g(x) \to 0$ as $x \to \infty$. However, in spite of its formal similarity to Theorem VIII, it relates not so much to classical Tauberian theory as to Wiener's theory of generalized harmonic analysis and Fourier spectra, and the proof depends heavily on this latter work [30a].

The other main Theorem XXV of the chapter is not closely related. The Tauberian element is confined to the equivalence of (22.59) and (22.60), depending on the theorem that

$$H^{-1} \int_0^H s(x)\,dx \to 0 \quad \text{as} \quad H \to \infty$$

$$\text{and} \quad \int_0^\infty \frac{\sin^2 \varepsilon u}{\varepsilon u^2} s(u)\, dx \to 0 \quad \text{as} \quad \varepsilon \to \infty$$

are equivalent for non-negative $s(x)$. This is a straightforward corollary of the general Tauberian theorem VIII, although there is an earlier proof using Littlewood's Tauberian theorem for Abel summation in Wiener's paper [27c].

Sonderdruck aus Mathematische Zeitschrift, Band 36 Heft 5.

A One-Sided Tauberian Theorem.

By

Norbert Wiener in Belmont (Mass. U. S. A.).

The present paper is the outcome of an attempt to apply the author's methods[1]) to a problem suggested to the author by Miss M. L. Cartwright. The theorem to be proved is the following:

Theorem I. *Let*

(1) $$\lim_{r\to 0}\sum_0^\infty a_n \exp(-n r e^{i\vartheta}) = s \qquad (-\pi/2 < \vartheta < \pi/2).$$

Let every quantity $n a_n$ *lie in the region*

(2) $$-\frac{\pi}{2} < -\alpha \leqq \arg(n a_n + B) \leqq \alpha < \frac{\pi}{2},$$

where α *and* B *are independent of* n. *Let*

(3) $$s_n = \sum_0^n a_n = O(n^m)$$

for some finite m. *Then*

(4) $$\sum_0^\infty a_n = s.$$

There is obviously no restriction in the assumption that $a_0 = 0$. We begin by writing

(5) $$s_{[u]} = s(u),$$

and putting (1) in the form

(6) $$\lim_{r\to 0} r e^{i\vartheta}\int_0^\infty s(u)\exp(-u r e^{i\vartheta})\,du = s,$$

which we can do because of (3). If (3) holds,

(7) $$\begin{aligned} & r\int_0^\infty |s(u)|\,|\exp(-u r e^{i\vartheta})|\,du \\ & \leqq K r\int_0^\infty u^m \exp(-u r\cos\vartheta)\,du \\ & = K m!/(\cos\vartheta)^{m+1} r^m. \end{aligned}$$

[1]) Tauberian Theorems, Annals of Mathematics **33** (1932), pp. 1—100.

Now let us introduce the auxiliary function

$$V(v) = \frac{1}{2\pi}\int_{-\infty}^{\infty} \frac{e^{-\xi^2/2}}{\Gamma(1-i\xi)} e^{\vartheta\xi} v^{-i\xi-1}\, d\xi. \tag{8}$$

It is easy to verify that for any k, $V(e^v)\, e^{kv}$ is the Fourier transform of a function of L_2, and hence belongs to L_2. This depends on the fact that $e^{-\xi^2/2}/\Gamma(1-i\xi)$ is entire and belongs to L_2 in the real part of ξ for every imaginary part of ξ. From this, it follows by the Schwarz inequality, since

$$\left|\int_a^{\pm\infty} V(e^v)\, e^{lv}\, dv\right| \leq \sqrt{\int_a^{\pm\infty} [V(e^v)\, e^{kv}]^2\, dv \int_a^{\pm\infty} e^{(2l-2k)v}\, dv} \tag{9}$$

that $V(e^v)\, e^{lv}$ belongs to L_1, from which we may deduce that $V(v)\, v^{\pm\nu}$ is always absolutely integrable over $(0,\infty)$. Hence, by (7), the double integral

$$v\int_0^\infty V(vr)\, dr\, r e^{i\vartheta} \int_0^\infty s(u) \exp(-ure^{i\vartheta})\, du \tag{10}$$

is absolutely convergent, and may be written

$$e^{i\vartheta}\int_0^\infty s(u)\, du \int_0^\infty vrV(vr) \exp(-ure^{i\vartheta})\, dr. \tag{11}$$

Thus by (6),

$$\lim_{v\to\infty} e^{i\vartheta}\int_0^\infty s(u)\, du \int_0^\infty vr V(vr) \exp(-ure^{i\vartheta})\, dr = s\int_0^\infty V(r)\, dr = s, \tag{12}$$

since

$$\int_0^\infty V(r)\, dr = \frac{1}{2\pi}\int_{-\infty}^{\infty} dy \int_{-\infty}^{\infty} \frac{e^{-\xi^2/2}}{\Gamma(1-i\xi)} e^{\vartheta\xi} e^{-iy\xi}\, d\xi = \frac{e^0}{\Gamma(1)} = 1 \tag{13}$$

by the Fourier integral theorem. Now,

$$\begin{aligned}
& e^{i\vartheta}\int_0^\infty vr\, V(vr) \exp(-ure^{i\vartheta})\, dr \\
&= \frac{e^{i\vartheta}}{2\pi}\int_0^\infty vr \exp(-ure^{i\vartheta})\, dr \int_{-\infty}^{\infty} \frac{e^{-\xi^2/2}}{\Gamma(1-i\xi)} e^{\vartheta\xi} (vr)^{-i\xi-1}\, d\xi \\
&= \frac{e^{i\vartheta}}{2\pi}\int_{-\infty}^{\infty} \frac{e^{-\xi^2/2}}{\Gamma(1-i\xi)} e^{\vartheta\xi}\, d\xi \int_0^\infty (vr)^{-i\xi} \exp(-ure^{i\vartheta})\, dr \\
&= \frac{1}{2\pi v}\int_{-\infty}^{\infty} \frac{e^{-\xi^2/2}}{\Gamma(1-i\xi)}\, d\xi \int_0^\infty \varrho^{-i\xi} e^{-\frac{u}{v}\varrho}\, d\varrho \\
&= \frac{1}{2\pi v}\int_{-\infty}^{\infty} \left(\frac{u}{v}\right)^{i\xi-1} e^{-\xi^2/2}\, d\xi = \frac{1}{2\pi u}\int_{-\infty}^{\infty} e^{i\xi\log\frac{u}{v} - \xi^2/2}\, d\xi \\
&= \frac{e^{-(\log\frac{u}{v})/2}}{u\sqrt{2\pi}} = \frac{\left(\frac{u}{v}\right)^{-1-\frac{\log u/v}{2}}}{v\sqrt{2\pi}}.
\end{aligned} \tag{14}$$

Thus (12) becomes:

$$(15)\qquad \lim_{v\to\infty}\frac{1}{\sqrt{2\pi v}}\int_0^\infty s(u)\left(\frac{u}{v}\right)^{-1-\frac{\log u/v}{2}}du = s.$$

This may again be written

$$(16)\qquad \lim_{\lambda\to 0}\sum a_n K(n\lambda) = s$$

where

$$(17)\qquad K(w) = \frac{1}{\sqrt{2\pi}}\int_{\log w}^\infty e^{-x^2/_2}dx.$$

Let us now write

$$(18)\qquad a_n = c_n e^{i\alpha} + d_n e^{-i\alpha};\qquad s = p e^{i\alpha} + q e^{-i\alpha},$$

where all the c_n's and d_n's are real, as well as p and q. It is easy to verify that $n c_n$ and $n d_n$ are bounded below, and that

$$(19)\qquad \lim_{\lambda\to 0}\sum c_n K(n\lambda) = q;\qquad \lim_{\lambda\to 0}\sum d_n K(n\lambda) = q.$$

Thus by Theorem XIII of the author's previous paper, in view of the fact that

$$(20)\qquad K(0) = 1$$

and

$$(21)\qquad \int_0^\infty K(w)\, w^{iu-1}dw = e^{-u^2/_2} \neq 0$$

for any real argument, it follows that

$$(22)\qquad \sum c_n = p;\qquad \sum d_n = q;$$

and by (18), that

$$(4)\qquad \sum_0^\infty a_n = s.$$

(Eingegangen am 28. Mai 1932.)

Druck von Oscar Brandstetter in Leipzig.
Printed in Germany.

Comments on [33b]

H. R. Pitt

The proof given for the theorem is by an application of the general Tauberian theorem, but it is not quite direct since the original kernel is transformed to a more tractable one by an auxiliary Abelian transformation. The argument is clearer after an exponential change of scale to the standard form. We then have

$$\int_{-\infty}^{\infty} k(x-y)s(e^y)\,dy \to 0, \quad k(x) = e^{-x+i\theta}\exp[-e^{-x+i\theta}],$$
$$K(\xi) = \int_{-\infty}^{\infty} e^{i\xi x}k(x)\,dx = e^{-\theta\xi}\Gamma(1-i\xi).$$

Then $q(x)[e^x V(e^x)$ in the text$]$ is defined by

$$q(x) = \frac{1}{2\pi}\int_{-\infty}^{\infty} e^{-i\xi x}e^{-\xi^2/2}e^{\theta\xi}[\Gamma(1-i\xi)]^{-1}\,d\xi$$

and is easily shown to be integrable in $(-\infty, \infty)$, while

$$Q(\xi) = \int_{-\infty}^{\infty} e^{i\xi x}q(x)\,dx = e^{-\xi^2/2}e^{\theta\xi}[\Gamma(1-i\xi)]^{-1}.$$

It then follows that $\int_{-\infty}^{\infty} h(x-y)s(e^y)\,dy \to 0$, where

$$h(x) = \int_{-\infty}^{\infty} q(x-y)k(y)\,dy = (2\pi)^{-1}e^{-x^2/2}, \quad H(\xi) = e^{-\xi^2/2}.$$

The remainder of the proof is an easy deduction from the general Tauberian theorem. If we had $s_n = O(1)$ instead of the weaker $s_n = O(n^m)$, the general theorem could be applied directly to the original kernel since $K(\xi) \neq 0$. In fact, this can still be done when $s_n = O(n^m)$, but the deduction of the boundedness of s_n is more intricate.

Commentary on the Tauberian-Theoretic Papers [27c, d], [28b], [29b], [32a, b], [33b]

J. Korevaar

1. The paper [27c]. Around 1925, Wiener was led to the following question in connection with his work on general harmonic analysis. Let $F(t)$ be integrable over every bounded interval $(0, T)$. Does the existence of one of the asymptotic averages

$$\lim_{T\to\infty}\frac{1}{T}\int_0^T F(t)\,dt, \tag{1.1}$$

$$\lim_{\varepsilon\downarrow 0}\frac{2}{\pi\varepsilon}\int_0^\infty F(t)\frac{\sin^2\varepsilon t}{t^2}\,dt \tag{1.2}$$

imply the existence of the other, and their equality? In the desired application, $F(t)$ is real and non-negative. (For background, cf. N. Levinson, *Wiener's life*,[15] in particular pp. 17–18.)

Discussing the question with A. E. Ingham in Göttingen in 1926, Wiener came to realize that his problem was related to the class of so-called Abelian and Tauberian theorems, in particular to the, then recent, work of G. H. Hardy and J. E. Littlewood. Specifically, Wiener's problem bore some resemblance to the following theorems relating Cesàro and Abel summability. If a series $\sum_1^\infty a_n$, with partial sums s_n, is Cesàro summable, that is, if the asymptotic average

$$\lim_{N\to\infty}\frac{1}{N}\sum_1^N s_n \tag{1.3}$$

exists, then the series is Abel summable, that is, the asymptotic average

$$\lim_{x\uparrow 1}\frac{\sum_1^\infty s_n x^n}{\sum_1^\infty x^n} \tag{1.4}$$

exists, and the two limits have the same value. This is an "easy" Abelian theorem. In the opposite direction, Hardy and Littlewood had proved the "difficult" Tauberian theorem that the existence of (1.4) implies that of (1.3) whenever the s_n satisfy a "Tauberian condition" such as boundedness, non-negativity, or one-sided boundedness.

In 1926, S. Bochner and G. H. Hardy published a paper in which they proved the "easy" implication (1.1) $\Rightarrow$ (1.2) under a mild condition on $F(t)$ (cf. Theorem I in [27c]). Soon thereafter, and this is the main content of [27c], Wiener established the "difficult" implication (1.2) $\Rightarrow$ (1.1) under the Tauberian condition $F(t) \geqslant 0$. Two aspects of the proof deserve special comment. One is that Wiener actually uses the theorem of Hardy and Littlewood stated above (more accurately, he uses an integral analog). The other is that this is the first time Wiener uses the method of trigonometric integrals to pass from one type of average to a different one. The transition is made through explicit representation of one kernel in terms of another (p. 121); in later work, where explicit representation is not possible, the transition would also require suitable approximation.

2. Papers [27d], [28b]. At the time when Wiener, through conversation with Ingham and correspondence with Hardy, became interested in Tauberian theorems, R. Schmidt had just published his work in the direction of a unified Tauberian theory. Using the theory of moments, Schmidt had succeeded in incorporating many of the varied results of Hardy and Littlewood in a general framework. However, the most difficult Tauberian theorem of all, a Tauberian theorem for Lambert series, had proved resistant to Schmidt's efforts. This Lambert Tauberian theorem implies the prime number theorem, but Hardy and Littlewood had only been able to prove it with the aid of number theoretic estimates similar to the prime number theorem.

It was at Schmidt's suggestion that Wiener enlarged the scope of his Tauberian investigations, and began working towards a powerful general theory based on Fourier methods, or "complex moments." Schmidt also proposed that the test for a successful theory should be that it handle the Lambert Tauberian theorem.

Thus Wiener's general Tauberian theory was born in papers [27d] and [28b]. The Comptes Rendus announcement [27d] is somewhat obscure, in part because of misprints, but paper [28b] gives a very readable account of the new method. In [28b] the "general" Tauberian theorem (see pp. 163–164) does not yet appear in the full generality that it would attain in [32a], but the form given suffices for most applications.

The proof depends on two simple ideas. Suppose f has an asymptotic average relative to the L^1 kernel N_1, that is,

$$\lim_{\eta \to -\infty} \int_{-\infty}^{\infty} f(\xi - \eta) N_1(\xi)\, d\xi \tag{2.1}$$

exists (p. 164). Then one can find other "admissible" kernels S, that is, L^1

functions S relative to which f has an asymptotic average, (i) by convolution with arbitrary L^1 functions R:

$$S = N_1 * R; \tag{2.2}$$

(ii) through approximation in L^1 (from below and above) by admissible kernels. Convolution is needed to manufacture sufficiently many new admissible kernels S, and since (2.2), by Fourier transformation, is equivalent to

$$\hat{S} = \hat{N}_1 \cdot \hat{R}, \tag{2.3}$$

the determination of $\hat{R}$ for a given candidate $\hat{S}$ requires division by $\hat{N}_1$. Thus Wiener's now famous hypothesis that the Fourier transform

$$\hat{N}_1(u) = \int_{-\infty}^{\infty} N_1(\xi)e^{iu\xi}\,d\xi \tag{2.4}$$

should be different from zero for all real u. In order to make R in (2.2) come out an L^1 function, it is convenient to require that $\hat{S}$ vanish outside a bounded interval and that $\hat{S}$ and $\hat{N}_1$ be continuously differentiable. The latter can be assured by imposing the growth conditions $\xi S(\xi) \in L^1$, $\xi N_1(\xi) \in L^1$. This is essentially what Wiener does in [28b]. A very nice and simple version of the proof appears in N. Levinson's essay *Wiener's life*,[15] pp. 18–21.

From his general Tauberian theorem, Wiener derives a very convenient and nowadays often neglected second fundamental theorem (p. 178) which is directly applicable to various classical series. He shows in particular that this second theorem contains the Lambert Tauberian theorem. In this application, the nonvanishing of the relevant Fourier transform follows from the nonvanishing of the Riemann zeta function $\zeta(s)$ on the line Re $s = 1$. Having received an independent proof at Wiener's hands, the Lambert Tauberian theorem could henceforth be used to prove the prime number theorem. Never before had so little information about the zeta function as the nonvanishing of $\zeta(1 + iy)$ been sufficient to obtain that result. (Since the advent of the "elementary" proof of the prime number theorem by A. Selberg and P. Erdös in 1949, one knows of course that the zeta function can be avoided altogether.)

Paper [28b] also contains a third fundamental theorem (pp. 180–181) that can be applied to Borel summability. Thus Wiener had succeeded in developing a general Tauberian theory so powerful that it could handle all the major Tauberian results known at the time.

3. Comments on the monograph [32a] and a brief remark on [29b]. In [32a] Wiener's basic general Tauberian theorem receives its definitive form (Theorem VIII, p. 25). There is no longer a growth condition on the kernel N_1 (now denoted by K_1); the nonvanishing of the Fourier transform $\hat{N}_1$ or $\hat{K}_1$ is the only requirement.

Taking the preceding section as point of departure, one may describe the new method of proof in the following way. It is necessary to refine the discussion of equation (2.3). To this end, Wiener shows (cf. p. 18) that if S is any L^1 function such that $\hat{S}$ has compact support, then

$$\hat{S}/\hat{N}_1 = \hat{R}, \quad \text{with } R \in L^1. \tag{3.1}$$

It will follow that $S = N_1 * R$ is admissible, and hence all L^1 functions N_2 or K_2 are admissible: they can be approximated, in L^1, by functions S whose Fourier transforms have compact support. To prove (3.1), Wiener shows that a function with compact support is the Fourier transform of an L^1 function R, if and only if it is locally given by absolutely convergent Fourier series (cf. Lemma 11f, p. 14). The result then follows from his now famous results on absolutely convergent Fourier series (Lemmas II a–e, pp. 10–14) about which more will be said in section 4.

In [32a], Wiener actually uses the above argument to prove his famous "closure of translates" theorem for L^1 (Theorem II, p. 9) which is equivalent to the basic Tauberian theorem. By the Hahn-Banach theorem, this closure theorem is also equivalent to the fact that the functional equation

$$N_1 * V = 0 \tag{3.2}$$

fails to have a nonzero bounded solution V, if and only if $\hat{N}_1(u) \neq 0$ for all u. Wiener does not make this observation, but several subsequent proofs of his general Tauberian theorem have been based on (3.2). The difficulty with (3.2) is that bounded functions V have no Fourier transform in the classical sense, hence a direct discussion of (3.2) requires some sort of generalized Fourier transform. In this context one may mention proofs by T. Carleman (1944),[4] A. Beurling (1947),[2] H. Pollard (1953)[19] and J. Korevaar (1965).[13]

The last paper contains a relatively simple discussion of (3.2) based on Schwartz's theory of distributions. It may be of interest to reproduce here the core of the argument which serves to replace N_1 in (3.2) by suitable test functions ϕ belonging to Schwartz's space $\mathscr{S}$. Suppose that $\hat{N}_1(c) \neq 0$. Let $\hat{\phi}(u)$ be a C^∞ function of the form $\hat{w}\{(u - c)/\varepsilon\}$ which is equal to 1 on $(c - \varepsilon, c + \varepsilon)$ and equal to 0 outside $(c - 2\varepsilon, c + 2\varepsilon)$. Simple considerations show that for small ε, the quotient

$$\frac{\hat{\phi}(u)}{\hat{N}_1(u)} = \frac{\hat{\phi}(u)/\hat{N}_1(c)}{1 - \{1 - \hat{N}_1(u)/\hat{N}_1(c)\}\hat{\phi}(\frac{1}{2}u + \frac{1}{2}c)}$$

(in which $\hat{\phi}(\frac{1}{2}u + \frac{1}{2}c)$ is equal to 1 wherever $\hat{\phi}(u) \neq 0$) is the Fourier transform $\hat{R}(u)$ of an L^1 function R. Thus by (3.2),

$$\phi * V = (R * N_1) * V = 0. \tag{3.3}$$

The function ϕ will belong to $\mathscr{S}$; hence one may take the Fourier transform of (3.3) to obtain the relation

$$\hat{\phi}(u)\,\hat{V}(u) = 0.$$

It follows in particular that $\hat{V}(u) = 0$ on $(c - \varepsilon, c + \varepsilon)$. Thus the nonvanishing of $\hat{N}_1(u)$ on $(-\infty, \infty)$ will imply that $\hat{V}(u) = 0$ for all u, and hence $V = 0$. In the other direction, it is easy to see that if $\hat{N}_1(c) = 0$, then (3.2) has a bounded exponential solution.

It may be observed that equation (3.2) can be dealt with in a classical manner if N_1 is a little better than just an L^1 function, cf. M. Kac (1965).[10]

Besides Theorem VIII, paper [32a] contains several general Tauberian theorems involving Stieltjes integrals (pp. 26–32) and designed to facilitate applications to infinite series. In practice, these results are not used as often as a very convenient theorem obtained later by H. R. Pitt, (1938).[17] The latter theorem is equivalent to Wiener's basic theorem, but the conclusion from (2.1) is that, under appropriate conditions, f has a limit rather than just an asymptotic average relative to other kernels N_2. For an account, cf. the books by Hardy[7] and Pitt.[18]

In [32a], Wiener discusses systematically how all Tauberian theorems known at the time can be derived from his general results. The paper contains a somewhat simpler derivation of the Lambert Tauberian theorem than was given in [28b]. Further simplification is possible when one uses Pitt's work referred to above. Wiener also shows in [32a] how the beautiful, and then very new, theorem of his student S. Ikehara (theorem XVII, p. 44) fits into the general framework. This "Landau-Ikehara-Wiener theorem" is now used in many books to provide a proof of the prime number theorem. A known Tauberian theorem for Borel summability and certain Tauberian theorems of Hardy and Littlewood for power series find a natural place in a chapter on Tauberian theorems involving kernels close to convolution type (pp. 62–

72). Work of Hardy and Littlewood on Cesàro summability of Fourier series and integrals which Wiener had already discussed in an ad-hoc manner in [29b] becomes more accessible in the setting of a chapter on quasi-Tauberian theorems (pp. 72–79). The final chapter (pp. 79–94) deals with various types of strong convergence and summability, some related to Wiener's general harmonic analysis.

4. Applications, extensions and refinements of the results in [32a]. *Absolutely convergent Fourier series.* Wiener's Lemma IIe on the reciprocal $1/f$ of a nonvanishing function f of class A (periodic f with absolutely convergent Fourier series) has received numerous extensions. In 1934 Lévy showed that if φ is analytic on the range of $f \in A$, then also $\varphi(f) \in A$. It is now known that real analyticity of φ is sufficient, and that no larger class of functions φ works for all f with range in a given set (H. Helson, J. -P. Kahane). There are corresponding results for Fourier-Stieltjes transforms and in the abstract framework of (infinite) locally compact Abelian groups. For an account, cf. W. Rudin's book,[20] Chapter 6.

An abstract version of Lemma IIe plays an important role in the theory of commutative complex Banach algebras with identity. I. M. Gelfand has shown that an element x of such an algebra has an inverse, if and only if its (abstract) Fourier transform ("Gelfand transform") is never zero. Cf. R. E. Edwards's book[5], in particular section 11.4.10.

Span of translates theorems. Let $\mathscr{F}$ be a family of functions $f \in L^1$, let $Z(\mathscr{F})$ denote the set of real zeros common to the Fourier transforms $\hat{f}(f \in \mathscr{F})$, and let $S(\mathscr{F})$ denote the closed span in L^1 of the translates of the functions $f \in \mathscr{F}$. One may also describe $S(\mathscr{F})$ as the closed ideal generated by $\mathscr{F}$, using convolution as multiplication. Wiener showed that $S(\mathscr{F}) = L^1$, if and only if $Z(\mathscr{F})$ is empty (Theorem IV, p. 19). But which functions g are approximable by linear combinations of translates of functions $f \in \mathscr{F}$ when $Z(\mathscr{F})$ is not empty? Wiener's Theorem III (p. 19) gives a partial answer to this question. Further contributions in the positive direction were made by a number of authors, including A. Beurling, S. Mandelbrojt and S. Agmon, and H. Pollard. However, for a long time it was not known if $S(\mathscr{F})$ always contains all $g \in L^1$ such that $Z(g) \supset Z(\mathscr{F})$. The latter question was finally resolved negatively through the work of L. Schwartz and P. Malliavin. For an account, and for generalizations to the case of locally compact Abelian groups, cf. Rudin's book,[20] Chapter 7.

There are also span of translates theorems for spaces other than L^1. Wiener himself dealt with the much simpler case of L^2 (Theorem I, p. 7). R. P. Boas, Jr. and S. Bochner considered the case of C_0 in 1938,[3] cf. also Korevaar, (1947).[12] A. Beurling considered another space of continuous functions in connection with a different proof of Wiener's Tauberian theorem (1945).[1] Mean periodic functions were defined by L. Schwartz (1947),[21] as functions whose translates do not span in a certain space. For an account, cf. Edwards's book,[5] Section 11.2.3.

Generalizations and refinements of the Tauberian theorem. Through the work of many mathematicians, including V. Ditkin, R. Godement, I. E. Segal, S. Mandelbrojt and S. Agmon, I. Kaplansky, G. W. Mackey, H. Helson, L. H. Loomis and H. J. Reiter, Wiener's theorem has been extended to arbitrary locally compact Abelian groups. In the process, it has become stronger also in the classical case. There is a good account in Rudin's book,[20] Chapter 7. For this and related more abstract results, cf. also the books by Edwards,[5] E. Hewitt and K. A. Ross,[8] Y. Katznelson[11] and L. H. Loomis.[16]

An extension of a different character may be found in A. Beurling, (1945)[1]. There is also a distributional form of Wiener's theorem: see T. H. Ganelius's book,[6] in particular p. 13.

There now exist very refined remainder estimates related to Wiener's theorem. Beurling proved the first such result; a large number of best possible results are due to Ganelius. An account may be found in Ganelius's book.

A few applications. Wiener's Tauberian theorems have been applied to problems in many areas. A complete account will not be attempted; only a few applications will be mentioned, mostly those initiated by Wiener himself. A number of these have been indicated already, notably, applications to general harmonic analysis and to the classical Tauberian theorems of Hardy and Littlewood. There have been very successful applications to the asymptotic behavior of entire functions, starting with Paley and Wiener [34d], Chapter 5. Then there are applications to Tauberian theorems for gap theorems, the so-called high indices theorems: see [35a] and N. Levinson's book,[14] Chapters 10, 11.[†] The result in Wiener's short paper [33b] should be compared with the classical Tauberian theorems of Littlewood, and Hardy and Littlewood for power series, where the Tauberian conditions are $|na_n| \leqslant K$ or $na_n \geqslant -K$. One may finally mention applications to ergodic theory; cf. the book by E. Hille and R. S. Phillips,[9] Chapter 18.

[†] *Editor's note.* See also the comprehensive commentary by Professor P. J. Koosis in this volume.

References

1. A. Beurling, Acta Math. 77(1945), 127–136.
2. A. Beurling, C. R. Acad. Sci., Paris 225(1947), 326–328.
3. R. P. Boas, Jr., and S. Bochner, Ann. of Math. 39(1938), 287–300.
4. T. Carleman, *L'intégrale de Fourier et questions qui s'y rattachent*, Uppsala, 1944.
5. R. E. Edwards, *Fourier series*, vol. 2, New York, 1967.
6. T. H. Ganelius, *Tauberian remainder theorems*, Berlin, 1971.
7. G. H. Hardy, *Divergent series*, Oxford, 1949.
8. E. Hewitt and K. A. Ross, *Abstract harmonic analysis*, vol. 2, Berlin, 1970.
9. E. Hille and R. S. Phillips, *Functional analysis and semigroups*, Providence, 1957.
10. M. Kac, Proc. Amer. Math. Soc. 16(1965), 1155–1157.
11. Y. Katznelson, *An introduction to harmonic analysis*, New York, 1958.
12. J. Korevaar, Duke Math. J. 14(1947), 31–50.
13. J. Korevaar, Proc. Amer. Math. Soc. 16(1965), 353–355.
14. N. Levinson, *Gap and density theorems*, New York, 1940.
15. N. Levinson, *Wiener's life*, Bull. Amer. Math. Soc. 72(1966), no. 1, part II, 1–32.
16. L. H. Loomis, *An introduction to abstract harmonic analysis*, Princeton, 1953.
17. H. R. Pitt, Proc. London Math. Soc. 44(1938), 243–288.
18. H. R. Pitt, *Tauberian theorems*, Oxford, 1958.
19. H. Pollard, Duke Math. J. 20(1953), 499–512.
20. W. Rudin, *Fourier analysis on groups*, Interscience, New York, 1962.
21. L. Schwartz, Ann. of Math. 48(1947), 857–929.

ANALYTIC PROPERTIES OF THE CHARACTERS OF INFINITE ABELIAN GROUPS

By N. WIENER, Cambridge, U. S. A. and R. E. A. C. PALEY, Cambridge, England

Haar has recently discussed the characters of denumerable Abelian groups. The authors give an explicit method for obtaining these characters without the use of matrices, and extend the notion of character to a special class of non-denumerable Abelian groups. The characters form a group with respect to multiplication, and this group may be regarded as in a certain sense dual to the original group. The original group is isomorphic with a certain group of functions in the space of the characters, which will be orthogonal in the denumerable case. We have the following theorem: *Let f (c) be any function defined over the space of the characters. Let it possess an absolutely convergent series in the orthogonal functions just mentioned, and let it vanish for no character. Then 1/f (c) possesses an absolutely convergent series in the same set of functions.* From this we may deduce theorems analogous to the general Tauberian theorems of one of the authors. In such analogies, the orthogonal functions defined in characteristic space will be the generalizations of the functions exp inx.

The space of the characters must of course possess an intrinsic definition of measure to make integration and orthogonality possible. This intrinsic measure is easy to establish. We have not only an intrinsic definition of measure but an intrinsic notion of limit and of interval, neighborhood, etc. It is thus possible to set up a complete analogue to the notion of almost periodic function, replacing translation numbers by operators of the group. As one would expect, an almost periodic function turns out to be one which may be approximated uniformly by a polynomial in the analogues of the trigonometric functions in the space of characters. The Wiener theory of generalized harmonic analysis likewise possesses a precise analogue in character space.

The Walsh orthogonal functions form an example of a set of our orthogonal functions in character space. We present many other similar examples.

Reprinted from the Proceedings of the NATIONAL ACADEMY OF SCIENCES,
Vol. 19, No. 2, pp. 253–257. February, 1933.

CHARACTERS OF ABELIAN GROUPS

BY R. E. A. C. PALEY[1] AND N. WIENER

DEPARTMENT OF MATHEMATICS, MASSACHUSETTS INSTITUTE OF TECHNOLOGY

Communicated January 14, 1933

1. In this note we give a preliminary account of some results which will be published in detail later.

The subject of group characters has been studied by Haar[2] who considers them in the case where the group is denumerable and by Peter and Weyl[3] who consider the characters of continuous groups. In this paper we are concerned only with Abelian groups, but, in this restricted field, we obtain results which are more complete than those of the two papers cited.

2. Let us consider an Abelian group G. We define a character of the group to be a function $\chi(A)$ of absolute value 1, defined on the elements A of G, such that, for any other transformation T of the group, we have

$$\chi(TA) = \chi(T)\chi(A) \tag{1}$$

We observe that the product of two functions satisfying (1) is another function also satisfying (1), so that the set of characters of the group form another group themselves. There is a certain sense in which it is possible to build up a duality between these two groups, and we shall discuss it later.

3. We begin our investigations by obtaining explicitly the functions which satisfy (1) in the case when the group G is denumerable. The simplest case is that in which the elements of G are all periodic, and, for that case, we will state our construction here, without proof. In the general case the procedure is essentially similar.

Suppose that the elements of G are written in a series

$$E, A_1, A_2, A_3, \ldots, A_n, \ldots, \tag{2}$$

where E denotes the identity transformation. We write $B_1 = A_1$, and suppose that n_1 is the least positive integer for which $A_1^{n_1} = E$. We strike out of the series (2) all the elements of the form $A_1^{m_1}$ $(1 \leqslant m_1 \leqslant n_1)$. Let B_2 be the first of the A's which remains. Let n_2 be the lowest integer for which $B_2^{n_2} = B_1^{k_{12}}$. We now strike out all the A's which are of the form $B_1^{m_1}B_2^{m_2}$ $(1 \leqslant m_1 \leqslant m_2,\ 1 \leqslant n_1 \leqslant n_2)$. If we continue indefinitely we have a series of equations of the form

$$\begin{aligned} B_1^{n_1} &= E, \\ B_2^{n_2} &= B_1^{k_{12}}, \\ B_3^{n_3} &= B_1^{k_{13}}\, B_2^{k_{23}}, \\ &\ldots\ldots \\ &\ldots\ldots \end{aligned}$$

and every member of G can be expressed as a finite product of B's.

We now construct a table in the following way. We first write 1 in a row by itself. Below it we write down all the solutions x_p of $x^{n_1} = 1$. Below each x_p we write down all the solutions $x_{p,q}$ of $x^{n_2} = x_p^{k_{12}}$. Below each $x_{p,q}$ we write all the solutions $x_{p,q,r}$ of $x^{n_3} = x_p^{k_{13}}x_{p,q}^{k_{23}}$, and so on.

We are now in a position to define our characters. For every infinite sequence $p, q, r, \ldots$ $(0 \leqslant p \leqslant n_1 - 1,\ 0 \leqslant q \leqslant n_2 - 1, \ldots)$ we define a character $\chi_{p,q,r}, \ldots$ to satisfy

$$\chi(B_1) = x_p,\ \chi(B_2) = x_{p,q},\ \chi\ (B_3) = x_{p,q,r}, \ldots$$

Then if

$$A = B_{m_1}^{\lambda_1}\, B_{m_2}^{\lambda_2} \ldots B_{m_k}^{\lambda k},$$

we write

$$\chi(A) = \chi^{\lambda_1}(B_{m_1})\, \chi^{\lambda_2}(B_{m_2}) \ldots \chi^{\lambda k}(B_{m_k}),$$

and we have defined χ for all A's.

4. Suppose now that G has more than $\aleph$ members. We show how to construct a set of denumerable Abelian groups $\{H\}$ such that every

member of G can be expressed uniquely as a finite product of members of the groups $\{H\}$. The simplest way to see this is the following. Suppose for simplicity that none of the elements of G are periodic and that they are arrayed in a well ordered series

$$E, A_1, A_2, \ldots \tag{3}$$

We write $B_1 = A_1$, and let H_1 consist of all the members of G which satisfy a relation of the form

$$A^m = B_1^n \mp E.$$

We delete from the series (3) all the members of H_1. Let B_2 be the first A which remains. H_2 is to consist now of all the A's which satisfy

$$A^m = B_2^n \mp E.$$

We delete from the series (3) all the members of G of the form

$$D_1^{n_1}\, D_2^{n_2}$$

where D_1 and D_2 belong to H_1 and H_2, respectively, and so on. We may proceed transfinitely, at each stage deleting all the finite products of members of groups H which we have already considered. If we now define a character χ for each of the groups $\{H\}$, we may define a character χ of G by means of the equation (1). Thus the set of characters of G can be completely determined.

We find ourself however in the position of having too many characters. Thus, for instance, if c denotes the number of the continuum and G has c members, then so has the set $\{H\}$. Each H has c characters, and thus G has $c^c > c$ characters. Our task therefore is to find some method of weeding out the superfluous characters. In certain cases we can do this.

5. Let us return to the denumerable case. Let G be the denumerable group and A its members, and let G^1 denote the group of characters χ. We observe that if A is fixed then $\chi(A)$ is a character of the group G^1 of elements χ, and from our point of view these ($\aleph$) are the only interesting ones.

Our procedure is as follows. We build up a system of measure (by no means the only reasonable one) on the set χ in such a way that the functions $\chi(A)$ are the only measurable characters of G^1. All we have to do is to make a correspondence between the character $\chi_{p,q,r},\ldots$ and the point

$$\frac{p}{n_1} + \frac{q}{n_1 n_2} + \frac{r}{n_1 n_2 n_3} + \ldots \tag{4}$$

of the line (0,1). This correspondence is (1 — 1) except for a set of measure zero, and we now define the measure of a set of $\chi_{p,q,r},\ldots$ to be that of the corresponding set of points (4). As we might expect, the system of characters $\chi(A)$ form a complete normalized orthogonal system for the space χ, so that, if A, A' are different, the integral

$$\int_{G^1} \chi(A)\chi(A')d\chi$$

vanishes.

6. As an example of the procedure we have just defined we may give the case where G is defined by the equations $B_1^2 = E$, $B_2^2 = B_1$, $B_3^2 = B_2$, . . ., and consists of all finite products of the B's. Then the table defined above is

$$\begin{array}{cccccccccccccccc} &&&&&&& 1 &&&&&&&& \\ &&& 1 &&&&&&&& -1 &&&& \\ & 1 &&&& -1 &&&& i &&&& -i && \\ 1 && -1 && i && -i && \omega && -\omega && i\omega && -i\omega & \\ &&&&&&& \cdots\cdots &&&&&&&& \\ &&&&&&& \cdots\cdots &&&&&&&& \end{array} \qquad (5)$$

where ω denotes $exp\left(\frac{1}{4}\pi i\right) = \sqrt{i}$. The group G^1 is then the dyadic group.

7. There are two directions in which we have succeeded in extending the ideas set out above. The first consists in extending tables of the form (5) so that they are unbounded to the right as well as downward. This enables us to define a system of measure on G^1 which is infinite instead of finite (we might, for instance, make a $(1 - 1)$ correspondence between the members of G^1 and the interval $(-\infty, \infty)$). The measurable characters G of G^1 are now no longer denumerable but of the order of the continuum. We can now build up a system of measure on G such that the elements of G^1 are the only measurable characters on G, and we have complete duality. The best known case is that in which both G and G^1 are the translation group on the infinite interval. One which is not continuous is that for which there is a correspondence between the elements of G (or G^1) and the finite binary numbers

$$\alpha_{-n}\, 2^n + \alpha_{-n+1}\, 2^{n-1} + \ldots + \alpha_0 + \alpha_1\, 2^{-1} + \alpha_2\, 2^{-2} + \ldots + \alpha_n\, 2^{-n} + \ldots, \qquad (6)$$

where $\alpha = 0$ or 1 (we shall have to regard, for example, 1 and $2^{-1} + 2^{-2} + \ldots$ as distinct), and we define the product of (6) and

$$\beta_{-n}\, 2^n + \beta_{-n+1}\, 2^{n-1} + \ldots + \beta_0 + \beta_1\, 2^{-1} + \beta_2\, 2^{-2} + \ldots + \beta_n\, 2^{-n} + \ldots$$

to be

$$\gamma_{-n}\, 2^n + \gamma_{-n+1}\, 2^{n-1} + \ldots + \gamma_0 + \gamma_1\, 2^{-1} + \gamma_2\, 2^{-2} + \ldots + \gamma_n\, 2^{-n} + \ldots,$$

where

$$\gamma_r = \alpha_r + \beta_r \text{ (mod. 2)} \quad (r = -n, -n+1, \ldots 0, 1, 2, \ldots n, \ldots).$$

8. Our other generalization is that in which G^1 instead of being too large is too small. Suppose, for instance, that G^1 denotes Hilbert space $\sum x_n^2 < \infty$, and that the product of $\{x_n\}$ and $\{y_n\}$ is $\{x_n + y_n\}$. The most general character which is continuous is

$$exp\left\{2\pi i\,(\Sigma\lambda_n x_n)\right\},$$

where $\sum\lambda_n^2$ must converge. And, in this case, G also will be a Hilbert space. We use a very valuable theorem due to Banach[4] to show that if G^1 denotes a separable metric space of type[5] B, and if the fundamental operation is that of addition, then we can find a denumerable set of characters of G^1 such that, at any rate formally, every continuous character can be expressed as a finite or infinite product of powers of these fundamental characters.

[1] INTERNATIONAL RESEARCH FELLOW.

[2] A. Haar, "Über unendliche komutative Gruppen," *Math. Zeit.*, **33**, 129–159 (1931).

[3] F. Peter and H. Weyl, "Die Vollständigkeit der primitiven Darstellungen einer geschlossenen kontinuierlichen Gruppe," *Math. Ann.*, **97**, 737–755 (1927).

[4] S. Banach, *Théorie des opérations linéaires*, Warszawa, 1932.

[5] See, e.g., Banach, loc. cit., Chapter V.

Comments on [32f], [33c]

E. Hewitt

These two notes are an isolated phenomenon in the development of the theory of duality for compact and locally compact Abelian groups. In the first announcement, the authors state the Wiener-Lévy theorem for the absolutely convergent Fourier series on some (it is not clear exactly which) compact Abelian groups. This theorem was proved later by Gel'fand[2] as an illustration of his theory of Banach algebras. In the second note, the authors make an unsuccessful effort to compute the characters of a countably infinite Abelian group. This was completely done a year later by J. W. Alexander.[1] The authors also mistakenly state that some of the characters of an uncountable Abelian group are "superfluous" for analysis on these groups. Just a year after the publication of the second note, Pontryagin[3] set down the entire theory of duality for locally compact second countable Abelian groups, followed very shortly by van Kampen's similar work without countability restrictions. There is no evidence that either Pontryagin or van Kampen was aware of the notes of Paley and Wiener. Apparently Wiener never returned to this topic after Paley's death, in 1933. One could hazard a guess that had Paley lived, he and Wiener would have worked out the proper setting for their analysis on the basis of the Pontryagin-van Kampen duality theorem, and no one can tell what else they might have discovered. But these two notes remain only a hint of what might have been.

References

1. J. W. Alexander, Ann. of Math. (2) 35(1934), 389–395.
2. I. M. Gel'fand, Mat. Sbornik (N. S.), (51) 9(1941), 51–66.
3. L. S. Pontryagin, Ann. of Math. (2) 35(1934), 361–388.
4. E. R. van Kampen, Ann. of Math. (2) 36(1935), 448–463.

THE TOTAL VARIATION OF $g(x+h)-g(x)$

BY

NORBERT WIENER and R. C. YOUNG*

1. There is a fundamental theorem in the theory of Lebesgue integration that if $f(x)$ be any function integrable Lebesgue, over the interval (a, b), then the integral

$$\int_a^b | f(x + h) - f(x) | \, dx \tag{1}$$

tends to zero with h. This theorem is usually proved by approximating to $f(x)$ in terms of continuous functions, for which the property is obvious.

The integral (1) is the total variation in (a, b) of the function $F(x+h)-F(x)$, where

$$F(x) = \int^x f(t)dt$$

and the quoted property is equivalent to the following statement:

(I). *If $F(x)$ be any absolutely continuous function in (a, b), the total variation of*

$$F(x + h) - F(x)$$

tends to zero with h.

This result no longer holds if we substitute for $F(x)$ any non-absolutely continuous function $G(x)$ of bounded variation. Indeed, it provides a necessary and sufficient condition for absolute continuity.† This fact may be first rendered plausible by taking the simplest case of a discontinuous function of bounded variation, and bearing in mind that a general continuous function of bounded variation is always a limit of simple discontinuous ones. If we assume for instance

$$g(x) = \alpha \text{ for } x \leqq c,$$
$$g(x) = \beta \text{ for } c < x,$$

* Presented to the Society, October 29, 1932; received by the editors April 21, 1932.

† This result was proved, by an entirely different method, by A. Plessner, Journal für Mathematik, vol. 160 (1929), pp. 26–32. A study of the total variation of $g(x+h)-g(x)$ when $g(x)$ is constant in the complementary intervals of Cantor's set is contained in an article by Hille and Tamarkin, American Mathematical Monthly, vol. 36 (1929), pp. 255–264. The existence of these two papers was called to our attention after the present paper had been completed. We notice also an erroneous statement in a paper by J. M. Whittaker, Proceedings of the Edinburgh Mathematical Society, (2), vol. 1 (1929), p. 232 (Lemma 1).

we have, for $h>0$,

$$g(x+h)-g(x)=0 \qquad \text{for } x \leqq c-h \text{ and } x>c,$$

$$g(x+h)-g(x)=\beta-\alpha \text{ for } c-h<x\leqq c$$

and over any interval (a, b) containing c and $c+h$ internally, the total variation of $g(x+h)-g(x)$ is the sum of the absolute values of its jumps, viz.

$$2|\beta-\alpha| = 2\left|\int_a^b dg(x)\right| = \left|\int_a^b dg(x)\right| + \left|\int_a^b dg(x+h)\right|.$$

In this special case we may then write, for every h, and every interval (a, b), $a\neq c$, $b\neq c$,

$$\int_a^b |d[g(x+h)-g(x)]| = \int_a^b |dg(x)| + \int_a^b |dg(x+h)|. \tag{2}$$

Actually this is the relation which we shall show (§3) to hold for any singular function $g(x)$ of bounded variation, if not for all h, at any rate for almost all. By a singular function we mean a function of bounded variation whose derivative is zero almost everywhere, of which the above discontinuous $g(x)$ provides a very special example.

An arbitrary function of bounded variation is the sum of an absolutely continuous function and a singular function of bounded variation, uniquely defined and called the singular part of the original function. Thus the proof of the above statement will carry with it, as an immediate consequence of I, the following result:

(II) *If $g(x)$ be any function of bounded variation with singular part $\gamma(x)$ continuous at the end points a, b,* we have*

$$\begin{aligned}\overline{\lim_{h\to 0}} \int_a^b |d[g(x+h)-g(x)]| &= \overline{\lim_{h\to 0}} \int_a^b |d[\gamma(x+h)-\gamma(x)]| \\ &= 2\int_a^b |d\gamma(x)|.\end{aligned} \tag{3}$$

This statement is of course only a rough corollary of the result for singular functions and absolutely continuous functions. More precise consequences to be born in mind are

(i) the left-hand equality in (3) holds not merely for h tending to 0 continuously, but for h tending to 0 through any discontinuous sequence;

* As an unessential restriction reducing $\int_{a+0}^{b+0}|d\gamma|$ to $\int_a^b|d\gamma|$.

(ii) the right-hand equality in (3) holds similarly for discontinuous approach of h to 0, if a certain set of measure 0 be avoided by h. To obtain any intermediary or the lower value of the considered limits, we have to take subsequences of this set of measure 0.

Examples will be constructed (§§4–6) to show what various possibilities exist with regard to the exceptional set of values of h. It will be seen that this set may be more than countable, or countable, or entirely absent. In particular therefore the limits in (3) may be unique limits. On the other hand, the examples will also show that the lower limits corresponding to the upper limits in (3) may in other cases have any lesser non-negative value. The case in which the lower limit is 0 would seem to have particular interest, and it might be useful to examine in greater detail the particular sequences of h_n tending to 0, for which this limit is obtained. It seems a question for instance whether a non-absolutely continuous function $g(x)$ exists for which $h_n=1/n$ would provide a sequence of this kind, or more generally, any other special sequence for which h_n/h_{n+1} is bounded.

2. We use the following simple lemmas.

LEMMA 1. *Given two functions $g_1(x)$, $g_2(x)$ of bounded variation, of which $g_1(x)$ has derivative zero, $g_1'(x)=0$, on the set complementary to a given set H_1 in (a, b). Then we have*

$$\int_a^b |d[g_1(x) - g_2(x)]| = \int_a^b |dg_1| + \int_a^b |dg_2| - 2\theta \int_{H_1} |dg_2|, \tag{4}$$

where $0 \leqq \theta \leqq 1$.*

We have

$$\int_{CH_1} |dg_2| - \int_{CH_1} |dg_1| \leqq \int_{CH_1} |d[g_2(x) - g_1(x)]|$$
$$\leqq \int_{CH_1} |dg_2| + \int_{CH_1} |dg_1|.$$

Since

$$\int_{CH_1} |dg_1| = 0,$$

we have

* This lemma has been stated initially in a less general form. The authors are indebted to Dr. S. Saks for the suggestion of extending their proof to the present, essentially more general, situation. We refer to de la Vallée Poussin, *Intégrales de Lebesgue, Fonctions d'Ensemble, Classes de Baire*, Paris, 1916, pp. 90–95, as to the notion of the total variation over a set, and as to some formulas which we are using here.

$$\int_{CH_1} |d[g_2 - g_1]| = \int_{CH_1} |dg_2| .$$

On the other hand, over the set H_1 itself,

$$-\int_{H_1} |dg_2| \leqq \int_{H_1} |d[g_2 - g_1]| - \int_{H_1} |dg_1| \leqq \int_{H_1} |dg_2| .$$

Hence, by addition,

$$\int_a^b |dg_2| - 2\int_{H_1} |dg_2| \leqq \int_a^b |d[g_2 - g_1]| - \int_a^b |dg_1| \leqq \int_a |dg_2|$$

giving (4) as required.

Lemma 2. *Let E_h be a variable set depending on a parameter h, and such that each point x belongs to E_h at most for a set of measure* 0 *of values of h*. Then for each function $g(x)$ of bounded variation,*

$$\int_{E_h} |dg| = 0$$

for almost all values of h. (More precisely, the exceptional values of h form a set of measure 0 *depending on g.)*

This is immediate from the theory of change of order of integration in a repeated Stieltjes integral.† Let $G(x)$ be the indefinite total variation of $g(x)$:

$$G(x) = \int^x |dg|$$

and $E(x, h)$ the characteristic function of E_h, equal to 1 in E_h, and 0 elsewhere, for each value of h. We have then

$$V(h) = \int_{E_h} |dg| = \int_{-\infty}^{\infty} E(x, h)dG(x).$$

Integrating this with respect to h, we get

$$\int_{-\infty}^{\infty} V(h)dh = \int_{-\infty}^{\infty} dh \int_{-\infty}^{\infty} E(x, h)dG(x) = \int_{-\infty}^{\infty} dG(x) \int_{-\infty}^{\infty} E(x, h)dh,$$

* In other words, E_h is the section y=constant h, of a plane set E whose sections x=constant all have measure 0. Such a set E has plane measure 0, and it is well known that E_h must then have measure 0 for almost all h. The present Lemma states that E_h has also measure 0 with respect to any function of bounded variation, for almost all h, and is an immediate adaptation of the classical result.

† See, e.g., L. C. Young, *The Theory of Integration*, Cambridge Tracts, No. 21, p. 41 (Theorem IV).

where, by the hypothesis,

$$\int_{-\infty}^{\infty} E(x, h)dh = 0 \text{ for each } x,$$

as the measure of the set of values of h for which x belongs to E_h. Thus

$$\int_{-\infty}^{\infty} V(h)dh = 0,$$

and hence the non-negative function $V(h)$ vanishes except at most in a set of measure 0.

In our application of this Lemma, the set E_h will be the h-translation of a fixed set E_0 of measure 0, that is, the set of points x such that $x-h$ belongs to E_0. It has the characteristic function

$$E_h(x) = E_0(x + h).$$

For fixed x this still represents a set of measure 0 in h, and so the hypothesis of the Lemma is fulfilled. Thus we have

COROLLARY. *Given any function of bounded variation* $g(x)$, *and any set* E_0 *of measure* 0, *the total variation of* g *over the* h*-translation of* E_0 *vanishes except at most for a set of values of* h *of measure* 0.

3. We have now immediately our

THEOREM. *If* $g(x)$ *be a function of bounded variation constant in the complementary intervals of a closed set* H *of measure* 0, *we have, for almost all* h,

$$\int_a^b |d[g(x+h) - g(x)]| = \int_a^b |dg(x+h)| + \int_a^b |dg(x)|. \tag{5}$$

For, $g(x+h)$ is then constant in the complementary intervals of the h-translation H_h of H, so by Lemma 1, for

$$g_2(x) = g(x+h), \quad g_1(x) = g(x),$$

the two sides of (5) differ by at most

$$2\int_{H_h} |dg|$$

and by Lemma 2, in the Corollary form, this vanishes for almost all h.

In constructing examples, it is simplest to consider periodic functions with interval of periodicity (a, b), and assume for instance $a=0$, $b=1$. Then (5) takes the form

$$\int_0^1 |d[g(x+h) - g(x)]| = 2\int_0^1 |dg(x)|. \tag{5'}$$

In that case also it suffices to consider only positive values of h, since $g(x+h)-g(x)$ is then still periodic with the same period as g, and so

$$\int_0^1 |\, d[g(x+h) - g(x)]\,| = \int_0^1 |\, d[g(x) - g(x-h)]\,| .$$

Furthermore we need only consider monotonic functions, with $\int_0^1 dg = 1$.

4. The classical example of a singular function of bounded variation is that of the monotone function constant in the complementary intervals of Cantor's typical ternary set* and representing an even mass distribution over this set, in the interval (0, 1). If x be expressed as a ternary fraction

$$x = \cdot\alpha_1\alpha_2 \cdots \alpha_i \cdots = \sum_{i=1}^{\infty}\alpha_i 3^{-i},\ \alpha_i = 0,\ 1 \text{ or } 2, \tag{6}$$

and α_n be the first (if any) of its digits equal to 1, the function is defined at x to have the value

$$g(x) = \sum_{i=1}^{n-1}\alpha_i 2^{-i-1} + \alpha_n 2^{-n}, \tag{7}$$

or if there be no digit 1 in (6), then

$$g(x) = \sum_{i=1}^{\infty}\alpha_i\, 2^{-i-1}. \tag{7'}$$

This function is a particular example of a class of functions possessing the following property; this is most easily described by introducing the expression "ternary interval of order n" to designate specifically the open intervals of length $1/3^n$ whose left-hand end points are the terminating ternary fractions of at most n digits, all even:

$$\cdot\alpha_1\alpha_2 \cdots \alpha_n\, 0\,0\,0 \cdots,\ \alpha_i = 0 \text{ or } 2.$$

Each such interval is contained in exactly one of lower order, and contains exactly 2^{r-n} intervals of higher order $r > n$. The intervals of given order are of course mutually exclusive and non-abutting.

PROPERTY (A). *For each ternary interval* (α, β) *of order* n, *and each ternary interval* (α', β') *of order* $(n+l)$ *contained in it,*

$$\left|\frac{g(\beta') - g(\alpha')}{g(\beta) - g(\alpha)}\right| \leqq \delta$$

where l *is a positive integer and* δ *a constant less than* 1, *independent of the particular interval.*

* The set of all non-terminating ternary fractions with even digits only, together with the terminating fractions whose last digit at most is odd.

Such functions share with the ordinary measuring function $m(x)=x$ a property relative to fractions with prescribed digits which we shall use in the form of

LEMMA 3. *If $g(x)$ be any function, constant in the complementary intervals of Cantor's ternary set H_0, and possessing the property* (A), *and if E be a subset of H_0 in which (in the ternary scale)*

$$x = \cdot\alpha_1\alpha_2 \cdots \alpha_i \cdots \tag{6}$$

has fixed prescribed digits (0 *or* 2) *for an infinity of indices,*

$$\alpha_{n_1}, \alpha_{n_2}, \cdots, \alpha_{n_i}, \cdots, n_1 < n_2 < \cdots \to \infty$$

(*the other digits being arbitrary,* 0 *or* 2), *then*

$$\int_E |dg| = 0.$$

Consider the ternary intervals of order n_k and left-hand end points

$\cdot\beta_1\beta_2 \cdots \beta_{n_1-1}\alpha_{n_1}\beta_{n_1+1} \cdots \beta_{n_2-1}\alpha_{n_2}\beta_{n_2+1} \cdots \beta_{n_k-1}\alpha_{n_k}$, α_{n_i} as above, $\beta_i = 0$ or 2,

and let σ_k denote their sum-set. Then $E \subset \sigma_k$ for each k (actually $E=\lim_{k\to\infty}\sigma_k$). Also

$$\int_{\sigma_k} |dg| \leqq \delta \int_{\sigma_{k-l}} |dg|\,. \tag{8}$$

This is because each interval of σ_k is in a different ternary interval of order $n_k - l$ contained in σ_{k-l}, and by property (A) contributes not more than δ times the variation over this interval to the total variation over σ_k. Thus if $[k/l]$ be the integral part of k/l,

$$\int_E |dg| \leqq \int_{\sigma_k} |dg| \leqq \delta^{[k/l]} \int_0^1 |dg| \to 0 \text{ with } 1/k$$

since $\delta < 1$.

This lemma is applied in conjunction with another relating purely to the ternary set H_0:

LEMMA 4. *To each non-terminating ternary fraction $h = \cdot a_1a_2 \cdots a_i \cdots$ there corresponds an infinite sequence of indices*

$$n_1 < n_2 < \cdots < n_i < \cdots \to \infty$$

and two specifications for the whole sequence of digits $\{c_{n_i}\}$ (0 *or* 2), *which a ternary fraction $x = \cdot c_1c_2 \cdots c_i \cdots$ must certainly satisfy if it is to belong to both H_0 and its h-translation H_h.*

This means that the common part H_0H_h is the sum of two sets of the kind considered in Lemma 3. What we prove precisely is as follows:

(i) If h has an infinite number of odd digits

$$a_{n_i} = 1,\ n_1 < n_2 < \cdots < n_i < \cdots \to \infty$$

then the corresponding digits of x must be alternately 0 and 2, i.e.

$$c_{n_1} = c_{n_3} = c_{n_5} = \cdots = 0 \text{ or } 2,$$
$$c_{n_2} = c_{n_4} = c_{n_6} = \cdots = 2 \text{ or } 0.$$

(ii) If h has only a finite number of odd digits,

$$a_i \neq 1 \text{ for } i > N,$$

and (r_i) are the indices for which $a_i=0$, (s_i) those for which $a_i=2$, after the Nth, then in x we must have either

$$c_{r_1} = c_{r_2} = \cdots = c_{r_i} = \cdots = 0$$

or

$$c_{s_1} = c_{s_2} = \cdots = c_{s_i} = \cdots = 2.$$

We deduce this from the following obvious facts, true for any h, terminating or not:

Given in the ternary scale

$$h = \cdot a_1a_2 \cdots a_i \cdots \qquad (a_i = 0, 1 \text{ or } 2),$$
$$x = \cdot b_1b_2 \cdots b_i \cdots \qquad (b_i = 0 \text{ or } 2),$$
$$x + h = \cdot c_1c_2 \cdots c_i \cdots \pmod 1 \qquad (c_i = 0 \text{ or } 2),$$

we have

(a) if $a_{n_1} = a_{n_2} = 1,\ a_i \neq 1$ for $n_1 < i < n_2$,

then either

$$b_{n_1} = 0, \quad b_{n_2} = 2 \ (\text{and } c_{n_1} = 2,\ c_{n_2} = 0),$$

or the same with n_1 and n_2 interchanged;

(b) if $a_r = b_r = 0\ (= c_r),\ a_s = b_s = 2\ (= c_s)$,

then $a_i = 1$ for some index between r and s (implying $|r-s|>1$).

In (a) we consider two consecutive odd digits in h, and affirm that the two corresponding digits in x (and $x+h$) cannot be both 0 or both 2. In fact $b_{n_2}=2$ would imply that we had in the formal addition $x+h$ to carry 1 from the n_2th place right back to the n_1th, where it would compound to 2 with a_{n_1}, and imply $b_{n_1}=0$ if c_{n_1} is to be even. And similarly $b_{n_2}=0$ implies $b_{n_1}=2$.

In (b) we consider two places in each of which h and x have the same digits, but different in the two places, and we conclude that between those two places h must have at least one digit $=1$. We may assume that between those two places no further coincidences occur, and we see at once that if the formal addition $x+h$ is to yield only even digits between those two places, and h had none $=1$ there, these digits in $x+h$ would all be 0 with 1 to carry or all 2, with nothing to carry, and a 1 would appear in the sum in the earlier of the two places r and s.

From Lemmas 3 and 4, we deduce that, *for any function* $g(x)$ *of period* 1, *constant in the complementary intervals of Cantor's set* H_0 *and possessing the above property* (A), *we must have, whenever* h *is a non-terminating ternary fraction,*

$$\int_{H_h} |dg| = \int_{H_0 \cdot H_h} |dg| = 0.$$

From this and Lemma 1, we deduce (as in the proof of our principal theorem), that *for all such functions* $g(x)$,

$$\int_0^1 |d[g(x+h) - g(x)]| = 2\int_0^1 |dg|$$

whenever h *is a non-terminating ternary fraction.*

The exceptional values of h thus belong to the countable set of terminating ternary fractions.

5. In the case of Cantor's function (7, 7′), property (A) holds with $l=1$ and $\delta=\frac{1}{2}$, and equality sign. This provides therefore an instance of a function $g(x)$ for which the exceptional values of h are at most countable. It may be seen moreover that in this case every terminating ternary fraction is an exceptional value of h. For this it suffices to remark that (5) can certainly not hold when $g(x+h)-g(x)$ is constant in an interval in which neither $g(x)$ nor $g(x+h)$ is constant. And if, when

$$h = \cdot a_1 a_2 \cdots a_i \cdots a_n,$$

we choose (as we can)

$$x_0 = \cdot b_1 b_2 \cdots b_i \cdots b_n, \quad \cdot b_i = 0 \text{ or } 2,$$

so that x_0+h has also only digits 0 or 2, then the ternary interval of order n and left-hand end point x_0 satisfies this condition.* If we go into the question more nearly, in order to investigate the lower limit of

$$\int_0^1 |d[g(x+h) - g(x)]| \text{ as } h \to 0,$$

* Cf. also Hille and Tamarkin, loc. cit., p. 261.

we find that for our present function, whereas

$$\int_0^1 |d[g(x+h)-g(x)]| = 2$$

for h non-terminating, we have for h terminating of exactly n digits,

$$1 \leqq \int_0^1 |d[g(x+h)-g(x)]| \leqq 2(1-2^{-n}), \tag{9}$$

with actual equality on either side for suitable values of h and each n. For let

$$h = \cdot a_1a_2 \cdots a_n, \quad a_i = 0, 1 \text{ or } 2, \ a_n \neq 0;$$

and suppose to fix the ideas that $a_n=2$. Then every ternary interval of order n and left-hand end point

$$\cdot b_1b_2 \cdots b_n, \quad b_i = 0 \text{ or } 2,$$

with $b_n=2$, translates through h into an interval in which g is constant, that is to say is itself an interval in which $g(x+h)$ is constant. Similarly if $a_n=1$ and we take $b_n=0$. So over half the ternary intervals of order n, the relation (5) certainly holds. Similarly it holds for all the intervals of length 2^{-n} and left-hand end points $\cdot b_1b_2 \cdots b_k$ with $b_k=1, k\leqq n$, over which $g(x)$ is constant. There only remain the other half of the ternary intervals of order n, over which $g(x)$ has variation 1/2, and $g(x+h)$ at most 1/2. So in this case the right-hand side of (5′) cannot exceed the left by more than 1.

The other inferences follow also very simply. For instance

$$a_n = a_{n-k_1} = \cdots = a_{n-k_r} = 1, \quad a_i = 0 \text{ or } 2 \text{ for } n-k_j < i < n-k_{j-1},$$
$$j = 1, 2, \cdots, r+1, \quad k_0 = 0, \ k_{r+1} = n,$$

according as j is odd or even, give some of the (everywhere dense set of) values of h for which equality occurs on the right of (9), while

$$a_i = 2 \text{ for } i < n, \ a_n = 1 \quad \text{or} \quad a_i = 0 \text{ for } i < n, \ a_n = 2$$

give the only values for which equality occurs on the left.

6. The classical monotone function just discussed expresses the total mass in $(0, x)$ due to mass unity, evenly distributed among the ternary intervals of order n, for each n, and its constancy in each complementary interval of the ternary set H_0 expresses the fact that there is no mass in these complementary intervals.

We now modify this construction so that the function is constant not only in the complementary intervals of H_0, but also in a set (everywhere dense on H_0) of the ternary intervals themselves, and show that we can choose these additional intervals of constancy (in a more than countable number of ways) so that the resulting functions satisfy (5) for every terminating ternary fraction h. The functions will all possess the property (A), and therefore satisfy (5) for non-terminating values of h, and so we shall have our example of functions for which h has no exceptional values, (5) being always true.

The definition is obtained inductively by supposing the mass distribution effected in the ternary intervals of order $2n$ and now dividing the mass of each such interval equally among *three* (instead of all) of the four ternary intervals of order $2(n+1)$ which it contains. The choice of these three intervals may be effected in four different ways, and as there will be exactly 3^n ternary intervals of order $2n$ actually containing mass, this means that we have $4\cdot 3^n$ different modes of distribution to choose from at each stage. In any case, each ternary interval of order $n+2$ will contain at most $1/3$ of the mass of the interval of order n containing it (viz. either 0 or exactly $1/3$ of it).

As the mass of any interval is the increment of the monotone function $g(x)$ representing the ultimate mass in $(0, x)$, we see that each function so obtained possesses the property (A) of page 332 with $l=2$, $\delta=1/3$.

The construction provides moreover a more than countable set of functions of this kind, for all of which (5) therefore holds whenever h is a non-terminating ternary fraction. By eliminating a certain minority of these functions, we may arrange so that (5) holds also for each of the remaining countable set of values of h. The simplest way of doing this is to associate with each of these values of h, i.e., terminating of say k digits, an infinite sequence of indices $>k$:

$$N_i = N_i^{(h)} \to \infty$$

such that

$$N_i^{(h)} \neq N_j^{(h')} \text{(all } i, j)$$

whenever $h' \neq h$ (this can always be done in any number of ways*), and to stipulate that for each $n = N_i^{(h)}$, two ternary intervals of order $2n$ which are h-translations one of the other (for that h) should never both be intervals devoid of mass. For instance if

$$h = \cdot a_1 a_2 \cdots a_k, \qquad a_i = 0, 1 \text{ or } 2, \ a_k \neq 0,$$

and

* For instance if $\{h_i\}$ be the considered set of values of h in any countable order, and $\{p_i\}$ the sequence of primes in increasing order, we can associate with each h_i the sequence of integers $p_i p_{i+n}$, $n = n_0, n_0+1, \cdots$.

$$x = \cdot b_1 b_2 \cdots b_k b_{k+1} \cdots b_{2n}, \quad b_i = 0 \text{ or } 2,$$

is the left-hand end point of a ternary interval of order $2n$ devoid of mass, and if $x+h$ is still a left-hand end point of a ternary interval of order $2n$, i.e.

$$x + h = \cdot c_1 c_2 \cdots c_k b_{k+1} \cdots b_{2n}, \; c_i = 0 \text{ or } 2,$$

then this latter interval must not be devoid of mass. Since $2n > k+1$, the two ternary intervals considered are necessarily in two different intervals of order $2(n-1)$, so that at the stage $n-1$ in our construction we can certainly make our choice (in at least 3^{3^n} different ways) so that the condition be fulfilled for that n. By the fact that to each index n that we have particularly to take into consideration corresponds only one h (with reference to which the restriction is made), a question of incompatibility for the different values of h cannot arise, and we are certainly left with a majority of the functions considered. For this residue, we see that for each h, (5) holds outside at most two of the ternary intervals of order $2n$ in each ternary interval of order $2(n-1)$ for each $n = N_i^{(h)}$. Thus if σ_i represent the intervals of this order $2n$ for which (5) may possibly not hold (more precisely in which neither $g(x)$ nor $g(x+h)$ is constant), we have as in the proof of Lemma 2 (inequality (8)),

$$\int_{\sigma_i} |dg| \leqq \delta \int_{\sigma_{i-1}} |dg|, \; \delta = \tfrac{2}{3},$$

and hence this variation tends to 0 with $1/i$. Thus (5) holds in the complementary intervals of a set over which $g(x)$ has total variation 0, that is, holds absolutely, for the arbitrary considered terminating h.

7. The object of rarifying the exceptional values of h as much as possible was achieved by an increased concentration of the unit mass distributed over our interval (0, 1), introducing additional intervals of constancy for $g(x)$. Conversely, we can multiply the exceptional set of values of h by breaking up the intervals of constancy, and diffusing the total mass more over the whole interval.

By this means we shall obtain a function with a more than countable set of exceptional h, and at the same time we shall find our example of a case in which the lower limit corresponding to (3) is 0.*

Let $n_1, n_2, \cdots$ be a sequence of integers increasing so rapidly that $\sum_1^\infty 1/n_k$ converges. Let x be any number in (0, 1). It is a theorem due to Cantor that x may always be expressed in the form

$$x = \frac{m_1}{n_1} + \frac{m_2}{n_1 n_2} + \frac{m_3}{n_1 n_2 n_3} + \cdots, \tag{10}$$

* The actual construction is adapted from a suggestion of Mr. A. S. Besicovitch.

where $m_1 < n_1$, $m_2 < n_2$, $\cdots$, and that this representation is unique, save for the ambiguity

$$\frac{1}{n_1 n_2 \cdots n_k} = \frac{n_{k+1} - 1}{n_1 \cdots n_{k+1}} + \frac{n_{k+2} - 1}{n_1 \cdots n_{k+2}} + \cdots,$$

which gives a non-terminating alternative to any terminating expression of a number. Let us assume that every n_k is even, equaling $2\nu_k$. If x is expressed as in (10), let

$$f(x) = \frac{m_1/2}{\nu_1} + \frac{m_2/2}{\nu_1\nu_2} + \frac{m_3/2}{\nu_1\nu_2\nu_3} + \cdots,$$

if every m_k is even, and

$$f(x) = \frac{m_1/2}{\nu_1} + \frac{m_2/2}{\nu_1\nu_2} + \cdots + \frac{m_{n-1}/2}{\nu_1\nu_2 \cdots \nu_{n-1}} + \frac{[m_n/2] + 1}{\nu_1\nu_2 \cdots \nu_n},$$

if m_n is the first odd m_k. Clearly $f(x)$ is a monotone function, constant over every interval

$$\left(\frac{2\mu_k + 1}{n_1 n_2 \cdots n_k}, \frac{2\mu_k + 2}{n_1 n_2 \cdots n_k}\right) \text{ where } \mu_k < \nu_k.$$

These intervals for any given value of k fill up half the line, and for $k = 1, 2, 3, \cdots, K$ fill up a set of intervals of measure $1 - 2^{-K}$.

Let h be defined as

$$\frac{a_1}{n_1} + \frac{a_2}{n_1 n_2} + \cdots + \frac{a_k}{n_1 n_2 \cdots n_k} + \cdots,$$

where every a_k is 0 or 2. We have

$$f(x + h) - f(x) = \frac{a_1}{2\nu_1} + \frac{a_2}{2\nu_1\nu_2} + \frac{a_3}{2\nu_1\nu_2\nu_3} + \cdots$$

whenever

$$2 \leqq m_r \leqq n_r - 2,\ 2 \leqq m_{r+1} \leqq n_{r+1} - 2, \cdots,$$

where r is the first index for which $a_k \neq 0$. The total variation of $f(x)$ over the set of points for which $m_k < 2$ does not however exceed $1/\nu_k$, and the same is true of the total variation of $f(x+h)$ over those intervals. A similar result applies to the range $n_k - m_k < 2$. Thus the total variation of $f(x+h) - f(x)$ over these intervals does not exceed

$$4(1/\nu_r + 1/\nu_{r+1} + \cdots + 1/\nu_{r+k} + \cdots) \tag{11}$$

which is hence an upper bound for the total variation of $f(x+h)-f(x)$ over $(0, 1)$. The total variation of $f(x)$ over the same interval is 1. Thus we have here a set of values of h of the power of the continuum including arbitrarily small values, for which

$$\int_0^1 | d[f(x+h) - f(x)] | \neq 2 \int_0^1 | df(x) | ,$$

and the total variation on the left-hand side tends to zero (since (11) does so) when h tends to zero through this set.

MASSACHUSETTS INSTITUTE OF TECHNOLOGY,
CAMBRIDGE, MASS.
GIRTON COLLEGE,
CAMBRIDGE, ENGLAND

ON ABSOLUTELY CONVERGENT FOURIER-STIELTJES TRANSFORMS

By N. Wiener and H. R. Pitt

(1.1) **Introduction.** Suppose that $f(x)$ is of bounded variation in $(-\infty, \infty)$ and let

$$F(x) = \int e^{-iyx}\, df(y).$$

(Here, and in what follows, integrals in which the limits are not specified are always taken over the range $(-\infty, \infty)$.)

We define $\mathfrak{A}$ to be the class of functions $F(x)$ which can be expressed in this form. If $F(x)$ is periodic, it belongs to $\mathfrak{A}$ only if its Fourier series is absolutely convergent, and Wiener[1] has shown that the condition $F(x) \neq 0$, or the equivalent condition

$$\underline{\text{Bd}}\, | F(x) | > 0, \tag{1.1.1}$$

is sufficient to make $[F(x)]^{-1}$ belong to $\mathfrak{A}$. The same result for almost periodic functions has been proved by Cameron[2] and Pitt.[3]

Our object here is to investigate how far this is true in the general case.

We can write

$$f(x) = h(x) + g(x) + s(x),$$

where $h(x)$ is a step function, $g(x)$ is absolutely continuous and $s(x)$ is singular in the sense defined by Lebesgue; that is, it is continuous, of bounded variation, not constant, and has derivative zero at almost all points. We can write

$$F(x) = H(x) + G(x) + S(x),$$

where $H(x)$, $G(x)$, $S(x)$ are the transforms of $h(x)$, $g(x)$, $s(x)$; and we use the letters H, G, S in this sense throughout the paper. If $F(x)$ belongs to $\mathfrak{A}$ and

$$F(x) = \int e^{-iyx}\, df(y),$$

we write

$$T\{F(x)\} = \int | df(y) |.$$

Received March 14, 1938.

[1] N. Wiener, *Tauberian theorems*, Annals of Mathematics, vol. 33(1932), pp. 1–100.

[2] R. H. Cameron, *Analytic functions of absolutely convergent generalized trigonometric sums*, this Journal, vol. 3(1937), pp. 682–688.

[3] H. R. Pitt, *A theorem on absolutely convergent trigonometrical series*, Journal of Math. and Phys., M. I. T., vol. 16(1938), pp. 191–195.

Our main conclusion is that the condition $\underline{\mathrm{Bd}} \mid F(x) \mid > 0$ is sufficient to make $[F(x)]^{-1}$ belong to $\mathfrak{A}$ provided that $T\{S(x)\}$ is "not too large", and that otherwise the conclusion is not true.

2. We require a number of elementary lemmas.

(2.1) LEMMA 1. *If $F_1(x)$, $F_2(x)$ belong to $\mathfrak{A}$, then so do $c_1F_1(x) + c_2F_2(x)$ and $F_1(x)F_2(x)$, c_1 and c_2 being any real or complex numbers. Moreover,*

(a) $$T\{c_1F_1(x) + c_2F_2(x)\} \leqq |c_1| T\{F_1(x)\} + |c_2| T\{F_2(x)\},$$

(b) $$T\{F_1(x)F_2(x)\} \leqq T\{F_1(x)\} T\{F_2(x)\}.$$

The first result is obvious. To prove the second, we write

$$F_1(x) = \int e^{-iyx} df_1(y), \qquad F_2(x) = \int e^{-iyx} df_2(y).$$

Then

$$\begin{aligned} F_1(x)F_2(x) &= \int e^{-iux} df_2(u) \int e^{-iyx} df_1(y) \\ &= \int df_2(u) \int e^{-iyx} df_1(y - u) \\ &= \int e^{-iyx} df(y), \end{aligned}$$

where

$$f(y) = \int f_1(y - u)\, df_2(u),$$

except possibly in an enumerable set of points. If $f(y)$ is defined appropriately at these points,

$$T\{F_1(x)F_2(x)\} = \int |df(y)| \leqq \int |df_1(y)| \int |df_2(y)| = T\{F_1(x)\}T\{F_2(x)\}.$$

(2.2) LEMMA 2. *If $F(x)$ belongs to $\mathfrak{A}$ and $T\{F(x)\} = \alpha < 1$, then $[1 + F(x)]^{-1}$ belongs to $\mathfrak{A}$ and*

$$T\{[1 + F(x)]^{-1}\} \leqq (1 - \alpha)^{-1}.$$

We have

(2.2.1) $$[1 + F(x)]^{-1} = \sum_{n=0}^{\infty} (-1)^n [F(x)]^n,$$

the series being absolutely and uniformly convergent for $-\infty < x < \infty$. It follows from Lemma 1(b) that $[F(x)]^n$ belongs to $\mathfrak{A}$ and that $T\{[F(x)]^n\} \leqq \alpha^n$. We can therefore write

$$[1 + F(x)]^{-1} = \sum_{n=0}^{\infty} \int e^{-iyx} df_n(y), \qquad \int | df_n(y) | \leqq \alpha^n,$$

$$[1 + F(x)]^{-1} = \int e^{-iyx} d\left[\sum_{n=0}^{\infty} f_n(y) \right],$$

$$T\{[1 + F(x)]^{-1}\} = \int \left| d\left[\sum_{n=0}^{\infty} f_n(y) \right] \right| \leqq \sum_{n=0}^{\infty} \int | df_n(y) | \leqq \sum_{n=0}^{\infty} \alpha^n = (1 - \alpha)^{-1}.$$

(2.3) We require certain auxiliary functions $\Omega(\epsilon, \lambda, x)$ and $\Gamma(x)$ which are defined as follows. If $\epsilon > 0$, $\lambda > 0$, $2\epsilon\lambda \leqq \pi$, we write

$$\Omega(\epsilon, \lambda, x) = \begin{cases} 1 & (0 \leqq | x | < \epsilon), \\ 2 - \dfrac{| x |}{\epsilon} & (\epsilon \leqq | x | < 2\epsilon), \\ 0 & (2\epsilon \leqq | x | \leqq \pi/\lambda), \end{cases}$$

$$\Omega(\epsilon, \lambda, x + 2\pi/\lambda) = \Omega(\epsilon, \lambda, x).$$

We write

$$\Gamma(x) = \begin{cases} \Omega(1, \tfrac{1}{2}\pi, x) & (| x | \leqq 2), \\ 0 & (| x | > 2). \end{cases}$$

LEMMA 3. (a) *$\Omega(\epsilon, \lambda, x)$ belongs to $\mathfrak{A}$ and*

$$T\{ \Omega(\epsilon, \lambda, x)\} \leqq C,$$

C being an absolute constant.

(b) *For any fixed λ,*

$$\lim_{\epsilon \to 0} T\{\Omega(\epsilon, \lambda, x)[1 - e^{-i\lambda x}]\} = 0.$$

(c) $$\Gamma(x) = \int e^{-iyx} \gamma(y)\, dy, \qquad \int | \gamma(y) | \, dy < \infty.$$

Let us first consider (a). An easy calculation shows that

$$\Omega(\epsilon, \lambda, x) = \sum_{-\infty}^{\infty} d_n e^{-i\lambda n x},$$

where

$$d_0 = \frac{3\lambda\epsilon}{2\pi}, \qquad d_n = \frac{2}{\pi\lambda\epsilon n^2} \sin \frac{n\lambda\epsilon}{2} \sin \frac{3n\lambda\epsilon}{2} \qquad (n \neq 0).$$

It is plain that $\Omega(\epsilon, \lambda, x)$ belongs to $\mathfrak{A}$. Further,

$$T\{\Omega(\epsilon, \lambda, x)\} \leqq \sum_{3\lambda\epsilon|n|\leqq\pi} |d_n| + \frac{2}{\pi\lambda\epsilon}\sum_{3\lambda\epsilon|n|>\pi}\frac{1}{n^2}$$

$$\leqq \frac{3\lambda\epsilon}{2\pi} + 1 + \frac{2}{\pi\lambda\epsilon}\sum_{3\lambda\epsilon|n|>\pi}\frac{1}{n^2} \leqq C,$$

since $2\epsilon\lambda \leqq \pi$.

Next consider (b). We have

$$\Omega(\epsilon, \lambda, x)[1 - e^{-i\lambda x}] = \sum (d_n - d_{n-1})e^{-i\lambda nx}.$$

It is easy to show that $\lim_{\epsilon\to 0}(d_n - d_{n-1}) = 0$ for each n and that

$$|d_n - d_{n-1}| \leqq \frac{A}{n^2} \qquad (n \neq 0, n \neq 1),$$

A being independent of ϵ. The conclusion then follows by uniform convergence.

(c) is obvious by direct calculation.

(2.4) LEMMA 4. *If $G(x)$ belongs to $\mathfrak{A}$ with $g(x)$ absolutely continuous, and if*

$$\underline{\text{Bd}}\,|1 + G(x)| > 0,$$

then $[1 + G(x)]^{-1}$ belongs to $\mathfrak{A}$.

Let N be a positive integer and let

$$3\epsilon^2 N = 4, \qquad x_n = 3\epsilon n - \frac{2}{\epsilon} \qquad (n = 0, 1, \cdots, N).$$

Then

$$\sum_{n=0}^{N}\Gamma[(x - x_n)/\epsilon] = 1 \qquad (|x| \leqq 2/\epsilon),$$

$$1 = \Gamma(\epsilon x) + 1 - \Gamma(\epsilon x),$$

and since $\Gamma(\epsilon x) = 0$ for $x \geqq 2/\epsilon$, we have

$$1 = \Gamma(\epsilon x)\sum_{n=0}^{N}\Gamma[(x - x_n)/\epsilon] + [1 - \Gamma(\epsilon x)],$$

$$\text{(2.4.1)} \qquad \frac{1}{1 + G(x)} = \Gamma(\epsilon x)\sum_{n=0}^{N}\frac{\Gamma[(x - x_n)/\epsilon]}{1 + G(x)} + \frac{1 - \Gamma(\epsilon x)}{1 + G(x)}.$$

We shall show that each term of this sum belongs to $\mathfrak{A}$ when N is sufficiently large.

We have

$$\frac{\Gamma[(x - x_n)/\epsilon]}{1 + G(x)} = \frac{\Gamma[(x - x_n)/\epsilon]}{1 + G(x_n) + [G(x) - G(x_n)]\Gamma[\frac{1}{2}(x - x_n)/\epsilon]},$$

since the denominators differ only at points where $\Gamma[(x - x_n)/\epsilon] = 0$. Writing $\gamma_\epsilon(y) = 2\epsilon\gamma(2\epsilon y)$, we have

$$[G(x) - G(x_n)]\,\Gamma[\tfrac{1}{2}(x - x_n)/\epsilon]$$

$$= \int e^{-itx} g(t)\,dt \int e^{-iyx+iyx_n} \gamma_\epsilon(y)\,dy - \int e^{-itx_n} g(t)\,dt \int e^{-iyx+iyx_n} \gamma_\epsilon(y)\,dy$$

$$= \int g(t)\,dt \int e^{-iyx+ix_n(y-t)} \gamma_\epsilon(y - t)\,dy - \int e^{-itx_n} g(t)\,dt \int e^{-iyx+iyx_n} \gamma_\epsilon(y)\,dy$$

$$= \int e^{-iyx}\,dy \int e^{-ix_n(y-t)} g(t)[\gamma_\epsilon(y - t) - \gamma_\epsilon(y)]\,dt.$$

Hence

$$T\{[G(x) - G(x_n)]\,\Gamma[\tfrac{1}{2}(x - x_n)/\epsilon]\} \leqq \int \left| \int g(t)[\gamma_\epsilon(y - t) - \gamma_\epsilon(y)]\,dt \right| dy$$

$$\leqq \int |\,g(t)\,|\,dt \int |\,\gamma_\epsilon(y - t) - \gamma_\epsilon(y)\,|\,dy.$$

But

$$\int |\,\gamma_\epsilon(y - t) - \gamma_\epsilon(y)\,|\,dy = \int |\,\gamma(y - 2\epsilon t) - \gamma(y)\,|\,dy,$$

and this is bounded for all t and ϵ and tends to zero, for any fixed t, as $\epsilon \to 0$.

Since $|\,1 + G(x_n)\,|$ exceeds a positive number independent of N, it follows that if N is sufficiently large,

$$T\{[G(x) - G(x_n)]\,\Gamma[\tfrac{1}{2}(x - x_n)/\epsilon]\} \leqq |\,1 + G(x_n)\,|$$

for $n = 0, 1, \cdots, N$. Hence, appealing to Lemmas 1 and 2, we see that each term of the first part of the right side of (2.4.1) belongs to $\mathfrak{A}$.

As for the last term on the right of (2.4.1), we can write

$$\frac{1 - \Gamma(\epsilon x)}{1 + G(x)} = \frac{1 - \Gamma(\epsilon x)}{1 + G(x)[1 - \Gamma(\tfrac{1}{2}\epsilon x)]}.$$

Now

$$\Gamma(\tfrac{1}{2}\epsilon x) = \frac{2}{\epsilon} \int e^{-iyx} \gamma(2y/\epsilon)\,dy,$$

$$G(x) = \int e^{-iyx} g(y)\,dy.$$

Therefore

$$G(x)\,\Gamma(\tfrac{1}{2}\epsilon x) = \frac{2}{\epsilon}\int e^{-iyx}\,dy \int g(y-t)\gamma(2t/\epsilon)\,dt,$$

$$G(x)[1-\Gamma(\tfrac{1}{2}\epsilon x)] = \int e^{-iyx}\Big[g(y) - \frac{2}{\epsilon}\int g(y-t)\gamma(2t/\epsilon)\,dt\Big]dy$$

$$= \frac{2}{\epsilon}\int e^{-iyx}\Big[\int [g(y)-g(y-t)]\gamma(2t/\epsilon)\,dt\Big]dy,$$

since

$$\frac{2}{\epsilon}\int \gamma(2t/\epsilon)\,dt = 1.$$

Hence

$$T\{G(x)[1-\Gamma(\tfrac{1}{2}\epsilon x)]\} = \frac{2}{\epsilon}\int\int |\,[g(y)-g(y-t)]\gamma(2t/\epsilon)\,|\,dt\,dy$$

$$= \int |\,\gamma(t)\,|\,dt \int |\,g(y) - g(y-\tfrac{1}{2}t\epsilon)\,|\,dy.$$

Since $g(y) \in L(-\infty, \infty)$, $\int |\,g(y) - g(y - \tfrac{1}{2}t\epsilon)\,|\,dy$ is bounded for $-\infty < t < \infty$ and tends to zero for each fixed t, as $\epsilon \to 0$. Hence

$$\lim_{\epsilon\to 0} T\{G(x)[1-\Gamma(\tfrac{1}{2}\epsilon x)]\} = 0,$$

and it follows from Lemmas 1 and 2 that $[1 - \Gamma(\epsilon x)]/[1 + G(x)]$ belongs to $\mathfrak{A}$ when ϵ is sufficiently small.

(2.5) LEMMA 5. *Let $G(x)$, $F(x)$, $[F(x)]^{-1}$ belong to $\mathfrak{A}$, $g(x)$ being absolutely continuous. Let F and G satisfy*

$$\underline{\mathrm{Bd}}\,|\,F(x) + G(x)\,| > 0.$$

Then $[F(x) + G(x)]^{-1}$ belongs to $\mathfrak{A}$.

We have

$$\frac{1}{F(x)+G(x)} = \frac{1}{F(x)}\,\frac{1}{\{1 + G(x)[F(x)]^{-1}\}}.$$

Let

$$[F(x)]^{-1} = \int e^{-iyx}\,df^*(y), \qquad \int |\,df^*(y)\,| < \infty.$$

Since $g(x)$ is absolutely continuous, we can write

$$G(x) = \int e^{-iyx} g'(y)\,dy, \qquad \int |\,g'(y)\,|\,dy < \infty.$$

Then

$$G(x)[F(x)]^{-1} = \int e^{-iyx}\,dy \int g'(y-t)\,df^*(t),$$

and since $F(x)$ is bounded,

$$\underline{\mathrm{Bd}}\,|\,1 + G(x)[F(x)]^{-1}\,| = \underline{\mathrm{Bd}}\,|\,[F(x) + G(x)][F(x)]^{-1}\,| > 0,$$

and the conclusion follows from Lemma 4.

(3.1) THEOREM 1. *Suppose that $F(x)$ belongs to $\mathfrak{A}$ and let $\underline{\mathrm{Bd}}\,|\,F(x)\,| > 0$. Suppose that $H(x)$, $S(x)$ are defined as in (1.1), and that*

$$T\{S(x)\} < \underline{\mathrm{Bd}}\,|\,H(x)\,|\,.$$

Then $[F(x)]^{-1}$ belongs to $\mathfrak{A}$.

In view of Lemma 5, it is sufficient to prove that $[H(x) + S(x)]^{-1}$ belongs to $\mathfrak{A}$. Let

$$\underline{\mathrm{Bd}}\,|\,H(x)\,| = \delta, \qquad 0 < 3\eta \leqq \delta - T\{S(x)\}. \tag{3.1.1}$$

Then

$$H(x) = H_1(x) + H_2(x),$$

where

$$H_1(x) = \sum_{n=1}^{N} a_n e^{-i\lambda_n x}, \qquad H_2(x) = \sum_{n=N+1}^{\infty} a_n e^{-i\lambda_n x},$$

and N is chosen so that

$$T\{H_2(x)\} = \sum_{n=N+1}^{\infty} |\,a_n\,| \leqq \eta, \tag{3.1.2}$$

$$|\,H_1(x)\,| \geqq \delta - \eta. \tag{3.1.3}$$

Let K be an integer greater than 1 and let

$$\epsilon_n = \frac{2\pi}{3K\lambda_n}, \qquad \Omega_n(x) = \Omega(\epsilon_n, \lambda_n, x)$$

for $n = 1, 2, \cdots, N$. Then

$$\sum_{k=1}^{K} \Omega_n(x - 3k\epsilon_n) = 1$$

for $n = 1, 2, \cdots, N$ and all x, and it follows that

$$\prod_{n=1}^{N} \sum_{k=1}^{K} \Omega_n(x - 3k\epsilon_n) = 1.$$

Hence

$$\sum \prod_{n=1}^{N} \Omega_n(x - 3k_n \epsilon_n) = 1,$$

where the summation extends over the K^N sets $(k_1, \cdots, k_N)$ formed from integers $1, 2, \cdots, K$.

We can therefore write

(3.1.4) $$\frac{1}{H(x) + S(x)} = \sum \frac{\prod_{n=1}^{N} \Omega_n(x - 3k_n \epsilon_n)}{H(x) + S(x)}.$$

By Lemma 1, it is sufficient to show that each term of the sum belongs to $\mathfrak{A}$. Let x_0 be a point at which

$$\Omega(x) = \prod_{n=1}^{N} \Omega_n(x - 3k_n \epsilon_n) \neq 0.$$

(If there is no such point, the relevant term in the right side of (3.1.4) vanishes.) Then

(3.1.5) $$\frac{\Omega(x)}{H(x) + S(x)} = \frac{\Omega(x)}{H_1(x_0) + Q(x)},$$

where

(3.1.6) $$Q(x) = H_2(x) + S(x) + [H_1(x) - H_1(x_0)] \prod_{n=1}^{N} \Omega(2\epsilon_n, \lambda_n, x - 3\epsilon_n k_n),$$

since the two denominators differ only at points where $\Omega(x) = 0$. Since

$$[H_1(x) - H_1(x_0)] \prod_{n=1}^{N} \Omega(2\epsilon_n, \lambda_n, x - 3\epsilon_n k_n)$$
$$= \sum_{\nu=1}^{N} a_\nu (e^{-i\lambda_\nu x} - e^{-i\lambda_\nu x_0}) \prod_{n=1}^{N} \Omega(2\epsilon_n, \lambda_n, x - 3\epsilon_n k_n),$$

we have

(3.1.7) $$T\left\{[H_1(x) - H_1(x_0)] \prod_{n=1}^{N} \Omega(2\epsilon_n, \lambda_n, x - 3\epsilon_n k_n)\right\}$$
$$\leqq \sum_{\nu=1}^{N} |a_\nu| C^{N-1} T\{[e^{-i\lambda_\nu x} - e^{-i\lambda_\nu x_0}] \Omega(2\epsilon_\nu, \lambda_\nu, x - 3\epsilon_\nu k_\nu)\},$$

by Lemmas 1 and 3. Since $\Omega(x_0) \neq 0$, we have

$$\Omega_\nu(x_0 - 3\epsilon_\nu k_\nu) \neq 0 \qquad (\nu = 1, 2, \cdots, N),$$

and therefore, by the definition of $\Omega_\nu(x)$,

(3.1.8) $$|x_0 - 3\epsilon_\nu k_\nu| < 2\epsilon_\nu, \qquad \mod (2\pi/\lambda_\nu),$$

for $\nu = 1, 2, \cdots, N$. Returning to (3.1.7), we see that

$$
\begin{aligned}
&T\{[e^{-i\lambda_\nu x} - e^{-i\lambda_\nu x_0}]\,\Omega(2\epsilon_\nu, \lambda_\nu, x - 3\epsilon_\nu k_\nu)\} \\
&\quad = T\{[e^{-i\lambda_\nu(x+3\epsilon_\nu k_\nu)} - e^{-i\lambda_\nu x_0}]\,\Omega(2\epsilon_\nu, \lambda_\nu, x)\} \\
(3.1.9)\qquad &\quad = T\{[e^{-i\lambda_\nu x} - e^{-i\lambda_\nu(x_0-3\epsilon_\nu k_\nu)}]\,\Omega(2\epsilon_\nu, \lambda_\nu, x)\} \\
&\quad \leqq T\{[1 - e^{-i\lambda_\nu x}]\Omega(2\epsilon_\nu, \lambda_\nu, x)\} + C\,|\,1 - e^{-i\lambda_\nu(x_0-3\epsilon_\nu k_\nu)}\,|.
\end{aligned}
$$

Hence, using (3.1.7), (3.1.8) and Lemma 1, we can choose K so that

$$
(3.1.10)\qquad T\left\{[H_1(x) - H_1(x_0)]\prod_{n=1}^{N}\Omega(2\epsilon_n, \lambda_n, x - 3\epsilon_n k_n)\right\} < \eta.
$$

Then, by (3.1.1), (3.1.2), (3.1.3) and (3.1.10), we have

$$
T\{Q(x)\} < \eta + \delta - 3\eta + \eta = \delta - \eta \leqq H_1(x_0),
$$

and the conclusion follows from (3.1.5) by using Lemma 2.

(4.1) We shall devote the remainder of this paper to showing that Theorem 1 is substantially the best possible result of its kind. In particular, we show that the conditions

$$
\underline{\mathrm{Bd}}\,|\,F(x)\,| > 0, \qquad F(x) \in \mathfrak{A}
$$

are not sufficient to make $[F(x)]^{-1} \in \mathfrak{A}$. We require Theorem 2 in the proof of Theorem 3, but it seems of some independent interest in itself.

THEOREM 2. *Let δ be > 0. Then we can define a function $S(x)$ of $\mathfrak{A}$ such that*

$$
(4.2.1)\qquad |\,S(x)\,| \leqq \delta,
$$

$$
(4.2.2)\qquad T\{S(x)\} = 1,
$$

$$
(4.2.3)\qquad T\left\{\sum_{k=1}^{\infty} H_k(x)[S(x)]^k\right\} = \sum_{k=1}^{\infty} T\{H_k(x)\}
$$

for any sequence of functions $H_k(x)$ defined by

$$
(4.2.4)\qquad H_k(x) = \sum_{j=1}^{\infty} a_{k,j}e^{-i\lambda_j x}, \qquad \sum_{k,j} |\,a_{k,j}\,| < \infty.
$$

We observe, first, that if $S(x)$ satisfies conditions (4.2.2), (4.2.3), then so does $[S(x)]^m$ for any integer m. Therefore, instead of (4.2.1), it is sufficient to show that

$$
(4.2.5)\qquad \overline{\mathrm{Bd}}\,|\,S(x)\,| < 1,
$$

for then $[S(x)]^m$ will be a function of the required type if m is sufficiently large.

In what follows we suppose that the function $s(x)$ is of bounded variation in $(-\infty, \infty)$ and define $S(x)$ by

$$
S(x) = \int e^{-iyx}\,ds(y).
$$

We denote by B (with or without a suffix) any array of the form

$$\cdot\ ---\ 00\ ---\ xxx\ ---\ 00-\ \cdots$$

in which there are arbitrarily long stretches of consecutive 0's, x's, and gaps. We shall define $s(x)$ by means of such arrays.

With any zero of B, at the r-th place say, we associate the set of 2^{r-1} intervals of length 2^{-r} whose binary representations have zeros in the r-th place, and suppose that $s(x)$ is constant in the complementary intervals. We suppose also that whenever x occurs in the r-th place of B, the variation (possibly zero) of $s(x)$ in each interval $(\nu 2^{-r}, (\nu + 1)2^{-r})$ $(\nu = 0, 1, \cdots, 2^r - 1)$ is divided equally between the two intervals $(\nu 2^{-r}, \nu 2^{-r} + 2^{-r-1})$ and $(\nu 2^{-r} + 2^{-r-1}, (\nu + 1) 2^{-r})$. In these circumstances we say that the function $e^{-i\lambda x}S(x)$, for any real λ, is of type B.

We say that two arrays are *disjoint* if there are arbitrarily long stretches of consecutive 0's in each which correspond to x's in the other. A set of arrays is disjoint if each pair is disjoint. We require the following result.

LEMMA 6. *Let $B_1, \cdots, B_K$ be disjoint, and let $S_1(x), \cdots, S_K(x)$ be of types $B_1, \cdots, B_K$, respectively. Let $H_k(x)$ $(k = 1, \cdots, K)$ be trigonometrical polynomials. Then*

$$T\left\{\sum_{k=1}^{K} S_k(x)H_k(x)\right\} = \sum_{k=1}^{K} T\{S_k(x)H_k(x)\}.$$

We may suppose without loss of generality that each of the polynomials $H_k(x)$ has K terms. Suppose that $S_\nu(x)$ is of type B_ν $(\nu = 1, \cdots, K)$, and suppose that B_1 has 0's and B_2 has x's from the N-th to the $(N + p)$-th place. We write

$$S_\nu(x)H_\nu(x) = \int e^{-iyx}\, d\sigma_\nu(y) \qquad (\nu = 1, \cdots, K).$$

Then the whole variation of $\sigma_1(y)$ is concentrated in K sets of 2^N intervals of the form

$$(\lambda_k + \nu 2^{-N}, \lambda_k + \nu 2^{-N} + 2^{-N-p}) \qquad (k = 1, \cdots, K; \nu = 0, \cdots, 2^N),$$

where λ_k is an exponent in $H_1(x)$. But for each k, this set of intervals cannot contain more than a fraction 2^{-p+1} of the variation of $s_2(x - \lambda)$, for any λ, since B_2 has x's between the N-th and $(N + p)$-th places. Therefore not more than a fraction $K \cdot 2^{-p+1}$ of the variation of $\sigma_2(x)$ is contained in these intervals. Then plainly

$$T\{S_1(x)H_1(x) + S_2(x)H_2(x)\} \geqq T\{S_1(x)H_1(x)\} + (1 - K2^{-p+1})T\{S_2(x)H_2(x)\},$$

and since p is arbitrarily large,

$$T\{S_1(x)H_1(x) + S_2(x)H_2(x)\} = T\{S_1(x)H_1(x)\} + T\{S_2(x)H_2(x)\}.$$

The extension of this argument to K functions is obvious.

We can now define $S(x)$. We suppose that $s(x)$ increases from 0 to $\frac{1}{2}$ in $(0, \frac{1}{2})$ and decreases from $\frac{1}{2}$ to 0 in $(\frac{1}{2}, 1)$. Let

$$S(x) = \int_0^1 e^{-iyx}\, ds(y).$$

If N is a positive integer, we can write

$$(4.2.6) \qquad S(x) = \sum_{\nu=0}^{2^N-1} S_\nu^N(x), \qquad S_\nu^N(x) = \int_{\nu 2^{-N}}^{(\nu+1)2^{-N}} e^{-iyx}\, ds(y).$$

If

$$\alpha_\nu^N = \int_{\nu 2^{-N}}^{(\nu+1)2^{-N}} |\, ds(y)\, |,$$

then

$$\sum_{\nu=0}^{2^N-1} \alpha_\nu^N = 1.$$

We proceed inductively (with respect to N). Suppose that $S_\nu^N(x)$ is of type B_ν^N, that the B_ν^N's have arbitrarily long stretches of gaps in common and that they have 0's or x's in the first N places. Moreover, we suppose that the B_ν^N's are such that if they are divided into two groups in any way, the arrays in each group have in common arbitrarily long stretches of 0's corresponding to x's in all the arrays of the other group.

Now consider the two groups of integers ν for which B_ν^N has x's or 0's, respectively, in the n-th place (these may not exhaust the integers $0, 1, \cdots, 2^N - 1$, as some of the B_ν^N's may have gaps). We suppose that for every n, the sum of α_ν^N's of the first group is not less than the sum of α_ν^N's of the second. In other words, we suppose that at least half of the variation of $s(x)$ is distributed by an x at each stage.

Now since we have only a finite number of B_ν^N's to define for each value of N, and we have arbitrarily long stretches of gaps in the arrays defined for integers less than N, it is plain that the B_ν^N's can be defined inductively so that the properties described above hold for every N. Since this process defines α_ν^N for all N and ν, the function $s(x)$ is defined completely.

Since $T\{[S(x)]^k\} \leqq 1$, it is sufficient to prove the theorem in the case in which there is a finite number of $H_k(x)$'s and each is a trigonometric polynomial. The result in the general case follows at once by a simple limit argument. Let ϵ be > 0, and let 2^N be $> k$. Then

$$(4.2.7) \qquad \begin{aligned} S(x) &= \sum_\nu S_\nu^N(x), \\ [S(x)]^k &= [\sum_\nu S_\nu^N(x)]^k = k! \sum S_{\nu_1}^N(x) S_{\nu_2}^N(x) \cdots S_{\nu_k}^N(x) + Q_k^N(x), \end{aligned}$$

where $\nu_1, \nu_2, \cdots, \nu_k$ are all different and $Q_k^N(x)$ is the sum of terms involving the square or higher power of at least one $S_\nu^N(x)$.

It is plain from the construction of $S(x)$ that

$$\lim_{N\to\infty} \max \alpha_\nu^N = 0,$$

and since $s(y)$ is monotonic in each interval $(\nu 2^{-N}, (\nu+1)2^{-N})$,

$$T\{S_{\nu_1}^N(x) \cdots S_{\nu_k}^N(x)\} = T\{S_{\nu_1}^N(x)\} \cdots T\{S_{\nu_k}^N(x)\} = \alpha_{\nu_1}^N \cdots \alpha_{\nu_k}^N.$$

Hence[4]

$$\text{(4.2.8)} \qquad \lim_{N\to\infty} T\{Q_k^N(x)\} = 0, \qquad \lim_{N\to\infty} k! \sum T\{S_{\nu_1}^N(x) \cdots S_{\nu_k}^N(x)\} = 1.$$

Now if two functions $S(x)$, $S'(x)$ are of types B, B', it is easy to show that $S(x)S'(x)$ is of a type obtained by adding B and B' according to the rule that in stretches of consecutive 0's or x's we write

$$0 + 0 = 0, \qquad x + 0 = x, \qquad x + x = x,$$

except possibly in the last place in any stretch. Thus, if

$$B = \cdot 0000 --- xxxx --- xxxx ---,$$

$$B' = \cdot 00000 -- 000 ---- xxxx ---,$$

then $S(x)S'(x)$ is of type

$$\cdot 000 ---- xx ------ xxx ----.$$

Then it follows from our definition of $S_\nu^N(x)$ that non-identical functions of the form

$$S_{\nu_1}^N(x) S_{\nu_2}^N(x) \cdots S_{\nu_k}^N(x)$$

are disjoint, so that, using (4.2.7) and (4.2.8), we can write

$$\text{(4.2.9)} \qquad [S(x)]^k = \sum_p S_{k,p}(x) + Q_k(x),$$

where all the $S_{k,p}(x)$'s are disjoint and

$$\text{(4.2.10)} \qquad \sum_{k,j} |a_{k,j}|\, T\{Q_k(x)\} \leqq \epsilon,$$

$$\text{(4.2.11)} \qquad \sum_p T\{S_{k,p}(x)\} \geqq 1 - \epsilon \qquad (k = 1, 2, \cdots, K).$$

[4] We use the fact that if $\alpha_\nu \geqq 0$, $\sum_{\nu=1}^{n} \alpha_\nu = 1$, then

$$(\alpha_1 + \alpha_2 + \cdots + \alpha_n)^k - k! \sum \alpha_{\nu_1}\alpha_{\nu_2} \cdots \alpha_{\nu_k} \to 0$$

as $\max \alpha_\nu \to 0$, the summation being over all sets of k different ν_j's.

Hence

$$\sum_k H_k(x)[S(x)]^k = \sum_k \sum_j a_{k,j} e^{-i\lambda_j x} [\sum_p S_{k,p}(x) + Q_k(x)],$$

$$T\{\sum_k H_k(x)[S(x)]^k\} \geqq T\{\sum_k \sum_p \sum_j a_{k,j} e^{-i\lambda_j x} S_{k,p}(x)\} - \epsilon$$

$$= \sum_{k,p} T\{\sum_j a_{k,j} e^{-i\lambda_j x} S_{k,p}(x)\} - \epsilon,$$

by Lemma 6. Now

$$S_{k,p}(x) = \int e^{-iyx} ds_{k,p}(y),$$

where $s_{k,p}(x)$ is constant except in an interval of length $k \cdot 2^{-N}$. Hence, if we choose N so that

$$k \cdot 2^{-N} \leqq \min_{j \neq j'} | \lambda_j - \lambda_{j'} |,$$

it is plain that

$$T\{\sum_j a_{k,j} e^{-i\lambda_j x} S_{k,p}(x)\} = \sum_j | a_{k,j} | T\{e^{-i\lambda_j x} S_{k,p}(x)\}$$

$$= \sum_j | a_{k,j} | T\{S_{k,p}(x)\}.$$

It follows that

$$T\{\sum_k H_k(x)[S(x)]^k\} \geqq \sum_{k,j} | a_{k,j} | \sum_p T\{S_{k,p}(x)\} - \epsilon$$

$$\geqq \sum_{k,j} | a_{k,j} | - \epsilon[1 + \sum_{k,j} | a_{k,j} |],$$

by (4.2.11). Since ϵ is arbitrary,

$$T\{\sum_k H_k(x)[S(x)]^k\} \geqq \sum_{k,j} | a_{k,j} | = \sum_k T\{H_k(x)\},$$

and since

$$T\{\sum_k H_k(x)[S(x)]^k\} \leqq \sum_k T\{H_k(x)\},$$

by Lemma 1, this completes the proof of (4.2.3).

Finally, we have to prove (4.2.5). Let x be $\geqq \frac{1}{2}\pi$, and choose an integer $N \geqq 1$ so that

(4.2.12) $$\pi \cdot 2^{N-2} \leqq x < \pi \cdot 2^{N-1}.$$

We write, as before,

$$S(x) = \sum_{\nu=0}^{2^N - 1} S_\nu^N(x).$$

Since at least a half of the total variation of $s(x)$ is divided equally between intervals $(\nu 2^{-N}, \nu 2^{-N} + 2^{-N-1})$ and $(\nu 2^{-N} + 2^{-N-1}, (\nu + 1)2^{-N})$, we can write

$$|S_\nu^N(x)| \leqq \beta_\nu + \sum_\nu \left| \int_0^{2^{-N}} e^{-iyx}\, d\sigma_\nu(y) \right|,$$

where $\beta_\nu \geqq 0$, $\sigma_\nu(y)$ increases from 0 to β'_ν in $(0, 2^{-N})$,

$$0 \leqq A \leqq \tfrac{1}{2}, \qquad \sum_\nu \beta_\nu = A, \qquad \sum_\nu \beta'_\nu = 1 - A.$$

Moreover, since the variation of $s(y)$ in the interval $(\nu 2^{-M}, \nu 2^{-M} + 2^{-M-k})$ is at least a fraction 2^{-k} of its variation in $(\nu 2^{-M}, (\nu + 1)2^{-M})$, we can suppose that

$$\sigma_\nu(2^{-N-3}) = \lambda_1 \beta'_\nu \geqq \tfrac{1}{8}\beta'_\nu,$$
$$\sigma_\nu(2^{-N-1} + 2^{-N-3}) - \sigma_\nu(2^{-N-1}) = \lambda_2 \beta'_\nu \geqq \tfrac{1}{8}\beta'_\nu.$$

Hence

$$\left| \int_0^{2^{-N}} e^{-iyx}\, d\sigma_\nu(y) \right| \leqq (1 - \lambda_1 - \lambda_2)\beta'_\nu + \left| \int_0^{2^{-N-3}} e^{-iyx}\, d\sigma_\nu(y) + \int_{2^{-N-1}}^{2^{-N-1}+2^{-N-3}} e^{-iyx}\, d\sigma_\nu(y) \right|.$$

Using (4.2.12), we see that

$$0 \leqq xy \leqq \tfrac{1}{16}\pi$$

in the first integral, while

$$\tfrac{1}{8}\pi \leqq xy \leqq \tfrac{5}{16}\pi$$

in the second. It follows that

$$\left| \int_0^{2^{-N}} e^{-iyx}\, d\sigma_\nu(y) \right| \leqq (1 - \lambda_1 - \lambda_2)\beta'_\nu + \beta'_\nu[\lambda_1 + \lambda_2 - \tfrac{1}{4} + \tfrac{1}{8}\,|1 + e^{i\theta_\nu}|]$$
$$= \beta'_\nu[\tfrac{3}{4} + \tfrac{1}{8}\,|1 + e^{i\theta_\nu}|],$$

where

$$\tfrac{1}{16}\pi \leqq \theta_\nu \leqq \tfrac{5}{16}\pi.$$

Hence

$$\left| \int_0^{2^{-N}} e^{-iyx}\, d\sigma_\nu(y) \right| \leqq C\beta'_\nu,$$

C being an absolute constant less than 1, and we have

$$|S(x)| \leqq A + (1 - A)C < 1$$

for $x \geqq \frac{1}{2}\pi$. If $x < \frac{1}{2}\pi$, we use the fact that $s(y)$ increases from 0 to $\frac{1}{2}$ in $(0, \frac{1}{2})$ and decreases from $\frac{1}{2}$ to 0 in $(\frac{1}{2}, 1)$. We have

$$S(x) = Ae^{i\theta} - Be^{i\varphi},$$

where

$$0 \leqq A \leqq \tfrac{1}{2}, \qquad 0 \leqq B \leqq \tfrac{1}{2},$$
$$0 \leqq \theta(x) \leqq \tfrac{1}{2}x, \qquad \tfrac{1}{2}x \leqq \varphi(x) \leqq x.$$

Then we have

$$0 \leqq \psi(x) = \varphi(x) - \theta(x) \leqq \tfrac{1}{2}\pi,$$
$$\begin{aligned} | S(x) | &\leqq | A - Be^{i\psi} | \\ &= | A - B\cos\psi - iB\sin\psi | \\ &\leqq | A - iB | \leqq 2^{-\frac{1}{2}} \end{aligned}$$

This completes the proof of Theorem 2.

(5.1) THEOREM 3. *Let δ, δ' satisfy $\delta > \delta' > 0$. Then there is a function $S(x)$ of $\mathfrak{A}$, with*

$$| S(x) | \leqq \delta', \qquad T\{S(x)\} = \delta,$$

such that, if $\underline{\mathrm{Bd}}\, | H(x) | = \delta$,

$$\frac{1}{F(x)} = \frac{1}{H(x) + S(x) + G(x)}$$

does not belong to $\mathfrak{A}$ for any $G(x)$ (even if $\underline{\mathrm{Bd}}\, | F(x) | > 0$).

If $| S(x) | \leqq \delta'$, then $| H(x) + S(x) | \geqq \delta - \delta' > 0$, so that it is sufficient, after Theorem 1, to show that $[H(x) + S(x)]^{-1}$ does not belong to $\mathfrak{A}$. We may plainly suppose that $\delta = 1$ and that

(5.1.1) $$\underline{\mathrm{Bd}}\, | H(x) - 1 | = 0.$$

Let η be > 0, and let

$$H(x) = \sum_{n=1}^{\infty} a_n e^{-i\lambda_n x},$$
$$H_1(x) = \sum_{n=1}^{N} a_n e^{-i\lambda_n x}, \qquad H_2(x) = \sum_{N+1}^{\infty} a_n e^{-i\lambda_n x},$$

where $N = N(\eta)$ is chosen so that

(5.1.2) $$T\{H_2(x)\} \leqq \tfrac{1}{4}\eta, \qquad | \underline{\mathrm{Bd}}\, | H_1(x) | - 1 | \leqq \tfrac{1}{4}\eta.$$

Now let $S(x)$ be the function defined in Theorem 2 (with δ' instead of δ). We assume that $[H(x) + S(x)]^{-1}$ belongs to $\mathfrak{A}$, and show that this leads to a contradiction. Let

(5.1.3) $$T\{[H(x) + S(x)]^{-1}\} = A < \infty.$$

Using (5.1.1) and the first inequality of (5.1.2), we can choose x_0 so that

(5.1.4) $$| H_1(x_0) - 1 | \leqq \tfrac{1}{2}\eta.$$

By Lemma 3(b), we can choose $\epsilon = \epsilon(N, \eta) = \epsilon(\eta)$ so that

$$T\{[H_1(x) - H_1(x_0)] \prod_{n=1}^{N} \Omega(2\epsilon, \lambda_n, x - x_0)\} < \tfrac{1}{4}\eta.$$

Then since

$$H(x) + S(x) = H_1(x_0) + H_2(x) + S(x) + [H_1(x) - H_1(x_0)] \prod_{n=1}^{N} \Omega(2\epsilon, \lambda_n, x - x_0)$$

at points where

$$\Omega(x) = \prod_{n=1}^{N} \Omega(\epsilon, \lambda_n, x - x_0) \neq 0,$$

we have, using (5.1.2), (5.1.4) and Lemma 1,

$$\frac{\Omega(x)}{H(x) + S(x)} = \frac{\Omega(x)}{1 + S(x) + P(x)}, \tag{5.1.5}$$

where

$$T\{P(x)\} \leqq \eta. \tag{5.1.6}$$

Now let M be a positive integer independent of η and let

$$M > 2A - 1. \tag{5.1.7}$$

We have

$$\begin{aligned} \frac{1}{1 + S(x) + P(x)} &= \sum_{\nu=0}^{M} (-1)^{\nu}[S(x) + P(x)]^{\nu} \\ &\quad + \frac{(-1)^{M+1}[S(x) + P(x)]^{M+1}}{1 + S(x) + P(x)} \\ &= \sum_{\nu=0}^{M} (-1)^{\nu}[S(x)]^{\nu} + Q(x), \end{aligned} \tag{5.1.8}$$

where

$$Q(x) = \frac{(-1)^{M+1}[S(x) + P(x)]^{M+1}}{1 + S(x) + P(x)} + \sum_{\nu=0}^{M} (-1)^{\nu}\{[S(x) + P(x)]^{\nu} - [S(x)]^{\nu}\}.$$

Using (5.1.6) and Lemma 1, we obtain

$$T\{S(x) + P(x)\} \leqq 1 + \eta,$$
$$T\{[S(x) + P(x)]^{\nu} - [S(x)]^{\nu}\} \leqq (1 + \eta)^{\nu} - 1.$$

Hence

$$\begin{aligned} T\{Q(x)\Omega(x)\} &\leqq (1 + \eta)^{M+1} T\{\Omega(x)[1 + S(x) + P(x)]^{-1}\} \\ &\quad + T\{\Omega(x)\} \sum_{\nu=0}^{M} [(1 + \eta)^{\nu} - 1] \\ &\leqq T\{\Omega(x)\} \left[(1 + \eta)^{M+1} A + \sum_{\nu=0}^{M} [(1 + \eta)^{\nu} - 1] \right], \end{aligned}$$

by (5.1.5) and (5.1.3), and it follows from this and (5.1.5) and (5.1.8) that

$$(5.1.9)\quad \begin{aligned} T\left\{\Omega(x)\sum_{\nu=0}^{M}(-1)^{\nu}[S(x)]^{\nu}\right\} &\leqq T\{\Omega(x)[H(x)+S(x)]^{-1}\} + T\{\Omega(x)Q(x)\} \\ &\leqq T\{\Omega(x)\}\left[A(1+(1+\eta)^{M+1}) + \sum_{\nu=0}^{M}[(1+\eta)^{\nu}-1]\right]. \end{aligned}$$

But by Theorem 2, we have

$$T\left\{\Omega(x)\sum_{\nu=0}^{M}(-1)^{\nu}[S(x)]^{\nu}\right\} = \sum_{\nu=0}^{M} T\{\Omega(x)(-1)^{\nu}\} = (M+1)T\{\Omega(x)\}.$$

Combining this with (5.1.9), and dividing by $T\{\Omega(x)\}$, we have

$$M + 1 \leqq A[1 + (1+\eta)^{M+1}] + \sum_{\nu=0}^{M}[(1+\eta)^{\nu} - 1].$$

Since M is independent of η we can let $\eta \to 0$ and obtain

$$M + 1 \leqq 2A.$$

This contradicts (5.1.7). Therefore $[H(x) + S(x)]^{-1}$ cannot belong to $\mathfrak{A}$.

MASSACHUSETTS INSTITUTE OF TECHNOLOGY AND HARVARD UNIVERSITY.

Comments on [33d], [38b]

W. Rudin

It may be worthwhile to restate the principal theorem of [33d] (p. 331) in modern measure-theoretic language, and to indicate a simple proof. Associate with each $\mu \in M(R)$ (the algebra of complex Borel measures on the real line R) and with each $h \in R$ the h-translate μ_h of μ, defined by

$$\mu_h(E) = \mu(E - h).$$

Let m be the Lebesgue measure of R; write $\mu \perp \lambda$ to mean that μ and λ are mutually singular (i.e., are concentrated on disjoint sets).

Theorem. *If $\mu \in M(R)$ and $\mu \perp m$, then $\mu \perp \mu_h$ for almost every h. (Here "a.e." refers to Lebesgue measure.)*

To prove this, let E be a Borel set in R with $m(E) = 0$ on which μ is concentrated. Apply Fubini's theorem to the product measure $m \times |\mu|$, where $|\mu|$ is the total variation of μ, and to the set $\{(x, y)\colon x + y \in E\} \subset R^2$, to conclude that

$$\int_R |\mu|(E - x)\, dm(x) = \int_R m(E - y)\, d|\mu|(y) = 0.$$

The last equality holds because $m(E - y) = 0$ for every y. It follows that $|\mu|(E - x) = 0$ for almost all x. For each such x, $\mu \perp \mu_x$, (QED).

The same theme—the existence of large collections of pairwise singular measures—also occurs in Theorem 2 of [38b]. In fact, formula (4.3.2) implies that if μ is the measure whose Fourier-Stieltjes transform is S, then any two of its convolution powers $\mu, \mu^2, \mu^3, \ldots$ are mutually singular.

It was later observed that if P is any independent perfect set and if μ is any continuous measure that is concentrated on $P \cup (-P)$, then μ has pairwise singular convolution powers. This has led to an easy proof of the so-called Wiener-Pitt phenomenon (Theorem 3 of [38b]—the best known result of this paper) in any nondiscrete locally compact Abelian group G. The same proof also yields the asymmetry of $M(G)$, for any such G. Details of this may be found in Section 5.3 of Rudin's book;[1] references to relevant papers by Sreider (1950), Hewitt (1955), and Williamson (1959) are also given there.

Reference

1. W. Rudin, *Fourier analysis on groups*, Interscience, New York, 1962.

Comments on [38b]

E. Hewitt

This paper provides still another example of the extraordinary seminal character of Wiener's work. This paper is the first to deal with the convolution algebra **M**(R) of all countably additive complex-valued Borel measures on the line R. The authors succeeded in proving a remarkable number of properties of this extremely complicated algebra. Let μ be a measure in **M**(R), and let μ_a, μ_s, and μ_d denote, respectively, the absolutely continuous, continuous singular, and purely discontinuous components of μ. The first main theorem of the paper states that if the norm of μ_s is less than the greatest lower bound of $|\hat{\mu}_a|$ on the dual line $\hat{R}$, then the measure μ has an inverse in **M**(R). This result is not difficult to prove with what is now known of the structure of multiplicative linear functionals on **M**(R), but it was a considerable achievement at that time.

The main result of the paper is that there exists a measure μ in **M**(R) whose Fourier-Stieltjes transform $\hat{\mu}$ is bounded in absolute value away from zero, and yet which has no inverse in **M**(R). Wiener and Pitt were led to inquire about the existence of such measures by noticing that nothing of the sort can happen for periodic functions with absolutely convergent Fourier series (this is the Wiener-Lévy theorem), or for almost periodic continuous functions on R (this theorem was proved by R. H. Cameron[1] and also by H. R. Pitt.[4] The basic idea of the proof of Wiener and Pitt is the construction of a continuous singular non-negative measure all of whose convolution powers are mutually singular. Much attention has been lavished on the algebra **M**(R) and the corresponding algebra **M**(G) for one or another class of locally compact groups G. All of this work has its origin in the paper of Wiener and Pitt. The next major contribution was by Yu. A. Šreĭder,[7] who succeeded in proving the Wiener-Pitt theorem in a quite different way, again for the group R. The theorem was extended to nondiscrete locally compact Abelian groups containing arbitrarily small elements of infinite order by Hewitt,[2] and to all nondiscrete locally compact Abelian groups by J. H. Williamson.[11] Williamson's proof is remarkable in that it transfers the original idea of Wiener and Pitt to a general group-theoretic setting. Williamson has also established the Wiener-Pitt theorem for the convolution algebra of measures on the non-negative real numbers.[12] An extreme form of the Wiener-Pitt phenomenon was identified by Hewitt and Kakutani,[3] in which it was shown that certain subspaces of **M**(G) have the property that all linear functionals of norm one on these subspaces admit multiplicative extensions over the entire algebra. Detailed study of $\mathbf{M}(G)$ continues. See, for example, J. L. Taylor[8–10] and also J. S. Pym.[5,6]

References

1. R. H. Cameron, Duke J. 3(1937), 682–688.
2. E. Hewitt, Michigan Math. J. 5(1958), 149–158.
3. E. Hewitt and S. Kakutani, Ill. J. Math. 4(1960), 553–574.
4. H. R. Pitt, J. Math and Phys., MIT 16(1938), 191–195.
5. J. S. Pym, Proc. Cambridge Philos. Soc. 65(1969), 597–599.
6. J. S. Pym, Proc. London Math. Soc. (3) 19(1969), 625–660.
7. Yu. A. Šreĭder, Mat. Sbornik (N. S.), (69) 27(1950), 297–318.
8. J. L. Taylor, Trans. Amer. Math. Soc. 119(1965), 150–166.
9. J. L. Taylor, Trans. Amer. Math. Soc. 128(1967), 257–263.
10. J. L. Taylor, Pacific J. Math. 31(1969), 809–826.
11. J. H. Williamson, Proc. Edinburgh Math. Soc. 11(1959), 195–206.
12. J. H. Williamson, Proc. Edinburgh Math. Soc. 13(1962), 37–38.

FOURIER-STIELTJES TRANSFORMS AND SINGULAR INFINITE CONVOLUTIONS.*

By Norbert Wiener and Aurel Wintner.

1. Let $F(x)$, $-\pi \leqq x \leqq \pi$, be a continuous monotone function and

$$(1) \qquad c_n = c_n(F) = \frac{1}{2\pi}\int_{-\pi}^{\pi} e^{-inx}dF(x) = \bar{c}_{-n} \qquad (n = 0, 1, 2, \cdots)$$

the sequence of its trigonometrical moments or Fourier-Stieltjes coefficients. If $F(x)$ is absolutely continuous, i. e., if

$$(2) \qquad F(x) = \int^{x} f(y)dy \qquad (-\pi \leqq x \leqq \pi)$$

holds for an L-integrable function f, then

$$(3) \qquad c_n = o(1)$$

by the Riemann-Lebesgue lemma. The converse is not true, since an example of Menchoff [8] shows that (3) may hold for a monotone function $F(x)$ which is singular, i. e., continuous and such that its derivative vanishes almost everywhere but $F(x) \neq$ const.

Let an absolute constant κ^* be defined by means of a Dedekind cut as follows: For a given number κ, there do or do not exist singular monotone functions $F(x)$ which satisfy

$$(4) \qquad c_n = o(n^{-\kappa})$$

according as $\kappa < \kappa^*$ or $\kappa > \kappa^*$. Menchoff's result, when expressed in terms of κ^*, does not state more than that κ^* cannot be negative, a fact which is implied by $c_n = O(1)$ and is, therefore, clear from (1). Also a result of Kershner [6], which has improved on Menchoff's example (3), does not give more than $\kappa^* \geqq 0$. However, Littlewood [7] has recently proved by an ingenious, though rather involved, construction that $\kappa^* \neq 0$. Littlewood's proof of $\kappa^* > 0$ is a pure existence proof, since it seems to be hardly possible to prove by his method that $\kappa^* > 10^{-10}$, say. At any rate, Littlewood's result $\kappa^* > 0$ implies the existence of a critical κ-range within which (4) is incapable

* Received May 9, 1938.

of deciding whether or not $F(x)$ is absolutely continuous. In fact, a well-known remark of Lebesgue shows that the Riemann-Lebesgue lemma, i. e. (3), cannot be improved for the class of all absolutely continuous F (not even if f in (2) is restricted to be continuous and periodic).

On the other hand, the critical κ-range is bounded, i. e., $\kappa^* < +\infty$. Actually, $\kappa^* \leqq \frac{1}{2}$. In fact, if F is such that (4) holds for a $\kappa > \frac{1}{2}$, then $\Sigma | c_n |^2 < +\infty$, and so the Fischer-Riesz theorem assures that (2) holds for a function f which is of class L, since it is of class L^2.

2. Accordingly, $0 < \kappa^* \leqq \frac{1}{2}$. No better result (e. g., $10^{-10} \leqq \kappa^* \leqq \frac{1}{2} - 10^{-10}$) is known to us from the existing literature. But we shall prove that $\kappa^* = \frac{1}{2}$. In other words, we shall show that there exists for every $\epsilon > 0$ a singular monotone function $F = F_\epsilon$ such that

$$c_n = O(n^{-\frac{1}{2}+\epsilon}). \tag{5}$$

It seems to be a reasonable guess that if $\gamma_1, \gamma_2, \cdots$ are given positive numbers for which $\Sigma \gamma_n^2 = +\infty$, then there exists a singular monotone function F such that $c_n = O(\gamma_n)$ or $c_n = o(\gamma_n)$.

It will remain undecided whether or not the width, $\kappa^* = \frac{1}{2}$, of the critical κ-range is directly connected with the critical widths $\frac{1}{2}$ in the theory of ordinary Dirichlet series (although the method will be such as to prove (5) also when n in (1) is allowed to vary continuously). Actually, it will be seen that the absolute constant κ^* is the "same" $\frac{1}{2}$ which occurs in the standard asymptotic formulae of the type

$$J_0(u) = O(| u |^{-\frac{1}{2}}), \qquad J_0(u) = \Omega(| u |^{-\frac{1}{2}}).$$

In fact, the exponent $-\frac{1}{2}$ of (5) will be introduced by using van der Corput's lemma,

$$\left| \int_a^b \cos g(x)\,dx \right| \leqq \text{Const.} \min_{a \leqq x \leqq b} | g''(x) |^{-\frac{1}{2}}, \tag{6}$$

where $g(x)$, $a \leqq x \leqq b$, is any real function with a non-vanishing continuous second derivative $g''(x)$, and Const. is independent of a, b and g (cf. Kershner [5]).

The proof of $\kappa^* = \frac{1}{2}$ will consist of four steps:

(i) It turns out (§ **5**) that, instead of proving the existence of a singular monotone $F(x) = F_\epsilon(x)$ for which the $c_n = c_n(F)$ are small in the sense of (5), it is sufficient to prove the existence of a singular monotone $G(x) = G_\epsilon(x)$ for which the $c_n = c_n(G)$ are small on the *average*, in the sense that

(7) $$\sum_{n=-N}^{N} | c_n | = O(N^{\epsilon}) \qquad (N \to +\infty).$$

(ii) It will be shown (§ **5**) that an $F(x)$, $-\pi \leqq x \leqq \pi$, which satisfies (5) can always be derived from a $G(x)$, $-\pi \leqq x \leqq \pi$, which satisfies (7), the passage from G to F consisting of an elementary one-to-one continuous transformation of the interval $-\pi \leqq x \leqq \pi$ into itself. It is only through this transformation that (5) becomes satisfied, since before this passage from G to F not even (3) is satisfied (cf. (29), § **7**). The effect of this transformation on the Fourier-Stieltjes coefficients may be estimated by means of the second mean value theorem, which introduces into (5) the exponent $-\frac{1}{2}$ of (6).

(iii) The theory of symmetric Bernoulli convolutions (cf. Jessen and Wintner [3], Kershner and Wintner [4], and Wintner [10], where further references are given) implies that the construction of a G which satisfies (7) can be based (§ **7**) on suitable estimations of the function

(8) $$L(u) = \prod_{k=1}^{\infty} \cos (3^{-k}u) \qquad (-\infty < u < +\infty)$$

for large $| u |$. Since $\cos v = 1 + O(v^2)$ as $v \to 0$, the product (8) defines an entire function of the exponential type. Actually, (8) is the Fourier-Stieltjes transform of the symmetric Bernoulli convolution which defines the classical Cantor function (cf. Carleman [1], pp. 223-226). Thus, while singular Bernoulli convolutions will occur, they will be used for quite another purpose than in the examples of Menchoff [8] and Kershner [6], examples which are in themselves singular Bernoulli convolutions and satisfy (3).

(iv) The remaining step, namely that estimate of (8) which was mentioned under (iii), will be made (§ **6**) by an adaptation of an enumeration process which is familiar in a similar, though easier, instance; cf. Wiener [9].

All of this will apply also for Fourier-Stieltjes transforms (§ **8**), instead of trigonometrical moments; so that in (1) the index n can be replaced by a continuous variable u.

3. Let $\xi = \xi(x)$, $-\pi \leqq x \leqq \pi$, be a strictly increasing continuous function such that $\xi(-\pi) = -\pi$, $\xi(\pi) = \pi$. Then, if $G(x)$, $-\pi \leqq x \leqq \pi$, is any continuous monotone function, so is the function $F(x)$, $-\pi \leqq x \leqq \pi$, which is defined by $G(x) = F(\xi(x))$, i. e., by $F(x) = G(\xi^{-1}(x))$. Clearly,

$$\int_{-\pi}^{\pi} e^{-imx} dF(x) = \int_{-\pi}^{\pi} e^{-imx} dG(\xi^{-1}(x)) = \int_{-\pi}^{\pi} e^{-im\xi(x)} dG(x).$$

Hence, on placing

(9) $$\lambda_{nm} = \frac{1}{2\pi}\int_{-\pi}^{\pi} e^{-im\xi(x)} e^{inx} dx, \qquad (m, n = 0, \pm 1, \cdots),$$

and developing the function $e^{-im\xi(x)}$, which is of bounded variation for every fixed m, into its Fourier series

$$e^{-im\xi(x)} = \sum_{n=-\infty}^{+\infty} \lambda_{nm} e^{-inx}, \qquad (m = 0, \pm 1, \cdots),$$

finally applying, for instance, a standard theorem of W. H. Young, one clearly obtains

(10) $$c_m(F) = \sum_{n=-\infty}^{+\infty} \lambda_{nm} c_n(G), \qquad (m = 0, \pm 1, \cdots),$$

if use is made of the notation (1).

Suppose, in particular, that the topological transformation $\xi = \xi(x)$ of the interval $[-\pi, \pi]$ into itself is such that the derivative of the function $\xi(x)$, or the distortion of the mapping $\xi = \xi(x)$, lies between two positive bounds. In this case it is clear that a monotone continuous function $G(x)$ is singular if and only if the same holds for the function $F(x) = G(\xi^{-1}(x))$.

4. Let

(11) $$\xi = \xi(x) \equiv \tfrac{1}{4}(3x + \tfrac{1}{\pi} x^2 \operatorname{sgn} x), \quad \text{where} \quad -\pi \leqq x \leqq \pi.$$

This $\xi(x)$ is continuous, strictly monotone and such that $\xi(-\pi) = -\pi$, $\xi(\pi) = \pi$. The coefficients (9) belonging to the mapping (11) are

(12) $$\lambda_{nm} = \frac{1}{\pi}\int_{0}^{\pi} \cos\{\tfrac{1}{4}m(3x + \tfrac{1}{\pi} x^2) - nx\} dx.$$

Hence, from (6),

(13) $$|\lambda_{nm}| < \text{Const.}\, |m|^{-\frac{1}{2}} \text{ for all } n, m.$$

It will be shown that

(14) $$|\lambda_{nm}| < \text{Const.}\, n^{-2} \text{ whenever } |n| > 2|m|.$$

First, it is easily verified by differentiation with respect to x that if $|n| > 2|m|$, the function $\{\ \}$ whose cosine is the integrand of (12) is strictly monotone for $0 \leqq x \leqq \pi$. Hence, on placing $\{\ \} = t$ and

(15) $$f_{nm}(t) = |(\tfrac{3}{4}m - n)^2 + \tfrac{1}{\pi} mt|^{-\frac{1}{2}},$$

the integral (12) can be written as

$$\lambda_{nm} = \pm\frac{1}{\pi}\int_0^{(m-n)\pi} f_{nm}(t)\cos t\,dt.$$

Consequently, on applying partial integration and noting that $\frac{1}{\pi} < 1$,

$$|\lambda_{nm}| \leqq \left|\int_0^{(m-n)\pi} f'_{nm}(t)\sin t\,dt\right|, \text{ where } ' = d/dt.$$

Now, it is easily verified from the assumption $|n| > 2|m|$ that the derivative, $f'_{nm}(t)$, of the function (15) is monotone between $t = 0$ and $t = (m-n)\pi$. Consequently, application of the second mean value theorem to the last integral gives

$$|\lambda_{nm}| \leqq 2|f'_{nm}(0)| + 2|f'_{nm}((m-n)\pi)|.$$

Hence, on calculating $f'_{nm}(0)$ and $f'_{nm}((m-n)\pi)$ from (15) and then using the assumption $|n| > 2|m|$, the appraisal (14) follows.

Incidentally, (13) also is a consequence of the second mean value theorem, since so is (6).

5. Writing (10) in the form

$$c_m(F) = \left\{\sum_{n=-\infty}^{-2|m|-1} + \sum_{n=-2|m|}^{2|m|} + \sum_{n=2|m|+1}^{+\infty}\right\}\lambda_{nm}c_n(G),$$

(13) and (14) imply that, as $m \to \pm\infty$,

$$c_m(F) = O\left(\sum_{n=-\infty}^{-2|m|} n^{-2}\right) + O\left(\sum_{n=-2|m|}^{2|m|} |c_n(G)|\, m^{-\frac{1}{2}}\right) + O\left(\sum_{n=2|m|}^{+\infty} n^{-2}\right),$$

since $|c_n(G)| \leqq c_0(G) = \text{const.}$, by (1). Accordingly,

$$c_m(F) = O(|m|^{-1}) + |m|^{-\frac{1}{2}}\, O\left(\sum_{n=-2|m|}^{2|m|} |c_n(G)|\right) \text{ as } m \to \pm\infty.$$

Hence, if $G(x)$ satisfies (7) for a fixed ϵ, then $F(x)$ satisfies (5) for the same ϵ. Furthermore, application of the last remark of § **3** to the mapping (11) shows that both monotone functions $F(x)$, $G(x)$ are singular if either of them is. This proves (i)-(ii), § **2**.

6. Next, (iv), § **2** will be established, by showing that there exists for every $\epsilon > 0$ a sufficiently large positive integer $p = p_\epsilon$ such that for the function (8) one has

$$\sum_{n=-N}^{N} |L(2\pi n)|^p = O(N^\epsilon) \text{ as } N \to +\infty, \tag{16}$$

and even

$$\sum_{n=-N}^{N} |L(\alpha n)|^p = O(N^\epsilon) \text{ uniformly for } \pi \leqq \alpha \leqq 2\pi, \text{ as } N \to +\infty. \tag{17}$$

First, if n is a positive integer, let $\nu = \nu_n$ denote the number of digits in the triadic representation of n; so that

$$n = \delta_{\nu+1}\delta_\nu \cdots \delta_k \cdots \delta_2\delta_1, \qquad (\nu = \nu_n), \tag{18}$$

where every δ_k is either 0 or 1 or 2, while $\delta_\nu \neq 0$ and $\delta_{\nu+1} = 0$. Clearly, the fractional part, $\beta - [\beta]$, of $\beta = 3^{-k}n$ is

$$0.\,\delta_k\delta_{k-1} \cdots \delta_2\delta_1 \qquad (k = 1, \cdots, \nu_n).$$

Hence, on considering the six possible pairs

$$\{\delta_k; \delta_{k-1}\} = \{0;1\},\ \{1;0\},\ \{1;2\},\ \{2;1\},\ \{2;0\},\ \{0;2\},$$

it is easily verified that the fractional part of $3^{-k}n$ either lies between $1/9$ and $4/9$ or between $5/9$ and $8/9$. Consequently, in both cases,

$$|\cos(3^{-k}2\pi n)| \leqq \cos(\pi/9) < \exp(-1/100).$$

Since this holds for $k = 1, \cdots, \nu_n$, and since every factor of the product (8) lies between -1 and 1, it follows that

$$|L(2\pi n)| < \exp(-\mu_n/100),$$

where μ_n denotes the number of changes of the digits δ in the finite sequence (18), i. e., the number of those k for which $\delta_{k+1} \neq \delta_k$. Thus, if $p > 0$,

$$|L(2\pi n)|^p < \exp(-p\mu_n/100). \tag{19}$$

Now, if m is a positive integer, the number of those among the first N positive integer n which have exactly m changes of digits in their triadic representation (18), i. e., the number of those n for which $1 \leqq n \leqq N$ and $\mu_n = m$, cannot exceed

$$(6\nu_N)^m/m!,$$

six being the number of distinct pairs of digits 0, 1, 2. It follows, therefore, from (19) that if the positive integer p is fixed and $N \to +\infty$, then

$$\sum_{n=1}^{N} |L(2\pi n)|^p = O\{\sum_{m=0}^{\infty} \exp(-pm/100) \cdot (6\nu_N)^m/m!\}.$$

Since (18) clearly implies that

$$\nu_n = O(\log n) \quad \text{as } n \to +\infty,$$

it follows that, if p is fixed and $N \to +\infty$,

$$\sum_{n=1}^{N} | L(2\pi n) |^p = \sum_{m=0}^{\infty} \{\exp(-p/100) \cdot O(\log N)\}^m/m!\,.$$

Consequently, there exists for every $\epsilon > 0$ a sufficiently large integer $p = p_\epsilon > 0$ such that, as $N \to +\infty$,

$$\sum_{n=1}^{N} | L(2\pi n) |^p = O\{\sum_{m=0}^{\infty} (\epsilon \log N)^m/m!\} = O\{\exp(\epsilon \log N)\},$$

and so, by (8),

$$\sum_{n=-N}^{N} | L(2\pi n) |^p = 1 + 2O\{\exp(\epsilon \log N)\} = O(N^\epsilon).$$

This proves (16). And (17) is proved in the same way, if, starting with an arbitrary positive n which need not be an integer, one applies (18) to the integral part $[n]$ of n, instead of to n. Then (19) remains valid if the subscript n of μ_n is replaced by $[n]$; so that the balance of the above proof can be repeated without change.

7. All that remains to be done is what is assigned by (iii), § **2**.

If $H = H(x)$, $-\infty < x < +\infty$, is a distribution function, (i.e., a monotone function such that $H(-\infty) = 0$, $H(+\infty) = 1$), let $\Lambda(u; H)$ denote its Fourier-Stieltjes transform,

$$\Lambda(u; H) = \int_{-\infty}^{+\infty} e^{iux} dH(x); \qquad -\infty < u < +\infty. \tag{20}$$

Then

$$\Lambda(u; H_{\mathrm{I}} * H_{\mathrm{II}}) = \Lambda(u; H_{\mathrm{I}})\Lambda(u; H_{\mathrm{II}}), \tag{21}$$

if $H_{\mathrm{I}} * H_{\mathrm{II}}$ denotes the convolution

$$\int_{-\infty}^{+\infty} H_{\mathrm{I}}(x - y) dH_{\mathrm{II}}(y), \tag{22}$$

which is a distribution function for any pair H_{I}, H_{II} of distribution functions. It is implied by (21) that

$$H_{\mathrm{I}} * H_{\mathrm{II}} = H_{\mathrm{II}} * H_{\mathrm{I}}, \qquad (H_{\mathrm{I}} * H_{\mathrm{II}}) * H_{\mathrm{III}} = H_{\mathrm{I}} * (H_{\mathrm{II}} * H_{\mathrm{III}}),$$

and that, for any positive integer p,

(23) $$\Lambda(u; H^{(p)}) = (\Lambda(u; H))^p,$$

if, starting with an H, one puts

(24) $$H^{(1)} = H, \quad H^{(2)} = H^{(1)} * H, \quad H^{(3)} = H^{(2)} * H, \cdots.$$

Let, in particular, $H(x)$ be the distribution function which is 0 if $x \leqq -\frac{1}{2}$ and 1 if $x \geqq \frac{1}{2}$, while if $|x| \leqq \frac{1}{2}$, then $H(x)$ is the classical Cantor function, obtained in the usual manner by successive trisections of the interval $-\frac{1}{2} \leqq x \leqq \frac{1}{2}$. Then $H(x)$ is, for $-\infty < x < +\infty$, the infinite convolution

$$H(x) = B(3x) * B(3^2x) * \cdots * B(3^kx) * \cdots,$$

where $B(x)$ is the distribution function which has a jump of $\frac{1}{2}$ at $x = -1$ and $x = 1$; so that $\Lambda(u; B) = \cos u$. Correspondingly,

(25) $$\Lambda(u; H) = \prod_{k=1}^{\infty} \cos(3^{-k}u).$$

Furthermore, from (24),

(26) $$H^{(p)}(x) = B(a_{1p}x) * B(a_{2p}x) * \cdots * B(a_{kp}x) * \cdots$$

where

(27) $$a_{kp} = 3^j, \text{ if } (j-1)p < k \leqq jp, \text{ where } j = 1, 2, \cdots.$$

Also, from (23), (25) and (8),

(28) $$\Lambda(u; H^{(p)}) = (L(u))^p.$$

Since it is clear from (8) that $L(3^h \cdot 2\pi)$ is independent of h for $h = 1, 2, \cdots$, and that $L(6\pi) > 0$, it follows from (28) that

(29) $$\Lambda(3^h \cdot 2\pi; H^{(p)}) = (L(6\pi))^p > 0 \text{ for } h = 1, 2, \cdots.$$

Now, the distribution function $H(x)$ is, if $-\frac{1}{2} \leqq x \leqq \frac{1}{2}$, the Cantor function, so that $H(-\frac{1}{2}) = 0$, $H(\frac{1}{2}) = 1$, and so $H(x)$ is continuous for $-\infty < x < +\infty$. Hence it is clear from the definition (20) of a convolution that each of the iterated functions (24) is continuous for $-\infty < x < +\infty$. On the other hand, $H^{(p)}(x)$ is, in view of (26), an infinite convolution of step functions. Consequently (cf. Jessen and Wintner [3], Theorem 35), the distribution function $H^{(p)}(x)$ either is singular or absolutely continuous. Since (29) assures that $\Lambda(u; H^{(p)}) \to 0$, $u \to \pm\infty$, is impossible for a fixed p, it follows that $H^{(p)}(x)$ is singular for every p.

Furthermore, $H(x)$ is constant outside the interval $-\frac{1}{2} \leqq x \leqq \frac{1}{2}$. Hence it is clear from (27) and from the definition (22) of a convolution that $H^{(p)}(x)$ is constant outside the interval $-\frac{1}{2}p \leqq x \leqq \frac{1}{2}p$. Accordingly, by (20),

$$\Lambda(u; H^{(p)}) = \int_{-\frac{1}{2}p}^{\frac{1}{2}p} e^{iux} dH^{(p)}(x).$$

It follows, therefore, from (28) and (17) that there exists for every $\epsilon > 0$ a sufficiently large $p = p_\epsilon$ such that, as $N \to +\infty$,

$$\sum_{n=-N}^{N} \left| \int_{-\frac{1}{2}p}^{\frac{1}{2}p} e^{ianx} dH^{(p)}(x) \right| = O(N^\epsilon)$$

holds uniformly for $\pi \leqq \alpha \leqq 2\pi$.

Thus, on keeping ϵ fixed (hence also $p = p_\epsilon$), and then changing the unit along the x-axis in such a way that the interval $-\frac{1}{2}p \leqq x \leqq \frac{1}{2}p$ becomes the interval $-\pi \leqq x \leqq \pi$, one has

$$(30) \qquad \sum_{n=-N}^{N} \left| \int_{-\pi}^{\pi} e^{inx} dG(x) \right| = O(N^\epsilon),$$

if $G = G_\epsilon$ denotes the monotone function which results from $H^{(p)}$, where $p = p_\epsilon$, upon the change of units. Since $H^{(p)}$ is singular, so is G. Hence, comparison of (30) and (1) assures the existence of a singular monotone G which satisfies (7).

This completes the proof of (5), i. e., of $\kappa^* = \frac{1}{2}$.

8. While the Cantor function $H(x)$ is constant almost everywhere, (27) implies that the singular monotone function (26) is, if $p > 1$, nowhere constant on an interval (cf. Kershner and Wintner [4], p. 547). Consequently, the above result implies one recently proved by Hartman and Kershner [2].

Actually, the above constructions can easily be extended to the case in which the sequence of Fourier-Stieltjes coefficients is replaced by the corresponding Fourier-Stieltjes transform. To this end, all that is needed is to replace in (10) the summation by an integration, and then, with the help of (17), to repeat the above proof with obvious modification. Then it follows that there exists for every $\epsilon > 0$ a singular monotone function $F(x)$, $-\pi \leqq x \leqq \pi$, such that

$$(31) \qquad \int_{-\pi}^{\pi} e^{iux} dF(x) = O(|u|^{-\frac{1}{2}+\epsilon}) \quad \text{as} \quad u \to \pm\infty,$$

where u is a continuous variable. It is also seen that (31) can be chosen as an even function, i. e., such that the curve $y = F(x)$ is symmetric with respect to the line $y = F(0)$.

MASSACHUSETTS INSTITUTE OF TECHNOLOGY,
THE JOHNS HOPKINS UNIVERSITY.

REFERENCES.

[1] T. Carleman, *Sur les équations intégrales singulières à noyau reél et symétrique*, Uppsala, 1923.

[2] P. Hartman and R. Kershner, "On the Fourier-Stieltjes transform of a singular function," *American Journal of Mathematics*, vol. 60 (1938), pp. 459-462.

[3] B. Jessen and A. Wintner, "Distribution functions and the Riemann zeta function," *Transactions of the American Mathematical Society*, vol. 38 (1935), pp. 48-88.

[4] R. Kershner and A. Wintner, "On symmetric Bernoulli convolutions," *American Journal of Mathematics*, vol. 57 (1935), pp. 541-548.

[5] R. Kershner, "Determination of a van der Corput-Landau absolute constant," *ibid.*, vol. 57 (1935), pp. 840-846.

[6] R. Kershner, "On singular Fourier-Stieltjes transforms," *ibid.*, vol. 58 (1936), pp. 450-452.

[7] J. E. Littlewood, "On the Fourier coefficients of functions of bounded variation," *Quarterly Journal of Mathematics*, vol. 7 (1936), pp. 219-226.

[8] D. Menchoff, "Sur l'unicité du développement trigonométrique," *Comptes Rendus*, vol. 163 (1916), pp. 433-436.

[9] N. Wiener, *The Fourier Integral and Certain of Its Applications*, Cambridge, 1933, p. 153.

[10] A. Wintner, "On a class of Fourier transforms," *American Journal of Mathematics*, vol. 58 (1936), pp. 45-90.

ON SINGULAR DISTRIBUTIONS

By Norbert Wiener and Aurel Wintner

Let $0 < a < 1$. Then $L_a(u) = \prod_{n=1}^{\infty} \cos(a^n u)$ is convergent for $-\infty < u < +\infty$. It is known that $L_a(u)$ is the Fourier-Stieltjes transform of a symmetric Bernoulli convolution $\sigma_a = \sigma_a(x)$, $-\infty < x < +\infty$; that the distribution function $\sigma_a(x)$ has no discontinuities and is constant for $|x| > a/(1-a)$, but not for $|x| > -\epsilon + a/(1-a)$; that $\sigma_a(x)$ is either nowhere or almost everywhere constant on the interval $|x| \leq a/(1-a)$ according as $a \geq \frac{1}{2}$ or $a < \frac{1}{2}$. It is not known whether $\sigma_a(x)$ is absolutely continuous for every $a \geq \frac{1}{2}$. All that is clear is that $\sigma_a(x)$ is absolutely continuous for $a = 2^{-1/\nu}$ $(\nu = 1, 2, \cdots)$, and that it acquires derivatives of arbitrarily high order as $\nu \to +\infty$. This follows from the multiplication rule of Fourier-Stieltjes transforms, since $L_{\frac{1}{2}}(u)$ is Vieta's product for $\sin u/u$, which is $O(|u|^{-1})$ as $u \to \pm\infty$.

Thus $\sigma_{\frac{1}{2}}(x)$ is linear $(= \frac{1}{2}x + \frac{1}{2})$ on its range $|x| \leq 1 = a/(1-a)$. Let $a < \frac{1}{2}$. Then[1] $\sigma_a(x)$ is of the Cantor type, $\sigma_a(x)$ being constant on a dense set of subintervals of $|x| \leq a/(1-a)$. In particular, $\sigma_{\frac{1}{3}}(x)$ is the classical Cantor function on $|x| \leq 2$. Kershner[2] has shown that, while the Fourier-Stieltjes transform $L_a(u)$ does not tend to 0 if $a = 1/q$, where $q = 3, 4, 5, \cdots$, we have $L_a(u) = O([\log |u|]^{-b})$, if a is a rational number p/q, not the reciprocal of an integer, while b is a positive function of p and q. Nothing is known about $L_a(u)$ if a is irrational. On the other hand, as $a \to \frac{1}{2} - 0$, the singular function $\sigma_a(x)$ tends to the linear function $\sigma_{\frac{1}{2}}(x)$.

We shall show in §1 that, no matter what the arithmetical structure of a may be, we shall have for every fixed $a < \frac{1}{2}$

$$\frac{1}{2u}\int_{-u}^{u} (L_a(v))^2\,dv = O\left(|u|^{\frac{\log 2}{\log a}}\right) \qquad (u \to \infty).$$

Thus the possible irregularities are smoothed out by averaging with respect to u.

Upon choosing a sufficiently near to $\frac{1}{2}$, it follows that there exists for

[1] R. Kershner and A. Wintner, Amer. Journ. Math. 57 (1935), 541–548.

[2] R. Kershner, Amer. Journ. Math. 58 (1936), 450–452.

every $\epsilon > 0$ a continuous distribution function which is constant almost everywhere and has a Fourier-Stieltjes transform which is $O(|u|^{\epsilon-1})$ on the average. Combining this with a device applied recently by the authors,[3] it follows that there exists for every $\epsilon > 0$ a monotone function $f(x)$, defined over the range $-\pi \leq x \leq \pi$, of the Cantor type, and with Fourier-Stieltjes constants $c_n = \bar{c}_{-n}$ such that $c_n = O(n^{-\frac{1}{2}+\epsilon})$. The functions of the previous paper, while singular, are not of the Cantor type, since they are not almost everywhere constant.

In §2, we shall apply the method of §1 to the weakly convergent products $\prod(1 + \cos q_n x)$, instead of to the Fourier-Stieltjes transforms $\prod \cos (a^n u)$. We shall suppose for the sake of simplicity that the q_n are positive integers. It is known[4] that if $q_{n+1}/q_n > \lambda$ holds for a fixed $\lambda > 3$,

$$\varphi_m(x) = -x + \int_0^x \prod_{n=1}^{m} (1 + \cos q_n y)\, dy$$

tends as $m \to \infty$ to a singular function $\varphi(x) = \varphi(x + 2\pi)$. We shall show that in order to assure the existence of a limit function $\varphi(x)$ which shall be continuous, of bounded variation, but not absolutely continuous, we may replace $\lambda > 3$ by $\lambda > 2$.

Corresponding to the fact that $L_a(u)$ becomes absolutely continuous for $a = \frac{1}{2}$, it is to be expected that the result $\lambda > 2$ is the best possible of its kind, so that $\lambda = 2$ is not admissible. This is actually the case. In fact, we shall verify in §3 that if $q_n = 2^n$, $\varphi(x)$ exists but is discontinuous.

In §4, we shall point out a parallelism between the results of §§2 and 3 and a celebrated Gegenbeispiel of Bohr.[5] Bohr's Gegenbeispiel is defined by a recursion formula tantamount to an infinite convolution, and hence its Fourier transform is defined by an infinite product, of which we can give the factors explicitly. This will be a weakly convergent infinite product not unlike those which we have discussed in the last two paragraphs.

§5 is devoted to the filling in of a lacuna in Wiener's general theory of spectral functions.[6] According to this theory, every autocorrelation of a function possessed of a spectrum, upon suitable choice of units,

[3] N. Wiener and A. Wintner, Amer. Journ. Math. 60 (1938), 513–532.

[4] Cf. p. 130 and p. 294 of A. Zygmund's Trigonometrical Series (1935), where further references are given.

[5] H. Bohr, Acta Math. 45 (1925), 112–117.

[6] N. Wiener, Acta Math. 55 (1931), 117–258, Chap. II.

is the Fourier-Stieltjes transform of a distribution function. It is quite obvious that if a distribution function is absolutely continuous, or if it is purely discontinuous, it may be obtained as the function of which some autocorrelation is the Fourier-Stieltjes transform. Up to the present, the only published example of an autocorrelation with singular Fourier-Stieltjes transform is one given by Mahler.[7] We shall show in this paragraph that the problem has nothing to do with the structure of particular singular distribution functions, since *every* given distribution function belongs to the spectrum analysis of the autocorrelation of a suitable chosen function. In this sense, the result of §5 is that Wiener's theory of spectrum analysis is identical with the theory of the Fourier-Stieltjes transforms of distribution functions. The proof will be based on a diagonal process which is made possible by the continuity theorem of P. Lévy.[8]

Finally, while the result of §5 makes the exhibition of specific singular spectra superfluous from the point of view of existence proofs, we shall nevertheless devote §6 to the construction of one special example, with a theory simpler than that of Mahler, and closely allied to the theory of infinite convolutions.

§1. Let $0 < a < \frac{1}{2}$ and $u \geq 0$. Then if $L_a(u) = \prod_{k=1}^{\infty} \cos(a^k u)$, we have for every n,

$$\int_0^u [L_a(v)]^2\, dv \leq \int_0^u \left[\prod_{k=1}^{n} \cos(a^k v)\right]^2 dv$$

$$= 2^{-2n} \int_0^u \left| \prod_{k=1}^{n} (e^{ia^k v} + e^{-ia^k v}) \right|^2 dv.$$

Hence

$$\int_0^u [L_a(v)]^2\, dv \leq 2^{-2n} \int_0^u |\sum \exp\{i(\pm a \pm a^2 \pm \cdots \pm a^n)v\}|^2\, dv.$$

Now write the integrand $|\sum|^2 = \sum\overline{\sum}$ as a linear combination of terms $e^{ib\nu}$, and count the number of terms which belong to the same frequency b. Let us notice that $\sum_{k+1}^{\infty} a^j = \dfrac{a^{k+1}}{1-a} = a^k \dfrac{a}{1-a} > a^k$. On

[7] K. Mahler, Journ. Math. Phys. Mass. Inst. Technology 6 (1927), 158–164.
[8] P. Lévy, C. R. 175 (1922), 854–856.

integrating between $v = 0$ and $v = u$, we obtain from e^{ibv} the divisor b or the factor u according as $b \neq 0$ or $b = 0$. Thus it is clear that

$$\int_0^u [L_a(v)]^2\,dv < \frac{\text{const.}}{2^{2n}}\left(2^n u + \sum_{k=0}^{n} \frac{2^{n+k}}{a^{n-k}}\right).$$

This may be written as

$$\int_0^u [L_a(v)]^2\,dv < \text{const.}\left(\frac{u}{2^n} + \frac{1}{(2a)^n} \sum_{k=0}^{n} (2a)^k\right).$$

Now let us again make use of the assumption that $a < \frac{1}{2}$. We have $\sum_{k=0}^{\infty} (2a)^k < +\infty$, and thus

$$\int_0^u [L_a(v)]^2\,dv < \text{const.}\left(\frac{u}{2^n} + \frac{1}{(2a)^n}\right)$$

for every $u \geq 1$ and $n \geq 1$. Choose $n = \left[\dfrac{\log u}{\log 1/a}\right]$ for every u. Then if we let $u \to \infty$, it follows that

$$\int_0^u [L_a(v)]^2\,dv = O\left(u^{\frac{\log 2a}{\log a}}\right) + O\left(u^{\frac{\log 2a}{\log a}}\right)$$

for every fixed $a < \frac{1}{2}$. However, $L_a(u) = L_a(-u)$, so that

$$\frac{1}{2u}\int_{-u}^u [L_a(v)]^2\,dv = O\left(u^{\frac{-\log 2}{\log 1/a}}\right),$$

as we have stated in the introduction.

Apparently O cannot be replaced by o. Actually $\dfrac{\log 2}{\log 1/a}$ is precisely[9] the Lipschitz exponent of the distribution function $\sigma_a(x)$ whose Fourier-Stieltjes transform is $L_a(u)$.

§2. Let $q_1, \cdots, q_n, \cdots$ be a sequence of positive integers. Put

$$\varphi_n(x) = -x + \int_0^x \prod_{k=1}^{n} (1 + \cos q_k y)\,dy,$$

so that $\varphi_n(x)$ is a trigonometrical polynomial and $\varphi_n(x) + x$ a strictly increasing function for $-\infty < x < \infty$. Clearly,

$$\varphi_{n+1}(x) - \varphi_n(x) = \int_0^x (\cos q_{n+1} y) \prod_{k=1}^{n} (1 + \cos q_k y)\,dy.$$

[9] F. Hausdorff, Math. Annalen 79 (1918), 157–179; cf. R. Kershner and A. Wintner, loc. cit. 1.

Now suppose that $q_{n+1}/q_n \geq \lambda$ for some fixed $\lambda > 2$. Then

$$\prod_{k=1}^{n} (1 + \cos q_k y) = \sum_{j=1}^{m} p_j \cos r_j y,$$

where m and r_j are non-negative integers and the p_j are non-negative numbers which depend on n in such a way that

$$\sum_{j=1}^{m} p_j = 2^n \quad \text{and} \quad r_j \leq \sum_{k=1}^{n} q_k \leq q_{n+1} \frac{1}{\lambda - 1},$$

$1/(\lambda - 1)$ being obtained by using the geometrical progression as majorant. Thus,

$$| \varphi_{n+1}(x) - \varphi_n(x) | < \text{const.} \sum_{j=1}^{m} \frac{p_j}{q_{n+1} - r_j},$$

where const. is independent of x and n. Furthermore, it is clear that

$$\sum_{j=1}^{m} \frac{p_j}{q_{n+1} - r_j} < \text{const} \sum_{n=1}^{m} \frac{p_n}{q_{n+1}} < \text{const} \left(\frac{2}{\lambda}\right)^n.$$

It follows, therefore, from $\lambda > 2$ that $\sum | \varphi_{n+1}(x) - \varphi_n(x) |$ is majorized by a fixed convergent geometrical progression; so that, as $n \to \infty$, $\varphi_n(x)$ tends uniformly to a limit function $\varphi(x)$.

This continuous $\varphi(x)$ has the period 2π and is of bounded variation. In fact, since every $\varphi_n(x) + x$ is continuous and monotone, while every $\varphi_n(x)$ has the period 2π, these properties carry over in the limit. Finally, $\varphi(x)$ cannot be absolutely continuous over $-\pi \leq x \leq \pi$. If it were, by the Riemann-Lebesgue lemma, its Fourier-Stieltjes coefficients would tend to 0. Actually,

$$\int_{-\pi}^{\pi} e^{ikx}\, d\varphi(x) = \lim_{n \to \infty} \int_{-\pi}^{\pi} e^{ikx}\, d\varphi_n(x)$$

for every k, while obviously

$$\frac{1}{\pi} \int_{-\pi}^{\pi} e^{ikx}\, d\varphi_n(x) \geq 1$$

whenever k is a q_j, for all large enough values of n.

§3. The assumption that the λ of the inequality $q_{n+1}/q_n \geq \lambda$ is greater than 2 can not be replaced by the milder assumption $\lambda > 2 - \epsilon$, nor even by $\lambda \geq 2$. As a Gegenbeispiel, let $q_n = 2^{n-1}$. Then, by complete induction, we may show that

$$\prod_{k=1}^{n} (1 + \cos q_k y) = \frac{\sin^2 (2^{n-1} y)}{2^n \sin^2 (y/2)}.$$

Thus, since $\int_0^{+\infty} \frac{\sin^2 y}{y^2}\, dy = \frac{\pi}{2}$, it may be proved (as in the Fejér theorem of which, in fact, this is a special case) that

$$\phi(x) = \lim_{n\to\infty} \phi_n(x) = -x + \lim_{n\to\infty} \int_0^x \prod_{k=1}^{n} (1 + \cos q_k y)\, dy$$

$$= -x + \lim_{n\to\infty} \int_0^x \frac{\sin^2 (2^{n-1}y)}{2^n \sin^2 (y/2)}\, dy = -x + \pi \operatorname{sgn} x \pmod{2\pi}.$$

The function $\varphi(x) + x$ is monotone, but is not continuous.

§4. Bohr[10] defines a "periodenartig" function as one which is connected with a positive absolute constant c in such a way that if ϵ is any positive number, there exists a translation number $\tau(\epsilon)$ pertaining to ϵ and exceeding c. He gives an example to show that such a function need not be almost periodic in his uniform sense. We shall show that his example is the Fourier-Stieltjes transform of a continuous but not absolutely continuous monotone function.

Bohr defines his example by the following recursion process: let $\tau_2, \tau_3, \cdots$ be a sequence of numbers such that

$$\tau_n > 2((n - 1)\tau_{n-1} + (n - 2)\tau_{n-2} + \cdots + 2\tau_2 + 1);\ \tau_2 > 2.$$

Put

$$f_1(x) = \begin{cases} 1 - |x| & \text{for } -1 \leqq x \leqq 1; \\ 0 & \text{for } |x| > 1; \end{cases}$$

$$f_n(x) = \sum_{m=n-1}^{1} \frac{n - m}{n} f_{n-1}(x + m\tau_n) + f_{n-1}(x) + \sum_{m=1}^{n-1} \frac{n - m}{n} f_{n-1}(x - m\tau_n).$$

Then the desired function is

$$f(x) = \lim_{n\to\infty} f_n(x),$$

which exists uniformly on every finite interval.

[10] H. Bohr, loc. cit. 5.

For simplicity, we shall assume that the numbers $\tau_2, \tau_3, \cdots$ are integers. Let us notice that if $\varphi_n(u)$ is the Fourier transform of $f_n(x)$,

$$\varphi_1(u) = \sqrt{\frac{2}{\pi}}\frac{1-\cos u}{u^2};$$

$$\varphi_n(u) = \varphi_{n-1}(u)\,\frac{\sin^2\dfrac{un\tau_n}{2}}{n\sin^2\dfrac{u\tau_n}{2}}.$$

Thus

$$\varphi_n(u) = \sqrt{\frac{2}{\pi}}\frac{1-\cos u}{u^2}\prod_{k=2}^{n}\frac{\sin^2\dfrac{uk\tau_k}{2}}{k\sin^2\dfrac{u\tau_k}{2}}.$$

The sum of the Fourier coefficients of $\sqrt{\dfrac{\pi}{2}}\dfrac{u^2\varphi_n(u)}{1-\cos u}$, which are all real and non-negative, will be

$$\prod_{k=2}^{n}\left(1+\frac{2(k-1)}{k}+\frac{2(k-2)}{k}+\cdots+\frac{2}{k}\right) = n!.$$

On the other hand, the coefficient of $e^{ium\tau_k}$ $(k \le n)$ will be $(n-m)/n$.

Let us put

$$\psi_n(u) = \int_0^u \sqrt{\frac{\pi}{2}}\frac{v^2\varphi_n(v)}{1-\cos v}\,dv.$$

We shall then have

$$\psi_{n+1}(u) - \psi_n(u) = \int_0^u \sqrt{\frac{\pi}{2}}\frac{v^2\varphi_n(v)}{1-\cos v}\left(\sum_{m=1}^{n-1}\frac{n-m}{n}\left(e^{ivm\tau_n}+e^{-ivm\tau_n}\right)\right)dv.$$

The recursive inequality between the τ_n's makes it clear that no term of this expression has a frequency less in absolute value than $\tau_n/2$. Since

$$\tau_n > 2^{n-1}(n-1)!$$

it follows that

$$|\psi_{n+1}(u) - \psi_n(u)| < \text{const.}\,\frac{(n+1)!}{2^{n-1}(n-1)!} < \text{const.}\, n^2 2^{-n}$$

This establishes the existence of

$$\psi(u) = \lim_{n\to\infty} \psi_n(u)$$

as a uniform limit over $-\infty < u < \infty$. The relations between the τ_n's also imply that over the range of frequencies between $-\tau_n$ and τ_n, the Fourier coefficients of $\psi_n(u)$ are the same as those of $\psi(u)$. Thus the average of the coefficients of $\psi(u)$ over this range is not greater than

$$\frac{n!}{2^n(n-1)!} = \frac{n}{2^n},$$

and the average of the coefficients over the infinite range can readily be shown to be 0. Consequently $\psi(u)$ is a continuous monotone function. Finally, this function cannot be absolutely continuous, since its Fourier coefficients are not $o(1/n)$. This reasoning follows the lines laid down in §2.

By the definition of $\varphi_n(u)$,

$$f_n(x) = \frac{1}{\sqrt{2\pi}} \int_{-\infty}^{\infty} \varphi_n(u) e^{-iux}\, du.$$

Hence, by the definition of $\psi_n(u)$,

$$f_n(x) = \frac{1}{2} \int_{-\infty}^{\infty} \frac{1 - \cos u}{u^2} e^{-iux}\, d\psi_n(u).$$

It follows from the continuity theorem of Lévy that

$$f(x) = \frac{1}{2} \int_{-\infty}^{\infty} \frac{1 - \cos u}{u^2} e^{-iux}\, d\psi(u).$$

If then we put

$$\Phi(u) = \sqrt{\frac{\pi}{2}} \int_{-\infty}^{u} \frac{1 - \cos v}{v^2}\, d\psi(v),$$

it is easy to show that $\Phi(u)$ is monotone and continuous, but not absolutely continuous. We shall have

$$f(x) = \frac{1}{\sqrt{2\pi}} \int_{-\infty}^{\infty} e^{-iux}\, d\Phi(u),$$

as stated in the introduction.

§5. One of the authors[11] has introduced the following theory of the spectrum as it occurs in optics: let $f(t)$ be a measurable function defined over the interval $-\infty < t < \infty$, and let

$$F(t) = \lim_{T\to\infty} \frac{1}{2T} \int_{-T}^{T} f(t + x)\bar{f}(x)\, dx$$

exist for every t, and be a continuous function of t. Then we shall have

$$F(t) = \int_{-\infty}^{\infty} e^{itu}\, d\sigma(u),$$

where $\sigma(u)$ is a monotone bounded function. This function $\sigma(u)$ is called the *spectrum* of $f(t)$. The question at issue is: what functions $\sigma(u)$ are spectra of suitable functions $f(t)$? We shall prove that every bounded non-increasing function is such a $\sigma(u)$.

First, if $\sigma(u)$ is the step-function increasing by a_n at the N points u_n, respectively, we can put

$$f(t) = \sum_1^N a_n^{\frac{1}{2}} e^{iu_n t}.$$

If $\sigma(u)$ is any bounded non-increasing function, which can be thought of as normalized by putting $\sigma(-\infty) = 0$, $\sigma(\infty) = 1$, let us write

$$\sigma_n(u) = \frac{1}{n}\, [n\sigma(u)].$$

This function $\sigma_n(u)$ is again monotone, lies between 0 and 1, and is in addition a step-function. Thus there exists a function $f_n(t)$ which has $\sigma_n(u)$ as its spectrum. Since $f_n(t)$ is almost periodic, there will then exist a quantity $A_n > 0$ such that if $B_n > A_n$, for all t

$$\left| \frac{1}{2B_n} \int_{-B_n}^{B_n} f_n(x + t)\bar{f}_n(x)\, dx - \lim_{T\to\infty} \frac{1}{2T} \int_{-T}^{T} f_n(x + t)\bar{f}_n(x)\, dx \right| < 2^{-n}.$$

Now let

$$f(t) = f_n(t) \quad \text{for} \quad \prod_1^{n-1} (A_k + k) \leq |t| < \prod_1^{n} (A_k + k)$$

and let

$$F_n(t) = \lim_{T\to\infty} \frac{1}{2T} \int_{-T}^{T} f_n(x + t)\bar{f}_n(x)\, dx$$

[11] Cf., e.g., Göttinger Nachrichten, 1925, 151–158; loc. cit. 6.

It is then an elementary result that

$$F(t) = \lim_{T\to\infty} \frac{1}{2T} \int_{-T}^{T} f(x+t)\bar{f}(x)\,dx$$

exists for every value of t, and is equal to

$$\lim_{n\to\infty} F_n(t) = \lim_{n\to\infty} \int_{-\infty}^{\infty} e^{itu'} d\sigma_n(u).$$

Since, by the definition of σ_n, $\sigma_n \to \sigma$ as $n \to \infty$, we have

$$F(t) = \int_{-\infty}^{\infty} e^{itu}\, d\sigma(u),$$

by the Lévy continuity theorem. This implies that $F(t)$ is continuous. This completes the proof.

§6. In this section, we shall give an example of a function $f(t)$ with a spectrum of the Cantor type. From the point of view of existence, this adds nothing to what has been proved in the last section, but the example is interesting on its own merits. In this example, we shall give in the first instance an array a_n with a spectrum of this sort. From the array theory, it is easy to derive a function $f(t)$ with a spectrum of the desired type, by putting

$$f(t) = a_{[t]}\,.$$

An array a_n $(-\infty < n < \infty)$ is said[12] to have a spectrum if

$$b_n = \lim_{N\to\infty} \frac{1}{2N+1} \sum_{m=-n}^{N} a_{m+n}\bar{a}_m$$

exists for every integer n, positive, negative, or zero. The spectrum will be given by the function

$$\sigma(u) = b_0 u + \sum_{-\infty}^{\infty}{}' \frac{b_n e^{inu}}{i_n},$$

which will be of limited total variation over every finite interval. In our case, we shall put

$$a_n = \prod_{v=2}^{\infty} 2^{\frac{1}{2}} \sin\left(\frac{\pi}{4} + \frac{2\pi n}{v!}\right).$$

[12] N. Wiener, Journ. Math. Phys. Mass. Inst. Technology 6 (1927), 145–157.

We shall show that $\{a_n\}$ has a spectrum $\sigma(u)$ which is continuous, and increases only over the set of points S given by the values of

$$2\pi\left(\frac{\alpha_2}{2!}+\frac{\alpha_3}{3!}+\cdots+\frac{\alpha_\nu}{\nu!}+\cdots\right),$$

$$\text{where}\qquad \alpha_2 = \pm 1, \alpha_3 = \pm 1, \cdots, \alpha_\nu = \pm 1, \cdots.$$

As this set is of the Cantor type and of zero measure, the desired result follows.

Not only does the product by which a_n is defined converge, but since

$$\sqrt{2}\sin\left(\frac{\pi}{4}+\frac{2n\pi}{\nu!}\right) = 1 + O(n/\nu!),$$

we have

$$|a_n| \le \prod_{\nu!<n} 2^{\frac{1}{2}} \prod_{\nu!\ge n} (1 + O(n/\nu!)) = O(1) \prod_{\nu!<n} 2^{\frac{1}{2}} = \exp O\left(\frac{\log n}{\log\log n}\right).$$

Let us put

$$a_{Mn} = \prod_{\nu=2}^{M} \sqrt{2}\sin\left(\frac{\pi}{4}+\frac{2n\pi}{\nu!}\right) = \prod_{\nu=2}^{M}\left(\frac{1-i}{2}e^{\frac{2\pi n i}{\nu!}} + \frac{1+i}{2}e^{\frac{-2\pi n i}{\nu!}}\right).$$

The frequencies of this expression are the numbers

$$B_{M\mu} = 2\pi\left(\frac{\alpha_2}{2!}+\frac{\alpha_3}{3!}+\cdots+\frac{\alpha_M}{M!}\right)$$

where each α_ν is ± 1, and the coefficient of $\exp(B_{M_\mu} in)$ is

$$A_{M\mu} = \left(\frac{1-i}{2^{\frac{1}{2}}}\right)^{\alpha_2+\cdots+\alpha_M} \cdot 2^{-M/2}.$$

Since

$$\frac{1-i}{2}+\frac{1+i}{2} = 1,$$

if μ_1 corresponds to the sequence $\alpha_1, \cdots, \alpha_{M-1}, \alpha_M$, and μ_2 to the sequence $\alpha_1, \cdots, \alpha_{M-1}, -\alpha_M$, while μ_3 corresponds to the sequence $\alpha_1, \cdots, \alpha_{M-1}$, we have

$$A_{M\mu_1} + A_{M\mu_2} = A_{M-1,\mu_3}.$$

Furthermore, the interval $(B_{M\mu_1}, B_{M\mu_2})$ contains no other $B_{M\mu}$. Thus the sum of the $A_{M\mu}$'s corresponding to the $B_{M\mu}$'s in the interval $(-\frac{1}{2}, x)$

contains not more than one term of absolute value $2^{-\frac{1}{2}}$, not more than one of absolute value $2^{-\frac{2}{2}}$, and so on; or in other words can not exceed $(1 - 2^{-\frac{1}{2}})^{-1}$ in modulus. An argument of exactly the same type will show that if we consider the interval (x, y), where

$$|x - y| \leq \frac{1}{\nu!} + \frac{1}{(\nu + 1)!} + \cdots,$$

the sum of the $A_{M\mu}$'s corresponding to the $B_{M\mu}$'s in this interval can not exceed

$$\frac{2^{\frac{\nu}{2}+1}}{2^{\frac{1}{2}} - 1}$$

in modulus.

Let it be noted that this bound does not depend on M. Let it also be noted that for all points x in any interval between intervals of the form

$$\frac{\alpha_2}{2!} + \frac{\alpha_3}{3!} + \cdots + \frac{\alpha_M}{M!} \pm \sum_{M+1}^{\infty} \frac{1}{\nu!},$$

the sum of the $A_{N\mu}$'s corresponding to $B_{N\mu}$'s in the interval $(-\frac{1}{2}, x)$ is independent of N for $N \geq M$. If we write $\varphi_N(x)$ for this sum, we thus see that it converges uniformly to a limit $\varphi(x)$ as $N \to \infty$. Thus for fixed ϵ, but uniformly in x,

$$\lim_{M \to \infty} \frac{1}{2\epsilon} \int_{-\frac{1}{2}}^{x} |\varphi_M(y + \epsilon) - \varphi_M(y - \epsilon)|^2 dy$$

$$= \frac{1}{2\epsilon} \int_{-\frac{1}{2}}^{x} |\varphi(y + \epsilon) - \varphi(y - \epsilon)|^2 dx.$$

Now let

$$\frac{1}{N!} - 2 \sum_{N+1}^{\infty} \frac{1}{\nu!} > 2\epsilon \geq \frac{1}{(N-1)!} - 2 \sum_{N}^{\infty} \frac{1}{\nu!}.$$

If we notice that

$$\frac{1-i}{2} + \frac{1+i}{2} = 1; \qquad \left(\frac{1-i}{2}\right)^2 + \left(\frac{1+i}{2}\right)^2 = 1;$$

we shall see that except over a set of points not exceeding

$$2^{N+4} \sum_{N+2}^{\infty} \frac{1}{\nu!} = O(2^{N+4}(N + 2)^{-N-2})$$

in measure, we shall have

$$| \varphi(x + \epsilon) - \varphi(x - \epsilon) |^2 = \sum | \varphi_N(A_{N\mu} + 0) - \varphi_N(A_{N\mu} - 0) |^2,$$

the sum being taken over all discontinuities of φ_N in $(x - \epsilon, x + \epsilon)$. Elsewhere our estimate of the sums of the $A_{M\mu}$'s may be used to give an upper estimate of $| \varphi(x + \epsilon) - \varphi(x - \epsilon) |$. Let us notice that

$$| \varphi_N(A_{N\mu} + 0) - \varphi_N(A_{N\mu} - 0) |^2 = 2^{-M}.$$

From this it follows that if $2\pi x$ lies in one of the intervals belonging to $\bar{S}$, then the double limit

$$\lim_{\substack{\epsilon \to 0 \\ M \to \infty}} \frac{1}{2\epsilon} \int_{-\frac{1}{2}}^{x} | \varphi_M(y + \epsilon) - \varphi_M(y - \epsilon) |^2 dy$$

exists. It is now not hard to prove that it will also exist for every x in $(-\frac{1}{2}, \frac{1}{2})$. Here we use the fact of the monotoneness of the approximations, and the fact that their maximum jumps tend to zero.

Now let us put

$$\sigma(u) = 2\pi \lim_{\substack{\epsilon \to 0 \\ M \to \infty}} \frac{1}{2\epsilon} \int_{-\frac{1}{2}}^{u/2\pi} | \varphi_M(y + \epsilon) - \varphi_M(y - \epsilon) |^2 dy$$

and let us take $\varphi_M(u)$ to be $2^{-M+1}\pi$ times the number of $A_{N\mu}$'s in $(-\frac{1}{2}, u/2\pi)$. We shall have

$$\sigma(u) = \lim_{M \to \infty} \sigma_M(u).$$

We shall easily be able to establish that

$$2\pi \lim_{\substack{\epsilon \to 0 \\ M \to \infty}} \frac{1}{2\epsilon} \int_{-\frac{1}{2}}^{\frac{1}{2}} | \varphi_M(y + \epsilon) - \varphi_M(y - \epsilon) |^2 \cos 2m\pi x \, dx = \int_{-\pi}^{\pi} \cos mu \, d\sigma(u).$$

Now,

$$a_{Mn} = \int_{-\frac{1}{2}}^{\frac{1}{2}} e^{2\pi n i x} \, d\varphi_N(x).$$

Thus

$$\int_{-\frac{1}{2}}^{\frac{1}{2}} e^{2\pi i n x} d\{\varphi_M(x + \epsilon) - \varphi_M(x - \epsilon)\} = a_{Mn}(e^{-2ni\pi\epsilon} - e^{2ni\pi\epsilon}),$$

if we define $\varphi(x)$ to be of period 1. Hence, on integrating by parts,

$$\int_{-\frac{1}{2}}^{\frac{1}{2}} (\varphi_M(x + \epsilon) - \varphi_M(x - \epsilon)) e^{2ni\pi x} dx = a_{Mn} \frac{\sin 2\pi n\epsilon}{n\pi}.$$

It follows that

$$\int_{-\pi}^{\pi} \cos mu \, d\sigma(u) = \lim_{\substack{M\to\infty \\ \epsilon\to 0}} \frac{\pi}{\epsilon} \sum_{-\infty}^{\infty} a_{M,m+n}\bar{a}_{Mn} \frac{\sin^2 2\pi n\epsilon}{n^2\pi^2}$$

$$= \lim_{\epsilon\to 0} \frac{\pi}{\epsilon} \sum_{-\infty}^{\infty} a_{m+n}\bar{a}_n \frac{\sin^2 2n\pi\epsilon}{n^2\pi^2}.$$

By a Tauberian theorem,[13] it follows that

$$\int_{-\pi}^{\pi} \cos mu \, d\sigma(u) = 2\pi \lim_{N\to\infty} \frac{1}{2N+1} \sum_{-N}^{N} a_{m+n}\bar{a}_n = 2\pi b_m,$$

and the array $\{a_n\}$ has a spectrum. We have

$$b_0 x + \sum_{-N}^{N}{}' \frac{b_n e^{inx}}{in} = b_0 x + 2\sum_{1}^{N} \frac{b_n}{n} \sin nx$$

$$= \frac{1}{2\pi} \int_{-\pi}^{\pi} \left\{ x + 2\sum_{1}^{N} \frac{\cos n\xi \sin nx}{n} \right\} d\sigma(\xi)$$

$$= \frac{1}{2\pi} \left\{ x + 2\sum_{1}^{N} \frac{\sin nx}{n} \right\} \{\sigma(\pi) - \sigma(-\pi)\}$$

$$+ \frac{1}{\pi} \int_{-\pi}^{\pi} \sigma(\xi) \sum_{1}^{N} \sin n\xi \sin nx \, d\xi.$$

The first part of this converges to 0, and the second part converges in the mean to $\sigma(u)$, which we have already seen to be a singular function.

MASSACHUSETTS INSTITUTE OF TECHNOLOGY,
THE JOHNS HOPKINS UNIVERSITY.

[13] N. Wiener, loc. cit. 6, §5.

Comments on [38c], [39c]

W. Rudin

These two papers deal with the following question: How rapidly can the Fourier coefficients $c_n = \hat{\mu}(n)$ of a singular measure μ on the unit circle tend to 0? That they can indeed tend to 0 was first shown by Menchoff (1916). In [38c] it is proved that for every $\varepsilon > 0$ there is a positive singular μ with

$$(*) \qquad c_n = O(n^{\varepsilon - 1/2}).$$

In [39c], a stronger result is claimed (p. 234): *for every $\varepsilon > 0$ there is a positive μ that satisfies $(*)$ and is of the "Cantor type"; i.e., μ is concentrated on a nowhere dense perfect set of Lebesgue measure* 0. The authors refer for this to the device used in [38c] ((i) and (ii) on pp. 514–515). This device consists in taking the image $\mu = dF$ of a measure $\lambda = dG$ carried by $[-\pi, \pi]$, by means of the homeomorphism ξ defined in (11), p. 516. The important property of ξ is that $\hat{\mu}(n) = O(n^{\varepsilon - 1/2})$, if

$$(7) \qquad \sum_{-N}^{N} |\hat{\lambda}(n)| = O(N^{\varepsilon}).$$

In [38c] the device is applied to $\lambda = \sigma * \sigma * \ldots * \sigma$, where σ is a Cantor-Lebesgue measure; i.e.,

$$\hat{\sigma}(n) = \prod_{j=1}^{\infty} \cos(a3^{-j}n).$$

This λ is no longer of the Cantor type. In [39c], Wiener and Wintner claim that it applies to the measure λ given by

$$\hat{\lambda}(u) = L_a(u) = \prod_{k=1}^{\infty} \cos a^k u$$

where $a < \frac{1}{2}$, because (p. 234) "the Fourier-Stieltjes transform is $O(|u|^{\varepsilon - 1})$ on the average." But what they actually prove (p. 236) is only that

$$\frac{1}{2u} \int_{-u}^{u} |L_a(v)|^2 \, dv = O(u^{\varepsilon - 1}),$$

hence that

$$\frac{1}{2u}\int_{-u}^{u}|\hat{\lambda}(v)|\,dv = O(u^{(\varepsilon-1)/2}),$$

which is very far from the needed relation (7). (This description of the error was communicated to the reviewer by J. P. Kahane.) The "Cantor type" result was proved later by Salem.[3]

In connection with the second paragraph on p. 233 of [39c], one should mention Salem's beautiful theorem: *$L_a(u)$ tends to 0 as $u \to \infty$, if and only if $1/a$ is not an algebraic integer all of whose conjugates have absolute value less than* 1. For a proof, see Salem's book[4] or the book by Kahane and Salem.[2] The latter contains a wealth of information on problems of this type.

The conjecture made on p. 514 of [38c] was proved by O. S. Ivasev-Musatov,[1] for sequences $\{\gamma_n\}$ that satisfy

$$\sum \gamma_n^2 = \infty, \quad n\gamma_n^2 \to 0, \quad n^{1+\varepsilon}\gamma_n^2 \to \infty, \quad \text{and} \quad n^k\gamma_n^2\uparrow,$$

for every $\varepsilon > 0$ and for some $k > 1/2$. Examples of such sequences are given by

$$\gamma_n = n^{-1/2}, \quad \gamma_n = (n\log n)^{-1/2}, \quad \gamma_n = (n\log n \operatorname{loglog} n)^{-1/2}, \ldots .$$

Without some regularity assumption on $\{\gamma_n\}$, the conjecture is actually false. For example, put $\gamma_n = (1 + |n|)^{-1}$ when $n \neq 2^k$ and put $\gamma_n = 1$ when $n = 2^k$. Then, $\Sigma\gamma_n^2 = \infty$. But if F is a function of bounded variation with $\hat{F}(n) = O(\gamma_n)$, then the measure μ induced by F satisfies

$$\hat{\mu}(n) = \hat{F}(n)/n = O(n^{-1}\gamma_n),$$

which implies that $\Sigma|\hat{\mu}(n)|^2 < \infty$. Thus F is the integral of an L^2-function, and is therefore absolutely continuous.

References

1. O. S. Ivasev-Musatov, Izv. Akad. Nauk SSSR, Ser. Mat. 21(1957), 559–578; AMS translation (2) 14(1960), 289–310.

2. J. P. Kahane and R. Salem, *Ensembles parfaits et séries trigonométriques*, Actualités Sci. Ind., Hermann, Paris, 1963.

3. R. Salem, J. Math. and Phys., MIT 21(1942), 69–82.

4. R. Salem, *Algebraic Numbers and Fourier Series*, Heath, Boston, 1963.

Comments on [38c]

E. Hewitt

This paper, remarkable for its era, has numerous ancestors and also numerous progeny. A general question, and an important one in harmonic analysis, is: How small can the Fourier-Stieltjes transform,

$$\hat{\mu}(n) = \int_{[-\pi,\pi]} \exp(-int)d\mu(t),$$

be for a continuous singular probability measure μ on the circle group $\mathbb{T}$? As early as 1916 Menchoff[11] showed that $\hat{\mu}(n)$ can be $O(\log(n)^{-1/2})$. Littlewood in 1936[10] showed that $\hat{\mu}(n)$ can be $o(n^{-\alpha})$ for some positive number α. (Littlewood's method could be used to show that α is larger than 1/5.) In the present paper Wiener and Wintner show that for every $\varepsilon > 0$ there is a μ such that $\hat{\mu}(n)$ is $o(n^{-1/2+\varepsilon})$. A simplified construction, with a single measure independent of ε, appears in Zygmund,[15] vol. 2, pp. 146–147. The paper gives a conjecture that $\hat{\mu}(n)$ can minorize any positive function on $\mathbb{Z}$ that is not in l_2. This was disproved by Hewitt and Ritter.[4]

The construction of Wiener and Wintner was later refined by Schaeffer,[13] by Wiener and Wintner themselves,[14] and by Salem in a long series of papers, see,[12] pp. 219–227, 239–251, 295–304, 316–333, 423–426, 481–493, 569–575.

A dramatic forward step was taken in 1952 by Ivašëv-Musatov.[7,8] He showed that $\hat{\mu}(n)$ can be very small indeed, subject to not being in l_2 and to certain regularity conditions. For example, $\hat{\mu}(n)$ can be $o((n \log(n) \log\log(n) \ldots \log_{(p)}(n))^{-1/2})$. The behavior of $\hat{\mu}(n)$ on the average was studied by Hewitt and Ritter,[3] continuing earlier work of Hewitt and Zuckerman.[5,6] Further refinements of the construction of Ivašëv-Musatov have been made by Körner.[9] Brown and Hewitt[1,2] recently extended the construction of Ivašëv-Musatov by different means to obtain still smaller transforms.

References

1. G. Brown and E. Hewitt, Proc. Nat. Acad. Sci. 75(1978), 5268–5269.

2. G. Brown and E. Hewitt, *Continuous singular measures with small Fourier-Stieltjes transforms* (forthcoming).

3. E. Hewitt and G. Ritter, Proc. Royal Irish Acad. Sci., sect. A, 77(1976), 265–287.

4. E. Hewitt and G. Ritter, Symposia Mathematica, Istituto Nazionale di Alta Matematica, Rome, 22(1977), 21–31.

5. E. Hewitt and H. S. Zuckerman, Proc. Cambridge Philos. Soc. 62(1966), 399–420.

6. E. Hewitt and H. S. Zuckerman, *Corrigendum*, Proc. Cambridge Philos. Soc. 63(1967), 367–368.

7. O. S. Ivašëv-Musatov, Doklady Akad. Nauk SSSR 82(1952), 9–11.

8. O. S. Ivašëv-Musatov, Izvestija Akad. Nauk SSSR, ser. mat. 21(1957), 559–578.

9. T. W. Körner, Ann. Inst. Fourier, Grenoble, 27(1977), 97–115.

10. J. E. Littlewood. Quarterly J. Math. (Oxford Series) 7(1936), 219–226.

11. D. E. Menchoff, C. R. Acad. Sci., Paris, 163(1916), 433–436.

12. R. Salem, *Oeuvres Mathématiques*, Hermann, Paris, 1967.

13. A. C. Schaeffer, Amer. J. Math. 61(1939), 934–940.

14. N. Wiener and A. Wintner, J. Math. Phys., MIT, 17(1938), 233–246.

15. A. Zygmund, *Trigonometric Series*, 2d ed., vols. 1, 2, Cambridge University Press, Cambridge, England, 1959.

CONVERGENCE PROPERTIES OF ANALYTIC FUNCTIONS OF FOURIER-STIELTJES TRANSFORMS*

BY

ROBERT H. CAMERON AND NORBERT WIENER

1. **Introduction.** Wiener and Pitt† have given conditions under which the reciprocal of an absolutely convergent Fourier-Stieltjes integral is again an absolutely convergent Fourier-Stieltjes integral. It is the purpose of this paper to generalize this result in two directions. We replace reciprocals by general analytic functions which may even be multiple-valued, and we replace absolute convergence by finiteness of certain more general norms. These norms are of two types, both of which depend on a parameter θ, $(0<\theta\leqq 1)$; and both reduce to total variation when $\theta=1$.

Following the notation of (WP), we let $f(x)$ denote a function of bounded variation in $(-\infty, \infty)$ for which $2f(x)=f(x+0)+f(x-0)$, and let

$$F(x) = \int_{-\infty}^{\infty} e^{-iyx}df(y).$$

We write

$$f(x) = h(x) + g(x) + s(x),$$

where $h(x)$ is a step-function, $g(x)$ is absolutely continuous, and $s(x)$ is continuous and has a zero derivative almost everywhere. We refer to $h(x)$, $g(x)$, $s(x)$ as the discrete, smooth, and singular parts of $f(x)$, and to their Fourier-Stieltjes transforms $H(x)$, $G(x)$, $S(x)$ as the almost periodic, transient, and unpredictable parts of $F(x)$; of course we have $F(x)=H(x)+G(x)+S(x)$. Moreover $h(x)$, $g(x)$, $s(x)$ are each of bounded variation and essentially uniquely determined by $f(x)$, while $H(x)$, $G(x)$, $S(x)$ are uniquely determined by $F(x)$.

We define for $0<\theta\leqq 1$,

$$T_\theta^*\{F(x)\} = 2\sum_{n=-\infty}^{\infty}\left[\int_n^{n+1} | df(y) |\right]^\theta$$

and‡

* Presented to the Society, December 28, 1938; received by the editors January 21, 1939.

† *On absolutely convergent Fourier-Stieltjes transforms*, Duke Mathematical Journal, vol. 4 (1938), pp. 420–436. This paper will be referred to as (WP).

‡ We use the symbol $\int |dh(y)|^\theta$ to mean the sum of the θ powers of the jumps of $h(y)$. By a jump we mean the whole jump $|h(y+0)-h(y-0)|$, not a half jump $|h(y+0)-h(y)|$.

$$T_\theta^{**}\{F(x)\} = 2\int_{-\infty}^{\infty} | dh(y) |^\theta + 2\sum_{n=-\infty}^{\infty}\left[\int_n^{n+1} | dg(y) + ds(y) |\right]^\theta;$$

and we say that $F(x) \varepsilon A_\theta^*$ *or* $F(x) \varepsilon A_\theta^{**}$ *if* $T_\theta^*[F(x)] < \infty$ *or* $T_\theta^{**}[F(x)] < \infty$. *We use the symbol* T_θ *to stand for* T_θ^* *or* T_θ^{**} *as* $\pm$ *stands for* $+$ *or* $-$; *and similarly we use* A_θ *for* A_θ^* *or* A_θ^{**}. Thus, the previous statement might have been written $F(x) \varepsilon A_\theta$ if $T_\theta[F(x)] < \infty$. We will also suppress the θ when no confusion will be caused.

We now state the main theorem of this paper:

THEOREM I. *Let* $F(x) \varepsilon A_\theta$, *let* R *be the closure of the set of values of* $F(x)$, *and let* R^* *be the set of complex numbers whose distance from* R *is not greater than* $\{T_\theta[S(x)]\}^{1/\theta}$. *Let* $\mathcal{J}(z)$ *be a multiple-valued function defined on an open set* $\mathcal{R}$ *containing* R^*; *and let* $\mathcal{J}(z)$ *consist of exactly* n *distinct nonintersecting analytic sheets in the neighborhood of each point of* $\mathcal{R}$. *Let the* n *continuous branches of* $\mathcal{J}[F(x)]$ *and* $\mathcal{J}[H(x)]$ *be denoted in some arbitrary order by* $[\mathcal{J}(F(x))]_j$, $(j=1, \cdots, n)$, *and by* $[\mathcal{J}(H(x))]_j$, $(j=1, \cdots, n)$. *Then there exist two permutations* $p_1, \cdots, p_n$ *and* $p_1', \cdots, p_n'$ *(each unique) of the numbers* $1, 2, \cdots, n$ *such that*

$$\lim_{N\to\infty} \frac{1}{N}\int_0^N |[\mathcal{J}(F(x))]_j - [\mathcal{J}(H(x))]_{p_j}|^2 dx = 0, \tag{1.1}$$

$$\lim_{N\to\infty} \frac{1}{N}\int_{-N}^0 |[\mathcal{J}(F(x))]_j - [\mathcal{J}(H(x))]_{p_j'}|^2 dx = 0. \tag{1.2}$$

Moreover if for any particular j *we have* $p_j = p_j'$, *then* $[\mathcal{J}(F(x))]_j \varepsilon A_\theta$.

2. **Properties of the norms.** We must first establish the fact that the norms $T_\theta\{F(x)\}$ satisfy the axioms

I. $T(F_1)+T(F_2) \geqq T(F_1+F_2)$;
II. $T(F_1)T(F_2) \geqq T(F_1F_2)$;
III. $|a|^\theta T_\theta(F) = T_\theta(aF)$.

The first of these relations follows immediately from the inequality $a^\theta + b^\theta \geqq (a+b)^\theta$, which holds whenever $a \geqq 0$, $b \geqq 0$, and $0 < \theta \leqq 1$ (as we can readily see by choosing $a > b$ and considering the function $(b/a)^\theta + 1 - (b/a+1)^\theta$ and its θ derivative). The third relation is obvious; and it therefore only remains to prove axiom II. Assuming therefore that $F(x) = F_1(x)F_2(x)$, we obtain

$$f(y) = \int_{-\infty}^{\infty} f_1(y-u)df_2(u)$$

except at a countable set of points. Then

$$\begin{aligned} T_\theta^*\{F_1(x)F_2(x)\} &= 2\sum_{n=-\infty}^{\infty}\left[\int_n^{n+1}\left|d_y\int_{-\infty}^{\infty}f_1(y-u)df_2(u)\right|\right]^\theta \\ &= 2\sum_{n=-\infty}^{\infty}\left[\int_n^{n+1}\left|d_y\sum_{m=-\infty}^{\infty}\int_m^{m+1}f_1(y-u)df_2(u)\right|\right]^\theta \\ &\leqq 2\sum_{n=-\infty}^{\infty}\left[\sum_{m=-\infty}^{\infty}\int_n^{n+1}\left|d_y\int_m^{m+1}f_1(y-u)df_2(u)\right|\right]^\theta \\ &\leqq 2\sum_{n=-\infty}^{\infty}\sum_{m=-\infty}^{\infty}\left[\int_m^{m+1}|df_2(u)|\int_{n-m-1}^{n-m+1}|df_1(u)|\right]^\theta \\ &= 2\sum_{m=-\infty}^{\infty}\left[\int_m^{m+1}|df_2(u)|\right]^\theta\sum_{n=-\infty}^{\infty}\left[\int_{n-1}^{n+1}|df_1(y)|\right]^\theta \\ &\leqq T_\theta^*[F_2(x)]\cdot T_\theta^*[F_1(x)]. \end{aligned}$$

Thus II holds for T_θ^*; and since T_θ^* and T_θ^{**} are identical for functions with zero almost periodic part, II holds for T_θ^{**} applied to functions of the form $F(x)=G(x)+S(x)$. Since I holds for T_θ^{**}, we need merely show that II holds if $F_1(x)=H_1(x)$ and $F_2(x)=H_2(x)$ and also holds if $F_1(x)=H_1(x)$ and $F_2(x)=G_2(x)+S_2(x)$. But $H(x)$ is merely an infinite sum of terms of the form $ae^{i\lambda x}$, and since I can be extended to infinite sums, we need merely show that II holds for products of the form $a_1e^{i\lambda_1 x}a_2e^{i\lambda_2 x}$ and $ae^{i\lambda x}[G(x)+S(x)]$. Direct substitution takes care of the first of these products; and the proof is completed by noting that if m is the greatest integer less than λ,

$$\begin{aligned} T_\theta^{**}\{ae^{i\lambda x}[G(x)+S(x)]\} &= T_\theta^{**}\left\{a\int_{-\infty}^{+\infty}e^{ixy}d[g(y-\lambda)+ds(y-\lambda)]\right\} \\ &= 2|a|^\theta\sum_{n=-\infty}^{\infty}\left[\int_n^{n+1}|dg(y-\lambda)+ds(y-\lambda)|\right]^\theta \\ &\leqq 2|a|^\theta\sum_{n=-\infty}^{\infty}\left[\int_{n-m-1}^{n-m+1}|dg(y)+ds(y)|\right]^\theta \\ &\leqq T_\theta^{**}\{ae^{i\lambda x}\}\cdot T_\theta^{**}\{G(x)+S(x)\}. \end{aligned}$$

3. **Functions of small norm.** We begin to prove Theorem I by first proving that the special case of it in which $\mathcal{J}(z)$ is single-valued and $F(x)$ is a constant plus a function of small norm.

LEMMA 1. *Let $F(x)\ \varepsilon\ A_\theta$, let $F(x)=\tau+F_1(x)$ where $T_\theta[F_1(x)]<K^\theta$, and let $\mathcal{J}(z)$ be analytic in a circle about τ of radius K. Then $\mathcal{J}\{F(x)\}\ \varepsilon\ A_\theta$.*

For $\mathcal{J}(z)$ has a Taylor's series $\mathcal{J}(z)=\sum_{n=0}^{\infty}a_n(z-\tau)^n$ converging when

$|z-\tau|<K$. Then if $T_\theta[F_1(x)]<K_1^\theta<K^\theta$, $\sum_{n=0}^\infty |a_n| K_1^n$ converges and $|a_n|K_1^n$ is bounded in n. Hence $\sum_{n=0}^\infty |a_n|^\theta W^n$ converges when $0<W<K_1^\theta$. Thus

$$T_\theta[\mathcal{J}(F(x))] = T_\theta\left[\sum_{n=0}^{\infty} a_n(F(x)-\tau)^n\right]$$

$$= T_\theta\left[\sum_{n=0}^{\infty} a_n(F_1(x))^n\right] \leqq \sum_{n=0}^{\infty} T_\theta[a_n(F_1(x))^n]$$

$$\leqq \sum_{n=0}^{\infty} |a_n|^\theta \{T_\theta[F_1(x)]\}^n,$$

and since $T_\theta[F_1(x)]<K_1^\theta$, it follows that the last sum is finite. Thus $\mathcal{J}(F(x)) \varepsilon A_\theta$.

4. **The space** $C\mathfrak{T}_n$. *Let $\mathfrak{T}_n$ be the set of points each of which consists of n ordered numbers, each reduced modulo 2π. Let C be the set of all real numbers, together with one special symbol ∞. Let $C\mathfrak{T}_n$ be the product space of C and $\mathfrak{T}_n$.*

The set of points $(x_1, \cdots, x_n)$ of $\mathfrak{T}_n$ which satisfy

$$|x_j - x_j'| < \epsilon \ (\mathrm{mod}\ 2\pi), \qquad j = 1, \cdots, n,$$

is called the ϵ-neighborhood of $(x_1', \cdots, x_n')$. The set of points x of C which satisfy $|x-x'|<\epsilon$ is called the ϵ-neighborhood of x. The set of finite points x which satisfy $|x|>1/\epsilon$ together with ∞, is called the ϵ-neighborhood of ∞. Product neighborhoods such as the ϵ-neighborhood of $(x_1, \cdots, x_n; x)$ or $(x_1, \cdots, x_n; \infty)$ are defined in the usual way. The ϵ-neighborhood of $(x_1, \cdots, x_n; \infty)$ will be called an infinite $C\mathfrak{T}_n$ neighborhood, and $(x_1, \cdots, x_n; \infty)$ will be called an infinite point of $C\mathfrak{T}_n$. It is obvious that the Heine-Borel theorem holds for the whole space.

5. **Finiteness of the norm a local property.** *A function $F(x)$ is called locally of finite norm in a finite C neighborhood N if there exists a function $F^*(x)$ which is of finite norm and equals $F(x)$ when x is in N. A function $F(x)$ is called locally of finite norm with respect to $\lambda_1, \cdots, \lambda_n$ in a $C\mathfrak{T}_n$ neighborhood N if there exists a function $F^*(x)$ which is of finite norm and equals $F(x)$ when $(\lambda_1 x, \cdots, \lambda_n x; x)$ is in N.*

LEMMA 2. *Let $\lambda_1, \cdots, \lambda_n$ be given. Then a necessary and sufficient condition that a function $f(x)$ be of finite norm is that it be locally of finite norm in a C neighborhood of each finite point of C and locally of finite norm with respect to $\lambda_1, \cdots, \lambda_n$ in a $C\mathfrak{T}_n$ neighborhood of each infinite point of $C\mathfrak{T}_n$.*

The necessity of the condition is obvious; so we need only prove sufficiency. We note at the outset that the hypothesis implies that $f(x)$ is locally of finite norm with respect to $\lambda_1, \cdots, \lambda_n$ in a $C\mathfrak{T}_n$ neighborhood of every point of $C\mathfrak{T}_n$. For a function which equals $f(x)$ in the ϵ-neighborhood of the point x_0 of C necessarily equals $f(x)$ when $(\lambda_1 x, \cdots, \lambda_n x; x)$ is in the ϵ-neighborhood of the point $(x_1, \cdots, x_n; x_0)$ of $C\mathfrak{T}_n$. Thus for each point P of $C\mathfrak{T}_n$ there is an $\epsilon_P > 0$ and a function $f_P(x)$ which is of finite norm and which equals $f(x)$ when $(\lambda_1 x, \cdots, \lambda_n x; x)$ is in the ϵ_P-neighborhood $\mathcal{N}_P$ of P. Then by the Heine-Borel theorem there is a finite number of points $P_1, \cdots, P_q$, such that the $\epsilon/2$-neighborhoods of P_j cover $C\mathfrak{T}_n$. Choose an integer

$$N > 2\pi[\min(\epsilon_{P_1}, \cdots, \epsilon_{P_q})]^{-1} \max[|\lambda_1|, \cdots, |\lambda_n|; 1].$$

Let $\Phi^*(x)$ be an even function which is zero on $|x| > 1$, unity at $x = 0$, and is continuous and has derivatives of all orders everywhere and satisfies $\Phi^*(x) + \Phi^*(1-x) = 1$ on $0 \leqq x \leqq 1$. Thus, to be specific, we may define

$$\Phi^*(x) = \begin{cases} \dfrac{1}{2} - \dfrac{\int_0^{2|x|-1} e^{(\xi^2-1)^{-1}} d\xi}{2\int_0^1 e^{(\xi^2-1)^{-1}} d\xi} & \text{if } 0 < |x| \leqq 1, \\ 0 & \text{if } 1 \leqq |x|. \end{cases}$$

We obviously have

$$\sum_{k=-N^2}^{N^2} \Phi(x-k) = \begin{cases} 1 & \text{if } |x| \leqq N^2, \\ 0 & \text{if } |x| > (N+1)^2. \end{cases}$$

Moreover if

$$\begin{cases} \Phi(x) = \Phi^*\left(\dfrac{Nx}{\pi}\right), & |x| < \pi, \\ \Phi(x) = \Phi(x + 2\pi), & \text{for all } x, \end{cases}$$

then

$$\sum_{k=1}^{2N} \Phi\left(x - \frac{\pi k}{N}\right) = 1, \quad \text{for all } x.$$

Thus

$$\sum_{k_1, \cdots, k_n = 1}^{2N} \Phi\left(\lambda_1 x - \frac{\pi}{N} k_1\right) \cdots \Phi\left(\lambda_n x - \frac{\pi}{N} k_n\right) = 1 \quad \text{for all } x,$$

and

$$\begin{aligned} f(x) &= \sum_{k=-N^2}^{N^2} \Phi^*(Nx - k) f(x) \\ &\quad + \left[1 - \sum_{k=-N^2}^{N^2} \Phi^*(Nx-k)\right] \sum_{k_1, \cdots, k_n=1}^{2N} \Phi\left(\lambda_1 x - \frac{\pi}{N} k_1\right) \\ &\quad \cdots \Phi\left(\lambda_n x - \frac{\pi}{N} k_n\right) f(x). \end{aligned}$$

for all x. Thus if we show that $\Phi^*(Nx-k)f(x)$ and

$$\left[1 - \sum_{k=-N^2}^{N^2} \Phi^*(Nx - k)\right] \Phi\left(\lambda_1 x - \frac{\pi}{N} k_1\right) \cdots \Phi\left(\lambda_n x - \frac{\pi}{N} k_n\right) f(x)$$

are of finite norm for all $k, k_1, \cdots, k_n$, it follows that $f(x)$ is of finite norm.

To show this for $\Phi^*(Nx-k)$, consider the point P_j whose $\epsilon_{P_j}/2$-neighborhood covers

$$\left(\frac{k}{N}\lambda_1, \cdots, \frac{k}{N}\lambda_n; \frac{k}{N}\right).$$

The ϵ_{P_j}-neighborhood $\mathcal{N}_j$ of this point covers the $\epsilon_{P_j}/2$-neighborhood of

$$\left(\frac{k}{N}\lambda_1, \cdots, \frac{k}{N}\lambda_n; \frac{k}{N}\right).$$

Thus when $|Nx-k|<1$, $(\lambda_1 x, \lambda_2 x, \cdots, \lambda_n x; x)$ is in $\mathcal{N}_j$, and when $\Phi^*(Nx-k)$ is not zero, $f_{P_j}(x)$ equals $f(x)$ and for all x

$$T^*(Nx - k)f(x) = T^*(Nx - k)f_{P_j}(x),$$

which is of finite norm.

Again, consider the point P_l whose $\epsilon_{P_l}/2$-neighborhood covers the point $(k_1\pi/N, \cdots, k_n\pi/N; \infty)$. The ϵ_{P_l}-neighborhood $\mathcal{N}_l$ of P_l covers the $\epsilon_{P_l}/2$-neighborhood of $(k_1\pi/N, \cdots, k_n\pi/N; \infty)$. Thus when

$$|\lambda_1 x - k_1\pi/N| < \pi/N, \cdots, |\lambda_n x - k_n\pi/N| < \pi/N, \qquad |x| > N,$$

$(\lambda_1 x, \lambda_2 x, \cdots, \lambda_n x, x)$ is in $\mathcal{N}_l$ and

$$\left[1 - \sum_{k=-N^2}^{N^2} \Phi^*(Nx - k)\right] \Phi\left(\lambda_1 x - \frac{\pi}{N} k_1\right) \cdots \Phi\left(\lambda_n x - \frac{\pi}{N} k_n\right) f(x)$$
$$= \left[1 - \sum_{k=-N^2}^{N^2} \Phi^*(Nx - k)\right] \Phi\left(\lambda_1 x - \frac{\pi}{N} k_1\right) \cdots \Phi\left(\lambda_n x - \frac{\pi}{N} k_n\right) f_{P_l}(x).$$

Thus $f(x)$ is of finite norm.

6. **Periodic functions with assigned derivatives and small norm.** Section 5 makes it necessary only to show that $\mathcal{Y}[F(x)]$ is locally of finite norm corresponding to all infinite points $C\mathfrak{T}_n$ and all finite points of C; and §3 shows that this will be established if we show how to replace functions by others locally equivalent but of sufficiently small norm. We begin by replacing exponential functions by locally equivalent functions of small norm.

The first step is to find functions of small norm having derivatives at the origin equal to those of $e^{ix}-1$.

Such a function is given by

$$\Psi_{m,p}(x) = e^{(i/p)\psi_m(px)} - 1$$

where

$$\psi_m(x) = x - \frac{2\pi\int_0^x \sin^{2m}\xi\, d\xi}{\int_0^{2\pi}\sin^{2m}\xi\, d\xi}; \tag{6.1}$$

for $\psi_m(x)$ is obviously periodic and hence is an exponential polynomial. Thus $\psi_m(px)$ has norms independent of p (for p an integer greater than 1) and the norms of $(1/p)\psi_m(px)=O(1/p^\theta)$. Thus for fixed m the norm of $\Psi_{m,p}(x)$ can be made arbitrarily small by making p sufficiently great; and since $\psi_m(x)=x+O(x^{2m+1})$ at the origin, it follows that $\Psi_{m,p}(x)=e^{ix}-1+O(x^{2m+1})$, and $\Psi_{m,p}(x)$ and $e^{ix}-1$ have the same first $2m$ derivatives at $x=0$ and points congruent (mod 2π).

7. **Locally exponential functions of small norm.** Now to obtain a function of small norm which is actually equal to $e^{ix}-1$ in the neighborhood of points congruent to zero (mod 2π) we introduce the function

$$\Omega(\epsilon, x) = \begin{cases} 1, & 0 \leqq |x| \leqq \epsilon \pmod{2\pi}, \\ \dfrac{1}{2} - \dfrac{\int_0^{2|x|/\epsilon-3} e^{(\xi^2-1)^{-1}}d\xi}{2\int_0^1 e^{(\xi^2-1)^{-1}}d\xi}, & \epsilon \leqq |x| \leqq 2\epsilon \pmod{2\pi}, \\ 0, & 2\epsilon \leqq |x| \leqq \pi \pmod{2\pi} \end{cases}$$

which is obviously of finite norm for all θ and of period 2π.

Let $H(x)$ be a periodic function of period 2π whose first m derivatives are continuous everywhere. Then if $H(0)=H'(0)=\cdots=H^{(m)}(0)=0$, and $1/m<\theta\leqq 1$, we shall show that

$$\lim_{\epsilon\to 0} T_\theta\{H(x)\Omega(\epsilon, x)\} = 0. \tag{7.1}$$

For if $H(x)\Omega(\epsilon, x)=P(\epsilon, x)=\sum_{n=-\infty}^{\infty}p_n(\epsilon)e^{inx}$, we have by integration by parts

$$p_n = \frac{1}{2\pi}\int_0^{2\pi} P(\epsilon, x)e^{-inx}dx = \frac{1}{2\pi(in)^m}\int_0^{2\pi} P^{(m)}(\epsilon, x)e^{-inx}dx \quad \text{for} \quad n \neq 0.$$

Thus if $L(\epsilon)$ is the greatest value taken on by either $|P(\epsilon, x)|$ or $|P^{(m)}(\epsilon, x)|$ for all x, we have $|p_n(\epsilon)|\leqq 2L(\epsilon)/(1+|n|^m)$ for all n and

$$T_\theta\{P(\epsilon, x)\} \leqq 2[L(\epsilon)]^\theta \sum_{n=-\infty}^{\infty}\frac{2}{(1+|n|^m)^\theta} \quad \text{for all } m.$$

Since $m\theta>1$, this sum converges, and we need merely show that $L(\epsilon)\to 0$ to establish (7.1). But $\max_x |P(\epsilon, x)| \to 0$ as $\epsilon \to 0$ since $H(x)$ is continuous and vanishes at zero and $\Omega(\epsilon, x)$ is bounded and is zero when $|x| \geqq 2\epsilon$ (mod 2π). Moreover

$$P^{(m)}(\epsilon, x) = \sum_{j=0}^{m} C_{m,j} H^{(m-j)}(x)\Omega^{(j)}(\epsilon, x),$$

and we shall show that each term of this sum approaches zero uniformly as $\epsilon \to 0$. In the interval $|x| < 2\epsilon$ the function $\Omega(\epsilon, x)$ is a function of x/ϵ alone, and hence its j'th derivative with respect to x is less than $C_j \epsilon^{-j}$, where C_j is independent of ϵ and x. Moreover $H^{(m-j)}(x)x^{-j} \to 0$ as $x \to 0$; so

$$\max_{|x| \leqq 2\epsilon} \frac{|H^{(m-j)}(x)|}{\epsilon^j} \to 0 \quad \text{as } \epsilon \to 0.$$

Thus

$$\begin{aligned} \max_x |H^{(m-j)}(x)\Omega^{(j)}(\epsilon, x)| &= \max_{|x| \leqq 2\epsilon} |H^{(m-j)}(x)\Omega^{(j)}(\epsilon, x)| \\ &\leqq \left[\max_{|x| \leqq 2\epsilon} \frac{|H^{(m-j)}(x)|}{\epsilon^j} \right] \cdot C_j \to 0 \end{aligned}$$

and $L(\epsilon) \to 0$ as $\epsilon \to 0$, and (7.1) is established.

We now define

$$E_{m,p}(\epsilon, x) = \Omega(\epsilon, x)\{e^{ix} - 1 - \Psi_{m,p}(x)\} + \Psi_{m,p}(x)$$

and have, for fixed m and θ such that $2m\theta > 1$,

$$\lim_{p \to \infty} \left[\lim_{\epsilon \to 0} T_\theta\{E_{m,p}(\epsilon, x)\} \right] = 0,$$

while for $|x| < \epsilon$ (mod 2π), $E_{m,p}(\epsilon, x) = e^{ix} - 1$.

8. ***G*-functions with assigned derivatives and small norm.** We now seek to replace a G-function by a function of small norm C-locally equivalent to it. We begin by finding a function of small norm having the same derivatives as the given function at the origin.

Let $G(x)$ be an entire function which vanishes at the origin and m a positive integer. Then $G[\psi_m(px)/p]$ (where $\psi_m(x)$ is the function defined in (6.1)) has its first $2m$ derivatives at $x=0$ equal to those of $G(x)$, and the norm of $G[\psi_m(px)/p]$ approaches zero as $p \to \infty$.

9. ***G*-functions locally of small norm.** *Let $G(x)$ be a G-function which is also an entire function having $G(0) = G'(0) = \cdots = G^{(m)}(0) = 0$. Let*

$$\Omega^*(\epsilon, x) = \begin{cases} \Omega(\epsilon, x), & 0 \leqq |x| \leqq \pi, \\ 0, & |x| \geqq \pi. \end{cases}$$

Then if $1/m<\theta\leqq 1$, *we shall show that*

$$\lim_{\epsilon\to 0} T_\theta\{\Omega^*(\epsilon, x)G(x)\} = 0. \tag{9.1}$$

This statement is proved in much the same way as the corresponding statement for H-functions.

Let

$$\Omega^*(\epsilon, x)G(x) = P^*(\epsilon, x) = \int_{-\infty}^{\infty} p^*(\epsilon, \xi)e^{-i\xi x}d\xi,$$

so that

$$p^*(\epsilon, \xi) = \frac{1}{2\pi}\int_{-\pi}^{\pi} P^*(\epsilon, x)e^{i\xi x}dx.$$

Integrating by parts we have

$$p^*(\epsilon, \xi) = \frac{1}{(-i\xi)^m 2\pi}\int_{-\pi}^{\pi} P^{*(m)}(\epsilon, x)e^{i\xi x}dx \quad \text{if} \quad \xi\neq 0.$$

Thus

$$|p^*(\epsilon, \xi)| < \frac{2}{|\xi|^m+1} L^*(\epsilon) \quad \text{for all } \xi,$$

where $L^*(\epsilon)$ is the greater of the upper bounds of $|P^*(\epsilon, x)|$ and $|P^{*(m)}(\epsilon, x)|$ on $|x|\leqq\pi$. Now as $\epsilon\to 0$, the bounds of $|P^*(\epsilon, x)|$ and $|P^{*(m)}(\epsilon, x)|$ approach zero for the same reason that $|P(\epsilon, x)|$ and $|P^{(m)}(\epsilon, x)|$ approached zero in §7. Thus $\lim_{\epsilon\to 0} L^*(\epsilon)=0$, and

$$\begin{aligned} T_\theta[P^*(\epsilon, x)] &= 2\sum_{n=-\infty}^{\infty}\left[\int_n^{n+1} |p^*(\epsilon, \xi)|\, d\xi\right]^\theta \\ &\leqq 2[L^*(\epsilon)]^\theta \sum_{n=-\infty}^{\infty}\left[\int_n^{n+1}\frac{2d\xi}{(1+|\xi|^m)}\right]^\theta; \end{aligned}$$

and if $m\theta>1$, (9.1) holds.

We now define for any G-function which is also an entire function and vanishes at the origin,

$$\Gamma_{m,p}(G|\,\epsilon, x) = \Omega^*(\epsilon, x)\{G(x) - G[\psi_m(px)/p]\} + G[\psi_m(px)/p],$$

and we have for fixed m and θ such that $2m\theta>1$,

$$\lim_{p\to\infty}\lim_{\epsilon\to 0} T_\theta\{\Gamma_{m,p}(G|\,\epsilon, x)\} = 0,$$

while for $|x|<\epsilon$, $\Gamma_{m,p}(G|\,\epsilon, x)=G(x)$.

10. ***G*-functions of small norm at** ∞. Turning now to the C-neighborhood of ∞, we seek to replace an entire G-function by a G-function of small norm equal to the given function near ∞. Let $G(x)$ be an entire G-function given by

$$G(x) = \int_{-A}^{A} e^{-iux} g(u)du$$

(where $g(u)=0$ if $|u|>A$), and let

$$G_\delta(x) = \int_{-\infty}^{\infty} e^{-iux} g_\delta(u)du$$

where

$$g_\delta(u) = \frac{1}{\delta^3}\int_u^{u+\delta}\int_{\xi_3}^{\xi_3+\delta}\int_{\xi_2}^{\xi_2+\delta} g(\xi_1)d\xi_1 d\xi_2 d\xi_3$$

is the triple smoothing of $g(u)$.

Then

$$T_\theta[G(x) - G_\delta(x)] = 2\sum_{n=-\infty}^{\infty}\left[\int_n^{n+1} | g(u) - g_\delta(u) | du\right]^\theta;$$

so

$$\lim_{\delta\to 0} T_\theta[G(x) - G_\delta(x)] = 0.$$

Now let

$$[1 - \Omega^*(N, x)]G_\delta(x) = P_\delta(N, x) = \int_{-\infty}^{\infty} p_\delta(N, \xi)e^{-i\xi x}d\xi.$$

Since $g_\delta(u)$ has two continuous derivatives, if $m>1/\theta$, $G_\delta^{(k)}(x)=o(1/x^2)$ and $P_\delta^{(k)}(N, x)=o(1/x^2)$ at $\pm\infty$ for $k=0, 1, \cdots, m$. But

$$\frac{1}{2\pi}\int_{-\infty}^{\infty} P_\delta(N, x)e^{iux}dx = p_\delta(N, u),$$

and integrating by parts, we have

$$p_\delta(N, u) = \frac{1}{(-iu)^m 2\pi}\int_{-\infty}^{\infty} e^{iux}P_\delta^{(m)}(N, x)dx.$$

Thus if $L_\delta(N)$ is the greater of the upper bounds of

$$|(x^2+1)P_\delta(N, x)|, \qquad |(x^2+1)P_\delta^{(m)}(N, x)|,$$

we have

$$|p_\delta(N, u)| \leqq \min\left(1, \frac{1}{|u|^m}\right)\frac{\pi}{2\pi} L_\delta(N) \leqq \frac{L_\delta(N)}{1+|u|^m};$$

so that

$$T_\theta[P_\delta(N, x)] = 2\sum_{n=-\infty}^{\infty}\left[\int_n^{n+1} | p_\delta(N, \xi) | \, du\right]^\theta$$
$$\leqq 2[L_\delta(N)]^\theta \sum_{n=-\infty}^{\infty}\left[\int_n^{n+1} \frac{du}{1+|u|^m}\right]^\theta,$$

where the latter sum converges if $m\theta>1$. But $1-\Omega^*(N, x)$ is never numerically greater than 1 and is zero when $|x|<N$; so $\overline{\text{Bd}}\; P_\delta(N, x)\to 0$ as $N\to\infty$. Moreover for $N>1$, each derivative of $1-\Omega^*(N, x)$ is bounded in N and x, and is zero when $|x|<N$. Thus each term of Leibnitz' expansion of

$$(1+x^2)\frac{d^m}{dx^m}[(1-\Omega^*(N, x))G_\delta(x)]$$

has its upper bounds approach zero as $N\to\infty$. Hence $L_\delta(N)\to 0$ as $N\to\infty$, and $\lim_{N\to\infty} T_\theta[P_\delta(N, x)]=0$.

Finally, we define

$$\Gamma^*_{N,\delta}(G\,|\,x) = G(x) - G_\delta(x) + P_\delta(N, x)$$

and have

$$\lim_{\delta\to 0}\lim_{N\to\infty} T_\theta[\Gamma^*_{N,\delta}(G\,|\,x)] = 0$$

and $\Gamma^*_{N,\delta}(G|x)=G(x)$ for $x\geqq 2N$.

11. **Proof of Theorem** I. Returning now to the proof of Theorem I, we note that the fact that $\mathcal{J}[F(x)]$ consists of n continuous functions is obvious, as is also the same fact for $\mathcal{J}(H(x))$, since each value of $H(x)$ belongs to R. Now the symmetric functions of the n values of $\mathcal{J}(H(x))$ are single-valued analytic functions of $\mathcal{J}(H(x))$ which are therefore almost periodic; and by Walther's theorem on algebraic functions of almost periodic functions, it follows that each branch of $\mathcal{J}(H(x))$ is almost periodic.

Since the closed set R^* is contained in the open set $\mathcal{R}$, there is a positive number η such that every number within a distance η of R^* is in $\mathcal{R}$. Choose L so great that $|G(x)|<\eta/2$ when $|x|\geqq L$. Then $|F(x)-H(x)|<\eta/2+|S(x)|$, so that if once chosen on the same branch, $F(x)$ and $H(x)$ remain on the same branch of $\mathcal{J}(z)$ when $x\geqq L$ and also when $x\leqq -L$. Thus given $\epsilon>0$, we can find $\delta>0$ such that if $x>L$ and $|F(x)-H(x)|<\delta$, then $|[\mathcal{J}(F(x))]_j - [\mathcal{J}(H(x))]_{p_j}|<\epsilon$, where p_j is so chosen that $[\mathcal{J}(F(x))]_j$ and $[\mathcal{J}(H(x))]_{p_j}$ are on the same branch for $x>L$. From this (1.1) and (1.2) readily follow. But two almost periodic functions never have their mean square difference zero, and hence p_j and p_j' are unique.

Now suppose that $p_j=p_j'$; and write simply $\mathcal{J}(F(x))$ and $\mathcal{J}(H(x))$ for

$[\mathcal{J}(F(x))]_j$ and $[\mathcal{J}(H(x))]_{p_j}$. Then whenever $|x| \geqq L$, $F(x)$ and $H(x)$ are to be taken on the same branch of $\mathcal{J}(z)$.

In the closed region $\mathcal{R}^*_{\eta/2}$ of points within $\eta/2$ of R^*, there is a minimum distance γ between the branches of $\mathcal{J}(z)$, so that for all z in $\mathcal{R}^*_{\eta/2}$, $|[\mathcal{J}(z)]_j - [\mathcal{J}(z)]_k| \geqq \gamma$ for all pairs of branches. Now if M is the least common module of $\mathcal{J}(H(x))$ and $H(x)$, we can find a finite number of elements $\mu_1, \cdots, \mu_q$ of M and a positive number ϵ_1 such that whenever

$$|\mu_j h| < \epsilon_1 \pmod{2\pi}, \qquad j = 1, \cdots, q,$$

then

$$|\mathcal{J}(H(x+h)) - \mathcal{J}(H(x))| < \gamma/2 \quad \text{for all } x,$$

and $|H(x+h) - H(x)|$ is uniformly so small that if $\mathcal{J}$ were taken on the same branch for both of these values, then $|\mathcal{J}(H(x+h)) - \mathcal{J}(H(x))|$ would also be less than $\gamma/2$. Thus $\mathcal{J}(H(x+h))$ and $\mathcal{J}(H(x))$ are on the same branch for all x.

Choose $\bar{\delta}$ so small that $[4\bar{\delta} + T_\theta[S(x)]]^{1/\theta} + (2\bar{\delta})^{1/\theta} < T_\theta[S(x)]^{1/\theta} + \eta/2$. Let $H(x) = \sum_{k=1}^{\infty} h_k e^{i\lambda_k x}$, and let q' be chosen so that if $H^*(x) = \sum_{k=q'+1}^{\infty} h_k e^{i\lambda_k x}$, then $T_\theta(H^*(x)) < \bar{\delta}$. We shall show that $\mathcal{J}(F(x))$ is locally of finite norm with respect to $\mu_1, \cdots, \mu_q, \lambda_1, \cdots, \lambda_{q'}$ in the $\mathcal{T}_{q+q'}C$ neighborhood of every infinite point of $\mathcal{T}_{q+q'}C$; and hence after a similar argument for finite C neighborhoods that $\mathcal{J}[F(x)]$ is of finite norm.

Let us consider the $\mathcal{T}C$ neighborhood of $(\mu_1\xi_1, \cdots, \mu_q\xi_q, \lambda_1\xi_1', \cdots, \lambda_{q'}\xi_{q'}'; \infty)$. Let

$$S(x) = \int_{-\infty}^{\infty} e^{-i\xi x} ds(\xi), \qquad G(x) = \int_{-\infty}^{\infty} e^{-i\xi x} g(\xi) d\xi;$$

and let $G^*(x) = \int_{-A}^{A} e^{-i\xi x} g(\xi) d\xi$, where A is so great that $T_\theta[G(x) - G^*(x)] < \bar{\delta}$. Let $\Gamma^*_{N',\delta'}[G^*|x] = G^{**}(x)$, where N', δ' are chosen so that $T_\theta[G^{**}(x)] < \bar{\delta}$. Then $G^{**}(x) = G^*(x)$ when $|x| > 2N'$. Let

$$H^{**}(x) = \sum_{k=1}^{q'} h_k E_{m,p_k}(\epsilon^{(k)}, \lambda_k x - \lambda_k \xi_k) e^{i\lambda_k \xi_k},$$

where m, p_k, $\epsilon^{(k)}$ are so chosen that

$$T_\theta[h_k E_{m,p_k}(\epsilon^{(k)}, \lambda_k x - \lambda_k \xi_k)] < \bar{\delta}/2q'.$$

Then $T_\theta[H^{**}(x)] < \bar{\delta}$, and

$$H^{**}(x) = H(x) - H^*(x) - \sum_{k=1}^{q'} h_k e^{i\lambda_k \xi_k}$$

whenever $|\lambda_k(x-\xi_k)|<\epsilon^{(k)}$ (mod 2π), $(k=1,\cdots,q')$. Let

$$\epsilon = \min(\epsilon_1/2,\ 1/2N',\ 1/L,\ \epsilon^{(1)},\cdots,\epsilon^{(q')}),$$

and consider the ϵ-neighborhood of $(\mu_1\xi_1,\cdots,\mu_q\xi_q,\lambda_1\xi_1',\cdots,\lambda_{q'}\xi_{q'}';\ \infty)$. We shall show that the function $\mathcal{J}(F(x))$ is locally of finite norm with respect to $\mu_1,\cdots,\mu_q,\lambda_1,\cdots,\lambda_{q'}$ in this neighborhood. We say $x\ \varepsilon\ \mathcal{N}$ if $(x\mu_1,\cdots,x\mu_q, x\lambda_1,\cdots,x\lambda_{q'},x)$ is in this neighborhood.

If $x\ \varepsilon\ \mathcal{N}$, we have $G^{**}(x)=G^*(x)$ and

$$H^{**}(x) = H(x) - H^*(x) - \tau \tag{11.1}$$

where $\tau=\sum_{k=1}^{q'}h_ke^{i\lambda_k\xi_k}$; and if x_1, x_2 are any two points in $\mathcal{N}$,

$$|(x_1-x_2)\mu_j| < \epsilon_1 \ (\mathrm{mod}\ 2\pi), \qquad j=1,\cdots,q,$$

and

$$|\mathcal{J}(H(x_1)) - \mathcal{J}(H(x_2))| < \gamma/2,$$

and $\mathcal{J}[H(x_1)]$ and $\mathcal{J}[H(x_2)]$ are on the same branch of $\mathcal{J}(z)$. Thus $\mathcal{J}(z)$ is a single-valued analytic function over the set of values of $H(x)$ for x in $\mathcal{N}$, and since there is no branch point or singularity within $[T_\theta(S(x))]^{1/\theta}+\eta/2$ of these points, it follows that for x in $\mathcal{N}$, $\mathcal{J}(z)$ is analytic and single-valued for all $z=F(x)$. Moreover $\mathcal{J}(z)$ has an analytic and single-valued branch for all points within $\eta/2+[T_\theta(S(x))]^{1/\theta}$ of points for which $z=F(x)$ with x in $\mathcal{N}$, and on this branch $\mathcal{J}[F(x)]$ has the values we have agreed to denote by $\mathcal{J}(F(x))$.

For x in $\mathcal{N}$, $F(x)=F^*(x)$, where

$$F^*(x) = \tau + H^*(x) + H^{**}(x) + [G(x) - G^*(x)] + G^{**}(x) + S(x).$$

Moreover

$$T_\theta\{F^*(x) - \tau\} < 4\bar{\delta} + T_\theta[S(x)], \tag{11.2}$$

and by (11.1) the point τ is within $(2\bar{\delta})^{1/\theta}$ of R; and by (11.2) and the definition of $\bar{\delta}$, we find that $\mathcal{J}(z)$ is analytic and single-valued throughout a circle of radius greater than $\{T_\theta[F^*(x)-\tau]\}^{1/\theta}$ about τ. Thus by Lemma 1, $\mathcal{J}[F^*(x)]\ \varepsilon\ A^*$, and since $\mathcal{J}[F(x)]=\mathcal{J}[F^*(x)]$ in $\mathcal{N}$, $\mathcal{J}[F(x)]$ is locally of finite norm with respect to $(\mu_1,\cdots,\mu_q,\lambda_1,\cdots,\lambda_{q'})$ in a $C\mathfrak{T}$ neighborhood of $(\mu_1\xi_1,\cdots,\mu_q\xi_q,\lambda_1\xi_1',\cdots,\lambda_{q'}\xi_{q'}';\ \infty)$. The simpler fact that $\mathcal{J}[F(x)]$ is locally of finite norm in the neighborhood of every finite point of C can be proved in a similar but simpler manner; and we therefore conclude that $\mathcal{J}[F(x)]\ \varepsilon\ A_\theta$.

MASSACHUSETTS INSTITUTE OF TECHNOLOGY,
CAMBRIDGE, MASS.

Comments on [39d]

W. Rudin

Wiener' work in harmonic analysis furnished much of the inspiration and provided many of the guidelines that later shaped the development of the theory of Banach algebras. Paper [39d] is a case in point. Its main result states (in modern terminology) that certain holomorphic functions operate in certain algebras A_θ of Fourier-Stieltjes transforms. A_1 is the usual convolution algebra of all complex Borel measures on the line; it is a Banach algebra. The algebras A_θ, with $\theta < 1$, are subalgebras of A_1 that are topological algebras but not Banach algebras.

It was of course known from paper [38b] that the domain of the operating function had (in certain cases) to be assumed to be quite a bit larger than the closure of the range of the Fourier-Stieltjes transform on which it was to operate. This accounts for part of the complication in the statement of Theorem I. Another complication is due to the fact that multiple-valued operating functions are admitted.

This "symbolic calculus" is now a standard feature of the Gelfand theory. Later work by Šilov[2] and by Arens and Calderon[1] has extended it to situations in which holomorphic functions of n complex variables operate on n-tuples of elements of a Banach algebra. Again, multiple-valued functions are admitted. These problems have also been studied in other topological algebras, for example, by L. Waelbroeck.[3]

References

1. R. F. Arens and A. -P. Calderon, Ann. Math. 62(1955), 204–216.
2. G. E. Šilov, Mat. Sbornik (N.S.) 32(1953), 353–364.
3. L. Waelbroeck, J. Math. Pures et Appl. 33(1954), 147–186.

Comments on [39d]

R. H. Cameron and I. Richards

This paper, of course, is the outgrowth of a long line of developments, beginning with Wiener's famous "Tauberian theorems" [32a]. R. H. Cameron proved the analogue of the Wiener-Lévy Theorem for absolutely convergent almost periodic functions,[1] and his result can be extended to sums of L^1 functions and discrete measures (see below). Then Wiener and Pitt [38b] showed that the extension to arbitrary Fourier-Stieltjes transforms does not hold. (Note the terms "almost periodic, transient, and unpredictable" used in the present paper.)[†] This paper "rescues" the theorem for general Fourier-Stieltjes transforms, but at the expense of requiring the analytic function $\mathscr{F}$ to be defined on the larger set R^* instead of the range R.

The size of R^* depends on the "unpredictable" part of the measure, and when this part vanishes, then $R^* = R$. Setting $\theta = 1$ gives the usual "L^1-norm" on the space of bounded variation functions, and taking $\mathscr{F}$ to be single-valued reduces the complexity of the statement. The multi-valued case has applications, however, e.g., to quadratic equations.[2]

Concerning methodology, Wiener made strong use of the fact that $A(R)$ (Fourier transform of the space $L^1(R)$) is a "local algebra," i.e., that a function which coincides with an element of $A(R)$ in a neighborhood of each point (including ∞) actually belongs to $A(R)$. This idea has been used down to the present day. The proof of Wiener's Tauberian theorem given in Rudin's book *Fourier Analysis on Groups*[5] does not differ substantially from Wiener's own (except that Wiener used explicit functions, whereas Rudin abstracts out the properties of these functions which are needed). Some things do change, however, and various approaches to the "analytic" side of Wiener's argument have since been found. We mention three lemmas which serve this purpose:

(1) If $\int_{-\infty}^{\infty} f(x)\,dx = 0, \quad g \in L^1, \quad \text{and } g_a(x) = a \cdot g(ax),$

then

$$\|f * g_a\|_1 \to 0 \quad \text{as } a \to 0 \quad \text{(Wiener);}$$

(2) $\|f\|_{sp} = \lim_{n \to \infty} [\|f^{(n)}\|_1]^{1/n}$ (Beurling);

[†] *Editor's note.* Their usage here bears only a tenuous relationship to their systematic usage by Wiener and others in prediction and filtering later on.

(3) Gelfand's theorem.

In the first method one uses power series in the "localized transform" $(f * g_a)^\wedge$, where $\hat{g}_a \equiv 1$ in a neighborhood of the origin. The second method is explained in the well-known book *Functional analysis* by Riesz and Nagy[4] (where Cameron's theorem, mentioned above, is extended to arbitrary sums of L^1 functions and discrete measures). A development of the third method, which gets maximum mileage out of it, is given in Dunford and Schwartz' treatise, *Linear operators*, Vol. 2.[3]

References

1. R. H. Cameron, Duke J. 3(1937), 682–688.
2. R. H. Cameron, J. Math. and Phys. 21(1942), 57–62.
3. N. Dunford and J. T. Schwartz, *Linear Operations*, vol. 2, Interscience, New York, 1963.
4. F. Riesz and B. von Sz.-Nagy, *Functional Analysis*, Ungar, New York, 1955.
5. W. Rudin, *Fourier Series on Groups*, Interscience, New York, 1962.

TAYLOR'S SERIES OF ENTIRE FUNCTIONS OF SMOOTH GROWTH

By N. Wiener and W. T. Martin

1. **Introduction.**[1] Let $a_n \geqq 0$ $(n = 1, 2, \cdots)$, and

$$\sum_1^\infty a_n x^n \sim sH(x) \qquad (x \to \infty). \tag{1}$$

We may rewrite (1) in the form

$$\sum_1^\infty a_n e^{n\xi - F(\xi)} \to s \qquad (\xi \to \infty), \tag{2}$$

where

$$H(x) = e^{F(\log x)}. \tag{3}$$

We shall derive the following Tauberian theorem.

Theorem 1. *Let $a_n \geqq 0$ $(n = 1, 2, \cdots)$, and let* (2) *hold, where*

(4) *F is four times continuously differentiable for $a \leqq x$ for some a;*

$$F''(x) \geqq \text{const.} > 0 \textit{ for } a \leqq x; \tag{5}$$

$$F'''(x) = o([F''(x)]^{3/2}) \qquad (x \to \infty); \tag{6}$$

$$F^{\text{iv}}(x) = o([F''(x)]^2) \qquad (x \to \infty). \tag{7}$$

Then for any positive λ,

$$\lim_{x \to \infty} \frac{(2\pi)^{1/2}}{\lambda} \sum_{x \leqq \psi(n) \leqq x+\lambda} a_n e^{nG(n) - F(G(n))} = s, \tag{8}$$

where $\psi(x)$ is defined by

$$\int_a^x [G'(x)]^{1/2}\, dx = \psi(x), \tag{9}$$

and $G(x)$ is the inverse function to $F'(x)$ for $a \leqq x$.

As a converse to this theorem we shall prove

Theorem 2. *Let $a_n \geqq 0$ $(n = 1, 2, \cdots)$, and let* (8) *hold for every positive value of λ, where F fulfills the conditions* (4), (5), (6) *and* (7) *and ψ and G are defined as in Theorem* 1. *Then* (2) *holds.*

Received January 19, 1937.

[1] The authors' attention was directed to problems of this type by Professor Vijayaraghavan. It has come to our attention that a similar group of problems has been recently attacked by Mr. Kales of Brown University by quite different methods. While neither direction of work is reducible to the other, Mr. Kales' priority in entering the field is clear.

In the next theorem we place some additional restrictions on F and we obtain certain inequalities related to ψ.

THEOREM 3. (i) *Let F fulfill the conditions* (4) *and* (5) *and the further condition*

$$\text{(10)} \quad \text{for some positive } \epsilon,\ F''(x) = O([F'(x)]^{1+\epsilon}) \qquad (x \to \infty).$$

Then for any positive ϵ,

$$\text{(11)} \qquad n^{-\frac{1}{2}-\epsilon} < \psi'(n)$$

and

$$\text{(12)} \qquad n^{\frac{1}{2}-\epsilon} < \psi(n),$$

for sufficiently large values of n. (ii) *Let F fulfill the conditions* (4) *and* (5) *and the additional condition*

(13) $F'(x)$ *is ultimately greater than any power of x.*

Then for any positive ϵ,

$$\text{(14)} \qquad \psi(n) < n^{\frac{1}{2}+\epsilon}$$

for sufficiently large values of n.

We shall obtain the following theorems on entire functions of smooth growth as immediate consequences of these Tauberian theorems.

THEOREM 4. *Let*

$$\text{(15)} \qquad f(x) = \sum_0^\infty b_n x^n$$

be an integral function, and let

$$\text{(16)} \qquad \frac{1}{2\pi}\int_0^{2\pi} |f(re^{i\theta})|^2 d\theta \sim s e^{F(2 \log r)} \qquad (r \to \infty),$$

where F fulfills the conditions (4), (5), (6) *and* (7). *Then for any positive* λ,

$$\text{(17)} \qquad \lim_{x\to\infty} \frac{(2\pi)^{\frac{1}{2}}}{\lambda} \sum_{x \leq \psi(n) \leq x+\lambda} |b_n|^2 e^{nG(n)-F(G(n))} = s,$$

where ψ *and G are defined as in Theorem* 1.

The converse to Theorem 4 is

THEOREM 5. *Let* (15) *be an entire function and let* (17) *hold for every positive value of* λ, *where F fulfills the conditions* (4), (5), (6) *and* (7), *and G and* ψ *are defined as in Theorem* 1. *Then* (16) *holds.*

If we place additional conditions on F, we have the following gap theorem.

THEOREM 6. *Let* (15) *be an entire function, and let* (16) *hold, where* $s \neq 0$ *and F satisfies the conditions* (4), (5), (6), (7), (10) *and* (13). *Then* (17) *holds where* ψ *and G are defined as in Theorem* 1. *Furthermore, for any positive* ϵ *we shall always have* (11) *and*

$$\text{(18)} \qquad n^{\frac{1}{2}-\epsilon} < \psi(n) < n^{\frac{1}{2}+\epsilon}$$

for sufficiently large values of n. Thus the function (15) *can have only a finite number of gaps of magnitude* $(\nu, \nu + \nu^{\frac{1}{2}+\epsilon})$, *that is, for any positive* ϵ, *the equations*

$$a_n = a_{n+1} = \cdots = a_{n+[n^{\frac{1}{2}+\epsilon}]} = 0$$

can hold for at most a finite number of values of n.

As a consequence of Theorems 1 and 2 we shall derive the following non-linear Tauberian theorem.

THEOREM 7. *Let* $a_n \geqq 0$ $(n = 1, 2, \cdots)$, *and let*

$$c_n = \sum_1^n a_m\, a_{n-m} \qquad (n = 1, 2, \cdots). \tag{19}$$

If F fulfills the conditions (4), (5), (6) *and* (7) *and if* ψ *and G are defined as in Theorem* 1, *then the following two statements are equivalent*: (i) *the relation* (8) *holds for every positive value of* λ; (ii) *the relation*

$$\lim_{x\to\infty} \frac{(2\pi)^{\frac{1}{2}}}{\lambda} \sum_{x \leqq \sqrt{2}\psi(\frac{1}{2}n) \leqq x+\lambda} c_n e^{nG(\frac{1}{2}n) - 2F(G(\frac{1}{2}n))} = s^2 \tag{20}$$

holds for every positive value of λ.

For the proof of Theorem 1 we shall put

$$\xi = \varphi(u + \psi(n)), \tag{21}$$

so that u occurs as the difference of two variables

$$u = \varphi^{-1}(\xi) - \psi(n). \tag{22}$$

Here ϕ and ψ are functions to be determined with regard to the following considerations. We shall develop the exponent $n\xi - F(\xi)$ in a Taylor's series with remainder and we shall determine ϕ and ψ in such a manner that the coefficient of u shall vanish and that the coefficient of $\frac{1}{2}u^2$ shall be negative unity. Then we shall use the conditions on F to show that the remainder in the Taylor's series approaches zero as $\xi \to \infty$; indeed, that it approaches zero in such a manner that (2) shall imply that

$$\sum_1^\infty a_n\, e^{nG(n) - F(G(n))}\, e^{-\frac{1}{2}u^2} \to s \qquad (\xi \to \infty). \tag{23}$$

We shall then show that the General Tauberian Theorem of Wiener[2] is applicable to (23). Here the kernel $K(u)$ is $e^{-\frac{1}{2}u^2}$ whose Fourier transform $e^{-\frac{1}{2}x^2}$ is non-vanishing, and, as we have noted, u occurs as the difference of two variables. The relation (8) will follow immediately.

Theorem 2, the converse of Theorem 1, will follow by an application of the converse portion of the General Tauberian Theorem just referred to. The other theorems will follow at once.

2. Let us put

$$\xi = \varphi(u + \psi(n)). \tag{21}$$

[2] N. Wiener, *Tauberian theorems*, Annals of Mathematics, (2), vol. 33 (1932), pp. 1–100.

We have, formally, on holding n fast and developing in powers of u,

$$(24)\quad n\xi - F(\xi) = n\varphi(u + \psi(n)) - F(\varphi(u + \psi(n)))$$
$$= n\varphi(\psi(n)) - F(\varphi(\psi(n)))$$
$$+ u[n - F'(\varphi(\psi(n)))]\varphi'(\psi(n))$$
$$+ \frac{u^2}{2}\{[n - F'(\varphi(\psi(n)))]\varphi''(\psi(n)) - F''(\varphi(\psi(n)))[\varphi'(\psi(n))]^2\}$$
$$+ \frac{u^3}{6} D_v^3[n\varphi(v) - F(\varphi(v))]_{v=\theta u+\psi(n)},$$

where $0 \leqq \theta = \theta(u, n) \leqq 1$.

We desire that the coefficient of u shall be zero, and the coefficient of $\frac{1}{2}u^2$ negative unity, whether or not n is an integer. Necessary conditions for this are, formally, that

$$(25)\qquad F'(\varphi(\psi(w))) = w,$$

$$(26)\qquad \varphi'(\psi(w)) = \psi'(w),$$

for, as regards the latter, the coefficient of $\frac{1}{2}u^2$ is

$$(27)\qquad \{D_w[w\varphi'(\psi(w)) - F'(\varphi(\psi(w)))\varphi'(\psi(w))] - \varphi'(\psi(w))\}/\psi'(w).$$

Denoting by $G(w)$ the inverse function to $F'(w)$, we see that (25) and (26) become

$$(28)\qquad \varphi(\psi(w)) = G(w),$$

$$(29)\qquad [\psi'(w)]^2 = G'(w).$$

For our functions φ, ψ we are thus led to take

$$(30)\qquad \psi(w) = \int_a^w [G'(t)]^{\frac{1}{2}}\,dt, \qquad \varphi(w) = G(\psi^{-1}(w)),$$

where a is so large that F is four times continuously differentiable for $w \geqq a$ and $0 <$ const. $\leqq F''(w)$. With ψ, φ so defined, (24), with w in place of n, is valid for w and $u + \psi(w)$ sufficiently large and (25), (26), (28) and (29) are valid for $w \geqq a$.

The last term of (24) may be written

$$(31)\quad \frac{u^3}{6}\{w\varphi'''(\theta u + \psi(w)) - F'''(\varphi(\theta u + \psi(w)))[\varphi'(\theta u + \psi(w))]^3$$
$$- 3F''(\varphi(\theta u + \psi(w)))\varphi'(\theta u + \psi(w))\varphi''(\theta u + \psi(w))$$
$$- F'(\varphi(\theta u + \psi(w)))\varphi'''(\theta u + \psi(w))\}.$$

Differentiating (25) and using (26), we have

$$F''(\varphi(\psi(w)))[\varphi'(\psi(w))]^2 = 1. \tag{32}$$

For a general argument z, (32) becomes

$$F''(\varphi(z))[\varphi'(z)]^2 = 1, \tag{33}$$

and yields on differentiation

$$F'''(\varphi(z))[\varphi'(z)]^3 + 2F''(\varphi(z))\varphi'(z)\varphi''(z) = 0. \tag{34}$$

Thus if $\theta u + \psi(w) = z$, (31) becomes

$$\begin{aligned} &\frac{u^3}{6}\{w\varphi'''(z) - F''(\varphi(z))\varphi'(z)\varphi''(z) - F'(\varphi(z))\varphi'''(z)\} \\ &\qquad = \frac{u^3}{6}\left\{[F'(\varphi(\psi(w))) - F'(\varphi(z))]\varphi'''(z) - \frac{\varphi''(z)}{\varphi'(z)}\right\}. \end{aligned} \tag{35}$$

Now, by the law of the mean, there will be some v in the interval between $\psi(w)$ and z such that

$$F'(\varphi(\psi(w))) - F'(\varphi(z)) = -\theta u F''(\varphi(v))\varphi'(v) = -\frac{\theta u}{\varphi'(v)}. \tag{36}$$

Expression (35) thus becomes

$$-\frac{u^3}{6}\left\{\theta u\frac{\varphi'''(z)}{\varphi'(v)} + \frac{\varphi''(z)}{\varphi'(z)}\right\}, \tag{37}$$

and

$$e^{w\xi - F(\xi)} = \exp\left\{w\varphi(\psi(w)) - F(\varphi(\psi(w))) - \tfrac{1}{2}u^2 - \frac{u^3}{6}\left[\theta u\frac{\varphi'''(z)}{\varphi'(v)} + \frac{\varphi''(z)}{\varphi'(z)}\right]\right\}. \tag{38}$$

We shall now prove two facts relating to the function

$$A(u, \xi) = w\xi - F(\xi) - wG(w) + F(G(w)), \tag{39}$$

where

$$\xi = \varphi(u + \psi(w)). \tag{40}$$

(A) *For any positive* U

$$\max_{|u|\leqq U} |A(u, \xi) + \tfrac{1}{2}u^2| \to 0 \qquad (\xi \to \infty); \tag{41}$$

(B) *there are constants* w_0 *and* ξ_0 *such that, if*

$$2 \leqq |u|, \qquad u \leqq \varphi^{-1}(\xi) - \psi(w_0), \qquad \xi_0 \leqq \xi, \tag{42}$$

then

$$A(u, \xi) \leqq -|u|/2. \tag{43}$$

For the proof of (A) let us differentiate

$$\varphi'(\psi(x)) = [G'(x)]^{\frac{1}{2}} \tag{44}$$

to obtain

$$\varphi''(\psi(x))\psi'(x) = \tfrac{1}{2}[G'(x)]^{-\frac{1}{2}}G''(x). \tag{45}$$

Thus

$$\varphi''(\psi(x)) = \frac{G''(x)}{2G'(x)}; \tag{46}$$

$$\varphi'''(\psi(x))\psi'(x) = \frac{G'(x)G'''(x) - [G''(x)]^2}{2[G'(x)]^2}; \tag{47}$$

$$\frac{\varphi''(\psi(x))}{\varphi'(\psi(x))} = \frac{G''(x)}{2[G'(x)]^{\frac{3}{2}}}; \tag{48}$$

$$\frac{\varphi'''(\psi(x))}{\varphi'(\psi(x))} = \frac{G'(x)G'''(x) - [G''(x)]^2}{2[G'(x)]^3}. \tag{49}$$

Translated into terms of G, (6) and (7) become

$$G''(x) = o([G'(x)]^{\frac{3}{2}}) \tag{50}$$

$$G'''(x) = o([G'(x)]^2). \qquad (x \to \infty) \tag{51}$$

In view of (50) and (51) it follows from (48) and (49) that

$$\frac{\varphi''(x)}{\varphi'(x)} \to 0 \qquad (x \to \infty) \tag{52}$$

and

$$\frac{\varphi'''(x)}{\varphi'(x)} \to 0 \qquad (x \to \infty). \tag{53}$$

Now if $|z - v| <$ const., then

$$\frac{\varphi'(z)}{\varphi'(v)} \to 1 \qquad (z \to \infty). \tag{54}$$

since

$$\log \frac{\varphi'(z)}{\varphi'(v)} = \int_v^z \frac{\varphi''(x)}{\varphi'(x)}\, dx \to 0 \qquad (z \to \infty). \tag{55}$$

The relations (52), (53) and (54) yield the result (A).

For the proof[3] of (B) let us differentiate $A(u, \xi)$ partially with respect to u,

$$\begin{aligned}\frac{\partial A}{\partial u} = -\frac{\partial A}{\partial \psi(w)} = \frac{G(w) - \xi}{\psi'(w)} &= \frac{\varphi(\psi(w)) - \varphi(u + \psi(w))}{\varphi'(\psi(w))} \\ &= -u\,\frac{\varphi'(\theta u + \psi(w))}{\varphi'(\psi(w))}.\end{aligned} \tag{56}$$

[3] This proof has been revised in accordance with a suggesti n of J. J. Gergen.

Now by (54),

$$\frac{\partial A}{\partial u} \to -u \qquad (\xi \to \infty) \tag{57}$$

uniformly for $| u | \leqq 2$. Next let ξ_0 be determined so that

$$-u - \tfrac{1}{2} \leqq \frac{\partial A}{\partial u} \leqq -u + \tfrac{1}{2}; \qquad A(u, \xi) \leqq -\tfrac{1}{2}u^2 + 1, \tag{58}$$

for $| u | \leqq 2$, $\xi_0 \leqq \xi$ and let w_0 be so determined that

$$| G''(w) | \leqq [G'(w)]^{\frac{3}{2}} \tag{59}$$

for $\psi(w) \geqq \psi(w_0)$. We shall now show that (B) holds with these values ξ_0, w_0. In fact, let $\xi_0 \leqq \xi$ and let $\bar{u}$ be any value of u such that

$$\bar{u} \leqq \varphi^{-1}(\xi) - \psi(w_0) \tag{60}$$

and

$$\left| \frac{\partial A}{\partial u} \right]_{u=\bar{u}} \bigg| \leqq 1. \tag{61}$$

Then, since

$$\frac{\partial^2 A}{\partial u^2} = -1 + \tfrac{1}{2} \frac{G''(w)}{[G'(w)]^{\frac{3}{2}}} \frac{\partial A}{\partial u}, \tag{62}$$

it follows that

$$-\tfrac{3}{2} \leqq \frac{\partial^2 A}{\partial u^2} \bigg]_{u=\bar{u}} \leqq -\tfrac{1}{2}. \tag{63}$$

In view of (56) it follows that

$$\frac{\partial A}{\partial u} \begin{array}{ll} > 0 & \text{if } u < 0, \\ < 0 & \text{if } u > 0, \end{array} \tag{64}$$

and hence

$$\left| \frac{\partial A}{\partial u} \right| < 1 \qquad \text{for } u \text{ on the open interval } (0, \bar{u}). \tag{65}$$

Since

$$\frac{\partial A}{\partial u} \bigg]_{u=\frac{3}{2}} \leqq -1, \qquad \frac{\partial A}{\partial u} \bigg]_{u=-\frac{3}{2}} \geqq 1, \tag{66}$$

it follows that

$$| \bar{u} | \leqq \tfrac{3}{2}. \tag{67}$$

Accordingly,

$$\frac{\partial A}{\partial u} < -1 \qquad \text{for } \xi_0 \leqq \xi, \quad 2 \leqq u \leqq \varphi^{-1}(\xi) - \psi(w_0); \tag{68}$$

$$\frac{\partial A}{\partial u} > 1 \qquad \text{for } \xi_0 \leqq \xi, \quad u \leqq -2, \; u \leqq \varphi^{-1}(\xi) - \psi(w_0). \tag{69}$$

(B) then follows.

Setting

$$p(\xi) = \varphi^{-1}(\xi) - \psi(w_0), \tag{70}$$

we see that (A) and (B) imply not only that for any positive λ

$$\sum_{\nu=-\infty}^{[p(\xi)]} \max_{\nu-\lambda \leqq u \leqq \nu} [e^{w\xi - F(\xi) - wG(w) + F(G(w))}] < M_\lambda < \infty \tag{71}$$

for all sufficiently large ξ, but also that for any positive λ

$$\lim_{\xi\to\infty} \sum_{\nu=-\infty}^{[p(\xi)]} \max_{\nu-\lambda \leqq u \leqq \nu} | e^{-\frac{1}{2}u^2} - e^{w\xi - F(\xi) - wG(w) + F(G(w))} | = 0, \tag{72}$$

for the ends of each function run down, as we have seen, faster than $e^{-\frac{1}{4}|u|}$ and any finite portion of

$$\exp \{w\xi - F(\xi) - wG(w) + F(G(w))\} \tag{73}$$

approaches $e^{-\frac{1}{2}u^2}$ as $\xi \to \infty$.

Let us consider

$$\sum_{w_0}^{\infty} a_n e^{n\xi - F(\xi)} = \sum_{w_0}^{\infty} a_n e^{nG(n) - F(G(n))} e^{n\xi - F(\xi) - nG(n) + F(G(n))}. \tag{74}$$

Since this approaches s as $\xi \to \infty$, it follows that the sum is $\leqq s + 1$ for $\xi \geqq \xi_2$ for some ξ_2. But $\exp \{n\xi - F(\xi) - nG(n) + F(G(n))\}$ is greater than some constant $h_\lambda > 0$ for $0 \leqq u \leqq \lambda$, $\xi \geqq \xi_3$, for some $\xi_3 \geqq \xi_2$, where $\xi = \varphi(u + \psi(n))$. Hence

$$\sum_{0 \leqq u \leqq \lambda} a_n e^{nG(n) - F(G(n))} \leqq \frac{s+1}{h_\lambda}, \qquad \xi \geqq \xi_3. \tag{75}$$

Now $0 \leqq u \leqq \lambda$ is equivalent to

$$\varphi^{-1}(\xi) - \lambda \leqq \psi(n) \leqq \varphi^{-1}(\xi) \tag{76}$$

and hence

$$\sum_{\nu \leqq \psi(n) \leqq \nu + \lambda} a_n e^{nG(n) - F(G(n))} < M'_\lambda \qquad (\nu = 0, 1, 2, \cdots). \tag{77}$$

From these facts we prove immediately that

$$\sum_{n=w_0}^{\infty} a_n e^{nG(n) - F(G(n)) - \frac{1}{2}u^2} \to s \qquad (\xi \to \infty); \tag{78}$$

indeed

$$\begin{aligned} \Big| \sum_{w_0}^{\infty} a_n e^{nG(n) - F(G(n))} \{e^{-\frac{1}{2}u^2} - e^{n\xi - F(\xi) - nG(n) + F(G(n))}\} \Big| \\ \leqq \sum_{\nu=-\infty}^{[p(\xi)]} \Big\{ \Big[\max_{\nu-1 \leqq u \leqq \nu} | e^{-\frac{1}{2}u^2} - e^{n\xi - F(\xi) - nG(n) + F(G(n))} | \Big] \\ \cdot \sum_{\nu-1 \leqq u \leqq \nu} a_n e^{nG(n) - F(G(n))} \Big\} \to 0 \qquad (\xi \to \infty), \end{aligned} \tag{79}$$

and this proves (78). If we define

$$(80)\qquad L(y) = \begin{cases} (2\pi)^{\frac{1}{2}} \sum_{\psi(w_0)\leqq\psi(n)<y} a_n e^{nG(n)-F(G(n))} & \text{for } y > \psi(w_0), \\ 0 & \text{for } y \leqq \psi(w_0), \end{cases}$$

then (78) is equivalent to

$$(81)\qquad \int_{-\infty}^{\infty} e^{-\frac{1}{2}(x-y)^2}\, dL(y) \to s \int_{-\infty}^{\infty} e^{-\frac{1}{2}u^2}\, du \qquad (x \to \infty).$$

Now let us apply the General Tauberian Theorem of Wiener (loc. cit., pp. 27, 28) to (81). The kernel $e^{-\frac{1}{2}u^2}$ has the non-vanishing Fourier transform $e^{-\frac{1}{2}x^2}$; the function $L(y)$ is a monotone function such that

$$(82)\qquad \int_{\nu}^{\nu+\lambda} dL(y) = (2\pi)^{\frac{1}{2}} \sum_{\nu\leqq\psi(n)\leqq\nu+\lambda} a_n e^{nG(n)-F(G(n))} < (2\pi)^{\frac{1}{2}} M'_\lambda$$

and hence if

$$(83)\qquad K_2(u) = \begin{cases} 1 & \text{for } 0 \leqq -u \leqq \lambda, \\ 0 & \text{otherwise}, \end{cases}$$

then

$$(84)\qquad \int_{-\infty}^{\infty} K_2(x-y)\,dL(y) \to s \int_{-\infty}^{\infty} K_2(u)\,du = \lambda s.$$

That is,

$$(85)\qquad \int_{x}^{x+\lambda} dL(y) \to \lambda s \qquad (x \to \infty),$$

or

$$(86)\qquad \frac{(2\pi)^{\frac{1}{2}}}{\lambda} \sum_{x\leqq\psi(n)\leqq x+\lambda} a_n e^{nG(n)-F(G(n))} \to s \qquad (x \to \infty).$$

This concludes the proof of Theorem 1.

3. For the proof of Theorem 2 we note that the class Σ of functions of the form (83) is such that there is no x for which for every function $K_2(u)$ of Σ

$$(87)\qquad \frac{1}{(2\pi)^{\frac{1}{2}}} \int_{-\infty}^{\infty} K_2(u) e^{iux}\, du = 0.$$

Thus (84) implies[4] (81) and this proves Theorem 2. This argument shows the validity of Theorem 2 even if (8) holds only for two values of λ whose ratio is irrational.

[4] See Wiener, loc. cit., p. 26.

4. For the proof of Theorem 3 let (4), (5) and (10) hold. Then for some positive ϵ

$$G'(F'(x)) = \frac{1}{F''(x)} > \frac{1}{[F'(x)]^{1+\epsilon}} \qquad x > x_\epsilon, \tag{88}$$

and hence for every positive ϵ

$$x^{-1-\epsilon} < G'(x); \tag{89}$$

$$x^{-\frac{1}{2}-\epsilon} < \psi'(x), \tag{90}$$

for sufficiently large values of x. From (90) it follows that for every positive ϵ,

$$x^{\frac{1}{2}-\epsilon} < \psi(x) \tag{91}$$

for sufficiently large values of x. This proves part (i) of Theorem 3.

For part (ii) let (4), (5) and (13) hold. In view of (5) and (13) it follows that $G(x)$ is $O(x^\epsilon)$ for every $\epsilon > 0$. Applying the Schwarz inequality to (38), we have

$$\begin{aligned} \psi(x) &\leqq \left[\int_a^x dx \int_a^x G'(x)dx\right]^{\frac{1}{2}} \\ &= O(x^{\frac{1}{2}}[G(x)]^{\frac{1}{2}}) \\ &= O(x^{\frac{1}{2}+\epsilon}) \qquad \epsilon > 0, \end{aligned} \tag{92}$$

which yields (14).

5. Let (15) be an integral function. The condition (16) gives

$$\frac{1}{2\pi}\int_0^{2\pi} |f(re^{i\theta})|^2 d\theta = \sum_0^\infty |b_n|^2 r^{2n} \sim se^{F(2\log r)} \qquad (r \to \infty). \tag{93}$$

If we set

$$a_n = |b_n|^2, \qquad \xi = \log r^2, \tag{94}$$

(93) becomes

$$\sum_0^\infty a_n e^{n\xi - F(\xi)} \to s \qquad (\xi \to \infty). \tag{95}$$

Applying Theorem 1 to this case, we see that the conclusions of Theorem 4 follow. Theorem 5 is an immediate consequence of Theorem 2. In order to obtain the result on gaps stated in Theorem 6, let us assume that (11), (12) and (14) hold for sufficiently large values of n and let us consider the large values of n for which

$$x \leqq \psi(n) \leqq x + \lambda. \tag{96}$$

Let $g(x)$ be the inverse function to $\psi(x)$. Then the values of n for which (96) holds are those for which

$$g(x) \leqq n \leqq g(x + \lambda). \tag{97}$$

But

$$g(x+\lambda) = g(x) + \lambda g'(x+\theta\lambda) \qquad 0 \leqq \theta \leqq 1$$

(98)
$$= g(x) + \lambda \frac{1}{\psi'(g(x+\theta\lambda))}$$
$$= g(x) + \lambda O([g(x+\theta\lambda)]^{\frac{1}{2}+\epsilon})$$
$$= g(x) + \lambda O([g(x)]^{\frac{1}{2}+\epsilon}).$$

Thus the values of n for which (96) holds are those for which

(99)
$$g(x) \leqq n \leqq g(x) + \lambda O([g(x)]^{\frac{1}{2}+\epsilon})$$

and hence if $s \neq 0$, the function (15) can have only a finite number of gaps of magnitude $(\nu, \nu + \nu^{\frac{1}{2}+\epsilon})$.

6. Let $a_n \geqq 0$ $(n = 1, 2, \cdots)$, and let

(100)
$$c_n = \sum_1^n a_m a_{n-m} \qquad (n = 1, 2, \cdots).$$

If (8) holds for every positive value of λ, in view of Theorem 2 it follows that (2) holds and hence

(101)
$$\left[\sum_1^\infty a_n e^{n\xi}\right]^2 e^{-2F(\xi)} \to s^2 \qquad (\xi \to \infty);$$

(102)
$$\sum_1^\infty c_n e^{n\xi - 2F(\xi)} \to s^2 \qquad (\xi \to \infty).$$

In view of Theorem 1, (102) implies that

(103)
$$\lim_{x\to\infty} \frac{(2\pi)^{\frac{1}{2}}}{\lambda} \sum_{x \leqq \psi(n) \leqq x+\lambda} c_n e^{nG_1(n) - F_1(G_1(n))} = s^2,$$

where

(104)
$$F_1(x) = 2F(x); \qquad G_1(x) = G(\tfrac{1}{2}x);$$
$$G_1'(x) = \tfrac{1}{2}G'(\tfrac{1}{2}x); \qquad \psi_1'(x) = \frac{1}{\sqrt{2}}\psi'(\tfrac{1}{2}x);$$
$$\psi_1(x) = \sqrt{2}\,\psi(\tfrac{1}{2}x) + \text{const}.$$

Thus (103) is equivalent to (20). Similarly, if (20) holds for every positive value of λ, (102) holds and hence (2) holds. This shows the complete equivalence of the two statements contained in Theorem 7.

MASSACHUSETTS INSTITUTE OF TECHNOLOGY.

TAYLOR'S SERIES OF FUNCTIONS OF SMOOTH GROWTH IN THE UNIT CIRCLE

By W. T. Martin and N. Wiener

1. **Introduction.** We have recently proved a series of results connecting the growth of an entire function with the growth of certain expressions depending on the coefficients.[1] There is a closely related series of results which concern, not the growth of an entire function, but the growth of a function as it approaches the circle of convergence. As will be seen in the formulation of our results, there is a precise methodological similarity between theorems of the two types, and, indeed, they differ merely by the fact that a parameter which is negative in one case is positive in the other. It is nevertheless worth while to give an explicit formulation of the theorems because they are of less familiar character than the other set of theorems and involve results which have only recently been obtained by Vijayaraghavan and Wiener, Avakumović and Karamata, and Pitt (see footnotes 4, 6) in contrast with the theorems of the entire type which are best exemplified by the classical theory of the Borel summation. The circle of convergence theorems are more directly applicable to a series of interesting problems in the analytical theory of numbers, and, in particular, they allow us to carry Tauberian methods in the problem of partitions much further than Hardy and Ramanujan at one time believed possible.[2] There is a considerable prospect of their further utility in unexplored portions of this field. It is true that the revolutionary work of Rademacher[3] overshadows all less perfect methods, whether Tauberian or not. We wish to call attention to the fact that it is possible by an artifice to eliminate the condition of positivity which is necessary for Tauberian theorems of the type considered, or, more accurately, to guarantee its satisfaction by a consideration not of a singularity of a power series at an individual point on the circle of convergence, but rather the behavior of the integral of the square of its modulus as we approach the circle of convergence.

There will be many places in which the detail of proof is so similar to that of our previous paper that anything like completeness is scarcely necessary. Where this is the case we shall consider ourselves at liberty to present our argument in a highly schematic form and to refer to our previous paper for technical details.

Received January 12, 1938.

[1] N. Wiener and W. T. Martin, *Taylor's series of entire functions of smooth growth*, this Journal, vol. 3(1937), pp. 213–223.

[2] G. H. Hardy and S. Ramanujan, *Asymptotic formulae in combinatory analysis*, Proc. of the London Math. Soc., vol. 17(1918), pp. 75–115. See especially pp. 89, 90.

[3] H. Rademacher, *On the partition function* $p(n)$, Proc. of the London Math. Soc., vol. **43** (1937), pp. 241–254.

Let $a_n \geqq 0$ $(n = 1, 2, \cdots)$, and let

$$\sum_1^\infty a_n x^n \sim sH(x) \qquad (x \to 1^-). \tag{1.1}$$

We may rewrite (1.1) in the form

$$\sum_1^\infty a_n e^{F(\xi) - n\xi} \to s \qquad (\xi \to 0^+), \tag{1.2}$$

where

$$F(\xi) = -\log H(e^{-\xi}). \tag{1.3}$$

We shall derive the following Tauberian theorem.

THEOREM 1. *Let* $a_n \geqq 0$ $(n = 1, 2, \cdots)$, *and let* (1.2) *hold, where*
(1.4) *F is four times continuously differentiable for* $0 < \xi \leqq \xi_0$ *for some positive* ξ_0 ;

$$F''(\xi) \leqq \text{const.} < 0 \quad \textit{for } 0 < \xi \leqq \xi_0\, ; \tag{1.5}$$

$$\int_\xi^{\xi_0} [-F''(t)]^{\frac{1}{2}}\, dt \to \infty \qquad (\xi \to 0^+); \tag{1.6}$$

$$F'''(\xi) = o([-F''(\xi)]^{\frac{3}{2}}) \qquad (\xi \to 0^+); \tag{1.7}$$

$$F^{\mathrm{iv}}(\xi) = o([F''(\xi)]^2) \qquad (\xi \to 0^+). \tag{1.8}$$

Then for every positive λ,

$$\lim_{x\to\infty} \frac{(2\pi)^{\frac{1}{2}}}{\lambda} \sum_{x \leqq \psi(n) < x+\lambda} a_n e^{F(G(n)) - nG(n)} = s, \tag{1.9}$$

where $\psi(x)$ *is defined by*

$$\int_a^x [-G'(x)]^{\frac{1}{2}}\, dx = \psi(x) \tag{1.10}$$

and $G(x)$ (*for* $x > a = F'(\xi_0)$) *is the inverse function to* $F'(\xi)$; *that is,* $G(F'(\xi)) = \xi$ *for* $0 < \xi \leqq \xi_0$.

Theorem 1 and Theorem 4, which we will deduce from Theorem 1, have important applications in the analytical theory of numbers. In a later paper we hope to present some results based upon them.

A special case of Theorem 1, due to Vijayaraghavan and Wiener (in an unpublished paper), is the following

THEOREM A. *Let* $a_n \geqq 0$ $(n = 1, 2, \cdots)$, *and let*

$$\sum_1^\infty a_n x^n \sim s \exp[(1-x)^{-1}] \qquad (x \to 1^-). \tag{1.11}$$

Then for every positive λ

$$\lim_{N\to\infty} \frac{(2\pi)^{\frac{1}{2}}}{\lambda} \sum_{N \leqq n < N+\lambda N^{\frac{3}{4}}} a_n \exp\left[-2n^{\frac{1}{2}}\right] = s\,\frac{e^{\frac{1}{2}}}{2^{\frac{1}{4}}}. \tag{1.12}$$

To derive Theorem A from Theorem 1 we write $x = e^{-\xi}$. Then (1.11) becomes

$$\sum a_n e^{-n\xi} \sim s \exp\left\{\frac{1}{1-e^{-\xi}}\right\} = se^{1/\xi} \exp\left\{\frac{1}{1-e^{-\xi}} - \frac{1}{\xi}\right\}. \tag{1.13}$$

Noting that

$$\exp\left\{\frac{1}{1-e^{-\xi}} - \frac{1}{\xi}\right\} \to e^{\frac{1}{2}} \quad \text{as } \xi \to 0^+,$$

we see that (1.13) becomes

$$\sum a_n \exp\left\{-\frac{1}{\xi} - n\xi\right\} \to se^{\frac{1}{2}} \qquad (\xi \to 0^+). \tag{1.14}$$

Thus for this case $F(\xi) = -\xi^{-1}$, and it is easily verified that this function satisfies conditions (1.4), $\cdots$, (1.8). Moreover

$$G(x) = x^{-\frac{1}{2}}, \qquad \psi(x) = 2^{\frac{1}{2}} x^{\frac{1}{4}}$$

and (1.9) becomes

$$\lim_{x\to\infty} \frac{(2\pi)^{\frac{1}{2}}}{\lambda} \sum_{x \leqq 2^{\frac{1}{2}} n^{\frac{1}{4}} < x+\lambda} a_n \exp\left[-2n^{\frac{1}{2}}\right] = se^{\frac{1}{2}} \tag{1.15}$$

from which (1.12) easily follows.

Avakumović and Karamata have proved the following[4]

THEOREM B. *Let* $A(y)$ *be* $\geqq 0$ *and non-decreasing for* $y \geqq 0$ *and let*

$$J(x) = \int_0^\infty e^{-y/x} A(y)\, dy$$

converge for $x > 0$. *Let*[5]

$$J(x) \asymp x^k e^x \qquad (x \to \infty) \tag{1.16}$$

for some real k. *Then*

$$\exp\left[2x^{\frac{1}{2}} - (2+\epsilon)x^{\frac{1}{4}}(\log x)^{\frac{1}{2}}\right] < A(x) < Mx^{\frac{1}{2}(k-1)} \exp\left[2x^{\frac{1}{2}}\right]. \tag{1.17}$$

H. R. Pitt has proved a theorem which generalizes Theorem B, at the same time furnishing more precise results.[6] His theorem in this particular case is

[4] Avakumović and J. Karamata, *Über einige Taubersche Sätze, deren Asymptotik von Exponentialcharacter ist*. I, Mathematische Zeitschrift, vol. 41(1936), pp. 345–356.

[5] Hardy writes $f(x) \asymp g(x)$ $(x \to \infty,\ g(x) > 0)$ if there are two constants $0 < \delta < \Delta < \infty$ such that $\delta < f(x)/g(x) < \Delta$ for $x > x_0$.

[6] Pitt discovered and proved his theorem independently of our results. In conversation with him we discussed the results and later he showed how the case $k = 0$ of his theorem is deducible from ours. In §3 we shall give the proof of this portion of Pitt's theorem as submitted to us by him.

THEOREM C. *Suppose that* $A(y)$ *is non-decreasing in* $(0, \infty)$ *and that*

$$J(x) = \int_0^\infty e^{-y/x}\, dA(y)$$

converges absolutely for $x > 0$. *Let* k *be any real number,*

$$\alpha > 0, \qquad B = \alpha^{-\alpha/(1+\alpha)}(1 + \alpha),$$
$$h(x) = e^{-Bx^2} x^{1-2k/\alpha} A(x^{2(1+\alpha)/\alpha}).$$

Let

$$J(x) \sim x^k \exp [x^\alpha] \qquad (x \to \infty).$$

Then

$$e^{-\epsilon(x)} \leqq h(x) \leqq \epsilon(x), \tag{1.18}$$

where

$$\lim_{x \to \infty} \frac{\epsilon(x)}{x} = 0.$$

As a special case of Theorem C Pitt has obtained the following extension of Avakumović's and Karamata's result.

THEOREM D. *If* $A(y)$ *is non-decreasing in* $(0, \infty)$ *and*

$$J(x) \sim x^k e^x \qquad (x \to \infty),$$

then

$$\exp [2x^{\frac{1}{2}} - o(x^{\frac{1}{2}})] \leqq A(x) \leqq o(x^{\frac{1}{2}k} \exp [2x^{\frac{1}{2}}]). \tag{1.19}$$

As a converse to Theorem 1 we shall prove

THEOREM 2. *Let* $a_n \geqq 0$ $(n = 1, 2, \cdots)$, *and let* (1.9) *hold for every positive value of* λ, *where* F *fulfills the conditions* (1.4), $\cdots$, (1.8) *and* ψ *and* G *are defined as in Theorem* 1. *Then* (1.2) *holds.*

In the next theorem we place some additional restrictions on F and we obtain certain inequalities related to ψ.

THEOREM 3. (i) *Let* F *fulfill conditions* (1.4), (1.5), (1.6) *and the additional condition:*

(1.20) *for some* p *in* $\frac{1}{2} < p < 1$ *and for all sufficiently small positive* ϵ,

$$-F''(\xi) = O([F'(\xi)]^{2p+\epsilon}) \qquad (\xi \to 0^+).$$

Then for every positive ϵ

$$n^{-p-\epsilon} < \psi'(n) \tag{1.21}$$

and

$$n^{1-p-\epsilon} < \psi(n) \tag{1.22}$$

for sufficiently large values of n.

(ii) *Let F fulfill conditions* (1.4), (1.5), (1.6) *and the additional condition*:
(1.23) *for some p in* $\frac{1}{2} < p < 1$ *and for all sufficiently small positive* ϵ,

$$[F'(\xi)]^{2p-\epsilon} < -F''(\xi)$$

for sufficiently small (positive) values of ξ.
Then for every positive ϵ,

$$\psi(n) < n^{1-p+\epsilon} \tag{1.24}$$

for sufficiently large values of n.

As immediate consequences of these theorems we shall obtain the following theorems on smoothly growing functions analytic in the unit circle.

THEOREM 4. *Let*

$$f(z) = \sum_0^\infty b_n z^n \tag{1.25}$$

be analytic for $|z| < 1$, *and let*

$$\frac{1}{2\pi}\int_0^{2\pi} |f(re^{i\vartheta})|^2 d\vartheta \sim se^{-F(-2\log r)} \qquad (r \to 1^-), \tag{1.26}$$

where F fulfills conditions (1.4), $\cdots$, (1.8). *Then for every positive* λ,

$$\lim_{x\to\infty} \frac{(2\pi)^{\frac{1}{2}}}{\lambda} \sum_{x \leq \psi(n) < x+\lambda} |b_n|^2 e^{F(G(n))-nG(n)} = s, \tag{1.27}$$

where ψ *and G are defined as in Theorem* 1.

The converse to Theorem 4 is

THEOREM 5. *Let* (1.25) *be analytic for* $|z| < 1$, *and let* (1.27) *hold for every positive value of* λ, *where F fulfills conditions* (1.4), $\cdots$, (1.8) *and* ψ *and G are defined as in Theorem* 1. *Then* (1.26) *holds.*

If we place the additional restrictions (1.20) and (1.23) on F, we have the following gap theorem.

THEOREM 6. *Let* (1.25) *be analytic for* $|z| < 1$, *and let* (1.26) *hold, where* $s \neq 0$ *and F satisfies conditions* (1.4), $\cdots$, (1.8), (1.20), (1.23). *Then* (1.27) *holds where* ψ *and G are defined as in Theorem* 1. *Furthermore, for every positive* ϵ *we shall always have* (1.21) *and*

$$n^{1-p-\epsilon} < \psi(n) < n^{1-p+\epsilon} \tag{1.28}$$

for sufficiently large values of n. Thus the function (1.25) *can have only a finite number of gaps of magnitude* $(\nu, \nu + \nu^{p+\epsilon})$, *that is, for any positive* ϵ, *the equations*

$$a_n = a_{n+1} = \cdots = a_{n+[n^{p+\epsilon}]} = 0$$

can hold for at most a finite number of values of n.

A slightly more general form of Theorem 1, which we shall first prove, is

THEOREM 7. *Let* $\{\lambda_n\}$ *be any sequence such that*

$$(1.29) \qquad 0 \leqq \lambda_0 < \lambda_1 < \lambda_2 < \cdots < \lambda_n < \cdots, \qquad \lim_{n\to\infty} \lambda_n = \infty;$$

let $a_n \geqq 0$ $(n = 1, 2, \cdots)$, *and let*

$$(1.30) \qquad \sum_1^\infty a_n e^{F(\xi)-\lambda_n \xi} \to s \qquad (\xi \to 0^+),$$

where F *satisfies conditions* (1.4), $\cdots$, (1.8). *Then for every positive* λ,

$$(1.31) \qquad \lim_{x\to\infty} \frac{(2\pi)^{\frac{1}{2}}}{\lambda} \sum_{x \leqq \psi(\lambda_n) < x+\lambda} a_n e^{F(G(\lambda_n))-\lambda_n G(\lambda_n)} = s,$$

where ψ *and* G *are defined as in Theorem* 1. *Conversely, if* $\{\lambda_n\}$ *is any sequence satisfying* (1.29), *if* $a_n \geqq 0$ $(n = 1, 2, \cdots)$, *and if* (1.31) *holds for every positive value of* λ, *then* (1.30) *holds.*

2. For the proof of Theorem 7 let us consider the exponent in (1.30) with a general argument w instead of λ_n: $F(\xi) - w\xi$. Let us place

$$(2.1) \qquad \xi = \varphi(u + \psi(w)).$$

Developing the exponent $F(\xi) - w\xi$ in a Taylor's series with remainder and determining φ and ψ in such a manner that the coefficient of u vanishes and that the coefficient of $\frac{1}{2}u^2$ is negative unity, we find

$$(2.2) \qquad \psi(w) = \int_a^w [-G'(x)]^{\frac{1}{2}}\, dx, \qquad \varphi(w) = G(\psi^{-1}(w)),$$

where $G(x)$ is the function inverse to $F'(\xi)$: $G(F'(\xi)) = \xi$. The details here are entirely analogous to those of our earlier paper (see reference in footnote 1). With these functions φ, ψ, using the conditions on F, we find that the remainder in the Taylor's series approaches zero; indeed, it approaches zero in such a manner that (1.31) implies that

$$(2.3) \qquad \sum_1^\infty a_n e^{F(G(\lambda_n))-\lambda_n G(\lambda_n)} \to s \qquad (\xi \to 0^+).$$

It is an easy step from this to apply Wiener's General Tauberian Theorem[7] to (2.3). Here the kernel $K(u)$ is $e^{-\frac{1}{2}u^2}$ whose Fourier transform $e^{-\frac{1}{2}x^2}$ is non-vanishing, and u occurs as the difference of two variables: $u = \varphi^{-1}(\xi) - \psi(w)$. The relation (1.31) follows at once.

The second part of Theorem 7 follows at once from the converse portion of Wiener's Tauberian Theorem (see reference in footnote 7, p. 26).

Theorems 1 and 2 are special cases of Theorem 7 for $\lambda_n = n$. Theorems 4 and 5 follow at once from these theorems.

[7] N. Wiener, *Tauberian theorems*, Ann. of Math., (2), vol. 33(1932), pp 1–100.

Using the definition (2.2) of $\psi(w)$ and the hypotheses on F in Theorem 3, one easily obtains the conclusions of Theorem 3. Theorem 6 follows immediately from Theorems 3 and 4.

3. In this section we shall show how the portion of Pitt's theorem (Theorem C of this paper) for which $k = 0$ is deducible from Theorem 1. The method used in showing this is essentially due to Pitt. Using his method, we first prove the following result from which we will deduce the desired result.

THEOREM E. *Let $A(y)$ be non-decreasing for $0 < y < \infty$, and let $F(\xi)$ satisfy conditions* (1.4), $\cdots$, (1.8). *If*

$$\int_0^\infty e^{-y\xi}\, dA(y) \sim se^{-F(\xi)} \qquad (\xi \to 0^+), \tag{3.1}$$

then for every positive λ there is an x_λ such that

$$(2\pi)^{-\frac{1}{2}}\tfrac{1}{2}s\lambda e^{-R(g(x-\lambda))} \leqq A(g(x)) \leqq h_\lambda(x-\lambda)e^{-R(g(x))} + H_\lambda \sum_{1\leqq n<x/\lambda} e^{-R(g(x-n\lambda))} \tag{3.2}$$

for $x \geqq x_\lambda$, where

$$R(x) = F(G(x)) - xG(x), \tag{3.3}$$

$$g(x) = \psi^{-1}(x) \qquad (\psi(g(x)) \equiv x), \tag{3.4}$$

$$h_\lambda(x) = \int_{g(x)}^{g(x+\lambda)} e^{R(y)}\, dA(y), \qquad H_\lambda = \operatorname*{l.u.b.}_{0\leqq x<\infty} h_\lambda(x), \tag{3.5}$$

and G and ψ are defined as in Theorem 1.

As a consequence of Theorem 1 we have for every positive λ

$$\lim_{x\to\infty} h_\lambda(x) = s(2\pi)^{-\frac{1}{2}}\lambda. \tag{3.6}$$

Furthermore, since

$$R'(y) = F'(G(y))G'(y) - yG'(y) - G(y) = -G(y) < 0,$$

it follows that $R(y)$ is decreasing. Since $R(y)$ is decreasing and $A(y)$ is non-decreasing, we have from (3.5)

$$e^{R(g(x+\lambda))}\int_{g(x)}^{g(x+\lambda)} dA(y) \leqq h_\lambda(x) \leqq e^{R(g(x))}\int_{g(x)}^{g(x+\lambda)} dA(y). \tag{3.7}$$

As a matter of convenience let us take $A(0) = 0$. Then from (3.7)

$$\begin{aligned} e^{R(g(x+\lambda))}[A(g(x+\lambda)) - A(g(x))] &\leqq h_\lambda(x) \\ &\leqq e^{R(g(x))}\int_0^{g(x+\lambda)} dA(y) = e^{R(g(x))}A(g(x+\lambda)). \end{aligned} \tag{3.8}$$

Using (3.6) and the second inequality in (3.8), we have

$$(3.9)\qquad \begin{aligned} A(g(x)) &\geqq h_\lambda(x-\lambda)e^{-R(g(x-\lambda))} \\ &\geqq \tfrac{1}{2}s(2\pi)^{-\frac{1}{2}}\lambda e^{-R(g(x-\lambda))} \end{aligned} \qquad (x \geqq x_\lambda).$$

The first inequality in (3.8), together with (3.6), gives

$$(3.10)\qquad \begin{aligned} A(g(x)) &\leqq \sum_{0\leqq n<x/\lambda} [A(g(x-n\lambda)) - A(g(x-\overline{n+1}\lambda))] \\ &\leqq \sum_{0\leqq n<x/\lambda} h_\lambda(x-\overline{n+1}\lambda)e^{-R(g(x-n\lambda))} \\ &\leqq h_\lambda(x-\lambda)e^{-R(g(x))} + \sum_{1\leqq n<x/\lambda} h_\lambda(x-\overline{n+1}\lambda)e^{-R(g(x-n\lambda))} \\ &\leqq h_\lambda(x-\lambda)e^{-R(g(x))} + H_\lambda \sum_{1\leqq n<x/\lambda} e^{-R(g(x-n\lambda))}. \end{aligned}$$

This concludes the proof of Theorem E.

We now interpret this result for the case

$$(3.11)\qquad F(\xi) = -\xi^{-\alpha}, \qquad \alpha > 0; \qquad s \neq 0.$$

Then

$$(3.12)\qquad \begin{aligned} F'(\xi) &= \alpha\xi^{-\alpha-1}, \qquad G(x) = \frac{\alpha^{1/(\alpha+1)}}{x^{1/(\alpha+1)}}, \\ \psi(x) &= 2(\alpha+1)^{\frac{1}{2}}\alpha^{-\frac{1}{2}}\alpha^{-\frac{1}{2}\alpha/(\alpha+1)}x^{\frac{1}{2}\alpha/(\alpha+1)} \\ &= 2\alpha^{-\frac{1}{2}}B^{\frac{1}{2}}x^{\frac{1}{2}\alpha/(\alpha+1)} \qquad (B = (\alpha+1)\alpha^{-\alpha/(\alpha+1)}), \\ R(x) &\equiv F(G(x)) - xG(x) = -Bx^{\alpha/(\alpha+1)}, \\ g(x) &= \left(\frac{\alpha x^2}{4B}\right)^{(\alpha+1)/\alpha} \end{aligned}$$

The relation (3.2) becomes

$$(3.13)\qquad \begin{aligned} &(2\pi)^{-\frac{1}{2}}\tfrac{1}{2}s\lambda \exp\left[B\frac{\alpha}{4B}(x-\lambda)^2\right] \leqq A\left(\left(\frac{\alpha x^2}{4B}\right)^{(\alpha+1)/\alpha}\right) \\ &\leqq h_\lambda(x-\lambda)\exp\left[B\frac{\alpha}{4B}x^2\right] + H_\lambda \sum_{1\leqq n<x/\lambda} \exp\left[B\frac{\alpha}{4B}(x^2-2n\lambda x+n^2\lambda^2)\right]. \end{aligned}$$

The first inequality gives

$$(3.14)\qquad A(x^{2(\alpha+1)/\alpha})e^{-Bx^2} \geqq (2\pi)^{-\frac{1}{2}}\tfrac{1}{2}s\lambda \exp[-\alpha^{\frac{1}{2}}B^{\frac{1}{2}}\lambda x],$$

$$(3.15)\qquad \liminf_{x\to\infty} x^{-1}[\log A(x^{2(\alpha+1)/\alpha}) - Bx^2] \geqq -B^{\frac{1}{2}}\lambda.$$

Since this is true for every positive λ,

$$(3.16)\qquad \liminf_{x\to\infty} x^{-1}[\log A(x^{2(\alpha+1)/\alpha}) - Bx^2] \geqq 0.$$

Hence

$$A(x^{2(\alpha+1)/\alpha})e^{-Bx^2} \geqq e^{-o(x)}. \tag{3.17}$$

The second inequality in (3.13) gives

$$\begin{aligned} \exp\left[-B\frac{\alpha}{4B}x^2\right]A\left(\left(\frac{\alpha x^2}{4B}\right)^{(\alpha+1)/\alpha}\right) &\leqq h_\lambda(x-\lambda) + H_\lambda \sum_{1\leqq n<x/\lambda} e^{-\frac{1}{4}\alpha n\lambda(2x-n\lambda)} \\ &\leqq h_\lambda(x-\lambda) + H_\lambda \sum_{1\leqq n<x/\lambda} e^{-\frac{1}{4}\alpha n\lambda x} \\ &\leqq h_\lambda(x-\lambda) + H_\lambda \frac{e^{-\frac{1}{4}\alpha\lambda x} - e^{-\frac{1}{4}\alpha\lambda x[x/\lambda]}}{1-e^{-\frac{1}{4}\alpha\lambda x}}. \end{aligned} \tag{3.18}$$

Hence

$$\limsup_{x\to\infty} \exp\left[-B\frac{\alpha}{4B}x^2\right]A\left(\left(\frac{\alpha x^2}{4B}\right)^{(\alpha+1)/\alpha}\right) \leqq s(2\pi)^{-\frac{1}{2}}\lambda. \tag{3.19}$$

Since λ is arbitrary,

$$\limsup_{x\to\infty} e^{-Bx^2}A(x^{2(\alpha+1)/\alpha}) \leqq 0. \tag{3.20}$$

Hence

$$e^{-Bx^2}A(x^{2(\alpha+1)/\alpha}) = o(1). \tag{3.21}$$

The relations (3.17) and (3.21) yield the conclusion of Pitt's theorem for $k = 0$.

MASSACHUSETTS INSTITUTE OF TECHNOLOGY.

Comments on [37a], [38d]

W. Fuchs

These two papers are rather isolated and have not led to further developments. They have connections to the Wiman-Valiron theory of the relation between maximum modulus and maximum term of a power series. For a special class of entire functions, the "admissible" functions, W. K. Hayman[1] proved the following:

Let $H(z) = \Sigma a_n z^n$ ($a_n \geqslant 0$) be admissible, write (with Wiener and Martin) $F(\log r) = \log H(r)$. Define $G(r)$ by $F'(G(r)) = r$. Then

$$a_n e^{nG(n)} \sim (2\pi F''(G(n))^{-1/2} e^{F(G(n))} \qquad (n \to \infty).$$

Since

$$F''(G(r)) = G'^{-1}(r),$$

we have

$$((2\pi)^{1/2}/\lambda) \sum_{n=n_1}^{n=n_2} a_n e^{nG(n)-F(G(n))} \sim (1/\lambda) \sum_{n_1}^{n_2} G'^{1/2}(n) \qquad (n_1 > n_2 \to \infty).$$

Assuming that G' is sufficiently smooth,

$$\sum_{n_1}^{n_2} G'^{1/2}(n) \sim \int_{n_1}^{n_2} G'^{1/2}(x)dx \sim \psi(n_2) - \psi(n_1),$$

which yields (8) of [37a]. The class of admissible functions of Hayman is not identical with the class of series considered in [37a], so that his paper does not provide an alternative approach to a proof of the results of [37a]. But the discussion shows that the statements of [37a] concern a fairly complicated expression whose behavior can be described in terms of quantities connected with the Wiman-Valiron theory.

Reference

1. W. K. Hayman, *A generalization of Stirling's formula*, J. f. r. angew. M. 196(1956), 67–95.

A GENERALIZATION OF IKEHARA'S THEOREM

BY N. WIENER AND H. R. PITT

1. Suppose that $s(y)$ is measurable, $s(y) \geq -K$, and that $s(y)e^{-\sigma y}$ is integrable in $(0, \infty)$ for any positive σ. Let $\omega = \sigma + it$,

$$S(\omega) = \int_0^\infty e^{-\omega y} s(y)\, dy. \tag{1.1}$$

Let $B > 0$. If $0 < \alpha \leq 1$, we denote by C_α the curve

$$| t | = B\sigma^\alpha, \qquad \sigma \geq 0, \tag{1.2}$$

and denote the y-axis by C_0.

One of the conditions in our main theorem in that $S(\omega)$ is integrable along C_α. We first explain what this involves. If $0 < \alpha \leq 1$, we suppose that $S(\omega)$ is absolutely integrable on C_α and that we have the usual relation

$$S(\omega) = \frac{1}{2\pi i} \int_{C_\alpha} \frac{S(z)\, dz}{z - \omega} \tag{1.3}$$

for points ω inside C_α. If $\alpha = 0$, we suppose that $S(\omega)$ can be defined for $\sigma = 0$ in such a way that for any positive A

$$S(\omega) = \frac{1}{2\pi i} \int_{C'} \frac{S(z)\, dz}{z - \omega}, \tag{1.4}$$

where C' consists of the three lines $t = \pm A$, $\sigma \geq 0$ and $| t | \leq A$, $\sigma = 0$.

Our main object is to prove the following result.

THEOREM 1. *Suppose that*

$$A > 0, \qquad 0 \leq \alpha < 1, \tag{1.5}$$

$$s(y) \geq -K, \tag{1.6}$$

and that $S(\omega)$, defined as above, is integrable on C_α. Then

$$\lim_{n \to \infty} x^{-\alpha} \int_{x - Ax^\alpha}^{x + Ax^\alpha} s(y)\, dy = 0.$$

247

We prove the theorem first in the extreme cases $\alpha = 0$, $\alpha = 1$, and then show how these are related to known results. In particular, we deduce Ikehara's theorem[1] from the case $\alpha = 0$.

We may plainly suppose that $s(y) = 0$ for $y \leq 1$. Then

$$
\begin{aligned}
(1.7) \qquad |S(\sigma + it)| &\leq \int_1^\infty e^{-\sigma y} |s(y)| \, dy \leq e^{-\frac{1}{2}\sigma} \int_1^\infty e^{-\frac{1}{2}\sigma y} s(y) \, dy \\
&= 0(e^{-\frac{1}{2}\sigma}) \quad \text{as} \quad \sigma \to \infty .
\end{aligned}
$$

This is enough to ensure that (1.3) and (1.4) are true, by Cauchy's theorem, when $S(\omega)$ is continuous on C_α at the origin.

2. LEMMA 1. *If $s(y) \geq -K$ and $S(\omega)$ is integrable on C_1, then*

$$\lim_{x \to \infty} x^{-1} \int_0^x s(y) \, dy = 0.$$

[Since $s(y) = 0$ for $y \leq 0$, this is plainly equivalent to the case $\alpha = 1$ of the theorem.]

We have, using (1.3),

$$S(\sigma) = \frac{1}{2\pi i} \int_{C_1} \frac{S(\omega) \, d\omega}{\omega - \sigma},$$

$$|\sigma S(\sigma)| \leq \frac{1}{2\pi} \int_{C_1} \left| \frac{S(\omega) \, d\omega}{\omega - \sigma} \right| = 0(1) \qquad (\sigma \to 0),$$

since $\sigma |\omega - \sigma|^{-1}$ is bounded on C_1 and tends to 0 as $\sigma \to 0$ at each point of C_1 except the origin. Since $s(y) \geq -K$, the conclusion follows from a well known theorem[2] of Hardy and Littlewood.

LEMMA 2. *If $A > 0$, $s(y) \geq -K$ and $S(\omega)$ is integrable on C_0, then*

$$\lim_{x \to \infty} \int_{x-A}^{x+A} s(y) \, dy = 0.$$

If $\lambda > 0$, we can write[3]

$$
\begin{aligned}
\int e^{-ity} h_\lambda(x - y) \, dy &= \left(1 - \frac{|t|}{\lambda}\right) e^{-itx} \qquad (|t| \leq \lambda), \\
&= 0 \qquad (|t| > \lambda),
\end{aligned}
$$

[1] S. Ikehara, Journal of Math. and Phys., M. I. T. (1931), 1–12. S. Bochner, "*Ein Satz von Landau und Ikehara.*" Math. Zeitschrift 37 (1933), 1–9.

[2] G. H. Hardy and J. E. Littlewood, "*Tauberian Theorems concerning power series and Dirichlet series whose coefficients are positive.*" Proc. London Math. Soc., 13 (1913), 174–198. We use an extension of Theorem 8 with $L(x) = 1$, $\alpha = 1$.

[3] Integrals in which the limits are omitted are over the range $(-\infty, \infty)$.

where

$$h_\lambda(x) = \frac{\sin^2 \lambda x}{\pi \lambda x^2}.$$

Also

$$\int e^{-ity} e^{-\sigma y} s(y)\, dy = S(\sigma + it),$$

and it follows from the Fourier transform analogue of Parseval's theorem that

$$(2.1) \quad \int h_\lambda(x - y) e^{-\sigma y} s(y)\, dy = \int_{-\lambda}^{\lambda} \left(1 - \frac{|t|}{\lambda}\right) e^{-itx} S(\sigma + it)\, dt.$$

We take C' to be the three lines $t = \pm 2\lambda$, $\sigma \geq 0$, and $|t| \leq 2\lambda$, $\sigma = 0$. Then it follows from (1.4) and the fact that $S(\omega)$ is absolutely integrable on C' that

$$\frac{1}{2\pi i} \int_{C'} \frac{S(z)\, dz}{z - \omega} = 0$$

if z is outside C'. Hence, if $|t| \leq \lambda$,

$$S(\sigma + it) = \frac{1}{2\pi i} \int_{C'} \frac{S(z)\, dz}{z - \sigma - it}, \qquad 0 = \frac{1}{2\pi i} \int_{C'} \frac{S(z)\, dz}{z + \sigma - it},$$

and we obtain by subtraction

$$S(\sigma + it) = \frac{1}{2\pi} \int_{-2\lambda}^{2\lambda} \frac{2\sigma S(iy)\, dy}{(y - t)^2 + \sigma^2} + \epsilon(\sigma, t),$$

where

$$|\epsilon(\sigma, t)| \leq \frac{2\sigma}{\pi(\lambda^2 - \sigma^2)} \int_{C'} |S(z)|\, dz.$$

It is easy to deduce from this that

$$\lim_{\sigma \to 0} \int_{-\lambda}^{\lambda} |S(\sigma + it) - S(it)|\, dt = 0,$$

and therefore, using the fact that $s(y) + K \geq 0$ in (2.1), we have

$$\int h_\lambda(x - y) s(y)\, dy = \int_{-\lambda}^{\lambda} \left(1 - \frac{|t|}{\lambda}\right) e^{-itx} S(it)\, dt.$$

By the Riemann-Lebesgue theorem, it follows that

$$(2.2) \quad \lim_{x \to \infty} \int h_\lambda(y) s(x - y)\, dy = 0.$$

Now let

$$\epsilon > 0, \qquad g_\epsilon(x) = \int_{x-\epsilon}^{x+\epsilon} s(y)\, dy.$$

Since (2.2) holds for any positive λ, and since $h_\lambda(x) \geq \lambda/\pi^2$ for $|x| \leq 1/\lambda$, it follows that $g_\epsilon(x)$ is bounded and

$$\text{(2.3)} \qquad \overline{\lim_{x\to\infty}} |g_\epsilon(x)| \leq \eta(\epsilon), \qquad \lim_{\epsilon\to 0} \eta(\epsilon) = 0.$$

On integrating (2.2) between $x - A$ and $x + A$, we have

$$\lim_{x\to\infty} \int h_\lambda(y) g_A(x - y)\, dy = 0.$$

Now $g_A(x)$ is bounded for fixed A, and

$$\lim_{x\to\infty} |g_A(x - y) - g_A(x)| \leq 2\eta(y).$$

Hence

$$\overline{\lim_{x\to\infty}} \left| g_A(x) - \int h_\lambda(y) g_A(x - y)\, dy \right| \leq \overline{\lim_{x\to\infty}} \int h_\lambda(y) |g_A(x - y) - g_A(x)|\, dy$$

$$\leq 2 \int h_\lambda(y)\eta(y)\, dy = 2 \int h_1(y)\eta(y/\lambda)\, dy.$$

On letting $\lambda \to \infty$, it follows from (2.3) that the left hand side is zero, and hence

$$\lim_{x\to\infty} g_A(x) = 0.$$

Ikehara's theorem can be stated as follows.

THEOREM 2. *Suppose that $k(y)$ is non-decreasing in $(0, \infty)$ and that*

$$f(\omega) = \int_0^\infty e^{-\omega y}\, dk(y)$$

converges absolutely for $\sigma > 1$. Suppose that $f(w) - (w - 1)^{-1}$ is continuous to the right of $\sigma = 1$. Then

$$\lim_{y\to\infty} e^{-y} k(y) = 1.$$

Let $s(y) = e^{-y}k(y) - 1$, so that $s(y) \geq -1$. We have, by partial integration,

$$\int_0^\infty e^{-\omega y} s(y)\, dy = \int_0^\infty e^{-(\omega\pi) y} k(y)\, dy - \int_0^\infty e^{-\omega y}\, dy = \frac{f(\omega\pi)}{\omega\pi} - \frac{1}{\omega},$$

and this is continuous to the right of the y-axis. Therefore, by Lemma 2,

$$\lim_{x\to\infty} \int_{x-A}^{x+A} e^{-y} k(y)\, dy = 2A$$

for any positive A. But since $k(y)$ is monotone,

$$e^{2A} k(x + A)e^{-(x+A)} \geq e^{-y} k(y) \geq e^{-2A} k(x - A)e^{-(x-A)}$$

for $| x - y | \leq 2A$. Therefore, as $x \to \infty$,

$$e^{2A} \varliminf k(x)e^{-x} \geq 1 \geq e^{-2A} \varlimsup k(x)e^{-x},$$

and we obtain the required conclusion by letting $A \to 0$.

3. *Proof of Theorem* 1 *for* $0 < \alpha < 1$.

We write C instead of C_α and put $\alpha^{-1} = a$. Our hypothesis implies that $S(| t |^a + it)$ is absolutely integrable in $(-\infty, \infty)$. We first show that

$$\text{(3.1)} \qquad \lim_{\epsilon\to 0} \int e^{ixt} S(\epsilon + | t |^a + it)\, dt = \int e^{ixt} S(| t |^a + it)\, dt.$$

We have, by (1.3)

$$S(\epsilon + | t |^a + it) = \frac{1}{2\pi i} \int_C \frac{S(\omega)\, d\omega}{\omega - \epsilon - | t |^a - it},$$

$$0 = \frac{1}{2\pi i} \int_C \frac{S(\omega)\, d\omega}{\omega + \epsilon - | t |^a - it},$$

and on subtracting we get

$$S(\epsilon + | t |^a + it) = \frac{\epsilon}{\pi i} \int_C \frac{S(\omega)\, d\omega}{(\omega - | t |^a - it)^2 - \epsilon^2}.$$

Let η be positive and independent of ϵ. Then

$$\text{(3.2)} \qquad \begin{aligned} \int_{-\eta}^{\eta} e^{ixt} S(\epsilon + | t |^a + it)\, dt &= \int_{-\eta}^{\eta} e^{ixt} dt \frac{\epsilon}{2\pi i} \int_C \frac{S(\omega)\, d\omega}{(\omega - | t |^a - it)^2 - \epsilon^2} \\ &= \frac{1}{2\pi i} \int_C S(\omega)\, d\omega \int_{-\eta}^{\eta} \frac{\epsilon e^{ixt} dt}{(\omega - | t |^a - it)^2 - \epsilon^2}. \end{aligned}$$

Since $\alpha < 1$, we can choose $\eta > 0$ so that

$$\text{(3.3)} \qquad (3\eta)^{2a-1} a^2 \leq \tfrac{1}{2}.$$

Then if $\omega = | \tau |^a + i\tau$, $| \tau | \leq 2\eta$, we have

$$|\, | \tau |^a - | t |^a \,| \leq |\, | \tau | - | t |\, |\, (3\eta)^{a-1} a \leq | \tau - t |\, (3\eta)^{a-1} a,$$

by the second mean-value theorem, and it follows that

$$| (\omega - | t |^a - it)^2 - \epsilon^2 | = | [(| \tau |^a - | t |^a) + i(\tau - t)]^2 - \epsilon^2 |$$
$$\geq (\tau - t)^2 + \epsilon^2 - a^2(3\eta)^{2a-2}(\tau - t)^2 \geq \tfrac{1}{2}(\tau - t)^2 + \epsilon^2.$$

Hence, if $| \tau | \leq 2\eta$,

$$\left| \int_{-\eta}^{\eta} \frac{\epsilon e^{ixt} dt}{(\omega - | t |^a - it)^2 - \epsilon^2} \right| \leq 2\epsilon \int \frac{dt}{(\tau - t)^2 + 2\epsilon^2} = 2 \int \frac{dt}{t^2 + 2} < 6. \tag{3.4}$$

If $| \tau | \geq 2\eta$, we have $| \tau - t | \geq \eta$, $| | \tau |^a - | t |^a | \geq \eta^a$, and

$$| (\omega - | t |^a - it)^2 - \epsilon^2 | \geq 2 | \tau - t | | | \tau |^a - | t |^a | \geq 2\eta^{1+a}.$$

For such values of τ,

$$\left| \int_{-\eta}^{\eta} \frac{\epsilon e^{ixt} dt}{(\omega - | t |^a - it)^2 - \epsilon^2} \right| \leq \epsilon\eta^{-a} = 0(1) \qquad (\epsilon \to 0). \tag{3.5}$$

Combining (3.2), (3.4) and (3.5), we obtain

$$\overline{\lim_{\epsilon \to 0}} \left| \int_{-\eta}^{\eta} e^{ixt} S(\epsilon + | t |^a + it)\, dt \right| \leq \frac{3}{\pi} \int_{-\eta}^{\eta} | S(| t |^a + it) |\, dt. \tag{3.6}$$

Now

$$\begin{aligned} &\overline{\lim_{\epsilon \to 0}} \left| \int e^{ixt} S(| t |^a + it)\, dt - \int e^{ixt} S(\epsilon + | t |^a + it)\, dt \right| \\ &\qquad \leq \overline{\lim_{\epsilon \to 0}} \left| \int_{-\infty}^{-\eta} + \int_{\eta}^{\infty} \right| S(| t |^a + it) - S(\epsilon + | t |^a + it) \Big| dt \\ &\qquad + \overline{\lim_{\epsilon \to 0}} \left| \int_{-\eta}^{\eta} e^{ixt} S(\epsilon + | t |^a + it)\, dt \right| + \int_{-\eta}^{\eta} | S(| t |^a + it) |\, dt. \end{aligned} \tag{3.7}$$

It follows from (1.7) that $S(\epsilon + | t |^a + it)$ is majorized in the range $| t | \geq \eta$ by an integrable function which is independent of ϵ, and since $S(\omega)$ is uniformly continuous for $\sigma \geq \eta^a$, the first term on the right of (3.7) vanishes. Using (3.6), it is plain that the sum of the remaining terms does not exceed

$$\left(1 + \frac{3}{\pi}\right) \int_{-2\eta}^{2\eta} | S(| t |^a + it) |\, dt.$$

Since this is arbitrarily small by choice of η, we have established (3.1).

Now using (3.1), we have

$$\int e^{ixt} S(|t|^a + it)\,dt = \lim_{\epsilon \to 0} \int e^{ixt} S(\epsilon + |t|^a + it)\,dt$$

$$(3.8) \qquad = \lim_{\epsilon \to 0} \int e^{ixt}\,dt \int s(y) e^{-(\epsilon+|t|^a+it)y}\,dy$$

$$= \lim_{\epsilon \to 0} \int s(y) e^{-\epsilon y}\,dy \int e^{-|t|^a y + i(x-y)t}\,dt$$

$$= \lim_{\epsilon \to 0} \int s(y) e^{-\epsilon y} y^{-a} F[(x-y)y^{-a}]\,dy,$$

where

$$(3.9) \qquad F(u) = \int e^{itu} e^{-|t|^a}\,dt = 2\int_0^\infty e^{-t^a} \cos ut\,dt.$$

We now consider two cases, according as a is not an even integer, or is an even integer. In the first case, if we put $\nu = [a]$ the function

$$e^{-|t|^a} + |t|^a e^{-|t|}$$

has $\nu + 2$ successive derivatives belonging to L over $(-\infty, \infty)$. We shall have

$$(3.10) \qquad \int_{-\infty}^{\infty} (e^{-|t|^a} + |t|^a e^{-|t|}) e^{iut}\,dt = O(u^{-\nu-2}) = o(u^{-a-1}),$$

and

$$(3.11) \qquad \int_{-\infty}^{\infty} |t|^a e^{-|t|} e^{iut}\,dt = \int_0^\infty t^a e^{-t(1+iu)}\,dt + \int_0^\infty t^a e^{-t(1-iu)}\,dt$$

$$= 2\Gamma(a+1) R[(1+iu)^{-a-1}].$$

Accordingly, unless a is an even integer,

$$(3.12) \qquad F(u) = \Omega(u^{-a-1}) = O(u^{-a-1}),$$

and $F(u)$ is free from zeros for $u >$ some u_0. Moreover, $F(u)$ is continuous and $O(u^{-2})$ for large values of u, and is therefore integrable in $(-\infty, \infty)$.

Since $s(y) + K \geq 0$ and $e^{-\epsilon y}$ increases as $\epsilon \to 0$, it follows that $s(y)y^{-a}F[(x-y)y^{-a}]$ is integrable, and by (3.8)

$$\int s(y) y^{-a} F[(x-y)y^{-a}]\,dy = \lim_{\epsilon \to 0} \int s(y) e^{-\epsilon y} y^{-a} F[(x-y)y^{-a}]\,dy$$

$$= \int e^{ixt} S(|t|^a + it)\,dt.$$

Therefore, by the Riemann-Lebesgue theorem,

$$(3.13)\qquad \lim_{x\to\infty} \int s(y)y^{-\alpha}F[(x-y)y^{-\alpha}]\,dy = 0.$$

In the case in which a is an even integer, it is a known fact[4] that there exists a $\theta > 0$, such that

$$F(u) = 2\int_0^\infty e^{-t^a}\cos ut\,dt = O\left(\exp\left(-\theta\,|\,u\,|^{\frac{a}{a-1}}\right)\right).$$

It follows that for each x,

$$F[(x-y)y^{-\alpha}] = O(\exp(-\vartheta y)),$$

and hence, by dominated convergence, that

$$\int s(y)y^{-\alpha}F[(x-y)y^{-\alpha}]\,dy = \lim_{\epsilon\to 0}\int s(y)e^{-\epsilon y}y^{-\alpha}F[(x-y)y^{-\alpha}]\,dy$$

Thus (3.13) is again valid, and if we put $\gamma = (1-\alpha)^{-1} = a/(a-1)$, we can change our variables x and y, and establish

$$(3.14)\qquad \lim_{x\to\infty}\int_0^\infty s(y^\gamma)F[(x^\gamma - y^\gamma)y^{1-\gamma}]\,dy = 0.$$

Let us notice that since $\gamma > 1$, if $x > 0$, $y > 0$, we have

$$|\,(x^\gamma - y^\gamma)y^{1-\gamma}\,| = |\,y\,|\left|\left(\frac{x}{y}\right)^\gamma - 1\right| \geq |\,y\,|\left|\frac{x}{y} - 1\right| = |\,x - y\,|.$$

Furthermore,

$$F[(x^\gamma - y^\gamma)y^{1-\gamma}] = F[(y^\gamma + \gamma y^{\gamma-1}(x-y) + O(x-y)^2y^{\gamma-2}y^{1-\gamma} - y]$$
$$= F\left[\gamma(x-y) + \frac{O(x-y)^2}{y}\right].$$

Thus there exists a function $F^*(x, y)$, defined over $(-\infty, \infty)$ in both arguments; with the properties

$$F^*(x, y) \leq \limsup_{u>|x-y|} F(u),$$

and for all B, we have uniformly in u over $-B < x < B$,

$$\lim F^*(x+u, u) = F(x);$$

[4] Titchmarsh, "Theory of Functions," 8-47.

and such that

$$\lim_{x\to\infty}\int_0^\infty s(y^\gamma)F^*(x, y)\,dy = 0.$$

In the case in which a is not an even integer, we have

$$F^*(x, y) = O(|x - y|^{-a-1})$$

and in the even integer case,

$$F^*(x, y) = O(\exp(-\vartheta\,|x - y|^\gamma).$$

In the first case,

$$\int_{-\infty}^\infty |F^*(x, y)|\,(A + |x - z|^{a+1})^{-1}dz = O(|x - y|^{-a-1}),$$

and in the second case,

$$\int_{-\infty}^\infty |F^*(x, y)|\exp(-\eta|x - z|^\gamma)\,dz = O\left(\exp\left(-\frac{\eta}{2^\gamma}|x - y|^\gamma\right)\right)$$

$$(0 < \eta < \gamma).$$

In either case, we shall thus have

$$\lim_{x\to\infty}\int_0^\infty s(y^\gamma)Q(x, y)\,dy = 0, \tag{3.15}$$

where $Q(x, y)$ is defined in the first case by

$$Q(x, y) = \int_{-\infty}^\infty F^*(z, y)(A + |x - z|^{a+1})^{-1}dz,$$

and in the second case by

$$Q(x, y) = \int_{-\infty}^\infty F^*(z, y)\exp(-\eta|x - z|^\gamma)\,dz.$$

We shall now show that if A is large enough, or η small enough, $Q(x, y)$ will be positive, will have an integral with respect to y which is bounded in x, and over the range $-\infty < y < \infty$, will exceed a fixed positive constant for large enough values of x.

$$Q(x, y) = Q_1(x, y) + Q_2(x, y) + Q_3(x, y)$$

where

$$Q_1(x, y) = \int_{-\infty}^{y-B} F^*(z, y)(A + |x - z|^{a+1})^{-1}dz;$$

$$Q_2(x, y) = \int_{y-B}^{y+B} F^*(z, y)(A + |x - z|^{a+1})^{-1}dz;$$

$$Q_3(x, y) = \int_{y+B}^{\infty} F^*(z, y)(A + |x - z|^{a+1})^{-1}dz$$

As x becomes positively infinite, we shall have uniformly in y,

$$Q_2(x, y) \sim R(x - y),$$

where

$$\begin{aligned} R(x) &= \int_{-B}^{B} F(\gamma z)(A + |x - z|^{a+1})^{-1}dz \\ &= (A + |x|^{a+1})^{-1} \int_{-B}^{B} F(\gamma z)\, dz \\ &\qquad + \int_{-B}^{B} F(\gamma z) \frac{|x|^{a+1} - |x - z|^{a+1}}{(A + |x|^{a+1})(A + |x - z|^{a+1})} dz \\ &= (A + |x|^{a+1})^{-1} \int_{-B}^{B} F(\gamma z)\, dz \\ &\qquad + \vartheta \frac{B^a}{A} (A + |x|^{a+1})^{-1} \qquad (|\vartheta| < \text{const.}). \end{aligned}$$

On the other hand,

$$|Q_1(x, y)| \leq \text{const.} \int_{-\infty}^{-B} |z|^{-a-1}(A + |(x - y) - z|^{a+1})^{-1}dz.$$

If we take the part of this integral corresponding to $|x - y - z|^{a+1} > A$, it can not exceed

$$\text{const.}(A + |x - y|^{a+1})^{-1}B^{-a}.$$

If on the other hand, we take the part for which $|x - y - z|^{a+1} < A$, it can not exceed

$$\text{const.}\ A^{-1-(a+1)^{-1}} \cdot \min (B^{-a-1}, (A^{-(A+1)} - |x - y|)^{-a-1}),$$

where $\min(u, v)$ signifies the smaller of the two quantities, u and v, or their common value if they coincide. If now $B = A^{\frac{1}{a+1}}$, we see that

$$|Q_1(x, y)| \leq \text{const.}(A + |x - y|^{a+1})^{-1}B^{-a}.$$

Similarly

$$| Q_3(x, y) | \leq \text{const.}(A + | x - y |^{a+1})^{-1} B^{-a}.$$

Let us notice that as A and consequently B tends to infinity,

$$B^a/A = A^{-1/(a+1)} \to 0,$$

and

$$\int_{-B}^{B} F(\gamma z)\, dz \to \int_{-\infty}^{\infty} F(\gamma z)\, dz = \frac{2\pi}{\gamma} > 0.$$

Hence, if we combine Q_1, Q_2, and Q_3, we see that for large enough A independent of x, for all sufficiently large x, we have for all y,

$$Q(x, y) > \left(\frac{2\pi}{\gamma} - \epsilon\right)(A + | x |^{a+1})^{-1}.$$

We also can prove at once that

$$\int_{-\infty}^{\infty} Q(x, y)\, dy < \text{const.}$$

In the case in which a is an even integer, we can give a similar proof that η may be selected once for all, so that $Q(x, y)$ will be positive for all y's, for sufficiently large values of x, and that the last formula will hold good. As in the first case, for large enough x, the lower bound of the value of $Q(x, y)$ over $x \leq y \leq x + 1$ will be positive. Remembering that $s(y)$ is bounded below, we shall have by (3.15), for some constant H,

$$\int_{x}^{x+1} s(y^\gamma)\, dy \leq H. \tag{3.16}$$

It is now easy to show that

$$\lim_{A \to \infty} \left\{ \sum_{n=-\infty}^{\infty} \overline{Bd}_{n \leq x \leq n+1} \, | F(x^\gamma - y^\gamma) y^{1-\gamma}] - F(\gamma(x - y)) | \right\} = 0.$$

and combining this with (3.14) and (3.16), we obtain

$$\lim_{x \to \infty} \int s(y^\gamma) F(\gamma(x - y))\, dy = 0.$$

It follows now from a known extension[5] of Wiener's general Tauberian theorem, since the Fourier transform of $F(x)$ is never zero, that

$$\lim_{x\to\infty} \int_{x-A}^{x+A} s(y^\gamma)\,dy = 0. \tag{3.17}$$

Finally, we have

$$\lim_{x\to\infty} x^{-\alpha} \int_{x-Ax^\alpha}^{x+Ax^\alpha} s(y)\,dy = \lim_{x\to\infty} \gamma x^{-\alpha} \int_{x^{1-\alpha}-A+0(1)}^{x^{1-\alpha}+A+0(1)} s(y^\gamma) y^{\gamma-1} dy$$

$$= \lim_{x\to\infty} \gamma \int_{x^{1-\alpha}-A+0(1)}^{x^{1-\alpha}+A+0(1)} s(y^\gamma)\,dy = \lim_{x\to\infty} \gamma \int_{x^{1-\alpha}-A}^{x^{1-\alpha}+A} s(y^\gamma)\,dy$$

since $y^{\gamma-1}x^{-\alpha} \to 1$ uniformly in the range of integration, and $s(y) \geq -K$. The conclusion now follows from (3.17).

MASSACHUSETTS INSTITUTE OF TECHNOLOGY AND HARVARD UNIVERSITY.

---

[5] H. R. Pitt, "*A Remark on Wiener's General Tauberian Theorem.*" Duke Mathematical Journal, 4 (1938).

Some prime-number consequences of the Ikehara theorem.

By NORBERT WIENER and LEONARD GELLERT in Cambridge, Mass.

SELBERG[1] and ERDŐS[2] have recently shown the Prime Number Theorem to be demonstrable by elementary methods[1]). The present paper is devoted, not to elementary proofs of this theorem, but to simple analytical considerations which may throw light on the reasons why such elementary, but not easy, proofs may be expected to function. Our fundamental tool is the Ikehara theorem[3]), to the effect that if for $\Re z > 1$,

$$\int_0^\infty u^{-z} dF(z) = \frac{a}{z-1} + G(z), \tag{1}$$

where $F(z)$ is monotone; and if over every finite range of x,

$$\lim_{x\to 1+0} G(x+iy) = H(y), \tag{2}$$

where $H(y) \in L$ over every finite range; then

$$\lim_{A\to\infty} \frac{1}{A} F(A) = a. \tag{3}$$

We shall consider the two Dirichlet series:

$$\int_0^\infty u^{-z} \log u \; d\bar{\omega}(u) = -\frac{\zeta'(z)}{\zeta(z)}, \tag{4}$$

and

$$-\int \frac{\zeta''(z)}{\zeta(z)} dz = -\frac{\zeta'(z)}{\zeta(z)} + \int \zeta'(z) d\left(\frac{1}{\zeta(z)}\right) = -\frac{\zeta'(z)}{\zeta(z)} - \int \left(\frac{\zeta'(z)}{\zeta(z)}\right)^2 dz =$$

$$= \int_0^\infty u^{-z} \log u \; d\bar{\omega}(u) - \int dz \left(\int_0^\infty u^{-z} \log u \; d\bar{\omega}(u)\right)^2 = \tag{5}$$

$$= \int_0^\infty u^{-z} \log u \; d\bar{\omega}(u) - \int dz \int_0^\infty u^{-z} \log u \; d\bar{\omega}(u) \int_0^\infty v^{-z} \log v \; d\bar{\omega}(v).$$

1) A. SELBERG, An elementary proof of the prime number theorem, *Annals of Math.*, (2) **50** (1949), pp. 305—313.

2) P. ERDŐS, On a new method in the elementary theory of numbers which leads to an elementary proof of the prime number theorem, *Proceedings National Academy of Sciences U.S.A.*, **35** (1949), pp. 374—384.

3) N. WIENER, Tauberian theorems, *Annals of Math.*, (2) **33** (1933), pp. 1 – 100.

Now,

$$-\int dz \int_0^\infty u^{-z} \log u \, d\tilde{\omega}(u) \int_0^\infty v^{-z} \log v \, d\tilde{\omega}(v) =$$

$$= -\int dz \int_0^\infty \log u \, d\tilde{\omega}(u) \int_0^\infty w^{-z} \log \frac{w}{u} \, d_w \tilde{\omega}\left(\frac{w}{u}\right) =$$

(6) $$= -\int dz \int_0^\infty \log u \, d\tilde{\omega}(u) \int_0^\infty w^{-z} d_w \int_0^{\frac{w}{u}} \log v \, d\tilde{\omega}(v) =$$

$$= -\int dz \int_0^\infty w^{-z} d_w \int_{u=0}^{u=\infty} \log u \, d\tilde{\omega}(u) \int_0^{\frac{w}{u}} \log v \, d\tilde{\omega}(v) =$$

$$= \text{const.} + \int_0^\infty w^{-z} \frac{1}{\log w} d_w \int_{u=0}^{u=\infty} \log u \, d\tilde{\omega}(u) \int_0^{\frac{w}{u}} \log v \, d\tilde{\omega}(v).$$

Hence,

(7) $$-\int \frac{\zeta''(z)}{\zeta(z)} dz = \text{const.} + \int_0^\infty u^{-z} \log u \, d\tilde{\omega}(u) +$$

$$+ \int_0^\infty u^{-z} \frac{1}{\log u} d_u \int_{w=0}^{w=\infty} \log w \, d\tilde{\omega}(w) \int_0^{\frac{u}{v}} \log v \, d\tilde{\omega}(v).$$

Let us now consider $\zeta(z)$ on the line $z = 1 + iy$. It is well known that $\zeta(z)$ is analytic on this line, except at $z = 1$, when it is of the form $\varphi(z) + \frac{1}{z-1}$, where $\varphi(z)$ is analytic. It consequently has on the line only zeros of finite order. Now, if ε is real and positive,

(8) $$\left| \frac{1}{\zeta(1+iy+\varepsilon)} \right| \leq \left| \sum \frac{\mu(n)}{n^{1+iy+\varepsilon}} \right| \leq \sum \frac{1}{n^{1+\varepsilon}} \leq \zeta(1+\varepsilon) = O\left(\frac{1}{\varepsilon}\right);$$

so that the zeta function cannot have a pole of higher order than 1 on the 1-line. It follows that $-\frac{\zeta'(z)}{\zeta(z)}$ cannot have a singularity on the 1-line that is not a pole of order 1 with residue -1, except for the singularity at $z = 1$, which is a pole of order 1 with residue 1. Thus either $-\frac{\zeta'(z)}{\zeta(z)}$ has no other singularity on the 1-line than the pole at $z = 1$, in which case

(9) $$\frac{1}{u} \int_0^u \log u \, d\tilde{\omega}(u) \to 1,$$

or it has a singularity at $1+i\lambda$. In the first case, we may write (9):

$$\frac{\log u\,\varpi(u)}{u}-\frac{1}{u}\int_0^u\frac{\varpi(u)\,du}{u}\to 1\,; \tag{10}$$

and since it is well known that $\varpi(u)=o(u)$, it follows that

$$\varpi(u)\sim\frac{u}{\log u}, \tag{11}$$

which is equivalent to the prime number theorem.

On the other hand, let

$$\zeta(1+i\lambda)=0. \tag{12}$$

Then

$$-\frac{\zeta'(z)}{\zeta(z)}-\frac{1}{2}\frac{\zeta'(z+i\lambda)}{\zeta(z+i\lambda)}-\frac{1}{2}\frac{\zeta'(z-i\lambda)}{\zeta(z-i\lambda)}= \tag{13}$$

$$=\int_0^\infty u^{-z}(1+\cos\lambda\log u)\log u\,d\varpi(u)$$

will be a Dirichlet series with non-negative coefficients with no singularity at $z=1$. It is easy to show that it can then have no singularity on $\Re z=1$. Thus by a very weak form of the Ikehara theorem,

$$\frac{1}{u}\int_0^u(1+\cos\lambda\log u)\log u\,d\varpi(u)\to 0. \tag{14}$$

This is to say that

$$\varpi(u)=\varpi_1(u)+\varpi_2(u), \tag{15}$$

where

$$\lim_{u\to\infty}\frac{1}{u}\int_0^u\log u\,d\varpi_1(u)=0, \tag{16}$$

and where $\varpi_2(u)$ only increases over the intervals

$$1+\cos\lambda\log u<\varepsilon, \tag{17}$$

or

$$|\lambda\log u-n\pi|<\varepsilon. \tag{18}$$

On the other hand, $-\int\frac{\zeta''(z)}{\zeta(z)}dz$ is a function which may be shown to behave like

$$\frac{\dfrac{2}{(z-1)^2}}{\dfrac{1}{z-1}}=\frac{2}{z-1} \tag{19}$$

at $z=1$, and to be analytic elsewhere in $x\geq 1$, except for possible logarithmic singularities where $\zeta(z)=0$. These do not interfere with Lebesgue integrability.

Hence the Ikehara theorem applies, and

$$(20)\qquad \frac{1}{u}\int_0^u \log u\, d\bar{\omega}(u)+\frac{1}{u}\int_0^u \frac{1}{\log u}\, d_u \int_{w=0}^{w=\infty} \log w\, d\bar{\omega}(w)\int_0^{\frac{u}{w}} \log v\, d\bar{\omega}(v) \to 2.$$

Hence

$$(21)\qquad \int_{|\lambda \log u - n\pi| < \varepsilon_1} \left(\log u\, d\bar{\omega}(u)+\frac{1}{\log u}\, d_u \int_{w=0}^{w=\infty} \log w\, d\bar{\omega}(w)\int_0^{\frac{u}{w}} \log v\, d\bar{\omega}(v)\right) \sim$$

$$=2e^{\frac{\pi n}{\lambda}}\left(e^{\frac{\varepsilon_1}{\lambda}}-e^{-\frac{\varepsilon_1}{\lambda}}\right),$$

so that

$$(22)\qquad \int_{|\lambda \log u - n\pi| < \varepsilon_1} \log u\, d\bar{\omega}(u) \leq 2e^{\frac{n\pi}{\lambda}}\left(e^{\frac{\varepsilon_1}{\lambda}}-e^{-\frac{\varepsilon_1}{\lambda}}\right).$$

Combining this with (16), (17), and (18), we see that for large u's

$$(23)\qquad \frac{1}{u}\int_0^u \log u\, d\bar{\omega}(u) \leq \frac{1}{u}\left(e^{\frac{\varepsilon_1}{\lambda}}-e^{-\frac{\varepsilon_1}{\lambda}}\right)\frac{2ue^{\varepsilon_1 \pi}}{1-e^{-\pi/\lambda}},$$

which is asymptotically not greater than

$$(24)\qquad \frac{2\left(e^{\frac{\varepsilon_1}{\pi}}-e^{-\frac{\varepsilon_1}{\pi}}\right)}{1-e^{-\pi/\lambda}}.$$

Since ε_1 is arbitrary, we see that

$$(25)\qquad \lim_{u\to\infty} \frac{1}{u}\int_0^u \log u\, d\bar{\omega}(u) = 0.$$

This is however inconsistent with the known elementary Chebychev theorem, to the effect that

$$(26)\qquad \varliminf_{u\to\infty} \frac{1}{u}\int_0^u \log u\, d\bar{\omega}(u) > 0,$$

so that we have succeeded in eliminating the hypothesis that there is a λ for which $\zeta(1+i\lambda)=0$, and have proved the prime number theorem.

(Received July 11, 1949.)

On a local L^2-variant of Ikehara's theorem

BY NORBERT WIENER AND AUREL WINTNER *

1. A fundamental theorem of Vivanti and Pringsheim states that if

$$a_n \geqq 0 \tag{1}$$

holds for all coefficients of a power series

$$p(z) = \sum_{n=0}^{\infty} a_n z^n \tag{2}$$

which converges for $|z| < 1$, then either the point $z = 1$ or no point of the circumference $|z| = 1$ is a singular point of the function $p(z)$. Clearly, the theorem becomes false if (1) is relaxed to

$$s_n \geqq 0, \quad \text{where} \quad s_n = \sum_{k=0}^{n} a_k \tag{3}$$

(for, by changing the value of a_0 alone, $|s_n| <$ const. can be reduced to (3), whereas trivial examples show $|s_n| <$ const. is certainly not sufficient for the alternative of Vivanti-Pringsheim). Still less is it possible to relax (1) to the existence of some Cesàro index $m(=1, 2, ...)$ satisfying, for the series $\sum a_n$, the condition

$$S_n^m \geqq 0 \quad (\text{as } n \to \infty, \text{ while } m \text{ is fixed}), \tag{4}$$

since (3) is equivalent to the case $m = 0$ of (4).

The Vivanti-Pringsheim theorem has a certain variant which, as an application of his Fourier methods in general Tauberian theory (cf., in particular, the proof of Ikehara's theorem in [2], § 19), one of us proved, but did not publish, some time ago. This theorem (quoted, but not proved, in [1], p. 242 and p. 250, item 12.6f) states that if (1) holds for the coefficients of a power series (2) which con-

* Received Sept. 5-1957.

verges for $|z|<1$ and is of class L^2 on *some* (no matter how narrow) sector containing the segment $0 \leqq z < 1$, then it is of class L^2 on the entire circle. By this is meant that if there exists on *some* arc $-\varepsilon \leqq \theta \leqq \varepsilon$ of the circumference $|z|=1$ a measurable function $p_0(\theta)$ satisfying

$$(5) \qquad \int_{-\varepsilon}^{\varepsilon} |p(re^{i\theta}) - p_0(e^{i\theta})|^2 d\theta \to 0 \text{ as } r \to 1,$$

with $z = re^{i\theta}$ (and $r < 1$) in (2), and if (1) is assumed, then $p_0(\theta)$ can be extended from $[-\varepsilon, \varepsilon]$ to $[-\pi, \pi]$ in such a way that

$$(6) \qquad \int_{-\pi}^{\pi} |p(re^{i\theta}) - p_0(e^{i\theta})|^2 d\theta \to 0 \text{ as } r \to 1$$

will hold (it being understood that $p_0(e^{i\theta})$ is of class L^2 on $[-\pi, \pi]$).

2. It will be shown in this paper that, in contrast with the Vivanti-Pringsheim theorem itself, *the* L^2*-variant can be generalized so as to relax* (1) *to* (3), *and even to* (4).

The theorem remains true if the power series (2) is replaced by a Laplace integral and (3) or (4) is adjusted to this general case. We shall give the proof for this more general case. This will have, among other things, the advantage of exhibiting the parallelism with the proof of Ikehara's theorem directly.

3. Let $f(s)$ be a function which is regular in the half-plane $\sigma > 1$, where $s = \sigma + it$, and suppose that there exists on a fixed t-interval $[-a, a]$ a function $f_0(t)$ of class L^2 satisfying

$$(7) \qquad \int_{-a}^{a} |f(\sigma + it) - f_0(t)|^2 dt \text{ as } \sigma \to 1$$

(by $\sigma \to 1$ is meant $\sigma \to 1 + 0$). Then $f(s)$ will be called *of class* $L^2(a)$. In view of the completeness of the space of the L^2-functions on a t-interval $[-a, a]$, condition (7) is equivalent to

$$(8) \qquad \int_{-a}^{a} |f(\sigma_1 + it) - f(\sigma_2 + it)|^2 dt \to 0 \text{ as } (\sigma_1, \sigma_2) \to (1, 1),$$

where $\sigma_1 (> 1)$ and $\sigma_2 (> 1)$ are independent variables.

In this terminology, the main theorem, to be proved, can be formulated as follows:

Let $\mu(u)$, where $0 \leqq u < \infty$, be of bounded variation on every finite interval $0 \leqq u \leqq U (< \infty)$, suppose that the Laplace-Stieltjes transform,

$$(9) \qquad f(s) = \int_0^\infty e^{-su}\, d\mu(u),$$

of μ is convergent* (but, possibly, not absolutely convergent) *on the half-plane $\sigma > 1$, where $s = \sigma + it$, and let

$$(10) \qquad \mu(0) = 0 \text{ and } \mu(u) \geqq 0, \text{ where } 0 < u < \infty$$

(but

$$(10 \text{ bis}) \qquad d\mu(u) \geqq 0, \quad \text{where} \quad 0 \leqq u < \infty,$$

is not assumed). *Then the function $f(s)$ cannot be of class $L^2(a)$ for any $a = \varepsilon$ unless it is of class $L^2(a)$ for every $a = N$.*

In contrast, the proof of Ikehara's theorem ([2], § 19), which assumes (10 bis), cannot be improved so as to relax (10 bis) to (10). In this connection, cf. [3].

4. Conditions (10) and (10 bis) correspond to (3) and (1) respectively, if (9) is identified with (2), by choosing $\mu(u)$ to be a step-function and placing $z = e^{-s}$ (except that the radius of convergence R of (2) then becomes $R = L/e$, instead of $R = 1$).

Correspondingly, the generalization (4) of (3) results if the assumption $\mu(u) \geqq 0$ is relaxed to the assumption that there exists some m satisfying $\mu_m(u) \geqq 0$ (for large u), where

$$\mu_m(u) = \int_0^u \mu_{m-1}(v)\, dv, \qquad \mu_1(u) = \mu(u).$$

But it will be clear from the proof that the case of an arbitrary m can be treated in the same way as the case $m = 1$, if the partial integration, leading from (9) to (13) and (14) below, is applied m times. For this reason, it will be sufficient to deal with the case of the assumption (10), where $m = 1$ (the assumption (10 bis) would correspond to $m = 0$).

5. In the proof of the theorem, it can be assumed that the given a-value, the value for which $f(s)$ is supposed to be of class $L^2(a)$, is $a = 1$ (this normalization is accomplished by a change of the unit of length on the t-axis). Then the assumption is that

$$(11)\quad \int_{-1}^{1} |f(1+\varepsilon+it) - f(1+\eta+it)|^2\, dt \to 0 \text{ as } (\varepsilon, \eta) \to (0, 0)$$

(where $1 + \varepsilon = \sigma_1$ and $1 + \eta = \sigma_2$ in the earlier notations), and the assertion is that

$$\int_{-a}^{a} |f(1+\varepsilon+it) - f(1+\eta+it)|^2\, dt \to 0$$

must hold for every fixed positive $a < \infty$.

To this end, it will be sufficient to show that

$$(12)\quad \int_{-\frac{1}{2}}^{\frac{1}{2}} |f(1+\varepsilon+it+ic) - f(1+\eta+it+ic)|^2\, dt \to 0$$

holds for every real c. In fact, (12) can be written in the form

$$\int_{-\frac{1}{2}+c}^{\frac{1}{2}+c} |f(1+\varepsilon+it) - f(1+\eta+it)|^2\, dt \to 0.$$

If this is applied to $c = 3/2$ and to $c = -3/2$, and if both of the resulting relations are added to (11), it follows that (11) remains true if its $[-1,1]$ is increased to $[-2,2]$.

Clearly, a repetition of this process leads to $[-N, N]$ for any fixed $N > 0$. This is the reason why it will be sufficient to prove that (11) leads to (12) for any fixed real c.

6. If the integral (9) is convergent for $\sigma > 1$, and if, without loss of generality, $\mu(0) = 0$ in (9), then a partial integration shows that

$$(13)\quad f(s) = sF(s),$$

where

$$(14)\quad F(s) = \int_0^{\infty} e^{-su} \mu(u)\, du,$$

the convergence of the integral (14) for $\sigma > 1$ being part of the statement. [It is worth emphasizing that, whereas the condition (10), which was not used here, and the convergence of (9) for $\sigma > 1$ do not imply the absolute convergence of (9) for $\sigma > 1$, they do imply the absolute convergence of (14) for $\sigma > 1$.]

Since both s and $1/s$ are regular and bounded on every $(\sigma + it)$-rectangle of the form $1 < \sigma < 2$, $-N < t < N$, it is clear from (13) that (11) and (12) are equivalent to

$$(15) \qquad \int_{-1}^{1} | F(1 + \varepsilon + it) - F(1 + \eta + it)|^2 \, dt \to 0$$

and

$$(16) \qquad \int_{-\frac{1}{2}}^{\frac{1}{2}} | F(1 + \varepsilon + it + ic) - F(1 + \eta + it + ic)|^2 \, dt \to 0$$

respectively. Hence the assertion of the theorem is equivalent to the statement that, *if* (10) *is assumed*, (15) leads to (16) for every real c.

Actually, (16) will prove to be true, not only for every fixed c, but *uniformly* on the *infinite* range $-\infty < c < \infty$. But this additional information is immaterial in the proof of the theorem.

7. The beginning of the proof is about the same as that of Ikehara's theorem; it proceeds as follows:

For any fixed $\sigma > 1$ and for any fixed real c, put

$$F_\sigma(t; c) = (1 - |t|) F(\sigma + it + ic) \text{ if } |t| \leqq 1, \quad F_\sigma(t; c) = 0 \text{ if } |t| \geqq 1,$$

where $-\infty < t < \infty$, and also put

$$G_\sigma(t; c) = e^{-\sigma u} \mu(u) e^{-icu} \text{ if } u \geqq 0, \quad G_\sigma(u; c) = 0 \text{ if } u \leqq 0,$$

where $-\infty < u < \infty$ (and $\mu(0) = 0$). Then, since the function

$$(17) \qquad S(u) = \left(\sin \frac{1}{2} u\right)^2 / \left(\frac{1}{2} u\right)^2, \qquad (S(0) = 1),$$

where $-\infty < u < \infty$, is a positive constant multiple of the Fourier transform of max $(0, 1 - |t|)$, where $-\infty < t < \infty$, it is readily verified from (14), where $\mu(u) \geqq 0$, that the Fourier transform of the function $F_\sigma(t; c)$ of t is the convolution $S(u) * G_\sigma(u; c)$ (if a normalizing constant factor, such as $(2\pi)^{-\frac{1}{2}}$, is disregarded). In view of the definitions of $F_\sigma(t;c)$ and $G_\sigma(u;c)$, this means that the identity

(18) $$\int_{-1}^{1} e^{ixt}(1-|t|)F(\sigma+it+ic)\,dt = H_\sigma(x;c),$$

where

(19) $$H_\sigma(x;c) = \int_0^\infty e^{-\sigma u}\mu(u)e^{-icu}S(x-u)\,du,$$

holds for $\sigma > 1$, $-\infty < x < \infty$, $-\infty < c < \infty$ (to a neglected positive constant factor); and that, by Plancherel's relation,

$$\int_{-1}^{1}(1-|t|)^2|F(\sigma+it+ic)|^2\,dt = \int_{-\infty}^{\infty}|H_\sigma(x;c)|^2\,dx$$

(to a constant factor, the square of the preceding constant factor).

If all of this is applied, not to the functions belonging to one $\sigma > 1$, but to the difference of the functions belonging to two σ values, $\sigma = 1+\varepsilon$ and $\sigma = 1+\eta$, where $\varepsilon > 0$ and $\eta > 0$, then what results is that the expression

(20) $$\int_{-1}^{1}(1-|t|)^2|F(1+\varepsilon+it+ic)-F(1+\eta+it+ic)|^2\,dt$$

is a constant multiple of the expression

(21) $$\int_{-\infty}^{\infty}|H_{1+\varepsilon}(x;c)-H_{1+\eta}(x;c)|^2\,dx.$$

8. Since $(1-|t|)^2$ has a positive minimum on the interval $-\frac{1}{2}\leq t\leq\frac{1}{2}$, it is clear that the integral on the left of the relation (16) is majorized by a constant multiple of the integral (20). Hence, in order to prove that the relation (16) holds for every fixed c, it is sufficient to show that the function (21) of $(\varepsilon, \eta; c)$ tends to 0 as $(\varepsilon, \eta) \to (0, 0)$. But this property of (21) can readily be concluded (as a matter of fact, *uniformly* for $-\infty < c < \infty$), as follows:

According to (10) and (17), both functions μ, S are non-negative. Hence it is clear from (19) that

$$| H_{1+\varepsilon}(x ; c) - H_{1+\eta}(x ; c) |$$
$$\leqq \left| \int_0^\infty (e^{-(1+\varepsilon)u} - e^{-(1+\eta)u}) \mu(u) S(x-u) | e^{-icu} | du \right|$$

(whether $\varepsilon > \eta$, $\varepsilon < \eta$ or $\varepsilon = \eta$). But since $| e^{-icu} | = 1$, it is also seen from (19) that the last inequality can be written in the form

$$| H_{1+\varepsilon}(x ; c) - H_{1+\eta}(x ; c) | \leqq | H_{1+\varepsilon}(x ; 0) - H_{1+\eta}(x ; 0) |.$$

Hence, if (ε, η) is fixed, the function (21) of c is majorized by the value attained by (21) at $c = 0$.

Consequently, if (ε, η) is fixed, the function (21) of c is majorized by a constant multiple of the value attained by (20) at $c = 0$, and the constant factor is independent of (ε, η). But since $(1 - |t|)^2 \leqq 1$ if $-1 \leqq t \leqq 1$, the value of (20) at $c = 0$ is majorized by the integral occurring in (15). Since the limit relation (15), where $(\varepsilon, \eta) \rightarrow (0, 0)$, is assumed this proves that the function (21) of $(\varepsilon, \eta ; c)$ tends to 0 as $(\varepsilon, \eta) \rightarrow (0, 0)$, uniformly for $-\infty < c < \infty$.

REFERENCES

[1] R. P. Boas, Jr., *Entire functions*, New York, 1954.

[2] N. Wiener, *The Fourier integral and certain of its applications*, Cambridge, 1933.

[3] A. Wintner, « On the Tauberian nature of Ikehara's theorem », *American Journal of Mathematics*, vol. 69 (1947), pp. 99-103.

Massachusetts Institute of Technology
The Johns Hopkins University.

NOTES ON PÓLYA'S AND TURÁN'S HYPOTHESES CONCERNING LIOUVILLE'S FACTOR

by

Norbert Wiener (Cambridge, Mass.) and **Aurel Wintner** (Baltimore, Md.)

1. It was observed by Pólya [8] that if $\lambda(n)$ denotes Liouville's factor (i.e., the completely multiplicative function defined by $\lambda(p) = -1$ or, equivalently, by

$$(1) \qquad \sum_{n=1}^{\infty} \lambda(n)/n^s = \Pi(1+p^{-s})^{-1} \equiv \zeta(2s)/\zeta(s),$$

where $\sigma > 1$), then, on the one hand, there is a considerable numerical evidence for the truth of the hypothesis that

$$(2) \qquad \sum_{n=1}^{m} \lambda(n) \leqq 0 \quad \text{for} \quad m > 1$$

(in this regard, cf. [4]) and that, on the other hand, a general function-theoretical theorem of Landau [7] readily leads to the following conclusion: The hypothesis (2) (or its weakened form, requiring only

$$(3) \qquad \sum_{n=1}^{m} \lambda(n) \leqq 0 \quad \text{for} \quad m > m'$$

if m' is large enough) implies Riemann's hypothesis,

$$(4) \qquad \zeta(s) \neq 0 \quad \text{for} \quad \sigma > \frac{1}{2}$$

[and, incidentally, the simplicity of all zeros of $\zeta(s)$ also; so that

$$(5) \qquad d\zeta(s)/ds \neq 0 \quad \text{at} \quad s = s_0 \quad \text{if} \quad \zeta(s_0) = 0,$$

where s_0 must be of the form $\frac{1}{2} + it_0$, by (4)].

A hypothesis similar to, but different from, (3) was proposed by Turán [9], p. 35, as follows:

$$\sum_{n=1}^{m} \lambda(n)/n \geqq 0 \quad \text{for} \quad m > m'', \tag{6}$$

if m'' is large enough. If (6) is true, then (4) and (5) can be concluded as before (in this regard, nothing would change if $\leqq 0$ and $\geqq 0$ in (3) were changed to $\geqq 0$ and $\leqq 0$ respectively). But what is alluring in this variant of (3) is that, owing to a general theorem of Bohr [2] (concerning the Diophantine irrelevance of « phase-shifts » for the zeros of absolutely convergent Dirichlet series on an open half-plane), the truth of (6) is implied by the following hypothesis (which, since

$$\zeta(s) \neq 0 \quad \text{for} \quad \sigma > 1 \tag{7}$$

is obvious from

$$\sum_{n=1}^{\infty} n^{-s} = \prod_{p} (1 - p^{-s})^{-1} = \zeta(s), \tag{8}$$

where $\sigma > 1$, appears to be comparatively innocent):

$$\sum_{n=1}^{m} n^{-s} \neq 0 \quad \text{for} \quad \sigma > 1 \quad \text{if} \quad m > m''. \tag{9}$$

The deduction of (6) from the hypothesis (9) is contained in the more general argument on pp. 14-15 in [9]; concerning (9) ➤ (4), cf. p. 4 *loc. cit.* (Theorem 1, the simple wording of which a footnote attributes to Jessen).

2. It is worth mentioning that if (9) is true for $m'' = 0$, then not only (4) and (5) hold but there also follow the assertions, say (4 bis) and (5 bis), which result if $\zeta(s)$ is replaced by $L(s)$ in (4) and (5), where $L(s) = L(s; \chi)$ denotes the Dirichlet series belonging to an arbitrarily *real* character. This can be concluded as follows:

A recent result of Showalter (published in [1]) implies that if (6) is true for $m'' = 0$, then

$$\sum_{n=1}^{m} \chi(n)/n \geqq \sum_{n=1}^{m} \lambda(n)/n$$

is true for every m (and for every real χ). Hence, if (6 bis) denotes the assertion which results when λ is replaced by χ in (6), then, since (9) implies (6), it follows that the case $m'' = 0$ of (9) implies the case $m'' = 0$ of (6 bis). But

(6 bis) (even with some $m'' > 0$ only) leads to (4 bis) and (5 bis) if Landau's function-theoretical theorem is applied in the same way as before.

On the other hand, if (2 bis) denotes the hypothesis which results if λ is replaced by χ in (2), then (2 bis) is false (for certain χ), whether (2) be true or false. In fact, not even the hypothesis (4), which is weaker than (2), is involved in the failure of (2 bis). For it was shown by Heilbronn [5] that not only (2 bis) is false (without any hypothesis) but not even a weakened form of (2 bis), proposed by Chowla [3], can hold.

Accordingly, (4 bis) cannot be concluded from (2) in the way in which (4 bis) was concluded from the case $m'' = 0$ of (6). Hence the case $m'' = 0$ of the hypothesis (9) appears to contain substantially more than what is contained in the hypothesis (2), the latter being adjusted only to (7), i.e., to the *principal* character χ alone. This curious situation is pointed out here because of the comments of Turán [9], p. 35, which are to the effect that (6), or at any rate a relaxed form of (6), might be more amenable than (3) is.

3. Turán's hypothesis (9) refers to the *partial sums* of $\zeta(s) = \sum n^{-s}$ for $\sigma > 1$. Can (4) [and (5)] be concluded if (9) is weakened to the corresponding hypothesis involving the Abel sums of $\zeta(s) = \sum n^{-s}$ for $\sigma > 1$? This question arises quite naturally. For, on the one hand, partial sums are generally more wobbly than Abel sums (the simplest instance of this remark being the content of Abel's own theorem, the converse of which is false) and, on the other hand, whether (3) or (6) is assumed, the deduction of (4) and (5) depends on an application of Landau's theorem to an integral representation of the function (1), a representation in which the partial sums of the numerical series of $\sum \lambda(n)$ or $\sum \lambda(n)/n$ form the integrand. There is, however, another integral representation of a Dirichlet series, a representation in which

(10) $$f(e^{-x}), \quad \text{where} \quad f(r) = \sum_{n=1}^{\infty} a_n r^n \quad \text{and} \quad 0 \leqq r < 1,$$

takes the place of

(11) $$\sum_{n \leqq x} a_n,$$

if $\sum a_n/n^s$ is the function to be represented by the integral (in both cases, the range of integration is the half-line $0 < x < \infty$, which can be reduced to $1 \leqq x < \infty$ in the second case). In particular, Riemann's proofs of the functional equation depend (with $f(r) = \sum r^n$ or $f(r) = \sum r^{n^2}$) on the integral representation of $\sum a_n/n^s = \zeta(s)$ in terms of (10), rather than of (11); and it was shown in [11] that even such Tauberian theorems as that of Ikehara (cf. [10], § 19) can be generalized so as to put the (Tauberian) assumption of *monotony* on the Abelian generator, rather than on the partial sum (as a function of x).

On the other hand, Landau's general theorem is applicable whether (11) or (10) is supposed to be of constant sign for large x (this, of course, is less than the corresponding assumption of monotony, just mentioned). Since Turán's way of applying Bohr's general theorem (cf. § 1 above) can also be adjusted to the replacement of (11) by (10), the answer to the question, raised at the beginning of this § 3, easily proves to be affirmative.

4. In order to see this, replace the partial sums of $\zeta(s) = \sum n^{-s}$, where $\sigma > 1$, by its Abelian generator *

$$\zeta_r(s) = \sum_{n=1}^{\infty} r^n/n^s, \tag{12}$$

where $0 \leqq r < 1$. Then what corresponds to the assumption of a sufficiently large m'' satisfying (9) is the assumption that there exists a sufficiently small ε having the property that

$$\zeta_r(s) \neq 0 \quad \text{for} \quad \sigma > 1 \quad \text{if} \quad 1 - \varepsilon < r < 1 \tag{13}$$

$(0 < \varepsilon < 1)$; a hypothesis which, in view of (12), goes over into the trivial fact (7) if $r \to 1 - 0$. Clearly, the refinement announced in § 3 for Turán's result can be formulated as follows:

If there exists a positive constant $\varepsilon(< 1)$ *for which the functions* (12) *satisfy* (13), *then* (4) *is true [as is, incidentally,* (5) *as well].*

REMARK. There are two comments to be made on this inference.

First, the converse inference can hardly be made; we know of nothing to prevent that (4) [even with (5)] be possibly true under the assumption that the « Abelian hypothesis » (13) is false. But we shall refer in § 8 to *another* « Abelian hypothesis », one which is *equivalent* to (4) [without (5)].

Secondly, the Abelian relaxation of (9) to (13) is an *actual* relaxation; for, lacking any conceivable Tauberian restriction (such as an appropriate condition of unilateral boundedness), we cannot infer (9) from (13). We shall mention in

* Actually, if r has a fixed value within the unit circle of the complex r-plane, then (12) is an entire function of s, since the series is uniformly convergent on every fixed half-plane $\sigma > -$ const.

§ 8 a striking parallel instance, one concerning (4) itself, in which precisely the Abelian relaxation is certainly not vacuous (unless (4) in false).

In view of § 3, the proof of the italicized implication requires no new ideas. But since the details are not lengthy, and since a little formal trick (the use of the parameter u in § 6 below) with the *subsequent* identification $s+u$ with s is involved, for the sake of completeness we shall give the details.

5. For a fixed positive $r<1$, write the Dirichlet series $\zeta_r(s)$ in the form

$$\zeta_r(s)=\sum_{n=1}^{\infty} r^n \exp(-s\log n), \qquad (\exp = n^{-s}), \tag{14}$$

confine s to the half-plane $\sigma>1$ (even though any half-plane $\sigma>c$ would be allowed), and apply to (14) Bohr's general theorem in [2], by using for the sequence $\log 1, \ldots, \log n, \ldots$ the basis $\log 2, \ldots, \log p, \ldots$, and for Bohr's « phase-shifting » of (14) into

$$\sum_{n=1}^{\infty} r^n \exp(i\varphi_n)\, n^{-s} \tag{15}$$

the angles defined (mod 2π) as follows: $\varphi_n=0$ or $\varphi_n=\pi$ according as the number of all (not necessarily distinct) prime divisors p of n is even or odd. In view of (1), this means that $\exp(i\varphi_n)$ is chosen to be $\lambda(n)$ for every n (also for $n=1$). Hence (15) goes over into

$$\zeta_r(s)=\sum_{n=1}^{\infty} r^n \lambda(n)/n^s. \tag{16}$$

Accordingly, Bohr's general theorem is applicable to the two functions (14), (16); its assertion reduces to the following: For a fixed r, the function $\zeta_r(s)$ attains on the half-plane $\sigma>1$ those and only those values which the function $\zeta_r(s)$, belonging to the same r as $\zeta_r(s)$, attains on the half-plane $\sigma>1$. In particular, the hypothesis (13) (for a fixed ε) is equivalent to

$$\zeta_r(s)\neq 0 \quad \text{for} \quad \sigma>1 \quad \text{if} \quad 1-\varepsilon<r<1 \tag{17}$$

(with the same ε).

Choose $s=\sigma$ in (17) and note that, according to (16), the function $\zeta_r(\sigma)$ is real-valued on the half-line $1<\sigma<\infty$ and tends to the limit $r^1\lambda(1)/1^s=r\lambda(1)=r>0$ as $\sigma\to\infty$. Consequently, the case $s=\sigma$ of (17) means that

$$f(r;\,u)\equiv\sum_{n=1}^{\infty} r^n\lambda(n)/n^u>0 \quad \text{for} \quad 1<u<\infty \quad \text{if} \quad 1-\varepsilon<r<1, \tag{18}$$

where, for later use, $s = \sigma$ is replaced by u (and the first of the relations (18) is the definition of f).

6. In order to pass from (18) to an appropriate application of Riemann's general identity (referred to in § 3), it will be sufficient to use that identity only in its following particular form: If an ordinary Dirichlet series $\sum a_n/n^s$ is absolutely convergent for $\sigma > 1$, then

$$\Gamma(s)\sum_{n=1}^{\infty} a_n/n^s = \int_0^{\infty} x^{s-1} f(e^{-x})\,dx \tag{19}$$

holds for $\sigma > 1$, if $f(r)$ is defined by (10). Choose $a_n = \lambda(n)/n^u$, where u is fixed on the region $1 < u < \infty$. Then $\sum a_n/n^s$ is absolutely convergent for $\sigma > 1$ (and even for $\sigma > 0$) and (19) reduces to

$$\Gamma(s)\zeta(2s + 2u)/\zeta(s + u) = \int_0^{\infty} x^{s-1} f(e^{-x};\ u)\,dx, \tag{20}$$

if use is made of the definitions, (**1**) and (18), of $\lambda(n)$ and f.

It is now possible to apply Landau's theorem, his extension [7] (to Laplace or Mellin integrals) of the Vivanti-Pringsheim theorem (on power series). In fact, let Landau's theorem be applied to the function (20) of s, *while u is kept fixed (but $u > 1$)*. Then it follows from (18) that, when the real part of s decreases (from $\sigma = \infty$), the function (of s) on the left of (20) will *first* become singular at a real s (unless it is an entire function-which it is not).

Finally, $\Gamma(s)$ has no zero, and is regular, for $\sigma > 0$, and it is well-known (from the functional equation of ζ) that $\zeta(z)$ has no zero on the half-line $0 < z < \infty$. Since $\zeta(2z)$ is regular on the half-line $\frac{1}{2} < z < \infty$, it follows that the function $\zeta(2s + 2u)/\zeta(s + u)$ of s is regular on the s-domain $\sigma + u > \frac{1}{2}$. Hence, the function $\zeta(2z)/\zeta(z)$ of the complex variable z is regular so long as the real part of z exceeds $\frac{1}{2}$. Thus (4) follows.

By keeping t fixed but letting $\sigma \to \frac{1}{2} + 0$ in $s = \sigma + it$, the additional information (5) follows from (20) and (18) in *precisely* the same way as when (9) is applied to the corresponding integral representation of $\zeta(2s)/\zeta(s)$ (that is to say, in *about* the same way in which (5) is usually concluded from the Mertens hypothesis).

7. No implicative connection between (3) and (6) or between (3) and (13) is known. It is therefore worth mentioning that, so far as (4) and (5) are concerned, Pólya's hypothesis can be relaxed in the same way in which the hypothesis (13) on the Abelian generator, (12), of (1) relaxes Turán's hypothesis.

In fact, let $f(r)$ now denote the case $a_n = \lambda(n)$ of (10) and, instead of (3), suppose merely that there exists a sufficiently small ε^* having the property that

$$f(r) \equiv \sum_{n=1}^{\infty} \lambda(n) r^n \leqq 0 \quad \text{for} \quad 1 - \varepsilon^* < r < 1 \tag{18*}$$

$(0 < \varepsilon^* < 1)$. Then, since (19) reduces for $a_n = \lambda(n)$ to

$$\Gamma(s)\zeta(2s)/\zeta(s) = \int_0^{\infty} x^{s-1} f(e^{-x})\, dx, \tag{20*}$$

(4) [and (5)] can be concluded as before (and, in one respect, more conveniently, since there is now no parameter u to deal with).

It is similarly seen that (4) [and (5)] can be concluded not only from the Mertens hypothesis, $M(x) = O(x^{\frac{1}{2}})$, but also from its Abelian relaxation,

$$g(r) = O(1-r)^{-\frac{1}{2}}, \tag{21}$$

where

$$M(x) = \sum_{n \leqq x} \mu(n) \quad \text{and} \quad g(r) = \sum_{n=1}^{\infty} \mu(n) r^n;$$

that either of the unilateral hypotheses $g(r) = O_L(1-r)^{-\frac{1}{2}}$, $g(r) = O_R(1-r)^{-\frac{1}{2}}$ would also suffice; finally, that not only the relation $M(x) = o(x^{\frac{1}{2}})$, claimed by Stieltjes, is false (Landau) but (for the same reason) not even its Abelian relaxation, $g(r) = o(1-r)^{-\frac{1}{2}}$, can possibly be true.

8. The Abelian reductions considered above, the reductions of Mertens' Pólya's and Turán's hypotheses to (21), (18) and (13) respectively, are more than mere *analytical* subtleties, since there is a *number-theoretical* parallel in which Abelian reduction is basic (under, and for, the truth of Riemann's hypothesis). This instance is the case $a_n = \Lambda(n) - 1$ of (10)-(11); so that

$$\sum_{n=1}^{\infty} a_n/n^s = -\zeta'/\zeta(s) - \zeta(s).$$

Let $h(r)$, $H(x)$ denote the corresponding functions (10), (11), i.e., put

$$h(r) = \sum_{n=1}^{\infty} \Lambda(n) r^n - (1-r)^{-1} \quad \text{and} \quad H(x) = \sum_{n \leqq x} \Lambda(n) - [x].$$

Then the situation is as follows:

(i) *Whether Riemann's hypothesis* (4) *be true or false,* it is certainly true that

$$H(x) \neq O(x^{\frac{1}{2}}), \tag{22}$$

and even

$$H(x) \neq O_R(x^{\frac{1}{2}}) \quad \text{and} \quad H(x) \neq O_L(x^{\frac{1}{2}}). \tag{22 bis}$$

(ii) Nevertheless, *under Riemann's hypothesis* (4), the Abelian relaxation,

$$h(r) = O(1-r)^{\frac{1}{2}}, \tag{23}$$

of the (certainly false) estimate $H(x) = O(x^{\frac{1}{2}})$ is true; in fact, (23) is equivalent to (4), since even

$$h(r) = O_R(1-r)^{\frac{1}{2}} \quad \text{or} \quad h(r) = O_L(1-r)^{\frac{1}{2}} \tag{23 bis}$$

implies (4).

In order to conclude (4) from (23), or just from either of the assumptions (23 bis), it is sufficient to apply (19) to $a_n = \Lambda(n) - 1$, and then to use Landau's theorem [7] in the same way as above. The corresponding (but unconditional) assertion, (22), of (i), and even the unilateral refinement, (22 bis), of (22) goes back to Phragmén (1891-1901; for later references, cf. [6], p. 139). On the other hand, Hardy and Littlewood computed the « Abelian relaxation » of the explicit formula of Riemann-Mangoldt (a relaxation which, incidentally, is much easier to prove than that formula itself), and concluded from it that (23), the Abelian relaxation of the (unconditionally) wrong relaxation $H(x) = O(x^{\frac{1}{2}})$ of (23), is implied by (4); cf. [6], pp. 134-138. Hence (23) and (4) are equivalent (and the whole of (ii) is now clear, since (23) is equivalent to the two relations (23 bis) together).

Tamworth, N. H., June 1957

REFERENCES

[1] P. T. Bateman and S. Chowla, « *The equivalence of two conjectures in the theory of numbers* », Journal of the Indian Mathematical Society, vol. 17 (1953), pp. 177-181.

[2] H. Bohr, « *Zur Theorie der allgemeinen Dirichletschen Reihen* », Mathematische Annalen, vol. 79 (1919), pp. 136-150.

[3] S. Chowla, « *Note on Dirichlet's L- functions* », Acta Arithmetica, vol. 1 (1936) pp. 113-114.

[4] H. Gupta, « *On a table of values of* $L_1(n)$ », Proceedings of the Indian Academy of Sciences, vol. 12 (1940), pp. 407-409.

[5] H. Heilbronn, « *On real characters* », Acta Arithmetica, vol. 2 (1937), pp. 212-213.

[6] G. H. Hardy and J. E. Littlewood, « *The Riemann Zeta-function and the theory of the distribution of primes* », Acta Mathematica, vol. 41 (1918), pp. 119-196.

[7] E. Landau, « *Ueber einen Satz von Tschebischeff* », Mathematische Annalen, vol. 61 (1905), pp. 527-550.

[8] G. Pólya, « *Verschiedene Bemerkungen zur Zahlentheorie* », Jahresbericht der Deutschen Mathematiker-Vereinigung, vol. 28 (1919), pp. 31-40.

[9] P. Turán, « *On some approximative Dirichlet-polynomials in the theory of the zeta-function of Riemann* », Danske Vid. Selsk. Mat. hyphen Fys. Medd., vol. 24 (1948), no. 17.

[10] N. Wiener, « *The Fourier integral and certain of its applications* », Cambridge University Press, 1933.

[11] A. Wintner, « *On Riemann's reduction of Dirichlet series to power series* », American Journal of Mathematics, vol. 69 (1947), pp. 769-789.

ON THE NON-VANISHING OF EULER PRODUCTS.*

By NORBERT WIENER and AUREL WINTNER.

Let a function $\chi = \chi_p$, defined on the set of all prime numbers p, be such as to satisfy the following two conditions:

$$(1) \qquad -1 \leqq \chi_p \leqq 1$$

(e.g., $\chi_p{}^2 = |\chi_p|$, i.e., $\chi_p = -1, 0, 1$) and

$$(2) \qquad f(s) \text{ is meromorphic on } \sigma = 1,$$

if $f(s)$ the Euler product

$$(3) \qquad f(s) = \Pi(1 - \chi_p/p^s)^{-1} \qquad (\sigma > 1).$$

By the condition (2) is meant the assumption that there exist an open domain D in the s-plane and a meromorphic function $f(s)$ on D in such a way that D contains the half-plane $\sigma \geqq 1$ (but possibly no half-plane $\sigma \geqq 1 - \epsilon$, where $\epsilon > 0$) and that $f(s)$ is identical on the half-plane $\sigma > 1$ with the function which is there represented by the product (3); a function which, in view of (1), is regular (and non-vanishing) for $\sigma > 1$. On the other hand, the condition (1) means that if χ is extended from primes to positive integers by the assignments $\chi_1 = 1$ and $\chi_n \chi_m = \chi_{nm}$ (so that (3) becomes identical with the absolutely convergent Dirichlet series $\Sigma \chi_n/n^s$ if $\sigma > 1$), then the resulting real-valued representation χ_n of the multiplication on the semi-group of the positive integers n is a bounded representation (in fact, lim sup $|\chi_n| = \infty$ unless $|\chi_n| \leqq 1$ for every n). The two extreme cases allowed by (1) and (2) result if $\chi_p = 1$ and $\chi_p = -1$ (for every p); in fact, since (3) then reduces to

$$(4) \qquad \Pi(1 - p^{-s})^{-1} = \zeta(s) \text{ and } \Pi(1 + p^{-s})^{-1} = \zeta(2s)/\zeta(s),$$

(2) is satisfied in both cases.

The following considerations were suggested by a point occurring in Hadamard's, and also in de la Vallée Poussin's, proof of

$$(5) \qquad \zeta(s) \neq 0 \text{ on } \sigma = 1.$$

* Received June 29, 1957.

The point in question is the following: Besides the case $f(s) = \zeta(s)$ of (2), it is assumed in the proof of (5) that $\zeta(s)$ *is regular (rather than just meromorphic) at every* $s = 1 + it \neq 0$. It is this fact from which the negation of (5) leads to a contradiction, since what is actually proved is the following: If $\zeta(s)$ would have a zero $s = 1 + it_0$, then it would also have some pole $s = 1 + it^0$ distinct from the pole $s = 1$ (in fact, it is shown that t^0 could be chosen to be $2t_0$; cf., e.g., [1], pp. 29-30). Accordingly, the proof of (5) fails to succeed if all that is known for the case $f(s) = \zeta(s)$ of (3) is the following pair of assumptions: (2) and

(6) $$f(s) \text{ has a pole at } s = 1.$$

There is quite another proof of (4), the function-theoretical proof adapted by Ingham from Landau's proof [2] of Dirichlet's theorem $L(1) \neq 0$ (in Landau's function-theoretical proof, $L(s)$ belongs to a real character); cf. [1], p. 89. But this proof is even more restrictive than the preceding one, since what is now assumed is the regularity of $f(s)$ at every point $s \neq 1$ of the half-plane $\sigma \geqq \frac{1}{2}$ (on the boundary line $\sigma = \frac{1}{2}$, less than regularity is needed).

Thus it is natural to ask whether

(7) $$f(s) \neq 0 \text{ on } \sigma = 1,$$

the generalization of (5) for an arbitrary case (1) of (3), must or need not be true if only (2) and (6) are assumed. For, as seen above, the classical argument applies only if (1) is retained but (2) is strengthened to the hypothesis that

(8) $$f(s) \text{ is regular on } \sigma = 1 \text{ if } s \neq 1.$$

The purpose of this note is to show that the answer to the question, just raised, is in the affirmative; in other words, that (7) remains true if (8) is relaxed to (2). Curiously enough, (8) proves to be a consequence of the other hypotheses; so that the situation is as follows:

(I) *In terms of a sequence* $\chi_2, \chi_3, \cdots$ *satisfying* (1), *define on the open half-plane* $\sigma > 1$ *a (regular, non-vanishing) function* $f(s)$ *by the corresponding Euler product* (3), *and suppose that, as* $\sigma \rightarrow 1 + 0$, *the function* $f(s)$ *behaves in such a way that conditions* (2) *and* (6) *are satisfied. Then both* (8) *and* (7) *hold.*

Actually, (I) is not the final result, since the assumption (6) of (I) proves to be superfluous. This is the content of the theorem at the end of this paper.

Although (I) contains (5), there will result no new proof of (5), since (5), along with

(9) $$1/\zeta(s) \neq 0 \text{ on } \sigma = 1 \text{ if } s \neq 1$$

(that is to say, the regularity and the non-vanishing of $\zeta(1+it)$ for $0 < t < \infty$); will be granted in the proof of (I). What will directly be proved is not (I) itself but the following variant of (I):

(II) *The assertions of* (I) *remain true if its assumption* (6) *is replaced by the assumption*

(10) $$f(1) = 0$$

(and all the other assumptions of (I) are retained).

Proof of (II). For $-\infty < t < \infty$, let j_t denote the index (the logarithmic residue) of $f(s)$ at $s = 1 + it$, and let J_t belong to

(11) $$F(s) = \zeta(s)f(s)$$

in the same way as j_t belongs to $f(s)$. Since (2) is assumed for $f(s)$ itself and for $f(s) = \zeta(s)$, it is clear from (5), (9) and (11) that $j_t = J_t$ for $t \neq 0$. Since the assertions, (8) and (7), of (II) are equivalent to $j_t = 0$ for $t \neq 0$, and since (10) and (11) imply that $J_0 = 0$ (for $\zeta(s)$ has a simple pole at $s = 1$), it follows that (II) will be proved if it is shown that

(12) $$|J_t| \leqq J_0 \qquad (J_0 = 0)$$

holds for every t. But the truth of (12) can be concluded by an elementary argument (cf. [3], where (8), rather than just (2), is assumed, (10) is retained, and (1) is generalized to complex-valued χ_p, with $|\chi_p| \leqq 1$).

In order to prove (12), note that, if $\sigma > 1$, logarithmic differentiation of (3) leads to

(13) $$-f'/f(s) = \sum_p \sum_{k=1}^{\infty} (\chi_p \log p)/p^{ks}.$$

If (13) is added to the case $\chi_p = 1$ of (13), it follows from (11) and from the first of the relations (4) that, if $\sigma > 1$,

(14) $$-F'/F(s) = \sum_p \sum_{k=1}^{\infty} a_p/p^{ks},$$

where

$$a_p = (1 + \chi_p)\log p. \tag{15}$$

But since (15) and (1) imply that $a_p \geqq 0$, it is clear from (11) that, if $\sigma > 1$,

$$|(\sigma - 1)F'/F(\sigma + it)| \leqq |(\sigma - 1)F'/F(\sigma)| \text{ for } -\infty < t < \infty. \tag{16}$$

If $\sigma \to 1 + 0$, then (16) goes over into $|J_t| = |J_0|$, and so (12) follows from the parenthetical equality in (12).

This completes the proof of (II).

Actually, the following extension of (II) was also proved:

(II*) *The assertions of* (II) *remain true if its assumption* (1) *is replaced by the unilateral restriction* $-1 \leqq \chi_p$, *provided that the Euler product* (3) *is uniformly convergent on every compact subset of the half-plane* $\sigma > 1$.

The latter proviso (which, since χ_p is bounded from below, is satisfied if, though not only if, $\chi_p \leqq$ Const. for some Const.) cannot of course be omitted. But then the restriction $-1 \leqq \chi_p$ will suffice in the preceding proof, since, in view of (15), this unilateral restriction is equivalent to $a_p \geqq 0$.

Much deeper lies the following fact:

(II bis) *Under the assumptions of* (II), *and even those of* (II*), *the function* $f(s)$ *possesses an analytic continuation which is meromorphic on some half-plane* $\sigma > 1 - \epsilon$, *where* $\epsilon = \epsilon_f > 0$, *and* $f(s)$ *does not vanish at any point* $s \neq 1$ *of this half-plane* $\sigma > 1 - \epsilon$.

[REMARK. *If it is true that* $\zeta(s)$ *has no zero on some open half-plane containing the line* $\sigma = 1$ (*for instance, if Riemann's hypothesis is true*) *then, in* (II bis), *the word "meromorphic" can be replaced by "regular."*]

In fact, since $\zeta(s)$ has a simple pole at $s = 1$, it is clear from (2), (10) and (11) that, if $\epsilon > 0$ is small enough, $F(s)$ is regular at every point of the half-line $s > 1 - \epsilon$, and that $F(s)$ can have a zero on such a half-line only if $s = 1$ is a multiple zero of $f(s)$ (in fact, (3) does not vanish for $s > 1$). But $f(s)$ cannot have a multiple zero at $s = 1$, since, if $s > 1$ and $-1 \leqq \chi_p$, it is clear from (3) that

$$f(s) \geqq \Pi(1 + p^{-s})^{-1} \sim \text{const.}\ (s - 1) \text{ as } s \to 1 - 0,$$

where const. $= \zeta(2) > 0$, by the second of the relations (4).

Accordingly, $F(s)$ is a regular and non-vanishing function, and so the logarithmic derivative of $F(s)$ is a regular function, on some half-line $s > 1 - \epsilon$. Since $a_p \geqq 0$, it now follows from Landau's theorem (on Dirichlet series $\Sigma c_n/n^s$ with $c_n \geqq 0$), a theorem referred to but not used above, that the Dirichlet series (14) must converge for $s > 1 - \epsilon$ and, therefore, for $\sigma > 1 - \epsilon$. Consequently, the logarithmic derivative of $F(s)$ is a regular function, and therefore $F(s)$ is a non-vanishing regular function, on the half-plane $\sigma > 1 - \epsilon$. In view of (11), this proves both (II bis) and the Remark following (II bis).

Proof of (I). If χ_p is replaced by $-\chi_p$ (for every p), then the assumption (1) remains unaltered but (3) becomes replaced by

(17) $$f^*(s) = \Pi(1 + \chi_p/p^s)^{-1}.$$

The connection between (3) and (17) is involutory. But while it is clear from (1) that (17) represents a non-vanishing regular function for $\sigma > 1$, it is not quite obvious that (2), too, is an involutory property; in other words, that

(18) $f^*(s)$ is meromorphic on $\sigma = 1$

if (2) is assumed. It turns out, however, that (18) follows from (2) and is, therefore, equivalent to (2).

First, since (2) remains true if $f(s)$ is replaced by $1/f(s)$, more than (18) will follow if it is ascertained that the quotient of $f^*(s)$ and $1/f(s)$ is a non-vanishing regular function on the line $\sigma = 1$. But (3) and (17) show that, if $\sigma > 1$, this quotient is identical with

(19) $$f(s)f^*(s) = \Pi(1 - \chi_p^2/p^{2s})^{-1}.$$

On the other hand, it is clear from (1) that (19) is a regular and non-vanishing function on the half-plane $\sigma > \frac{1}{2}$ and, therefore, on the line $\sigma = 1$. [This conclusion is curious, since, by placing either $\chi_p = 1$ or $\chi_p = 0$, and making the alternative choice in an appropriate ("lacunary") manner, it is easy to obtain a pair of functions (1), (17), the factors on the left of (19), which are regular on the half-plane $\sigma > \theta$ but become singular at every point of the line $\sigma = \theta$, where $\frac{1}{2} < \theta < 1$, rather than $\theta = \frac{1}{2}$.]

Let (1*), (2*), $\cdots$ denote what results if $f(s)$ is replaced by $f^*(s)$, and χ_p by $-\chi_p$, in (1), (2), $\cdots$ respectively. Then, as just shown, (1) and (2) together are equivalent to (1*) and (2*) together. Since (II) has already been proved, it follows that (1*), (2*) and (10*) imply both (8*) and (7*).

But the quotient of $f^*(s)$ and $1/f(s)$ and, hence, the quotient of $1/f(s)$ and $f^*(s)$, was seen to be a non-vanishing regular function on the line $\sigma = 1$. Hence (8*) and (7*) are equivalent to (7) and (8) respectively and so, since (10) goes over into (6) if $f(s)$ is replaced by $1/f(s)$, the proof of (I) is now complete.

* *

*

If $f(s)$ is an L-function of Dirichlet belonging to a real non-principal character (so that, in particular, (1) and (3) hold for a certain χ_p which is capable of the three values -1, 0, 1 only), and if the meromorphic (rather than the regular) behavior of $L(s)$ on $\sigma = 1$ is assumed, then, since $L(1) \neq 0$ (Dirichlet) but $L(1) \neq \infty$, neither the non-vanishing nor the regularity of $L(s)$ on $\sigma = 1$ can be concluded from (I) and (II) together. Hence there arises the question whether or not (7) and (8) remain true if (1) and (2) are retained for (3) but (6) and (10) are replaced by

$$0 \neq f(1) \neq \infty \tag{20}$$

(or, what in view of (1) and (3), where $\sigma > 1$, is the same thing, by $0 < f(1) < \infty$). It will be shown that the answer to this question is affirmative:

(III) *The assertions of* (I) *remain true if* (6) *is replaced by* (20).

In order to prove (III), it will be sufficient to show that, under the assumptions of (III), the line $\sigma = 1$ cannot contain a zero of $f(s)$. For if this is granted, then the remaining assertion of (III), according to which $f(s)$ cannot have a pole on $\sigma = 1$, will follow in the same way in which (I) and (II) were proved to be equivalent (that is, by applying (17) and (19) in the same way as above). Hence it is sufficient to assume that $f(s)$ has a zero $s = 1 + ia$ on $\sigma = 1$ and to show that the existence of at least one such real number a leads to a contradiction.

First, $a \neq 0$, since $f(1) \neq 0$. Next, since $f(s + ia) = 0$ at $s = 1$, and since (1) and (3) imply that $f(1 + it)$ and $f(1 - it)$ are complex conjugate values, it is clear that the function

$$G(s) = \zeta^2(s) f(s + ia) f(s - ia) \tag{21}$$

(which, in view of (2), is meromorphic on $\sigma = 1$) will have at $s = 1$ neither a zero nor a pole if the zero $s = 1 + ia$ of $f(s)$ is a simple zero (the factor $\zeta^2(s)$ of (21) has a double pole at $s = 1$). But the argument applied in the proof of (II bis) also shows that $f(s)$ cannot have a multiple zero on $\sigma=1$ (simply because (1) and (3) imply that, if $\sigma>1$,

$$|f(\sigma+it)|\geqq\Pi(1+p^{-})^{-1}=\zeta(2\sigma)/\zeta(\sigma)>\text{Const.}(\sigma-1)$$

holds for every t and for a certain Const. >0). Accordingly, if K_t denotes the logarithmic residue of $G(s)$ at the point $s=1+it$, then $K_0=0$.

It will be shown that

$$(22) \qquad |K_t|\leqq|K_0|$$

holds for every t. In view of $K_0=0$, this will prove that $K_t=0$ holds for every t. It will therefore follow from (5), (9) and (21) that $f(s+ia)f(s-ia)$ is of index 0, and so a non-vanishing regular function, at every point $s\neq1$ of the line $\sigma=1$. Since $f(1+it)$ and $f(1-it)$ are complex conjugates, and since $a\neq0$, it now follows that $f(s)$ is regular (rather than just meromorphic) on $\sigma=1$. But since $s=1+ia$ and $s=1-ia$ are two distinct zeros of $f(s)$, this contains a contradiction. In fact, it was shown in [3] that if (1) (or, in the complex case, just $|\chi_p|\leqq1$) holds in (3), where $\sigma>1$, and if the function $f(s)$ remains regular at every point of $\sigma=1$, then $f(s)$ cannot have more than one zero on $\sigma=1$.

In view of this contradiction, the proof of (III) will be complete if the truth of (22) is proved (for every t). But (22) can be proved, along the lines of the proof of (12), as follows:

If $s=\sigma+it$, where $\sigma>1$, is replaced in (13) once by $s+ia$ and once by $s-ia$, it follows, by addition, that the logarithmic derivative of $f(s+ia)f(s-ia)$ is

$$(23) \qquad -\sum_p\sum_{k=1}^{\infty}(\lambda_{p,k}\log p)/p^{ks}, \text{ where } \lambda_{p,k}=2\chi_p\cos(ak\log p).$$

On the other hand, it follows from the first of the relations (4) that the logarithmic derivative of $\zeta^2(s)$ results from (23) if χ_p and a are replaced by 1 and 0 respectively. If the resulting series is added to (23) itself, then the definition, (21), of $G(s)$ shows that

$$(24) \qquad -G'/G(s)=\sum_p\sum_{k=1}^{\infty}b_{p,k}/p^{ks} \text{ for } \sigma>1,$$

where

$$(25) \qquad b_{p,k}=2\mu_{p,k}\log p, \qquad \mu_{p,k}=1+\chi_p\cos(ak\log p).$$

But it is seen from (25) and (1) that $b_{p,k}\geqq0$. Hence it is clear from (24) that (16) remains true if F is replaced by G. Consequently, (22) follows by letting $\sigma - 1 \rightarrow +0$.

Since this completes the proof of (III), the final result is as follows:

THEOREM. *If there is assigned to every prime p a value χ_p satisfying $-1 \leqq \chi_p \leqq 1$, and if the function $f(s)$, defined for $\sigma > 1$ by the Euler product $f(s) = \Pi(1 - \chi_p/p^s)^{-1}$, remains meromorphic at every point of the line $\sigma = 1$, then $f(s)$ is a regular and non-vanishing function at every point $s \neq 1$ of the line $\sigma = 1$.*

This theorem, which is equivalent to (I), (II) and (III) together, can be completed by the further information supplied by (II bis) and (II*) [and also by the hypothetical information contained in the Remark formulated after (II bis)].

MASSACHUSETTS INSTITUTE OF TECHNOLOGY,
THE JOHNS HOPKINS UNIVERSITY.

REFERENCES.

[1] A. E. Ingham, *The distribution of primes*, Cambridge Tracts, no. 30, 1932.

[2] E. Landau, "Ueber einen Satz von Tschebyscheff," *Mathematische Annalen*, vol. 61 (1905), pp. 527-550.

[3] A. Wintner, "On the non-vanishing of certain Dirichlet series," *American Journal of Mathematics*, vol. 74 (1952), pp. 723-725.

Commentary on the Number-Theoretic Papers [39e], [50a], [56a], [57b, c]

P. T. Bateman and H. G. Diamond

Comments on [39e]

The main theorem of the present paper is a Tauberian theorem which is "intermediate" between that of Wiener and Ikehara on the one hand, and a theorem of Hardy and Littlewood on the other. In the notation of the paper, these are the cases $\alpha = 0$ and $\alpha = 1$, respectively. A result of this type had been conjectured by J. Karamata and a special case (with $\alpha = 1/2$) had been given by V. G. Avakumović.[1]

This theorem is also given in the book *Tauberian Theorems* by H. R. Pitt,[6] pp. 135–138; but the proof there assumes that $F(u) > 0$ for all real u, where

$$F(u) = F_a(u) = \int_{-\infty}^{\infty} e^{-|t|^a} e^{iut}\, dt.$$

This assumption is valid if and only if $a \leqslant 2$, however, since $F_a(u)$ is known to have real zeros for $a > 2$ (cf. G. Pólya[7]), and to have no real zeros for $0 < a \leqslant 2$ (cf. P. Lévy[5]).

Theorems related to the Wiener-Pitt result have been given by Subhankulov,[8] along with some applications. See also T. Ganelius, *Tauberian Remainder Theorems.*[3] The present paper of Wiener and Pitt, however, does not appear to have been well-known.

In proving Lemma 1 the authors use a Tauberian theorem of the Hardy-Littlewood type for the Laplace transform. The original theorem of Hardy and Littlewood applied to power series. The theorem used by Wiener and Pitt is as follows: *If* $s(y) \geqslant -k$ *and if*

$$S(\sigma) = \int_0^{\infty} s(y)e^{-\sigma y}\, dy = o(1/\sigma) \quad as \quad \sigma \to 0+,$$

then

$$\frac{1}{x}\int_0^{x} s(y)\, dy = o(1) \quad as \quad x \to \infty.$$

The proof given is based on a technique of Karamata.[4] The hypotheses imply that for any polynomial P and hence for any Riemann integrable function P that

$$\int_0^\infty (s(y) + k)e^{-\sigma y} P(e^{-\sigma y})dy = \frac{k}{\sigma} \int_0^1 P(x)\,dx + o\left(\frac{1}{\sigma}\right)$$

as $\sigma \to 0^+$. The result follows by taking $P(x) = 1/x$ when $1/e \leqslant x \leqslant 1$ and zero otherwise.

The O-estimate for $F(u)$ given on line 5 of page 254 is correct, but the reference is not, since Titchmarsh gives only a cruder estimate for $\max\limits_{|z|=R} |F(z)|$ for z in the complex plane. The estimate for large real u may be obtained along the following lines. Since a is an even integer, $F(u)$ may be expressed as the Fourier integral of e^{-t^a}. Integration by parts N times yields

$$F(u) = \frac{1}{(iu)^N} \int_{-\infty}^{\infty} e^{-itu} \left(\frac{d}{dt}\right)^N \{e^{-t^a}\}\,dt.$$

The Nth derivative of $\exp(-t^a)$ is estimated by the Cauchy inequality, using, e.g., a circle centered at t with radius

$$R = \exp\{(\log |u| - c - 1)/(a - 1)\}$$

where

$$c^{1/a} = 1 + \csc\{\pi/(4a)\}.$$

For $|t| > R \csc\{\pi/(4a)\}$, we exploit the fact that

$$\mathrm{Re}(\{t + Re^{i\theta}\}^a) \geqslant (|t| - R)^a/\sqrt{2},$$

while the opposite case is handled by a trivial estimate. Finally, taking N to be about

$$\exp\{a(\log |u| - c - 1)/(a - 1)\},$$

we obtain the desired estimate for $F(u)$.

We have recently discovered a simpler proof of the main theorem involving a Bochner-Ikehara type argument along line segments $\sigma + it$, $-\frac{B}{2}\sigma^\alpha \leqslant t \leqslant \frac{B}{2}\sigma^\alpha$ where $\sigma \to 0+$ with $1/x$. See H. G. Diamond.[2]

Errata and Specific Remarks

p. 247, line 6:	(In the proof the authors take $B = 1$.) The B occurring on p. 254 and the B occurring on pp. 256–257 have nothing to do with the B here.
p. 247, line 9:	Replace the second "in" by "is."
p. 247, line 4, from bottom:	Replace $0 \leqslant \alpha < 1$ by $0 \leqslant \alpha \leqslant 1$.
p. 247, last line:	Replace $n \to \infty$ by $x \to \infty$.
p. 248, lines 7–8:	This remark as stated applies primarily to (1.3) and the case $\alpha > 0$. In order to get (1.4) when $\alpha = 0$, it would be appropriate to assume that $S(\omega)$ is defined on $\sigma = 0$ by continuity, or as an L_1 limit on each finite stretch of C_0.
p. 248, line 15:	There should be a factor σ under the integral sign. Also, $O(1)$ should be replaced by $o(1)$.
p. 249, line 2:	Replace $\sin^2 \lambda x$ by $2 \sin^2 \frac{1}{2}\lambda x$.
p. 249, formula (2.1):	There should be a factor $1/2\pi$ on the right and e^{-itx} should be replaced by e^{itx}.
p. 249, line 12:	Replace "if z is" by "if ω is." Also, replace "if $\|t\| \leqslant \lambda$" by "if $\|t\| \leqslant \lambda$ and $0 < \sigma < \lambda$."
p. 249, line 3 from bottom:	A factor $1/2\pi$ is missing on the right, and e^{-itx} should be replaced by e^{itx}.
p. 250, line 9:	Replace lim by $\overline{\lim}$.
p. 250, line 5 from bottom:	Replace "to the right of $\sigma = 1$" by "in the closed half-plane $\sigma \geqslant 1$."
p. 250, line 3 from bottom:	We may assume that $k(0) = 0$.
p. 250, last line:	Replace $\omega\pi$ by $\omega + 1$, three times.
p. 251, line 5:	Replace $2A$ by A.
p. 251, formula (3.2):	Replace $2\pi i$ by πi, twice.
p. 251, formula (3.3):	Replace $2a - 1$ by $2a - 2$.
p. 252, line 1:	Omit "second."
p. 252, formula (3.5):	Replace $O(1)$ by $o(1)$.
p. 252, formula (3.6):	Right-hand side should be $\dfrac{6}{\pi}\displaystyle\int_{-2\eta}^{2\eta}$.
p. 252, line 2 from bottom:	Replace $1 + 3/\pi$ by $1 + 6/\pi$.
p. 253:	The asymptotic formula for F implied by (3.10) and (3.11)

when a is not an even integer was derived by contour integration in the paper of Pólya mentioned earlier.

p. 254, line 7 from bottom: Insert) after $y^{\gamma-2}$. This estimate is valid for $0 < \frac{1}{2}y < x < \frac{3}{2}y$, say.

p. 254, line 5 from bottom: Presumably,

$$F^*(x, y) = F((x^\gamma - y^\gamma)y^{1-\gamma})$$

if $x > 0$, $y > 0$ and is otherwise chosen so that

$$|F^*(x, y)| < \sup_{u > |x-y|} |F(u)|.$$

p. 254, line 3 from bottom: Omit "lim". Put absolute value sign about $F^*(x, y)$ and $F(u)$.

p. 254, line 2 from bottom: Replace u by x.

p. 254, last line: The limit is as $u \to \infty$. Also, the limit should be $F(\gamma x)$.

p. 255, lines 8, 10: Replace $F^*(x, y)$ by $F^*(z, y)$.

p. 255, line 11: Replace $0 < \eta < \gamma$ by $0 < \eta < \vartheta$.

p. 255, line 8 from bottom: The A here is not that of the theorem. It will have to be taken rather large.

p. 256, line 5: Interchange x and y.

p. 256, line 12 from bottom: We use the fact that

$$\int_{-\infty}^{\infty} |zF(z)|\, dz < \infty.$$

p. 256, last 11 lines: In estimating Q_1, it is helpful to take $B = A^{1/(a+1)}$ from the start and consider separately the three cases $y - x < 0$, $0 \leqslant y - x \leqslant 10B$, $y - x > 10B$, say,

p. 257, line 8: Interchange x and y in the last half of the line. Thus, we should have "for all sufficiently large y, we have for all x."

p. 257, line 9: Replace $|x|^{a+1}$ by $|x - y|^{a+1}$.

p. 257, equation (3.16): Replace $s(y^\gamma)$ by $|s(y^\gamma)|$.

p. 257, line 3 from bottom: The first part of the line should be

$$\lim_{x\to\infty} \sum_{n=0}^{\infty} \overline{\underset{n \leqslant y \leqslant n+1}{Bd}}$$

p. 258, first sentence:	Actually the authors use the technique of p. 439 of Pitt's paper, rather than Pitt's theorem as such. In fact Pitt's theorem is not applicable, since F is not non-negative everywhere if $a > 2$.
p. 258, lines 5–6;	Replace $x^{1-\alpha} \pm A + O(1)$ by $x^{1-\alpha} \pm (1 - \alpha)A + o(1)$.
p. 258, line 6	Replace $x^{1-\alpha} \pm A$ by $x^{1-\alpha} \pm (1 - \alpha) A$.

References

1. V. G. Avakumović, C. R. Acad. Sci., Paris, 204(1937), 224–226.
2. H. G. Diamond, Proc. Amer. Math. Soc. 31(1972), 152–158.
3. T. Ganelius, *Tauberian remainder Theorems*, Springer Lecture Notes in Mathematics, Vol. 232.
4. J. Karamata, Math. Zeit. 32(1930), 319–320.
5. P. Lévy, C. R. Acad. Sci., Paris, 176(1923), 1118–1120.
6. H. R. Pitt, *Tauberian theorems*, Oxford, 1958.
7. G. Pólya, Messenger of Math. 52(1923), 185–188.
8. M. A. Subhankulov, AMS Transl. (2) 26(1963), 311–338.

Comments on [50a]

In proving the prime number theorem by means of the Tauberian theorem of Wiener and Ikehara, one must first establish that $\zeta(s) \neq 0$ for $\sigma = \operatorname{Re} s = 1$. This fact is usually established by means of a trigonometric inequality.

In the present paper, Wiener and Geller set out to show that the non-vanishing of ζ could be deduced from the identity $\zeta''/\zeta = (\zeta'/\zeta)' + (\zeta'/\zeta)^2$ by still more applications of the Wiener-Ikehara theorem. The identity just quoted lies behind the formula of Selberg which leads to an elementary proof of the prime number theorem.

A. E. Ingham[1] in his review of the present article noted that the paper could be shortened drastically (and the dependence on the Wiener-Ikehara theorem eliminated) by an argument of the following type. The function

$$g(s) = -\frac{\zeta'(s)}{\zeta(s)} - \frac{1}{2}\frac{\zeta'(s+i\lambda)}{\zeta(s+i\lambda)} - \frac{1}{2}\frac{\zeta'(s-i\lambda)}{\zeta(s-i\lambda)}$$

has a Dirichlet series with non-negative coefficients and has no singularity at $s = 1$. Thus for $\sigma > 1$

$$-\operatorname{Re} g(\sigma + i\lambda) \leqslant g(\sigma) \qquad (= O(1)),$$

and the supposed zero of ζ at $1 + i\lambda$ is seen to be incompatible with the known regularity of ζ at $1 + 2i\lambda$. However, this proof is just a disguised version of the usual proof which is based on the trigonometrical inequality

$$3 + 4\cos\theta + \cos 2\theta \geqslant 0,$$

since

$$2\{g(\sigma) + \operatorname{Re} g(\sigma + i\lambda)\} = -3\frac{\zeta'(\sigma)}{\zeta(\sigma)} - 4\operatorname{Re}\frac{\zeta'(\sigma + i\lambda)}{\zeta(\sigma + i\lambda)} - \operatorname{Re}\frac{\zeta'(\sigma + 2i\lambda)}{\zeta(\sigma + 2i\lambda)}.$$

The argument of Wiener and Geller is rather intricate compared with the usual proofs. However, it does give a method of exploiting the idea that a zero of ζ at $1 + i\lambda$ would entail an untenable concentration of primes in intervals of the form

$$[e^{(2n\pi - \pi - \varepsilon_1)/\lambda}, e^{(2n\pi - \pi + \varepsilon_1)/\lambda}].$$

Errata and Specific Remarks

p. 25, line 2: Replace "Gellert" by "Geller".

p. 25, formula (4): It is more customary to use Π instead of $\tilde{\omega}$.

pp. 25–26, formulas (5), (6), and (7): Some readers may prefer to replace these formulas by

$$-\int_{\infty}^{s} \frac{\zeta''(z)}{\zeta(z)}\,dz = \sum_{k=2}^{\infty} \frac{\Lambda(k)}{k^s} + \sum_{k=2}^{\infty} \frac{a_k}{k^s},$$

where Λ denotes the von Mangoldt function, and $a_k = \frac{1}{\log k} \sum_{mn=k} \Lambda(m)\Lambda(n)$. The only fact used about a_k in the sequel is that $a_k \geqslant 0$.

p. 26, line 3: Replace d_v by d_w.

p. 26, line 12 from bottom: Replace u/v by u/w as the upper limit of the last integral.

p. 26, line 6 from bottom: Replace "pole" by "zero".

p. 27, formula (18): The formula should read $|\lambda \log u - (2n - 1)\pi| < \varepsilon_1$.
p. 28, formula (21): Replace n by $2n - 1$ twice.
p. 28, formula (22): The formula should read

$$\int_{|\lambda \log u - (2n-1)\pi| \leqslant \varepsilon_1} \log u \, d\tilde{\omega}(u) \leqslant 2e^{\frac{(2n-1)\pi}{\lambda}} \{e^{\varepsilon_1/\lambda} - e^{-\varepsilon_1/\lambda} + o(1)\}.$$

p. 28, formula (23): The formula should read

$$\frac{1}{u}\int_0^u \log u \, d\tilde{\omega}(u) \leqslant \frac{1}{u}(e^{\varepsilon_1/\lambda} - e^{-\varepsilon_1/\lambda})\frac{2ue^{\varepsilon_1/\lambda}}{1 - e^{-2\pi/\lambda}} + o(1).$$

p. 28, formula (24): The formula should read

$$\frac{2(e^{\varepsilon_1/\lambda} - e^{-\varepsilon_1/\lambda})e^{\varepsilon_1/\lambda}}{1 - e^{-2\pi/\lambda}}$$

Reference

1. A. E. Ingham, Math. Rev. 11(1950), 644–645.

Comments on [56a]

The authors establish an L^2 analogue of Landau's theorem on the convergence of Dirichlet series with positive coefficients. They prove that if F is the Laplace transform of a non-negative function, convergent for Re $s > 1$, then for any $\sigma > 1$, $a > 0$, $\varepsilon > 0$ an inequality of the type

$$(*) \qquad \int_{-a}^{a} |F(\sigma + it)|^2 \, dt \leqslant \left(8\left[\frac{a}{\varepsilon}\right] + 1\right) \int_{-\varepsilon}^{\varepsilon} |F(\sigma + it)|^2 \, dt$$

is valid.

The main idea of the proof, which occurs in §§7–8, is the following: if h and ϕ are both Fourier transforms of non-negative functions in $L_1 \cap L_2$, then the Plancherel theorem gives

$$\int_{-\infty}^{\infty} |h(t + c)\phi(t)|^2\,dt \leqslant \int_{-\infty}^{\infty} |h(t)\varphi(t)|^2\,dt$$

for any real c. Taking $h(t) = \max(1 - |t|/\varepsilon, 0)$ and making the obvious estimates, we obtain

$$\int_{c-\varepsilon/2}^{c+\varepsilon/2} |\phi(t)|^2\,dt \leqslant 4 \int_{-\varepsilon}^{\varepsilon} |\phi(t)|^2\,dt.$$

We then set $\phi(t) = F(\sigma + it)$ and deduce (*) by superposition.

S. Wainger[3] has remarked that an analogous L^p result can be deduced from the L^2 result when p is a positive even integer. He showed that the analogous result fails when $1 \leqslant p < 2$. H. S. Shapiro[2] has shown that the L^p analogue fails for every $p > 2$, $p \neq 2, 4, 6, \ldots, \infty$.

The "variant of the Vivanti-Pringsheim theorem", mentioned in §1, follows immediately from a lemma of Erdös and Fuchs[1] to the effect that under condition (1) we have for any $r < 1$

$$\int_{-\pi}^{\pi} |p(re^{i\theta})|^2\,d\theta \leqslant \frac{3\pi}{\varepsilon} \int_{-\varepsilon}^{\varepsilon} |p(re^{i\theta})|^2\,d\theta.$$

It is interesting to note that the argument of the present paper establishes this Erdös-Fuchs lemma (up to a constant factor), but that the method of Erdös and Fuchs, which utilizes the inequality $\Sigma|c_n|^2 \leqslant (\Sigma|c_n|)^2$, does not appear to carry over to integrals.

In §2 the authors suggest a generalization of the L^2 variant of the Vivanti-Pringsheim theorem. As the example $a_n = (-1)^n$ shows, the italicized statement is not correct for radius of convergence 1. The power series version of the main theorem of the paper should read as follows:

Theorem. *Suppose $\Sigma_{n=0}^{\infty} a_n z^n$ has positive radius of convergence $r_0 < 1$. Suppose that for some fixed m and all $n \geqslant 0$ the inequalities $S_n^{(m)} \geqslant 0$ hold, where*

$$S_n^{(0)} = a_0 + a_1 + \ldots + a_n,$$
$$S_n^{(m)} = S_0^{(m-1)} + S_1^{(m-1)} + \ldots + S_n^{(m-1)} \qquad (m = 1, 2, \ldots).$$

Finally, assume that for some $\varepsilon > 0$

$$\lim_{r,r' \to r_0 -} \int_{-\varepsilon}^{\varepsilon} |p(re^{i\theta}) - p(r'e^{i\theta})|^2 \, d\theta = 0.$$

Then

$$\lim_{r,r' \to r_0 -} \int_{-\pi}^{\pi} |p(re^{i\theta}) - p(r'e^{i\theta})|^2 \, d\theta = 0.$$

The reason that the radius of convergence must be strictly less than 1 is easily seen by passing to Laplace-Stieltjes transforms and considering the argument of Wiener and Wintner for abscissa of convergence 0. Their proof is not valid for this case because the functions f and F in (13) behave quite differently near 0.

Errata and Specific Remarks

p. 54, formula (7):	Insert $\to 0$ before "as."
p. 55, fourth line of §4:	Replace L by 1.
p. 55, last paragraph:	The m used here differs by one from the m used on 53.
p. 56, first sentence:	The expression "change of the unit of length on the t axis" is a little unfortunate in the present circumstances. On the one hand, such operations are not valid for analytic functions such as f; on the other hand, the authors treat $f(1 + \varepsilon + it)$ as a function of two real variables, for which the stretching would be legitimate. In any case the simplifications produced by taking $a = 1$ are negligible indeed.
p. 58, line 9:	A factor 2π is needed on the left-hand side in order to make the equality exact. The constant factors occurring in the theory of Fourier integrals are treated rather whimsically in this section. In any case, their values are largely irrelevant for the purpose of the present paper.

References

1. P. Erdös and W. H. J. Fuchs, J. London Math. Soc. 31(1956), 67–73.
2. H. S. Shapiro, Quart. J. Math. Oxford Ser. (2), 26(1975), 9–18.
3. S. Wainger, Proc. Amer. Math. Soc. 20(1969), 16–18.

Comments on [57b]

This paper considers various conjectures, most of which imply the truth of the Riemann hypothesis (RH). Some of these hypotheses, e.g. (3) and (6), have since been shown false (cf. C. B. Haselgrove,[1] R. S. Lehman[2, 3] and R. Spira[4]) while others of these hypotheses are very much open to doubt.

The main theorem of the paper asserts that the RH is a consequence of the following hypothesis:

(*) *There exists an absolute constant* $\varepsilon > 0$ *such that* $\Sigma_{n=1}^{\infty} r^n n^{-s} \neq 0$ *for all pairs* r, s *with* $\operatorname{Re} s > 1$ and $1 - \varepsilon < r < 1$.

The preceding theorem and its proof are modeled on the following result contained in a paper of Turán ([9] in the authors' bibliography): If we assume the hypothesis

(**) *for each* $N > N_0$, $\Sigma_{n=1}^{N} n^{-s} \neq 0$ *for all* s *with* $\operatorname{Re} s > 1$,

then (6) holds and hence the RH is true. However, (6) is now known to be false, and so the hypothesis (**) cannot hold.

The authors apparently believed that hypothesis (*) is weaker than hypothesis (**). However, there appears to be no justification (except the *a posteriori* fact that (**) is false) for the authors' assertion that (*) constitutes an "Abelian relaxation" of (**). Indeed, on the one hand we see no obvious way of concluding hypothesis (*) from hypothesis (**), and on the other hand there are simple examples of infinite series whose partial sums are all non-zero and whose Abel means have real zeros smaller than but arbitrarily close to 1. An example is the series whose Abel means are

$$F(r) = 1 + \sum_{n=1}^{\infty} (-1)^{a(n)} 2^n r^{b(n)} = \prod_{n=1}^{\infty} (1 - 2^{2^n} r^{4^n}) \qquad (0 < r < 1),$$

where if

$$n = 2^{m_1} + 2^{m_2} + \ldots + 2^{m_r}, \quad 0 \leqslant m_1 < m_2 < \ldots < m_r,$$

then $a(n) = r$ and

$$b(n) = 4^{m_1} + 4^{m_2} + \ldots + 4^{m_r}.$$

It is clear that $F(r)$ has zeros arbitrarily close to 1, and, on the other hand, by the uniqueness of binary expansions, that

$$1 + \sum_{n=1}^{N} (-1)^{a(n)} 2^n \neq 0.$$

Errata and Specific Remarks

p. 241, line 5:	Presumably "(3)" should be replaced by "(3) and (6)". However, it is easy to see that these reversed inequalities could not possibly hold.
p. 241, line 4 from bottom:	Replace $\chi(n)$ by $\lambda(n)\,\chi(n)$.
p. 241, line 2 from bottom:	Replace χ by $\lambda\chi$.
p. 244, lines 4, 6, 9, and 12 from bottom:	Replace ζ by Z.
p. 246, line 4:	Replace (1) by (9).
p. 247, line 2:	Replace $(1 - r)^{-1}$ by $r(1 - r)^{-1}$.
p. 247, (23) and (23 bis):	Replace $(1 - r)^{1/2}$ by $(1 - r)^{-1/2}$.

References

1. C. B. Haselgrove, Mathematika 5(1958), 141–145.
2. R. S. Lehman, Math. of Computation 14(1960), 311–320.
3. R. S. Lehman, Acta Arith. 11(1966), 397–410.
4. R. Spira, Math. of Computation 22(1968), 163–173.

Comments on [57c]

The authors consider functions f defined for $\sigma = \text{Re}\, s > 1$ by

$$f(s) = \prod_p (1 - \chi_p p^{-s})^{-1},$$

where χ_p is real and $-1 \leqslant \chi_p \leqslant 1$. They prove that if f can be continued as a meromorphic function into some domain containing the line $\sigma = 1$, then $f(s)$ is regular and nonvanishing at each point of the line $\sigma = 1$, except possibly at $s = 1$.

The proof treats separately the three cases in which $f(1)$ is zero, infinite, or neither zero nor infinite. The basic information used is that $-\zeta'(s)/\zeta(s) - f'(s)/f(s)$ has a Dirichlet series with non-negative coefficients and that $\zeta(s)$ has neither zeros nor poles on the line $\sigma = 1$, $s \neq 1$.

The theorem offers no new proof of the non-vanishing of $\zeta(s)$ on $\sigma = 1$. It is interesting to note that an *a priori* knowledge of regularity on $\sigma = 1$, which is important for Hadamard's proof that $\zeta(s) \neq 0$ on $\sigma = 1$, is not needed in the present context. However, we know of no application of the theorem in which regularity on $\sigma = 1$, $s \neq 1$, is not known in advance.

Errata and Specific Remarks

p. 802, line 3: Replace 0 by 1.

p. 802, line 11: Replace (4) by (5).

p. 803, line 12 from bottom: To show that $J_0 = 0$ and not merely $J_0 \geqslant 0$ we need the fact that f cannot have a multiple zero at 1. This follows either from the argument at the bottom of p. 804, or from the fact that the coefficients on the right-hand side of (14) are non-negative.

p. 803, line 4 from bottom: Replace χ_p by χ_p^k. Thus in what follows a_k must be replaced by $a_{p,k}$, where $a_{p,k} = 1 + \chi_p^k \log p$.

p. 804, line 5: Replace $|J_t| = |J_0|$ by $|J_t| \leqslant |J_0|$.

p. 804, statement of (II*): Presumably the authors meant to require $\chi_p < p$ for all p, since the case $\chi_p > p$ leads to poles with real part greater than 1 and the hypothesis $f(1) = 0$ is impossible to achieve if $\chi_p = p$ for some p.

p. 804, line 2 from bottom: Replace $1 - 0$ by $1 + 0$.

p. 807, line 3: Replace $1 + p^{-}$ by $1 + p^{-\sigma}$.

p. 807, lines 11–17: We do not understand the argument here and so offer the following substitute: If we take $s = 1 + i\alpha$, we see that $f(1 + 2i\alpha)$ is finite, since $f(1) \neq 0$. But then we get a contradiction by letting σ approach 1 from above in the inequality: $\{\zeta(\sigma)f(\sigma)\}^3|\zeta(\sigma + i\alpha)f(\sigma + i\alpha)|^4\ |\zeta(\sigma + 2i\alpha)f(\sigma + 2i\alpha)| \geqslant 1$.

p. 807, Formulas (23) and (25): Replace χ_p by χ_p^k.

THE HISTORICAL BACKGROUND OF HARMONIC ANALYSIS

BY

NORBERT WIENER

This address deals with certain aspects of the theory of Fourier series and integrals during the nineteenth century. The subject of Fourier series has been a central one both in the region of pure mathematical analysis and in mathematical physics. It is just because this subject faces two ways that its history is of especial interest. At each period in the history of mathematics, the relation between mathematics and physics has been viewed from an aspect somewhat peculiar to the period; and in tracing the history of this subject, we shall have in miniature a history of the larger relations between the two disciplines.

The first beginnings of Fourier series theory may be regarded as very ancient, for the musical theories of Pythagoras have already in them elements which later investigation has shown to be of trigonometric nature. The whole classical theory of cycles and epicycles as found in Ptolemaic astronomy is also something which a modern point of view would interpret as harmonic analysis. These two motives, that of the acoustical-optical wave theory, and that of astronomical and geophysical periodicities, united by Plato in the music of the spheres, have ever since furnished the chief stimuli of the natural sciences to the study of harmonic analysis. The motions of the planets, the tides, and the irregular recurrences of the weather, with their hidden periodicities, form a counterpart to the vibrating string and the phenomena of light.

However, it is probably a mere picturesque feat of the imagination to push harmonic analysis further back than Huygens. In Huygens' principle, which by the way was not correctly used by Huygens himself, we resolve a wave front into a set of centers of instantaneous disturbances, and by continuing these disturbances over a small interval of time we are in some way able to determine the new wave front.

While we can not extract from Huygens the Huygens principle in rigorous modern form, we at least find in his work a preoccupation with phenomena continuous over space and with their transportation from point to point by a process of expansion. It is not fanciful to see an extension of this point of view in the philosophy of Leibniz, who is under a heavy intellectual debt to Huygens, here and elsewhere. Leibniz regards space as a plenum, and considers that disturbances are transported from one region to another by a point to point transfer of energy, and not by an actual action at a distance.

The tool for the discussion of such plena was also formulated by Leibniz, and is the differential calculus. It was indeed the Leibnizian school, rather than Leibniz himself, which developed the theory of partial differential equations, or, as we should put it in the jargon of the present day, field theory. In this we must mention above all other names those of the Bernoulli family and of Euler. It is with Clairaut, d'Alembert, and Daniel Bernoulli that trigonometric expansions in the strict sense first come into prominence. Clairaut introduced them in the astronomical theory of the perturbing function, but their most famous application lies elsewhere. Among the simplest of partial differential equations is that of the vibrating string. It is one-dimensional and conserves energy, and it was observed by d'Alembert and Euler to possess a solution in the form of the pulse moving with a certain fixed velocity either to the right or to the left. This holds for the infinite string; in the case of the finite string the fastening at the ends serves to reflect an arriving pulse either with or without reversal of phase according to the nature of the constraint, and send it back along the line to meet and cross a similar pulse reflected from the other end. The solution of the string problem, given the initial displacement in velocity, is thus possible by means of extraordinary simplicity.

Besides the wave solution of the problem of the vibrating string, it was early seen that another solution of what may be called a vibration type is possible. Instead of looking for solutions whose geometrical character remains unchanged by bodily motion along the string, we look for solutions which remain unchanged except in scale by the passage of time. These represent what are called stationary vibrations of the string. Their analytic representation is that of a function of position alone multiplied by a function of time alone. Both functions are trigonometric in character. It will be found that neither the one nor the other is completely arbitrary, but on the other hand, that each one can assume only one of a discrete set of admissible frequencies, the so-called characteristic frequencies of the string.

The problem thus arises of synthesizing a solution of general type from these particular solutions. As the differential equation is linear, the natural means of synthesis is addition. It was realized very early—by Daniel Bernoulli, in 1753, for instance, as well as indirectly by d'Alembert—that solutions additively compounded from stationary solutions have a high degree of generality, and that they may be used in many cases interchangeably with the formally simpler wave solutions of the string problem.

However, the precise relations between these two types of solution still stood in need of elucidation. Series developments of particular numbers date well back into the seventeenth century, and the classical series development of a function—the power series of Taylor and Maclaurin—was known to Newton early in the eighteenth century. It was rediscovered about 1715 and soon was added to the repertory of any working mathematician. It is not surprising that certain highly special features of that development were taken as typical for series developments in general. In particular, the coefficients of a Taylor series are all obtained by successive differentiation in the neighborhood of a single point. It is thus a part of the very nature of such an expansion that it can represent only a function which is determined everywhere by its behavior in any single such neighborhood. The ordinary elementary functions of algebra and trigonometry, together with the logarithm and its inverse, the exponential function, are of this type. It was only natural that many mathematicians should expect that the series of stationary vibrations arising from the string problem might display a similar property.

On the other hand, the wave solution of the string problem is manifestly not confined to solutions with such a property of analyticity. A disturbance of any form, at least if it is smooth enough to have a slope, may be transported bodily along a string, reflected at the ends, transported back again, and so on; and the motion thus obtained is sufficiently perspicuous to make the verification of the fact that it satisfies the differential equation of the vibrating string a matter of trivial difficulty.

The early days of the Bernoulli vibration solution of the string problem were naturally stormy, even though neither Euler nor d'Alembert fully realized the degree of arbitrariness of the solution of the string problem which they had given, and in particular, did not believe the functions entering into their problem might coincide over one region and differ over another; they were not ready to believe that Bernoulli series could represent a motion of the degree of irregularity which fitted into their own solutions. Obviously no thoroughly satisfactory answer to this dilemma was possible until it was realized just how the coefficients of Bernoulli's trigonometric series depended on the functions represented. Lagrange made an attempt to arrive at this representation by a study of the vibrating string with equal concentrated masses distributed uniformly along it, and then proceeded to the limit as the points became sown more and more thickly. But his limit argument left much to be desired in rigor. The final formal representation of the coefficients of a trigonometric series is due to Fourier and, possibly independently, to Laplace.

Fourier's problem was closely analogous to that of the vibrating string, being, like it, the study of a partial differential equation with given initial conditions, although his equation was that of the flow of heat and was what we would now call of the parabolic type, as opposed to the equation of the string, which is of the hyperbolic type. Fourier represented his initial conditions by a trigonometric series in which the coefficients were given as definite integrals. This same representation of a function was employed by Laplace in connection with the use of the method of generating functions in his theory of probability. The moment this explicit representation was given, it became obvious that the method was formally applicable to a function made up of several discrete and independent pieces with angles or jumps between them. Thus a completely new idea of the nature of a function was originated, and perhaps the greatest step was taken in the theory which is now known as that of the functions of real variables.

We have mentioned that Laplace's study of trigonometric series arises in connection with the theory of generating functions. Generating functions were already known to Euler and applied by him to the theory of partitions and other similar combinatory problems. The basis of this whole theory is the formula which derives the coefficients of the product of two Taylor series from the coefficients of each series taken separately. It will be seen that the product coefficient is the sum of products of the coefficients from the two factor series taken in such a way that the sum of the exponents is fixed. Such a sum is very nearly what is known in modern terminology as the convolution or "Faltung" of the coefficients of the two series. This convolution is merely changed by a relabeling of its terms when one of the series is changed by multiplication by an integral power of a parametric variable. From another point of view, if we translate the coefficients of one series as a whole with respect to the coefficients of the other, we merely introduce a translation of the coefficients of the product series.

If we have a number of independent phenomena which combine additively, the distribution of the quantity representing the combined phenomena is derived by the successive convolution of the distribution functions of the components. This is the reason for introducing generating functions into the theory of probability. These functions represent a vital part of Laplace's contribution to the theory and dominate the theory at the present time, although the variables which we raise to a power in the series are generally now taken to be complex. In the light of this interpretation of the series a statement in a recent edition of the Encyclopedia Britannica, to the effect that Laplace's method of generating functions is obsolete, is not justifiable.

While the method of generating functions can be formulated for real Taylor or Laurent series, the problem of determining the coefficients of such a series is not physically natural enough to make this the most effective way of introducing generating functions. Laplace already knew that generating functions could be treated very effectively in the case in which the parametric variable is of absolute value 1. He was also aware of the advantages of replacing a discrete set of powers of the parametric variable combined in an ordinary sum by a continuum of such values combined in an integral. The so-called Laplace transform and its complex analogue now called the Fourier transform were both known to Laplace as well as to Fourier, who used them in connection with the problem of the conduction of heat in the half-plane, and so on.

Thus, at the very beginning of the theory of Fourier series, ideas belonging to a large number of related fields made their appearance. We have vibration problems, problems for the theory of the conduction of heat, combinatorial and other theory of number problems, and problems of probability. Nevertheless, the orthodox line of development of Fourier series theory went through the work of Fourier and, in so far as it was practically applied, it was chiefly to the boundary-value problems of the theory of partial differential equations as they occur in mathematical physics. Fourier himself, following along lines already laid out by Laplace, studied developments in spherical harmonics and Bessel functions and laid the groundwork for the general theory of orthogonal functions. The immediate followers of Fourier, such as Poisson, Sturm, Liouville, on the one hand, and Hamilton and Dirichlet on the other, partly devoted their efforts to a further exploration of this direction and partly to a more rigorous justification of the formulas which Fourier had attacked in a rough heuristic manner. This was a period when the exuberant fertility of the eighteenth century had already begun to give way to a more critical spirit, and when the influence of Cauchy had led mathematicians to make a much more careful examination of the convergence of their expansions. Thus it was no more than natural that an overwhelming part of the efforts of the students of Fourier series during the last century should have been devoted to questions of convergence.

We have already mentioned Cauchy. Besides his indirect influence on the theory of Fourier series from the point of view of rigor, his study of functions of a complex variable led to a better understanding of the relation between the Taylor series and the differential method of determining its coefficients on the one hand, and the Fourier series and the integral method of determining its coefficients on the other. It was seen that these two theories, widely discrepant as they seem to be when we remain in the real domain, form part of the same theory in the complex plane. The real part of a function of a complex variable is a harmonic function; this fact also led to a study of the relation between Fourier series and harmonic functions. This, it is true, may equally well be studied from the point of view of Fourier himself and the boundary-value problem on the circle. Both of these theories have a definite relation to the work of Poisson.

At the same time that the Fourier series theory was receiving this development either from the point of view of pure mathematics or from that of a rather rigidly formalized branch of applied mathematics in the case of boundary-value problems, other physical problems were familiarizing the physicists with the notion of trigonometric developments in a totally different direction. The old string problem received an enormously broad extension with the increasing popularity of wave theories, both in the domain of sound and in that of optics. This was particularly due to Young and to Fresnel. These theories borrowed much from the older string theory, but in view of the greater number of dimensions to which they applied, went far beyond it. The idea of white light as the additive combination of monochromatic vibrations was, of course, along the same lines as the theory of the Fourier integral, but there seems to be no evidence that any of the earlier writers saw that it demanded far more than had been developed explicitly by Fourier and Laplace.

To this same order of ideas belong questions of coherency and interference, of the different types of polarization, and so on. The astronomical motive in Fourier analysis, which had already appeared in the formal investigations of stability by Laplace, Lagrange, and Poisson, became the foundation of a new lunar theory in the hands of Hill and Brown. In all of this work, one can only wonder at the ability of the physicist working with the crudest of mathematical tools and total lack of rigor, to develop an intuition which, almost without exception, saves him from the apparently inevitable consequences of his mathematical sins.

It is a real tragedy of the mathematics of the nineteenth century that, on the one hand, its physical stimulus was continually schematized, reduced in breadth, thrown into the background, or even forgotten, while on the other the experimental work of the physicists gradually sundered itself from the fertilizing influence of the development of mathematical theory. For a long period the theory of Fourier series tended to contract itself to the scope of a mathematical game whose applications, whatever they might have been, were regarded as of little interest outside of mathematics. On the other hand, the harmonic analysis of the physicist gropingly developed its own technique as if the physicist had believed that the mathematical difficulties of principle had all been swept away somewhere obscurely in the literature by the pure mathematician, and that all he needed to do was to recreate crudely his own working formulas.

Within the limited field which the mathematicians have taken for their own, the first efforts of Poisson, Dirichlet, and their followers were directed to the purpose of finding some field, no matter how limited, in which the theory of the Fourier development could be carried out rigorously. The theory of functions of limited total variation and of uniform convergence are to a large extent by-products of this effort. Dirichlet's conditions represent the first important satisfactory achievement of the sort. The works of Riemann on the unicity of trigonometric developments attach themselves closely to this order of ideas, and foreshadow in certain respects the later theory of summable series.

On the other hand, once Fourier developments had been established on a sound basis in any region, no matter how narrow, mathematicians took upon themselves the problem of delimiting the exact bounds of this region. Inasmuch as the Fourier coefficients are defined by integration, this demanded a reinvestigation and a revision of the notions of measure and integral. It was soon seen that functions much more general than those of limited total variation could be given at least formal Fourier development, and what is much more important, that sets of coefficients much more general than those of functions satisfying the Dirichlet conditions determined, in some sense or other, the Fourier series of well defined functions.

In the course of time these activities led to numerous by-products. Parseval's relation between the sum of the squares of the coefficients of a Fourier series and the integral of the square of the function represented was a powerful stimulus to the investigation of the general question: under what circumstances can a function be determined by an arbitrary set of coefficients such that the sum of their squares converges. Far more recently, Hurwitz also studied this region of ideas. Again, the theory of the summability of series owes much of the interest which it has excited to its Fourier series applications. Here the work of Fejér is to be mentioned as the fountainhead of an important school of research. It became quite clear that the convergence theory, useful as it was in giving rigor to a limited region in Fourier analysis, was quite incapable of delimiting a sharp and well defined region in the theory of such series, and a great deal of Euler's work which had been buried under the wave of rigorism in the time of Cauchy was brought back to the light of day. We must mention Borel as one of the pioneers in this direction, as in so many that have helped to form the modern theory of harmonic analysis.

The ultimate triumph of this movement is to be found in the Riesz-Fischer theorem. Before we come to this theorem in detail, however, it is necessary to go back to the domain of mathematical physics and see what some of the leading ideas of statistical mechanics were in the last half of the nineteenth century. We have already seen the early connection of Fourier analysis and statistical theory in the work of Laplace. This connection is intimate and reappears on several different levels. Besides Laplace's problem of the composition of probabilities, the theory of light brought in the germs of a statistical theory of harmonic analysis itself. No one has ever seen an oscillograph of a ray of visible light. The evidence that this light is of a wave character is and can be only such evidence as is applicable to a statistical assemblage of vibrations, rather than to a single pure vibration. Thus the founders of wave optics were driven despite themselves, to invent a statistical harmonic analysis. Young, Fresnel, Goüy, and Schuster all did important work in this direction, as well as in the related field of partially polarized light. In the hands of Schuster, this range of ideas, together with ideas from meteorology and astronomy, led to the discovery of the periodogram, a theory for which the pure mathematical basis was defective for many decades. It was only in 1924 that a paper by Wiener established the precise connection between this and the rigorous modern theory of the Fourier integral. Another spectrum theory, earlier than that of Wiener and covering radically different aspects of singular distributions, was developed by Hellinger in order to analyze the unitary invariants of Hermitian transformations of Hilbert space.

Besides these two relations between harmonic analysis and statistics, the autonomous development of the theory of statistical mechanics has introduced yet a third, and this the most important. Statistical mechanics had its inception in Maxwell's theory of the distribution of the velocities of the particles in a gas. Its more mature form is due to Gibbs and concerns, instead of the distribution of quantities attached to the different particles of a single system, the distribution of all systems where the laws of force are given and only the initial conditions are allowed to vary. In any case, statistical mechanics makes no assertion about the individual particle or the individual system, but merely about what will happen except for a fraction of the particles or systems, which is so small that in some sense or other it may be neglected. If the Maxwell gas is taken as filling an infinite space, or if the Gibbs ensemble of systems is really taken as complete, these sets of negligible likelihood must be taken as in some sense or other of zero likelihood. We thus need for the complete justification of statistical mechanics a rigorous theory of measure in which sets of zero measure have a legitimate place. It is, of course, clear that a theory of sets of zero measure in rigorous form was not reached in a day. The fundamental assumption of Gibbs, for example, that an ensemble of dynamic systems in some way traces in the course of time a distribution of parameters which is identical with the distribution of parameters of all systems at a given time, was first stated in a form not merely inadequate, but impossible. This is the famous ergodic hypothesis. It is one of the greatest triumphs of recent mathematics in America, or elsewhere, that the correct formulation of the ergodic hypothesis and the proof of the theorem on which it depends have both been found by the elder Birkhoff of Harvard.

However, long before a really adequate formulation of the ergodic hypothesis was found, the notion of "almost all" had become an accepted part of the equipment of every physicist. It was under the influence of ideas belonging to this domain that Poincaré at the end of the last century developed the philosophy of questions in the theory of probability which marked the first really great progress in that theory since the days of Laplace. The ideas of statistical randomness and phenomena of zero probability were current among the physicists and mathematicians in Paris around 1900, and it was in a medium heavily ionized by these ideas that Borel and Lebesgue solved the mathematical problem of measure.

On the mathematical side, the problem of measure was absolutely vital to the foundation of a symmetrical theory of Fourier series. The formulas connecting a function and its Fourier coefficients are so symmetrical that it was early obvious that the problem of determining the coefficients of a function and the problem of determining a function from its coefficients were closely parallel. In any earlier theories of measure and integration, on the other hand, the problem of proceeding to a function from a set of coefficients enjoying some simple and very general property had no natural solution. The question of putting the theory of integration on a broad enough basis to allow a solution of these problems was attacked simultaneously by many mathematicians in many places. Besides Borel and Lebesgue in Paris and W. H. Young in England, we must mention the anticipatory work of Osgood and Pierpont in America which, if it did not lead completely to the goal of the new integration theory, at least did much to point to the direction in which that goal would be found. After the Lebesgue integral had been discovered, Daniell gave it an abstract form independent of dimensionality.

The Lebesgue theory of measure and integration produced an immediate and explosive expansion in the theory of harmonic analysis. The chief result was the Riesz-Fischer theorem, to the effect that there exists a perfect equivalence between the class of Lebesgue measurable functions with Lebesgue integrable square moduli, and the class of functions which can be represented by Fourier series for which the sum of the squares of the moduli of the Fourier coefficients converges. It is this theory which gives the motivation of the infinitely-many dimensional space of Hilbert. To this realm of ideas belong Weyl's lemma, to the effect that a sequence of functions of this class converging in the mean converges in the mean to a limit defined almost everywhere, and the Plancherel theorem as to the existence of Fourier transforms for an analogous class.

Both in the case of Fourier series and in that of the Fourier integral, there is an L^p theory analogous to the L^2 theory but less perfect. Here the chief names are Hölder, Hausdorff, Minkowski, Titchmarsh, and Paley.

Another block of related theorems includes the Lebesgue theorem that an integrable function is almost everywhere the derivative of its integral, and Lebesgue's other theorem that the Fourier series of an integrable function is almost everywhere Cesàro summable of the first order to the function. The whole theory of the convergence and summability of Fourier series was recast by a group of mathematicians working largely in England, among whom must be mentioned W. H. Young, Hobson, Hardy, Littlewood, and Chapman, together with the brothers Riesz, Szász, and Zygmund on the continent. The theory of summability in connection with trigonometric series, while anticipated by Abel and Poisson, was first consciously employed by Riemann. The study of convergence factors and exponents of convergence for Fourier series and even for general orthogonal series developed a literature of its own, particularly in the hands of the Polish school. Although no natural necessary and sufficient condition for the convergence of a Fourier series was forthcoming, such a condition was found by Hardy and Littlewood for summability of unspecified order.

It is impossible to follow the classical theory of Fourier series into all its ramifications, nor is it any easier to do so with the related theory of the Lebesgue integral. This latter theory, it is important to note, soon established relations with allied statistical theories. The work of Einstein and Smoluchowski on the Brownian motion had led in 1909 to the important investigations of Perrin in Paris, and it was not long before he was consulting his colleague Borel on the matter. Besides his work on measure, Borel was known for his investigations in the theory of probability, and was in close touch with the ideas of Poincaré. Perrin had noticed that the paths of particles subjected to the Brownian motion were suggestive of the non-differentiable continuous curves of Weierstrass. It is perhaps not too far-fetched to see in this suggestion a strong reinforcement of the motives which led Borel to apply the methods of the theory of probability—that is, of measure—to analytical problems. This is his theory of "probabilités denombrables," later studied by Steinhaus, Rademacher, Wiener, and others. In the hands of Paley and Zygmund, it became the nucleus of a fertile method for the devising of Gegenbeispiele in the theory of Fourier series. Wintner, first by himself and then in collaboration with Jessen, developed a parallel theory which yielded much information on the distribution of the values of the Riemann zeta function and other related functions. The Wiener theory of differential space is a direct development of the mathematical ideas behind the theory of the Brownian motion. It has led to a very general theory of continuous randomness, and is related to the work of Khintchine and Kolmogoroff, and to the theories of stochastic processes developed by these authors, Cramér, Lévy, and others. Through the work of Cramér, ideas of this order have been introduced into certain generalizations of number theory. Lastly, it now seems likely that a systematic study of such notions may lead to a really constructive systematization of the study of random ensembles in statistical mechanics, as, for example, in the theory of turbulence or in that of the statistical mechanics of liquids.

Besides the classical theories of Fourier series and integrals, a number of investigations have been made of trigonometric developments not falling strictly under either of these heads. Non-harmonic Fourier series are at least as old as Fourier, and the Dirichlet series of the student of analytic number theory give rise to trigonometric series, where the frequencies do not form an arithmetic progression. It was these number-theoretic series on the one hand, and the dynamical nonharmonic Fourier series of Bohl, on the other, which led Harold Bohr, in the early twenties, to formulate the notion of the almost periodic function. His original papers became the nucleus for a considerable literature. Bohr's fundamental theorems, to the effect that the Parseval theorem holds in a certain "mean" form for almost periodic functions, and that these functions are precisely those which admit a uniform approximation by trigonometric polynomials, were proved in widely different ways by Weyl, de la Vallée Poussin, and Wiener. The theory of Wiener was sufficiently general in scope to cover periodograms and the theory of white light. The class of almost periodic functions was generalized by Stepanoff, Weyl, Besicovitch, and Wiener. The definition of almost periodic functions was recast by Bochner into the form, "the class of all continuous functions, such that among any infinite sequence of translations of one of them, a uniformly convergent subset can be chosen." Bohr himself studied almost periodic functions in the complex domain, and was followed by Jessen, Favard, and Cameron in the investigation of the distribution of the values of an analytic almost periodic function in strips.

The more general question of the harmonic analysis of the orbits of particles in a dynamical system owes much to the work of Birkhoff. Birkhoff's work represents an essential refinement of the ideas of Poincaré on Poisson stability. Here we must mention the related physical theory of adiabatic invariants, as developed by Ehrenfest and Levi-Civita. Birkhoff's ergodic theory furnishes the necessary *point d'appui* of Wiener's generalized harmonic analysis. Indeed, it was for the purpose of such a harmonic analysis that the ergodic theorem was first applied. A concrete achievement in this direction is the theory of mean planetary motion of Wintner, von Kampen, Weyl, and others. Other forms of generalized harmonic analysis go back to the work of Hahn, and concern themselves with what should now be called the theory of Stieltjes-Fourier transforms being pursued so successfully by Wintner and his collaborators. Other writers who have contributed to the theory of generalized harmonic analysis are Bochner and Berry.

Almost periodic functions are characterized by a certain property of their translations, which is a property invariant under translation. This is related to the fact that the functions e^{iux} are the characteristic functions of the translation group. Harmonic analysis is thus the appropriate tool for the study of linear problems formally invariant under this group. As most physical phenomena are not attached to a fixed origin in time, and as indeed the existence of such an attachment would make it impossible to repeat an experiment, and to develop a consistent physics, harmonic analysis is indicated over a wide domain of practical problems. These group-theoretical aspects of harmonic analysis are foreshadowed in the work of Lagrange and Laplace on ordinary difference and partial differential equations with constant coefficients. The calculus of operations associated with the names of Arbogast and Boole, as well as the work of Heaviside on the electric circuit, represent further developments of this point of view. The example of Heaviside shows how directly it has remained in contact with physical and engineering practice. After Heaviside, obscure and intuitive as he was, an extensive interpretative literature developed, in which the names of Bromwich, Doetsch, Jeffries, and others are to be mentioned. The methods of Heaviside, as well as the alternative approach of Carson, have been pushed into the background by the conscious, explicit use of the Fourier integral, which has been much facilitated by the recent tables of Fourier transforms published by Campbell and Foster, of the Bell Telephone Laboratories.

The considerations which lie back of the operational calculus of Heaviside are clearly of group-theoretical nature, as we have seen. Other groups besides the translation-group possess characteristic functions, and a theory of expansions in terms of these functions. The characteristic functions belonging to the denumerable abelian group, all of whose operators are of period two, are the so-called Walsh functions, and they have as a basis the well known Rademacher functions. The functions of Haar and Franklin, while not characteristic, are closely related. It was a heuristic remark by Paley that the Walsh functions have expansion properties very nearly identical with those of the trigonometric functions, of which they form a good working model. This suggested to Paley and the author the formation of a bridge between the theory of Fourier developments and the theory of group measure of Haar and Pontrjagin. The work of von Neumann on almost periodic functions on groups represents a further development of this line of research, and in particular, the extension to the non-abelian case is an important advance. It has its inspiration in Weyl's approach to the theory of almost periodic functions, and in the earlier work of Hilbert and E. Schmidt.

While the theory of Tauberian theorems is not a Fourier theory in its origins, and while the original method of Hardy and Littlewood has nothing to do with trigonometric considerations, an early use of Tauberian methods was to establish the Poisson approach to Fourier series on a rigorous basis. However, there are other more intimate relations between the Tauberian theorems and harmonic analysis. Many of the most common theorems of Tauberian type or, in other words, inverse theorems concerning limit processes, deal with averages of a function which have a certain sort of invariance with respect to a multiplicative change in scale. The logarithmic transformation reduces this multiplicative change of scale to an additive change, and hence transforms Tauberian problems into a form suitable for the application of the methods of harmonic analysis. I repeat, the base of this application consists in the group-theoretical properties of the multiplication group. The earlier work of Hardy and Littlewood on these problems had no such group-theoretical justification, and it is in a paper of Wiener that a rational basis was found for introducing Fourier methods. Later on the author saw that the key theorem was one concerning the reciprocal of an absolutely convergent Fourier series. With the aid of these methods the solution of Tauberian problems has been reduced to a routine. The author's work was stimulated by the earlier work and personal suggestions of Robert Schmidt. The Toeplitz-Schmidt theory of gestrahlte Mittelbildungen bears the same relation to the author's work that the Laplace integral does to the Fourier transform.

Another closely related piece of work is that of Karamata, which, however, did not have in its inception any Fourier aspect whatever. With the aid of Tauberian theorems the author's pupil Ikehara and he have introduced simplifications into the proof of the prime number theorem, and these simplifications have led to subsequent work by Heilbronn, Landau, and others. This Tauberian work has found direct application in the theory of Fourier series and in other branches of harmonic analysis, and has also been turned by Paley and the author to the study of the behavior of entire functions of exponential type in the complex domain. Here their work joins with that of Pólya, Miss Cartwright, and others, and in its application to the study of gap series, with that of Mandelbrojt. A closely related field of harmonic analysis work is the theory of quasi-analytic functions, where the original introduction of trigonometric methods is due to de la Vallée Poussin, and where Paley, Mandelbrojt, and the author have carried the theory further. The recent very refined work of Levinson develops the theory in all these directions.

It is impossible in the time allotted me to take up all branches of harmonic analysis, and the motives which have played a rôle in forming the theories. The theory of the Laplace integral has been developed by Widder, Doetsch, S. Bernstein, Hille, Tamarkin, and others. In Japan, Izumi and his school have done much noteworthy work. The general theory of Watson transforms is also to be mentioned here. It will be seen that recent work in harmonic analysis is highly international, and that no inconsiderable part of it has been performed in this country.

While the historical facts in any concrete situation rarely point a clear-cut moral, it is worth while noting that the recent fertility of harmonic analysis has followed a refertilization of the field with physical ideas. It is a falsification of the history of mathematics to represent pure mathematics as a self-contained science drawing inspiration from itself alone and morally taking in its own washing. Even the most abstract ideas of the present time have something of a physical history. It is quite a tenable point of view to urge this even in such fields as that of the calculus of assemblages, whose exponents, Cantor and Zermelo, have been deeply interested in problems of statistical mechanics. Not even the influence of this theory on the theory of integration, and indirectly on the theory of Fourier series, is entirely foreign to physics. The somewhat snobbish point of view of the purely abstract mathematician would draw but little support from mathematical history. On the other hand, whenever applied mathematics has been merely a technical employment of methods already traditional and jejune, it has been very poor applied mathematics. The desideratum in mathematical as well as physical work is an attitude which is not indifferent to the extremely instructive nature of actual physical situations, yet which is not dominated by these to the dwarfing and paralyzing of its intellectual originality. Viewed as a whole, the theory of harmonic analysis has a very fine record of this sort. It is not a young theory, but neither is it yet in its dotage. There is much more to be learned and much more to be proved.

Massachusetts Institute of Technology,
Cambridge, Mass.

Comments on [38e]

P. Masani

This 12-page American Mathematical Society semi-centennial address, delivered in 1938, traces certain salient aspects of the evolution of harmonic analysis, and in the light of this suggests a broadening of its scope and the adoption of a dialectical attitude towards it. It offers a valuable blueprint for a more thorough and up-to-date history of the subject, since it stresses the factors that many a historian of harmonic analysis could easily overlook. The article is interpretative rather than narrative: it contains no equations or references. Nor from the standpoint of scholarship, is it comparable to the classic reviews of Riemann[7] and Burkhardt.[1,2,3] The reader might find it illuminating to fall back on these sources while reading Wiener's account. Even so he will need an empathic attitude in order to make certain diffuse passages come out true.

According to Wiener the *prehistory* of the subject commences with *Huygen's Principle*, which emphasized contiguous determinism and stimulated the formulation of the partial differential equations of field physics. The *history* itself begins with attempts to solve these equations under boundary conditions with the aid of sinusoids: the work of d'Alembert, Euler, Bernoulli and Lagrange on the vibrating string, of Fourier on steady-state heat conduction, etc. It should be added, of course, that harmonic analysis becomes relevant only when the contiguous deterministic mechanism is assumed to be "temporarily homogeneous" (i.e. its controlling parameters are deemed to be time-independent) and only when its "small" oscillations are investigated. For only then do we get *linear* partial differential operators in our equations having *sinusoidal* eigenfunctions. The propagation of these sinusoids being known from a spectral analysis of the operator, the Principle of Superposition tells us how any linear combination of sinusoids will propagate. The question of expansion of an arbitrary function in terms of sinusoids thus becomes vital. Wiener traces this development, pointing out its tenuous similarities with the earlier Taylor expansion, how unlike the former even discontinuous functions could now be expanded, and how this led to the abandonment of certain illusions and forced on reluctant minds a widening of the function concept.

Wiener then turns to the concept of *moment-generating function* of a distribution (i.e. its Laplace-Stieltjes transform), which Laplace introduced into his analytic theory of probabilities (1812). This does not exist for all distributions; but once the fear of i is lost, and we insert it in the exponent, i.e. take the Fourier-Stieltjes transform, we get the so-called *characteristic function*, which always exists. The great importance of these functions stems from the fact that the distribution of a sum of independent random variables is a convolution of the individual distributions, and our transforms convert convolutions into multiplications. This explains why harmonic analysis plays such a vital role in the calculus of probabilities, why, for instance, the characteristic function looms large in all proofs of the Central Limit Theorem. This aspect of the development could have been missed by a historian, less mathematically perceptive than Wiener.

Wiener then turns to the next stage, which saw a considerable widening of the range of applications of Fourier analysis, and the establishment of contact with Cauchy's theory of holomorphic functions f, via harmonic functions: the real part of the Taylor series of f is the Fourier series of Re f. This phase was dominated by the *movement towards rigor* initiated by Cauchy. Poisson, Dirichlet and their followers strived to keep the theory within a framework in which the expansions could be rigorously established. They were thus led to seek specific and rather narrow classes of functions, such as those of bounded variation.

But while the field of the purely mathematical theory of Fourier series was being thus constricted, new and greater demands were being made on harmonic analysis by the flourishing *wave theory of light* of Fresnel and Young. This theory had emerged from Newton's epoch-making experiments on the dispersion and interference of light, and the discoveries of diffraction and polarization. These phenomena are statistical. As Wiener says,

> No one has ever seen an oscillograph of a ray of visible light. The evidence that this light is of a wave character is and can be only such evidence as is applicable to a *statistical assemblage of vibrations*, rather than to a single pure vibration. Thus the founders of wave optics were driven despite themselves, to invent a *statistical* harmonic analysis. [38e, p. 62, my emphasis]

The physicists who undertook this task loved harmonic analysis. ("A great mathematical poem" is how Lord Kelvin had described Fourier's classic.) But pure mathematicians had so restricted their horizons as to the scope of this field that they were impotent to offer advice. A similar numbness greeted the remarkable work of A. W. Hill on lunar theory.

Wiener calls this schism a "tragedy" for 19th-century mathematics. But the tragedy turned out to be the middle step of a remarkable dialectical development. For as Wiener narrates, the systematic pursuit of rigor was not without its creative side. It forced a clarification of the number concept, gave rise to the set theory of Cantor, the theories of measure and integration of Borel and Lebesgue, and following in their wake came L_2 and the Riesz-

Fischer and Fourier-Plancherel theorems. These proved to be just the theories required for the rigorous formulation of the remarkable ideas advanced by the great physicists. The 19th-century "tragedy" was thus instrumental in ushering in, around 1920, a new movement towards a broadened harmonic analysis, from which the statistical physicists could draw.

Wiener then describes the salient features of this *new movement*. There were two (interacting) tendencies, one stochastic, the other not directly so, and many stimuli were at work. On the stochastic side, the main achievements were the theories of the Brownian movement, of random Fourier series, of stationary stochastic processes in the time and spectral domains, the Birkhoff Ergodic Theorem and the theory of mean-motions of Wintner and Weyl—the last two hailing from the (non-stochastic) theory of dynamical systems. On the non-stochastic side were the Bohr theory of almost periodic functions stemming from researches in Dirichlet series, Wiener's generalized harmonic analysis and Tauberian theory, and the Haar, Radamacher and Walsh systems. To this list of Wiener should be added Bochner's theorem on positive-definite functions and the von Neumann and Stone theorems on the unitary representations of $\mathbb{N}$ and $\mathbb{R}$.

This volume and about half of the first are testimony to Wiener's great contribution to this movement. The central idea governing much of his work is *translation invariance* or *stationarity*: under the additive group $\mathbb{R}$ for harmonic analysis, and under the multiplicative group $\mathbb{R}_+$ for Tauberian theory.

We cannot of course in this review continue this history beyond 1938. For the evolution of *abstract harmonic analysis* and the theory of group representations, stemming from the researches of Haar, Andre Weil, Gelfand and Beurling and their schools, W. F. Eberlein and G. W. Mackey, we must refer to Hewitt and Ross, vols. I, II, (especially pp. 518–522 and pp. 547–552), and to G. W. Mackey.[5,6] For the development of the field from the distributional standpoint, we would refer to Schwartz's classic work.[8] For the major application centering around the *prediction and filtering* of stationary stochastic processes, causal, anti-causal factorization and Hopf-Wiener equations, we would refer to the contributions in the Wiener Memorial Issue of the Bull. Amer. Math. Soc. 72(1966) No. 1, Part II, and the forthcoming Volume III of these collected works. Its potentialities in the area of *turbulence and statistical* mechanics are explored in the commentary by Drs. B. McMillan and G. Deem in Volume I, pp. 654–671 of these works, and in my commentary on [30a], §D, in this volume. For the remarkable progress in the *celestial mechanics of mean motion* by pursuit of these and related ideas, we refer to the book of S. Stenberg,[9] and the references therein. We would have liked to refer to a work on the application of generalized harmonic analysis to the theory of music, but we know of none.

Music and celestial mechanics—they remind us that harmonic analysis began in a way with Pythagoras and Ptolemy. Its remarkable evolution is a memorable chapter in the intellectual ascent of man.

References

1. H. Burkhardt, *Entwicklungen nach oscillirenden Functionen*, Jahresber. der Math. Ver., 10(1902), 228–342.

2. H. Burkhardt, *Trigonometrische Interpolation* (1904), Enc. der math. Wiss., Band II, Teil 1, 642–694, Leipzig, 1899–1916.

3. H. Burkhardt, *Trigonometrische Riehen und Integrale bis eta* (1850, 1904), Enc. der math. Wiss. Band II, Teil 1, 819–1354, Leipzig, 1899–1916.

4. E. Hewitt and K. Ross, *Abstract harmonic analysis*, vols. 1, 2, Springer Verlag, 1963, 1970.

5. G. W. Mackey, *Induced representations of groups and quantum mechanics*, Benjamin, New York, 1918.

6. G. W. Mackey, *Harmonic analysis as the exploration of symmetry—a historical survey*, in History of Analysis, ed. by R. J. Stanton and R. O. Wells, Rice Univ. Press, Houston, 1978.

7. B. Riemann, *Ueber die Darstellbarkeit einer Function durch eine trigonometrische Riehe* (1867), in *Gesammelte mathematische Werke*, Leipzig, 1892.

8. L. Schwartz, *Théorie des distributions*, vols. 1, 2 Hermann, Paris, 1966.

9. S. Stenberg, *Celestial mechanics*, Benjamin, New York, 1969.

IE
Classical Harmonic and Complex Analysis

THE EQUIVALENCE OF EXPANSIONS IN TERMS OF ORTHOGONAL FUNCTIONS[1]

By J. L. WALSH, OF HARVARD UNIVERSITY, AND N. WIENER.

Introduction. Suppose we are considering the nth partial sum of the formal expansion of an arbitrary function $f(x)$ in terms of the set of functions $\{\phi_n(x)\}$ normal and orthogonal on the interval $0 \leqq x \leqq 1$. This sum is

$$\sum_{i=1}^{n} \phi_i(x) \int_0^1 f(y)\phi_i(y)dy = \int_0^1 f(y) \left\{ \sum_{i=1}^{n} \phi_i(x)\phi_i(y) \right\} dy. \tag{1}$$

If we are considering the nth partial sum of the expansion of $f(x)$ in terms of the functions $\phi_n(x)$ which together with the functions $\psi_n(x)$ form a biorthogonal set normalized over the interval (0,1), we find that this sum is

$$\sum_{i=1}^{n} \phi_i(x) \int_0^1 f(y)\psi_i(y)dy = \int_0^1 f(y) \left\{ \sum_{i=1}^{n} \phi_i(x)\psi_i(y) \right\} dy. \tag{2}$$

If we replace the partial sums (1) and (2) by the analogous Cesàro sums of order 1, we get

$$\int_0^1 f(y) \left\{ \sum_{i=1}^{n} \frac{n-i+1}{n} \phi_i(x)\phi_i(y) \right\} dy \tag{3}$$

$$\text{and } \int_0^1 f(y) \left\{ \sum_{i=1}^{n} \frac{n-i+1}{n} \phi_i(x)\psi_i(y) \right\} dy,$$

respectively. Similar expressions may easily be obtained for Cesàro sums of higher orders. Again, the Fourier integral representation for $f(x)$ is

$$\lim_{n\to\infty} \frac{1}{\pi} \int_{-\infty}^{\infty} f(y) \frac{\sin n(y-x)}{y-x} dy. \tag{4}$$

All of the expressions (1) – (4) have at their nth stage, as their

[1] Presented to the American Mathematical Society, December 28, 1921

nth approximation to the arbitrary function $f(x)$, an expression of the form

$$\int_a^b f(y)\, K_n(x,y)dy. \tag{5}$$

In the expansions which have been mentioned, this represents $f(x)$ in some sense or other, either by convergence (as $n\to\infty$), or convergence in the mean, or perhaps purely formally. For certain well-known sequences $\{K_n(x,y)\}$ the nature of this mode of representation has been ascertained with more or less completeness. The question naturally arises whether for other sequences $\{K_n\}$ the problem of the mode of representation of f may not be reducible to that of better-known sequences $\{K_n\}$. This is the chief problem attacked in the present paper; the principal result is stated in Theorem I. We shall consider also the representation of a function by a given sequence $\{K_n\}$ at different points of the interval considered; the results found are stated in Theorems II and III.

It is clear that the formal expansions of a function $f(x)$ in terms of the two sequences $\{K_n\}$ and $\{K^1{}_n\}$ will have the same properties in all that concerns convergence, convergence in the mean, uniform convergence, Gibbs's Phenomenon, etc., if merely

$$\lim_{n\to\infty}\int_a^b f(y)\left\{ K_n(x,y) - K^1{}_n(x,y)\right\} dy = 0 \tag{6}$$

uniformly in x. If this is true we shall say that the formal expansions of $f(x)$ *have the same convergence properties* or are *equivalent*. If this is true for all functions $f(x)$ of summable square[2] we shall say that not only the expansions of $f(x)$ are equivalent but that the two sequences $\{K_n\}$ and $\{K_n^1\}$ are equivalent. These definitions presuppose a range of values for x which is frequently $a\leqq x\leqq b$ and will be considered to be this entire interval unless otherwise specified. The range may be specified, however, as any sub-set of points on this interval.

[2] Here and throughout this paper, when we postulate the summability of the square of a function we tacitly assume that the function is measurable; it follows that the function itself is also summable.

The notion of equivalence has been employed by Haar[3] and by Young[4] in specific cases. A similar notion involving the absoluteness of the convergence of the series

$$\sum \int_a^b f(y) \left\{ K_{n+1}(x,y) - K_{n+1}^1(x,y) - K_n(x,y) + K_n^1(x,y) \right\} dy, \quad (7)$$

has been discussed in some detail by Walsh,[5] and sufficient conditions have been found for its applicability. To the knowledge of the authors, however, no necessary and sufficient conditions for equivalence have been found. The first part of the present paper is devoted to a proof of **Theorem I**, which can be expressed in a simplified form as follows:

A necessary and sufficient condition that the sequences $\{ K_n \}$ *and* $\{ K_n^1 \}$ *be equivalent is that*

(a) *the expansions of all functions of a closed set*[6] *in terms of* K_n *and* K_n^1 *have the same convergence properties and that*

(b) *there exist a finite A such that for all n and x*

$$\int_a^b \{ K_n(x,y) - K_n^1(x,y) \}^2 dy < A. \quad (8)$$

Haar, in the second article cited, makes use of the sufficiency of this condition in the particular case of the equivalence of the Fourier and Legendre developments, and in a footnote mentions a mode of proving its sufficiency that is of perfectly general application, employing Hilbert's theory of bilinear forms in infinitely many variables.

Theorem I is applied in the first part of the present paper to the demonstration of the equivalence of the sine and cosine series.

[3] Mathematische Annalen, Vol. 69 (1910), pp. 331–371; Vol. 78 (1917–1918), pp. 121–136. The former of these papers refers to the equivalence of the Sturm-Liouville and Fourier cosine developments, the latter refers to the equivalence of the Legendre polynomial and Fourier cosine developments.

[4] Proceedings of the London Mathematical Society, (2) Vol. XVIII, pp. 141–162, 163–200, especially pp. 156, 150. These papers consider the equivalence of the Legendre polynomial and Bessel developments to the better-known Fourier developments.

[5] Transactions of the American Mathematical Society, Vol. 22 (1921), pp. 230–239. See also a forthcoming paper in the Annals of Mathematics.

[6] A set of functions is said to be *closed* or *complete* if each function of the set is of summable square and if there is no non-null function of summable square orthogonal to every function of the set.

In the second part of this paper we discuss certain problems of the interval equivalence of expansions. Thus the Fourier sine development of an arbitrary function has in the interior of a period the following important properties:

(*a*) The convergence of the development of $f(x)$ for $x=x_1$ and the nature of the approach of the approximating functions (the partial sums) to the limit function in the neighborhood of x_1 depend merely on the values of $f(x)$ for arguments in the neighborhood of x_1;

(*b*) The nature of the approach of the approximating functions to the limit function in the neighborhood of x_1 depends in no way on the actual position of x_1 in the entire interval; in other words, the difference between the sine series for $f(x)$ in the neighborhood of x_1 and the sine series for $f(x+h)$ in the neighborhood of x_1-h converges uniformly to zero in x and h.

In order to make these properties more tangible, we formulate the following definitions:

(*a*) the sequence $\{K_n\}$ is said to be ***regular*** if, whenever $f(x) \equiv g(x)$ over the interval (c,d) $[a<c<d<b]$, we have

$$\lim_{n\to\infty}\int_a^b [f(y)-g(y)]\,K_n(x,y)dy = 0 \tag{9}$$

uniformly over any closed interval entirely interior to (c,d);

(*b*) the sequence $\{K_n\}$ is said to be ***uniform*** if it is regular and if whenever x and $x+h$ lie in the interval $(a+\epsilon,\ b-\epsilon)$, where ϵ is an arbitrary positive quantity, we have

$$\lim_{n\to\infty}\int_{x-\epsilon}^{x+\epsilon} f(y)\,\{K_n(x,y)-K_n(x+h,y+h)\}\,dy=0 \tag{10}$$

uniformly in x and h for every $f(y)$ of summable square; this is the condition already referred to, but it will be convenient to require also that

$$\lim_{n\to\infty}\int_a^b\left\{g(x)-\int_a^b K_n(x,y)g(y)dy\right\}^2dx=0 \tag{11}$$

shall be significant and valid for every $g(x)$ of summable square.

That condition (11) is equivalent to property (*b*) follows from

Theorem II. *If the sequence* $\{K_n\}$ *is regular, then for every* ϕ *of summable square*

$$\lim_{n\to\infty}\int_a^{x-\epsilon}\phi(y)\,K_n(x,y)dy=\lim_{n\to\infty}\int_{x+\epsilon}^{b}\phi(y)\,K_n(x,y)dy=0 \qquad (12)$$

uniformly in x *over the interval* $(a+\epsilon,\ b-\epsilon)$.

We shall prove that if $\{K_n\}$ is uniform, it is equivalent over $(a+\epsilon,\ b-\epsilon)$ to a sequence of the form $K_n(y-x)$. From this there will follow

Theorem III. *If the sequence* $\{K_n\}$ *is uniform and if* $F(x)$ *can be written in the form,*

$$\int_a^x \Phi(u)du, \qquad (13)$$

where $\Phi(u)$ *is of summable square in* (a,b), *then the formal expansion of* $F(x)$ *in terms of the sequence* $\{K_n\}$ *converges uniformly over* $(a+\epsilon,\ b-\epsilon)$ *to the value* $F(x)$.

I. **The equivalence of expansions.**

Let $\{f_n(x)\}$ and $\{\phi_n(x)\}$ be two closed sets of functions, normal and orthogonal, on the interval $0\leqq x\leqq 1$. A condition that the sequences $\{K_n\}$ corresponding be equivalent is that the sequence of functions

$$\sum_{j=1}^{n}\{f_j(x)f_j(y)-\phi_j(x)\phi_j(y)\}=Q_n(x,y) \qquad (14)$$

be equivalent to zero. If this is true we shall say that the sets $\{f_n\}$ and $\{\phi_n\}$ are equivalent; the condition for this equivalence thus reduces to the condition that for every $F(x)$ of summable square we have

$$\lim_{n\to\infty}\int_0^1 F(y)Q_n(x,y)dy=0 \qquad (15)$$

uniformly in x.

We shall now prove

Theorem I. *A necessary and sufficient condition that a sequence of functions* $\{Q_n(x,y)\}$ *of summable square in* y *should have the* *property that for every F of summable square* (15) *should hold uniformly in x over a point set S contained in* (0.1) *is that there exist a number A such that for all n and for all x in S,*

$$\int_0^1 [Q_n(x,y)]^2 dy < A, \tag{16}$$

and that for all the functions $\{\psi_m\}$ *of a set closed on* (0, 1),

$$\lim_{n\to\infty} \int_0^1 \psi_m(y) Q_n(x,y) dy = 0 \tag{17}$$

uniformly in x over S.

The sufficiency of (16) and (17) is readily demonstrated.[7] Let $\{\psi_m\}$ be taken, as it may without essential restriction, as a closed normal orthogonal set. Then by the Riesz-Fischer theorem, a necessary and sufficient condition that a function $F(x)$ be of summable square is that we may write

$$F(x) \sim \sum_{j=1}^{\infty} a_j \psi_j(x),\ a_j = \int_0^1 F(x) \psi_j(x) dx, \tag{18}$$

where the symbol $\sim$ is to be interpreted as convergence in the mean. Moreover, by the Riesz-Fischer theory,[8] we have

$$\int_0^1 F(y) Q_n(x,y) dy = \sum_{j=1}^{\infty} a_j \int_0^1 \psi_j(y) Q_n(x,y) dy$$

$$= \sum_{j=1}^{m} a_j \int_0^1 \psi_j(y) Q_n(x,y) dy + \sum_{j=m+1}^{\infty} a_j \int_0^1 \psi_j(y) Q_n(x,y) dy. \tag{19}$$

Then we have from the general inequality,

$$\sum_{j=1}^{k} \alpha_j \beta_j \leqq \left\{ \sum_{j=1}^{k} \alpha_j^2 \sum_{j=1}^{k} \beta_j^2 \right\}^{\frac{1}{2}} \tag{20}$$

which holds whatever may be the quantities α_j and β_j, we have

[7] Cf. Hilbert, Integralgleichungen, p. 148.
[8] This follows immediately from the Schwarz inequality.

$$\left|\sum_{j=m+1}^{\infty} a_j \int_0^1 \psi_j(y) Q_n(x,y)\,dy\right|$$
$$\leqq \left\{ \sum_{i=m+1}^{\infty} a_i^2 \sum_{j=m+1}^{\infty} \left[\int_0^1 \psi_j(y) Q_n(x,y)\,dy\right]^2 \right\}^{\frac{1}{2}}$$
$$\leqq \left\{ \sum_{j=m+1}^{\infty} a_j^2 \sum_{j=1}^{\infty} \left[\int_0^1 \psi_j(y) Q_n(x,y)\,dy\right]^2 \right\}^{\frac{1}{2}} \tag{21}$$
$$\leqq \left\{ \sum_{j=m+1}^{\infty} a_j^2 \int_0^1 \left[Q_n(x,y)\right]^2 dy \right\}^{\frac{1}{2}} \leqq \left\{ \sum_{j=m+1}^{\infty} a_j^2 \right\}^{\frac{1}{2}} A^{\frac{1}{2}}.$$

Let the function $F(x)$ be given. We may choose m so that the last term in (21) is less than $\epsilon/2$, by the convergence of Σa_j^2. We choose N, so that for $n > N$, $1 \leqq j \leqq m$ we have

$$\left| a_j \sum_{i=1}^{m} a_i^2 \int_0^1 \psi_j(y) Q_n(x,y)\,dy \right| \leqq \frac{\epsilon a_j^2}{2}, \tag{22}$$

for all x in S. Then it follows from (19) that

$$\left| \int_0^1 F(y)\, Q_n(x,y)\,dy \right| < \epsilon \tag{23}$$

uniformly for all x in S. This completes the proof of the sufficiency of (16) and (17).

It remains to prove the necessity of (16) and (17); in fact, merely the necessity of (16), for that of (17) is obvious. Suppose (16) not to be true. Then there are two infinite sequences of numbers n_k and x_k, such that

$$\int_0^1 \left[Q_{n_k}(x_k,y)\right]^2 dy > 2^{2k}. \tag{24}$$

Set $C_1(y) = Q_{n_1}(x_1,y)$. Let $C_2(y)$ be the first $Q_{n_k}(x_k,y)$, such that

$$\left| \int_0^1 C_2(y)\, Q_{n_1}(x,y)\,dy \right| < 1 \tag{25}$$

for all x in S; such a $Q_{n_k}(x_k,y)$ will exist because of (15). Similarly, let $C_k(y)$ be the first $Q_{n_k}(x_k,y)$, such that we have

$$\left| \int_0^1 C_k(y)\, C_j(y)\,dy \right| < 1 \tag{26}$$

for all x in S and for every $j < k$.

We now consider the series

$$\sum_{n=1}^{\infty} \frac{C_n(y)}{2^n \left\{ \int_0^1 [C_n(y)]^2 dy \right\}^{\frac{1}{2}}}, \tag{27}$$

and shall prove that this series converges in the mean. That is, we are to prove

$$\lim_{n\to\infty} \int_0^1 \left\{ \sum_{j=n}^{n+p} \frac{C_j(y)}{2^j \left\{ \int_0^1 [C_j(y)]^2 dy \right\}^{\frac{1}{2}}} \right\}^2 dy = 0; \tag{28}$$

or, what is the same thing,

$$\lim_{n\to\infty} \sum_{k=n}^{n+p} \sum_{j=n}^{n+p} \frac{\int_0^1 C_j(y)\, C_k(y)\, dy}{2^{j+k} \left\{ \int_0^1 [C_j(y)]^2 dy \int_0^1 [C_k(y)]^2 dy \right\}^{\frac{1}{2}}} = 0. \tag{29}$$

We find at once from the Schwarz inequality that

$$\int_0^1 C_j(y)\, C_k(y) dy \leqq \left\{ \int_0^1 [C_j(y)]^2 dy \int_0^1 [C_k(y)]^2 dy \right\}^{\frac{1}{2}}, \tag{30}$$

so that (29) will follow if merely

$$\lim_{n\to\infty} \sum_{k=n}^{\infty} \sum_{j=n}^{\infty} 2^{-(j+k)} = 0\,. \tag{31}$$

This is, however, immediately true, since

$$\sum_{k=n}^{\infty} \sum_{j=n}^{\infty} 2^{-(j+k)} = \left\{ \sum_{j=n}^{\infty} 2^{-j} \right\}^2 = 2^{2-2n} \tag{32}$$

Hence, there is a function $\Phi(x)$ to which the series (27) converges in the mean.[9] The limit function $\Phi(x)$ is summable and of summable square, and (27) can be integrated term by term.

Let us now consider $\int_0^1 \Phi(x) C_n(x) dx.$

We have

$$\int_0^1 \Phi(x) C_n(x) dx = \sum_{m=1}^{\infty} \frac{\int_0^1 C_n(x)\, C_m(x) dx}{2^m \left\{ \int_0^1 [C_m(x)]^2 dx \right\}^{\frac{1}{2}}} \tag{33}$$

[9] See, for example, Plancherel, Rendiconti di Palermo, Vol. 30 (1910), p. 292.

We find from (24)

$$\frac{\int_0^1 [C_n(x)]^2 dx}{2^n \left\{ \int_0^1 [C_n(x)]^2 dx \right\}^{\frac{1}{2}}} > 1; \tag{34}$$

and we find from (24) and (26)

$$\frac{\left| \int_0^1 C_n(x)\, C_m(x) dx \right|}{2^m \left\{ \int_0^1 [C_m(x)]^2 dx \right\}^{\frac{1}{2}}} < 2^{-2m}. \tag{35}$$

Then we have finally,

$$\int_0^1 \Phi(x) C_n(x) dx > 1 - \sum_{m=1}^{\infty} 2^{-2m} = \frac{2}{3}, \tag{36}$$

which is inconsistent with (15). This completes the proof of the necessity of (16) and the proof of Theorem I.

In particular, we have proved that a necessary and sufficient condition that the normal orthogonal sets $\{f_n\}$ and $\{\phi_n\}$ be equivalent is that [in the notation of (14)]

$$\int_0^1 [Q_n(x,y)]^2 dy < A \tag{37}$$

for all n and for all x on the interval (0, 1), while [$\{f_n\}$ being assumed closed]

$$\lim_{n\to\infty} \int_0^1 f_j(y)\, Q_n(x,y) dy = 0 \tag{38}$$

for every j uniformly in x.

Formulæ (37) and (38) are susceptible of a good deal of transformation. We may substitute for (37)

$$\sum_{j=1}^{n} [f_j(x)]^2 + \sum_{j=1}^{n} [\phi_j(x)]^2 - 2 \sum_{j=1}^{n} \sum_{k=1}^{n} f_j(x) \phi_k(x) \int_0^1 f_j(y) \phi_k(y)\, dy < A, \tag{39}$$

and for (38)

$$\lim_{n\to\infty} \left\{ f_j(x) - \sum_{k=1}^{n} \phi_k(x) \int_0^1 \phi_k(y) f_j(y) dy \right\} = 0, \tag{40}$$

uniformly in x for every j. Formula (40) simply states that the ϕ series for each f_j converges uniformly to the value f_j.

Equivalence as defined for these sets $\{f_n\}$ and $\{\phi_n\}$ refers to the identity of the convergence properties of the two series formed as in (18) from the two sets of functions. This equivalence includes equivalence of the two new series found by summing those two former series by the Cesàro mean of the first order. There is, however, an equivalence for the new series without any reference to the old series, or, what is the same thing, which involves the identity of the convergence properties of the original series so far as concerns summability or uniform summability of the first or of higher orders. The definition of this equivalence is that for every function F of summable square we must have

$$\lim_{n\to\infty} \frac{1}{n}\int_0^1 F(y) \sum_{j=1}^{n} Q_j(x,y)dy = 0 \tag{41}$$

uniformly in x. For this equivalence it is necessary and sufficient that there be an A such that for all n and x

$$A > \frac{1}{n^2}\int_0^1 \left\{ \sum_{j=1}^{n} Q_j(x,y) \right\}^2 dy = \frac{1}{n^2}\int_0^1 \left\{ \sum_{j=1}^{n} (n+1-j)[f_j(x) f_j(y) - \phi_j(x)\phi_j(y)] \right\}^2 dy$$

$$= \frac{1}{n^2}\left\{ \sum_{j=1}^{n} (n+1-j)^2 \left\{ [f_j(x)]^2 + [\phi_j(x)]^2 \right\} - 2\sum_{j=1}^{n}\sum_{k=1}^{n} (n+1-j)(n+1-k) \left\{ f_j(x)\phi_k(x) \int_0^1 f_j(y)\phi_k(y)dy \right\} \right\}, \tag{42}$$

while

$$\lim_{n\to\infty} \frac{1}{n}\int_0^1 f_j(y) \sum_{k=1}^{n} Q_k(x,y)dy = 0 \tag{43}$$

uniformly in x for all j. Condition (43) may be written

$$\lim_{n\to\infty}\left[\frac{n+1-j}{n} f_j(x) - \sum_{1}^{n} \frac{(n+1-k)\phi_k(x)\int_0^1 \phi_k(y) f_j(y)dy}{n} \right] = 0, \tag{44}$$

which merely asserts that every f_j is the uniform Cesàro sum of its ϕ series. It is of course assumed here that the set $\{f_j\}$ is closed.

Every type of summation has a type of equivalence appropriate to it, and every such type of equivalence admits of a treatment similar to the one just given for Cesàro summability of the first kind. We shall have a boundedness condition analogous to (42) and a set of summability conditions not unlike (44).

In many cases of equivalence of the sets $\{f_n\}$ and $\{\phi_n\}$, the two sets of functions are so related that we have, for all n and x,

$$\sum_{j=1}^{n} \left\{ f_j(x) - \phi_j(x) \right\}^2 < A. \tag{45}$$

In this case it is possible to replace (39) by an equivalent condition to the effect that,

$$\left| \sum_{j=1}^{n} \sum_{k=1}^{n} A_{jk} f_j(x) \phi_k(x) \right| < B, \tag{46}$$

where

$$A_{jk} = \int_0^1 f_j(x)\phi_k(x)dx - \delta_{jk}, \tag{47}$$

δ_{jk} being the Kronecker symbol that is zero or unity according as $j \neq k$ or $j = k$.

The special case of Theorem I, where the range S is a single point, deserves some detailed consideration. Thus we have the theorem concerning a sequence $\{Q_n\}$ of functions of summable square:

In order that for all functions $F(y)$ of summable square we have

$$\lim_{n\to\infty} \int_0^1 F(y)Q_n(y)dy = 0, \tag{48}$$

it is necessary and sufficient that (48) *be true for a closed set of F's and that there be an A such that for all n*

$$\int_0^1 [Q_n(y)]^2 dy < A. \tag{49}$$

We shall now prove the following theorem:

Let there be a sequence $\{Q_n(y)\}$ such that (49) *obtains. Then a* *necessary and sufficient condition that there exist a function $Q(y)$ such that*

$$\lim_{n\to\infty}\int_0^1 F(y)Q_n(y)dy = \int_0^1 F(y)Q(y)dy \tag{50}$$

for all functions $F(y)$ of summable square is that, for all functions of a closed set $\{F_m(y)\}$,

$$\lim_{n\to\infty}\int_0^1 F_m(y)Q_n(y)dy$$

exist.

The necessity of the condition is obvious; we proceed to prove its sufficiency. We clearly have the following relation, if the closed set $\{F_m(y)\}$ is chosen (as may be done with no loss of generality) as a normal orthogonal set:

$$\sum_{m=1}^{\infty}\left\{\int_0^1 F_m(y)Q_n(y)dy\right\}^2 < A. \tag{51}$$

It therefore follows that,

$$\sum_{m=1}^{\infty}\left\{\lim_{n\to\infty}\int_0^1 F_m(y)Q_n(y)dy\right\}^2 \leqq A. \tag{52}$$

Then, by the Riesz-Fischer Theorem, there exists a function $Q(y)$ of summable square such that,

$$Q(y) \frown\!\smile \sum_{m=1}^{\infty} F_m(y)\left\{\lim_{n\to\infty}\int_0^1 F_m(y)Q_n(y)\,dy\right\}, \tag{53}$$

where the sign $\frown\!\smile$ has the same significance as in (18).

Then we have

$$\int_0^1 Q(y)F_m(y)dy = \lim_{n\to\infty}\int_0^1 F_m(y)Q_n(y)dy, \tag{54}$$

$$\lim_{n\to\infty}\int_0^1 F_m(y)\left\{Q_n(y)-Q(y)\right\}dy=0. \tag{55}$$

We have also

$$\int_0^1\left\{Q_n(y)-Q(y)^2\right\}dy \leqq 2\int_0^1 [Q_n(y)]^2dy+2\int_0^1 [Q(y)]^2dy<4A, \tag{56}$$

so that as in (48) we have

$$\lim_{n\to\infty}\int_0^1 F(y)\left\{Q_n(y)-Q(y)\right\}dy=0, \tag{57}$$

and our proof is complete.

It may also be remarked that if (50) is assumed, we may prove successively (55), (54), (53), (52), (49), under the assumption that the integrals that occur in (49) all exist.

The relation between $Q_n(x)$ and $Q(x)$ in (50) is analogous to convergence in the mean, and is perhaps sufficiently important to deserve a special name. The authors suggest, as a possible verbal equivalent of (50), the statement that $\{Q_n\}$ is *quasi-convergent* to Q.

The equivalence of the sine and cosine series. The nth partial sum of the sine series for $f(x)$ over the interval $(0,\pi)$ is

$$\frac{1}{2\pi}\int_0^\pi\left\{f(y)\frac{\sin(2n+1)\frac{y-x}{2}}{\sin\frac{y-x}{2}}-f(y)\frac{\sin(2n+1)\frac{y+x}{2}}{\sin\frac{y+x}{2}}\right\}dy. \tag{58}$$

The corresponding nth partial sum of the cosine series is

$$\frac{1}{2\pi}\int_0^\pi\left\{f(y)\frac{\sin(2n+1)\frac{y-x}{2}}{\sin\frac{y-x}{2}}+f(y)\frac{\sin(2n+1)\frac{y+x}{2}}{\sin\frac{y+x}{2}}\right\}dy. \tag{59}$$

The difference between these nth partial sums is

$$\frac{1}{\pi}\int_0^\pi f(y)\frac{\sin(2n+1)\frac{x+y}{2}}{\sin\frac{y+x}{2}}dy. \tag{60}$$

Accordingly, the condition analogous to (16) is the uniform boundedness of

$$\int_0^\pi \frac{\sin^2(2n+1)\frac{y+x}{2}}{\sin^2\frac{y+x}{2}}\,dy < \int_0^\pi \csc^2\frac{y+x}{2}\,dy$$
$$= 2\int_{\frac{x}{2}}^{\frac{\pi+x}{2}} \csc^2 u\,du = 2\cot\frac{x}{2} - 2\cot\frac{\pi+x}{2}. \tag{61}$$

But the extreme right-hand member of (61) is uniformly bounded over any interval $(\epsilon, \pi-\epsilon)$ if $\epsilon > 0$. Furthermore, the cosine expansions of the sine functions converge uniformly over $(0, \pi)$. Hence, by Theorem I, the sine and cosine expansions of an arbitrary function (summable and of summable square) have the same convergence properties over the interval $(\epsilon,\pi-\epsilon)$.

II. **Regular and uniform sequences.**

Let the sequence $\{K_n\}$ be regular over the interval (0,1) according to the definition already given. This is equivalent to saying that the expansion in terms of $\{K_n\}$ of a function which is of summable square and is zero throughout an interval (a,b) [which is part of the interval (0,1)] converges uniformly to zero over any closed interval (a^1,b^1) entirely interior to (a,b). That is, if ϕ is any function of summable square, we must have

$$\lim_{n\to\infty}\int_0^a \phi(y)\,K_n(x,y)dy = 0,$$
$$\lim_{n\to\infty}\int_b^1 \phi(y)\,K_n(x,y)dy = 0, \tag{62}$$

uniformly over (a^1,b^1). It follows at once that there exists a constant A such that

$$\int_0^a [K_n(x,y)]^2\,dy < A,$$
$$\int_b^1 [K_n(x,y)]\,dy < A, \tag{63}$$

for all n and for all x in (a^1, b^1). The first of inequalities (63) must hold for every $x > a$, and the second for every $x < b$.

Consider the following diagram of the range of variation of x and y in $K_n(x,y)$:

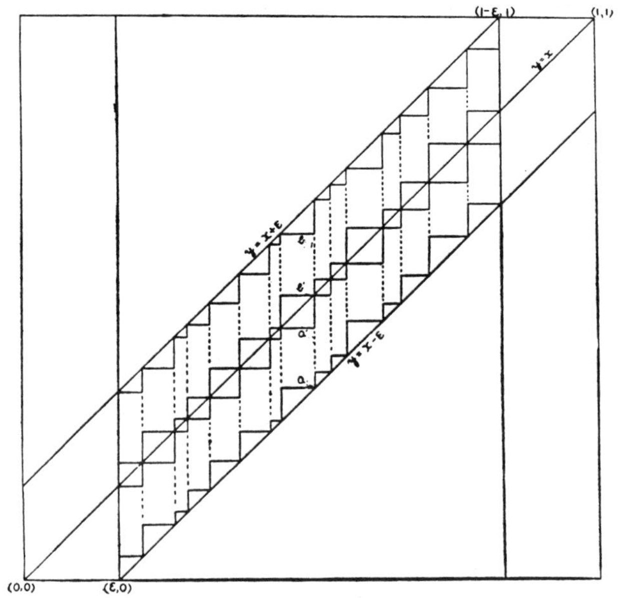

The portion of the figure formed by the lines $x=0$, $x=\epsilon$, $x=1-\epsilon$, $x=1$, $y=0$, $y=1$, $y=x$, $y=x-\epsilon$ and $y=x+\epsilon$ needs no explanation. The parallelogram bounded by the lines $x=\epsilon$, $x=1-\epsilon$, $y=x-\epsilon$, $y=x+\epsilon$ is divided by a finite number of verticals at horizontal distances each less than $\epsilon/2$. The resulting subdivisions of this parallelogram are numbered from the left, $1, 2, \ldots n \ldots$. In the nth subdivision four horizontal lines are drawn. The line $y=b_n$ passes through the left-hand upper corner of the subdivision, that is, passes through the intersection of the left-hand bounding vertical line, and $y=x+\epsilon$. Similarly, the line $y=a_n$ passes through the right-hand lower corner, the intersection of the right-hand bounding vertical and $y=x-\epsilon$. The lines $y=a_n^1$ and $y=b_n^1$ pass through the intersection of the line $y=x$ with the left-hand and right-hand bounding verticals, respectively. Obviously we have $a_n<a_n^1<b_n^1<b_n$.

It follows immediately from (62) that we have

$$\lim_{m\to\infty}\int_0^{a_n}\phi(y)\,K_m(x,y)dy=0,$$
$$\lim_{m\to\infty}\int_{b_n}^{1}\phi(y)\,K_m(x,y)dy=0, \tag{64}$$

uniformly over (a_n^1,b_n^1). We shall use these relations to establish Theorem II.

Let us apply the Schwarz inequality to $\int_{x-\epsilon}^{\alpha}\phi\,(y)\,K_m(x,y)\,dy$, where $x-\epsilon\leqq\alpha\leqq a_n$; we have

$$\int_{x-\epsilon}^{\alpha}\phi(y)\,K_m(x,y)dy\Big|\leqq\sqrt{\int_0^{\alpha}\left\{K_m(x,y)\right\}^2dy\int_{x-\epsilon}^{\alpha}\left\{\phi(y)\right\}^2dy}. \tag{65}$$

Hence, so long as $a_n^1\leqq x\leqq b_n^1$ it follows from (63) that there is a constant A_n independent of x, m and α, such that

$$\left|\int_{x-\epsilon}^{\alpha}\phi(y)\,K_m(x,y)dy\right|\leqq\sqrt{A_n\int_{x-\epsilon}^{\alpha}\left\{\phi(y)\right\}^2dy}. \tag{66}$$

We now subdivide each region of the x-axis between $x=a_n^1$ and $x=b_n^1$ by verticals $x=c_1^{(n)}=a_n^1$, $x=c_2^{(n)}$, . . . $x=c_p^{(n)}=b_n^1$. These lines determine a succession of triangles much like the triangles shown in the figure. Each of the new triangles is bounded by the line $y=x-\epsilon$, by a vertical line $x=c_q^{(n)}$, and by a horizontal line $y=c_{q+1}^{(n)}-\epsilon$. If we increase indefinitely the divisions in such a manner that the maximum value of $c_{q+1}^{(n)}-c_q^{(n)}$ approaches zero, and if α is always chosen so as to be the $c_{q+1}^{(n)}-\epsilon$ next greater than x, it follows from the uniform continuity of $\int_0^{\zeta}\left\{\phi(y)\right\}^2dy$, and the fact that the A_n's are constant, that

$$\lim_{\max(c_{q+1}^{(n)}-c_q^{(n)})\to 0}\left|\int_{x-\epsilon}^{\alpha}\phi(y)\,K_n(x,y)dy\right|=0, \tag{67}$$

uniformly for all such integrals.

Let the process of subdivision just described be carried so far that the uniform upper bound of the integrals in (67) is $\eta/2$.

It is then possible to make m so large, by (64), that

$$\left|\int_0^a \phi(y)\,K_m(x,y)dy\right| < \frac{\eta}{2}, \tag{68}$$

where the appropriate a is chosen for every x, uniformly over each of the intervals $a_n^1 \leqq x \leqq b_n^1$ and uniformly for all the intervals. Thus we have

$$\left|\int_0^{x-\epsilon} \phi(y)\,K_m(x,y)dy\right| \leqq \left|\int_0^a \phi(y)\,K_m(x,y)dy\right| + \left|\int_{x-\epsilon}^a \phi(y)\,K_m(x,y)dy\right| < \eta \tag{69}$$

uniformly for each of the intervals $a_n^1 \leqq x \leqq b_n^1$ and uniformly for all those intervals.

This means that

$$\lim_{m\to\infty}\int_0^{x-\epsilon} \phi(y)\,K_m(x,y)dy = 0 \tag{70}$$

uniformly in x for $\epsilon \leqq x \leqq 1-\epsilon$. This establishes the first part of Theorem II, and the second part may be established in precisely the same way.

Uniform expansions. If the sequence $\{K_n\}$ is uniform over (0,1), then by Theorem II and (10) $\{K_n\}$ is equivalent over the sub-interval $(\epsilon, 1-\epsilon)$ to the sequence whose nth term is $K_n(\epsilon, y-x+\epsilon)$ for $x-\epsilon \leqq y \leqq x+\epsilon$ and zero elsewhere. We shall denote this latter function by $K_n(y-x)$.

Then we see from (11) that

$$\lim_{n\to\infty}\int_\epsilon^{1-\epsilon}\left\{g(x) - \int_\epsilon^{1-\epsilon} K_n(y-x)g(y)dy\right\}^2 dx = 0. \tag{71}$$

Hence, it is true uniformly for the whole class C of functions such that $\int_0^1 \{f(x)\}^2 dx < A$, that

$$\lim_{n\to\infty}\int_\epsilon^{1-\epsilon} f(x)dx \int_\epsilon^{1-\epsilon} K_n(y-x)\,g(y)dy = \int_\epsilon^{1-\epsilon} f(x)g(x)dx. \tag{72}$$

The definition of $K_n(u)$ and (11) assure the existence of $\int_{-1}^{1}\{K_n(u)\}^2du$. Then the Fourier series $\frac{a_1}{2}+\sum_1^\infty(a_i\cos\pi u+b_i\sin\pi u)$ of $K_n(u)$ converges in the mean to the value $K_n(u)$. Denote by $S_m(u)$ the mth partial sum of this series and by $R_m(u)$ the corresponding remainder. We clearly have

$$\int_\epsilon^{1-\epsilon}f(x)dx\int_\epsilon^{1-\epsilon}S_m(y-x)g(y)dy=\int_\epsilon^{1-\epsilon}g(y)dy\int_\epsilon^{1-\epsilon}S_m(y-x)f(x)dx,$$

for S_m is the sum of a finite number of products of functions of y by functions of x. Moreover, we have, by the Schwarz inequality,

$$\begin{aligned}&\left[\int_\epsilon^{1-\epsilon}f(x)dx\int_\epsilon^{1-\epsilon}R_m(y-x)g(y)dy\right]^2\\ &\quad\leqq\int_\epsilon^{1-\epsilon}\{f(x)\}^2dx\int_\epsilon^{1-\epsilon}\left\{\int_\epsilon^{1-\epsilon}R_m(y-x)g(y)dy\right\}^2dx\\ &\quad\leqq\int_\epsilon^{1-\epsilon}\{f(x)\}^2dx\int_\epsilon^{1-\epsilon}dx\left\{\int_\epsilon^{1-\epsilon}\{R_m(y-x)\}^2dy\int_\epsilon^{1-\epsilon}\{g(y)\}^2dy\right\}\\ &\quad\leqq\int_\epsilon^{1-\epsilon}\{f(x)\}^2dx\int_{-1}^{1}\{R_m(u)\}^2du\int_\epsilon^{1-\epsilon}\{g(y)\}^2dy.\end{aligned}$$

Hence, since a similar inequality holds if the integration is performed in the reverse order, we have

$$\begin{aligned}&\int_\epsilon^{1-\epsilon}f(x)dx\int_\epsilon^{1-\epsilon}K_n(y-x)g(y)dy\\ &\qquad=\lim_{m\to\infty}\int_\epsilon^{1-\epsilon}f(x)dx\int_\epsilon^{1-\epsilon}S_m(y-x)g(y)dy\\ &\qquad=\lim_{m\to\infty}\int_\epsilon^{1-\epsilon}g(y)dy\int_\epsilon^{1-\epsilon}S_m(y-x)f(x)dx\\ &\qquad=\int_\epsilon^{1-\epsilon}g(y)dy\int_\epsilon^{1-\epsilon}K_n(y-x)f(x)dx.\end{aligned}$$

Thus we find, from (72),

$$\lim_{n\to\infty}\int_\epsilon^{1-\epsilon}g(y)dy\int_\epsilon^{1-\epsilon}K_n(y-x)f(x)dx=\int_\epsilon^{1-\epsilon}f(x)g(x)dx. \quad (73)$$

Let us set $g(x)=0$ outside of the interval $(2\epsilon, 1-2\epsilon)$, so that we have

$$\begin{aligned}\int_{2\epsilon}^{1-2\epsilon} f(x)g(x)dx &= \lim_{n\to\infty}\int_{2\epsilon}^{1-2\epsilon} g(y)dy\int_{\epsilon}^{1-\epsilon} K_n(y-x)f(x)dx \\ &= \lim_{n\to\infty}\int_{2\epsilon}^{1-2\epsilon} g(y)dy\int_{y-\epsilon}^{y+\epsilon} K_n(y-x)f(x)dx \\ &= \lim_{n\to\infty}\int_{2\epsilon}^{1-2\epsilon} g(y)dy\int_{-\epsilon}^{\epsilon} K_n(u)f(y-u)du \\ &= \lim_{n\to\infty}\int_{-\epsilon}^{\epsilon} K_n(u)du\int_{2\epsilon}^{1-2\epsilon} g(y)f(y-u)dy \\ &= \lim_{n\to\infty}\int_{-\epsilon+\zeta}^{\epsilon+\zeta} K_n(\eta-\zeta)d\eta\int_{2\epsilon}^{1-2\epsilon} g(y)\, f(y+\zeta-\eta)dy \\ &= \lim_{n\to\infty}\int_{0}^{1} K_n(\eta-\zeta)d\eta\int_{2\epsilon}^{1-2\epsilon} g(y)f(y+\zeta-\eta)dy,\end{aligned} \tag{74}$$

uniformly for all f's in C and for all values of ζ on the interval $(\epsilon, 1-\epsilon)$. Let us set $F(\eta)=f_\zeta(\zeta-\eta)$, so that we have $f_\zeta(x)= F(\zeta-x)$, and hence

$$\int_0^1\left\{f_\zeta(x)\right\}^2 dx=\int_0^1\left\{F(\zeta-x)\right\}^2 dx=\int_{\zeta-1}^{\zeta}\left\{F(u)\right\}^2 du. \tag{75}$$

Then if $F(u)$ is defined over $(2\epsilon-1, 1-2\epsilon)$, and if $\int_{2\epsilon-1}^{1-2\epsilon}\{F(u)\}^2du$ exists, we have

$$\lim_{n\to\infty}\int_0^1 K_n(\eta-\zeta)d\eta\int_{2\epsilon}^{1-2\epsilon} F(\eta-y)g(y)dy=\int_{2\epsilon}^{1-2\epsilon} f_\zeta(x)g(x)dx \tag{76}$$

uniformly for $\epsilon\leqq\zeta\leqq 1-\epsilon$. That is, any function $\Phi(x)$ that can be expressed in the form

$$\int_{2\epsilon}^{1-2\epsilon} F(x-y)g(y)dy,$$

is the uniform limit of its expansion in terms of $\{K_n\}$.

Among the functions of the form

$$\int_{2\epsilon}^{1-2\epsilon} F(x-y)g(y)dy$$

are those for which $g(x) \equiv 1$. For such a function we have

$$\Phi(x) = \int_{2\epsilon}^{1-2\epsilon} F(x-y)dy = \int_{x+2\epsilon-1}^{x-2\epsilon} F(u)du = \int_{x+4\epsilon-1}^{x} F(\nu-2\epsilon)d\nu. \tag{77}$$

If we set $F(\nu-2\epsilon) = G(\nu)$ and let $G(\nu) = 0$ for $\nu < 0$, we ν find

$$\Phi(x) = \int_0^x G(\nu)d\nu \tag{78}$$

for $x \leqq 1-4\epsilon$. Thus, if we use (78) as the definition of $\Phi(x)$, the $\{K_n(x,y)\}$ expansion of Φ will converge uniformly to the value Φ over the interval $(4\epsilon, 1-4\epsilon)$. This completes the proof of Theorem III.

The property expressed in Theorem III is of course familiar in the case of the Fourier sine expansion.[10]

[10] Since receiving page proof of this paper, it has come to the attention of the authors that a necessary and sufficient condition for (48) including (49) as part of its enunciation has been given by Lebesque (*Annales de la Faculté de Toulouse*, (3), Vol. I (1909), p. 55).

Comments on [22d]

J. L. Walsh

This paper is a contribution to the study of the equivalence of expansions, a topic primarily studied by Haar,[3] Young[4] and Walsh,[5] these citations being the same as those on p. 105 of [22d]. (The second cited paper by Walsh appeared subsequently in Ann. of Math., 24(1922), 109–120.) The memoir by Lebesgue[10] [22d, p. 122] on singular integrals was not known to the authors when this paper was written; it contains (p. 55) condition (49) and (p. 72) condition (16). Lebesgue considers Fourier series in particular, but gives no special detailed study of expansions in terms of orthogonal functions in general.

Of course, numerous later studies of the equivalence of expansions have been made, for instance on Birkhoff series, by M. H. Stone.[1]

Reference

1. M. H. Stone, Trans. Amer. Math. Soc. 28(1926), 695–761.

NOTE ON QUASI-ANALYTIC FUNCTIONS[1]

By Norbert Wiener

Let $f(x)$ be an even function with period 2π. Let its Fourier series be

$$f_0/2+f_1 \cos x+ \ldots +f_n \cos nx+ \ldots$$

Let us write,

$$f|g=\sum_0^\infty A_k^2 f_k g_k \qquad [A_k \neq 0].$$

Let A be the class of functions $f(x)$ for which $f|f$ is bounded. We shall term A *infinitely differentiable* if

$$\frac{1}{A_k}=o(k^{-h})$$

for all positive h, and *quasi-analytic* if it is infinitely differentiable, and if whenever $f(x)$ belongs to A and

$$f(0)=f''(0)= \ldots =f^{(2n)}(0)= \ldots =0$$

then $f(x)=0$.

The following theorems are proved without difficulty.

Theorem I. *If A is infinitely differentiable and*

$$\sum_0^\infty A_k^2 a_k^2$$

converges, there is a function $a(x)$ of A such that

$$a(x)=a_0/2+a_1 \cos x+ \ldots +a_n \cos nx+ \ldots$$

[1] This paper owes its inception to a course of lectures delivered by Prof. Charles de la Vallée Poussin at the Massachusetts Institute of Technology in November, 1924. These lectures have appeared in the *Bulletin de la Société Mathématique de France*, Vol LII, section 1. The use of Fourier series for the study of quasi-analytic functions is of course due to Professor de la Vallée Poussin. The use of quadratic forms for the characterization of quasi-analytic functions follows out an idea of Carleman in several recent papers in the *Comptes Rendus*, and the expansion here developed is essentially that of Carleman.

Proof. Clearly $A_k a_k$ is bounded. Hence

$$a_k = o(k^{-h}).$$

Therefore $\Sigma |a_k|$ converges, so that

$$a_0/2 + a_1 \cos x + \ . \ . \ . \ + a_n \cos nx + \ . \ . \ .$$

converges uniformly.

Theorem II. *If A is infinitely differentiable, every function of A possesses derivatives of all orders, and vice versa.*

Proof. As above, if A is infinitely differentiable, and f belongs to A, then

$$f_k = o(k^{-h}).$$

As to the converse theorem, let A not be infinitely differentiable. Form

$$\Sigma \delta_k \cos kx$$

where

$$\delta_k = \frac{1}{k^2 A_k}; \qquad [\,|A_k| \geq 1]$$

$$\delta_k = \frac{1}{k^2} \qquad [\,|A_k| < 1].$$

Clearly $\Sigma\ \delta_k \cos kx$ converges to a function belonging to A; clearly, moreover, this function cannot be differentiated more than $2+h$ times, where h is the greatest number for which

$$\frac{1}{A_k} = O(k^{-h}).$$

Theorem III. *Let A be quasi-analytic, and let $B_k = A_k/k^2$. Let B be determined in terms of B_k, in the same way in which A is determined in terms of A_k. Then B is quasi-analytic.*

Proof. The functions of A are the second derivatives of the functions of B. Given a function belonging to A, the corresponding function belonging to B is determined except for a term $h+kx$. The k is determined by the periodicity of the functions of B. The h is uniquely determined when the value of the function at the origin is known.

Corollary. *If A is quasi-analytic,*

$$\mathbf{A}(x)=\frac{1}{2A_0}+\frac{\cos x}{A_1}+\ . \ . \ . \ +\frac{\cos nx}{A_n}+\ . \ . \ .$$

belongs to a quasi-analytic class, and conversely if $\mathbf{A}(x)$ belongs to a quasi-analytic class, A is quasi-analytic.

Proof. As for the direct theorem, the quasi-analytic class in question is B itself. As for the inverse theorem, if $\mathbf{A}(x)$ belongs to a quasi-analytic class B, and f belongs to A, then

$$f_k=o\left(\frac{1}{A_k}\right),$$

and f belongs to B. It then results from the definition of quasi-analyticity that A is likewise quasi-analytic. It is here and subsequently understood that no $A_k=0$.

Theorem IV. *An infinitely differentiable class A is a quasi-analytic class when and only when the class of functions $\mathbf{A}^{(2n)}(x)$ is closed over $(0, \pi)$.*

Proof. By Theorem I, a necessary and sufficient condition that $f(x)$ belong to A is that

$$\phi(x)\sim A_0f_0+A_1f_1\cos x+\ . \ . \ . \ +A_nf_n\cos nx+\ . \ . \ .$$

be summable and of summable square. Furthermore,

$$\begin{aligned} f^{(2n)}(0) &= (-1)^n[f_1+2^{2n}f_2+3^{2n}f_3+\ . \ . \ .] \\ &= \frac{2}{\pi}\int_0^\pi \phi(x)\mathbf{A}^{(2n)}(x)dx. \end{aligned}$$

Hence the quasi-analyticity of A means that the functions $\mathbf{A}^{(2n)}(x)$ $[n=0, 1, \ . \ . \ .]$ admit no summable function of summable square orthogonal to them all.

Corollary. *If $f(x)$ is a member of a quasi-analytic class, and no f_n vanishes, then f and its even derivatives form a closed set over $(0, \pi)$. Under any circumstances, if f is a member of a quasi-analytic class, there is a function $g(x)$ such that if*

$$\int_0^\pi f(x)\cos nx\neq 0$$

then
$$\int_0^\pi [f(x)-g(x)] \cos nx = 0,$$

while g and its derivatives form a closed set over $(0, \pi)$. *Conversely, if* $f(x)$ *is an even function of period* 2π, *possessing at all points derivatives of all orders, and if with its even derivatives it forms a closed set over* $(0, \pi)$, *it is itself a member of a quasi-analytic class.*

Theorem V. *Let A be a quasi-analytic class of functions. Let* $\{\phi_k(x)\}$ *be a set of normal and orthogonal functions closed over* $(0, \pi)$. *Let*

$$\phi_k(x) \sim a_{k0} + a_{k1} \cos x + \ldots + a_{kn} \cos nx + \ldots \quad [k=1, 2, \ldots]$$

Put

$$\Phi_k(x) = \frac{a_{k0}}{A_0} + \frac{a_{k1}}{A_1} \cos x + \ldots + \frac{a_{kn}}{A_n} \cos nx + \ldots$$

Then the functions of A are precisely those which can be represented in the form

$$f(x) = a_1\Phi_1(x) + a_2\Phi_2(x) + \ldots + a_n\Phi_n(x) + \ldots,$$

where $\sum_1^\infty a_n^2$ *converges. We shall have*

$$a_n = f \mid \Phi_n.$$

Proof. Let
$$f_n(x) = a_1\Phi_1(x) + \ldots + a_n\Phi_n(x),$$
and let
$$g_n(x) = a_1\phi_1(x) + \ldots + a_n\phi_n(x).$$
We have
$$\Phi_k \mid \Phi_l = \frac{2}{\pi}\int_0^\pi \phi_k(x)\phi_l(x)dx = 0.$$
Hence
$$(f_m - f_n) \mid (f_m - f_n) = \frac{2}{\pi}\int_0^\pi [g_m(x) - g_n(x)]^2 dx$$
$$= \frac{2}{\pi} \sum_{m+1}^{n} a_k^2$$
Thus
$$\lim_{m, n \to \infty} (f_m - f_n) \mid (f_m - f_n) = 0.$$

Consequently
$$\lim_{m,n\to\infty}\int_0^\pi [f_m^{(k)}(x)-f_n^{(k)}(x)]^2\,dx=0.$$
From this we deduce that $f(x)$ exists, and that $f_n(x)$ tends uniformly towards $f(x)$ with all its derivatives.

If we write
$$f(x)=\Sigma b_k \cos kx,$$
we obtain formally
$$b_k=\sum_j \frac{a_j\alpha_{jk}}{A_k}\,.$$
The convergence of $\Sigma a_j\alpha_{jk}$ may be deduced with the aid of a form of the Schwartz inequality, and *a fortiori* the convergence of the series for b. It is easy to verify with the aid of the classical theory of orthogonal transformations that
$$\sum_k(\sum_j a_j\alpha_{jk})^2$$
converges. From this it results at once that $f(x)$ belongs to A. Indeed, the classical theory of orthogonal transformations yields us at once the result that
$$\sum_k(\sum_j a_j\alpha_{jk})^2$$
converges when and only when
$$\sum_k a_k^2$$
converges. Hence the functions of A are precisely those which can be represented in the form
$$\Sigma a_k\Phi_k(x),$$
where $\Sigma\, a_k^2$ converges.

In particular, we may choose for our functions $\phi_n(x)$ the set we obtain by normalizing $\mathbf{A}^{(2n)}(x)$. Let us have
$$\phi_n(x)=A_{n0}\,\mathbf{A}\,(x)+A_{n1}\,\mathbf{A}''(x)+\;.\;.\;.\;+A_{n,\,n-1}\,\mathbf{A}^{(2n-2)}(x).$$
We shall then have
$$f\,\Big|\,\Phi_n=\sum_0^{n-1}(-1)^jA_{nj}f(x)\Big|\left[\frac{d^{2j}1/dx^{2j}}{2A^2_0}+\frac{\cos x}{A_1^2}+.\;.\;.+\frac{k^{2j}\cos kx}{A^2_k}+.\;.\;.\right]$$
$$=\sum_0^{n-1}(-1)^jA_{nj}\left[f_0\frac{d^{2j}1}{dx^{2j}}+f_1+2^{2j}f_2+.\;.\;.\right]$$
$$=\sum_0^{n-1}A_{nj}f^{(2j)}(0).$$

It is hence necessary that

$$\sum_{n=1}^{\infty}\left\{\sum_{j=0}^{n-1} A_{nj} f^{(2j)}(0)\right\}^2$$

should converge if f is to belong to A.

Let us now start with the supposition that

$$\sum_{1}^{\infty}\left\{\sum_{j=0}^{n-1} A_{nj}\alpha_j\right\}^2$$

converges. Form

$$f(x)=\sum_{n=1}^{\infty}\Phi_n(x)\sum_{j=0}^{n-1} A_{nj}\alpha_j$$

This will belong to A. As before, we shall have

$$f\,|\,\Phi_n=\sum_{0}^{n-1} A_{nj} f^{(2j)}(0)$$

However, by Theorem V, we have

$$f\,|\,\Phi_n=\sum_{0}^{n-1} A_{nj}\alpha_j.$$

Hence

$$\sum_{0}^{n-1} A_{nj}[f^{(2j)}(0)-\alpha_j]=0. \qquad [n=1,2,\ldots]$$

Unless therefore there is some relation of linear dependence between the functions $\mathbf{A}^{(2n)}_{(x)}$, we must have identically

$$f^{(2j)}(0)=\alpha_j$$

Such a relation of linear dependence would mean that the function f satisfies a differential equation of finite order with constant coefficients, and hence has a terminating Fourier series. This, however, we have excluded by definition. We thus obtain:

Theorem VI. *A necessary and sufficient condition that there be a function $f(x)$ of the quasi-analytic class A such that*

$$f^{(2n)}(0)=\alpha_n \qquad [n=0,1,2,\ldots]$$

is that

$$\sum_{1}^{\infty}\left\{\sum_{j=0}^{n-1} A_{nj}\alpha_j\right\}^2$$

converge. The function is then

$$\sum_{n=1}^{\infty}\Phi_n(x)\sum_{j=0}^{n-1} A_{nj}\alpha_j.$$

As was pointed out to the author by Prof. de la Vallée Poussin, this theorem is at least in appearance very surprising. The A's can be so chosen as to make A consist entirely of analytic functions — for example, we may take $A_k = e^k$. In this case, any change in the value of a single α_j, excepting the first, will destroy the possibility of finding a *periodic* analytic function satisfying the assigned initial conditions. It is hence quite paradoxical to find the possibility of finding a quasi-analytic function of an assigned class with given initial values for the function and its derivatives dependent on the convergence of a series involving the initial values, and on this alone. Such, however, appears to be the case.

Comments on [25e]

J. W. Neuberger

In this paper Wiener defines the term "infinitely differentiable" (for certain cosine series) rather indirectly, but shows that his definition implies the expected property. An infinitely differentiable collection A of cosine series is said to be *quasi-analytic* provided it is true that if f belongs to A and $f(0) = f''(0) = f^{(4)}(0) = \cdots = 0$, then $f(x) = 0$ for all x in $[0, 2\pi]$. It is then shown that quasi-analyticity holds for a certain collection of cosine series, if and only if an associated collection of functions is closed (in the sense that there is no non-zero summable function of summable square that is orthogonal to all members of the collection). This very interesting connection between closure and quasi-analyticity does not seem to have been investigated further.

PROCEEDINGS

OF THE

NATIONAL ACADEMY OF SCIENCES

Volume 13 February 15, 1927 Number 2

ON THE CLOSURE OF CERTAIN ASSEMBLAGES OF TRIGONOMETRICAL FUNCTIONS

By Norbert Wiener

Department of Mathematics, Massachusetts Institute of Technology

Communicated December 2, 1926

THEOREM: *Let*

$$\cos \lambda_n x \ \ [0 \leq \lambda_1 < \lambda_2 < \ldots, \lambda_{k+1} - \lambda_k \geq A]$$

be a closed set of functions over $(0,\pi)$. *If*

$$P(x) = \sum_1^n a_k \cos \lambda_k x,$$

let

$$\sum_1^n a_k^2 \leq C \int_0^\pi [P(x)]^2 dx, \tag{1}$$

C being dependent only on the sequence $\{\lambda_k\}$. *Let*

$$|\lambda_k - \mu_k| \leq D < \sqrt{\frac{2A}{3c\pi^3} + \left(A + \frac{1}{2}\right)^2} - A - \frac{1}{2}. \ \ [k = 1, 2, 3, \ldots].$$

Then the sequence $\cos \mu_n x$ *is closed over* $(0,\pi)$.

Proof. Let $\varphi(x)$ be any summable function of summable square over $(0, \pi)$. Let

$$\int_0^\pi \left[\varphi(x) - \sum_1^n a_k \cos \lambda_k x\right]^2 dx < e.$$

The a_k's may be so chosen because of the closure of $\{\cos \lambda_k x\}$. Let us consider

$$\psi(x) = \sum_1^n a_k(\cos \lambda_k x - \cos \mu_k x).$$

We have

$$2\psi(x)\frac{\sin x/2}{x} = \int_0^\infty F(u) \cos ux \, du,$$

where

$$F(u) = \sum_1^n a_k F_k(u) \cdot$$

$$F_k(u) = \operatorname{sgn}\left(u-\lambda_k+\frac{1}{2}\right)-\operatorname{sgn}\left(u-\lambda_k-\frac{1}{2}\right)-\operatorname{sgn}\left(u-\mu_k+\frac{1}{2}\right) + \operatorname{sgn}\left(u-\mu_k-\frac{1}{2}\right).^1$$

It follows that

$$\int_0^\infty 4[\psi(x)]^2\frac{\sin^2 x/2}{x^2}dx = \frac{\pi}{2}\int_0^\infty [F(u)]^2du$$
$$\leq 2\pi\left(2+\frac{1+D}{A}\right)D\sum_1^n a_k^2,$$

from which we conclude that

$$\int_0^\pi \Big[\psi\,(x)\Big]^2 dx \leq \frac{\pi^3}{2}\left(2+\frac{1+D}{A}\right)D\sum_1^n a_k^2.$$

Hence

$$\int_0^\pi\left[\varphi\,(x)-\sum_1^n a_k\cos\mu_k x\right]^2 dx$$
$$\leq 3\epsilon+\frac{3\pi^3}{2}\left(2+\frac{1+D}{A}\right)D\sum_1^n a_k^2$$
$$\leq 3\epsilon+\frac{3\pi^3}{2}\left(2+\frac{1+D}{A}\right)DC\int_0^\pi\left[\sum_1^n a_k\cos\mu_k x\right]^2 dx$$
$$\leq 3\epsilon+\frac{3\pi^3}{2}\left(2+\frac{1+D}{A}\right)DC\left[\int_0^\pi [\varphi(x)]^2dx+2\sqrt{\epsilon}\sqrt{\int_0^\pi [\varphi(x)]^2\,dx}+\epsilon\right]$$
$$\leq \eta+\vartheta\int_0^\pi [\varphi(x)]^2dx,$$

where $\vartheta < 1$, and n can be made as small as is desired by taking ϵ small enough. We can hence choose our a_k's in such a way that

$$\int_0^\pi\left[\varphi(x)-\sum_1^n a_k\cos\mu_k x\right]^2 dx\leq\vartheta_1\int_0^\pi [\varphi(x)]^2dx,$$

where $\vartheta_1 < 1$ and is independent of φ.

We may now choose a set of numbers b_k in such a way that

$$\int_0^\pi\left[\varphi(x)-\sum_1^n a_k\cos\mu_k x-\sum^{n_1} b_k\cos\mu_k x\right]^2 dx$$
$$\leq\vartheta_1\int_0^\pi\left[\varphi(x)-\sum_1^n a_k\cos\mu_k x\right]^2 dx\leq\vartheta_1^2\int_0^\pi [\varphi(x)]^2dx.$$

By repeating this process, we obtain a trigonometric polynomial $P_n(x)$ in the functions $\cos\mu_k x$ such that

$$\int_0^{\pi} [\varphi(x) - P_n(x)]^2 dx \leq \vartheta_1^n \int_0^{\pi} [\varphi(x)]^2 dx.$$

This proves the closure of the set cos $\mu_k x$.

As a special case, let $\lambda_k = k-1$, $C = 2/\pi$. Then D can be any number less than $\sqrt{\frac{1}{3\pi^2} + \frac{9}{4}} - \frac{3}{2}\cdot$ In particular, *if*

$$\nu < \sqrt{\frac{1}{3\pi^2} + \frac{9}{4}} - \frac{3}{2}$$

the sequence

$$\cos \nu x, \cos (1 + \nu)x, \cos (2 + \nu)x, \ldots.$$

is closed over $(0, \pi)$.

[1] The terms at the beginning may have to be modified slightly because $F_k(u)$ is to vanish for negative u. This is allowed for in the following estimate.

A CLASS OF GAP THEOREMS

by NORBERT WIENER (Cambridge, Mass.).

The present paper is devoted to the study of the following:

THEOREM I ([1]). - *Let the Maclaurin series* $f(z)=\sum_{0}^{\infty} a_n z^{\lambda_n}$ *have the unit circle for its circle of convergence, and let there exist a function* $\Phi(\theta)$ *belonging to* L_2 *such that over the arc* (α, β),

$$\lim_{r\to 1-0} \int_{\alpha}^{\beta} |f(re^{i\theta}) - \Phi(\theta)|^2 d\theta = 0.$$

Let the λ_n*'s be arranged in increasing order, and let* $\beta > \alpha$. *Let*

$$\lambda_{n+1} - \lambda_n \geqslant L = \frac{16\pi}{\beta - \alpha}.$$

Then there exists a function $\Psi(\theta)$ *belonging to* L_2 *over* $(0, 2\pi)$ *and coinciding with* $\Phi(\theta)$ *over* (α, β) *such that*

$$\lim_{r\to 1-0} \int_{0}^{2\pi} |f(re^{i\theta}) - \Psi(\theta)|^2 d\theta = 0,$$

and

$$\int_{0}^{2\pi} |\Psi(\theta)|^2 d\theta \leqslant \frac{16\pi}{\beta-\alpha} \int_{\alpha}^{\beta} |\Psi(\theta)|^2 d\theta.$$

To prove this let us consider the polynomial $P(\theta) = \sum_{1}^{N} a_n e^{i\lambda_n\theta}$. We have for the FOURIER transform of $\dfrac{e^{i\lambda_n\theta}\sin\frac{L\theta}{2}}{\theta}$ the functions

$$\frac{1}{\sqrt{2\pi}} \int_{-\infty}^{\infty} e^{iu\theta} e^{i\lambda_n\theta} \frac{\sin\frac{L\theta}{2}}{\theta} d\theta = \begin{cases} \sqrt{\frac{\pi}{2}} & \text{if } |\lambda_n + u| < \frac{L}{2}; \\ 0 & \text{if } |\lambda_n + u| > \frac{L}{2}. \end{cases}$$

([1]) Theorem I follows the argument of Theorem XL in the forthcoming book of PALEY and WIENER: *Fourier Transforms in the Complex Domain.*

These differ from zero over non-overlapping ranges and are orthogonal. By PLANCHEREL's theorem,

$$(1) \qquad \int_{-\infty}^{\infty} |P(\theta)|^2 \frac{\sin^2 \frac{L\theta}{2}}{\theta^2} d\theta = \sum_1^N |a_n|^2 \frac{\pi L}{2}.$$

Hence

$$\int_{-\frac{\pi}{L}}^{\frac{\pi}{L}} |P(\theta)|^2 d\theta \leqslant \frac{\pi^3}{2L} \sum_1^N |a_n|^2$$

and in general

$$(2) \qquad \int_{-\frac{\pi}{L}+y}^{\frac{\pi}{L}+y} |P(\theta)|^2 d\theta \leqslant \frac{\pi^3}{2L} \sum_1^N |a_n|^2.$$

Again, by (2), if $A > \frac{\pi}{L}$,

$$\int_{-A}^{A} |P(\theta)|^2 d\theta \leqslant \pi^2 A \sum_1^N |a_n|^2.$$

Thus

$$\left[\int_A^\infty + \int_{-\infty}^{-A}\right] \frac{|P(\theta)|^2}{\theta^2} d\theta = \int_A^\infty \frac{1}{\theta^2} d \int_{-\theta}^{\theta} |P(\theta)|^2 d\theta =$$

$$= \left| -\frac{1}{A^2} \int_{-A}^{A} |P(\theta)|^2 d\theta + 2 \int_A^\infty \frac{d\theta}{\theta^2} \frac{1}{\theta} \int_{-\theta}^{\theta} |P(x)|^2 dx \right| \leqslant$$

$$\leqslant 2\pi^2 \sum_1^N |a_n|^2 \int_A^\infty \frac{dx}{x^2} = \frac{2\pi^2}{A} \sum_1^N |a_n|^2.$$

Thus again by (1)

$$\int_{-A}^{A} |P(\theta)|^2 \frac{\sin^2 \frac{L\theta}{2}}{\theta^2} d\theta \geqslant \left[\frac{\pi L}{2} - \frac{2\pi^2}{A}\right] \sum_1^N |a_n|^2.$$

Now let $A = \frac{4\pi^2}{\pi L - 2C}$ where C is some sufficiently small positive constant. Then

$$\int_{-\frac{4\pi^2}{\pi L - 2C}}^{\frac{4\pi^2}{\pi L - 2C}} |P(\theta)|^2 d\theta \geqslant \frac{4}{L^2} \int_{-\frac{4\pi^2}{\pi L - 2C}}^{\frac{4\pi^2}{\pi L - 2C}} |P(\theta)|^2 \frac{\sin^2 \frac{L\theta}{2}}{\theta^2} d\theta \geqslant \frac{4C}{L^2} \sum_1^N |a_n|^2.$$

In particular, if $C=\frac{\pi L}{2}$,

$$\int_{-\frac{8\pi}{L}}^{\frac{8\pi}{L}} |P(\theta)|^2 d\theta \geqq \frac{2\pi}{L} \sum_1^N |a_n|^2.$$

If $\beta-\alpha=\frac{16\pi}{L}$ it follows that

$$\int_\alpha^\beta |P(\theta)|^2 d\theta \geqq \frac{\beta-\alpha}{8} \sum_1^N |a_n|^2 = \frac{\beta-\alpha}{16\pi} \int_0^{2\pi} |P(\theta)|^2 d\theta.$$

Now let us consider $f(re^{i\theta})$ where $r<1$. We have over $(0, 2\pi)$

$$f(re^{i\theta}) = \underset{N\to\infty}{\text{limit in mean}} \sum_1 a_n r^{\lambda_n} e^{i\lambda_n\theta} = \underset{N\to\infty}{\text{limit in mean}}\, f_N(re^{i\theta}).$$

We have just shown that

$$\int_0^{2\pi} |f_N(re^{i\theta})|^2 d\theta \leqq \frac{16\pi}{\beta-\alpha} \int_\alpha^\beta |f_N(re^{i\theta})|^2 d\theta,$$

and it follows at once that

$$\int_0^{2\pi} |f(re^{i\theta})|^2 d\theta \leqq \frac{16\pi}{\beta-\alpha} \int_\alpha^\beta |f(re^{i\theta})|^2 d\theta.$$

Thus

$$\overline{\lim_{r\to 1}} \int_0^{2\pi} |f(re^{i\theta})|^2 d\theta \leqq \frac{16\pi}{\beta-\alpha} \int_\alpha^\beta |\varphi(\theta)|^2 d\theta.$$

That is

$$\overline{\lim_{r\to 1}} \sum_1^\infty r^{2\lambda_n} |a_n|^2 \leqq \frac{8}{\beta-\alpha} \int_\alpha^\beta |\varphi(\theta)|^2 d\theta,$$

whence we find that

$$\sum_1^\infty |a_n|^2 \leqq \frac{8}{\beta-\alpha} \int_\alpha^\beta |\varphi(\theta)|^2 d\theta,$$

and if we put

$$\psi(\theta) \sim \sum a_n e^{i\lambda_n\theta},$$

then

$$\int_0^{2\pi} |\psi(\theta)|^2 d\theta \leqq \frac{16\pi}{\beta-\alpha} \int_\alpha^\beta |\psi(\theta)|^2 d\theta$$

and using BESSEL equality for FOURIER Series we have

$$\lim_{r\to 1-0} \int_0^{2\pi} |f(re^{i\theta}) - \psi(\theta)|^2 d\theta = 0.$$

An intermediate corollary of Theorem I is:

THEOREM II. - *On the hypothesis of Theorem I, if $\psi(\theta)$ is the n-times repeated integral of $\psi^{(n)}(\theta)$ over (α, β), it is also the n-times repeated integral $\psi^{(n)}(\theta)$ over $(0, 2\pi)$, $\psi^{n}(\theta)$ is of period 2π and*

$$\int_0^{2\pi} |\psi^{(n)}(\theta)|^2 d\theta \leq \frac{16\pi}{\beta - \alpha} \int_\alpha^\beta |\psi^{(n)}(\theta)|^2 d\theta.$$

Let it be noted that the constant in this formula is independent of n. The theorem is obtained by applying Theorem I to $\sum (i\lambda_k)^n a_k z^{\lambda_k}$.

A further corollary is:

THEOREM III. - *On the hypothesis of Theorem I, if $\psi(\theta)$ is infinitely often differentiable, and*

$$\text{(a)} \qquad \int_\alpha^\beta |\psi^{(n)}(\theta)|^2 d\theta < A_n^2 B^n$$

where A_n is given and B is some positive quantity, then

$$\text{(b)} \qquad \int_0^{2\pi} |\psi^{(n)}(\theta)|^2 d\theta \leq A_n^2 C^n,$$

where C is some positive quantity. If over (α, β), (a) is true, which is always the case if

$$\text{(c)} \qquad |\psi^{(n)}(\theta)| < A_n B_1^n$$

for some B_1, then over the entire circle,

$$\text{(d)} \qquad |\psi^{(n-1)}\theta| < A_n C_1^n$$

for some C_1. In particular, if ψ is analytic over (α, β) it is analytic over the entire circle, and 1 is not the radius of convergence of the circle, which contradicts our assumptions. Thus the circle $|r| = 1$ is a natural boundary.

The only parts of Theorem III which need discussion are the transition from (b) to (d), and the analytic case. As to the transition from (b) to (d), let it be noted that as ψ is periodic, the real and the imaginary parts of $\psi^{(n)}(\theta)$ must vanish at places θ_1 and θ_2 (not necessarily identical) between 0 and 2π. Thus by the SCHWARZ inequality

$$|\psi^{(n-1)}(\theta)| \leq \left| \int_{\theta_1}^{\theta} R\psi^{(n)}(w)dw \right| + \left| \int_{\theta_2}^{\theta} I\psi^{(n)}(w)dw \right| \leq$$

$$\leq 2 \int_0^{2\pi} |\psi^{(n)}(w)|\, dw \leq 2 \sqrt{\int_0^{2\pi} dw \int_0^{2\pi} |\psi^{(n)}(w)|^2 dw} \leq$$

$$\leq 2\sqrt{2\pi}\, A_n C^{\frac{n}{2}} \leq A_n C_1^n$$

for some C_1.

As to the analytic case, let us notice that $\psi(\theta)$ is analytic over (α, β) when and only when there exists a C such that over (α, β)

$$\psi^{(n)}(\theta) \leq n!\, C^n.$$

Then over $(0, 2\pi)$ we shall have a B such that

$$\psi^{(n)}(\theta) \leq (n+1)!\, C_1^n \leq n!\, n C_1^n \leq n!\, C_2^n,$$

Thus $\psi(\theta)$ is analytic for all θ, and about each point of the circle of convergence, we may describe a small circle within which $f(z)$ may be continued analytically. By the HEINE-BOREL theorem, we may cover the entire unit circle with a finite number of overlapping circles of this sort, which together with the unit circle will contain a large circle concentric with the unit circle. The circle of convergence is hence not the unit circle.

The class of all functions satisfying (c) or (d) is quasi-analytic over its respective interval if

$$\int_0^\infty \log\left(\sum_{\nu=0}^\infty \frac{x^{2\nu}}{A_\nu^2}\right)\frac{dx}{1+x^2}$$

diverges. This statement is precisely equivalent to that of the divergence

$$\int_0^\infty \log\left(\sum_{\nu=0}^\infty \frac{x^{2\nu+2}}{A_\nu^2}\right)\frac{dx}{1+x^2}.$$

Thus we obtain

THEOREM IV. - *On the hypothesis of Theorem I, if $f(re^{i\theta})$ belongs to a quasi-analytic class along the sector (α, β) of the circle of convergence, it belongs to a related quasi-analytic class over the entire circle of convergence.*

The culminating theorem of this paper is

THEOREM V. - *Let $f(z) = \sum_0^\infty a_n z^{\lambda_n}$ have the unit circle as its circle of convergence, and let the λ_n's be a set of integers arranged in increasing order. Let $\lambda_{n+1} - \lambda_n \to \infty$. Then the unit circle is a natural boundary. Further, let $f(z)$ tend over some sector to continuous boundary values as $|z| \to 1$. Then if these boundary values are quasi-analytic over any sector, as in* (c), *with $A_n \geq n!$ they exist and belong to a related quasi-analytic class over the entire unit circle.*

Exactly similar results hold for the general DIRICHLET series $\sum a_n e^{\lambda_n z}$ in the abscissa of convergence, where λ_n are real but not necessarily integers, and are proved in exactly the same way.

To prove Theorem V let us notice that if we subtract an appropriate polynomial from $f(z)$, we can make the gaps between the remaining successive λ_n's all as large as we please. We therefore may apply Theorems III and IV to the remainder. We have already so defined the quasi-analytic class of (c) that the functions contained in it remain in it after the addition of an analytic function.

It will be observed that our gap-theorems include the classical HADAMARD gap-theorem, but not the more powerful gap-theorem of FABRY. It is still an open question whether the latter theorem can be established by methods of this type, which essentially go back to the fact that a lacunary FOURIER series represents a function with a certain type of generalized periodicity of arbitrary small period.

Reprinted from the
Science Reports of the National Tsinghua University
Ser. A, Vol. III, No. 3, December 1935

FABRY'S GAP THEOREM

BY NORBERT WIENER*

Departments of Mathematics and Electrical Engineering

We wish to prove Fabry's[1] theorem, to the effect that if

(1) $0<\lambda_1<\ldots<\lambda_n<\ldots;\ 1/\lambda_n=o(1/n);\lambda_{n+1}-\lambda_n>C(n=1,2,\ldots)$;

then the function

$$f(x)=\sum_1^\infty a_n e^{-\lambda_n x} \tag{2}$$

has its ordinate of convergence (if any) as a natural boundary. We follow in many respects lines laid down by Pólya[2], but the details of our proof are considerably simpler.

We introduce the auxiliary function:

$$\varphi(z)=\prod_1^\infty\left(1-\frac{z^2}{\lambda_n^2}\right)=o(e^{\alpha|z|}) \text{ for all } \alpha>0. \tag{3}$$

We put

$$\begin{cases} \mu_n=\dfrac{nC}{3} \text{ if } \left(\dfrac{nC}{3},\dfrac{(n+1)C}{3}\right) \text{ contains no } \lambda_m; \\ \mu_n=\lambda_m \text{ if } \left(\dfrac{nC}{3},\dfrac{(n+1)C}{3}\right) \text{ contains } \lambda_m; \\ \nu_n=\dfrac{2nC}{3}-\mu_n;\ \dfrac{nC}{3}-\mu_n=\epsilon_n; \end{cases} \tag{4}$$

---

* Professor of Mathematics, Massachusetts Institute of Technology, U.S.A., on leave as Research Professor at Tsing-hua University, Peiping, China, 1935-1936.

1 E. Fabry, Acta Mathematica 22(1898), pp.65-87.

2 G. Pólya, Untersuchungen über Lücken und Singularitäten von Potenzreihen. Mathematische Zeitschrift 29(1929), pp.549-640. This article contains a very complete account of the literature.

and

$$(5)\qquad \psi(z)=z^2\prod_1^\infty\left(1-\frac{z^2}{\mu_n^2}\right)\left(1-\frac{z^2}{\nu_n^2}\right)$$

$$=z^2\prod_1^\infty\frac{\left(\left(\frac{nc}{3}-\epsilon_n\right)^2-z^2\right)\left(\left(\frac{nc}{3}+\epsilon_n\right)^2-z^2\right)}{\left(\frac{n^2c^2}{9}-\epsilon_n^2\right)^2}$$

$$=z^2\prod_1^\infty\frac{\frac{n^4c^4}{81}+\epsilon_n^4+z^4-2z^2\frac{n^2c^2}{9}-2\epsilon_n^2\frac{n^2c^2}{9}-2z^2\epsilon_n^2}{\frac{n^4c^4}{81}-2\epsilon_n^2\frac{n^2c^2}{9}+\epsilon_n^4}$$

$$\sim \text{const.}\ z^2\prod_1^\infty\left(1-\frac{9z^2}{n^2c^2}\right)^2\left\{1+\frac{\epsilon_n^4-2\epsilon_n^2\frac{n^2c^2}{9}-2z^2\epsilon_n^2}{\left(\frac{n^2c^2}{9}-z^2\right)^2}\right\}$$

$$\sim \text{const.}\ \sin^2\frac{3\pi z}{c}\prod_1^\infty\left\{1+\frac{O(1)}{z^2-\frac{n^2c^2}{9}}+\frac{O(n^2)}{\left(z^2-\frac{n^2c^2}{9}\right)^2}\right\}$$

$$=O(1)\ \text{if}\ |Iz|=1$$

Thus by the Phragmén-Lindelöf theorem, as x goes to $\pm\infty$ along the real axis,

$$(6)\qquad \psi(x)=O(1)$$

Furthermore,

$$(7)\qquad \varphi(z)=\psi(z)/\chi(z)$$

where $\chi(z)$ is a function of the form

$$(8)\qquad \chi(z)=z^2\prod_1^\infty\left(1-\frac{z^2}{\sigma_n^2}\right)$$

and the σ_n's have a definite density. We have

$$\frac{1}{x}\log\left|\chi(x(1+i\epsilon)\right| = o(1) + \frac{1}{x}\sum_{1}^{\infty}\log\left|1 - \frac{x^2(1-{}^2\epsilon+2i\epsilon)}{\sigma_n{}^2}\right|$$

$$= o(1) + \frac{1}{2x}\sum_{1}^{\infty}\log\left(1 - 2(1-\epsilon^2)\frac{x^2}{\sigma_n{}^2} + (1+\epsilon^2)^2\frac{x^4}{\sigma_n{}^4}\right)$$

$$= o(1) + \frac{1}{2x}\int_0^\infty \log\left(1-2(1-\epsilon^2)\frac{x^2}{u^2}+(1+\epsilon^2)^2\frac{x^4}{u^4}\right)d\sum_1^n \sigma_n$$

$$= o(1) + \frac{1}{2x}\int_0^\infty \sum_1^n \sigma_n \frac{-4(1-\epsilon^2)\frac{x^2}{u^3}+4(1+\epsilon^2)^2\frac{x^4}{u^5}}{1-2(1-\epsilon^2)\frac{x^2}{u^2}+(1+\epsilon^2)^2\frac{x^4}{u^4}}du$$

$$\sim \frac{1}{2}\lim_{n\to\infty}\frac{1}{n}\sum_1^n \sigma_k \int_0^\infty \frac{-4(1-\epsilon^2)\frac{1}{u^2}+4(1+\epsilon^2)^2\frac{1}{u^4}}{1-2(1-\epsilon^2)\frac{1}{u^2}+(1+\epsilon^2)^2\frac{1}{u^4}}du$$

$$\sim \frac{1}{2}\lim_{n\to\infty}\frac{1}{n}\sum_1^n \sigma_k \int_0^\infty \frac{-4u^2(1-\epsilon^2)+(1+\epsilon^2)^2}{u^4-2u^2(1-\epsilon^2)+(1+\epsilon^2)^2}du$$

$$= \pi\,\xi \lim_{n\to\infty}\frac{1}{n}\sum_1^n \sigma_k.$$

Hence if x is real,

(9) $$\left|\chi(x)\right| \leq \left|\chi(x(1+\epsilon))\right| = o\left(\exp(2\pi\epsilon|x|\lim_{n\to\infty}\frac{1}{n}\sum_1^n \sigma_k)\right)$$

This also holds along any abscissa. Thus by (5) and (7),

(10) $$\frac{1}{\varphi(x+iA)} = o\left(e^{\alpha|x|}\right) \qquad (\alpha > 0)$$

Again,

$$\log \varphi(z) = \sum_{1}^{\infty} \log \left(1 - \frac{z^2}{\lambda_n^2}\right), \tag{11}$$

whence

$$\frac{\varphi'(z)}{\varphi(z)} = \sum_{1}^{\infty} - \frac{2z}{\lambda_n^2 - z^2} \tag{12}$$

It follows that for x real, if λ_m is the nearest λ_n to x,

$$\frac{(x-\lambda_m)\varphi'(x)}{\varphi(z)} = O(\log x). \tag{13}$$

Hence, by (10),

$$\log |\varphi'(x)| < o(x), \tag{14}$$

and

$$\frac{\varphi(x)}{\varphi'(\lambda_n)(x-\lambda_n)} < \text{const. } e^{\alpha[|x|+|\lambda_n|]} \tag{15}$$

for every positive α. Therefore

$$\varphi_n(x) = \frac{e^{\epsilon|\lambda_n|} e^{-\epsilon|x|}\varphi(x)}{\varphi'(\lambda_n)(x-\lambda_n)} \tag{16}$$

belongs to L_2 for every positive ϵ, and if

$$G_n(u) = \frac{1}{2\pi}\int_{-\infty}^{\infty} e^{-inx}\varphi_n(x)\,dx \tag{17}$$

then

$$\int_{-\infty}^{\infty} G_n(u)\, e^{i\lambda_m u}\, du = \delta_{mn}. \tag{18}$$

It is easily verified that $G_n(u) = O(u^{-2})$ at $\pm\infty$. Thus if $f(x)$ is a periodic function belonging to L_2, of the form $\sum_0^\infty a_n e^{i\lambda_n x}$, we have

$$(19) \qquad \int_{-\infty}^{\infty} f(u)\, G_n(u)\, du = a_n.$$

Now[3],

$$(20) \qquad G_n(u) = \frac{1}{2\pi}\int_0^{\infty} \frac{e^{\epsilon|\lambda_n|}\varphi(x)}{\varphi'(\lambda_n)(x-\lambda_n)} e^{x(-\epsilon-iu)}\, dx$$

$$- \frac{1}{2\pi}\int_0^{\infty} \frac{e^{\epsilon|\lambda_n|}\varphi(x)}{\varphi'(\lambda_n)(x-\lambda_n)} e^{x(\epsilon-iu)}\, dx$$

where the ∞ in both cases is chosen in such a direction as to make the integrals significant. The expression

$$(21) \qquad H_n(w) = \frac{1}{2\pi}\int_0^{\infty} \frac{e^{\epsilon|\lambda_n|}\varphi(x_n)}{\varphi'(\lambda_n)(x-\lambda_n)} e^{wx}\, dx$$

is a single-valued function of w, with no singularity other than the origin. It is $O(1/w)$ at ∞. We have

$$(22) \qquad G_n(u) = H_n(-\epsilon - iu) - H_n(\epsilon - iu).$$

Thus (19) becomes

$$(23) \qquad a_n = \int_{-\infty}^{\infty} [H_n(-\epsilon - iu) - H_n(\epsilon - iu)]\, f(u)\, du,$$

and if $f(z)$ is analytic in the half-place $Iz > -\xi_1 > -2\xi$, this becomes

$$(24) \qquad a_n = \int_C [H_n(-\epsilon - iu) - H_n(\epsilon - iu)]\, f(u)\, du,$$

3 Pólya, loc. cit., p. 580.

where C' consists of the real axis from $-A$ to A, together with a semicircle in the upper half plane from A to $-A$. This gives

$$(25)\qquad a_n=\int_C H_n(-iu)\, f(u-i\epsilon)\, du,$$

where C is a circle of radius ϵ about the origin. By (16).

$$(26)\qquad |H_n(w)|\leq\frac{1}{2\pi}\int_0^\infty e^{\alpha|\lambda_n|}e^{(\alpha-|w|)x}\,dx=O(e^{\beta\lambda_n})$$

for $|w|=\epsilon$ and every $\beta>0$. Thus

$$(27)\qquad |a_n|=O(e^{\beta\lambda_n})\left[\int_C|f(u-i\epsilon)|^2\,du\right]^{\frac{1}{2}}$$

Furthermore, (25) gives

$$(28)\qquad a_n=e^{\lambda_n y}\int_C H_n(-iu)\, f(u+iy-i\epsilon)\, du$$

for $y<0$. This is however an analytic function of y for $y>-A$ if f is analytic for arguments with imaginary part greater than $-A-2\epsilon$ and real part less in modulus than 2ϵ. Thus (28) and (27) remain true under these hypotheses. Under these conditions,

$$(29)\qquad |a_n|=O(e^{-\lambda_n A+\lambda_n\beta})$$

and the ordinate of convergence of the Dirichlet series of $f(z)$ is at most $-A+\beta$. Thus if there is any interval over which $f(z)$ is analytic for $Iz>-A-2\epsilon$, the ordinate of convergence is $>-A+\beta$. Since ϵ and β, this ordinate of convergence, if it exists, must be a natural boundary.

Now let $f(z)$ be an entire function. We have by (27)

$$(30)\ \left|f(z)\right| = \left|\sum_0^\infty a_n e^{i\lambda_n Rz} e^{-\lambda_n Iz}\right| \leq \left[\int_C |f(z-2i\epsilon)|^2 dz\right]^{\frac{1}{2}} \sum_0^\infty e^{-\epsilon\lambda_n/2}$$

$$\leq \text{const.} \max_{\substack{|R\omega| \leq 2\epsilon \\ 0 \leq I\omega \leq 3\epsilon}} \left| f(z-w) \right|$$

With the aid of this fact, we may readily prove Pólya's theorem, to the effect that over every interval of Rz, $f(z)$ has the same order, whether in the ordinary sense or in that of Lindelöf, and the same type in that order[4].

The methods of this paper are also available to prove Pólya's theorems concerning functions where the maximum density of the coefficients, though not zero, is less than 1.

4 Loc. cit., p. 624, Theorem VI.

ANALYSE MATHÉMATIQUE. — *Sur les séries de Fourier lacunaires. Théorèmes directs.* Note de MM. **S. Mandelbrojt** et **Norbert Wiener**, présentée par M. Émile Borel.

Mandelbrojt a démontré les théorèmes suivants :

Théorème I. — 1° *Si* $f(t)$ *est intégrable* L *dans l'intervalle* $(0, 2\pi)$, *et si*

$$\overline{\lim_{\alpha \to +0}} \left\{ -\left(\frac{1}{\log \alpha}\right) \log \left[-\log \int_{t_0}^{t_0+\alpha} |f(t)|\, dt \right] \right\} = \delta > 0 \qquad (0 \leqq t_0 < 2\pi);$$

2° *Si*

$$f(t) \sim \sum_{\nu=1}^{\infty} (a_{n_\nu} \cos n_\nu t + b_{n_\nu} \sin n_\nu t),$$

où

$$\Sigma n_\nu^{-\sigma-\varepsilon} < \infty, \qquad \Sigma n_\nu^{-\sigma+\varepsilon} = \infty \quad (\textit{pour tout } \varepsilon > 0) \qquad (\sigma < 1);$$

3° *Si*

$$\delta > \sigma(1-\sigma)^{-1},$$

la fonction $f(t)$ *est nulle presque partout* ([3]).

Théorème II. — 1° *Si* $f(t)$ *possède la propriété* 2° *du théorème I;*

2° *Si* $f(t)$ *est indéfiniment dérivable, ses dérivées vérifiant les inégalités*

$$|f^{(n)}(t)| < \mathrm{K}^n m_n \qquad (0 \leqq t \leqq 2\pi)$$

où

$$\frac{1}{\sigma} > \varlimsup_{n \to \infty} \frac{\log m_n}{n \log n};$$

3° *Si*

$$f^{(n)}(0) = 0 \qquad (n = 0, 1, 2, \ldots)$$

$f(t)$ *est identiquement nulle* (Mandelbrojt, *loc. cit.*).

---

(1) *Bull. Ac. R. Belg.*, Cl. des Sc., **21**, 1935, p. 385.

(2) *Annals of Mathematics*, **36**, 1935, p. 607, ainsi que *Physical Review*, **46**, 1934, p. 505.

(3) Mandelbrojt, *Séries de Fourier et classes quasi-analytiques de fonctions*, Paris, 1935, p. 101 et 128.

La méthode de Mandelbrojt relève essentiellement de la possibilité de construire une famille de fonctions $\varphi_\alpha(t)\,(t \geqq 0)$, nulles pour $t > \alpha$, leurs maxima $M_\alpha = \operatorname{Max}|\varphi_\alpha(t)|$ ne croissant pas trop vite lorsque $\alpha \to 0$, leurs coefficients de Fourier étant tous nuls pour les indices $n = n_\nu$ $(\nu = 1, 2, \ldots, p-1, p+1, \ldots)$ sauf pour $\nu = p$, les coefficients d'indice n_p ne décroissant pas trop vite avec α. M. Mandelbrojt a réalisé cette construction en partant de la théorie des fonctions quasi-analytiques.

Or Wiener et Palay ont démontré qui si $\Phi(z)$ est une fonction entière telle que $\Phi(z) = o\,(e^{\alpha|z|})$, $\int_{-\infty}^{\infty} |\Phi(x)|^2\,dx < \infty$, la fonction $g(t)$, dont $\Phi(z)$ est la transformée de Fourier, est liée à celle-ci par la relation $\Phi(z) = \int_{-\alpha}^{\alpha} g(t)\,e^{itz}\,dt$ (1). Les méthodes de Wiener et Palay permettent alors de construire des fonctions $g_\alpha(t)$ pouvant intervenir dans les démonstrations des théorèmes de Mandelbrojt au même titre que les fonctions $\varphi_\alpha(t)$ de ce dernier. La théorie des transformées de Fourier fournit ainsi un instrument utile dans les considérations qui précèdent. En remarquant que les théorèmes de Denjoy-Carleman peuvent également être démontrés par la théorie des transformées de Fourier (Palay and Wiener, *loc. cit.*), on pourra comparer facilement les deux méthodes dont il est question dans cette Note. (Étant donné la relation qui existe entre la théorie des fonctions quasi-analytiques et celle des transformées de Fourier.) Voici les grandes lignes de la démonstration du théorème I.

Soient $z = x + iy$, $\vartheta = 1 + \varepsilon$ $(\varepsilon > 0)$, $N > 2^{\frac{1}{\varepsilon}} - 1$, et posons

$$F(z) = \frac{\sin \pi z}{z \prod\limits_{n<N} \left(1 - \dfrac{z^2}{[n^\vartheta]^2}\right)}, \qquad \Phi_\alpha(z) = \prod_{n>1} \left(1 - \frac{z^2}{n_\nu^2}\right) F(\alpha z + i).$$

On a pour tout $\sigma' > \sigma$ $(1/\vartheta > \sigma')$, A et $B^{\sigma'}$ étant des constantes > 0,

$$(1) \qquad |\Phi_\alpha(x)| < B_{\sigma'} e^{-A(|x|\alpha)^{\frac{1}{\vartheta}}} + x^{\sigma'},$$

$$(2) \qquad \Phi_\alpha(z) = o(e^{K\alpha|z|}) \qquad (k > \pi).$$

Il résulte de la théorie des transformées de Fourier

$$\Phi_\alpha(z) = \int_{-K\alpha}^{K\alpha} g_\alpha(t)\,e^{itz}\,dt; \qquad \int_{-\infty}^{\infty} |x\Phi_\alpha(x)|^2\,dx = \frac{1}{2\pi} \int_{-K\alpha}^{K\alpha} |g'_\alpha(t)|^2\,dt,$$

(1) Palay and Wiener, *American Math. Soc. Colloquium Publication*, **19**, 1934, p. 12 et 14.

d'où l'on tire, en vertu de (1),

$$\int_{-K\alpha}^{K\alpha} |g'_\alpha(t)|^2\, dt = O\left(e^{\alpha\frac{-\sigma'}{1-\sigma'\vartheta}}\right),$$

$$\underset{|t|\leqq K\alpha}{\operatorname{Max}} |g_\alpha(t)| = O\left(e^{\alpha\frac{-\sigma'}{1-\sigma'\vartheta}}\right),$$

Posons $t_0 = 0$, $n_1 = 0$; on a, pour α assez petit,

$$a_0 = \int_0^{2k\alpha} f(t)\, g(t-k\alpha)\, dt \left[\int_0^{2k\alpha} g(t-k\alpha)\, dt\right]^{-1} = \frac{1}{\operatorname{sh}\pi}\int_0^{2k\alpha} f(t)\, g(t-k\alpha)\, dt.$$

D'où il résulte

$$|a_0| \leqq \underset{|t|\leqq k\alpha}{\operatorname{Max}} |g_\alpha(t)| \int_0^{2k\alpha} |f(t)|\, dt = \int_0^{2k\alpha} |f(t)|\, dt . O\left(e^{\alpha - \frac{\sigma'}{\sigma'\vartheta}}\right).$$

En donnant à α les valeurs α_n telles que

$$\int_0^{2k\alpha_n} |f(t)|\, dt < e^{-(2k\alpha_n)^{-\delta-\eta}} \qquad (\eta > 0, \text{ petit}),$$

on voit, en posant $\alpha_n \to 0$, que $a_0 = 0$. Pour démontrer que a_{n_p} est égal à zéro on remplacera $f(t)$ par $f(t)\cos n_p t$; de même pour b_{n_p}, on considérera $f(t)\sin n_p t$.

Si l'on suppose que $f(t)$ est une fonction paire (et si $t_0 = \pi$) les démonstrations se simplifient, car on peut alors n'employer qu'une seule fonction $\Phi(z)$, au lieu d'une famille Φ_α.

Quant au théorème II, Mandelbrojt l'a tiré du théorème I, mais on peut aussi le démontrer directement.

ANALYSE MATHÉMATIQUE. — *Séries de Fourier lacunaires. Théorèmes inverses.* Note de MM. **Norbert Wiener** et **Szolem Mandelbrojt**, présentée par M. Émile Borel.

Les inégalités qui interviennent dans les théorèmes I et II de Mandelbrojt, théorèmes dont une nouvelle démonstration a été indiquée dans une Note précédente ([1]), ne peuvent pas être améliorées. Étant donné que les hypothèses du théorème II entraînent celle du théorème I, il suffit de démontrer notre affirmation pour le théorème II. Nous pouvons en effet démontrer le théorème suivant :

Quel que soit $\sigma < 1$, *on peut construire une fonction* $f(t)$, *non identiquement nulle, indéfiniment dérivable (périodique de période* 2π), *telle que*

$$f(t) = \Sigma a_{n_\nu} \cos n_\nu t$$

avec $\Sigma n_\nu^{-\sigma-\varepsilon} < \infty$, $\Sigma n_\nu^{-\sigma+\varepsilon} = \infty$ $(\varepsilon > 0)$, *et telle que* $|f^{(n)}(x)| < m_n$ $(n = 0, 1, \ldots)$, *où*

$$\lim \frac{\log m_n}{n \log n} = \frac{1}{\sigma},$$

et enfin, telle que $f^{(n)}(-\pi) = f^{(n)}(\pi) = 0$ $(n = 0, 1, \ldots)$.

Ce théorème améliore d'une certaine manière un théorème de Mandelbrojt, qui ne porte que sur une suite infinie de valeurs de σ.

La démonstration du théorème que nous venons d'énoncer va encore être basée sur la théorie des transformées de Fourier (de Wiener et Paley); et tout comme dans la Note précédente, ce sont les éléments de cette théorie qui interviennent dans les démonstrations de la théorie de la quasi-analyticité qui permettront d'éviter cette dernière.

Passons maintenant à la construction de notre fonction $f(t)$; posons

$$\Phi(z) = \sin z \left[z \prod_{N}^{\infty} \left(1 - \frac{z^2}{\left[\nu^{\frac{1}{\sigma}}\right]^2} \right) \right]^{-1},$$

($[x]$ désignant la partie entière de x), N étant suffisamment grand pour que les quantités $\left[\nu^{\frac{1}{\sigma}}\right]$ $(\nu \geqq N)$ soient toutes différentes, et considérons la fonction

$$f(t) = \frac{1}{2\pi} \int_{-\infty}^{\infty} \Phi(z)\, e^{itz}\, dz.$$

([1]) Mandelbrojt et Norbert Wiener, *Comptes rendus*, 203, 1936, p. 34.

Comme on a sur deux droites $y = \pm a$ $(a > 0)$

$$\Phi(z) = O(e^{-k|z|^\sigma}) \qquad (k > 0),$$

la même inégalité est valable, d'après le principe de Phragmèn-Lindelöf, sur l'axe réel lui-même.

On peut, par conséquent, écrire

$$\int_{-\infty}^{\infty} \left[\sum_{n=1}^{\infty} \frac{x^n}{\left(\frac{2n}{ek}\right)^{\frac{n}{\sigma}}} \Phi(x) \right]^2 dx < \infty$$

et

$$\int_{-\infty}^{\infty} |x^n \Phi(x)|^2 \, du = o(\mathrm{A}n)^{\frac{n}{\sigma}}.$$

Mais il résulte de la théorie des transformées de Fourier que

$$\int_{-\pi}^{\pi} |f^{(n)}(t)|^2 \, dt = \frac{1}{2\pi} \int_{-\infty}^{\infty} |\Phi(x) x^n|^2 \, dx,$$

d'où l'on conclut que

$$|f^{(n-1)}(t)| = \left| \int_{-\pi}^{t} f^{(n)}(t) \, dt \right| \leqq \int_{-\pi}^{\pi} |f^{(n)}(t)| \, dt = o(\mathrm{A}n)^{\frac{n}{\sigma}}.$$

On vérifie alors immédiatement que $f(t)$ possède toutes les propriétés précisées.

GAP THEOREMS

By Norbert Wiener, Cambridge, Mass.

In general, a gap theorem is the assertion that a function $f(x)$ with the Fourier or Dirichlet series

(1) $$f(x)=\sum a_n e^{i\lambda_n x},$$

or the Taylor series

(2) $$f(x)=\sum a_n x^{\lambda_n},$$

has certain properties which depend merely on the distribution of the λ_n.[1] In this sense, it is a gap theorem to assert that the function with the Fourier series

(3) $$f(x)=\sum_{-\infty}^{\infty} a_n e^{inx},$$

is of period 2π.

Closely connected with the very notion of a gap theorem is the problem of determining the coefficients a_n in terms of the function $f(x)$. The determinations which are of most interest are the integral determinations of the form

(4) $$a_n=\int_C f(x) g_n(x) dx.$$

The path C of integration is a matter of great importance. It is very often the doubly infinite axis of reals. In case (3), we have

(5) $$a_n=\frac{1}{2\pi}\int_{-\pi}^{\pi} f(x) e^{-inx} dx,$$

and the path of integration is a segment of the axis of reals, of length 2π. Let it be noted that c and g_n are not in general uniquely determined — that we may write, in fact,

(6) $$a_n=\frac{1}{2\pi}\int_{A-\pi}^{A+\pi} f(x)\, e^{-inx} dx$$

in place of (5), for any real A.

[1] Paley and Wiener. Am. Math. Soc. Colloquium Pub. Vol. XIX. Chapters V—VIII. — N. Levinson, On a problem of Pólya, Am. Journal of Math., Vol. LVIII, p. 791. By the same author, Non Harmonic Fourier Series, Ann. of Math. Vol. 37, p. 919.

If we consider formula (4) as applied to case (1), we see that

(7)
$$\int_C e^{i\lambda_m x} g_n(x)\,dx = \begin{cases} 1 \text{ if } m=n; \\ 0 \text{ if } m \neq n. \end{cases}$$

Thus if we put

(8)
$$\int_C e^{iux} g_n(x)\,dx = G_n(u),$$

we have

(9)
$$G_n(\lambda_m) = \begin{cases} 1 \text{ if } m=n; \\ 0 \text{ if } m \neq n. \end{cases}$$

In general, if

(10)
$$G(u) = G_n(u)(u-\lambda_n),$$

we shall have

(11)
$$G(\lambda_n) = 0$$

for all values of n. If we now put $G_n(u)$ for the Lagrange interpolation function

(12)
$$G_n(u) = \frac{G(u)}{(u-\lambda_n)\,G'(\lambda_n)},$$

or for the limiting value of this function, in case both numerator and denominator tend to zero for a particular value of u, (9) will be satisfied. Hence at least formally, we have reduced the problem of determining all the functions $G_n(u)$ to the determination of the single function $G(u)$.

If C is a curve of finite diameter (or a set of points of finite diameter in the complex plane), $G_n(u)$ and hence $G(u)$ are integral functions of the exponential type. As the λ_n are zeros of such a function, the Weierstrass product

(13)
$$P(u) = \prod \left(1 - \frac{u}{\lambda_n}\right) e^{\frac{u}{\lambda_n}}$$

converges, and we may write

(14)
$$G(u) = P(u)\,Q(u),$$

where $Q(u)$ is an entire function of not more than exponential type. For a given C and a given set of λ_n, the problem of the determination of the g_n reduces to the determination of the entire functions $Q(u)$ such that we may write

(15)
$$\frac{P(u)\,Q(u)}{(u-\lambda_n)\left[\frac{d}{du}(P(u)\,Q(u))\right]_{u=\lambda_n}} = \int_C e^{iux} g_n(x)\,dx.$$

To solve this, we must be able to recognize the functions of the form

$$\int_C e^{iux} g(x)\, dx. \tag{16}$$

There are certain cases where the relevant theorems are known. If $g(x)$ belongs to L^2, and C is the segment of the axis of reals from $-A$ to A, the functions (16) are integral functions dominated by $e^{A|u|}$, and belonging to L^2 along the doubly infinite axis of reals. Conversely, every integral function subject to these conditions has a representation (16), where C is the real interval $(-A, A)$ and $g(x)$ belongs to L^2 over this interval. This theorem is due to the late R. E. A. C. Paley and myself.[1]

Closely connected with this theorem is another, due to Titchmarsh, Pólya, Paley, myself, and Levinson,[2] to the effect that in an integral function of this sort, the zeros are located (with the exception of a set, the number of which, in a circle of radius R about the origin, is $o(R)$), in the sectors $-\varepsilon<\vartheta<\varepsilon$, $\pi-\varepsilon<\vartheta<\pi+\varepsilon$, and that in each of these sectors, the number of zeros with modulus $<R$ is

$$\frac{RA}{\pi} + o(R). \tag{17}$$

This theorem allows us to conclude that if the Pólya maximal density of the zeros of $\varphi(u)$ either to the right or to the left of the origin exceeds $\frac{A}{\pi}$, it is impossible to put

$$\varphi(u)=\int_{-A}^{A} e^{iux} g(x)\, dx, \tag{18}$$

where $g(x)$ belongs to L^2 over $(-A, A)$.

Another result from the Pólya theory of the indicatrix is that if $\varphi(u)$ is an integral function dominated by $e^{A|u|}$, then it may be written in form (16), where C is any circle about the origin of radius $>A$, and $g(x)$ is analytic on this circle. Conversely, every function of form (16), where $g(x)$ is bounded, and C a circle of radius A about the origin, is dominated by $e^{A|u|}$ for large values of $|u|$. The number of zeros of such a function within a circle of radius R, divided by R, has an upper bound of oscillation not exceeding $\frac{2A}{\pi}$. An integral function of order 1, for which this upper bound does not exceed $\frac{2A}{\pi}$, is dominated by $e^{(A+\varepsilon)|u|}$.

[1] Paley and Wiener, loc. cit. chapter V.

[2] N. Levinson, loc. cit.

This latter class of results leads directly to the Fabry gap theorem. Let the λ_n be a set of zero density, and let

$$G(u) = \prod \left(1 - \frac{u^2}{\lambda_n^2}\right). \tag{19}$$

This function has less than exponential growth. By the theory of the indicatrix,

$$G(u) = \int_C g(z)\, e^{iuz}\, dz \tag{20}$$

where C is an arbitrarily small circle about the origin, on which $g(x)$ is bounded. Furthermore,

$$g(z) = \frac{1}{2\pi} \int_0^\infty G(u)\, e^{-iuz}\, du, \tag{21}$$

where the integral is taken in such a sense as to make it converge. Similarly, if we put

$$g_n(z) = \frac{1}{2\pi} \int_0^\infty e^{-izu} \frac{G(u)}{(u-\lambda_n)\, G'(\lambda_n)}\, du, \tag{22}$$

we shall have

$$\int_C g_n(z)\, e^{iuz}\, dz = \frac{G(u)}{(u-\lambda_n)\, G'(\lambda_n)} = \begin{cases} 0 \text{ if } u = \lambda_m\, (m \neq n); \\ 1 \text{ if } u = \lambda_n. \end{cases} \tag{23}$$

It follows that

$$e^{\varepsilon|\lambda_n|}\, e^{-\varepsilon|u|} \int_C g_n(z)\, e^{iuz}\, dz = \begin{cases} 0 \text{ if } |u| = |\lambda_m|\, (m \neq n);\ u = -\lambda_n; \\ 1 \text{ if } u = \lambda_n. \end{cases} \tag{24}$$

Now,

$$\begin{aligned} &\frac{1}{2\pi} \int_{-\infty}^{\infty} e^{-iuz} \frac{G(u)}{(u-\lambda_n)\, G'(\lambda_n)}\, e^{\varepsilon|\lambda_n|}\, e^{-\varepsilon|u|}\, du \\ &= \frac{1}{2\pi} \int_0^{\infty} e^{-iuz} \frac{G(u)}{(u-\lambda_n)\, G'(\lambda_n)}\, e^{\varepsilon|\lambda_n|}\, e^{-\varepsilon u}\, du \\ &\quad - \frac{1}{2\pi} \int_0^{-\infty} e^{-iuz} \frac{G(u)}{(u-\lambda_n)\, G'(\lambda_n)}\, e^{\varepsilon|\lambda_n|}\, e^{\varepsilon u}\, du \\ &= \{g_n(z - i\varepsilon) - g_n(z + i\varepsilon)\}\, e^{\varepsilon|\lambda_n|}. \end{aligned} \tag{25}$$

Thus

$$(26)\qquad \int_{-\infty}^{\infty} \{g_n(z-i\varepsilon)-g_n(z+i\varepsilon)\}\, e^{-iuz}\, dz = \begin{cases} e^{-\varepsilon|\lambda_n|} \text{ if } u=\lambda_n; \\ 0 \text{ if } |u| = |\lambda_m|\,(m \neq n);\; u=-\lambda_n. \end{cases}$$

Hence if $f(x)$ is defined by (1), we have formally

$$(27)\qquad a_n = \int_{-\infty}^{\infty} e^{\varepsilon|\lambda_n|} f(z)\, \{g_n(z-i\varepsilon)-g_n(z+i\varepsilon)\}\, dz.$$

It is not difficult to prove that this is actually true if the distance between two λ_n's has a positive lower bound, and $\Sigma|a_n|^2$ converges.

Now let this be the case, and let

$$(28)\qquad F(z) = \sum_{1}^{\infty} a_n e^{-\lambda_n z}.$$

Then

$$(29)\qquad a_n = \int_{-\infty}^{\infty} \{g_n(z-i\varepsilon)-g_n(z+i\varepsilon)\}\, e^{\varepsilon|\lambda_n|}\, F(iz)\, dz.$$

Now, $F(z)$ has no singularities in the half-plane $Rz>0$, and tends to zero uniformly as $Rz\to\infty$. The function $g_n(z)$ has a singularity at 0, and nowhere else. Thus

$$(30)\qquad \int_{-\infty}^{\infty} g_n(z+i\varepsilon)\, F(iz)\, dz = 0,$$

and

$$(31)\qquad a_n = \int_{C} g_n(z)\, e^{\varepsilon|\lambda_n|}\, F(iz-\varepsilon)\, dz,$$

where C is a small circle about the origin. It is understood, of course, that $F(iz-\varepsilon)$ has a series (28) uniformly and absolutely convergent for all values of z on C.

It is easy to show that on a circle C of radius η,

$$(32)\qquad g_n(z) = O(e^{\beta\lambda_n})$$

for every $\beta>0$. From this and (31), it results that

$$(33)\qquad |a_n| \leqq \limsup_{C}\, |F(iz-\varepsilon)|\, e^{\varepsilon|\lambda_n|}\ \text{const.}$$

It follows readily that

$$(34)\qquad \begin{aligned} |F(x+iy+x_1)| &\leqq \Sigma |a_n|\, e^{-\lambda_n x} e^{-\lambda_n x_1} \\ &\leqq \limsup_{C}\, |F(iz-x-\varepsilon)|\, \Sigma e^{\lambda_n(\varepsilon-x_1)}\ \text{const.} \\ &\leqq \text{const.} \limsup_{C}\, |F(iz-x-\varepsilon)|. \end{aligned}$$

Thus the value of $|F(z)|$ on an ordinate more than a constant distance to the right of a small circle of fixed radius does not exceed a constant multiple of the maximum value of the expression on the circle, provided, at any rate, that the circle is to the right of the ordinate of convergence of the Dirichlet series for $F(z)$. It is not difficult to frame an argument from analytical continuation to show that this remains true provided merely that $F(z)$ is analytic in a region including the small circle and all points to the right. From this it follows at once that the boundary of convergence of the Dirichlet series of F must be a line made up of singularities. Similarly, we may establish Pólya's theorem, that over every interval of Iz, $F(z)$ has the same order, whether in the ordinary sense or in that of Lindelöf, and the same type in that order.[1]

In this theorem we have assumed that the density of the λ_n's is zero and that $\lambda_{n+1}-\lambda_n$ has a positive lower bound. The methods are available to prove Pólya's theorems concerning functions where the maximum density of the λ_n's, though not zero, is <1.

There is another important class of gap theorems dealing with the case where $\lambda_{n+1}-\lambda_n \to \infty$. In this case, a more direct approach than that through the a_n's may be obtained by considering the properties of the function $f(x)$ analogous to periodicity.[2] It may be shown that there is a quasi-periodicity property characterizing almost periodic functions for which there is a minimum distance between two proper frequencies. Let $f(x)$ be an almost periodic function in the Stepanoff class S_2. Let C be the class of all functions of the form

$$(35) \qquad \sum_1^N a_n f(x+b_n) \qquad (a_n \text{ complex, } b_n \text{ real}).$$

Let A and B be real numbers such that for all x_1 and x_2 and for all functions $g(x)$ in C,

$$(36) \qquad \int_{x_1}^{x_1+A} |g(x)|^2 dx \Big/ \int_{x_2}^{x_2+A} |g(x)|^2 dx < B.$$

Let D be the lower bound of the A's for which B is finite. Then if D is finite, $f(x)$ is said to be pseudoperiodic, with pseudoperiod D. The class of pseudoperiodic functions is the same as the class of the functions $f(x)$ of the form

$$(37) \qquad \sum_{-\infty}^{\infty} a_n e^{i\lambda_n x} (\lambda_{n+1} > \lambda_n;\ \text{lim. inf. } \lambda_{n+1}-\lambda_n > 0,\ \Sigma |a_n|^2 < \infty).$$

---

[1] Pólya, Untersuchungen über Lücken und Singularitäten von Potenzreihen, Math. Zeitschrift, Band 29, 1929.

[2] Paley and Wiener, loc. cit. chapter VII.

Furthermore, if E is the lower limit of $\lambda_{n+1}-\lambda_n$, we have

$$DE \leqq \text{absolute constant}. \tag{38}$$

Hence if

$$f(x) \sim \sum_{-\infty}^{\infty} a_n e^{i\lambda_n x} \left(\Sigma |a_n|^2 < \infty,\ \lambda_{n+1} > \lambda_n,\ \lim_{n \to \pm\infty} \lambda_{n+1} - \lambda_n = \infty\right) \tag{39}$$

$f(x)$ is the sum of a trigonometrical polynomial and a function for which $D < D_1$. It results at once, from this and the fact that D depends only on the λ_n's, not on the a_n's, that any series of inequalities on the integrals of the squares of the moduli of f and its derivatives, if it is invariant under the addition of an analytic function, holds over every interval of x provided the corresponding inequalities hold over any given interval of x. Crudely stated, if f satisfies the conditions for a type of quasianalyticity over any interval, however small, it satisfies the conditions for a related type of quasianalyticity over every interval, however large. This is closely related to certain theorems of Mandelbrojt.[1]

Mandelbrojt's more recent work has been to develop a theory intermediate between the Carleman theory of quasi-analytic functions and the standard gap theory. He is chiefly concerned with functions such as

$$f(x) = \sum_{-\infty}^{\infty} a_n e^{i\lambda_n x}, \tag{40}$$

where the λ_n's are integers, $\lambda_{-1} = -\lambda_n$, and

$$\sum_{1}^{\infty} \lambda_n^{-\sigma-\varepsilon} < \infty, \quad \sum_{1}^{\infty} \lambda_n^{-\sigma+\varepsilon} = \infty. \quad (0 < \sigma < 1). \tag{41}$$

He discusses two problems which are partial inverses of one another:

(a) The construction for a given value σ of non-null functions $f(x)$ with derivatives satisfying certain inequalities of the form

$$|f^{(n)}_{(x)}| \leqq m_n a^n; \tag{42}$$

(b) The proof that it is impossible for a given value of σ to construct non-null functions $f(x)$ for which

$$\int_a^{a+\varepsilon} |f(x)|\, dx \tag{43}$$

tends to zero more rapidly than certain given functions of ε.

---

[1] S. Mandelbrojt, Séries Lacunaires, Actualités scientifique et Industrielles, Paris 1936.

In problem (a), as it is only σ that plays a rôle, and not the λ_n, we shall put

(44) $$\lambda_n = [n^{\frac{1}{\sigma}}].$$

We shall then consider the integral function

(45) $$\varphi(u) = \prod_1^{\infty} \left(1 - \frac{u^2}{\lambda_n^2}\right).$$

It may be proved that except in the neighborhood of the λ_n's,

(46) $$|\varphi(u)| = \Omega(|u|^{-R} e^{|u|^{\sigma}}) \qquad (k \text{ fixed}).$$

Now let

(47) $$\psi(u) = \frac{\sin \pi u}{\varphi(u)}.$$

It is possible to show that on the real axis

(48) $$|\psi(u)| = O(|u|^{k_1} e^{-|u|^{\sigma}}) \qquad (k_1 \text{ fixed}).$$

Thus ψ belongs to L^2. The Fourier transform of ψ differs from 0 only over $(-\pi, \pi)$, by theorems we have quoted already. If we take it as the $f(x)$ of (40), we see at once that in the Fourier series of f will have as coefficients

(49) $$\frac{1}{2\pi} \int_{-\pi}^{\pi} f(x) e^{-inx} dx,$$

which will vanish unless n is of the form $\pm \lambda_k$.

By a change involving not more than a finite number of λ_n's, it is thus possible to obtain a function $\psi_1(u)$, integral and dominated by $e^{\pi|u|}$, which is $O(e^{-|u|^{1/\sigma}})$ on the real axis, and which vanishes for every integral argument, *except* a set of numbers $\pm \mu_n$ such that

(50) $$\sum \mu_n^{-\sigma} = \infty, \quad \sum \mu_n^{-(\sigma+\varepsilon)} < \infty.$$

The function

(51) $$f(x) = \int_{-\infty}^{\infty} \psi_1(u) e^{iux} du$$

will then have a series of the form

(52) $$f(x) = \sum_{-\infty}^{\infty} a_n e^{i\mu_n x}.$$

Moreover, f will vanish outside of $(-\pi, \pi)$. We shall have

(53) $$f^{(n)}(x) = \int_{-\infty}^{\infty} (iu)^n \psi_1(u) e^{iux} du.$$

Hence

$$\int_{-\pi}^{\pi}|f^{(n)}(x)|^2\,dx=\int_{-\infty}^{\infty}|f^{(n)}(x)|^2\,dx=2\pi\int_{-\infty}^{\infty}u^{2n}|\psi_1(u)|^2\,dx$$

(54) $$\leqq \text{const.}\int_0^\infty u^{2n}e^{-u^\sigma}\,du$$

$$\leqq \text{const.}\int_0^\infty w^{\frac{2n}{\sigma}}e^{-w}\sigma\, w^{\frac{1}{\sigma}-1}\,dw$$

$$=\text{const.}\,\sigma\,\Gamma((2n+1)\sigma).$$

Hence, as f is differentiable with all derivatives over $(-\infty, \infty)$, and vanishes outside $(-\pi, \pi)$

$$|f^{(n)}(x)|=\Big|\int_{-\pi}^{x}f^{(n-1)}(y)\,dy\Big|$$

$$\leqq\Big\{\int_{-\pi}^{x}dy\int_{-\pi}^{x}|f^{(n-1)}(y)|^2\,dx\Big\}^{\frac{1}{2}}$$

$$\leqq \text{const.}\left\{\frac{1}{\sigma}\,\Gamma((2n-1)/\sigma)\right\}^{\frac{1}{2}}$$

(55) $$\leqq \text{const.}\left\{\frac{1}{\sigma}\left\{\frac{2n-2}{\sigma}\right\}^{\frac{2n-2}{\sigma}+\frac{1}{2}}e^{-\frac{2n-2}{\sigma}}\right\}^{\frac{1}{2}}$$

$$=\text{const.}\,\sigma^{-\frac{n-1}{\sigma}+\frac{3}{4}}(2n-1)^{\frac{n-1}{\sigma}+\frac{1}{4}}e^{-\frac{n-1}{\sigma}}$$

$$\leqq \text{const.}\,a^n n^{\frac{n}{\sigma}},$$

where a does not depend on n. If we put $n^{\frac{n}{\sigma}}=m_n$, we have for all n

(56) $$\frac{\log m_n}{n\log n}=\frac{1}{\sigma}>1\,,\ \frac{1}{\sigma}<\infty.$$

The possibility of forming such a function $f(x)$ has been proved by Mandelbrojt and myself,[1] and is an extension of a result of Mandelbrojt.

Case (b) may also be reduced, and in essence has already been reduced by Mandelbrojt, to the problem of forming certain functions with known Fourier transforms, and given by infinite products. We return to the

---

[1] Mandelbrojt and Wiener, Comptes Rendus 203, 1936 p. 34 and 203, 1936, p. 233. — Mandelbrojt, Séries de Fourier et classes quasi-analytique de fonctions. Paris 1935, p. 101 and 108.

beginning of this lecture, and to formula 4, where C is the interval $(-\varepsilon, \varepsilon)$ on the real axis. Then in (8),

$$\int_{-\varepsilon}^{\varepsilon} e^{iux} g_n(x)\,dx = G_n(u). \tag{57}$$

This means that $G_n(u)$ is an integral function, dominated by $e^{|n|}$, and belonging to L^2 along the real axis. The density of its zeros is $\frac{\varepsilon}{\pi}$, both to the right and to the left of the origin. The function $G(u)$ of (10) will have the same density.

This suggests a way of obtaining the function $G(u)$. Let us form the function

$$P_1(u) = \prod_1^\infty \left(1 - \frac{u^2}{\lambda_n^2}\right). \tag{58}$$

This is an integral function of order σ, whenever

$$\sum \lambda_n^{-\sigma-\varepsilon_1} < \infty, \quad \sum \lambda_n^{-\sigma+\varepsilon_1} = \infty. \tag{59}$$

Then

$$|P_1(u)| = O(e^{|u|^{\sigma+\varepsilon_1}}). \tag{60}$$

Now let

$$\nu_n = \left[n^{\frac{1}{\sigma+\varepsilon_2}} \frac{\varepsilon}{\pi} \right] \frac{\pi}{\varepsilon}. \tag{61}$$

Let N be the least integer that is so large that from $n = N$ on, no two ν_n's are the same. Let

$$\varphi_2(u) = \prod_N^\infty \left(1 - \frac{u^2}{\nu_n^2}\right) \tag{62}$$

and

$$\psi_2(u) = \frac{\sin \varepsilon u}{\varphi_2(u)}. \tag{63}$$

It may be shown that if u is real

$$\psi_2(u) = O(e^{-|u|^{\sigma+\varepsilon_2-\varepsilon_3}}). \qquad (\varepsilon_3 > 0). \tag{64}$$

Thus

$$u^{-1}\psi_2(u) P_1(u) = O(e^{-|u|^{\sigma+\varepsilon_2-\varepsilon_3}}). \tag{65}$$

This function furthermore has zeros at the points $\pm\lambda_n$. Thus

$$f_2(x) = \int_{-\infty}^{\infty} u^{-1}\psi_2(u) P_1(u) e^{iux}\,du \tag{66}$$

is a function differing from 0 only over $(-\varepsilon, \varepsilon)$, and with a Fourier series

$$\alpha_0 + \sum_1^\infty (\alpha_n \cos nx + \beta_n \sin nx), \tag{67}$$

where

(68) $$\alpha_{\lambda_n} = \beta_{\lambda_n} = 0; \qquad (n = 1, 2, 3, \cdots)$$

$$\alpha_0 \neq 0.$$

We have

$$\int_{-\varepsilon}^{\varepsilon} |f_2(x)|^2 \, dx = 2\pi \int_{-\infty}^{\infty} u^{-2} |\psi_2(u) P_1(u)|^2 \, du$$

$$= O(\varepsilon^2) \int_{-\infty+i}^{\infty+i} e^{2|u|^{\sigma+\varepsilon_1}} \frac{1}{\left| \prod_N^{\infty} \left(1 - \frac{u^2}{\nu_n^2}\right) \right|^2} du$$

(69) $$= O(\varepsilon^2) \int_i^{\infty+i} e^{2|u|^{\sigma+\varepsilon_1}} \left| \prod_1^{N-1} \left(- \frac{u^2 \pi^2}{n^{\frac{2}{\sigma+\varepsilon_2}} \varepsilon^2} \right) \right| \prod_N^{\infty} \left| \frac{1 - \frac{u^2 \pi^2}{n^{\frac{2}{\sigma+\varepsilon_2}} \varepsilon^2}}{1 - \frac{u^2}{\nu_n^2}} \right|^2 \left\{ \prod_1^{\infty} \left| 1 - \frac{u^2 \pi^2}{n^{\frac{2}{\sigma+\varepsilon_2}} \varepsilon^2} \right|^2 \right\}^{-1} du$$

$$= O(\varepsilon^2) \int_0^{\infty} e^{-u^{\frac{\sigma+\varepsilon_2}{2}}} \frac{u^{4N} \pi^{4N}}{\varepsilon^{4N} (\Gamma(N))^{\frac{4}{\sigma+\varepsilon_2}}} du.$$

Now,

(70) $$N = O\left(\frac{(\sigma+\varepsilon_2)\pi}{\varepsilon}\right)^{\frac{1}{\sigma+\varepsilon_2}-1}.$$

Thus

$$\int_{-\varepsilon}^{\varepsilon} |f_2(x)|^2 \, dx = O(\varepsilon^2) \frac{\pi^{4N}}{\varepsilon^{4N} (\Gamma(N))^{\frac{4}{\sigma+\varepsilon_2}}} \int_0^{\infty} e^{-u^{\frac{\sigma+\varepsilon_2}{2}}} u^{4N} \, du$$

(71) $$= \frac{O(\pi^{4N})}{\varepsilon^{4N} (\Gamma(N))^{\frac{4}{\sigma+\varepsilon_2}}} \Gamma\left(4N\left(\frac{\sigma+2\varepsilon_2}{3}\right)\right)$$

$$= O\left(e^{\varepsilon^{-\left(\frac{1}{\sigma+\varepsilon_2} - 1 + \varepsilon_3\right)}}\right).$$

Again, by (14),

$$\int_{-\varepsilon}^{\varepsilon} f_2(x)\,dx = \frac{\varepsilon}{2\pi}. \tag{72}$$

Now, let $f(x)$ belong to L^2, and let

$$f(x) \sim a_0 + \sum_{1}^{\infty} (a_n \cos \lambda_n x + b_n \sin \lambda_n x). \tag{73}$$

Then

$$\int_{-\varepsilon}^{\varepsilon} f(x+x_1) f_2(x)\,dx = a_0 \varepsilon. \tag{74}$$

Thus

$$\begin{aligned} a_0 &= \frac{1}{\varepsilon} \int_{-\varepsilon}^{\varepsilon} f(x+x_1) f_2(x)\,dx \\ &\leq \frac{\text{const.}}{\varepsilon} \left\{ \int_{-\varepsilon}^{\varepsilon} |f(x+x_1)|^2\,dx \right\}^{\frac{1}{2}} O\left(e^{\varepsilon^{-\left(\frac{1}{\sigma+\varepsilon_2}-1+\varepsilon_3\right)}}\right) \end{aligned} \tag{75}$$

for all $\varepsilon_3 > 0$.

The quantity ε_2 is also arbitrary. Thus if

$$\int_{-\varepsilon}^{\varepsilon} |f(x+x_1)|^2\,dx = O\left(e^{-\varepsilon^{1-\frac{1}{\sigma}+\varepsilon_4}}\right), \tag{76}$$

we have $a_0 \to 0$ with ε, and hence $a_0 = 0$. Here x_1 may depend on ε.

A similar argument, based on the estimate of max $|f_2(x)|$ instead of $\int_{-\varepsilon}^{\varepsilon} |f_2(x)|^2\,dx$, will show that if

$$\int_{-\varepsilon}^{\varepsilon} |f(x+x_1)|\,dx = O\left(e^{-\varepsilon^{1-\frac{1}{\sigma}+\varepsilon_4}}\right) \tag{77}$$

we have $a_0 = 0$, and in general, $a_n = b_n = 0$. This establishes Mandelbrojt's theorem: *if*

$$\overline{\lim_{a \to +0}} \frac{\log\left(-\log \int_{x_0}^{x_0+a} |f(x)|\,dx\right)}{-\log a} = \varrho > 0, \tag{78}$$

if

$$\Sigma n_i^{-\sigma-\varepsilon} < \infty, \quad \Sigma n_i^{-\sigma+\varepsilon} = \infty; \quad (\varepsilon > 0), \tag{79}$$

if

(80) $$f(x) \sim \frac{a_0}{2} + \sum_{i=1}^{\infty} (a_{n_i} \cos n_i x + b_{n_i} \sin n_i x),$$

and if

(81) $$\varrho > \frac{\sigma}{1-\sigma}$$

then $f(x)$ *is equivalent to* 0. The results in problem (b) lead readily to a proof that this result cannot be extended to the case where $\varrho = \frac{\sigma}{1-\sigma}$, and conversely, this result may be used to indicate that the results of problem (a) cannot be extended to the case where

(82) $$\underline{\lim} \frac{\log m_n}{n \log n} > \frac{1}{\sigma}.$$

A TAUBERIAN GAP THEOREM OF HARDY AND LITTLEWOOD

BY NORBERT WIENER[1]

The theorem here discussed resembles typical Tauberian theorems in being a partial inverse of Abel's theorem on power series, but differs in the nature of the auxiliary hypothesis imposed to permit the inversion. The classical Tauberian theorem of Hardy and Littlewood reads[2]: *Let* $f(x)=\sum_1^\infty a_n e^{-nx}$ *converge for* $\Re x>0$, *and let* $\lim_{x\to 0} f(x)=s$. *Then if*

$$a_n = O(1/n), \tag{1}$$

we shall have

$$\sum_1^\infty a_n = s.$$

Here the auxiliary hypothesis is (1), and concerns the order of magnitude of the coefficients of $\sum a_n e^{-nx}$. In another type of theorem also due to Hardy and Littlewood, the auxiliary hypothesis concerns the exponents. The following theorem was conjectured by Littlewood in 1913 and proved by Hardy and Littlewood in 1924[3].

THEOREM I. *Let*

$$0<\lambda_1<\lambda_2<\cdots<\lambda_n<\cdots;\ \lambda_{n+1}/\lambda_n \geq \vartheta > 1 \quad (n=1,2,\cdots) \tag{2}$$

Let

(1) Professor of Mathematics, Massachusetts Institute of Technology, U.S.A., on leave as Research Professor at Tsing-hua University, Peiping, China, 1935-1936.

(2) J. E. Littewood, On the converse of Abel's theorem on Power Series. *Proc. Lond. Math. Soc.*, (2) **9** (1910), 434-448.

(3) Abel's Theorem and its converse II, *Proc. Lond. Math. Soc.* (2) **22** (1924), 254-269.

(3) $$\sum_1^\infty a_n e^{-\lambda_n x} = f(x)$$

converge for $\Re x > 0$, *and let*

(4) $$\lim_{x \to +0} f(x) = s$$

Then

(5) $$\sum_1^\infty a_n = s$$

Up to the present, this theorem has stood alone in the literature. In this paper, I give an alternative proof of the theorem, and bring it into connection with a lemma from the theory of interpolation.

Let it be noted that Theorem I may easily be reduced to a theorem on convergent series. Let us suppose that we have established the following lemma:

LEMMA I. *Let* $\{\lambda_n\}$ and $f(x)$ *be defined as in Theorem* I, *and let* $\sum_1^\infty a_n$ *converge. Then there exists a sequence* $\{\varphi_n(x)\}$ *of measurable functions with the properties:*

(6) $$\sum_1^n a_n = \int_0^\infty f(x)\, \varphi_n(x)\, dx$$

(7) $$\int_0^\infty \varphi_n(x)\, dx = 1; \qquad (n = 1, 2, \cdots)$$

(8) $$\int_0^\infty |\varphi_n(x)|\, dx < \text{const.}; \qquad (n = 1, 2, \cdots)$$

(9) $$\lim_{n \to \infty} \int_\varepsilon^\infty |\varphi_n(x)|\, dx = 0 \quad \textit{for all } \varepsilon > 0.$$

Assuming Lemma I, we proceed to Theorem I as follows: if $\{a_n\}$ and $f(x)$ are taken as in Theorem I, $\sum_1^\infty a_n e^{\lambda_n \eta}$ converges for $\eta > 0$. Thus by Lemma I.

(10) $$\sum_1^n a_k e^{-\lambda_k \eta} = \int_0^\infty f(x + \eta)\, \varphi_n(x)\, dx$$

Since

$$\lim_{\eta \to 0} f(x+\eta) = f(x) \tag{11}$$

uniformly over $(0, \infty)$, it follows from (8) that

$$\sum_{1}^{n} a_k = \int_0^\infty f(x)\, \varphi_n(x)\, dx. \tag{12}$$

Thus by (6) and (7),

$$| \sum_1^n a_k - s | \leqq \int_0^\infty | f(x) - s | \cdot | \varphi_n(x) | \, dx \tag{13}$$

$$\leqq \int_0^\infty | \varphi_n(x) | \, dx \max_{x \leqq \varepsilon} | f(x) - s | + 2 \max_{x < \infty} | f(x) | \int_\varepsilon^\infty | \varphi_n(x) | \, dx.$$

Since $f(x)$ is continuous, for any ε_1, we may choose ε so small that

$$\max_{x \leqq \varepsilon} | f(x) - s | \leqq \varepsilon_1 / (2 \operatorname*{l.u.b.}_n \int_0^\infty | \varphi_n(x) | \, dx) \tag{14}$$

Having fixed ε, we may then choose η so large that

$$\int_\varepsilon^\infty | \varphi_n(x) | \, dx \leqq \varepsilon_1 / (4 \max | f(x) |) \tag{15}$$

in accordance with (9). Thus for all sufficiently large n,

$$| \sum_1^n a_k - s | \leqq \varepsilon_1 / 2 \tag{16}$$

This establishes (5).

Lemma I itself depends on:

LEMMA II. *Let*

$$\cdots < \mu_{-n} < \cdots < \mu_{-1} < \mu_0 < \mu_1 < \cdots < \mu_n < \cdots; \tag{17}$$

$$D < | \mu_{k+1} - \mu_k | < 2D \qquad (k = \cdots, -1, 0, 1, \cdots). \tag{18}$$

Then there exists a sequence $\{ k_n(\xi) \}$ *of functions measurable over* $(-\infty, \infty)$, *such that*

$$\int_{-\infty}^\infty | h_n(\xi) | \, d\xi < \text{const.} \qquad (n = 1, 2, \cdots) \tag{19}$$

$$\lim_{n \to \infty} \int_{-\infty}^B | h_n(\xi) | \, d\xi = 0 \qquad \textit{for all real } B. \tag{20}$$

and that there exists an A independent of n, such that

$$(21)\qquad \frac{1}{\pi}\int_{\mu_n}^{\mu_{n+1}} d\xi \int_{-\infty}^{\infty} \frac{\operatorname{Sin}^2 A(\eta-\xi)}{A(\eta-\xi)^2}\, h_m(\eta)\, d\eta = \begin{cases} 0 & \text{if } m \neq n, \\ 1 & \text{if } m = n. \end{cases}$$

In deducing Lemma I from Lemma II, we make use of the auxiliary function

$$(22)\qquad K(\eta) = \frac{1}{2\pi}\int_{-2A}^{2A} \frac{1-\frac{|z|}{2A}}{\Gamma(iz+1)}\, e^{-i\eta z}\, dz = O(1/\eta^2)$$

and of the functions

$$(23)\qquad f_n(\eta) = \int_{-\infty}^{\infty} h_n(\xi)\, K(\eta-\xi)\, d\xi$$

we have

$$(24)\qquad \int_{-\infty}^{\infty} K(\eta-\xi)\, e^{-e^{\xi}} e^{\xi}\, d\xi = \int_{-\infty}^{\infty} \frac{d\xi}{2\pi} \int_{-2A}^{2A} \frac{1-\frac{|z|}{2A}}{\Gamma(iz+1)}\, e^{-i(\eta-\xi)z}\, dz\, e^{-e^{\xi}} e^{\xi}$$

$$= \frac{1}{2\pi}\int_{-2A}^{2A} \left(1-\frac{|z|}{2A}\right) e^{-i\eta z}\, dz$$

$$= \frac{1}{\pi}\,\frac{\sin^2 A\eta}{A\eta^2}.$$

Thus

$$(25)\qquad \int_{-\infty}^{\infty} f_n(\eta)\, e^{-e^{\xi-\eta}} e^{\xi-\eta}\, d\eta = \frac{1}{\pi}\int_{-\infty}^{\infty} h_n(\eta)\, \frac{\sin^2 A(\eta-\xi)}{A(\eta-\xi)^2}\, d\eta.$$

This gives us

$$(26)\qquad \int_{\mu_n}^{\mu_{n+1}} d\xi \int_{-\infty}^{\infty} f_m(\eta)\, e^{-e^{\xi-\eta}} e^{\xi-\eta}\, d\eta = \begin{cases} 0 & \text{if } m \neq n, \\ 1 & \text{if } m = n. \end{cases}$$

We now introduce the transformation,

$$(27)\qquad e^{\mu_n} = \nu_n; \qquad \varphi_n(x) = \frac{1}{x} f_n(-\log x).$$

Formula (26) now takes the form

(28) $$\int_{\nu_n}^{\nu_{n+1}} dx \int_0^\infty y\, \varphi_m(y)\, e^{-xy}\, dy = \begin{cases} 0 & \text{if } m \neq n, \\ 1 & \text{if } m = n. \end{cases}$$

Thus if

(29) $$f(x) = \sum_{-\infty}^{\infty} x\, s_n \int_{\nu_n}^{\nu_{n-1}} e^{-xy}\, dy$$

converges absolutely, we have

(30) $$\int_0^\infty \varphi_m(y)\, f(y)\, dy = s_n.$$

Furthermore,

(31) $$\sum_{-\infty}^{\infty} x\, s_n \int_{\nu_n}^{\nu_{n+1}} e^{-xy}\, dy = \sum_{-\infty}^{\infty} (s_n - s_{n-1})\, e^{-\nu_n x}.$$

We have thus eslatlished (6) if we replace our λ_n by ν_n. The only difference between the λ_n and the ν_n it that ν_{n+1}/ν_n has an upper limit as well as a lower limit. We may introduce such on upper limit in the case of the set $\{\lambda_n\}$ by introducing dummy λ_m's, with vanishing a_n's.

Formula (8) results from (23) and (22), and (7) results from (25) if we integrate over $(-\infty, \infty)$ with respect to ξ, by virtue of (21). Formula (9) may be derived from (19) and (20) as follows:

(32) $$\int_0^\infty |\varphi_n(x)|\, dx = \int_{-\infty}^{-\log \varepsilon} |f_n(\eta)|\, d\eta \leqq \int_{-\infty}^{-\log \varepsilon} d\eta \int_{-\infty}^{\infty} |h_n(\xi)|\, |k(\eta - \xi)|\, d\xi$$

$$= \int_{-\infty}^{-\log \varepsilon} d\eta \int_{-\infty}^{-2\log \varepsilon} |h_n(\xi)|\, |k(\eta - \xi)|\, d\xi + \int_{-\infty}^{-\log \varepsilon} d\eta \int_{-2\log \varepsilon}^{\infty} |h_n(\xi)|\, |k(\eta - \xi)|\, d\xi \quad [\varepsilon < 1]$$

$$\leqq \int_{-\infty}^{\infty} |k(\eta)|\, d\eta \int_{-\infty}^{-2\log \varepsilon} |h_n(\xi)|\, d\xi + \int_{-\infty}^{\infty} |h_n(\xi)|\, d\xi \int_{-\infty}^{\log \varepsilon} |k(\eta)|\, d\eta.$$

We have thus completed the transition from Lemma II to Lemma I.

We now come to the proof of Lemma II. We introduce the step functions:

(33) $$h_n^0(x) = 1 \text{ over } (\mu_n, \mu_{n+1}) \text{ and } 0 \text{ elsewhere.}$$

It is easy to show that we may choose A once and for all so large that

(34) $$\int_{-\infty}^{\infty} \left| h_n^{(0)}(x) - \frac{1}{\pi} \int_{\mu_n}^{\mu_{n+1}} \frac{\sin^2 A(\eta - x)}{A(\eta - x)^2} d\eta \right| dx \leqq D/2.$$

Hence if

(35) $$l(x) = \sum_{-\infty}^{\infty} a_n h_n^0(x), \qquad \sum_{-\infty}^{\infty} | a_n | < \infty,$$

we get

(36) $$\int_{-\infty}^{\infty} \left| l(x) - \frac{1}{\pi} \int_{-\infty}^{\infty} l(\eta) \frac{\sin^2 A(\eta - x)}{A(\eta - x)^2} d\eta \right| dx \leqq \tfrac{1}{2} \int_{-\infty}^{\infty} | l(x) | \, dx.$$

Now let us define by mathematical induction:

(37) $$g_n^k(x) = \frac{1}{\pi} \int_{-\infty}^{\infty} h_n^{(k)}(x) \frac{\sin^2 A(\eta - x)}{A(\eta - x)^2} d\eta;$$

(38) $$h_n^{(k)}(x) = \frac{1}{\mu_{m+1} - \mu_m} \int_{\mu_m}^{\mu_{m+1}} [h_n^{(k-1)}(\xi) - g_n^{(k-1)}(\xi)] \, d\xi \qquad (\mu_m < x < \mu_{m+1});$$

(39) $$h_n(x) = \sum_{n=0}^{\infty} h_n^k(x),$$

(40) $$g_n(x) = \sum_{n=0}^{\infty} g_n^k(x).$$

By (37),

(41) $$\int_{-\infty}^{\infty} | g_n^k(x) | \, dx \leqq \int_{-\infty}^{\infty} | h_n^{(k)}(x) | \, dx,$$

and by (38) and (36),

(42) $$\int_{-\infty}^{\infty} | h_n(x) | \, dx \leqq \int_{-\infty}^{\infty} | h_n^{(k-1)}(x) - g_n^{(k-1)}(x) | \, dx \leqq \tfrac{1}{2} \int_{-\infty}^{\infty} | h_n^{(k-1)}(x) | \, dx$$
$$\leqq 2^{-k} \int_{-\infty}^{\infty} | h_n^0(x) | \, dx \leqq 2^{1-k} D.$$

Thus the series (39) converges in the mean L_1 to a function of L_1, and the same may readily be proved of (40). Moreover, by (38),

$$\int_{\mu_m}^{\mu_{m+1}} h_n^{(k)}(x)\,dx = \int_{\mu_m}^{\mu_{m+1}} h_n^{(k-1)}(x)\,dx - \int_{\mu_m}^{\mu_{m+1}} g_n^{(k-1)}(x)\,dx; \tag{43}$$

and if we sum up these equalities, we get:

$$\int_{\mu_m}^{\mu_{m+1}} [h_n(x) - h_n^0(x)]\,dx = \int_{\mu_m}^{\mu_{m+1}} [h_n(x) - g_n(x)]\,dx. \tag{44}$$

From this it follows that:

$$\int_{\mu_m}^{\mu_{m+1}} g_n(x)\,dx = \begin{cases} 1 & \text{if} \quad m = n \\ 0 & \text{if} \quad m \neq n. \end{cases} \tag{45}$$

This estalishes (21). By (42), together with (18),

$$\int_{-\infty}^{\infty} |h_n(x)|\,dx \leqq 4D. \tag{46}$$

Thus (19) holds. As to (20), let us notice that:

$$\int_{-\infty}^{\infty} \frac{d\eta}{(1 + (x - \eta)^2)(1 + (\eta - \mu_n)^2)} \leqq \frac{2\pi}{1 + (x - \mu_n)^2} \tag{47}$$

$$\frac{1}{\mu_{m+1} - \mu_m} \int_{\mu_m}^{\mu_{m+1}} \frac{dx}{1 - (x - a)^2} = \frac{1}{1 + (\sigma_m - a)^2} \quad (\mu_m \leqq \sigma_m \leqq \mu_{m+1}) \tag{48}$$

$$\leqq \frac{4}{1 + (x - a)^2} \quad \text{for any } x \text{ in } (\mu_m, \mu_{m+1}).$$

Thus since

$$|h_n^0(x)| < \frac{\text{const.}}{1 + (x - \mu_n)^2}; \tag{49}$$

We get

$$|h_n^{(k)}(x)| < \frac{C_k}{1 + (x - \mu_n)^2}. \tag{50}$$

Hence

$$\left| h_n(x) - \sum_k^{\infty} h_n^{(k)}(x) \right| < \frac{\text{const.}}{1 + (x - \mu_n)^2} \tag{51}$$

From this follows

$$\int_{-\infty}^{-A} | h_n(x) | \, dx \leqq \sum_k^{\infty} \int_{-\infty}^{\infty} | h_n^{(k)}(x) | \, dx + O\left(\frac{1}{A + \mu_n} \right) \tag{52}$$

$$\leqq 2^{2-k} D + O\left(\frac{1}{A + \mu_n} \right).$$

Thus for every k,

$$\overline{\lim_{n \to \infty}} \int_{-\infty}^{-A} | h_n(x) | \, dx \leqq 2^{2-k} D, \tag{53}$$

which yields

$$\lim_{n \to \infty} \int_{-\infty}^{-A} | h_n(x) | \, dx = 0. \tag{54}$$

Thus gives us (20), and completes the proof of Lemma II.

Tsing Hua University
April, 1936

(Received 2, May, 1936)

Reprinted from
The Science Reports of National Tsing Hua University
Series A, Vol. III, Nos. 4 & 5, July, 1936

A THEOREM OF CARLEMAN

BY NORBERT WIENER*

Departments of Mathematics and Electrical Engineering

Carleman[1] has recently stated the following theorem:

THEOREM I. *Let*

(1) $a_n > 0 \quad (n=1,2,\ldots);\ 0 < \lambda_1 < \lambda_2 < \ldots;$

and let

(2) $$f(x) = \sum_{\nu=1}^{\infty} \frac{a_\nu}{\lambda_\nu + x}.$$

Let

(3) $$xf(x) = A \log x + B + 0(e^{-\alpha x^{1/2}}) \qquad (\alpha > 0).$$

Then if $0 < \delta < 1/2$,

(4) $$\sum_{x < \lambda_n < x + x^{1/2+\delta}} a_n \sim A x^{\delta - 1/2} \quad as\ x \to \infty.$$

We shall here give a proof of this theorem which is based on considerations apparently applicable to a large class of similar theorems. We have

(5) $$-xf(x) + x\int_x^\infty f(y)\frac{dy}{y} = A + O(e^{-\alpha x^{1/2}}).$$

* Professor of Mathematics, Massachusetts Institute of Technology, U.S.A., on leave as Research Professor at Tsing Hua University, Peiping, China, 1935-1936.

1. Proceedings of the Scandinavian Mathematical Congress, 1934.

That is,

$$(6)\quad x\sum_{v=1}^{\infty} a_n\left(\frac{1}{\lambda_n}\log(1+\lambda_n/x)-\frac{1}{\lambda_n+x}\right)=A+O(e^{-\alpha x^{1/2}}).$$

If we put

$$(7)\qquad \varphi(u)=\sum_{\lambda_n\leq u} a_n\,\lambda_n,$$

we may write this

$$(8)\qquad \frac{1}{x}\int_0^\infty\left(\frac{x^2}{u^2}\log\left(1+\frac{u}{x}\right)-\frac{1}{\left(\frac{u}{x}\frac{u}{x}+1\right)}\right)d\varphi(u)=A+O(e^{-\alpha x^{1/2}}).$$

We have

$$(9)\qquad \int_0^\infty u^{iz}\left(\frac{1}{u^2}\log(1+u)-\frac{1}{u(u+1)}\right)du$$

$$=\frac{-\pi z}{(z+i)\sin\pi iz}.$$

Hence

$$(10)\qquad \lim_{z\to 0}\frac{-\pi z}{(z+i)\sin\pi iz}=1;\quad \frac{-\pi x}{(x+i)\sin\pi ix}\neq 0\ (x\text{ real}).$$

We may thus apply a theorem of the author[2]; from which we obtain

$$(11)\qquad \varphi(u)\sim Au.$$

Indeed, for this we only need the weaker assumption

$$(12)\qquad xf(x)=A\log x+B+0\,(1)$$

in place of (3).

We now introduce the auxiliary function

$$(13)\qquad F(x)=\prod_1^\infty\left(1-\frac{z^2}{n^{2/(1+\varepsilon)}}\right)\qquad (0<\varepsilon<1).$$

[2] Cf. The Fourier Integral and Certain of its Applications, Cambridge, 1933, chapter on Tauberian Theorems.

It is easy to establish that if y is real.

$$(14)\ y^{-1-\varepsilon}\log F(iy) = y^{-1-\varepsilon}\sum_{1}^{\infty}\log\left(1+\frac{y^2}{n^{2/(1+\varepsilon)}}\right)$$

$$= y^{-1-\epsilon}\int_{0}^{\infty}\log\left(1+\frac{y^2}{n^2}\right)d[u^{1+\varepsilon}]$$

$$= \int_{0}^{\infty}\frac{2y^2}{u^3}\frac{[u^{1+\epsilon}]}{y^{1+\epsilon}(1+\frac{y^2}{u^2})}du$$

$$< \int_{0}^{\infty}\frac{2u_{\varepsilon}}{1+u^2}du = const.$$

Similarly, if x is real,

$$(15)\ x^{-1-\varepsilon}\log|F(x)| = x^{-1-\varepsilon}\sum_{1}^{\infty}\log\left|1-\frac{x^2}{n^{2(1+\varepsilon)}}\right|$$

$$= x^{-1-\varepsilon}\int_{0}^{\infty}\log\left|1-\frac{x^2}{u^2}\right|d\,[u^{1+\varepsilon}]$$

$$\sim x^{-1-\varepsilon}\int_{0}^{\infty}\log\left|1-\frac{x^2}{u^2}\right|(1+\varepsilon)u^{\varepsilon}\,du$$

$$= (1+\epsilon)\int_{0}^{\infty}\log\left|1-\frac{1}{u^2}\right|u^{\epsilon}\,du$$

$$= (1+\epsilon)\int_{0}^{\infty}\log\left|1-\frac{1}{u^2}\right|u^{\epsilon}du + (1+\epsilon)\int_{2^{-\frac{1}{2}}}^{\infty}\log\left|1-\frac{1}{u^2}\right|u^{\epsilon}du$$

$$< (1+\varepsilon)\,2^{-\epsilon/2}\int_{0}^{\infty}\log\left|1-\frac{1}{u^2}\right|du$$

$$= (1+\epsilon)2^{-\epsilon/2}\left\{\lim_{\eta\to 0}\left\{\int_0^{1-\eta}\log(1-u^2)\,du\right.\right.$$

$$\left.+\int_{1+\eta}^{\infty}\log\left(1-\frac{1}{u^2}\right)du-2\int_0^1\log u\,du\right\}=0.$$

Thus $\log|F(x)|$ is asymptotically a negative constant multiplied by $x^{1+\epsilon}$. Combining this with (14), we see that there exist a constant $K>0$ such that

$$(16)\qquad \log\left|F(z)\,\epsilon^{-K(-iz)^{1+\epsilon}}\right|$$

is bounded on the real axis and the positive imaginary axis. Thus by the Phragmén-Lindelof theorem, it is bounded in the upper half-plane. That is, there exist a $C_1>0$ and a C_2 such if $y/|x|<C_2$;

$$(17)\qquad \log|F(x+iy)|\leq K+|x|^{1+\epsilon}R(-i+C_2)^{1+\epsilon}$$

$$\leq K-C_1|x|^{1-\epsilon}.$$

Similarly, if $y/|x|>C_2$, there exists a constant C_2 such that

$$(18)\qquad \log|F(x+iy)|\leq K+C_3|y|^{1-\epsilon}.$$

Combining (17) and (18), in all cases a constant C_4 and positive constants C_5 and C_6 exist, for which

$$(19)\qquad \log|F(x+iy)|\leq C_4-C_5|x|^{1+\epsilon}+C_6|y|^{1+\epsilon}$$

in both half-planes. That is,

$$(20)\qquad |F(x+iy)|^2< const.\ e^{-2C_5|x|^{1+\epsilon}+2C_6|y|^{1+\epsilon}}.$$

Let

$$(21)\qquad \{F(x+iy)\}^2=\int_{-\infty}^{\infty}G(u)e^{iux-uy}du.$$

Then

$$(22)\quad \left|\int_{-\infty}^{\infty} G(\tfrac{u}{N})\,\frac{\sinh \pi u(u+i)}{\pi i u}\, e^{iux}\,du\right|$$

$$=\left|\frac{N}{2\pi}\int_{-\infty}^{\infty}(G(u)(e^{(iux+u\pi)N}+e^{i(ux-u\pi)N})(\frac{1}{i}+\frac{1}{\sqrt{u}})du\right|$$

$$=\left|\frac{N}{2\pi i}\{(F(N(x-\pi i)))^2+(F(N(x+\pi i)))^2\}\right.$$

$$\left.+\frac{N}{2}\int_{-\pi}^{\pi}\{F(N(x-yi))\}^2\,dy\right|$$

$$< N \text{ const } e^{-2C_5\{N|x|\}^{1+\epsilon}+2C_6\{N\pi\}^{1+\epsilon}}$$

Let us notice that (8) and (9) lead to

$$(23)\quad \int_{-\infty}^{\infty}\{A\,d\eta-e^{-\eta}d\varphi(e^{\eta})\}\int_{-\infty}^{\infty}e^{iz(\xi-\eta)}\left(\frac{\sin h_\pi z(z+i)}{\pi i z}\right)^{-1}dz$$

$$=O(e^{-\alpha e^{\xi/2}}).$$

We also have

$$(24)\quad \frac{1}{2\pi}\int_{-\infty}^{\infty}dx\int_{-\infty}^{\infty}G(\tfrac{u}{N})\,\frac{\sinh \pi u(u+i)}{\pi i u}\,e^{uix}\,du$$

$$\times\int_{-\infty}^{\infty}e^{iu(y-x)}\left(\frac{\sinh \pi u(u+i)}{\pi\, iu}\right)^{-1}du$$

$$=\int_{-\infty}^{\infty}G(\tfrac{u}{N})e^{uiy}du=N\{F(Ny)\}^2$$

By an exchange of the order of an absolutely convergent double integral, (23) thus leads to

$$(25)\quad \left|\int_{-\infty}^{\infty}N\{F(N(\xi-\eta))\}^2\{A\,d\eta-e^{-\eta}d\varphi(e^{\eta})\}\right|$$

$$< \left[\int_0^{\xi(1-\epsilon_1)} + \int_{\xi(1-\epsilon)_1}^{\infty} \right] N \text{const.}\, exp\{-2C_5\{N|\xi-\eta|\}^{1+\epsilon}$$

$$+ 2C_6(N\pi)^{1+\epsilon}\} O(e^{-\alpha e^{\frac{\eta}{2}}}) du$$

$$< N \text{ const. } exp\{-2C_5\{N|\epsilon_1\xi|\}^{1+\epsilon} + 2C_6(N\pi)^{1+\epsilon}\}$$

$$+ \text{ const. } exp\ \{2C_6(N\pi)\}^{1+\epsilon} - \alpha e^{\xi(\frac{1-\epsilon_1}{2})}$$

If now N is a function of ξ not exceeding

(26) $$C_7\ exp\ (\xi \frac{1-\epsilon_1}{2(1+\epsilon)}),$$

where

(27) $$C_7 < \frac{1}{\pi} \left(\frac{\alpha}{2C_6}\right)^{(1+\epsilon)^{-1}},$$

(25) yields us

(28) $$\left| \int_{-\infty}^{\infty} N\{F(N(\xi-\eta))\}^2 \{A d\eta - e^{-\eta}\ d\varphi(e^{\eta}))\} \right.$$

$$< \text{ const. } exp(-C_8\xi^{1+\epsilon}) \quad (C_8>0).$$

If we now take N as a function of ξ alone, we have

(29) $$\frac{1}{\xi_1} \int_{\xi}^{\xi+\xi_1} d\xi \int_{-\infty}^{\infty} N\{F(N(\xi-\eta))\}^2 \{A d\eta - e^{-\eta} d\varphi(e^{\eta})\}$$

$$= O(exp(-C_9\ \xi^{1+\epsilon})) \qquad (C_9>0),$$

where

(30) $$\xi_1 = exp\{(\delta-\tfrac{1}{2})\xi\} \qquad \left(\delta > \frac{\epsilon+\epsilon_1}{2(1+\epsilon)}\right).$$

It then results directly from (29) that

$$(31)\ \int_{-\infty}^{\infty}\{F(x)\}^2dx\,\frac{1}{\xi_1}\int_{\xi}^{\xi+\xi_1}(A\,d\eta-e^{-\eta}\,d\varphi(e^{\eta}))$$

$$\leq O(exp(-C_9\xi^{1+\varepsilon}))+\frac{1}{\xi_1}\int_{-\infty}^{\infty}[\,e^{\xi-\eta}d\varphi(e^{\eta-\xi})$$

$$+e^{\xi+\xi_1-\eta}d\varphi(e^{\eta-\xi-\xi_1})+2A]exp(-2C_5|N\eta|^{1+\varepsilon})(1)O\,d\eta$$

$$=O(1)+\frac{O(1)}{N\xi_1}\int_{-\infty}^{\infty}[\,e^{\xi-\eta}d\varphi(e^{\eta-\xi})$$

$$+\,e^{\xi+\xi_1-\eta}\,d\varphi(e^{\eta-\xi-\xi_1})]N exp(-2C_5|N\eta|^{1+\varepsilon})\,d\eta\,.$$

By (8),

$$(32)\ \int_{-\infty}^{\infty}[e^{\xi-\eta}\log(1+e^{\eta-\xi})-(1+e^{\eta-\xi})^{-1}]e^{-\eta}\,d\varphi(e)^{\eta}$$

$$=O(1).$$

By (28), and (26), if ϵ_2 is small enough,

$$(33)\ N\int_{BN^{\epsilon_2}-1}^{BN^{\epsilon_2}-1}\{F(N(\xi-\eta))\}^2\{e^{-\eta}\,d\varphi(e^{\eta})=O(1).$$

Combining these, it is easy to see that

$$(34)\ \int_{-\infty}^{\infty}N\ exp\,(-2C_5(N|\xi-\eta|)^{1+\varepsilon})\,e^{-\eta}\,d\varphi(e^{\eta})=O(1).$$

Thus by (31),

$$(35)\quad \frac{1}{\xi_1}\int_{\xi}^{\xi+\xi_1}(A\,d\eta-e^{-\eta}\,d\varphi(e^{\eta}))\rightarrow 0.$$

A change of scale converts this into

$$(36)\quad \int_x^{x+x^{1/2+\delta}+O(x^{2\delta})} \frac{d\varphi(u)}{u} = \sum_{x<\lambda_n \leq x+x^{1/2+\delta}+O(x^{2\delta})} a_n$$

$$\sim A\, x^{\delta - 1/2},$$

which is trivially equivalent to Carleman's theorem.

Commentary on the Gap-Theoretic Papers [27a], [34b], [35a] and [36a–e]†

P. J. Koosis

It is convenient to divide the following set of eight papers on gap theorems into five groups. In the first two groups, various properties of power series $\Sigma_0^\infty a_n z^{\lambda_n}$ or trigonometric series $\Sigma_{-\infty}^{\infty} a_n e^{i\lambda_n\theta}$ which depend on some separation condition for the exponents λ_n (series with such a separation condition are called *gap series*) are investigated, using two different methods. One is essentially a real variable method; the other involves complex variables. There then follows a longer expository article on the work covered in the first two groups of papers. It is placed in a group by itself. The remaining two groups consist of papers of different character.

There follow commentaries for the papers in each group. Reference to a formula in one of the papers will be by the *arabic* number of that formula, as found in that particular paper. Formulas occurring in the commentary itself will be designated by *lower case roman* numerals.

Group I: [27e], [34b]

Paper [27e] deals with the comparison of two systems of functions $\{\cos \lambda_k x\}$, $\{\cos \mu_k x\}$, the object being to show that if the first system is complete in $L_2(0, \pi)$ and has the Bessel property (1), then the second system is also complete in $L_2(0, \pi)$ provided that the μ_k do not differ too much from the λ_k. Wiener's idea here is to use $\int_0^\infty |f(x)|^2 \frac{\sin^2 x/2}{x^2}\, dx$ as an estimate for $\int_0^\pi |f(x)|^2\, dx$ and to evaluate the first quantity with the help of Fourier integrals.

What really lies behind this paper is the problem of deducing completeness, on a given interval, of a system of exponentials $e^{i\mu_k x}$ from that of some related system $\{e^{i\lambda_k x}\}$. After publishing [27e] Paley and Wiener found a simpler real-variable comparison method which works in case the second system is orthogonal on the interval in question (e.g., $\lambda_k = k$ and an interval of length 2π). Their work is found in Chapter VII of Paley-Wiener [34d]. The method was improved by Duffin and Eachus,[7] whose work is reproduced on p. 208 in the book of Riesz-Nagy.[31] Redheffer found another real-variable comparison method, based on successive integration by parts. His method works for general (even complex) exponents λ_k and μ_k, and is explained in §4 of Alexander-Redheffer's paper.[1]

† These reviews were written in 1970 and therefore do not contain references to work published after that year.

Sharper results, including some of the best possible for the special case where $\lambda_k = k$ and the interval has length 2π are given in Chapters I and IV of Levinson's book.[20] Levinson uses complex variable methods, in which entire functions of exponential type play a prominent role. The above cited paper of Alexander and Redheffer[1] gives a complex-variable method for comparing the completeness properties of two general systems of exponentials on a given interval. Further discussion of the use of complex-variable methods is postponed to the commentaries on Group II and Group III. [34b] takes up the general question of estimating

$$\int_I |\sum_k a_k e^{i\lambda_k x}|^2 \, dx$$

in terms of $\Sigma_k |a_k|^2$ for a given interval I under the assumption that $|\lambda_{k+1} - \lambda_k| \geqslant$ some given L. The kind of result obtained says that the two quantities are equivalent if I is long enough, and the problem is to find, in terms of L, a lower bound for the lengths of the I for which

$$\text{(i)} \qquad A_I \sum_k |a_k|^2 \leqslant \int_I |\sum_k a_k e^{i\lambda_k x}|^2 \, dx \leqslant B_I \sum_k |a_k|^2.$$

With the help of this double inequality it is possible to show, for functions of the form $\Sigma_k a_k e^{i\lambda_k x}$, that many properties hold on the whole real line as soon as they are known to hold on the interval I.

Wiener's idea here, an outgrowth of the one in [27e], is to estimate $\int_{-A}^{A} |P(x)|^2 \, dx$ from above *and* below by working out $\int_{-\infty}^{\infty} |P(x)|^2 \frac{\sin^2 Lx/2}{x^2} dx$. In this way, he shows that (i) holds whenever length $I \geqslant 16\pi/L$. Actually, in the argument on the top of the third page of [34b], one may take $C > 0$ to be arbitrarily small, and thus prove that length $I > 8\pi/L$ is sufficient for (i). The formula immediately following (2) is derived therefrom by putting $y = 2\pi/L, -2\pi/L, 4\pi/L, -4\pi/L$, etc., and counting. A better result is obtained if we simply observe that $|\lambda_{k+1} - \lambda_k| \geqslant L$ implies $|\lambda_{k+1} - \lambda_k| \geqslant \delta$ for $0 < \delta \leqslant L$, so that, for $A = \pi/\delta \geqslant \pi/L$, we can substitute δ for L in (2), and get

$$\int_{-A}^{A} |P(x)|^2 \, dx \leqslant \frac{\pi^2 A}{2} \sum_k |a_k|^2.$$

Using this in the computation following equation (2), we see that (i) holds as soon as length $I > 4\pi/L$.

The method used in [34b] was improved by Ingham,[10] who replaced $\frac{\sin^2 Lx/2}{x^2}$ by a function vanishing identically outside $[-A, A]$. This led to the (obviously) best possible lower bound $2\pi/L$ for length I. Ingham's argument is given in Zygmund's book,[34] Chapter V, §9.

If $\{\lambda_k\}$ is a real sequence wfth $|\lambda_{k+1} - \lambda_k| \geqslant$ some $\delta > 0$ (such sequences are now called *regular*), let Λ denote the infimum of the lengths of the I for which (i) holds. There is a general problem to express Λ in terms of some geometric property of the sequence $\{\lambda_k\}$. The solution is in terms of a certain *density* Δ for the sequence, discovered by Kahane. (Redfeffer[28] already had this density, but it is there used for a different problem.) If we let $n(t)$ sgn t be the number of λ_k between 0 and t,

$$\Delta = \lim_{r \to \infty} \limsup_{t \to \pm\infty} \frac{n(t + r) - n(t)}{r}.$$

Kahane showed that $\Lambda = 2\pi\Delta$. His first proofs of this result (starting in 1953–54[11] and given more completely in 1957[12]) leaned heavily on the theory of entire functions of exponential type. Later in 1962[13] he was able to show how to base the proof mainly on Ingham's simple real-variable method which is directly related, as we have seen, to the idea in [34b]. The treatment in Kahane's paper[13] is very exhaustive, covering systems of exponentials in several variables. Landau[18] must also be cited in this connection.

The idea initiated in these two papers of Wiener (more particularly, in the second one) is very simple and natural. It has, nevertheless, led finally to rather complete and definitive results on gap series.

Group II: [35a], [36b, c]

In each of these papers, essential use is made of entire functions of exponential type, constructed expressly for the problem at hand as canonical products

$$\text{(ii)} \qquad \prod_n \left(1 - \frac{z^2}{\lambda_n^2}\right).$$

These canonical products are employed together with various instruments of complex-variable theory, for instance, Lagrange's interpolation formula or the Paley-Wiener Theorem, and an essential part of the work consists of making various estimates on them in order that these other tools may be used effectively.

[35a] is concerned with the way the density of the sequence $\{\lambda_n\}$ is

related to the distribution of singularities of

$$\text{(iii)} \qquad \sum_n a_n e^{-\lambda_n s}$$

on that series' line of convergence. As Wiener himself remarks at the beginning of this paper, the idea of using the product (ii) to study the properties of (iii) goes back to Pólya.[27] It is not completely clear to what kind of Dirichlet series Wiener is applying his reasoning. At one point (sentence following (18)), he seems to be restricting himself to Taylor series. A simpler version of the work in this paper, valid for series (iii) with separated exponents (i.e., with $|\lambda_j - \lambda_l| \geqslant$ some $\delta > 0$ whenever $j \neq l$) is contained in §12.7 of Boas's book.[6]

Actually, a much better investigation of general Dirichlet series (iii) whose exponents λ_n have finite density was already in print for some two years when [35a] appeared. We are referring to the book by V. Bernstein.[3] Bernstein's investigation runs along the same lines as that in the present paper, but goes much deeper because he applies Lagrange's interpolation formula in its expression with a contour integral, according to the method of residues. By proper choices of contour he is able to exploit the minimum modulus theorems for entire functions of the form (ii), and thus study the cases where the λ_n in (iii) get very close together, take into account the effect of grouping terms on the convergence of (iii), and so forth.

There is more on Dirichlet series in Levinson's book, [20] in the papers of L. Schwartz[33] (who made extensive use of Bernstein's method described above), and Bejar-Alamo,[2] who studies series (iii) with complex λ_n. A particularly simple and elegant presentation is found in Chapter II of Kahane's thesis.[11] Further references will be given with the commentary on Group III.

In reading [35a], the reader should replace the phrase "by (10)", occuring after equation (13), with the words "by (3) and (10)".

[36b] contains a simplified proof of a theorem in Mandelbrojt.[23] The idea here is to get functions enjoying certain desired properties by first constructing their Fourier transforms as products of the form (ii), and then applying the Paley-Wiener Theorem. The paper has some misprints, and a clearer and more complete version is given in [36d] in group III. The statement following the last formula in this paper is incorrect, because $f(t)\cos n_p t$ is not necessarily of the form

$$\sum_p (a_p \cos n_p t + b_p \sin n_p t)$$

when $f(t)$ is. The only way out is to repeat, for each p, the argument given by Wiener for $p = 0$. This necessitates a modification, depending on p, in the definition of the products $\Phi_\alpha(z)$, and a small amount of additional calculation.

A simple explanation of the method used in [36b] is found on pp. 60–61 of Kahane,[11] where further applications are given.

[36c] contains an example to show that one of the results in [36b] is the best possible. Again, an entire function satisfying certain estimates is constructed as a product of form (ii), and the Paley-Wiener theorem is then applied. Pp. 65–69 of Kahane[11] (see especially p. 68) contain further interesting developments of the method used in [36c].

Group III: [36d]

This is an expository article about the material in [34b], [35a], [36b, c]. Wiener has substantially clarified the ideas of those papers, and the present one is much easier to read.

The explanation of what goes on in [35a] is fairly good in [36d]. Wiener does not reproduce here all of the estimations carried out in [35a]. He is still not sure about the precise class of Dirichlet series to which his argument applies, and the restriction following formula (27) to series (iii) having $\Sigma_n |a_n|^2 < \infty$ is unnecessary. If the distance between successive λ_n is bounded away from zero, $\Sigma_n a_n e^{-\lambda_n s}$ is actually bounded in any half-plane properly contained in its half-plane of convergence. As noted above, a good explanation of all these matters was already available in V. Bernstein's book.[3]

[36b], and [36c] are also explained, and the estimates used in those papers are here worked out in considerable detail. This is especially useful in the case of [36b], whose misprints are cleaned up in [36d].

An important part of the work of Wiener thus far discussed is really about various properties associated with systems of exponentials $e^{i\lambda_n z}$. We saw that the papers of Group I had in them a simple real-variable idea which has led finally to the establishment of definitive results about quasi-analyticity properties of series $\Sigma_n a_n e^{i\lambda_n x}$.

At bottom of the papers in Group II lies the notion, not really original with Wiener, of getting at various properties of the system $\{e^{i\lambda_n z}\}$ through investigation of the canonical product $\prod_n \left(1 - \frac{z^2}{\lambda_n^2}\right)$. This idea has been extremely fruitful in the further development of the subject and has, along with that development, led to considerable extension of our knowledge about entire functions of exponential type. It is clear that Wiener himself saw the importance of such an approach, for a good part of Paley-Wiener [34d] revolves around it. Many of the later developments find their origin in that book; the immediate outgrowth was Levinson's book,[20] which was for many years the definitive one on such questions. Earlier work is contained in Mandelbrojt[23] and in the book of V. Bernstein[3] already cited. More recent material is found in L. Schwartz,[33] Kahane,[11] Levin,[19] Mandelbrojt,[24] Malliavin,[21] and Malliavin and Rubel.[22] Mandelbrojt[25] should also be cited in connection with Dirichlet series.

[27e] was involved with the important question of deciding whether or not a given system of exponentials $e^{i\lambda_n x}$ is complete on a real interval of given length. The subject was taken up again in Chapters V and VI of Paley-Wiener [34d], where it is treated by both real and complex methods. It turns out that the important results in this connection have all been obtained by a complex method, based partly on the use of $\prod_n \left(1 - \frac{z^2}{\lambda_n^2}\right)$, and partly on that of Jensen's formula or some variant thereof.

Let $0 < \lambda_1 < \lambda_2 < \ldots$, and let $n(t)$ denote the number of λ_k between 0 and t. Chapter V of Paley-Wiener [34d] has the result that $\{e^{i\lambda_k x}\}$ is complete on any real interval of length $< 2\pi\dot{D}$, where

$$\dot{D} = \limsup_{t \to \infty} \frac{n(t)}{t}.$$

Further and more refined results of this kind, some best possible, are contained in Levinson.[20] L. Schwartz[33] observed that there is, associated with each sequence $\{\lambda_k\}$, a number R, $0 \leqslant R \leqslant \infty$, such that the $e^{i\lambda_k x}$ are simultaneously complete in all the spaces $L_p(-A, A)$, $1 \leqslant p < \infty$, $\mathscr{C}(-A, A)$, whenever $A < R$, and not complete in any of them if $A > R$. He called R the radius of completeness of the sequence $\{\lambda_k\}$, and saw that the main problem in the study of completeness was to express R in terms of some geometric property associated with sequences $\{\lambda_k\}$. It was already clear from the work in Paley-Wiener [34d] and Levinson[20] that R is determined by some *density* for the sequence $\{\lambda_k\}$. The problem was to find the right density.

The above mentioned result of Paley and Wiener shows that if $\frac{n(t)}{t}$ tends to a limit D as $t \to \infty$, then $R \geqslant \pi D$. For a long time it was thought that R must *equal* πD in that case. This was disproved by Kahane;[14] later, another counterexample was given by Koosis,[17] and his procedure was greatly extended by Redheffer.[30] (Most of the work in this last paper was actually done 6 or 7 years before the paper was published.) In the meantime, various upper bounds on R were being obtained (Redheffer[28–30] and Koosis[16]).

The correct density for expressing R was discovered by Beurling and Malliavin some time in 1960 or 1961. Their work became known, partly through the circulation of mimeographed material, soon after it was completed, but its full publication took some time (Beurling and Malliavin).[4,5] It is the second of their papers which contains the actual solution of the completeness problem. That paper is difficult to read, and it is helpful to study the expository article of Kahane[15] before tackling it. (The proof of the lemma in §5 of Kahane[15] is marred by an error, but that difficulty can be straightened out. Malliavin, in an early mimeographed report, had avoided that trouble by making the preliminary change of variable $x = \log \xi$ before doing the work covered in §5 of Kahane.[15])

Here is the result of Beurling and Malliavin. Take the graph of $n(t)$ vs. t, with its *vertical gaps*, coinciding with the jumps of $n(t)$, *filled in*, and, for a given $D > 0$, consider the abscissae t having the property that $(t + \tau, n(t) + D\tau)$ is *never on that graph* for $\tau > 0$. These t fill a certain open set on the horizontal axis. The *complement* of that open set consists of a (possibly finite) sequence of disjoint closed intervals $I_p(D)$, one of which may have infinite length. All the λ_k lie in the union of the $I_p(D)$. (The reader should make a diagram at this point. If one imagines, in the plane of the paper, parallel rays of light with slope D shining down on the graph from the upper right-hand corner, the $I_p(D)$ will lie under those portions of the graph that are in shadow.) Let $\mathscr{D}$ be the infimum of the set of D for which

$$\sum_p \left[\int_{I_p(D)} \frac{dx}{x} \right]^2 < \infty.$$

The theorem of Beurling and Malliavin says that R, the radius of completeness of $\{\lambda_k\}$, is equal to $\pi\mathscr{D}$. Even the system $\{e^{\pm i\lambda_k x}; k = 1, 2, 3, \ldots\}$ will be incomplete on any interval of length $> 2\pi\mathscr{D}$, whereas the functions $e^{i\lambda_k x}$, $k = 1, 2, 3, \ldots$ alone are complete on any interval of length $< 2\pi\mathscr{D}$.

Here, the inequality $R \geqslant \pi\mathscr{D}$ is fairly easy, and Redheffer had almost found it independently in 1961 (see Redheffer[30]). What is difficult is the establishment of the reverse inequality, and it is there that a really deep study of the canonical product $\prod_k \left(1 - \frac{z^2}{\lambda_k^2}\right)$ is required. The essentially elementary estimation methods employed for this purpose by Paley-Wiener [34d], Levinson,[20] Redheffer,[28-30] and Koosis[16] were simply not sufficient, and the decisive step was the application of potential theory to the study of such products by Beurling and Malliavin[4].

Besides the two methods used by Wiener for studying gap series, there is a third one, based on Jensen's formula or one of its variants. This is also the method used to get lower bounds for R in the completeness problem just discussed. Insofar as the distribution of singularities is concerned, the method is more useful for the study of Taylor series than of general Dirichlet series. One example of such an application is given in §12.6 of Boas's book.[6] Here is another, especially simple and elegant, which is due to Kahane. He told me about it 20 years ago, but it has never before been published.

Suppose $f(z) = \Sigma_0^\infty a_n z^n$, and that the radius of convergence of the series is 1. Then, for some sequence of n tending to ∞, we must have $\frac{\log |a_n|}{n} \to 0$. *Let S be any such sequence. We are going to show that if the power series has many gaps near the terms corresponding to the $n \in S$, then $f(z)$ must have many singularities on the circle of convergence.* More precisely, let Λ be the set of n for which $a_n = 0$, and let $\Lambda(t)$ denote the number of points of Λ between 0 and t. Let $0 < \alpha \leqslant \pi$, and suppose there is some $\beta > 0$ such that

$$\text{(iv)} \qquad \limsup_{\delta \to 0} \liminf_{\substack{t \to \infty \\ |t - S| \leqslant \beta t}} \frac{\Lambda(t + \delta t) - \Lambda(t)}{\delta t} > \frac{\pi - \alpha}{\pi}.$$

(By $|t - S|$ is meant the distance from t to the nearest point of S.)

Theorem. *If* (iv) *holds, any arc of* $|z| = 1$ *having length* 2α *has on it at least one singularity of* $f(z)$.

Proof. Take any fixed $\gamma > 0$ small in comparison with β, and choose a $\delta > 0$ much smaller than γ such that

$$\text{(v)} \qquad \frac{\Lambda(t + \delta t) - \Lambda(t)}{\delta t} \geqslant \text{ say } D > \frac{\pi - \alpha}{\pi},$$

whenever $|t - S| \leqslant \beta t$ and t is sufficiently large. Taking a large $n \in S$ and putting $t = (1 + \delta)^k n$ with $k = 0, 1, \ldots \left[\frac{\log(1 + \beta)}{\log(1 + \delta)}\right]$ in (v), it is easy to verify, by addition, that $\Lambda(n + r) - \Lambda(n) \geqslant D(1 - \delta)r - D\delta n$ for $0 \leqslant r \leqslant \beta n$. In particular,

$$\frac{1}{\gamma n} \int_0^{\gamma n} \frac{\Lambda(n + r) - \Lambda(n)}{r} dr \geqslant \frac{1}{\gamma n} \int_{\delta n}^{\gamma n} \frac{\Lambda(n + r) - \Lambda(n)}{r} dr$$
$$\geqslant D(1 - \delta)\left(1 - \frac{\delta}{\gamma}\right) - D\frac{\delta}{\gamma} \log \frac{\gamma}{\delta}$$

if $n \in S$ is large; and if $\delta > 0$ is very small in comparison with γ, we get

(vi) $$\frac{1}{\gamma n}\int_0^{\gamma n} \frac{\Lambda(n + r) - \Lambda(n)}{r}\,dr \geqslant \quad \text{say} \quad D' > \frac{\pi - \alpha}{\pi}$$

for large $n \in S$. In the same way, we see that

(vii) $$\frac{1}{\gamma n}\int_0^{\gamma n} \frac{\Lambda(n) - \Lambda(n - r)}{r}\,dr \geqslant D'$$

for large $n \in S$. From (vi) and (vii), we get

(viii) $$\frac{1}{\gamma n}\int_0^{\gamma n} \frac{\Lambda(n + r) - \Lambda(n - r)}{r}\,dr \geqslant 2D' > 2\frac{\pi - \alpha}{\pi},$$

valid for sufficiently large $n \in S$.

Suppose $f(z)$ can be continued analytically across some closed arc of length 2α on $|z| = 1$. There is no loss of generality in assuming that arc to be symmetric about the point -1, and in that case, Cauchy's formula for the coefficients of $f(z) = \Sigma_0^\infty a_n z^n$ takes the form

(ix) $$a_n = \frac{1}{2\pi i}\int_{\Gamma_r} f(e^\zeta)e^{-n\zeta}\,d\zeta$$

after making the change of variable $z = e^\zeta$. Here, r is any positive number <1, and Γ_r is the path, oriented upwards, which is composed of the three vertical segments

$$\begin{aligned} &\mathscr{R}\zeta = \log r, |\mathscr{I}\zeta| \leqslant \pi - \alpha, \\ &\mathscr{R}\zeta = p, \; -\pi \leqslant \mathscr{I}\zeta \leqslant -(\pi - \alpha), \\ &\mathscr{R}\zeta = p, \pi - \alpha \leqslant \mathscr{I}\zeta \leqslant \pi, \end{aligned}$$

and the two horizontal ones

$$\begin{aligned} &\log r \leqslant \mathscr{R}\zeta \leqslant p, \mathscr{I}\zeta = -(\pi - \alpha), \\ &\log r \leqslant \mathscr{R}\zeta \leqslant p, \mathscr{I}\zeta = \pi - \alpha. \end{aligned}$$

The number p is some fixed quantity > 0, depending on how far beyond the arc in question $f(z)$ can be analytically continued.

Let

$$A(w) = \frac{1}{2\pi i} \int_{\Gamma_r} f(e^{\zeta}) e^{-w\zeta} \, d\zeta. \tag{x}$$

Then $A(w)$ is an entire function of exponential type, and $A(n) = a_n$ by (ix). It is easy to see by drawing a figure that $\mathscr{R} w\zeta \geqslant -(\pi - \alpha)|\mathscr{I} w| - |\mathscr{R} w| \log \frac{1}{r}$ for every ζ on Γ_r, provided that $|\arg w| < \arctan \frac{1}{\alpha}\left(p + \log \frac{1}{r}\right)$. Using this in (x) with r chosen very close to 1, we get, for $|\arg w| \leqslant \arctan (p/\alpha)$,

$$|A(w)| \leqslant M_{\varepsilon} \exp\{(\pi - \alpha)|\mathscr{I} w| + \varepsilon |w|\}, \tag{xi}$$

where $\varepsilon > 0$ can be taken as small as we please.

Now S was chosen to be a sequence such that

$$\frac{\log |A(n)|}{n} = \frac{\log |a_n|}{n} \to 0$$

for $n \to \infty$ in S, and Λ was a sequence of n satisfying $A(n) = 0$. Take a fixed $\gamma > 0$, small in comparison with p/α, and to the function $A(w)$ let us apply Jensen's formula for a circle of radius γn about some large $n \in S$. We will get

$$\int_0^{\gamma n} \frac{\Lambda(n + r) - \Lambda(n - r)}{r} \, dr \tag{xii}$$

$$\leqslant \frac{1}{2\pi} \int_{-\pi}^{\pi} \log |A(n + \gamma n e^{i\theta})| \, d\theta - \log |A(n)|.$$

According to the choice of γ, we may use (xi) to estimate the right side of (xii) and we find after an easy calculation that

$$\frac{1}{\gamma n} \int_0^{\gamma n} \frac{\Lambda(n + r) - \Lambda(n - r)}{r} \, dr$$

$$\leqslant 2 \frac{\pi - \alpha}{\pi} + \frac{\varepsilon(1 + \gamma)}{\gamma} - \frac{\log |A(n)|}{\gamma n} + \frac{\log M_{\varepsilon}}{\gamma n}.$$

Since $\frac{\log |A(n)|}{n} \to 0$ for $n \to \infty$ in S, this contradicts (viii) if we use (xi) with a small enough $\varepsilon > 0$. The proof is complete.

The theorem just established includes Pólya's (and hence Fabry's) gap theorem, and shows much more, namely, that it is the sparsity of terms in $\Sigma_0^\infty a_n z^n$ *near the ones with large* $|a_n|$ that causes the function represented by that series to have a certain distribution of singularities on the circle of convergence. Kahane's idea here was to apply Jensen's theorem *to a circle with moving centre*. This simple trick enabled Koosis[17] to obtain a decisive increase in the lower bound for R, the radius of completeness of a sequence (Thus, Kahane's contribution to that paper is finally acknowledged!), and a slight refinement of Koosis's procedure led to the inequality $R \geqslant \pi\mathscr{D}$, $\mathscr{D}$ being the density of Beurling and Malliavin. (As mentioned earlier, Redheffer, independently, had almost obtained that inequality in this way.) The hard part was recognizing (and proving) that $R \leqslant \pi\mathscr{D}$.

Results like the ones in [36b], discussed again in [36d], can also be arrived at by using Jensen's formula instead of constructing canonical products having prescribed properties. The method in question is due to Levin and Lifschutz, and is presented in Appendix II of Levin's book.[19] It bears somewhat the same relation to the one employed in [36b] that Kahane's argument, reproduced above, bears to Wiener's in [35a]. Chapter I of Kahane's thesis[11] contains a systematic discussion of the method. Instead of Jensen's formula, others of like character (e.g., Carleman's) may be used.

Group IV: [36e]

The heart of [36e] is in Lemma II. Here is the idea behind that lemma. The paper is about the exponentials $e^{-\lambda_k x}$, and making the change of variable $x = e^{\xi}$ in these exponentials leads to the consideration of translates of $e^{-e^{\xi}} e^{\xi}$. The function $\Gamma(iz + 1)$ then comes in as the Fourier transform of $e^{-e^{\xi}} e^{\xi}$. Instead of (22), one *would like* to write

(xiii) $$K(\eta) = \int_{-\infty}^{\infty} \frac{e^{-i\eta z}}{\Gamma(iz + 1)}\, dz.$$

Then we could use the functions $\frac{h_n^{(0)}(\xi)}{\mu_{n+1} - \mu_n}$ in place of the $h_n(\xi)$ in (23); and (26) would hold with the functions $f_n(\eta)$ thus defined. Instead of (21), one would just use the obvious orthogonality of the $h_n^{(0)}(\xi)$. ($h_n^{(0)}$, defined in (33), is merely the characteristic function of the interval $[\mu_n, \mu_{n+1}]$.)

The trouble is that $\frac{1}{\Gamma(iz+1)}$ increases exponentially for real $z \to \pm\infty$ making it impossible for us to use (xiii). Formula (22) is thus used as a substitute for (xiii), and this modification causes the Fejer kernel $\frac{\sin^2 A\xi}{\pi A\xi^2}$ to come into the picture. It is the presence of the latter that compels Wiener to construct functions h_n satisfying (21) instead of mere biorthogonality to the $h_n^{(0)}$, so that, by means of (23), he may get functions f_n which will do the job.

Wiener observed that if A is large, (21) is *almost* a requirement that the h_n be biorthogonal to the $h_n^{(0)}$. Since the $h_n^{(0)}$ *are* biorthogonal to themselves, it is reasonable to expect that the h_n could be constructed from the $h_n^{(0)}$ by a process of successive approximation. And that is exactly the way Wiener gets them. His construction is reminiscent of the one found at the beginning of Chapter VII in Paley-Wiener [34d].

Now let us clarify some of the details in the proof of Lemma II. Formula (36) follows from (34) because, if $\ell(x)$ is given by (35),

$$\int_{-\infty}^{\infty} |\ell(x)|\, dx \geqslant D \sum_n |a_n|.$$

The indices of summation in (39) and (40) should be k, and not n.

According to the recursive definition (38), the functions $h_n^{(k-1)}(x)$ are of the form (35), so, in (42),

$$\int_{-\infty}^{\infty} |h_n^{(k-1)}(x) - g_n^{(k-1)}(x)|\, dx$$

has the form of the left side of (36), making (42) a consequence of (36). In the extreme left-hand member of (42) one should read $h_n^{(k)}(x)$ for $h_n(x)$.

The last three chapters of Levinson's book[20] are based on the idea of IV. Therein, the gap Tauberian theorem proved here is greatly extended and generalized. Levinson replaces Wiener's real-variable proof of Lemma II by various constructions involving canonical products. The difference between the two approaches is somewhat like that between the method of Group II and that of Group I. (A better comparison is with the difference between Chapter IV of Levinson[20] and Chapter VII of Paley-Wiener [34d].)

Chapter 7 of Hardy's book[9] contains a simple classical proof of the Hardy-Littlewood higher indices theorem. Rudin[32] observed, by the way, (and this not until 1966!) that the Hadamard separation condition on the

λ_n cannot be relaxed. A different treatment of gap Tauberian theorems is given by Pitt.[26] Gaier[8] uses an interesting complex-variable approach, based on the Phragmén-Lindelöf theorem.

Group V: [36a]

Strictly speaking, this paper has nothing to do with gap theorems. Estimations of canonical products do, however, play a prominent role in it. The article is very difficult to read, and is heavily burdened with misprints. China was at war at the time, and Peking, where the article was published, passed under Japanese occupation a few months after its appearance.

This reviewer is not sure that the main result of the paper is really proven. The following is an explanation of the details that he has been able to clarify.

The first two pages of [36a] are given over to a proof of Carleman's theorem (related to the main one of the paper, but simpler) to the effect that (12) implies (11). In (12), read $o(1)$ instead of $O(1)$.

Come we now to the main part of the paper. The tilde of asymptotic equality in (15) is incorrect ($F(x)$ has real zeros!) and should be replaced by an inequality. This makes no difference as far as the validity of (20) is concerned, but it is possible that Wiener used the asymptotic equality in one of the later steps, left out of this printed version. More about this later.

Here is a justification of (21). Let

$$G(u) = \frac{1}{2\pi}\int_{-\infty}^{\infty} [F(x)]^2 e^{-iux}\,dx.$$

Since $F(z)$ is entire, Cauchy's theorem yields, with the help of (20),

$$G(u) = \frac{1}{2\pi}\int_{-\infty}^{\infty} [F(x+iy)]^2 e^{-iu(x+iy)}\,dx$$

for each fixed y. That is,

$$G(u)e^{-yu} = \frac{1}{2\pi}\int_{-\infty}^{\infty} [F(x+iy)]^2 e^{-iux}\,dx;$$

and (21) follows from this by Fourier's inversion formula.

In (22), read $\dfrac{(u+i)\sinh \pi u}{\pi i u}$ for $\dfrac{\sinh \pi u(u+i)}{\pi i u}$. A similar rectification should be made in (23) and (24). In the second member of (22), the $+$ sign between the two exponentials should be replaced by a $-$ sign, and the term $1/\sqrt{u}$ by $1/u$. In the third member of (22), one of the $+$ signs between the terms should be a $-$ sign. The last member of (22) is correct.

In the second member of (25), the lower limit for the second integral sign should be $\xi(1 - \varepsilon_1)$. The last member of (25) has a set of parentheses missing in its second term.

The ξ in the *integrand* in (29) should be replaced by a dummy variable, say ζ, and the $d\xi$ replaced by $d\zeta$.

Formula (30) for ξ_1 comes from the statement of the theorem Wiener is trying to prove. N is taken to depend on ξ in the way specified by (26) so that $N\xi_1$ will tend rapidly to ∞ as $\xi \to \infty$. This property of $N\xi_1$ is needed later, when the last member of (31) is looked at.

Formula (31) is obtained as follows. First rewrite the left side of (29) as

$$\text{(xiv)} \qquad \int_{-\infty}^{\infty} \left\{ \frac{1}{\xi_1} \int_{\xi-\eta}^{\xi+\xi_1-\eta} N[F(N\tau)]^2 \, d\tau \right\} \{A\,d\eta - e^{-\eta}\,d\varphi(e^{\eta})\}.$$

For $\xi < \eta < \xi + \xi_1$,

$$\text{(xv)} \qquad \int_{\xi-\eta}^{\xi+\xi_1-\eta} N[F(N\tau)]^2 \, d\tau$$

differs from $\int_{-\infty}^{\infty} [F(x)]^2 \, dx$ by at most a constant multiple of

$$\text{(xvi)} \qquad N \left\{ \int_{|\eta-\xi|}^{\infty} + \int_{|\xi+\xi_1-\eta|}^{\infty} \right\} e^{-2C_5(N\tau)^{1+\varepsilon}} \, d\tau$$

in view of (20). For $\eta \notin (\xi, \xi + \xi_1)$, (xv) is bounded, in absolute value, by a constant multiple of (xvi), for the same reason. We estimate the integrals occuring in (xvi). They are both of the form

$$\text{(xvii)} \qquad N \int_{u}^{\infty} e^{-k(N\tau)^{1+\varepsilon}} \, d\tau,$$

with $u \geqslant 0$. For $Nu \geqslant 1$, (xvii) equals

$$e^{-k(Nu)^{1+\varepsilon}} \int_{Nu}^{\infty} e^{-k[s^{1+\varepsilon} - (Nu)^{1+\varepsilon}]} \, ds$$

$$\leqslant e^{-k(Nu)^{1+\varepsilon}} \int_{Nu}^{\infty} e^{-(1+\varepsilon)(Nu)^{\varepsilon} k(s-Nu)}\, ds \leqslant \text{const. } e^{-k(Nu)^{1+\varepsilon}}.$$

On the other hand, for $0 \leqslant Nu \leqslant 1$, (xvii) is *trivially* $\leqslant$ const. $e^{-k(Nu)^{1+\varepsilon}}$. So for $u \geqslant 0$,

$$\text{(xviii)} \qquad N \int_{u}^{\infty} e^{-k(N\tau)^{1+\varepsilon}}\, d\tau \leqslant \text{const. } e^{-k(Nu)^{1+\varepsilon}}.$$

One now uses (xviii) in (xvi) in order to estimate (xv) for $\eta \in (\xi, \xi + \xi_1)$, as well as for $\eta \notin (\xi, \xi + \xi_1)$. The result is plugged into (xiv), which is then used as the left side of (29). This yields the required inequality between the first two members of (31), after making a change of variable.

There are, however, some misprints in (31). In its second member, the $d\eta$ should go with the $2A$, and $(1)O$ should be replaced by $O(1)$, which should stand in front of the integral sign. In the third member of (31), the first $O(1)$ should be replaced by $o(1)$, and the $d\eta$ following the integral sign should be eliminated.

In (32), replace $d\varphi(e)^{\eta}$ by $d\varphi(e^{\eta})$.

Now this is as far as I have been able to go! I cannot figure out what Wiener meant to write for formula (33). According to (28), equation (33) is true even if the integral is taken from $-\infty$ to ∞, and I cannot see how that would help to prove (34).

It is possible that Wiener got from (32) to (34) by using the incorrect relation

$$\log |F(x)| \sim -K |x|^{1+\varepsilon}$$

instead of (20). (See (15) and especially the remark immediately following same). In any event, I cannot see how (34) follows, and am thus not sure that the proof of (36), i.e., of (4), and hence, of the main theorem in this paper, is correct.

I have looked for references to [36a]. The *Zentralblatt* and the *Jahresberichte über die Fortschritte der Mathematik* both mention the result Wiener claims to have established, but do not discuss his proof. The paper is mentioned neither in Pitt's book,[26] nor in Pitt and Wiener [39d], which was published some two years after [36a] and deals with somewhat similar material.

It would be of interest to see whether or not the argument given in

[36a] can in fact be carried through. Perhaps this reviewer has just missed some simple point.

References

1. W. Alexander and R. Redheffer, *The excess of sets of complex exponentials*, Duke Math. J. 34(1967), 59–72.

2. J. Bejar-Alamo, *Funciones definidas por series de Dirichlet con exponentes complejos*, Memorias de mat. del instituto Jorge Juan, Madrid, 1946.

3. V. Bernstein, *Leçons sur les progrès récents de la théorie des séries de Dirichlet*, Gauthier-Villars, Paris, 1933.

4. A. Beurling and P. Malliavin, *On Fourier transforms of measures with compact support*, Acta Math. 107(1962), 291–309.

5. A. Beurling and P. Malliavin, *On the closure of characters and zeros of entire functions*, Acta Math. 118(1967), 80–93.

6. R. Boas, *Entire Functions*, Academic Press, New York, 1954.

7. R. Duffin and J. Eachus, *Some notes on an expansion theorem of Paley and Wiener*, Bull. Amer. Math. Soc. 48(1942), 850–855.

8. D. Gaier, *Complex variable proofs of Tauberian theorems*, Matscience Rep. 56, Inst. of Math. Sci., Madras, no date.

9. G. Hardy, *Divergent series*, Oxford, 1949.

10. A. Ingham, *Some trigonometric inequalities with applications to the theory of series*, Math. Zeit. 41(1936), 367–379.

11. J. -P. Kahane, *Sur quelques problèmes d'unicité et de prolongement* Ann. Inst. Fourier Grenoble, V(1953–54), 39–130.

12. J. -P. Kahane, *Sur les fonctions moyenne-périodiques bornées*, Ann. Inst. Fourier Grenoble, VII(1957), 293–314.

13. J. -P. Kahane, *Pseudo-périodicité et séries de Fourier lacunaires*, Ann. Ecole Norm. Sup., Paris, (3) 79(1962), 93–150.

14. J. -P. Kahane, *Sur la totalité des suites d'exponentielles imaginaires*, Ann. Inst. Fourier Grenoble, VIII(1959), 273–275.

15. J. -P. Kahane, *Travaux de Beurling et Malliavin*, Séminaire Bourbaki, 1961–62, W. A. Benjamin, New York, 1966, pp. 225-01 to 225-13.

16. P. Koosis, *Sur la non-totalité de certaines suites d'exponentielles*, Ann. Ecole Norm. Sup., Paris, 75(1958), 125–152.

17. P. Koosis, *Sur la totalité des systèmes d'exponentielles imaginaires*, C. R. Acad. Sci., Paris, 250(1960), 2102–2103.

18. H. Landau, *Necessary density conditions for sampling and interpolation of certain entire functions*, Acta Math., 117(1967), 37–52.

19. B. Levin, *Distribution of Zeros of Entire Functions*. Russian edition pub. 1956, Moscow. English transl. Amer. Math. Soc. (1964).

20. N. Levinson, *Gap and density theorems*. Amer. Math. Sec. Colloq. Pubs. XXVI, New York, 1940.

21. P. Malliavin, *Sur la croissance radiale d'une fonction méromorphe*, Ill. J. Math. 1(1957), 259–296.

22. P. Malliavin and L. Rubel, *On small entire functions of exponential type with given zeros*, Bull. Soc. Math. de France, 89(1961), 175–206.

23. S. Mandelbrojt, *Séries de Fourier et classes quasi-analytiques de fonctions*, Gauthier-Villars, Paris, 1935.

24. S. Mandelbrojt, *Séries adhérentes, régularisation des suites, applications*, Gauthier-Villars, Paris, 1952.

25. S. Mandelbrojt, *Dirichlet series*, Rice Inst. Pamphlet XXXI, no. 4, Houston, 1944.

26. H. Pitt, *Tauberian theorems*, Oxford, Bombay, 1958.

27. G. Pólya, *Untersuchungen über Lücken und Singularitäten von Potenzreihen*. Math. Zeit., 29(1929), 549–640.

28. R. Redheffer, *Remarks on incompleteness*, Proc. Amer. Math. Soc. 2(1951), 365–369.

29. R. Redheffer, *On even entire functions with zeros having a density*, Trans. Amer. Math. Soc. 77(1954), 32–61.

30. R. Redheffer, *Elementary remarks on completeness*, Duke Math. J., 35(1968), 103–116.

31. F. Riesz and N. Nagy, *Leçons d'analyse fonctionnelle*. 2d. ed., Budapest, 1953.

32. W. Rudin, *A converse to the high indices theorem*, Proc. Amer. Math. Soc. 17(1966), 434–435.

33. L. Schwartz, *Etude des sommes d'exponentielles*. 2d. ed., Hermann, Paris, 1959.

34. A. Zygmund, *Trigonometric Series I, II*. 2d. ed., Cambridge, 1959.

ANALYSE MATHÉMATIQUE. — *Sur les fonctions indéfiniment dérivables sur une demi-droite*. Note [1] de MM. **Szolem Mandelbrojt** et **Norbert Wiener**, présentée par M. Paul Montel.

Cette Note donne une nouvelle méthode permettant d'établir des théorèmes du genre de ceux obtenus par S. Mandelbrojt, concernant les fonctions indéfiniment dérivables sur l'intervalle I ($x \geqq 0$).

Soit $f(x)$ une fonction indéfiniment dérivable sur I, avec

$$\overline{\text{borne}}\, |f^{n}(x)| = m_n < \infty \qquad (n \geqq 0,\ f^{(\nu_n)}(0) = 0),$$

où $\{\nu_n\}$ est une suite d'entiers non négatifs croissants. S. Mandelbrojt a indiqué des relations entre les suites $\{m_n\}$ et $\{\nu_n\}$ permettant d'affirmer que $f(x) \equiv 0$ [2].

---

(1) Séance du 10 novembre 1947.

(2) S. Mandelbrojt, *Sur une inégalité fondamentale* [*Annales de l'École Normale Supérieure*, (3), 63, fasc. 4, pp. 352–378].

L'idée essentielle de cette Note réside dans l'emploi d'une fonction auxiliaire, holomorphe dans un demi-plan, dont la variable indépendante correspond à l'ordre de différentiation et dont les valeurs *sont données* par celles des dérivées de la fonction donnée à l'origine. La fonction donnée est supposée impaire, ce qui ne constitue pas une vraie restriction.

Les classes considérées ici ne contiennent que des fonctions analytiques. Nous démontrerons, dans une Note ultérieure, un théorème couvrant des classes bien plus vastes. Il est d'ailleurs difficile de comparer les conditions présentes à celles des théorèmes de Mandelbrojt, même en se bornant aux classes analytiques.

Soit $\{p_n\}$ une suite d'entiers non négatifs et

$$P(z)=\prod_{n=1}^{\infty}\frac{2p_n+z}{2p_n-z}\,e^{-\frac{z}{p_n}}.$$

Si $\{m_n\}$ est une suite non négative avec $\lim \log m_n/n=\infty$, nous désignerons par $\log \overline{m}(x)$ l'ordonnée du polygone de Newton des points $(n,\ \log m_n)$. Si $\underline{\lim}\log m_n/n=c<\infty$, on posera $\log\overline{m}(x)=c$, $\overline{m}_n=\overline{m}(n)$.

Théorème. — *Soit $f(x)$ une fonction indéfiniment dérivable sur* I, *avec*

$$m_n=\overline{\text{borne}}\,|f^{(n)}(x)|<\infty\ (n\geqq 0),\qquad f^{(2n)}(0)=0,\qquad f^{(2p_k+1)}(0)=0,$$

où $\{p_k\}$ est une suite d'entiers positifs.

S'il existe une fonction croissante $N(\sigma)$ *et une fonction à variation bornée*

$$G(\sigma)\qquad (a<\sigma<\infty)\qquad \text{avec}\qquad 0<l<G(\sigma)\leqq\frac{\pi}{2},$$

telles que

$$\overline{m}_{2n+1}\,|P(2n+1)|=O(1),$$
$$\log\{\overline{m}[e^{\sigma}\cos G(\sigma)]\,|P[e^{\sigma+iG(\sigma)}]|\}<-N(\sigma),$$
$$\int^{\infty}N(\sigma)e^{-\frac{\pi}{2}\int^{\sigma}\frac{du}{G(u)}}d\sigma=\infty,$$

la fonction $f(x)$ est identiquement nulle.

D'après un théorème de A. Gorny (³), il existe une constante A telle que $m_n\leqq A^n\overline{m}_n$ $(n\geqq 1)$. Posons $g(t)=f(t)/t$ et

$$H(\omega)=\underset{B\to\infty}{\text{l.i.m.}}\,(2\pi)^{-\frac{1}{2}}\int_{-B}^{B}g(t)\,e^{it\omega}d\omega.$$

Comme on a

$$g^{(\nu)}(t)=t^{-\nu-1}\int_0^t f^{(\nu+1)}(u)\,u^{\nu}\,du,$$

(³) *Contribution à l'étude des fonctions dérivables d'une variable réelle* (*Acta Mathematica*, **71**, 1939, pp. 317-358).

on peut écrire

$$\begin{aligned}\int_0^\infty |H(\omega)|^2\omega^{2\nu}d\omega &= \int_0^\infty |g^{(\nu)}(t)|^2 dt \\ &\leq \int_0^1 \left| t^{-\nu-1}\int_0^t m_{\nu+1}u^\nu du\right|^2 dt \\ &\quad + \int_1^\infty \left| t^{-\nu-1}(f^{(\nu)}(t)\,t^\nu - \nu\int_0^t f^{(\nu)}(u)\,u^{\nu-1}du\right|^2 dt \\ &\leq m_{\nu+1}^2(\nu+1)^{-2} + 4\,m_\nu^2 \leq C^\nu \overline{m}_{\nu+1}^2,\end{aligned}$$

où C est une constante. Posons maintenant

$$F(z) = \int_0^\infty H(\omega)\, e^{z\log\omega}\, d\omega,$$

avec $\log\omega$ réel. La fonction $F(z)$ est holomorphe pour $\mathcal{R}z > 0$, et

$$F(2p_n) = \int_0^\infty H(\omega)\,\omega^{2p_n}\,d\omega = 0 \qquad (n \geq 1).$$

En choisissant convenablement $q > 0$, $B > 0$, on démontre aisément que

$$\Phi(z) = F(z)\,P(z)\,(1+z)^{-q}\,e^{-Bz}$$

est holomorphe pour $\mathcal{R}z > 0$, et, dans ce demi-plan, l'inégalité suivante a lieu

$$|\Phi(z)| \leq 2\,\overline{m}\left(2\left[\frac{x}{2}\right]+5\right)\left|P\left(2\left[\frac{x}{2}\right]+5\right)\right|$$

($[c]$ désigne la partie entière de c, $x = \mathcal{R}z$).

Un théorème établi par S. Mandelbrojt et G. Mac-Lane concernant les fonctions dans une région-bande (4) permet alors de prouver que $\Phi(z) \equiv 0$, et par suite $f(x) \equiv 0$.

Comments on [36b, c], [47a]

S. Mandelbrojt

In the first note, [36b], the authors established a new proof of a theorem published by the commentator some years earlier. It relates the rapidity of decrease of the mean value of the modulus of an L function, say, in $[0, \alpha]$, when α decreases to zero, to the "lacunarity" of it's Fourier series—exactly to the "exponent of convergence", σ, of n_ν^{-1}, if the series admits only terms of the form $(a_\nu \cos n_\nu x + b_\nu \sin n_\nu x)$. The, now famous, Paley-Wiener's theorem, which is used here for one of the very first times, permits one to construct a family of functions $g_\alpha(t)$ of support $[0, \alpha]$, the maximum of the modulus of these functions increasing not too rapidly, and this family playing a role similar to that played by the family of infinitely differentiable functions φ_α introduced in the proof of the commentator. Thus, observing that the Paley-Wiener's ideas on the Fourier transform have already served these two writers for their new proof of the classical Denjoy-Carleman's theorem on quasi-analyticity, we may notice the deep relationship between these ideas in the two fields: different aspects of quasi-analyticity, on the one hand, and Paley-Wiener's theorem, on the other hand.

In the second note, [36c], it is shown that the established relationship between the decrease of the mean value of the function and the exponent of convergence σ of n_ν^{-1} is sharp, for every $\sigma(0 < \sigma < 1)$. In the earlier commentator's work this was shown only for a sequence of σ tending to zero.

The third note, [47a], gives new types of "generalized quasi-analyticity". This amounts to the affirmation that an infinitely differentiable function of which the growth of the maxima of the absolute values of the derivatives is given, cannot have *too many* derivatives zero at a point, without being a constant. The theorem proved in this note, cannot, as it is pointed out by the authors, be compared in a simple way to the corresponding older theorem of the commentator. In the process of proof indications are given on the repartition of the zeros on the real axis of a function F, holomorphic in the half-plane $x > 0$, the maximum of the modulus of F on parallels to the y-axis being of a certain growth when x tends to infinity.

Results similar to those of the first two notes were since often mentioned, generalized, and used by such writers as Hirshman and Jankins,[1] Levin,[3] and J. P. Kahane.[2] The ideas of the third note were largely used in a significant way in Malliavin's Thesis.[4]

References

1. I. I. Hirshman and J. A. Jenkins, Proc. Amer. Math. Soc. 1(1950), 390–393; 512–517.
2. J. -P. Kahane, Ann. Inst. Fourier 5(1955), 39–130.
3. B. Levin, Dokl. USSR 9(51), 3(1941), 693–710.
4. P. Malliavin, Acta Math. 93(1955), 179–255.

HERMITIAN POLYNOMIALS AND FOURIER ANALYSIS

By Norbert Wiener

It is the purpose of the present paper to bring out the relation of the development of a function in a series of Hermite orthogonal functions to its Fourier development, and to extend certain results of Hermann Weyl[1] leading to Fourier developments of fractional order.

As is well known,

$$e^{-t^2+2tx} = \sum_{n=0}^{\infty} \frac{H_n(x)}{n!} t^n.$$

We wish to solve the following integral equation:

$$e^{-\omega^2 t^2+2t\omega x-(x^2/2)} = \int_{-\infty}^{\infty} e^{t^2+2ity-(y^2/2)} K(x, y)dy. \tag{1}$$

We shall clearly have

$$\omega^n H_n(x) e^{-(x^2/2)} = \int_{-\infty}^{\infty} i^n H_n(y) e^{-(y^2/2)} K(x, y)dy,$$

so that the transformation

$$\varphi(x) = \int_{-\infty}^{\infty} K(x, y) f(y)dy$$

will replace the nth Hermite function by the same function, multiplied by the factor $(\omega/i)^n$. To solve equation (1), let us put

$$K(x, y)e^{x^2/2-y^2/2} = Q(x, y).$$

Then

$$\int_{-\infty}^{\infty} Q(x, y)e^{2ity}dy = e^{-t^2(\omega^2+1)+2t\omega x}.$$

[1] H. Weyl, Quantenmechanik und Gruppentheorie, *Ztsch. f. Physik*, vol. 46 (1927), pp. 1–47. Especially note pp. 22, 33.

70

In case $R(\omega^2+1) > 0$, the Fourier transformation yields us

$$Q(x, y) = \frac{1}{\pi}\int_{-\infty}^{\infty} e^{-t^2(\omega^2+1)+2t(\omega x-iy)}dt$$

$$= \frac{1}{\pi}\int_{-\infty}^{\infty} e^{-\left[t\sqrt{\omega^2+1}-\frac{\omega x-iy}{\sqrt{\omega^2+1}}\right]^2} e^{\left[\frac{\omega x-iy}{\sqrt{\omega^2+1}}\right]^2} dt$$

$$= \frac{e^{\frac{\omega^2x^2-2i\omega xy-y^2}{\omega^2+1}}}{\sqrt{\pi(\omega^2+1)}},$$

so that

$$K(x, y) = \frac{e^{(x^2+y^2)\frac{\omega^2-1}{2(\omega^2+1)}-\frac{2i\omega xy}{\omega^2+1}}}{\sqrt{\pi(\omega^2+1)}}.$$

Let us put $\omega = e^{i\theta}$. Then

$$K(x, y) = \frac{e^{i\left[\frac{x^2+y^2}{2}\tan\theta - xy\sec\theta-\theta/2\right]}}{\sqrt{2\pi\sec\theta}}.$$

In the particular case where $\theta = 0$, this reduces to

$$K(x, y) = e^{-ixy}/\sqrt{2\pi}.$$

Let us suppose that $f(x)$ is quadratically summable over $(-\infty, \infty)$. The operator

$$\int_{-A}^{A} K(x, y)f(y)dy$$

is obtained from $f(y)$ by the following stages:

(1). We mutilate $f(y)$ into the function

$$f_A(y) = f(y), \quad [|y| < A]$$
$$f_A(y) = 0. \quad [|y| \geq A].$$

(2). We multiply $f_A(y)$ by $e^{i(y^2\tan\theta/2)-i(\theta/2)}$. In case θ is real, this does not affect $|f_A(y)|$.

(3). We make the Fourier transform of the function obtained from (2) as a function of z. This leaves the integral of the absolute value of the square invariant.

(4). We put $z = y\sec\theta$, and divide the whole function by $\sqrt{\sec\theta}$, to keep the integral of the square invariant.

(5). We multiply by $e^{i(x^2 \tan \vartheta)/2}$ If θ is real, it results from these steps that

$$\int_{-\infty}^{\infty} dx \left| \int_{-A}^{A} K(x, y) f(y) dy \right|^2 = \int_{-A}^{A} | f(y) |^2 dy$$

and that

$$\varphi(x) = \underset{A \to \infty}{\text{l. m.}} \int_{-A}^{A} K(x, y) f(y) dy$$

exists, and is such that

$$\int_{-\infty}^{\infty} | \varphi(x) |^2 dx = \int_{-\infty}^{\infty} | f(y) |^2 dy.$$

Since:

(1). The Hermite functions constitute a closed orthogonal set;

(2). The transformation

$$\varphi(x) = \underset{A \to \infty}{\text{l. m.}} \int_{-A}^{A} K(x, y) f(y) dy \tag{2}$$

is "length-preserving";

$$\omega^n H_n(x) e^{-(x^2/2)} = \int_{-\infty}^{\infty} i^n H_n(y) e^{-(y^2/2)} K(x, y) dy, \tag{3}$$

the transformation (2) simply multiplies the coefficient of $H_n(y)e^{-y^2/2}$ in the Hermite development of the quadratically summable function $f(y)$ by $(\omega/i)^n$, at least when $|\omega| = 1$. If $(\omega/i)^{2\nu} = 1$, it follows that

$$\begin{aligned} f(-x) = {} & \underset{A_1 \to \infty}{\text{l. m.}} \int_{-A_1}^{A_1} K(x, \xi_1) d\xi_1 \\ & \underset{A_2 \to \infty}{\text{l. m.}} \int_{-A_2}^{A_2} K(\xi_1, \xi_2) d\xi_2 \cdots \underset{A_n \to \infty}{\text{l. m.}} \int_{-A_n}^{A_n} K(\xi_{n-1}, \xi_n) f(\xi_n) d\xi_n, \end{aligned} \tag{3}$$

as the changing of the sign of the odd terms in the Hermite development of a function $f(x)$ transforms it into $f(-x)$. Where $\nu = 2$, (3) yields the ordinary Fourier repeated integral, so that it may be regarded as a generalized Fourier integral expansion. Its convergency properties—as opposed to those of convergence in the mean—seem to merit some study.

It is a matter of some interest to see how transformation (2) acts on $iyf(y)$. We have

$$\frac{\partial K(x, y)}{\partial x} = \left[x \left(\frac{\omega^2 - 1}{\omega^2 + 1} \right) - \frac{2i\omega y}{\omega^2 + 1} \right] K(x, y).$$

Hence

$$yK(x, y) = \frac{-(\omega^2 + 1)\dfrac{\partial K}{\partial x} + (\omega^2 - 1)xK}{2i\omega}$$

$$= i \cos \theta \frac{\partial K}{\partial x} + x \sin \theta K.$$

It thus follows formally from (2) that

$$(4) \quad i \cos \theta \frac{d\varphi(x)}{dx} + x \sin \theta \, \varphi(x) = \underset{A \to \infty}{\text{l. m.}} \int_{-A}^{A} K(x, y) y f(y) dy.$$

In case both $f(y)$ and $yf(y)$ are quadratically summable, (4) is easily demonstrable, as we may show by actually proceeding to the limit in the mean. Thus the transformation (2) transforms the operator x into the operator $i \cos \theta(d/dx) + \sin \theta x$.

If we apply to $f(y)$ the transformation $e^{iy\mu}$, we multiply $K(x, y)$ by $e^{iy\mu}$. Now

$$K(x + \lambda, y) = \frac{e^{i\left[\frac{(x^2+y^2)}{2} \tan \theta + x\lambda \tan \theta + (\lambda^2/2) \tan \theta - xy \sec \theta - \lambda y \sec \theta - (\theta/2)\right]}}{\sqrt{2\pi \sec \theta}}$$

$$= K(x, y) e^{i(x\lambda \tan \theta - y\lambda \sec \theta)} e^{i(\lambda^2 \tan \vartheta)/2}.$$

If we let $\lambda = -\mu \cos \theta$, we have

$$K(x - \mu \cos \theta, y) e^{ix\mu \sin \theta} = K(x, y) e^{iy\mu} e^{i(\mu^2 \sin 2\vartheta)/4}.$$

Thus the transformation $e^{iy\mu}$ on $f(y)$ corresponds to the transformation which turns $\varphi(x)$ into $e^{-i[\mu x \sin \theta - (\mu^2/4) \sin 2\theta]} \varphi(x - \mu \cos \theta)$. Comparing this with our definition of the transform of $yf(y)$, we obtain

$$e^{-(\mu^2/4) i \sin 2\theta} e^{i\mu x \sin \theta} \varphi(x - \mu \cos \theta)$$

as a natural definition of

$$e^{[ix \sin \theta - \cos \theta (d/dx)]\mu} \varphi(x).$$

This is in agreement with the definition of Weyl.

Comments on [29d]

T. Hida

This paper seems to have been written for the purpose of establishing the Fourier transform of fractional order so as to be in agreement with H. Weyl's work on Quantum Mechanics. However, we now find a lot of important connections with many fields in mathematics.

Let us put $\omega = ie^{i\theta}$, instead of $\omega = e^{i\theta}$ as done by Wiener, in the expression of the kernel $K(x, y)$ on p. 71. Then, the kernel gives a unitary transformation, call it F_θ, acting on $L^2(R^1)$, and we get a one-parameter, periodic unitary group $\{F_\theta;\ \theta \text{ real}\}$ with F_0 = identity and with $F_{\theta+2\pi} = F_\theta$. This group plays an important role not only in Fourier analysis but also in the investigations of the heat equation, the complex Brownian motion, quantum mechanics and several others. The infinitesimal generator of F_θ is given by $-\frac{1}{2}i\left(\frac{d^2}{dx^2} - x^2 + 1\right)$, and the eigenfunctions are $H_n(x) \exp(-x^2/2)$, $n \geqslant 0$, as is described in this paper.

The assertion in the last paragraph of the paper may be interpreted as a statement of the commutation relation between $\{F_\theta\}$ and the one-parameter unitary group of multiplications by $e^{ix\mu}$, $-\infty < \mu < \infty$.

Errata

Page 70,

line 3 from bottom: $K(x, y)e^{x^2 - y^2/2}$ should read $K(x, y)e^{(x^2 - y^2)/2}$.

Page 71,

line 8: (the denominator of the right side) $\sqrt{2\pi \sec \theta}$ should read $\sqrt{2\pi \cos \theta}$.

line 7 from bottom: $e^{i(y^2 \tan \theta/2) - i(\theta/2)}$ should read $e^{i(y^2 \tan \theta)/2 - i(\theta/2)}$.

line 2 from bottom: $y \sec \theta$ should read $x \sec \theta$.

line 1 from bottom: $\sqrt{\sec \theta}$ should read $\sqrt{\cos \theta}$.

Page 72,

line 10 from bottom: $(\omega/i)^{2\nu} = 1$ should read $(\omega/i)^\nu = -1$.

line 13: The number (3) should be deleted.

line 8 from bottom: $\ldots \operatorname*{l.m.}_{A_n \to \infty} \int_{-A_n}^{A_n} K(\xi_{n-1}, \xi_n) f(\xi_n)\, d\xi_n$ should read

$$\ldots \operatorname*{l.m.}_{A_\nu \to \infty} \int_{-A_\nu}^{A_\nu} K(\xi_{\nu-1}, \xi_\nu) f(\xi_\nu)\, d\xi_\nu.$$

Page 73,

line 12 from bottom: (the denominator of the right side) $\sqrt{2\pi \sec \theta}$ should read $\sqrt{2\pi \cos \theta}$.

line 6 from bottom: $e^{-i[\mu x \sin \theta - (\mu^2/4) \sin 2\theta]}$ should read $e^{i[\mu x \sin \theta - (\mu^2/4) \sin 2\theta]}$.

NOTES ON THE THEORY AND APPLICATION OF FOURIER TRANSFORMS. III, IV, V, VI, VII*

BY

R. E. A. C. PALEY AND N. WIENER

III. On Müntz's theorem

1. We give a proof of the following theorem which is Szász's† generalization of Müntz's‡ theorem. The method is similar to one employed by Carleman§ to prove Müntz's theorem. Incidentally we give a theorem concerning the distribution of the zeros of functions analytic in a half-plane, which is analogous to, and in some respects more general than, another theorem of Carleman's paper.

2. We recall the following well known

THEOREM I. *Let*

$$\zeta_n = \rho_n e^{i\theta_n},\ 0 \leqq \rho_n < 1 \qquad (n = 1, 2, \cdots) \tag{2.1}$$

be a set of points in the unit circle $|\zeta| < 1$. *Suppose that an analytic function* $f_1(\zeta)$ *is regular in* $|\zeta| < 1$, *satisfies*

$$\int_{-\pi}^{\pi} | f_1(\rho e^{i\theta}) |^2 d\theta \leqq B,\ \rho < 1, \tag{2.2}$$

where B is a constant which depends only on f_1, *and has zeros at the points* ζ_n. *Then*

$$\sum_{n=1}^{\infty}(1 - \rho_n) < \infty. \tag{2.3}$$

Conversely, if the series (2.3) *converges, then there exists a bounded function* $f_1(\zeta)$ *which has zeros at the points* ζ_n.

Suppose we invert the interior of the unit circle into the half-plane $\mathfrak{J}(z) > 0$, by means of the substitution

* Presented to the Society, October 28, 1933; received by the editors March 16, 1933. Notes I and II of the Series have appeared in this volume of these Transactions, pp. 348–355.

† O. Szász, *Über die Approximation stetiger Funktionen durch lineare Aggregate von Potenzen*, Mathematische Annalen, vol. 77 (1916), pp. 482–496.

‡ C. H. Müntz, *Über den Approximationssatz von Weierstrass*, Schwarz's Festschrift, Berlin, 1914, pp. 303–312.

§ T. Carleman, *Über die Approximation analytischer Funktionen durch lineare Aggregate von vorgegebenen Potenzen*, Arkiv för Matematik, Astronomi och Fysik, vol. 17 (1922–23), No. 9, pp. 1–30.

(2.4) $$\zeta = \frac{1 + iz}{1 - iz}.$$

Suppose that the point $\zeta = \rho e^{i\theta}$ inverts into $z = x + iy$. Then it is a matter of elementary algebra to verify that the convergence of the series (2.3) is equivalent to that of the series

$$\sum_{n=1}^{\infty} \frac{y_n}{1 + x_n^2 + y_n^2},$$

$z_n = x_n + iy_n$, $y_n > 0$, being the inverse of $\zeta_n = \rho_n e^{i\theta_n}$.

We are now in a position to prove our next theorem, which is a generalization of Carleman's:

THEOREM II. *Let*

(2.5) $$z_n = x_n + iy_n \qquad (n = 1, 2, \cdots)$$

denote a sequence of points in the half plane $\Im(z) > 0$, *and let* $f(z)$ *be a function regular in* $\Im(z) > 0$ *which satisfies*

(2.6) $$\int_{-\infty}^{\infty} |f(x + iy)|^2 dx < 1$$

and has zeros at the points z_n. *Then*

(2.7) $$\sum_{n=1}^{\infty} \frac{y_n}{1 + x_n^2 + y_n^2} < \infty.$$

Conversely, if the series (2.7) *converges, then there exists a bounded function* $f(z)$ *regular in* $\Im(z) > 0$, *satisfying* (2.6) *and having zeros at the points* z_n.

For the second half of the theorem we have only to observe, by Theorem I, that there exists a function $g(z)$ with zeros at the points z_n, analytic in $\Im(z) > 0$, and less than 1 in absolute value. We have now only to write

$$f(z) = \tfrac{1}{2}(z + i)^{-1} g(z)$$

and condition (2.6) is satisfied.

To prove the first part of the Theorem it will be sufficient to show that the function $f(z)$ can be represented by a Cauchy integral

(2.8) $$f(z) = (2\pi i)^{-1} \int_{-\infty}^{\infty} \frac{f^*(x')}{x' - z} dx', \ \Im(z) > 0,$$

where the function $f^*(x')$ satisfies

(2.9) $$\int_{-\infty}^{\infty} |f^*(x')|^2 dx' \leqq 1.$$

Indeed, it is readily seen that the substitution (2.4) transforms the integral of the right-hand member of (2.8) into

$$(2\pi i)^{-1}\int_{-\pi}^{\pi}\frac{f_1^*(\zeta')d\zeta'}{\zeta'-\zeta}+C,\quad \zeta'=e^{i\phi},$$

where $f_1^*(\zeta')$ corresponds to $f^*(x')$ and

$$C=(2\pi i)^{-1}\int_{-\infty}^{\infty}\frac{f^*(x')}{x'+i}dx'.$$

It follows that, on putting $f_1(\zeta)\equiv f(z)$,

$$\int_{-\pi}^{\pi}|f_1(\rho e^{i\theta})|^2dt\leqq B$$

and we may apply Theorem I.

Now suppose that

$$0<\epsilon<y<y_0,\quad 2|x|<N_0<N.$$

Then

$$2\pi if(z)=\int_{-N}^{N}\frac{f(x'+i\epsilon)}{x'+i\epsilon-z}dx'-\int_{-N}^{N}\frac{f(x'+iy_0)}{x'+iy-z}dx'$$
$$+i\int_{\epsilon}^{y_0}\frac{f(N+iy')}{N+iy'-z}dy'-i\int_{\epsilon}^{y_0}\frac{f(-N+iy')}{-N+iy'-z}dy'.$$

But

$$\frac{1}{N_0}\int_{N_0}^{2N_0}dN\int_{\epsilon}^{y_0}\frac{|f(N+iy')|}{|N+iy'-z|}dy'\leqq\frac{2}{N_0^2}\int_{\epsilon}^{y_0}dy'\int_{N_0}^{2N_0}|f(N+iy')|\,dN$$
$$\leqq\frac{2}{N_0^2}\int_{\epsilon}^{y_0}dy'N_0^{1/2}\left[\int_{N_0}^{2N_0}|f(N+iy')|^2dN\right]^{1/2}\leqq 2(y_0-\epsilon)N_0^{-3/2}$$

and tends to zero as $N_0\to\infty$. Similar analysis applies to the last term. Hence

$$2\pi if(z)=\lim_{N_0\to\infty}\frac{1}{N_0}\int_{N_0}^{2N_0}dN\left[\int_{-N}^{N}\frac{f(x'+i\epsilon)}{x'+i\epsilon-z}dx'-\int_{-N}^{N}\frac{f(x'+iy_0)}{x'+iy_0-z}dx'\right]$$
$$=\int_{-\infty}^{\infty}\frac{f(x'+i\epsilon)}{x'+i\epsilon-z}dx'-\int_{-\infty}^{\infty}\frac{f(x'+iy_0)}{x'+iy_0-z}dx',$$

since the last two integrals converge absolutely. Now make y_0 tend to infinity. The second integral does not exceed

$$\left[\int_{-\infty}^{\infty} |f(x'+iy_0)|^2 dx'\right]^{1/2}\left[\int_{-\infty}^{\infty} \frac{dx'}{|x'+iy_0-z|^2}\right]^{1/2},$$

and so tends to zero.

Since condition (2.6) is satisfied, by a classical argument of F. Riesz† there exists a sequence $\{\epsilon_k \downarrow 0\}$ such that $f(x+i\epsilon_k)$ converges weakly to a function $f^*(x)$ satisfying (2.9). Hence

$$2\pi i f(z) = \int_{-\infty}^{\infty} \frac{f(x'+i\epsilon_k)}{x'-z}\,dx' \to \int_{-\infty}^{\infty} \frac{f^*(x')}{x'-z}\,dx' \text{ as } \epsilon_k \to 0,$$

which is the desired result.

3. We proceed now to our main theorem.

THEOREM III. *A necessary and sufficient condition for the closure* L^2 *of the functions* t^{λ_n}, $\Re(\lambda_n) > -\frac{1}{2}$, *in the interval* $(0, 1)$, *is the divergence of the series*

$$\sum_{n=1}^{\infty} \frac{1+2\Re(\lambda_n)}{1+|\lambda_n|^2}. \tag{3.1}$$

We observe first that if the functions t^{λ_n} are not closed L^2 then there exists a function $\phi(t)$ of integrable square on (0, 1) which is orthogonal to them all, so that we have

$$0 < \int_0^1 |\phi(t)|^2 dt < \infty, \tag{3.2}$$

$$\int_0^1 \phi(t) t^{\bar{\lambda}_n} dt = 0 \qquad (n = 1, 2, 3, \cdots). \tag{3.3}$$

Conversely if the system $\{t^{\lambda_n}\}$ is closed then there exists no function $\phi(t)$ not identically zero satisfying (3.2) and (3.3). Now let us write $t=e^x$; the conditions (3.2) and (3.3) become

$$0 < \int_{-\infty}^{0} |\phi(e^x)|^2 e^x dx < \infty,$$

$$\int_{-\infty}^{0} \phi(e^x) \exp[x(1+\bar{\lambda}_n)]\, dx = 0 \qquad (n = 1, 2, \cdots).$$

Upon writing $\phi(e^x)e^{x/2} = \Phi(x)$, we transform the last two formulas into

† *Untersuchungen über Systeme integrierbarer Funktionen*, Mathematische Annalen, vol. 69 (1910), pp. 449–497; pp. 466–468.

(3.4) $$0 < \int_{-\infty}^{0} |\Phi(x)|^2 dx < \infty ,$$

(3.5) $$\int_{-\infty}^{0} \Phi(x) \exp [x(\tfrac{1}{2} + \bar{\lambda}_n)] dx = 0 \qquad (n = 1, 2, \cdots).$$

Thus the closure or non-closure L^2 of the functions t^{λ_n} on (0, 1) is equivalent to the non-existence or existence of a function satisfying (3.4) and (3.5), which is equivalent to the closure or non-closure of the functions $\exp [x(\frac{1}{2}+\lambda_n)]$ on the interval $(-\infty, 0)$.

4. Proceeding now to the proof of the theorem we observe first that, if there exists a function satisfying (3.4) and (3.5), then the function

$$f(z) = \int_{-\infty}^{0} \Phi(x') e^{-izx'} dx$$

exists for $\mathfrak{J}(z) > 0$, and defines an analytic function in that half-plane. Further, by (3.5), it vanishes at the points $(\frac{1}{2}+\bar{\lambda}_n)i$. Since, by Plancherel's theorem,

$$\int_{-\infty}^{\infty} | f(x + iy) |^2 dx = 2\pi \int_{-\infty}^{0} | \Phi(x') |^2 e^{2yx'} dx',$$

the series (3.1) converges by Theorem II. Thus the non-closure of the functions $\{x^{\lambda_n}\}$ implies the convergence of (3.1).

To obtain the converse we have to show that when (3.1) converges then we can find a function $\Phi(x)$ satisfying (3.4) and (3.5). Now in virtue of Theorem II we can find a function $f(z)$ which is analytic for $\mathfrak{J}(z) > 0$, is uniformly bounded in this half-plane, and vanishes at the points $(\frac{1}{2}+\bar{\lambda}_n)i$, with the integral

$$\int_{-\infty}^{\infty} | f(x + iy) |^2 dx$$

uniformly bounded. Let $g_y(\xi)$ denote the Fourier transform

$$(2\pi)^{-1/2} \int_{-\infty}^{\infty} f(x + iy) e^{ix\xi} dx$$

of $f(x+iy)$. Then the argument given in detail in the first note of this series (Theorem II) shows that $g_y(\xi)$ is of the form $G(\xi)e^{y\xi}$ for $\xi < 0$ and vanishes for $\xi > 0$, where

$$\int_{-\infty}^{0} | G(\xi) |^2 d\xi < \infty .$$

Now

$$(2\pi)^{-1/2}\int_{-\infty}^{0} G(\xi)\exp\left[\xi(\tfrac{1}{2}+\bar{\lambda}_n)\right]d\xi$$

$$= (2\pi)^{-1/2}\int_{-\infty}^{0} G(\xi)\exp\left[\xi\Re(\tfrac{1}{2}+\lambda_n)\right]\exp\left[-i\xi\Im(\lambda_n)\right]d\xi$$

$$= (2\pi)^{-1/2}\int_{-\infty}^{0} g_y(\xi)\exp\left[-i\xi\Im(\lambda_n)\right]d\xi \qquad (y = \Re(\tfrac{1}{2}+\lambda_n))$$

$$= f\{\Im(\lambda_n) + i\Re(\tfrac{1}{2}+\lambda_n)\} = f\{(\tfrac{1}{2}+\bar{\lambda}_n)i\} = 0.$$

Thus, for all n,

$$\int_{-\infty}^{0} G(\xi)\exp\left[\xi(\tfrac{1}{2}+\bar{\lambda}_n)\right]d\xi = 0.$$

We have only to identify $\Phi(x)$ with $G(x)$ and our theorem is proved.

5. The problem of the closure in L^2 of functions $e^{\lambda_n x}$ on a finite interval is much more difficult than the corresponding one for an interval which is infinite in one direction. We have obtained a number of theorems in this direction but nothing like a complete answer to the problem. In the case however where all the numbers λ_n are real (we need no longer assume that λ_n is positive or negative) a necessary and sufficient condition for the closure of the functions $e^{\lambda_n x}$ on a finite interval is the divergence of the series

$$\sum_{n=1}^{\infty}\frac{1}{1+|\lambda_n|}.$$

IV. A THEOREM ON CLOSURE

1. The present note is devoted to the proof of the following theorem:

THEOREM I. *The set of functions $\{e^{-\pi|x|/2}e^{i\lambda_n x}\}$ is closed L^2 over $(-\infty, \infty)$ when and only when*

$$\sum_{n=1}^{\infty}\frac{\cos\Im(\lambda_n)}{\cosh\Re(\lambda_n)} = \infty, \quad -\tfrac{1}{2}\pi < \Im(\lambda_n) < \tfrac{1}{2}\pi. \tag{1.1}$$

The condition (1.1) should be contrasted with the condition

$$\sum_{n=1}^{\infty}\frac{\Im(\lambda_n)-\frac{1}{2}\pi}{1+|\lambda_n|^2} = \infty, \quad \Im(\lambda_n) > -\tfrac{1}{2}\pi,$$

which is a necessary and sufficient condition for the closure L^2 of the functions $\{e^{-\pi|x|/2}e^{i\lambda_n x}\}$ on the interval $(0, \infty)$ (see e.g. the preceding note of this series). Thus if for instance all the numbers λ_n are real, the conditions for closure on the intervals $(0, \infty)$ and $(-\infty, \infty)$ are

$$\sum_{n=1}^{\infty} \frac{1}{1+|\lambda_n|^2} = \infty , \quad \sum_{n=1}^{\infty} e^{-|\lambda_n|} = \infty$$

respectively.

2. We shall prove the theorem by the following chain of lemmas:

LEMMA 1. *As* $|x|\to\infty$ *in either direction along the real axis,*

$$|\Gamma(ix+\tfrac{1}{2})| \sim (2\pi)^{1/2}e^{-\pi|x|/2}.$$

This is an immediate consequence of Stirling's formula.

LEMMA 2. *The set of functions* $\{e^{-\pi|x|/2}e^{i\lambda_n x}\}$ *is closed* L^2 *when and only when the set* $\{\Gamma(ix+\frac{1}{2})e^{i\lambda_n x}\}$ *is closed* L^2.

For let

$$\int_{-\infty}^{\infty} |f(x)|^2 dx < \infty \tag{2.1}$$

and

$$\int_{-\infty}^{\infty} f(x)\Gamma(ix+\tfrac{1}{2})e^{i\lambda_n x}dx = 0 \qquad (n = 1, 2, \cdots). \tag{2.2}$$

Let

$$g(x) = f(x)\Gamma(ix+\tfrac{1}{2})e^{\pi|x|/2}.$$

Then, by Lemma 1,

$$\int_{-\infty}^{\infty} |g(x)|^2 dx < \infty \tag{2.3}$$

and

$$\int_{-\infty}^{\infty} g(x)e^{-\pi|x|/2}e^{i\lambda_n x}dx = 0 \qquad (n = 1, 2, \cdots). \tag{2.4}$$

Similarly, (2.1) and (2.2) follow from (2.3) and (2.4). Thus there exists a function $f(x)$ orthogonal to the set $\{\Gamma(ix+\frac{1}{2})e^{i\lambda_n x}\}$ when and only when there exists a function $g(x)$ orthogonal to the set $\{e^{-\pi|x|/2}e^{i\lambda_n x}\}$, and one set is closed L^2 when and only when the other set is closed.

LEMMA 3. *The set of functions*

$$e^{v/2}\exp(-e^{v-\lambda_n}) \qquad (n = 1, 2, \cdots) \tag{2.5}$$

is closed L^2 *over* $(-\infty, \infty)$ *when and only when the set of functions* $\{\Gamma(ix+\frac{1}{2})e^{i\lambda_n x}\}$ *is closed* L^2 *over* $(-\infty, \infty)$.

This follows from the fact that the Fourier transform of (2.5), apart from a constant factor, is $\Gamma(ix+\frac{1}{2})e^{i\lambda_n x}$, and that, by Plancherel's theorem, L^2-closure is invariant under a Fourier transformation.

LEMMA 4. *The set of functions* $\{x^{\mu_n}\}$, $\mu_n = e^{-\lambda_n} - \frac{1}{2}$, *is closed* L^2 *over* (0, 1) *when and only when the set of functions* (2.5) *is closed* L^2 *over* $(-\infty, \infty)$.

For the pair of statements

$$\int_0^1 |f(x)|^2 dx < \infty$$

and

$$\int_0^1 f(x) x^{\mu_n} dx = 0$$

is equivalent to the pair of statements

$$\int_{-\infty}^{\infty} |g(v)|^2 dv = 0$$

and

$$\int_{-\infty}^{\infty} g(v) e^{v/2} \exp(-e^{v-\lambda_n}) dv = 0,$$

where

$$g(v) = f[\exp(-e^v)] e^{v/2} \exp(-e^v/2),$$

and we may again apply the argument of Lemma 2.

LEMMA 5. *The set of functions* $\{x^{\mu_n}\}$, $\Re(\mu_n) > -\frac{1}{2}$, *is closed* L^2 *over* (0, 1) *when and only when*

$$\sum_{n=1}^{\infty} \frac{2\Re(\mu_n) + 1}{1 + |\mu_n|^2} = \infty \tag{2.6}$$

This is a well known theorem of Szász.†

Condition (2.6) is equivalent to (1.1). Theorem I now follows by combining Lemmas 2, 3, 4, and 5.

V. ON ENTIRE FUNCTIONS

1. Let $f(z)$ be an entire function, $f(0)=1$, and $\{z_\nu\}$ the sequence of zeros of $f(z)$. We denote by $M_f(r)$ and $m_f(r)$ respectively the maximum and minimum of $|f(z)|$ on the circle $|z| = r$, and by $n_f(r)$ the number of zeros of $f(z)$

† Loc. cit. See also our preceding note.

contained in the interior of this circle. The purpose of this note is to prove the following two theorems:

THEOREM I. *Let*

$$\log M_f(r) = O(r^{1/2}) \tag{1.1}$$

and

$$\int_0^\infty \log^+ m_f(r) r^{-3/2} dr < \infty . \tag{1.2}$$

Then

$$n_f(r) \sim A r^{1/2} \tag{1.3}$$

where the constant A is determined by

$$A = -\pi^{-2} \int_0^\infty \log \prod_{\nu=1}^\infty \left| \left(1 - \frac{x}{|z_\nu|}\right) \right| x^{-3/2} dx. \tag{1.4}$$

THEOREM II. *Let $f(z)$ be an entire function of order not exceeding $\frac{1}{2}$. If the conditions*

$$n_f(r) \sim B r^{1/2}, \tag{1.5}$$

$$B = -\pi^{-2} \int_0^\infty \log | f(x) | x^{-3/2} dx, \quad f(0) = 1, \tag{1.6}$$

are satisfied, then all roots of $f(z)$ are positive.

The proofs of these theorems are based upon a lemma which is of independent interest. This lemma is discussed in the next §2. In §3 we give proofs of Theorems I and II. In §4 we give a modification of the lemma of §2 and discuss its application to the theory of the Riemann zeta-function. In the last §5 we give proof of some results analogous to those of §2. They are in part contained in a paper by Titchmarsh,† and in part represent extensions of his results.

2. Let $\{\lambda_\nu\}$ be a monotone sequence of positive numbers such that the series $\sum_1^\infty \lambda_\nu^{-2}$ converges. We set

$$\phi(z) = \prod_{\nu=1}^\infty \left(1 - \frac{z^2}{\lambda_\nu^2}\right). \tag{2.1}$$

† E. C. Titchmarsh, *On integral functions with real negative zeros*, Proceedings of the London Mathematical Society, (2), vol. 26 (1927), pp. 185–200.

LEMMA 2.1. *If* $\sum_1^\infty \lambda_\nu^{-2}$ *converges then the statements*

$$\log \phi(iy) \sim \pi A |y| \text{ as } |y| \to \infty, \tag{2.2}$$

$$\int_{-\infty}^{\infty} \log |\phi(x)| x^{-2} dx = -\pi^2 A, \tag{2.3}$$

are completely equivalent.

We have, assuming $y>0$,

$$\begin{aligned} (\pi y)^{-1} \log \phi(iy) &= (\pi y)^{-1} \sum_{\nu=1}^{\infty} \log\left(1 + \frac{y^2}{\lambda_\nu^2}\right) \\ &= (\pi y)^{-1} \int_0^\infty \log\left(1 + \frac{y^2}{t^2}\right) d\Lambda(t) \end{aligned} \tag{2.4}$$

where $\Lambda(t)$ is the number of λ_ν's not exceeding t. Similarly

$$\begin{aligned} -\pi^{-2}\int_{-y}^{y} \log|\phi(x)| x^{-2}dx &= -2\pi^{-2}\int_0^y x^{-2}dx \int_0^\infty \log\left|1 - \frac{x^2}{t^2}\right| d\Lambda(t) \\ &= -2\pi^{-2}\int_0^\infty d\Lambda(t) \int_0^y \log\left|1 - \frac{x^2}{t^2}\right| x^{-2}dx \\ &= -2\pi^{-2}y^{-1}\int_0^\infty d\Lambda(t) \frac{y}{t} \int_0^{y/t} \log|1 - s^2| s^{-2}ds. \end{aligned} \tag{2.5}$$

Expressions (2.4), (2.5) are both of the form

$$\frac{1}{y}\int_0^\infty N\left(\frac{t}{y}\right) d\Lambda(t) \tag{2.6}$$

where $\Lambda(t)$ is a monotone increasing function. In (2.4) we have

$$N(\lambda) = N_1(\lambda) = \frac{1}{\pi} \log\left(1 + \frac{1}{\lambda^2}\right), \tag{2.7}$$

while, in (2.5),

$$\begin{aligned} N(\lambda) = N_2(\lambda) &= -\frac{2}{\pi^2 \lambda}\int_0^{1/\lambda} \log|1 - x^2| x^{-2}dx \\ &= \frac{2}{\pi^2}\left\{ \log\left|1 - \frac{1}{\lambda^2}\right| + \frac{1}{\lambda}\log\left|\frac{1+\lambda}{1-\lambda}\right|\right\}. \end{aligned} \tag{2.8}$$

The function $N_1(\lambda)$ is positive and monotone decreasing, the same being also true of $N_2(\lambda)$ since

(2.9)
$$N_2'(\lambda) = -\frac{2}{\pi^2\lambda^2}\log\left|\frac{1+\lambda}{1-\lambda}\right|$$

If we write $N(\lambda)$ for either of $N_1(\lambda)$, $N_2(\lambda)$, the following properties are easily established:

(2.10)
$$N(\lambda) = \begin{cases} O\left(\log\frac{1}{\lambda}\right) & \text{as } \lambda \to 0, \\ O\left(\frac{1}{\lambda^2}\right) & \text{as } \lambda \to \infty, \end{cases}$$

(2.11)
$$\sum_{k=-\infty}^{\infty} \max_{2^k \leq \lambda \leq 2^{k+1}} \lambda N(\lambda) < \infty, \quad N(\lambda) > 0,$$

$$\int_0^\infty N_1(\lambda)\lambda^{it}d\lambda = \frac{1}{\pi(it+1)}\int_0^\infty \frac{\mu^{(it-1)/2}d\mu}{1+\mu} = \frac{1}{(it+1)\cosh\frac{\pi t}{2}},$$

$$\int_0^\infty N_2(\lambda)\lambda^{it}d\lambda = \frac{2}{\pi^2(it+1)}\int_0^\infty \lambda^{it-1}\log\left|\frac{1+\lambda}{1-\lambda}\right|d\lambda$$

$$= \frac{2}{\pi^2(it+1)}\sum_{k=-\infty}^{\infty}\frac{1}{\left(\frac{t}{2}\right)^2+\left(k+\frac{1}{2}\right)^2} = \frac{2\tan\frac{\pi it}{2}}{\pi it(it+1)}.$$

It follows that

(2.12)
$$\int_0^\infty N(\lambda)\lambda^{it}d\lambda \neq 0 \text{ when } t \text{ is real},$$

(2.13)
$$\int_0^\infty N(\lambda)d\lambda = 1.$$

We observe finally that the expressions

$$\frac{1}{y}\int_0^\infty N_1\left(\frac{t}{y}\right)dA(t) = (\pi y)^{-1}\log\phi(iy),$$

$$\frac{1}{y}\int_0^\infty N_2\left(\frac{t}{y}\right)dA(t) = -\pi^{-2}\int_{-y}^{y}\log|\phi(x)|\,x^{-2}dx$$

$\to 0$ as $y \to 0$. Hence either of the statements

(2.14)
$$\frac{1}{y}\int_0^\infty N_i\left(\frac{t}{y}\right)dA(t) \to A \text{ as } y \to \infty, \quad i = 1, 2,$$

implies the boundedness of the corresponding integral

$$\frac{1}{y}\int_0^\infty N_i\left(\frac{t}{y}\right)d\Lambda(t)$$

over the range $(0, \infty)$. A direct application of a Tauberian theorem of Wiener† shows that statements (2.14) are completely equivalent, which is precisely the result of our Lemma 2.1.

3. We now proceed to the proofs of Theorems 1 and 2.

Proof of Theorem I. We first observe that by a known theorem the assertions $\log M_f(r)=O(r^{1/2})$ and $n_f(r)=O(r^{1/2})$ are equivalent. It follows that if we replace each zero by another zero with the same absolute value but situated on the positive part of the real axis, changing in effect

$$\prod_1^\infty\left(1-\frac{z}{z_\nu}\right) \quad \text{into} \quad \prod_1^\infty\left(1-\frac{z}{|z_\nu|}\right),$$

we certainly do not affect the truth of (1.1). Secondly this process will decrease $m_f(r)$ for every value of r, so that we do not affect the truth of (1.2) either. Thus it is legitimate to assume that all the zeros of $f(z)$ are real and positive. Let them be

$$\lambda_1^2,\ \lambda_2^2,\ \cdots;\quad 0<\lambda_1\leqq\lambda_2\leqq\cdots;\quad \sum_1^\infty\lambda_\nu^{-2}<\infty\ .$$

Our next observation is that, by some theorems of Titchmarsh‡, for a function of this special type the assertions

$$n_f(r)\sim Ar^{1/2} \text{ and } \log M_f(r)\sim\pi Ar^{1/2}$$

are equivalent, so that we may leave $n_f(r)$ and confine our attention to $M_f(r)$. We write

$$z^{1/2} = w = u+iv$$

so that $f(z)$ is transformed into the new function

$$f(w^2) = \phi(w) = \prod_{\nu=1}^\infty\left(1-\frac{w^2}{\lambda_\nu^2}\right)$$

which satisfies

$$\log^+|\phi(w)| = O(|w|) \tag{3.1}$$

and

† N. Wiener, *Tauberian theorems*, Annals of Mathematics, (2), vol. 33 (1932), pp. 1–100; Theorem XI′, p. 30.

‡ Loc. cit., Theorems I and II.

(3.2)
$$\int_{-\infty}^{\infty} \log^+ |\phi(u)| u^{-2}du < \infty .$$

We have to show that

(3.3)
$$\log \phi(iv) \sim \pi A |v| .$$

Since the series $\sum_1^\infty \lambda_\nu^{-2}$ converges, by Lemma 2.1 it will be sufficient to establish the convergence of the integral

(3.4)
$$\int_{-\infty}^{\infty} \log |\phi(u)| u^{-2}du = -\pi^2 A .$$

It follows from (3.1) that the ratio $\Lambda(t)/t$ is bounded. Hence, by (2.5) and (2.9),

$$\begin{aligned}\int_{-v}^{v} \log |\phi(u)| u^{-2}du &= -\frac{\pi^2}{y}\int_0^\infty N_2\left(\frac{t}{y}\right)d\Lambda(t) \\ &= \frac{\pi^2}{y}\int_0^\infty \Lambda(t)d_tN_2\left(\frac{t}{y}\right) = -2\int_0^\infty t^{-2}\Lambda(t)\log\left|\frac{y+t}{y-t}\right|dt \\ &= O\left\{\int_0^v \frac{dt}{t}\log\left|\frac{1+t/y}{1-t/y}\right| + \int_v^\infty \frac{dt}{t}\log\left|\frac{1+y/t}{1-y/t}\right|\right\} = O(1).\end{aligned}$$

Being combined with (3.2) this shows that

$$\int_{-v}^{v} \log^- |\phi(u)| u^{-2}du = O(1).$$

Hence the integral

$$\int_{-\infty}^{\infty} \log^- |\phi(u)| u^{-2}du$$

converges, whence, again by (3.2), the convergence of the integral (3.4) follows. Expression (1.2) for the constant A of Theorem I is now immediately obtained.

Another, non-Tauberian, proof of Theorem 1 proceeds as follows. We have shown in the above discussion that the integral

$$\int_{-\infty}^{\infty} |\log |\phi(u)|| u^{-2}du$$

converges. Hence the harmonic function

$$F(u, v) = \log |\phi(u+iv)| - \frac{1}{\pi}\int_{-\infty}^{\infty} \frac{v\log|\phi(u')|}{(u'-u)^2+v^2}du'$$

exists in the upper half-plane $v>0$, and vanishes for $v=0$ (except at the zeros of $\phi(w)$, $u=\lambda_n^2$). It may be extended by reflection to the lower half-plane. The resulting harmonic function will be continuous everywhere, even at the zeros of $\phi(w)$. Indeed, $F(u, v)$ vanishes along the segment of the line $v=0$ through such a point, and thus cannot have a logarithmic singularity there, while the order of singularity cannot be greater than logarithmic. Thus $F(u, v)$ is the real part of an entire function.

By (1.1),

$$\log|\phi(z)| = O(|z|).$$

Now,

$$\int_{-\infty}^{\infty} \frac{v \log|\phi(u')|}{(u'-u)^2+v^2} du' = \int_{-\infty}^{-2u} + \int_{-2u}^{2u} + \int_{2u}^{\infty}$$

$$\leqq 8v \int_{-\infty}^{\infty} |\log|\phi(u')|| \, u'^{-2} du' + \text{const} \int_{-2u}^{2u} \frac{u'v}{(u'-u)^2+v^2} du'$$

$$= O(|z|).$$

Thus we must have

$$F(u, v) = \pi A v, \quad v > 0,$$

for some A. Thus, as $v \to \infty$,

$$\log \phi(iv) = \log|\phi(iv)| = \pi A v + \frac{1}{\pi} \int_{-\infty}^{\infty} \frac{v \log|\phi(u')|}{u'^2+v^2} du' = \pi A v + o(v)$$

which is the desired result.

Proof of Theorem II. If $\{z_\nu\}$ is the sequence of zeros of $f(z)$ we have

$$f(z) = \prod_{\nu=1}^{\infty} \left(1 - \frac{z}{z_\nu}\right).$$

We set

$$\psi(w) = f(w^2) = \prod_{\nu=1}^{\infty} \left(1 - \frac{w^2}{z_\nu}\right),$$

$$f^*(z) = \prod_{\nu=1}^{\infty} \left(1 - \frac{z}{\lambda_\nu^2}\right), \quad |z_\nu| = \lambda_\nu^2,$$

$$\phi(w) = f^*(z) = \prod_{\nu=1}^{\infty} \left(1 - \frac{w^2}{\lambda_\nu^2}\right), \; z = w^2.$$

Then, since $\sum_1^\infty \lambda_\nu^{-2} < \infty$,

$$n_f(r^2) = n_{f^*}(r^2) \sim \pi^{-1} \log M_{f^*}(r^2) = \pi^{-1} \log M_\phi(r) = \pi^{-1} \log \phi(ir).$$

Hence, by hypothesis (1.5) of Theorem II,

$$\log \phi(ir) \sim \pi B,$$

and, by Lemma 2.1,

$$\int_{-\infty}^{\infty} \log |\phi(u)| u^{-2} du = -\pi^2 B.$$

By hypothesis (1.4),

$$-\pi^2 B = \int_0^{\infty} \log |f(x)| x^{-3/2} dx = \int_{-\infty}^{\infty} \log |\psi(u)| u^{-2} du.$$

Thus we must have

$$\int_{-\infty}^{\infty} [\log |\psi(u)| - \log |\phi(u)|] u^{-2} du = 0. \tag{3.5}$$

On the other hand,

$$\log |\psi(u)| - \log |\phi(u)| = \sum_{\nu=1}^{\infty} \log \frac{|z_\nu - u^2|}{||z_\nu| - u^2|} > 0,$$

unless all roots z_ν are real and positive. Thus relation (3.5) implies that all roots z_ν are positive, and Theorem II is proved.

4. In this paragraph we use the notation of §2, but make a slightly different assumption concerning the asymptotic behavior of $\phi(iy)$.

LEMMA 4.1. *If the series* $\sum_1^{\infty} \lambda_\nu^{-2}$ *converges, then the statements*

$$\log \phi(iy) \sim \pi A |y| \log |y| \text{ as } |y| \to \infty \tag{4.1}$$

and

$$\int_{-y}^{y} \log |\phi(x)| x^{-2} dx \sim -\pi^2 A \log |y| \tag{4.2}$$

are completely equivalent.

We assume $y>0$. Using the kernels $N_1(\lambda)$, $N_2(\lambda)$ of §2 we may replace (4.1), (4.2) respectively by

$$(y \log y)^{-1} \int_0^{\infty} N\left(\frac{t}{y}\right) d\Lambda(t) \to A, \quad N(\lambda) = N_1(\lambda), N_2(\lambda). \tag{4.3}$$

We now observe that either of the statements (4.3) implies

$$\Lambda(y) = O(y \log y). \tag{4.4}$$

Indeed if (4.3) is satisfied with $N = N_1$ or $N = N_2$, then

$$O(1) \geqq (y \log y)^{-1} \int_0^y N\left(\frac{t}{y}\right) d\Lambda(t)$$

$$= N(1)(y \log y)^{-1}\Lambda(y) - (y \log y)^{-1} \int_0^y \Lambda(t) d_t N\left(\frac{t}{y}\right)$$

$$> N(1)\Lambda(y)(y \log y)^{-1},$$

since $N(\lambda)$ is positive and decreasing. Next we prove that, under the condition (4.4), (4.3) is equivalent to

$$\frac{1}{y}\int_0^\infty N\left(\frac{t}{y}\right) d\Lambda^*(t) \to A, \quad N(\lambda) = N_1(\lambda),\ N_2(\lambda), \tag{4.5}$$

where

$$\Lambda^*(y) = \int_0^y (\log t)^{-1} d\Lambda(t). \tag{4.6}$$

It is readily seen from (4.4) and (4.6) that $\Lambda^*(y)$ vanishes for sufficiently small y, while

$$\Lambda^*(y) = O(y) \text{ as } y \to \infty .$$

Now the difference of the left-hand members of (4.5) and (4.3) is equal to

$$I(y) = \frac{1}{y}\int_0^\infty N\left(\frac{t}{y}\right)\left(\frac{1}{\log t} - \frac{1}{\log y}\right) d\Lambda(t)$$

$$= (y \log y)^{-1} \int_0^\infty N\left(\frac{t}{y}\right) \log \frac{y}{t} d\Lambda^*(t)$$

$$= -(y \log y)^{-1} \int_0^\infty \Lambda^*(t) d_t\left[N\left(\frac{t}{y}\right) \log \frac{y}{t}\right]$$

$$= O\left\{(\log y)^{-1} \int_0^\infty t\frac{d}{dt}\left[N(t) \log \frac{1}{t}\right] dt\right\} = O\left(\frac{1}{\log y}\right),$$

and $\to 0$ as $y \to \infty$ or $y \to 0$. The same theorem of Wiener which was applied in the proof of Lemma 2.1 shows immediately the equivalence of the two statements (4.5); hence the two statements (4.3), and consequently (4.1) and (4.2), are also equivalent.

In order to apply Lemma 4.1 to the theory of the Riemann zeta-function we introduce

$$\Xi(z) = \xi\left(\frac{1}{2} + iz\right)$$

(4.7)

$$= \frac{1}{2}\left(\frac{1}{2} + iz\right)\left(\frac{1}{2} - iz\right)\pi^{-1/4-iz/2}\Gamma\left(\frac{1}{4} + \frac{iz}{2}\right)\zeta\left(\frac{1}{2} + iz\right).$$

It is known that $\Xi(z)$ is an entire function, is even and has all zeros in the strip $|\mathfrak{I}(z)| < \frac{1}{2}$. Moreover

$$\log \Xi(iy) = O(y) + \log \Gamma(y/2) \sim \tfrac{1}{2} y \log y,$$

$$\Xi(z) = c \prod_{\nu=1}^{\infty}\left(1 - \frac{z^2}{z_\nu^2}\right), \quad \sum_1^{\infty} |z_\nu|^{-2} < \infty, \; c = \Xi(0).$$

We set

$$z_\nu = z_\nu' + iz_\nu'', \; z_\nu'' > 0, \; |z_\nu''| < \tfrac{1}{2}; \; |z_\nu| = \lambda_\nu.$$

Let us put

$$\mathrm{H}(z) = c \prod_{\nu=1}^{\infty}\left(1 - \frac{z^2}{\lambda_\nu^2}\right).$$

We have outside the strip $|\mathfrak{I}(z)| \leqq 1$,

$$\log\left|\frac{\mathrm{H}(z)}{\Xi(z)}\right| = -\sum_{\nu=1}^{\infty} \log\left|\frac{z_\nu^2 - z^2}{\lambda_\nu^2 - z^2}\right| = -\sum_{\lambda_\nu \leqq |2z|} - \sum_{\lambda_\nu > |2z|}$$

$$= \sum_{\lambda_\nu < |2z|} O\left(\frac{|z|}{\lambda_\nu^2}\right) + \sum_{\lambda_\nu > |2z|} O\left(\frac{|z|^2}{\lambda_\nu^2}\right) = O(|z|).$$

Thus, assuming $y > 0$,

$$\log \mathrm{H}(iy) \sim \tfrac{1}{2} y \log y,$$

and, by Lemma 4.1,

$$(4.8) \qquad (\log y)^{-1} \int_{-y}^{y} \log |c^{-1}\mathrm{H}(x)| \, x^{-2} dx \to -\frac{\pi}{2} \text{ as } y \to \infty .$$

Again, on the real axis,

$$\left|1 - \frac{x^2}{z_\nu^2}\right| \geqq \left|1 - \frac{x^2}{\lambda_\nu^2}\right|,$$

whence

$$\log(|1 - x^2/z_\nu^2| / |1 - x^2/\lambda_\nu^2|) \geqq 0.$$

Furthermore

$$0 \leqq I_\nu = \int_{-\infty}^{\infty} \log \frac{\left|1 - \frac{x^2}{z_\nu^2}\right|}{\left|1 - \frac{x^2}{\lambda_\nu^2}\right|} x^{-2} dx$$

$$\leqq \frac{2}{\lambda_\nu^2} \int_0^\infty \log \left|\frac{1+t^2}{1-t^2}\right| t^{-2} dt = O\left(\frac{1}{\lambda_\nu^2}\right),$$

hence we can integrate term-wise and obtain

$$0 < \int_{-\infty}^{\infty} \log \left|\frac{\Xi(x)}{\mathrm{H}(x)}\right| x^{-2} dx = \sum_{\nu=1}^{\infty} I_\nu < \infty .$$

Then, by (4.8),

$$(\log y)^{-1} \int_{-y}^{y} \log \left| c^{-1}\Xi(x) \right| x^{-2} dx \to -\frac{\pi}{2} \cdot$$

If we return to the zeta-function using (4.7), this gives our final result

$$\int_1^y \frac{\log \left| \zeta(\tfrac{1}{2} + ix) \right|}{x^2} dx = o(\log y). \tag{4.9}$$

5. Titchmarsh has† discussed asymptotic properties of entire functions with real negative zeros. In this paragraph we indicate some results which overlap with those of Titchmarsh. The method used in deriving these results is closely analogous to that used in proving Lemma 2.1; therefore we shall give here only a brief outline of the proof leaving the details to the reader.

Let

$$f(z) = \prod_{\nu=1}^{\infty} \left(1 + \frac{z}{a_\nu}\right)$$

be an entire function all of whose zeros $\{-a_\nu\}$ are negative. It will be assumed that

$$0 < a_1 \leqq a_2 \leqq \cdots, \quad \sum_{\nu=1}^{\infty} a_\nu^{-1} < \infty . \tag{5.1}$$

For simplicity we shall use the symbol $n(r)$ instead of $n_f(r)$ of the preceding paragraphs. The letter x will designate a real positive variable which tends to infinity.

† Loc. cit.

LEMMA 5.1. *Let λ, ρ, θ be fixed numbers such that*

$$\lambda > 0, \quad 0 < \rho < 1, \quad |\theta| < \pi. \tag{5.2}$$

Then the statements

(i) $$n(x) \sim \lambda x^{\rho},$$

(ii) $$\log f(x) \sim \pi\lambda \operatorname{cosec} \pi\rho \, x^{\rho},$$

(iii) $$\log | f(xe^{i\theta}) | \sim \pi\lambda \operatorname{cosec} \pi\rho \cos \theta\rho \, x^{\rho},$$

(iv) $$\int_0^x r^{-1-\pi/(2\theta)} \log | f(re^{i\theta}) | \, dr \sim - \frac{\pi\lambda \operatorname{cosec} \pi\rho \cos \theta\rho}{\rho - \dfrac{\pi}{2\theta}} x^{\rho-\pi/(2\theta)}$$

are all equivalent. In the last statement (iv) *the right-hand member in the case $\theta = \pi/(2\theta)$ should be replaced by its limiting value as $\rho \to \pi/(2\theta)$.*

We first observe that the convergence of the series $\sum_1^\infty a_\nu^{-1}$ implies

$$n(x) = o(x). \tag{5.3}$$

Next let us put

$$\omega(x) = x^{-\rho} n(x). \tag{5.4}$$

In view of the fact that $n(x)$ is monotone increasing it is readily seen that the statements (i), which can be written as

$$\omega(x) \to \lambda, \tag{5.5}$$

and

$$\int_0^x \omega(r) dr \sim \lambda x, \tag{5.6}$$

are equivalent.†

Our next step is to transform the left-hand members of (ii–iv) in such a way as to allow an immediate application of Wiener's Tauberian theorems. We have

$$x^{-\rho} \log f(x) = x^{-\rho} \int_0^\infty \log\left(1 + \frac{x}{t}\right) dn(t)$$

$$= x^{-\rho} \int_0^\infty n(t) \frac{x}{t(t+x)} dt = \frac{1}{x} \int_0^\infty \omega(t) \frac{\left(\dfrac{t}{x}\right)^{\rho-1}}{1 + \dfrac{t}{x}} dt,$$

† This is readily proved directly or derived from a theorem of Wiener, loc. cit., Theorem XIII, pp. 34–35; it also follows from a well known theorem of Landau, *Beiträge zur analytischen Zahlentheorie*, Rendiconti del Circolo Matematico di Palermo, vol. 26 (1908), pp. 169–302; p. 218.

$$x^{-\rho}\log|f(xe^{i\theta})| = \frac{1}{2}x^{-\rho}\int_0^\infty \log\left|1+\frac{x}{t}e^{i\theta}\right|^2 dn(t)$$

$$= \frac{1}{x}\int_0^\infty \omega(t)\left(\frac{t}{x}\right)^{\rho-1}\frac{1+\frac{t}{x}\cos\theta}{1+\frac{2t}{x}\cos\theta+\frac{t^2}{x^2}}dt,$$

$$x^{\pi/(2\theta)-\rho}\int_0^x r^{-1-\pi/(2\theta)}\log|f(re^{i\theta})|\,dr$$

$$= x^{\pi/(2\theta)-\rho}\int_0^x r^{-1-\pi/(2\theta)}dr\, r^{\rho-1}\int_0^\infty \omega(t)\left(\frac{t}{r}\right)^{\rho-1}\frac{1+\frac{t}{r}\cos\theta}{1+\frac{2t}{r}\cos\theta+\frac{t^2}{r^2}}dt$$

$$= -\frac{1}{x}\int_0^\infty \omega(t)dt\left(\frac{x}{t}\right)^{1-\rho+\pi/(2\theta)}\int_{x/t}^\infty \frac{1+\frac{1}{r}\cos\theta}{1+\frac{2}{r}\cos\theta+\frac{1}{r^2}}r^{-1-\pi/(2\theta)}dr$$

(see (5.9) below, for $u=0$, $\theta\rho=\pi/2$). Thus all the statements (ii–iv) are expressible in the form

$$\frac{1}{x}\int_0^\infty \omega(t)N\left(\frac{t}{x}\right)dt \to \lambda \text{ as } x\to\infty, \tag{5.7}$$

where $N(y)$ stands, respectively, for

$$N_3(y) = \frac{1}{\pi\operatorname{cosec}\pi\rho}\frac{y^{\rho-1}}{1+y},$$

$$N_4(y) = \frac{1}{\pi\operatorname{cosec}\pi\rho\cos\theta\rho}y^{\rho-1}\frac{1+y\cos\theta}{1+2y\cos\theta+y^2},$$

$$N_5(y) = -\frac{\rho-\frac{\pi}{2\theta}}{\pi\operatorname{cosec}\pi\rho\cos\theta\rho}y^{\rho-1-\pi/(2\theta)}\int_{1/y}^\infty \frac{1+\frac{1}{r}\cos\theta}{1+\frac{2}{r}\cos\theta+\frac{1}{r^2}}r^{-1-\pi/(2\theta)}dr.$$

A direct computation yields

(5.8)
$$\int_0^\infty \frac{y^{iu+\rho-1}}{1+y}dy = \pi \operatorname{cosec} \pi(iu+\rho),$$

$$\int_0^\infty \frac{y^{iu+\rho-1}(1+y\cos\theta)}{1+2y\cos\theta+y^2}dy$$

(5.9)
$$= \frac{1}{2}\left[\int_0^\infty \frac{y^{iu+\rho-1}}{1+ye^{i\theta}}dy + \int_0^\infty \frac{y^{iu+\rho-1}}{1+ye^{-i\theta}}dy\right]$$

$$= \int_0^\infty \frac{y^{iu+\rho-1}}{1+y}\,\frac{1}{2}\left[(e^{i\theta})^{-iu-\rho} + (e^{-i\theta})^{-iu-\rho}\right]dy$$

$$= \pi \operatorname{cosec} \pi(iu+\rho)\cos\theta(iu+\rho),$$

$$\int_0^\infty y^{iu+\rho-1-\pi/(2\theta)}dy\int_{1/y}^\infty r^{-1-\pi/(2\theta)}\frac{1+\dfrac{1}{r}\cos\theta}{1+\dfrac{2}{r}\cos\theta+\dfrac{1}{r^2}}dr$$

(5.10)
$$= \int_0^\infty r^{-1+\pi/(2\theta)}\frac{1+r\cos\theta}{1+2r\cos\theta+r^2}dr\int_r^\infty y^{iu+\rho-1-\pi/(2\theta)}dy$$

$$= -\frac{\pi \operatorname{cosec} \pi(iu+\rho)\cos\theta(iu+\rho)}{iu+\rho-\dfrac{\pi}{2\theta}}.$$

The last result is first derived in the case where $0<\rho<1+\pi/(2\theta)$, but is easily extended to the general case $0<\rho<1$ by analytic continuation. It is an easy matter to verify that the kernels $N_3(y)$, $N_4(y)$ when $|\theta|<\pi/2$, and $N_5(y)$ are possessed of all the properties of the kernel $N(y)$ stated in the proof of Lemma 2.1. We set $\Lambda(t)=\int_0^t\omega(t)dt$. Since $\omega(t)\geqq 0$, $\Lambda(t)$ is monotone increasing. Hence Wiener's theorem used in §2 may be applied here with the result that the statements (ii), (iii) when $|\theta|<\pi/2$, and (iv) are equivalent, while either of (ii) or (iv) implies (iii) when $\pi/2<|\theta|<\pi$. It should be observed that the kernel $N_4(y)$ is not positive when $|\theta|>\pi/2$, while $N_5(y)$ is positive over the whole range $|\theta|<\pi$. The introduction of this kernel was necessitated by the lack of positiveness of $N_4(y)$ when $|\theta|>\pi/2$. Another theorem of Wiener† will show that either of the statements (ii), (iii) when $|\theta|<\pi/2$, and (iv) implies (5.6), hence (5.5) which is the same as (i). On the other hand it may be proved directly‡ that (i) implies (ii), hence also (iii) and (iv). This completes the proof of Lemma 5.1.

† N. Wiener, loc. cit., Theorem XI'', pp. 31–32.

‡ Titchmarsh, loc. cit., Theorem I.

VI. On two problems of Pólya

1. Pólya† has set the following problem: *Let the real numbers* $m_1, m_2, \cdots$ *have the properties* $0<m_1<m_2<\cdots$ *and*

$$\underline{\lim}_{n\to\infty} \frac{n}{m_n} > \frac{b-a}{2\pi} > 0. \tag{1.1}$$

Furthermore, let $f(x)$ *be continuous in the closed interval* $[a, b]$. *Then it will follow from*

$$\int_a^b f(x)\cos m_n x\,dx = \int_a^b f(x)\sin m_n x\,dx = 0 \tag{1.2}$$

that $f(x)$ *vanishes identically.*

There is no restriction in supposing $b=-a=\pi$. We shall prove the following more general theorem:

Theorem I. *Let* $0<m_1<m_2<\cdots$ *and let*

$$\overline{\lim}_{n\to\infty} \frac{n}{m_n} > 1. \tag{1.3}$$

Then if $f(x)$ *belongs to* L^2 *and*

$$\int_{-\pi}^{\pi} f(x)e^{\pm i m_n x}dx = 0 \qquad (n = 1, 2, 3, \cdots), \tag{1.4}$$

$f(x)$ *vanishes except over a set of measure zero.*

It is very important that we have replaced $\underline{\lim}$ by $\overline{\lim}$. This yields us a much deeper theorem.

Since (1.4) is satisfied with $f(x)$ replaced by $f(x)\pm f(-x)$, it suffices to consider only the cases of $f(x)$ even or odd. We shall give the discussion of the case $f(x)$ even, under the additional assumption that $\int_{-\pi}^{\pi} f(t)dt\neq 0$. The case where this assumption is not satisfied as well as the case of $f(x)$ odd will require but slight modifications which may be left to the reader. We set

$$\phi(u) = \int_{-\pi}^{\pi} f(t)e^{iut}dt, \tag{1.5}$$

where the entire function $\phi(u)$ is even and where we may assume without loss of generality that $\phi(0)=1$. We observe that, on setting $u=\sigma+i\tau$, we have

† Jahresbericht der Deutschen Mathematiker-Vereinigung, vol. 40 (1931), Abteilung 2, p. 81, Problem 108.

$$(1.6)\quad |\phi(u)| = |\phi(\sigma + i\tau)| \leqq \left\{\int_{-\pi}^{\pi} |f(t)|^2 dt\right\}^{1/2} e^{\pi|\tau|} = O(e^{\pi|\tau|}).$$

On the other hand by the theory of Fourier transforms we know that $\phi(\sigma) \subset L^2$ over $(-\infty, \infty)$ and that the Fourier transform of $\phi(\sigma)$ vanishes outside $(-\pi, \pi)$. Hence by Theorem II of our Note I,†

$$(1.7)\qquad \int_{-\infty}^{\infty} \frac{|\log|\phi(\sigma)||}{1+\sigma^2} d\sigma < \infty.$$

By the change of variable $u^2 = z$ we obtain a function $\psi(z) = \phi(u)$ which satisfies the conditions of Theorem I of our preceding Note V. It follows at once that the limit

$$(1.8)\qquad \lim_{r\to\infty} \frac{n_\phi(r)}{r} = A$$

exists. Let $\{u_\nu\}$ be the sequence of zeros of $\phi(u)$. It is clear that $\{\pm m_\nu\}$ is a subsequence of $\{u_\nu\}$. Hence, by (1.3),

$$(1.9)\qquad 2 < \overline{\lim_{n\to\infty}} \frac{2n}{m_n} \leqq \lim_{r\to\infty} \frac{n_\phi(r)}{r} = A.$$

However, by Jensen's theorem, in view of (1.6),

$$(1.10)\qquad \begin{aligned} \frac{1}{r}\int_0^r \frac{n_\phi(t)}{t} dt &= \frac{1}{2\pi r}\int_0^{2\pi} \log|\phi(re^{i\theta})|\, d\theta \\ &\leqq \frac{1}{2\pi r}\int_0^{2\pi} \pi r |\sin\theta|\, d\theta + O\left(\frac{1}{r}\right) = 2 + O\left(\frac{1}{r}\right). \end{aligned}$$

Hence

$$(1.11)\qquad A = \lim_{r\to\infty} \frac{1}{r}\int_0^r \frac{n_\phi(t)}{t} dt \leqq 2.$$

The resulting contradiction shows that $f(x)$ must vanish except for a set of measure zero.

2. Pólya‡ has also set the following problem: *Let $f(z)$ be an entire function which is bounded for the integral arguments $z = 0, \pm 1, \pm 2, \cdots, \pm n, \cdots$. Let*

$$(2.1)\qquad M_f(r) = o(r).$$

Then $f(z)$ reduces to a constant.

† The present volume of these Transactions, p. 349.

‡ Jahresbericht der Deutschen Mathematiker-Vereinigung, vol. 40 (1931), Abteilung 2, p. 80, Problem 105.

It is clearly sufficient to prove the theorem for an even $f(z)$, for if $f(z)$ is odd, we need only consider $f(z)/z$, which will be even, and will hence reduce to a constant which can only be zero. The general function may then be treated by reducing it to the sum of an odd and an even part.

If $f(z)$ is even,

$$g(z) = [f(z) - f(0)]z^{-2} \tag{2.2}$$

will be entire. Thus

$$\sum_{n=-\infty}^{\infty} | g(n) | < \infty . \tag{2.3}$$

Let us form

$$G(z) = \sum_{n=-\infty}^{\infty} g(n) \frac{\sin \pi(n - z)}{\pi(n - z)} . \tag{2.4}$$

Clearly

$$G(x + iy) = O(y^{-1}e^{\pi|y|}). \tag{2.5}$$

Let us now form the entire function

$$H(z) = [g(z) - G(z)] \operatorname{cosec} \pi z. \tag{2.6}$$

For all values of z and all integral values of n we shall have

$$\begin{aligned} H[(n + \tfrac{1}{2}) + iy] &= O\{\exp (\epsilon[(n + \tfrac{1}{2})^2 + y^2]^{1/2} - \pi | y |)\} \\ &\quad + O(1/|y|) \\ &= e^{\epsilon|n+1/2|}O(e^{(\epsilon-\pi)|y|}) + O(1/|y|) \end{aligned} \tag{2.7}$$

uniformly in n. We have here employed (2.1) and (2.5). Hence

$$\int_{-\infty}^{\infty} | H(n + \tfrac{1}{2} + iy) |^2 dy = O(e^{2\epsilon|n|}), \tag{2.8}$$

for all ϵ.

Let us put

$$x_1 = [x + \tfrac{1}{2}], \; x_2 = [x - \tfrac{1}{2}].$$

Then, by Cauchy's theorem,

$$\begin{aligned} H(x + iy) &= (2\pi i)^{-1} \int_{-\infty}^{\infty} \frac{H(x_1 + \frac{1}{2} + iy_1)}{x_1 + \frac{1}{2} + iy_1 - x - iy} dy_1 \\ &\quad - (2\pi i)^{-1} \int_{-\infty}^{\infty} \frac{H(x_2 - \frac{1}{2} + iy_1)}{x_2 - \frac{1}{2} + iy_1 - x - iy} dy_1 . \end{aligned} \tag{2.9}$$

Hence

$$(2.10)\quad \int_{-\infty}^{\infty} H(x+iy)e^{iuv}dy$$
$$= (2\pi i)^{-1}\int_{-\infty}^{\infty} H\left(x_1+\frac{1}{2}+iy_1\right)e^{iuv_1}dy_1\int_{-\infty}^{\infty}\frac{e^{iuv}dy}{x_1+\frac{1}{2}-x-iy}$$
$$-(2\pi i)^{-1}\int_{-\infty}^{\infty} H\left(x_2-\frac{1}{2}+iy_1\right)e^{iuv_1}dy_1\int_{-\infty}^{\infty}\frac{e^{iuv}}{x_2-\frac{1}{2}-x-iy},$$

and, by the Plancherel theorem and the Schwarz unequality,

$$\int_{-\infty}^{\infty}|H(x+iy)|^2dy \leqq \text{const}\left\{\int_{-\infty}^{\infty}|H(x_1+\tfrac{1}{2}+iy)|^2dy + \int_{-\infty}^{\infty}|H(x_2-\tfrac{1}{2}+iy)|^2dy\right\}.$$

Thus, by (2.8),

$$(2.11)\quad \int_{-\infty}^{\infty}|H(x+iy)|^2dy = O(e^{2\epsilon|x|}).$$

By an application of Cauchy's theorem,

$$(2.12)\quad \int_{-\infty}^{\infty}H(x+iy)e^{iuv}dy = e^{-ux}\int_{-\infty}^{\infty}H(iy)e^{iuv}dy \equiv e^{-ux}\phi(y).$$

Thus by the Plancherel theorem,

$$(2.13)\quad \int_{-\infty}^{\infty}|\phi(u)|^2e^{-2ux}du = O(e^{2\epsilon|x|}).$$

This is however only possible if $\phi(u)$ vanishes almost everywhere for $|u|>\epsilon$. Since ϵ is arbitrarily small, $\phi(u)$ must be equivalent to zero. Thus $H(z)$ vanishes, and $g(z)=G(z)$. On the other hand,

$$(2.14)\quad M_G(r) \sim \pi r,$$

unless every $g(n)$ is zero. This yields

$$(2.15)\quad G(z) = g(z) = 0,\ f(z) = f(0),$$

which is the desired result.

VII. On the Volterra equation

1. A theorem of Mercer† asserts that *if* $0<\alpha<1$, *and if*

† J. Mercer, *On the limits of real variants*, Proceedings of the London Mathematical Society, (2), vol. 5 (1907), pp. 206–224.

(1.1) $$\alpha s_n + (1-\alpha)\frac{1}{n}\sum_1^n s_\nu \to s,$$

then

(1.2) $$s_n \to s.$$

This theorem possesses generalizations of a non-trivial nature. The continuous analogue asserts that if $0<\alpha<1$ and if

(1.3) $$\alpha s(x) + \frac{1-\alpha}{x}\int_1^x s(y)dy \to s \text{ as } x\to\infty,$$

then

(1.4) $$s(x) \to s.$$

By a change of independent variable, this asserts that if

(1.5) $$\alpha S(\xi) + (1-\alpha)\int_0^\xi e^{\eta-\xi}S(\eta)d\eta \to s,$$

then

(1.6) $$S(\xi) \to s.$$

This statement is a particular case of the following theorem:

THEOREM I. *Let $F(x)$ be measurable and bounded over every finite range $(0, A)$. Let $K(x) \subset L$ over $(0, \infty)$, that is,*

(1.7) $$\int_0^\infty |K(\xi)|\, d\xi < \infty .$$

Let

(1.8) $$F(x) + \int_0^x K(x-\xi)F(\xi)d\xi \to s \text{ as } x\to\infty .$$

Then if

(1.9) $$\int_0^\infty K(\xi)e^{-w\xi}d\xi \neq -1,\ \Re(w) \geqq 0,$$

we shall have

(1.10) $$F(x) \to s\left[1 + \int_0^\infty K(\xi)d\xi\right].$$

Conversely, let $K(x) \subset L$, $\int_0^\infty K(\xi)d\xi \neq -1$, and let (1.8) *imply* (1.10) *for every $F(x)$ satisfying our conditions. Then* (1.9) *must be true.*

2. The second part of this theorem may be proved by reductio ad absurdum by putting

$$F(x) = e^{w_0 x},$$

where

$$\int_0^\infty K(\xi)e^{-w_0\xi}d\xi = -1, \; \Re(w_0) \geqq 0, \; w_0 \neq 0.$$

Then

$$\left| F(x) + \int_0^x K(x-\xi)F(\xi)d\xi \right| = \left| e^{w_0 x} \int_x^\infty K(\xi)e^{-w_0\xi}d\xi \right|$$

$$\leqq \int_x^\infty |K(\xi)| d\xi \to 0.$$

As (1.10) is obviously false, the second part of the theorem is proved.

3. The first part of Theorem I will appear as a corollary to a theorem concerning the Volterra integral equation of the closed cycle.

We shall use the symbol

$$A \bigstar B(x)$$

to indicate the "Faltung" of the two functions $A(x)$, $B(x)$,

$$A \bigstar B(x) = \int_0^x A(\xi)B(x-\xi)d\xi = \int_0^x A(x-\xi)B(\xi)d\xi = B \bigstar A(x).$$

It is well known that the (bounded and measurable) solution of the Volterra integral equation

(3.1) $$G(x) = F(x) + K \bigstar F(x)$$

is uniquely determined and given by

(3.2) $$F(x) = G(x) + Q \bigstar G(x)$$

where the resolvent kernel $Q(x)$ itself is determined from

(3.3) $$Q(x) + K(x) = -K \bigstar Q(x) = -Q \bigstar K(x),$$

or else, by

(3.4) $$Q(x) = \sum_{n=1}^{\infty}(-1)^n K^n(x), \; K^n(x) = K \bigstar K \bigstar \cdots \bigstar K(x).$$

We observe that the solution of (3.3) is easily obtained by using Laplace transforms.† Let us designate by

† See, e.g., S. Bochner, *Vorlesungen über Fouriersche Integrale*, Leipzig, 1932, chapter VII. Other references are also found there.

$$k(w) = \int_0^\infty K(\xi)e^{-w\xi}d\xi, \tag{3.5}$$

$$q(w) = \int_0^\infty Q(\xi)e^{-w\xi}d\xi, \tag{3.6}$$

the Laplace transforms of $K(x)$, $Q(x)$. Equation (3.4) reduces then to

$$q(w) = \frac{-k(w)}{1 + k(w)}, \tag{3.7}$$

and $Q(x)$ will be found by inversion of a Laplace integral.

The theorem in question may now be stated as follows:

THEOREM II. *A necessary and sufficient condition that* $Q(x) \subset L$ *over* $(0, \infty)$, *that is,*

$$\int_0^\infty |Q(\xi)| d\xi < \infty, \tag{3.8}$$

is that

$$k(w) = \int_0^\infty K(\xi)e^{-w\xi}d\xi \neq -1, \quad \Re(w) \geqq 0. \tag{3.9}$$

If this theorem holds true, the first part of Theorem I is immediately derived. Indeed, under the assumptions made we have

$$F(x) = G(x) + \int_0^x G(x - \xi)Q(\xi)d\xi.$$

Here $G(\xi)$ is bounded over every finite range and $\to s$ as $\xi \to \infty$. Hence $G(\xi)$ is bounded over the whole range $(0, \infty)$. Since $Q(\xi)$ is integrable over $(0, \infty)$ we may pass to limit as $x \to \infty$ under the integral sign, with the result

$$F(x) \to s + s\int_0^\infty Q(\xi)d\xi = s[1 + q(0)] = s[1 + k(0)]^{-1}$$

which is precisely the desired formula (1.10).

4. To prove the necessity of (3.9) we observe that if (3.8) holds then $q(w)$ as well as $k(w)$ are analytic in the half-plane $\Re(w) > 0$ and continuous up to and including the boundary $\Re(w) = 0$. This implies that the denominator in the right-hand member of (3.7) does not vanish for $\Re(w) \geqq 0$, so that (3.9) holds.

The proof that (3.9) is sufficient is more difficult. We introduce the auxiliary functions

$$\phi_A(u) = \begin{cases} 1, \ |u| < A, \\ 2 - |u|/A, \ A \leqq |u| \leqq 2A, \\ 0, \ |u| > 2A, \end{cases} \tag{4.1}$$

and put

$$q^*(w) \equiv \frac{-k(w)}{1 + k(w)}, \tag{4.2}$$

$$q^*(iu) = q_1(u) + q_2(u), \tag{4.3}$$

$$q_1(u) = \phi_A(u)q^*(iu), \ q_2(u) = [1 - \phi_A(u)]q^*(u). \tag{4.4}$$

We wish to prove that if A is sufficiently large, $q_1(u)$ and $q_2(u)$ are both Fourier transforms of functions of L.

To begin with,

$$q_1(u) = \begin{cases} \dfrac{-\phi_A(u)k(iu)}{\phi_{2A}(u) + \phi_{2A}(u)k(iu)} \text{ when } |u| < 2A, \\ 0 \text{ when } |u| \geqq 2A. \end{cases}$$

Thus $q_1(u)$ is the quotient of two functions, each the Fourier transform of a function of L, each vanishing outside a finite range, and such that the denominator function only vanishes in points interior to the region of vanishing of the numerator function. We may then appeal to a theory due to Wiener† to show that $q_1(u)$ is the Fourier transform of a function of L.

We have

$$q_2(u) = [1 - \phi_A(u)]\frac{-k(iu)[1 - \phi_{A/2}(u)]}{1 + k(iu)[1 - \phi_{A/2}(u)]}.$$

It is easy to show that this is the Fourier transform of a function of L when the same is true of

$$-k(iu)[1 - \phi_{A/2}(u)]\{1 + k(iu)[1 - \phi_{A/2}(u)]\}^{-1} = \sum_{n=1}^{\infty}(-1)^n\{k(iu)[1 - \phi_{A/2}(u)]\}^n.$$

Now

$$\{k(iu)[1 - \phi_{A/2}(u)]\}^n$$

is the Fourier transform of a function $h_n(x)$ for which

---

† N. Wiener, *The Fourier Integral and Certain of its Applications*, Cambridge, 1933; Lemmas 6_7, 6_{10}, 6_{18}.

$$\int_{-\infty}^{\infty} | h_n(\xi) | \, d\xi \leqq \left[\int_{-\infty}^{\infty} | h_1(\xi) | \, d\xi \right]^n$$

$$= \left[\int_{-\infty}^{\infty} d\xi \left| K(\xi) - \frac{1}{\pi A} \int_0^{\infty} K(\eta) \frac{\cos \frac{A}{2}(\xi - \eta) - \cos A(\xi - \eta)}{(\xi - \eta)^2} d\eta \right| \right]^n .$$

An argument of the familiar Fejér type will show that we may choose A so large that the integral in brackets is less than any given number λ, $0<\lambda<1$. It will follow at once that $q_2(u)$ is the Fourier transform of a function $F_2(x)$ for which

$$\int_{-\infty}^{\infty} | F_2(\xi) | \, d\xi \leqq \frac{\lambda}{1 - \lambda} \cdot$$

Combining this with the similar result for $q_1(u)$, we see that we may write

$$(4.5) \qquad q^*(iu) = \frac{- k(iu)}{1 + k(iu)} = \int_{-\infty}^{\infty} F(\xi) e^{-iu\xi} d\xi, \quad F(\xi) \subset L.$$

We may rewrite (4.5) as

$$(4.6) \qquad \int_{-\infty}^{0} F(\xi) e^{-iu\xi} d\xi = - \int_{0}^{\infty} F(\xi) e^{-iu\xi} d\xi + q^*(iu).$$

Now, it is readily seen that $k(w) \to 0$ as $|w| \to \infty$, uniformly in the half-plane $\Re(w) \geqq 0$. Since, by hypothesis, $1+k(w) \neq 0$ for $\Re(w) \geqq 0$, there exists a positive constant c such that

$$| 1 + k(w) | \geqq c > 0.$$

Thus

$$- \int_0^{\infty} F(\xi) e^{-w\xi} d\xi + q^*(w)$$

is a function of w analytic and bounded in the right half-plane, and continuous up to and including the imaginary axis. Similarly,

$$\int_{-\infty}^{0} F(\xi) e^{-w\xi} d\xi$$

is a function of w analytic and bounded in the left half-plane, and continuous up to and including the imaginary axis. Furthermore, the two functions are identical on the imaginary axis. By the classical argument of Riemann-Painlevé it readily results that they are parts of the same analytic function, which is thus entire and bounded. It hence reduces to a constant, and since

(4.7)
$$\int_{-\infty}^{0} F(\xi)e^{-w\xi}d\xi \to 0 \text{ as } w \to -\infty,$$

this constant can only be 0. Thus

(4.8)
$$q^*(w) = \frac{-k(w)}{1+k(w)} = \int_0^{\infty} F(\xi)e^{-w\xi}d\xi.$$

On the other hand, it follows readily from (3.4) that there exists a $w_0 > 0$ such that

(4.9)
$$\frac{-k(w)}{1+k(w)} = \int_0^{\infty} Q(\xi)e^{-w\xi}d\xi, \ \Re(w) > w_0,$$

and

(4.10)
$$\int_0^{\infty} |Q(\xi)| e^{-w_0\xi}d\xi < \infty .$$

By the uniqueness theorem for Laplace transforms we conclude that $F(x)e^{-wx}$ and $Q(x)e^{-wx}$ coincide almost everywhere, whence $Q(x) \subset L$.

5. In this proof, we have used the theorem† of Wiener that if a function has an absolutely convergent Fourier series and does not vanish, its reciprocal has an absolutely convergent Fourier series. P. Lévy‡ has pointed out that the same methods suffice for the following theorem: if a function $f(x)$ has an absolutely convergent Fourier series, and $\Phi(u)$ is analytic over the range of values of $f(x)$, then $\Phi[f(x)]$ has an absolutely convergent Fourier series. By methods not essentially different from those of this paper, we may extend this theorem as follows: if $f(x)$ is the Fourier transform of a function of L, and $\Phi(u)$ is analytic over the range of values of $f(x)$, including 0, then

$$\Phi[f(x)]$$

is the Fourier transform of a function of L.

---

† Loc. cit., Lemma 6_{16}.

‡ P. Lévy, *Sur la convergence absolue des séries de Fourier*, Paris Comptes Rendus, vol. 196 (1933), p. 463.

Massachusetts Institute of Technology,
Cambridge, Mass.

Comments on [33e, Part I]

J. W. Neuberger

Part I of [33e] concerns a proof of a theorem (Theorem I) of Carleman. Theorem I is proved by using Theorem II, a generalization of a result due to de la Vallée Poussin. Theorem II involves Fourier transforms whereas the result of de la Vallée Poussin is for Fourier series. This marks an early use of Fourier transforms in the study of quasi-analyticity.

Quasi-analytic collections can be specified by fixing a certain adequate rate of decrease that a function may have in the directions of plus and minus infinity and then taking for the quasi-analytic collection the collection of all Fourier transforms of such functions. De la Vallée Poussin, in the paper Wiener cites, was concerned with the simplier case of Fourier *series* where one specifies a rate of dècrease of Fourier coefficients. The two ideas can be combined, of course, using Fourier-Stieltjes transforms. The best results in this direction known to this writer are contained in A. Beurling's lecture notes.[1] Beurling uses Fourier-Stieltjes transforms of certain classes of (possibly unbounded) measures and still obtains (Theorem IV of Lecture 3) quasi-analytic collections in the manner described above. Beurling also obtains very comprehensive gap theorems from his theory (Corollary 4.1 of lecture 3).

Reference

1. A. Beurling, *On quasi-analyticity and general distributions*, Multigraphed Lectures, Stanford Univ., 1961.

Comments on [33e, Part II]

R. P. Gosselin

Part II of [33e] is an elegant example of the use of analytic methods in Fourier theory. Analytic methods in the theory of conjugate functions had already been used by M. Riesz in his fundamental work concerning the boundedness of the conjugate operator in the L^p norm.[2] The authors' result contrasts with those for which similar hypotheses are made about the Fourier coefficients. Thus there is a well-developed but relatively elementary theory about the integrability of sine and cosine series with monotone decreasing coefficients (cf. A. Zygmund[3]). The one theorem of the paper is too special, apparently, to have had much aftermath. In fact, real variable methods have largely taken over the theory of singular integrals, particularly in the work of Calderon and Zygmund.[1] Although the hypothesis of oddness of the function cannot be eliminated entirely, as the authors point out, there are useful variants of the theorem of the paper which can be obtained from it by consideration of the kernel of the conjugate operator.

References

1. A. -P. Calderon and A. Zygmund, Acta Math. 88(1952), 85–139.
2. M. Riesz, Math. Zeit. 27(1927), 104–113.
3. A. Zygmund, *Trigonometric Series*, Cambridge, 1959.

REMARKS ON THE CLASSICAL INVERSION FORMULA FOR THE LAPLACE INTEGRAL

D. V. WIDDER AND N. WIENER

If a function $f(s)=f(\sigma+i\tau)$ is defined for $\sigma>0$ by the Laplace integral

$$f(s) = \int_0^\infty e^{-st}\phi(t)dt, \tag{1}$$

then the classical inversion formula is

$$\phi(t) = \frac{1}{2\pi i}\int_{c-i\infty}^{c+i\infty} f(s)e^{st}ds, \qquad c > 0, t > 0. \tag{2}$$

Conditions for the validity of this formula have frequently been discussed. However, the authors know of no adequate treatment* of the case when $\phi(t)$ belongs to L^2 in $(0, \infty)$:

$$\int_0^\infty |\phi(t)|^2dt < \infty. \tag{3}$$

We employ here the usual notation,

$$\underset{a\to\infty}{\text{l.i.m.}}\ \phi_a(t) = \phi(t),$$

to mean that $\phi_a(t)$ and $\phi(t)$ belong to L^2 in $(-\infty, \infty)$ and that

$$\lim_{a\to\infty}\int_{-\infty}^\infty |\phi_a(t) - \phi(t)|^2dt = 0.$$

It is clear first that if (3) holds then (1) converges absolutely for $\sigma>0$, since

$$|f(\sigma + i\tau)|^2 = \left|\int_0^\infty e^{-st}\phi(t)dt\right|^2 \leqq \int_0^\infty e^{-2\sigma t}dt\int_0^\infty |\phi(t)|^2dt.$$

Moreover, by the Plancherel theorem regarding Fourier transforms,

$$\underset{a\to\infty}{\text{l.i.m.}}\int_0^a e^{-i\tau t}\phi(t)dt$$

* But compare G. Doetsch, *Bedingungen für die Darstellbarkeit einer Funktion als Laplace Integral und eine Umkehrformel für die Laplace-Transformation*, Mathematische Zeitschrift, vol. 42 (1936), p. 272, Theorem 1.

exists. We denote it by $f(i\tau)$. The same theorem gives us at once that

$$\underset{a\to\infty}{\text{l.i.m.}}\ \frac{1}{2\pi}\int_{-a}^{a} f(i\tau)e^{i\tau t}d\tau = \begin{cases}\phi(t), & t>0,\\ 0, & t<0,\end{cases}$$

or

$$\underset{a\to\infty}{\text{l.i.m.}}\ \frac{1}{2\pi i}\int_{-ia}^{ia} f(s)e^{st}ds = \begin{cases}\phi(t), & t>0,\\ 0, & t<0.\end{cases} \tag{4}$$

Hence (2) with $c=0$ is valid in the sense of (4). However, if $c>0$, (2) is no longer valid in this sense.

If $c>0$, it is again clear from the Plancherel theorem that

$$\underset{a\to\infty}{\text{l.i.m.}}\ \frac{1}{2\pi}\int_{-a}^{a} f(c+i\tau)e^{i\tau t}d\tau = \begin{cases}e^{-ct}\phi(t), & t>0,\\ 0, & t<0.\end{cases}$$

But this does not imply that

$$\underset{a\to\infty}{\text{l.i.m.}}\ \int_{c-ia}^{c+ia} f(s)e^{st}ds = \begin{cases}\phi(t), & t>0,\\ 0, & t<0,\end{cases}$$

unless $f(s)$ is identically zero. For, set $\phi_a(t)=\int_{-a}^{a}f(c+i\tau)e^{i\tau t}d\tau$. It will be sufficient to show that

$$\int_{-\infty}^{\infty} e^{2ct}\,|\,\phi_a(t)\,|^2dt = \infty \tag{5}$$

for some a. Choose a so that

$$f(c+ia) \neq f(c-ia). \tag{6}$$

This is possible, for otherwise we should have by use of (1) that

$$\int_0^{\infty} e^{-ct}\phi(t)\sin at\,dt = 0$$

for all a. By the uniqueness theorem for the Fourier sine transform this would imply that $\phi(t)$ is equivalent to zero and that $f(s)$ is identically zero. In fact we see that (6) may be satisfied for some a in every interval however small.

An integration by parts of the integral defining $\phi_a(t)$ gives

$$\begin{aligned}\phi_a(t) &= \frac{f(c+ia)e^{iat}-f(c-ia)e^{-iat}}{it} - \frac{1}{t}\int_{-a}^{a} e^{i\tau t}f'(c+i\tau)d\tau\\ &= \frac{f(c+ia)e^{iat}-f(c-ia)e^{-iat}}{it} + o\left(\frac{1}{|t|}\right), \qquad |t|\to\infty,\end{aligned}$$

$$= \frac{e^{iat}}{it}\left[\{f(c+ia)-f(c-ia)\}+f(c-ia)(1-e^{-2iat})\right]$$

$$+ o\left(\frac{1}{|t|}\right).$$

Hence

$$t^2|\phi_a(t)|^2 \geqq \left[\frac{k}{2} - 2l\,|\sin at|\right]^2, \qquad t > t_0,$$

where t_0 is a sufficiently large positive number and

$$k = |f(c+ia) - f(c-ia)| \neq 0, \qquad l = |f(c-ia)|.$$

Since $2l|\sin at| < k/4$ in an interval of length δ, say, about $t=0$ and in intervals congruent to this one, modulo π/a, it is clear that the integrand of (5) exceeds $k^2e^{2ct}/16t^2$ in infinitely many intervals of length δ. This is sufficient to insure the divergence of the integral.

We collect our results in the following form:

THEOREM. *If $\phi(t)$ belongs to L^2 in $(0, \infty)$, then it has a Laplace transform $f(\sigma+i\tau)$ defined for $\sigma>0$ by the absolutely convergent integral*

$$f(\sigma+i\tau) = \int_0^\infty e^{-(\sigma+i\tau)t}\phi(t)dt,$$

and for $\sigma=0$ by

$$f(i\tau) = \underset{a\to\infty}{\text{l.i.m.}} \int_0^a e^{-i\tau t}\phi(t)dt.$$

The inversion formula

$$\underset{a\to\infty}{\text{l.i.m.}}\ \frac{1}{2\pi i}\int_{c-ia}^{c+ia} f(s)e^{st}ds = \begin{cases}\phi(t), & t>0,\\ 0, & t<0,\end{cases}$$

is false ($f(s)\not\equiv 0$) for $c>0$ and valid for $c=0$. For all $c \geqq 0$

$$\underset{a\to\infty}{\text{l.i.m.}}\ \frac{1}{2\pi}\int_{-a}^{a} f(c+i\tau)e^{i\tau t}d\tau = \begin{cases}e^{-ct}\phi(t), & t>0,\\ 0, & t<0.\end{cases}$$

HARVARD UNIVERSITY AND
MASSACHUSETTS INSTITUTE OF TECHNOLOGY

Comments on [38f]

D. V. Widder

This note concerns the validity of the classical inversion of the Laplace integral,

$$f(s) = \int_0^{\infty} e^{-st} \varphi(t)\, dt,$$

case $\varphi(t)$ belongs to $L^2(0, \infty)$. It is ordinarily written as

$$\varphi(t) = \underset{a \to \infty}{\text{l.i.m.}} \frac{e^{ct}}{2\pi} \int_{-a}^{a} e^{ity} f(c + iy)\, dy, \quad c \geqq 0.$$

The authors show that, although this formula is valid for $c = 0$, it need not be true for $c > 0$. In this sense the note is a clarification and correction of earlier work by G. Doetsch [Math. Zeit. 42(1936), p. 272, thm 1].

Reprinted from the
TRANSACTIONS OF THE AMERICAN MATHEMATICAL SOCIETY
Vol. 52, No. 2, pp. 249-256
September, 1942

ON THE OSCILLATION OF THE DERIVATIVES OF A PERIODIC FUNCTION

BY
GEORGE PÓLYA AND NORBERT WIENER

1. *Let $f(x)$ be a real valued periodic function of period 2π defined for all real values of x and possessing derivatives of all orders. Let N_k denote the number of changes of sign of $f^{(k)}(x)$ in a period. We consider the order of magnitude of N_k as $k \to \infty$.*

(I) *If $N_k = O(1)$, $f(x)$ is a trigonometric polynomial.*

(II) *If $N_k = O(k^\delta)$ where δ is fixed, $0 < \delta < 1/2$, $f(x)$ is an entire function of finite order not exceeding $(1-\delta)/(1-2\delta)$.*

(III) *If $N_k = o(k^{1/2})$, $f(x)$ is an entire function.*

We prove this theorem by consideration of the Fourier series of $f(x)$

$$f(x) = \sum c_n e^{inx}, \tag{1}$$

$c_{-n} = \bar{c}_n$ $(n = 0, 1, 2, \cdots)$. Here, as in what follows, the sign $\sum$ without explicitly stated limits means a summation from $-\infty$ to ∞. Under the present conditions, the series (1) is absolutely and uniformly convergent for real x, and so are the Fourier series of $f'(x), f''(x), \cdots$, obtained from (1) by term by term differentiation. If we focus our attention on the Fourier series, we may express the general trend of our theorem by saying that *a small amount of oscillation in the higher derivatives implies a rapid decrease in the coefficients,* this decrease being so extreme in case (I) that all coefficients from a certain point onward vanish.

The theorem we have to prove and a few analogous facts(1) point towards a general principle which cannot yet be stated in precise terms but which is not entirely unsuitably expressed by saying that *a small amount of oscillation in the higher derivatives indicates a great amount of simplicity in the analytic nature of the function.*

An analogous theorem may be formulated for almost periodic functions. As in other theorems of this kind, the number of changes of sign in a period is replaced by the density of these changes over the infinite line and a trigonometric polynomial is replaced by an entire function of exponential type. The extension of case (I) of our theorem offers the least difficulty.

Presented to the Society, May 2, 1941; received by the editors August 8, 1941.

(1) See S. Bernstein, *Leçons sur les Propriétés Extrémales*, Paris, 1926, pp. 190–197 and Communications de la Société Mathématique de Kharkow, (4), vol. 2 (1928), pp. 1–11; R. P. Boas and G. Pólya, Proceedings of the National Academy of Sciences, vol. 27 (1941), pp. 323–325.

2. We start with a few preliminary remarks on changes of sign. We consider first a real-valued function $f(x)$ which is defined in an interval $a \leqq x \leqq b$. We say that this function has N changes of sign in this interval if it is possible to find $N+1$, and no more, abscissae $x_0, x_1, \cdots, x_N$ in the interval such that

$$x_0 < x_1 < \cdots < x_{N-1} < x_N, \tag{2}$$

$$f(x_{\nu-1})f(x_\nu) < 0, \qquad \nu = 1, 2, 3, \cdots, N. \tag{3}$$

If a function has N changes of sign in an interval, its derivative has there at least $N-1$ changes of sign. This variant of Rolle's theorem is easily proved by considering ξ_ν, such that

$$f(x_\nu) - f(x_{\nu-1}) = (x_\nu - x_{\nu-1})f'(\xi_\nu), \qquad x_{\nu-1} < \xi_\nu < x_\nu,$$

and observing that $f(x_\nu)f'(\xi_\nu) > 0$ and that therefore

$$f'(\xi_{\nu-1})f'(\xi_\nu) < 0, \qquad \nu = 2, 3, \cdots, N.$$

Applying this to the function $e^{ax}f(x)$ (where a is a real constant) and its derivative $[e^{ax}f(x)]' = e^{ax}[af(x) + f'(x)]$, we see that *the number of changes of sign of*

$$(a + D)f(x)$$

(where D is the symbol of differentiation) *is not inferior to $N-1$, N being the number of changes of sign of $f(x)$.*

Now let $f(x)$ be periodic with the period 2π. We say that the number of changes of sign of $f(x)$ *in a period* is N, if it is possible to find just $N+1$, and no more, abscissae $x_0, x_1, \cdots, x_N$ such that

$$x_N = x_0 + 2\pi,$$

and (2), (3) hold. Observe that $f(x_N) = f(x_0)$ and that, therefore, N is necessarily *even*. Hence it follows that the number of changes of sign of $(a+D)f(x)$ in a period is not inferior to that of $f(x)$. We defined N_k in our initial statement; now we see that

$$N_0 \leqq N_1 \leqq N_2 \leqq \cdots \leqq N_{k-1} \leqq N_k \leqq \cdots. \tag{4}$$

Observing that

$$(a + D)(a - D) \sum \frac{c_n e^{inx}}{a^2 + n^2} = \sum c_n e^{inx},$$

we obtain:

LEMMA I. *The number of changes of sign of the function* (1) *in a period is not inferior to that of*

$$\sum \frac{c_n e^{inx}}{a^2 + n^2}.$$

3. The series (1) represents a trigonometric polynomial of order m if $c_n=0$ for $n=m+1, m+2, \cdots$. If $f(x)$ is a trigonometric polynomial of order m, it cannot have more than $2m$ roots in a period; this is well known. Observe that for large k, $f^{(k)}(x)$ has actually $2m$ changes of sign, because as $k\to\infty$, $(im)^{-k}f^{(k)}(x)$ approaches the first or the second of the two expressions

$$(5) \qquad -c_{-m}e^{-imx}+c_m e^{imx}, \qquad c_{-m}e^{-imx}+c_m e^{imx},$$

according as k is odd or even. The second of these expressions is of the form $2|c_m| \cos(mx-\gamma)$, with a certain real γ and the first is of the same form except for a factor i.

The case (I) of our theorem characterizes the trigonometric polynomials and can be stated as follows: *A real-valued periodic function $f(x)$ possessing derivatives of all orders is a trigonometric polydomial if and only if the number of changes of sign of $f^{(k)}(x)$ remains bounded for $k\to\infty$.*

In order to prove this we consider (1). We have to show that some $f^{(k)}(x)$ have an arbitrarily great number of changes of sign if there are $c_n\neq 0$ with arbitrarily large subscripts n. More precisely we shall show this:

If $m>0$ and $c_m\neq 0$, then all derivatives of (1), *from a certain stage onward, have not less than $2m$ changes of sign.*

In fact, by repeated application of Lemma I, we ascertain that

$$(6) \qquad f^{(k)}(x) = i^k \sum n^k c_n e^{inx}$$

does not have fewer changes of sign than

$$(7) \qquad i^k \sum \left(\frac{2mn}{m^2+n^2}\right)^k c_n e^{inx}.$$

But since it is given that $c_m\neq 0$ and that $\sum c_n$ is absolutely convergent, we have from a certain k onward

$$(8) \qquad |c_m| > \left(\sum_{n=1}^{m-1}+\sum_{n=m+1}^{\infty}\right)\left(\frac{2mn}{m^2+n^2}\right)^k |c_n|.$$

Indeed we have for $n>0$, $n\neq m$, $0<2mn<m^2+n^2$, and therefore, each term tends to 0 on the right-hand side of (8) for $k\to\infty$.

But if (8) holds for a certain even k, the sum in (7) has the same sign as the second expression (5) in all those real points x in which this latter reaches $2|c_m|$, the maximum of its absolute value. This maximum is reached with alternating signs, in equidistant points, the distance of two consecutive points being π/m. Therefore (7) has not less than $2m$ changes of sign. We have proved this for even k but the same is true and the proof is nearly the same for odd k. Then, by Lemma I, (6) has not less than $2m$ changes of sign, and case (I) of our theorem is proved.

4. We consider the case (III) of our theorem before case (II).

If the periodic function $f(x)$ is analytic along the whole real axis, it is analytic in a certain horizontal strip bisected by the real axis and the Fourier series (1), which is a Laurent series in

$$z = e^{ix},$$

converges in the interior of the strip. Hence, by examining

$$\limsup_{n\to\infty} |c_n|^{1/n} \tag{9}$$

we can distinguish the following three cases:

If $f(x)$ is not analytic along the whole real axis, (9) has the value 1.

If $f(x)$ is analytic in a certain horizontal strip of width $2h$ bisected by the real axis, but in no wider horizontal strip, (9) has the value e^{-h}.

If $f(x)$ is an entire function, (9) has the value 0.

In order to prove case (III) of our theorem, we have to show that in the first two cases $N_k = o(k^{1/2})$ is excluded. We prove the following statement.

If there exists a positive number γ *such that*

$$\limsup_{n\to\infty} |c_n| e^{n\gamma} = \infty, \tag{10}$$

then there exists a positive number g such that $f^{(k)}(x)$ *has, for an infinity of values of* k, *not less than* $(k/g)^{1/2}$ *changes of sign.*

By the considerations of the foregoing section, $f^{(k)}(x)$ has certainly not less than $2m$ changes of sign if (8) holds. Using (10), we have to find an arbitrarily large m and a corresponding k such that (8) holds. We shall succeed in finding such an m by applying the following known lemma(²).

LEMMA II. *We consider two infinite sequences* $l_1, l_2, \cdots, l_n, \cdots$ *and* $s_1, s_2, \cdots, s_n, \cdots$, *and suppose that*

$$l_n \geqq 0, \qquad n = 1, 2, 3, \cdots, \tag{11}$$

$$0 < s_1 < s_2 < s_3 < \cdots, \tag{12}$$

$$\lim_{n\to\infty} l_n = 0, \tag{13}$$

$$\limsup_{n\to\infty} l_n s_n = \infty. \tag{14}$$

Then there exists an infinity of integers m such that

$$l_m \geqq l_{m+\mu}, \qquad \mu = 1, 2, 3, \cdots,$$

$$l_m s_m \geqq l_{m-\mu} s_{m-\mu}, \qquad \mu = 1, 2, \cdots, m-1.$$

We put

(²) See G. Pólya and G. Szegö, *Aufgaben und Lehrsätze aus der Analysis*, Berlin, 1925, vol. 1, pp. 18 and 173, Problem 109.

$$|c_n| = l_n, \qquad e^{n\gamma} = s_n.$$

This choice satisfies (11), (12), (13), (14); in fact, (13) is satisfied because (1) is convergent, and (14) is satisfied because we have supposed (10). Thus we obtain an infinity of m such that

$$|c_{m+\mu}| \leqq |c_m|, \qquad \mu = 1, 2, 3, \cdots,$$
$$|c_{m-\mu}| e^{(m-\mu)\gamma} \leqq |c_m| e^{m\gamma}, \qquad \mu = 1, 2, \cdots, m-1.$$

This we use to estimate the following sum. (Our ultimate aim is to prove (8).)

$$\tag{15} \begin{aligned} &\sum_{\mu=1}^{m-1}\left(\frac{2m(m-\mu)}{m^2+(m-\mu)^2}\right)^k \left|\frac{c_{m-\mu}}{c_m}\right| + \sum_{\mu=1}^{\infty}\left(\frac{2m(m+\mu)}{m^2+(m+\mu)^2}\right)^k \left|\frac{c_{m+\mu}}{c_m}\right| \\ &< \sum_{\mu=1}^{m-1}\left(\frac{2m(m-\mu)}{m^2+(m-\mu)^2}\right)^k e^{\gamma\mu} + \sum_{\mu=1}^{\infty}\left(\frac{2m(m+\mu)}{m^2+(m+\mu)^2}\right)^k \\ &= S_1 + S_2. \end{aligned}$$

We introduced the abbreviations

$$\tag{16} S_1 = \sum_{\mu=1}^{m-1} e^{\gamma\mu}\left(1 + \frac{\mu^2}{2m(m-\mu)}\right)^{-k},$$

$$\tag{17} S_2 = m\sum_{\mu=1}^{\infty}\left(\frac{2(1+\mu/m)}{1+(1+\mu/m)^2}\right)^k \frac{1}{m},$$

and we shall consider S_1 and S_2 in turn.

(1) Split the sum S_1 in two parts, μ being less than or equal to $m/2$ in the first part and greater than $m/2$ in the second. Using the fact that

$$(1+x)^{-1} < e^{-x/2}, \qquad 0 < x < 1,$$

we obtain

$$\begin{aligned} S_1 &< \sum_{1}^{m/2} e^{\gamma\mu - k\mu^2/4m^2} + \sum_{m/2}^{m-1} e^{\gamma\mu}(4/5)^k \\ &< \sum_{1}^{m/2} e^{-[(k/4m^2)-\gamma]\mu} + me^{\gamma m}(4/5)^k \\ &< e^{-(g-\gamma)}/(1-e^{-(g-\gamma)}) + me^{\gamma m}(4/5)^{4gm^2}. \end{aligned}$$

We put

$$\tag{18} k = 4gm^2$$

where g is a positive integer, $g>\gamma$. We choose a fixed g such that for sufficiently great m

$$\tag{19} S_1 < 1/2.$$

(2) The function $2x(1+x^2)^{-1}$ decreases for $x>1$. Therefore by (17)

$$S_2 < m\int_1^\infty \left(\frac{2x}{1+x^2}\right)^k dx \sim \frac{m}{2}\left(\frac{2\pi}{k}\right)^{1/2} = \frac{1}{2}\left(\frac{\pi}{2g}\right)^{1/2}.$$

We used a well known asymptotic evaluation of definite integrals(³) and (18). If $g \geqq 2$, which we assume, we obtain for sufficiently great m

$$S_2 < 1/2. \tag{20}$$

But (15), (19), (20) show that (8) is true so that $f^{(k)}(x)$ has not fewer changes of sign than

$$2m = (k/g)^{1/2}.$$

5. We now proceed to the proof of case (II).

LEMMA III. *The Fourier series* (1) *represents an entire function of the finite order* λ, $\lambda>1$, *if and only if*

$$\liminf_{n\to\infty} \frac{\log\log(1/|c_n|)}{\log n} = \frac{\lambda}{\lambda-1}. \tag{21}$$

The proof consists of two parts. Both parts follow familiar lines; so we do not give all the details.

(1) Assume that $f(x)$ is entire and of order λ. Then for a fixed positive ϵ and for sufficiently great $|x|$,

$$|f(x)| < e^{|x|^{\lambda+\epsilon}}. \tag{22}$$

If we evaluate c_n and shift the line of integration (periodicity and Cauchy's formula), we obtain as a result that if r is any positive number,

$$\begin{aligned} c_n &= \frac{1}{2\pi}\int_{-ir-\pi}^{-ir+\pi} f(x)e^{-inx}dx \\ &= \frac{1}{2\pi}e^{-nr}\int_{-\pi}^{\pi} f(-ir+u)e^{-inu}du, \end{aligned} \tag{23}$$

$$|c_n| \leqq e^{(r+\pi)^{\lambda+\epsilon}-nr}.$$

Here we use (22). We choose r, for given n, so that this right-hand side of (23) shall be a minimum. It follows by straight-forward calculation that (21) holds with "$\geqq$" instead of "$=$."

(2) Assume that

$$\liminf_{n\to\infty} \frac{\log\log(1/|c_n|)}{\log n} = \kappa > 0.$$

(³) See, for example, G. Pólya and G. Szegö, loc. cit., vol. 1, pp. 78 and 244, Problem 201.

Therefore we have, for a given positive ϵ and all sufficiently great n

$$|c_n e^{inx}| < e^{-n^{\kappa-\epsilon}+|x|n}.$$

We choose n, for a given x, so that the right-hand side is a maximum. This maximum gives the right order of magnitude because the terms of (1) whose index surpasses a certain multiple of the index of the maximum term, yield a negligible contribution. We find that the order λ of $f(x)$ satisfies the inequality

$$\lambda \leqq \frac{\kappa}{\kappa - 1}\cdot$$

This gives (21) with "$\leqq$" instead of "$=$."

6. Now we are prepared to prove case (II) of our theorem. We have to show that *if the entire function $f(x)$ is of order λ, and $\epsilon > 0$ then, for an infinity of k,*

$$N_k > k^{(\lambda-1)/(2\lambda-1)-\epsilon}.$$

Put $\lambda/(\lambda-1)+\eta=\gamma$, η being positive and small. By Lemma III, the fact we have to show can be stated as follows:

If there exists a positive number γ, $\gamma > 1$, such that

$$\limsup_{n\to\infty} |c_n| e^{n^\gamma} = \infty,$$

then there exists a positive g such that $f^{(k)}(x)$ has, for an infinity of k, more than $(k/g)^{1/(\gamma+1)}$ changes of sign.

We apply Lemma II, whose conditions are satisfied by

$$l_n = |c_n|, \qquad s_n = e^{n^\gamma}.$$

We obtain the result that for an infinity of m,

$$|c_m| \geqq |c_{m+\mu}|, \qquad \mu = 1\ 2, \cdots,$$
$$|c_m| e^{m^\gamma} \geqq |c_{m-\mu}| e^{(m-\mu)^\gamma}, \qquad \mu = 1, 2, \cdots, m-1.$$

Hence, using the fact that $\gamma > 1$, we obtain

$$\begin{aligned} &\sum_{\mu=1}^{m-1}\left(\frac{2m(m-\mu)}{m^2+(m-\mu)^2}\right)^k\left|\frac{c_{m-\mu}}{c_m}\right| + \sum_{\mu=1}^{\infty}\left(\frac{2m(m+\mu)}{m^2+(m+\mu)^2}\right)^k\left|\frac{c_{m+\mu}}{c_m}\right| \\ &< \sum_{\mu=1}^{m-1}\left(\frac{2m(m-\mu)}{m^2+(m-\mu)^2}\right)^k e^{\gamma m^{\gamma-1}\mu} + \sum_{\mu=1}^{\infty}\left(\frac{2m(m+\mu)}{m^2+(m+\mu)^2}\right)^k \qquad (24) \\ &= S_1' + S_2. \end{aligned}$$

S_2 has the same meaning as before (see (17)), and

$$S_1' = \sum_{\mu=1}^{m-1} e^{\gamma m^{\gamma-1}\mu}\left(1+\frac{\mu^2}{2m(m-\mu)}\right)^{-k}$$

$$(25) \qquad < \sum_{\mu=1}^{m/2} e^{\gamma m^{\gamma-1}\mu - k\mu^2/4m^2} + me^{\gamma m^{\gamma}}(4/5)^k$$

$$= \sum_{\mu=1}^{m/2} e^{-m^{\gamma-1}(g\mu^2-\gamma\mu)} + me^{\gamma m^{\gamma}}(4/5)^{4gm^{\gamma+1}}.$$

We put

$$k = 4gm^{\gamma+1},$$

and we choose g so that

$$\gamma + 1 \leqq g < \gamma + 2,$$

and so that k is an integer. This choice assures that

$$S_1' \to 0, \qquad S_2 \to 0$$

for $m \to \infty$ (see (25) and the considerations preceding (20)). Therefore, by (24), (8) is true and $f^{(k)}(x)$ has not fewer changes of sign than

$$2m > 2[k/4(\gamma+2)]^{1/(\gamma+1)}.$$

BROWN UNIVERSITY,
PROVIDENCE, R. I.
MASSACHUSETTS INSTITUTE OF TECHNOLOGY,
CAMBRIDGE, MASS.

Comments on [42a]

R. Askey

To place this paper in a historical perspective, we recall that a C^∞ function whose derivatives do not change sign on an interval (a, b) has an extension as an analystic function to an open set in the complex plane which contains (a, b) by a theorem of S. Bernstein.[1] Extensions of this theorem have been found regularly. A recent survey was given by R. P. Boas.[2] In 1940, D. V. Widder proved the surprising theorem that $(-1)^n f^{(2n)}(x) \geqslant 0$, $0 < x < 1$, implies that $f(x)$ is the restriction to $(0, 1)$ of an entire function of exponential type at most π. The strong conclusion from the seemingly weak hypothesis caught the fancy of the mathematical community, and many papers were written on this subject in the early 1940's.

The first paper to weaken the assumptions on the derivatives to allow each of them to change sign was this paper of Pólya and Wiener published in 1942. It was followed almost immediately by a series of four papers, "On the oscillation of differential transforms, I, II, III, IV" by Hille, Schaeffer, and Szegö. A summary of these results was given by Pólya, "On the zeros of the derivatives of a function and its analytic character."[3] The last paragraph in this review is worth quoting:

> Looking back at the new results and new analogies that have been added quite recently to our knowledge of the subject, we may obtain the impression of grasping the outlines of a well balanced theory leading to further results.

This has not happened, for interests changed and the only more recent result I could find is in "Proof of a conjecture of I. J. Schoenberg" by I. I. Hirschman.[4] He proved that the only C^∞ function which vanishes for $|t| \geqslant a$ for which $f^{(n)}(t)$ has at most n sign changes, $n = 0, 1, \ldots$, is identically zero. This can be used to show that the only 2π periodic, C^∞ function with $N_k(f) \leqslant k$, $k = 0, 2, \ldots$, which vanishes with all its derivatives at a point is identically zero. Thus the condition $N_k(f) \leqslant k$, which it is probably not sufficiently strong to imply analyticity, is a type of quasi-analyticity condition.

Let us hope that publication of these collected works will rekindle interest in these problems. A well-developed theory should be worked out.

References

1. S. Bernstein, Math. Anal. 75(1914), 449–468.

2. R. P. Boas, Amer. Math. Monthly 78(1971), 1085–1093.

3. G. Pólya, *On the zeros of the derivatives of a function and its analytic character*, Bull. Amer. Math. Soc. 49(1943), 178–191.

4. I. I. Hirschman, Proc. Amer. Math. Soc. 1(1950), 63–65.

Index of Papers and Commentaries